INTERNAL COMBUSTION ENGINES

(Including Air Compressors, Gas Turbines and Jet Propulsion)

INTERNAL COMBUSTION ENGINES

(Including Air Compressors, Gas Turbines and Jet Propulsion)

[A textbook for the students preparing for B.Tech, B.E., AMIE, UPSC (Engg. Services) and GATE Examinations]

SI UNITS

By

Er. R.K. RAJPUT

M.E. (Hons.), Gold Medallist, Grad. (Mech. Engg. & Elect. Engg.); M.I.E. (*India*); M.S.E.S.I.; M.I.S.T.E.; C.E. (*India*)

Recipient of:

"Best Teacher (Academic) Award"

"Distinguished Author Award"

"Jawahar Lal Nehru Memorial Gold Medal"
for an outstanding research paper
(Institution of Engineers–India)

Principal (*Formerly*):

- Thapar Polytechnic College;
- Punjab College of Information Technology,

PATIALA

LAXMI PUBLICATIONS (P) LTD

(An ISO 9001:2008 Company)

BENGALURU • CHENNAI • COCHIN • GUWAHATI • HYDERABAD
JALANDHAR • KOLKATA • LUCKNOW • MUMBAI • RANCHI • NEW DELHI
BOSTON (USA) • ACCRA (GHANA) • NAIROBI (KENYA)

A TEXTBOOK OF INTERNAL COMBUSTION ENGINES

Printed and bound in India
Typeset at Goswami Associates, Delhi
First Edition: 2005, Second Edition : 2007, Third Edition : 2016
ISBN 81-318-0066-0

Branches

✆	Bengaluru	080-26 75 69 30	
✆	Chennai	044-24 34 47 26,	24 35 95 07
✆	Cochin	0484-237 70 04,	405 13 03
✆	Guwahati	0361-254 36 69,	251 38 81
✆	Hyderabad	040-27 55 53 83,	27 55 53 93
✆	Jalandhar	0181-222 12 72	
✆	Kolkata	033-22 27 43 84	
✆	Lucknow	0522-220 99 16	
✆	Mumbai	022-24 91 54 15,	24 92 78 69
✆	Ranchi	0651-220 44 64	

PUBLISHED IN INDIA BY

Laxmi Publications (P) Ltd.
(An ISO 9001:2008 Company)
113, GOLDEN HOUSE, DARYAGANJ,
NEW DELHI - 110002, INDIA
Telephone : 91-11-4353 2500, 4353 2501
Fax : 91-11-2325 2572, 4353 2528
www.laxmipublications.com info@laxmipublications.com

C—
Printed at:

CONTENTS

Chapter *Pages*

Chapter *Pages*

PREFACE TO THE THIRD EDITION

In this edition, besides revising the book, **Chapter 22** has been thoroughly updated/modified and a **new Chapter 23** has been added, with the following *contents*:

Chapter 22: **Universities Examinations' Questions (Latest-selected) with Answers/Solutions**

A. Short Answer Questions.

B. Conventional Questions with Answers/Solutions.

Chapter 23: **"GATE" and "UPSC" Examinations Questions (Latest-selected) with Answers/Solutions**

A. Conventional Questions with Answers/Solutions.

B. Multiple-Choice Questions with Answers and "Explanations".

Any constructive suggestions for the improvement of this book are most welcome.

Er. R.K.Rajput

(Author)

PREFACE TO THE FIRST EDITION

This treatise on **"Internal Combustion Engines"** (*Including gas turbines*) contains comprehensive treatment of the subject matter in a simple, lucid and direct language. It envelops a large number of solved problems properly graded including typical worked examples from examination point of view.

The book comprises 21 chapters. All chapters are saturated with much needed text, supported by simple and self explanatory figures. At the end of each chapter—**Highlights, Objective Type Questions, Theoretical Questions** and **Unsolved Examples** have been added ; besides this a **"Question Bank"** containing **"Additional Objective Type Questions (with Answers and Solution—Comments)"**, **"Theoretical Questions with Answers"** and **"Additional Typical Examples** (*Including Universities and Competitive Examination Questions*)" have been included to make the book a comprehensive and a complete unit in all respects.

The book will prove to be a boon to the students preparing for engineering undergraduate, A.M.I.E., post graduate, U.P.S.C. and other competitive examinations.

The author's thanks are due to his wife Ramesh Rajput for extending all co-operation during preparation of the manuscript and proof reading.

In the end the author wishes to express his gratitude to Shri R.K. Gupta, Chairman, Sh. Saurabh Gupta, Managing Director, Laxmi Publications Pvt. Ltd., New Delhi for taking a lot of pains in bringing out the book with very good presentation in a short span of time.

Although every care has been taken to make the book free of errors both in text as well as in solved examples, yet the author shall feel obliged if any errors present are brought to his notice. Constructive criticism of the book will be warmly received.

Er. R.K.Rajput

(Author)

1

Basic Concepts of Thermodynamics

1.1. DEFINITION OF THERMODYNAMICS

Thermodynamics may be defined as follows :

Thermodynamics is an axiomatic science which deals with the relations among heat, work and properties of system which are in equilibrium. It describes state and changes in state of physical systems.

Or

Thermodynamics is *the science of the regularities governing processes of energy conversion.*

Or

Thermodynamics is *the science that deals with the interaction between energy and material systems.*

Thermodynamics, basically entails *four laws* or axioms known as Zeroth, First, Second and Third law of thermodynamics.

— the *First law* throws light on *concept of internal energy.*

— the *Zeroth law* deals with *thermal equilibrium* and establishes a *concept of temperature.*

— the *Second law* indicates the limit of *converting heat into work* and introduces the *principle of increase of entropy.*

— third law defines the *absolute zero of entropy.*

These laws are based on experimental observations and have no *mathematical proof.* Like all physical laws, these laws are based on *logical reasoning.*

1.2. THERMODYNAMIC SYSTEMS

1.2.1. System, Boundary and Surroundings

System. A system is *a finite quantity of matter or a prescribed region of space* (Refer Fig. 1.1)

Boundary. The *actual or hypothetical envelope enclosing the system* is the boundary of the system. The boundary may be fixed or it may move, as and when a system containing a gas is compressed or expanded. The boundary may be *real* or *imaginary.* It is not difficult to envisage a real boundary but an example of imaginary boundary would be one drawn around a system consisting of the fresh mixture about to enter the cylinder of an I.C. engine together with the remanants of the last cylinder charge after the exhaust process (Refer Fig. 1.2).

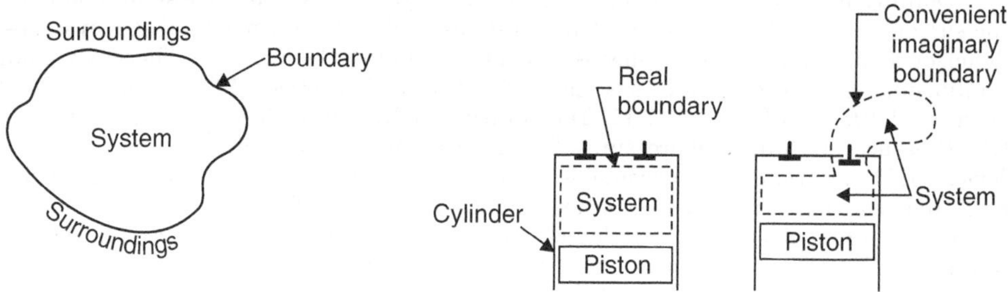

Fig. 1.1. The system.

Fig. 1.2. The real and imaginary boundaries.

1.2.2. Closed System

Refer Fig. 1.3. If the boundary of the system is impervious to the flow of matter, it is called a *closed system.* An example of this system is mass of gas or vapour contained in an engine cylinder, the boundary of which is drawn by the cylinder walls, the cylinder head and piston crown. Here the *boundary is continuous and no matter may enter or leave.*

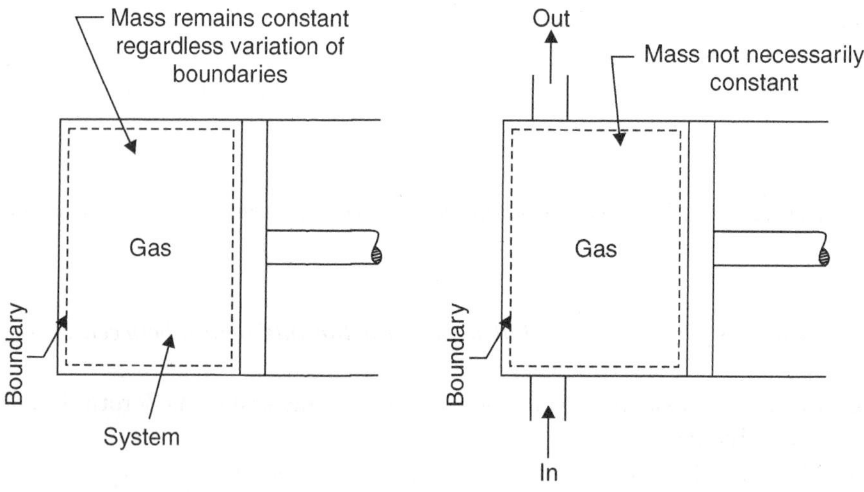

Fig. 1.3. Closed system.

Fig. 1.4. Open system.

1.2.3. Open System

Refer Fig. 1.4. An open system is one in which *matter flows into or out of the system.* Most of the engineering systems are open.

1.2.4. Isolated System

An isolated system is that system *which exchanges neither energy nor matter with any other system or with environment.*

1.2.5. Adiabatic System

An adiabatic system is one *which is thermally insulated from its surroundings.* It can, however, *exchange work with its surroundings.* If it does not, it becomes an isolated system.

Phase. A phase is a quantity of matter which is homogeneous throughout in chemical composition and physical structure.

1.2.6. Homogeneous System

A system which consists of a single phase is termed as *homogeneous system.* Examples : Mixture of air and water vapour, water plus nitric acid and octane plus heptane.

1.2.7. Heterogeneous System

A system which consists of two or more phases is called a *heterogeneous system.* Examples : Water plus steam, ice plus water and water plus oil.

1.3. PURE SUBSTANCE

A pure substance is one that has a homogeneous and invariable chemical composition even though there is a change of phase. In other words, it is a system which is (*a*) homogeneous in composition, (*b*) homogeneous in chemical aggregation. Examples : Liquid, water, mixture of liquid water and steam, mixture of ice and water. The mixture of liquid air and gaseous air is not a pure substance.

1.4. THERMODYNAMIC EQUILIBRIUM

A system is in *thermodynamic equilibrium* if the temperature and pressure at all points are same ; there should be no velocity gradient ; the chemical equilibrium is also necessary. Systems under temperature and pressure equilibrium but not under chemical equilibrium are sometimes said to be in metastable equilibrium conditions. *It is only under thermodynamic equilibrium conditions that the properties of a system can be fixed.*

Thus for attaining a state of *thermodynamic equilibrium* the following three types of equilibrium states must be achieved :

1. **Thermal equilibrium.** The temperature of the system does not change with time and has same value at all points of the system.

2. **Mechanical equilibrium.** There are no unbalanced forces within the system or between the surroundings. The pressure in the system is same at all points and does not change with respect to time.

3. **Chemical equilibrium.** No chemical reaction takes place in the system and the chemical composition which is same throughout the system does not vary with time.

1.5. PROPERTIES OF SYSTEMS

A property of a system is a characteristic of the system which depends upon its state, but not upon how the state is reached. There are two sorts of property :

1. **Intensive properties.** These properties *do not depend on the mass of the system. Examples* : Temperature and pressure.

2. **Extensive properties.** These properties *depend on the mass of the system. Example* : Volume. Extensive properties are often divided by mass associated with them to obtain the intensive properties. For example, if the volume of a system of mass m is V, then the specific volume of matter within the system is $\frac{V}{m} = v$ which is an intensive property.

1.6. STATE

***State** is the condition of the system at an instant of time as described or measured by its properties. Or each unique condition of a system is called a state.*

It follows from the definition of state that each property has a single value at each state. Stated differently, all properties are *state* or *point functions.* Therefore, all properties are identical for identical states.

On the basis of the above discussion, we can determine if a given variable is *property* or not by applying the following *tests* :

— *A variable is a property, if and only if, it has a single value at each equilibrium state.*

— *A variable is a property, if and only if, the change in its value between any two prescribed equilibrium states is single-valued.*

Therefore, *any variable whose change is fixed by the end states is a property.*

1.7. PROCESS

A process occurs when the system undergoes a change in a state or an energy transfer at a steady state. A process may be *non-flow* in which a fixed mass within the defined boundary is undergoing a change of state. Example : a substance which is being heated in a closed cylinder undergoes a non-flow process (Fig. 1.3). *Closed systems undergo non-flow processes.* A process may be a flow process in which mass is entering and leaving through the boundary of an open system. In a steady flow process (Fig. 1.4) mass is crossing the boundary from surroundings at entry, and an equal mass is crossing the boundary at the exit so that the total mass of the system remains constant. In an open system it is necessary to take account of the work delivered from the surroundings to the system at entry to cause the mass to enter, and also of the work delivered from the system at surroundings to cause the mass to leave, as well as any heat or work crossing the boundary of the system.

Quasi-static process. Quasi means 'almost'. A quasi-static process is also called a *reversible process. This process is a succession of equilibrium states and infinite slowness is its characteristic feature.*

1.8. CYCLE

Any process or series of processes whose end states are identical is termed a **cycle**. The processes through which the system has passed can be shown on a state diagram, but a complete section of the path requires in addition a statement of the heat and work crossing the boundary of the system. Fig. 1.5 shows such a cycle in which a system commencing at condition '1' changes in pressure and volume through a path 123 and returns to its initial condition '1'.

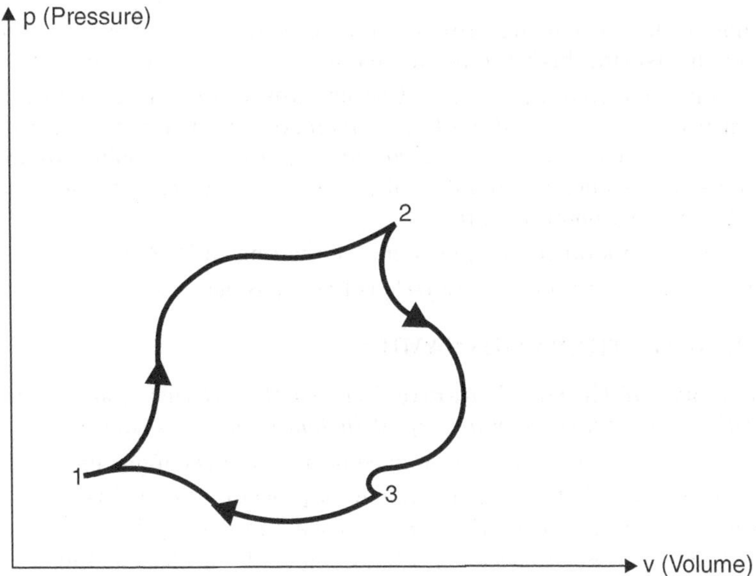

Fig. 1.5. Cycle of operations.

1.9. POINT FUNCTION

When two properties locate a *point* on the graph (co-ordinate axes) then those properties are called as *point function.*

Examples. Pressure, temperature, volume etc.

$$\int_1^2 dV = V_2 - V_1 \text{ (an } \textit{exact differential}\text{)}.$$

1.10. PATH FUNCTION

There are certain quantities which cannot be located on a graph by a *point* but are given by the *area* or so, on that graph. In that case, the area on the graph, pertaining to the particular process, is a function of the path of the process. Such quantities are called *path functions.*

Examples. Heat, work etc.

Heat and work are *inexact differentials.* Their change cannot be written as difference between their end states.

Thus $\int_1^2 \delta Q \neq Q_2 - Q_1$ and is shown as ${}_1Q_2$ or Q_{1-2}

Similarly $\int_1^2 \delta W \neq W_2 - W_1$, and is shown as ${}_1W_2$ or W_{1-2}

Note. The operator δ is used to denote inexact differentials and operator d is used to denote exact differentials.

1.11. TEMPERATURE

- *The temperature is a thermal state of a body which distinguishes a hot body from a cold body.* The temperature of a body is *proportional to the stored molecular energy i.e.* the average molecular kinetic energy of the molecules in a system. (A particular molecule does not have a temperature, it has energy. The gas as a system has temperature).

- Instruments for measuring ordinary temperatures are known as "*thermometers*" and those for measuring high temperatures are known as "*pyrometers*".
- It has been found that a gas will not occupy any volume at a certain temperature. This temperature is known as *absolute zero temperature*. The temperatures measured with absolute zero as basis are called *absolute temperatures*. Absolute temperature is stated in degrees centigrade. The point of absolute temperature is found to occur at 273.15°C below the freezing point of water.

Then : Absolute temperature = Thermometer reading in °C + 273.15.

Absolute temperature in degree centigrade is known as degrees kelvin, denoted by K (SI unit).

1.12. ZEROTH LAW OF THERMODYNAMICS

- **'Zeroth law of thermodynamics'** *states that if two systems are each equal in temperature to a third, they are equal in temperature to each other.*

Example. Refer Fig. 1.6. System '1' may consist of a mass of gas enclosed in a rigid vessel fitted with a pressure gauge. If there is no change of pressure when this system is brought into contact with system '2' a block of iron, then the two systems are equal in temperature (assuming that the systems 1 and 2 do not react each other chemically or electrically). Experiment reveals that if system '1' is brought into contact with a third system '3' again with no change of properties then systems '2' and '3' will show no change in their properties when brought into contact provided they do not react with each other chemically or electrically. Therefore, '2' and '3' must be in equilibrium.

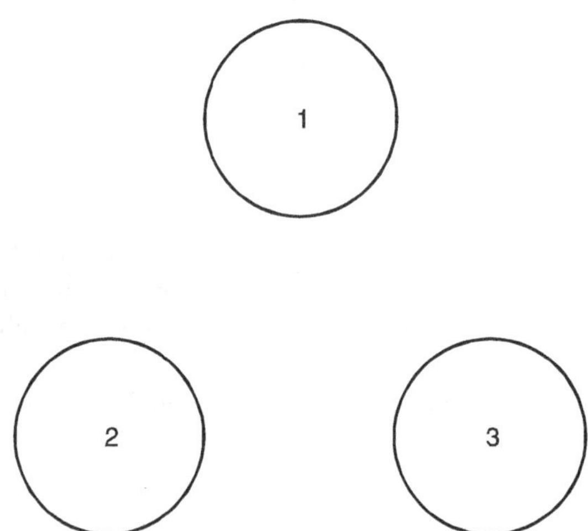

Fig. 1.6. Zeroth law of thermodynamics.

- This law was enunciated by R.H. Fowler in the year 1931. However, since the first and second laws already existed at that time, it was designated as *zeroth law* so that it *precedes* the first and second laws *to form a logical sequence.*

1.13. PRESSURE

1.13.1. Definition of Pressure

Pressure is defined as a *force per unit area*. Pressures are exerted by gases, vapours and liquids. The instruments that we generally use, however, record pressure as the difference between two pressures. Thus, it is the *difference between the pressure exerted by a fluid of interest and the ambient atmospheric pressure.* Such devices indicate the pressure either above or below that of the atmosphere. When it is *above the atmospheric pressure,* it is termed *gauge pressure* and is *positive.* When it is *below* atmospheric, it is *negative* and is known as *vacuum.* Vacuum readings are given in millimetres of mercury or millimetres of water below the atmosphere.

It is necessary to establish an absolute pressure scale which is independent of the changes in atmospheric pressure. A pressure of absolute zero can exist only in complete vacuum. *Any pressure measured above the absolute zero of pressure* is termed an *'absolute pressure'.*

A schematic diagram showing the *gauge pressure, vacuum pressure* and the *absolute pressure* is given in Fig. 1.7.

Mathematically :

(*i*) Absolute pressure = Atmospheric pressure + Gauge pressure

$$p_{abs.} = p_{atm.} + p_{gauge}.$$

(*ii*) Vacuum pressure = Atmospheric pressure – Absolute pressure.

Vacuum is defined as the *absence of pressure.* A *perfect vacuum* is obtained when *absolute pressure is zero,* at this instant *molecular momentum is zero.*

Atmospheric pressure is measured with the help of barometer.

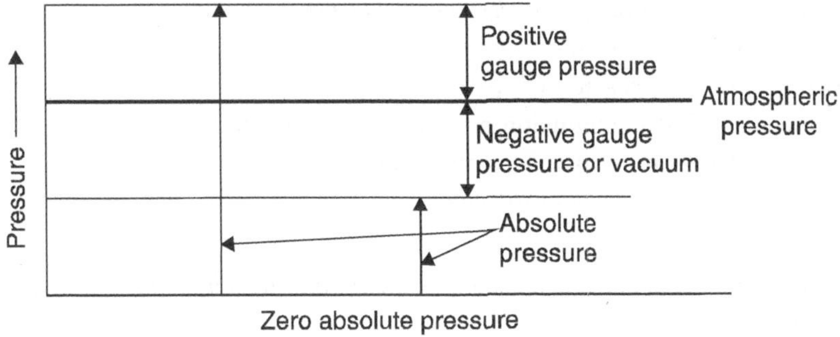

Fig. 1.7. Schematic diagram showing gauge, vacuum and absolute pressures.

1.13.2. Unit for Pressure

The fundamental SI unit of pressure is N/m^2 (sometimes called *pascal,* Pa) or bar. 1 bar $= 10^5$ $N/m^2 = 10^5$ Pa. Standard atmospheric pressure = 1.01325 bar = 0.76 m Hg.

Low pressures are often expressed in terms of mm of water or mm of mercury. This is an abbreviated way of saying that the pressure is such that which will support a liquid column of stated height.

1.13.3. Types of Pressure Measurement Devices

The pressure may be measured by means of indicating gauges or recorders. These instruments may be mechanical, electro-mechanical, electrical or electronic in operation.

1. **Mechanical instruments.** These instruments may be classified into following two groups :

— The *first group* includes those instruments in which the *pressure* measurement is made by *balancing an unknown force with a known force.*

— The *second group* includes those employing *quantitative deformation of an elastic member for pressure measurement.*

2. **Electro-mechanical instruments.** These instruments usually *employ a mechanical means for detecting the pressure and electrical means for indicating or recording the detected pressure.*

3. **Electronic instruments.** Electronic pressure measuring instruments normally depend on some physical change that can be detected and indicated or recorded electronically.

1.14. REVERSIBLE AND IRREVERSIBLE PROCESSES

Reversible process. *A reversible process (also sometimes known as quasi-static process) is one which can be stopped at any stage and reversed so that the system and surroundings are exactly restored to their initial states.*

This process has the following *characteristics* :

1. It must pass through the same states on the reversed path as were initially visited on the forward path.

2. This process when undone will leave no history of events in the surroundings.

3. It must pass through a continuous series of equilibrium states.

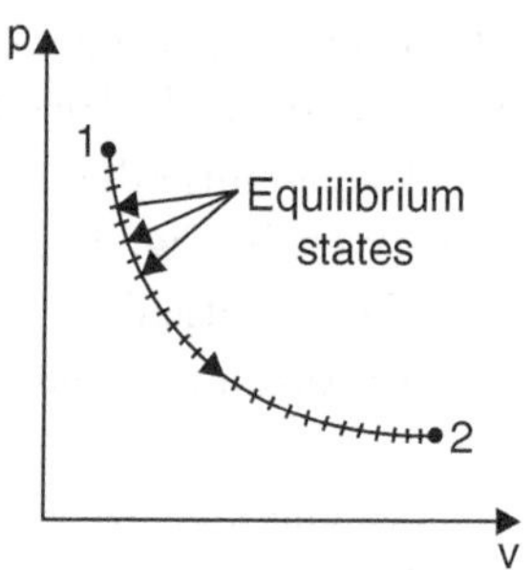

Fig. 1.8. Reversible process.

No real process is truely reversible but some processes may approach reversibility, to close approximation.

Examples. Some *examples of nearly reversible processes* are :

(*i*) Frictionless relative motion.

(*ii*) Expansion and compression of spring.

(*iii*) Frictionless adiabatic expansion or compression of fluid.

(*iv*) Polytropic expansion or compression of fluid.

(*v*) Isothermal expansion or compression.

(*vi*) Electrolysis.

Irreversible process. An *irreversible process is one in which heat is transferred through a finite temperature.*

Examples :

(*i*) Relative motion with friction
(*ii*) Combustion
(*iii*) Diffusion
(*iv*) Free expansion
(*v*) Throttling
(*vi*) Electricity flow through a resistance
(*vii*) Heat transfer
(*viii*) Plastic deformation.

An irreversible process is usually represented by a dotted (or discontinuous) line joining the end states to indicate that the intermediate states are indeterminate (Fig. 1.9).

Irreversibilities are of *two types* :

1. **External irreversibilities.** These are associated with *dissipating effects outside the working fluid.*

Example. *Mechanical friction occurring during a process due to some external source.*

2. **Internal irreversibilities.** These are associated with *dissipating effects within the working fluid.*

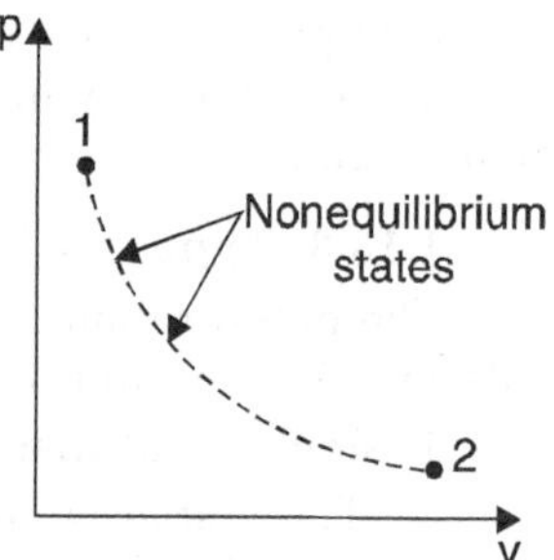

Fig. 1.9. Irreversible process.

Example. *Unrestricted expansion of gas, viscosity and inertia of the gas.*

1.15. ENERGY, WORK AND HEAT

1.15.1. Energy

Energy is a general term embracing *energy in transition* and *stored energy.* The stored energy of a substance may be in the forms of *mechanical energy* and *internal energy* (other forms of stored energy may be chemical energy and electrical energy). Part of the stored energy may take the form of either potential energy (which is the gravitational energy due to height above a chosen datum line) or kinetic energy due to velocity. The balance part of the energy is known as *internal energy.* In a *non-flow process* usually there is no change of potential or kinetic energy and hence change of mechanical energy will not enter the calculations. In a *flow process,* however, there may be changes in both potential and kinetic energy and these must be taken into account while considering the changes of stored energy. ***Heat and work*** are *the forms of energy in transition.* These are the only forms in which energy can cross the boundaries of a system. *Neither heat nor work can exist as stored energy.*

1.15.2. Work and Heat

Work

Work is said to be done when a *force moves through a distance.* If a part of the boundary of a system undergoes a displacement under the action of a pressure, the work done W is the product of the force (pressure × area), and the distance it moves in the direction of the force. Fig. 1.10 (*a*) illustrates this with the conventional piston and cylinder arrangement, the heavy line defining the boundary of the system. Fig. 1.10 (*b*) illustrates another way in which work might be applied to a system. A force is exerted by the paddle as it changes the momentum of the fluid, and since this force moves during rotation of the paddle room work is done.

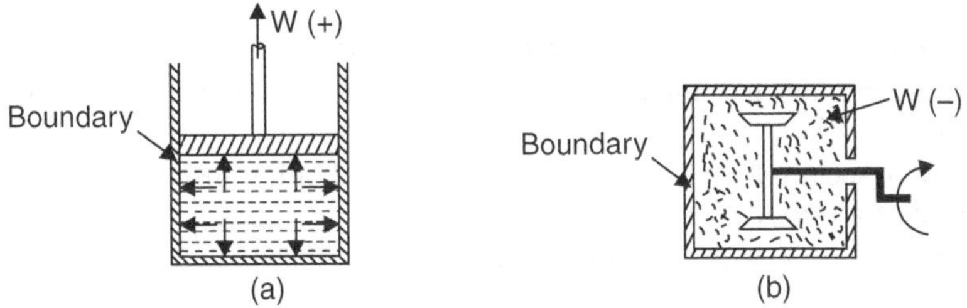

Fig. 1.10

"Work" is a transient quantity which only appears at the boundary while a change of state is taking place within a system. "Work" is 'something' which appears at the boundary when a system changes its state due to the movement of a part of the boundary under the action of a force.

Sign convention :

- If the work is done *by* the system *on* the surroundings, *e.g.* when a fluid expands pushing a piston outwards, the work is said to be *positive.*

i.e., *Work output of the system* = + W

- If the work is done *on* the system *by* the surroundings, *e.g.* when a force is applied to a rotating handle, or to a piston to compress a fluid, the work is said to be *negative.*

i.e., *Work input to system* = – W

Heat

Heat (denoted by the symbol Q), may be, defined in an analogous way to work as follows :

"Heat is 'something' which appears at the boundary when a system changes its state due to a difference in temperature between the system and its surroundings".

Heat, like work, is a *transient quantity* which only appears at the boundary while a change is taking place within the system.

It is apparent that neither δW or δQ are exact differentials and therefore any integration of the elemental quantities of work or heat which appear during a change from state 1 to state 2 must be written as

$$\int_1^2 \delta W = W_{1-2} \text{ or } {}_1W_2 \text{ (or } W\text{), and}$$

$$\int_1^2 \delta Q = Q_{1-2} \text{ or } {}_1Q_2 \text{ (or } Q\text{)}$$

Sign convention :

If the heat flows *into* a system *from* the surroundings, the quantity is said to be *positive* and, conversely, if heat flows *from* the system to the surroundings it is said to be *negative.*

In other words :

Heat received by the system $= + Q$

Heat rejected or given up by the system $= - Q$.

Comparison of Work and Heat

Similarities :

(*i*) Both are *path functions and inexact differentials.*

(*ii*) Both are boundary phenomenon *i.e.,* both are recognized at the boundaries of the system as they cross them.

(*iii*) Both are associated with a process, not a state. Unlike properties, work or heat has no meaning at a state.

(*iv*) Systems possess energy, but not work or heat.

Dissimilarities :

(*i*) In heat transfer temperature difference is required.

(*ii*) In a stable system there cannot be work transfer, however, there is no restriction for the transfer of heat.

(*iii*) The sole effect external to the system could be reduced to rise of a weight but in the case of a heat transfer other effects are also observed.

1.16. FIRST LAW OF THERMODYNAMICS

It is observed that when a system is made to undergo a complete cycle then net work is done *on* or *by* the system. Consider a cycle in which net work is done by the system. Since energy cannot be created, this mechanical energy must have been supplied from some source of energy. Now the system has been returned to its initial state : Therefore, its *intrinsic* energy is unchanged, and hence the mechanical energy has not been provided by the system itself. The only other energy involved in the cycle is the heat which was supplied and rejected in various processes. Hence, by the law of conservation of energy, the net work done by the system is equal to the net heat supplied to the system. The First Law of Thermodynamics can, therefore, be stated as follows :

"When a system undergoes a thermodynamic cycle then the net heat supplied to the system from the surroundings is equal to net work done by the system on its surroundings.

or
$$\oint dQ = \oint dW$$

where $\oint$ represents the sum for a complete cycle.

The first law of Thermodynamics *cannot be proved analytically, but experimental evidence has repeatedly confirmed its validity*, and since no phenomenon has been shown to contradict it, the first law is accepted as a *law of nature*. It may be remarked that no restriction was imposed which limited the application of first law to reversible energy transformation. Hence the first law applies to reversible as well as irreversible transformations : For non-cyclic process, a more general formulation of first law of thermodynamics is required. A new concept which involves a term called *internal energy* fulfills this need.

— The First Law of Thermodynamics may also be stated as follows :

"Heat and work are mutually convertible but since energy can neither be created nor destroyed, the total energy associated with an energy conversion remains constant".

Or

— **"No machine can produce energy without corresponding expenditure of energy, *i.e.*, it is impossible to construct a perpetual motion machine of first kind".**

1.17. THE PERFECT GAS

1.17.1. The Characteristic Equation of State

— At temperatures that are considerably in excess of critical temperature of a fluid, and also at very low pressure, the vapour of fluid tends to obey the equation

$$\frac{pv}{T} = \text{constant} = R$$

In practice, no gas obeys this law rigidly, but many gases tend towards it.

An imaginary ideal gas which obeys this law is called a *perfect gas*, and the equation $\frac{pv}{T} = R$, is called the *characteristic equation of a state of a perfect gas*. The constant R is called the *gas constant*. Each perfect gas has a different gas constant.

Units of R are Nm/kg K or kJ/kg K.

Usually, the characteristic equation is written as

$$pv = RT \qquad ...(1.1)$$

or for m kg, occupying V m^3

$$pV = mRT \qquad ...(1.2)$$

— The characteristic equation in *another form*, can be derived by using kilogram-mole as a unit.

The *kilogram-mole* is defined as a quantity of a gas equivalent to M kg of the gas, where M is the molecular weight of the gas (*e.g.* since the molecular weight of oxygen is 32, then 1 kg mole of oxygen is equivalent to 32 kg of oxygen).

As per definition of the kilogram-mole, for m kg of a gas, we have

$$m = nM \qquad ...(1.3)$$

where n = Number of moles.

Note. Since the standard of mass is the kg, kilogram-mole will be written simply as mole.

Substituting for m from Eqn. (1.3) in Eqn. (1.2) gives

$$pV = nMRT$$

or

$$MR = \frac{pV}{nT}$$

According to *Avogadro's hypothesis* the volume of 1 mole of any gas is the same as the volume of 1 mole of any other gas, when the gases are at the same temperature and pressure. Therefore, $\frac{V}{n}$ is the same for all gases at the same value of p and T. That is the quantity $\frac{pV}{nT}$ is a *constant* for all gases. This constant is called *universal gas constant*, and is given the symbol, R_0.

i.e.,

$$MR = R_0 = \frac{pV}{nT}$$

or

$$pV = nR_0T \qquad ...(1.4)$$

Since $MR = R_0$, then

$$R = \frac{R_0}{M} \qquad ...(1.5)$$

It has been found experimentally that the volume of 1 mole of any prefect gas at 1 bar and 0°C is approximately 22.71 m^3.

Therefore from Eqn. (1.4),

$$R_0 = \frac{pV}{nT} = \frac{1 \times 10^5 \times 22.71}{1 \times 273.15}$$

$$= 8314.3 \text{ Nm/mole K}$$

Using Eqn. (1.5), the gas constant for any gas can be found when the molecular weight is known.

Example. For oxygen which has a molecular weight of 32, the gas constant

$$R = \frac{R_0}{M} = \frac{8314}{32} = 259.8 \text{ Nm/kg K}.$$

1.17.2. Specific Heats

— The specific heat of a solid or liquid is usually defined as the *heat required to raise unit mass through one degree temperature rise.*

— For small quantities, we have

$$dQ = mcdT$$

where m = Mass

c = Specific heat

dT = Temperature rise.

For a gas there are an infinite number of ways in which heat may be added between any two temperatures, and hence a *gas could have an infinite number of specific heats.* However, only two specific heats for gases are defined.

Specific heat at constant volume, c_v

and *Specific heat at constant pressure,* c_p.

We have

$dQ = m\, c_p\, dT$ For a reversible non-flow process at *constant pressure* ...(1.6)

and $dQ = m\, c_v\, dT$ For a reversible non-flow process at *constant volume* ...(1.7)

The values of c_p and c_v, for a perfect gas, are constant for any one gas at all pressures and temperatures. Hence, integrating Eqns. (1.6) and (1.7), we have

Flow of heat in a reversible constant pressure process

$$= mc_p (T_2 - T_1) \quad ...(1.8)$$

Flow of heat in a reversible constant volume process

$$= mc_v (T_2 - T_1) \quad ...(1.9)$$

In case of *real gases*, c_p and c_v *vary with temperature*, but a suitable *average value* may be used for most practical purposes.

1.17.3. Joule's Law

Joule's law states as follows :

"The internal energy of a perfect gas is a function of the absolute temperature only."

i.e., $u = f(T)$

To evaluate this function let 1 kg of a perfect gas be heated at constant volume.

According to non-flow energy equation,

$$dQ = du + dW$$

$dW = 0$, since volume remains constant

$$\therefore \quad dQ = du$$

At constant volume for a perfect gas, from Eqn. (1.7), for 1 kg

$$dQ = c_v dT$$

$$\therefore \quad dQ = du = c_v dT$$

and integrating $u = c_v\, T + K$, K being constant.

According to Joule's law $u = f(T)$, which means that internal energy varies linearly with absolute temperature. Internal energy can be made zero at any arbitrary reference temperature. For a perfect gas it can be assumed that $u = 0$ when $T = 0$, hence constant K is zero.

i.e. Internal energy, $u = c_v\, T$ for a perfect gas ...(1.10)

or For mass m, of a prefect gas

Internal energy, $U = mc_v\, T$ ...(1.11)

For a perfect gas, in any process between states 1 and 2, we have from Eqn. (1.11)

Gain in internal energy,

$$U_2 - U_1 = mc_v (T_2 - T_1) \quad ...(1.12)$$

Eqn. (1.12) gives the gains of internal energy for a perfect gas between two states *for any process, reversible* or *irreversible.*

1.17.4. Relationship Between Two Specific Heats

Consider a perfect gas being heated at constant pressure from T_1 to T_2.

According to non-flow equation,

$$Q = (U_2 - U_1) + W$$

Also for a perfect gas,

$$U_2 - U_1 = mc_v (T_2 - T_1)$$

$$Q = mc_v (T_2 - T_1) + W$$

In a constant pressure process, the work done by the fluid,

$$W = p(V_2 - V_1)$$

$$= mR(T_2 - T_1) \qquad \left[\begin{array}{l} \because \; p_1V_1 = mRT_1 \\ p_2V_2 = mRT_2 \\ p_1 = p_2 = p \text{ in this case} \end{array}\right]$$

On substituting $Q = mc_v (T_2 - T_1) + mR (T_2 - T_1) = m(c_v + R) (T_2 - T_1)$

But for a constant pressure process,

$$Q = mc_p (T_2 - T_1)$$

By equating the two expressions, we have

$$m(c_v + R)(T_2 - T_1) = mc_p(T_2 - T_1)$$

$$\therefore \quad c_v + R = c_p$$

or $$c_p - c_v = R \qquad ...(1.13)$$

Dividing both sides by c_v, we get

$$\frac{c_p}{c_v} - 1 = \frac{R}{c_v}$$

$$\therefore \quad c_v = \frac{R}{\gamma - 1} \qquad ...[1.13\ (a)]$$

(where $\gamma = c_p/c_v$)

Similarly, dividing both sides by c_p, we get

$$c_p = \frac{\gamma R}{\gamma - 1} \qquad ...[1.13\ (b)]$$

$$\left[\begin{array}{l} \text{In M.K.S. units} : c_p - c_v = \frac{R}{J}; c_v = \frac{R}{J(\gamma - 1)}, c_p = \frac{\gamma R}{(\gamma - 1)J} \\ \text{In SI units the value of } J \text{ is unity.} \end{array}\right]$$

1.17.5. Enthalpy

— One of the fundamental quantities which occur invariably in thermodynamics is the sum of internal energy (u) and pressure volume product (pv). This sum is called **Enthalpy** (h).

i.e., $$h = u + pv \qquad ...(1.14)$$

— The enthalpy of a fluid is the property of the fluid, since it consists of the sum of a property and the product of the two properties. Since enthalpy is a property like internal energy, pressure, specific volume and temperature, it can be introduced into any problem whether the process is a flow or a non-flow process.

The total enthalpy of mass, m, of a fluid can be

$$H = U + pV, \text{ where } H = mh.$$

For a *perfect gas,*

Referring equation (1.14), $h = u + pv$

$$= c_vT + RT \qquad [\because \; pv = RT]$$

$$= (c_v + R)T$$

$$= c_pT \qquad [\because \; c_p = c_v + R]$$

i.e., $$h = c_pT$$

and $$H = mc_pT.$$

(Note that, since it has been assumed that $u = 0$ at $T = 0$, then $h = 0$ at $T = 0$).

1.17.6. Ratio of Specific Heats

The ratio of specific heat at constant pressure to the specific heat at constant volume is given by the symbol γ (gamma).

i.e., $$\gamma = \frac{c_p}{c_v} \quad ...(1.15)$$

Since $c_p = c_v + R$, it is clear that c_p must be *greater* than c_v for any perfect gas. It follows, therefore, that the ratio, $\frac{c_p}{c_v} = \gamma$ is *always greater than unity.*

In general, the approximate values of γ are as follows :

For *monoatomic* gases such as *argon, helium* = 1.6.

For *diatomic* gases such as *carbon monoxide, hydrogen, nitrogen* and *oxygen* = 1.4.

For triatomic gases such as *carbondioxide* and *sulphur dioxide* = 1.3.

For some hydro-carbons the value of γ is quite low.

[*e.g.*, for ethane γ = 1.22, and for isobutane γ = 1.11]

Table 1.1. Summary of Processes for Perfect Gas (Unit mass)

Process	*Index* n	*Heat added*	$\int_1^2 pdv$	*p, v, T relations*	*Specific heat, c*
Constant pressure	$n = 0$	$c_p(T_2 - T_1)$	$p(v_2 - v_1)$	$\frac{T_2}{T_1} = \frac{v_2}{v_1}$	c_p
Constant volume	$n = \infty$	$c_v(T_2 - T_1)$	0	$\frac{T_1}{T_2} = \frac{p_1}{p_2}$	c_v
Constant temperature	$n = 1$	$p_1v_1 \log_e \frac{v_2}{v_1}$	$p_1v_1 \log_e \frac{v_2}{v_1}$	$p_1v_1 = p_2v_2$	∞
Reversible adiabatic	$n = \gamma$	0	$\frac{p_1v_1 - p_2v_2}{\gamma - 1}$	$p_1v_1^{\gamma} = p_2v_2^{\gamma}$ $\frac{T_2}{T_1} = \left(\frac{v_1}{v_2}\right)^{\gamma - 1}$ $= \left(\frac{p_2}{p_1}\right)^{\frac{\gamma - 1}{\gamma}}$	0
Polytropic	$n = n$	$c_n(T_2 - T_1)$ $= c_v\left(\frac{\gamma - n}{1 - n}\right) \times (T_2 - T_1)$ $= \frac{\gamma - n}{\gamma - 1} \times$ work done (non–flow)	$\frac{p_1v_1 - p_2v_2}{n - 1}$	$p_1v_1^{n} = p_2v_2^{n}$ $\frac{T_2}{T_1} = \left(\frac{v_1}{v_2}\right)^{n - 1}$ $= \left(\frac{p_2}{p_1}\right)^{\frac{n - 1}{n}}$	$c_n = c_v\left(\frac{\gamma - n}{1 - n}\right)$

Note. Equations must be used keeping dimensional consistence.

1.18. STEADY FLOW ENERGY EQUATION (S.F.E.E.)

In many practical problems, the rate at which the fluid flows through a machine or piece of apparatus is constant. This type of flow is called *steady flow*.

Assumptions :

The following *assumptions* are made in the system analysis :

(*i*) The mass flow through the system remains constant.

(*ii*) Fluid is uniform in composition.

(*iii*) The only interaction between the system and surroundings are work and heat.

(*iv*) The state of fluid at any point remains constant with time.

(*v*) In the analysis only potential, kinetic and flow energies are considered.

Fig. 1.11 shows a schematic flow process for an open system. An open system is one in which both mass and energy may cross the boundaries. A wide interchange of energy may take place within an open system. Let the system be an automatic engine with the inlet manifold at the first state point and exhaust pipe as the second point. There would be an interchange of chemical energy in the fuel, kinetic energy of moving particles, internal energy of gas and heat transferred and shaft work within the system. From Fig. 1.11 it is obvious that if there is no variation of flow of mass or energy with time across the boundaries of the system the steady flow will prevail. The conditions may pass through the cyclic or non-cyclic changes within the system. As a result the mass entering the system equals the mass leaving, also energy entering the system equals energy leaving.

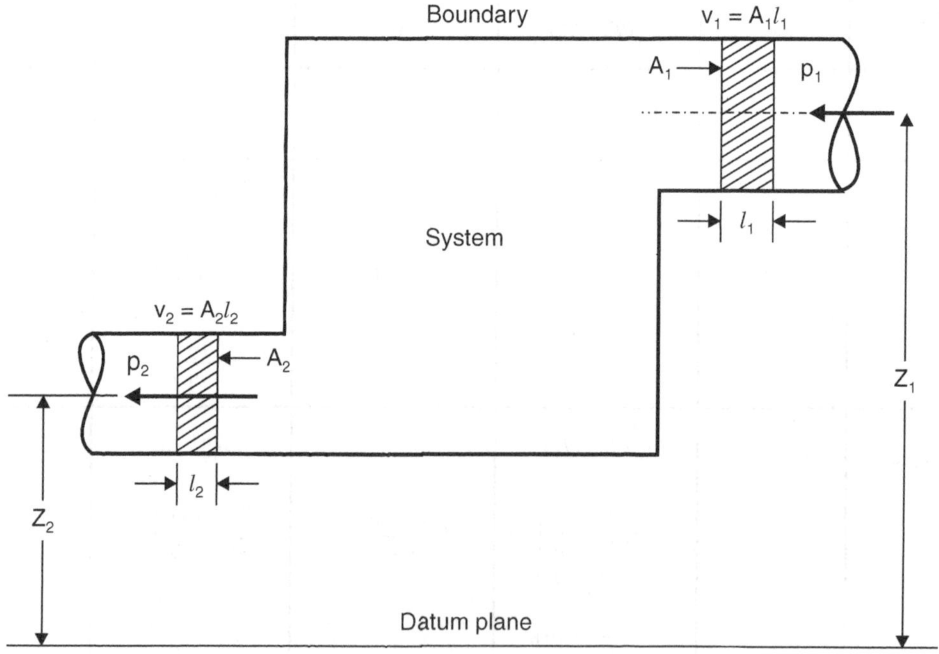

Fig. 1.11

The steady flow equation can be expressed as follows :

$$u_1 + \frac{C_1^2}{2} + Z_1 g + p_1 v_1 + Q = u_2 + \frac{C_2^2}{2} + Z_2 g + p_2 v_2 + W \quad ...(1.16)$$

$$(u_1 + p_1 v_1) + \frac{C_1^2}{2} + Z_1 g + Q = (u_2 + p_2 v_2) + \frac{C_2^2}{2} + Z_2 g + W$$

$$h_1 + \frac{C_1^2}{2} + Z_1 g + Q = h_2 + \frac{C_2^2}{2} + Z_2 g + W \quad [\because h = u + pv]$$

If Z_1 and Z_2 are neglected, we get

$$h_1 + \frac{C_1^2}{2} + Q = h_2 + \frac{C_2^2}{2} + W \quad ...[1.16\ (a)]$$

where Q = Heat supplied (or entering the boundary) per kg of fluid ;

W = Work done by (or work coming out of the boundary) 1 kg of fluid ;

C = Velocity of fluid ;

Z = Height above datum ;

p = Pressure of the fluid ;

u = Internal energy per kg of fluid ;

pv = Energy required for 1 kg of fluid.

This equation is applicable to any medium in any steady flow. It is applicable not only to rotary machines such as centrifugal fans, pumps and compressors but also to reciprocating machines such as steam engines.

In a steady flow the rate of mass flow of fluid at any section is the same as at any other section. Consider any section of cross-sectional area A, where the fluid velocity is C, the rate of volume flow past the section is CA. Also, since mass flow is volume flow divided by specific volume,

$$\text{Mass flow rate, } \dot{m} = \frac{CA}{v} \quad ...(1.17)$$

(where v = *specific volume* at the section)

This equation is known as the **continuity of mass equation**.

With reference to Fig. 1.11.

$$\therefore \quad \dot{m} = \frac{C_1 A_1}{v_1} = \frac{C_2 A_2}{v_2} \quad ...[1.17\ (a)]$$

1.18.1. Energy Relations for Flow Process

The energy equation (m kg of fluid) for a steady flow system is given as follows :

$$m\left(u_1 + \frac{C_1^2}{2} + Z_1 g + p_1 v_1\right) + Q = m\left(u_2 + \frac{C_2^2}{2} + Z_2 g + p_2 v_2\right) + W$$

i.e.,

$$Q = m\left[(u_2 - u_1) + (Z_2 g - Z_1 g) + \left(\frac{C_2^2}{2} - \frac{C_1^2}{2}\right) + (p_2 v_2 - p_1 v_1)\right] + W$$

i.e.,

$$Q = m\left[(u_2 - u_1) + g(Z_2 - Z_1) + \left(\frac{C_2^2 - C_1^2}{2}\right) + (p_2 v_2 - p_1 v_1)\right] + W$$

$$= \Delta U + \Delta PE + \Delta KE + \Delta (pv) + W$$

where $\Delta U = m\,(u_2 - u_1)$

$$\Delta PE = mg\,(Z_2 - Z_1)$$

$$\Delta KE = m\left(\frac{C_2^2 - C_1^2}{2}\right)$$

$$\Delta pv = m(p_2v_2 - p_1v_1)$$

$$\therefore \quad Q - \Delta U = [\Delta PE + \Delta KE + \Delta(pv) + W] \quad ...(1.18)$$

For non-flow process, $Q = \Delta U + W = \Delta U + \int_1^2 pdV$

i.e.,
$$Q - \Delta U = \int_1^2 pdV \quad ...(1.19)$$

1.19. LIMITATIONS OF FIRST LAW OF THERMODYNAMICS

It has been observed that *energy can flow* from a system in the form of *heat* or *work*. The first law of thermodynamics sets no limit to the amount of the total energy of a system which can be caused to flow out as work. A limit is imposed, however, as a result of the principle enunciated in the second law of thermodynamics which states that heat will flow naturally from one energy reservoir to another at a lower temperature, but not in opposite direction without assistance. This is very important because a heat engine operates between two energy reservoirs at different temperatures.

Further the first law of thermodynamics *establishes equivalence between the quantity of heat used and the mechanical work but does not specify the conditions under which conversion of heat into work is possible, neither the direction in which heat transfer can take place.* This gap has been *bridged* by the second law of thermodynamics.

1.20. PERFORMANCE OF HEAT ENGINE AND REVERSED HEAT ENGINE

Refer Fig. 1.12 (*a*). A *heat engine* is used to produce the maximum work transfer from a given positive heat transfer. The measure of success is called the *thermal efficiency* of the engine and is defined by the ratio :

$$\text{Thermal efficiency, } \eta_{th} = \frac{W}{Q_1} \quad ...(1.20)$$

where W = Net work transfer from the engine, and

Q_1 = Heat transfer to engine.

For a *reversed heat engine* [Fig. 1.12 (*b*)] acting as a *refrigerator* when the purpose is to achieve the maximum heat transfer from the cold reservoir, the measure of success is called the co-efficient of performance (C.O.P.). It is defined by the ratio :

$$\text{Co-efficient of performance, } (\text{C.O.P.})_{ref.} = \frac{Q_2}{W} \quad ...(1.21)$$

where Q_2 = Heat transfer *from cold reservoir*

W = The net work transfer to the refrigerator.

For a **reversed heat engine** [Fig. 1.12 (*b*)] acting as a *heat pump*, the measure of success is again called the *co-efficient of performance*. It is defined by the ratio :

Co-efficient of performance, $(\text{C.O.P.})_{heat\ pump} = \dfrac{Q_1}{W}$ ...(1.22)

where Q_1 = Heat transfer *to hot reservoir*

W = Net work transfer to the heat pump.

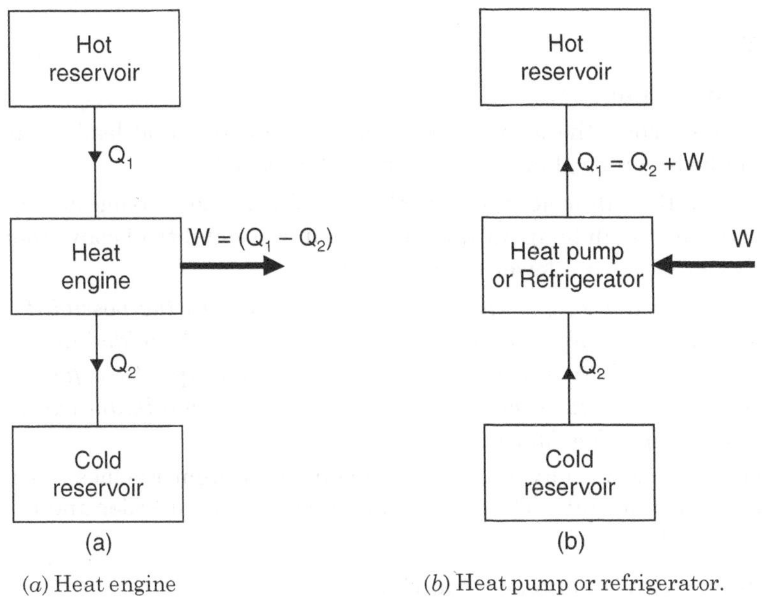

(*a*) Heat engine (*b*) Heat pump or refrigerator.

Fig. 1.12

In all the above three cases application of the first law gives the relation $Q_1 - Q_2 = W$, and this can be used to rewrite the expressions for thermal efficiency and co-efficient of performance solely in terms of the heat transfers.

$$\eta_{th} = \frac{Q_1 - Q_2}{Q_1} \quad ...(1.23)$$

$$(C.O.P.)_{ref} = \frac{Q_2}{Q_1 - Q_2} \quad ...(1.24)$$

$$(C.O.P.)_{heat\ pump} = \frac{Q_1}{Q_1 - Q_2} \quad ...(1.25)$$

It may be seen that η_{th} is *always less than unity* and $(C.O.P.)_{heat\ pump}$ is *always greater than unity*.

1.21. STATEMENTS OF SECOND LAW OF THERMODYNAMICS

The second law of thermodynamics has been enunciated meticulously by Clausius, Kelvin and Planck in slightly different words although both statements are basically identical. Each statement is based on an *irreversible process*. The *first considers transformation of heat between two thermal reservoirs* while the *second considers the transformation of heat into work.*

1.21.1. Clausius Statement

"It is impossible for a self acting machine working in a cyclic process unaided by any external agency, to convey heat from a body at a lower temperature to a body at a higher temperature".

In other words, heat of, itself, cannot flow from a colder to a hotter body.

1.21.2. Kelvin-Planck Statement

"It is impossible to construct an engine, which while operating in a cycle produces no other effect except to extract heat from a single reservoir and do equivalent amount of work".

Although the Clausius and Kelvin-Planck statements appear to be different, they are really equivalent in the sense that a *violation of either statement implies violation of other.*

1.22. ENTROPY

1.22.1. Introduction

In heat engine theory, the term entropy plays a vital role and leads to important results which by other methods can be obtained much more laboriously.

It may be noted that all heat is not equally valuable for converting into work. Heat that is supplied to a substance at high temperature has a greater possibility of conversion into work than heat supplied to a substance at a lower temperature.

*"**Entropy** is a function of a quantity of heat which shows the possibility of conversion of that heat into work. The increase in entropy is small when heat is added at a high temperature and is greater when heat addition is made at a lower temperature. Thus for maximum entropy, there is minimum availability for conversion into work and for minimum entropy there is maximum availability for conversion into work."*

The entropy attains its maximum value when the system reaches a stable equilibrium state from a non-equilibrium state. This is the state of maximum disorder and is one of maximum thermodynamic probability.

1.22.2. Temperature-Entropy Diagram

If entropy is plotted horizontally and absolute temperature vertically, the diagram so obtained is called *temperature entropy* (*T-s*) diagram. Such a diagram is shown in Fig. 1.13. If working

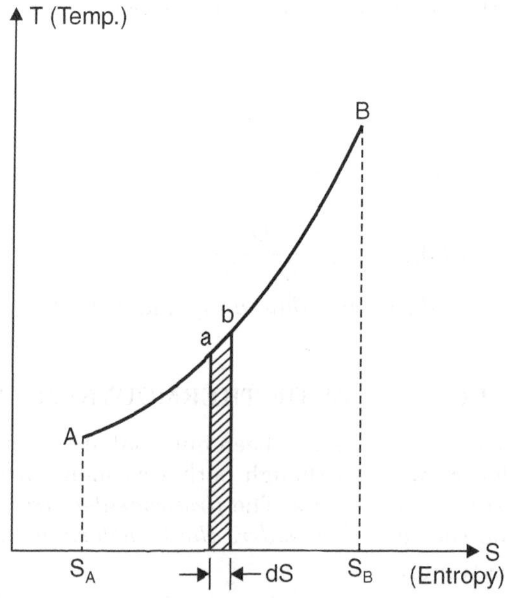

Fig. 1.13. Temperature-entropy diagram.

fluid receives a small amount of heat dQ in an elementary portion ab of an operation AB when temperature is T, and if dQ is represented by the shaded area of which T is the mean ordinate, the width of the figure must be $\frac{dQ}{T}$. This is called *'increment of entropy'* and is denoted by dS. The total heat received by the operation will be given by the area under the curve AB and $(S_B - S_A)$ will be corresponding increase of entropy.

From above we conclude that :

Entropy change, $$dS = \frac{\text{Heat Change }(Q)}{\text{Absolute temperature }(T)}.$$

"***Entropy** may also be defined as the thermal property of a substance which remains constant when substance is expanded or compressed adiabatically in a cylinder*".

Note. 's' stands for specific entropy whereas 'S' means total entropy (*i.e.*, $S = ms$).

1.22.3. Characteristics of Entropy

The *characteristics of entropy* in a summarised form are given below :

1. It increases when heat is supplied irrespective of the fact whether temperature changes or not.
2. It decreases when heat is removed whether temperature changes or not.
3. It remains unchanged in all adiabatic frictionless processes.
4. It increases if temperature of heat is lowered without work being done as in a throttling process.

Table 1.2. Summary of Formulae

S. No.	*Process*	*Change of entropy (per kg)*
1.	General case	(*i*) $c_v \log_e \frac{T_2}{T_1} + R \log_e \frac{v_2}{v_1}$ (in terms of T and v)
		(*ii*) $c_v \log_e \frac{p_2}{p_1} + c_v \log_e \frac{v_2}{v_1}$ (in terms of p and v)
		(*iii*) $c_p \log_e \frac{T_2}{T_1} - R \log_e \frac{p_2}{p_1}$ (in terms of T and p)
2.	Constant volume	$c_v \log_e \frac{T_2}{T_1}$
3.	Constant pressure	$c_p \log_e \frac{T_2}{T_1}$
4.	Isothermal	$R \log_e \frac{v_2}{v_1}$
5.	Adiabatic	Zero
6.	Polytropic	$c_v \left(\frac{n-\gamma}{n-1}\right) \log_e \frac{T_2}{T_1}$

1.23. THE THIRD LAW OF THERMODYNAMICS

The third law of thermodynamics is stated as follows :

"The entropy of all perfect crystalline solids is zero at absolute zero temperature".

The third law of thermodynamics, often referred to as **Nernst law,** provides the basis for the calculations of absolute entropies of substances.

According to this law, if the entropy is zero at $T = 0$, the absolute entropy $s_{ab.}$ of a substance at any temperature T and pressure p is expressed by the expression

$$s_{ab.} = \int_0^{T_s = T_{f_1}} c_{ps} \frac{dT}{T} + \frac{h_{sf}}{T_s} + \int_{T_s}^{T_{f_2} = T_g} c_{pf} \frac{dT}{T} + \frac{h_{fg}}{T_g} \int_{T_g}^{T} c_{pg} \frac{dT}{T} \quad ...(1.26)$$

where $T_s = T_{f_1} = T_{sf} = T_{sat}$ ...for fusion,

$T_{f_2} = T_g = T_{fg} = T_{sat}$ ...for vaporisation,

c_{ps}, c_{pf}, c_{pg} = Constant pressure specific heats for solids, liquids and gas, and

h_{sf}, h_{fg} = Latent heats of fusion and vaporisation.

Thus by putting $s = 0$ at $T = 0$, one may integrate zero kelvin and standard state of 298.15 K and 1 atm., and find the entropy difference.

Further, it can be shown that the entropy of a crystalline substance at $T = 0$ is not a function of pressure, *viz.*, $\left(\frac{\partial s}{\partial p}\right)_{T=0} = 0$.

However, at temperature above absolute zero, the entropy is a function of pressure also.

1.24. AVAILABLE AND UNAVAILABLE ENERGY

There are many forms in which an energy can exist. But even under ideal conditions all these forms cannot be converted completely into work. This indicates that energy has two parts :

— Available part.

— Unavailable part.

'Available energy' is the *maximum portion of energy which could be converted into useful work by ideal processes which reduce the system to a dead state* (a state in equilibrium with the earth and its atmosphere). Because there can be only one value for maximum work which the system alone could do while descending to its dead state, it follows immediately that *'Available energy' is a property.*

A system which has a pressure difference from that of surroundings, work can be obtained from an expansion process, and if the system has a different temperature, heat can be transferred to a cycle and work can be obtained. But when the temperature and pressure becomes equal to that of the earth, transfer of energy ceases, and although the system contains internal energy, this energy is *unavailable.*

Summarily available energy denote, the latent capability of energy to do work, and in this sense it can be applied to energy in the system or in the surroundings.

The theoretical maximum amount of work which can be obtained from a system at any state p_1 and T_1 when operating with a reservoir at the constant pressure and temperature p_0 and T_0 is called **'availability'**.

HIGHLIGHTS

1. *Thermodynamics* is an axiomatic science which deals with the relations among heat, work and properties of systems which are in equilibrium. It basically entails four laws or axioms known as *Zeroth, First, Second* and *Third* law of thermodynamics.
2. A *system* is a finite quantity of matter or a prescribed region of space.

   A system may be a *closed, open* or *isolated* system.
3. A *phase* is a quantity of matter which is homogeneous throughout in chemical composition and physical structure.
4. A *homogeneous system* is one which consists of a *single phase.*
5. A *heterogeneous system* is one which consists of *two or more phases.*
6. A *pure substance* is one that has a homogeneous and invariable chemical composition even though there is a change of phase.
7. A system is in *thermodynamic equilibrium* if temperature and pressure at all points are same ; there should be no *velocity gradient.*
8. A *property of a system* is a characteristic of the system which depends upon its state, but not upon how the state is reached.

   Intensive properties do not depend on the mass of the system.

   Extensive properties depend on the mass of the system.
9. *State* is the condition of the system at an instant of time as described or measured by its properties. Or each unique condition of a system is called a state.
10. A *process* occurs when the system undergoes a change in state or an energy transfer takes place at a steady state.
11. Any process or series of processes whose end states are identical is termed a *cycle.*
12. The *pressure* of a system is the force exerted by the system on unit area of boundaries. Vacuum is defined as the absence of pressure.
13. A *reversible process* is one which can be stopped at any stage and reversed so that the system and surroundings are exactly restored to their initial states.

    An *irreversible process* is one in which heat is transferred through a finite temperature.
14. Zeroth law of termodynamics states that if two systems are each equal in temperature to a third, they are equal in temperature to each other.
15. Infinite slowness is the characteristic feature of a quasi-static process. A quasi-static process is a succession of equilibrium states. It is also called a reversible process.
16. *Internal energy* is the heat energy stored in a gas. The internal energy of a perfect gas is a function of *temperature* only.
17. First law of thermodynamics states :

    — Heat and work are mutually convertible but since energy can neither be created nor destroyed, the total energy associated with an energy conversion remains constant.

    Or

    — No machine can produce energy without corresponding expenditure of energy, *i.e.* it is impossible to construct a perpetual motion machine of first kind.

    First law can be expressed as follows :

    $$Q = \Delta E + W$$

    $$Q = \Delta U + W$$ ... if electric, magnetic, chemical energies are absent and changes in potential and kinetic energies are neglected.

18. There can be no machine which would continuously supply mechanical work without some form of energy disappearing simultaneously. Such a fictitious machine is called a perpetual motion machine of the first kind, or in brief, PMM1. A PMM1 is thus impossible.

19. The energy of an isolated system is always constant.

20. In case of

(*i*) **Reversible constant volume process** (v = constant)

$$\Delta u = c_v(T_2 - T_1)\ ;\ W = 0\ ;\ Q = c_v\,(T_2 - T_1)$$

(*ii*) **Reversible constant pressure process** (p = constant)

$$\Delta u = c_v(T_2 - T_1)\ ;\ W = p(v_2 - v_1)\ ;\ Q = c_p\,(T_2 - T_1)$$

(*iii*) **Reversible temperature or isothermal process** (pv = constant)

$$\Delta u = 0,\ W = p_1 V_1 \log_e r,\ Q = W$$

where r = expansion or compression ratio.

(*iv*) **Reversible adiabatic process** (pv^γ = constant)

$$\pm\,\Delta u = \mp\ W = \frac{R(T_1 - T_2)}{\gamma - 1}\ ;\ Q = 0\ ;\ \frac{T_2}{T_1} = \left(\frac{v_1}{v_2}\right)^{\gamma - 1} = \left(\frac{p_2}{p_1}\right)^{\frac{\gamma - 1}{\gamma}}$$

(*v*) **Polytropic reversible process** (pv^n = constant)

$$\Delta u = c_v\,(T_2 - T_1)\ ;\ W = \frac{R(T_1 - T_2)}{n - 1}\ ;\ Q = \Delta u + W\ ;$$

and $$\frac{T_2}{T_1} = \left(\frac{v_1}{v_2}\right)^{n - 1} = \left(\frac{p_2}{p_1}\right)^{\frac{n - 1}{n}}$$ and $$Q = \left(\frac{\gamma - n}{\gamma - 1}\right) \times W.$$

21. Steady flow equation can be expressed as follows :

$$u_1 + \frac{C_1^2}{2} + Z_1 g + p_1 v_1 + Q = u_2 + \frac{C_2^2}{2} + Z_2 g + p_2 v_2 + W \quad ...(i)$$

or $$h_1 + \frac{C_1^2}{2} + Q = h_2 + \frac{C_2^2}{2} + W, \text{ neglecting } Z_1 \text{ and } Z_2 \quad ...(ii)$$

where Q = Heat supplied per kg of fluid ; W = Work done by 1 kg of fluid ;
C = Velocity of fluid ; Z = Height above datum ;
p = Pressure of the fluid ; u = Internal energy per kg of fluid ;
pv = Energy required per kg of fluid.

This equation is applicable to any medium in any steady flow.

22. Clausius statement :

"It is impossible for a self-acting machine working in a cyclic process, unaided by any external agency, to convey heat from a body at a lower temperature to a body at a higher temperature."

Kelvin-Planck statement :

"It is impossible to construct an engine, which while operating in a cycle produces no other effect except to extract heat from a single reservoir and do equivalent amount of work".

Although above statements of second law of thermodynamics appear to be different, they are really equivalent in the sense that violation of either statement implies violation of other.

23. Clausius inequality is given by,

$$\sum_{\text{Cycle}} \left(\frac{\delta Q}{T}\right) \le 0$$

"When a system performs a reversible cycle, then

$$\sum_{\text{Cycle}} \left(\frac{\delta Q}{T}\right) = 0,$$

but when the cycle is not reversible

$$\sum_{\text{Cycle}} \left(\frac{\delta Q}{T}\right) < 0."$$

24. **'Entropy'** is a function of a quantity of heat which shows the possibility of conversion of that heat into work. The increase in entropy is small when heat is added at a high temperature and is greater when heat addition is made at lower temperature. Thus for maximum entropy, there is a minimum availability for conversion into work and for minimum entropy there is maximum availability for conversion into work.
25. **The third law of thermodynamics** is stated as follows :
"The entropy of all perfect crystalline solids is zero at absolute zero temperature".

OBJECTIVE TYPE QUESTIONS

Choose the correct answer :

1. A definite area or space where some thermodynamic process takes place is known as
(*a*) thermodynamic system (*b*) thermodynamic cycle
(*c*) thermodynamic process (*d*) thermodynamic law.
2. An open system is one in which
(*a*) heat and work cross the boundary of the system, but the mass of the working substance does not
(*b*) mass of working substance crosses the boundary of the system but the heat and work do not
(*c*) both the heat and work as well as mass of the working substances cross the boundary of the system
(*d*) neither the heat and work nor the mass of the working substances cross the boundary of the system.
3. An isolated system
(*a*) is a specified region where transfer of energy and/or mass take place
(*b*) is a region of constant mass and only energy is allowed to cross the boundaries
(*c*) cannot transfer either energy or mass to or from the surroundings
(*d*) is one in which mass within the system is not necessarily constant
(*e*) none of the above.
4. In an extensive property of a thermodynamic system
(*a*) extensive heat is transferred (*b*) extensive work is done
(*c*) extensive energy is utilised (*d*) all of the above
(*e*) none of the above.
5. Which of the following is an intensive property of a thermodynamic system ?
(*a*) Volume (*b*) Temperature
(*c*) Mass (*d*) Energy.
6. Which of the following is the extensive property of a thermodynamic system ?
(*a*) Pressure (*b*) Volume
(*c*) Temperature (*d*) Density.
7. When two bodies are in thermal equilibrium with a third body they are also in thermal equilibrium with each other. This statement is called
(*a*) Zeroth law of thermodynamics (*b*) First law of thermodynamics
(*c*) Second law of thermodynamics (*d*) Kelvin-Planck's law.

8. The temperature at which the volume of a gas becomes zero is called
   (*a*) absolute scale of temperature (*b*) absolute zero temperature
   (*c*) absolute temperature (*d*) none of the above.
9. The value of one bar (in SI units) is equal to
   (*a*) 100 N/m^2 (*b*) 1000 N/m^2 (*c*) 1×10^4 N/m^2
   (*d*) 1×10^5 N/m^2 (*e*) 1×10^6 N/m^2.
10. The absolute zero pressure will be
   (*a*) when molecular momentum of the system becomes zero
   (*b*) at sea level (*c*) at the temperature of – 273 K
   (*d*) under vacuum conditions (*e*) at the centre of the earth.
11. Absolute zero temperature is taken as
   (*a*) – 273°C (*b*) 273°C
   (*c*) 237°C (*d*) – 373°C.
12. Which of the following is correct ?
   (*a*) Absolute pressure = gauge pressure + atmospheric pressure
   (*b*) Gauge pressure = absolute pressure + atmospheric pressure
   (*c*) Atmospheric pressure = absolute pressure + gauge pressure
   (*d*) Absolute pressure = gauge pressure – atmospheric pressure.
13. The unit of energy in SI units is
   (*a*) Joule (J) (*b*) Joule metre (Jm)
   (*c*) Watt (W) (*d*) Joule/metre (J/m).
14. One watt is equal to
   (*a*) 1 Nm/s (*b*) 1 N/min (*c*) 10 N/s
   (*d*) 100 Nm/s (*e*) 100 Nm/m.
15. One joule (J) is equal to
   (*a*) 1 Nm (*b*) kNm
   (*c*) 10 Nm/s (*d*) 10 kNm/s.
16. The amount of heat required to raise the temperature of 1 kg of water through 1°C is called
   (*a*) specific heat at constant volume (*b*) specific heat at constant pressure
   (*c*) kilo calorie (*d*) none of the above.
17. The heating and expanding of a gas is called
   (*a*) thermodynamic system (*b*) thermodynamic cycle
   (*c*) thermodynamic process (*d*) thermodynamic law.
18. A series of operations, which take place in a certain order and restore the initial condition is known as
   (*a*) reversible cycle (*b*) irreversible cycle
   (*c*) thermodynamic cycle (*d*) none of the above.
19. The condition for the reversibility of a cycle is
   (*a*) the pressure and temperature of the working substance must not differ, appreciably, from those of the surroundings at any stage in the process
   (*b*) all the processes, taking place in the cycle of operation, must be extremely slow
   (*c*) the working parts of the engine must be friction free
   (*d*) there should be no loss of energy during the cycle of operation
   (*e*) all of the above (*f*) none of the above.
20. In an irreversible process, there is a
   (*a*) loss of heat (*b*) no loss of heat
   (*c*) gain of heat (*d*) no gain of heat.

21. The main cause of the irreversibility is
(*a*) mechanical and fluid friction (*b*) unrestricted expansion
(*c*) heat transfer with a finite temperature difference
(*d*) all of the above (*e*) none of the above.

22. According to kinetic theory of heat
(*a*) temperature should rise during boiling (*b*) temperature should fall during freezing
(*c*) at low temperature all bodies are in solid state
(*d*) at absolute zero there is absolutely no vibration of molecules
(*e*) none of the above.

23. A system comprising a single phase is called a
(*a*) closed system (*b*) open system (*c*) isolated system
(*d*) homogeneous system (*e*) heterogeneous system.

24. If all the variables of a stream are independent of time it is said to be in
(*a*) steady flow (*b*) unsteady flow (*c*) uniform flow
(*d*) closed flow (*e*) constant flow.

25. A control volume refers to
(*a*) a fixed region in space (*b*) a specified mass
(*c*) an isolated system (*d*) a reversible process only
(*e*) a closed system.

26. Internal energy of a perfect gas depends on
(*a*) temperature, specific heats and pressure (*b*) temperature, specific heats and enthalpy
(*c*) temperature, specific heats and entropy (*d*) temperature only.

27. In reversible polytropic process
(*a*) true heat transfer occurs (*b*) the entropy remains constant
(*c*) the enthalpy remains constant (*d*) the internal energy remains constant
(*e*) the temperature remains constant.

28. An isentropic process is always
(*a*) irreversible and adiabatic (*b*) reversible and isothermal
(*c*) frictionless and irreversible (*d*) reversible and adiabatic
(*e*) none of the above.

29. The net work done per kg of gas in a polytropic process is equal to

(*a*) $p_1 v_1 \log_e \frac{v_2}{v_1}$ (*b*) $p_1 (v_1 - v_2)$ (*c*) $p_2 \left(v_2 - \frac{v_1}{v_2} \right)$

(*d*) $\frac{p_1 v_1 - p_2 v_2}{n - 1}$ (*e*) $\frac{p_2 v_1 - p_2 v_2}{n - 1}$.

30. Steady flow occurs when
(*a*) conditions do not change with time at any point
(*b*) conditions are the same at adjacent points at any instant
(*c*) conditions change steadily with the time
(*d*) $\left(\frac{\partial v}{\partial t} \right)$ is constant.

31. A reversible process requires that
(*a*) there be no heat transfer (*b*) newton's law of viscosity be satisfied
(*c*) temperature of system and surroundings be equal
(*d*) there be no viscous or coloumb friction in the system
(*e*) heat transfer occurs from surroundings to system only.

32. The first law of thermodynamics for steady flow
(*a*) accounts for all energy entering and leaving a control volume
(*b*) is an energy balance for a specified mass of fluid
(*c*) is an expression of the conservation of linear momentum
(*d*) is primarily concerned with heat transfer.
(*e*) is restricted in its application to perfect gases.

33. The characteristic equation of gases $pV = mRT$ holds good for
(*a*) monoatomic gases (*b*) diatomic gas (*c*) real gases
(*d*) ideal gases (*e*) mixture of gases.

34. A gas which obeys kinetic theory perfectly is known as
(*a*) monoatomic gas (*b*) diatomic gas (*c*) real gas
(*d*) pure gas (*e*) perfect gas.

35. Work done in a free expansion process is
(*a*) zero (*b*) minimum (*c*) maximum
(*d*) positive (*e*) negative.

36. Which of the following is not a property of the system ?
(*a*) Temperature (*b*) Pressure (*c*) Specific volume
(*d*) Heat (*e*) None of the above.

37. In the polytropic process equation pv^n = constant, if $n = 0$, the process is termed as
(*a*) constant volume (*b*) constant pressure (*c*) constant temperature
(*d*) adiabatic (*e*) isothermal.

38. In the polytropic process equation pv^n = constant, if n is infinitely large, the process is termed as
(*a*) constant volume (*b*) constant pressure (*c*) constant temperature
(*d*) adiabatic (*e*) isothermal.

39. The processes or systems that do not involve heat are called
(*a*) isothermal processes (*b*) equilibrium processes (*c*) thermal processes
(*d*) steady processes (*e*) adiabatic processes.

40. In a reversible adiabatic process the ratio (T_1/T_2) is equal to

(*a*) $\left(\frac{p_1}{p_2}\right)^{\frac{\gamma-1}{\gamma}}$ (*b*) $\left(\frac{v_1}{v_2}\right)^{\frac{\gamma-1}{\gamma}}$

(*c*) $(v_1 v_2)^{\frac{\gamma-1}{2\gamma}}$. (*d*) $\left(\frac{v_2}{v_1}\right)^{\gamma}$.

41. In isothermal process
(*a*) temperature increases gradually (*b*) volume remains constant
(*c*) pressure remains constant (*d*) enthalpy change is maximum
(*e*) change in internal energy is zero.

42. During throttling process
(*a*) internal energy does not change (*b*) pressure does not change
(*c*) entropy does not change (*d*) enthalpy does not change
(*e*) volume change is negligible.

43. When a gas is to be stored, the type of compression that would be ideal is
(*a*) isothermal (*b*) adiabatic (*c*) polytropic
(*d*) constant volume (*e*) none of the above.

44. If a process can be stopped at any stage and reversed so that the system and surroundings are exactly restored to their initial states, it is known as
(*a*) adiabatic process (*b*) isothermal process (*c*) ideal process
(*d*) frictionless process (*e*) energyless process.

45. The state of a substance whose evaporation from its liquid state is complete, is known as
(*a*) vapour (*b*) perfect gas
(*c*) air (*d*) steam.

46. In SI units, the value of the universal gas constant is
(*a*) 0.8314 J/mole/K (*b*) 8.314 J/mole/K
(*c*) 83.14 J/mole/K (*d*) 831.4 J/mole/K
(*e*) 8314 J/mole/K.

47. When the gas is heated at constant pressure, the heat supplied
(*a*) increases the internal energy of the gas (*b*) increases the temperature of the gas
(*c*) does some external work during expansion (*d*) both (*b*) and (*c*)
(*e*) none of the above.

48. The gas constant (R) is equal to the
(*a*) sum of two specific heats (*b*) difference of two specific heats
(*c*) product of two specific heats (*d*) ratio of two specific heats.

49. The heat absorbed or rejected during a polytropic process is
(*a*) $\left(\frac{\gamma - n}{\gamma - 1}\right) \times$ work done (*b*) $\left(\frac{\gamma - n}{\gamma - 1}\right)^2 \times$ work done
(*c*) $\left(\frac{\gamma - n}{\gamma - 1}\right)^{1/2} \times$ work done (*d*) $\left(\frac{\gamma - n}{\gamma - 1}\right)^3 \times$ work done.

50. Second law of thermodynamics defines
(*a*) heat (*b*) work (*c*) enthalpy
(*d*) entropy (*e*) internal energy.

51. For a reversible adiabatic process, the change in entropy is
(*a*) zero (*b*) minimum (*c*) maximum
(*d*) infinite (*e*) unity.

52. For any reversible process, the change in entropy of the system and surroundings is
(*a*) zero (*b*) unity (*c*) negative
(*d*) positive (*e*) infinite.

53. For any irreversible process the net entropy change is
(*a*) zero (*b*) positive (*c*) negative
(*d*) infinite (*e*) unity.

54. The processes of a Carnot cycle are
(*a*) two adiabatic and two constant volume
(*b*) one constant volume and one constant pressure and two isentropics
(*c*) two adiabatics and two isothermals (*d*) two constant volumes and two isothermals
(*e*) two isothermals and two isentropics.

55. Isentropic flow is
(*a*) irreversible adiabatic flow (*b*) ideal fluid flow (*c*) perfect gas flow
(*d*) frictionless reversible flow (*e*) reversible adiabatic flow.

56. In a Carnot engine, when the working substance gives heat to the sink
(*a*) the temperature of the sink increases (*b*) the temperature of the sink remains the same
(*c*) the temperature of the source decreases
(*d*) the temperatures of both the sink and the source decrease
(*e*) changes depend on the operating conditions.

57. If the temperature of the source is increased, the efficiency of the Carnot engine

(*a*) decreases
(*b*) increases
(*c*) does not change
(*d*) will be equal to the efficiency of a practical engine
(*e*) depends on other factors.

58. The efficiency of an ideal Carnot engine depends on

(*a*) working substance
(*b*) on the temperature of the source only
(*c*) on the temperature of the sink only
(*d*) on the temperatures of both the source and the sink
(*e*) on the construction of engine.

59. The efficiency of a Carnot engine using an ideal gas as the working substance is

(*a*) $\frac{T_1 - T_2}{T_1}$
(*b*) $\frac{T_1}{T_1 - T_2}$
(*c*) $\frac{T_1 T_2}{T_1 - T_2}$
(*d*) $\frac{T_1 - T_2}{T_1 T_2}$
(*e*) $\frac{T_2(T_1 - T_2)}{T_1(T_1 + T_2)}$.

60. In a reversible cycle, the entropy of the system

(*a*) increases
(*b*) decreases
(*c*) does not change
(*d*) first increases and then decreases
(*e*) depends on the properties of working substance.

61. A frictionless heat engine can be 100% efficient only if its exhaust temperature is

(*a*) equal to its input temperature
(*b*) less than its input temperature
(*c*) 0°C
(*d*) 0°K
(*e*) – 100°C.

62. Kelvin-Planck's law deals with

(*a*) conservation of energy
(*b*) conservation of heat
(*c*) conservation of mass
(*d*) conversion of heat into work
(*e*) conversion of work into heat.

63. Which of the following statements is correct according to Clausius statement of second law of thermodynamics ?

(*a*) It is impossible to transfer heat from a body at a lower temperature to a body at a higher temperature
(*b*) It is impossible to transfer heat from a body at a lower temperature to a body at a higher temperature, without the aid of an external source.
(*c*) It is possible to transfer heat from a body at a lower temperature to a body at a higher temperature by using refrigeration cycle
(*d*) None of the above.

64. According to Kelvin-Planck's statement of second law of thermodynamics

(*a*) It is impossible to construct an engine working on a cyclic process, whose sole purpose is to convert heat energy into work
(*b*) It is possible to construct an engine working on a cyclic process, whose sole purpose is to convert the heat energy into work
(*c*) It is impossible to construct a device which while working in a cyclic process produces no effect other than the transfer of heat from a colder body to a hotter body
(*d*) When two dissimilar metals are heated at one end and cooled at the other, the e.m.f. developed is proportional to the difference of their temperatures at the two end.
(*e*) None of the above.

65. The property of a working substance which increases or decreases as the heat is supplied or removed in a reversible manner is known as

(*a*) enthalpy
(*b*) internal energy
(*c*) entropy
(*d*) external energy.

66. The entropy may be expressed as a function of

(*a*) pressure and temperature (*b*) temperature and volume

(*c*) heat and work (*d*) all of the above

(*e*) none of the above.

67. The change of entropy, when heat is absorbed by the gas is

(*a*) positive (*b*) negative (*c*) positive or negative.

68. Which of the following statements is *correct* ?

(*a*) The increase in entropy is obtained from a given quantity of heat at a low temperature

(*b*) The change in entropy may be regarded as a measure of the rate of the availability of heat for transformation into work

(*c*) The entropy represents the maximum amount of work obtainable per degree drop in temperature

(*d*) All of the above.

69. The condition for the reversibility of a cycle is

(*a*) the pressure and temperature of working substance must not differ, appreciably from those of the surroundings at any stage in the process

(*b*) all the processes taking place in the cycle of operation, must be extremely slow

(*c*) the working parts of the engine must be friction free

(*d*) there should be no loss of energy during the cycle of operation

(*e*) all of the above.

70. In an irreversible process there is a

(*a*) loss of heat (*b*) no loss of work

(*c*) gain of heat (*d*) no gain of heat.

71. The main cause for the irreversibility is

(*a*) mechanical and fluid friction (*b*) unrestricted expansion

(*c*) heat transfer with a finite temperature difference

(*d*) all of the above.

72. The efficiency of the Carnot cycle may be increased by

(*a*) increasing the highest temperature (*b*) decreasing the highest temperature

(*c*) increasing the lowest temperature (*d*) decreasing the lowest temperature

(*e*) keeping the lowest temperature constant.

73. Which of the following is the *correct* statement ?

(*a*) All the reversible engines have the same efficiency

(*b*) All the reversible and irreversible engines have the same efficiency

(*c*) Irreversible engines have maximum efficiency

(*d*) All engines are designed as reversible in order to obtain maximum efficiency.

ANSWERS

1. (*a*)	**2.** (*c*)	**3.** (*c*)	**4.** (*e*)	**5.** (*b*)	**6.** (*b*)	**7.** (*a*)
8. (*b*)	**9.** (*d*)	**10.** (*a*)	**11.** (*a*)	**12.** (*a*)	**13.** (*a*)	**14.** (*a*)
15. (*a*)	**16.** (*c*)	**17.** (*b*)	**18.** (*c*)	**19.** (*e*)	**20.** (*a*)	**21.** (*d*)
22. (*d*)	**23.** (*d*)	**24.** (*a*)	**25.** (*a*)	**26.** (*d*)	**27.** (*a*)	**28.** (*d*)
29. (*d*)	**30.** (*a*)	**31.** (*d*)	**32.** (*a*)	**33.** (*c*)	**34.** (*e*)	**35.** (*a*)
36. (*d*)	**37.** (*b*)	**38.** (*a*)	**39.** (*e*)	**40.** (*a*)	**41.** (*e*)	**42.** (*d*)
43. (*a*)	**44.** (*c*)	**45.** (*b*)	**46.** (*e*)	**47.** (*d*)	**48.** (*b*)	**49.** (*a*)
50. (*d*)	**51.** (*a*)	**52.** (*a*)	**53.** (*b*)	**54.** (*e*)	**55.** (*e*)	**56.** (*b*)
57. (*b*)	**58.** (*d*)	**59.** (*a*)	**60.** (*c*)	**61.** (*d*)	**62.** (*d*)	**63.** (*b*)
64. (*e*)	**65.** (*c*)	**66.** (*a*)	**67.** (*a*)	**68.** (*d*)	**69.** (*e*)	**70.** (*a*)
71. (*d*)	**72.** (*d*)	**73.** (*a*).				

THEORETICAL QUESTIONS

1. Define a thermodynamic system. Differentiate between open system, closed system and an isolated system.
2. How does a homogeneous system differ from a heterogeneous system ?
3. What do you mean by a pure substance ?
4. Explain the following terms :
   (*i*) State, (*ii*) Process, and (*iii*) Cycle.
5. Explain briefly zeroth law of thermodynamics.
6. What is a quasi-static process ?
7. What do you mean by 'reversible work' ?
8. Define 'internal energy' and prove that it is a property of a system.
9. Explain the First Law of Thermodynamics as referred to closed systems undergoing a cyclic change.
10. State the First Law of Thermodynamics and prove that for a non-flow process, it leads to the energy equation $Q = \Delta U + W$.
11. What is the mechanical equivalent of heat ? Write down its value when heat is expressed in kJ and work is expressed in N-m.
12. What do you mean by "Perpetual motion machine of first kind-PMM 1" ?
13. Why only in constant pressure non-flow process, the enthalpy change is equal to heat transfer ?
14. Prove that the rate of change of heat interchange per unit change of volume when gas is compressed or expanded is given by $\frac{\gamma - n}{\gamma - 1} \times \frac{pdv}{J}$.
15. Write down the general energy equation for steady flow system and simplify when applied for the following systems :
    (*i*) Centrifugal water pump (*ii*) Reciprocating air compressor
    (*iii*) Steam nozzle (*iv*) Steam turbine
    (*v*) Gas turbine.
16. Explain clearly the difference between a non-flow and a steady flow process.
17. State the limitations of first law of thermodynamics.
18. What is the difference between a heat engine and a reversed heat engine ?
19. Enumerate the conditions which must be fulfilled by a reversible process. Give some examples of ideal reversible processes.
20. What is an irreversible process ? Give some examples of irreversible processes.
21. Give the following statements of second law of thermodynamics.
    (*i*) Clausius statement
    (*ii*) Kelvin-Planck statement.
22. Define heat engine, refrigerator and heat pump.
23. What is the perpetual motion machine of the second kind ?
24. What do you mean by 'Thermodynamic temperature' ?
25. What do you mean by 'Clausius inequality' ?
26. Describe the working of a Carnot cycle.
27. Derive an expression for the efficiency of the reversible heat engine.
28. What do you mean by the term 'Entropy' ?

2

Introduction to Internal Combustion Engines

2.1. HEAT ENGINES

Any type of engine or machine which derives heat energy from the combustion of fuel or any other source and converts this energy into mechanical work is termed as a ***heat engine****.*

Heat engines may be *classified* into two main classes as follows :

1. External Combustion Engines.
2. Internal Combustion Engines.

1. External combustion engines (E.C. engines)

In this case, *combustion of fuel takes place outside the cylinder* as in case of *steam engines* where the heat of combustion is employed to generate steam which is used to move a piston in a cylinder. Other examples of external combustion engines are *hot air engines, steam turbine* and *closed cycle gas turbine*. These engines are generally used for driving locomotives, ships, generation of electric power etc.

2. Internal combustion engines (I.C. engines)

In this case, *combustion of the fuel with oxygen of the air occurs within the cylinder* of the engine. The internal combustion engines group includes engines employing mixtures of combustible gases and air, known as *gas engines*, those *using lighter liquid fuel* or spirit known as *petrol engines* and those using heavier liquid fuels, known as *oil compression ignition* or *diesel engines*.

The detailed classification of heat engines is given in Fig. 2.1.

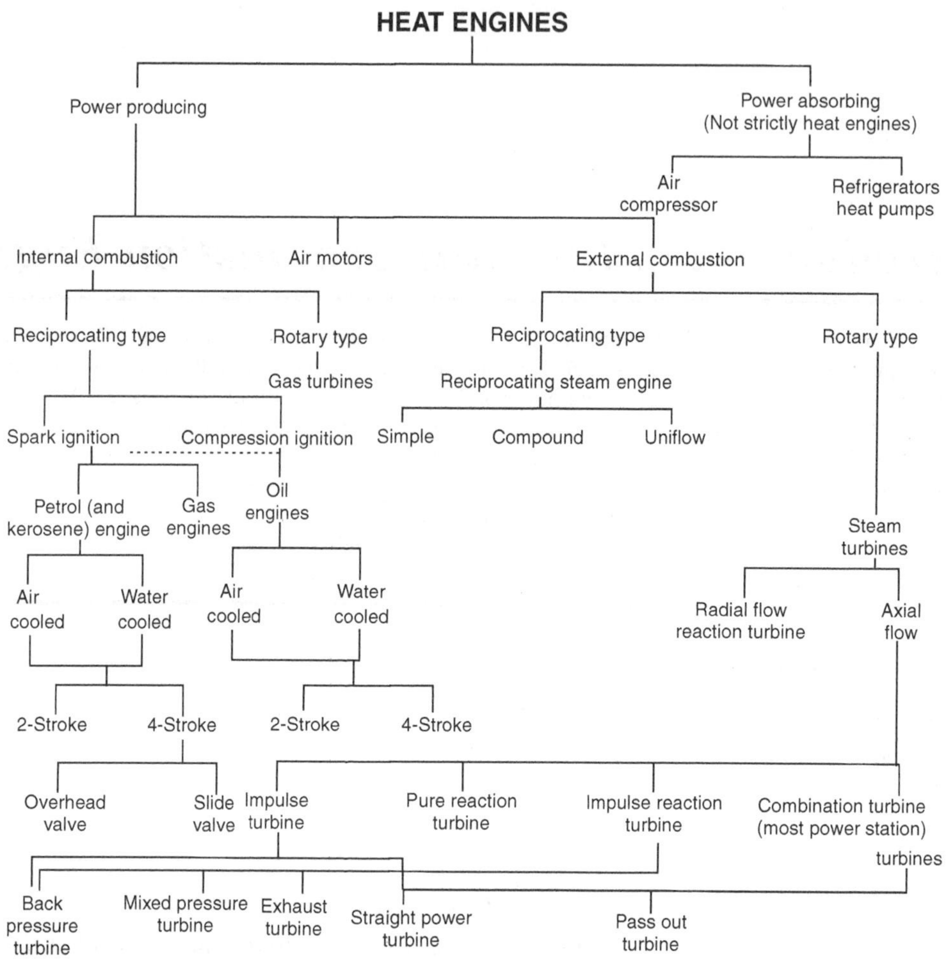

Fig. 2.1. Classification of heat engines.

Advantages of reciprocating internal combustion engines over external combustion engines :

Reciprocating internal combustion engines offer the following *advantages* over external combustion engines :

1. Overall efficiency is high.
2. Greater mechanical simplicity.
3. Weight to power ratio is generally low.
4. Generally lower initial cost.
5. Easy starting from cold conditions.
6. These units are compact and thus require less space.

Advantages of the external combustion engines over internal combustion engines :

The external combustion engines claim the following *advantages* over internal combustion engines :

1. Starting torque is generally high.
2. Because of external combustion of fuel, cheaper fuels can be used. Even solid fuels can be used advantageously.
3. Due to external combustion of fuel it is possible to have flexibility in arrangement.
4. These units are self-starting with the working fluid whereas in case of internal combustion engines, some additional equipment or device is used for starting the engines.

2.2. DEVELOPMENT OF I.C. ENGINES

Brief early history of development of I.C. engines is as follows :

- Many different styles of internal combustion engines were built and tested during the seond half of the 19th century.
- The first fairly practical engine was invented by J.J.E. Lenoir which appeared on the scene about 1860. During the next decade, several hundred of these engines were built with power upto about 4.5 kW and mechanical efficiency upto 5%.
- The Otto-Langen engine with efficiency improved to about 11% was first introduced in 1867 and several thousands of these were produced during the next decade. This was a type of atmospheric engine with the power stroke propelled by atmospheric pressure acting against a vacuum.
- Although many people were working on four-stroke cycle design, Otto was given credit when his prototype engine was built in 1876.
- In the 1880s, the internal combustion engines first appeared in automobiles. Also in this decade the two-stroke cycle engine became practical and was manufactured in large number.
- Rudolf Diesel, by 1892, had perfected his compression ignition engine into basically the same diesel engine known today. This was after years of development work which included the use of solid fuel in his early experimental engines.
- *Early compression engines were noisy, large, slow, single cylinder engines. They were, however, generally more efficient than spark ignition engines.*
- It wasn't until the 1920s that multicylinder compression ignition engines were made small enough to be used with automobile and trucks.
- *Wakdle's first rotary engine* was tested at NSV, Germany in 1957.
- The practical *stirling engines* in small number are being produced since 1965.
  - These engines require costly material and advanced technology for manufacture.
  - Thermal efficiencies higher than 30% have been obtained.
  - The *advantages* of stirling engine *are low exhaust emission and multi-fuel capability.*

2.3. CLASSIFICATION OF I.C. ENGINES

Internal combustion engines may be classified as given below :

1. **According to cycle of operation :**

(*i*) Two stroke cycle engines

(*ii*) Four stroke cycle engines.

2. **According to cycle of combustion :**

(*i*) Otto cycle engine (combustion at constant volume)

(*ii*) Diesel cycle engine (combustion at constant pressure) (*iii*) Dual-combustion or Semi-Diesel cycle engine (combustion partly at constant volume and partly at constant pressure).

3. **According to arrangement of cylinder :** Refer Fig. 2.2.

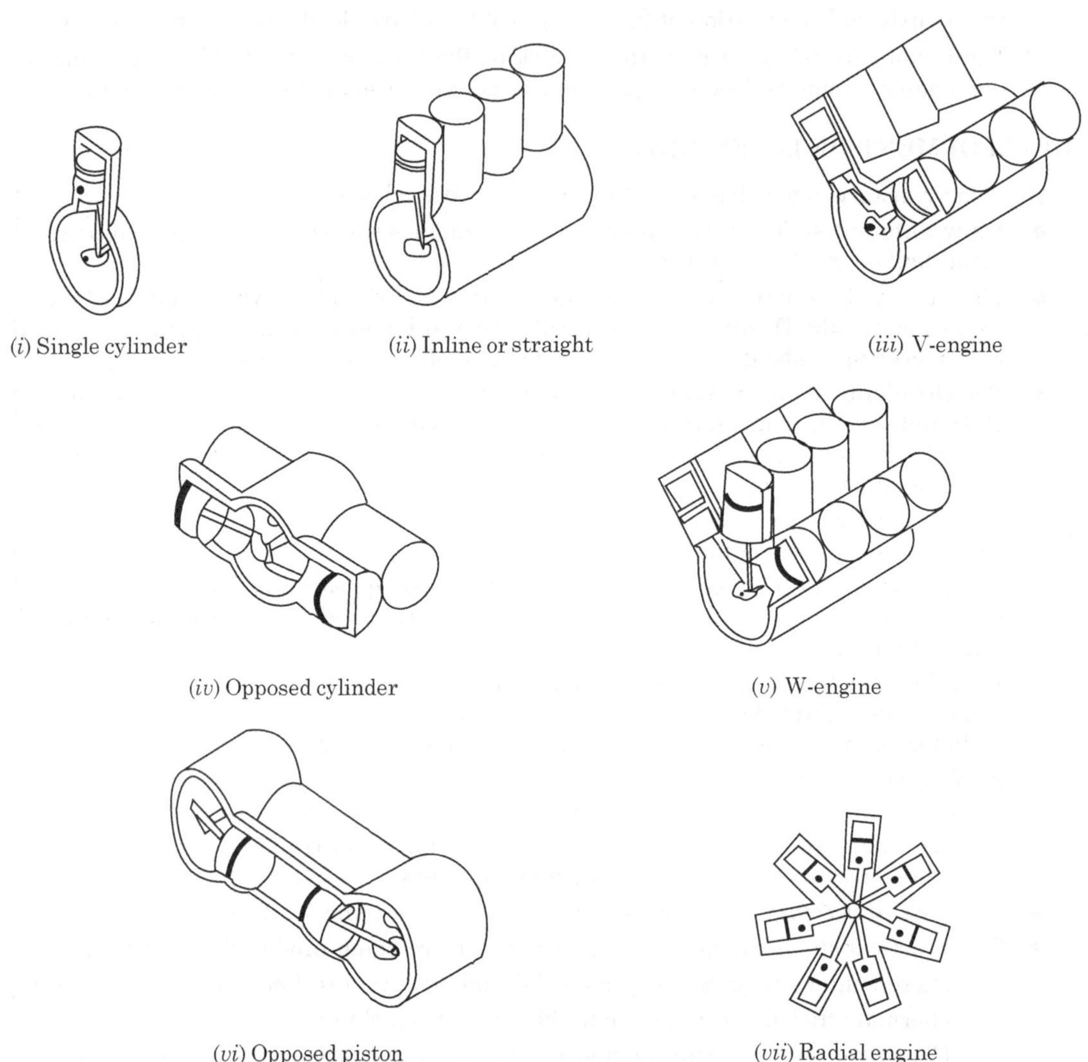

Fig. 2.2. Engine classification by cylinder arrangement.

(*i*) **Single cylinder engine.** Engine has one cylinder and piston connected to the crankshaft.

(*ii*) **In-line or straight engines.** Cylinders are positioned in a straight line one behind the other along the length of the crankshaft.

(*iii*) **V-engine**

- An engine with two cylinder banks (*i.e.*, two-in-line engines) inclined at an angle to each other and with one crankshaft.
- Most of the bigger automobiles use the 8-cylinder V-engine (4-cylinder in-line on each side of V).

(*iv*) **Opposed cylinder engine**

- Two banks of cylinders opposite each other on a single crankshaft (a V-engine with 180° V).
- These are common on small aircraft and some automobiles with even number of cylinders from two to eight or more.

(*v*) **W-engine**

- Same as V-engine except with three banks of cylinders on the same crankshaft.
- Not common, but some have been developed for racing automobiles.

(*vi*) **Opposed piston engine**

- In this type of engine there are two pistons in each cylinder with the combustion chamber in the centre between the pistons.
- A single combustion process causes two power strokes, at the same time, with each piston being pushed away from the centre and delivering power to a separate crankshaft at each end of this cylinder.

(*vii*) **Radial engine**

- It is an engine with pistons positioned in a circular plane around the central crankshaft. The connecting rods of the pistons are connected to a master rod which, in turn, is connected to the crankshaft.
- In a radial engine the bank of cylinders always has an *odd* number of cylinders ranging from 3 to 13 or more.
- Operating on a four-stroke cycle, every other cylinder fires and has a power stroke as the crankshaft rotates, giving a smooth operation.
- Many medium and large size propeller-driven aircraft use radial engines. For large aircraft two or more banks of cylinders are mounted together, one behind the other on a single crankshaft, making one powerful smooth engine.
- Very large ship engines exist with upto 54 cylinders, six banks of 9 cylinder each.

4. **According to their uses :**

(*i*) Stationary engine (*ii*) Portable engine
(*iii*) Marine engine (*iv*) Automobile engine
(*v*) Aero engine etc.

5. **According to the speed of the engine :**

(*i*) Low speed engine (*ii*) Medium speed engine
(*iii*) High speed engine.

6. **According to method of ignition :**

(*i*) Spark-ignition engine (*ii*) Compression-ignition engine.

7. **According to method of cooling the cylinder :**

(*i*) Air-cooled engine (*ii*) Water-cooled engine.

8. **According to method of governing :**

(*i*) Hit and miss governed engine (*ii*) Quality governed engine
(*iii*) Quantity governed engine.

9. **According to valve arrangement :**

(*i*) Over head valve engine (*ii*) *L*-head type engine
(*iii*) *T*-head type engine (*iv*) *F*-head type engine.

10. **According to number of cylinders :**

(*i*) Single-cylinder engine (*ii*) Multi-cylinder engine.

11. **According to air intake process :**

(*i*) *Naturally aspirated.* No intake air pressure boost system.

(*ii*) *Supercharged.* Intake air pressure increased with the compressor driven off the engine crankshaft.

(*iii*) *Turbocharged.* Intake air pressure increased with the turbine-compressor driven by the engine exhaust gases.

(*iv*) *Crankcase compressed.* Two stroke-cycle engine which uses the crankcase as the intake air compressor. Limited development work has also been done on the design and construction of four-stroke cycle engines with crankcase compression.

12. **According to fuel employed :**

(*i*) Oil engine (*ii*) Petrol engine

(*iii*) Gas engine (*iv*) Kerosene engine

(*v*) LPG engine (*vi*) Alcohol-ethyl, methyl engine

(*vii*) Duel fuel engine (*viii*) Gasohol (90% gasoline and 10% alcohol).

13. **Method of fuel input for S.I. engines :**

(*i*) *Carburetted*

(*ii*) *Multipoint port fuel injection.* One or more injectors at each cylinder intake.

(*iii*) *Throttle body fuel injection.* Injectors upstream in intake manifold.

2.4. APPLICATION OF I.C. ENGINES

The I.C. engines are generally used for :

(*i*) Road vehicles (*e.g.*, scooter, motorcycle, buses etc.)

(*ii*) Aircraft

(*iii*) Locomotives

(*iv*) Construction in civil engineering equipment such as bull-dozer, scraper, power shovels etc.

(*v*) Pumping sets

(*vi*) Cinemas

(*vii*) Hospital

(*viii*) Several industrial applications.

The *applications of various engines separately* are listed below :

1. **Small two-stroke petrol engines :**

- These engines are employed where *simplicity and the low cost of the prime mover are primary considerations.*
- The 50 c.c. engines develops maximum brake power (B.P.) of 1.5 kW at 5000 r.p.m. and is *used in mopeds.*
- The 100 c.c. engine developing maximum brake power of about 3 kW at 5000 r.p.m. is *used in scooters.* The 150 c.c. engine develops maximum brake power of about 5 kW at 5000 r.p.m.
- The 250 c.c. engine developing a maximum brake power of about 9 kW at 4500 r.p.m. is generally *used in motor cycles.*

- These engines also find applications in very small electric generating sets, pumping sets etc.

2. **Small four-stroke petrol engines :**

- These engines are primarily used in automobiles.
- These are also used in pumping sets and mobile electric generating sets.

These days diesel engines are taking them over, in the above mentioned applications.

3. **Four stroke diesel engines :**

- The four-stroke diesel engine (a versatile prime mover) is manufactured in diameter ranging from 50 mm to 600 mm with speeds ranging from 100 to 4400 r.p.m., the power delivered per cylinder varying from 1 to 1000 kW.
- Diesel engine is employed for the following :
  - — Pumping sets
  - — Construction machinery
  - — Air compressors and drilling jigs
  - — Tractors
  - — Jeeps, cars and taxies
  - — Mobile and stationary electric generating plant
  - — Diesel-electric locomotive
  - — Boats and ships.

4. **Two stroke diesel engines :**

- These engines having very high power are usually employed for *ship propulsion* and generally have bores above 60 cm, uniflow with exhaust valves or loop scavenged.

Example. Nordberg, *2 stroke, 12-cylinder 80 cm bore and 155 cm stroke, diesel engine develops 20000 kW at 120 r.p.m.*

5. **Radial piston engine in small aircraft propulsion :**

- Radial four stroke petrol engines having power range from 300 kW to 4000 kW have been used in small aircrafts.
- In modern large aircrafts, instead of these engines, gas turbine plant as turboprop engine or turbojet engine and gas turbine engines are used.

2.5. ENGINE CYCLE-ENERGY BALANCE

Refer Fig. 2.3. It shows the energy flow through the reciprocating engine. The analysis is based on the first law of thermodynamics which states that energy can neither be created nor destroyed, it can be converted from one form to other.

- In an I.C. engine fuel is fed to the combustion chamber where it burns in the presence of air and its chemical energy is converted into heat. All this energy is not available for driving the piston since a portion of this energy is lost through exhaust, coolant and radiation. The remaining energy is converted to power and is called indicated energy or *indicated power* (I.P.). The ratio of this energy to the input fuel energy is called indicated thermal efficiency $[\eta_{th.(I)}]$.

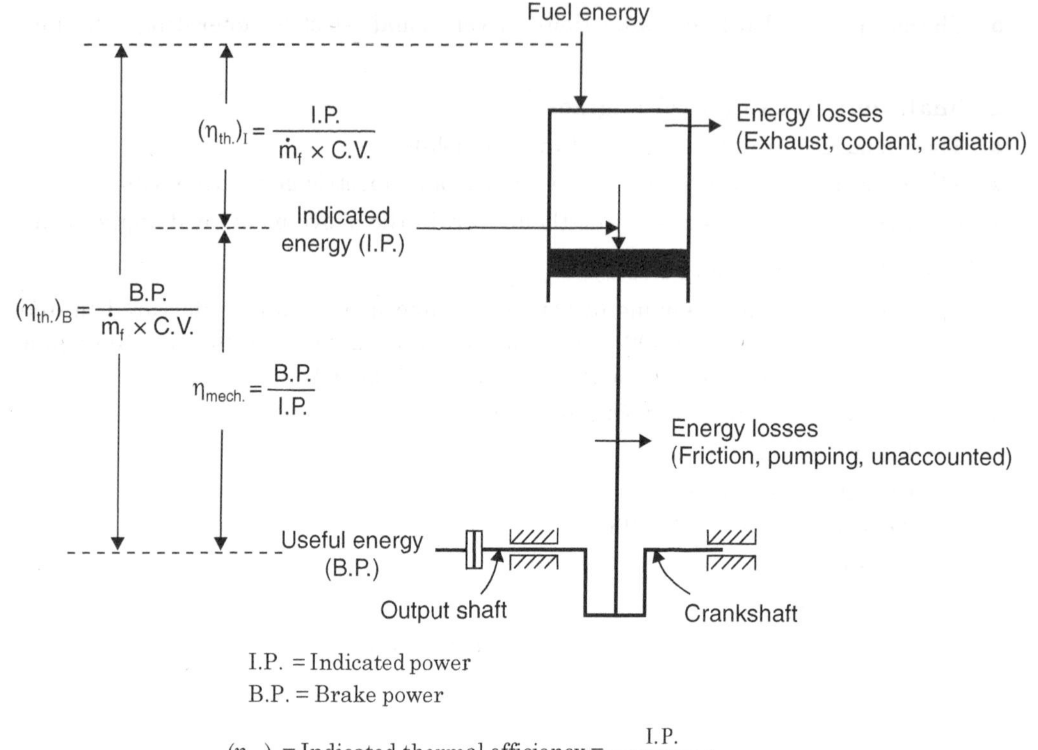

I.P. = Indicated power

B.P. = Brake power

$(\eta_{th.})_I$ = Indicated thermal efficiency = $\frac{I.P.}{\dot{m}_f \times C.V.}$

(where $\dot{m}_f$ = mass of fuel in kg/s and C.V. = calorific value)

$(\eta_{th.})_B$ = Brake thermal efficiency.

Fig. 2.3. The energy flow through the reciprocating engine.

- The energy available at the piston passes through the connecting rod to the crankshaft. In this transmission of energy/power there are losses due to friction, pumping, etc. The sum of all these losses, converted to power, is termed as **friction power (F.P.).** The remaining energy is the *useful mechanical energy and is termed as* **shaft energy** or **brake power (B.P.).** The *ratio of energy at shaft to fuel input energy is called* **brake thermal efficiency** $[\eta_{th.(B)}]$.
- The *ratio of shaft energy to the energy available at the piston is called* **mechanical efficiency** $(\eta_{mech.})$.

2.6. BASIC IDEA OF I.C. ENGINES

The basic idea of internal combustion engine is shown in Fig. 2.4. The cylinder which is closed at one end is filled with a mixture of fuel and air. As the crankshaft turns it pushes cylinder. The piston is forced up and compresses the mixture in the top of the cylinder. The mixture is set alight and, as it burns, it creates a gas pressure on the piston, forcing it down the cylinder. This motion is shown by arrow '1'. The piston pushes on the rod which pushes on the crank. The crank is given rotary (turning) motion as shown by the arrow '2'. The fly wheel fitted on the end of the crankshaft stores energy and keeps the crank turning steadily.

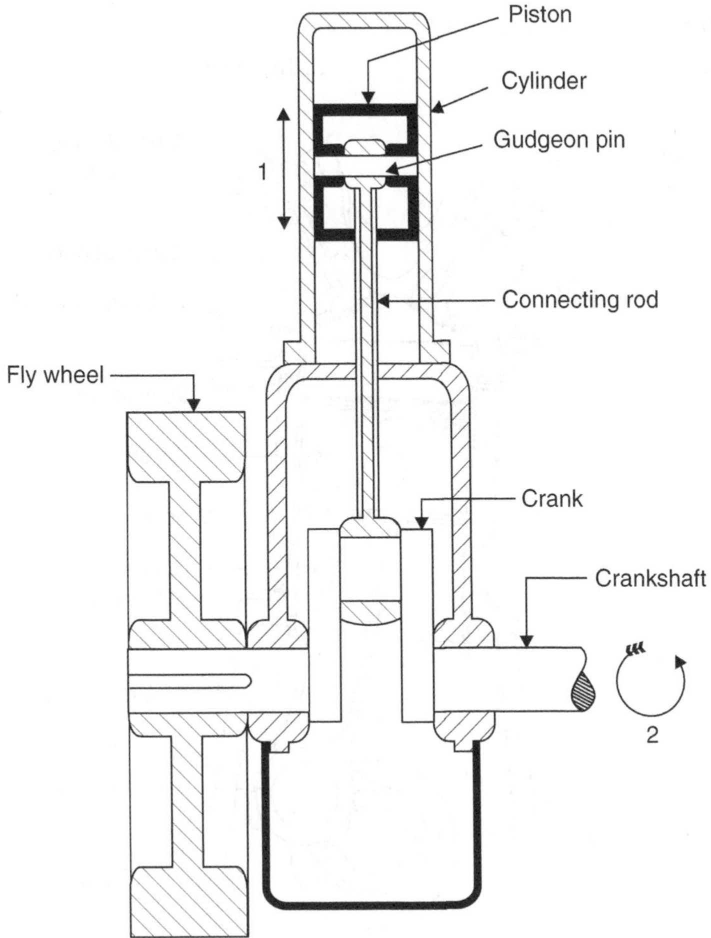

Fig. 2.4. Basic idea of I.C. engine.

2.7. DIFFERENT PARTS OF I.C. ENGINES

Here follows the detail of the various parts of an internal combustion engine.

A cross-section of an air-cooled I.C. engine with principal parts is shown in Fig. 2.5.

A. Parts common to both petrol and diesel engine :

1. Cylinder
2. Cylinder head
3. Piston
4. Piston rings
5. Gudgeon pin
6. Connecting rod
7. Crankshaft
8. Crank
9. Engine bearing
10. Crankcase
11. Flywheel
12. Governor
13. Valves and valve operating mechanisms.

B. Parts for petrol engines only :

1. Spark plugs
2. Carburettor
3. Fuel pump.

C. Parts for Diesel engine only :

1. Fuel pump.
2. Injector.

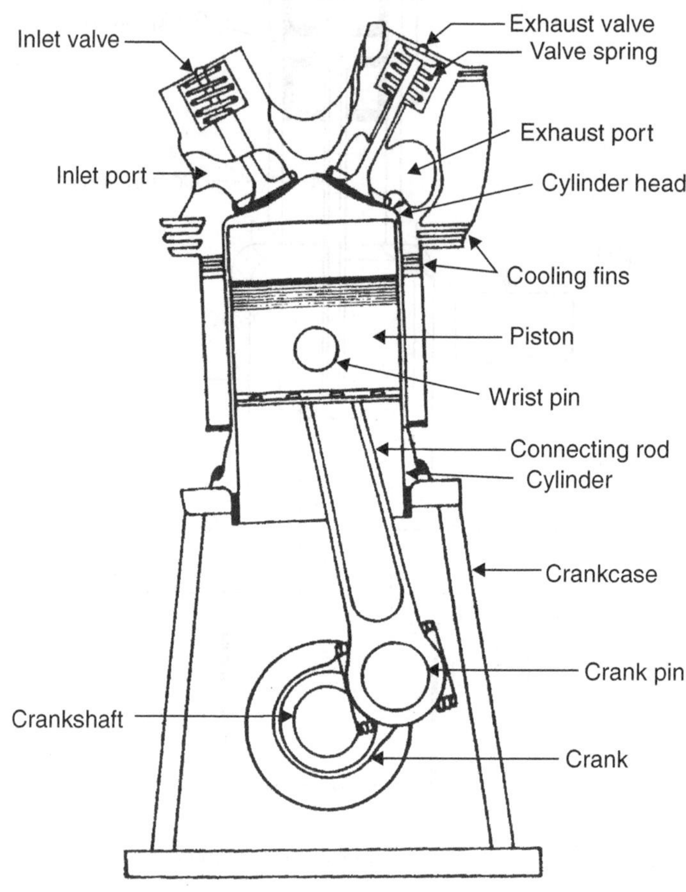

Fig. 2.5. Air-cooled I.C. engine.

A. Parts common to both petrol and diesel engines :

1. Cylinder

The cylinder contains gas under pressure and guides the piston. It is in direct contact with the products of combustion and it must be cooled. The ideal form consists of a plain cylindrical barrel in which the piston slides. The movement of the piston or stroke being in most cases, longer than the bore. This is known as the *"stroke bore ratio"*. The upper end consists of a combustion or clearance space in which the ignition and combustion of the charge takes place. In practice, it is necessary to depart from the ideal hemispherical slope in order to accommodate the valves, sparking plugs etc. and to control the combustion. Sections of an air-cooled cylinder and a water-cooled cylinder are shown in Fig. 2.6 and 2.7 respectively. *The cylinder is made of hard grade cast iron and is usually, cast in one piece.*

2. Cylinder head

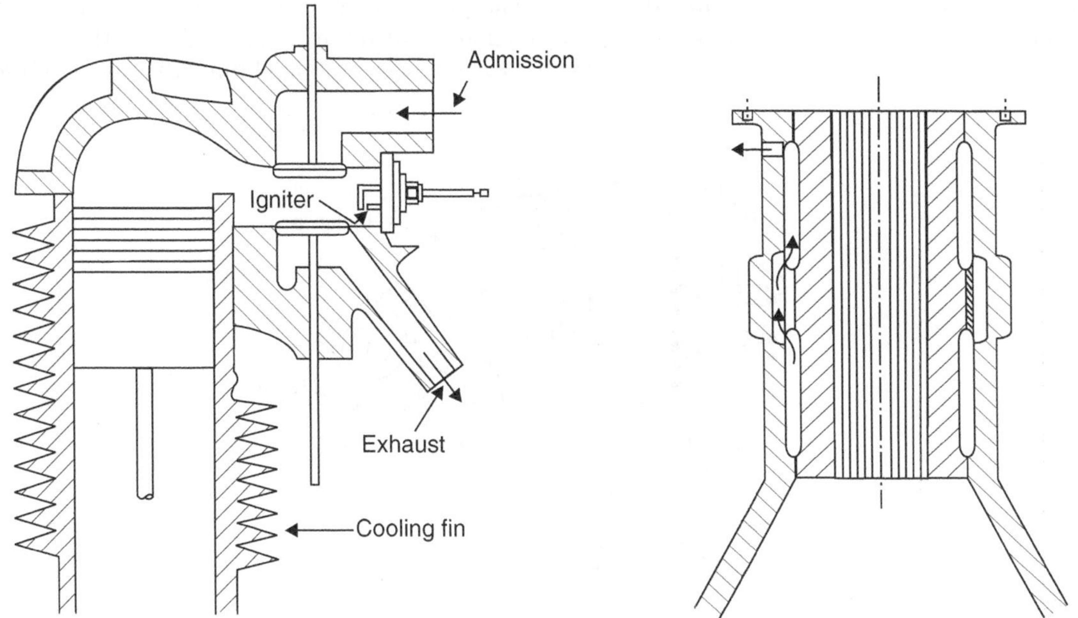

Fig. 2.6. Air-cooled cylinder.

Fig. 2.7. Water-cooled cylinder.

One end of the cylinder is closed by means of a *removable cylinder head* (Fig. 2.6) which usually contains the inlet or admission valve [Fig. 2.8 (*a*)] for admitting the mixture of air and

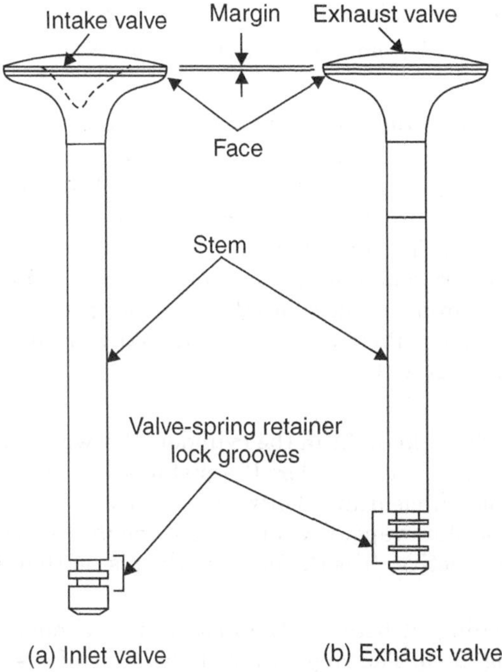

Fig. 2.8

fuel and exhaust valve [Fig. 2.8 (*b*)] for discharging the product of combustion. Two valves are kept closed, by means of cams (Fig. 2.9) geared to the engine shaft. The passage in the cylinder head leading to and from the valves are called *ports*. The pipes which connect the inlet ports of the various cylinders to a common intake pipe for the engine is called the inlet *manifold*. If the exhaust ports are similarly connected to a common exhaust system, this system of piping is called *exhaust manifold*.

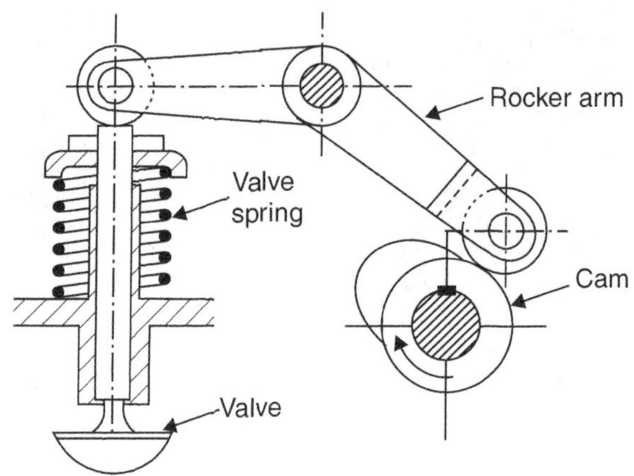

Fig. 2.9. Cam and rocker arm.

The main purpose of the cylinder head is to seal the working ends of the cylinders and not to permit entry and exit of gases on cover head valve engines. The inside cavity of head is called the *combustion chamber*, into which the mixture is compressed for firing. Its *shape controls the direction and rate of combustion*. Heads are drilled and tapped with correct thread to take the ignition spark plug. All the combustion chambers in an engine must be of same shape and size. The shape may be in part controlled by the piston shape.

The cylinder head is usually made of cast iron or aluminium.

3. **Piston**

A piston is fitted to each cylinder as a face to receive gas pressure and transmit the thrust to the connecting rod.

The piston must (*i*) give gas tight seal to the cylinder through bore, (*ii*) slide freely, (*iii*) be light and (*iv*) be strong. The thrust on the piston on the power stroke tries to tilt the piston as the connecting rod swings, side ways. The piston wall, called the skirt must be strong enough to stand upto this side thrust. *Pistons are made of cast iron or aluminium alloy for lightness.* Light alloy pistons expand more than cast iron one therefore they need large clearances to the bore, when cold, or special provision for expansion. Pistons may be solid skirt or split skirt. A section through a split skirt piston is shown in Fig. 2.10.

4. **Piston rings**

The piston must be a fairly loose fit in the cylinder. If it were a tight fit, it would expand as it got hot and might stick tight in the cylinder. If a piston sticks it could ruin the engine. On the other hand, if there is too much clearance between the piston and cylinder walls, much of the pressure from the burning gasoline vapour will leak past the piston. This means, that the push on the piston will be much less effective. It is the push on the piston that delivers the power from the engines.

To provide a good sealing fit between the piston and cylinder, pistons are equipped with piston rings, as shown in Fig. 2.10. The rings are usually made of cast iron of fine grain and high elasticity which is not affected by the working heat. Some rings are of alloy spring steel. They are split at one point so that they can be expanded and slipped over the end of the piston and into ring grooves which have been cut in the piston. When the piston is installed in the cylinder the rings

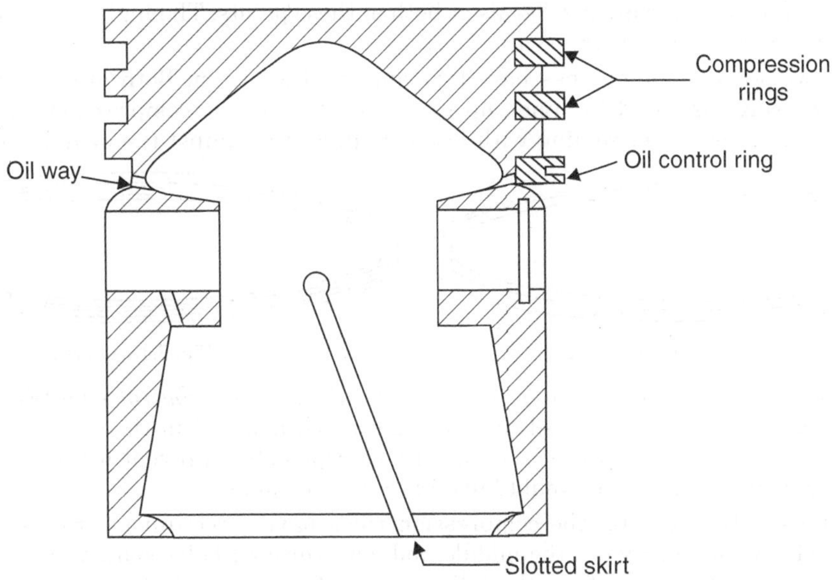

Fig. 2.10. Section through a split skirt piston.

are compressed into ring grooves which have been cut in the piston. When the piston is installed in the cylinder, the rings are compressed into the ring grooves so that the split ends come almost together. The rings fit tightly against the cylinder wall and against the sides of the ring grooves in the piston. Thus, *they form a good seal between the piston and the cylinder wall.* The rings can expand or contact as they heat and cool and still make a good deal. Thus they are free to slide up and down the cylinder wall.

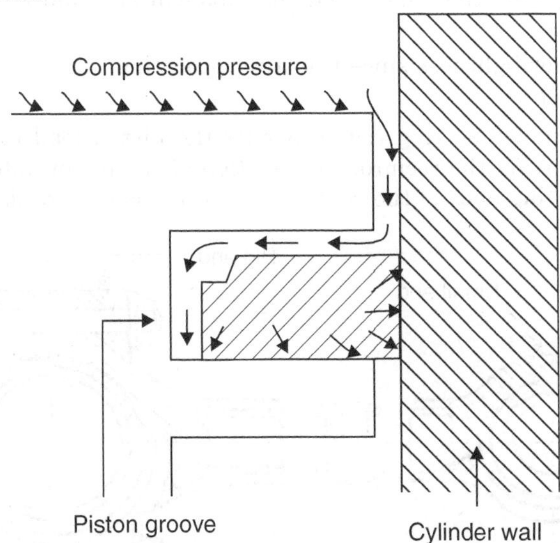

Fig. 2.11. Working of a piston ring.

Fig. 2.11 shows how the piston ring works to hold in the compression and combustion pressure. The arrows show the pressure above the piston passing through clearance between the piston and the cylinder wall. It presses down against the top and against the back of the piston rings as shown by the arrows. This pushes the piston ring firmly against the bottom of the piston ring groove. As a result there are good seals at both of these points. The higher the pressure in the combustion chamber, the better the seal.

Small two stroke cycle engines have two rings on the piston. Both are compression rings (Fig. 2.12). Two rings are used to divide up the job of holding the compression and combustion pressure. This produces better sealing with less ring pressure against the cylinder wall.

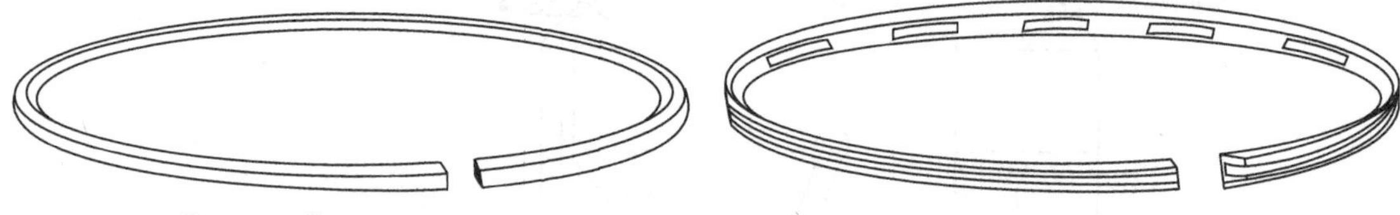

Fig. 2.12. Compression ring. Fig. 2.13. Oil ring.

Four stroke cycle engines have an extra ring, called the oil control ring (Fig. 2.13). Four stroke cycle engines are so constructed that they get much more oil in the cylinder wall than do two stroke cycle engines. This additional oil must be scraped off to prevent it from getting up into the combustion chamber, where it would burn and cause trouble.

Refer Figs. 2.12 and 2.13, the compression rings have a rectilinear cross-section and oil rings are provided with a groove in the middle and with through holes spaced at certain interval from each other. The oil collected from the cylinder walls flows through these holes into the piston groove whence through the holes in the body of the piston and down its inner walls into the engine crankcase.

5. **Gudgeon pin** (or wrist pin or piston pin)

These are *hardened steel parallel spindles* fitted through the piston bosses and the small end bushes or eyes to allow the connecting rods to swivel. Gudgeon pins are a press fit in the piston bosses of light alloy pistons when cold. For removal or fitting, the piston should be dipped in hot water or hot oil, this expands the bosses and the pins can be removed or fitted freely without damage.

It is made hollow for lightness since it is a reciprocating part.

6. **Connecting rod**

Refer Fig. 2.14. The connecting rod transmits the piston load to the crank, causing the latter to turn, thus converting the reciprocating motion of the piston into a rotary motion of the crankshaft. The lower or "big end" of the connecting rod turns on "crank pins".

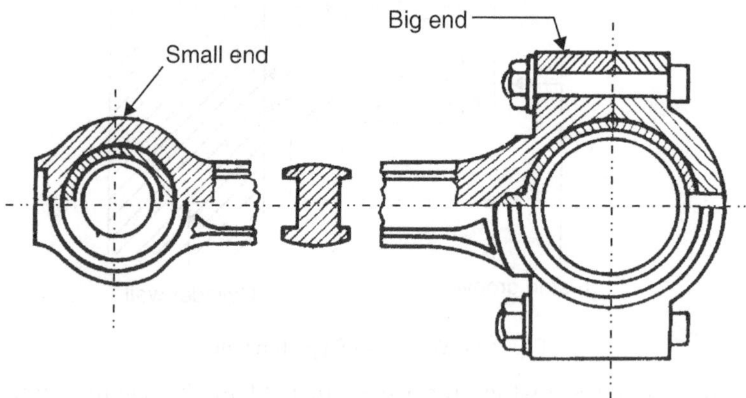

Fig. 2.14. Connecting rod.

The connecting rods are made of nickle, chrome and chrome vandium steels. For small engines the material may be aluminium.

7. **Crank**

The piston moves up and down in the cylinder. This up and down motion is called *reciprocating motion*. The piston moves in a straight line. The straight line motion must be changed to rotary, or turning motion, in most machines, before it can do any good. That is rotary motion is required to make wheels turn, a cutting blade spin or a pulley rotate. To change the reciprocating motion to rotary motion a crank and connecting rod are used. (Figs. 2.15 and 2.16). The connecting rod connects the piston to the crank.

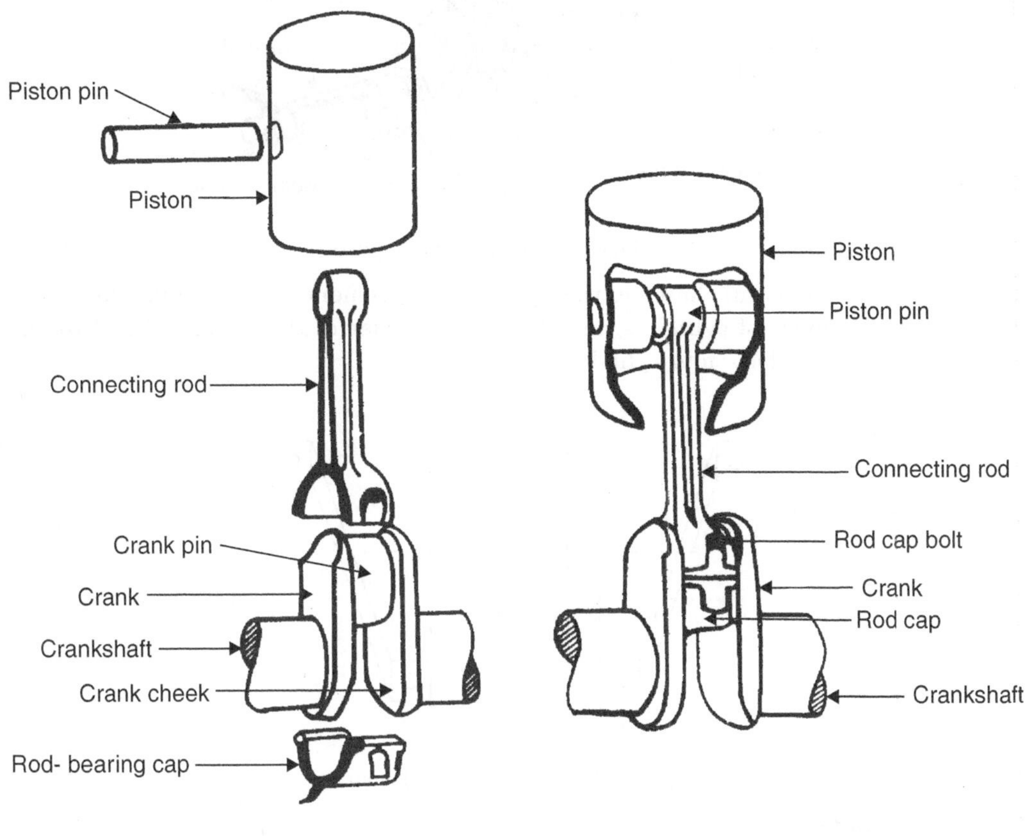

Fig. 2.15 Fig. 2.16

Note. The crank end of the connecting rod is called rod *"big end"*. The piston end of the connecting rod is called the rod *"small end"*.

8. **Crankshaft**

The crank is part of the crankshaft. The crankshaft of an internal combustion engine receives via its cranks the efforts supplied by the pistons to the connecting rods. All the engines auxiliary mechanisms with mechanical transmission are geared in one way or the another to the crankshaft. *It is usually a steel forging, but some makers use special types of cast iron such as spheroidal graphitic or nickel alloy castings* which are cheaper to produce and have good service life. Refer Fig. 2.17. The crankshaft converts the reciprocating motion to rotary motion. The crank shaft mounts in bearings which, encircle the journals so it can rotate freely.

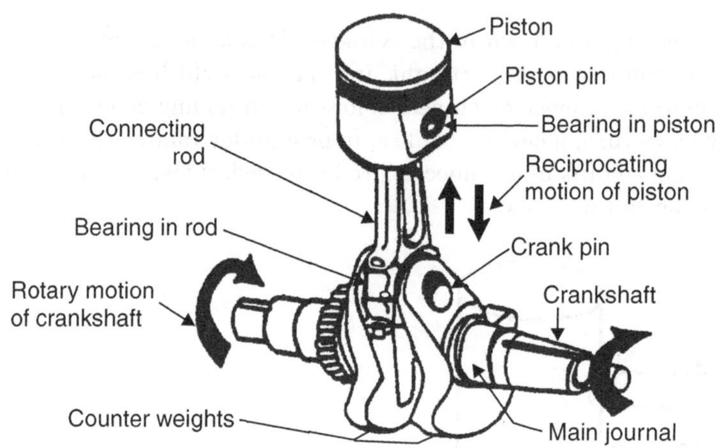

Fig. 2.17. Crank shaft and other parts.

The shape of the crankshaft *i.e.* the mutual arrangement of the cranks depend on the number and arrangement of cylinders and the turning order of the engine. Fig. 2.18 shows a typical crankshaft layout for a four cylinder engine.

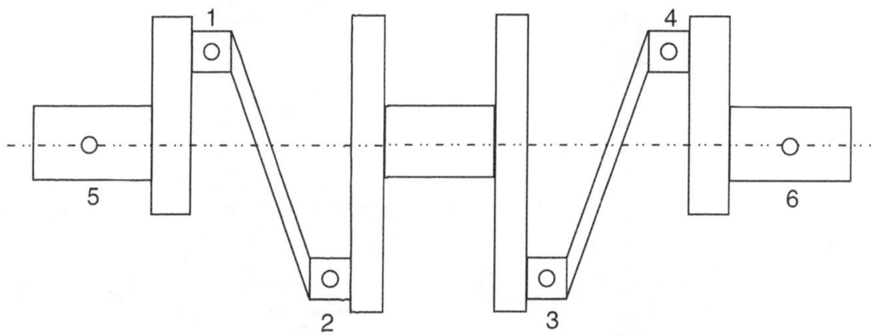

Fig. 2.18. Typical crankshaft layout.

9. Engine bearing

The crankshaft is supported by bearing. The connecting rod big end is attached to the crank pin on the crank of the crankshaft by a bearing. A piston pin at the rod small end is used to attach the rod to the piston. The piston pin rides in bearings. Every where there is rotary action in the engine, bearings are used to support the moving parts. The purpose of bearing is to reduce the friction and allow the parts to move easily. Bearings are lubricated with oil to make the relative motion easier.

Bearings used in engines are of two types : *sliding* or *rolling* (Fig. 2.19).

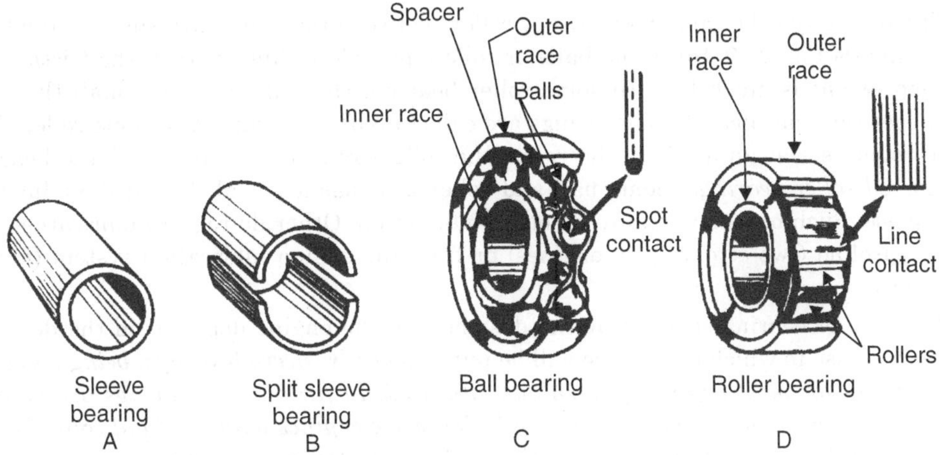

Fig. 2.19. Bearings.

The sliding type of bearings are sometimes called *bushings or sleeve bearings* because they are in the shape of a sleeve that fits around the rotating journal or shaft. The sleeve-type connecting rod big end bearings usually called simply rod bearings and the crankshaft supporting bearings called the main bearings are of the split sleeve type. They must be split in order to permit their assembly into the engine. In the rod bearing, the upper half of the bearing is installed in the rod, the lower half is installed in the rod bearing cap. When the rod cap is fastened to the rod shown in Fig. 2.16 a complete sleeve bearing is formed. Likewise, the upper halves of the main bearings are assembled in the engine and then the main bearing caps, with the lower bearing halves are attached to the engine to complete the sleeve bearings supporting the crankshaft.

The typical bearing half is made of steel or bronze back to which a lining of relatively soft bearing material is applied. Refer Fig. 2.20. This relatively soft bearing material, which is made of several materials such as copper, lead, tin and other metals, has the ability to conform to slight irregularities of the shaft rotating against it. If wear does take place, it is the bearing that wears and the bearing can be replaced instead of much more expansive crankshaft or other engine part.

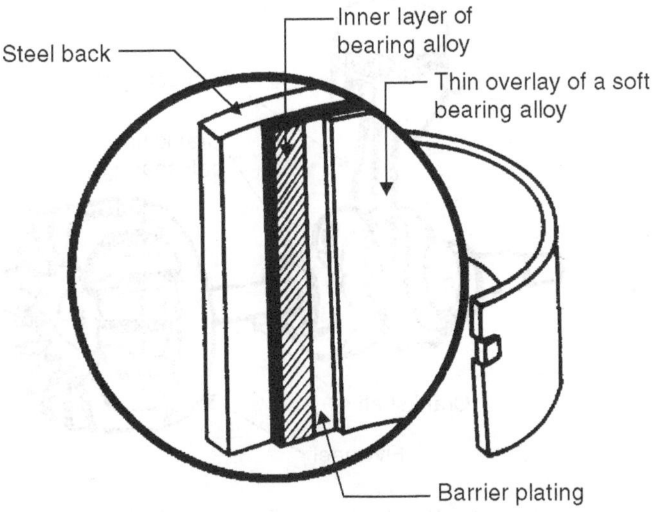

Fig. 2.20. Bearing half (details).

The rolling-type bearing uses balls or rollers between the stationary support and the rotating shaft. Refers Fig. 2.19. Since the balls or rollers provide rolling contact, the frictional resistance to movement is much less. In some roller bearing, the rollers are so small that they are hardly bigger than needles. These bearings are called *needle bearings.* Also some rollers bearings have the rollers set at an angle to the races, the rollers roll in are tapered. These bearings are called *tapered roller bearings.* Some ball and roller bearings are sealed with their lubricant already in place. Such bearings require no other lubrication. Other do require lubrication from the oil in the gasoline (two stroke cycle engines) or from the engine lubrication system (four stroke cycle engines).

The type of bearing selected by the designers of the engine depends on the design of the engine and the use to which the engine will be put. *Generally, sleeve bearings, being less expensive and satisfactory for most engine applications, are used. In fact sleeve bearings are used almost universally in automobile engines. But you will find some engines with ball and roller bearings to support the crankshaft and for the connecting rod and piston-pin bearings.*

10. **Crankcase**

The main body of the engine to which the cylinders are attached and which contains the crankshaft and crankshaft bearing is called *crankcase.* This member also holds other parts in alignment and resists the explosion and inertia forces. It also protects the parts from dirt etc. and serves as a part of lubricating system.

11. **Flywheel**

Refer Figs. 2.4 and 2.21. A flywheel (steel or cast iron disc) secured on the crank shaft performs the following *functions* :

(*a*) Brings the mechanism out of dead centres.

(*b*) Stores energy required to rotate the shaft during preparatory strokes.

(*c*) Makes crankshaft rotation more uniform.

(*d*) Facilitates the starting of the engine and overcoming of short time over loads as, for example, when the machine is started from rest.

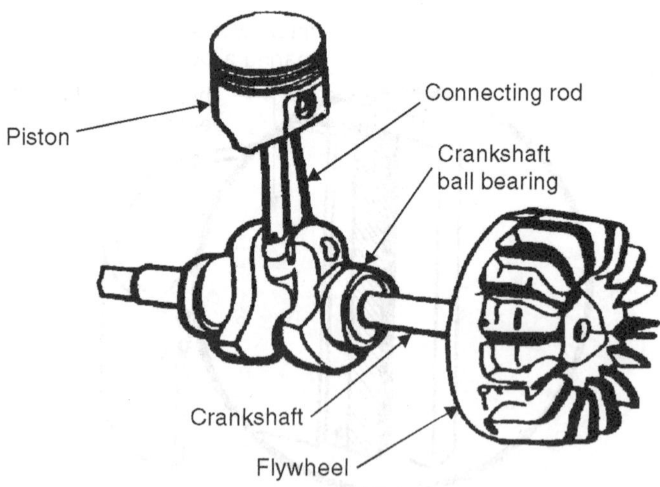

Fig. 2.21. Flywheel secured on crankshaft.

The weight of the flywheel depends upon the nature of variation of the pressure. The flywheel for a double-acting steam engine is lighter than that of a single-acting one. Similarly, the flywheel for a two-stroke cycle engine is lighter than a flywheel used for a four-stroke cycle engine. *Lighter flywheels are used for multi-cylinder engines.*

12. Governor

A governor may be defined *as a device for regulating automatically output of a machine by regulating the supply of working fluid.* When the speed decreases due to increase in load the supply valve is opened by mechanism operated by the governor and the engine therefore speeds up again to its original speed. If the speed increases due to a decrease of load the governor mechanism closes the supply valve sufficiently to slow the engine to its original speed. *Thus the function of a governor is to control the fluctuations of engine speed due to changes of load.*

Comparison between a Flywheel and a Governor

	Flywheel	*Governor*
1.	It is provided on engines and fabricating machines *viz.*, rolling mills, punching machines ; shear machines, presses etc.	It is provided on prime movers such as engines and turbines.
2.	Its function is to store the available mechanical energy when it is in excess of the load requirement and to part with the same when the available energy is less than that required by the load.	Its function is to regulate the supply of driving fluid producing energy, according to the load requirement so that at different loads almost a constant speed is maintained.
3.	It works continuously from cycle to cycle.	It works intermittently *i.e.* only when there is change in load.
4.	In engines it takes care of fluctuations of speed during thermodynamic cycle.	It takes care of fluctuations of speed due to variation of load over long range of working engines and turbines.
5.	In fabrication machines it is very economical to use it in that it reduces capital investment on prime movers and their running expenses.	But for governor, there would have been unnecessarily more consumption of driving fluid. Thus it economies its consumption.

Types of governor :

Governors are classified as follows :

1. Centrifugal governor

(*i*) *Gravity controlled,* in which the centrifugal force due to the revolving masses is largely balanced by gravity.

(*ii*) *Spring controlled,* in which the centrifugal force is largely balanced by springs.

2. Inertia and flywheel governors

(*i*) Centrifugal type, in which centrifugal forces play the major part in the regulating action.

(*ii*) Inertia governor, in which the inertia effect predominates.

The *inertia type* governors are fitted to the crankshaft or flywheel of an engine and so differ radically in appearance from the centrifugal governors. The balls are so arranged that the inertia force caused by an angular acceleration or retardation of the shaft tends to alter their positions. The amount of displacement of governor balls is controlled by suitable springs and through the governor mechanism, alters the fuel supply to the engine. The inertia governor is more sensitive than centrifugal but it becomes very difficult to balance the revolving parts. For this reason *centrifugal governors are more frequently used.* We shall discuss centrifugal governors only.

Important centrifugal governors are :

1. Watt governor
2. Porter governor
3. Proell governor
4. Hartnell governor.

1. Watt governor

It is the primitive governor as used by Watt on some of his early steam engines. It is used for a very slow speed engine and this is why it has now become obsolete.

Refer Fig. 2.22. Two arms are hinged at the top of the spindle and two revolving balls are fitted on the other ends of the arms. One end of each of the links are hinged with the arms, while the other ends are hinged with the sleeve, which may slide over the spindle. The speed of the crankshaft is transmitted to the spindle through a pair of bevel gears by means of a suitable arrangement. So the rotation of the spindle of the governor causes the weights to move away from the centre due to the centrifugal force. This makes the sleeve to move in the upward direction. This movement of the sleeve is transmitted by the lever to the throttle valve which partially closes or opens the steam pipe and reduces or increases the supply of steam to the engine. So the engine speed may be adjusted to a normal limit.

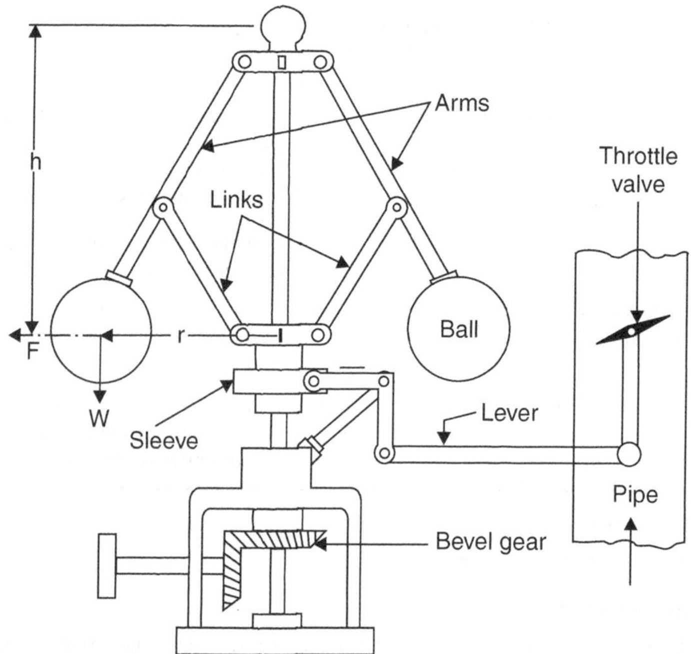

Fig. 2.22. Watt governor.

2. Porter governor

Fig. 2.23 shows diagrammatically a Porter governor where two or more masses called the governor balls rotate about the axis of the governor shaft which is driven through suitable gearing from the engine crankshaft. The governor balls are attached to the arms. The lower arms are attached to the *sleeve which acts as a central weight.* If the speed of the rotation of the balls increases owing to a decrease of load on the engine, the governor balls fly outwards and the sleeve moves upwards thus closing the fuel passage till the engine speed comes back to its designed speed. If the engine speed decreases owing to an increase of load, the governor balls fly inwards and the sleeve moves downwards thus opening the fuel passage more for oil till the engine speed comes back to its designed speed. The engine is said to be running at its designed speed when the outward inertia or centrifugal force is just balanced by the inward controlling force.

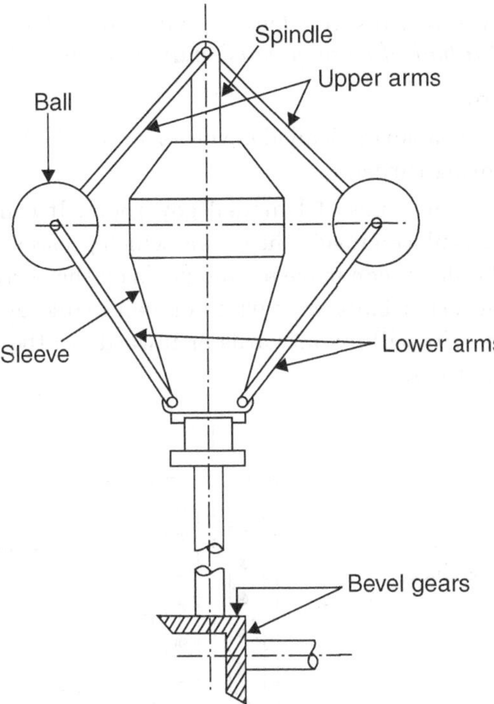

Fig. 2.23. Porter governor.

3. **Proell governor**

Refer Fig. 2.24. It is a modification of porter governor. The governor balls are carried on an *extension of the lower arms*. For given values of weight of the ball, weight of the sleeve and height

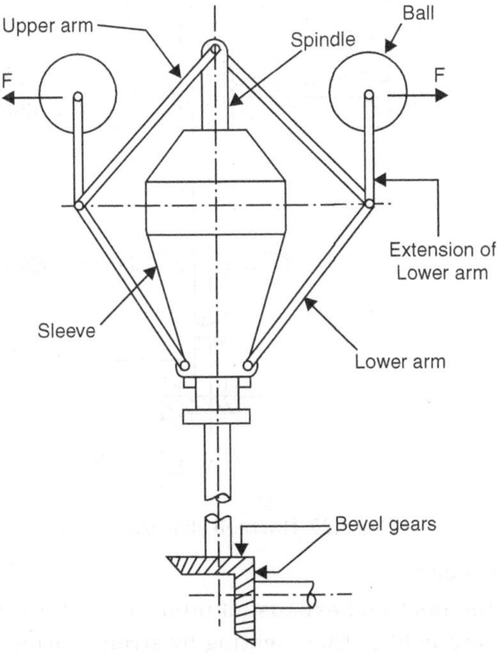

Fig. 2.24. Proell governor.

of the governor, a Proell governor runs at a *lower speed* than a Porter governor. *In order to give the same equilibrium speed a ball of smaller mass may be used in Proell governor.*

4. Hartnell governor

The Hartnell governor is a spring loaded governor in which the controlling force, to a great extent, is provided by the spring thrust.

Fig. 2.25 shows one of the types of Hartnell governors. It consists of casing fixed to the spindle. A compressed spring is placed inside the casing which presses against the top of the casing and on adjustable collars. The sleeve can move up and down on the vertical spindle depending upon the speed of the governor. Governor balls are carried on bell crank lever which are pivoted on the lower end of the casing. The balls will fly outwards or inwards as the speed of the governor shaft increases or decreases respectively.

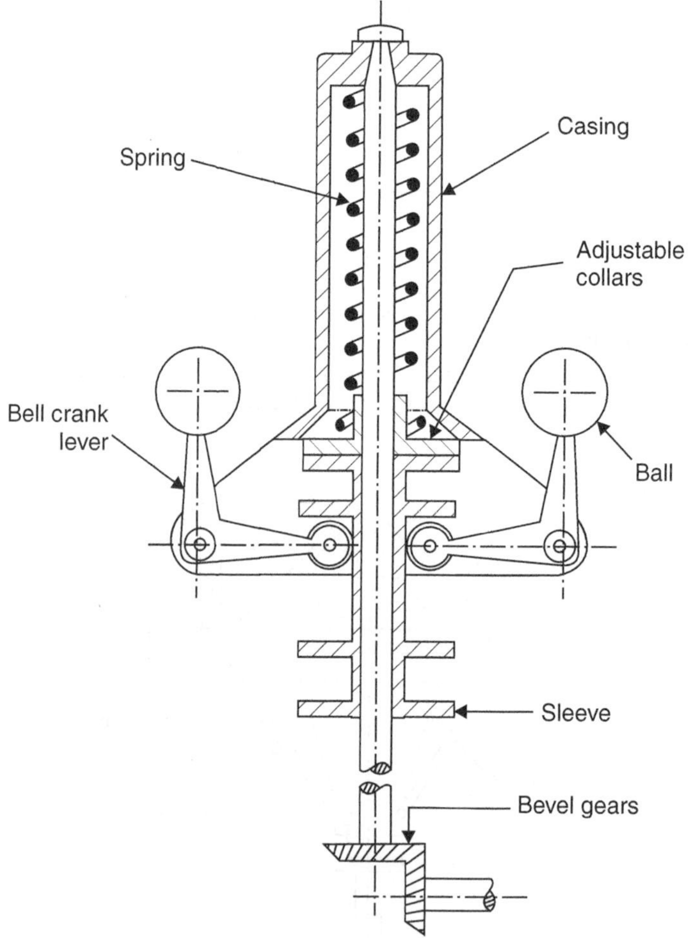

Fig. 2.25. Hartnell gbovernor.

5. Valves and valve gears

With few exceptions the inlet and exhaust of internal combustion engines are controlled by poppet valves. These valves are held to their seating by strong springs, and as the valves usually open inwards, the pressure in the cylinder helps to keep them closed. The valves are lifted from their seats and the ports opened either by cams having projecting portion designed to give the period of opening required or by eccentrics operating through link-work. Of these two methods the cam gear is more commonly used, but in either case it is necessary that the valve gear shaft of an engine should rotate but once from beginning to end of a complete cycle, however many strokes may be involved in the completion of that cycle. This is necessary to secure a continuous regulation of the valve gear as required. For this purpose the cams or eccentrics of four-stroke engines are mounted on shafts driven by gearing at half the speed of the crankshaft. The curves used for the acting faces of the cams depend on the speed of the engine and rapidity of valve opening desired.

Fig. 2.26 shows a valve gear for I.C. engine. It consists of poppet valve, the steam bushing or guide, valve spring, spring retainer, lifter or push rod, camshaft and half speed gear for a four-

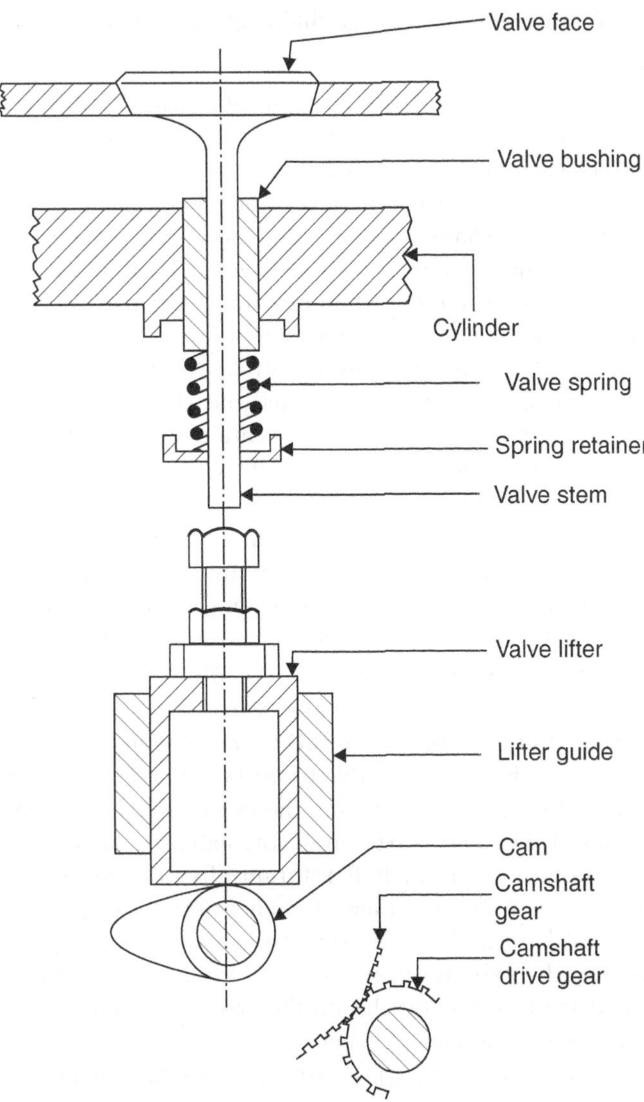

Fig. 2.26. Valve gear for I.C. engine.

stroke engine. The poppet valve, in spite of its shortcomings of noise and difficulties of cooling is commonly used due to its simplicity and capacity for effective sealing under all operating conditions. The valve is subjected to very heavy duty. It holds in combustion chamber and is exposed to high temperatures of burning gases. Exhaust valve itself may attain a high temperature while external cooling is not available. Special heat resisting alloys are therefore used in the construction of the exhaust valve and it may sometimes have a hollow construction filled with mineral salts to provide for heat dissipation. The salts become liquid when valve is working and transfer heat from the head to the stem from which it is carried through the stem guide to the cylinder block.

The timing of the valves *i.e.* their opening and closing with respect to the travel of the piston is very important thing for efficient working of the engine. The drive of the camshaft is arranged through gears or chain and sprocket (called timing gear or timing chain). Any wearing of the gears or chain and sprocket would result in disturbing the precise timing of the valves. It is desirable, therefore, to avoid use of multiple gears of long chains in the camshaft drive.

Valve timing

Theoretically the valves open and close at top dead centre (T.D.C.) or at bottom dead centre (B.D.C.) but practically they do so some time before or after the piston reaches the upper or lower limit of travel. There is a reason for this. Look at the inlet valve, for example. It normally opens several degrees of crankshaft-rotation before T.D.C. on the exhaust stroke. That is the intake valve begins to open before the exhaust stroke is finished. This gives the valve enough time to reach the fully open position before the intake stroke begins. Then, when the intake stroke starts, the intake valve is already wide open and air fuel mixture can start to enter the cylinder, immediately. Likewise the intake valve remains open for quite a few degrees of crankshaft rotation after the piston has passed B.D.C. at the end of the intake stroke. This allows additional time for air fuel mixture to continue to flow into the cylinder. The fact that the piston has already passed B.D.C. and is moving up or the compression stroke while the intake valve is still open does not effect the movement of air fuel mixture into the cylinder. Actually air fuel mixture is still flowing in as the intake valve starts to close.

This is due to the fact that air-fuel mixture has inertia. That is, it attempts to keep on flowing after it once starts through the carburettor and into the engine cylinder. The momentum of the mixture then keeps it flowing into the cylinder even though the piston has started up on the compression stroke. This packs more air-fuel mixture into the cylinder and results in a stronger power stroke. In other words, this improves *volumetric efficiency.*

For a some what similar reason, the exhaust valve opens well before the piston reaches B.D.C. on the power stroke. As the piston nears B.D.C., most of the push on the piston has ended and nothing is lost by opening the exhaust valve towards the end of the power stroke. This gives the exhaust gases additional time to start leaving the cylinder so that exhaust is well started by the time the piston passes B.D.C. and starts up on the exhaust stroke. The exhaust valve then starts opening for some degrees of crankshaft rotation after the piston has passed T.D.C. and intake stroke has started. This makes good use of momentum of exhaust gases. They are moving rapidly towards the exhaust port, and leaving the exhaust valve open for a few degrees after the intake stroke starts giving the exhaust gases some additional time to leave the cylinder. This allows more air-fuel mixture to enter on the intake stroke so that the stronger power stroke results. That is, it improves volumetric efficiency.

The actual timing of the valves varies with different four stroke cycle engines, but the typical example for an engine is shown in Fig. 2.27. Note that the inlet valve opens 15° of crankshaft rotation before T.D.C. on the exhaust stroke and stays open until 50° of crankshaft rotation after B.D.C. on the compression stroke. The exhaust valve opens 50° before B.D.C. on the power stroke and stays open 15° after T.D.C. on the inlet stroke. This gives the two valves an overlap of 30° at the end of exhaust stroke and beginning of the *compression stroke.*

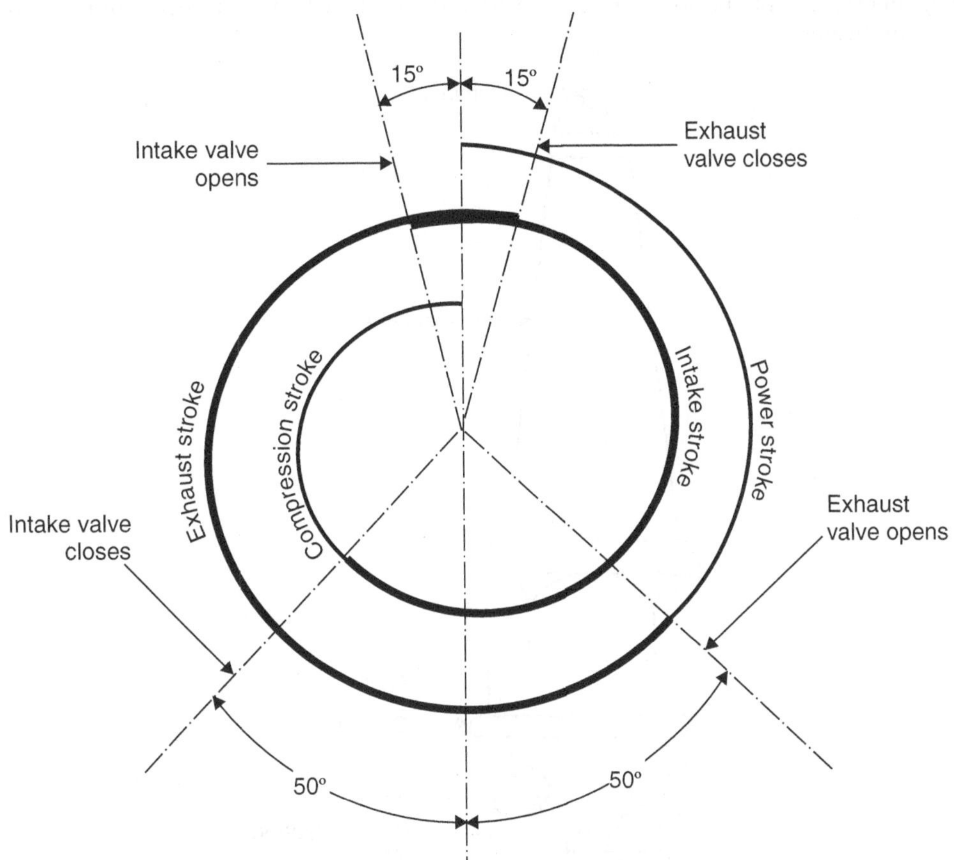

Fig. 2.27. Typical valve timing diagram.

B. Parts common to petrol engine only :

Spark-plug

The main function of a spark-plug is to conduct the high potential from the ignition system into the combustion chamber. It provides the proper gap across which spark is produced by applying high voltage, to ignite the combustion chamber.

A spark-plug entails the following requirements :

(*i*) It must withstand peak pressures up to atleast 55 bar.

(*ii*) It must provide suitable insulation between two electrodes to prevent short circuiting.

(*iii*) It must be capable of withstanding high temperatures to the tune of 2000°C to 2500°C over long periods of operation.

(*iv*) It must offer maximum resistance to erosion burning away of the spark points irrespective of the nature of fuel used.

(*v*) It must possess a high heat resistance so that the electrodes do not become sufficiently hot to cause the preignition of the charge within the engine cylinder.

(*vi*) The insulating material must withstand satisfactorily the chemical reaction effects of the fuel and hot products of combustion.

(*vii*) Gas tight joints between the insulator and metal parts are essential under all operating conditions.

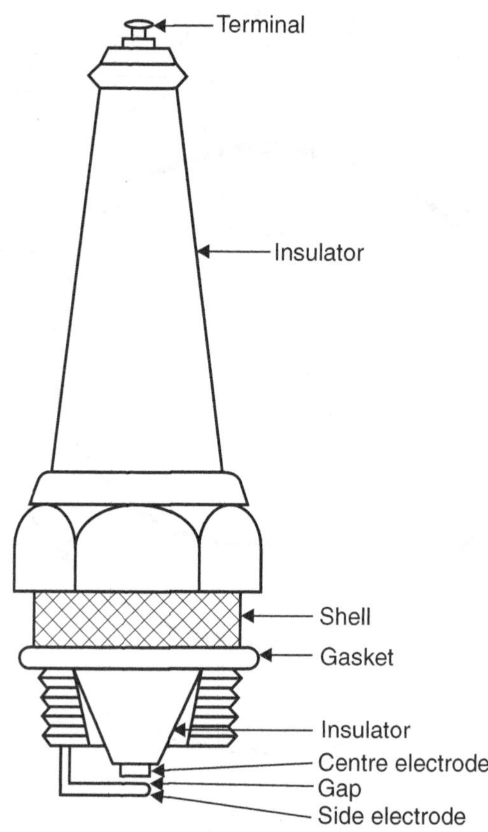

Fig. 2.28. Spark-plug.

Refer Fig. 2.28. The spark-plug consists of a metal shell having two electrodes which are insulated from each other with an air gap. High tension current jumping from the supply electrode produces the necessary spark. Plugs are sometimes identified by the heat range or the relative temperature obtained during operation. The correct type of plug with correct width of gap between the electrodes are important factors. The spark-plug gap can be easily checked by means of a feeler gauge and set as per manufacturer's specifications. It is most important that while adjusting the spark plug it is the outer earthed electrode *i.e.*, tip which is moved in or out gradually for proper setting of the gap. No bending force should be applied on the centre-electrode for adjusting the gap as this can cause crack and fracture of insulation and the plug may become absolutely useless.

Porcelain is commonly used as insulating material in spark-plugs, as it is cheap and easy to manufacture. Mica can also be used as insulating material for spark-plugs. Mica, however, cannot withstand high temperatures successfully.

- **Operating Heat Range :**
  - A spark-plug heat range is a measure of the plug's ability to transfer heat from the central electrode and insulator nose to the cylinder-head and cooling system.
  - When the heat absorbed by the plug's central electrode and insulator nose exceeds the capability of the plug to dissipate this heat in the same time, then the plug will *overheat* and the central electrode temperature will rise above its safe operating limit of about 900 to 950°C. *Above the plug's upper working temperature-limit, the central electrode will glow and ignite the air-fuel mixture before the timed spark actually occurs.* This condition is *known as* **auto-ignition** as it *automatically starts the combustion process independently of the controlled ignition spark.* The danger of this occuring is in the fact that it may take place relatively early in the compression stroke. Consequently, the pressure generated in the particular cylinder suffering from auto-ignition will oppose the upward movement of the piston. Excessive mechanical stresses will be produced in the reciprocating and rotating components and an abnormal rise in the cylinder temperature would, if allowed to continue, damage the engine.
  - If the plug's ability to transfer heat away from the central electrode and insulator tip *exceeds* that of the input heat from combustion, over the same time span, then the plug's central electrode and insulator nose *would operate at such a low temperature as to permit the formation of carbon deposits around the central nose of the plug.* This *critical lower temperature region is usually between 350°C and 400°C* and, at temperatures below this, carbon or oil deposits will *foul the insulation,* creating conducting shunts to the inside of the metal casing of the plug. Consequently, if *deposits are permitted to form, a proportion of the ignition spark energy will bypass the plug gap so that there will be insufficient energy left to ionize the electrode with the result that misfiring will result.* Establishing a heat balance between the plug's input and output heat flow, so that the *plug's temperature remains just in excess of 400°C, provides a self cleaning action on both the surfaces of the electrodes and insulator.*
  - A good spark-plug design tries to match the heat flowing from the plug to the heat flowing into it, caused by combustion under all working conditions, so that the plug operates below the upper temperature limit at full load, but never drops below the lower limit when idling or running under light-load conditions.

- **Firing Voltage :**

  A certain *minimum voltage* is necessary to make the spark jump the electrode air gap, the actual magnitude of the voltage required will depend upon the following *factors* :

  (*i*) Compression pressure (*ii*) Mixture strength

  (*iii*) Electrode gap (*iv*) Electrode tip temperature.

- **Tightness of Spark-plug :**
  - The seat joint tightness is essential for good heat dissipation.
  - Spark-plugs should not be over tightened otherwise the plug metal casing may become distorted, causing the central electrode insulator to break its seal and become loose. *Combustion gas may then escape through the plug with the result that it overheats.*

— An under-tightened plug may work itself loose and cause combustion gas to escape between the plug and cylinder-head plug hole threads to the atmosphere, again this will *result in overheating and rapid deterioration of the electrode tips.*

It is best to torque the plug to a definite degree of tightness.

Simple carburettor

The function of a carburettor is to atomise and metre the liquid fuel and mix it with the air as it enters the induction system of the engine, maintaining under all conditions of operation fuel-air proportions appropriate to those conditions.

All modern carburettors are based upon Bernoulli's theorem,

$$C^2 = 2gh$$

where C is the velocity in metres/sec, g is the acceleration due to gravity in metre/sec^2 and h is the head causing the flow expressed in metres of height of a column of the fluid.

The equation of mass rate of flow is given by,

$$m = \rho A\sqrt{2gh}$$

where ρ is the density of the fluid and A is the cross-sectional area of fluid stream.

In Fig. 2.29 is shown simple carburettor. L is the float chamber for the storage of fuel. The fuel supplied under gravity action or by fuel pump enters the float chamber through the filter F.

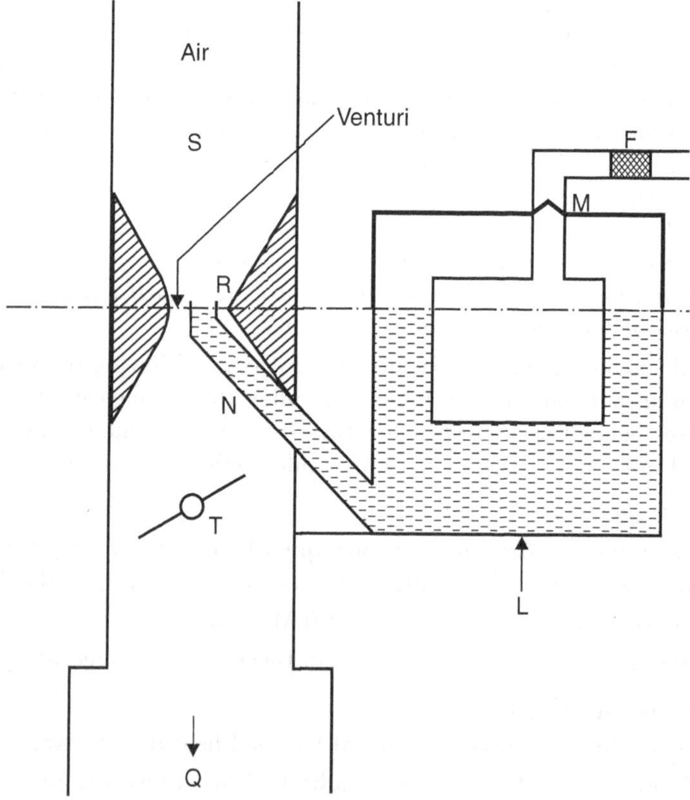

Fig. 2.29. Simple carburettor.

The arrangement is such that when the oil reaches a particular level the float valve *M* blocks the inlet passage and thus cuts off the fuel oil supply. On the fall of oil level, the float descends down, consequently intake passage opens and again the chamber is filled with oil. Then the float and the float valve maintains a constant fuel oil level in the float chamber. *N* is the jet from which the fuel is sprayed into the air stream as it enters the carburettor at the inlet *S* and passes through the throat or venturi *R*. The fuel level is slightly below the outlet of the jet when the carburettor is inoperative.

As the piston moves down in the engine cylinder, suction is produced in the cylinder as well as in the induction manifold *Q* as a result of which air flows through the Carburettor. The velocity of air increases as it passes through the construction at the venturi *R* and pressure decreases due to conversion of a portion of pressure head into kinetic energy. Due to decreased pressure at the venturi and hence by virtue of difference in pressure (between the float chamber and the venturi) the jet issues fuel oil into air stream. Since the jet has a very fine bore, the oil issuing from the jet is in the form of fine spray ; it vapourises quickly and mixes with the air. This air fuel mixture enters the engine cylinder ; its quantity being controlled by varying the position of the throttle valve *T*.

Limitations :

(*i*) Although theoretically the air fuel ratio supplied by a simple (single jet) carburettor should remain constant as the throttle goes on opening, actually it provides increasingly richer mixture as the throttle is opened. This is because of the reason that the density of air tends to decrease as the rate of flow increases.

(*ii*) During idling, however, the nearly closed throttle causes a reduction in the mass of air flowing through the venturi. At such low rates of air flow, the pressure difference between the float chamber and the fuel discharge nozzle becomes very small. It is sufficient to cause fuel to flow through the jet.

(*iii*) Carburettor does not have arrangement for providing rich mixture during starting and warm up.

In order to *correct for faults* :

(*i*) number of compensating devices are used for (*ii*) an idling jet is used which helps in running the engine during idling. For (*iii*) choke arrangement is used.

Fuel pump (for carburettor-petrol engine).

Refer Fig. 2.30. This type of pump is used in petrol engine for supply of fuel to the carburettor. Due to rotation of the crankshaft the cam pushes the lever in the upward direction. One end of the lever is hinged while the other end pulls the diaphragm rod with the *diaphragm*. So the diaphragm comes in the downward direction against the compression of the spring and thus a vacuum is produced in the pump chamber. This causes the fuel to enter into the pump chamber from the *glass bowl* through the *strainer* and the inlet valve, the impurities of the fuel ; if there is any, deposit at the bottom of the glass bowl. On the return stroke the spring pushes the diaphragm in the upward direction forcing the fuel from the pump chamber into the carburettor through the *outlet valve*.

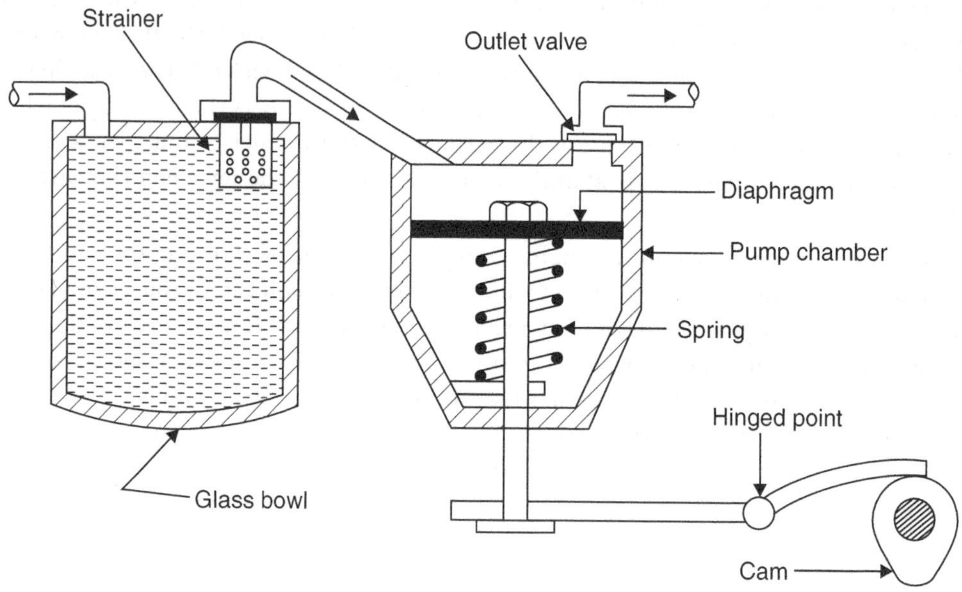

Fig. 2.30. Fuel pump for carburettor.

Parts for Diesel engine only :

FUEL PUMP

Refer Fig. 2.31. *L* is the plunger which is driven by a cam and tappet mechanism at the bottom (not shown in the figure) *B* is the barrel in which the plunger reciprocates. There is the rectangular vertical groove in the plunger which extends from top to another helical groove. *V* is the delivery valve which lifts off its seat under the liquid fuel pressure and against the spring force (*S*). The fuel pump is connected to fuel atomiser through the passage *P*, *SP* and *Y* are the spill and supply ports respectively. When the plunger is at its bottom stroke the ports *SP* and *Y* are uncovered (as shown in the Fig. 2.31) and oil from low pressure pump (not shown) after being filtered is forced into the barrel. When the plunger moves up due to cam and tappet mechanism, a stage reaches when both the ports *SP* and *Y* are closed and with the further upward movement of the plunger the fuel gets compressed. The high pressure thus developed lifts the delivery valve off its seat and fuel flows to atomiser through the passage *P*. With further rise of the plunger, at a certain moment, the port *SP* is connected to the fuel in the upper part of the plunger through the rectangular vertical groove by the helical groove ; as a result of which a sudden drop in pressure occurs and the delivery valve falls back and occupies its seat against the spring force. The plunger is rotated by the rack *R* which is moved in or out by the governor. *By changing the angular position of the helical groove (by rotating the plunger) of the plunger relative to the supply port, the length of stroke during which the oil is delivered can be varied and thereby quantity of fuel delivered to the engine is also varied accordingly.*

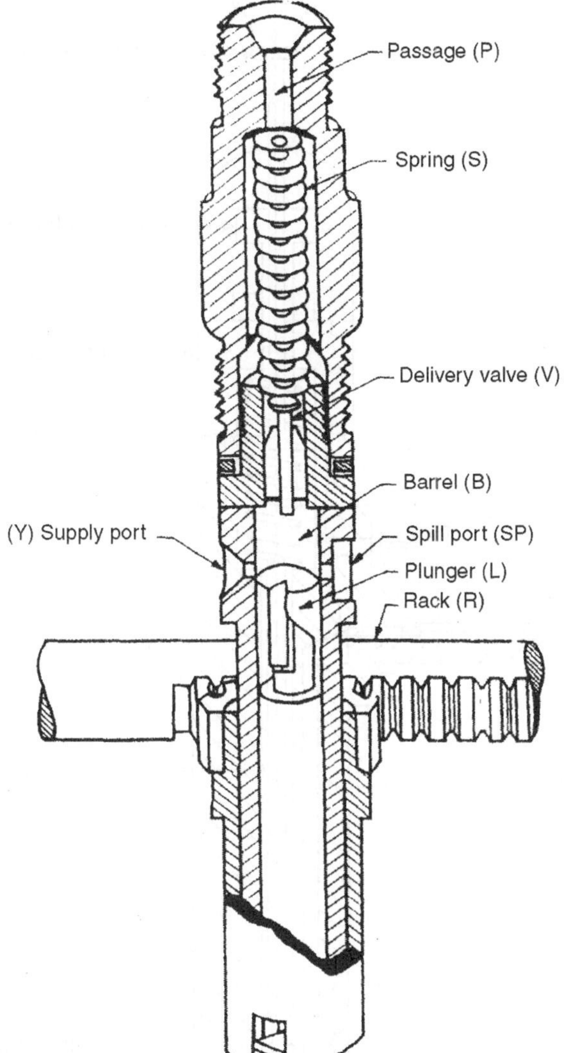

Fig. 2.31. Fuel pump.

Fuel atomiser or injector

Refer Fig. 2.32. It consists of a nozzle valve (*NV*) fitted in the nozzle body (*NB*). The nozzle valve is held on its seat by a spring '*S*' which exerts pressure through the spindle *E*. '*AS*' is the adjusting screw by which the nozzle valve lift can be adjusted. Usually the nozzle valve is set to lift at 135 to 170 bar pressure. *FP* is the feeling pin which indicates whether valve is working properly or not. The oil under pressure from the fuel pump enters the injector through the passages *B* and *C* and lifts the nozzle valve. The fuel travels down the nozzle *N* and injected into the engine cylinder in the form of fine sprays. When the pressure of the oil falls, the nozzle valve occupies its seat under the spring force and fuel supply is cut off. Any leakage of fuel accumulated above the valve is led to the fuel tank through the passage *A*. The leakage occurs when the nozzle valve is *worn out*.

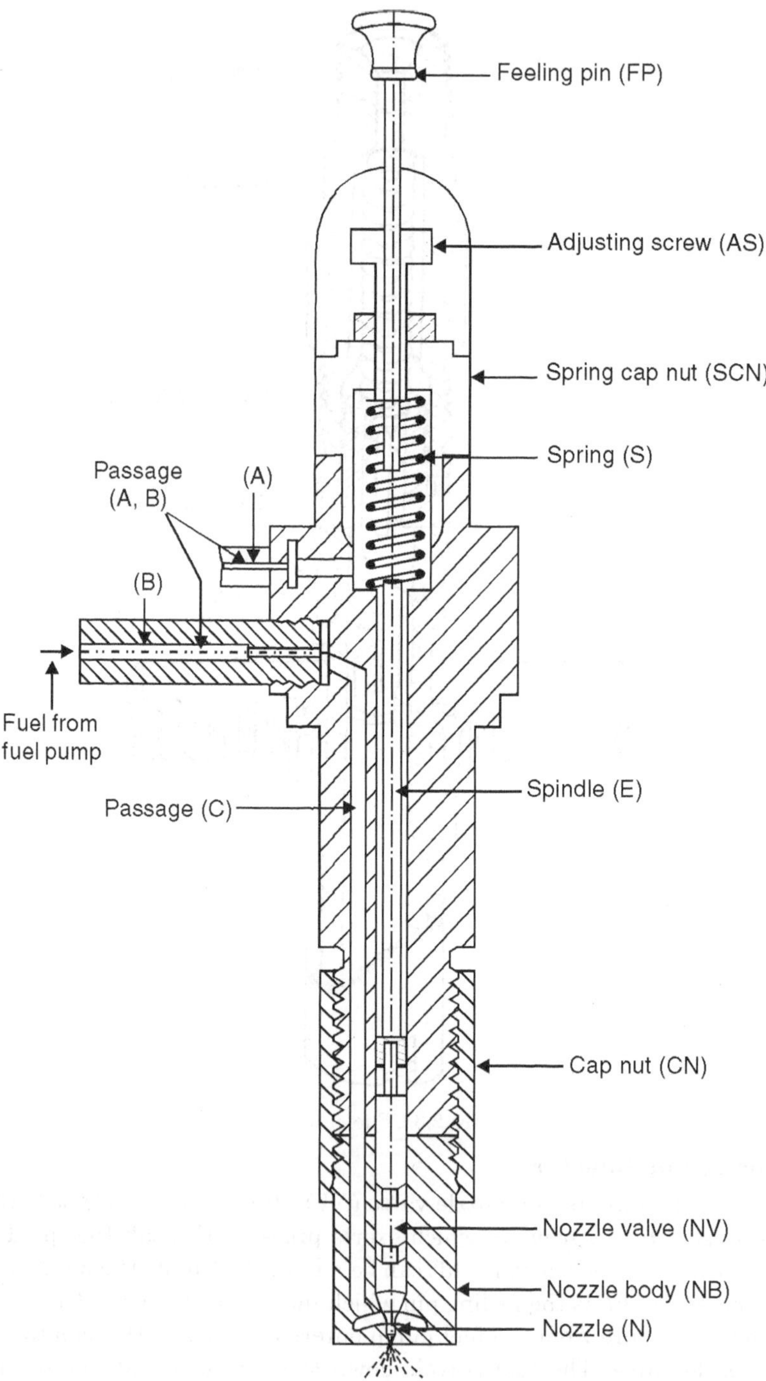

Fig. 2.32. Fuel atomiser or injector.

List of engine parts, materials, method of manufacture and functions :

	Name of the part	*Material*	*Function*	*Method of manufacture*
1.	*Cylinder*	Hard grade cast iron	Contains gas under pressure and guides the piston.	Casting
2.	*Cylinder head*	Cast iron or aluminium	Main function is to seal the working end of the cylinder and not to permit entry and exit of gases on overhead valve engines.	Casting, forging
3.	*Piston*	Cast iron or aluminium alloy	It acts as a face to receive gas pressure and transmits the thrust to the connecting rod.	Casting, forging
4.	*Piston rings*	Cast iron	Their main function is to provide a good sealing fit between the piston and cylinder.	Casting
5.	*Gudgeon pin*	Hardened steel	It supports and allows the connecting rod to swivel.	Forging
6.	*Connecting rod*	Alloy steel ; for small engines the material may be aluminium	It transmits the piston load to the crank, causing the latter to turn, thus converting the reciprocating motion of the piston into rotary motion of the crankshaft.	Forging
7.	*Crankshaft*	In general the crankshaft is made from a high tensile forging, but special cast irons are sometimes used to produce a light weight crank shaft that does not require a lot of machining.	It converts the reciprocating motion of the piston into the rotary motion.	Forging
8.	*Main bearings*	The typical bearing half is made of steel or bronze back to which a lining of relatively soft bearing material is applied.	The function of bearing is to reduce the friction and allow the parts to move easily.	Casting
9.	*Flywheel*	Steel or cast iron.	In engines it takes care of fluctuations of speed during thermodynamic cycle.	Casting
10.	*Inlet valve*	Silicon chrome steel with about 3% carbon.	Admits the air or mixture of air and fuel into engine cylinder.	Forging
11.	*Exhaust valve*	Austenitic steel	Discharges the product of combustion.	Forging

2.8. TERMS CONNECTED WITH I.C. ENGINES

Refer Fig. 2.33.

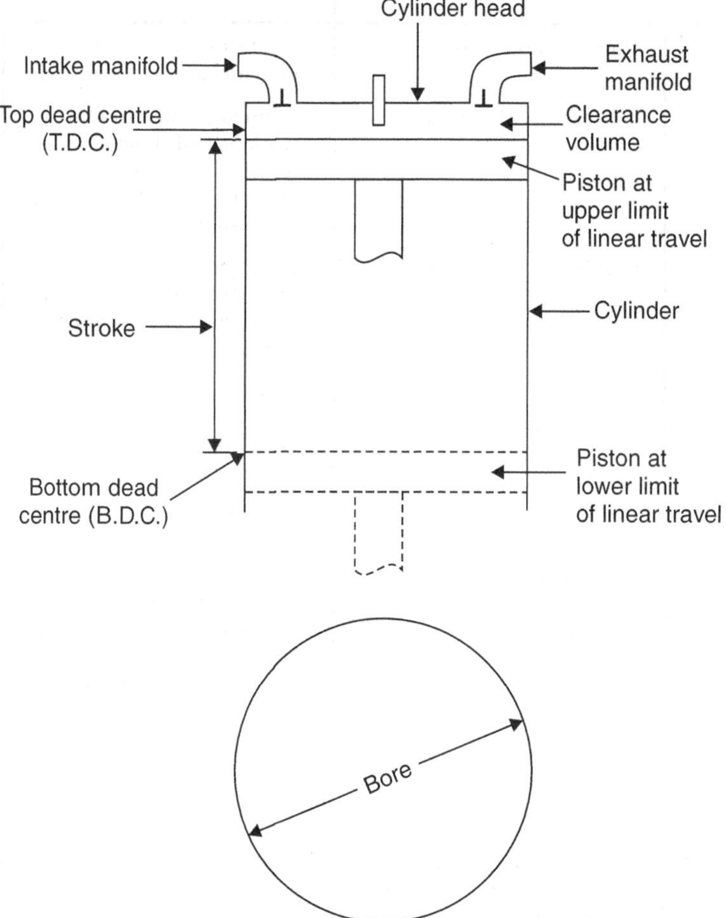

Fig. 2.33. Terms relating I.C. engines.

Bore. *The inside diameter of the cylinder is called "bore".*

Stroke. As the piston reciprocates inside the engine cylinder, it has got limiting upper and lower positions beyond which it cannot move and reversal of motion takes place at these limiting positions.

The linear distance along the cylinder axis between two limiting positions, is called "stroke".

Top Dead Centre (T.D.C.). *The top most position of the piston towards cover end side of the cylinder is called "top dead centre".* In case of horizontal engines, this is known as *inner dead centre.*

Bottom Dead Centre (B.D.C.). *The lowest position of the piston towards the crank end side of the cylinder is called "bottom dead centre".* In case of horizontal engines it is called *outer dead centre.*

Clearance volume. *The volume contained in the cylinder above the top of the piston, when the piston is at top dead centre, is called the "clearance volume".*

- Bore sizes of engines range from 0.5 m down to 0.5 cm. The ratio of bore of stroke D/L, for small engines is usually from 0.8 to 1.2.

— An engine with $L = D$ is often called a **square** *engine* ;

— If $L > D$ the engine is **under square ;**

— If $L < D$ the engine is **over square.**

Very large engines are *always under square,* with *stroke lengths up to four times bore diameter.*

Swept volume. *The volume swept through by the piston in moving between top dead centre and bottom dead centre, is, called "swept volume or piston displacement".* Thus, when piston is at bottom dead centre, total volume = swept volume + clearance volume.

- Typical values for engine displacement range from 0.1 cm^3 for small model airplanes to about 8 litres for large automobiles to much large number for large ship engines. The displacement of a modern average automobile engine is about two to three litres.
- For a given displacement volume, a longer stroke allows for a smaller bore (under square), resulting in less surface area in the combustion chamber and correspondingly less heat loss. This increases thermal efficiency within the combustion chamber. However, the longer stroke results in higher piston speed and higher friction losses that reduce the output power which can be obtained off the crankshaft. If the stroke is shortened, the bore must be increased and the engine will be over square. This decreases friction losses but increases heat transfer losses. *Most modern automobile engines are near square, with some slightly over square and some slightly under square.*

Compression ratio. *It is ratio of total cylinder volume to clearance volume.*

Refer Fig. 2.33. Compression ratio (r) is given by

$$r = \frac{V_s + V_c}{V_c}$$

where V_s = Swept volume, V_c = Clearance volume.

The compression ratio varies from 5 : 1 to 11 : 1 (average value 7 : 1 to 9 : 1) in *S.I. engines* and from 12 : 1 to 24 : 1 (average value 15 : 1 to 18 : 1) in *C.I. engines.*

- Modern spark ignition (S.I.) engines have compression ratios of 8 to 11, while compression ignition (C.I.) engines have compression ratios in the range 12 to 24. *Engines with superchargers or turbochargers usually have lower compression ratios than naturally aspirated engines.*
- Various attempts have been made to develop engines with a *variable compression ratio.* One such system uses a *split piston that expands due to changing hydraulic pressure caused by engine speed and load.* Some two-stroke cycle engines have been built which have a sleeve-type valve that changes the slot opening on the exhaust port. The piston where the exhaust port is fully closed can be adjusted by several degrees of engine rotation. This changes the *effective compression ratio* of the engine.

Piston speed. *The average speed of the piston is called "piston speed".*

Piston speed = 2 LN

where L = Length of the stroke, and

N = Speed of the engine in r.p.m.

- Average engine speed for all engines will normally be in the range of 5 to 15 m/s with large diesel engines on the low end and high performance automobile engines on the high end. There are following *two reasons* why engines operate in this range :
  — *First,* this is about the *safe limit which can be tolerated by material strength of the engine components.*
  — The *second* reason why maximum average piston speed is limited is *because of the gas flow into and out of cylinders.* Piston speed determines the instantaneous flow

rate of air-fuel into the cylinder during intake and exhaust flow out of the cylinder during the exhaust stroke. *Higher piston speeds would require larger valves to allow for higher flow rates.* In most engines, valves are at a maximum size with no room for enlargement.

Some Other Terms :

Direct Injection (D.I.). *Fuel injection into the main combustion chamber of an engine.* Engines have either one main combustion chamber (open chamber) or a divided combustion chamber made up of a main chamber and a smaller connected secondary chamber.

Indirect Injection (I.D.I.). *Fuel injection into the secondary chamber of an engine with a divided combustion chamber.*

Smart Engine. *Engine with computer controls that regulate operating characteristics* such as *air-fuel ratio, ignition timing, valve timing, exhaust control, intake tuning* etc.

Engine Management System (E.M.S.). Computer and electronics used to control smart engines.

Wide Open Throttle (W.O.T.). Engine operated with *throttle valve fully open when maximum power and/or speed is desired.*

Ignition Delay (I.D.). It is the *time interval between ignition initiation and the actual start of combustion.*

Air-Fuel Ratio (A/F). It is the *ratio of the air to mass of fuel input into engine.*

Fuel-Air Ratio (F/A). It is the *ratio of fuel to mass of air input into engine.*

2.9. WORKING CYCLES

An internal combustion engine can work on any one of the following cycles :

(*a*) Constant volume or Otto cycle

(*b*) Constant pressure or Diesel cycle

(*c*) Dual combustion cycle.

These may be either *four stroke cycle or two stroke cycle engines.*

(*a*) **Constant volume or Otto cycle.** The cycle is so called because heat is supplied at constant volume. Petrol, gas, light oil engines work on this cycle. In the case of a petrol engine the proper mixing of petrol and air takes place in the carburettor which is situated outside the engine cylinder. The proportionate mixture is drawn into the cylinder during the suction stroke. In a gas engine also, air and gas is mixed outside the engine cylinder and this mixture enters the cylinder during the suction stroke. In light oil engines the fuel is converted to vapours by a vapouriser which receives heat from the exhaust gases of the engine and their mixture flows towards engine cylinder during suction stroke.

(*b*) **Constant pressure or diesel cycle.** In this cycle only air is drawn in the engine cylinder during the suction stroke, this air gets compressed during the compression stroke and its pressure and temperature increase by a considerable amount. Just before the end of the stroke a metered quantity of fuel under pressure adequately more than that developed in the engine cylinder is injected in the form of fine sprays by means of a fuel injector. Due to very high pressure and temperature of the air the fuel ignites and hot gases thus produced throw the piston downwards and work is obtained. *Heavy oil engines make use of this cycle.*

(*c*) **Dual combustion cycle.** This cycle is also called *semi-diesel cycle.* It is so named because heat is added *partly at constant volume and partly at constant pressure.* In this cycle only air is drawn in the engine cylinder during suction stroke. The air is then compressed in hot combustion chamber at the end of the cylinder during the compression stroke to a pressure of about 26 bar. The heat of compressed air together with heat of combustion chamber ignites the fuel. The fuel is injected into the cylinder just before the end of compression stroke where it ignites immediately. The fuel injection is continued until the point of cut off is reached. The burning of fuel at first takes place at constant volume and continues to burn at constant pressure during the first part of expansion or working stroke. The field of application of this cycle is heavy oil engines.

2.10. INDICATOR DIAGRAM

An indicator diagram is a graph between pressure and volume ; the former being taken on vertical axis and the latter on the horizontal axis. This is obtained by an instrument known as *indicator*. The indicator diagrams are of two types : (*a*) Theoretical or hypothetical, (*b*) Actual. The theoretical or hypothetical indicator diagram is always longer in size as compared to the actual one, since in the former losses are neglected. The ratio of the area of the actual indicator diagram to the theoretical one is called *diagram factor*.

2.11. FOUR STROKE CYCLE ENGINES

Here follows the description of the four stroke otto and diesel-cycle engines.

Otto engines. The Otto four stroke-cycle refers to its use in petrol engines, gas engines, light oil engines in which the mixture of air and fuel are drawn in the engine cylinder. Since ignition in these engines is due to a spark, therefore they are also called *spark ignition engines.*

The various strokes of a four stroke (Otto) cycle engine are detailed below.

Refer Fig. 2.34.

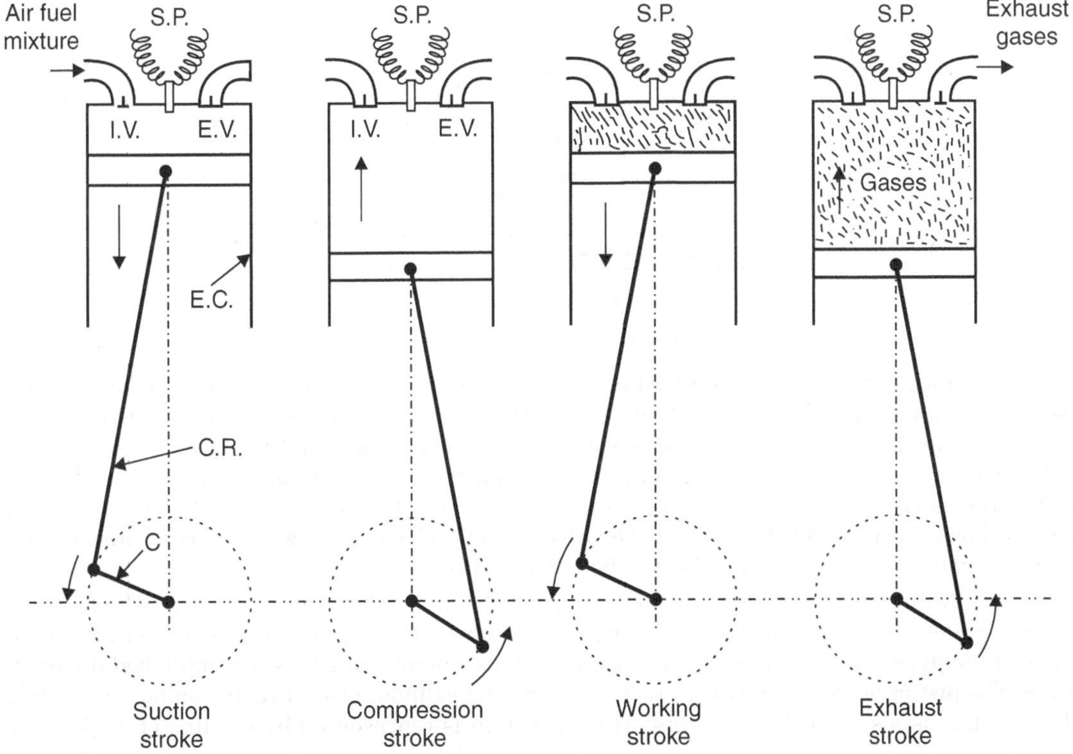

Fig. 2.34. Four stroke Otto cycle engine.

1. **Suction stroke.** During this stroke (also known as induction stroke) the piston moves from top dead centre (T.D.C.) to bottom dead centre (B.D.C.) ; the inlet valve opens and proportionate fuel air mixture is sucked in the engine cylinder. This operation is represented by the line 5—1 (Fig. 2.32). The exhaust valve remains closed throughout the stroke.

2. **Compression stroke.** In this stroke, the piston moves (1—2) towards (T.D.C.) and compresses the enclosed fuel air mixture drawn in the engine cylinder during suction. The pressure of the mixture rises in the cylinder to a value of about 8 bar. Just before the end of this stroke the operating-plug initiates a spark which ignites the mixture and combustion takes place at constant volume (line 2—3) (Fig. 2.35). Both the inlet and exhaust valves remain closed during the stroke.

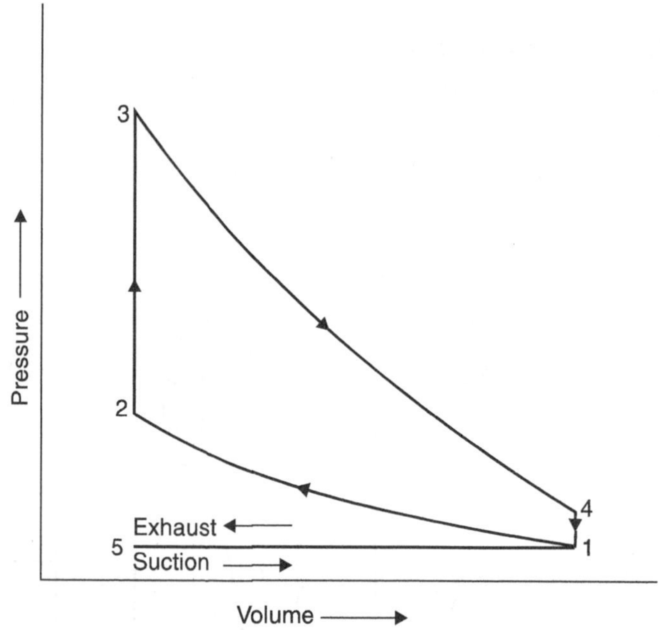

Fig. 2.35. Theoretical p-V diagram of a four stroke Otto cycle engine.

3. **Expansion or working stroke.** When the mixture is ignited by the spark plug the hot gases are produced which drive or throw the piston from T.D.C. to B.D.C. and thus the work is obtained in this stroke. It is during this stroke when we get work from the engine ; the other three strokes namely suction, compression and exhaust being idle. *The flywheel mounted on the engine shaft stores energy during this stroke and supplies it during the idle strokes.* The expansion of the gases is shown by 3-4. (Fig. 2.35). Both the valves remain closed during the start of this stroke but when the piston just reaches the B.D.C. the exhaust valve opens.

4. **Exhaust stroke.** This is the last stroke of the cycle. Here the gases from which the work has been collected become useless after the completion of the expansion stroke and are made to escape through exhaust valve to the atmosphere. This removal of gas is accomplished during this stroke. The piston moves from B.D.C. to T.D.C. and the exhaust gases are driven out of the engine cylinder ; this is also called *scavenging.* This operation is represented by the line (1-5) (Fig. 2.35).

Fig. 2.36 shows the actual indicator diagram of four stroke Otto cycle engine. It may be noted that line 5-1 is below the atmospheric pressure line. This is due to the fact that owing to restricted area of the inlet passages the entering fuel air mixture cannot cope with the speed of the piston. The exhaust line 4-5 is slightly above the atmospheric pressure line. This is due to restricted exhaust passages which do not allow the exhaust gases to leave the engine-cylinder quickly.

The loop which has area 4-5-1 is called *negative loop* ; it gives the pumping loss due to admission of fuel air mixture and removal of exhaust gases. The area 1-2-3-4 is the total or gross work obtained from the piston and network can be obtained by subtracting area 451 from the area 1-2-3-4.

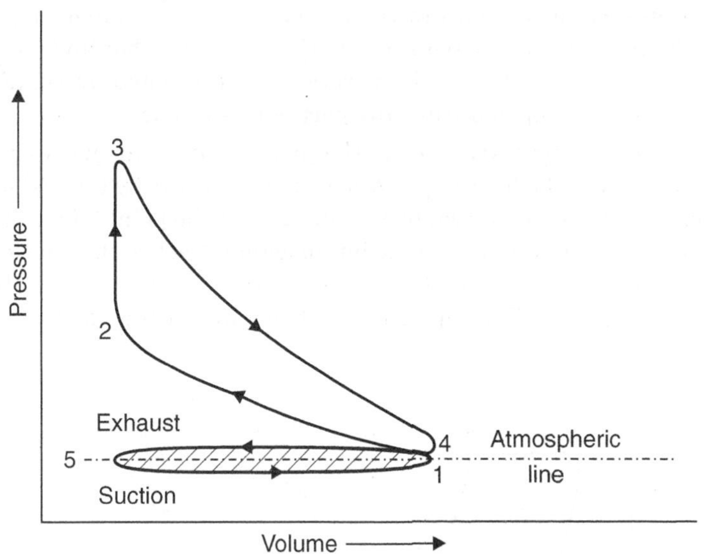

Fig. 2.36. Actual p-V diagram of a four stroke Otto cycle engine.

Diesel engines (four stroke cycle). As is the case of Otto four stroke ; this cycle too is completed in four strokes as follows. (Refer Fig. 2.37).

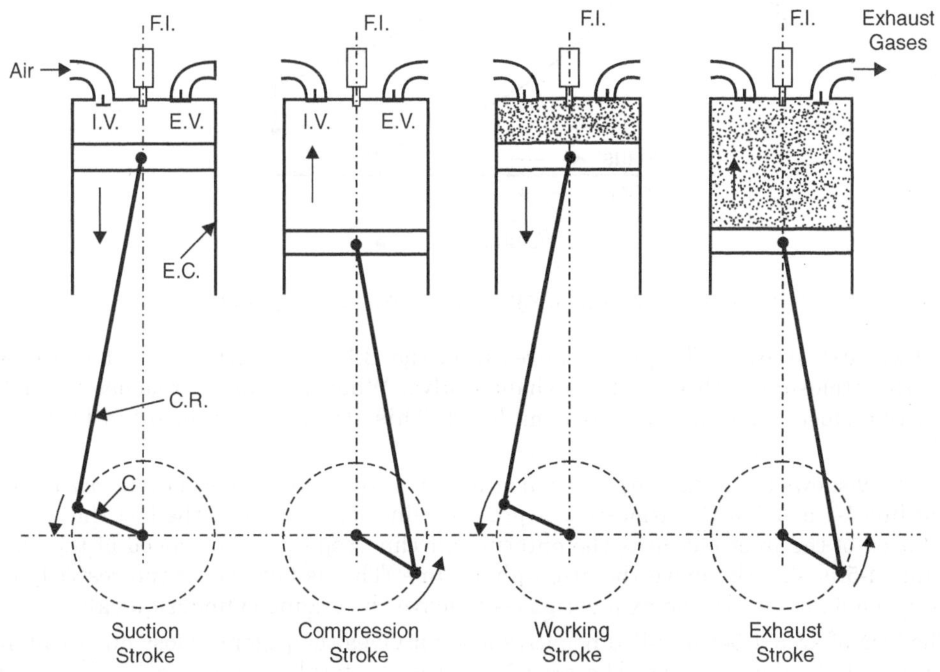

Fig. 2.37. Four stroke Diesel cycle engine.

1. **Suction stroke.** With the movement of the piston from T.D.C. to B.D.C. during this stroke, the inlet valve opens and the air at atmospheric pressure is drawn inside the engine cylinder ; the exhaust valve however remains closed. This operation is represented by the line 5-1 (Fig. 2.38).

2. **Compression stroke.** The air drawn at atmospheric pressure during the suction stroke is compressed to high pressure and temperature (to the value of 35 bar and 600°C respectively) as the piston moves from B.D.C. to T.D.C. This operation is represented by 1-2 (Fig. 2.38). Both the inlet and exhaust valves do not open during any part of this stroke.

3. **Expansion or working stroke.** As the piston starts moving from T.D.C. a matered quantity of fuel is injected into the hot compressed air in fine sprays by the fuel injector and it (fuel) starts burning at constant pressure shown by the line 2-3. At the point 3 fuel supply is cut off. The fuel is injected at the end of compression stroke but in actual practice the ignition of the fuel starts before the end of the compression stroke. The hot gases of the cylinder expand adiabatically to point 4, thus doing work on the piston. The expansion is shown by 3-4 (Fig. 2.38).

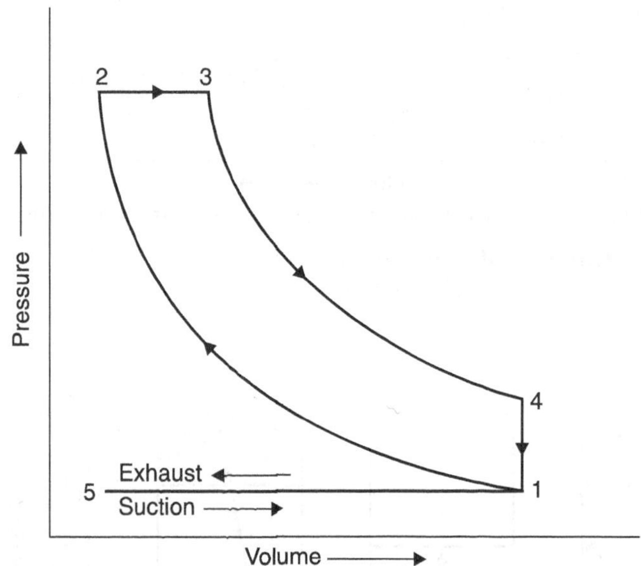

Fig. 2.38. Theoretical p-V diagram of a four stroke Diesel cycle.

4. **Exhaust stroke.** The piston moves from the B.D.C. to T.D.C. and the exhaust gases escape to the atmosphere through the exhaust valve. When the piston reaches the T.D.C. the exhaust valve closes and the cycle is completed. This stroke is represented by the line 1-5 (Fig. 2.38).

Fig. 2.39 shows the actual indicator diagram for a four-stroke Diesel cycle engine. It may be noted that line 5-1 is below the atmospheric pressure line. This is due to the fact that owing to the restricted area of the inlet passages the entering air can't cope with the speed of the piston. The exhaust line 4-5 is slightly above the atmospheric line. This is because of the restricted exhaust passages which do not allow the exhaust gases to leave the engine cylinder quickly.

The loop of area 4-5-1 is called negative loop ; it gives the pumping loss due to admission of air and removal of exhaust gases. The area 1-2-3-4 is the total or gross work obtained from the piston and net work can be obtained by subtracting area 4-5-1 from area 1-2-3-4.

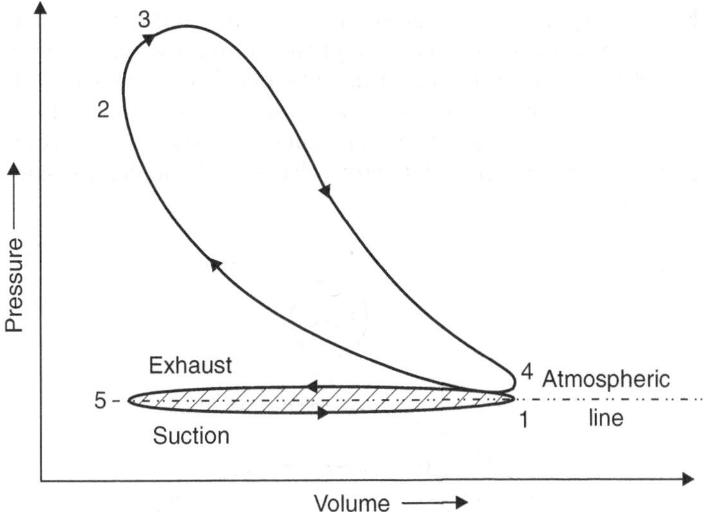

Fig. 2.39. Actual *p-V* diagram of four stroke Diesel cycle.

Valve Timing Diagrams (Otto and Diesel engines)

1. **Otto engine.** Fig. 2.40 shows a theoretical valve timing diagram for *four stroke "Otto*

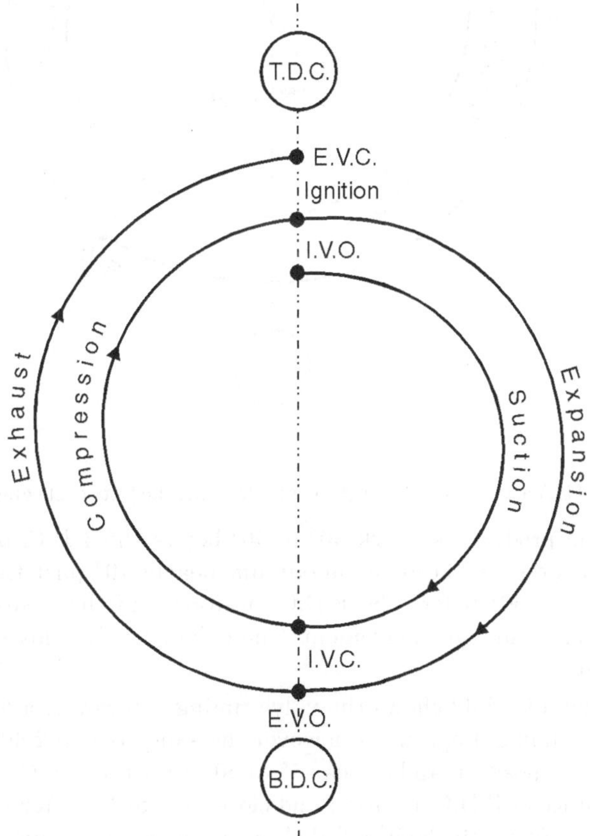

Fig. 2.40. Theoretical valve timing diagram (four stroke Otto cycle engine).

cycle" engines which is self-explanatory. In actual practice, it is difficult to open and close the valve instantaneously ; so as to get better performance of the engine the valve timings are modified. In Fig. 2.41 is shown an actual valve timing diagram. The inlet valve is opened 10° to 30° in advance of the T.D.C. position to enable the fresh charge to enter the cylinder and to help the burnt gases at the same time, to escape to the atmosphere. The suction of the mixture continues up to 30°-40° or even 60° after B.D.C. position. The inlet valve closes and the compression of the entrapped mixture starts.

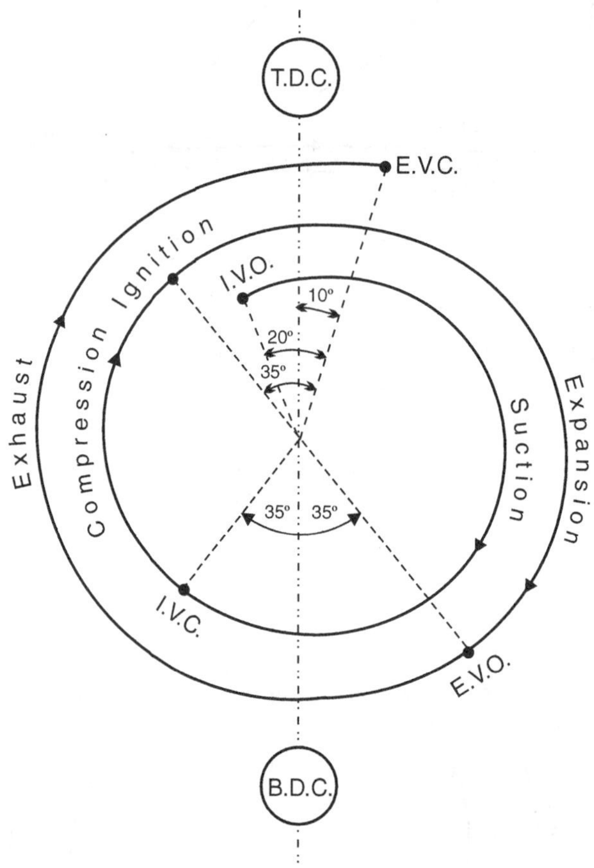

Fig. 2.41. Actual valve timing diagram (four Stroke Otto cycle engines).

The sparking plug produces a spark 30° to 40° before the T.D.C. position ; thus fuel gets more time to burn. The pressure becomes maximum nearly 10° past the T.D.C. position. The exhaust valve opens 30° to 60° before the B.D.C. position and the gases are driven out of the cylinder by piston during its upward movement. The exhaust valve closes when piston is nearly 10° past T.D.C. position.

2. **Diesel engines.** Fig. 2.42 shows the valve timing diagram of a *four stroke "Diesel cycle" engine* (theoretical valve timing diagram, is however the same as Fig. 2.40). Inlet valve opens 10° to 25° in advance of T.D.C. position and closes 25° to 50° after the B.D.C. position. Exhaust valve opens 30° to 50° in advance of B.D.C. position and closes 10° to 15° after the T.D.C. position. The fuel injection takes place 5° to 10° before T.D.C. position and continues up to 15° to 25° near T.D.C. position.

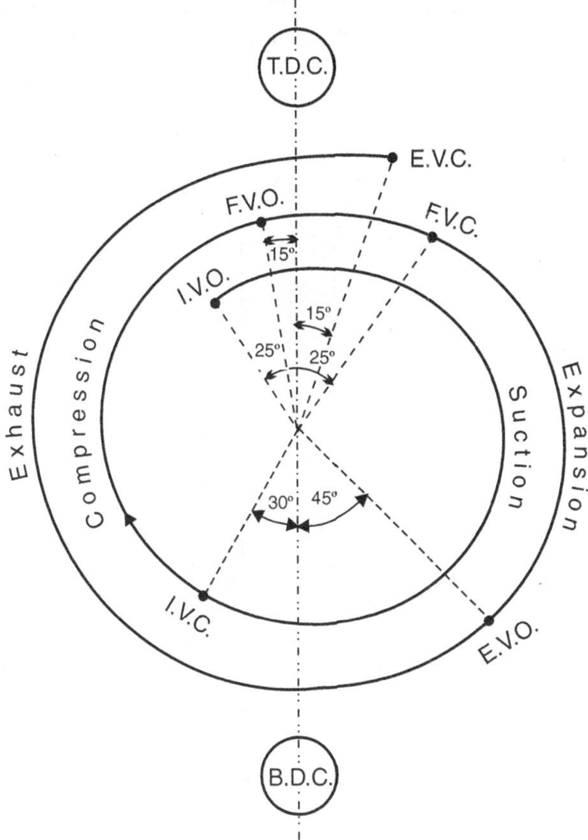

Fig. 2.42. Actual valve timing diagram (four stroke Diesel cycle engines).

2.12. TWO STROKE CYCLE ENGINES

In 1878, Dugald-clerk, a British engineer introduced a cycle which could be completed in two *strokes of piston rather than four strokes* as is the case with the four stroke cycle engines. The engines using this cycle were called two stroke cycle engines. In this engine suction and exhaust strokes are eliminated. Here *instead of valves, ports are used. The exhaust gases are driven out from engine cylinder by the fresh charge of fuel entering the cylinder nearly at the end of the working stroke.*

Fig. 2.43 shows a two stroke petrol engine (used in scooters, motor cycles etc.). The cylinder L is connected to a closed crank chamber C.C. During the upward stroke of the piston M, the gases in L are compressed and at the same time fresh air and fuel (petrol) mixture enters the crank chamber through the valve V. When the piston moves downwards, V closes and the mixture in the crank chamber is compressed. Refer Fig. 2.43 (*i*), the piston is moving upwards and is compressing an explosive change which has previously been supplied to L. Ignition takes place at the end of the stroke. The piston then travels downwards due to expansion of the gases (Fig. 2.43 (*ii*)) and near the end of this stroke the piston uncovers the exhaust port (E.P.) and the burnt exhaust gases escape through this port (Fig. 2.43 (*iii*)). The transfer port (T.P.) then is uncovered immediately, and the compressed charge from the crank chamber flows into the cylinder and is deflected upwards by the hump provided on the head of the piston. It may be noted that the incoming air petrol mixture helps the removal of gases from the engine-cylinder ; if, in case these exhaust gases do not leave the cylinder, the fresh charge gets diluted and efficiency of the engine will decrease. The piston then again starts moving from B.D.C. to T.D.C. and the charge gets compressed when E.P. (exhaust port) and T.P. are covered by the piston ; thus the cycle is repeated.

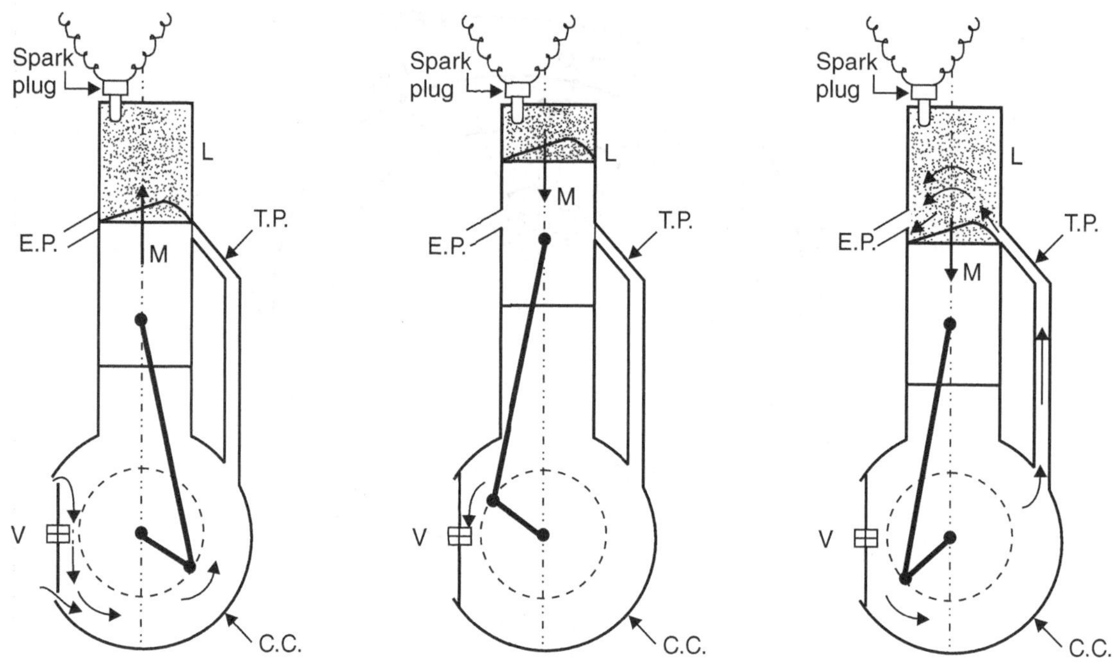

Fig. 2.43. Two stroke cycle engine.

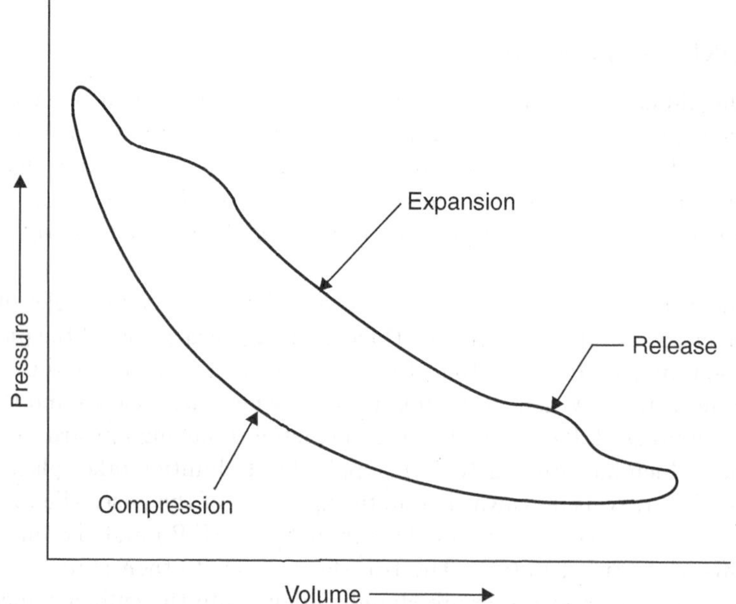

Fig. 2.44. p-V diagram for a two stroke cycle engine.

Fig. 2.44 shows the *p-V* diagram for a two stroke cycle engine. It is only for the main cylinder or the top side of the piston. Fig. 2.45 shows self-explanatory port timing diagram for a two stroke cycle engine.

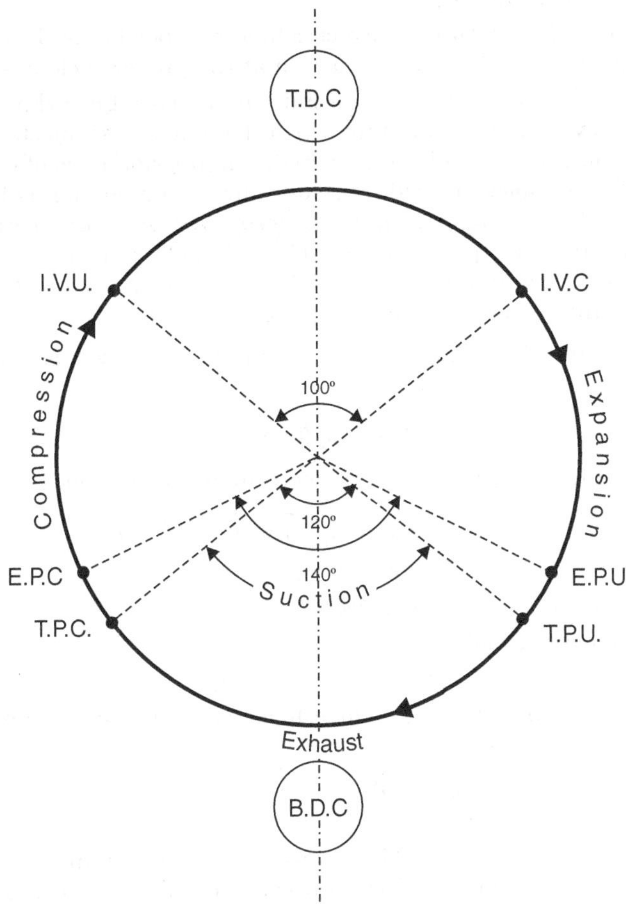

Fig. 2.45. Port timing diagram.

In a two stroke Diesel cycle engine all the operations are the same as in the spark ignition (Otto cycle) engine with the differences ; firstly in this case, only air is admitted into cylinder instead of air fuel mixture and secondly fuel injector is fitted to supply the fuel instead of a sparking plug.

2.13. INTAKE FOR COMPRESSION IGNITION ENGINES

- *The compression ignition (C.I.) engines are operated unthrottled with engine speed and power controlled by the amount of fuel injected during each cycle.* This allows for high volumetric efficiency at all speeds, *with the intake system designed for very little flow restriction of the incoming air.* Further raising the volumetric efficiency is the fact that no fuel is added until late in compression stroke, after air intake is fully completed. In addition many C.I. engines are *turbocharged, which enhances air intake even more.*

- The addition of fuel is made late in the compression stroke, starting somewhere around 20° before T.D.C. Injectors mounted in the cylinder head inject directly into the combustion chamber, where self ignition occurs due to the high temperature of the air caused by compression heating.
- It is important that fuel with the correct cetane number be used in an engine so that self-ignition initiates the start of combustion at the proper cycle position.
- For C.I. engines, the injection pressure must be much higher than that required for S.I. engines. The cylinder pressure into which the fuel is first injected is very high near the end of the compression stroke, due to high compression ratio of C.I. engines. By the time the final fuel is injected, peak pressure during combustion is being experienced. *Pressure must be high enough so that fuel spray will penetrate across the entire combustion chamber.* Injection pressures of **200** bar to **2000** bar are common *with average fuel droplet size generally decreasing with increasing pressure.* Orifice hole size of injectors is typically in the range of **0.2** to **1.0** mm diameter.
- The mass flow rate of fuel ($\dot{m}_f$) through an injector, during injection, is given by the relation :

$$\dot{m}_f = C_D A_n \sqrt{2\rho_f \Delta p} \qquad ...(2.1)$$

The total mass of fuel (m_f) injected into one cylinder during one cycle is given as :

$$m_f = C_D A_n \sqrt{2\rho_f \Delta p}\ (\Delta\theta/360\ N) \qquad ...(2.2)$$

where, C_D = Discharge co-efficient of injector,
A_n = Flow area of nozzle orifice(s),
ρ_f = Density of fuel,
Δp = Pressure differential across injector,
$\Delta\theta$ = Crank angle through which injection takes place (in degrees), and
N = Engine speed.

Again, $$p_{inj.} \simeq \Delta p \qquad ...(2.3)$$
and, $$p_{inj.} \simeq N^2 \qquad ...(2.4)$$

(To ensure that the crank angle of rotation through which injection takes place is almost constant for all speeds)

— *Large engines must have very high injection pressure and high spray velocity.*

— *For optimum fuel viscosity and spary penetration, it is important to have fuel at the correct temperature.*

Often engines are equipped with temperature sensors and means of heating or cooling the incoming fuel. *Many large truck engines are equipped with heated fuel filters. This allows the use of cheaper fuel that has less viscosity control.*

— *In small engines more costly, lower viscosity fuel is required.*

2.14. COMPARISON OF FOUR STROKE AND TWO STROKE CYCLE ENGINES

S.No.	*Aspects*	*Four Stroke Cycle Engines*	*Two Stroke Cycle Engines*
1.	*Completion of cycle*	The cycle is completed in *four strokes of the piston* or in *two revolutions of the crankshaft.* Thus one power stroke is obtained in every two revolutions of the crankshaft.	The cycle is completed in *two strokes of the piston* or in *one revolution of the crankshaft.* Thus one power stroke is obtained in each revolution of the crankshaft.

S.No.	*Aspects*	*Four Stroke Cycle Engines*	*Two Stroke Cycle Engines*
2.	*Flywheel required -heavier or lighter*	Because of the above turning-movement is not so uniform and hence *heavier* flywheel is needed.	More uniform turning movement and hence *lighter* flywheel is needed.
3.	*Power produced for same size of engine*	Again because of one power stroke for two revolutions, power produced for same size of engine is *small* or for the same power the engine is heavy and bulky.	Because of one power stroke for one revolution, power produced for same size of engine is *more* (theoretically twice, actually about 1.3 times) or for the same power the engine is light and compact.
4.	*Cooling and lubrication requirements*	Because of one power stroke in two revolutions *lesser* cooling and lubrication requirements. Lesser rate of wear and tear.	Because of one power stroke in one revolution *greater* cooling and lubrication requirement. Great rate of wear and tear.
5.	*Value and valve mechanism*	The four stroke engine *contains* valve and valve mechanism.	Two stroke engines have *no* valves but only ports (some two stroke engines are fitted with conventional exhaust valves).
6.	*Initial cost*	Because of the heavy weight and complication of valve mechanism, *higher* is the initial cost.	Because of light weight and simplicity due to absence of valve mechanism, *cheaper* in initial cost.
7.	*Volumetric efficiency*	Volumetric efficiency *more* due to more time of induction.	Volumetric efficiency *less* due to lesser time for induction.
8.	*Thermal and part-load efficiencies*	Thermal efficiency higher, part load efficiency better than two stroke cycle engine.	Thermal efficiency lower, part load efficiency lesser than four stroke cycle engine.
9.	*Applications*	Used where efficiency is important ; in *cars, buses, trucks, tractors, industrial engines, aeroplane, power generators etc.*	In two stroke petrol engine some fuel is exhausted during scavenging. Used where (*a*) *low cost*, and (*b*) *compactness and light weight important.* Two stroke (air cooled) petrol engines used in very small sizes only, *lawn movers, scooters motor cycles* (lubricating oil mixed with petrol). Two stroke diesel engines used in *very large sizes* more than 60 cm bore, for *ship propulsion* because of low weight and compactness.

2.15. COMPARISON OF SPARK IGNITION (S.I.) AND COMPRESSION IGNITION (C.I.) ENGINES

S.No.	*Aspects*	*S.I. engines*	*C.I. engines*
1.	*Thermodynamic cycle*	Otto cycle	Diesel cycle ...... For slow speed engines Dual cycle ...... For high speed engines
2.	*Fuel used*	Petrol	Diesel.

S.No.	Aspects	S.I. engines	C.I. engines
3.	*Air-fuel ratio*	10 : 1 to 20 : 1	18 : 1 to 100 : 1.
4.	*Compression ratio*	upto 11 ; Average value 7 to 9 ; Upper limit of compression ratio fixed by *anti-knock quality of fuel.*	12 to 24 ; Average value 15 to 18 ; Upper limit of compression ratio is limited by *thermal and mechanical stresses.*
5.	*Combustion*	Spark ignition	Compression ignition.
6.	*Fuel supply*	By carburettor ...... cheap method	By injection ...... expensive method.
7.	*Operating pressure* (*i*) Compression pressure (*ii*) Maximum pressure	7 bar to 15 bar 45 bar to 60 bar	30 bar to 50 bar 60 bar to 120 bar.
8.	*Operating speed*	High speed : 2000 to 6000 r.p.m.	Low speed : 400 r.p.m. Medium speed : 400 to 1200 r.p.m. High speed : 1200 to 3500 r.p.m.
9.	*Control of power*	Quantity governing ...... by throttle	Quality governing ...... by rack.
10.	*Calorific value*	44 MJ/kg	42 MJ/kg.
11.	*Cost of running*	high	low.
12.	*Maintenance cost*	Minor maintenance required	Major overall required but less frequently.
13.	*Supercharging*	Limited by *detonation*. Used only in *aircraft engines.*	Limited by *blower power and mechanical and thermal stresses. Widely used.*
14.	*Two stroke operation*	*Less suitable*, fuel loss in scavenging. But small two stroke engines are used in mopeds, scooters and motorcycles due to their *simplicity and low cost.*	No fuel loss in scavenging. *More suitable.*
15.	*High powers*	No	Yes.
16.	*Distribution of fuel*	A/F ratio is not optimum in multi-cylinder engines.	Excellent distribution of fuel in multi-cylinder engines.
17.	*Starting*	Easy, low cranking effort.	Difficult, high cranking effort.
18.	*Exhaust gas temperature*	High, due to low thermal efficiency.	Low, due to high thermal efficiency.
19.	*Weight per unit power*	Low (0.5 to 4.5 kg/kW).	High (3.3 to 13.5 kg/kW).
20.	*Initial capital cost*	Low	High due to heavy weight and study construction, costly construction, 1.25-1.5 times.
21.	*Noise and vibration*	Less	More idle noise problem.
22.	*Uses*	Mopeds, scooters, motorcycles, simple engine passenger cars, aircrafts etc.	Buses, trucks locomotives, tractors, earth moving machinery and stationary generating plants.

2.16. COMPARISON BETWEEN A PETROL ENGINE AND A DIESEL ENGINE

S.No.	*Petrol engine*	*Diesel engine*
1.	Air petrol mixture is sucked in the engine cylinder during suction stroke.	Only air is sucked during suction stroke.
2.	Spark plug is used.	Employs an injector.
3.	Power is produced by spark ignition.	Power is produced by compression ignition.
4.	Thermal efficiency up to 25%.	Thermal efficiency up to 40%.
5.	Occupies less space.	Occupies more space.
6.	More running cost.	Less running cost.
7.	Light in weight.	Heavy in weight.
8.	Fuel (Petrol) costlier.	Fuel (Diesel) cheaper.
9.	Petrol being volatile is dangerous.	Diesel is non-dangerous as it is non-volatile.
10.	Pre-ignition possible.	Pre-ignition not possible.
11.	Works on Otto cycle.	Works on Diesel cycle.
12.	Less dependable.	More dependable.
13.	Used in *cars* and *motor cycles.*	Used in heavy duty vehicles like *trucks, buses* and *heavy machinery.*

2.17. HOW TO TELL A TWO STROKE CYCLE ENGINE FROM A FOUR STROKE CYCLE ENGINE ?

S.No.	*Distinguishing features*	*Four stroke cycle engine*	*Two stroke cycle engine*
1.	*Oil sump and oil-filter plug*	It has an oil sump and oil-filter plug.	It does not have oil sump and oil-filter plug.
2.	*Oil drains etc.*	It requires oil drains and refills periodically, just an automobile do.	In this type of engine, the oil is added to the gasoline so that a mixture of gasoline and oil passes through the carburettor and enters the crankcase with the air.
3.	*Location of muffler (exhaust silencer)*	It is installed at the head end of the cylinder at the exhaust valve location.	It is installed towards the middle of the cylinder, at the exhaust port location.
4.	Name plate	If the name plate mentions the type of oil and the crankcase capacity, or similar data, it is a four stroke cycle engine.	If the name plate tells to mix oil with the gasoline, it is a two stroke cycle engine.

HIGHLIGHTS

1. Any type of engine or machine which derives heat energy from the combustion of fuel or any other source and converts this energy into mechanical work is termed as a **heat engine**.
2. The function of a carburettor is to atomise and meter the liquid fuel and mix it with air as it enters the injection system of the engine maintaining under all conditions of operation fuel air proportion approximate to those conditions.
3. The two basic ignition systems in current use are :
   (*i*) Battery or coil ignition system
   (*ii*) Magneto ignition system.
4. Following are the methods of governing I.C. engines :
   (*i*) Hit and miss method (*ii*) Quality governing
   (*iii*) Quantity governing.
5. Pre-ignition is the premature combustion which starts before the application of spark. Overheated spark plugs and exhaust valves which are the main causes of pre-ignition should be carefully avoided in engines.
6. A very sudden rise to pressure during combustion accompanied by metallic hammer like sound is called **detonation**. The region in which detonation occurs is farthest removed from the sparking plug, and is named the 'detonation zone' and even with severe detonation this zone is rarely more than that one quarter the clearance volume.
7. The **octane number** is the percentage of octane in the mixture [of iso-octane (high rating) and normal heptane (low rating), by volume] which knocks under the same conditions as the fuel.
8. **Delay period or ignition lag** is the time immediately following injection of fuel during which the ignition process is being initiated and the pressure does not rise beyond the value it would have due to compression of air.
9. Higher the cetane rating of the fuel lesser is the propensity for diesel knock. In general a high octane value implies a low cetane value.
10. The purpose of **supercharging** is to raise the volumetric efficiency above that value that which can be obtained by normal aspiration. Supercharging of petrol engines, because of its poor fuel economy, is not very popular and is used only when a large amount of power is needed or when more power is needed to compensate altitude loss.
11. **Dissociation** refers to disintegration of burnt gases at high temperatures. It is a reversible process and increases with temperature. Dissociation, in general, causes a loss of power and efficiency.
12. **Performance of I.C. engines**. Some important relations :

   (*i*) Indicated power (I.P.) $= \frac{np_{mi}LANk \times 10}{6}$ kW

   (*ii*) Brake Power (B.P.) $= \frac{(W - S)\,\pi\,(D_b + d)N}{60 \times 1000}$ kW or $\left(= \frac{2\pi NT}{60 \times 1000} \text{ kW}\right)$

   (*iii*) Mechanical efficiency, $\eta_{mech.} = \frac{\text{B.P.}}{\text{I.P.}}$

   (*iv*) Thermal efficiency (indicated), $\eta_{th.(I)} = \frac{\text{I.P.}}{\dot{m}_f \times C}$

   and thermal efficiency (brake), $\eta_{th.(B)} = \frac{\text{B.P.}}{\dot{m}_f \times C}$

   where $\dot{m}_f$ = mass of fuel used in kg/sec.

   (*v*) $\eta_{relative} = \frac{\eta_{thermal}}{\eta_{air\text{-}standard}}$

(*vi*) Measurement of air consumption by **air box method** :

Volume of air passing through the orifice, $V_a = 840\, AC_d \sqrt{\dfrac{h_w}{\rho_a}}$

and mass of air passing through the orifice,

$$m_a = 0.066\, C_d \times d^2 \sqrt{h_w \rho_a} \text{ kg/min}$$

where, A = Area of orifice, m^2

d = Diameter of orifice, cm

h_w = Head of water in 'cm' causing the flow

ρ_a = Density of air in kg/m^3 under atmospheric conditions.

OBJECTIVE TYPE QUESTIONS

Choose the correct answer :

1. In a four stroke cycle engine, the four operations namely suction, compression, expansion and exhaust are completed in the number of revolutions of crank shaft equal to
   (*a*) four (*b*) three
   (*c*) two (*d*) one.
2. In a two stroke cycle engine, the operations namely suction, compression, expansion and exhaust are completed in the number of revolutions of crank shaft equal to
   (*a*) four (*b*) three
   (*c*) two (*d*) one.
3. In a four stroke cycle S.I. engine the cam shaft runs
   (*a*) at the same speed as crank shaft (*b*) at half the speed of crank shaft
   (*c*) at twice the speed of crank shaft (*d*) at any speed irrespective of crank shaft speed.
4. The following is an S.I. engine
   (*a*) Diesel engine (*b*) Petrol engine
   (*c*) Gas engine (*d*) none of the above.
5. The following is C.I. engine
   (*a*) Diesel engine (*b*) Petrol engine
   (*c*) Gas engine (*d*) none of the above.
6. In a four stroke cycle petrol engine, during suction stroke
   (*a*) only air is sucked in (*b*) only petrol is sucked in
   (*c*) mixture of petrol and air is sucked in (*d*) none of the above.
7. In a four stroke cycle diesel engine, during suction stroke
   (*a*) only air is sucked in (*b*) only fuel is sucked in
   (*c*) mixture of fuel and air is sucked in (*d*) none on the above.
8. The two stroke cycle engine has
   (*a*) one suction valve and one exhaust valve operated by one cam
   (*b*) one suction valve and one exhaust valve operated by two cams
   (*c*) only ports covered and uncovered by piston to effect charging and exhausting
   (*d*) none of the above.
9. For same output, same speed and same compression ratio the thermal efficiency of a two stroke cycle petrol engine as compared to that for four stroke cycle petrol engine is
   (*a*) more (*b*) less
   (*c*) same as long as compression ratio is same (*d*) same as long as output is same.
10. The ratio of brake power to indicated power of an I.C. engine is called
    (*a*) mechanical efficiency (*b*) thermal efficiency
    (*c*) volumetric efficiency (*d*) relative efficiency.

ANSWERS

1. (*c*)	**2.** (*d*)	**3.** (*b*)	**4.** (*b*)	**5.** (*a*)	**6.** (*c*)	**7.** (*a*)
8. (*c*)	**9.** (*b*)	**10.** (*a*).				

THEORETICAL QUESTIONS

1. Name the two general classes of combustion engines and state how do they basically differ in principle ?

2. Discuss the relative advantages and disadvantages of internal combustion and external combustion engines.

3. What are the two basic types of internal combustion engines ? What are the fundamental differences between the two ?

4. What is the function of a governor ? Enumerate the types of governors and discuss with a neat sketch the Porter governor.

5. Differentiate between a flywheel and a governor.

6. (*a*) State the function of a carburettor in a petrol engine.
(*b*) Describe a simple carburettor with a neat sketch and also state its limitations.

7. Explain with neat sketches the construction and working of the following :
(*i*) Fuel pump (*ii*) Injector.

8. Explain the following terms as applied to I.C. engines :
Bore, stroke, T.D.C., B.D.C., clearance volume, swept volume, compression ratio and piston speed.

9. Explain with suitable sketches the working of a four stroke otto engine.

10. Discuss the difference between ideal and actual valve timing diagrams of a petrol engine.

11. In what respects four stroke diesel cycle (compression ignition) engine differs from four stroke cycle spark ignition engine ?

12. Discuss the difference between theoretical and actual valve timing diagrams of a diesel engine ?

13. What promotes the development of two stroke engines ? What are the two main types of two stoke engines ?

14. Describe with a suitable sketch the two stroke cycle spark ignition (SI) engine. How its indicator diagram differs from that of four stroke cycle engine ?

15. Compare the relative advantages and disadvantages of four stroke and two stroke cycle engines.

3

Air Standard Cycles

3.1. Definition of a cycle. 3.2. Air standard efficiency. 3.3. The Carnot cycle. 3.4. Constant volume or Otto cycle. 3.5. Constant pressure or Diesel cycle. 3.6. Dual combustion cycle. 3.7. Comparison of Otto, Diesel and Dual combustion cycles—Efficiency versus compression ratio—For the same compression ratio and the same heat input—For constant maximum pressure and heat supplied. 3.8. Atkinson cycle. 3.9. Ericsson cycle. 3.10. Brayton cycle. 3.11. Stirling cycle. 3.12. Miller cycle. 3.13. Lenoir cycle—Highlights—Objective Type Questions—Theoretical Questions—Unsolved Examples.

3.1. DEFINITION OF A CYCLE

A **cycle** is defined as a *repeated series of operations occuring in a certain order.* It may be repeated by repeating the processes in the same order. The cycle may be of imaginary perfect engine or actual engine. The former is called **ideal cycle** and the latter **actual cycle**. In ideal cycle all accidental heat losses are prevented and the working substance is assumed to behave like a perfect working substance.

3.2. AIR STANDARD EFFICIENCY

To compare the effects of different cycles, it is of paramount importance that the effect of the calorific value of the fuel is altogether eliminated and this can be achieved by considering air (which is assumed to behave as a perfect gas) as the working substance in the engine cylinder. *The efficiency of engine using air as the working medium is known as an* "**Air standard efficiency**". This efficiency is oftenly called **ideal efficiency**.

The actual efficiency of a cycle is always *less* than the air-standard efficiency of that cycle under ideal conditions. This is taken into account by introducing a new term "**Relative efficiency**" which is defined as :

$$\eta_{relative} = \frac{\text{Actual thermal efficiency}}{\text{Air standard efficiency}} \quad ...(3.1)$$

The analysis of all air standard cycles is based upon the following *assumptions* :

Assumptions :

1. The gas in the engine cylinder is a perfect gas *i.e.*, it obeys the gas laws and has constant specific heats.
2. The physical constants of the gas in the cylinder are the same as those of air at moderate temperatures *i.e.*, the molecular weight of cylinder gas is 29.

$$c_p = 1.005 \text{ kJ/kg-K}, \ c_p = 0.718 \text{ kJ/kg-K}.$$

3. The compression and expansion processes are adiabatic and they take place without internal friction, *i.e.*, these processes are isentropic.
4. No chemical reaction takes place in the cylinder. Heat is supplied or rejected by bringing a hot body or a cold body in contact with cylinder at appropriate points during the process.

5. The cycle is considered closed with the same 'air' always remaining in the cylinder to repeat the cycle.

3.3. THE CARNOT CYCLE

This cycle has the *highest possible efficiency* and consists of four simple operations namely,

(*a*) Isothermal expansion

(*b*) Adiabatic expansion

(*c*) Isothermal compression

(*d*) Adiabatic compression.

The condition of the Carnot cycle may be imagined to occur in the following way :

One kg of a air is enclosed in the cylinder which (except at the end) is made of perfect non-conducting material. A source of heat '*H*' is supposed to provide unlimited quantity of heat, non-conducting cover '*C*' and a sump '*S*' which is of infinite capacity so that its temperature remains unchanged irrespective of the fact how much heat is supplied to it. The temperature of source *H* is T_1 and the same is of the working substance. The working substance while rejecting heat to sump '*S*' has the temperature T_2 *i.e.*, the same as that of sump *S*.

Following are the *four stages* of the Carnot cycle. Refer Fig. 3.1 (*a*).

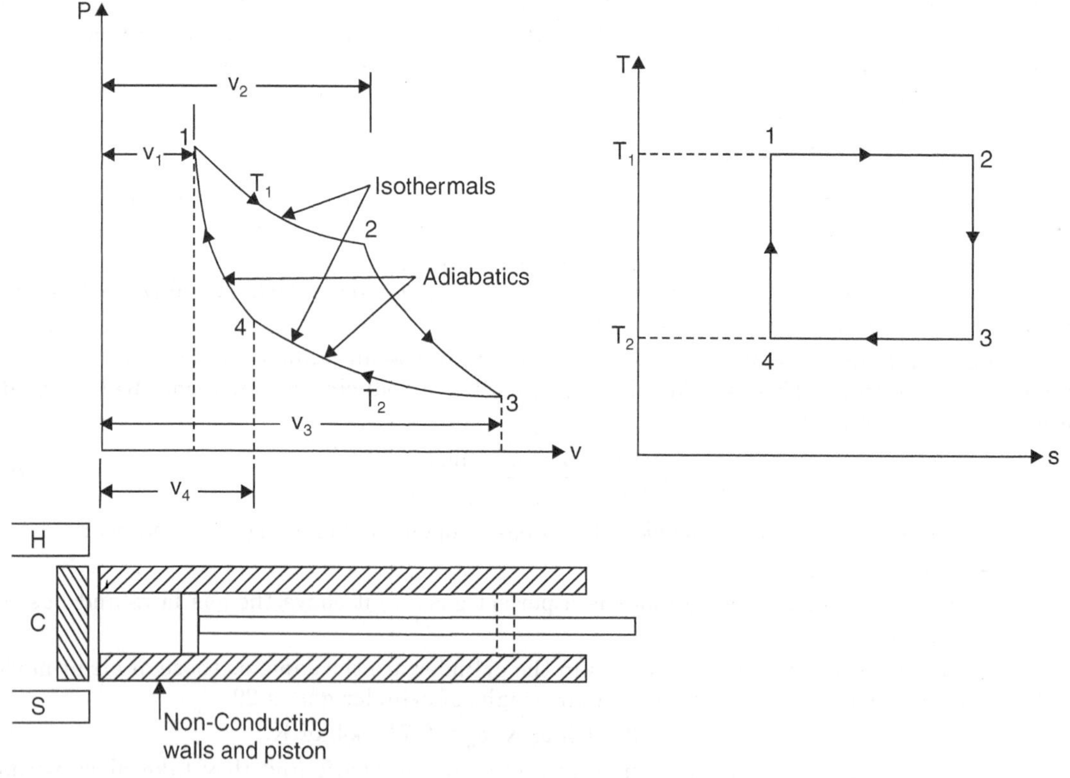

(*a*) Four stages of carnot cycle

(*b*) T-s diagram

Fig. 3.1

Stage (1). Line 1-2 [Fig. 3.1 (*a*)] represents the isothermal expansion which takes place at temperature T_1 when source of heat H is applied to the end of cylinder. Heat supplied in this case is given by $RT_1 \log_e r$ and where r is the ratio of expansion.

Stage (2). Line 2-3 represents the application of non-conducting cover to the end of the cylinder. This is followed by the adiabatic expansion and the temperature falls from T_1 to T_2.

Stage (3). Line 3-4 represents the isothermal compression which takes place when sump '*S*' is applied to the end of cylinder. Heat is rejected during this operation whose value is given by $RT_2 \log_e r$ where r is the ratio of compression.

Stage (4). Line 4-1 represents repeated application of non-conducting cover and adiabatic compression due to which temperature increases from T_2 to T_1.

It may be noted that ratio of expansion during isothermal 1-2 and ratio of compression during isothermal 3-4 must be equal to get a closed cycle.

Fig. 3.1 (*b*) represents the Carnot cycle on *T-s* coordinates.

Now according to law of conservation of energy,

Heat supplied = Work done + Heat rejected

Work done = Heat supplied – Heat rejected

$$= RT_1 \,.\, \log_e r - RT_2 \log_e r$$

$$\text{Efficiency of cycle} = \frac{\text{Work done}}{\text{Heat supplied}} = \frac{R \log_e r\,(T_1 - T_2)}{RT_1 \,.\, \log_e r}$$

$$= \frac{T_1 - T_2}{T_1} \qquad \text{...(3.2)}$$

From this equation, it is quite obvious that if temperature T_2 decreases, efficiency increases and it becomes 100% if T_2 becomes absolute zero which, of course is impossible to attain. Further more *it is not possible to produce an engine that should work on Carnot's cycle as it would necessitate the piston to travel very slowly during first portion of the forward stroke (isothermal expansion) and to travel more quickly during the remainder of the stroke (adiabatic expansion) which however is not practicable.*

Example 3.1. *A Carnot engine working between 400°C and 40°C produces 130 kJ of work. Determine :*

(*i*) *The engine thermal efficiency.*

(*ii*) *The heat added.*

(*iii*) *The entropy changes during heat rejection process.*

Solution. Temperature, $T_1 = T_2 = 400 + 273 = 673$ K

Temperature, $T_3 = T_4 = 40 + 273 = 313$ K

Work produced, $W = 130$ kJ.

(*i*) **Engine thermal efficiency, η_{th} :**

$$\eta_{th.} = \frac{673 - 313}{673} = \mathbf{0.535 \text{ or } 53.5\%.} \quad \textbf{(Ans.)}$$

(*ii*) **Heat added :**

$$\eta_{th.} = \frac{\text{Work done}}{\text{Heat added}}$$

i.e.,

$$0.535 = \frac{130}{\text{Heat added}}$$

$$\therefore \quad \text{Heat added} = \frac{130}{0.535} = \mathbf{243 \text{ kJ.}} \quad \textbf{(Ans.)}$$

(*iii*) **Entropy change during the heat rejection process, $(S_3 - S_4)$:**

Heat rejected = Heat added – Work done

= 243 – 130 = 113 kJ

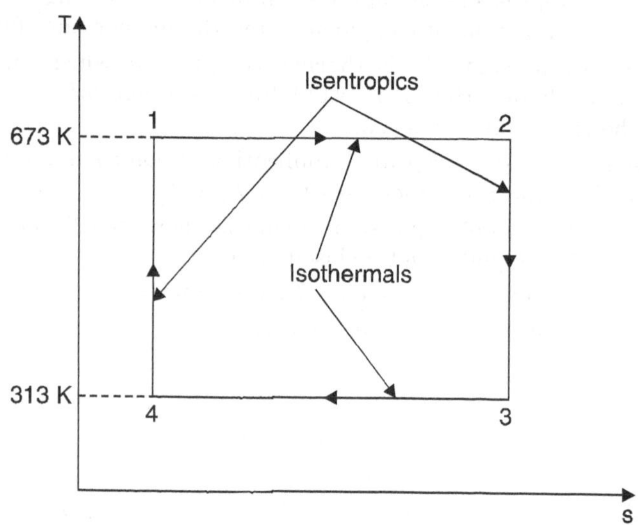

Fig. 3.2

Heat rejected $= T_3 (S_3 - S_4) = 113$

$$\therefore \quad (S_3 - S_4) = \frac{113}{T_3} = \frac{113}{313} = \mathbf{0.361\ kJ/K.} \quad \textbf{(Ans.)}$$

Example 3.2. *0.5 kg of air (ideal gas) executes a Carnot power cycle having a thermal efficiency of 50 per cent. The heat transfer to the air during the isothermal expansion is 40 kJ. At the beginning of the isothermal expansion the pressure is 7 bar and the volume is 0.12 m³. Determine :*

(*i*) *The maximum and minimum temperatures for the cycle in K ;*

(*ii*) *The volume at the end of isothermal expansion in m³ ;*

(*iii*) *The heat transfer for each of the four processes in kJ.*

For air c_v = 0.721 kJ/kg K, and c_p = 1.008 kJ/kg K. **(U.P.S.C. 1993)**

Solution. Refer Fig. 3.3. *Given* : m = 0.5 kg ; η_{th} = 50% ; Heat transfered during isothermal expansion = 40 kJ ; p_1 = 7 bar, V_1 = 0.12 m³ ; c_v = 0.721 kJ/kg K ; c_p = 1.008 kJ/kg K.

(*i*) **The maximum and minimum temperatures, T_1, T_2 :**

$$p_1 V_1 = mRT_1$$

$$7 \times 10^5 \times 0.12 = 0.5 \times 287 \times T_1$$

$$\therefore \quad \text{Maximum temperature, } T_1 = \frac{7 \times 10^5 \times 0.12}{0.5 \times 287} = \mathbf{585.4\ K.} \quad \textbf{Ans.}$$

$$\eta_{cycle} = \frac{T_1 - T_2}{T_1} \Rightarrow 0.5 = \frac{585.4 - T_2}{585.4}$$

$$\therefore \quad \text{Minimum temperature, } T_2 = 585.4 - 0.5 \times 585.4 = \mathbf{22.7\ K.} \quad \textbf{(Ans.)}$$

(*ii*) **The volume at the end of isothermal expansion V_2 :**

Heat transferred during isothermal expansion

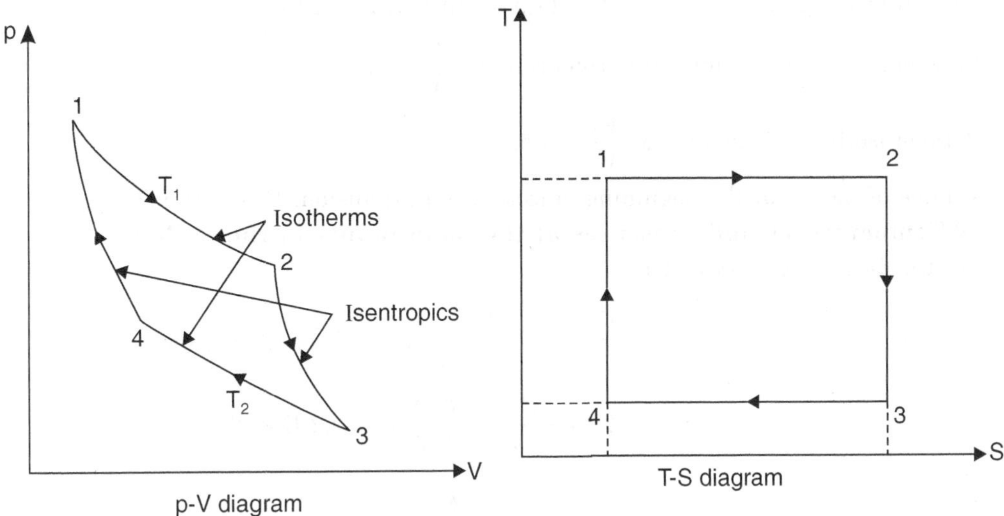

Fig. 3.3. Carnot cycle.

$$= p_1V_1 \ln(r) = mRT_1 \ln\left(\frac{V_2}{V_1}\right) = 40 \times 10^3 \qquad \text{......(Given)}$$

or $$0.5 \times 287 \times 585.4 \ln\left(\frac{V_2}{0.12}\right) = 40 \times 10^3$$

or $$\ln\left(\frac{V_2}{0.12}\right) = \frac{40 \times 10^3}{0.5 \times 287 \times 585.4} = 0.476$$

or $$V_2 = 0.12 \times (e)^{0.476} = \mathbf{0.193\ m^3}. \quad \textbf{(Ans.)}$$

(*iii*) **The heat transfer for each of the four processes :**

Process	*Classification*	*Heat transfer*
1—2	Isothermal expansion	40 kJ
2—3	Adiabatic reversible expansion	zero
3—4	Isothermal compression	– 40 kJ
4—1	Adiabatic reversible compression	zero. **(Ans.)**

☞**Example 3.3.** *In a Carnot cycle, the maximum pressure and temperature are limited to 18 bar and 410°C. The ratio of isentropic compression is 6 and isothermal expansion is 1.5. Assuming the volume of the air at the beginning of isothermal expansion as 0.18 m*^{3}*, determine :*

(*i*) *The temperature and pressures at main points in the cycle.*

(*ii*) *Change in entropy during isothermal expansion.*

(*iii*) *Mean thermal efficiency of the cycle.*

(*iv*) *Mean effective pressure of the cycle.*

(*v*) *The theoretical power if there are 210 working cycles per minute.*

Solution. Refer Fig. 3.4.

Maximum pressure, $p_1 = 18$ bar

Maximum temperature, $T_1 = (T_2) = 410 + 273 = 683$ K

Ratio of isentropic (or adiabatic) compression, $\frac{V_4}{V_1} = 6$

Ratio of isothermal expansion, $\frac{V_2}{V_1} = 1.5.$

Volume of the air at the beginning of isothermal expansion, $V_1 = 0.18$ m³.

(*i*) **Temperatures and pressures at the main points in the cycle :**

For the *isentropic process* 4-1

$$\frac{T_1}{T_4} = \left(\frac{V_4}{V_1}\right)^{\gamma - 1} = (6)^{1.4 - 1} = (6)^{0.4} = 2.05$$

$$\therefore \qquad T_4 = \frac{T_1}{2.05} = \frac{683}{2.05} = 333.2 \text{ K} = T_3$$

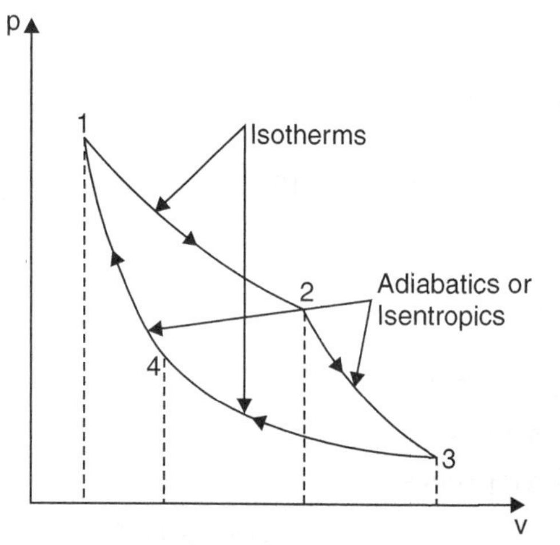

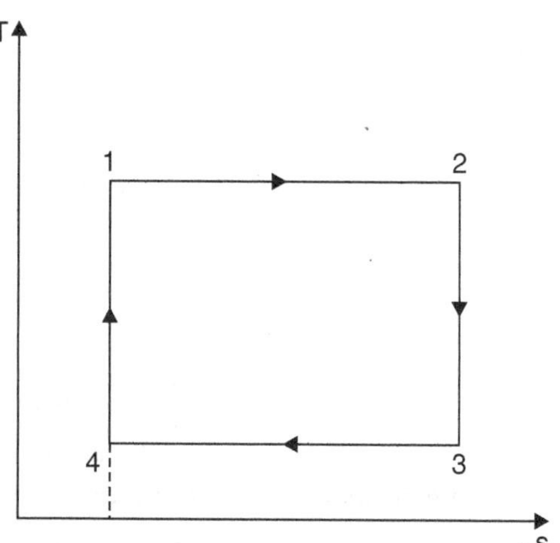

Fig. 3.4

$$\text{Also,} \qquad \frac{p_1}{p_4} = \left(\frac{V_4}{V_1}\right)^{\gamma} = (6)^{1.4} = 12.29$$

$$\therefore \qquad p_4 = \frac{p_1}{12.29} = \frac{18}{12.29} = 1.46 \text{ bar}$$

For the *isothermal process* 1-2

$$p_1V_1 = p_2V_2$$

$$p_2 = \frac{p_1V_1}{V_2} = \frac{18}{1.5} = 12 \text{ bar}$$

For *isentropic process* 2-3, we have

$$p_2V_2^{\gamma} = p_3V_3^{\gamma}$$

$$p_3 = p_2 \times \left(\frac{V_2}{V_3}\right)^{\gamma} = 12 \times \left(\frac{V_1}{V_4}\right)^{\gamma} \qquad \left[\because \frac{V_4}{V_1} = \frac{V_3}{V_2}\right]$$

$$= 12 \times \left(\frac{1}{6}\right)^{1.4} = \mathbf{0.97\ bar.}\ \ \textbf{(Ans.)}$$

Hence

$$\left.\begin{aligned} p_1 &= \mathbf{18\ bar} \quad T_1 = T_2 = \mathbf{683\ K} \\ p_2 &= \mathbf{12\ bar} \\ p_3 &= \mathbf{0.97\ bar} \quad T_3 = T_4 = \mathbf{333.2\ K} \\ p_4 &= \mathbf{1.46\ bar} \end{aligned}\right\} \textbf{(Ans.)}$$

(*ii*) **Change in entropy :**

Change in entropy during isothermal expansion,

$$S_2 - S_1 = mR \log_e \left(\frac{V_2}{V_1}\right) = \frac{p_1V_1}{T_1} \log_e \left(\frac{V_2}{V_1}\right) \qquad \left[\because pV = mRT \text{ or } mR = \frac{pV}{T}\right]$$

$$= \frac{18 \times 10^5 \times 0.18}{10^3 \times 683} \log_e (1.5) = \mathbf{0.192\ kJ/K.}\ \ \textbf{(Ans.)}$$

(*iii*) **Mean thermal efficiency of the cycle :**

Heat supplied,

$$Q_s = p_1V_1 \log_e \left(\frac{V_2}{V_1}\right)$$

$$= T_1 (S_2 - S_1)$$

$$= 683 \times 0.192 = 131.1 \text{ kJ}$$

Heat rejected,

$$Q_r = p_4V_4 \log_e \left(\frac{V_3}{V_4}\right)$$

$= T_4 (S_3 - S_4)$ because increase in entropy during heat addition is equal to decrease in entropy during heat rejection.

$\therefore$ $$Q_r = 333.2 \times 0.192 = 63.97 \text{ kJ}$$

$\therefore$ *Efficiency,*

$$\eta = \frac{Q_s - Q_r}{Q_s} = 1 - \frac{Q_r}{Q_s}$$

$$= 1 - \frac{63.97}{131.1} = \mathbf{0.512\ \ or\ \ 51.2\%.}\ \ \textbf{(Ans.)}$$

(*iv*) **Mean effective pressure of the cycle, $\mathbf{p_m}$:**

The mean effective pressure of the cycle is given by

$$p_m = \frac{\text{Work done per cycle}}{\text{Stroke volume}}$$

$$\frac{V_3}{V_1} = 6 \times 1.5 = 9$$

Stroke volume,

$$V_s = V_3 - V_1 = 9V_1 - V_1 = 8V_1 = 8 \times 0.18 = 1.44 \text{ m}^3$$

$\therefore$ $$p_m = \frac{(Q_s - Q_r) \times J}{V_s} = \frac{(Q_s - Q_r) \times 1}{V_s} \qquad (\because J = 1)$$

$$= \frac{(131.1 - 63.97) \times 10^3}{1.44 \times 10^5} = \mathbf{0.466\ bar.}\ \ \textbf{(Ans.)}$$

(*v*) **Power of the engine, P :**

Power of the engine working on this cycle is given by

$$P = (131.1 - 63.97) \times (210/60) = \mathbf{234.9\ kW.} \quad \textbf{(Ans.)}$$

Example 3.4. *A reversible engine converts one-sixth of the heat input into work. When the temperature of the sink is reduced by 70°C, its efficiency is doubled. Find the temperature of the source and the sink.*

Solution. Let T_1 = Temperature of the source (*K*), and

T_2 = Temperature of the sink (*K*).

First case :

$$\frac{T_1 - T_2}{T_1} = \frac{1}{6}$$

i.e., $6T_1 - 6T_2 = T_1$

or $5T_1 = 6T_2$ or $T_1 = 1.2T_2$ ...(*i*)

Second case :

$$\frac{T_1 - [T_2 - (70 + 273)]}{T_1} = \frac{1}{3}$$

$$\frac{T_1 - T_2 + 343}{T_1} = \frac{1}{3}$$

$$3T_1 - 3T_2 + 1029 = T_1$$

$$2T_1 = 3T_2 - 1029$$

$$2 \times (1.2T_2) = 3T_2 - 1029 \qquad (\because \; T_1 = 1.2\ T_2)$$

$$2.4T_2 = 3T_2 - 1029$$

or $0.6T_2 = 1029$

$\therefore$ $T_2 = \dfrac{1029}{0.6}$ = **1715 K or 1442°C. (Ans.)**

and $T_1 = 1.2 \times 1715$ = **2058 K or 1785°C. (Ans.)**

Example 3.5. *An inventor claims that a new heat cycle will develop 0.4 kW for a heat addition of 32.5 kJ/min. The temperature of heat source is 1990 K and that of sink is 850 K. Is his claim possible ?*

Solution. Temperature of heat source, T_1 = 1990 K

Temperature of sink, T_2 = 850 K

Heat supplied, = 32.5 kJ/min

Power developed by the engine, P = 0.4 kW

The most efficient engine is one that works on Carnot cycle

$$\eta_{carnot} = \frac{T_1 - T_2}{T_1} = \frac{1990 - 850}{1990} = 0.573 \text{ or } 57.3\%$$

Also, thermal efficiency of the engine,

$$\eta_{th} = \frac{\text{Work done}}{\text{Heat supplied}} = \frac{0.4}{(32.5/60)} = \frac{0.4 \times 60}{32.5}$$

$$= 0.738 \text{ or } 73.8\%$$

which is not feasible as no engine can be more efficient than that working on Carnot cycle.

Hence claims of the inventor is **not true. (Ans.)**

Example 3.6. *An ideal engine operates on the Carnot cycle using a perfects gas as the working fluid. The ratio of the greatest to the least volume is fixed and is x : 1, the lower temperature of the cycle is also fixed, but the volume compression ratio 'r' of the reversible adiabatic compression is variable. The ratio of the specific heats is γ.*

Show that if the work done in the cycle is a maximum then,

$$(\gamma - 1) \log_e \frac{x}{r} + \frac{1}{r^{\gamma - 1}} - 1 = 0.$$

Solution. Refer Fig. 3.1.

$$\frac{V_3}{V_1} = x \; ; \; \frac{V_4}{V_1} = r$$

During isotherms, since compression ratio = expansion ratio

$$\therefore \quad \frac{V_3}{V_4} = \frac{V_2}{V_1}$$

Also
$$\frac{V_3}{V_4} = \frac{V_3}{V_1} \times \frac{V_1}{V_4} = x \times \frac{1}{r} = \frac{x}{r}$$

Work done per kg of the gas

$$= \text{Heat supplied} - \text{Heat rejected} = RT_1 \log_e \frac{x}{r} - RT_2 \log_e \frac{x}{r}$$

$$= R(T_1 - T_2) \log_e \frac{x}{r} = RT_2 \left(\frac{T_1}{T_2} - 1 \right) \log_e \frac{x}{r}$$

But
$$\frac{T_1}{T_2} = \left(\frac{V_4}{V_1} \right)^{\gamma - 1} = (r)^{\gamma - 1}$$

∴ Work done per kg of the gas,

$$W = RT_2 \, (r^{\gamma - 1} - 1) \log_e \frac{x}{r}$$

Differentiating W w.r.t. 'r' and equating to zero

$$\frac{dW}{dr} = RT_2 \left[(r^{\gamma - 1} - 1) \left\{ \frac{r}{x} \times (-xr^{-2}) \right\} + \log_e \frac{x}{r} \{ (\gamma - 1) r^{\gamma - 2} \} \right] = 0$$

or
$$(r^{\gamma - 1} - 1) \left(-\frac{1}{r} \right) + (\gamma - 1) \times r^{\gamma - 2} \log_e \frac{x}{r} = 0$$

or
$$- r^{\gamma - 2} + \frac{1}{r} + r^{\gamma - 2} (\gamma - 1) \log_e \frac{x}{r} = 0$$

or
$$r^{\gamma - 2} \left\{ -1 + \frac{1}{r \, . \, r^{\gamma - 2}} + (\gamma - 1) \log_e \frac{x}{r} \right\} = 0$$

or
$$-1 + \frac{1}{r \, . \, r^{\gamma - 2}} + (\gamma - 1) \log_e \frac{x}{r} = 0$$

$$(\gamma - 1) \log_e \frac{x}{r} + \frac{1}{r^{\gamma - 1}} - 1 = 0. \quad \textbf{Proved.}$$

3.4. CONSTANT VOLUME OR OTTO CYCLE

This cycle is so named as it was conceived by 'Otto'. On this cycle, petrol, gas and many types of oil engines work. *It is the standard of comparison for internal combustion engines.*

Fig. 3.5 (*a*) and (*b*) shows the theoretical p-V diagram and T-s diagrams of this cycle respectively.

The point 1 represents that cylinder is full of air with volume V_1, pressure p_1 and absolute temperature T_1.

Line 1-2 represents the adiabatic compression of air due to which p_1, V_1 and T_1 change to p_2, V_2 and T_2 respectively.

Line 2-3 shows the supply of heat to the air at constant volume so that p_2 and T_2 change to p_3 and T_3 (V_3 being the same as V_2).

Line 3-4 represents the adiabatic expansion of the air. During expansion p_3, V_3 and T_3 change to a final value of p_4, V_4 or V_1 and T_4 respectively.

Line 4-1 shows the rejection of heat by air at constant volume till original state (point 1) reaches.

Consider ***1 kg of air*** (working substance) :

Heat supplied at constant volume $= c_v(T_3 - T_2)$.

Heat rejected at constant volume $= c_v(T_4 - T_1)$.

But, work done = heat supplied – heat rejected

$$= c_v(T_3 - T_2) - c_v(T_4 - T_1)$$

$$\therefore \quad \text{Efficiency} = \frac{\text{Work done}}{\text{Heat supplied}} = \frac{c_v(T_3 - T_2) - c_v(T_4 - T_1)}{c_v(T_3 - T_2)}$$

$$= 1 - \frac{T_4 - T_1}{T_3 - T_2} \quad \text{...(i)}$$

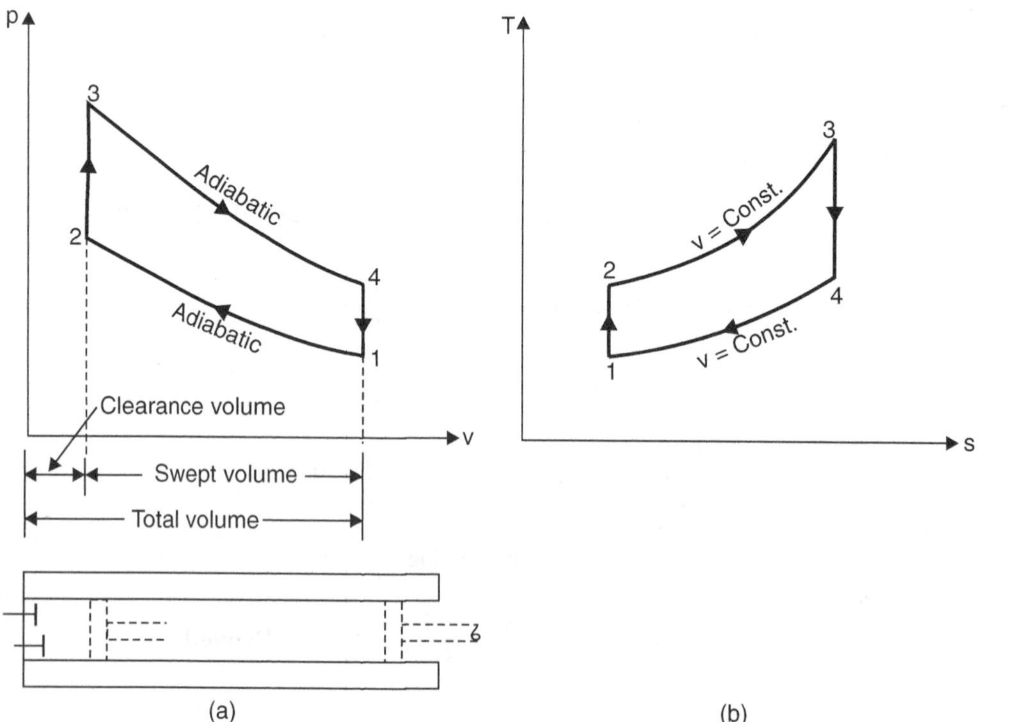

Fig. 3.5

Let compression ratio, $r_c\ (= r) = \dfrac{v_1}{v_2}$

and expansion ratio, $r_e\ (= r) = \dfrac{v_4}{v_3}$

(These *two ratios are same* in this cycle)

As $$\frac{T_2}{T_1} = \left(\frac{v_1}{v_2}\right)^{\gamma - 1}$$

Then, $$T_2 = T_1 \,.\, (r)^{\gamma - 1}$$

Similarly, $$\frac{T_3}{T_4} = \left(\frac{v_4}{v_3}\right)^{\gamma - 1}$$

or $$T_3 = T_4 \,.\, (r)^{\gamma - 1}$$

Inserting the values of T_2 and T_3 in equation (*i*), we get

$$\eta_{otto} = 1 - \frac{T_4 - T_1}{T_4 \,.\, (r)^{\gamma - 1} - T_1 \,.\, (r)^{\gamma - 1}} = 1 - \frac{T_4 - T_1}{r^{\gamma - 1}(T_4 - T_1)}$$

$$= 1 - \frac{1}{(r)^{\gamma - 1}} \quad \text{...(3.3)}$$

This expression is known as the ***air standard efficiency of the Otto cycle.***

It is clear from the above expression that efficiency increases with the increase in the value of r, which means we can have maximum efficiency by increasing r to a considerable extent, but *due to practical difficulties its value is limited to about 9.*

The *net work done per kg* in the Otto cycle can also be expressed in terms of p, v. If p is expressed in bar *i.e.* 10^5 N/m^2, then work done

$$W = \left(\frac{p_3v_3 - p_4v_4}{\gamma - 1} - \frac{p_2v_2 - p_1v_1}{\gamma - 1}\right) \times 10^2 \text{ kJ} \quad \text{...(3.4)}$$

Also $$\frac{p_3}{p_4} = r^{\gamma} = \frac{p_2}{p_1}$$

$\therefore$ $$\frac{p_3}{p_2} = \frac{p_4}{p_1} = r_p$$

where r_p stands for *pressure ratio.*

and $$v_1 = rv_2 = v_4 = rv_3 \qquad \left[\because \frac{v_1}{v_2} = \frac{v_4}{v_3} = r\right]$$

$\therefore$ $$W = \frac{1}{\gamma - 1}\left[p_4v_4\left(\frac{p_3v_3}{p_4v_4} - 1\right) - p_1v_1\left(\frac{p_2v_2}{p_1v_1} - 1\right)\right]$$

$$= \frac{1}{\gamma - 1}\left[p_4v_4\left(\frac{p_3}{p_4r} - 1\right) - p_1v_1\left(\frac{p_2}{p_1r} - 1\right)\right]$$

$$= \frac{v_1}{\gamma - 1}\left[p_4\,(r^{\gamma - 1} - 1) - p_1\,(r^{\gamma - 1} - 1)\right]$$

$$= \frac{v_1}{\gamma - 1}\left[(r^{\gamma - 1} - 1)(p_4 - p_1)\right]$$

$$= \frac{p_1 v_1}{\gamma - 1}\left[(r^{\gamma - 1} - 1)(r_p - 1)\right] \quad \text{...[3.4 (a)]}$$

Mean effective pressure ($\mathbf{p_m}$) is given by :

$$p_m = \left[\left(\frac{p_3 v_3 - p_4 v_4}{\gamma - 1} - \frac{p_2 v_2 - p_1 v_1}{\gamma - 1}\right) \div (v_1 - v_2)\right] \text{bar} \quad \text{...(3.5)}$$

Also
$$p_m = \frac{\left[\frac{p_1 v_1}{\gamma - 1}(r^{\gamma - 1} - 1)(r_p - 1)\right]}{(v_1 - v_2)}$$

$$= \frac{\frac{p_1 v_1}{\gamma - 1}[(r^{\gamma - 1} - 1)(r_p - 1)]}{v_1 - \frac{v_1}{r}}$$

$$= \frac{\frac{p_1 v_1}{\gamma - 1}[(r^{\gamma - 1} - 1)(r_p - 1)]}{v_1\left(\frac{r - 1}{r}\right)}$$

i.e.,
$$\mathbf{p_m} = \frac{p_1 r[(r^{\gamma - 1} - 1)(r_p - 1)]}{(\gamma - 1)(r - 1)} \quad \text{...(3.6)}$$

Example 3.7. *The efficiency of an Otto cycle is 60% and $\gamma = 1.5$. What is the compression ratio ?*

Solution. Efficiency of Otto cycle, $\eta = 60\%$

Ratio of specific heats, $\gamma = 1.5$

Compression ratio, $r = ?$

Efficiency of Otto cycle is given by

$$\eta_{Otto} = 1 - \frac{1}{(r)^{\gamma - 1}}$$

$$0.6 = 1 - \frac{1}{(r)^{1.5 - 1}}$$

or
$$\frac{1}{(r)^{0.5}} = 0.4 \quad \text{or} \quad (r)^{0.5} = \frac{1}{0.4} = 2.5 \quad \text{or} \quad r = 6.25$$

Hence, *compression ratio* $=$ **6.25. (Ans.)**

Example 3.8. *An engine of 250 mm bore and 375 mm stroke works on Otto cycle. The clearance volume is 0.00263 m³. The initial pressure and temperature are 1 bar and 50°C. If the maximum pressure is limited to 25 bar, find the following :*

(i) The air standard efficiency of the cycle.

(ii) The mean effective pressure for the cycle.

Assume the ideal conditions.

Solution. Bore of the engine, $D = 250 \text{ mm} = 0.25 \text{ m}$

Stroke of the engine, $L = 375 \text{ mm} = 0.375 \text{ m}$

Clearance volume, $V_c = 0.00263 \text{ m}^3$

Initial pressure, $p_1 = 1 \text{ bar}$

Initial temperature, $T_1 = 50 + 273 = 323 \text{ K}$

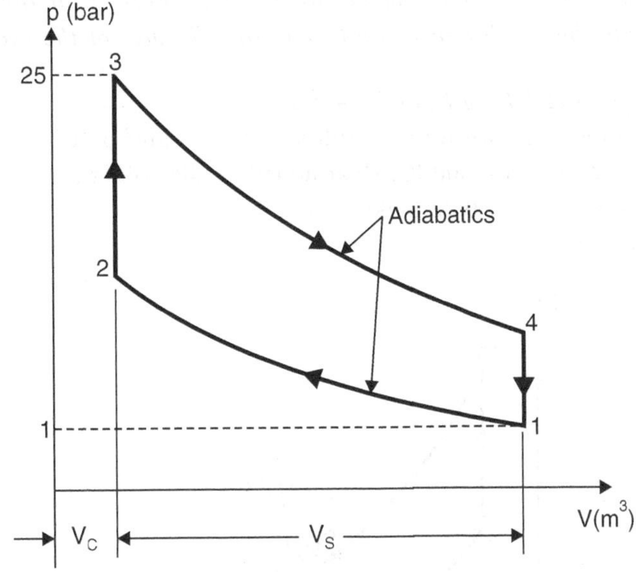

Fig. 3.6

Maximum pressure, $p_3 = 25 \text{ bar}$

Swept volume, $V_s = \pi/4\ D^2L = \pi/4 \times 0.25^2 \times 0.375 = 0.0184 \text{ m}^3$

Compression ratio, $r = \dfrac{V_s + V_c}{V_c} = \dfrac{0.0184 + 0.00263}{0.00263} = 8.$

(*i*) **Air standard efficiency :**

The air standard efficiency of Otto cycle is given by

$$\eta_{\text{Otto}} = 1 - \frac{1}{(r)^{\gamma - 1}} = 1 - \frac{1}{(8)^{1.4-1}} = 1 - \frac{1}{(8)^{0.4}}$$

$$= 1 - 0.435 = \mathbf{0.565} \quad \textbf{or} \quad \mathbf{56.5\%.} \quad \textbf{(Ans.)}$$

(*ii*) **Mean effective pressure, p_m :**

For adiabatic (or isentropic) process 1-2

$$p_1V_1^{\gamma} = p_2V_2^{\gamma}$$

or

$$p_2 = p_1 \left(\frac{V_1}{V_2}\right)^{\gamma} = 1 \times (r)^{1.4} = 1 \times (8)^{1.4} = 18.38 \text{ bar}$$

$\therefore$ Pressure ratio, $r_p = \dfrac{p_3}{p_2} = \dfrac{25}{18.38} = 1.36$

The mean effective pressure is given by

$$p_m = \frac{p_1 r[(r^{\gamma - 1} - 1)(r_p - 1)]}{(\gamma - 1)(r - 1)} = \frac{1 \times 8\,[\{(8)^{1.4-1} - 1\}\,(1.36 - 1)]}{(1.4 - 1)\,(8 - 1)} \qquad \text{...[Eqn. (3.6)]}$$

$$= \frac{8(2.297 - 1)(0.36)}{0.4 \times 7} = 1.334 \text{ bar}$$

Hence *mean effective pressure* = **1.334 bar. (Ans.)**

Example 3.9. *The minimum pressure and temperature in an Otto cycle are 100 kPa and 27°C. The amount of heat added to the air per cycle is 1500 kJ/kg.*

(i) Determine the pressures and temperatures at all points of the air standard Otto cycle.

(ii) Also calculate the specific work and thermal efficiency of the cycle for a compression ratio of 8 : 1.

Take for air : $c_v = 0.72$ *kJ/kg K, and* $\gamma = 1.4$. **(GATE, 1998)**

Solution. Refer Fig. 3.7. *Given :* $p_1 = 100$ kPa $= 10^5$ N/m^2 or 1 bar ;

$T_1 = 27 + 273 = 300$ K ; Heat added = 1500 kJ/kg ;

$r = 8 : 1$; $c_v = 0.72$ kJ/kg ; $\gamma = 1.4$.

Consider 1 kg of air.

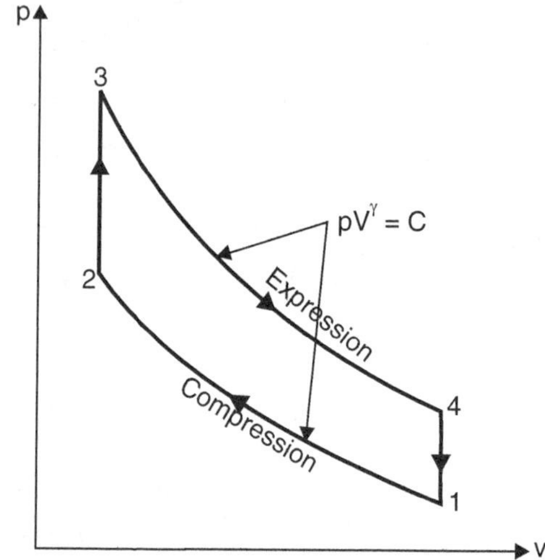

Fig. 3.7

(*i*) **Pressures and temperatures at all points :**

Adiabatic Compression process 1-2 :

$$\frac{T_2}{T_1} = \left(\frac{v_1}{v_2}\right)^{\gamma - 1} = (r)^{\gamma - 1} = (8)^{1.4 - 1} = 2.297$$

$\therefore$ $T_2 = 300 \times 2.297 =$ **689.1 K. (Ans.)**

Also $p_1 v_1^{\gamma} = p_2 v_2^{\gamma}$

or $$\frac{p_2}{p_1} = \left(\frac{v_1}{v_2}\right)^{\gamma} = (8)^{1.4} = 18.379$$

$\therefore$ $p_2 = 1 \times 18.379 =$ **18.379 bar. (Ans.)**

Constant volume process 2-3 :

Heat added during the process,

$$c_v (T_3 - T_2) = 1500$$

or $\quad 0.72\,(T_3 - 689.1) = 1500$

or $$T_3 = \frac{1500}{0.72} + 689.1 = \mathbf{2772.4\ K.}\quad \textbf{(Ans.)}$$

Also, $$\frac{p_2}{T_2} = \frac{p_3}{T_3} \Rightarrow p_3 = \frac{p_2T_3}{T_2} = \frac{18.379 \times 2772.4}{689.1} = \mathbf{73.94\ bar.}\quad \textbf{(Ans.)}$$

Adiabatic Expansion process 3-4 :

$$\frac{T_3}{T_4} = \left(\frac{v_4}{v_3}\right)^{\gamma - 1} = (r)^{\gamma - 1} = (8)^{1.4 - 1} = 2.297$$

$\therefore$ $$T_4 = \frac{T_3}{2.279} = \frac{2772.4}{2.297} = \mathbf{1206.9\ K.}\quad \textbf{(Ans.)}$$

Also, $$p_3v_3^{\gamma} = p_4v_4^{\gamma} \Rightarrow p_4 = p_3 \times \left(\frac{v_3}{v_4}\right)^{\gamma} = 73.94 \times \left(\frac{1}{8}\right)^{1.4} = \mathbf{4.023\ bar.}\quad \textbf{(Ans.)}$$

(*ii*) **Specific work and thermal efficiency :**

Specfic work = Heat added – Heat rejected

$$= c_v\,(T_3 - T_2) - c_v(T_4 - T_1) = c_v\,[(T_3 - T_2) - (T_4 - T_1)]$$

$$= 0.72\,[(2772.4 - 689.1) - (1206.9 - 300)] = \mathbf{847\ kJ/kg.}\quad \textbf{(Ans.)}$$

Thermal efficiency, $$\eta_{th} = 1 - \frac{1}{(r)^{\gamma - 1}}$$

$$= 1 - \frac{1}{(8)^{1.4 - 1}} = 0.5647 \text{ or } \mathbf{56.47\%.}\quad \textbf{(Ans.)}$$

Example 3.10. *An air standard Otto cycle has a volumetric compression ratio of 6, the lowest cycle pressure of 0.1 MPa and operates between temperature limits of 27°C and 1569°C.*

(i) Calculate the temperature and pressure after the isentropic expansion (ratio of specific heats = 1.4).

(ii) Since it is observed that values in (i) are well above the lowest cycle operating conditions, the expansion process was allowed to continue down to a pressure of 0.1 MPa. Which process is required to complete the cycle ? Name the cycle so obtained.

(iii) Determine by what percentage the cycle efficiency has been improved. **(GATE, 1994)**

Solution. Refer Fig. 3.8. *Given :* $\frac{v_1}{v_2} = \frac{v_4}{v_3} = r = 6$; $p_1 = 0.1$ MPa = 1 bar ; $T_1 = 27 + 273 = 300$ K ; $T_3 = 1569 + 273 = 1842$ K ; $\gamma = 1.4$.

(*i*) **Temperature and pressure after the isentropic expansion, T_4, p_4 :**

Consider 1 kg of air :

For *the compression process 1-2 :*

$$p_1v_1^{\gamma} = p_2v_2^{\gamma} \Rightarrow p_2 = p_1 \times \left(\frac{v_1}{v_2}\right)^{\gamma} = 1 \times (6)^{1.4} = 12.3 \text{ bar}$$

Also $$\frac{T_2}{T_1} = \left(\frac{v_1}{v_2}\right)^{\gamma - 1} = (6)^{1.4 - 1} = 2.048$$

$\therefore$ $$T_2 = 300 \times 2.048 = 614.4 \text{ K}$$

For *the constant volume process 2-3 :*

$$\frac{p_2}{T_2} = \frac{p_3}{T_3} \Rightarrow p_3 = \frac{p_2 T_3}{T_2} = 12.3 \times \frac{1842}{614.4} = 36.9 \text{ bar}$$

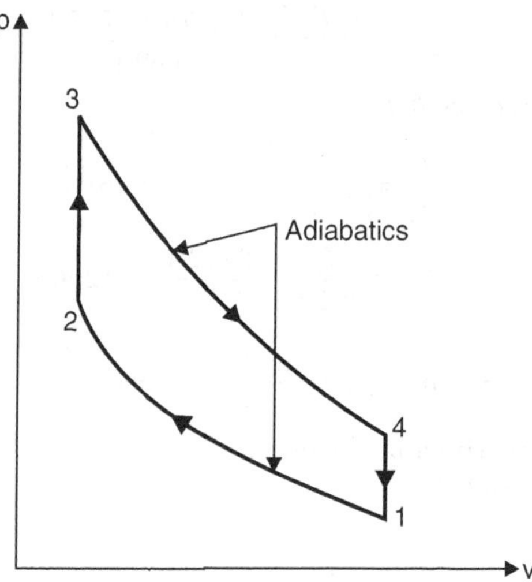

Fig. 3.8

For *the expansion process 3-4 :*

$$\frac{T_3}{T_4} = \left(\frac{v_4}{v_3}\right)^{\gamma - 1} = (6)^{1.4 - 1} = 2.048$$

$$\therefore \qquad T_4 = \frac{T_3}{2.048} = \frac{1842}{2.048} = \mathbf{900\ K.} \quad \textbf{(Ans.)}$$

Also
$$p_3 v_3^{\gamma} = p_4 v_4^{\gamma} \Rightarrow p_4 = p_3 \times \left(\frac{v_3}{v_4}\right)^{\gamma}$$

or
$$p_4 = 36.9 \times \left(\frac{1}{6}\right)^{1.4} = \mathbf{3\ bar.} \quad \textbf{(Ans.)}$$

(*ii*) **Process required to complete the cycle :**

Process required to complete the cycle is the *constant pressure* scavenging.

The cycle is called **Atkinson cycle** (Refer Fig. 3.9).

(*iii*) **Percentage improvement/increase in efficiency :**

$$\eta_{\text{Otto}} = 1 - \frac{1}{(r)^{\gamma - 1}} = 1 - \frac{1}{(6)^{1.4 - 1}} = 0.5116 \text{ or } \mathbf{51.16\%.} \quad \textbf{(Ans.)}$$

$$\eta_{\text{Atkinson}} = \frac{\text{Work done}}{\text{Heat supplied}} = \frac{\text{Heat supplied} - \text{Heat rejected}}{\text{Heat supplied}}$$

$$= \frac{c_v(T_3 - T_2) - c_p(T_5 - T_1)}{c_v(T_3 - T_2)} = 1 - \frac{c_p(T_5 - T_1)}{c_v(T_3 - T_2)} = 1 - \frac{\gamma(T_5 - T_1)}{(T_3 - T_2)}$$

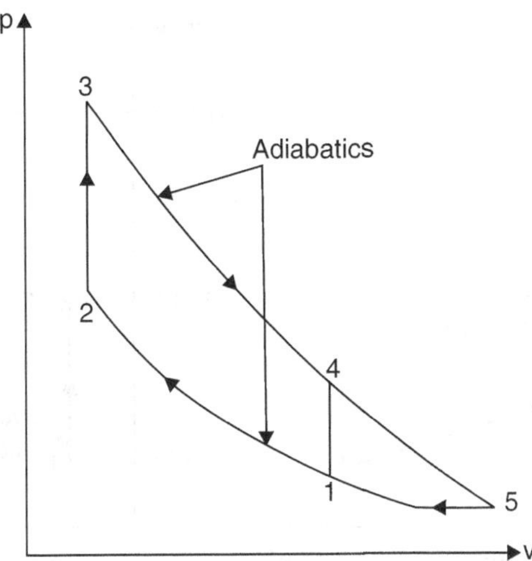

Fig. 3.9. Atkinson cycle.

Now, $$\frac{T_5}{T_3} = \left(\frac{p_5}{p_3}\right)^{\frac{\gamma-1}{\gamma}} \quad \text{or} \quad T_5 = 1842 \times \left(\frac{1.0}{36.9}\right)^{\frac{1.4-1}{1.4}} = 657 \text{ K}$$

$\therefore$ $$\eta_{\text{Atkinson}} = 1 - \frac{1.4(657-300)}{(1842-614.4)} = 0.5929 \quad \text{or} \quad \mathbf{59.29\%}.$$

$\therefore$ **Improvement in efficiency** = 59.29 – 51.16 = **8.13%. (Ans.)**

Example 3.11. *A certain quantity of air at a pressure of 1 bar and temperature of 70°C is compressed adiabatically until the pressure is 7 bar in Otto cycle engine. 465 kJ of heat per kg of air is now added at constant volume. Determine :*

(*i*) *Compression ratio of the engine.*

(*ii*) *Temperature at the end of compression.*

(*iii*) *Temperature at the end of heat addition.*

Take for air c_p *= 1.0 kJ/kg K,* c_v *= 0.706 kJ/kg K.*

Show each operation on p-V and T-s diagrams.

Solution. Refer Fig 3.10.

Initial pressure, $p_1 = 1$ bar

Initial temperature, $T_1 = 70 + 273 = 343$ K

Pressure after adiabatic compression, $p_2 = 7$ bar

Heat addition at constant volume, $Q_s = 465$ kJ/kg of air

Specific heat at constant pressure, $c_p = 1.0$ kJ/kg K

Specific heat at constant volume, $c_v = 0.706$ kJ/kg K

$\therefore$ $$\gamma = \frac{c_p}{c_v} = \frac{1.0}{0.706} = 1.41$$

(*i*) **Compression ratio of engine, r :**

According to *adiabatic compression* 1-2

$$p_1 V_1^{\gamma} = p_2 V_2^{\gamma}$$

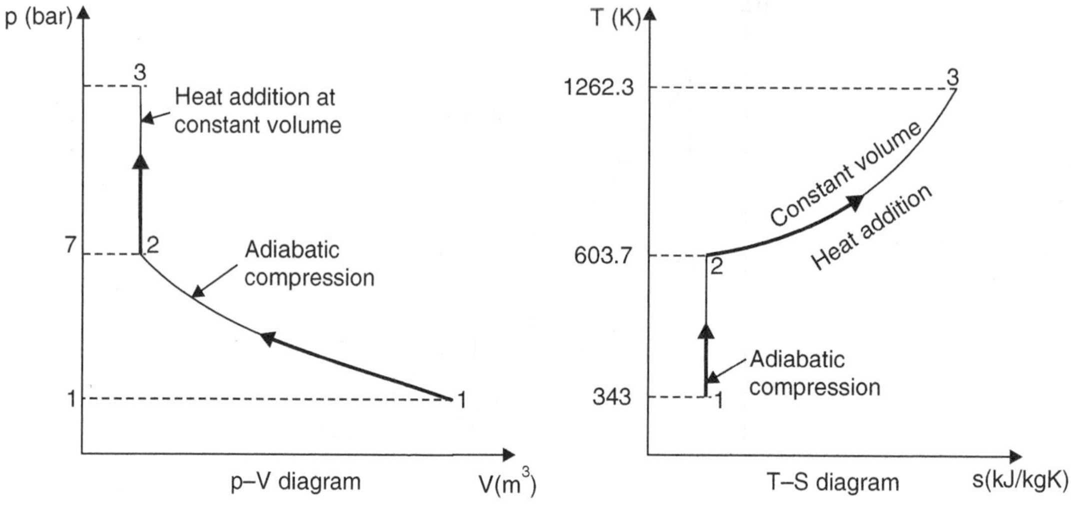

Fig. 3.10

or $$\left(\frac{V_1}{V_2}\right)^{\gamma} = \frac{p_2}{p_1}$$

or $$(r)^{\gamma} = \frac{p_2}{p_1} \qquad \left(\because \frac{V_1}{V_2} = r\right)$$

or $$r = \left(\frac{p_2}{p_1}\right)^{\frac{1}{\gamma}} = \left(\frac{7}{1}\right)^{\frac{1}{1.41}} = (7)^{0.709} = 3.97$$

Hence *compression ratio of the engine* = **3.97. (Ans.)**

(*ii*) **Temperature at the end of compression, T_2 :**

In case of *adiabatic compression 1-2,*

$$\frac{T_2}{T_1} = \left(\frac{V_1}{V_2}\right)^{\gamma - 1} = (3.97)^{1.41 - 1} = 1.76$$

$$\therefore \quad T_2 = 1.76\ T_1 = 1.76 \times 343 = 603.7 \text{ K or } 330.7°\text{C}$$

Hence *temperature at the end of compression* = **330.7°C. (Ans.)**

(*iii*) **Temperature at the end of heat addition, T_3 :**

According to *constant volume heating operation 2-3*

$$Q_s = c_v\ (T_3 - T_2) = 465$$

$$0.706\ (T_3 - 603.7) = 465$$

or $$T_3 - 603.7 = \frac{465}{0.706}$$

or $$T_3 = \frac{465}{0.706} + 603.7 = 1262.3 \text{ K or } 989.3°\text{C}$$

Hence *temperature at the end of heat addition* = **989.3°C. (Ans.)**

Example 3.12. *In a constant volume 'Otto cycle', the pressure at the end of compression is 15 times that at the start, the temperature of air at the beginning of compression is 38°C and maximum temperature attained in the cycle is 1950°C. Determine :*

(*i*) *Compression ratio.*
(*ii*) *Thermal efficiency of the cycle.*
(*iii*) *Work done.*
Take γ *for air = 1.4.*

Solution. Refer Fig. 3.11.

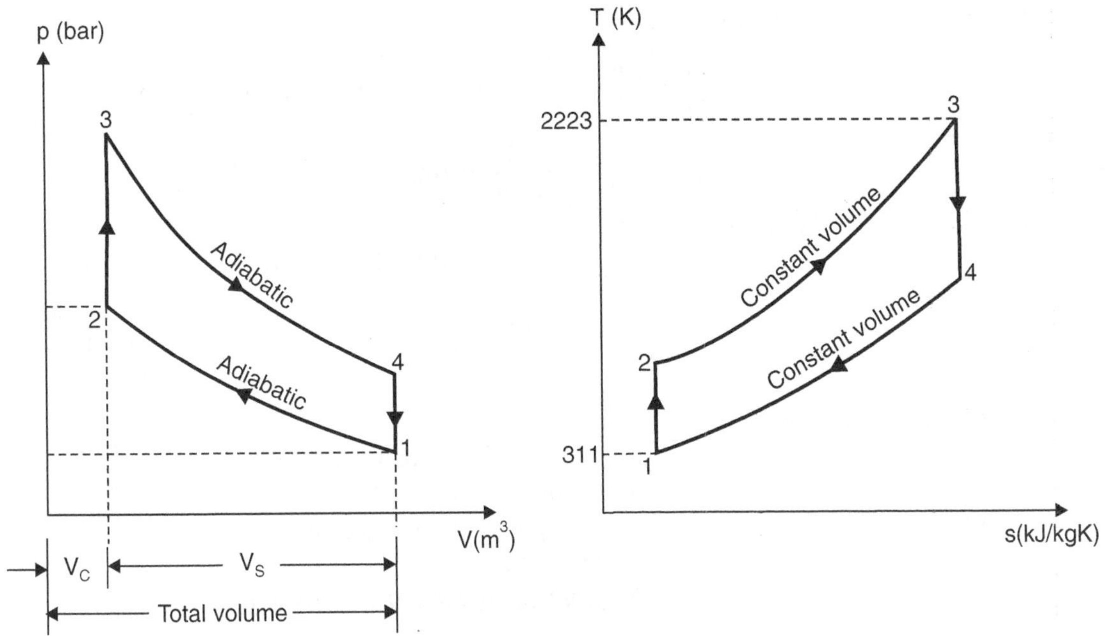

Fig. 3.11

Initial temperature, $T_1 = 38 + 273 = 311$ K

Maximum temperature, $T_3 = 1950 + 273 = 2223$ K.

(*i*) **Compression ratio, r :**

For *adiabatic compression 1-2,*

$$p_1 V_1^{\gamma} = p_2 V_2^{\gamma}$$

or

$$\left(\frac{V_1}{V_2}\right)^{\gamma} = \frac{p_2}{p_1}$$

But

$$\frac{p_2}{p_1} = 15 \qquad \text{...(Given)}$$

$\therefore$

$$(r)^{\gamma} = 15 \qquad \left[\because\ r = \frac{V_1}{V_2}\right]$$

or

$$(r)^{1.4} = 15$$

or

$$r = (15)^{\frac{1}{1.4}} = (15)^{0.714} = 6.9$$

Hence *compression ratio* = **6.9. (Ans.)**

(*ii*) **Thermal efficiency :**

Thermal efficiency, $\eta_{th} = 1 - \dfrac{1}{(r)^{\gamma - 1}} = 1 - \dfrac{1}{(6.9)^{1.4 - 1}}$ = **0.538 or 53.8%. (Ans.)**

(*iii*) **Work done :**

Again, for *adiabatic compression 1-2,*

$$\frac{T_2}{T_1} = \left(\frac{V_1}{V_2}\right)^{\gamma - 1} = (r)^{\gamma - 1} = (6.9)^{1.4 - 1} = (6.9)^{0.4} = 2.16$$

or $$T_2 = T_1 \times 2.16 = 311 \times 2.16 = 671.7 \text{ K or } 398.7°\text{C}$$

For *adiabatic expansion process 3-4,*

$$\frac{T_3}{T_4} = \left(\frac{V_4}{V_3}\right)^{\gamma - 1} = (r)^{\gamma - 1} = (6.9)^{0.4} = 2.16$$

or $$T_4 = \frac{T_3}{2.16} = \frac{2223}{2.16} = 1029 \text{ K or } 756°\text{C}$$

Heat supplied *per kg of air*

$$= c_v(T_3 - T_2) = 0.717(2223 - 671.7)$$

$$= 1112.3 \text{ kJ/kg of air} \qquad \left[c_v = \frac{R}{\gamma - 1} = \frac{0.287}{1.4 - 1} = 0.717 \text{ kJ/ kg K}\right]$$

Heat rejected per kg of air

$$= c_v(T_4 - T_1) = 0.717(1029 - 311)$$

$$= 514.8 \text{ kJ/kg of air}$$

∴ *Work done* = Heat supplied – Heat rejected

= 1112.3 – 514.8

= **597.5 kJ or 597500 N-m. (Ans.)**

☞**Example 3.13.** *An engine working on Otto cycle has a volume of 0.45 m*^{3}*, pressure 1 bar and temperature 30°C at the beginning of compression stroke. At the end of compression stroke, the pressure is 11 bar. 210 kJ of heat is added at constant volume. Determine :*

(*i*) *Pressures, temperatures and volumes at salient points in the cycle.*

(*ii*) *Percentage clearance.*

(*iii*) *Efficiency.*

(*iv*) *Net work per cycle.*

(*v*) *Mean effective pressure.*

(*vi*) *Ideal power developed by the engine if the number of working cycles per minute is 210.*

Assume the cycle is reversible.

Solution. Refer Fig. 3.12

Volume, $V_1 = 0.45 \text{ m}^3$

Initial pressure, $p_1 = 1$ bar

Initial temperature, $T_1 = 30 + 273 = 303$ K

Pressure at the end of compression stroke, $p_2 = 11$ bar

Heat added at constant volume = 210 kJ

Number of working cycles/min. = 210.

(*i*) **Pressures, temperatures and volumes at salient points :**

For *adiabatic compression 1-2,*

$$p_1 V_1^{\gamma} = p_2 V_2^{\gamma}$$

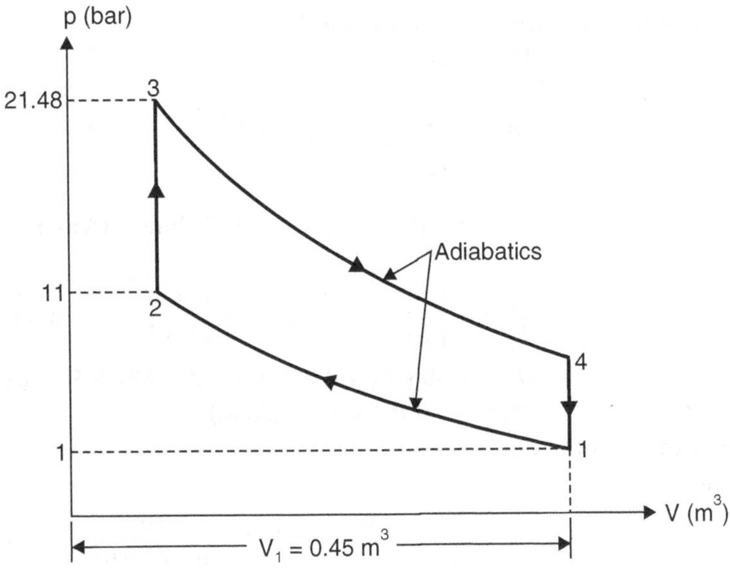

Fig. 3.12

or $$\frac{p_2}{p_1} = \left(\frac{V_1}{V_2}\right)^{\gamma} = (r)^{\gamma} \quad \text{or} \quad r = \left(\frac{p_2}{p_1}\right)^{\frac{1}{\gamma}} = \left(\frac{11}{1}\right)^{\frac{1}{1.4}} = (11)^{0.714} = 5.5$$

Also $$\frac{T_2}{T_1} = \left(\frac{V_1}{V_2}\right)^{\gamma - 1} = (r)^{\gamma - 1} = (5.5)^{1.4 - 1} = 1.977 \simeq 1.98$$

$\therefore$ $T_2 = T_1 \times 1.98 = 303 \times 1.98 =$ **600 K. (Ans.)**

Applying gas laws to points 1 and 2, we have

$$\frac{p_1V_1}{T_1} = \frac{p_2V_2}{T_2}$$

$\therefore$ $V_2 = \frac{T_2}{T_1} \times \frac{p_1}{p_2} \times V_1 = \frac{600 \times 1 \times 0.45}{303 \times 11} =$ **0.081 m³. (Ans.)**

The heat supplied during the *process 2-3* is given by :

$$Q_s = m\, c_v\, (T_3 - T_2)$$

where $$m = \frac{p_1V_1}{RT_1} = \frac{1 \times 10^5 \times 0.45}{287 \times 303} = 0.517 \text{ kg}$$

$\therefore$ $$210 = 0.517 \times 0.71\, (T_3 - 600)$$

or $T_3 = \frac{210}{0.517 \times 0.71} + 600 =$ **1172 K. (Ans.)**

For the *constant volume process 2-3,*

$$\frac{p_3}{T_3} = \frac{p_2}{T_2}$$

$\therefore$ $p_3 = \frac{T_3}{T_2} \times p_2 = \frac{1172}{600} \times 11 =$ **21.48 bar. (Ans.)**

$V_3 = V_2 =$ **0.081 m³. (Ans.)**

For the *adiabatic (or isentropic) process 3-4,*

$$p_3V_3^{\gamma} = p_4V_4^{\gamma}$$

$$p_4 = p_3 \times \left(\frac{V_3}{V_4}\right)^{\gamma} = p_3 \times \left(\frac{1}{r}\right)^{\gamma}$$

$$= 21.48 \times \left(\frac{1}{5.5}\right)^{1.4} = \textbf{1.97 bar.} \quad \textbf{(Ans.)}$$

Also
$$\frac{T_4}{T_3} = \left(\frac{V_3}{V_4}\right)^{\gamma - 1} = \left(\frac{1}{r}\right)^{\gamma - 1} = \left(\frac{1}{5.5}\right)^{1.4 - 1} = 0.505$$

$\therefore$
$$T_4 = 0.505\ T_3 = 0.505 \times 1172 = \textbf{591.8 K.} \quad \textbf{(Ans.)}$$

$$V_4 = V_1 = \textbf{0.45 m}^3. \quad \textbf{(Ans.)}$$

(*ii*) **Percentage clearance :**

Percentage clearance

$$= \frac{V_c}{V_s} = \frac{V_2}{V_1 - V_2} \times 100 = \frac{0.081}{0.45 - 0.081} \times 100$$

$$= \textbf{21.95\%.} \quad \textbf{(Ans.)}$$

(*iii*) **Efficiency :**

The heat rejected per cycle is given by

$$Q_r = mc_v(T_4 - T_1)$$

$$= 0.517 \times 0.71\ (591.8 - 303) = 106 \text{ kJ}$$

The air-standard efficiency of the cycle is given by

$$\eta_{otto} = \frac{Q_s - Q_r}{Q_s} = \frac{210 - 106}{210} = \textbf{0.495 or 49.5\%.} \quad \textbf{(Ans.)}$$

[Alternatively :

$$\eta_{otto} = 1 - \frac{1}{(r)^{\gamma - 1}} = 1 - \frac{1}{(5.5)^{1.4 - 1}} = \textbf{0.495 or 49.5\%.} \quad \textbf{(Ans.)}$$]

(*iv*) **Mean effective pressure, p_m :**

The mean effective pressure is given by

$$p_m = \frac{W \text{ (work done)}}{V_s \text{(swept volume)}} = \frac{Q_s - Q_r}{(V_1 - V_2)}$$

$$= \frac{(210 - 106) \times 10^3}{(0.45 - 0.081) \times 10^5} = \textbf{2.818 bar.} \quad \textbf{(Ans.)}$$

(*v*) **Power developed, P :**

Power developed, P = Work done per second

= Work done per cycle × Number of cycles per second

= (210 – 106) × (210/60) = **364 kW.** **(Ans.)**

Example 3.14. (*a*) *Show that the compression ratio for the maximum work to be done per kg of air in an Otto cycle between upper and lower limits of absolute temperatures T_3 and T_1 is given by*

$$r = \left(\frac{T_3}{T_1}\right)^{1/2(\gamma - 1)}$$

(*b*) *Determine the air-standard efficiency of the cycle when the cycle develops maximum work with the temperature limits of 310 K and 1220 K and working fluid is air. What will be the percentage change in efficiency if helium is used as working fluid instead of air ? The cycle operates between the same temperature limits for maximum work development.*

Consider that all conditions are ideal.

Solution. Refer Fig. 3.13.

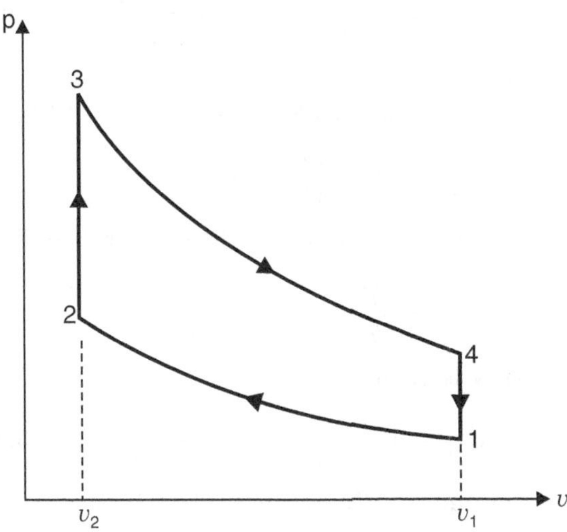

Fig. 3.13

(*a*) The work done per kg of fluid in the cycle is given by

$$W = Q_s - Q_r = c_v\,(T_3 - T_2) - c_v\,(T_4 - T_1)$$

But $$\frac{T_2}{T_1} = \left(\frac{v_1}{v_2}\right)^{\gamma - 1} = (r)^{\gamma - 1}$$

$\therefore$ $$T_2 = T_1\,.\,(r)^{\gamma - 1} \quad ...(i)$$

Similarly, $$T_3 = T_4\,.\,(r)^{\gamma - 1} \quad ...(ii)$$

$\therefore$ $$W = c_v\left[T_3 - T_1\,.\,(r)^{\gamma - 1} - \frac{T_3}{(r)^{\gamma - 1}} + T_1\right] \quad ...(iii)$$

This expression is a function of r when T_3 and T_1 are fixed. The value of W will be maximum when

$$\frac{dW}{dr} = 0.$$

$\therefore$ $$\frac{dW}{dr} = -\,T_1\,.\,(\gamma - 1)\,(r)^{\gamma - 2} - T_3\,(1 - \gamma)\,(r)^{-\gamma} = 0$$

or $$T_3\,(r)^{-\gamma} = T_1\,(r)^{\gamma - 2}$$

or $$\frac{T_3}{T_1} = (r)^{2(\gamma - 1)}$$

$\therefore$ $$r = \left(\frac{T_3}{T_1}\right)^{1/2(\gamma - 1)}$$ **Proved.**

(*b*) **Change in efficiency :**

For *air* $\gamma = 1.4$

$$\therefore \quad r = \left(\frac{T_3}{T_1}\right)^{1/2(1.4-1)} = \left(\frac{1220}{310}\right)^{1/0.8} = 5.54$$

The air-standard efficiency is given by

$$\eta_{otto} = 1 - \frac{1}{(r)^{\gamma-1}} = 1 - \frac{1}{(5.54)^{1.4-1}} = \mathbf{0.495 \text{ or } 49.5\%.} \quad \textbf{(Ans.)}$$

If *helium* is used, then the values of

$$c_p = 5.22 \text{ kJ/kgK} \quad \text{and} \quad c_v = 3.13 \text{ kJ/kg K}$$

$$\therefore \quad \gamma = \frac{c_p}{c_v} = \frac{5.22}{3.13} = 1.67$$

The compression ratio for maximum work for the temperature limits T_1 and T_3 is given by

$$r = \left(\frac{T_3}{T_1}\right)^{1/2(\gamma-1)} = \left(\frac{1220}{310}\right)^{1/2(1.67-1)} = 2.77$$

The air-standard efficiency is given by

$$\eta_{otto} = 1 - \frac{1}{(r)^{\gamma-1}} = 1 - \frac{1}{(2.77)^{1.67-1}} = \mathbf{0.495 \quad or \quad 49.5\%.}$$

Hence change in efficiency is nil. (Ans.)

Example 3.15. (*a*) *An engine working on Otto cycle, in which the salient points are 1, 2, 3 and 4, has upper and lower temperature limits T_3 and T_1. If the maximum work per kg of air is to be done, show that the intermediate temperature is given by*

$$T_2 = T_4 = \sqrt{T_1 T_3}\,.$$

(*b*) *If an engine works on Otto cycle between temperature limits 1450 K and 310 K, find the maximum power developed by the engine assuming the circulation of air per minute as 0.38 kg.*

Solution. (*a*) Refer Fig. 3.13 (Example 3.14).

Using the equation (*iii*) of example 3.14.

$$W = c_v \left[T_3 - T_1 \,.\, (r)^{\gamma-1} - \frac{T_3}{(r)^{\gamma-1}} + T_1\right]$$

and differentiating W w.r.t. r and equating to zero

$$r = \left(\frac{T_3}{T_1}\right)^{1/2(\gamma-1)}$$

$$T_2 = T_1 (r)^{\gamma-1} \quad \text{and} \quad T_4 = T_3/(r)^{\gamma-1}$$

Substituting the value of r in the above equation

$$T_2 = T_1 \left[\left(\frac{T_3}{T_1}\right)^{1/2(\gamma-1)}\right]^{\gamma-1} = T_1 \left(\frac{T_3}{T_1}\right)^{1/2} = \sqrt{T_1 T_3}$$

Similarly,

$$T_4 = \frac{T_3}{\left[\left(\frac{T_3}{T_1}\right)^{1/2(\gamma-1)}\right]^{\gamma-1}} = \frac{T_3}{\left(\frac{T_3}{T_1}\right)^{1/2}} = \sqrt{T_3 T_1}$$

$\therefore \qquad T_2 = T_4 = \sqrt{T_1 T_3}$. **Proved.**

(*b*) **Power developed, P :**

$$\left.\begin{aligned} T_1 &= 310 \text{ K} \\ T_3 &= 1450 \text{ K} \\ m &= 0.38 \text{ kg} \end{aligned}\right\} \quad \text{....(Given)}$$

Work done $\qquad W = c_v\,[(T_3 - T_2) - (T_4 - T_1)]$

$$T_2 = T_4 = \sqrt{T_1 T_3} = \sqrt{310 \times 1450} = 670.4 \text{ K}$$

$\therefore \qquad W = 0.71\,[(1450 - 670.4) - (670.4 - 310)]$

$\qquad = 0.71\,(779.6 - 360.4) = 297.6$ kJ/kg

Work done per second $\qquad = 297.6 \times (0.38/60) = 1.88$ kJ/s

Hence **power developed, P = 1.88 kW. (Ans.)**

Example 3.16. *For the same compression ratio, show that the efficiency of Otto cycle is greater than that of Diesel cycle.*

Solution. Refer Fig. 3.14.

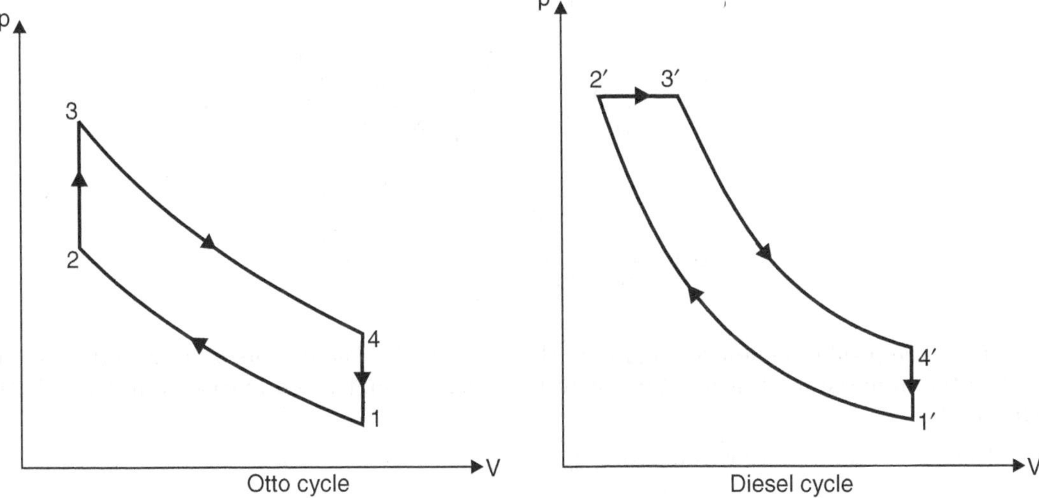

Fig. 3.14

We know that

$$\eta_{otto} = 1 - \frac{1}{(r)^{\gamma - 1}}$$

and

$$\eta_{diesel} = 1 - \frac{1}{(r)^{\gamma - 1}} \times \frac{1}{\gamma}\left\{\frac{\rho^{\gamma} - 1}{\rho - 1}\right\}$$

As the compression ratio is same,

$$\frac{V_1}{V_2} = \frac{V_1'}{V_2'} = r$$

If $\dfrac{V_4'}{V_3'} = r_1$, then cut off ratio, $\rho = \dfrac{V_3'}{V_2'} = \dfrac{r}{r_1}$

Putting the value of ρ in η_{diesel}, we get

$$\eta_{\text{diesel}} = 1 - \frac{1}{(r)^{\gamma-1}} \times \frac{1}{\gamma}\left[\frac{\left(\frac{r}{r_1}\right)^{\gamma} - 1}{\frac{r}{r_1} - 1}\right]$$

From above equation, we observe

$$\frac{r}{r_1} > 1$$

Let $r_1 = r - \delta$, where δ is a small quantity.

Then $$\frac{r}{r_1} = \frac{r}{r-\delta} = \frac{r}{r\left(1-\frac{\delta}{r}\right)} = \left(1-\frac{\delta}{r}\right)^{-1} = 1 + \frac{\delta}{r} + \frac{\delta^2}{r^2} + \frac{\delta^3}{r^3} + \ldots\ldots$$

and $$\left(\frac{r}{r_1}\right)^{\gamma} = \frac{r^{\gamma}}{r^{\gamma}\left(1-\frac{\delta}{r}\right)^{\gamma}} = \left(1-\frac{\delta}{r}\right)^{-\gamma} = 1 + \frac{\gamma\delta}{r} + \frac{\gamma(\gamma+1)}{2!} \cdot \frac{\delta^2}{r^2} + \ldots\ldots$$

$\therefore$ $$\eta_{\text{diesel}} = 1 - \frac{1}{(r)^{\gamma-1}} \times \frac{1}{\gamma}\left[\frac{\frac{\gamma \cdot \delta}{r} + \frac{\gamma(\gamma+1)}{2!} \cdot \frac{\delta^2}{r^2} + \ldots\ldots}{\frac{\delta}{r} + \frac{\delta^2}{r^2} + \ldots\ldots}\right]$$

$$= 1 - \frac{1}{(r)^{\gamma-1}}\left[\frac{\frac{\delta}{r} + \frac{\gamma+1}{2} \cdot \frac{\delta^2}{r^2} + \ldots\ldots}{\frac{\delta}{r} + \frac{\delta^2}{r^2} + \ldots\ldots}\right]$$

The ratio inside the bracket is greater than 1 since the co-efficients of terms δ^2/r^2 is greater than 1 in the numerator. It means that something more is subtracted in case of diesel cycle than in Otto cycle.

Hence, *for same compression ratio* $\eta_{otto} > \eta_{diesel}$.

3.5. CONSTANT PRESSURE OR DIESEL CYCLE

This cycle was introduced by Dr. R. Diesel in 1897. It differs from Otto cycle, in, that heat is supplied at constant pressure instead of at constant volume. Fig. 3.15 (*a* and *b*) shows the *p-V* and *T-s* diagrams of this cycle respectively.

This cycle comprises of the following ***operations*** :

(*i*) 1-2......*Adiabatic compression.*

(*ii*) 2-3......*Addition of heat at constant pressure.*

(*iii*) 3-4......*Adiabatic expansion.*

(*iv*) 4-1......*Rejection of heat at constant volume.*

Point 1 represents that the cylinder is full of air. Let p_1, V_1 and T_1 be the corresponding pressure, volume and absolute temperature. The piston then compresses the air adiabatically (*i.e.* pV^{γ} = constant) till the values become p_2, V_2 and T_2 respectively (at the end of the stroke) at point 2. Heat is then added from a hot body at a constant pressure. During this addition of heat let

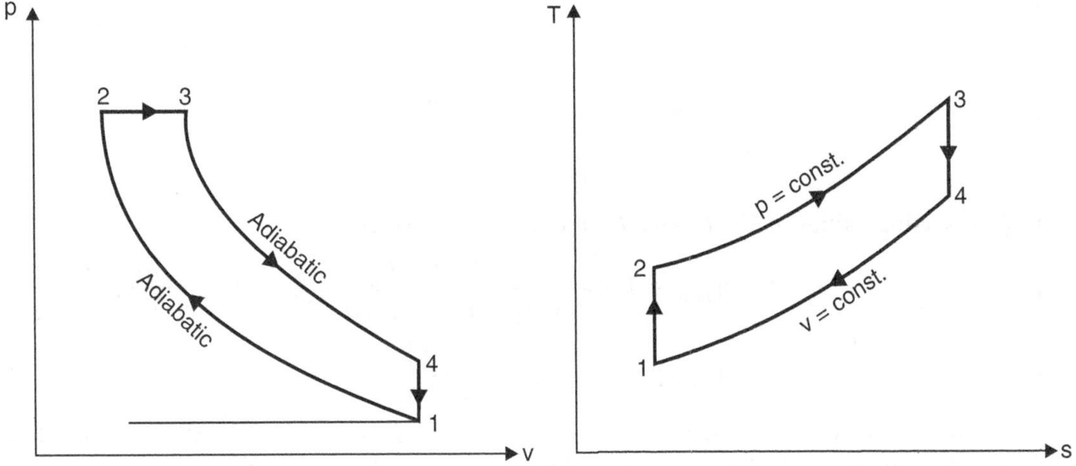

Fig. 3.15

volume increases from V_2 to V_3 and temperature T_2 to T_3, corresponding to point 3. This point (3) is called the *point of cut off*. The air then expands adiabatically to the conditions p_4, V_4 and T_4 respectively corresponding to point 4. Finally, the air rejects the heat to the cold body at constant volume till the point 1 where it returns to its original state.

Consider 1 kg of air.

Heat supplied at constant pressure $= c_p(T_3 - T_2)$

Heat rejected at constant volume $= c_v(T_4 - T_1)$

Work done $=$ Heat supplied – Heat rejected

$$= c_p(T_3 - T_2) - c_v(T_4 - T_1)$$

$$\therefore \quad \eta_{\text{diesel}} = \frac{\text{Work done}}{\text{Heat supplied}}$$

$$= \frac{c_p(T_3 - T_2) - c_v(T_4 - T_1)}{c_p(T_3 - T_2)}$$

$$= 1 - \frac{(T_4 - T_1)}{\gamma(T_3 - T_2)} \quad \text{...(i)} \left[\because \frac{c_p}{c_v} = \gamma\right]$$

Let compression ratio, $r = \dfrac{v_1}{v_2}$ and cut off ratio, $\rho = \dfrac{v_3}{v_2}$ *i.e.* $\dfrac{\text{Volume at cut-off}}{\text{Clearance volume}}$

Now, during *adiabatic compression 1-2,*

$$\frac{T_2}{T_1} = \left(\frac{v_1}{v_2}\right)^{\gamma - 1} = (r)^{\gamma - 1} \quad \text{or} \quad T_2 = T_1 \,.\, (r)^{\gamma - 1}$$

During *constant pressure process 2-3,*

$$\frac{T_3}{T_2} = \frac{v_3}{v_2} = \rho \quad \text{or} \quad T_3 = \rho \,.\, T_2 = \rho \,.\, T_1 \,.\, (r)^{\gamma - 1}$$

During *adiabatic expansion 3-4,*

$$\frac{T_3}{T_4} = \left(\frac{v_4}{v_3}\right)^{\gamma - 1}$$

$$= \left(\frac{r}{\rho}\right)^{\gamma-1} \qquad \left(\because \frac{v_4}{v_3} = \frac{v_1}{v_3} = \frac{v_1}{v_2} \times \frac{v_2}{v_3} = \frac{r}{\rho}\right)$$

$$\therefore \quad T_4 = \frac{T_3}{\left(\frac{r}{\rho}\right)^{\gamma-1}} = \frac{\rho \,.\, T_1 (r)^{\gamma-1}}{\left(\frac{r}{\rho}\right)^{\gamma-1}} = T_1 \,.\, \rho^{\gamma}$$

By inserting values of T_2, T_3 and T_4 in equation (*i*), we get

$$\eta_{\text{diesel}} = 1 - \frac{(T_1 \,.\, \rho^{\gamma} - T_1)}{\gamma(\rho \,.\, T_1 \,.\, (r)^{\gamma-1} - T_1 \,.\, (r)^{\gamma-1})} = 1 - \frac{(\rho^{\gamma} - 1)}{\gamma (r)^{\gamma-1} (\rho - 1)}$$

or

$$\eta_{\text{diesel}} = 1 - \frac{1}{\gamma (r)^{\gamma-1}} \left[\frac{\rho^{\gamma} - 1}{\rho - 1}\right] \qquad ...(3.7)$$

It may be observed that equation (3.7) for efficiency of diesel cycle is different from that of the Otto cycle only in bracketed factor. This factor is always greater than unity, because $\rho > 1$. Hence *for a given compression ratio, the Otto cycle is more efficient.*

The *net work* for diesel cycle can be expressed in terms of pv as follows :

$$W = p_2(v_3 - v_2) + \frac{p_3 v_3 - p_4 v_4}{\gamma - 1} - \frac{p_2 v_2 - p_1 v_1}{\gamma - 1}$$

$$= p_2 (\rho v_2 - v_2) + \frac{p_3 \rho v_2 - p_4 r v_2}{\gamma - 1} - \frac{p_2 v_2 - p_1 r v_1}{\gamma - 1}$$

$$\left[\begin{array}{l} \because \ \frac{v_3}{v_2} = \rho \quad \therefore \ v_3 = \rho v_2 \ \text{and} \ \frac{v_1}{v_2} = r \quad \therefore \ v_1 = r v_2 \\ \text{But } v_4 = v_1 \quad \therefore \ v_4 = r v_2 \end{array}\right]$$

$$= p_2 v_2 (\rho - 1) + \frac{p_3 \rho v_2 - p_4 r v_2}{\gamma - 1} - \frac{p_2 v_2 - p_1 r v_2}{\gamma - 1}$$

$$= \frac{v_2 [p_2 (\rho - 1)(\gamma - 1) + p_3 \rho - p_4 r - (p_2 - p_1 r)]}{\gamma - 1}$$

$$= \frac{v_2 \left[p_2 (\rho - 1)(\gamma - 1) + p_3 \left(\rho - \frac{p_4 r}{p_3}\right) - p_2 \left(1 - \frac{p_1 r}{p_2}\right)\right]}{\gamma - 1}$$

$$= \frac{p_2 v_2 [(\rho - 1)(\gamma - 1) + \rho - \rho^{\gamma} \,.\, r^{1-\gamma} - (1 - r^{1-\gamma})}{\gamma - 1}$$

$$\left[\because \ \frac{p_4}{p_3} = \left(\frac{v_3}{v_4}\right)^{\gamma} = \left(\frac{\rho}{r}\right)^{\gamma} = \rho^{\gamma} r^{-\gamma}\right]$$

$$= \frac{p_1 v_1 r^{\gamma-1} [(\rho - 1)(\gamma - 1) + \rho - \rho^{\gamma} r^{1-\gamma} - (1 - r^{1-\gamma})]}{\gamma - 1}$$

$$\left[\because \ \frac{p_2}{p_1} = \left(\frac{v_1}{v_2}\right)^{\gamma} \ \text{or} \ p_2 = p_1 \,.\, r^{\gamma} \ \text{and} \ \frac{v_1}{v_2} = r \ \text{or} \ v_2 = v_1 r^{-1}\right]$$

$$= \frac{p_1 v_1 r^{\gamma-1} [\gamma(\rho - 1) - r^{1-\gamma} (\rho^{\gamma} - 1)]}{(\gamma - 1)} \qquad ...(3.8)$$

Mean effective pressure p_m is given by :

$$p_m = \frac{p_1 v_1 r^{\gamma-1}[\gamma(\rho-1) - r^{1-\gamma}(\rho^\gamma - 1)]}{(\gamma-1)v_1\left(\frac{r-1}{r}\right)}$$

or

$$\mathbf{p_m} = \frac{p_1 r^\gamma [\gamma(\rho-1) - r^{1-\gamma}(\rho^\gamma - 1)]}{(\gamma-1)(r-1)} \quad ...(3.9)$$

Example 3.17. *A diesel engine has a compression ratio of 15 and heat addition at constant pressure takes place at 6% of stroke. Find the air standard efficiency of the engine.*

Take γ for air as 1.4.

Solution. Refer Fig. 3.16.

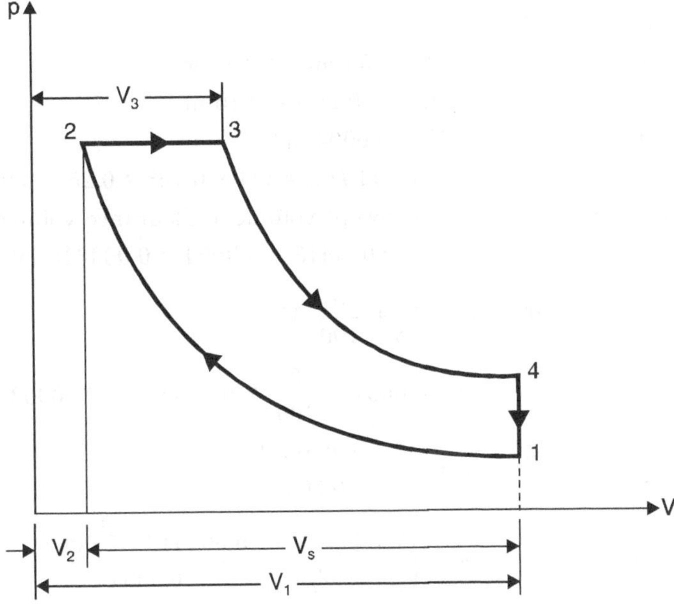

Fig. 3.16

Compression ratio, $r \left(= \frac{V_1}{V_2}\right) = 15$

γ for air = 1.4

Air standard efficiency of diesel cycle is given by

$$\eta_{diesel} = 1 - \frac{1}{\gamma (r)^{\gamma-1}}\left[\frac{\rho^\gamma - 1}{\rho - 1}\right] \quad ...(i)$$

where ρ = cut-off ratio = $\frac{V_3}{V_2}$

But $\quad V_3 - V_2 = \frac{6}{100} V_s \quad (V_s = \text{stroke volume})$

$$= 0.06\,(V_1 - V_2) = 0.06\,(15\,V_2 - V_2)$$
$$= 0.84\,V_2 \quad \text{or} \quad V_3 = 1.84\,V_2$$

$$\therefore \qquad \rho = \frac{V_3}{V_2} = \frac{1.84\,V_2}{V_2} = 1.84$$

Putting the value in eqn. (*i*), we get

$$\eta_{diesel} = 1 - \frac{1}{1.4\,(15)^{1.4-1}}\left[\frac{(1.84)^{1.4} - 1}{1.84 - 1}\right]$$

$$= 1 - 0.2417 \times 1.605 = \mathbf{0.612 \text{ or } 61.2\%}. \quad \textbf{(Ans.)}$$

Example 3.18. *The stroke and cylinder diameter of a compression ignition engine are 250 mm and 150 mm respectively. If the clearance volume is 0.0004 m³ and fuel injection takes place at constant pressure for 5 per cent of the stroke determine the efficiency of the engine. Assume the engine working on the diesel cycle.*

Solution. Refer Fig. 3.16.

Length of stroke, $L = 250 \text{ mm} = 0.25 \text{ m}$

Diameter of cylinder, $D = 150 \text{ mm} = 0.15 \text{ m}$

Clearance volume, $V_2 = 0.0004 \text{ m}^3$

Swept volume, $V_s = \pi/4\, D^2L = \pi/4 \times 0.15^2 \times 0.25 = 0.004418 \text{ m}^3$

Total cylinder volume

$$= \text{Swept volume} + \text{Clearance volume}$$
$$= 0.004418 + 0.0004 = 0.004818 \text{ m}^3$$

Volume at point of cut-off, $V_3 = V_2 + \frac{5}{100} V_s$

$$= 0.0004 + \frac{5}{100} \times 0.004418 = 0.000621 \text{ m}^3$$

$\therefore$ Cut-off ratio, $\rho = \frac{V_3}{V_2} = \frac{0.000621}{0.0004} = 1.55$

Compression ratio, $r = \frac{V_1}{V_2} = \frac{V_s + V_2}{V_2} = \frac{0.004418 + 0.0004}{0.0004} = 12.04$

Hence,

$$\eta_{diesel} = 1 - \frac{1}{\gamma\,(r)^{\gamma-1}}\left[\frac{\rho^\gamma - 1}{\rho - 1}\right] = 1 - \frac{1}{1.4 \times (12.04)^{1.4-1}}\left[\frac{(1.55)^{1.4} - 1}{1.55 - 1}\right]$$

$$= 1 - 0.264 \times 1.54 = \mathbf{0.593 \text{ or } 59.3\%}. \quad \textbf{(Ans.)}$$

Example 3.19. *Calculate the percentage loss in the ideal efficiency of a diesel engine with compression ratio 14 if the fuel cut-off is delayed from 5% to 8%.*

Solution. Let the clearance volume (V_2) be unity.

Then, compression ratio, $r = 14$

Now, when the fuel is cut-off at 5%, we have

$$\frac{\rho - 1}{r - 1} = \frac{5}{100} \quad \text{or} \quad \frac{\rho - 1}{14 - 1} = 0.05 \quad \text{or} \quad \rho - 1 = 13 \times 0.05 = 0.65$$

$$\therefore \qquad \rho = 1.65$$

$$\eta_{diesel} = 1 - \frac{1}{\gamma (r)^{\gamma - 1}} \left[\frac{\rho^\gamma - 1}{\rho - 1} \right] = 1 - \frac{1}{1.4 \times (14)^{1.4 - 1}} \left[\frac{(1.65)^{1.4} - 1}{1.65 - 1} \right]$$

$$= 1 - 0.248 \times 1.563 = 0.612 \quad \text{or} \quad 61.2\%$$

When the fuel is cut-off at 8%, we have

$$\frac{\rho - 1}{r - 1} = \frac{8}{100} \quad \text{or} \quad \frac{\rho - 1}{14 - 1} = \frac{8}{100} = 0.08$$

$\therefore$ $$\rho = 1 + 1.04 = 2.04$$

$$\eta_{diesel} = 1 - \frac{1}{\gamma (r)^{\gamma - 1}} \left[\frac{\rho^\gamma - 1}{\rho - 1} \right] = 1 - \frac{1}{1.4 \times (14)^{1.4 - 1}} \left[\frac{(2.04)^{1.4} - 1}{2.04 - 1} \right]$$

$$= 1 - 0.248 \times 1.647 = \mathbf{0.591} \quad \textbf{or} \quad \mathbf{59.1\%.} \quad \textbf{(Ans.)}$$

Hence percentage loss in efficiency due to delay in fuel cut-off

$$= 61.2 - 59.1 = \mathbf{2.1\%.} \quad \textbf{(Ans.)}$$

Example 3.20. *The mean effective pressure of a Diesel cycle is 7.5 bar and compression ratio is 12.5. Find the percentage cut-off of the cycle if its initial pressure is 1 bar.*

Solution. Mean effective pressure, $p_m = 7.5$ bar

Compression ratio, $r = 12.5$

Initial pressure, $p_1 = 1$ bar

Refer Fig. 3.15.

The mean effective pressure is given by

$$p_m = \frac{p_1 r^\gamma [\gamma (\rho - 1) - r^{1-\gamma} (\rho^\gamma - 1)]}{(\gamma - 1)(r - 1)} \quad \text{...[Eqn. (3.9)]}$$

$$7.5 = \frac{1 \times (12.5)^{1.4} [1.4 (\rho - 1) - (12.5)^{1-1.4} (\rho^{1.4} - 1)]}{(1.4 - 1)(12.5 - 1)}$$

$$7.5 = \frac{34.33[1.4 \rho - 1.4 - 0.364\rho^{1.4} + 0.364]}{4.6}$$

$$7.5 = 7.46 (1.4 \rho - 1.036 - 0.364 \rho^{1.4})$$

$$1.005 = 1.4 \rho - 1.036 - 0.364 \rho^{1.4}$$

or $$2.04 = 1.4 \rho - 0.364 \rho^{1.4} \quad \text{or} \quad 0.346 \rho^{1.4} - 1.4 \rho + 2.04 = 0$$

Solving by trial and error method, we get

$$\rho = 2.24$$

$\therefore$ $$\% \; cut\text{-}off = \frac{\rho - 1}{r - 1} \times 100 = \frac{2.24 - 1}{12.5 - 1} \times 100 = \mathbf{10.78\%.} \quad \textbf{(Ans.)}$$

☞**Example 3.21.** *An engine with 200 mm cylinder diameter and 300 mm stroke works on theoretical Diesel cycle. The initial pressure and temperature of air used are 1 bar and 27°C. The cut-off is 8% of the stroke. Determine :*

(i) Pressures and temperatures at all salient points.

(ii) Theoretical air standard efficiency.

(iii) Mean effective pressure.

(iv) Power of the engine if the working cycles per minute are 380.

Assume that compression ratio is 15 and working fluid is air.

Consider all conditions to be ideal.

Solution. Refer Fig. 3.17.

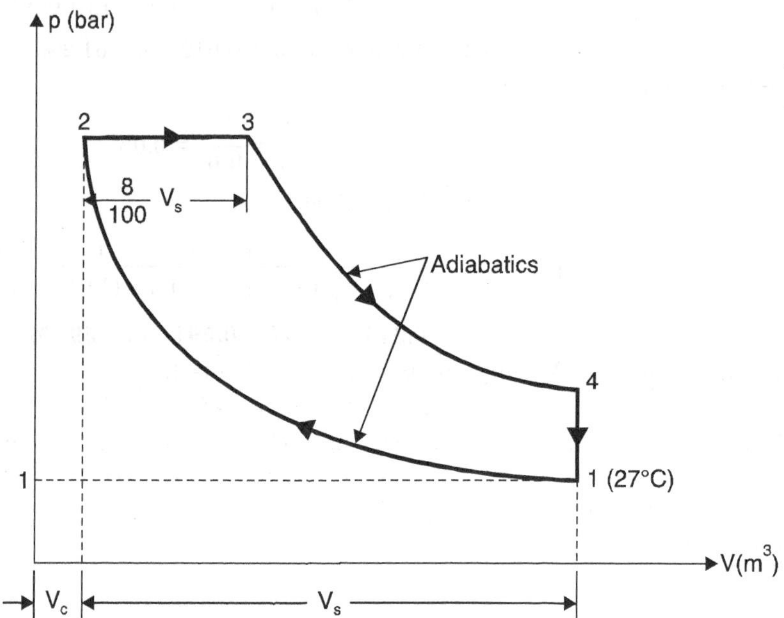

Fig. 3.17

Cylinder diameter, $D = 200$ mm or 0.2 m

Stroke length, $L = 300$ mm or 0.3 m

Initial pressure, $p_1 = 1.0$ bar

Initial temperature, $T_1 = 27 + 273 = 300$ K

Cut-off $= \frac{8}{100} V_s = 0.08\ V_s$

(*i*) **Pressures and temperatures at salient points :**

Now, stroke volume, $V_s = \pi/4\ D^2L = \pi/4 \times 0.2^2 \times 0.3 = 0.00942\ m^3$

$$V_1 = V_s + V_c = V_s + \frac{V_s}{r-1} \qquad \left[\because\ V_c = \frac{V_s}{r-1}\right]$$

$$= V_s\left(1 + \frac{1}{r-1}\right) = \frac{r}{r-1} \times V_s$$

i.e.,

$$V_1 = \frac{15}{15-1} \times V_s = \frac{15}{14} \times 0.00942 = \mathbf{0.0101\ m^3}. \quad \textbf{(Ans.)}$$

Mass of the air in the cylinder can be calculated by using the gas equation,

$$p_1V_1 = mRT_1$$

$$m = \frac{p_1V_1}{RT_1} = \frac{1 \times 10^5 \times 0.0101}{287 \times 300} = 0.0117 \text{ kg/cycle}$$

For the *adiabatic (or isentropic) process 1-2,*

$$p_1 V_1^{\gamma} = p_2 V_2^{\gamma} \quad \text{or} \quad \frac{p_2}{p_1} = \left(\frac{V_1}{V_2}\right)^{\gamma} = (r)^{\gamma}$$

$\therefore$ $\mathbf{p_2} = p_1 \cdot (r)^{\gamma} = 1 \times (15)^{1.4} = \mathbf{44.31\ bar.}$ **(Ans.)**

Also, $$\frac{T_2}{T_1} = \left(\frac{V_1}{V_2}\right)^{\gamma - 1} = (r)^{\gamma - 1} = (15)^{1.4 - 1} = 2.954$$

$\therefore$ $\mathbf{T_2} = T_1 \times 2.954 = 300 \times 2.954 = \mathbf{886.2\ K.}$ **(Ans.)**

$$\mathbf{V_2} = V_c = \frac{V_s}{r - 1} = \frac{0.00942}{15 - 1} = \mathbf{0.0006728\ m^3.}\ \textbf{(Ans.)}$$

$\mathbf{p_2 = p_3 = 44.31\ bar.}$ **(Ans.)**

% cut-off ratio $$= \frac{\rho - 1}{r - 1}$$

$$\frac{8}{100} = \frac{\rho - 1}{15 - 1}$$

i.e., $\rho = 0.08 \times 14 + 1 = 2.12$

$\therefore$ $\mathbf{V_3} = \rho\, V_2 = 2.12 \times 0.0006728 = \mathbf{0.001426\ m^3.}$ **(Ans.)**

$$\left[\begin{array}{l} V_3 \text{ can also be calculated as follows:} \\ V_3 = 0.08 V_s + V_c = 0.08 \times 0.00942 + 0.0006728 = 0.001426\ \text{m}^3 \end{array}\right]$$

For the *constant pressure process 2-3,*

$$\frac{V_3}{T_3} = \frac{V_2}{T_2}$$

$\therefore$ $$\mathbf{T_3} = T_2 \times \frac{V_3}{V_2} = 886.2 \times \frac{0.001426}{0.0006728} = \mathbf{1878.3\ K.}\ \textbf{(Ans.)}$$

For the *isentropic process 3-4*

$$p_3 V_3^{\gamma} = p_4 V_4^{\gamma}$$

$$\mathbf{p_4} = p_3 \times \left(\frac{V_3}{V_4}\right)^{\gamma} = p_3 \times \frac{1}{(7.07)^{1.4}}$$

$$= \frac{44.31}{(7.07)^{1.4}} = \mathbf{2.866\ bar.}\ \textbf{(Ans.)}$$

$$\left[\because \frac{V_4}{V_3} = \frac{V_4}{V_2} \times \frac{V_2}{V_3} = \frac{V_1}{V_2} \times \frac{V_2}{V_3} = \frac{r}{\rho}, \quad \because V_4 = V_1 = \frac{15}{2.12} = 7.07\right]$$

Also, $$\frac{T_4}{T_3} = \left(\frac{V_3}{V_4}\right)^{\gamma - 1} = \left(\frac{1}{7.07}\right)^{1.4 - 1} = 0.457$$

$\therefore$ $\mathbf{T_4} = T_3 \times 0.457 = 1878.3 \times 0.457 = \mathbf{858.38\ K.}$ **(Ans.)**

$\mathbf{V_4} = V_1 = \mathbf{0.0101\ m^3.}$ **(Ans.)**

(*ii*) **Theoretical air standard efficiency :**

$$\eta_{diesel} = 1 - \frac{1}{\gamma (r)^{\gamma - 1}}\left[\frac{\rho^{\gamma} - 1}{\rho - 1}\right] = 1 - \frac{1}{1.4\,(15)^{1.4 - 1}}\left[\frac{(2.12)^{1.4} - 1}{2.12 - 1}\right]$$

$$= 1 - 0.2418 \times 1.663 = \mathbf{0.598\ or\ 59.8\%.}\ \textbf{(Ans.)}$$

(*iii*) **Mean effective pressure, p_m :**

Mean effective pressure of Diesel cycle is given by

$$p_m = \frac{p_1(r)^{\gamma}[\gamma(\rho - 1) - r^{1-\gamma}(\rho^{\gamma} - 1)}{(\gamma - 1)(r - 1)}$$

$$= \frac{1 \times (15)^{1.4}[1.4(2.12 - 1) - (15)^{1-1.4}(2.12^{1.4} - 1)]}{(1.4 - 1)(15 - 1)}$$

$$= \frac{44.31[1.568 - 0.338 \times 1.863]}{0.4 \times 14} = \mathbf{7.424\ bar.}\quad \textbf{(Ans.)}$$

(*iv*) **Power of the engine, P :**

Work done per cycle $= p_m V_s = \dfrac{7.424 \times 10^5 \times 0.00942}{10^3} = 6.99$ kJ/cycle

Work done per second = Work done per cycle × No. of cycles per second

= 6.99 × 380/60 = 44.27 kJ/s = 44.27 kW

Hence *power of the engine* = **44.27 kW. (Ans.)**

Example 3.22. *The volume ratios of compression and expansion for a diesel engine as measured from an indicator diagram are 15.3 and 7.5 respectively. The pressure and temperature at the beginning of the compression are 1 bar and 27°C.*

Assuming an ideal engine, determine the mean effiective pressure, the ratio of maximum pressure to mean effective pressure and cycle efficiency.

Also find the fuel comsumption per kWh if the indicated thermal efficiency is 0.5 of ideal efficiency, mechanical efficiency is 0.8 and the calorific value of oil 42000 kJ/kg.

Assume for air : $c_p = 1.005$ *kJ/kg K ;* $c_v = 0.718$ *kJ/kg K,* $\gamma = 1.4$. **(U.P.S.C. 1996)**

Solution. Refer Fig. 3.18. *Given :* $\dfrac{V_1}{V_2} = 15.3$; $\dfrac{V_4}{V_3} = 7.5$

$p_1 = 1$ bar ; $T_1 = 27 + 273 = 300$ K ; $\eta_{th(I)} = 0.5 \times \eta_{air\text{-}standard}$; $\eta_{mech.} = 0.8$; $C = 42000$ kJ/kg.

The cycle is shown in the Fig. 3.18, the subscripts denote the respective points in the cycle.

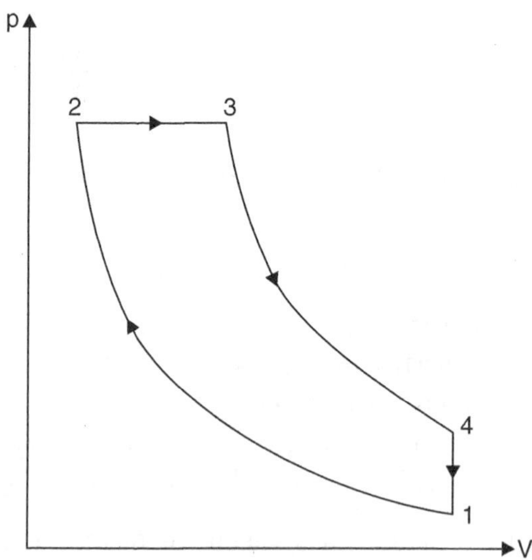

Fig. 3.18. Diesel cycle.

Mean effective pressure, p_m :

$$p_m = \frac{\text{Work done by the cycle}}{\text{Swept volume}}$$

Work done = Heat added – Heat rejected

Heat added $= mc_p\,(T_3 - T_2)$, and

Heat rejected $= mc_v\,(T_4 - T_1)$

Now assume air as a perfect gas and mass of oil in the air-fuel mixture is negligible and is not taken into account.

Process 1-2 is an *adiabatic compression process,* thus

$$\frac{T_2}{T_1} = \left(\frac{V_1}{V_2}\right)^{\gamma - 1} \quad \text{or} \quad T_2 = T_1 \times \left(\frac{V_1}{V_2}\right)^{1.4 - 1} \quad (\text{since } \gamma = 1.4)$$

or $$T_2 = 300 \times (15.3)^{0.4} = 893.3 \text{ K}$$

Also, $$p_1 V_1^{\gamma} = p_2 V_2^{\gamma} \Rightarrow p_2 = p_1 \times \left(\frac{V_1}{V_2}\right)^{\gamma} = 1 \times (15.3)^{1.4} = 45.56 \text{ bar}$$

Process 2-3 is a *constant pressure process,* hence

$$\frac{V_2}{T_2} = \frac{V_3}{T_3} \Rightarrow T_3 = \frac{V_3 T_2}{V_2} = 2.04 \times 893.3 = 1822.3 \text{ K}$$

Assume that the volume at point 2 (V_2) is 1 m^3. Thus the mass of air involved in the precess,

$$m = \frac{p_2 V_2}{RT_2} = \frac{45.56 \times 10^5 \times 1}{287 \times 893.3} = 17.77 \text{ kg}$$

$$\left[\because \frac{V_4}{V_3} = \frac{V_1}{V_3} = \frac{V_1}{V_2} \times \frac{V_2}{V_3} \quad \text{or} \quad \frac{V_3}{V_2} = \frac{V_1}{V_2} \times \frac{V_3}{V_4} = \frac{15.3}{7.5} = 2.04\right]$$

Process 3-4 is an *adiabatic expansion process,* thus

$$\frac{T_4}{T_3} = \left(\frac{V_3}{V_4}\right)^{\gamma - 1} = \left(\frac{1}{7.5}\right)^{1.4 - 1} = 0.4466$$

or $$T_4 = 1822.3 \times 0.4466 = 813.8 \text{ K}$$

$\therefore$ Work done $= mc_p\,(T_3 - T_2) - mc_v\,(T_4 - T_1)$

$= 17.77\,[1.005\,(1822.3 - 893.3) - 0.718\,(813.8 - 300)] = 10035$ kJ

$$\therefore \quad p_m = \frac{\text{Work done}}{\text{Swept volume}} = \frac{10035}{(V_1 - V_2)} = \frac{10035}{(15.3V_2 - V_2)} = \frac{10035}{14.3}$$

$= 701.7$ kN/m^2 = **7.017 bar. (Ans.)**

($\because$ $V_2 = 1$ m^3 assumed)

Ratio of maximum pressure to mean effective pressure

$$= \frac{p_2}{p_m} = \frac{45.56}{7.017} = 6.49. \quad \textbf{(Ans.)}$$

Cycle efficiency, η_{cycle} :

$$\eta_{cycle} = \frac{\text{Work done}}{\text{Heat supplied}}$$

$$= \frac{10035}{mc_p\,(T_3 - T_2)} = \frac{10035}{17.77 \times 1.005\,(1822.3 - 897.3)} = \mathbf{0.6048 \quad or \quad 60.48\%.} \quad \textbf{(Ans.)}$$

Fuel consumption per kWh ; $\mathbf{m_f}$

$\eta_{th(I)} = 0.5\ \eta_{cycle} = 0.5 \times 0.6048 = 0.3024$ or 30.24%

$$\eta_{th(B)} = 0.3024 \times 0.8 = 0.242$$

Also,
$$\eta_{th(B)} = \frac{\text{B.P.}}{m_f \times C} = \frac{1}{\frac{m_f}{3600} \times 42000} = \frac{3600}{m_f \times 42000}$$

or
$$0.242 = \frac{3600}{m_f \times 42000}$$

or
$$m_f = \frac{3600}{0.242 \times 42000} = \mathbf{0.354\ kg/kWh.} \quad \textbf{(Ans.)}$$

3.6. DUAL COMBUSTION CYCLE

This cycle (also called the *limited pressure cycle or mixed cycle*) is a combination of Otto and Diesel cycles, in a way, that h*eat is added partly at constant volume and partly at constant pressure ; the advantage of which is that more time is available to fuel (which is injected into the engine cylinder before the end of compression stroke) for combustion. Because of lagging characteristics of fuel this cycle is invariably used for diesel and hot spot ignition engines.*

The dual combustion cycle (Fig. 3.19) consists of the following ***operations*** :

(*i*) 1-2—*Adiabatic compression*

(*ii*) 2-3—*Addition of heat at constant volume*

(*iii*) 3-4—*Addition of heat at constant pressure*

(*iv*) 4-5—*Adiabatic expansion*

(*v*) 5-1—*Rejection of heat at constant volume.*

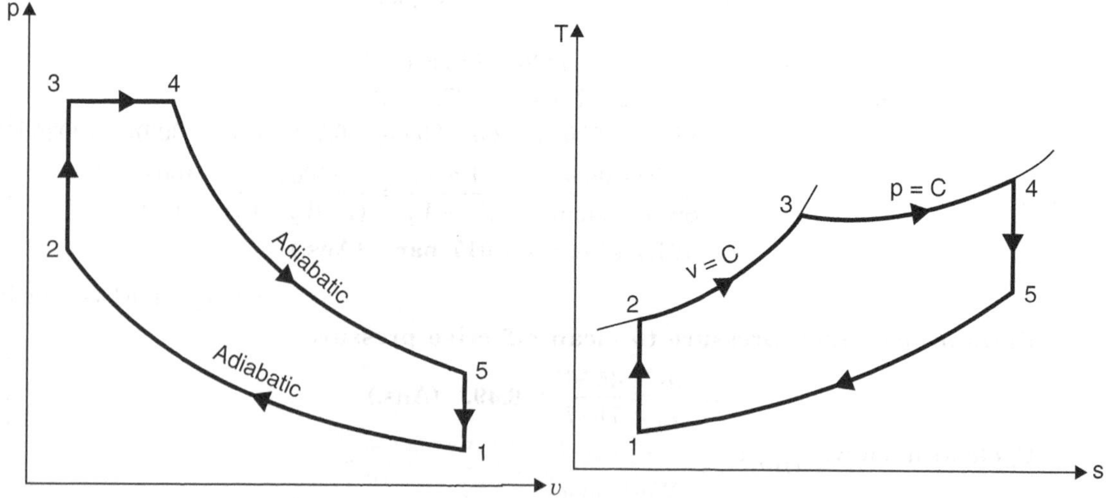

Fig. 3.19

Consider 1 kg of air.

Total heat supplied = Heat supplied during the operation 2-3 + Heat supplied during the operation 3-4

$$= c_v(T_3 - T_2) + c_p(T_4 - T_3)$$

Heat rejected during operation 5-1 $= c_v(T_5 - T_1)$

Work done = Heat supplied – Heat rejected

$$= c_v(T_3 - T_2) + c_p(T_4 - T_3) - c_v(T_5 - T_1)$$

$$\eta_{dual} = \frac{\text{Work done}}{\text{Heat supplied}} = \frac{c_v (T_3 - T_2) + c_p (T_4 - T_3) - c_v (T_5 - T_1)}{c_v (T_3 - T_2) + c_p (T_4 - T_3)}$$

$$= 1 - \frac{c_v (T_5 - T_1)}{c_v (T_3 - T_2) + c_p (T_4 - T_3)}$$

$$= 1 - \frac{c_v (T_5 - T_1)}{(T_3 - T_2) + \gamma (T_4 - T_3)} \quad ...(i) \quad \left(\because \gamma = \frac{c_p}{c_v}\right)$$

Compression ratio, $r = \frac{v_1}{v_2}$

During *adiabatic compression process 1-2,*

$$\frac{T_2}{T_1} = \left(\frac{v_1}{v_2}\right)^{\gamma - 1} = (r)^{\gamma - 1} \quad ...(ii)$$

During *constant volume heating process,*

$$\frac{p_3}{T_3} = \frac{p_2}{T_2}$$

or $\frac{T_3}{T_2} = \frac{p_3}{v_2} = \beta$, where β is known as ***pressure or explosion ratio.***

or

$$T_2 = \frac{T_3}{\beta} \quad ...(iii)$$

During *adiabatic expansion process 4-5,*

$$\frac{T_4}{T_5} = \left(\frac{v_5}{v_4}\right)^{\gamma - 1}$$

$$= \left(\frac{r}{\rho}\right)^{\gamma - 1} \quad ...(iv)$$

$$\left(\because \frac{v_5}{v_4} = \frac{v_1}{v_4} = \frac{v_1}{v_2} \times \frac{v_2}{v_4} = \frac{v_1}{v_2} \times \frac{v_3}{v_4} = \frac{r}{\rho}, \rho \text{ being the cut-off ratio}\right)$$

During *constant pressure heating process 3-4,*

$$\frac{v_3}{T_3} = \frac{v_4}{T_4}$$

$$T_4 = T_3 \frac{v_4}{v_3} = \rho\, T_3 \quad ...(v)$$

Putting the value of T_4 in the equation (*iv*), we get

$$\frac{\rho T_3}{T_5} = \left(\frac{r}{\rho}\right)^{\gamma - 1} \quad \text{or} \quad T_5 = \rho \,.\, T_3 \,.\, \left(\frac{\rho}{\gamma}\right)^{\gamma - 1}$$

Putting the value of T_2 in equation (*ii*), we get

$$\frac{\frac{T_3}{\beta}}{T_1} = (r)^{\gamma - 1}$$

$$T_1 = \frac{T_3}{\beta} \cdot \frac{1}{(r)^{\gamma - 1}}$$

Now inserting the values of T_1, T_2, T_4 and T_5 in equation (*i*), we get

$$\eta_{dual} = 1 - \frac{\left[\rho . T_3 \left(\frac{\rho}{r}\right)^{\gamma - 1} - \frac{T_3}{\beta} \cdot \frac{1}{(r)^{\gamma - 1}}\right]}{\left[\left(T_3 - \frac{T_3}{\beta}\right) + \gamma(\rho T_3 - T_3)\right]} = 1 - \frac{\frac{1}{(r)^{\gamma - 1}}\left(\rho^{\gamma} - \frac{1}{\beta}\right)}{\left(1 - \frac{1}{\beta}\right) + \gamma(\rho - 1)}$$

i.e.,

$$\eta_{dual} = 1 - \frac{1}{(r)^{\gamma - 1}} \cdot \frac{(\beta . \rho^{\gamma} - 1)}{[(\beta - 1) + \beta\gamma(\rho - 1)]} \quad \text{...(3.10)}$$

Work done is given by,

$$W = p_3(v_4 - v_3) + \frac{p_4 v_4 - p_5 v_5}{\gamma - 1} - \frac{p_2 v_2 - p_1 v_1}{\gamma - 1}$$

$$= p_3 v_3(\rho - 1) + \frac{(p_4 \rho v_3 - p_5 r v_3) - (p_2 v_3 - p_1 r v_3)}{\gamma - 1}$$

$$= \frac{p_3 v_3 (\rho - 1)(\gamma - 1) + p_4 v_3 \left(\rho - \frac{p_5}{p_4} r\right) - p_2 v_3 \left(1 - \frac{p_1}{p_2} r\right)}{\gamma - 1}$$

Also $\qquad \frac{p_5}{p_4} = \left(\frac{v_4}{v_5}\right)^{\gamma} = \left(\frac{\rho}{r}\right)^{\gamma}$ and $\frac{p_2}{p_1} = \left(\frac{v_1}{v_2}\right)^{\gamma} = r^{\gamma}$

also, $\qquad p_3 = p_4,\ v_2 = v_3,\ v_5 = v_1$

$$\therefore \quad W = \frac{v_3[p_3(\rho - 1)(\gamma - 1) + p_3(\rho - \rho^{\gamma} r^{1-\gamma}) - p_2(1 - r^{1-\gamma})]}{(\gamma - 1)}$$

$$= \frac{p_2 v_2[\beta(\rho - 1)(\gamma - 1) + \beta(\rho - \rho^{\gamma} r^{1-\gamma}) - (1 - r^{1-\gamma})]}{(\gamma - 1)}$$

$$= \frac{p_1(r)^{\gamma} v_1/r[\beta\gamma(\rho - 1) + (\beta - 1) - r^{1-\gamma}(\beta\rho^{\gamma} - 1)]}{\gamma - 1}$$

$$= \frac{p_1 v_1 r^{\gamma - 1}[\beta\gamma(\rho - 1) + (\beta - 1) - r^{\gamma - 1}(\beta\rho^{\gamma} - 1)]}{\gamma - 1} \quad \text{...(3.11)}$$

Mean effective pressure ($\mathbf{p_m}$) is given by,

$$p_m = \frac{W}{v_1 - v_2} = \frac{W}{v_1\left(\frac{r - 1}{r}\right)} = \frac{p_1 v_1[r^{1-\gamma}\beta\gamma(\rho - 1) + (\beta - 1) - r^{1-\gamma}(\beta\rho^{\gamma} - 1)]}{(\gamma - 1)v_1\left(\frac{r - 1}{r}\right)}$$

$$\mathbf{p_m} = \frac{p_1(r)^{\gamma}[\beta(\rho - 1) + (\beta - 1) - r^{1-\gamma}(\beta\rho^{\gamma} - 1)]}{(\gamma - 1)(r - 1)} \quad \text{...(3.12)}$$

Example 3.23. *The swept volume of a diesel engine working on dual cycle is 0.0053 m^3 and clearance volume is 0.00035 m^3. The maximum pressure is 65 bar. Fuel injection ends at 5 per cent of the stroke. The temperature and pressure at the start of the compression are 80°C and 0.9 bar. Determine the air standard efficiency of the cycle. Take γ for air = 1.4.*

Solution. Refer Fig. 3.20.

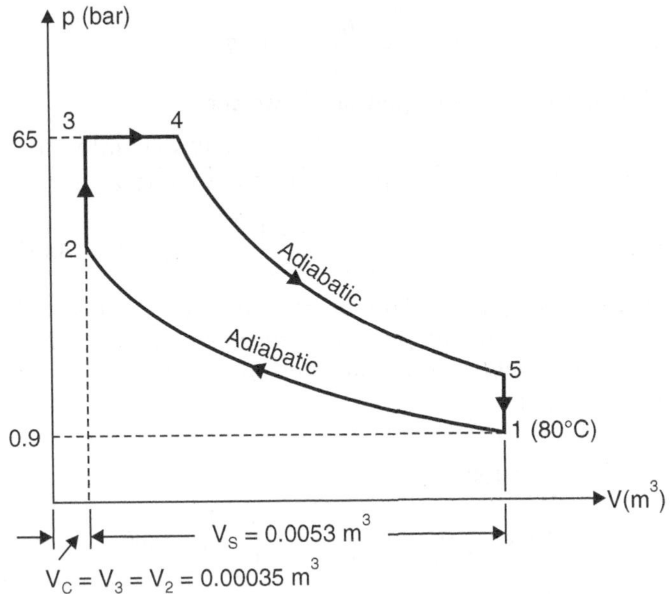

Fig. 3.20

Swept volume, $V_s = 0.0053 \text{ m}^3$

Clearance volume, $V_c = V_3 = V_2 = 0.00035 \text{ m}^3$

Maximum pressure, $p_3 = p_4 = 65$ bar

Initial temperature, $T_1 = 80 + 273 = 353$ K

Initial pressure, $p_1 = 0.9$ bar

$\eta_{\textbf{dual}} = ?$

The efficiency of a dual combustion cycle is given by

$$\eta_{\text{dual}} = 1 - \frac{1}{(r)^{\gamma-1}}\left[\frac{\beta \,.\, \rho^{\gamma} - 1}{(\beta - 1) + \beta\gamma(\rho - 1)}\right] \qquad ...(i)$$

Compression ratio, $r = \dfrac{V_1}{V_2} = \dfrac{V_s + V_c}{V_c} = \dfrac{0.0053 + 0.00035}{0.00035} = 16.14$

[∵ $V_2 = V_c$ = Clearance volume]

Cut-off ratio, $\rho = \dfrac{V_4}{V_3} = \dfrac{\frac{5}{100}V_s + V_3}{V_3} = \dfrac{0.05V_s + V_c}{V_c}$ (∵ $V_2 = V_3 = V_c$)

$$= \frac{0.05 \times 0.0053 + 0.00035}{0.00035} = 1.757 \text{ say } 1.76$$

Also during the *compression operation 1-2*

$$p_1 V_1^{\gamma} = p_2 V_2^{\gamma}$$

or $$\frac{p_2}{p_1} = \left(\frac{V_1}{V_2}\right)^{\gamma} = (16.14)^{1.4} = 49.14$$

or $$p_2 = p_1 \times 49.14 = 0.9 \times 49.14 = 44.22 \text{ bar}$$

Pressure or explosion ratio, $\beta = \frac{p_3}{p_2} = \frac{65}{44.22} = 1.47$

Putting the value of r, ρ and β in equation (*i*), we get

$$\eta_{dual} = 1 - \frac{1}{(16.14)^{1.4-1}}\left[\frac{1.47 \times (1.76)^{1.4} - 1}{(1.47-1) + 1.47 \times 1.4\,(1.76-1)}\right]$$

$$= 7 - 0.328\left[\frac{3.243 - 1}{0.47 + 1.564}\right] = \mathbf{0.6383 \text{ or } 63.83\%.} \quad \textbf{(Ans.)}$$

Example 3.24. *An oil engine working on the dual combustion cycle has a compression ratio 14 and the explosion ratio obtained from an indicator card is 1.4. If the cut-off occurs at 6 per cent of stroke, find the ideal efficiency. Take γ for air = 1.4.*

Solution. Refer Fig. 3.19.

Compression ratio, $r = 14$

Explosion ratio, $\beta = 1.4$

If ρ is the cut-off ratio, then $\frac{\rho - 1}{r - 1} = \frac{6}{100}$ or $\frac{\rho - 1}{14 - 1} = 0.06$

$\therefore$ $\rho = 1.78$

Ideal efficiency is given by

$$\eta_{ideal\ or\ dual} = 1 - \frac{1}{(r)^{\gamma-1}}\left[\frac{(\beta\rho^{\gamma} - 1)}{(\beta - 1) + \beta\gamma(\rho - 1)}\right]$$

$$= 1 - \frac{1}{(14)^{1.4-1}}\left[\frac{1.4 \times (1.78)^{1.4} - 1}{(1.4-1) + 1.4 \times 1.4(1.78-1)}\right]$$

$$= 1 - 0.348\left[\frac{3.138 - 1}{0.4 + 1.528}\right] = \mathbf{0.614 \text{ or } 61.4\%.} \quad \textbf{(Ans.)}$$

Example 3.25. *The compression ratio for a single-cylinder engine operating on dual cycle is 9. The maximum pressure in the cylinder is limited to 60 bar. The pressure and temperature of the air at the beginning of the cycle are 1 bar and 30°C. Heat is added during constant pressure process upto 4 per cent of the stroke. Assuming the cylinder diameter and stroke length as 250 mm and 300 mm respectively, determine :*

(*i*) *The air standard efficiency of the cycle.*

(*ii*) *The power developed if the number of working cycles are 3 per second.*

Take for air c_v = 0.71 kJ/kg K and : c_p = 1.0 kJ/kgK.

Solution. Refer Fig. 3.21.

Cylinder diameter, $D = 250$ mm $= 0.25$ m

Compression ratio, $r = 9$

Stroke length, $L = 300$ mm $= 0.3$ m

Initial pressure, $p_1 = 1$ bar
Initial temperature, $T_1 = 30 + 273 = 303$ K
Maximum pressure, $p_3 = p_4 = 60$ bar
Cut-off = 4% of stroke volume
Number of working cycles/sec. = 3.

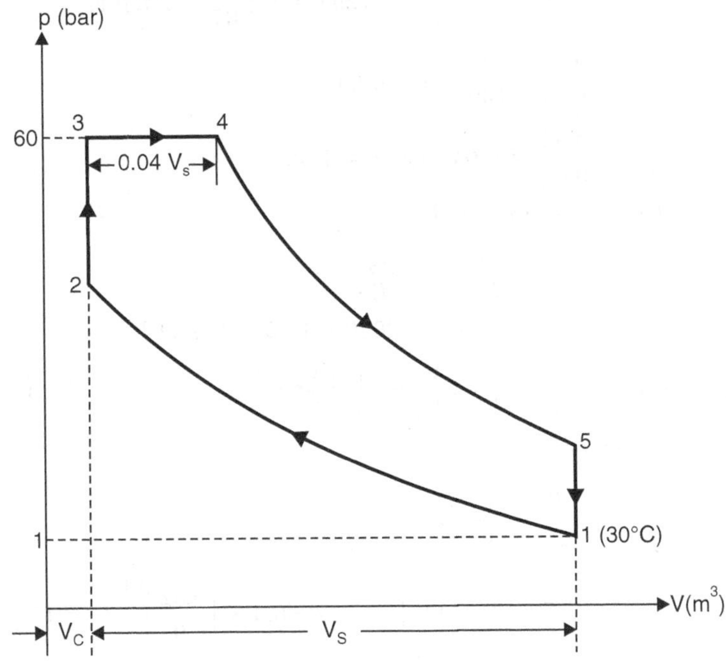

Fig. 3.21

(*i*) **Air standard efficiency :**

Now, swept volume, $V_s = \pi/4\ D^2L = \pi/4 \times 0.25^2 \times 0.3$
$= 0.0147 \text{ m}^3$

Also, compression ratio, $r = \dfrac{V_s + V_c}{V_c}$

i.e.,
$$9 = \frac{0.0147 + V_c}{V_c}$$

$$\therefore \quad V_c = \frac{0.0147}{8} = 0.0018 \text{ m}^3$$

$$\therefore \quad V_1 = V_s + V_c = 0.0147 + 0.0018 = 0.0165 \text{ m}^3$$

For the *adiabatic (or isentropic) process 1-2,*

$$p_1V_1^{\gamma} = p_2V_2^{\gamma}$$

$$p_2 = p_1 \times \left(\frac{V_1}{V_2}\right)^{\gamma} = 1 \times (r)^{\gamma} = 1 \times (9)^{1.4} = 21.67 \text{ bar}$$

Also,
$$\frac{T_2}{T_1} = \left(\frac{V_1}{V_2}\right)^{\gamma - 1} = (r)^{\gamma - 1} = (9)^{1.4 - 1} = (9)^{0.4} = 2.408$$

$\therefore \quad T_2 = T_1 \times 2.408 = 303 \times 2.408 = 729.6 \text{ K}$

For the *constant volume process 2-3,*

$$\frac{T_3}{p_3} = \frac{T_2}{p_2}$$

$$\therefore \quad T_3 = T_2 \cdot \frac{p_3}{p_2} = 729.6 \times \frac{60}{21.67} = 2020 \text{ K}$$

Also,
$$\frac{\rho - 1}{r - 1} = \frac{4}{100} \quad \text{or} \quad 0.04$$

$$\therefore \quad \frac{\rho - 1}{9 - 1} = 0.04 \quad \text{or} \quad \rho = 1.32$$

For the *constant pressure process 3-4,*

$$\frac{V_4}{T_4} = \frac{V_3}{T_3} \quad \text{or} \quad \frac{T_4}{T_3} = \frac{V_4}{V_3} = \rho$$

$$\therefore \quad T_4 = T_3 \times \rho = 2020 \times 1.32 = 2666.4 \text{ K}$$

Also expansion ratio,
$$\frac{V_5}{V_4} = \frac{V_5}{V_2} \times \frac{V_2}{V_4} = \frac{V_1}{V_2} \times \frac{V_3}{V_4} = \frac{r}{\rho} \qquad [\because V_5 = V_1 \text{ and } V_2 = V_3]$$

For *adiabatic process 4-5,*

$$\frac{T_5}{T_4} = \left(\frac{V_4}{V_5}\right)^{\gamma - 1} = \left(\frac{\rho}{r}\right)^{\gamma - 1}$$

$$T_5 = T_4 \times \left(\frac{\rho}{r}\right)^{\gamma - 1} = 2666.4 \times \left(\frac{1.32}{9}\right)^{1.4 - 1} = 1237 \text{ K}$$

Also
$$p_4 V_4^{\gamma} = p_5 V_5^{\gamma}$$

$$p_5 = p_4 \cdot \left(\frac{V_4}{V_5}\right)^{\gamma} = 60 \times \left(\frac{r}{\rho}\right)^{\gamma} = 60 \times \left(\frac{1.32}{9}\right)^{1.4} = 4.08 \text{ bar}$$

Heat supplied,
$$Q_s = c_v(T_3 - T_2) + c_p(T_4 - T_3)$$
$$= 0.71\ (2020 - 729.6) + 1.0\ (2666.4 - 2020) = 1562.58 \text{ kJ/kg}$$

Heat rejected,
$$Q_r = c_v\ (T_5 - T_1)$$
$$= 0.71\ (1237 - 303) = 663.14 \text{ kJ/kg}$$

$$\eta_{\text{air-standard}} = \frac{Q_s - Q_r}{Q_s} = \frac{1562.85 - 663.14}{1562.58} = \mathbf{0.5756 \text{ or } 57.56\%.} \quad \textbf{(Ans.)}$$

(*ii*) **Power developed by the engine, P :**

Mass of air in the cycle is given by

$$m = \frac{p_1 V_1}{R T_1} = \frac{1 \times 10^5 \times 0.0165}{287 \times 303} = 0.0189 \text{ kg}$$

$\therefore$ Work done per cycle $= m(Q_s - Q_r)$

$= 0.0189\ (1562.58 - 663.14) = 16.999$ kJ

Power developed = Work done per cycle × No. of cycles per second

$= 16.999 \times 3 =$ **50.99 say 51 kW. (Ans.)**

☞**Example 3.26.** *In an engine working on Dual cycle, the temperature and pressure at the beginning of the cycle are 90°C and 1 bar respectively. The compression ratio is 9. The maximum pressure is limited to 68 bar and total heat supplied per kg of air is 1750 kJ. Determine :*

(*i*) *Pressure and temperatures at all salient points*

(*ii*) *Air standard efficiency*

(*iii*) *Mean effective pressure.*

Solution. Refer Fig. 3.22.

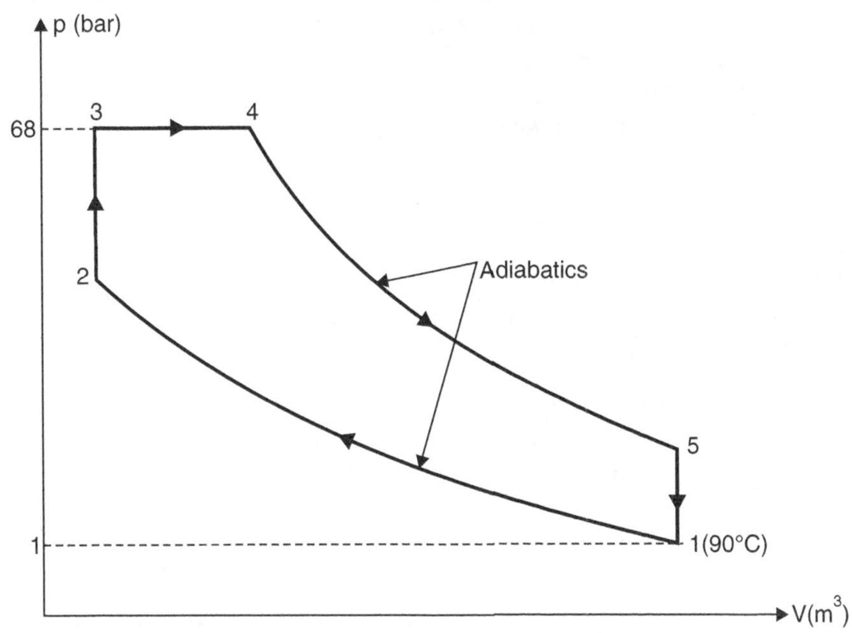

Fig. 3.22

Initial pressure, $p_1 = 1$ bar

Initial temperature, $T_1 = 90 + 273 = 363$ K

Compression ratio, $r = 9$

Maximum pressure, $p_3 = p_4 = 68$ bar

Total heat supplied $= 1750$ kJ/kg

(*i*) **Pressures and temperatures at salient points :**

For the *isentropic process* 1-2,

$$p_1V_1^\gamma = p_2V_2^\gamma$$

$$\mathbf{p_2} = p_1 \times \left(\frac{V_1}{V_2}\right)^\gamma = 1 \times (r)^\gamma = 1 \times (9)^{1.4} = \mathbf{21.67\ bar.} \quad \textbf{(Ans.)}$$

Also, $$\frac{T_2}{T_1} = \left(\frac{V_1}{V_2}\right)^{\gamma - 1} = (r)^{\gamma - 1} = (9)^{1.4 - 1} = 2.408$$

$\therefore$ $$\mathbf{T_2} = T_1 \times 2.408 = 363 \times 2.408 = \mathbf{874.1\ K.} \quad \textbf{(Ans.)}$$

$$\mathbf{p_3 = p_4 = 68\ bar.} \quad \textbf{(Ans.)}$$

For the *constant volume process* 2-3,

$$\frac{p_2}{T_2} = \frac{p_3}{T_3}$$

$\therefore$ $$\mathbf{T_3} = T_2 \times \frac{p_3}{p_2} = 874.1 \times \frac{68}{21.67} = \mathbf{2742.9\ K.}\quad \textbf{(Ans.)}$$

Heat added at constant volume

$$= c_v (T_3 - T_2) = 0.71\ (2742.9 - 874.1) = 1326.8 \text{ kJ/kg}$$

$\therefore$ Heat added at constant pressure

$$= \text{Total heat added} - \text{Heat added at constant volume}$$
$$= 1750 - 1326.8 = 423.2 \text{ kJ/kg}$$

$\therefore$ $$c_p(T_4 - T_3) = 423.2$$

or $$1.0(T_4 - 2742.9) = 423.2$$

$\therefore$ $$\mathbf{T_4 = 3166\ K.}\quad \textbf{(Ans.)}$$

For *constant pressure process* 3-4,

$$\rho = \frac{V_4}{V_3} = \frac{T_4}{T_3} = \frac{3166}{2742.9} = 1.15$$

For *adiabatic (or isentropic) process* 4-5,

$$\frac{V_5}{V_4} = \frac{V_5}{V_2} \times \frac{V_2}{V_4} = \frac{V_1}{V_2} \times \frac{V_3}{V_4} = \frac{r}{\rho} \qquad \left(\because\ \rho = \frac{V_4}{V_3}\right)$$

Also $$p_4 V_4^{\gamma} = p_5 V_5^{\gamma}$$

$\therefore$ $$\mathbf{p_5} = p_4 \times \left(\frac{V_4}{V_5}\right)^{\gamma} = 68 \times \left(\frac{\rho}{r}\right)^{\gamma} = 68 \times \left(\frac{1.15}{9}\right)^{1.4} = \mathbf{3.81\ bar.}\quad \textbf{(Ans.)}$$

Again, $$\frac{T_5}{T_4} = \left(\frac{V_4}{V_5}\right)^{\gamma - 1} = \left(\frac{\rho}{r}\right)^{\gamma - 1} = \left(\frac{1.15}{9}\right)^{1.4 - 1} = 0.439$$

$\therefore$ $$\mathbf{T_5} = T_4 \times 0.439 = 3166 \times 0.439 = \mathbf{1389.8\ K.}\quad \textbf{(Ans.)}$$

(*ii*) **Air standard efficiency :**

Heat rejected during constant volume process 5-1,

$$Q_r = C_v(T_5 - T_1) = 0.71(1389.8 - 363) = 729 \text{ kJ/kg}$$

$\therefore$ $$\eta_{\text{air-standard}} = \frac{\text{Work done}}{\text{Heat supplied}} = \frac{Q_s - Q_r}{Q_s}$$

$$= \frac{1750 - 729}{1750} = \mathbf{0.5834\ or\ 58.34\%.}\quad \textbf{(Ans.)}$$

(*iii*) **Mean effective pressure, $\mathbf{p_m}$:**

Mean effective pressure is given by

$$p_m = \frac{\text{Work done per cycle}}{\text{Stroke volume}}$$

or $$p_m = \frac{1}{V_s}\left[p_3 (V_4 - V_3) + \frac{p_4 V_4 - p_5 V_5}{\gamma - 1} - \frac{p_2 V_2 - p_1 V_1}{\gamma - 1}\right]$$

$$V_1 = V_5 = r\,V_c,\ V_2 = V_3 = V_c,\ V_4 = \rho V_c,\ V_s = (r - 1)\,V_c \qquad \left[\because\ r = \frac{V_s + V_c}{V_c} = 1 + \frac{V_s}{V_c}\ \ \therefore\ V_s = (r - 1)\,V_c\right]$$

$$\therefore \quad p_m = \frac{1}{(r-1)V_c}\left[p_3(\rho V_c - V_c) + \frac{p_4\rho V_c - p_5 \times rV_c}{\gamma - 1} - \frac{p_2V_c - p_1rV_c}{\gamma - 1}\right]$$

$r = 9,\ \rho = 1.15,\ \gamma = 1.4$

$p_1 = 1$ bar, $p_2 = 21.67$ bar, $p_3 = p_4 = 68$ bar, $p_5 = 3.81$ bar

Substituting the above values in the above equation, we get

$$p_m = \frac{1}{(9-1)}\left[68(1.15-1) + \frac{68 \times 1.15 - 3.81 \times 9}{1.4 - 1} - \frac{21.67 - 9}{1.4 - 1}\right]$$

$$= \tfrac{1}{8}\ (10.2 + 109.77 - 31.67) = 11.04 \text{ bar}$$

Hence, *mean effective pressure* = **11.04 bar. (Ans.)**

Example 3.27. *An I.C. engine operating on the dual cycle (limited pressure cycle) the temperature of the working fluid (air) at the beginning of compression is 27°C. The ratio of the maximum and minimum pressures of the cycle is 70 and compression ratio is 15. The amounts of heat added at constant volume and at constant pressure are equal. Compute the air standard thermal efficiency of the cycle. State three main reasons why the actual thermal efficiency is different from the theoretical value.*

Take γ for air = 1.4. **(U.P.S.C. 1997)**

Solution. Refer Fig. 3.23. *Given :* $T_1 = 27 + 273 = 300$ K ; $\frac{p_3}{p_1} = 70,\ \frac{v_1}{v_2} = \frac{v_1}{v_3} = 15$

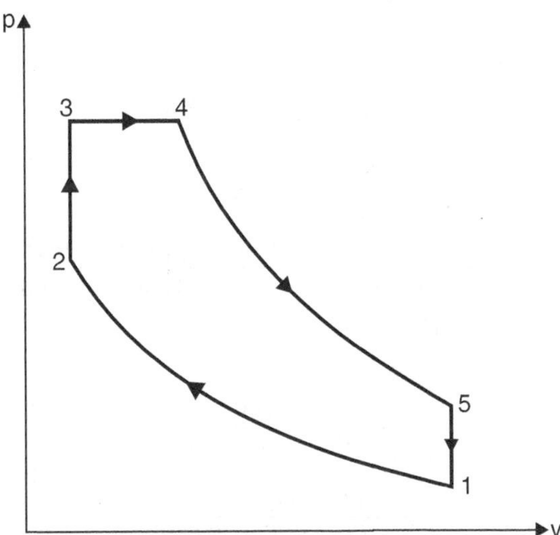

Fig. 3.23. Dual cycle.

Air standard efficiency, $\eta_{\text{air-standard}}$ **:**

Consider *1 kg of air.*

Adiabatic compression process 1–2 :

$$\frac{T_2}{T_1} = \left(\frac{v_1}{v_2}\right)^{\gamma - 1} = (15)^{1.4 - 1} = 2.954$$

$\therefore \quad T_2 = 300 \times 2.954 = 886.2 \text{ K}$

$$\frac{p_2}{p_1} = \left(\frac{v_1}{v_2}\right)^{\gamma} = (15)^{1.4} \quad \Rightarrow \quad p_2 = 44.3\, p_1$$

Constant pressure process 2-3 :

$$\frac{p_2}{T_2} = \frac{p_3}{T_3}$$

or $$T_3 = T_2 \times \frac{p_3}{p_2} = 886.2 \times \frac{70 p_1}{44.3 p_1} = 1400 \text{ K}$$

Also, Heat added at constant volume = Heat added at constant pressure ...(Given)

or $$c_v\,(T_3 - T_2) = c_p\,(T_4 - T_3)$$

or $$T_3 - T_2 = \gamma\,(T_4 - T_3)$$

or $$T_4 = T_3 + \frac{T_3 - T_2}{\gamma} = 1400 + \frac{1400 - 886.2}{1.4} = 1767 \text{ K.}$$

Constant volume process 3-4 :

$$\frac{v_3}{T_3} = \frac{v_4}{T_4} \quad \Rightarrow \quad \frac{v_4}{v_3} = \frac{T_4}{T_3} = \frac{1767}{1400} = 1.26$$

Also, $$\frac{v_4}{v_3} = \frac{v_4}{(v_1/15)} = 1.26 \quad \text{or} \quad v_4 = 0.084\, v_1$$

Also, $$v_5 = v_1$$

Adiabatic expansion process 4-5 :

$$\frac{T_4}{T_5} = \left(\frac{v_5}{v_4}\right)^{\gamma - 1} = \left(\frac{v_1}{0.084 v_1}\right)^{1.4 - 1} = 2.69$$

$$\therefore \quad T_5 = \frac{T_4}{2.69} = \frac{1767}{2.69} = 656.9 \text{ K}$$

$$\therefore \quad \eta_{\text{air-standard}} = \frac{\text{Work done}}{\text{Heat supplied}} = \frac{\text{Heat supplied} - \text{Heat rejected}}{\text{Heat supplied}}$$

$$= 1 - \frac{\text{Heat rejected}}{\text{Heat supplied}}$$

$$= 1 - \frac{c_v(T_5 - T_1)}{c_v(T_3 - T_2) + c_p(T_4 - T_3)}$$

$$= 1 - \frac{(T_5 - T_1)}{(T_3 - T_2) + \gamma(T_4 - T_3)}$$

$$= 1 - \frac{(656.9 - 300)}{(1400 - 886.2) + 1.4(1767 - 1400)} = \mathbf{0.653 \text{ or } 65.3\%.\ (Ans.)}$$

Reasons for actual thermal efficiency being different from the theoretical value :

1. In theoretical cycle working substance is taken *air* whereas in actual cycle *air with fuel acts as working substance.*

2. The fuel combustion phenomenon and associated problems like dissociation of gases, dilution of charge during suction stroke, etc. have *not* been taken into account.

3. Effect of variable specific heat, heat loss through cylinder walls, inlet and exhaust velocities of air/gas etc. have *not* been taken into account.

☞**Example 3.28.** *A Diesel engine working on a dual combustion cycle has a stroke volume of 0.0085 m^3 and a compression ratio 15 : 1. The fuel has a calorific value of 43890 kJ/kg. At the end of suction, the air is at 1 bar and 100°C. The maximum pressure in the cycle is 65 bar and air fuel ratio is 21 : 1. Find for ideal cycle the thermal efficiency. Assume c_p = 1.0 kJ/kg K and c_v = 0.71 kJ/kg K.*

Solution. Refer Fig. 3.24.

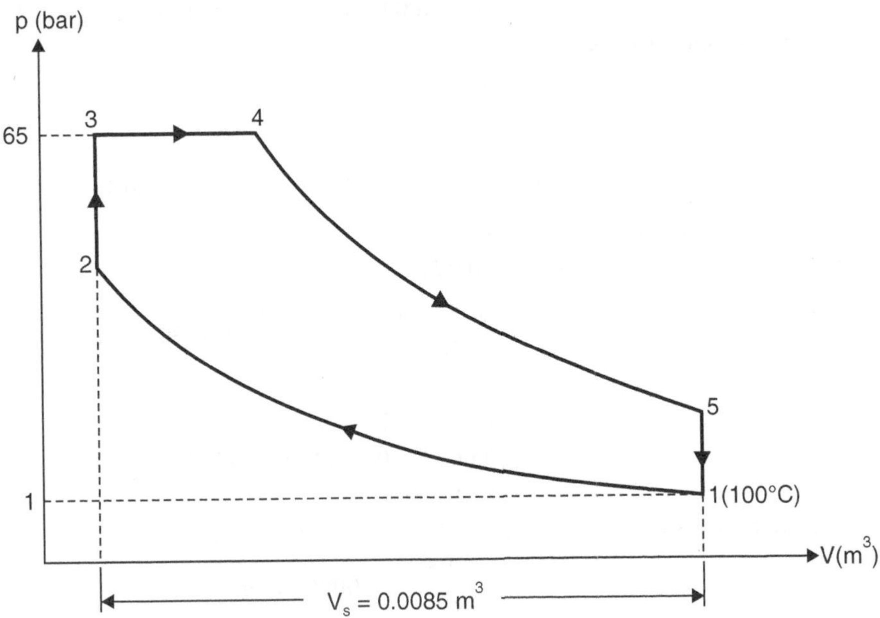

Fig. 3.24

Initial temperature, $T_1 = 100 + 273 = 373$ K

Initial pressure, $p_1 = 1$ bar

Maximum pressure in the cycle, $p_3 = p_4 = 65$ bar

Stroke volume, $V_s = 0.0085$ m^3

Air-fuel ratio = 21 : 1

Compression ratio, $r = 15 : 1$

Calorific value of fuel, $C = 43890$ kJ/kg

$c_p = 1.0$ kJ/kg K, $c_v = 0.71$ kJ/kg K

Thermal efficiency :

$$V_s = V_1 - V_2 = 0.0085 \text{ m}^3$$

and as $$r = \frac{V_1}{V_2} = 15, \text{ then } V_1 = 15V_2$$

∴ $$15V_2 - V_2 = 0.0085$$

or $$14V_2 = 0.0085$$

or $$V_2 = V_3 = V_c = \frac{0.0085}{14} = 0.0006 \text{ m}^3$$

or $$V_1 = 15V_2 = 15 \times 0.0006 = 0.009 \text{ m}^3$$

For *adiabatic compression process 1-2,*

$$p_1V_1^{\gamma} = p_2V_2^{\gamma}$$

or $$p_2 = p_1 \cdot \left(\frac{V_1}{V_2}\right)^{\gamma} = 1 \times (15)^{1.41} \qquad \left[\gamma = \frac{c_p}{c_v} = \frac{1.0}{0.71} = 1.41\right]$$

$$= 45.5 \text{ bar}$$

Also, $$\frac{T_2}{T_1} = \left(\frac{V_1}{V_2}\right)^{\gamma-1} = (r)^{\gamma-1} = (15)^{1.41-1} = 3.04$$

$\therefore$ $$T_2 = T_1 \times 3.04 = 373 \times 3.04 = 1134 \text{ K or } 861°\text{C}$$

For *constant volume process 2-3,*

$$\frac{p_2}{T_2} = \frac{p_3}{T_3}$$

or $$T_3 = T_2 \times \frac{p_3}{p_2} = 1134 \times \frac{65}{45.5} = 1620 \text{ K or } 1347°\text{C}$$

According to characteristic equation of gas,

$$p_1V_1 = mRT_1$$

$\therefore$ $$m = \frac{p_1V_1}{RT_1} = \frac{1 \times 10^5 \times 0.009}{287 \times 373} = 0.0084 \text{ kg (air)}$$

Heat added during constant volume process 2-3,

$$= m \times c_v (T_3 - T_2)$$

$$= 0.0084 \times 0.71 (1620 - 1134)$$

$$= 2.898 \text{ kJ}$$

Amount of fuel added during the *constant volume process 2-3,*

$$= \frac{2.898}{43890} = 0.000066 \text{ kg}$$

Also as air-fuel ratio is 21 : 1.

$\therefore$ Total amount of fuel added $$= \frac{0.0084}{21} = 0.0004 \text{ kg}$$

Quantity of fuel added during the process 3-4,

$$= 0.0004 - 0.000066 = 0.000334 \text{ kg}$$

$\therefore$ *Heat added during the constant pressure operation 3-4*

$$= 0.000334 \times 43890 = 14.66 \text{ kJ}$$

But $(0.0084 + 0.0004)\, c_p (T_4 - T_3) = 14.66$

or $0.0088 \times 1.0 (T_4 - 1620) = 14.66$

$\therefore$ $$T_4 = \frac{14.66}{0.0088} + 1620 = 3286 \text{ K or } 3013°\text{C}$$

Again for operation 3-4,

$$\frac{V_3}{T_3} = \frac{V_4}{T_4} \quad \text{or} \quad V_4 = \frac{V_3T_4}{T_3} = \frac{0.0006 \times 3286}{1620} = 0.001217 \text{ m}^3$$

For adiabatic expansion operation 4-5,

$$\frac{T_4}{T_5} = \left(\frac{V_5}{V_4}\right)^{\gamma-1} = \left(\frac{0.009}{0.001217}\right)^{1.41-1} = 2.27$$

or $$T_5 = \frac{T_4}{2.27} = \frac{3286}{2.27} = 1447.5 \text{ K or } 1174.5°\text{C}$$

Heat rejected during constant volume process 5-1,

$$= m\, c_v\, (T_5 - T_1)$$
$$= (0.00854 + 0.0004) \times 0.71\, (1447.5 - 373) = 6.713 \text{ kJ}$$

Work done $= \text{Heat supplied} - \text{Heat rejected}$
$= (2.898 + 14.66) - 6.713 = 10.845 \text{ kJ}$

∴ *Thermal efficiency,*

$$\eta_{th} = \frac{\text{Work done}}{\text{Heat supplied}} = \frac{10.845}{(2.898 + 14.66)} = \mathbf{0.6176 \text{ or } 61.76\%.} \quad \textbf{(Ans.)}$$

☞**Example 3.29.** *The compression ratio and expansion ratio of an oil engine working on the dual cycle are 9 and 5 respectively. The initial pressure and temperature of the air are 1 bar and 30°C. The heat liberated at constant pressure is twice the heat liberated at constant volume. The expansion and compression follow the law $pV^{1.25}$ = constant. Determine :*

(i) Pressures and temperatures at all salient points.

(ii) Mean effective pressure of the cycle.

(iii) Efficiency of the cycle.

(iv) Power of the engine if working cycles per second are 8.

Assume : Cylinder bore = 250 mm and stroke length = 400 mm.

Solution. Refer Fig. 3.25.

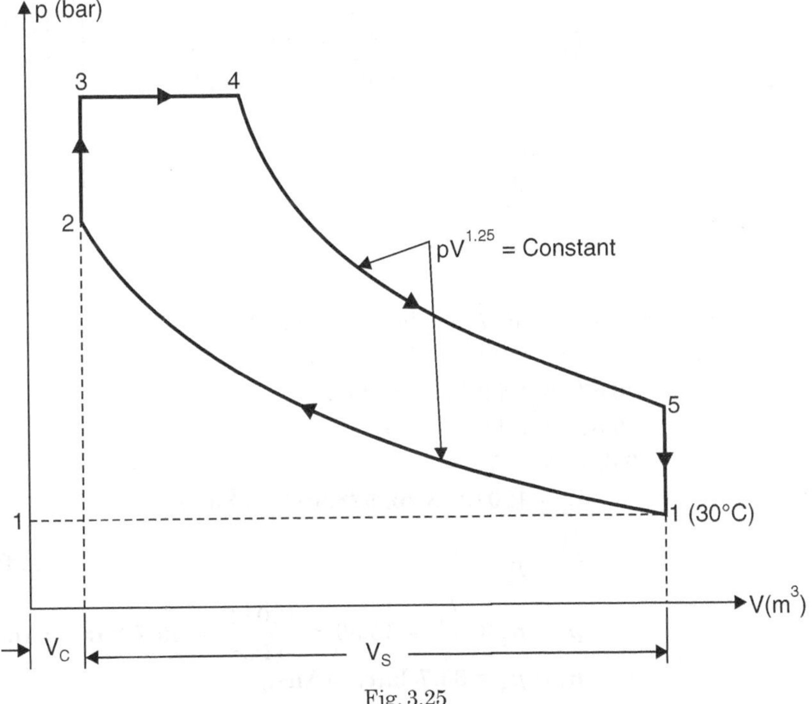

Fig. 3.25

Initial temperature, $T_1 = 30 + 273 = 303$ K

Initial pressure, $p_1 = 1$ bar

Compression and expansion law,

$$pV^{1.25} = \text{constant}$$

Compression ratio, $r_c = 9$
Expansion ratio, $r_e = 5$
Number of cycles/sec. $= 8$
Cylinder diameter, $D = 250 \text{ mm} = 0.25 \text{ m}$
Stroke length, $L = 400 \text{ mm} = 0.4 \text{ m}$
Heat liberated at constant pressure
$= 2 \times$ heat liberated at constant volume

(*i*) **Pressure and temperatures at all salient points :**

For *compression process 1-2,*

$$p_1 V_1^n = p_2 V_2^n$$

$$\therefore \quad \mathbf{p_2} = p_1 \times \left(\frac{V_1}{V_2}\right)^n = 1 \times (9)^{1.25} = \mathbf{15.59 \text{ bar.}} \quad \textbf{(Ans.)}$$

Also,
$$\frac{T_2}{T_1} = \left(\frac{V_1}{V_2}\right)^{n-1} = (9)^{1.25-1} = 1.732$$

$$\therefore \quad \mathbf{T_2} = T_1 \times 1.732 = 303 \times 1.732$$
$$= \mathbf{524.8 \text{ K or } 251.8°C.} \quad \textbf{(Ans.)}$$

Also, $c_p(T_4 - T_3) = 2 \times c_v(T_3 - T_2)$ (given) ...(*i*)

For constant pressure process 3-4,

$$\frac{T_4}{T_3} = \frac{V_4}{V_3} = \rho = \frac{\text{Compression ratio } (r_c)}{\text{Expansion ratio } (r_e)}$$
$$= \frac{9}{5} = 1.8$$
$$T_4 = 1.8\, T_3$$

$$\left[\begin{aligned} \frac{V_5}{V_4}(r_e) &= \frac{V_5}{V_3} \times \frac{V_3}{V_4} \\ &= \frac{V_1}{V_3} \times \frac{1}{\rho} \\ &= \frac{V_1}{V_2} \times \frac{1}{\rho} = \frac{r_c}{\rho} \\ \therefore \quad \rho &= \frac{r_c}{\frac{V_5}{V_4}} = \frac{r_c}{r_e} \end{aligned}\right]$$

Substituting the values of T_2 and T_4 in the equation (*i*), we get

$$1.0(1.8T_3 - T_3) = 2 \times 0.71(T_3 - 524.8)$$
$$0.8T_3 = 1.42(T_3 - 524.8)$$
$$0.8T_3 = 1.42T_3 - 745.2$$
$$\therefore \quad 0.62T_3 = 745.2$$
$$\mathbf{T_3} = \mathbf{1201.9 \text{ K or } 928.9°C.} \quad \textbf{(Ans.)}$$

Also,
$$\frac{p_3}{T_3} = \frac{p_2}{T_2}$$ ......for process 3-2

$$\therefore \quad \mathbf{p_3} = p_2 \times \frac{T_3}{T_2} = 15.59 \times \frac{1201.9}{524.8} = \mathbf{35.7 \text{ bar.}} \quad \textbf{(Ans.)}$$
$$\mathbf{p_4} = p_3 = \mathbf{35.7 \text{ bar.}} \quad \textbf{(Ans.)}$$
$$\mathbf{T_4} = 1.8T_3 = 1.8 \times 1201.9 = \mathbf{2163.4 \text{ K or } 1890.4°C.} \quad \textbf{(Ans.)}$$

For *expansion process 4-5,*

$$p_4 V_4^n = p_5 V_5^n$$

$$p_5 = p_4 \times \left(\frac{V_4}{V_5}\right)^n = p_4 \times \frac{1}{(r_e)^n} = \frac{35.7}{(5)^{1.25}} = \textbf{4.77 bar. (Ans.)}$$

Also
$$\frac{T_5}{T_4} = \left(\frac{V_4}{V_5}\right)^{n-1} = \frac{1}{(r_e)^{n-1}} = \frac{1}{(5)^{1.25-1}} = 0.668$$

$\therefore$ $T_5 = T_4 \times 0.668 = 2163.4 \times 0.668 =$ **1445 K or 1172°C. (Ans.)**

(*ii*) **Mean effective pressure, p_m :**

Mean effective pressure is given by

$$p_m = \frac{1}{V_s}\left[p_3(V_4 - V_3) + \frac{p_4V_4 - p_5V_5}{n-1} - \frac{p_2V_2 - p_1V_1}{n-1}\right]$$

$$= \frac{1}{(r_c - 1)}\left[p_3(\rho - 1) + \frac{p_4\rho - p_5r_c}{n-1} - \frac{p_2 - p_1r_c}{n-1}\right]$$

Now, $r_c = \rho$, $\rho = 1.8$, $n = 1.25$, $p_1 = 1$ bar, $p_2 = 15.59$ bar, $p_3 = 35.7$ bar, $p_4 = 35.7$ bar, $p_5 = 4.77$ bar

$$\therefore \quad p_m = \frac{1}{(9-1)}\left[35.7\,(1.8 - 1) + \frac{35.7 \times 1.8 - 4.77 \times 9}{1.25 - 1} - \frac{15.59 - 1 \times 9}{1.25 - 1}\right]$$

$$= \frac{1}{8}\,[28.56 + 85.32 - 26.36] = 10.94 \text{ bar}$$

Hence *mean effective pressure* = **10.94 bar. (Ans.)**

(*iii*) **Efficiency of the cycle :**

Work done per cycle is given by $W = p_mV_s$

here $V_s = \pi/4 D^2L = \pi/4 \times 0.25^2 \times 0.4 = 0.0196 \text{ m}^3$

$$\therefore \quad W = \frac{10.94 \times 10^5 \times 0.0196}{1000} \text{ kJ/cycle} = 21.44 \text{ kJ/cycle}$$

Heat supplied per cycle $= mQ_s$

where m is the mass of air per cycle which is given by

$$m = \frac{p_1V_1}{RT_1} \quad \text{where } V_1 = V_s + V_c = \frac{r_c}{r_c - 1}\,V_s$$

$$\left[\begin{aligned} r_c &= \frac{V_s + V_c}{V_c} = 1 + \frac{V_s}{V_c} \quad \text{or} \quad V_c = \frac{V_s}{r_c - 1} \\ \therefore \quad V_1 &= V_s + \frac{V_s}{r_c - 1} = V_s\left(1 + \frac{1}{r_c - 1}\right) = \frac{r_c}{r_c - 1}\,V_s \end{aligned}\right]$$

$$= \frac{9}{9-1} \times 0.0196 = 0.02205 \text{ m}^3$$

$$\therefore \quad m = \frac{1 \times 10^5 \times 0.02205}{287 \times 303} = 0.02535 \text{ kg/cycle}$$

$\therefore$ Heat supplied per cycle

$$= mQ_s = 0.02535[c_v(T_3 - T_2) + c_p(T_4 - T_3)]$$

$$= 0.02535[0.71(1201.9 - 524.8) + 1.0(2163.4 - 1201.9)]$$

$$= 36.56 \text{ kJ/cycle}$$

$$\textit{Efficiency} = \frac{\text{Work done per cycle}}{\text{Heat supplied per cycle}} = \frac{21.44}{36.56}$$

= **0.5864 or 58.64%. (Ans.)**

(*iv*) **Power of the engine, P :**

Power of the engine, P = Work done per second

= Work done per cycle × no. of cycles/sec.

= 21.44 × 8 = **171.52 kW. (Ans.)**

3.7. COMPARISON OF OTTO, DIESEL AND DUAL COMBUSTION CYCLES

Following are the important variable factors which are used as a basis for comparison of the cycles :

- Compression ratio.
- Maximum pressure
- Heat supplied
- Heat rejected
- Net work.

Some of the above mentioned variables are fixed when the performance of Otto, Diesel and dual combustion cycles is to be compared.

3.7.1. Efficiency Versus Compression Ratio

Fig. 3.26 shows the comparison for the air standard efficiencies of the Otto, Diesel and Dual combustion cycles at various compression ratios and with given cut off ratio for the Diesel and Dual combustion cycles. It is evident from the Fig. 3.26 that the air standard efficiencies increase with the increase in the compression ratio. *For a given compression ratio otto cycle is the most efficient while the Diesel cycle is the least efficient.* ($\eta_{otto} > \eta_{dual} > \eta_{diesel}$).

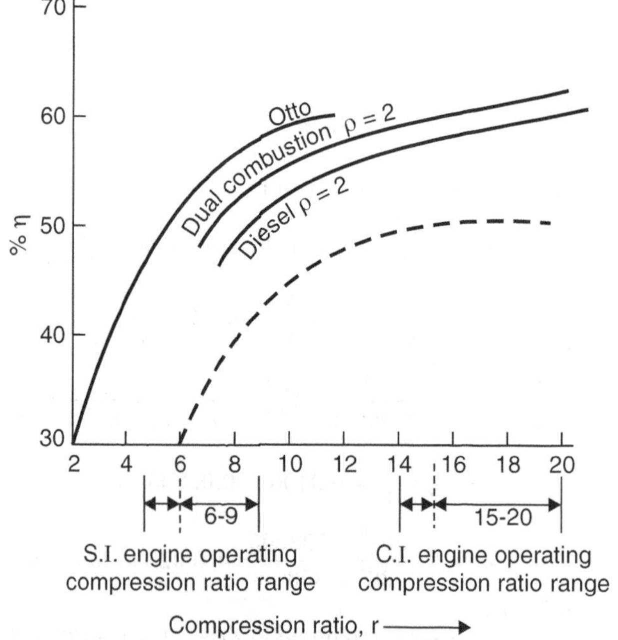

Fig. 3.26. Comparison of efficiency at various compression ratio.

Note. The maximum compression ratio for the petrol engine is limited by detonation. In their respective ratio ranges, the Diesel cycle is more efficient than the Otto cycle.

3.7.2. For the Same Compression Ratio and the Same Heat Input

A comparison of the cycles (Otto, Diesel and Dual) on the *p-v* and *T-s* diagrams for the *same compression ratio and heat supplied* is shown in the Fig. 3.27.

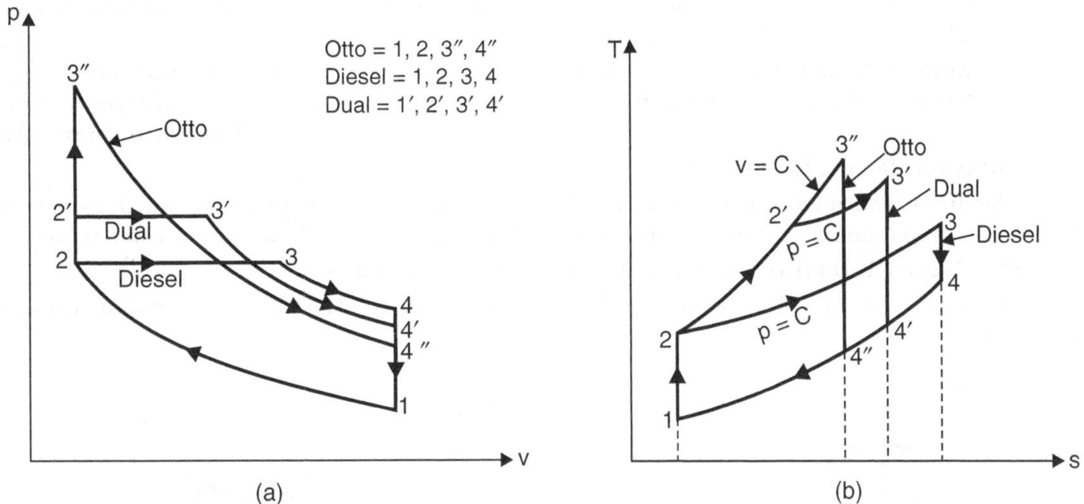

Fig. 3.27. (*a*) *p-v* diagram, (*b*) *T-s* diagram.

We know that, $$\eta = 1 - \frac{\text{Heat rejected}}{\text{Heat supplied}} \qquad ...(3.13)$$

Since all the cycles reject their heat at the same specific volume, process line from state 4 to 1, the quantity of *heat rejected from each cycle is represented by the appropriate area under the line 4 to 1 on the T-s diagram*. As is evident from the eqn. (3.13) the cycle which has the least heat rejected will have the highest efficiency. Thus, Otto cycle is the most efficient and Diesel cycle is the least efficient of the three cycles.

i.e., $$\eta_{otto} > \eta_{dual} > \eta_{diesel} .$$

3.7.3. For Constant Maximum Pressure and Heat Supplied

Fig. 3.28 shows the Otto and Diesel cycles on *p-v* and *T-s* diagrams for constant maximum pressure and heat input respectively.

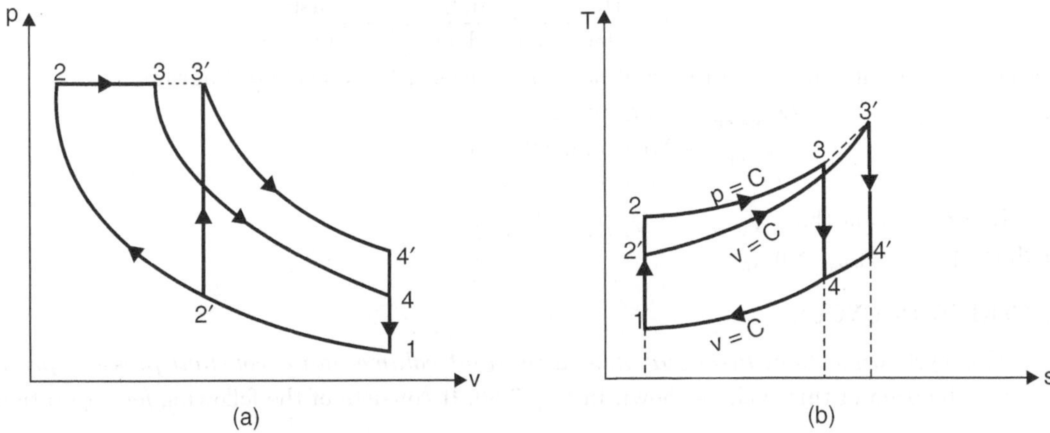

Fig. 3.28. (*a*) *p-v* diagram, (*b*) *T-s* diagram.

— For the maximum pressure the points 3 and 3′ must lie on a constant pressure line.

— On *T-s* diagram the heat rejected from the Diesel cycle is represented by the area under the line 4 to 1 and this area is less than the Otto cycle area under the curve 4′ to 1 ; hence the *Diesel cycle is more efficient than the Otto cycle for the condition of maximum pressure and heat supplied.*

Example 3.30. *(a) With the help of p-v and T-s diagram compare the cold air standard otto, diesel and dual combustion cycles for same maximum pressure and maximum temperature.* **(AMIE Summer, 1998)**

Solution. Refer Fig. 3.29. (*a, b*).

The air-standard Otto, Dual and Diesel cycles are drawn on common *p-v* and *T-s* diagrams for the same maximum pressure and maximum temperature, for the purpose of comparison.

Otto 1-2-3-4-1, Dual 1-2′-3′-3-4-1, Diesel 1-2″-3-4-1 (Fig 3.29 (*a*)).

Slope of constant volume lines on *T-s* diagram is higher than that of constant pressure lines. (Fig. 3.29 (*b*)).

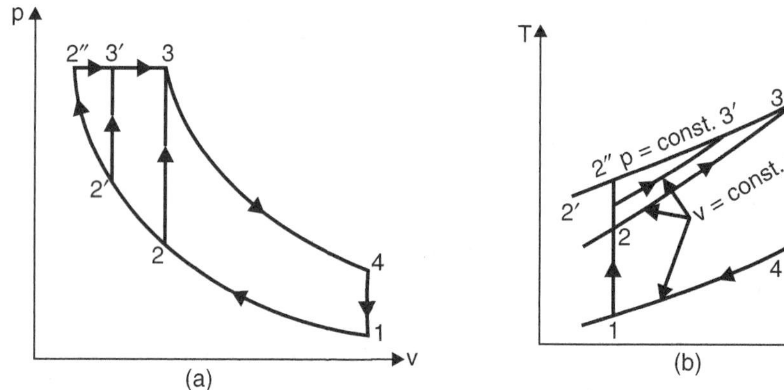

Fig. 3.29

Here the otto cycle must be limited to a low compression ratio (*r*) to fulfill the condition that point 3 (same maximum pressure and temperature) is to be a common state for all the three cycles.

The construction of cycles on *T-s* diagram proves that for the given conditions the heat rejected is same for all the three cycles (area under process line 4-1). Since, by definition,

$$\eta = 1 - \frac{\text{Heat rejected, } Q_r}{\text{Heat supplied, } Q_s} = 1 - \frac{\text{Const.}}{Q_s}$$

the cycle, with greater heat addition will be more efficient. From the *T-s* diagram,

$$Q_{s(\text{diesel})} = \text{Area under 2″-3}$$

$$Q_{s(\text{dual})} = \text{Area under 2′-3′-3}$$

$$Q_{s(\text{otto})} = \text{Area under 2-3.}$$

It can be seem that, $Q_{s(\text{diesel})} > Q_{s(\text{dual})} > Q_{s(\text{otto})}$

and thus, $\eta_{\text{diesel}} > \eta_{\text{dual}} > \eta_{\text{otto}}$.

3.8. ATKINSON CYCLE

This cycle consists of *two adiabatics*, a *constant volume and a constant pressure process*. *p-V* diagram of this cycle is shown in Fig. 3.30. It consists of the following *four* operations :

(*i*) 1-2—*Heat rejection at constant pressure*

(*ii*) 2-3—*Adiabatic compression*

(*iii*) 3-4—*Addition of heat at constant volume*

(*iv*) 4-1—*Adiabatic expansion.*

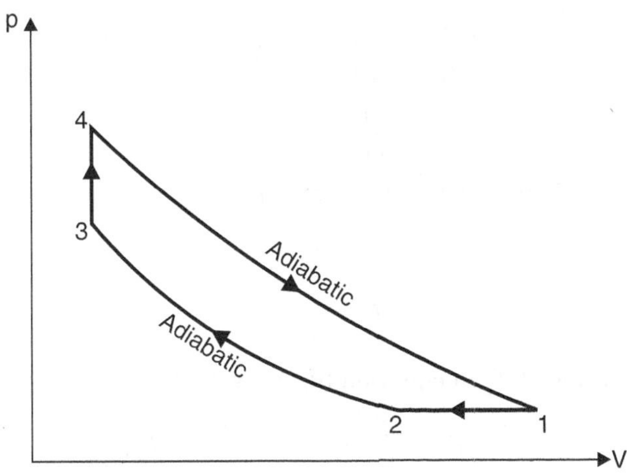

Fig. 3.30

Consider *1 kg of air*

Compression ratio $= \dfrac{v_2}{v_3} = \alpha$

Expansion ratio $= \dfrac{v_1}{v_4} = r$

Heat supplied at constant volume $= c_v(T_4 - T_3)$

Heat rejected $= c_v(T_1 - T_2)$

Work done = Heat supplied – Heat rejected

$= c_v(T_4 - T_3) - c_v(T_1 - T_2)$

$$\eta = \frac{\text{Work done}}{\text{Heat supplied}} = \frac{c_v\,(T_4 - T_3) - c_p(T_1 - T_2)}{c_v\,(T_4 - T_3)}$$

$$= 1 - \gamma \,.\, \frac{(T_1 - T_2)}{(T_4 - T_3)} \qquad ...(i)$$

During *adiabatic compression 2-3,*

$$\frac{T_3}{T_2} = \left(\frac{v_2}{v_3}\right)^{\gamma - 1} = (\alpha)^{\gamma - 1}$$

or $$T_3 = T_2\;(\alpha)^{\gamma - 1} \qquad ...(ii)$$

During constant pressure operation 1-2

$$\frac{v_1}{T_1} = \frac{v_2}{T_2}$$

or
$$\frac{T_2}{T_1} = \frac{v_2}{v_1} = \frac{\alpha}{r} \quad ...(iii)$$

$$\left(\frac{v_2}{v_1} = \frac{v_2}{v_3} \times \frac{v_3}{v_1} = \frac{v_2}{v_3} \times \frac{v_4}{v_1} = \frac{\alpha}{r}\right)$$

During *adiabatic expansion 4-1,*

$$\frac{T_4}{T_1} = \left(\frac{v_1}{v_4}\right)^{\gamma - 1} = (r)^{\gamma - 1}$$

$$T_1 = \frac{T_4}{(r)^{\gamma - 1}} \quad ...(iv)$$

Putting the value of T_1 in equation (*iii*), we get

$$T_2 = \frac{T_4}{(r)^{\gamma - 1}} \cdot \frac{\alpha}{r}$$

$$= \frac{\alpha T_4}{r^{\gamma}} \quad ...(v)$$

Substituting the value of T_2 in equation (*ii*), we get

$$T_3 = \frac{\alpha T_4}{(r)^{\gamma}} (\alpha)^{\gamma - 1} = \left(\frac{\alpha}{r}\right)^{\gamma} \cdot T_4$$

Finally putting the values of T_1, T_2 and T_3 in equation (*i*), we get

$$\eta = 1 - \gamma \left(\frac{\dfrac{T_4}{r^{\gamma - 1}} - \dfrac{\alpha T_4}{(r)^{\gamma}}}{T_4 - \left(\dfrac{\alpha}{r}\right)^{\gamma} . T_4} \right) = 1 - \gamma \left(\frac{r - \alpha}{r^{\gamma} - \alpha^{\gamma}}\right)$$

Hence, *air standard efficiency* $= 1 - \gamma . \left(\dfrac{r - \alpha}{r^{\gamma} - \alpha^{\gamma}}\right)$ ...(3.14)

Example 3.31. *A perfect gas undergoes a cycle which consists of the following processes taken in order :*

(*a*) *Heat rejection at constant pressure.*

(*b*) *Adiabatic compression from 1 bar and 27°C to 4 bar.*

(*c*) *Heat addition at constant volume to a final pressure of 16 bar.*

(*d*) *Adiabatic expansion to 1 bar.*

Calculate : (*i*) *Work done/kg of gas.*

(*ii*) *Efficiency of the cycle.*

Take : $c_p = 0.92$ *kJ/kg K,* $c_v = 0.75$ *kJ/kg K.*

Solution. Refer Fig. 3.31.

Pressure, $p_2 = p_1 = 1$ bar

Temperature, $T_2 = 27 + 273 = 300$ K

Pressure after adiabatic compression, $p_3 = 4$ bar

Final pressure after heat addition, $p_4 = 16$ bar

For *adiabatic compression 2-3,*

$$\frac{T_3}{T_2} = \left(\frac{p_3}{p_2}\right)^{\frac{\gamma - 1}{\gamma}} = \left(\frac{4}{1}\right)^{\frac{1.22 - 1}{1.22}} = 1.284 \quad \left[\gamma = \frac{c_p}{c_v} = \frac{0.92}{0.75} = 1.22\right]$$

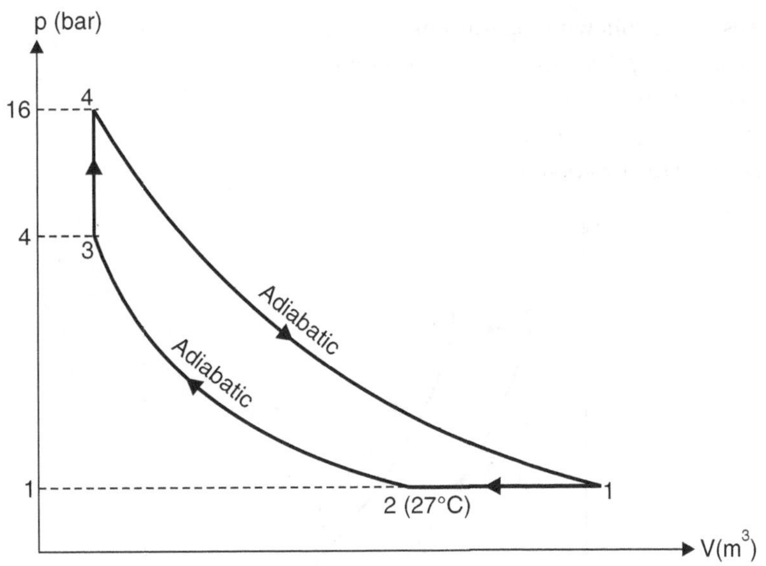

Fig. 3.31

$\therefore \qquad T_3 = T_2 \times 1.284 = 300 \times 1.284 = 385.2 \text{ K or } 112.2°\text{C}$

— For *constant volume process 3-4,*

$$\frac{p_4}{T_4} = \frac{p_3}{T_3}$$

$$T_4 = \frac{p_4 T_3}{p_3} = \frac{16 \times 385.2}{4} = 1540.8 \text{ K or } 1267.8°\text{C}$$

— For *adiabatic expansion process 4-1,*

$$\frac{T_4}{T_1} = \left(\frac{p_4}{p_1}\right)^{\frac{\gamma - 1}{\gamma}} = \left(\frac{16}{1}\right)^{\frac{1.22 - 1}{1.22}} = 1.648$$

or
$$T_1 = \frac{T_4}{1.648} = \frac{1540.8}{1.648} = 934.9 \text{ K or } 661.9°\text{C}.$$

(*i*) **Work done per kg of gas, W :**

Heat supplied $= c_v (T_4 - T_3)$

$= 0.75 (1540.8 - 385.2) = 866.7$ kJ/kg

Heat rejected $= c_p (T_1 - T_2) = 0.92(934.9 - 300) = 584.1$ kJ/kg

Work done/kg of gas, W = Heat supplied – Heat rejected

$= 866.7 - 584.1 = 282.6$ kJ/kg = **282600 N-m/kg. (Ans.)**

(*ii*) **Efficiency of the cycle :**

Efficiency, $\eta = \frac{\text{Work done}}{\text{Heat supplied}} = \frac{282.6}{866.7}$ = **0.326** or **32.6%. (Ans.)**

3.9. ERICSSON CYCLE

It is so named as it was invented by Ericsson. Fig. 3.32 shows *p-V* diagram of this cycle.

It comprises of the following operations :

(*i*) 1-2—*Rejection of heat at constant pressure*

(*ii*) 2-3—*Isothermal compression*

(*iii*) 3-4—*Addition of heat at constant pressure*

(*iv*) 4-1—*Isothermal expansion.*

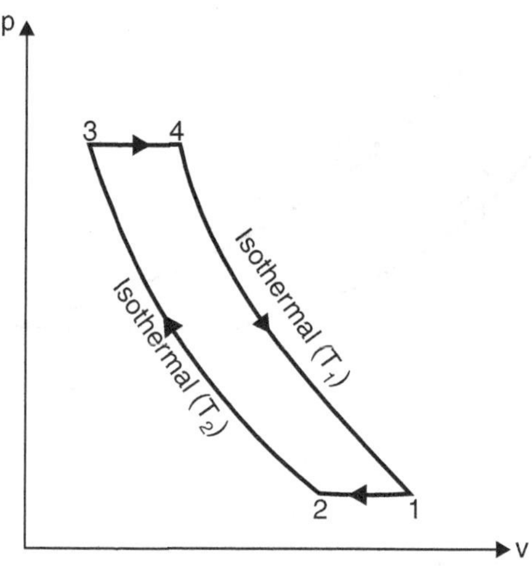

Fig. 3.32

Consider *1 kg of air.*

Volume ratio, $r = \frac{v_2}{v_3} = \frac{v_1}{v_4}$

Heat supplied to air from an external source

$= \text{Heat supplied during the isothermal expansion 4-1}$

$= RT_1 \log_e r$

Heat rejected by air to an external source

$= RT_2 \,.\, \log_e r$

Work done $= \text{Heat supplied} - \text{Heat rejected}$

$= RT_1 \,.\, \log_e r - RT_2 \,.\, \log_e r = R \log_e r\, (T_1 - T_2)$

$$\eta = \frac{\text{Work done}}{\text{Heat supplied}} = \frac{R \log_e r\, (T_1 - T_2)}{RT_1 \,.\, \log_e r}$$

$$= \frac{T_1 - T_2}{T_1} \quad \text{...(3.15)}$$

which is the same as Carnot cycle.

3.10. BRAYTON CYCLE

Brayton cycle is a constant pressure cycle for a perfect gas. It is also called **Joule cycle.** The heat transfers are achieved in reversible constant pressure heat exchangers. An ideal gas turbine plant would perform the processes that make up a Brayton cycle. The cycle is shown in the Fig. 3.33 (*a*) and it is represented on *p-v* and *T-s* diagrams as shown in Fig. 3.33 (*b, c*).

The various operations are as follows :

Operation 1-2. The air is compressed isentropically from the lower pressure p_1 to the upper pressure p_2, the temperature rising from T_1 to T_2. No heat flow occurs.

Operation 2-3. Heat flows into the system increasing the volume from V_2 to V_3 and temperature from T_2 to T_3 whilst the pressure remains constant at p_2. Heat received $= mc_p\,(T_3 - T_2)$.

Operation 3-4. The air is expanded isentropically from p_2 to p_1, the temperature falling from T_3 to T_4. No heat flow occurs.

Operation 4-1. Heat is rejected from the system as the volume decreases from V_4 to V_1 and the temperature from T_4 to T_1 whilst the pressure remains constant at p_1. Heat rejected $= mc_p$ $(T_4 - T_1)$.

$$\eta_{\text{air-standard}} = \frac{\text{Work done}}{\text{Heat received}}$$

$$= \frac{\text{Heat received/cycle} - \text{Heat rejected/cycle}}{\text{Heat received/cycle}}$$

$$= \frac{mc_p\,(T_3 - T_2) - mc_p\,(T_4 - T_1)}{mc_p\,(T_3 - T_2)} = 1 - \frac{T_4 - T_1}{T_3 - T_2}$$

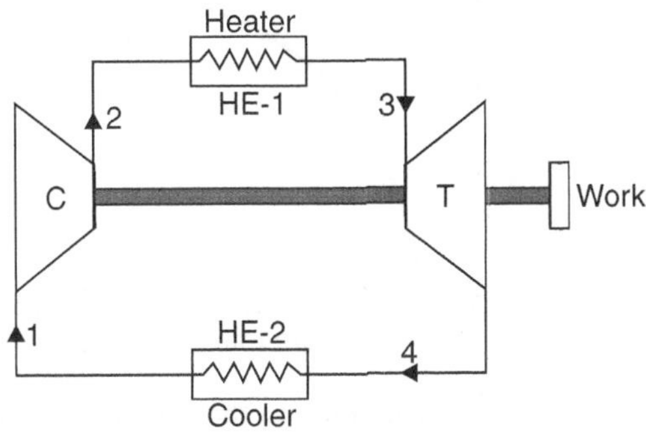

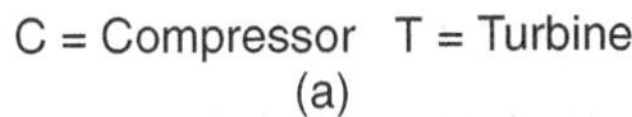
C = Compressor T = Turbine

(a)

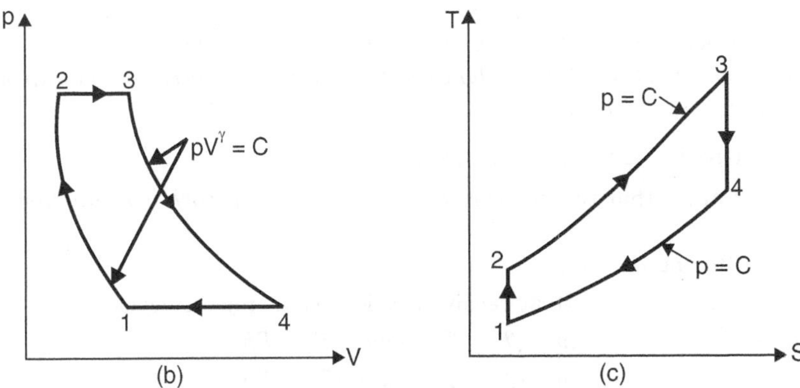

Fig. 3.33. Brayton cycle : (*a*) Basic components of a gas turbine power plant

(*b*) *p-V* diagram (*c*) *T-S* diagram.

Now, from isentropic expansion

$$\frac{T_2}{T_1} = \left(\frac{p_2}{p_1}\right)^{\frac{\gamma-1}{\gamma}}$$

$$T_2 = T_1 \, (r_p)^{\frac{\gamma-1}{\gamma}} \text{ , where } r_p = \text{Pressure ratio.}$$

Similarly $$\frac{T_3}{T_4} = \left(\frac{p_2}{p_1}\right)^{\frac{\gamma-1}{\gamma}} \quad \text{or} \quad T_3 = T_4 \, (r_p)^{\frac{\gamma-1}{\gamma}}$$

$$\therefore \quad \eta_{\text{air-standard}} = 1 - \frac{T_4 - T_1}{T_4 (r_p)^{\frac{\gamma-1}{\gamma}} - T_1 (r_p)^{\frac{\gamma-1}{\gamma}}} = 1 - \frac{1}{(r_p)^{\frac{\gamma-1}{\gamma}}} \qquad \text{...(3.16)}$$

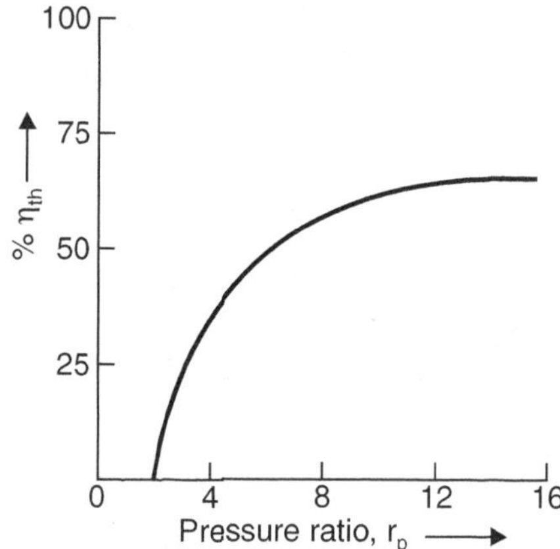

Fig. 3.34. Effect of pressure ratio on the efficiency of Brayton cycle.

The eqn. (3.16) shows that the *efficiency of the ideal joule cycle increases with the pressure ratio. The absolute limit of upper pressure is determined by the limiting temperature of the material of the turbine at the point at which this temperature is reached by the compression process alone, no further heating of the gas in the combustion chamber would be permissible and the work of expansion would ideally just balance the work of compression so that no excess work would be available for external use.*

Pressure ratio for maximum work :

Now we shall prove that the *pressure ratio for maximum work is a function of the limiting temperature ratio.*

Work output during the cycle

$$= \text{Heat received/cycle} - \text{Heat rejected/cycle}$$

$$= mc_p \, (T_3 - T_2) - mc_p \, (T_4 - T_1)$$

$$= mc_p \, (T_3 - T_4) - mc_p \, (T_2 - T_1)$$

$$= mc_p \, T_3 \left(1 - \frac{T_4}{T_3}\right) - T_1 \left(\frac{T_2}{T_1} - 1\right)$$

In case of a given turbine the minimum temperature T_1 and the maximum temperature T_3 are prescribed, T_1 being the temperature of the atmosphere and T_3 the maximum temperature which the metals of turbine would withstand. Consider the specific heat at constant pressure c_p to be constant. Then,

Since, $$\frac{T_3}{T_4} = (r_p)^{\frac{\gamma-1}{\gamma}} = \frac{T_2}{T_1}$$

Using the constant $'z' = \frac{\gamma - 1}{\gamma}$,

we have, work output/cycle

$$W = K\left[T_3\left(1 - \frac{1}{r_p^{\,z}}\right) - T_1\,(r_p^{\,z} - 1)\right]$$

Differentiating with respect to r_p

$$\frac{dW}{dr_p} = K\left[T_3 \times \frac{z}{r_p(z+1)} - T_1\, z r_p^{\,(z-1)}\right] = 0 \text{ for a maximum}$$

$$\therefore \quad \frac{zT_3}{r_p^{\,(z+1)}} = T_1 z (r_p)^{(z-1)}$$

$$\therefore \quad r_p^{\,2z} = \frac{T_3}{T_1}$$

$$\therefore \quad r_p = (T_3/T_1)^{1/2z} \quad i.e. \quad r_p = (T_3/T_1)^{\frac{\gamma}{2(\gamma-1)}} \qquad \text{...(3.17)}$$

Thus, the *pressure ratio for maximum work is a function of the limiting temperature ratio.*

Work Ratio

Work ratio is defined as the *ratio of net work output to the work done by the turbine.*

$$\therefore \quad \text{Work ratio} = \frac{W_T - W_C}{W_T}$$

$$\left[\begin{array}{ll}\text{where } W_T = \text{Work obtained from this turbine,} \\ \text{and } \quad W_C = \text{Work supplied to the compressor}\end{array}\right]$$

$$= \frac{mc_p(T_3 - T_4) - mc_p(T_2 - T_1)}{mc_p\,(T_3 - T_4)} = 1 - \frac{T_2 - T_1}{T_3 - T_4}$$

$$= 1 - \frac{T_1}{T_3}\left[\frac{(r_p)^{\frac{\gamma-1}{\gamma}} - 1}{1 - \frac{1}{(r_p)^{\frac{\gamma-1}{\gamma}}}}\right] = 1 - \frac{T_1}{T_3}(r_p)^{\frac{\gamma-1}{\gamma}}. \qquad \text{...(3.18)}$$

Example 3.32. *Air enters the compressor of a gas turbine plant operating on Brayton cycle at 101.325 kPa, 27°C. The pressure ratio in the cycle is 6. Calculate the maximum temperature in the cycle and the cycle efficiency. Assume $W_T = 2.5\ W_C$ where W_T and W_C are the turbine and the compressor work respectively. Take $\gamma = 1.4$.* **(P.U.)**

Solution. Pressure of intake air, $p_1 = 101.325$ kPa
Temperature of intake air, $T_1 = 27 + 273 = 300$ K
The pressure ratio in the cycle, $r_p = 6$

(*i*) **Maximum temperature in the cycle, T_3 :**

Refer Fig 3.35.

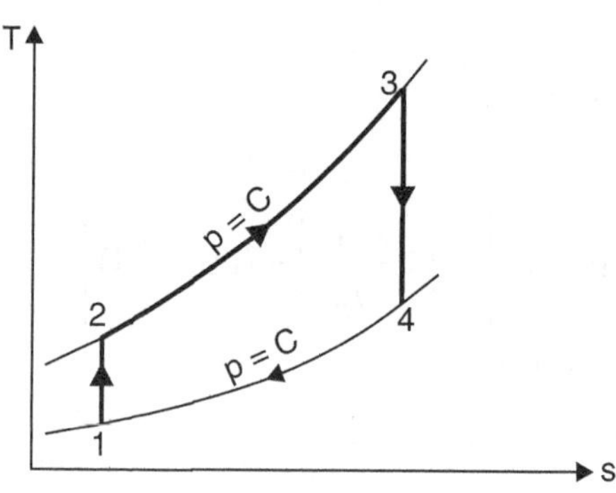

Fig. 3.35

$$\frac{T_2}{T_1} = \left(\frac{p_2}{p_1}\right)^{\frac{\gamma-1}{\gamma}} = (r_p)^{\frac{\gamma-1}{\gamma}} = (6)^{\frac{1.4-1}{1.4}} = 1.668$$

$\therefore$ $$T_2 = 1.668\ T_1 = 1.668 \times 300 = 500.4 \text{ K}$$

Also, $$\frac{T_3}{T_4} = (r_p)^{\frac{\gamma-1}{\gamma}} = (6)^{\frac{1.4-1}{1.4}} = 1.668$$

$\therefore$ $$T_4 = \frac{T_3}{1.668}$$

But $$W_T = 2.5\ W_C \quad \text{(Given)}$$

$\therefore$ $$mc_p\,(T_3 - T_4) = 2.5\ mc_p\,(T_2 - T_1)$$

or $$T_3 - \frac{T_3}{1.668} = 2.5\,(500.4 - 300) = 501 \text{ or } T_3\left(1 - \frac{1}{1.668}\right) = 501$$

$\therefore$ $$\mathbf{T_3} = \frac{501}{\left(1 - \frac{1}{1.668}\right)} = 1251 \text{ K or } \mathbf{978°C.} \quad \textbf{(Ans.)}$$

(*ii*) **Cycle efficiency, η_{cycle} :**

Now, $$T_4 = \frac{T_3}{1.668} = \frac{1251}{1.668} = 750 \text{ K}$$

$\therefore$ $$\eta_{cycle} = \frac{\text{Net work}}{\text{Heat added}} = \frac{mc_p\,(T_3 - T_4) - mc_p\,(T_2 - T_1)}{mc_p(T_3 - T_2)}$$

$$= \frac{(1251 - 750) - (500.4 - 300)}{(1251 - 500.4)} = 0.4 \text{ or } \mathbf{40\%.} \quad \textbf{(Ans.)}$$

$$\left[\textbf{Check}: \eta_{cycle} = 1 - \frac{1}{(r_p)^{\left(\frac{\gamma-1}{\gamma}\right)}} = 1 - \frac{1}{(6)^{\frac{1.4-1}{1.4}}} = 0.4 \text{ or } \mathbf{40\%.} \quad \textbf{(Ans.)}\right]$$

Example 3.33. *A gas turbine is supplied with gas at 5 bar and 1000 K and expands it adiabatially to 1 bar. The mean specific heat at constant pressure and constant volume are 1.0425 kJ/kg K and 0.7662 kJ/kg K respectively.*

(i) Draw the temperature-entropy diagram to represent the processes of the simple gas turbine system.

(ii) Calculate the power developed in kW per kg of gas per second and the exhaust gas temperature. **(GATE, 1995)**

Solution. *Given* : p_1 = 1 bar ; p_2 = 5 bar ; T_3 = 1000 K ; c_p = 1.0425 kJ/kg K ; c_v = 0.7662 kJ/kg K.

$$\gamma = \frac{c_p}{c_v} = \frac{1.0425}{0.7662} = 1.36$$

(*i*) **Temperature-entropy diagram :**

Temperature-entropy diagram representing the processes of the simple gas turbine system is shown in Fig. 3.36.

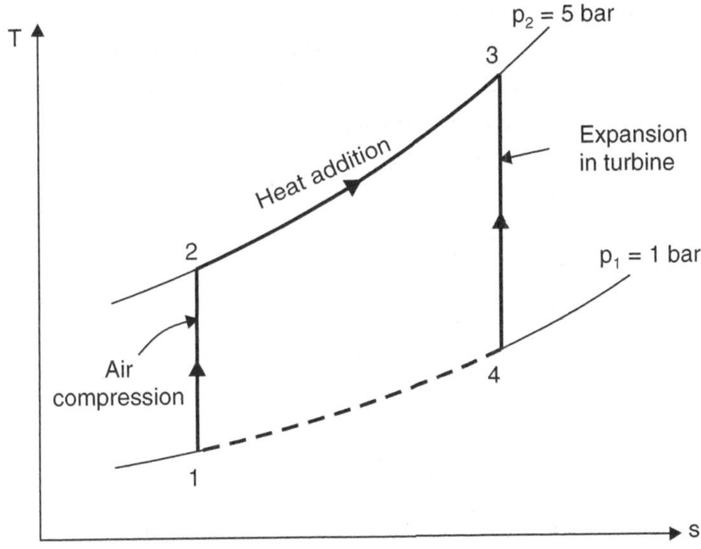

Fig. 3.36

(*ii*) **Power required :**

$$\frac{T_4}{T_3} = \left(\frac{p_1}{p_2}\right)^{\frac{\gamma-1}{\gamma}} = \left(\frac{1}{5}\right)^{\frac{1.36-1}{1.36}} = 0.653$$

$\therefore$ T_4 = 1000 × 0.653 = 653 K

Power developed per kg of gas per second

$$= c_p\,(T_3 - T_4)$$

= 1.0425 (1000 – 653) = **361.7 kW.** **(Ans.)**

Example 3.34. *An isentropic air turbine is used to supply 0.1 kg/s of air at 0.1 MN/m^2 and at 285 K to a cabin. The pressure at inlet to the turbine is 0.4 MN/m^2. Determine the temperature at turbine inlet and the power developed by the turbine. Assume c_p= 1.0 kJ/kg K.* **(GATE, 1999)**

Solution. *Given :* $\dot{m}_a = 0.1$ kg/s ; $p_1 = 0.1$ MN/m^2 = 1 bar, $T_4 = 285$ K ; $p_2 = 0.4$ MN/m^2 = 4 bar ; $c_p = 1.0$ kJ/kg K.

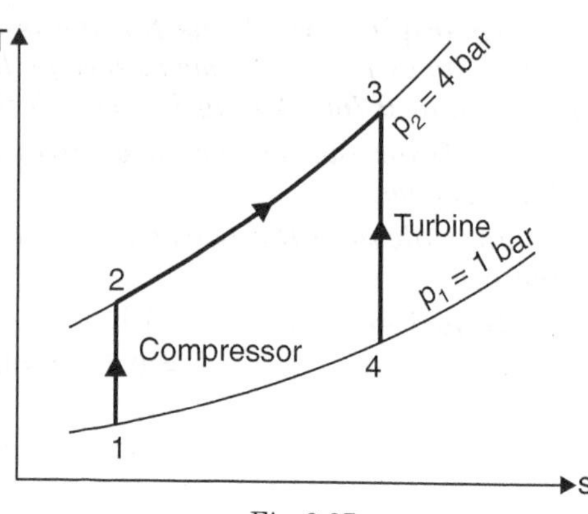

Fig. 3.37

Temperature at turbine inlet, T_3 :

$$\frac{T_3}{T_4} = \left(\frac{p_2}{p_1}\right)^{\frac{\gamma-1}{\gamma}} = \left(\frac{4}{1}\right)^{\frac{1.4-1}{1.4}} = 1.486$$

$\therefore$ $\mathbf{T_3 = 285 \times 1.486 = 423.5\ K.}$ **(Ans.)**

Power developed, P :

$$P = \dot{m}_a c_p (T_3 - T_4)$$
$$= 0.1 \times 1.0\ (423.5 - 285)$$
$$= \mathbf{13.85\ kW.}\ \textbf{(Ans.)}$$

Example 3.35. *Consider an air standard cycle in which the air enters the compressor at 1.0 bar and 20°C. The pressure of air leaving the compressor is 3.5 bar and the temperature at turbine inlet is 600°C. Determine per kg of air :*

(*i*) *Efficiency of the cycle,* (*ii*) *Heat supplied to air,*

(*iii*) *Work available at the shaft,* (*iv*) *Heat rejected in the cooler, and*

(*v*) *Temperature of air leaving the turbine.*

For air $\gamma = 1.4$ and $c_p = 1.005$ kJ/kg K.

Solution. Refer Fig. 3.35.

Pressure of air entering the compressor, $p_1 = 1.0$ bar

Temperature at the inlet of compressor, $T_1 = 20 + 273 = 293$ K

Pressure of air leaving the compressor, $p_2 = 3.5$ bar

Temperature of air at turbine inlet, $T_3 = 600 + 273 = 873$ K

(*i*) **Efficiency of the cycle, η_{cycle} :**

$$\eta_{cycle} = 1 - \frac{1}{(r_p)^{\frac{\gamma-1}{\gamma}}} = 1 - \frac{1}{(3.5)^{\frac{1.4-1}{1.4}}} = 0.30 \text{ or } \mathbf{30\%.}\ \textbf{(Ans.)} \qquad \left(\because\ r_p = \frac{p_2}{p_1} = \frac{3.5}{1.0} = 3.5\right)$$

(*ii*) **Heat supplied to air :**

For *compression process 1-2*, we have

$$\frac{T_2}{T_1} = \left(\frac{p_2}{p_1}\right)^{\frac{\gamma-1}{\gamma}} = \left(\frac{3.5}{1}\right)^{\frac{1.4-1}{1.4}} = 1.43$$

$\therefore$ $T_2 = T_1 \times 1.43 = 293 \times 1.43 \simeq 419$ K

$\therefore$ Heat supplied to air, $Q_1 = C_p (T_3 - T_2) = 1.005\ (873 - 419) = \mathbf{456.27\ kJ/kg.}$ **(Ans.)**

(*iii*) **Work available at the shaft, W :**

We know that, $$\eta_{cycle} = \frac{\text{Work output } (W)}{\text{Heat input } (Q_1)}$$

or $$0.30 = \frac{W}{456.27} \text{ or } W = 0.3 \times 456.27 = 136.88 \text{ kJ/kg}$$

(*iv*) **Heat rejected in the cooler, Q_2 :**

Work output (W) = Heat supplied (Q_1) – Heat rejected (Q_2)

$\therefore$ $Q_2 = Q_1 - W = 456.27 - 136.88 =$ **319.39 kJ/kg. (Ans.)**

(*v*) **Temperature of air leaving the turbine, T_4 :**

For *expansion (isentropic) process 3-4*, we have

$$\frac{T_3}{T_4} = (r_p)^{\frac{\gamma-1}{\gamma}} = (3.5)^{\frac{1.4-1}{1.4}} = 1.43$$

$$\therefore \quad T_4 = \frac{T_3}{1.43} = \frac{873}{1.43} = \textbf{610.5 K.} \quad \textbf{(Ans.)}$$

[**Check** : Heat rejected in the air cooler at constant pressure during the process 4-1 can also be calculated as : Heat rejected = $m \times c_p (T_4 - T_1) = 1 \times 1.005 \times (610.5 - 293) = 319.1$ kJ/kg]

Example 3.36. *Air enters the compressor of a gas turbine plant operating on Brayton cycle at 1 bar, 27°C. The pressure ratio in the cycle is 6. If $W_T = 2.5\ W_C$, where W_T and W_C are the turbine and compressor work respectively, calculate the maximum temperature and the cycle efficiency.* **(GATE, 1996)**

Solution. *Given* : $p_1 = 1$ bar ; $T_1 = 27 + 273 = 300$ K ; $\frac{p_2}{p_1} = 6$; $W_T = 2.5\ W_C$

Maximum temperature, T_3 :

$$\text{Now, } \frac{T_2}{T_1} = \left(\frac{p_2}{p_1}\right)^{\frac{\gamma-1}{\gamma}} = (6)^{\frac{1.4-1}{1.4}} = (6)^{\frac{0.4}{1.4}} = 1.668$$

$\therefore$ $T_2 = 300 \times 1.668 = 500.4$ K

$$\text{Also, } \frac{T_3}{T_4} = \left(\frac{p_2}{p_1}\right)^{\frac{\gamma-1}{\gamma}} = (6)^{\frac{0.4}{1.4}} = 1.668$$

or $T_3 = 1.668\ T_4$

Now, compressor work

$W_C = mc_p\ (T_2 - T_1)$,

and turbine work,

$W_T = mc_p\ (T_3 - T_4)$

Since $W_T = 2.5\ W_C$ ......(Given)

$\therefore$ $mc_p(T_3 - T_4) = 2.5 \times mc_p(T_2 - T_1)$

$$\left(T_3 - \frac{T_3}{1.668}\right) = 2.5\ (500.4 - 300)$$

$$\text{or} \quad T_3\left(1 - \frac{1}{1.668}\right) = 501$$

$\therefore$ $T_3 =$ **1251 K. (Ans.)**

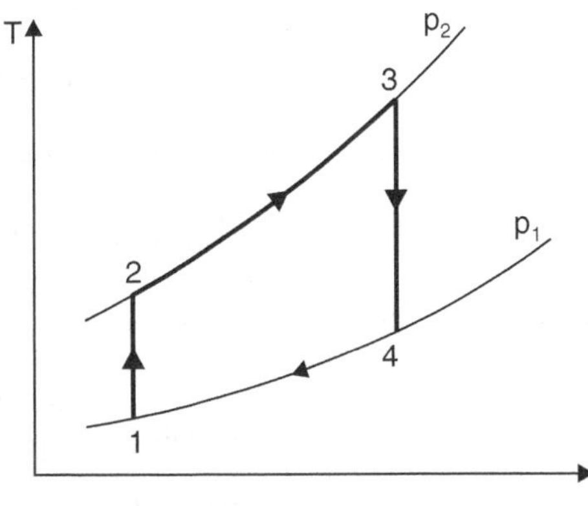

Fig. 3.38

Cycle efficiency, η_{cycle} :

$$\eta_{cycle} = \frac{W_T - W_C}{mc_p(T_3 - T_2)} = \frac{mc_p(T_3 - T_4) - mc_p(T_2 - T_1)}{mc_p(T_3 - T_2)}$$

$$= \frac{\left(1251 - \frac{1251}{1.668}\right) - (500.4 - 300)}{(1251 - 500.4)}$$

$$= \frac{501 - 200.4}{750.6} = \mathbf{0.40 \quad or \quad 40\%.} \quad \textbf{(Ans.)}$$

Example 3.37. *A closed cycle ideal gas turbine plant operates between temperature limits of 800°C and 30°C and produces a power of 100 kW. The plant is designed such that there is no need for a regenerator. A fuel of calorific 45000 kJ/kg is used. Calculate the mass flow rate of air through the plant and rate of fuel consumption.*

Assume c_p = 1 kJ/kg K and γ = 1.4. **(GATE, 1998)**

Solution. *Given* : $T_1 = 30 + 273 = 303$ K ; $T_3 = 800 + 273 = 1073$ K ; $C = 45000$ kJ/kg ; $c_p = 1$ kJ/kg K ; $\gamma = 1.4$; $W_{turbine} - W_{compressor} = 100$ kW.

$\dot{m}_a, \dot{m}_f$ **:**

Since no regenerator is used we can assume the turbine expands the gases upto T_4 in such a way that the exhaust gas temperature from the turbine is equal to the temperature of air coming out of the compressor *i.e.*, $T_2 = T_4$

$$\frac{p_2}{p_1} = \frac{p_3}{p_4}, \quad \frac{p_2}{p_1} = \left(\frac{T_2}{T_1}\right)^{\frac{\gamma}{\gamma - 1}} \text{ and } \frac{p_3}{p_4} = \left(\frac{T_3}{T_4}\right)^{\frac{\gamma}{\gamma - 1}}$$

$$\therefore \quad \frac{T_2}{T_1} = \frac{T_3}{T_4} = \frac{T_3}{T_2}$$

$$(\because \; T_2 = T_4 \; \text{......assumed})$$

or $$T_2^2 = T_1 T_3 \quad \text{or} \quad T_2 = \sqrt{T_1 T_3}$$

or $$T_2 = \sqrt{303 \times 1073} = 570.2 \text{ K}$$

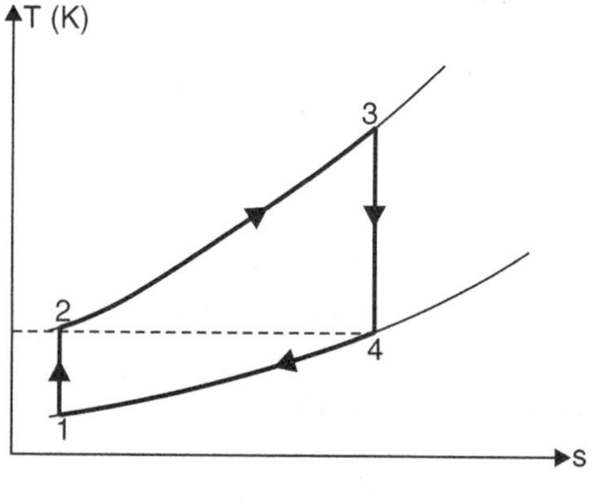

Now, $$W_{turbine} - W_{compressor} = \dot{m}_f \times C \times \eta$$

or $$100 = \dot{m}_f \times 45000 \times \left[1 - \frac{T_4 - T_1}{T_3 - T_2}\right]$$

$$= \dot{m}_f \times 45000 \left[1 - \frac{570.2 - 303}{1073 - 570.2}\right]$$

$$= \dot{m}_f \times 21085.9$$

or $$\dot{m}_f = \frac{100}{21085.9} = \mathbf{4.74 \times 10^{-3} \; kg/s.} \quad \textbf{(Ans.)}$$

Again, $$W_{turbine} - W_{compressor} = 100 \text{ kW}$$

$$(\dot{m}_a + \dot{m}_f)(T_3 - T_4) - \dot{m}_a \times 1 \times (T_2 - T_1) = 100$$

or $$(\dot{m}_a + 0.00474)(1073 - 570.2) - \dot{m}_a (570.2 - 303) = 100$$

or $$(\dot{m}_a + 0.00474) \times 502.8 - 267.2 \; \dot{m}_a = 100$$

or $$502.8 \; \dot{m}_a + 2.383 - 267.2 \; \dot{m}_a = 100$$

or $$235.6 \; \dot{m}_a = 97.617$$

$$\therefore \quad \dot{m}_a = \mathbf{0.414 \; kg/s.} \quad \textbf{(Ans.)}$$

Fig. 3.39

Example 3.38. *In a gas turbine plant working on Brayton cycle, the air at inlet is 27°C, 0.1 MP$_a$. The pressure ratio is 6.25 and the maximum temperature is 800°C. The turbine and compressor efficiencies are each 80%. Find compressor work, turbine work, heat supplied, cycle efficiency and turbine exhaust temperature. Mass of air may be considered as 1 kg. Draw T-s diagram.* **(AMIE Summer, 2000)**

Solution. Refer Fig. 3.40.

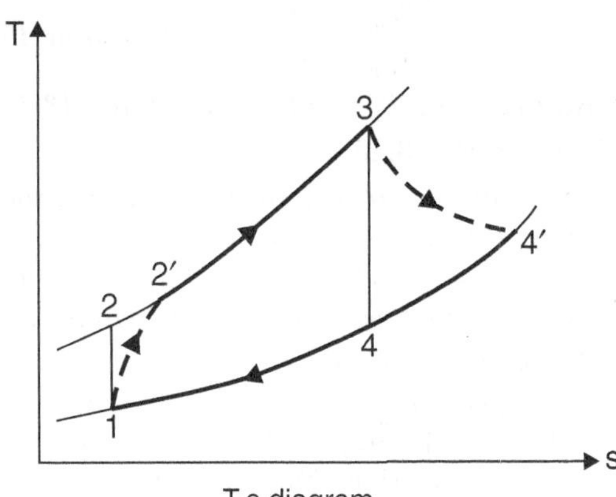

Fig. 3.40

Given : $T_1 = 27 + 273 = 300$ K ; $p_1 = 0.1$ MP$_a$; $r_p = 6.25$, $T_3 = 800 + 273 = 1073$ K

$\eta_{comp.} = \eta_{turbine} = 0.8.$

For the *compression process 1-2,* we have

$$\frac{T_2}{T_1} = \left(\frac{p_2}{p_1}\right)^{\frac{\gamma-1}{\gamma}} = (r_p)^{\frac{\gamma-1}{\gamma}} = (6.25)^{\frac{1.4-1}{1.4}} = 1.688$$

or $$T_2 = 300 \times 1.688 = 506.4 \text{ K}$$

Also, $$\eta_{comp.} = \frac{T_2 - T_1}{T_2' - T_1} \quad \text{or} \quad 0.8 = \frac{506.4 - 300}{T_2' - 300}$$

or $$T_2' = \frac{506.4 - 300}{0.8} + 300 = 558 \text{ K}$$

∴ **Compressor work,** $W_{comp.} = 1 \times c_p \times (T_2' - T_1)$

$= 1 \times 1.005 (558 - 300) =$ **259.29 kJ/kg. (Ans.)**

For *expansion process 3-4,* we have

$$\frac{T_3}{T_4} = \left(\frac{p_3}{p_4}\right)^{\frac{\gamma-1}{\gamma}} = (r_p)^{\frac{\gamma-1}{\gamma}} = (6.25)^{\frac{1.4-1}{1.4}} = 1.688$$

or $$T_4 = \frac{T_3}{1.688} = \frac{1073}{1.688} = 635.66 \text{ K}$$

Also, $$\eta_{turbine} = \frac{T_3 - T_4'}{T_3 - T_4} \quad \text{or} \quad 0.8 = \frac{1073 - T_4'}{1073 - 635.66}$$

or $$T_4' = 1073 - 0.8\ (1073 - 635.66) = 723.13\ \text{K}$$

$\therefore$ **Turbine work,** $W_{turbine} = 1 \times c_p \times (T_3 - T_4')$ (neglecting fuel mass)

$= 1 \times 1.005\ (1073 - 723.13) =$ **351.6 kJ/kg. (Ans.)**

Net work output, $W_{net} = W_{turbine} - W_{comp.} = 351.6 - 259.29 = 92.31$ kJ/kg

Heat supplied, $Q_{in} = 1 \times c_p \times (T_3 - T_2')$

$= 1 \times 1.005 \times (1073 - 558) =$ **517.57 kJ/kg. (Ans.)**

Cycle efficiency, $\eta_{cycle} = \dfrac{W_{net}}{Q_{in}} = \dfrac{92.31}{517.57} = 0.1783$ or **17.83%. (Ans.)**

Turbine exhaust temperature, $T_4' = 723.13$ K or **450.13°C. (Ans.)**

The *T-s* diagram is shown in Fig. 3.40.

Example 3.39. *Find the required air-fuel ratio in a gas turbine whose turbine and compressor efficiencies are 85% and 80%, respectively. Maximum cycle temperature is 875°C. The working fluid can be taken as air (c_p = 1.0 kJ/kg K, γ = 1.4) which enters the compressor at 1 bar and 27°C. The pressure ratio is 4. The fuel used has calorific value of 42000 kJ/kg. There is a loss of 10% of calorific value in the combustion chamber.* **(GATE, 1998)**

Solution. *Given :* $\eta_{turbine}$ = 85% ; $\eta_{compressor}$ = 80% ; $T_3 = 273 + 875 = 1148$ K, $T_1 = 27 + 273 = 300$ K ; c_p = 1.0 kJ/kg K ; γ = 1.4, p_1 = 1 bar, p_2 = 4 bar (since pressure ratio is 4) ; C = 42000 kJ/kg K ; η_{cc} = 90% (since loss in the combustion chamber is 10%)

For isentropric compression 1-2 :

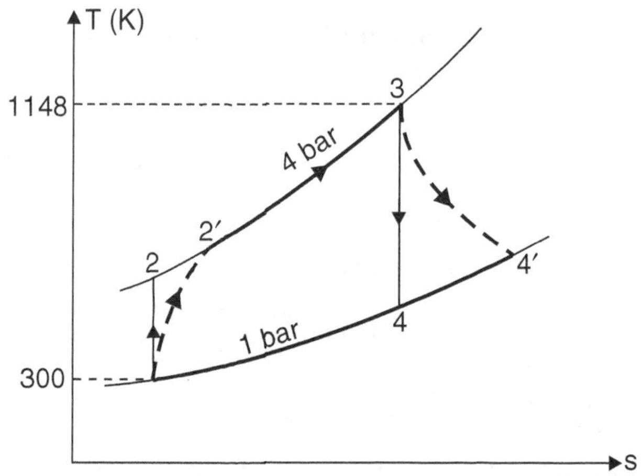

Fig. 3.41

$$\frac{T_2}{T_1} = \left(\frac{p_2}{p_1}\right)^{\frac{\gamma-1}{\gamma}} = (4)^{\frac{1.4-1}{1.4}} = 1.486$$

$\therefore$ $$T_2 = 300 \times 1.486 = 445.8\ \text{K}$$

$$\eta_{compressor} = \frac{T_2 - T_1}{T_2' - T_1}$$

or $$0.8 = \frac{445.8 - 300}{T_2' - 300}$$

or $$T_2' = \frac{445.8 - 300}{0.8} + 300 = 482.2 \text{ K}$$

Now, Heat supplied by the fuel = Heat taken by the burning gases

$$0.9 \times m_f \times C = (m_a + m_f) \times c_p \times (T_3 - T_2')$$

$$\therefore \quad C = \left(\frac{m_a + m_f}{m_f}\right) \times \frac{c_p(T_3 - T_2')}{0.9} = \left(\frac{m_a}{m_f} + 1\right) \times \frac{c_p(T_3 - T_2')}{0.9}$$

or $$42000 = \left(\frac{m_a}{m_f} + 1\right) \times \frac{1.00(1148 - 482.27)}{0.9} = 739.78\left(\frac{m_a}{m_f} + 1\right)$$

$$\therefore \quad \frac{m_a}{m_f} = \frac{42000}{739.78} - 1 = 55.77 \text{ say } 56$$

$\therefore$ A/F ratio = **56 : 1. (Ans.)**

3.11. STIRLING CYCLE

Fig. 3.42 shows a stirling cycle (1827) on *p-v* and *T-s* diagrams. It consists of *two isotherms and two constant volume processes.*

Process 1-2 is the *isothermal compression* with *heat rejection* Q_L to the surroundings at temperature T_L.

Process 3-4 is the *isothermal expansion* with *heat addition* Q_H from a source at temperature T_H.

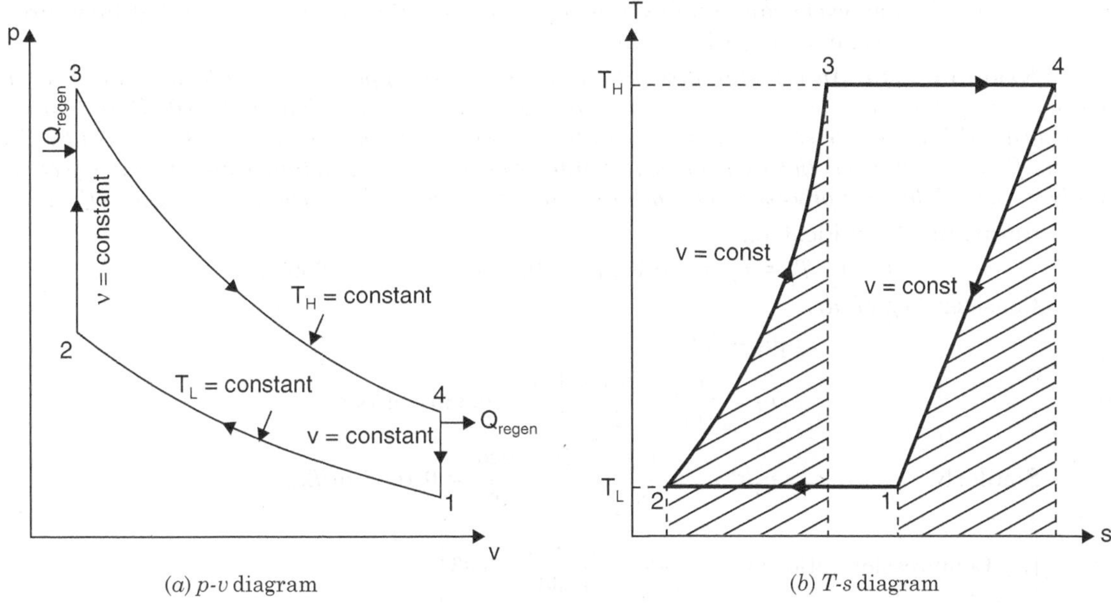

(*a*) *p-v* diagram (*b*) *T-s* diagram

Fig. 3.42. Stirling cycle.

Process 2-3 and **4-1** are *constant volume heat transfer processes.*

For *1 kg of ideal gas,*

$$Q_{1\text{-}2} \text{ (Heat rejected)} = W_{1\text{-}2} = -RT_L \ln \frac{v_1}{v_2} \text{ (compression)}$$

$$Q_{2\text{-}3} = c_v\,(T_H - T_L)\ ;\ W_{2\text{-}3} = 0 \text{ (since } v = \text{const.)}$$

$$Q_{3\text{-}4} \text{ (Heat supplied)} = W_{3\text{-}4} = RT_H \ln \frac{v_4}{v_3} = RT_H \ln \frac{v_1}{v_2} \text{ (Expansion)}$$

$$(\because\ v_4 = v_1 \text{ and } v_3 = v_2)$$

$$Q_{4\text{-}1} = -\,c_v(T_L - T_H) \text{ or } c_v\,(T_H - T_L)\ ;\ W_{4\text{-}1} = 0 \text{ (since } v = \text{const.)}$$

The efficiency of the Stirling cycle is *less* than that of the Carnot cycle *due to heat transfers at constant volume processes.* However, if a **regenerative** *arrangement* is used such that $Q_{4\text{-}1} = Q_{2\text{-}3}$ *i.e.* area under 1-4 is equal to area under 2-3, then the cycle efficiency becomes

$$\eta = \frac{RT_H \ln \frac{v_1}{v_2} - RT_L \ln \frac{v_1}{v_2}}{RT_H \ln \frac{v_1}{v_2}} = \frac{T_H - T_L}{T_H} \quad ...(3.19)$$

which means the *regenerative Stirling cycle* has *same efficiency as the Carnot cycle.*

The following points are worth noting :

- As far as the impractibility of accomplishing isothermal compression and expansion processes with a gas is concerned, the Stirling cycle *suffers from the limitation of the Carnot cycle.* But, it does *not* suffer from other drawbacks of the Carnot cycle, *viz,* very low m.e.p., the narrow *p-v* diagram and great susceptibility to the internal efficiencies of the compressor and the expander.
- The mean effective pressure (*m.e.p.*) of the Stirling cycle is *much* greater than that of the Carnot cycle.
- A reversed Stirling cycle with regeneration can similarly attain Carnot C.O.P.
- The Stirling cycle can suitably replace Otto cycle (having two constant volume processes) in reciprocating I.C. engines.

Example 3.40. *An air standard Stirling cycle is equipped with a 100 percent efficient regenerator system. The isothermal compression commences from 1 bar and 310 K, and subsequent heat addition at constant volume raises the pressure and temperature to 16 bar 930 K. The cycle is finally completed through an isothermal expansion and constant volume heat rejection. Analyse each of the four processes for work and heat transfer and determine the engine efficiency.*

Solution. Refer Fig.3.42.

Given : $p_1 = 1$ bar ; $T_1 = T_L = 310$ K ; $p_3 = 16$ bar ; $T_3 = T_H = 930$ K.

Consider *1 kg of air.*

$$p_1 v_1 = RT_1$$

or

$$v_1 = \frac{RT_1}{p_1} = \frac{287 \times 310}{1 \times 10^5} = 0.8897 \text{ m}^3/\text{kg}$$

Similarly

$$v_2 = v_3 = \frac{RT_3}{p_3} = \frac{287 \times 930}{16 \times 10^5} = 0.1668 \text{ m}^3/\text{kg}$$

$$\left[\therefore \text{ Compression ratio, } r = \frac{v_1}{v_2} = \left(\frac{v_4}{v_3}\right) = \frac{0.8897}{0.1668} = 5.33\right]$$

Process 1-2 : $T_1 = T_2 = 310 \text{ K} = T_L$

Heat rejected, $Q_{1\text{-}2} = W_{1\text{-}2} = RT_L \ln \left(\frac{v_1}{v_2}\right) = RT_L \ln (r)$

$= 287 \times 310 \ln (5.33) = 148878$ J/kg $= 148.88$ kJ/kg

Process 2-3 : $Q_{2\text{-}3} = W_{2\text{-}3} = 0$ (since volume is constant)

Process 3-4 : Heat supplied, $Q_{3\text{-}4} = W_{3\text{-}4} = RT_H \ln(r)$

$= 287 \times 930 \ln (5.33) = 446634$ J/kg or 446.63 kJ/kg

Process 4-1 : $Q_{4\text{-}1} = -c_v (T_L - T_H)$ or $c_v(T_H - T_L)$

Heat supplied during the process 2-3 = Heat rejected during the process 4-1.

∴ Work done = Net heat exchange during the isotherms

$= 446.63 - 148.88 = 297.75$ kJ/kg

$$\therefore \text{ Thermal efficiency, } \eta_{th} = \frac{\text{Work done}}{\text{Heat supplied}} = \frac{297.75}{446.63} = 0.6676 \text{ or } \mathbf{66.7\%}.$$

Since Stirling cycle is *completely reversible*, its efficiency is also given as,

$$\eta = \frac{T_H - T_L}{T_H} = \frac{930 - 310}{930} = 0.667 \text{ or } \mathbf{66.7\%.} \ \mathbf{(Ans.)}$$

3.12. MILLER CYCLE

The Miller cycle (named after R. H. Miller) is a modern modification of the Atkinson cycle and has an *expansion ratio greater than the compression ratio,* which is accomplished, however, in much a different way. *Whereas a complicated mechanical linkage system of some kind is required for an engine designed to operate on the Atkinson cycle, a Miller cycle engine uses unique valve timing to obtain the same desired results.*

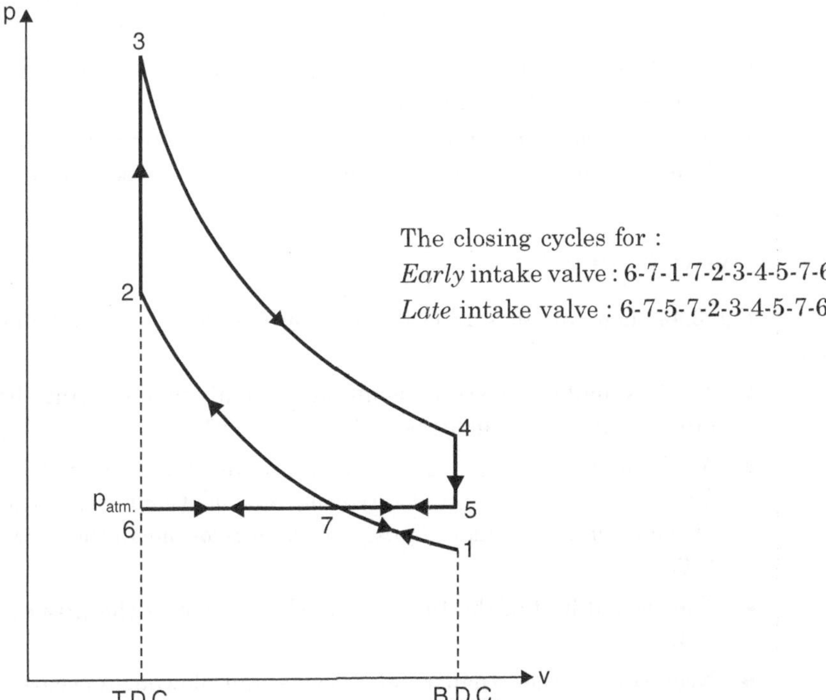

Fig. 3.43. Air-standard Miller cycle for unthrottled a naturally aspirated four-stroke cycle S.I. engine.

In a Miller cycle air intake is *unthrottled.* The quantity of air ingested into each cylinder is then *controlled by closing the intake valve at the proper time,* long before B.D.C. (point 7 in Fig. 3.43).

With the movement of the piston towards B.D.C. during the later part of the intake stroke, the cylinder pressure is reduced along the process 7-1.

When the piston reaches B.D.C. and starts back towards the T.D.C., the pressure again increases along the process 1-7.

Then, the resulting cycle is : 6-7-1-7-2-3-4-5-6.

— The work produced during 6-7 (intake process) is cancelled by 7-6 (exhaust process).

— The process 7-1 is cancelled by 1-7.

Hence net indicated work = Area within the loop 7-2-3-4-5-7 ; there being no pump work.

The compression ratio $= \dfrac{v_7}{v_2}$ ...(3.20)

The larger compression ratio $= \dfrac{v_1}{v_2} = \dfrac{v_4}{v_3}$ ...(3.21)

- A greater net indicated work per cycle is obtained as a result of the shorter compression stroke which absorbs work, combined with the longer expansion stroke which produces work.
- Further, by permitting air to flow through the intake system *unthrottled*, a *major loss* experienced by most S.I. engines is *eliminated.*
- Due to the absence of pump work, the Miller cycle engine has a higher thermal efficiency.
- Miller cycle engines are usually supercharged or turbocharged with peak intake manifold pressures of 150-200 kPa.
- Automobiles with Miller cycle engines were first marketed in the latter half of the 1990s. A typical value of the compression ratio is about 8 : 1, with an expansion ratio of about 10 : 1.

3.13. LENOIR CYCLE

Fig. 3.44 shows an air standard approximation for a historic Lenoir engine cycle, 1-2-3-4-5-2-1.

- The first half of the stroke is intake, with air-fuel entering the cylinder at atmospheric pressure ($p_{atm.}$) — **process 1–2.**
- At about halfway through the first stroke, the intake valve is *closed* and the air-fuel mixture is *ignited without any compression.* Combustion raises the temperature and pressure in the cylinder *almost at constant-volume* in the slow-moving engine—**process 2–3.**
- The second half of the first stroke then becomes the **power or expansion process 3–4.**
- Near B.D.C., the exhaust valve opens and blowdown occurs—**process 4–5.**
- Then follows the **exhaust stroke 5–1**, thus completing the two stroke cycle.

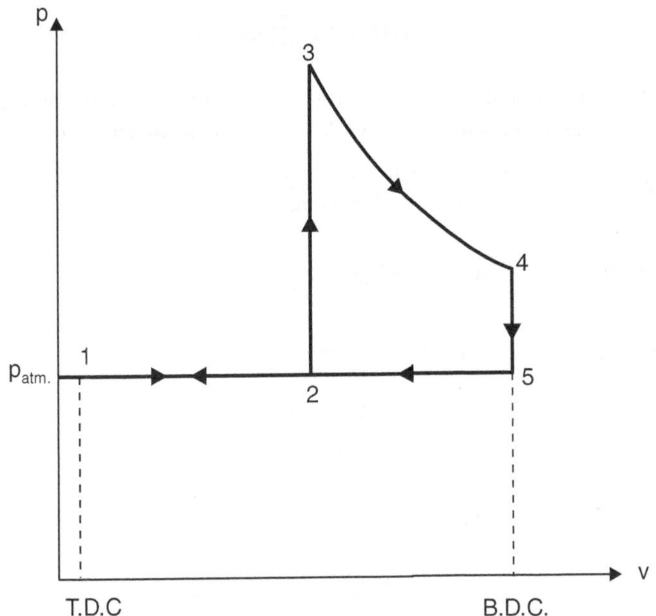

Fig. 3.44. Air standard approximation for historic Lenoir engine cycle, 1-2-3-4-5-2-1.

Thermodynamic analysis :

Consider the clearance volume to be essentially *nil*. Cancelling the intake process 1-2 and latter half of the stroke 2-1 thermodynamically on *p-v* coordinates, the cycle then becomes 2-3-4-5-2.

Process 2-3. *Constant volume heat input (combustion)*; All valves closed :

$$p_2 = p_1 = p_{atm.} \; ; \; v_3 = v_2 \; ; \; W_{2\text{-}3} = 0, \; Q_{2\text{-}3} = Q_{in} = c_v(T_3 - T_2) = (u_3 - u_2)$$

Process 3-4. *Isentropic power or expansion stroke* ; All valves closed :

$$Q_{3\text{-}4} = 0 \; ; \; T_4 = T_3(v_3/v_4)^{\gamma-1} \; ; \; p_4 = p_3\,(v_3/v_4)^{\gamma} \; ;$$

$$W_{3\text{-}4} = (p_4v_4 - p_3v_3)/(1 - \gamma) = R(T_4 - T_3)/(1 - \gamma)$$

Process 4-5. *Constant volume heat rejection (exhaust blow down)* ; Exhaust valve open and intake valve closed ;

$$v_5 = v_4 = v_{B.D.C.} \; ; \; W_{4\text{-}5} = 0, \; Q_{4\text{-}5} = Q_{out} = c_v(T_4 - T_5) = v_4 - v_5$$

Process 5-2. *Constant pressure exhaust stroke at* $p_{atm.}$; Exhaust valve open, and intake valve closed.

$$p_5 = p_2 = p_1 = p_{atm.} \; ; \; W_{5\text{-}2} = p_{atm.}(v_5 - v_2) \; ; \; Q_{5\text{-}2} = Q_{out} = (h_5 - h_2)$$

$$= c_p(T_5 - T_2).$$

Thus, thermal efficiency of Lenoir cycle,

$$(\eta_{th})_{\text{Lenoir}} = \frac{W_{\text{net}}}{Q_{\text{in}}} = \frac{Q_{\text{in}} - Q_{\text{out}}}{Q_{\text{in}}} = 1 - \frac{Q_{\text{out}}}{Q_{\text{in}}}$$

$$= 1 - \frac{[c_v\,(T_4 - T_5) + c_p\,(T_5 - T_2)]}{c_v(T_3 - T_2)}$$

$$= 1 - \frac{[(T_4 - T_5) + \gamma(T_5 - T_2)]}{T_3 - T_2} \qquad ...(3.22)$$

HIGHLIGHTS

1. **A cycle** is defined as a repeated series of operations occurring in a certain order.
2. The efficiency of an engine using air as the working medium is known as an '**Air standard efficiency**'.
3. Relative efficiency, $\eta_{\text{relative}} = \dfrac{\text{Actual thermal efficiency}}{\text{Air standard efficiency}}$.
4. **Carnot cycle** efficiency, $\eta_{\text{Carnot}} = \dfrac{T_1 - T_2}{T_1}$.
5. **Otto cycle** efficiency, $\eta_{\text{Otto}} = 1 - \dfrac{1}{(r)^{\gamma - 1}}$.

   Mean effective pressure, $p_{m(\text{Otto})} = \dfrac{p_1 r[(r^{\gamma-1} - 1)(r_p - 1)]}{(\gamma - 1)(r - 1)}$.
6. **Diesel cycle** efficiency, $\eta_{\text{Diesel}} = 1 - \dfrac{1}{\gamma(r)^{\gamma-1}}\left[\dfrac{\rho^{\gamma} - 1}{\rho - 1}\right]$

   Mean effective pressure, $p_{m(\text{Diesel})} = \dfrac{p_1 r^{\gamma}[\gamma(\rho - 1) - r^{1-\gamma}(\rho^{\gamma} - 1)]}{(\gamma - 1)(r - 1)}$.
7. **Dual cycle** efficiency, $\eta_{\text{Dual}} = 1 - \dfrac{1}{(r)^{\gamma-1}}\left[\dfrac{(\beta.\rho^{\gamma} - 1)}{(\beta - 1) + \beta\gamma(\rho - 1)}\right]$

   Mean effective pressure, $p_{m(\text{Dual})} = \dfrac{p_1 r^{\gamma}[\beta(\rho - 1) + (\beta - 1) - r^{1-\gamma}(\beta\rho^{\gamma} - 1)]}{(\gamma - 1)(r - 1)}$.
8. Atkinson cycle efficiency, $\eta_{\text{Atkinson}} = 1 - \gamma \cdot \dfrac{(r - \alpha)}{r^{\gamma} - \alpha^{\gamma}}$

   where α = compression ratio, r = expansion ratio.
9. Brayton cycle, $\eta_{\text{Brayton}} = 1 - \dfrac{1}{(r_p)^{\frac{\gamma-1}{\gamma}}}$, where r_p = pressure ratio.
10. Stirling cycle, $\eta_{stirling} = \dfrac{T_H - T_L}{T_H}$.
11. Miller cycle engines are usually supercharged or turbocharged with peak intake manifold pressures of 150 – 200 kPa.
12. Lenoir cycle, $(\eta_{th})_{\text{Lenoir}} = 1 - \dfrac{(T_4 - T_5) + \gamma(T_5 - T_2)}{T_3 - T_2}$

OBJECTIVE TYPE QUESTIONS

Choose the correct answer :

1. The air standard Otto cycle comprises
   (*a*) two constant pressure processes and two constant volume processes
   (*b*) two constant pressure and two constant entropy processes
   (*c*) two constant volume processes and two constant entropy processes
   (*d*) none of the above.

2. The air standard efficiency of Otto cycle is given by

(a) $\eta = 1 + \frac{1}{(r)^{\gamma+1}}$ (b) $\eta = 1 - \frac{1}{(r)^{\gamma-1}}$

(c) $\eta = 1 - \frac{1}{(r)^{\gamma+1}}$ (d) $\eta = 2 - \frac{1}{(r)^{\gamma-1}}$.

3. The thermal efficiency of theoretical Otto cycle
   (a) increases with increase in compression ratio
   (b) increases with increase in isentropic index γ
   (c) does not depend upon the pressure ratio
   (d) follows all the above.
4. The work output of theoretical Otto cycle
   (a) increases with increase in compression ratio
   (b) increases with increase in pressure ratio
   (c) increases with increase in adiabatic index γ
   (d) follows all the above.
5. For same compression ratio
   (a) thermal efficiency of Otto cycle is greater than that of Diesel cycle
   (b) thermal efficiency of Otto cycle is less than that of Diesel cycle
   (c) thermal efficiency of Otto cycle is same as that for Diesel cycle
   (d) thermal efficiency of Otto cycle cannot be predicted.
6. In air standard Diesel cycle, at fixed compression ratio and fixed value of adiabatic index (γ)
   (a) thermal efficiency increases with increase in heat addition cut off ratio
   (b) thermal efficiency decreases with increase in heat addition cut off ratio
   (c) thermal efficiency remains same with increase in heat addition cut off ratio
   (d) none of the above.

ANSWERS

1. (b) **2.** (b) **3.** (d) **4.** (d) **5.** (a) **6.** (b).

THEORETICAL QUESTIONS

1. What is a cycle ? What is the difference between an ideal and actual cycle ?
2. What is an air-standard efficiency ?
3. What is relative efficiency ?
4. Derive expressions of efficiency in the following cases :
   (i) Carnot cycle (ii) Diesel cycle (iii) Dual combustion cycle.
5. Explain "Air standard analysis" which has been adopted for I.C. engine cycles. State the assumptions made for air standard cycles.
6. Derive an expression for 'Atkinson cycle'.
7. Derive an expression for the thermal efficiency of Stirling cycle.
8. Explain the following cycles briefly and derive expressions of efficiency.
   (i) Miller cycle
   (ii) Lenoir cycle.

UNSOLVED EXAMPLES

1. A Carnot engine working between 377°C and 37°C produces 120 kJ of work. Determine :
   (*i*) The heat added in kJ. (*ii*) The entropy change during heat rejection process.
   (*iii*) The engine thermal efficiency. **[Ans.** (*i*) 229.5 kJ ; (*ii*) 0.353 kJ/K (*iii*) 52.3%]
2. Find the thermal efficiency of a Carnot engine whose hot and cold bodies have temperatures of 154°C and 15°C respectively. **[Ans.** 32.55%]
3. Derive an expression for change in efficiency for a change in compression ratio. If the compression ratio is increased from 6 to 8, what will be the percentage increase in efficiency ? **[Ans.** 8%]
4. The efficiency of an Otto cycle is 50% and γ is 1.5. What is the compression ratio ? **[Ans.** 4]
5. An engine working on Otto cycle has a volume of 0.5 m^3, pressure 1 bar and temperature 27°C at the commencement of compression stroke. At the end of compression stroke, the pressure is 10 bar. Heat added during the constant volume process is 200 kJ. Determine :
   (*i*) Percentage clearance (*ii*) Air standard efficiency
   (*iii*) Mean effective pressure
   (*iv*) Ideal power developed by the engine if the engine runs at 400 r.p.m. so that there are 200 complete cycles per minutes. **[Ans.** (*i*) 23.76% ; (*ii*) 47.2% ; (*iii*) 2.37 bar (*iv*) 321 kW]
6. The compression ratio in an air-standard Otto cycle is 8. At the beginning of compression process, the pressure is 1 bar and the temperature is 300 K. The heat transfer to the air per cycle is 1900 kJ/kg of air. Calculate :
   (*i*) Thermal efficiency (*ii*) The mean effective pressure.
   [Ans. (*i*) 56.47% ; (*ii*) 14.24 bar]
7. An engine 200 mm bore and 300 mm stroke works on Otto cycle. The clearance volume is 0.0016 m^3. The initial pressure and temperature are 1 bar and 60°C. If the maximum pressure is limited to 24 bar, find :
   (*i*) The air-standard efficiency of the cycle (*ii*) The mean effective pressure for the cycle.
   Assume ideal conditions. **[Ans.** (*i*) 54.08% ; (*ii*) 1.972 bar]
8. Calculate the air standard efficiency of a four stroke Otto cycle engine with the following data :
   Piston diameter (bore) = 137 mm ; Length of stroke = 130 mm ;
   Clearance volume = 0.00028 m^3.
   Express clearance as a percentage of swept volume. **[Ans.** 56.1% ; 14.6%]
9. In an ideal Diesel cycle, the temperatures at the beginning of compression, at the end of compression and at the end of the heat addition are 97°C, 789°C and 1839°C. Find the efficiency of the cycle.
   [Ans. 59.6%]
10. An air-standard Diesel cycle has a compression ratio of 18, and the heat transferred to the working fluid per cycle is 1800 kJ/kg. At the beginning of the compression stroke, the pressure is 1 bar and the temperature is 300 K. Calculate : (*i*) Thermal efficiency, (*ii*) The mean effective pressure.
    [Ans. (*i*) 61% ; (*ii*) 13.58 bar]
11. 1 kg of air is taken through a Diesel cycle. Initially the air is at 15°C and 1 ata. The compression ratio is 15 and the heat added is 1850 kJ. Calculate : (*i*) The ideal cycle efficiency, (*ii*) The mean effective pressure. **[Ans.** (*i*) 55.1% ; (*ii*) 13.4 bar]
12. What will be loss in the ideal efficiency of a Diesel engine with compression ratio 14 if the fuel cut off is delayed from 6% to 9% ? **[Ans.** 2.1%]
13. The pressures on the compression curve of a diesel engine are at $\frac{1}{8}$ th stroke 1.4 bar and at $\frac{7}{8}$ th stroke 14 bar. Estimate the compression ratio. Calculate the air standard efficiency of the engine if the cut off occurs at $\frac{1}{15}$ of the stroke. **[Ans.** 18.54 ; 63.7%]

14. A compression ignition engine has a stroke 270 mm, and a cylinder diameter of 165 mm. The clearance volume is 0.000434 m^3 and the fuel ignition takes place at constant pressure for 4.5 per cent of the stroke. Find the efficiency of the engine assuming it works on the Diesel cycle. **[Ans.** 61.7%]

15. The following data belong to a Diesel cycle :

Compression ratio = 16 : 1 ; Heat added = 2500 kJ/kg ; Lowest pressure in the cycle = 1 bar ; Lowest temperature in the cycle = 27°C.

Determine :

(*i*) Thermal efficiency of the cycle. (*ii*) Mean effective pressure.

[Ans. (*i*) 45% ; (*ii*) 16.8 bar]

16. The compression ratio of an air-standard Dual cycle is 12 and the maximum pressure in the cycle is limited to 70 bar. The pressure and temperature of cycle at the beginning of compression process are 1 bar and 300 K. Calculate : (*i*) Thermal efficiency, (*ii*) Mean effective pressure.

Assume : cylinder bore = 250 mm, stroke length = 300 mm, c_p = 1.005, c_v = 0.718 and γ = 1.4.

[Ans. (*i*) 61.92% ; (*ii*) 9.847 bar]

17. The compression ratio of a Dual cycle is 10. The temperature and pressure at the beginning of the cycle are 1 bar and 27°C. The maximum pressure of the cycle is limited to 70 bar and heat supplied is limited to 675 kJ/kg of air. Find the thermal efficiency of the cycle. **[Ans.** 59.5%]

18. An air standard Dual cycle has a compression ratio of 16, and compression begins at 1 bar, 50°C. The maximum pressure is 70 bar. The heat transferred to air at constant pressure is equal to that at constant volume. Determine :

(*i*) The cycle efficiency. (*ii*) The mean effective pressure of the cycle.

Take : c_p = 1.005 kJ/kg-K, c_v = 0.718 kJ/kg-K. **[Ans.** (*i*) 66.5% ; (*ii*) 4.76 bar]

19. Compute the air standard efficiency of a Brayton cycle operating between a pressure of 1 bar and a final pressure of 12 bar. Take γ = 1.4. **[Ans.** 50.8%]

4

Fuel-Air and Actual Cycles

4.1. FUEL-AIR CYCLES

4.1.1. Introduction

- In air standard cycles analysis highly simplified approximations are made. The air standard theory gives an estimate of engine performance which is much greater than the actual performance. This large variation is partly due to the non-instantaneous burning and valve operation incomplete combustion etc. ; the *major reason being over-simplication in using the values of the properties of the working fluid for cycle analysis.*
- In air cycle approximation it is assumed that air is a perfect gas having constant specific heats. *In actual engine the working fluid is not air but a mixture of air, fuel and exhaust gases. Furthermore, the specific heats of the working fluid are not constant but increase with rise in temperature, and at high temperature the combustion products are subjected to dissociation.*
- *The theoretical cycle based on the actual properties of the cylinder gases is called the* **Fuel-air cycle approximation** ; it provides a rough idea for comparison with the actual performance.

4.1.2. Factors considered for Fuel-air Cycle Calculations

The following *factors* are taken into considerations while making fuel-air cycle calculations :

1. The actual composition of cylinder gases (consisting of fuel, air, water vapour in air and residual gas). During the operation of engine the fuel-air ratio changes due to which the relative amounts of CO_2, water vapour etc. also change.
2. Increase of specific heats of gases (except monoatomic gases) with temperature increase, subsequently the value of γ also changes.
3. Since the fuel-air mixture does not completely combine chemically at high temperatures (above 1600 K), therefore, at equilibrium condition gases like CO_2, H_2, and O_2 may be present.

4. The variation in the number of molecules present in the cylinder as the temperature and pressure change.

4.1.3. Assumptions made for Fuel-Air Cycle Analysis

Beside considering above factors, the following ***assumptions*** *are made for the analysis of fuel-air cycle :*

1. Prior to combustion there is no chemical change in either fuel or air.
2. Subsequent to combustion, the change is always in chemical equilibrium.
3. The processes are *adiabatic* (*i.e.,* there is no exchange of heat between the gases and cylinder walls in any process). In addition, the expansion and compression processes are *frictionless.*
4. The velocities are *negligibly small in case of reciprocating engines.*

Furthermore, in case of a *consant volume fuel-air cycle* the following assumptions are made :

- The fuel is completely vaporised and perfectly mixed with the air ;
- The burning takes place instantaneously at T.D.C. (at constant volume).

4.1.4. Importance of Fuel-Air Cycle

- Whereas the air standard cycle exhibits the general effect of compression ratio on efficiency of the engine, the fuel-air cycle may be calculated for various fuel-air ratios, inlet temperatures and pressures (It is worth noting that fuel-air ratio and compression ratio are much more important parameters in comparison to inlet conditions).
- With the help of fuel-air cycle analysis a *very good estimate of power to be expected from the actual engine can be made.* Furthermore, it is possible *to approximate very closely peak pressures and exhaust temperatures on which design and engine structure depend.*

4.1.5. Variable Specific Heats

4.1.5.1. General aspects

The specific heat of any substance is the ratio of the heat required to raise the temperature of a unit mass of the substance through one degree centigrade. Different substances have different values of specific heat. In case of gases, the temperature can be raised in two ways, *e.g.* either at *constant pressure* or *constant volume.* Accordingly we have two specific heats c_p and c_v. It is often convenient to use specific heats for the mol of a substance. A mol is M kilograms, called kilogram-mol abbreviated as kg mol. Here M is the molecular weight of the substance.

Thus *molar specific heat*

$$C = M.\ c \text{ kJ/mol K}$$

Similarly, $$C_p = M.\ c_p \text{ kJ/mol K}$$

and $$C_v = M.\ c_v \text{ kJ/mol K}$$

In general, the specific heats are *not constant. The specific heat varies largely with temperature but not very significantly with pressure except at very high pressure.* Thus in simple calculations, the variation in specific heat with pressure is neglected.

The specific heats of gases increase with the rise in temperature since the vibrational energy of the molecules increases with temperature. The effect of variable specific heats on the engine performance at higher temperature is considerable and it is, therefore, necessary to study these effects.

It is generally assumed that the specific heat is a linear function of temperature and the following relations hold good.

$$c_p = a + KT \quad ...(4.1)$$
$$c_v = b + KT \quad ...(4.2)$$

where a, b and K are constant.

$$R = (c_p - c_v) = (a + KT) - (b + KT) = a - b \quad ...(4.3)$$

This linear relationship holds good upto about 1500 K. Above 1500 K, the increase is more rapid and the specific heats are given by :

$$c_p = a + K_1T + K_2T^2 \quad ...(4.4)$$
$$c_v = b + K_1T + K_2T^2 \quad ...(4.5)$$

If the internal energy of vibration is ignored the specific heats of gases may be written as :

$$R = c_p - c_v \text{ kJ/kg K or kcal/kg °K}.$$

For monoatomic gases :

$$c_v = \frac{3R}{2} \text{ kJ/kg K or kcal/kg °K}$$

$$Mc_v = \frac{3MR}{2} \text{ kJ/mol K or kcal/mol °K}$$

where R = The characteristic gas constant expressed in kJ/kg K or kcal/kg °K ;

and MR = Molecular weight × characteristic gas constant
= Universal gas constant
= 8.314 kJ/kg mol K or 1.986 kcal/kg mol °K

c_v is expressed in kJ/kg K or kcal/kg °K

Also $$c_p = R + c_v = R + \frac{3R}{2} = \frac{5R}{2}$$

$$Mc_p = \frac{5}{2} MR$$

Thus Mc_p (monoatomic) = 20.787 kJ/kg mol K or 4.965 kcal/kg mol °K

For diatomic gases :

$$c_v = \frac{5}{2} R \text{ or } Mc_v = \frac{5}{2} MR$$

and $$c_p = \frac{5}{2} R + R = \frac{7}{2} R$$

$\therefore$ Mc_p (diatomic) = 29.1 kJ/kg mol K or 6.95 kcal/kg mol °K.

For polyatomic gases :

(Approx.) $c_v = 3R$ or $M_{cv} = 3MR$

$c_p = 4R$ or $M_{cp} = 4M_{cp}$

Thus M_{cp} (polyatomic) = 33.26 kJ/kg mol K or 7.944 kcal/kg mol °K.

It can be shown, from above relations, that the value of ratio $\frac{c_p}{c_v}$ for the gases is :

Gases	c_p/c_v
Monoatomic	5/3
Diatomic	7/5
Polyatomic	4/3

All gases, except monoatomic gases, show an increase in specific heat at high temperature as shown in Fig. 4.1.

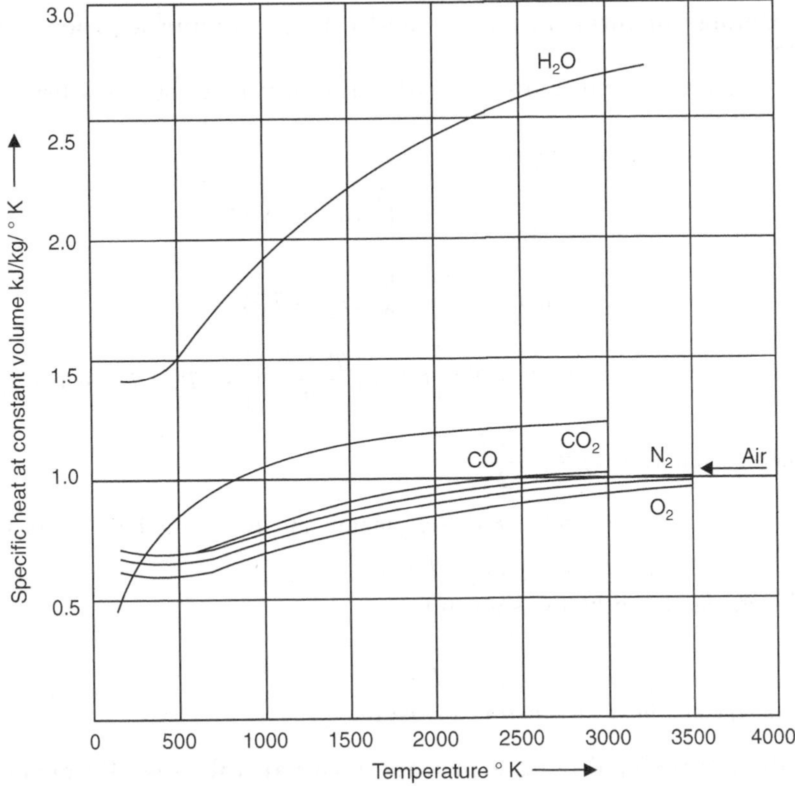

Fig. 4.1. Increase of specific heat with temperature.

Fig. 4.2 shows the effect of humidity on properties of air. R is the characteristic gas constant ; c_p and c_v are the specific heats at constant pressure and constant volume respectively, γ = ratio of specific heats.

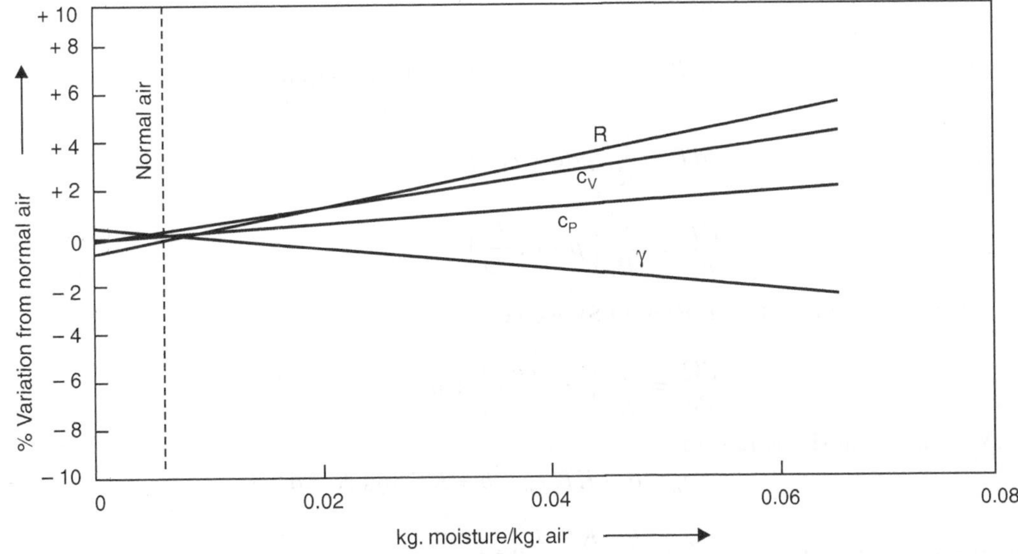

Fig. 4.2. Effect of humidity on properties of air.

4.1.5.2. Change of internal energy and ethalpy during a process with variable specific heats

The small change in internal energy (du) of a *unit mass* of a gas for small change in temperature (dT) is given by :

$$du = c_v\, dT$$

$$\therefore \quad u_2 - u_1 = \int_{T_1}^{T_2} c_v\, dT = \int_{T_1}^{T_2} (b + KT)\, dT$$

$$= b(T_2 - T_1) + \frac{K}{2}(T_2^2 - T_1^2)$$

$$= (T_2 - T_1)\left[b + K\left(\frac{T_1 + T_2}{2}\right)\right] = (T_2 - T_1)(b + KT_m)$$

where, T_m (mean temperature) $= \dfrac{T_1 + T_2}{2}$

$c_{vm} = b + KT_m$ (c_{vm} means mean specific heat at constant volume)

$$\therefore \quad u_2 - u_1 = c_{vm}(T_2 - T_1) \qquad ...(4.6)$$

Similarly, change in enthalpy is given by :

$$dh = c_p\, dT$$

$$\therefore \quad h_2 - h_1 = c_{pm}(T_2 - T_1) \qquad ...(4.7)$$

where, c_{pm} means specific heat at constant pressure.

4.1.5.3. Heat transfer during a process with variable specific heats

For a closed system, the heat-flow is expressed as :

$dQ = du + dW$ (considering 1 kg of gas)

or $\quad dQ = c_v\, dT + pdv$

$$\therefore \quad \frac{dQ}{dv} = c_v \cdot \frac{dT}{dv} + p \qquad ...(4.8)$$

Also, $\quad T = \dfrac{p \times v}{R}$ (According to gas equation)

$$\therefore \quad dT = \frac{1}{R}(pdv + vdp)$$

$$\therefore \quad \frac{dT}{dv} = \frac{1}{R}\left(p + v \cdot \frac{dp}{dv}\right)$$

Substituting this value in Eqn. (4.8), we get

$$\frac{dQ}{dv} = \frac{c_v}{R}\left(p + \frac{vdp}{dv}\right) + p$$

Now inserting the values of

$$c_p = a + KT,\ c_v = b + KT \text{ and } R = a - b$$

in the above equation, we get $\dfrac{dQ}{dv} = \dfrac{b + KT}{(a - b)}\left(p + \dfrac{vdp}{dv}\right) + p$

$$= \frac{1}{(a-b)}\left[bp + \frac{bvdp}{dv} + KTp + \frac{KTvdp}{dv} + ap - bp\right]$$

$$= \frac{b}{(a-b)}\left[v.\frac{dp}{dv} + \frac{KTp}{b} + \frac{KTv}{b}.\frac{dp}{dv} + \frac{a}{b}p\right]$$

$$= \frac{1}{\left(\frac{a}{b}-1\right)}\left(\frac{a}{b}p + v.\frac{dp}{dv}\right) + \frac{KT}{(a-b)}\left(p + v.\frac{dp}{dv}\right)$$

$$\frac{a}{b} = \frac{c_p - KT}{c_v - KT} = \gamma'$$

$$\therefore \quad \frac{dQ}{dv} = \frac{1}{(\gamma'-1)}\left(\gamma' p + v.\frac{dp}{dv}\right) + \frac{KT}{(a-b)}\left(p + v.\frac{dp}{dv}\right) \qquad ...(4.9)$$

If the expansion or compression is polytropic, then

pv^n = constant, where $n < \gamma'$.

Differentiating the above equation,

$$p\,.\,nv^{n-1}\,.\,dv + v^n\,.\,dp = 0$$

$$v\,.\,\frac{dp}{dv} = -p\,.\,n$$

Inserting this value in eqn. (4.9), the heat exchange in a polytropic process is given by

$$\frac{dQ}{dv} = \frac{1}{(\gamma'-1)}(\gamma' p - p\,.\,n) + \frac{KT}{(a-b)}(p - p\,.\,n)$$

$$= \frac{(\gamma'-n)}{(\gamma-1)}\,.\,p + \frac{KT}{(a-b)}(1-n)\,p = \left[\frac{\gamma'-n}{\gamma'-1} - \frac{(n-1)}{(a-b)}KT\right]p$$

$$\therefore \quad dQ = \left[\frac{(\gamma'-n)}{(\gamma'-1)} - \frac{(n-1)}{(a-b)}KT\right]pdv$$

$$= \left[\left(\frac{\gamma'-n}{\gamma'-1}\right) - \left(\frac{n-1}{a-b}\right)KT\right]dW \qquad ...(4.10)$$

If $K = 0$

$$dQ = \frac{\gamma'-n}{\gamma'-1}\,.\,dW.$$

4.1.5.4. Isentropic expansion with variable specific heats

The heat transfer to a system is expressed as :

$$dQ = du + dW$$

or $dQ = c_v\,dT + pdv$ (considering one kg of air)

For isentropic process, $dQ = 0$

$$\therefore \quad c_v\,dT + pdv = 0 \text{ or } c_v\frac{dT}{T} + \frac{pdv}{T} = 0$$

$\therefore \quad c_v \dfrac{dT}{T} + R \cdot \dfrac{dv}{v} = 0 \qquad \left(\because \dfrac{p}{T} = \dfrac{R}{v}\right)$

Inserting the value of R and c_v in the above equation, we get

$$(b + KT) \frac{dT}{T} + (a - b) \cdot \frac{dv}{v} = 0$$

$$\therefore \quad (b + KT) \frac{dT}{T} + (a - b) \frac{dv}{v} = 0$$

Integrating both sides, we get

$$b \log_e T + KT + (a - b) \log_e v = \text{constant}$$

or $\quad \log_e (T)^b + \log_e (e)^{KT} + \log_e (v)^{a-b} = \text{constant}$

or $\quad (T)^b (e)^{KT} (v)^{a-b} = \text{constant}$

or $\quad T.(e)^{\frac{K}{b}.T} \cdot (v)^{\left(\frac{a}{b} - 1\right)} = \text{constant} \qquad ...(4.11)$

$$\therefore \quad \frac{T}{v} \cdot (e)^{\frac{K}{b}T} . (v)^{\frac{a}{b}} = \text{constant}$$

But, $\quad \dfrac{T}{v} = \dfrac{p}{R} = \dfrac{p}{(a - b)}$

Inserting the value of $\dfrac{T}{v} = \dfrac{p}{R}$ in the above equation, we get

$$p(e)^{\frac{KT}{b}} . (v)^{\frac{a}{b}} = \text{constant}$$

or $\quad p(v)^{a/b} . e^{KT/b} = \text{constant} \qquad ...(4.12)$

4.1.5.5. 'Entropy change' during a process with variable specific heats

Let p_1, v_1, T_1 and p_2, v_2, T_2 be the initial and final conditions of the gas ; then change in entropy is given by (specific heats changing linearly with the temperature),

$$ds = \frac{dQ}{T} = \frac{pdv}{T} + \frac{c_v dT}{T}$$

$$= R \cdot \frac{dv}{v} + c_v \cdot \frac{dT}{T} \qquad \left(\because \frac{p}{T} = \frac{R}{v}\right)$$

$$= (a - b) \frac{dv}{v} + (b + KT) \frac{dT}{T}$$

Integrating between the limits, we get

$$s_2 - s_1 = (a - b) \log_e (v_2/v_1) + b \log_e (T_2/T_1) + K(T_2 - T_1) \qquad ...(4.13)$$

To express the equation in terms of p and v

$$\frac{p_1 v_1}{T_1} = \frac{p_2 v_2}{T_2} \text{ or } \frac{T_2}{T_1} = \frac{p_2 v_2}{p_1 v_1}$$

or $\quad \log_e T_2/T_1 = \log_e (p_2/p_1) + \log_e (v_2/v_1)$

Inserting this value in equation (4.13), we get

$$s_2 - s_1 = (a - b) \log_e (v_2/v_1) + b[\log_e (p_2/p_1) + \log_e (v_2/v_1)] + K(T_2 - T_1)$$

$$= a \log_e v_2/v_1 - b \log_e v_2/v_1 + b \log_e p_2/p_1 + b \log_e (v_2/v_1) + K(T_2 - T_1)$$

$$\therefore \quad s_2 - s_1 = a \log_e (v_2/v_1) + b \log_e (p_2/p_1) + K(T_2 - T_1) \quad ...(4.14)$$

To express the equation in terms of p and T

$$\log_e (v_2/v_1) = \log_e (T_2/T_1) - \log_e (p_2/p_1)$$

Inserting this value in eqn. (4.14), we get

$$s_2 - s_1 = a \log_e (T_2 - T_1) - (a - b) \log_e (p_2/p_1) + K (T_2 - T_1) \quad ...(4.15)$$

4.1.5.6. Effects of variable specific heats on air standard efficiencies of Otto and Diesel cycles

1. Otto cycle. The air standard efficiency of otto cycle is given by

$$\eta = 1 - \frac{1}{(r)^{\gamma - 1}}$$

$$= 1 - (r)^{-\frac{R}{c_v}} \qquad \left(\because \gamma - 1 = \frac{R}{c_v}\right)$$

$$\therefore \quad 1 - \eta = (r)^{-\frac{R}{c_v}}$$

Taking log on both sides, we get

$$\log (1 - \eta) = -\frac{R}{c_v} \log_e r$$

Differentiating the above equation, we have

$$-\frac{1}{1-\eta} \cdot d\eta = -R \log_e r \left(-\frac{1}{c_v^2} dc_v\right) = R \cdot \frac{dc_v}{c_v^2} \log_e r$$

The change in the efficiency with variation in specific heat is expressed as :

$$\frac{d\eta}{\eta} = -\frac{1-\eta}{\eta} \cdot \frac{R}{c_v} \cdot \frac{dc_v}{c_v} \cdot \log_e r$$

$$\therefore \quad \frac{d\eta}{\eta} = -\frac{1-\eta}{\eta} \cdot (\gamma - 1) \cdot \log_e r \cdot \frac{dc_v}{c_v} \quad ...(4.16)$$

The negative sign indicates the decrease in efficiency with increase in c_v.

The equation (4.16) gives *the percentage variation in air standard efficiency of Otto cycle on account of percentage variation in* c_v.

2. Diesel cycle. The air standard efficiency of the diesel cycle is expressed as :

$$\eta = 1 - \frac{1}{(r)^{\gamma - 1}} \left[\frac{\rho^{\gamma} - 1}{\gamma(\rho - 1)}\right]$$

$$\therefore \quad 1 - \eta = \frac{1}{(r)^{\gamma - 1}} \left[\frac{\rho^{\gamma} - 1}{\gamma(\rho - 1)}\right]$$

Taking log on both sides, we get

$$\log_e (1 - \eta) = -(\gamma - 1) \log_e r + \log_e (\rho^\gamma - 1) - \log_e \gamma - \log_e (\rho - 1)$$

Differentiating the above equation with respect to γ

$$-\frac{1}{1-\eta} \cdot \frac{d\eta}{d\gamma} = -\log_e r + \frac{\rho^\gamma \cdot \log_e \rho}{\rho^\gamma - 1} - \frac{1}{\gamma}$$

$$\therefore \qquad \frac{d\eta}{d\gamma} = (1 - \eta) \left[\log_e r - \frac{\rho^\gamma \cdot \log_e \rho}{\rho^\gamma - 1} + \frac{1}{\gamma} \right]$$

Multiplying the above equation by $\frac{d\gamma}{\eta}$

$$\frac{d\eta}{\eta} = \frac{1-\eta}{\eta} \cdot d\gamma \left[\log_e r - \frac{\rho^\gamma \cdot \log_e \rho}{\rho^\gamma - 1} + \frac{1}{\gamma} \right] \qquad ...(4.17)$$

But $\qquad c_p - c_v = R$

When R remains constant as the changes in c_p and c_v are considered with the same rate,

$$\therefore \qquad \gamma - 1 = \frac{R}{c_v}$$

Differentiating the above equation

$$d\gamma = -\frac{R}{{c_v}^2} \cdot dc_v = -\frac{R}{c_v} \cdot \frac{dc_v}{c_v} = -(\gamma - 1) \frac{dc_v}{c_v}$$

i.e.,

$$d\gamma = -(\gamma - 1) \frac{dc_v}{c_v} \qquad ...(4.18)$$

Inserting the value of $d\gamma$ from eqn. (4.18) into eqn. (4.17), we get

$$\frac{d\eta}{\eta} = -\frac{1-\eta}{\eta} (\gamma - 1) \left[\log_e r - \frac{\rho^\gamma \cdot \log_e \rho}{\rho^\gamma - 1} + \frac{1}{\gamma} \right] \frac{dc_v}{c_v} \qquad ...(4.19)$$

The eqn. (4.19) gives the percentage variation in air standard efficiency of Diesel cycle on account of percentage variation in c_v.

4.1.6. Effect of Variation of Specific Heats

The specific heats of gases increase with increase of temperature. Since the difference between c_p and c_v is constant, the *value of γ decreases as temperature increases.*

- Refer Fig. 4.3. During the *compression stroke,* if the variation of specific heats is taken in account, the final temperature and pressure would be lower than if constant specific heats are used. With variable specific heat the point at the end of compression is slightly lower, 2′ instead of 2.
- At the *end of combustion,* the pressure and temperature will be lower, represented by 3′ instead of 3. It is because of the following *reasons* :
  - — The temperature rise due to given heat release decreases as c_p increases, and
  - — The temperature at 2′ is lower than 2.

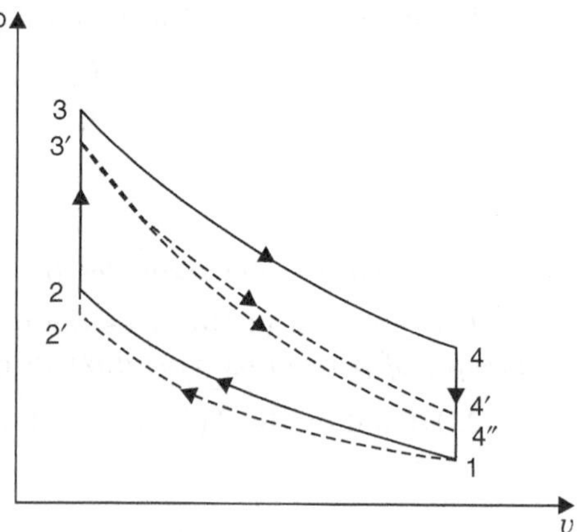

Fig. 4.3. Effect of variation of specific heats.

- The *reversible adiabatic expansion* from 3′ would be 3′-4″, but the expansion taking variable specific heat into account is above 3′-4″ and is represented by 3′-4′. The point 4′ will be below point 4 (the ideal expansion process starting from point 3 being along 3-4).

  Thus, it is seen that the *effect of variation of specific heats is to lower temperatures and pressures at point 2 and 3* and *hence to deliver less work than the corresponding cycle with constant specific heats.*

4.1.7. Dissociation

- **Dissociation (*or chemical equilibrium loss*)** *refers to disintegration of burnt gases at high temperature. It is a reversible process and increases with temperature.*
- During dissociation a considerable amount of heat is absorbed ; this heat will be liberated when the elements recombine as the temperature falls. Thus the general effect of dissociation is suppression of a part of the heat during the combustion period and liberation of it as expansion proceeds, a condition which is really identical with the effects produced by the change in specific heat. However, *the effect of dissociation is much smaller than that of change of specific heat.*
- *The dissociation, in general, lowers the temperature and, consequently, the pressures at the beginning of the stroke, this causes a loss of power and efficiency.*
- The dissociation is mainly of CO_2 into CO and O_2 ;

  $$2CO + O_2 \rightleftharpoons 2CO_2 + \text{heat}$$

  The dissociation of CO_2 commences at about 1000°C and at 1500°C it amounts to 1 per cent.

  There is very little dissociation of H_2O ;

  $$2H_2 + O_2 \rightleftharpoons 2H_2O + \text{heat}$$

- *Dissociation is more severe in the chemically correct mixture.* If the mixture is *weaker*, it *gives temperatures lower than those required for dissociation to take place* while if it is *richer*, during combustion it will *give out CO and O_2 both of which suppress the dissociation of CO_2.*

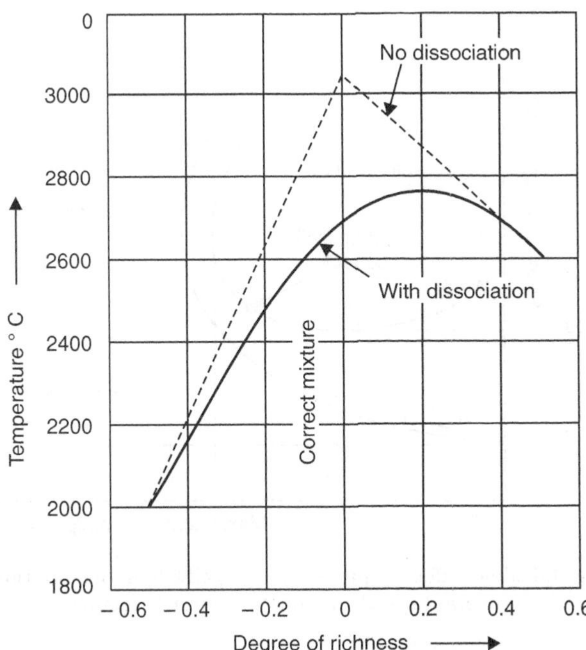

Fig. 4.4. Effect of dissociation on temperature at different mixture strength.

- *The dissociation has a more pronounced effect in* **S.I. engines.** In **C.I. engines,** *the heterogeneous mixture of air and fuel tend to lower the temperature and hence the dissociation.*

Fig. 4.4 shows the effect of dissociation on temperature at different mixture strength.

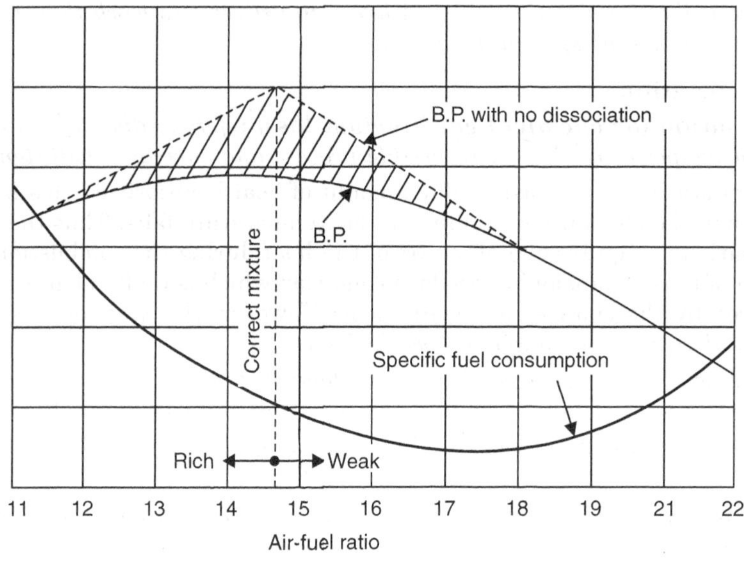

Fig. 4.5. Effect of dissociation on power.

Fig. 4.5 shows the effect of dissociation on power.

4.1.8. Thermal Efficiency and Fuel Consumption

Whereas air standard theory (simple) predicts no variation of thermal efficiency with mixture strength, *fuel-air analysis,* however, *suggests that the thermal efficiency will deteriorate as*

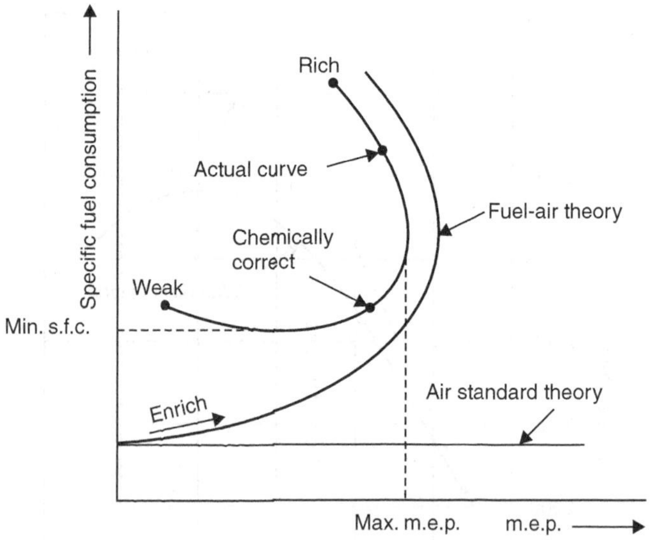

Fig. 4.6. Mean effective pressure *vs.* specific fuel consumption at constant speed and constant throttle setting.

the mixture supplied to an engine is enriched. This is due to the *increasing losses owing to variable specific heats and dissociation* as the engine temperatures are raised by enrichment towards the chemically correct ratio. If the enrichment continues beyond the chemically correct ratio it results in the supply of unusable excess fuel leading to rapid fall in the thermal efficiency. It would, therefore, appear that thermal efficiency would increase as the mixture is weakened. However, beyond a certain weakening the combustion becomes erratic which results in loss of efficiency. Thus the maximum efficiency lies within the weak zone near chemically correct ratio. Fig. 4.6 shows a plot between mean effective pressure (m.e.p.) *vs.* specific fuel consumption (s.f.c.) at constant speed and constant throttle setting.

4.1.9. Effect of Common Engine Variables

The effect of common engine variables on the temperature and pressure within the engine cylinder can be clearly understood by fuel-air analysis, as discussed below.

1. **Compression ratio :**

- The fuel-air cycle efficiency increases with compression ratio in the same manner as the air standard efficiency, this is shown in Fig. 4.7.

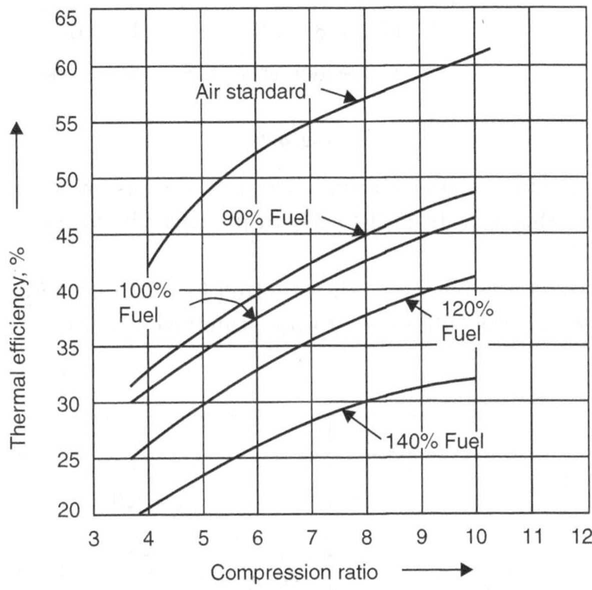

Fig. 4.7. Effect of compression ratio and mixture strength on thermal efficiency.

— It may be observed from Fig. 4.7 that the curve for **90%** of the theoretical fuel is higher than that for **100%** theoretical fuel, and with still higher percentage of fuel, the thermal efficiency drops. It also shows that *maximum efficiency is obtained with leaner mixtures.*

- Fig. 4.8 very clearly depicts the effect of mixture strength on thermal efficiency for various compression ratios (*r*).

— It shows that $\eta_{th.\ (I)}$ at a particular compression ratio, for the fuel-air cycle is higher for lean mixtures and *falls constantly as the mixture becomes rich,* till chemically correct fuel-air ratios are reached, and further falls more rapidly with the further enriching of the fuel beyond chemically correct mixture ratio.

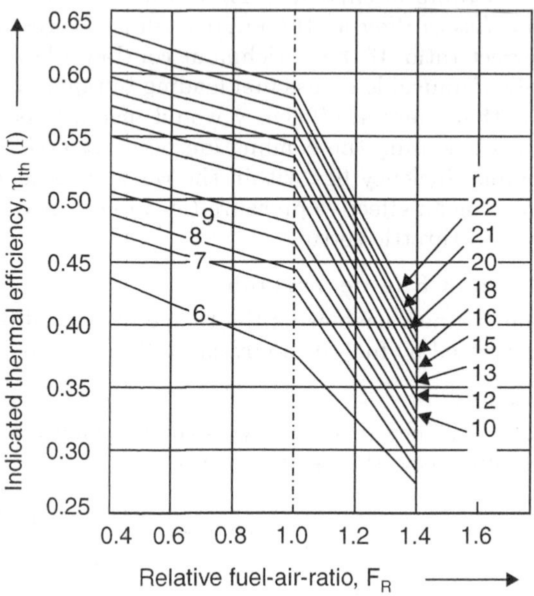

Fig. 4.8

— The $\eta_{th.\ (I)}$ is increased with increase in compression ratio at a fixed fuel-air ratio. The range shown is from 6 to 24 (In practice the value of *r* in S.I. engine hardly exceeds 10).

- As shown in Fig. 4.9, when a graph is plotted for *F/A* ratio (F_R) against ratio of fuel-air cycle indicated thermal efficiency to air standard efficiency, most revealing result is obtained.

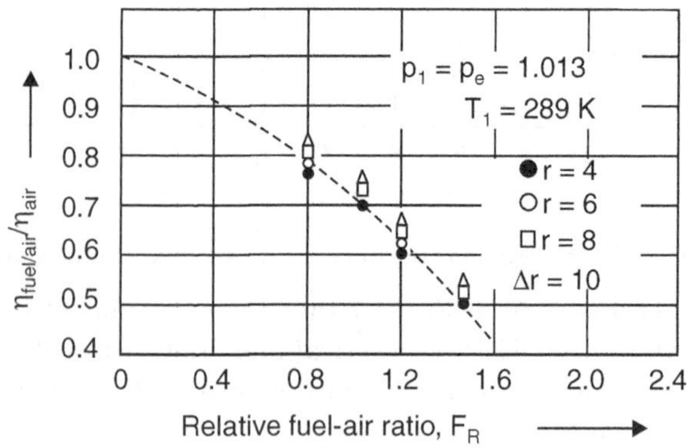

Fig. 4.9

— It is worth noting that for a particular value of F_R, the ratio of fuel-air cycle indicated thermal efficiency and air standard efficiency there is no variation of specific heat and no dissociation and therefore when $F_R = 1$, the maximum temperature must reach. But *by fuel-air cycle concept the maximum temperature is shifted to richer values of relative fuel-air ratios.*

In view of the above, the air standard cycle concept for predicting the performance of S.I. engines is misleading, whereas the fuel-air cycle concept seems to be very reliable.

2. **Fuel-air ratio :**

(i) Efficiency

- It has been experimentally evaluated that the $\eta_{th.(I)}$ is highest at lean fuel-air mixtures of the order of $F_R = 0.85$ (As the mixture is made lean, due to less energy input the temperature rise during the combustion will be less which results in lower specific heat and eventually lower chemical equilibrium losses. This results in higher efficiency and as the fuel-air ratio is reduced, it approaches the air-cycle efficiency as illustrated in Fig. 4.10).
- As shown in Fig. 4.10, in the range of the mixture ratios of operation, for S.I. engines, usually $F_R = 0.6$ to 1.4, the fuel-air cycle very closely resembles the actual curve (experimental). The air-standard cycle concept miserably fails, *not influenced by the fuel-air ratio.*

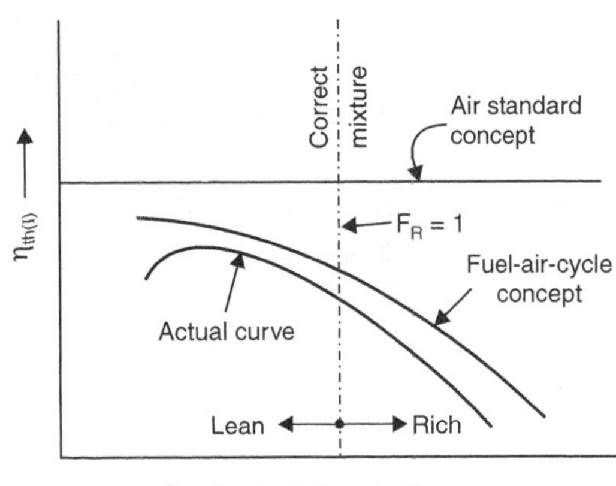

Fig. 4.10. Effect of mixture strength of $\eta_{th(I)}$ at a given compression ratio.

(ii) Maximum power :

Fig. 4.11 shows the effect of mixture strength on cycle power.

- According to air-standard theory maximum power is at chemically correct mixture whereas by fuel-air theory maximum power is obtained when the mixture is about 10 per cent rich. The *efficiency drops rapidly as the mixture becomes enriched* ; this is due to the following *reasons* :
  — Losses due to higher specific heats ;
  — Chemical equilibrium losses ;
  — Insufficient air which will result in formation of CO and H_2 in combustion, representing direct fuel wastage.

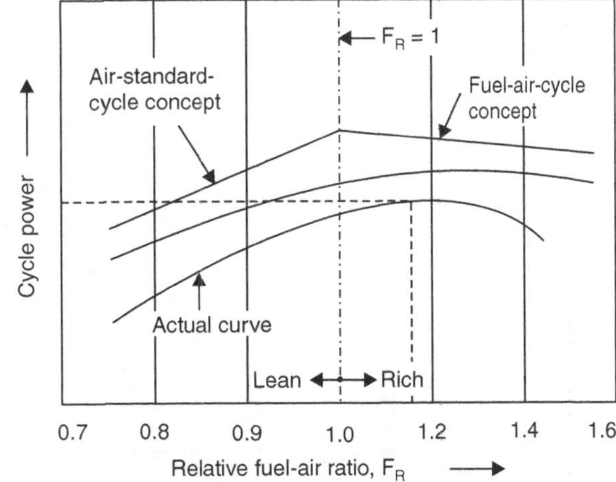

Fig. 4.11. Effect of mixture strength on cycle power.

(iii) Maximum temperature :

Fig. 4.12 shows the effect of F_R on maximum cycle temperature $T_3(K)$ at different compression ratios.

— The maximum temperature at a given compression ratio is reached when the mixture is slightly rich, 6% or so as shown in Fig. 4.12.

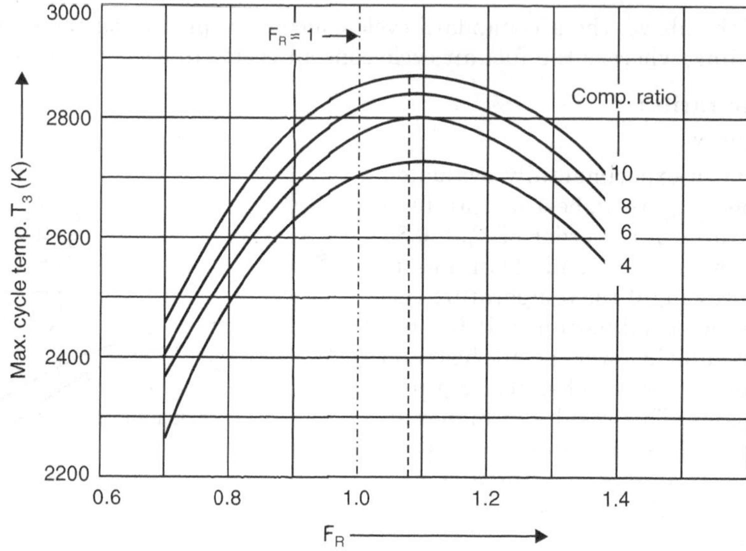

Fig. 4.12. Effect of F_R on maximum cycle temperature T_3 (K) at different compression ratios.

— In case of a chemically correct ratio there is still some oxygen present at the point 3 because of chemical equilibrium effect and hence rich mixture will cause more fuel to combine with oxygen at point 3, raising the temperature T_3. However, at richer mixtures more formation of CO overcomes the effect of more combustion and eventually reduces T_3.

(iv) Maximum pressure :

Fig. 4.13 shows the effect of F_R on maximum cycle pressure p_3 (bar) at different compression ratios.

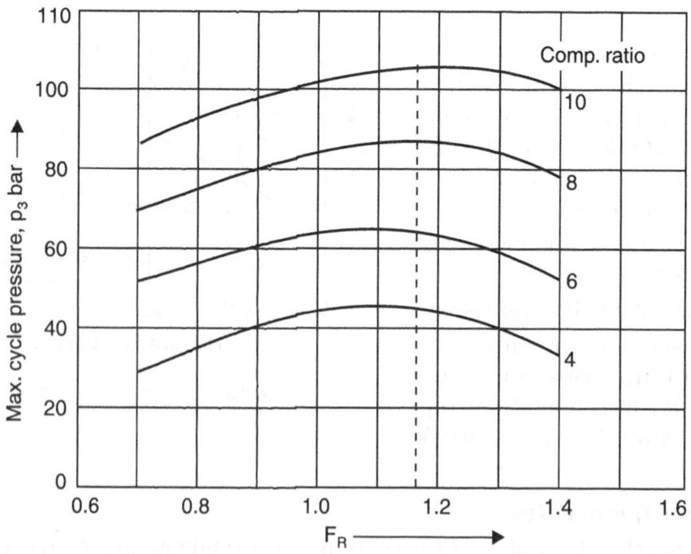

Fig. 4.13. Effect of F_R on maximum cycle pressure p_3 (bar) at different compression ratios.

— It follows more or less the same pattern as the maximum temperature versus F_R. *But the maximum occurs at slightly higher value of relative fuel-air ratios as compared to that of temperature.* This is *owing to molecular expansion.* There is increase of the mole of products after combustion.

(v) Exhaust temperature :

Fig. 4.14 shows effect of F_R on cycle exhaust gas temperature, T_4. *The exhaust gas temperature is maximum at the chemically correct mixture.*

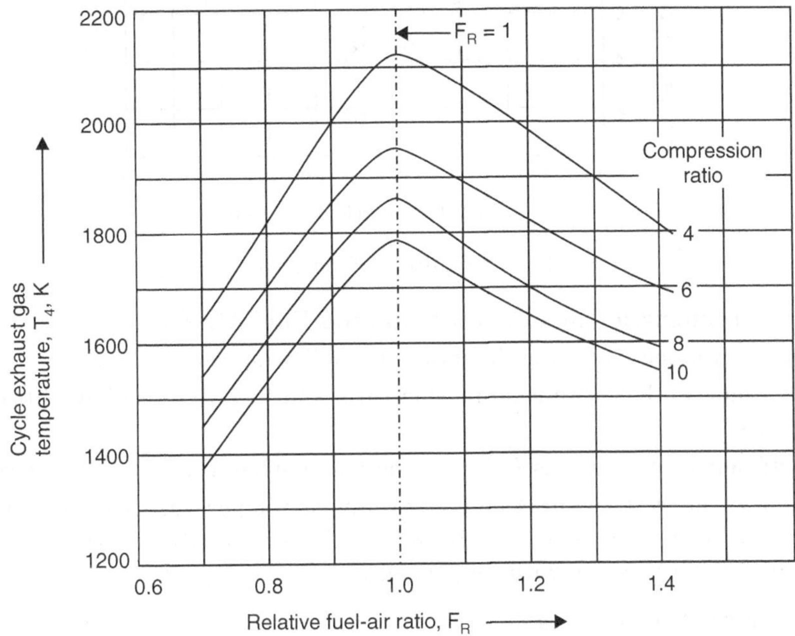

Fig. 4.14. Effect of F_R on T_4.

- The variation in exhaust temperatures at a fixed value of F_R shows that it decreases with increase in compression ratios. This is due to the fact *that increased expansion ratio causes the gases to do more work on the piston leaving less heat rejection at the end of the stroke.*
- The results are similar, for variation of compression ratio, for the air-standard cycle as well.

(vi) Mean effective pressure :

Fig. 4.15 shows the effect of F_R on m.e.p.

- m.e.p. *increases with compression* ratio as efficiency increases.
- It follows closely the maximum pressure curve, but the m.e.p. occurs at slightly less richer mixture ratios as compared to that for pressure.

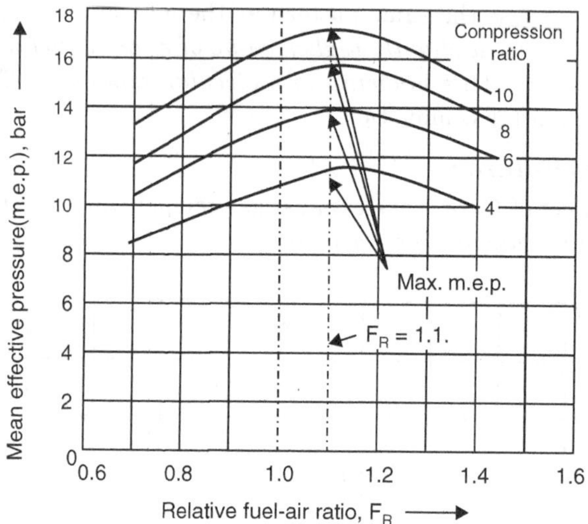

Fig. 4.15. Effect of F_R on m.e.p.

4.1.10. Characteristics of Constant Volume Fuel-Air Cycle

The constant volume fuel-air cycle entails the following *characteristics* :

1. The variables other than compression ratio and fuel-air ratio have little effect on the efficiency.
2. The efficiency decreases as *F/A* ratio increases, when the latter (*F/A* ratio) is variable.
3. When F_R is above 1.1 approximately, expansion temperature decreases with increasing *F/A* ratio, but the combustion is incomplete because mixture is rich and the net result is decrease in efficiency with increasing F_R.

4.1.11. Combustion Charts

- The thermodynamic charts embodying characteristics of cylinder gases are employed for computing fuel-air cycles, avoiding laborious calculations.
- *There are separate charts for fuel-air mixture and for products of combustion.* The **products of combustion charts** *are for different fuel-air ratio* (say F_R = 0.8 to 1.2) and are *used for calculation of pressure, temperature, volume and energy at various points in the cycle after burning has taken place.*
- The **chart for unburned mixture** is employed for *calculation of pressure, temperature, and energy of the cylinder contents before burning takes places.*

4.1.12. Gas Tables

The problems involving variable specific heats can be solved by the following *methods* :

1. By *integration of specific heat equation.* This method is tedious and time consuming.
2. By *enthalpy-entropy charts.* The major shortcoming of this method is large enough charts which can yield fairly accurate results are not available.
3. By **gas tables.** (Gas tables for air are given in Appendix)

— The gas tables which take into consideration the variation of specific heat with temperature, give the enthalpy, internal energy and energy function and have been computed and compiled for many gases and mixture of gases, including air. These tables are constructed assuming enthalpy and internal energy to be zero at absolute zero temperature and integrating zero pressure specific heat equations from 0(K) to the given temperature T (K).

— The most important *assumption* is the validity of the equation $pv = RT$, where p, v, R and T denote pressure, specific volume, the characteristic gas constant and the absolute thermodynamic temperature respectively. This is *true when the gas has a critical temperature very low as compared to the range of temperature* met with in engineering applications. For air at 0°C and 20 bar the deviation is only 1 per cent ; and at 0°C and 1 bar, the deviation is 0.1 per cent.

— The enthalpy and internal energy are function of temperature and, therefore, their values can be computed with *single variable property, i.e., temperature.*

Thus enthalpy h and internal energy u at any temperature T (K), are given by :

$$h = \int_0^T c_p \, dT$$

$$u = \int_0^T c_v \, dT$$

The entropy change involves both variables namely pressure and temperature.

In gas tables, h, p_r, u, v_r and ϕ are recorded for different values of temperature T (K).

Relative pressure, p_r :

$$T\,ds = dh - v\,dp = c_p\,dT - v\,dp \qquad \text{...(4.20)}$$

$(\because \; dh = c_p dT)$

For isentropic process, $ds = 0$, we get

$$0 = c_p\,dT - v\,dp \quad \text{or} \quad v\,dp = c_p\,dT$$

Dividing by $pv = RT$, we get

$$\frac{v\,dp}{pv} = \frac{c_p\,dT}{RT} \quad \text{or} \quad \frac{dp}{p} = \frac{c_p}{R} \cdot \frac{dT}{T}$$

or

$$\ln\left(\frac{p}{p_0}\right) = \frac{1}{R}\int_{T_0}^{T} c_p \cdot \frac{dT}{T} = \ln(p_r) \qquad \text{...(4.21)}$$

where T_0 is selected as a base temperature. It is seen that ratio $\frac{p}{p_0}$ is a function of temperature only, and is independent of the value of entropy.

— From eqn. (4.21) p_r can be calculated in terms of temperature.

On an isentropic path, for two states 1 and 2, we have

$$\frac{p_{r_1}}{p_{r_2}} = \frac{p_1/p_0}{p_2/p_0} = \frac{p_1}{p_2} \qquad \text{...(4.22)}$$

Thus, the ratio of relative pressure for two states having the same entropy is equal to the ratio of the absolute pressures for the same two states ; p_0 is chosen as *unity* for computed values.

Relative volume, v_r :

$$T\,ds = du + pdv \qquad \text{...(4.23)}$$

For an isentropic process, we have

$$0 = du + pdv$$

or

$$pdv = -\,du = -\,c_v dT$$

Dividing by $pv = RT$, we get

$$\frac{pdv}{pv} = -\frac{c_v dT}{RT}$$

$$\ln\left(\frac{v}{v_0}\right) = -\frac{1}{R}\int_{T_0}^{T} c_v \cdot \frac{dT}{T} = \ln(v_r) \quad ...(4.24)$$

v_r (designated as relative volume) is a function of temperature only.

Alternatively, $$v_r = \frac{RT}{p_r} \quad ...(4.25)$$

Thus for any two states on the isentropic,

$$\frac{v_{r_1}}{v_{r_2}} = \frac{v_1}{v_2} \quad ...(4.26)$$

Entropy function, ϕ :

$$Tds = c_p\, dT - v\, dp$$

$$\therefore \quad ds = \frac{c_p\, dT}{T} - \frac{vdp}{T} \quad \text{and} \quad \frac{v}{T} = \frac{R}{p}$$

$$\therefore \quad ds = c_p \cdot \frac{dT}{T} - R \cdot \frac{dp}{p}$$

If the zero for entropy is taken at T_0 temperature and p_0 pressure, where p_0 = unity, we have

$$s = \int_{T_0}^{T} c_p \cdot \frac{dT}{T} - R\ln(p)$$

$$= \phi - R\ln(p) \quad ...(4.27)$$

where $$\phi = \int_{T_0}^{T} c_p \frac{dT}{T} \quad ...(4.28)$$

ϕ is, therefore, a function of temperature only.

$$s_2 - s_1 = (\phi_2 - \phi_1) - R\ln\left(\frac{p_2}{p_1}\right) \quad ...[4.29\ (a)]$$

$$= \phi_2 - \phi_1 - R\ln\left(\frac{p_{r_2}}{p_{r_1}}\right) \quad ...[4.29\ (b)]$$

The properties of air are computed by assuming the following properties of air :

Gas	*Molecular weight*	*Percentage by volume*
Oxygen	32.00	20.99
Nitrogen	28.016	78.03
Argon	39.95	0.98

The molecular weight of air and gas constant of air work out to be 28.970 and 0.287 kJ/kg K respectively.

The enthalpy, relative pressure, and ϕ of air are computed by combining the corresponding values of the three constituents of air according to the law

$$h = \Sigma\, x_1 h_1 \quad ...(4.30)$$

$$\phi = \Sigma\, x_1\, \phi_1 \quad ...(4.31)$$

$$\ln (p_r) = \Sigma\, x_1 \ln (p_{r_1}) \quad ...(4.32)$$

where x_1 denotes the mol. fraction of gas 1 and the summation is carried over all constituent gases.

The internal energy is calculated by using the following formula :

$$h = u + pv = u + RT$$

or $$u = h - RT = h - 0.287\, T.$$

4.2. ACTUAL CYCLES

4.2.1. Introduction

In actual engine operation the following *losses occur,* due to which actual cycle efficiency is much lower than the air standard efficiency :

1. Dissociation losses
2. Losses due to variation of specific heats with temperature
3. Time losses
4. Losses due to incomplete combustion
5. Direct heat losses
6. Exhaust blowdown losses
7. Pumping losses.

- *If losses due to variable specific heats and dissociation are subtracted from the "air standard cycle", we get "fuel-air cycle analysis".*
- Furthermore, if *other losses are further subtracted from "fuel-air cycle analysis" we can very closely approximate the "actual cycle".*

4.2.2. Causes of Deviation of Actual Cycles from Fuel-Air Cycles

Important *causes of deviation of actual cycles from fuel-air cycles are* :

1. The progressive combustion rather than the instantaneous combustion.
2. The heat transfer to and from the working medium during compression and expansion.
3. Loss of work on the expansion stroke due to early opening of the exhaust valve, and exhaust blow down.
4. Gas leakage, fluid friction etc.

4.2.3. Real Fuel-Air Engine Cycles

The actual cycle which an I.C. engine experiences is not a thermodynamic cycle, in a true sense. *An ideal air-standard thermodynamic cycle occurs on a closed system of constant composition.* This is not what actually happens in an I.C. engine, and for this reason air-standard analysis gives, at best, only approximation to actual conditions and outputs.

Major differences are listed below :

1. *Real engines operate on an open cycle with changing composition.* Not only does the inlet gas composition differ from what exits, but often the mass flow rate is not the same.

— During combustion, total mass remains about the same but molar quantity changes.

— There is a *loss of mass* during the cycle due to *crevice flow and blowby past the pistons.* Most of the crevice flow is temporary loss of mass from the cylinder, but because it is greatest at the start of power stroke some output work is lost during expansion. Blowby can decrease the amount of mass in the cylinders by as much as 1% during compression and combustion.

2. Air-standard analysis treats the fluid flow through the entire engine as air and *approximates air as an ideal gas. In a real engine inlet flow may be all air, or it may be mixed up with 7% fuel, either gaseous or as liquid droplets, or both.*

— In air-standard analysis, even if all fluid in a engine cycle were air, *some error would be introduced by assuming it to be an ideal gas with constant specific heats. At the low pressures of inlet and exhaust, air can accurately be treated as an ideal gas, but at the higher pressures during combustion, air will deviate from ideal gas behaviour.* A more serious error is introduced by assuming constant specific heats for the analysis. Specific heats of a gas have a fairly *strong dependency on temperature and can vary as much as 30% in the temperature range of an engine.*

3. During the cycle of a real engine there are heat losses which are *neglected* in air-standard analysis.

— Loss of heat during combustion lowers actual peak temperature and pressure from what is predicted. The actual power stroke, therefore, starts at a lower pressure, and work output during expansion is decreased.

— Heat transfer continues during expansion, and this lowers the temperature and pressure below the ideal isentropic process towards the end of the power stroke. The result of heat transfer is a lower indicated thermal efficiency than predicted by air standard analysis.

— Heat transfer is also present during compression, which deviates the process from isentropic. However, this is less than during the expansion stroke due to the lower temperatures at this time.

4. Combustion requires a short but finite time to occur, and heat addition is not instantaneous at T.D.C.

— S.I. and C.I. engines generally have combustion efficiencies of about 95% and 98% respectively.

5. The blowdown process requires a finite real time and a finite cycle time, and does not occur at constant volume as in air-standard analysis. For this reason, the exhaust valve must open 40° to 60° before B.D.C., and output work at the latter end of expansion is lost.

6. The intake valve, in an actual engine, is not closed until after B.D.C. at the end of the intake stroke. Because of the flow restriction of the valve, air is still entering the cylinder at B.D.C., and volumetric efficiency would be lower if the valve is closed here. Because of this, however, actual compression does not start at B.D.C. but only after the inlet valve closes. With ignition then occuring before T.D.C., temperature and pressure rise before combustion is *less* than predicted by air-standard cycles.

7. Engine valves require a finite time to actuate.

4.2.4. Difference between Real Cycle and Fuel-Air Cycle

Assuming cycle to consist of compression and expansion strokes only, the differences between a real cycle and its equivalent fuel-air cycle are due to following *factors* :

1. Time losses (Including combustion loss)
2. Direct heat loss
3. Exhaust blowdown loss
4. Pumping loss
5. Rubbing friction loss.

1. **Time losses :**

Time losses may be *burning time loss* and *spark timings loss.*

(*a*) ***Burning time loss***

The burning time loss or merely *time loss is defined as the loss of power due to time required for mixing the fuel with air and for complete combustion.*

- In theoretical cycles the burning is *assumed to be instantaneous,* whereas in actual cycles the *burning process is completed in a finite interval of time.* The time required depends upon :
  (*i*) Fuel-air ratio ;
  (*ii*) Fuel chemical structure and its ignition temperature ;
  (*iii*) The flame velocity and the distance from the ignition point to the opposite side of the combustion chamber.
- The time required for combustion is such that under all circumstances some increase in volume takes place. The time interval between the passage of spark and completion of flame travel across the charge is approximately 40° crank rotation.

Fig. 4.16 shows the losses between real cycle and its equivalent fuel-air cycle.

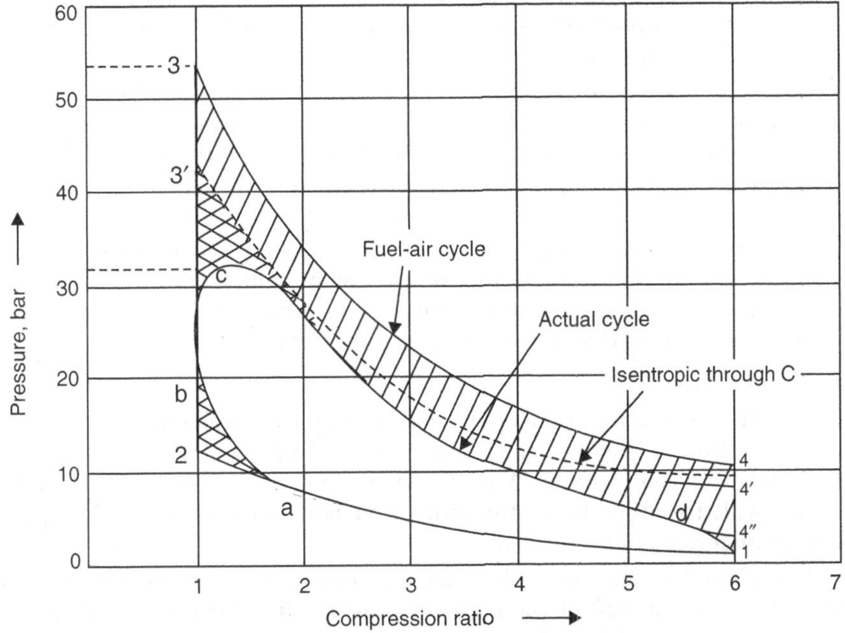

Fig. 4.16. The effect of time losses on p-v diagram.

— The effect of finite time being required for combustion is that the maximum pressure is not produced when the volume is minimum, as is expected. It is produced *sometime after T.D.C.* The pressure therefore rises in the first part of the working stroke from b to c, as shown in Fig. 4.16. The point 3 represents the maximum pressure had the combustion been instantaneous.

— The difference in area of actual cycle and fuel-air cycle shows the loss of power (the hatched-area).

(*b*) ***Spark timing loss***

After generation of spark in the cylinder, a definite time is required to start the burning of fuel. The effect of this is that maximum pressure is not reached at T.D.C. and it reaches late in the expansion stroke. *The time at which burning starts varies by varying the angle of advance (spark advance).*

(*i*) If the spark is given at T.D.C., the maximum pressure is low due to expansion of gases.

(*ii*) If the spark is advanced by 40° to start combustion at T.D.C., the combustion takes place at T.D.C. But the heat loss and the exhaust loss may be higher and again work obtained is not optimum.

In the above two cases, the work area is less, and, therefore, *power developed per cycle and efficiency are lower.*

Thus for *getting maximum work output, a moderate spark advance of 15° to 25° is the best.*

Incomplete combustion losses

— The loss due to incomplete combustion is included in time loss. It is not possible to get homogeneous fuel-air mixture inside the engine cylinder as fuel, air and residual gases are present in the engine cylinder before the start of ignition. There may be excess oxygen in one part and excess fuel in another part of the cylinder. Therefore, *incomplete combustion takes places in the region of excess fuel, and CO and O_2 both will appear in the exhaust gases.*

— *It is observed that energy release in S.I. engine is only about 95 per cent of the energy release when complete combustion would take place with near stoichiometric fuel-air ratio.* In actual engine, energy release is about 90 to 93 per cent of fuel energy input.

— It is always preferable to use a lean mixture to eliminate fuel waste, while a rich mixture is required to utilize all the oxygen. *Slightly lean mixtures give maximum efficiency, but too lean mixture will burn slowly, increase burning loss or may not burn causing total fuel loss.* In rich mixture some fuel is definitely wasted as adequate amount of oxygen is not available. *The flame speed in rich mixture is low and causes burning time loss leading to lowering of efficiency.*

2. **Direct heat loss :**

- When combustion of fuel takes place followed by the expansion stroke, the flow of heat takes place from cylinder gases through the cylinder walls and cylinder head into the water jacket or cooling fins. A part of heat enters the piston head and flows to the cylinder wall through the piston rings and is carried away by the engine oil which splashes underside of the piston.
- The loss of heat which takes place during combustion has the maximum effect, while that lost before the end of the expansion stroke has little effect, since it can do very small amount of useful work.
- During combustion and expansion, about 15% of the total heat is lost. Out of this, however, much is lost too late in the cycle to have done any useful work.
- In case all heat loss is recovered, about 20 per cent of it may appear as useful work.

3. **Exhaust blowdown loss :**

At the end of exhaust stroke, the cylinder pressure is about 7 bar. If the exhaust valve is opened at B.D.C., the piston has to do work against high cylinder pressure costing part of the exhaust stroke. When the exhaust valve is opened too early entire part of the expansion stroke is lost. Thus, best compromise is that exhaust valve be opened 40° to 70° before B.D.C., thereby, reducing the cylinder pressure to halfway to atmosphere before the start of the exhaust stroke.

4. **Pumping losses :**

- *The pumping loss is due to pumping gas from low inlet pressure to higher exhaust pressure.*
- The pumping loss *increases at part throttle* because throttling causes reduction in suction pressure.
- *Pumping loss increases with increase in speed.*

5. **Rubbing friction loss :**

The rubbing friction losses are caused due to :

(*i*) Friction between pistion and cylinder walls ;

(*ii*) Friction in various bearings ;

(*iii*) Friction in auxiliary equipment such as pumps and fans.

— The *piston friction increases rapidly with engine speed and to small extent by increases in m.e.p.*

— The bearing and auxiliary friction also increase with engine speed.

The engine efficiency is maximum at full load and reduces with the decrease in load. It is due to the fact that direct heat loss, pumping loss and rubbing friction loss increase at lower loads.

4.2.5. Comparison of Operations and Working Media for 'Air cycle', 'Fuel-air Cycle' and 'Actual Cycle' of S.I. Engines

1. **Air cycle :**

- The working medium is air throughout the cycle. It is assumed to be an ideal gas with constant properties.
- The working medium does not leave the system, and performs cyclic processes.
- There are not inlet and exhaust strokes.
- The compression and expansion processes are isentropic.
- The heat addition and rejection are instantaneous at T.D.C. and B.D.C. respectively, at constant volume.

2. **Fuel-air cycle :**

- The cylinder gases contain fuel, air, water vapour and residual gases.
- The fuel-air ratio changes during the operation of the engine which changes the relative amounts of CO_2, water vapour etc.
- The variations in the values of specific heat and γ with temperature, the effects of dissociation, and the variations in the number of molecules before and after combustion are considered.

Besides taking the above factors into consideration, the following *assumptions* are commonly made for the operation :

(*i*) No chemical change prior to combustion.

(*ii*) Charge is always in equilibrium after combustion.

(*iii*) Compression and expansion processes are frictionless, adiabatic.

(*iv*) Fuel completely vaporised and mixed with air.

(*v*) Burning takes place instantaneously, at constant volume, at T.D.C.

The fuel air cycle gives a very good estimate of the actual engine with regards to efficiency, power output, peak pressure, exhaust temperature etc.

3. **Actual cycle :**

- The working substance is a mixture of air and fuel vapour, with the products of combustion left from the previous cycle.
- The working substance undergoes change in the chemical composition.
- Variation in specific heats take place. Also the temperature and composition changes due to residual gases occur.
- The combustion is progressive rather than instantaneous.
- Heat transfer to and from the working medium to the cylinder walls take place.
- Exhaust blowdown losses *i.e.* loss of work due to early opening of the exhaust valves take place.
- Gas leakage and fluid friction are present.

WORKED EXAMPLES

Example 4.1. *Deduce an expression for the change in entropy of 1 kg of gas in terms of the temperature T_1 and T_2 before and after compression, if the law of compression is pv^n = constant and the specific heats are of the form,*

$$c_p = a + KT$$
$$c_v = b + KT.$$

Solution. We know that, $dQ = du + dW$

$$= c_v\, dT + pdv$$

Dividing both sides by T, $\dfrac{dQ}{T} = c_v \cdot \dfrac{dT}{T} + \dfrac{pdv}{T}$

or
$$ds = c_v \frac{dT}{T} + \frac{P}{T} \cdot dv$$

Inserting the value of c_v, $\quad ds = b\dfrac{dT}{T} + KdT + \dfrac{p}{T}\, dv$ ...(*i*)

Also,
$$pv = RT$$
$\therefore$
$$pdv + vdp = RdT$$
and
$$pv^n = C \quad ...(ii)$$
$\therefore$
$$p \cdot n(v)^{n-1} \cdot dv + v^n \cdot dp = 0$$
$$npdv = -\, vdp \quad ...(iii)$$

From (*ii*) and (*iii*), we get

$$pdv - npdv = RdT$$
$$pdv = \frac{RdT}{1-n}$$

Inserting in (*i*), we get $\quad ds = b\dfrac{dT}{T} + KdT + \dfrac{R}{(1-n)} \cdot \dfrac{dT}{T} = \left[b - \dfrac{R}{(n-1)}\right]\dfrac{dT}{T} + KdT$

Integrating, we get $\quad s_2 - s_1 = \left[b - \dfrac{R}{(n-1)}\right] \log_e \dfrac{T_2}{T_1} + K(T_2 - T_1)$

Substituting, $\quad R = a - b,$

or
$$s_2 - s_1 = \left[\mathbf{b} - \frac{\mathbf{a - b}}{\mathbf{(n-1)}}\right] \mathbf{log}_e \frac{\mathbf{T_2}}{\mathbf{T_1}} + \mathbf{K(T_2 - T_1)}. \quad \textbf{(Ans.)}$$

Example 4.2. *Determine the effect of percentage change in the efficiency of Otto cycle having a compression ratio of 8, if the specific heat at constant volume increases by 1.1 percent.*

Solution. *Given* : Compression ratio, $r = 8$;

Increase in specific heat at constant volume, $\frac{dc_v}{c_v} = 1.1\%$

Percentage change in Otto cycle efficiency, $\frac{d\eta}{\eta}$ **:**

The Otto cycle efficiency (η) is given by :

$$\eta = 1 - \frac{1}{(r)^{\gamma - 1}}$$

Now, $$c_p - c_v = R, \qquad \gamma - 1 = \frac{R}{c_v}$$

$\therefore$ $$\eta = 1 - (r)^{-(\gamma - 1)} = 1 - (r)^{-R/c_v}$$

or $$(1 - \eta) = (r)^{-R/c_v}$$

or $$\ln (1 - \eta) = -\frac{R}{c_v} \ln (r)$$

Differentiating both sides, we have

$$-\frac{1}{1-\eta} \cdot d\eta = \frac{R}{c_v^2} \ln (r) \times dc_v$$

or $$d\eta = -\frac{(1-\eta)R \ln (r)}{c_v^2} dc_v$$

or $$d\eta = -\frac{(1-\eta)(\gamma - 1)c_v \times \ln (r)}{c_v^2} \cdot dc_v$$

$$\frac{d\eta}{\eta} = -\frac{(1-\eta)(\gamma - 1) \ln (r)}{\eta} \cdot \frac{dc_v}{c_v}$$

Now $$\eta = 1 - \frac{1}{(8)^{1.4 - 1}} = 0.565$$

$\therefore$ $$\frac{d\eta}{\eta} = -\frac{(1 - 0.565)(1.4 - 1) \ln (8)}{0.565} \times \frac{1.1}{100}$$

$$= -\ \mathbf{0.704\%\ (decreased).}\quad \mathbf{(Ans.)}$$

Example 4.3. *Find the percentage change in efficiency of an Otto cycle for a compression ratio of 7 to 1 if the specific heat at constant volume increases by 3%.*

Solution. The change in efficiency with variation in specific heat is given by :

$$\frac{d\eta}{\eta} = -\frac{1-\eta}{\eta} (\gamma - 1) \log_e r \frac{dc_v}{c_v} \qquad \text{...}(i)$$

Also, $$\eta = 1 - \frac{1}{(r)^{\gamma - 1}} = 1 - \frac{1}{(7)^{1.4 - 1}} = 0.541 \text{ or } 54.1\%$$

Inserting this value in the equation (*i*), we get

$$\frac{d\eta}{\eta} = -\left(\frac{1-0.541}{0.541}\right)(1.4-1)\log_e 7 \times 0.03$$

$$= -\mathbf{0.0198 \text{ or } -1.98\%.} \quad \textbf{(Ans.)}$$

Negative sign indicates *decrease* in efficiency.

Example 4.4. *The following particulars relate to a Diesel cycle :*

Compression ratio = 18, cut-off = 5% of stroke, mean specific heat c_v for cycle = 0.71 kJ/kg K, characteristic gas constant = 0.285 kJ/kg K.

Knowing that the specific heat increases with temperature, if the mean specific heat for the air standard cycle increases by 2 per cent determine the percentage change in the air standard efficiency.

Solution. Compression ratio, $r = 18$

Cut-off = 5% of stroke

Mean specific heat for cycle, $c_v = 0.71$ kJ/kg K

Characteristic gas constant, $R = 0.285$ kJ/kg K

Increase in c_v = 2%.

Change in air-standard efficiency :

The percentage variation in the air standard efficiency on account of the percentage variation in the value of c_v is given by

$$\frac{d\eta}{\eta} = -\left(\frac{1-\eta}{\eta}\right)(\gamma-1)\left[\log_e r - \frac{\rho^\gamma \log_e \rho}{\rho^\gamma - 1} + \frac{1}{\gamma}\right]\frac{dc_v}{c_v} \quad ...(i)$$

[Article 4.1.5.6, Eqn. (4.19)]

Now, $$\rho = \frac{\%\text{ cut-off}}{100}(r-1)+1 = \frac{5}{100}(18-1)+1 = 1.85$$

and $$\gamma = 1 + \frac{R}{c_v} = 1 + \frac{0.285}{0.71} = 1.4$$

Also, $$\frac{dc_v}{c_v} = \frac{2}{100}$$

Efficiency of Diesel cycle is given by

$$\eta = 1 - \frac{1}{\gamma\,.\,(r)^{\gamma-1}}\left[\frac{\rho^\gamma - 1}{\rho - 1}\right] = 1 - \frac{1}{1.4\,(18)^{1.4-1}}\left[\frac{(1.85)^{1.4}-1}{1.85-1}\right]$$

$$= 1 - 0.2248\left(\frac{1.366}{0.85}\right) = 0.6387 \text{ or } 63.87\%$$

Now substituting the values of η and other quantities in the eqn. (*i*), we get

$$\frac{d\eta}{\eta} = -\frac{(1-0.6387)}{0.6387} \times (1.4-1)\left[\log_e 18 - \frac{(1.85)^{1.4}\log_e 1.85}{(1.85)^{1.4}-1} + \frac{1}{1.4}\right] \times \frac{2}{100}$$

$$= -0.5657 \times 0.4\left[2.89 - \frac{1.455}{1.366} + 0.714\right] \times \frac{2}{100}$$

$$= -\ 0.226\ (2.89 - 1.065 + 0.714) \times \frac{2}{100} = -\ 0.01147 = -\ 1.147\%$$

Negative sign means **decrease.**

Hence *percentage decrease in efficiency* = **1.147%. (Ans.)**

Example 4.5. *The following data relate to a petrol engine :*

Compression ratio = 7

Calorific value of fuel used = 44 MJ/kg

The air-fuel ratio = 15 : 1

The temperature and pressure of the charge at the end of stroke = 65°C, 1 bar

Index of compression = 1.33

The specific heat at constant volume, $(c_v) = 0.71 + 20 \times 10^{-5}\ T$ *kJ/kg K, when T is in K.*

Determine the maximum pressure in the cylinder. Compare this value with that of constant specific heat $c_v = 0.71$ *kJ/kg K.*

Solution. *Given* : $r = 7$; $C = 44 \times 10^3$ kJ/kg ; A/F ratio = 15 : 1 ; $T_1 = 65 + 273 = 338$ K ; $p_1 = 1$ bar ; $n = 1.34$, $c_v = 0.71 + 20 \times 10^{-5}$ T kJ/kg K.

Refer Fig. 4.17.

Maximum pressure in the cylinder, p_3 :

Consider *compression process 1-2* :

$$p_1 v_1^{\ n} = p_2 v_2^{\ n}$$

$$p_2 = p_1 \left(\frac{v_1}{v_2}\right)^n = 1 \times (7)^{1.33}$$

$$= 13.3 \text{ bar}$$

Now, $\dfrac{p_1 v_1}{T_1} = \dfrac{p_2 v_2}{T_2}$

$$\therefore \quad T_2 = T_1 \left(\frac{p_2 v_2}{p_1 v_1}\right)$$

$$= 338 \times \left(\frac{13.3}{1} \times \frac{1}{7}\right) = 642.2 \text{ K}$$

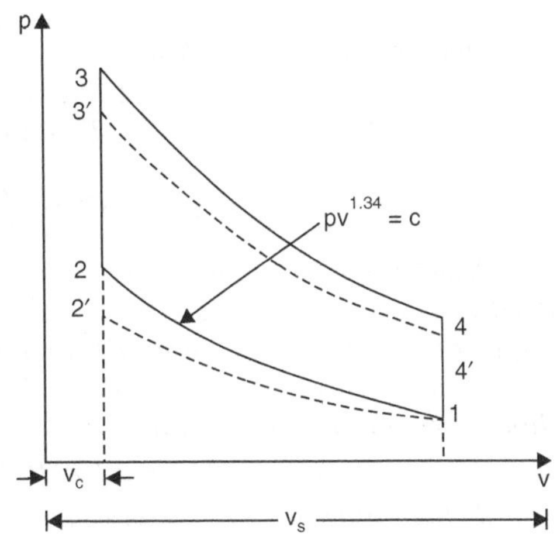

Fig. 4.17

Average temperature during combustion of charge = $\left(\dfrac{T_3 + T_2}{2}\right)$

Mean specific heat of product during combustion,

$$c_v \text{ mean} = 0.71 + 20 \times 10^{-5} \left[\frac{T_3 + T_2}{2}\right]$$

Assume 1 kg of air in the cylinder.

$$\text{Heat added per kg of charge} = \frac{44 \times 10^3}{16} = 2750 \text{ kJ/kg of air.}$$

$$Q = \text{mass of charge} \times c_v \times (T_3 - T_2)$$

$$2750 = 1 \times \left[0.71 + 20 \times 10^{-5} \left(\frac{T_3 + 642.2}{2}\right)\right] (T_3 - 642.2)$$

$$2750 = 0.71\,(T_3 - 642.2) + 10 \times 10^{-5}\,(T_3 + 642.2)(T_3 - 642.2)$$

$$2750 = 0.71\,T_3 - 456 + 10^{-4}\,(T_3^2 - 412421)$$

$$(2750 + 456) = 0.71\,T_3 + 10^{-4}T_3^2 - 41.24$$

or $$10^{-4}\,T_3^2 + 0.71T_3 = 3247.24$$

or $$T_3^2 + 7100\,T_3 - 3.247 \times 10^7 = 0$$

or $$T_3 = \frac{-7100 \pm \sqrt{7100^2 + 4 \times 3.247 \times 10^7}}{2}$$

$$= \frac{-7100 \pm 13427}{2} = 3163.5 \text{ K}$$

Consider *constant volume process 2-3.*

$$\frac{p_2}{T_2} = \frac{p_3}{T_3}$$

$\therefore$ $$p_3 = \frac{p_2 T_3}{T_2} = 13.3 \times \frac{3163.5}{642.2} = \mathbf{65.5\ bar.} \quad \textbf{(Ans.)}$$

For constant specific heat :

$$2750 = 0.71 \times (T_3 - 642.2)$$

$\therefore$ $$T_3 = 4515.4 \text{ K}$$

$\therefore$ $$p_3 = 13.3 \times \frac{4515.4}{642.2} = \mathbf{93.5\ bar.} \quad \textbf{(Ans.)}$$

Example 4.6. *An Otto cycle engine with a compression ratio of 10 uses a petroleum fuel of calorific value 48000 kJ/kg. The air fuel ratio is 15 : 1. The temperature and pressure of the charge at the end of suction are 57°C and 1 bar respectively. Determine the maximum pressure in the cycle with mean index of compression as 1.36 and the value of specific heat at constant volume heat addition expressed as*

$$c_v = 0.7117 + 2.1 \times 10^{-4} \text{ kJ/kg K where } T \text{ is the mean temperature.}$$

If the value of c_v remains constant at 0.7117 kJ/kg K, and also compression index is unaltered, how will the maximum pressure be affected ?

Solution. Compression ratio, $r = 10$

Calorific value of petroleum fuel = 48000 kJ/kg

Air-fuel ratio = 15 : 1

Temperature at the end of suction, $T_1 = 57 + 273 = 330$ K

Pressure at the end of suction, $p_1 = 1$ bar

Mean index of compression, $n = 1.36.$

We have $$p_1 V_1^{1.36} = p_2 V_2^{1.36}$$

$\therefore$ $$p_2 = p_1 \times \left(\frac{V_1}{V_2}\right)^{1.36} = 1 \times (10)^{1.36} = 22.91 \text{ bar}$$

Also, $$\frac{T_2}{T_1} = \left(\frac{p_2}{p_1}\right)^{\frac{n-1}{n}} = \left(\frac{22.91}{1}\right)^{\frac{1.36-1}{1.36}} = 2.29$$

$\therefore$ $$T_2 = 330 \times 2.29 = 756 \text{ K}$$

Mean specific heat during constant volume heat addition

$$c_{v_{(mean)}} = 0.7117 + 2.1 \times 10^{-4} \left(\frac{T_3 + T_2}{2}\right)$$

Thus heat added at constant volume/kg of charge

$$= \frac{48000}{16} = 3000 \text{ kJ/kg of charge}$$

And
$$3000 = \left[0.7117 + 2.1 \times 10^{-4} \left(\frac{T_3 + 756}{2}\right)\right] \times (T_3 - 756)$$

$\therefore$
$$3000 = 0.7117\ T_3 + \frac{2.1 \times 10^{-4}}{2} T_3^2 + \frac{2.1 \times 10^{-4} \times 756}{2} T_3 - 0.7117 \times 756$$
$$- \frac{2.1 \times 10^{-4} \times 756}{2} T_3 - \frac{2.1 \times 10^{-4}}{2} \times 756 \times 756$$

$$= 0.7117\ T_3 + 0.000105\ T_3^2 - 598$$
$$= 0.000105\ T_3^2 + 0.7117\ T_3 - 598$$

or
$$0.000105\ T_3^2 + 0.7117\ T_3 - 3598 = 0$$

Solving for T_3, $T_3 = 3370$ K

$\therefore$ *Maximum pressure in the cycle,*

$$p_3 = p_2 \times \frac{T_3}{T_2} = 22.91 \times \frac{3370}{756} = \textbf{102.1 bar.} \quad \textbf{(Ans.)}$$

When c_v remains constant at 0.7117 kJ/kg K ;

$$3000 = 0.7117\ (T_3 - 756)$$

or
$$T_3 = \frac{3000}{0.7117} + 756 = 4971 \text{ K}$$

and
$$p_3 = p_2 \frac{T_3}{T_2} = 22.91 \times \frac{4971}{756} = \textbf{150.6 bar.} \quad \textbf{(Ans.)}$$

Example 4.7. *Combustion in a diesel engine is assumed to begin at inner dead centre and to be at constant pressure. The air-fuel ratio is 27 : 1, the calorific value of the fuel is 43000 kJ/kg, and the specific heat of the products of combustion is given by :*

$c_v = 0.71 + 20 \times 10^{-5}\ T$; R for the products = 0.287 kJ/kg K.

If the compression ratio is 15 : 1, and the temperature at the end of compression 870 K, find at what percentage of the stroke combustion is completed.

Solution. *Given* : Air-fuel ratio = 27 : 1

Calorific value of fuel, c = 43000 kJ/kg

Specific heat of product of combustion : $c_v = 0.71 + 20 \times 10^{-5}$ T

R for products = 0.287 kJ/kg K

Compression ratio, r = 15 : 1

Temperature at the end of compression, T_2 = 870 K

Percentage of the stroke when combustion is completed :

For 1 kg of fuel the charge is 28 kg and the heating value is 43000 kJ/kg

$$dQ = m \int_2^3 c_p \, dT$$

and $$c_p = c_v + R$$
$$= 0.71 + 20 \times 10^{-5} T + 0.287$$
$$= 0.997 + 20 \times 10^{-5} T$$

$$\therefore \quad \frac{43000}{28} = \int_{T_2}^{T_3} (0.997 + 20 \times 10^{-5} \, T) \, dT$$

$$= \int_{T_2=870}^{T_3} 0.997 \, dT + (20 \times 10^{-5} \, T \, dT)$$

$$1535.7 = \left[0.997 \, T + 20 \times 10^{-5} \cdot \frac{T^2}{2} \right]_{T_2=870}^{T_3}$$

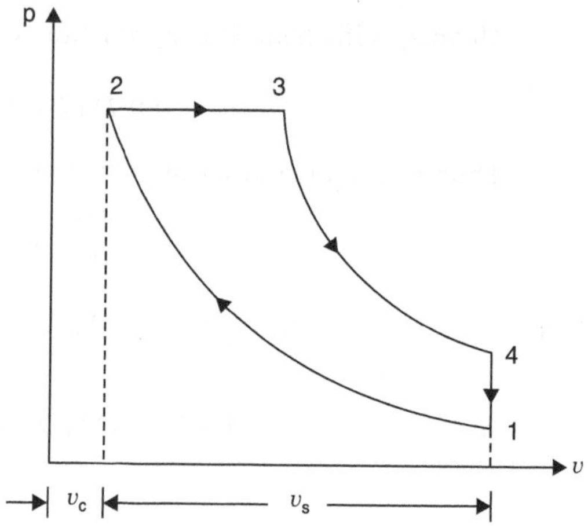

Fig. 4.18

or $$1535.7 = 0.997 (T_3 - 870) + 10 \times 10^{-5} (T_3^2 - 870^2)$$

or $$1535.7 = 0.997 \, T_3 - 867.4 + 10^{-4} \, T_3^2 - 75.69$$

or $$10^{-4} T_3^2 + 0.997 \, T_3 - 2478.8$$

or $$T_3^2 + 9970 \, T_3 - 2.4788 \times 10^7 = 0$$

$$T_3 = \frac{-9970 \pm \sqrt{(9970)^2 + 4 \times 2.4788 \times 10^7}}{2}$$

$$= \frac{-9970 \pm 14091}{2} = 2060 \text{ K}$$

Now, for *constant pressure process 2-3*, we have

$$\frac{v_2}{T_2} = \frac{v_3}{T_3}$$

or $$\frac{v_3}{v_2} = \frac{T_3}{T_2} = \frac{2060}{870} = 2.37$$

or $$v_3 = 2.37 \, v_2$$

$\therefore$ Combustion occupies $= \dfrac{2.37v_2 - v_2}{v_s} = \dfrac{1.37v_2}{14v_2} = 0.0978$ or **9.78% Stroke. (Ans.)**

$$\left[\because \; v_s = v_1 - v_c = v_1 - v_2 = 15v_2 - v_2 = 14v_2 \right]$$

Example 4.8. *In an oil engine working on dual combustion cycle the temperature and pressure at the beginning of compression are 87°C and 1 bar respectively. The compression ratio is 14 : 1. The heat supplied per kg of air is 1700 kJ, half of which is supplied at constant volume and half at constant pressure. Calculate :*

(i) The maximum pressure in the cycle.

(ii) The percentage of stroke at which cut-off occurs.

Take : γ for compression = 1.4 ; R = 0.287 kJ/kg K and

c_v for products of combustion = $0.71 + 20 \times 10^{-5}$ T.

Solution. *Given :* $T_1 = 87 + 273 = 360$ K ; $p_1 = 1$ bar ; $r = 14$;

$$Q_{2\text{-}3} = \frac{1700}{2} = 850 \text{ kJ/kg of air} \; ; \; Q_{3\text{-}4} = 850 \text{ kJ/kg of air} \; ;$$

γ for compression = 1.4 ; $R = 0.287$ kJ/kg K ; $c_v = 0.71 + 20 \times 10^{-5}\, T$

Consider *compression process 1-2 :*

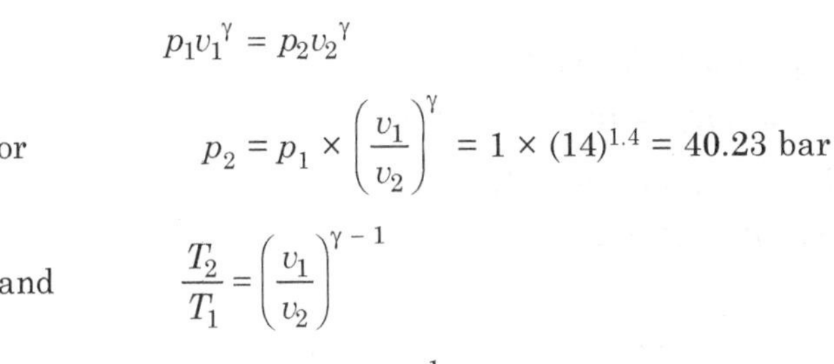

$$p_1 v_1^{\gamma} = p_2 v_2^{\gamma}$$

or $$p_2 = p_1 \times \left(\frac{v_1}{v_2}\right)^{\gamma} = 1 \times (14)^{1.4} = 40.23 \text{ bar}$$

and $$\frac{T_2}{T_1} = \left(\frac{v_1}{v_2}\right)^{\gamma - 1}$$

or $$T_2 = T_1 \left(\frac{v_1}{v_2}\right)^{\gamma - 1} = 360 \times (14)^{1.4 - 1} = 1034.6 \text{ K}$$

Fig. 4.19

Consider *constant volume process 2-3 :*

$$Q_{1-2} = m \int_{T_1}^{T_2} c_v \, dT$$

$$= 1 \int_{T_2}^{T_3} (0.71 + 20 \times 10^{-5}\, T)\, dT$$

$$850 = \left[0.71\, T + 20 \times 10^{-5} \frac{T^2}{2} \right]_{T_2}^{T_3}$$

or $$850 = 0.71(T_3 - T_2) + 10 \times 10^{-5}\, (T_3^2 - T_2^2)$$

or $$850 = 0.71\,(T_3 - 1034.6) + 10^{-4}\, [T_3^2 - (1034.6)^2]$$

or $$850 = 0.71\, T_3 - 734.6 + 10^{-4}\, T_3^2 - 107.04$$

or $$10^{-4}\, T_3^2 + 0.71\, T_3 - 1691.64 = 0$$

or $$T_3^2 + 7100\, T_3 - 1691.64 \times 10^4 = 0$$

or $$T_3 = \frac{-7100 \pm \sqrt{(7100)^2 + 4 \times 1691.64 \times 10^4}}{2}$$

$$= \frac{-7100 \pm 10866.3}{2} = 1883 \text{ K}$$

Further, $$\frac{p_2}{T_2} = \frac{p_3}{T_3} \quad \text{or} \quad p_3 = \frac{p_2 T_3}{T_2} = \frac{40.23 \times 1883}{1034.6} = \mathbf{73.22 \text{ bar.}} \quad \textbf{(Ans.)}$$

(*ii*) **Percentage of stroke at which cut-off occurs :**

$$c_p = c_v + R = 0.71 + 20 \times 10^{-5}\, T + 0.287$$

$$= 0.997 + 20 \times 10^{-5}\, T$$

Consider *constant pressure process 3-4 :*

$$Q_{3-4} = m \int_{T_3}^{T_4} c_p \, dT = 1 \int_{T_3}^{T_4} (0.997 + 20 \times 10^{-5} T) \, dT$$

$$850 = \left[\left(0.997 \, T + 20 \times 10^{-5} \frac{T^2}{2} \right) \right]_{T_3}^{T_4}$$

$$= 0.997 (T_4 - T_3) + 10^{-4} (T_4^2 - T_3^2)$$

$$= 0.997 (T_4 - 1883) + 10^{-4} (T_4^2 - 1883^2)$$

or $$850 = 0.997 \, T_4 - 1877.35 + 10^{-4} \, T_4^2 - 354.57$$

or $$10^{-4} \, T_4^2 + 0.997 \, T_4 - 3082 = 0$$

or $$T_4^2 + 9970 \, T_4 - 3082 \times 10^4 = 0$$

or $$T_4 = \frac{-9970 \pm \sqrt{(9970)^2 + 4 \times 3082 \times 10^4}}{2}$$

$$= \frac{-9970 + 14922.5}{2} = 2476 \text{ K}$$

Further, $$\frac{v_3}{T_3} = \frac{v_4}{T_4} \quad \text{or} \quad v_4 = v_3 \times \frac{T_4}{T_3} = v_3 \times \frac{2476}{1883} = 1.315 \, v_3$$

$\therefore$ $$\text{Cut-off} = \frac{v_4 - v_3}{v_1 - v_3} = \frac{1.315 \, v_3 - v_3}{14 v_3 - v_3} = \frac{0.315}{13}$$

$$= 0.0242 \quad \text{or} \quad \textbf{2.42\% of stroke. (Ans.)} \qquad \left(\because \; \frac{v_1}{v_2} = \frac{v_1}{v_3} = 14 \right)$$

Example 4.9. *An engine working on Otto cycle has a compression ratio 8. It uses hexane (C_6H_{14}) as the fuel which has a calorific value of 44 MJ/kg. The air-fuel ratio of the mixture is 13.8 : 1. The temperature and pressure of the mixture at the beginning of compression is 70°C and 1 bar respectively. If $c_v = 0.716$ and compression follows the law $pv^{1.35} = C$, determine the maximum temperature and pressure kJ/kg K reached in the cycle :*

(*i*) *Without considering the molecular contraction ;*

(*ii*) *Considering molecular contraction.*

Solution. *Given :* $r = 8$, $C = 44$ MJ/kg ; A/F ratio = 13.8 : 1, $T_1 = 70 + 273 = 343$ K ;

$p_1 = 1$ bar, $c_v = 0.716$ kJ/kg K, Law of compression : $pv^{1.35} = C$.

Maximum temperature (T_3), maximum pressure (p_3) :

(*i*) **Without considering the molecular contraction :**

The stoichiometric equation can be written as,

$$C_6H_{14} + xO_2 = yCO_2 + zH_2O$$

Equating atoms of the same element before and after combustion, we get

$$y = 6, \; z = 7 \text{ and } x = 9.5$$

$\therefore$ $$C_6H_{14} + 9.5O_2 = 6CO_2 + 7H_2O$$

$$\text{Gravemetric air-fuel } (A/F) \text{ ratio} = \frac{(9.5 \times 32)(100/23)}{12 \times 6 + 1 \times 14} = 15.37$$

The actual mixture strength (F/A ratio) expressed in terms of the chemically correct value is

$$\frac{15.37}{13.8} \times 100 = 111.4\%$$

i.e. the mixture used is 11.4% *rich in fuel*. The *combustion is, therefore, incomplete and hence CO will be formed.*

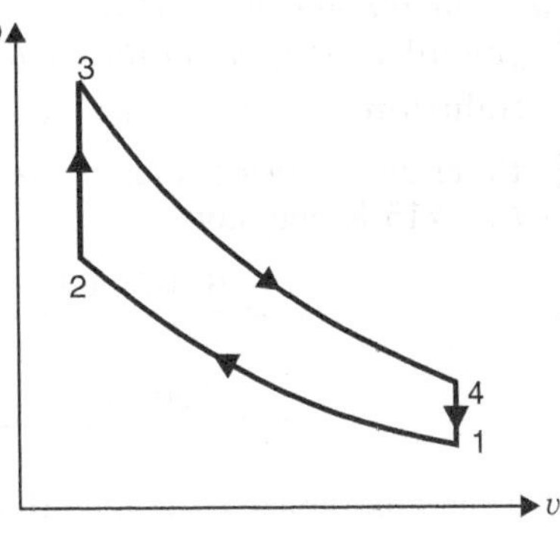

Fig. 4.20

$$1.114\ C_6H_{14} + 9.5O_2 = aCO_2 + bCO + cH_2O$$

Equating atoms of the same element before and after combustion, we get

$1.114 \times 6 = a + b$; $1.114 \times 14 = 2c$;

$9.5 \times 2 = 2a + b + c$

or $a + b = 6.684$; $c = 7.798$, $2a + b + c = 19$

Solving, we have : $a = 4.52$, $b = 2.16$, $c \sim 7.8$

By adding nitrogen on both sides we get the actual combustion equation as given below :

$$1.114\ C_6H_{14} + 9.5\ O_2 + 9.5 \times (79/21)\ N_2 = 4.52\ CO_2 + 2.16\ CO + 7.8\ H_2O + 9.5 \times (79/21)\ N_2$$

$\therefore$ Moles before combustion = 1.114 + 9.5 + 35.74 = 46.354 say 46.35

Moles after combustion = 4.52 + 2.16 + 7.8 + 35.74 = 50.22

$$\left[\because \text{ Molecular expansion} = \frac{50.22 - 46.35}{46.35} = 0.0835 \text{ or } 8.35\%\right]$$

From *compression process 1-2,* we have :

$$\frac{T_2}{T_1} = \left(\frac{v_1}{v_2}\right)^{n-1}$$

or $$\frac{T_2}{343} = (8)^{1.35-1} = 2.07 \qquad \therefore \quad T_2 = 710 \text{ K}$$

Also $$Q_{2-3} = c_v(T_3 - T_2) \text{ for 1 kg of mixture}$$

$$\frac{44 \times 10^3}{14.8} = 0.716\ (T_3 - 710) \quad \text{or} \quad T_3 = 4862 \text{ K}$$

Ignoring molecular expansion,

$$\frac{p_1v_1}{T_1} = \frac{p_3v_3}{T_3}$$

or $$\mathbf{p_3} = p_1 \times \frac{v_1}{v_3} \times \frac{T_3}{T_1} = 1 \times 8 \times \frac{4862}{343} = \mathbf{113.4\ bar.} \quad \textbf{(Ans.)}$$

(*ii*) **Considering molecular contraction :**

Since mass of the reactants and products is same and specific heats are assumed same, the *temperature of the products with* molecular expansion will *remain same* as without molecular expansion ; *only the pressure will change*

$pv = nRT$, where n is the number of moles

$\therefore$ $p \propto n$

$\therefore$ Pressure with molecular expansion $= 113.4 \times \dfrac{50.22}{46.35} = \mathbf{122.87\ bar.}$ **(Ans.)**

Example 4.10. *An ideal Otto cycle operating on air has a compression ratio 7. The temperatures at the end of isentropic compression is 442°C and at the end of expansion is 1347°C. Using gas tables determine the cycle work and efficiency.*

Solution. *Given* : $r = 7$; $T_2 = 442 + 273 = 715$ K ; $T_4 = 1337 + 273 = 1610$ K

From gas tables (Refer appendix), corresponding to $T_2 = 715$ K, we have

$$v_{r_2} = \frac{67.07 + 64.53}{2} = 65.8$$

$$u_2 = \frac{520.23 + 528.14}{2} = 524.2 \text{ kJ/kg}$$

From gas tables, corresponding to $T_4 = 1610$ K, we have

$$v_{r_4} = \frac{5.804 + 5.574}{2} \simeq 5.69$$

$$u_4 = \frac{1298.30 + 1316.96}{2} = 1307.63 \text{ kJ/kg}$$

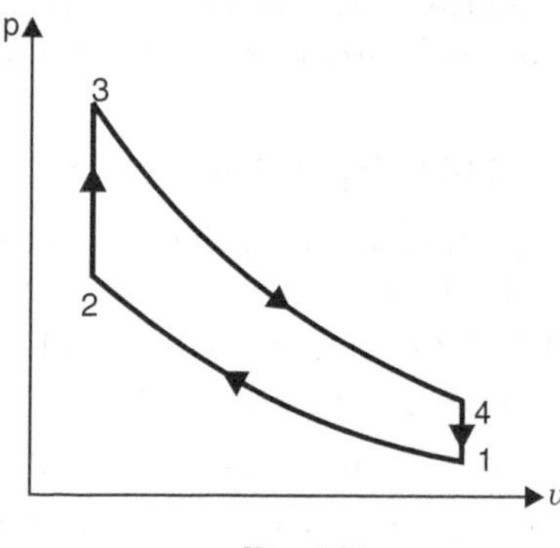

Fig. 4.21

Volume compression ratio $= \frac{v_1}{v_2} = \frac{v_{r_1}}{v_{r_2}} = 7$

$\therefore$ $v_{r_1} = 7 \times v_{r_2} = 7 \times 65.8 = 460.6$

From gas tables, corresponding to $v_{r_1} = 460.6$, we have

By interpolation : $T_1 = 340 - \frac{(340 - 330)}{(454.1 - 489.4)} \times (454.1 - 460.6) \simeq 338$ K

$$u_1 = 235.61 + \frac{242.82 - 235.61}{(340 - 330)} \times (338 - 330) = 241.38 \text{ kJ/kg}$$

Also, we have $\frac{v_4}{v_3} = \frac{v_{r_4}}{v_{r_3}} = 7$

$\therefore$ $v_{r_3} = \frac{v_{r_4}}{7} = \frac{5.69}{7} = 0.813$

From gas tables, corresponding to $v_{r_3} = 0.813$, we have

$$T_3 \simeq 2800 \text{ K} \quad \text{and} \quad u_3 \simeq 2462.5 \text{ kJ/kg}$$

The work done = Heat added – Heat rejected

$= (u_3 - u_2) - (u_4 - u_1)$

$= (2462.5 - 524.2) - (1307.63 - 241.38) =$ **872.05 kJ/kg. (Ans.)**

Thermal efficiency, $\eta_{th} = \frac{\text{Work done}}{\text{Heat added}} = \frac{872.05}{(2462.5 - 524.2)}$

$\simeq 0.45$ or **45%. (Ans.)**

HIGHLIGHTS

1. The specific heat varies largely with temperature but not very significantly with pressure except at very high pressure.
2. The specific heats of gases increase with rise in temperature since the vibrational energy of molecules increases with temperature.
3. Change of *internal energy* (per kg) during a process with variable specific heats,

$$u_2 - u_1 = c_{vm}(T_2 - T_1)$$

Change of enthalpy, $h_2 - h_1 = c_{pm}(T_2 - T_1)$.

4. Heat transfer during a process with variable specific heats,

$$dQ = \left[\left(\frac{\gamma' - n}{\gamma' - 1}\right) - \left(\frac{n-1}{a-b}\right) KT\right] dW.$$

5. In case of isentropic with variable specific heats

$$p(v)^{a/b} \,.\, e^{KT/b} = \text{constant}.$$

6. 'Entropy change' during a process with variable specific heats,

$$s_2 - s_1 = a \log_e\left(\frac{T_2}{T_1}\right) - (a-b)\log_e\left(\frac{p_2}{p_1}\right) + K(T_2 - T_1).$$

7. Percentage variation in air standard efficiency on account of percentage variation in c_v in case of :

(*i*) **Otto cycle :** $\dfrac{d\eta}{\eta} = -\dfrac{1-\eta}{\eta}(\gamma - 1)\log_e r \times \dfrac{dc_v}{c_v}$.

(*ii*) **Diesel cycle :** $\dfrac{d\eta}{\eta} = -\dfrac{1-\eta}{\eta}(\gamma - 1)\left[\log_e r - \dfrac{\rho^\gamma \,.\, \log_e \rho}{\rho^\gamma - 1} + \dfrac{1}{\gamma}\right]\dfrac{dc_v}{c_v}$.

8. *Dissociation* refers to disintegration of burnt gases at high temperature. It is a reversible process and increases with temperature.
9. Maximum efficiency is obtained with linear mixture.
10. The exhaust gas temperature is maximum at the chemically correct mixture.
11. If losses due to variable specific heats and dissociation are subtracted from the "air standard cycle", we get "fuel-air cycle analysis". If losses are further subtracted from "fuel-air cycle analysis" we can very closely approximate the "actual cycle".

OBJECTIVE TYPE QUESTIONS

Fill in the Blanks or Say "Yes" or "No" :

1. The value of c_v of a real gas ...... with increase in temperature.
2. The value of c_p of a real gas ...... with increase in temperature.
3. The value of c_p ...... with increase in moisture content.
4. The value of γ for air ...... with increase in moisture content in air.
5. The value of γ for a real gas ...... with increase in temperature.
6. The change of internal energy during a process with variable specific heats is equal to $c_{pm}(T_2 - T_1)$.
7. The change of enthalpy during a process with variable specific heats is equal to $c_{pm}(T_2 - T_1)$.
8. The theoretical cycle based on the actual properties of the cylinder gases is called fuel-air cycle approximation.

9. The ...... of any substance is the ratio of the heat required to raise the temperature of a unit mass of substance through one degree centigrade.
10. ...... refers to disintegration of burnt gases at high temperature.
11. Dissociation is ...... process and increases with temperature.
12. The effect of dissociation is much ...... than that of change of specific heat.
13. The dissociation of CO_2 commences at about 1000°C and at 1500°C it amounts to 5 percent.
14. Dissociation is ...... severe in the chemically correct mixture.
15. The dissociation has a more pronounced effect in ...... engines.
16. The fuel-air cycle efficiency increases with compression ratio in the same manner as the air standard efficiency.
17. Maximum efficiency is obtained with ...... mixtures.
18. The exhaust gas temperature is maximum with the chemically correct mixture.
19. The mean effective pressure increases with compression ratio.
20. In S.I. engines the combustion is ....
21. In C.I. engines the combustion is heterogeneous.
22. The ...... time loss is defined as loss of power due to time required for mixing the fuel with air and for complete combustion.

ANSWERS

1. increases	**2.** increases	**3.** increases	**4.** decreases	**5.** decreases
6. No	**7.** Yes	**8.** Yes	**9.** specific heat	**10.** Dissociation
11. reversible	**12.** smaller	**13.** No	**14.** more	**15.** S.I.
16. Yes	**17.** leaner	**18.** Yes	**19.** Yes	**20.** homogeneous
21. Yes	**22.** burning.			

THEORETICAL QUESTIONS

1. State the factors which should be taken into consideration while making fuel-air cycle calculations.
2. Enlist the assumptions which are made for fuel-air cycle analysis.
3. State the importance of fuel-air cycle.
4. What are molar specific heats ?
5. Derive expressions for change of internal energy and enthalpy during a process with variable specific heats.
6. Prove that in case of isentropic expansion with variable specific heats,

$$p(v)^{a/b} \cdot e^{KT/b} = \text{constant}.$$

7. Derive an expression for 'entropy change' during a process with variable specific heats.
8. Taking variation in specific heat into account, derive the following expressions :

For *Otto cycle :* $$\frac{d\eta}{\eta} = \left[-\frac{1-\eta}{\eta} \times (\gamma - 1) \times \log_e r\right] \frac{dc_v}{c_v}$$

For *Diesel cycle :* $$\frac{d\eta}{\eta} = -\frac{1-\eta}{\eta}(\gamma - 1)\left[\log_e r - \frac{\rho^\gamma \log_e \rho}{\rho^\gamma - 1} + \frac{1}{\gamma}\right]\frac{dc_v}{c_v}.$$

9. What is dissociation ? How does it affect power developed by the engine ?
10. Describe briefly the effect of dissociation on temperature at different mixture strength.
11. Explain the phenomenon of dissociation.
12. Are dissociation effects equally pronounced in S.I. and C.I. engines ? Explain.

13. Explain clearly the effect of compression ratio and mixture strength on thermal efficiency.
14. What is the effect of mixture strength on thermal efficiency at a given compression ratio.
15. What is the effect of mixture strength on cycle power ?
16. State the effect of F_R on maximum cycle temperature and pressure at different compression ratios.
17. State the characteristics of constant volume fuel-air cycle.
18. Discuss briefly "combustion charts".
19. What are combustion charts ? Where these are used and why ?
20. Write a short note on gas tables.
21. Discuss the effect of the following variables on pressure and temperature at salient points of Otto cycle on the basis of fuel-air cycle.
    (*i*) Compression ratio (*ii*) Fuel-air ratio.
22. What is the difference between air cycle and fuel-air cycle ? What are the assumptions in fuel-air cycle ?
23. What is the use of fuel-air cycle ?
24. What is the difference between air standard cycles and fuel-air cycles.
25. Make a comparative statement of operations and working media for air cycle, fuel-air cycle and actual cycle of S.I. engines.
26. Explain why a S.I. engine fails to operate if the air-fuel ratio is more than 20 : 1 while a C.I. engine can operate on an air-fuel ratio of even 50 : 1.
27. Explain how (*i*) time losses and (*ii*) incomplete combustion losses are accounted for in the real-cycle analysis.
28. "Air-fuel ratio in a S.I. engine varies from 8 to 16 approximately while such variation in a C.I. engine is from 100 at no-load to 20 at full load". Explain.

UNSOLVED EXAMPLES

1. Find the change in efficiency of an Otto cycle for a compression ratio of 7, if the specific heat at constant volume increases by 1 percent. **[Ans.** – 0.663%]
2. The following data relate to a petrol engine :
   Compression ratio = 6
   Calorific value of fuel used = 44000 kJ/kg
   The air-fuel ratio = 15 : 1
   The temperature and pressure of the charge at the end of the stroke = 60°C, 1 bar
   Index of compression = 1.32
   The specific heat at constant volume, $c_v = 0.71 + 20 \times 10^{-5}\, T$ kJ/kg K where T is in K.
   Determine the maximum pressure in the cylinder. Compare this value with that of constant specific heat $c_v = 0.71$ kJ/kg K. **[Ans.** 56.6 bar ; 80.5 bar]
3. The combustion in a diesel engine is assumed to begin at inner dead centre and to be at constant pressure. The air-fuel ratio is 28 : 1, the calorific value of the fuel is 42 MJ/kg, and the specific heat of the products of combustion is given by :
   $c_v = 0.71 + 20 \times 10^{-5}\, T$; R for the products = 0.287 kJ/kg K.
   If the compression ratio is 14 : 1, and the temperature at the end of compression is 800 K, find at what percentage of the stroke combustion is completed. **[Ans.** 10.96% stroke]
4. In an oil engine working on dual combustion cycle the temperature and pressure at the beginning of compression are 90°C and 1 bar respectively. The compression ratio is 13 : 1. The heat supplied per kg of air is 1675 J, half of which is supplied at constant volume and half at constant pressure. Calculate :
   (*i*) The maximum pressure in the cycle ;
   (*ii*) The percentage of stroke at which cut-off occurs.
   Take : γ for compression = 1.4 ; R = 0.287 kJ/kg K and c_v for products of combustion = $0.71 + 20 \times 10^{-5}\, T$.
   [Ans. 66.2 bar ; 2.64% of stroke]

5. An engine working on the Otto cycle, having compression ratio 7, uses hexane (C_6H_{14}) as the fuel. The calorific value of the fuel is 44 MJ/kg. The air-fuel ratio of the mixture is 13.67 : 1. Determine the maximum temperature and pressure reached in the cycle :

   (*i*) without considering the molecular contraction ;

   (*ii*) considering molecular contraction.

   Assume $c_v = 0.718$ kJ/kg K, compression follows the law $pv^{1.3} = C$. The temperature and pressure of the mixture at the beginning of compression are 67°C and 1 bar respectively.

   [**Ans.** (*i*) 4787 K ; 99.06 bar, (*ii*) 4787 K (no charge) ; 107.1 bar]

6. For an ideal Otto engine operating on air, the temperature at the end of isentropic compression is 452°C and at the end of expansion is 1347°C. The compression ratio is 7.5. Using gas tables find the cycle work and efficiency.

   [**Ans.** 979.4 kJ/kg ; 47.6%]

5

Combustion in S.I. Engines

5.1. INTRODUCTION

5.1.1. Definition of Combustion

Combustion *may be defined as a relatively rapid chemical combination of hydrogen and carbon in the fuel with the oxygen in the air, resulting in liberation of energy in the form of heat.*

Following conditions are necessary for combustion to take place :

1. A combustible mixture.
2. Some means to initiate combustion.
3. Stabilization and propagation of flame in the combustion chamber.

In spark ignition (S.I.) engines, a carburettor generally supplies a combustible mixture and the electric spark from a spark plug initiates the combustion.

5.1.2. Ignition Limits

- It has been observed through experiments that ignition of charge is only possible within certain limits of fuel-air ratio.

  The 'ignition limits' correspond approximately to those mixture ratios, at lean and rich ends of the scale, where the heat released by the spark is no longer sufficient to initiate combustion in the neighbouring unburnt mixture.
- Fig. 5.1 shows the ignition limits for hydrocarbons.

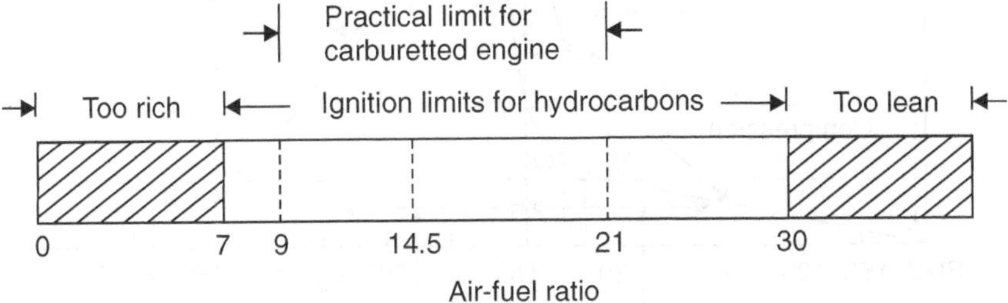

Fig. 5.1. Ignition limits for hydrocarbons.

— The ignition limits are wider at increased temperatures because of *higher rates of reaction and higher thermal diffusivity coefficients of the mixture.*
— The lower and upper limits of ignition of the mixture depend upon the *temperature and mixture ratio.*
— In case of hydrocarbon fuel the stoichiometric fuel-air ratio is about 1 : 15 and the fuel-air ratio lies between about 1 : 30 and 1 : 7.

5.2. COMBUSTION PHENOMENON

5.2.1. Normal Combustion

In a S.I. engine a single intensely high temperature spark passes across the electrodes, leaving behind a thin thread of flame. From this thin thread, combustion spreads to the envelope of mixture immediately surrounding it at a rate which *depends primarily upon the temperature of the flame front itself and to a secondary degree, upon both the temperature and the density of the surrounding envelope.* In the actual engine cylinder, the mixture is not at rest but is in highly turbulent condition. The *turbulence breaks the filament of a flame into a ragged front, thus presenting a far greater area of surface from which heat is being radiated ; hence its advance is speeded up enormously.*

According to Ricardo, the combustion process can be imagined as if developing in the following *two stages* :

(*i*) The growth and development of a self-propagating nucleus of flame (*ignition lag*). This is a *chemical process* and depends upon the following :
— The nature of fuel ;
— The temperature and pressure ;
— The proportion of the exhaust gas ;
— The temperature co-efficient of the fuel *i.e.,* the relationship between temperature and rate of acceleration of oxidation or burning.

(*ii*) The spread of the flame throughout the combustion chamber.

Fig. 5.2 shows the p-θ diagram of a petrol engine :

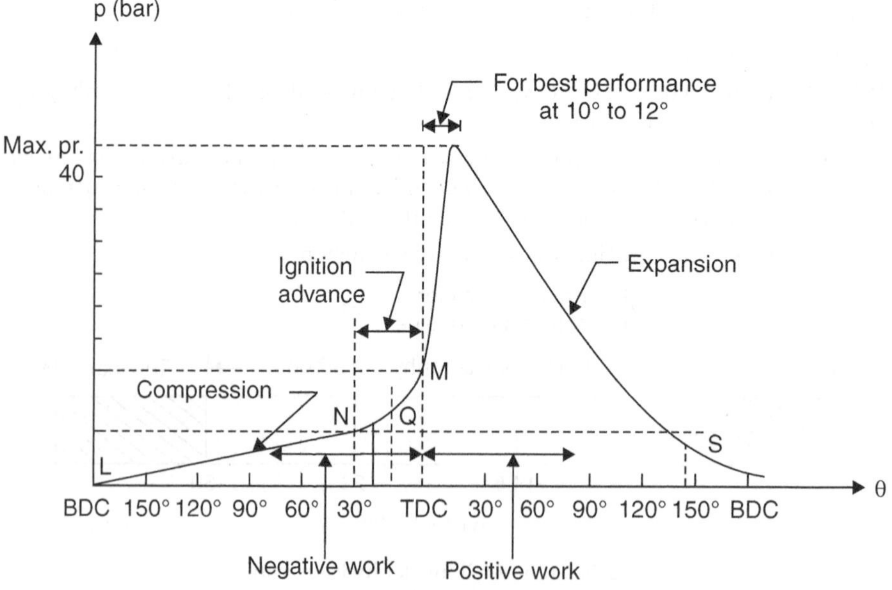

Fig. 5.2. Pressure-crank angle diagram of a petrol engine.

LNQM assumes *compression curve* having *no ignition.*

- First stage of combustion, the *ignition lag,* starts from this point and *no pressure rise is noticeable.*
- *Q* is the point where the *pressure rise can be detected.* From this point it deviates from the simple compression (motoring) curve.
- The time lag between first igniting of fuel and the commencement of the main phase of combustion is called the *period of incubation* or is also known as **ignition lag.** The time is normally about 0.0015 seconds. The maximum pressure is reached at about 12° after top dead centre point. Although the point of maximum pressure marks the completion of flame travel, it does not mean that at this point the whole of the heat of fuel has been liberated, for even after the passage of the flame, some further chemical adjustments due to reassociation, etc., will continue to a greater or less degree throughout the expansion stroke. This is known as ***after burning.***

Effect of engine variables on flame propagation :

1. ***Fuel-air ratio.*** When the *mixture is made leaner or is enriched* and *still more, the velocity of flame diminishes.*

2. ***Compression ratio.*** The *speed of combustion increases with increase of compression ratio.* The increase in compression ratio results in increase in temperature which increases the tendency of the engine to detonate.

3. ***Intake temperature and pressure.*** *Increase in intake temperature and pressure increases the flame speed.*

4. ***Engine load.*** As the *load on the engine increases,* the cycle pressures increase and hence the flame *speed increases.*

5. ***Turbulence.*** The flame speed is very low in non-turbulent mixture. A turbulent motion of the mixture intensifies the processes of heat transfer and mixing of the burned and unburned portions in the flame front. These two factors cause the *velocity of turbulent flame to increase practically in proportion to the turbulent velocity.*

6. ***Engine speed.*** The *flame speed increases almost linearly with engine speed.* The *crank angle required for flame propagation, which is the main phase of combustion, will remain almost constant at all speeds.*

7. ***Engine size.*** The number of crank degrees required for flame travel will be about the same irrespective of engine size, provided the engines are similar.

5.2.1.1. Factors affecting normal combustions in S.I. engines.

The factors which affect normal combustion in S.I. engines are briefly discussed below :

1. ***Induction pressure.*** As the pressure falls delay period increases and the ignition must be earlier at low pressures. A *vacuum control* may be incorporated.
2. ***Engine speed.*** As speed increases the constant time delay period needs more crank angle and ignition must be earlier. A *centrifugal control* may be employed.
3. ***Ignition timing.*** If ignition is too early the peak pressure will occur too early and work transfer falls. If ignition is too late the peak pressure will be low and work transfer falls. Combustion may not be complete by the time the exhaust valve opens and the valve may burn.
4. ***Mixture strength.*** Although the stoichiometric ratio should give the best results, the effect of dissociation shown in Fig. 5.3 is to make a slightly rich mixture necessary for maximum work transfer.
5. ***Compression ratio.*** An increase in compression ratio increases the maximum pressure and the work transfer.

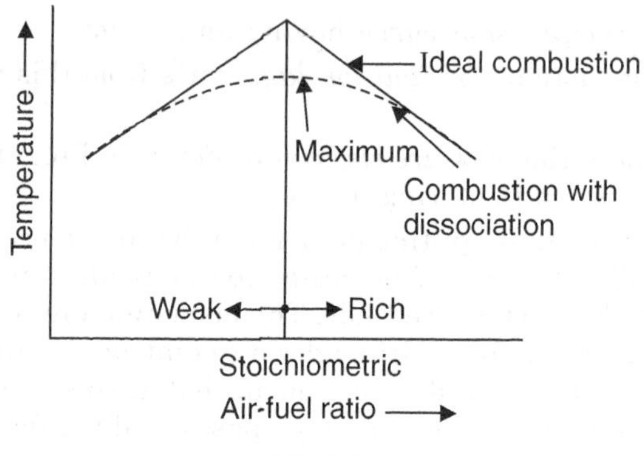

Fig. 5.3

6. **Combustion chamber.** The combustion chamber should be designed to *give a short flame path to avoid knock and it should promote optimum turbulence.*
7. **Fuel choice.**
   - The induction period of the fuel will affect the delay period.
   - The calorific value and the enthalpy of vaporisation will affect the temperatures achieved.

5.2.2. Abnormal Combustion

Due to excessively weak mixtures combustion may be slow or may be mis-timed. These are however obvious.

There are *two combustion abnormalities,* which are less obvious :

- The *first* of these is **pre** or **post ignition** of the mixture by *incandescent carbon particles in the chamber.* This will have the *effect of reducing the work transfer.*
- The *second* abnormality is generally known as *knock* and is a complex condition with many facets. A simple explanation shows that knock occurs when the unburnt portion of the gas in the combustion chamber is heated by combustion and radiation so that *its temperature becomes greater than the self ignition temperature.* If *normal progressive combustion is not completed before the end of the induction period* then a simultaneous explosion of the unburnt gas will occur. This explosion is accompanied by a detonation (pressure) wave which will be repeatedly reflected from the cylinder walls setting up a high frequency resonance which gives an audible noise. The detonation wave causes excessive stress and also destroys the thermal boundary layer at the cylinder walls causing overheating.

[**Note.** Refer articles 5.5 and 5.6 for details of pre-ignition and detonation respectively.]

5.3. EFFECT OF ENGINE VARIABLES ON IGNITION LAG

Ignition lag (the time lag between first igniting of fuel and the commencement of the main phase of combustion) is not a period of inactivity but is a chemical process. The ignition lag in terms of crank angles is 10° to 20° and in terms of time, 0.0015 second or so.

The duration of ignition lag depends on the following factors :

1. **Fuel.** Ignition lag depends on chemical nature of fuel. The *higher the self ignition temperature (S.I.T.) of fuel, longer the ignition lag.*
2. **Mixture ratio.** Ignition lag is the *smallest for the mixture ratio which gives the maximum temperature.* This mixture ratio is *somewhat richer than stoichiometric ratio.*

3. **Initial temperature and pressure.** Ignition lag is *reduced if the initial temperature and pressure are increased* (and these can be increased by increasing the compression ratio).
4. **Turbulence.** *Ignition lag is not much affected by the turbulence.*

5.4. SPARK ADVANCE AND FACTORS AFFECTING IGNITION TIMING

Spark advance. In order to obtain maximum power from an engine, the compressed mixture must deliver its maximum pressure at a time when the piston is about to commence its outward stroke and is nearest to T.D.C. Since there is a time lag between the occurrence of spark and the burning of the mixture, the *spark must take place before the piston reaches T.D.C. on its compression stroke, i.e.,* the *spark timing is advanced.* Usually the spark should occur at about 15° before T.D.C.

The correct instant for the introduction of spark is mainly determined by the **"ignition lag"**. The *factors affecting the ignition timings* are discussed below :

1. **Engine speed.** Suppose an engine has an ignition advance of θ degrees and operating speed in n r.p.s. Then time available for initiation of combustion is $\frac{\theta}{360n}$ seconds. Now if the engine speed is increased to $2n$ r.p.s. then in order to have the same time available for combustion, an ignition advance for 2θ degrees is required. Thus as the engine speed is *increased,* it will be necessary to advance the ignition progressively.

2. **Mixture strength.** In general *rich mixtures burn faster.* Hence, if the engine is operating with rich mixtures the optimum spark timings must be *retarded, i.e.,* the number of crank angle before T.D.C. at the time of ignition is *decreased* and the spark occurs *closer to T.D.C.*

3. **Part-load operation.** Part-load operation of a spark-ignition engine is affected by throttling the incoming charge. Due to throttling a small amount of charge enters the cylinder, and the dilution due to residual gases is also greater. In order to overcome the problem of exhaust gas dilution and the low charge density, at part-load operation the *spark advance must be increased.*

4. **Type of fuel.** Ignition delay will depend upon the type of fuel used in the engine. For *maximum power and economy a slow burning fuel needs a higher spark advance than a fast burning fuel.*

5.5. PRE-IGNITION

Refer Fig. 5.4.

- **Pre-ignition** is *the ignition of the homogeneous mixture in the cylinder, before the timed ignition spark occurs, caused by the local overheating of the combustible mixture.* For premature ignition of any local hot-spot to occur in advance of the timed spark on the combustion stroke it must attain a minimum temperature of something like 700–800°C.
- Pre-ignition is *initiated by some overheated projecting part* such as the *sparking plug electrodes, exhaust valve head, metal corners in the combustion chamber, carbon deposits or protruding cylinder head gasket rim etc.*
  — However, *pre-ignition is also caused by persistent detonating pressure shockwaves* scoring away the stagnant gases which normally protect the combustion chamber walls. The resulting increased heat flow through the walls, raises the surface temperature of any protruding poorly cooled part of the chamber, and this therefore provides a focal point for pre-ignition.
- The initiation of ignition and the propagation of the flame front from the heated hot-spot is similar to that produced by the spark-plug when it fires, the only difference

between the hot-spot and spark plug is their respective *instant of ignition.* Thus, the sparking plugs provides a timed and controlled moment of ignition whereas the heated surface forming the hot-spot builds upto the ignition temperature during each compression stroke and therefore the *actual instant of ignition is unpredicable.*

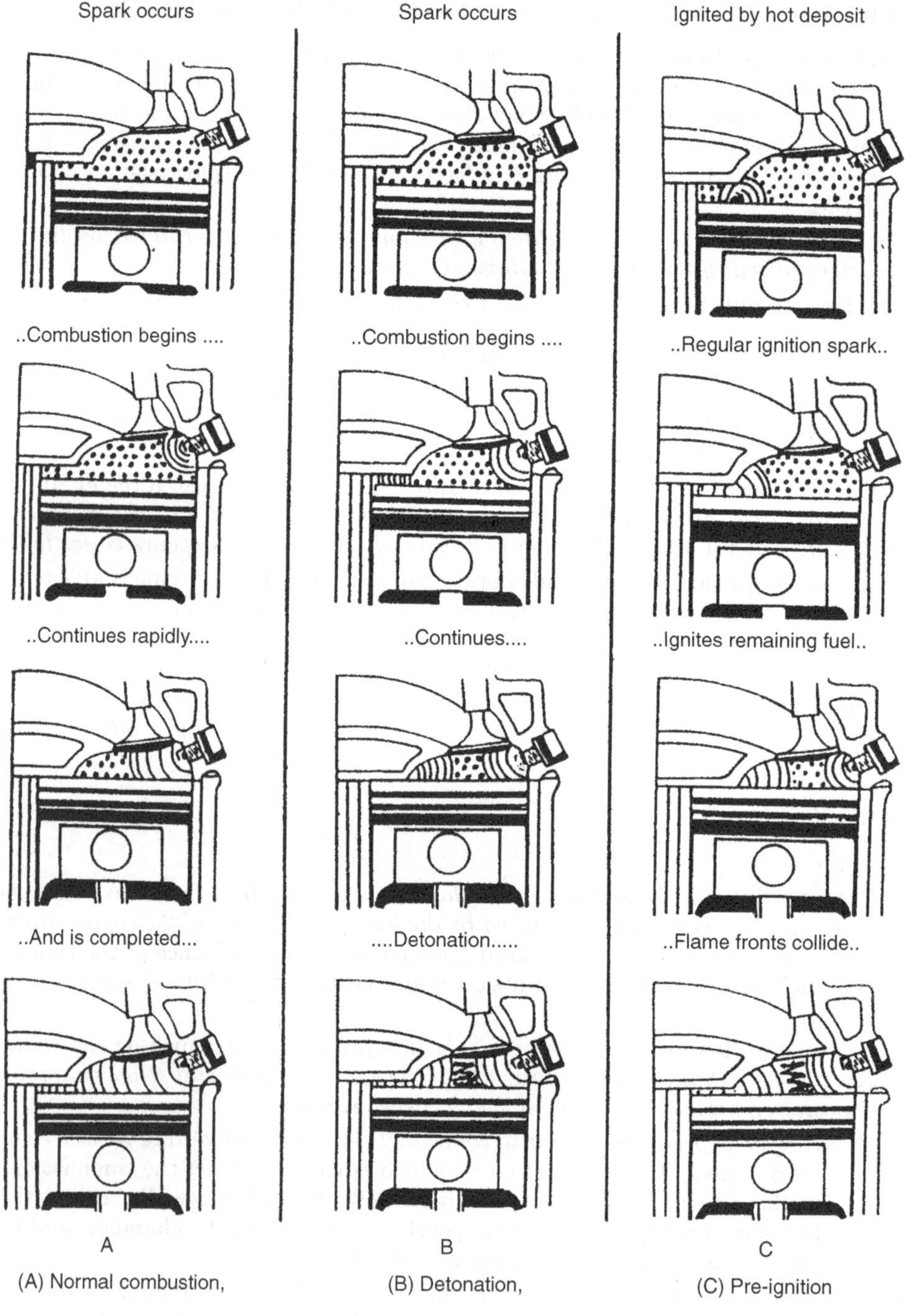

(A) Normal combustion, (B) Detonation, (C) Pre-ignition

Fig. 5.4

- The early ignition created by pre-ignition extends the total time and the burnt gases remain in the cylinder and therefore *increases the heat transfer on the chamber walls, as a result, the self-ignition temperature will occur earlier and earlier on each successive compression stroke.* Consequently, the peak cylinder pressure (which normally occurs at its optimum position of 10°–15° after T.D.C.) will progressively *advance* its position to T.D.C. where the cylinder pressure and temperature will be maximised.
  — The accumulated effects of an extended combustion time and rising peak cylinder pressure and temperature cause the self-ignition temperature to creep further and further ahead of T.D.C., and with it, peak cylinder pressure, which will now take place before T.D.C. so that **negative work** *will be done in compressing the combustion products* (Fig. 5.5).

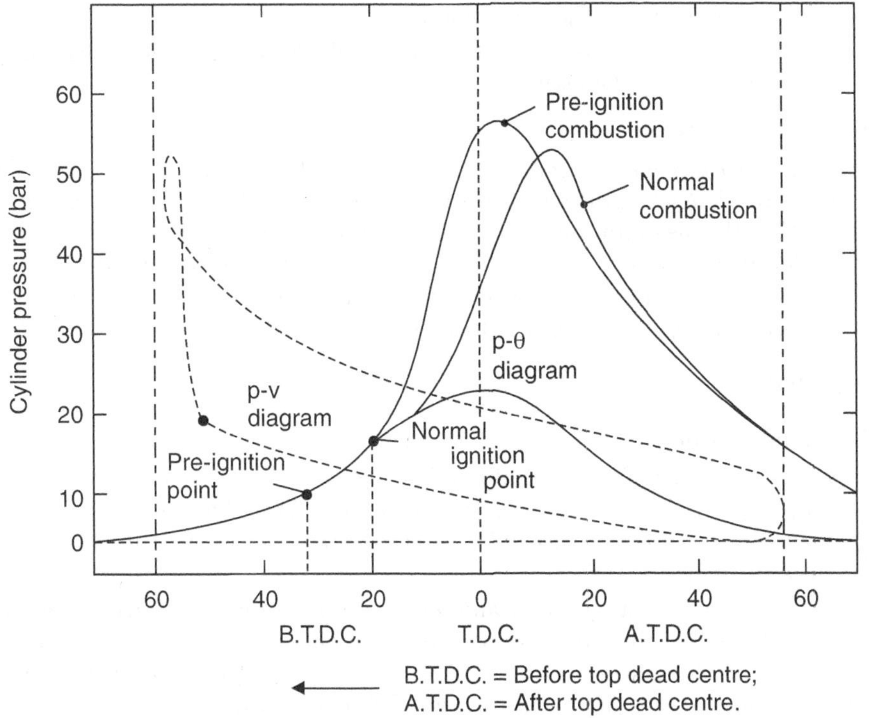

Fig. 5.5. Cylinder pressure variation when pre-ignition occurs.

Effects of pre-ignition :

1. It increases the tendency of detonation in the engines.

2. Pre-ignition is a serious type of abnormal combustion. It increases the heat transfer to the cylinder walls because high temperature gases remain in contact with the cylinder for a longer period. The load on the crankshaft during compression is abnormally high. This may *cause crank failure.*

3. Pre-ignition in a single-cylinder engine will result in a *steady reduction in speed and power output.*

4. The *real undesirable effects of pre-ignition* are *when it occurs only in one or more cylinders in a multi-cylinder engine.* Under these conditions, when the engine is driven hard, the unaffected cylinders will continue to develop their full power and speed, and so will drag the other piston or pistons, which are experiencing pre-ignition and are producing negative work, to and fro until *eventually the increased heat generated makes the pre-igniting cylinders' pistons and rings sieze.*

— Thus, the danger of the majority of cylinders operating efficiently while one or more cylinders are subjected to excessive pre-ignition is that the driver will only be aware of a loss in speed and power and therefore may try to work the engine harder to compensate for this, which only *intensifies the pre-ignition situation until siezure occurs.*

The following points are *worthnoting* :

1. Pre-ignition is not responsible for abnormally high cylinder pressure, but there can be a slight pressure rise above the normal due to the ignition point and, therefore, the peak pressure creeping forward to the T.D.C. position where maximum pressure occurs.

2. If pre-ignition occurs at the same time as the timed sparking plug fires, combustion will appear as normal. Therefore, if ignition is switched-off, the engine would continue to operate at the same speed as if it were controlled by the conventional timed spark, provided the self-ignition temperature continues to occur at the same point.

3. *Over-heated spark plugs and exhaust valves which are the main causes of pre-ignition should be carefully avoided in the engines.*

Tests for pre-ignition

- *The standard test for pre-ignition is to shut off the ignition. If the engine still fires, it is assumed that pre-ignition was taking place when the ignition was on.* Experience shows that this assumption is not always valid. *Sudden loss of power with no evidence of mechanical malfunctioning is fairly good evidence of pre-ignition.*
- The *best proof of pre-ignition is the appearance of an indicator card taken with a high speed indicator of the balanced-pressure type.*

5.6. DETONATION

5.6.1. Introduction

At present the amount of power that can be developed in the cylinder of a petrol engine is *fixed by the liability of a fuel to detonate, i.e.* just before the flame has completed its course across the combustion chamber and remaining unburnt charge fires throughout its mass spontaneously without external assistance.

The result is a tremendously rapid and local increase in pressure which sets up pressure waves that hit the cylinder walls with such violence that the walls emit a sound like a '*ping*'. It is the ping that manifests detonation. *Thus a very sudden rise of pressure during combustion accompanied by metallic hammer like sound is called* **detonation.**

The region in which detonation occurs is farthest removed from the sparking plug, and is named the "detonation zone" and even with severe detonation this zone is rarely more than one quarter the clearance volume.

5.6.2. Process of Detonation or Knocking

- The process/phenomenon of detonation or knocking may be explained by referring to the Fig. 5.6, which shows the cross-section of the combustion chamber with flame advancing from the spark plug location *A*. The advancing flame front compresses the end charge *BB′D* farthest from the spark plug, thus raising its temperature. The temperature of the end charge also increases due to heat transfer from the hot advancing flame front. Also some preflame oxidation may take place in end charge leading to further increase in its temperature.

  If the end charge *BB′D* reaches its auto-ignition temperature and remains for some time to complete the preflame reactions, the charge will autoignite leading to *knocking*

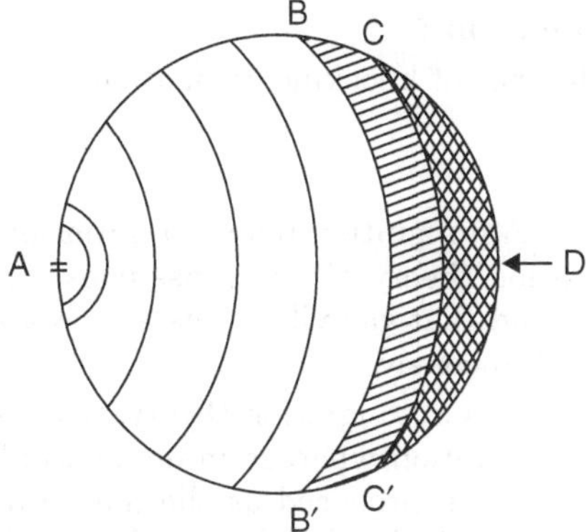

Fig. 5.6. Combustion with knocking.

combustion. During the preflame reaction period the flame front could move from *BB′* to *CC′*, and the knock occurs due to auto-ignition of the charge ahead of *CC′*. Here we have combustion unaccompanied by flame, producing a very high rate of pressure rise.

- The pressure-time diagram of detonating combustion in S.I. engines is drawn and labelled below :

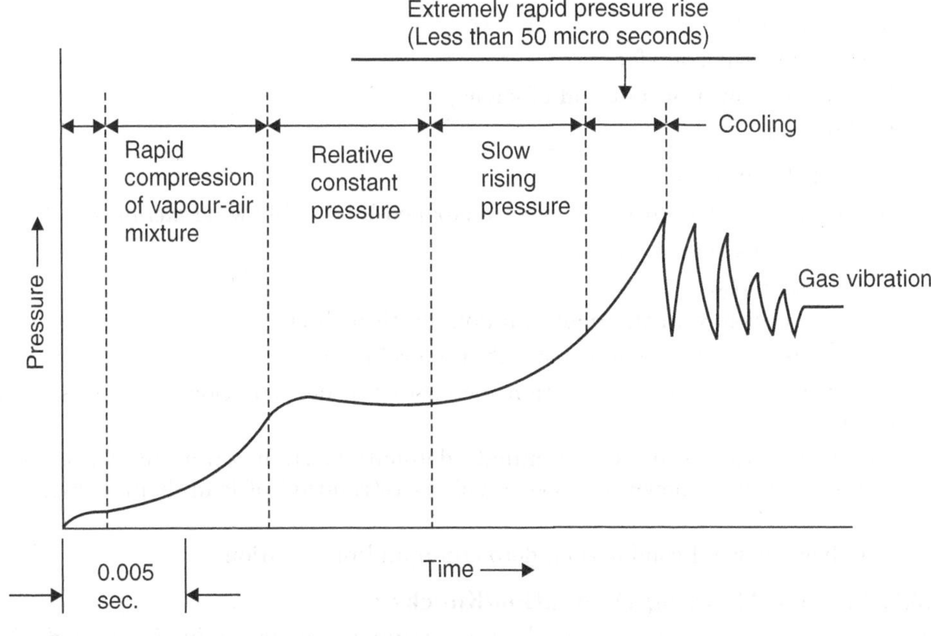

Fig. 5.7

- The *'intensity of detonation'* will depend mainly upon the *amount of energy contained in the 'end-mixture' and the rate of chemical reaction which releases it in the form of heat and a high intensity pressure-wave.* Thus, the earlier in the combustion process the detonation commences, the more unburnt end-mixture will be available to intensify the detonation. As little as 5 per cent of the total mixture charge when spontaneously ignited will be sufficient to produce a very violent knock.

5.6.3. Theories of Detonation :

There are two general theories of knocking/detonation :

(*i*) *The auto-ignition theory*

(*ii*) *The detonation theory.*

(*i*) **Auto-ignition theory.** Auto-ignition refers to initiation of combustion without the necessity of a flame. The auto-ignition theory of knock assumes that the flame velocity is *normal* before the onset-of auto-ignition and that gas vibrations are created by a number of end-gas elements auto-igniting almost simultaneously.

(*ii*) **Detonation theory.** In the auto-ignition theory, it is assumed that the flame velocity is normal before the onset of auto-ignition whereas in detonation theory a true *detonating wave* formed by preflame reactions has been proposed as the mechanism for explosive auto-ignition. Such a shock wave would travel through the chamber at about twice the sonic velocity and would compress the gases to pressures and temperatures where the reaction should be practically instantaneous.

In fact knocking or detonation is a complex phenomenon and no single explanation may be sufficient to explain it fully.

5.6.4. Effects of Detonation

1. Noise and roughness
2. Mechanical damage
3. Carbon deposits
4. Increase in heat transfer
5. Decrease in power output and efficiency
6. Pre-ignition.

Control of detonation :

The detonation can be controlled or even stopped by the following *methods* :

1. Increasing engine r.p.m.
2. Retarding spark.
3. Reducing pressure in the inlet manifold by throttling.
4. Making the ratio too lean or too rich, preferably latter.
5. ***Water injection.*** Water injection increases the delay period as well as reduces the flame temperature.
6. Use of high octane fuel can eliminate detonation. High octane fuels are obtained by adding additives known as dopes (such as tetra-ethyl of lead, benzol, xylene etc.), to petrol.

Fig. 5.4 shows normal combustion, detonation and pre-ignition.

5.6.5. Factors Affecting Detonation/Knocks :

The likelihood of knock is increased by any reduction in the induction period of combustion and any reduction in the progressive explosion flame velocity. Particular factors are listed below :

1. **Fuel choice.** A low self-ignition temperature promotes knock.

2. **Induction pressure.** Increase of pressure decreases the self-ignition temperature and the induction period. Knock will tend to occur *at full throttle.*

3. **Engine speed.** Low engine speeds will give low turbulence and low flame velocities (combustion period is constant in angle) and knock may occur at low speed.

4. **Ignition timing.** *Advanced ignition timing* increases peak pressures and *promotes* knock.

5. **Mixture strength.** *Optimum mixture strength* gives high pressures and *promotes* knock.

6. **Compression ratio.** *High compression ratios* increase the cylinder pressures and *promotes* knock.

7. **Combustion chamber design.** Poor design gives long flame paths, poor turbulence and insufficient cooling all of which *promote* knock.

8. **Cylinder cooling.** *Poor cooling* raises the mixture temperature and *promotes* knock.

5.7. PERFORMANCE NUMBER (PN)

Performance number is a *useful measure of detonation tendency.* It has been developed from the conception of knock limited indicated mean effective pressure (*klimep*), when inlet pressure is used as the dependent variable.

$$\text{Performance number (PN)} = \frac{klimep \text{ of test fuel}}{klimep \text{ of iso-octane}}$$

The performance number is *obtained on specified engine, under specified set of conditions by varying the inlet pressure.*

5.8. HIGHEST USEFUL COMPRESSION RATIO (HUCR)

The **highest useful compression ratio** *is the highest compression ratio employed at which a fuel can be used in a specified engine under specified set of operating conditions, at which detonation first becomes audible with both the ignition and mixture strength adjusted to give the highest efficiency.*

5.9. COMBUSTION CHAMBER DESIGN—S.I. ENGINES

Engine torque, power output and fuel consumption are profoundly influenced by the following :

(*i*) Engine compression ratio ;

(*ii*) Combustion chamber and piston crown shape ;

(*iii*) The number and size of the inlet and exhaust valves ;

(*iv*) The position of the sparking plug.

The following are the objects of good combustion chamber design :

1. *To optimize the filling and emptying of the cylinder with fresh unburnt charge respectively over the engine's operating speed range ; and*

2. *To create the condition in the cylinder for the air and fuel to be thoroughly mixed and then excited into a highly turbulent state so that the burning of the charge will be completed in the shortest possible time.*

3. To prevent the possibility of detonation at all times, as far as possible in order to achieve these fundamental requirements it is imperative to be aware of the *factors that contribute towards inducing the charge to enter the cylinder, to mix intimately, to burn both rapidly and smoothly and to expel the burnt gases.*

5.9.1. Induction Swirl

Refer Fig. 5.8.

- ***Swirl** is the rotational flow of charge within the cylinder about* its axis.
- *Swirl* is generated by constructing the intake system to give a tangential component to the intake flow as it enters the cylinder. This is done by *shaping and contouring the intake manifold, valve ports and even the piston face.*
- *Swirl greatly enhances the mixing of air and fuel* to give a homogeneous mixture in the very short time available for this in modern high speed engines. It is also a main mechanism for spreading of the flame front during the combustion process.

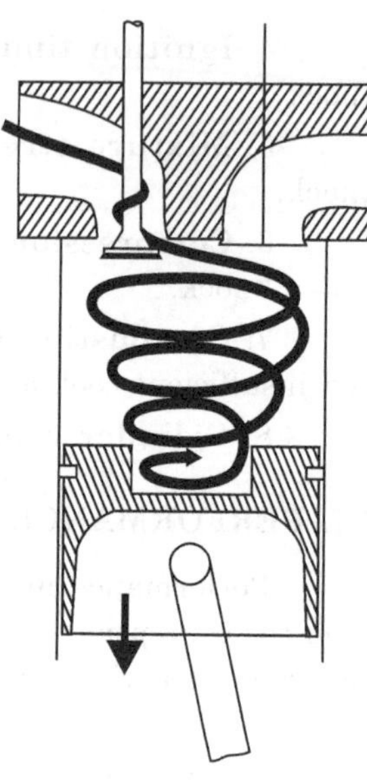

Fig. 5.8. Induction swirl.

The induction ports are classified as follows : Refer Fig. 5.9.

1. *Direct straight port.*
2. *Deflector wall port.*
3. *Masked valve port.*
4. *Helical port.* The intensity of swirl is influenced by the steepness of the port helix and the mean diameter of the spiral flow path about the valve axis.

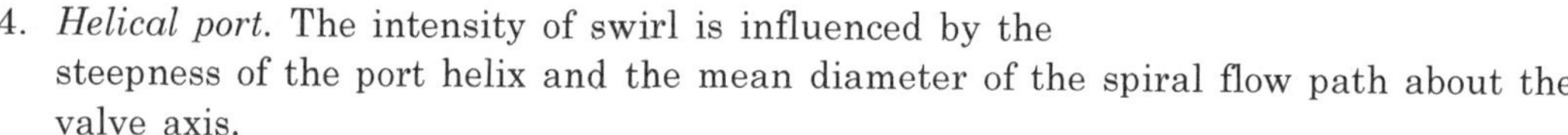

— Helical ports usually provide *higher flow discharges for equivalent levels of swirl compared with directed ports because* the whole periphery of the valve opening area can be fully utilized, and, as a result, higher volumetric efficiencies can be obtained in the low-to-mid speed range of the engine.

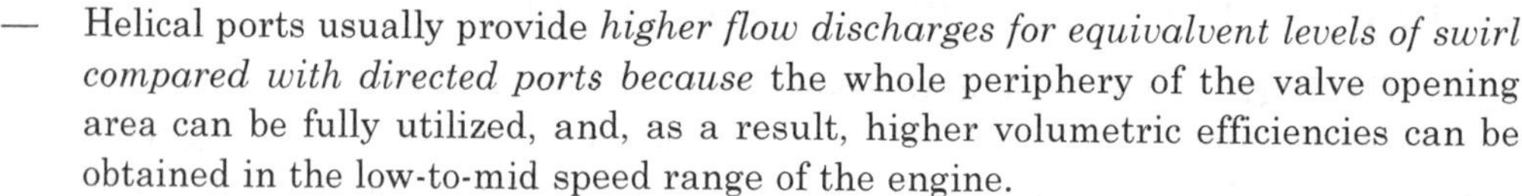

— These ports are *less sensitive to their position relative to the cylinder axis* since the swirl generated depends mainly on the port geometry above the valve and not how it enters the cylinder. Generally, the magnitude of swirl rises with increased valve lift.

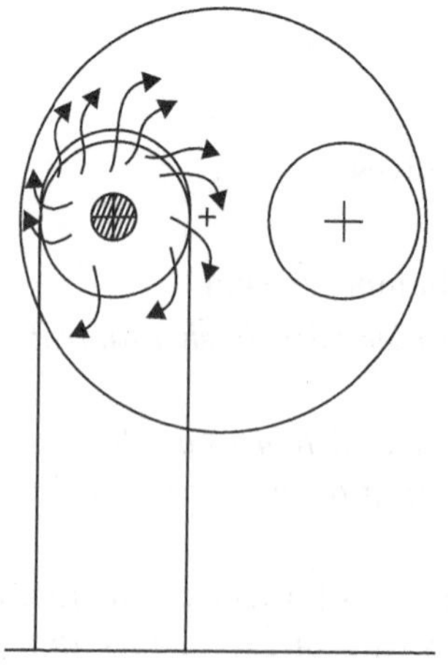

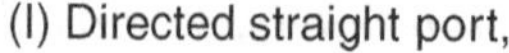

(I) Directed straight port,

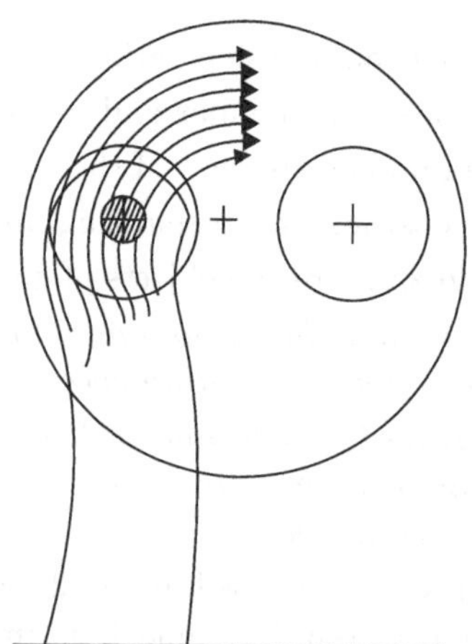

(ii) Deflector wall port,

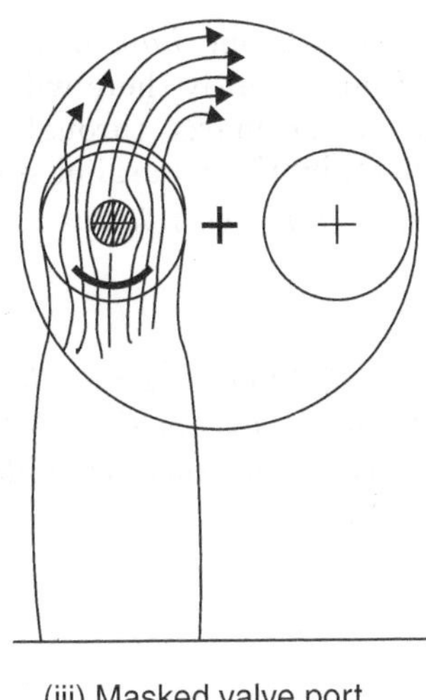
(iii) Masked valve port

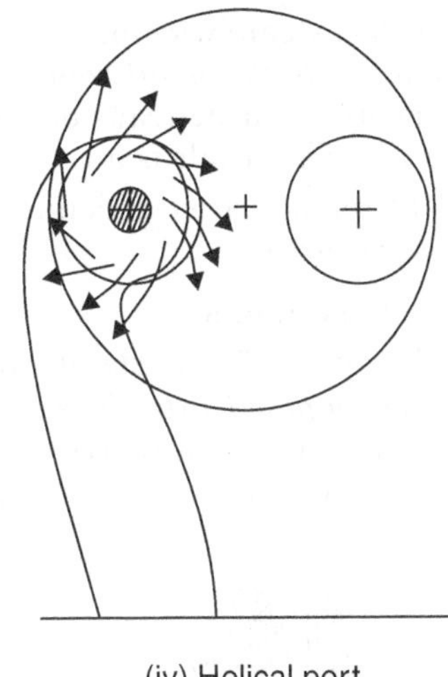
(iv) Helical port

Fig. 5.9. Induction ports.

— These ports, however, *suffer from a loss of volumetric efficiency in the upper speed range of the order 5 to 10%.*

- In **chamber wall deflected induction swirl,** the downward and circular movement of the mixture *generates an expanding, and then a contracting special swirl about the cylinder axis during both the induction and compression strokes, respectively.*

Methods of Intensifying the Rate of Burning

5.9.2. Squish and Tumble

- As the piston approaches T.D.C. at the end of compression stroke, the volume around the outlet edges of the combustion chamber is suddenly reduced to a very small value. Many modern combustion chamber designs have most of the clearance volume near the centreline of the cylinder. As the piston approaches T.D.C. the gas mixture occupying the volume *at the outer radius of the cylinder is forced radially inward as this outer volume is reduced to near zero. This radial inward motion of the gas mixture is called 'squish'.* It adds to other mass motions within the cylinder to mix the air and fuel and to quickly spread the flame front. Fig. 5.10 shows a typical compression squish.
- As the piston nears T.D.C. squish motion generates a *secondary rotational flow called* **"tumble"**. *This rotation occurs about a circumferential axis near the outer edge of the piston bowl.*

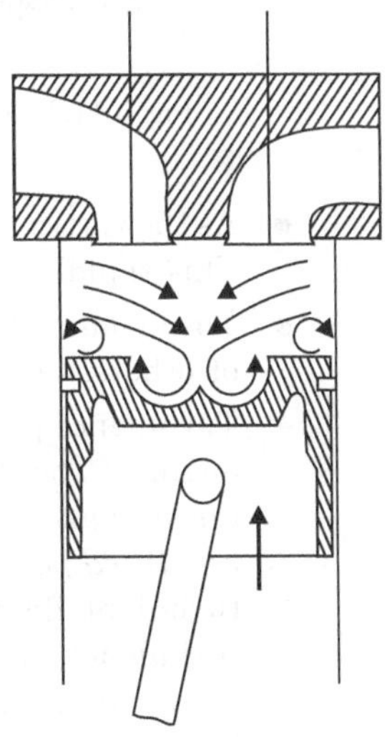
Fig. 5.10. Compression squish.

5.9.3. Quench Area

— The quench area is defined by the *parallel portion of the piston and cylinder head which almost touch each other as the piston approaches T.D.C.* These opposing flat surfaces sandwiching a thin lamina of charge between them, *have a large surface relative to the small volume trapped between them.* Consequently there will be a large amount of heat transferred from this thin lamina of hot charge through the metal walls. The result is a rapid cooling or quenching effect, by these parallel surfaces.

— The quench area is defined as *percentage of opposing flat area relative to the piston crown area.*

5.9.4. Turbulence

- *"Turbulence" consists of randomly dispersed vortices of different sizes which become superimposed into the air, or air and petrol mixtureflow stream* (Fig. 5.11). These vortices, which are carried along with the flow stream, represent small irregular breaks-ways that take on a concentric spiral motion (Fig. 5.12).

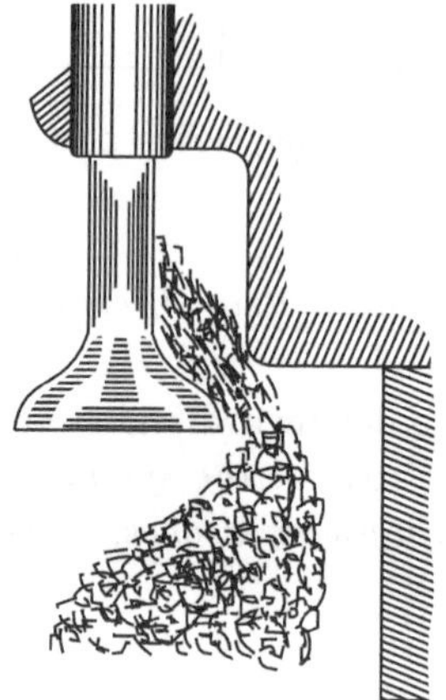

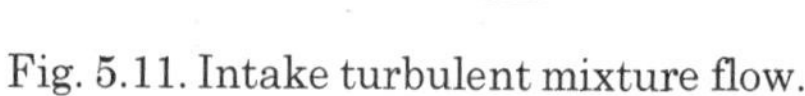

Fig. 5.11. Intake turbulent mixture flow.

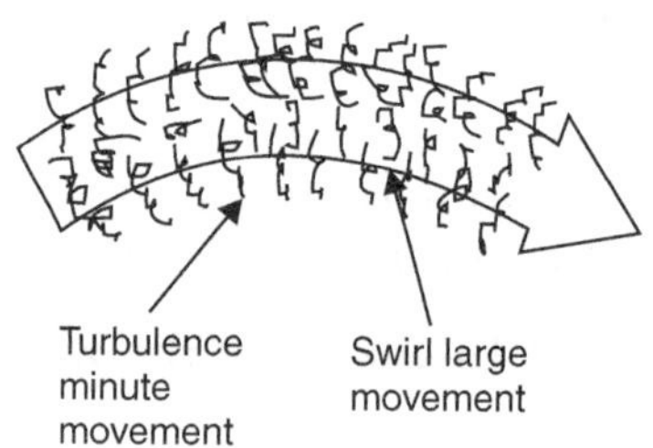

Fig. 5.12. Illustration of turbulence superimposed on mixture swirl.

- As the vortices whirl they will contact adjacent vortices causing viscous shear interaction. This rapidly speeds up the rate of heat transfer and fuel mixing.
- The amount of vortex activity, that is the formation of new vortices and the disintegration of others, *increases the turbulent flow with rising engine speed.*

— Turbulence plays a very important role in combustion phenomenon in S.I. (as well C.I.) engines. The flame speed is very low in non-turbulent mixtures. *A turbulent motion of the mixture intensifies the processes of heat transfer and mixing of the burned and unburned portions in the flame front* (diffusion). These two factors cause the velocity of turbulent flame to increase practically in proportion to the turbulent velocity. The turbulence of the mixture is due to admission of fuel-air mixture through comparatively narrow sections of the intake pipe, valves etc., in the suction stroke. The turbulence can be increased at the end of the compression stroke by suitable design of combustion chamber which involves the geometry of cylinder head and piston crown.

— The degree of turbulence increases *directly* with the piston speed.

The *effects of turbulence* can be summed up as follows :

1. Turbulence accelerates chemical action by intimate mixing of fuel and oxygen. Thus weak mixtures can be burnt.

2. The increase of flame speed due to turbulence reduces the combustion time and hence minimises the tendency to detonate.
3. Turbulence increase the heat flow to the cylinder wall and in the limit excessive turbulence may extinguish the flame.
4. Excessive turbulence results in the more rapid pressure rise (though maximum pressure may be lowered) and the high pressure rise causes the crankshaft to spring and rest of the engine to vibrate with high periodicity, resulting in rough and noisy running of the engine.

5.9.5. Flame Propagation

- Typical flame propagation velocities range from something like 15 to 70 m/s. This would relate to the combuston flame velocity increasing from about 15 m/s at an idle speed of about 1000 r.p.m. to roughly 70 m/s at a maximum speed of 6000 r.p.m.
- *When ignition occurs the nucleus of the flame spreads with the whirling or rotating vortices in the form of ragged burning crust from the initial spark plug ignition site.*
- *The speed of the flame propagation is roughly proportional to the velocity at the periphery of the vortices.*

5.9.6. Swirl Ratio

- Induction swirl can be generated by tangentially directing the air movement into the cylinder either by creating a preswirl in the induction port or by combining the tangential-directed flows with a preswirl helical port. *"Cylinder air swirl"* is defined *as the angular rotational speed about the cylinder axis.*
- ***Swirl ratio*** is defined as the *ratio of air rotational speed to crankshaft rotational speed.*
  — Helical ports can achieve swirl ratio of 3 to 5 at T.D.C. with a flat piston crown. However, if a bowl in the piston chamber is used, the swirl ratio can be increased to about 15 at T.D.C.

5.9.7. Surface-to-Volume Ratio

- In order to *minimise* the heat losses and formation of hydrocarbons within the combustion

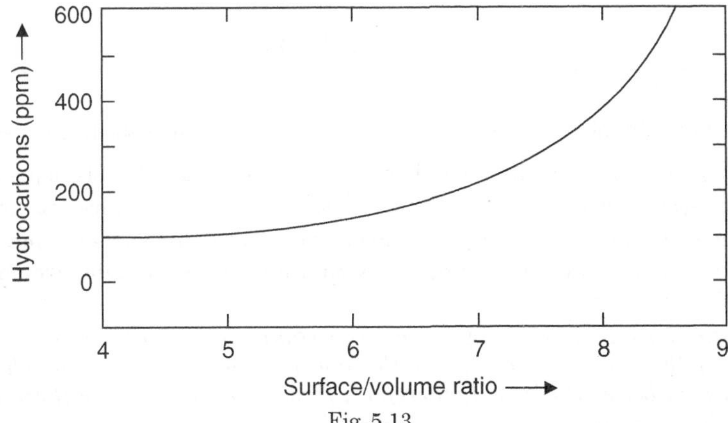

Fig. 5.13

chamber, the chamber volume should be maximised relative to its surface area, that is, the *chamber's surface area should be as small as possible relative to the volume occupied by the combustion chamber* (Fig. 5.13). The surface-to-volume ratio is the *ratio of the combustion surface area to that of its volume.*

- The surface-to-volume ratio *increases linearly with rising compression ratio.*

5.9.8. Stroke-to-Bore Ratio

- For various engines the stroke-to-bore ($L : D$) ratio can range from 0.6 : 1 to 1.4 : 1.
  — When $L = D$, the $L : D$ ratio is said to be *square* ;
  — When $L < D$, the $L : D$ ratio is said to be *oversquare* ;
  — When $L > D$, the engine is said to be *undersquare*.
- *"Oversquare" engines are more suitable for saloon car petrol engines,* whereas *"undersquare" engines are better utilised for large diesel engines.*

5.9.9. Compression Ratio (C.R.)

- When compression ratio increases from 5 : 1 to 10 : 1 the cylinder's compression pressure increases from 8.0 bar to 19.0 bar respectively (Fig. 5.14). Correspondingly, the maximum cylinder pressure increases from 32 bar to 82 bar and b.m.e.p. generated also increases from 9.4 bar to 11.8 bar over the same compression range respectively.
- The effect of higher cylinder pressure is to cause a corresponding rise in cylinder temperature from 360°C to 520°C over the same compression ratio rise. Raising the cylinder temperature reduces the ignition delay period for one set engine speed (Fig. 5.15). Thus, for an engine running in its mid-speed range, the ignition timing would be reduced from 37.5° to 12.5° before T.D.C. if its compression ratio is increased from 5 : 1 to 10 : 1.

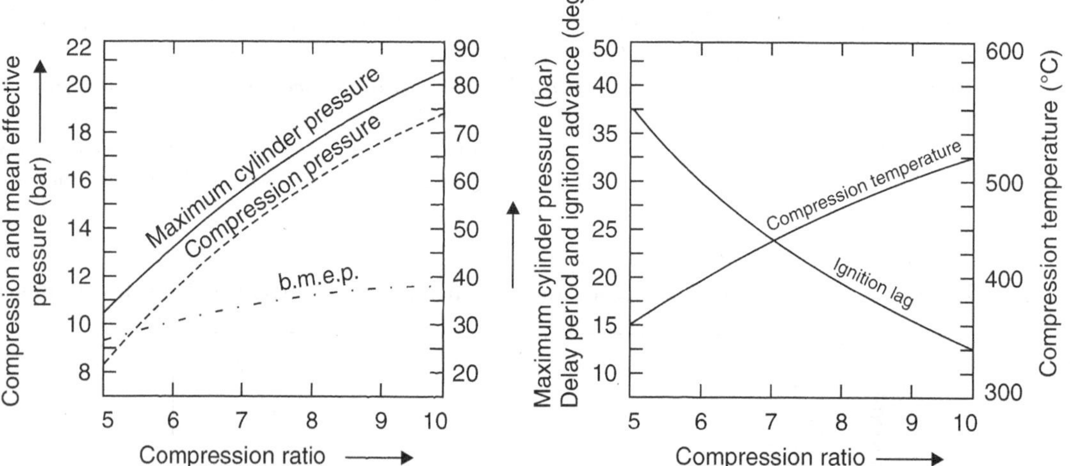

Fig. 5.14. Effect of compression ratio on the b.m.e.p. compression and maximum cylinder pressures.

Fig. 5.15. Effect of compression ratio on the air temperature and ignition lag.

- The effects of compression ratio on the characteristic pressure-volume diagram and the characteristic pressure-crank angle diagram for a petrol engine are shown in Fig. 5.14 and 5.15 respectively.
- *The main reason for raising the engine compression ratio is due to the increased density of the air-fuel mixture at the point of ignition, so that when the energy is released it is better utilized. It therefore, raises both the engine thermal efficiency and the developed power.*
- Out of the *major unwanted side effects* of raising the compression ratio is that there will be a corresponding *increase in cylinder pressure which,* in turn, *increases the piston-ring to cylinder-wall friction and compression and expansion heat losses.* Consequently, the *higher compression ratio produces a reduction in the mechanical efficiency.* Subsequently, increasing the compression ratio produces an *increase in thermal efficiency but at the expense of a falling mechanical efficiency.*

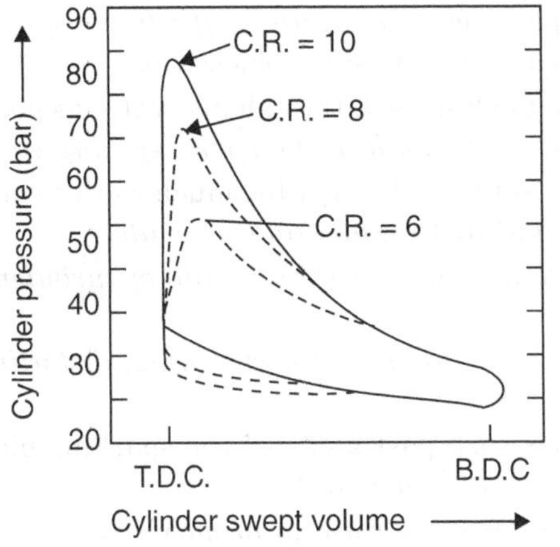

Fig. 5.16. Effect of compression ratio on the characteristic pressure-volume diagram for a petrol engine.

Fig. 5.17. Effect of compression ratio on the characteristic pressure-crank-angle diagram for a petrol engine.

- The *merits* and *limitations* of raising the compression ratio with regards to thermal efficiency and mechanical efficiency are shown in Fig. 5.18, whereas Fig. 5.19 shows the *benefits* of increased power and reduced specific fuel consumption with rising compression ratio.

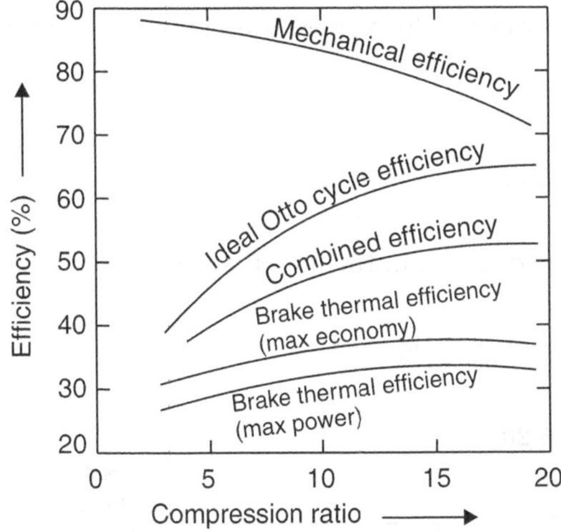

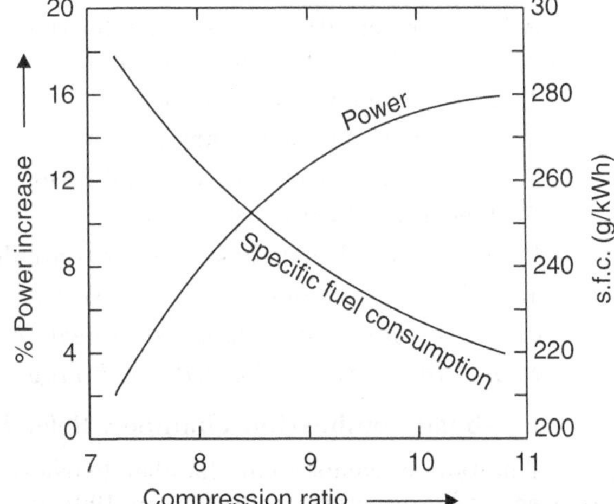

Fig. 5.18. Effect of compression ratio on an engine's thermal and mechanical efficiency.

Fig. 5.19. Effect of compression on engine power and specific fuel consumption.

For S.I. engines' combustion design practice, summarily, the following are required :

1. The smallest ratio of chamber surface-area to chamber volume as possible ... *to minimise heat losses to the cooling system.*
2. The shortest flame-front travel distance as possible ... *to minimise the combustion period.*
3. The provision for quenching the mixture farthest from the sparking plug ... *to prevent the end-gas overheating.* (However, it must not be excessive as this would prevent the end-gases burning and, therefore, it would cause a high level of hydrocarbons to be expelled to the exhaust).

4. The most central sparking plug position possible ... *to minimise the flame spread path* (or, alternately, *twin plugs* can be used to achieve the same objective).
5. The location of the sparking plug should be as close as to the exhaust valve as possible ... *to maximise the temperature of the mixture surrounding the sparking plug electrodes.*
6. The incoming mixture must have adequate squirrel (but not too much as this could lead to excessive heat losses) ... *to mix the air and fuel rapidly and intimately.*
7. The provision for squish zones ... *to excite the mixture into a state of turbulence just before the combustion occurs.*
8. The provision for cooling of the exhaust valve ... *to prevent overheating, distorting, and burning occurring.*
9. The provision for incoming fresh charge to sweep past and cool the sparking plug electrodes ... *to avoid pre-ignition under wide throttle opening.*
10. The utilisation of the highest possible compression ratio ... *to maximise the engine's thermal efficiency without promoting detonation.*
11. The inlet and exhaust valve sizes and numbers should be adequate ... *to expel the exhaust-gases and to fill the cylinder with the maximum mass of fresh charge in the upper speed range.*
12. The degree of turbulence created should be controlled ... *to prevent excessively high rates of burning and, correspondingly, limit very high rates of pressure rise which would cause rough and noisy running.*

5.10. SOME TYPES OF COMBUSTION CHAMBERS

A few representative types of combustion chambers of which there are many more variations are enumerated and discussed below :

1. *T*-head combustion chamber.
2. *L*-head combustion chamber.
3. *I*-head (or overhead valve) combustion chamber.
4. *F*-head combustion chamber.

It may be noted that these chambers are designed to obtain the objectives namely :

- A *high combustion rate at the start.*
- A *high surface-to-volume ratio near the end of burning.*
- A *rather centrally located spark plug.*

1. **T-head combustion chamber.** Refer Fig. 5.20.

This type of combustion chamber (earliest type) was used by Ford-motor corporation in 1908 in its famous model '*T*'.

The *T*-head design has the following *disadvantages* :

(*i*) Requires two cam shafts (for actuating the inlet valve and exhaust valve separately) by two cams mounted on the two cam shafts.

(*ii*) Very prone to detonation. There was violent detonation even at a compression ratio of 4 (with a fuel of octane number of 50).

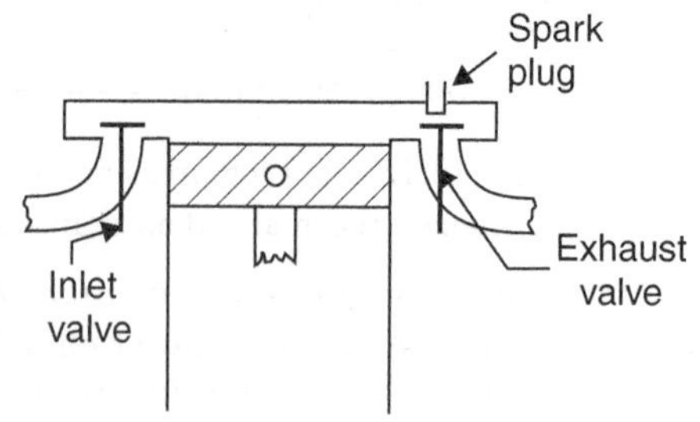

Fig. 5.20. *T*-head combustion chamber.

2. **L-head combustion chamber.** Refer Fig. 5.21.

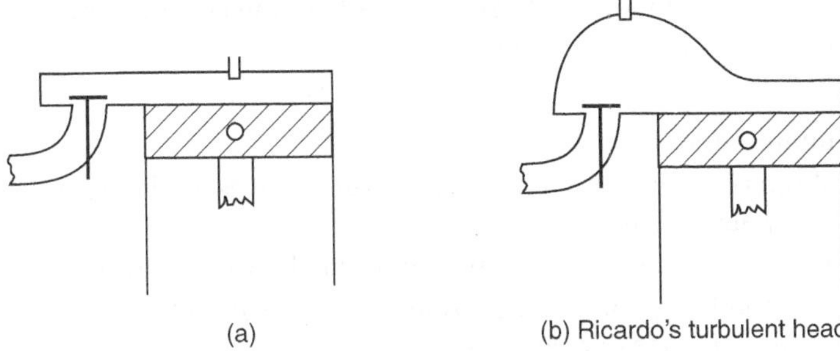

Fig. 5.21. *L*-head combustion chamber.

- It is a modification of the *T*-head type of combustion chamber. It provides the *two valves on the same side of the cylinder,* and the valves are operated through tappet by a *single camshaft.*
- Fig. 5.21 (*a*) and (*b*) shows two types of this side-valve engine. In these types *it is easy to lubricate the valve mechanism* with the detachable head, it may be noticed that *the cylinder head can be removed for cleaning or decarburising without disturbing valve-gear etc.*
  - — In Fig. 5.21 (*a*), the air flow has to take two right-angled turns to enter the cylinder. This causes a loss of velocity head, and a loss in turbulence level, *resulting in slow combustion process.*
  - — Fig. 5.21 (*b*) is the ***Ricardo's turbulent head design.*** The main body of the combustion chamber is concentrated over the valves leaving a slightly restricted passage communicating with the cylinder, thereby *creating additional turbulence during the compression stroke.* This design *reduces the knocking tendency by shortening the effective flame travel length* by bringing that portion of the head which lay over the further side of the piston into as close a contact as possible with the piston crown, forming a *quench space.* The thin layer of mixture (entrapped between the relatively cool piston and also cooled head) loses its heat rapidly, thereby avoiding knock. By placing the spark plug in the centre of the effective combustion space but with slight bias towards the exhaust valve, the flame travel length is reduced.

Advantages :

(*i*) Valve mechanism simple and easy to lubricate.

(*ii*) Detachable head easy to remove for cleaning and decarburising.

(*iii*) Valves of larger sizes can be provided.

Disadvantages :

(*i*) More surface-to-volume ratio and therefore *more heat loss.*

(*ii*) Longer length of flame travel.

(*iii*) Valve size restricted.

(*iv*) Thermal failure in cylinder block also. In *I*-head engine the thermal failure is confined to cylinder head only.

3. **I-head (or overhead valve) combustion chamber.** Refer Fig. 5.22.

This type of combustion chamber has both the inlet valve and the exhaust valve located in the cylinder head. An overhead engine is superior to side valve engine at high compression ratios.

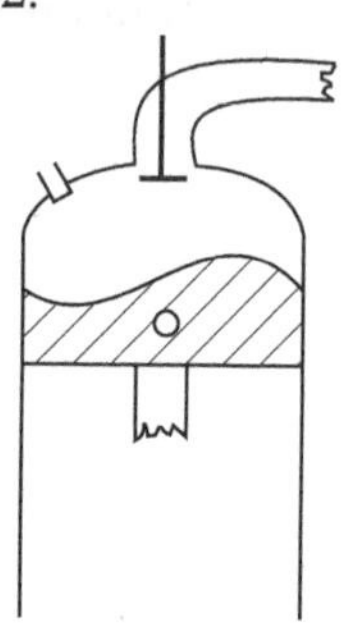

Fig. 5.22. *I*-head combustion chamber.

Advantages :

(*i*) Reduced pumping losses.

(*ii*) Higher volumetric efficiency (since the larger valves and larger lifts can be accommodated).

(*iii*) Less prone to detonation (since the path of flame travel is reduced).

(*iv*) Less force on the head bolts and therefore less possibility of leakage of compression gases or jacket water.

(*v*) Lower surface-volume ratio and, therefore, less heat loss and less air pollution.

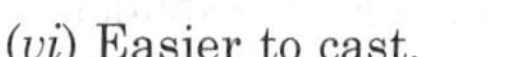

(*vi*) Easier to cast.

4. **F-head combustion chamber**

- In such a combustion chamber one valve is in head and other in the block. This design is a compromise between *L*-head and *I*-head combustion chambers.
- One of the most perfect *F*-head engines (wedge type) is the one used by the Rover company for several years. Its *advantages* are :

(*i*) High volumetric efficiency ;

(*ii*) Maximum compression ratio for fuel of given octane rating ;

(*iii*) High thermal efficiency ;

(*iv*) It can operate on leaner air-fuel ratios without misfiring.

The *drawback* of this design is the *complex mechanism for operation of valves and expensive special shaped piston.*

— Another successful design of this type of chamber is that used in Willys jeep.

Note 1. Some modern engines have divided chamber. These offer *high volumetric efficiency, good fuel economy,* and *cycle operation flexibility.* Their main *disadvantages* are *greater heat loss,* due to high surface area, and *high cost* and *difficulty in manufacturing.*

2. Very large engines are almost always C.I. engines. Because of their large combustion chambers and corresponding long flame travel distance, combined with slow engine speed, they would require very high octane fuel and very low compression ratio if operated as an S.I. engine. With the very long real time of combustion in the cylinder, it would be impossible to avoid serious knock problems.

5.10.1. Divided Combustion Chambers

- Some engines have divided combustion chambers, usually with about *80 percent* of the clearance volume in the **main chamber** above the piston and about *20 percent* of the volume as a **secondary chamber** *connected through a small orifice* (Fig. 5.23). *Combustion is started in the small secondary chamber, and the flame then passes through the orifice, where it ignites the main chamber.* Intake swirl is not as important in the main chamber of this type of engine, so the intake system can be designed for greater volumetric efficiency. It is desirable to have very high swirl in the secondary chamber, and the orifice between the chambers is shaped to supply this often, the *secondary chamber is called a "swirl chamber".* As the gases in the secondary chamber are consumed by combustion, the pressure rises and flaming gas expands back through the orifice and acts as a *torch ignition* for the main chamber. The expanding gas rushing back through the orifice creates a large secondary swirl in the main chamber,

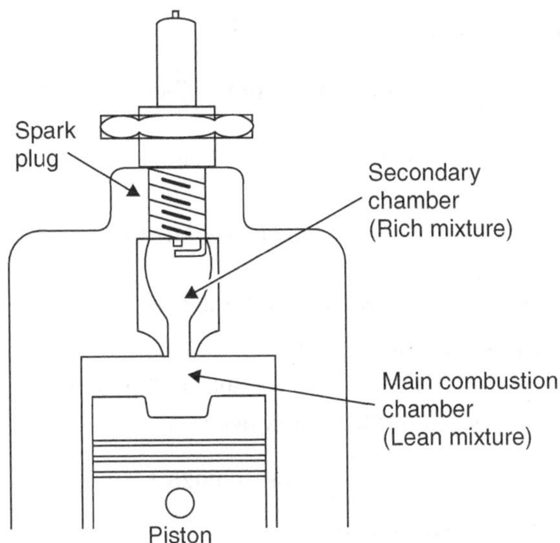

Fig. 5.23. Divided combustion chamber.

which enhances the combustion there. Creating an orifice that can do all this is a major design challenge.

- A divided chamber engine, oftenly, will also be a stratified charge engine. *The intake system is designed to supply a rich mixture in the secondary chamber and a lean mixture in the main chamber.* The rich mixture with very high swirl in the secondary chamber will ignite readily and combust very quickly. The flaming gases expanding back through the orifice will then ignite the lean mixture in the main chamber, a mixture often so lean that it would be difficult to ignite with a spark plug alone. The net result is *an engine that has good ignition and combustion, yet operates mostly lean to give good fuel economy.* Placement and timing of intake valves injectors to supply the proper air and fuel to all parts of this engine are extremely important.

Note. A variation of this type of combustion chamber on some C.I. engines is one with a totally passive secondary chamber, with all valves and injectors located in the main chamber. When combustion occurs in the main chamber, high pressure forces gas through the very small orifice and raises the pressure in the secondary chamber also. *When the pressure in the main chamber is reduced during the power stroke, the high pressure gases in the secondary chamber flow back into the main chamber. This holds the pressure in the main chamber to a higher level for a short time and gives a smooth, slightly greater force off the piston during the power stroke.* This kind of secondary chamber usually consists of about 5-10 percent of the clearance volume.

WORKED EXAMPLES

Example 5.1. *A S.I. engine operating at 1200 r.p.m. has a 10.2 cm bore with spark plug offset by 6 mm from the centre. The spark plug is fired at 20°C before T.D.C. It takes 6.5° of engine rotation for combustion to develop and get into flame propagation mode, where the average flame speed is 15.8 m/s. Calculate :*

(i) Time of one combustion process (i.e. time for flame front to reach the farthest cylinder wall) in sec. ;

(ii) Crank angle position at the end of combustion. **(Madras University)**

Solution. Maximum distance of flame travel

$$= \frac{1}{2} \text{ bore + spark plug offset}$$

$$= \frac{1}{2} \times 10.2 + \frac{6}{10} = 5.7 \text{ cm}$$

Time of flame travel $= \dfrac{5.7 \times 10^{-2}}{15.8} = 3.6076 \times 10^{-3}$ s

1200 r.p.m. $= \dfrac{1200}{60} \times 360 = 7200$ deg./s

$\therefore$ Crank angle for flame travel

$$= 3.6076 \times 10^{-3} \times 7200 = 25.975 \text{ deg.}$$

Time for combustion to develop = 6.5 crank degrees

$$= \frac{6.5}{7200} \quad \text{or} \quad 0.9028 \times 10^{-3} \text{ s}$$

(*i*) **Time for one combustion process**

= Time to develop + Time for propagation

$= 0.9028 \times 10^{-3}$ s $+ 3.6076 \times 10^{-3}$

$= \mathbf{4.5104 \times 10^{-3}}$. **(Ans.)**

(*ii*) **Total crank rotation**

= 6.5 + 26.975 = 33.48 degrees of crank rotation.

Since spark is fired at 20° before T.D.C, the crank position will be (33.48 – 20) or **13.48 degrees after T.D.C. (Ans.)**

Example 5.2. *In a trial on S.I. engine at full speed full power (i.e., fully open throttle) the spark occurred 26° bT.D.C (before top dead centre) and delay ended 4° bT.D.C. Assuming that the combustion period should finish 13° aT.D.C. (after top dead centre) for maximum power and that the effect of half closing the throttle at constant speed is to increase the delay period by 14% of the valve at full throttle, estimate the optimum spark timing for maximum power under following conditions :*

(*i*) *Under full throttle conditions when the engine is operated at half the maximum speed ;*

(*ii*) *When the engine is operated at conditions of half the maximum speed and the throttle half open.*

State how these alterations in optimum spark timing may be achieved in practice.

(Bombay University)

Solution.

• *The delay period, at constant throttle, is constant in time and thus increases in angle with the speed.*

• The *combustion period is constant in crank angle.*

The delay period = From 26° T.D.C. to 4° *b*T.D.C. *i.e.,* 22°

The combustion period = From 4° T.D.C. to 13° *a*T.D.C. *i.e.,* 17°.

(*i*) Full throttle half speed will result in delay angle being reduced to $\dfrac{22}{2}$ = 11° for the same time thus ignition timing should be arranged so that the total of 11 + 17 = 28°, ends 13° *a*T.D.C.

$\therefore$ **Time of spark** = 28 – 13 = **15° *b*T.D.C. (Ans.)**

A *centrifugal device* is used to accomplish this task.

(*ii*) Half throttle half speed will result in an increase of 14% in delay time over that at full throttle half speed *i.e.* by

$$\frac{14}{100} \times 11 = 1.54°$$

∴ Delay angle = 11 + 1.54 = 12.54°

Combustion period remains same as 17°

∴ Total period = 12.54 + 17 = 29.54° ; end is 13° *a*T.D.C.

∴ **Time of spark** = 29.54 – 13 = **16.54° *b*T.D.C. (Ans.)**

This is accomplished by *a vacuum device connected to the inlet manifold.*

HIGHLIGHTS

1. **Combustion** may be defined as a relatively rapid chemical combination of hydrogen and carbon in the fuel with the oxygen in the air, resulting in liberation of energy in the form of heat.
2. **Ignition lag** is the time lag between the first igniting of fuel and commencement of the main phase of combustion.
3. **Pre-ignition** is the ignition of the homogeneous mixture in the cylinder, before the timed ignition spark occurs, caused by the local overheating of the combustible mixture. The standard test for pre-ignition is to shut-off the ignition. If the engine still fires, it is assumed that pre-ignition was taking place when the ignition was on.
4. A very sudden rise of pressure during combustion accompanied by metallic hammer like sound is called **detonation.**
5. **Performance number** (PN) is a useful measure of detonation tendency,

$$PN = \frac{\text{klimep of test fuel}}{\text{klimep of iso-octane}}.$$

6. The *highest useful compression ratio* (HUCR) is the highest compression ratio employed at which a fuel can be used in a specified engine under a specified set of operating conditions, at which detonation first becomes audible with both the ignition and mixture strength adjusted to give the highest efficiency.
7. **Swirl** is rotational flow of charge within the cylinder about its axis. It is generated by constructing the intake system to give a tangential component to the intake flow as it enters the cylinder.
8. **Squish** is the radial inward motion of the gas mixture. As the piston near T.D.C. squish motion generates a secondary rotational flow called "*tumble*". This rotation occurs about a circumferential axis near the outer edge of the piston bowl.
9. **Quench area** is defined by the parallel portion of the piston and cylinder head which almost touch each other as the piston approaches T.D.C. It is defined as percentage of opposing flat area relative to the piston crown area.
10. **Turbulence** consists of randomly dispersed vortices of different sizes which become superimposed into the air, or air and petrol mixture flow stream.
11. The **speed of the flame propagation** is roughly proportional to the velocity at the periphery of the vortices.
12. **Cylinder air swirl** is defined as the ratio of angular rotational speed about the cylinder axis.
13. **Swirl ratio** is defined as the ratio of air rotational speed to crankshaft rotational speed.
14. The **surface to volume ratio is** the ratio of the combustion surface area to that of its volume. It increases linearly with rising compression ratio.
15. **Over-square** ($L < D$) engines are more suitable for saloon car petrol engines, whereas *under square* ($L > D$) engines are better utilised for large diesel engines.

OBJECTIVE TYPE QUESTIONS

Fill in the Blanks or Say 'Yes' or 'No' :

1. ... may be defined as a relatively rapid chemical combination of hydrogen and carbon in the fuel with the oxygen in the air resulting in liberation of energy in the form of heat.
2. The lower and upper limits of ignition of the mixture depend upon the temperature and mixture ratio.
3. The time lag between first igniting of fuel and commencement of the main phase of combustion is called the period of ...
4. The ignition lag is a chemical process.
5. An increase in compression ratio decreases the maximum pressure and the work transfer.
6. The higher the self ignition temperature of fuel, ... the ignition lag.
7. Ignition lag is the smallest for the mixture ratio which gives the maximum temperature.
8. Ignition lag is ... if the initial temperature and pressure are increased.
9. Ignition lag is much affected by turbulence.
10. Usually the spark should occur at about 15° before T.D.C.
11. The correct instant for the introduction of spark is mainly determined by the ... .
12. In general rich mixtures burn faster.
13. For maximum power and economy a slow burning fuel needs a higher spark advance than a fast burning fuel.
14. ... is the ignition of the homogeneous mixture in the cylinder, before the timed ignition spark occurs, caused by the local overheating of the combustible mixture.
15. Pre-ignition increases the tendency of detonation in the engines.
16. A very sudden rise of pressure during combustion accompanied by metallic hammer like sound is called ... .
17. ... number is a useful measure of detonation tendency.
18. ... is the rotational flow of charge within the cylinder about its axis.
19. The ... area is defined by the parallel portion of the piston and cylinder head which almost touch each other as the piston approaches T.D.C.
20. ... consists of randomly dispersed vortices of different sizes which becomes superimposed into the air, or air and petrol mixture flow steam.
21. 'Cylinder air swirl' is defined as the angular rotational speed about the cylinder axis.
22. ... ratio is defined as the ratio of air rotational speed to crankshaft rotational speed.
23. The degree of turbulence increases directly with the piston speed.
24. When $L < D$, the $L : D$ ratio is said to be ... .
25. Divided combustion chambers offer high volumetric efficiency, good fuel economy, and cycle operation flexibility.

ANSWERS

1. Combustion	**2.** Yes	**3.** incubation	**4.** Yes	**5.** No
6. Longer	**7.** Yes	**8.** reduced	**9.** No	**10.** Yes
11. ignition lag	**12.** Yes	**13.** Yes	**14.** Pre-ignition	**15.** Yes
16. detonation	**17.** Performance.	**18.** Swirl	**19.** quench	**20.** Turbulence
21. Yes	**22.** Swirl	**23.** Yes	**24.** oversquare	**25.** Yes.

THEORETICAL QUESTIONS

1. Define 'combustion'. State the general conditions necessary for combustion.
2. Discuss the ignition limits of hydrocarbon fuels.
3. Explain briefly combustion phenomenon in S.I. engines.
4. What do you mean by pre-ignition ? How can it be detected ?
5. Explain the difference between (*i*) pre-ignition, (*ii*) auto-ignition, and (*iii*) detonation.
6. Explain the phenomenon of auto-ignition. Explain how auto-ignition is responsible for knocking in S.I. engines.
7. Explain the phenomenon of knocking in S.I. engines. What are the different factors which influence the knocking ? Describe the methods used to suppress it.
8. Explain the main factors that influence the flame speed.
9. What is performance number ?
10. What are the factors that limit the compression ratio that can be used in petrol engines ?
11. "Abnormal combustion knock produced by surface ignition in S.I. engines is more harmful than normal combustion knock". Justify the statement.
12. How tetraethyl lead (T.E.L) improves the quality of fuel for S.I. engine ?
13. What do you mean by octane number of 85 and octane number of 75 ? What is H.U.C.R. ?
14. Space of the clearance volume controls the detonation in case of S.I. engine. Comment.
15. What is ignition lag ? Discuss the effect of engine variables on ignition lag.
16. Discuss the effects of the following variables on engine heat transfer :
    (*i*) Spark advance ; (*ii*) Engine output ;
    (*iii*) Pre-ignition and knocking.
17. "The highest compression ratio that can be used in a S.I. engine is limited by the detonation characteristics of the available fuel". Justify the statement.
18. "The retarding of spark timing in a S.I. engine will reduce detonation". Justify the statement.
19. What action can be taken with regard to the following variables, in order to reduce the possibility of detonation in a S.I. engine ? Justify your answers by reasons.
    (*i*) Compression ratio ; (*ii*) Mass of charge induced ;
    (*iii*) Mixture inlet temperature ; (*iv*) Engine speed ;
    (*v*) Distance of flame travel.
20. Discuss the effect of the following engine variables on flame propagation :
    (*i*) Compression ratio ; (*ii*) Fuel-air ratio ;
    (*iii*) Turbulence ; (*iv*) Engine load ;
    (*v*) Engine speed.
21. "Auto-ignition is the cause of detonation". Justify the statement.
22. "Compressed natural gas (CNG) is preferable in S.I. engine than C.I. engine" ? Justify the statement.
23. Why is spark advance required ? Discuss the factor that affect ignition timing.
24. On what basis are S.I. engines fuels compared when they are better than iso-octane in anti-knock characteristics ?
25. Discuss the three basic requirements of a good S.I. engine combustion chamber.
26. Discuss the general principles of S.I. engine combustion chamber design.
27. What are the advantages of overhead valve combustion chamber over side valve combustion chamber ?

6

Combustion in C.I. Engines

6.1. INTRODUCTION

The compression ignition (C.I.) engine was developed by Dr. Rudol Diesel, he got a patent of his engine in 1892.

It is a very important prime mover these days and is finding wide applications in :

— buses trucks, tractors ;

— locomotives,

— pumping sets ;

— stationary industrial applications ;

— small and medium electric power generation ;

— marine propulsion.

The following points are worth noting about C.I. engines :

- *Its thermal efficiency is higher than S.I. engines.*
- C.I. engine fuels (diesel oils) are less expensive than S.I. engine fuels (petrol or gasoline). Furthermore, since C.I. engines fuels have a higher specific gravity than petrol, and since fuel is sold on the volume basis (litres) and not on mass basis (kg), more kg of fuel per litre are obtained in purchasing C.I. engine fuel.

Due to the above mentioned factors the *running cost of C.I. engines is much less than S.I. engines* and as a consequence these engines find wide application in industrial, transport and other applications.

- A *C.I. engine is not much favoured in passenger cars due to the following reasons :*

  (*i*) Heavier weight ;

  (*ii*) Noise and vibration ;

  (*iii*) Smoke ;

  (*iv*) Odour.

  — In view of the utilisation of heavier compression ratios (12 : 1 to 22 : 1 compared to 6 : 1 to 11 : 1 of S.I. engines) the heavy forces act on the parts of the engine and therefore heavier parts are required.

— Also, because of heterogeneous mixture, lean mixture is used.

These factors make the engine heavier.

— The *incomplete combustion of heterogeneous mixture, and droplet combustion result in the smoke and odour.*

- C.I. engines are manufactured in the following range of speeds, speeds and power outputs :

Particulars	*Range*
1. Piston diameters	50 mm to 900 mm
2. Speeds	100 r.p.m. to 4400 r.p.m
3. Power output	2 B.P. to 40000 B.P.

6.2. COMBUSTION PHENOMENON IN C.I. ENGINES

- The process of combustion in the compression ignition (C.I.) engine is fundamentally different from that in a spark-ignition engine. In C.I. engine combustion occurs by the high temperature produced by the compression of the air, *i.e.* it is an *auto-ignition.* For this a minimum compression ratio of 12 is required. The efficiency of the cycle increases with higher values of compression ratio but the maximum pressure reached in the cylinder also increases. This requires heavier construction. The upper limit of compression ratio in a C.I. engine is due to mechanical factor and is a compromise between high efficiency and low weight and cost. The normal compression ratios are in the range of 14 to 17, but may be upto 23. The air-fuel ratios used in the C.I. engine lie between 18 and 25 as against about 14 in the S.I. engine, and *hence C.I. engines are bigger and heavier for the same power than S.I. engines.*
- In the C.I. engine, the intake is air alone and the fuel is injected at high pressure in the form of fine droplets near the end of compression. This leads to delay period in the C.I. engine, is greater than that in the S.I. engine. The *exact phenomenon of combustion in the C.I. engine* is explained below.
  - — Each minute droplet of fuel as it enters the highly heated air of engine cylinder is quickly surrounded by an envelope of its own vapour and this, in turn and at an appreciable interval is inflamed at the surface of the envelope. To evaporate the liquid, latent heat is abstracted from the surrounding air which reduces the temperature of the thin layer of air surrounding the droplet, and some time must elapse before this temperature can be raised again by abstracting heat from the main bulk of air in this vicinity. As soon as this vapour and the air in actual contact with it reach a certain temperature, ignition will take place. Once ignition has been started and a flame established the heat required for further evaporation will be supplied from that released by combustion. The vapour would be burning as fast as it can find fresh oxygen, *i.e., it will depend upon the rate at which it is moving through the air or the air is moving past it.*
  - — In the C.I. engine, the fuel is not fed in at once but is spread over a definite period. The first arrivals meet air whose temperature is only a little above their self-ignition temperature and the delay is more or less prolonged. The later arrivals find air already heated to a far higher temperature by the burning of their predecessors and therefore light up much more quickly, almost as they issue from the injector nozzle, but their subsequent progress is handicapped for there is less oxygen to find.
  - — If the air within the cylinder were motionless, only a small proportion of the fuel would find sufficient oxygen, for it is impossible to distribute the droplets uniformly throughout the combustion space. Therefore some air movement is absolutely essential, as in the S.I. engine. But there is a fundamental difference between the

air movements in the two types of engines. In the S.I. engine we call it *turbulence* and mean a confusion of whirls and eddies with no general direction of flow, (to break up the surface of the flame front, and to distribute the shreds of flame throughout an externally prepared combustible mixture). In the C.I. engine we call it *air swirl* and mean an orderly movement of the whole body of the air, with or without some eddying or turbulence, so as to bring a continuous supply of fresh air to each burning droplet and sweep away the products of combustion which otherwise tend to suffocate it.

Three phases of C.I. engine combustion :

In the C.I. engine, combustion may be considered in three distinct stages as shown in Fig. 6.1.

1. Ignition delay period.
2. Period of rapid or uncontrolled combustion.
3. Period of controlled combustion.

The *third phase is followed by* **after burning** (or burning on the expansion stroke), which may be called the *fourth phase of combustion.*

1. **Ignition delay period**

- The delay period is counted from the start of injection to the point where the p-θ combustion curve departs from air compression (or no ignition or motoring) curve.
- The delay period can be roughly sub-divided into ***physical delay*** and ***chemical delay.*** The period of ***physical delay is the time between the beginning of injection and the attainment of chemical reaction conditions.*** In the physical delay period, the fuel is atomized, vaporized, mixed with air, and raised in temperature. In the *chemical delay period reaction starts slowly and then accelerates until inflammation or ignition takes place* (it may be noted that the ignition delay in the S.I. engine is essentially equivalent to the chemical delay in the C.I. engine).

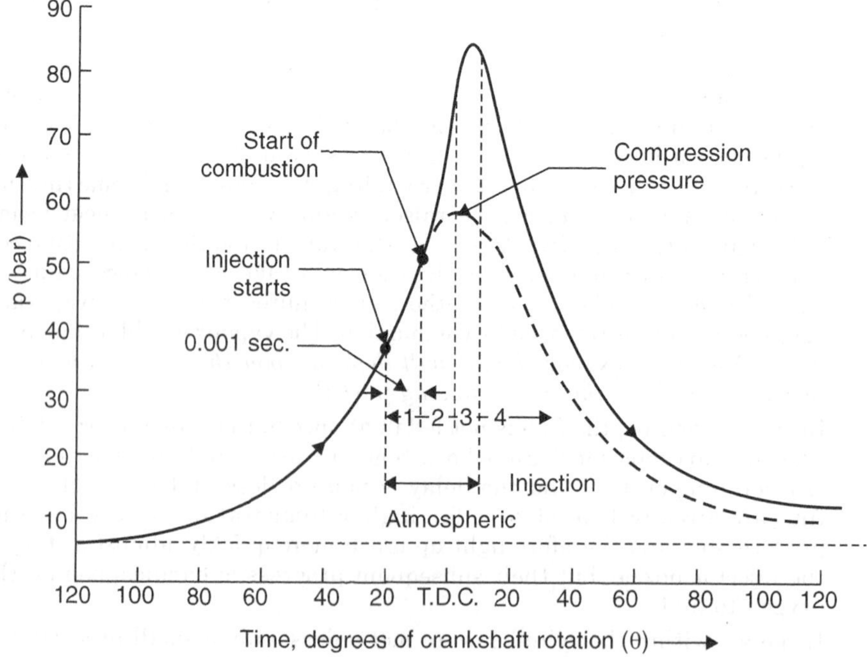

Fig. 6.1. Combustion phenomenon of C.I. engine.

- The delay period exerts a great influence in the C.I. engine combustion phenomenon. It is clear that the pressure reached during the second stage will depend upon the duration of the delay period ; *the longer the delay, the more rapid and higher the pressure rise,* since more fuel will be present in the cylinder before the rate of burning comes under control. This causes rough running and may cause *diesel knock.* Therefore we must *aim to keep the delay period as short as possible, both for the sake of smooth running and in order to maintain control over the pressure changes. But some delay period is necessary otherwise the droplet would not be dispersed in the air for complete combustion.* However, the delay period imposed upon is greater than what is needed and the designer's efforts are to shorten it as much as possible.

2. **Period of rapid or uncontrolled combustion.** The second stage of combustion in C.I. engines, after the delay period, is the *period of rapid or uncontrolled combustion.* This period is *counted from the end of the delay period to the point of maximum pressure on the indicator diagram.* In this second stage of combustion, the rise of pressure is rapid because during the delay period the droplets of fuel have had time to spread themselves out over a wide area and they have fresh air all around them. About one-third of heat is evolved during this process.

The rate of pressure rise depends on the amount of fuel present at the end of delay period, degree of turbulence, fineness of atomization and spray pattern.

3. **Period of controlled combustion.** At the end of second stage of combustion, the temperature and pressure, are so high that the fuel droplets injected in the third stage burn almost as they enter and any further pressure rise can be controlled by purely mechanical means, *i.e.* by the injection rate. The period of controlled combustion is assumed to end at maximum cycle temperature.The heat evolved by the end of controlled combustion is about 70 to 80 per cent.

4. **After burning**

- The combustion continues even after the fuel injection is over, because of poor distribution of fuel particles. This burning may continue in the expansion stroke upto 70° to 80° of crank travel from T.D.C. This continued burning, called the *after burning,* may be considered as the *fourth stage of the combustion.* The total heat evolved by the end of entire combustion process is 95 to 97% ; 3 to 5% of heat goes as unburned fuel in exhaust.
- In the *p-V* diagram, the stages of combustion are not seen because of little movement of piston with crank angle at the end and reversal of stroke. So for studying the combustion stages, therefore, a pressure-crank angle or time, *p*-θ or *p-t* diagram is invariably used. In the actual diagram, the various stages of combustion look merged, yet the individual stage is distinguishable.

Factors affecting combustion in C.I. engine :

The factors affecting combustion in C.I. engine are as follows :

(1) Ignition quality of fuel (cetane number)

(2) Injection pressure of droplet size

(3) Injection advance angle

(4) Compression ratio

(5) Intake temperature

(6) Jacket water temperature

(7) Intake pressure, supercharging

(8) Engine speed

(9) Load and Air to fuel ratio

(10) Engine size

(11) Type of combustion chamber.

6.3. FUNDAMENTALS OF THE COMBUSTION PROCESS IN DIESEL ENGINES

Effect of Compression Ratio and Engine Speed on Cylinder Pressure and Temperature

- The power output of a *diesel engine* is controlled by *varying the amount of fuel spray injected into a cylinder filled with compressed and heated air* whereas the *petrol engine* is controlled by *throttling the pre-mixed charge entering the cylinder.*
- The *pressure and temperature reached at the end of the compression stroke will depend primarily upon the compression ratio, intake temperature and speed of the engine.*
  — It has been observed that injection usually commences 15° to 20° before T.D.C. when both cylinder pressures and temperatures are much lower. As an *example*, a 15 : 1 compression ratio engine would have something like 600°C maximum temperature at T.D.C. but at 15° before T.D.C. this would only amount to 530°C.
  — Further it can be seen that the pressure and temperature rise in the cylinder with increased speed is largely *due to the reduced time available for compressed air to escape past the piston rings and heat to be lost through the cylinder walls and head.*

Diesel Engine Heterogeneous Charge Mixing

The air-fuel mixture formation, in the diesel engine, is of a *heterogeneous* natures, that is, it is locally concentrated at various sites and is therefore *unevenly distributed throughout the cylinder and combustion chamber.*

— Injected fuel spray penetrates the highly compressed and heated air mass where it is pulverised into many very small droplets in a localised formation. The mixing of the localised spray of fuel droplets in the hot air charge causes stoichiometeric (14.7 : 1 by weight) air-fuel ratio combustion zones to be established which are completely surrounded by pure air only. Thus the overall (averaged out) air-fuel mixture ratio range may vary from a rich, full load, 20 : 1, to a weak no-load, 100 : 1, air-fuel ratio.

— Most engines *operate with at least* **20% excess air** *due to difficulty of introducing sufficient exposed oxygen to the fuel vapour in the given time available so that the combustion process can be completed before the exhaust valve opens.* If the oxygen supply is *partially prevented from getting to the fuel vapour early enough during the power stroke then incomplete combustion, polluted exhaust gas and dark smoke will result.*

Diesel Engine Injected Spray Combustion Process

- Towards the end of the compression stroke when injection of the fuel into the combustion chamber commences, the quantity of fuel discharged is spread out over a predetermined period.
- The fuel spray enters the hot combustion chamber but does not immediately ignite, instead it breaks up into very small droplets (Fig. 6.2) and once these liquid droplets are formed, *their outer surfaces will immediately start to evaporate so there will be a liquid core surrounded with a layer of vapour.* At this point it should be explained that the burning of a hydrocarbon fuel in air is purely an *oxidation process.* Thus, initially, heat liberated from the oxidation of the fuel vapour is *less than the rate at which heat is extracted by convection and conduction,* but eventually a *critical temperature* is reached when the *rate of heat generated by oxidation exceeds the heat being dissipated by convection and radiation.* As a result, the temperature rises which, in turn, speeds up the oxidation process thus further increasing the heat released until a *flame site or sites are established,* this being known as the **ignition** and the temperature at which it occurs is called the **self-ignition temperature** of the fuel under these condition. *The heat required for further evaporation of the fuel droplets will thus be provided from heat released by the oxidation process, which is referred to as* **combustion.**

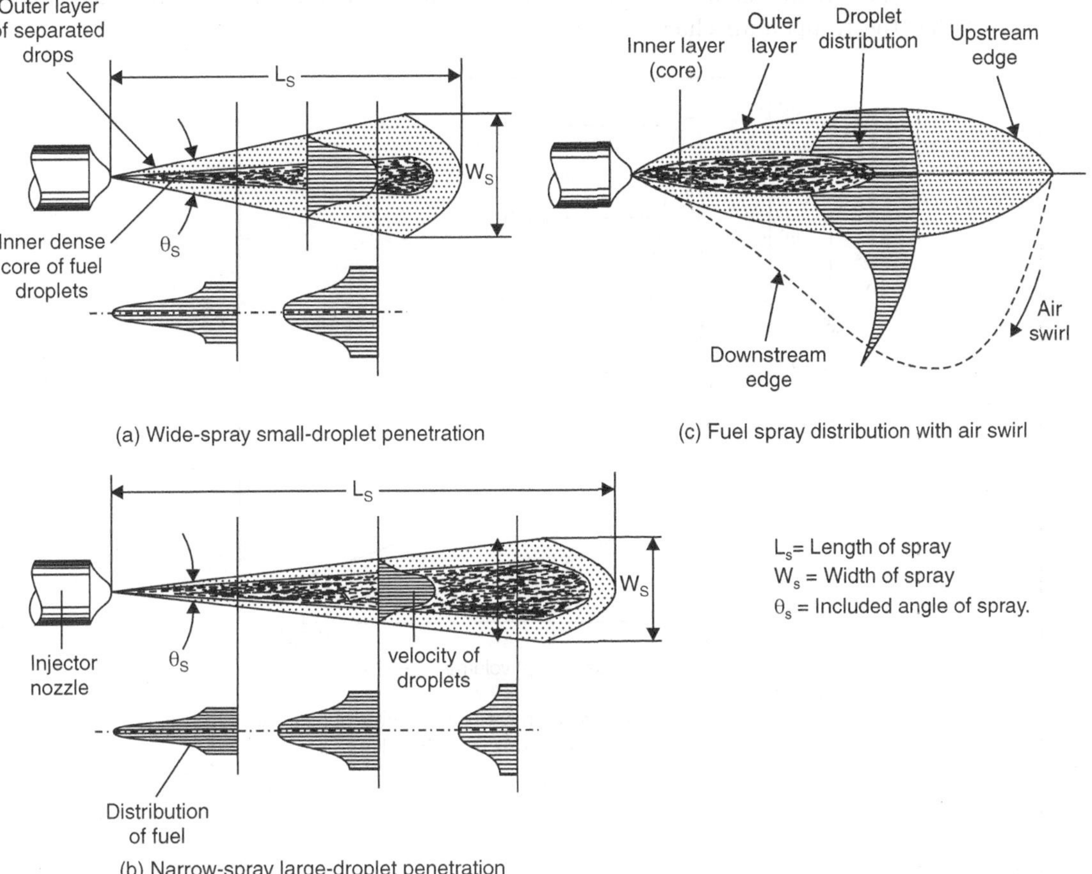

Fig. 6.2. Injected fuel spray characteristics.

- The liquid core, now surrounded by layers of heated vapour, oxidises burns as fast as it can ; that is it finds fresh oxygen to keep the chemical reaction going on.
- When the ***physical delay*** to convert the fuel spray into tiny droplets and the ***chemical reaction delay*** to establish ignition from the initial oxidation process are over, *the rate of burning is dependent on the speed at which the droplets are moving through the air or the air is moving past the droplets.*

Compression Ratio (r) :

Increase in compression ratio exercises the following *effects* :

- The cylinder compression pressure and temperature *increase* ; the ignition time lag between the point of injection to the instant when ignition first commences *reduces.*
- The density and turbulence of the charge *increase*, and this *increases the rate of burning and, accordingly the rate of pressure rise and the magnitude of the peak cylinder pressure reached.* The characteristics of the pressure rise relative to the piston stroke or crank-angle movement is illustrated in Fig. 6.3 and Fig. 6.4.
- Thermal efficiency and the specific fuel consumption are *improved* (Fig. 6.5)
- Raising compression ratio *results in reduction in the mechanical efficiency* as shown in Fig. 6.6 (since the higher cylinder pressures increase the pumping losses, friction

losses and compression and expansion losses as more work is done in squeezing together the trapped air charge).

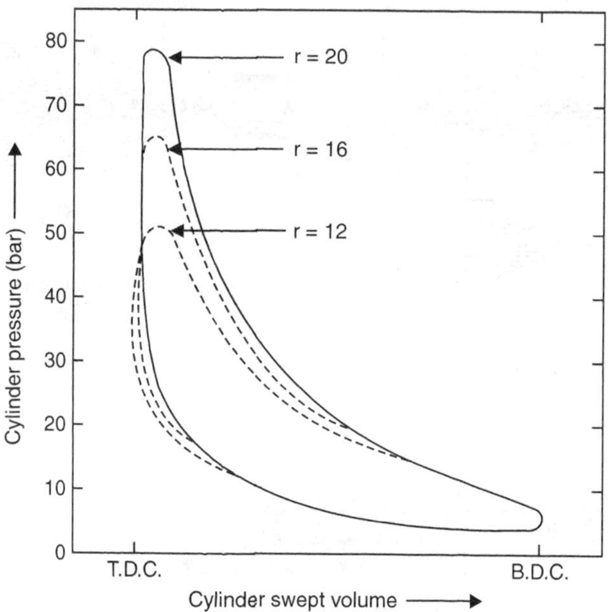

Fig. 6.3. Effect of compression ratio on the characteristic pressure—volume diagrams for a diesel engine.

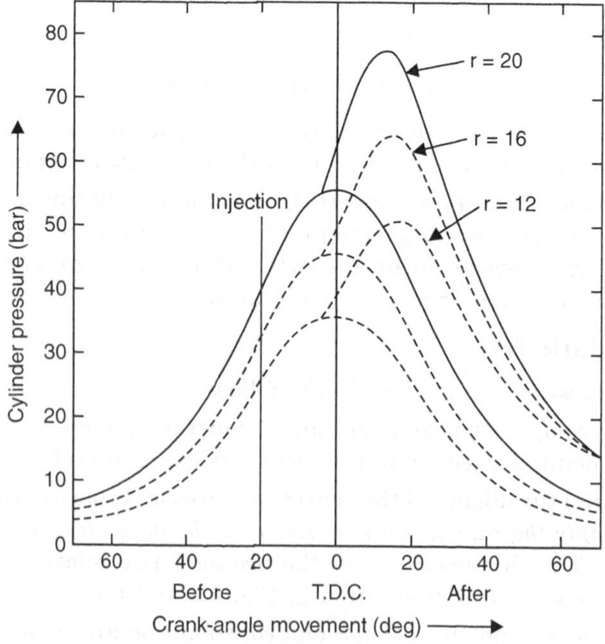

Fig. 6.4. Effect of compression ratio on the characteristic pressure –crank-angle movement diagrams for a diesel engine.

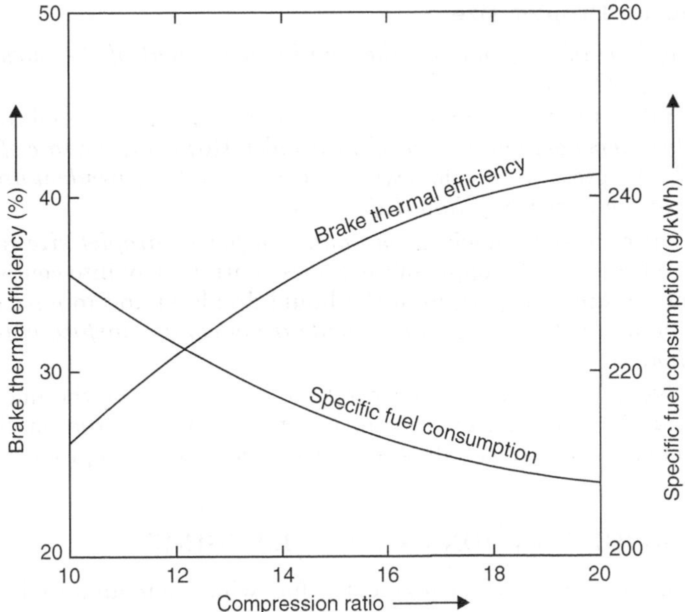

Fig. 6.5. Effect of compression ratio on the thermal efficiency and specific fuel consumption.

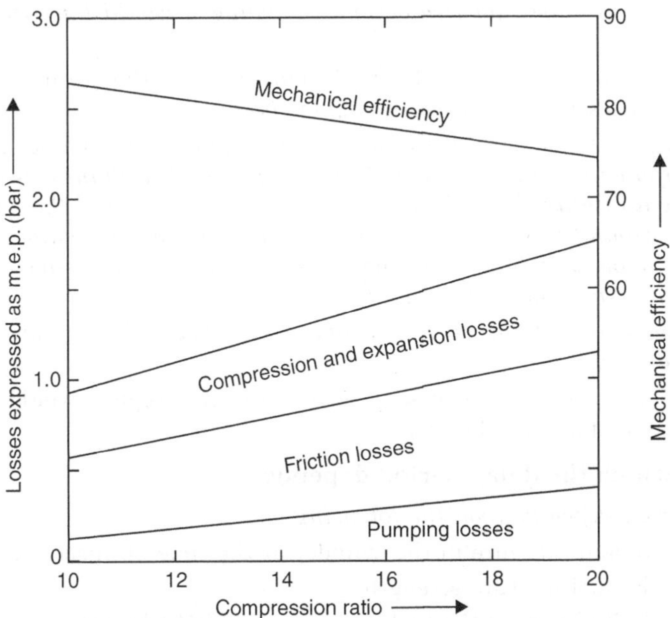

Fig. 6.6. Effect of compression ratio on the pumping, friction, compression and expansion losses and the resultant mechanical efficiency.

Injection Spray Droplet size

- *The rate of burning depends on the relative movement of the burning droplets to the surrounding air charge.*
- The time taken to establish and ignite a film of vapour surrounding a liquid droplet is *practically independent of the size of the droplet.* However, the *rate of burning* and correspondingly the pressure rise following ignition, will be *dependent upon the exposed surface area of the vaporising liquid droplets.*
- A compromise must be made to *maintain sufficient droplet size* (and, therefore, momentum so that a fresh supply of air comes continuously into contact with the shrinking size of the unburnt portion of the liquid droplets) and to *have available sufficient numbers of small droplets which provide an adequate surface vapour area for rapid combustion.*
- It is possible, to some extent, to control the droplet size by the injection needle spring closing load. Generally the greater the injector spring load, the smaller and finer will be the droplet size, whereas a light spring needle load tends to produce coarse liquid droplets.

6.4. DELAY PERIOD (OR IGNITION LAG) IN C.I. ENGINES

- In C.I. (compression ignition) engine, the fuel which is in atomised form is considerably colder than the hot compressed air in the cylinder. Although the actual ignition is almost instantaneous, *an appreciable time elapses before the combustion is in full progress. This time occupied is called the* **delay period** or **ignition lag**. *It is the time immediately following injection of the fuel during which the ignition process is being initiated and the pressure does not rise beyond the value it would have due to compression of air.*
- The delay period extends for about 13°, movement of the crank. The time for which it occurs *decreases with increase in engine speed.*
- In C.I. engine, the length of the delay period plays a vital role. *This period serves a useful purpose in that it allows the fuel jet to penetrate well into the combustion space. If there were* ***no delay*** *the fuel would burn at the injector and there would be an oxygen deficiency around the injector resulting in incomplete combustion. If the* ***delay is too long*** *the amount of fuel available for simultaneous explosion is too great and the resulting pressure rise is too rapid.*
  - The delay period affects the rate of pressure rise and hence knocking. It also affects *startability.*
  - Some delay period is necessary otherwise the droplets would not be dispersed in air for complete combustion.

Factors on which the delay period depends :

The *delay period depends upon the following :*

(*i*) Temperature and pressure in the cylinder at the time of injection.

(*ii*) Nature of the fuel mixture strength.

(*iii*) Relative velocity between the fuel injection and air turbulence.

(*iv*) Presence of residual gases.

(*v*) Rate of fuel injection.

(*vi*) To small extent the finess of the fuel spray.

The delay period increases with load but is not much affected by injection pressure.

- *The delay period should be as short as possible since a long delay period gives a more rapid rise in pressure and thus causes knocking.*

Effects of Various Factors on Delay Period :

Effects of various factors such as *fuel properties, intake temperature,* compression ratio, engine speed, type of combustion chamber, and injection advance are discussed below :

1. **Fuel properties :**

- The *self-ignition temperature (S.I.T.)* is the most important property of the fuel which affects the delay period.
  - — *A lower S.I.T. means a wide margin between it and the temperature of compressed air and hence lower delay period.*
  - — *Higher cetane number means a lower delay period and smoother engine operation.* Cetane number depends on the *chemical composition of fuel.* The *more paraffinic hydrocarbons are contained in fuel, higher will be the cetane number.*
- The other fuel properties which affect delay period are :

(*i*) Volatility ;

(*ii*) Latent heat ;

(*iii*) Viscosity ;

(*iv*) Surface tension.

— *Volatility and latent* heat affect the *time taken to form an envelope of vapour.*

— The *viscosity and surface tension* influence the *fineness of atomisation.*

2. **Intake temperature :**

- *Increase in intake temperature* would result in increase in compressed air temperature which would *reduce the delay period.*

3. **Compression ratio :**

- *Increase in compression ratio reduces delay period as it raises both temperature and density.*
  - — With increase in compression ratio, temperature of air increases. At the same time the minimum auto-ignition temperature decreases due to increased density of compressed air resulting in closer contact of molecules which thereby reduces the time of reaction when fuel is injected.
  - — As *the difference between compressed air temperature and minimum auto-ignition temperature increases, the delay period decreases.*

4. **Engine speed :**

- Delay period can be given either in terms of *absolute time* (in milliseconds) *or crank angle rotation.*
  - — At *constant speed,* delay period *is proportional to the delay angle.*
  - — In *variable speed operation,* delay period may *decrease in terms of milliseconds but increase in terms of crank angles.*

5. **Type of combustion chamber :**

- *A pre-combustion chamber gives shorter delay compared to an open type of combustion chamber.*

6. **Injection advance :**

- *Delay period increases with increase in injection advance angle.* The reason for increase in delay period with increase in injection advance angle is that pressures and temperatures are lower when injection begins.

- When injection advance angles are small, delay period reduces and operation of engine is smoother but power is reduced because large amount of fuel burns during expansion.

Abnormal Combustion in C.I. engines :

In C.I. engines, abnormal combustion is not a great problem as in S.I. engines. The only abnormality is ***"diesel knock"***. *This occurs when the delay period is excessively long so that there is a large amount of fuel in the cylinder for the simultaneous explosion phase.* The rate of pressure rise per degree of crank angle is then so great that an *audible knocking sound occurs.* Running is rough and if allowed to become extreme the *increase in mechanical and thermal stresses may damage the engine.* ***Knock is*** *thus a function of the fuel chosen and may be avoided by choosing a fuel with characteristics that do not give too long a delay period.*

6.5. DIESEL KNOCK

- *Diesel knock is the sound produced by the very rapid rate of pressure rise during the early part of the uncontrolled second phase of combustion.* The primary cause of an excessively high pressure rise is due to a *prolonged delay period* (Fig. 6.7). An extensive delay period can be due to the following *factors* :
  - (*i*) A low design compression ratio permitting only a marginal self-ignition temperature to be reached ;
  - (*ii*) A low combustion pressure due to worn piston rings or badly seating valves ;
  - (*iii*) Poor fuel ignition quality ; that is a low cetane number fuel ;
  - (*iv*) A poorly atomized fuel spray preventing early ignition to be established ;
  - (*v*) An inadequate injector needle spring load producing coarse droplet formation ;
  - (*vi*) A very low air intake temperature in cold wintry weather and during cold starting.

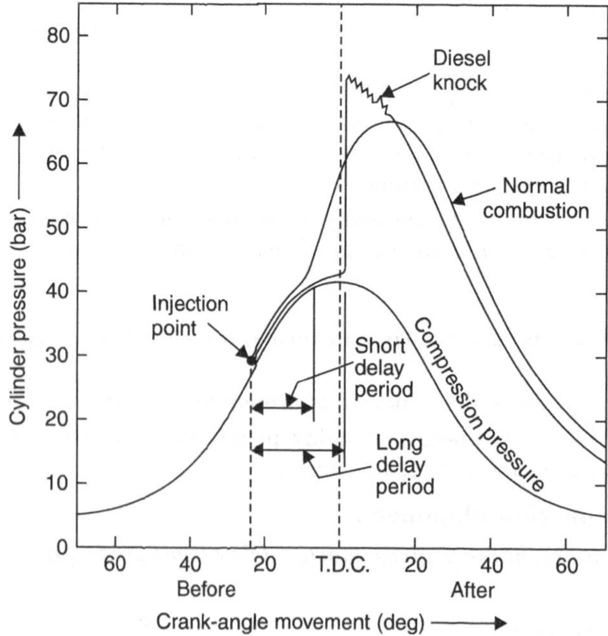

Fig. 6.7. Effect of short and long delay period on the characteristic $P—\theta$ diagram.

- A *very long ignition lag after injection causes a large proportion of the fuel discharge to enter the cylinder and to atomise before ignition and the propagation of burning actually occurs. Accordingly, when combustion does commence a relative amount of heat energy will be released almost immediately, this correspondingly produces the abnormally high rate of pressure rise, which is mainly responsible for rough and noisy combustion process under these condition (Fig. 6.7).*
- It has been observed generally, that provided the rate of pressure increase does not exceed 3 bar *per degree of crank-angle movement,* combustion will be relatively smooth, whereas *between a 3 and 4 bar pressure rise there is a tendency to knock and, above this rate of pressure rise, diesel knock will be prominent.*

Differences in the knocking phenomenon of the S.I. and C.I. Engines :

The following are the *differences in the knocking phenomena of the S.I. and C.I. engines* :

1. In the *S.I. engine,* the detonation occurs *near the end of combustion* whereas in the *C.I.* engine detonation occurs *near the beginning of combustion.*
2. The detonation in the S.I. engine is of a homogeneous charge causing very high rate of pressure rise and very high maximum pressure. In the C.I. engine, the fuel and air are imperfectly mixed and hence the rate of pressure rise is *normally lower than that in the detonating part of the charge in the S.I. engine.*
3. In the C.I. engine the fuel is injected into the cylinder only at the end of the compression stroke, *there is no question of pre-ignition as in S.I. engine.*
4. *In the S.I. engine, it is relatively easy to distinguish between knocking and non-knocking operation as the human ear easily finds the distinction.*
5. *Factors that tend to reduce detonation in the S.I. engine increase knocking in the C.I. engine.*

Methods of controlling diesel knock (Reducing delay period) :

The *diesel knock can be controlled by reducing delay period.* The delay is reduced by the following :

(*i*) High charge temperature.

(*ii*) High fuel temperature.

(*iii*) Good turbulence.

(*iv*) A fuel with a short induction period.

6.6. C.I. ENGINE COMBUSTION CHAMBERS

6.6.1. Primary Considerations in the Design of Combustion Chambers for C.I. Engines

In C.I. engines fuel is injected into the combustion chamber at about 15°C before T.D.C. during the compression stroke. For the best efficiency the combustion must complete within 15° to 20° of crank rotation after T.D.C. in the working stroke. Thus it is clear that injection and combustion both must complete in the short time. For best combustion mixing should be completed in the short time.

- In S.I. engine mixing takes place in carburettor, however in C.I. engines this has to be done in the combustion chamber. To achieve this requirement in a short period is an extremely difficult job particularly in high speed C.I. engines.
- From combustion phenomenon of C.I. engines it is evident that fuel-air contact must be limited during the delay period in order to limit $\frac{dp}{dt}$, the rate of pressure rise in the

second stage of combustion. This result can be obtained by *shortening the delay time. To achieve high efficiency and power the combustion must be completed when the piston is nearer to T.D.C., it is necessary to have rapid mixing of fuel and air during the third stage of combustion.*

- The design of combustion chamber for C.I. engines *must also take consideration of fuel injection system and nozzles to be used.*

The considerations can be summarized as follows :

1. High thermal efficiency.
2. Ability to use less expensive fuel (multi-fuel).
3. Ease of starting.
4. Ability to handle variations in speed.
5. Smoothness of operation *i.e.* avoidance of diesel knock and noise.
6. Low exhaust emission.
7. Nozzle design.
8. High volumetric efficiency.
9. High brake mean effective pressure.

6.6.2. Basic Methods of generating Air Swirl in C.I. Engines Combustion Chambers

There are *there basic methods of generating swirl in a C.I. engine combustion chamber,* which are mentioned below :

1. By *directing the flow of air during its entry to the cylinder known as* ***induction swirl.*** This method is *used in open combustion chamber.*

2. By *forcing the air through a tangential passage into a separate swirl chamber during compression stroke,* known as ***compression swirl.*** This method is used in *swirl chambers.*

3. By *use of initial pressure rise due to partial combustion to create swirl turbulence,* known as ***combustion induced swirl.*** This method is used in *pre-combustion chambers and air-cell chambers.*

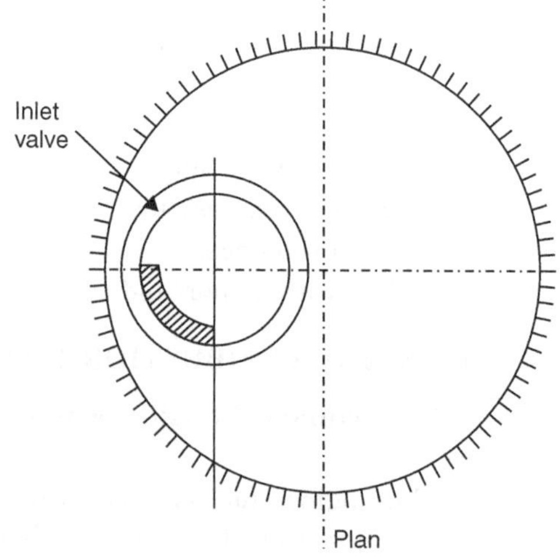

Fig. 6.8. Induction swirl by masking the inlet valve.

Induction swirl :

- In a *four stroke engine* induction swirl can be obtained either by careful formation of air intake passages or masking or shrouding a portion of circumference of inlet valve. The angle of mask is from 90° to 140° of the circumference.
- In *two stroke engine,* induction swirl is *created by suitable inlet port forms.*
- The induction swirl generated by air intake passages is very weak. If a masked inlet valve is used, it provides an obstruction in the passage which reduces volumetric efficiency. Therefore *swirl generated is weak* even with this method. With a weak swirl, a single orifice injection cannot provide the desired air fuel mixing. Therefore, with induction swirl, *we have to use a multiple-orifice injector.*

Advantages of induction swirl :

1. Easier starting (due to low intensity of swirl).
2. High excess air (low temperature), low turbulence (less heat loss), therefore indicated thermal efficiency is high.
3. Production of swirl requires no additional work.
4. Used with low speeds, therefore low quality of fuel can be used.

Disadvantages :

1. Shrouded valves, smaller valves, low volumetric efficiency.
2. Weak swirl, low air utilisation (60%), lower m.e.p. and large size (costly) engine.
3. Weak swirl, multi-orifice nozzle, high induction pressure, clogging of holes, high maintenance.
4. Swirl not proportional to speed ; efficiency not maintained at variable speed engine.
5. Influence minimum quantity of fuel. Complication at high loads and idling.

Compresssion swirl :

- The second method of generating swirl is by compression swirl in what is known as *swirl chamber*. A swirl chamber is a *divided chamber*. A divided combustion chamber is defined as one in which combustion space is divided into two or more distinct compartments, between which there are restrictions or throats small enough so that considerable pressure differences occur between them during combustion process.
- This swirl is maximum at about 15° before T.D.C. *i.e.* close to the time of injection. The fuel is injected into the swirl chamber and ignition and bulk of combustion takes place therein. A considerable amount of heat is lost when products of combustion pass back through the some throat and this loss of heat is reduced by employing a heat insulated chamber. Thus, it serves as a thermal regenerator receiving heat during combustion and expansion and returning the heat to air during compression stroke. However the loss of heat to surface of combustion chamber is greater than induction swirl.

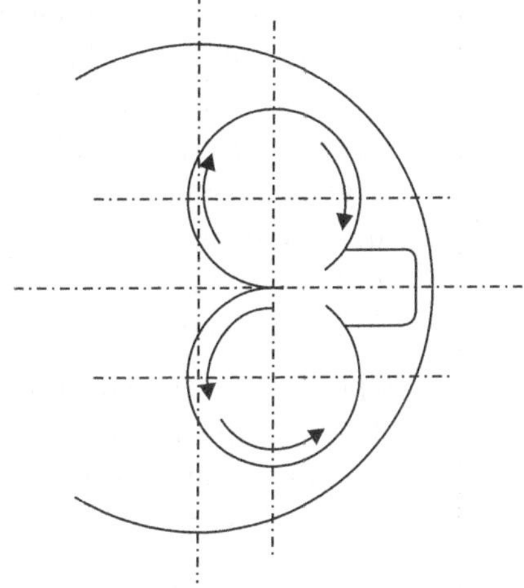

Fig. 6.9. Compression swirl.

- In *combustion swirl, a very strong swirl which increases with speed is generated.*

Advantage of compression swirl :

1. Large valves, high volumetric efficiency.
2. Single injector, pintle type (self cleaning), less maintenance.
3. Smooth engine operation.
4. Greater air utilization due to strong swirl. Smaller (cheaper) engine.
5. Swirl proportional to speed, suitable for variable speed operation.

Disadvantages :

1. Cold starting trouble due to high loss due to strong swirl, mechanical efficiency lower.
2. Less excess air ; lower indicated efficiency ; 5 to 8% more fuel consumption ; decreased exhaust valve life.
3. Cylinder more expensive in construction.
4. Work absorbed in producing swirl, mechanical efficiency lower.

Combustion induce swirl :

- This type of *swirl is induced by use of initial pressure rise due to partial combustion.*
- The combustion chambers which use this type of *swirl are not much favoured these days.*

6.6.3. Types of Combustion Chambers

In C.I. engines several types of combustion chambers are used. Each of these has its own peculiarities, and desirable, as well as undesirable features. Any one of these combustion chambers may produce good results in one field of application, but less desirable, or even poor results in another. No one combustion chamber design has yet been developed which will produce the best result is all types of engines. The particular design chosen, then, must be that which accomplishes the best performance for the application desired.

Four specific designs which find wide use in C.I. engines are discussed below :

A. The non-turbulent type
   (*i*) Open or direct combustion chamber.

B. The turbulent type
   (*i*) Turbulent chamber
   (*ii*) Pre-combustion chamber
   (*iii*) Energy cell.

1. **Open or direct combustion chamber :**

- Fig. 6.10 illustrates the usual design of *open combustion chamber*, which is representative of non-turbulent type.
- The fuel is injected directly into the upper portion of the cylinder, which acts as the combustion chamber. This type depends little on turbulence to perform the mixing. Consequently, the heat loss to the chamber walls is relatively low, and *easier starting results.* In order to obtain proper penetration and dispersal of the fuel necessary for mixing with the air, however, high injection pressures and multi-orifice nozzles are required. This necessitates small nozzle openings and results in more frequent clogging or diversion of the fuel spray by accumulated carbon particles, with consequent higher maintenance costs.

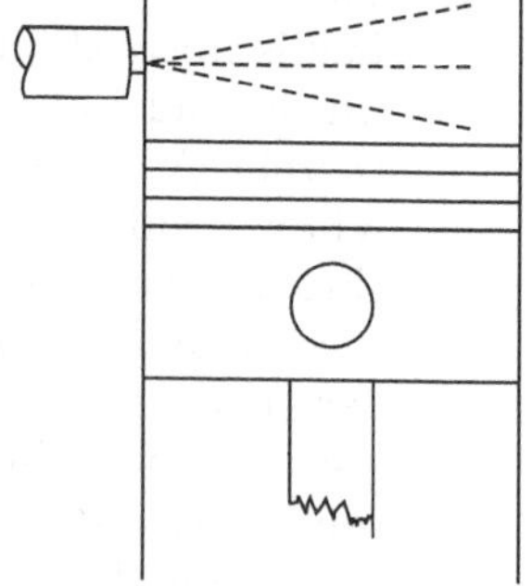

Fig. 6.10. Open or direct combustion chamber.

- This type of chamber is ordinarily used on *low speed engines*, where injection is spread through a greater period of time and thus ignition delay is a relatively less important factor. Consequently, *less costly fuels with longer ignition delay may be used.*
- Many attempts were made to improve the air motion in open chambers, the important are :
   (*a*) *by shrouding the inlet valve,* Refer Fig. 6.11 (*a*)
   (*b*) *by providing squish,* Refer Fig. 6.11 (*b*)

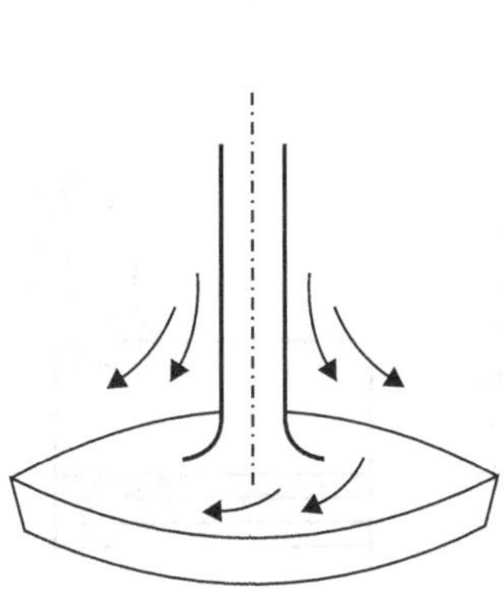

Fig. 6.11. (*a*) Air motion by shrouding the inlet valve.

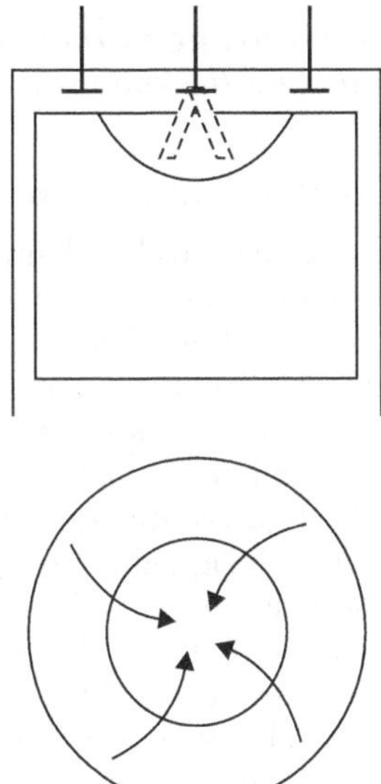

Fig. 6.11. (*b*) Squish air motion inside cylinder.

— By shrouding the inlet valve swirl motion is given to the air entering the cylinder which is believed to persist during compression stroke and the time of injection. This system gives better performance at low speeds, however volumetric efficiency reduces on account of reduction in inlet area due to shroud.

— Squish is provided by pushing the air at the end of the compression stroke in the space whose diameter is smaller than the cylinder bore. Because of the small clearance between the head and piston top when at T.D.C. air is pushed into combustion space providing air movement known as *squish*. The squish helps in mixing of fuel and air.

2. **Turbulent chamber.** Refer Fig. 6.12

In the '*turbulent chamber*' (Fig. 6.12) the upward moving piston forces all the air (or 70–80% of all air) at a greater velocity into a small antechamber, thus imparting a rotary motion to the air passing the pintle type nozzle. As the fuel is injected into the rotating air, it is partially mixed with this air, and commences to burn. The pressure built up in the antechamber by the expanding burning gases force the burning and unburned fuel and air mixtures back into the main chamber, again imparting high turbulence and further assisting combustion.

Advantages :

(*i*) The insulated or hot running combustion chamber shortens the delay period and limits the rate of pressure rise, resulting in smoother running.

(*ii*) The turbulence is responsible for rapid mixing and burning of fuel during the third stage of combustion.

(*iii*) Suitable for high speeds as the amount of turbulence is proportional to piston or engine speed. The burning in the third stage will be completed early without resulting in late burning.

(*iv*) The demands on the fuel injection system are not severe as it is not to be depended upon for mixing, distribution, etc.

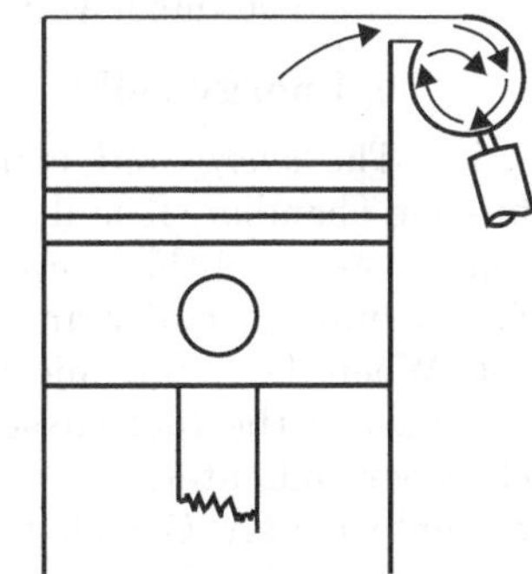

Fig. 6.12. Tubulent chamber.

The ***disadvantage*** is *that cold starting is difficult since air lose heat to combustion chamber walls during the compression stroke. The combustion chamber is relatively cool at the starting.*

3. **Pre-combustion chamber.** Refer Fig. 6.13.

Here the combustion chamber is separated into *two chambers.* The smaller one of the chambers occupy about *30 per cent of total combustion space.* The communication between two chambers is a narrow restricted passage or a number of small holes. The air is forced into the pre-combustion chamber by piston during the compression stroke. Fuel is injected into the pre-combustion chamber. The chamber is designed to run hot and this results in shortening the delay period of fuel which is highly desirable. The products from this chamber rushes into main combustion space through restricted passages, creating violent air motion. This violent air motion helps in rapid mixing and burning in the main combustion space. The fuel reaching the main combustion space has practically no delay period as the temperature is already high due to combustion in pre-combustion chamber and combustion in main chamber is rapid and complete (*i.e.* third stage of combustion) due to violent air motion.

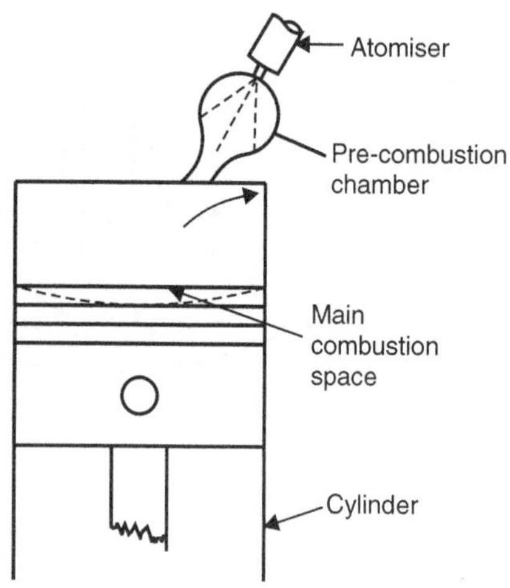

Fig. 6.13. Pre-combustion chamber.

Advantages :

(*i*) Due to short or practically no delay period for the fuel entering the main combustion space, *tendency to knock is minimum,* and as *such running is smooth.*

(*ii*) The combustion in the third stage is *rapid.*

(*iii*) As the mixing of fuel and air is thorough due to vialent projection of combustion products from pre-chamber, the *fuel injection system design need not be critical.*

Disadvantages :

(*i*) The velocity of burning mixture is too high during the passage from prechambers, so the heat loss is very high. This causes reduction in the thermal efficiency, which can be offset by increasing the compression ratio.

(*ii*) Cold starting will be difficult as the air loses heat to chamber walls during compression.

4. **Energy cell :**

The '*energy cell*' is more complex than the pre-combustion chamber. It is illustrated in Fig. 6.14. As the piston moves up on the compression stroke, some of the air is forced into the major and minor chambers of the energy cell. When the fuel is injected through the pintle type nozzle, part of the fuel passes across the main combustion chamber and enters the minor cell, where it is mixed with the entering air. Combustion first commences in the main combustion chamber where the temperature is higher, but the rate of burning is slower in this location, due to insufficient mixing of the fuel and air. The burning in the minor cell is slower at the start, but due to better mixing,

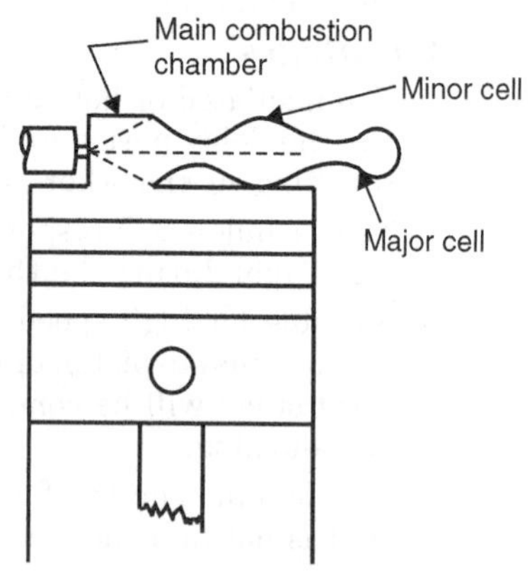

Fig. 6.14. Energy cell.

progresses at a more rapid rate. The pressures built up in the minor cell, therefore, force the burning gases out into the main combustion chamber, thereby creating added turbulence and producing better combustion in this chamber. In the mean time, pressure is built up in the major cell, which then prolongs the action of the jet stream entering the main chamber, thus continuing to induce turbulence in the main chamber.

5. **M. Combustion chamber :**

- After twenty years of research in 1954, Dr. Meuner of M.A.N., Germany developed M-process engine which ran without typical diesel combustion noise and hence it was named *'whisper engine'*.
- Fig. 6.15 shows a combustion chamber developed for small high speed engines. It differs from the other open combustion chamber engines in the respect that *fuel spray impinges tangentially on, and spreads over, the surface of a spherical space in the piston.* There is always some impingement of spray on the combustion chamber walls in all successful diesel engine designs. This impingement was not considered desirable till M.A.N. combustion system was experimented.
- The M.A.N. system's theory is that *enough of spray will ignite before impingement so that delay period will be normal while most of the fuel spray will evaporate from the hemispherical combustion space in piston prior to combustion.* Thus the *second stage of combustion is slowed down avoiding excessive rate of pressure rise.* Shrouded inlet valve is used to give air swirl in direction of arrow.

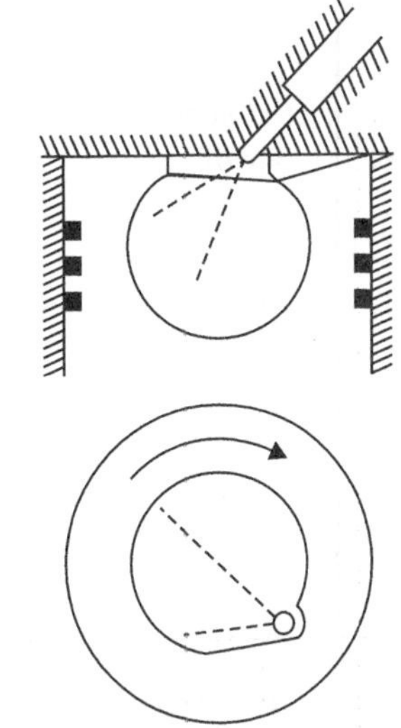

Fig. 6.15. M.A.N. 'M' combustion chamber.

Advantages :

'M-chamber' claims the following *advantages :*

(*i*) Low peak pressure.

(*ii*) Low rate of pressure rise.

(*iii*) Low smoke level.

(*iv*) Ability to operate on a wide range of liquid fuels (multi-fuel capability).

Disadvantages :

(*i*) Low volumetric efficiency.

(*ii*) Since fuel vaporisation depends upon the surface temperature of the combustion chamber, cold starting requires certain aids.

(*iii*) At starting and idling conditions hydrocarbon emissions may occur.

Table 6.1 gives *comparison between open combustion chambers and divided combustion chambers.*

S. No.	*Aspects*	*Open Combustion Chamber*	*Divided Combustion Chamber*
1.	*Fuel used*	Can consume fuels of good ignition quality, *i.e.* of shorter ignition delay or higher cetane number.	Can consume fuels of poor ignition quality *i.e.*, larger ignition delay. or lower cetane number.
2.	*Type of injection nozzle used*	Requires multiple hole injection nozzles for proper mixing of fuel	It is able to use single hole injection nozzles and moderate injection

		and air, and also higher injection pressures.	pressures. It can tolerate greater degree of nozzle fouling.
3.	*Sensitivity to fuel spray characteristic*	Sensitive.	Insensitive.
4.	*Mixing of fuel and air*	Mixing of fuel and air is not so efficient and thus high fuel/air ratios are not feasible without smoke.	Ability to use higher fuel/air ratios without smoke, due to proper mixing and consequent high air utilization factor.
5.	*Cylinder construction*	Cylinder construction is simple.	More expensive cylinder construction.
6.	*Starting*	Easy cold starting.	Difficult cold starting because of greater heat loss through the throat.
7.	*Thermal efficiency*	Open combustion chambers are thermally more efficient.	Divided combustion chambers suffer from irreversibilities like throttling through the throat during the compression and expansion ; thus leading to pressure losses and available heat losses. Therefore, these engines are thermally less efficient comparatively.

Summarily it may be said that a particular combustion chamber design must be chosen to perform a given job. No one combustion chamber can produce an ultimate of performance in all tasks. As most engineering work, the design of the chamber must be based on a compromise, after full considerations of the following factors :

(*i*) Heat lost to combustion chamber walls.

(*ii*) Injection pressure.

(*iii*) Nozzle design.

(*iv*) Maintenance.

(*v*) Ease of starting.

(*vi*) Fuel requirement.

(*vii*) Utilisation of air.

(*viii*) Weight relation of engine to power output.

(*ix*) Capacity for variable speed operation.

6.7. COLD STARTING OF C.I. ENGINES

- The important requirement of a C.I. engine is its easy starting from cold. To fulfil this requirement frequently compression ratios higher than necessary are used. Cold work, even so, may become difficult under the following conditions :

— When the cylinder liner is heavily worn ;

— When the valves are leaky ;

— Extreme cold climate (like Himalyan region).

Therefore, sometimes, it is necessary to provide some electrical aid for cold starting.

- *Open chamber direct injection engines are easiest to cold start because of the following reasons :*
  - (*i*) They have smallest surface to volume (S/V) ratio, as a consequence heat loss is minimum.
  - (*ii*) They have lowest intensity of swirl, due to which stagnant gas film remains on the cylinder walls which reduces heat transfer.

Cold starting aids for C.I. engines :

Several methods have been used in the past to achieve easy cold starting. Few of them are listed below :

1. *Preheating the engine cylinder by warm water.*
2. *Injection of a small quantity of lubricating oil or fuel oil.* This method temporarily raises the compression ratio, and seals the piston rings and valves.
3. *Provision of cartridges.*
4. *Modifying valve timings for starting.*
5. *Starting as petrol engine* by providing a carburettor and a spark plug. At starting compression ratio is reduced by providing an auxiliary chamber.

Modern starting aids of high speed engines :

The following basic three types of starting aids are used on modern high speed diesel engines :

1. *Electric glow plugs* (in the combustion chamber)
2. *Manifold heaters* (which ignite a small feed of fuel)
3. Injection of ether.

HIGHLIGHTS

1. The three phases of C.I. engine combustion are :
   (*i*) Ignition delay period (*ii*) Period of rapid or uncontrolled combustion
   (*iii*) Period of controlled combustion.
   The third phase is followed by after burning, which may be called the fourth phase of combustion.
2. The *period of physical delay* is the time between the beginning of injection and attainment of chemical reaction conditions.
   In the *chemical delay period* reaction starts slowly and then accelerates until inflammation or ignition takes place.
3. The *delay period* is the time immediately following injection of the fuel during which the ignition process is being initiated and the pressure does not rise beyond the value it would have due to compression of air.
4. The delay period should be as short as possible since a long delay period gives a more rapid rise in pressure and thus causes knocking.
5. *Diesel knock* is the sound produced by the very rapid rate of pressure rise during the early part of the uncontrolled second phase of combustion.
6. Four specific designs which find wide use in C.I. engines are :

A. The non-turbulent type :
   (*i*) Open combustion chamber

B. The turbulent type:
   (*i*) Turbulent chamber
   (*ii*) Pre-combustion chamber
   (*iii*) Energy cell.

OBJECTIVE TYPE QUESTIONS

Fill in the blanks or Say 'Yes' or 'No' :

1. The compression ignition engine was developed by ......
2. The thermal efficiency of C.I. engine is ...... than S.I. engines.
3. In C.I. engines the incomplete combustion of heterogeneous mixture, and droplet combustion result in smoke and odour.
4. The period of ...... delay is the time between the beginning of injection and attainment of chemical reaction condition.
5. In the ...... delay period, reaction starts slowly and then accelerates until inflammation or ignition takes place.
6. The second stage of combustion in C.I. engines after the delay period, is the period of ...... combustion.
7. The air-fuel mixture formation, in the diesel engine, is of a ...... mixture.
8. The delay period should be as ...... as possible.
9. Increase in intake temperature would result in increase in compressed air temperature which would increase the delay period.
10. Increase in compression ratio reduces delay period.
11. At constant speed, delay period is proportional to the delay angle.
12. Delay period decreases with increase in advance angle.
13. A pre-combustion chamber gives shorter delay compared to an open type of combustion chamber.
14. ...... is the sound produced by the very rapid rate of pressure rise during the early part of the uncontrolled second phase of combustion.
15. Factors that tend to reduce detonation in S.I. engine increase knocking in the C.I. engine.
16. Induction swirl results in easier starting of the C.I. engine.
17. 'M-process' engine, developed in 1954, was named 'whisper engine'.

ANSWERS

1. Dr. Rudol Diesel	**2.** higher	**3.** Yes	**4.** physical	**5.** chemical
6. uncontrolled	**7.** heterogeneous	**8.** short	**9.** No	**10.** Yes
11. Yes	**12.** No	**13.** Yes	**14.** Diesel knock	**15.** Yes
16. Yes	**17.** Yes.			

THEORETICAL QUESTIONS

1. When was C.I. engine developed and by whom ?
2. State the applications of C.I. engines.
3. Enlist the reasons for which C.I. engine is not much favoured in passenger cars.
4. Explain briefly the combustion phenomenon in C.I. engine.
5. Describe briefly various phases of C-I. engine combustion.
6. State the various factors which affect combustion in C.I. engine.
7. Explain briefly diesel engine injected spray combustion process.
8. What is delay period in C.I. engines ?
9. What is the difference between physical delay and chemical delay ?
10. State the factors on which delay period depends.
11. Explain the effect of the following factors on delay period :

    (*i*) Fuel properties (*ii*) Intake temperature

    (*iii*) Compression ratio (*iv*) Engine speed

    (*v*) Type of combustion chamber (*vi*) Injection advance.

12. Explain briefly the phenomenon of "Diesel knock".

13. State the differences in the knocking phenomena of S.I. and C.I. engines.

14. Enlist various methods of controlling diesel knock.

15. What should be the primary considerations in the design of combustion chambers for C.I. engines ?

16. Explain briefly basic methods of generating air swirl in C.I. engines combustion chambers.

17. Enlist the advantages and disadvantages of induction swirl.

18. State the advantages and disadvantages of compression swirl.

19. Explain briefly any two of the following combustion chambers :

(*i*) Open or direct combustion chamber (*ii*) Turbulent chamber

(*iii*) Pre-combustion chamber (*iv*) Energy cell.

20. Give the comparison between open combustion chambers and divided combustion chambers.

21. Write short note on cold starting of C.I. engines.

22. Explain briefly cold starting aids for C.I. engines.

23. Explain the phenomenon of knock in C.I. engines and compare it with S.I. engine knock.

24. How does the mixture composition in combustion chamber of a C.I. engine differ from that of a S.I. engine ?

25. "The factors that tend to increase detonation in S.I. engine tend to reduce knocking in C.I. engine".

Discuss the above statement with reference to the following influencing factors :

(*i*) Compression ratio ; (*ii*) Inlet temperature ;

(*iii*) Inlet pressure ; (*iv*) Self-ignition temperature of fuel ;

(*v*) Time lag of ignition of fuel ; (*vi*) r.p.m. ;

(*vii*) Combustion chamber wall temperature.

26. Why does rate of pressure rise during combustion is limited to a certain value ?

27. Discuss the influence of ignition delay on combustion processes in S.I. and C.I. engines. Explain how the presence of a knock inhibitor in fuel oil helps to change the ignition delay in C.I. engines.

28. "The requirement of air motion and swirl in a C.I. engine combustion chamber is more strigent than in a S.I. engine". Justify the statement.

29. "The induction swril in a C.I. engine helps in increasing indicated thermal efficiency". Justify the statement.

30. How are C.I. engine combustion classified ? What type of swirl is used in these chambers ?

31. "In agriculture field, it is better to use C.I. engine than S.I. engine". Justify the statement.

32. How can a diesel engine be converted to CNG engine ?

33. "The maximum substitution of diesel engine by CNG in a C.I. engine is limited by the cetane characteristics of the available fuel". Justify the statement.

34. Write a short note on aids for starting C.I. engines under extreme cold climate.

35. Describe the M-combustion system and discuss its relative merits with respect to D.I. chambers.

7

Air Capacity of Four Stroke Engines

7.1. INTRODUCTION

- **Air capacity (actual)** is *defined as the mass flow of fresh air through the engine per unit time.* The engine output depends on this parameter.
- The indicated power output of an engine may be expressed as follows :

Indicated power, I.P. $= (m_f \times C)\, \eta_{th\,(I)}$ ...(7.1)

where, m_f = Mass of fuel per unit time,

C = Calorific value of fuel, and

$\eta_{th\,(I)}$ = Indicated thermal efficiency.

From eqn. (7.1) it may be observed that power developed by the engine depends on the amount of fuel supplied (m_f). The amount of fuel which may be burnt usefully depends, however, on the amount of oxygen available (for its combustion) which in turn depends on the quantity of air supplied to the engine. The fuel-air ratio is defined as $F/A = m_f/m_a$. The eqn. (7.1) then becomes

$$\text{I.P.} = m_a\,(F/A \times C) \times \eta_{th\,(I)} \quad ...(7.2)$$

- The indicated thermal efficiency ($\eta_{th\,(I)}$) of an engine depends on :
  - (*i*) Compression ratio.
  - (*ii*) Ignition timing.
  - (*iii*) F/A (Fuel-air) ratio.
  - — With *fixed compression ratio and optimum ignition timing, $\eta_{th\,(I)}$ depends only on F/A ratio.* When F/A varies closely within the range of 0.075 and 0.085, efficiency decreases as F/A increases, making the product $[F/A \times C \times \eta_{th\,(I)}]$ approximately constant over this range; eventually *I.P. becomes proportional to m_a*. Thus the power developed by the engine, for a given valve of F/A ratio and $\eta_{th\,(I)}$ will be proportional to the amount of air the engine can take in per unit time, as illustrated in Fig. 7.1.

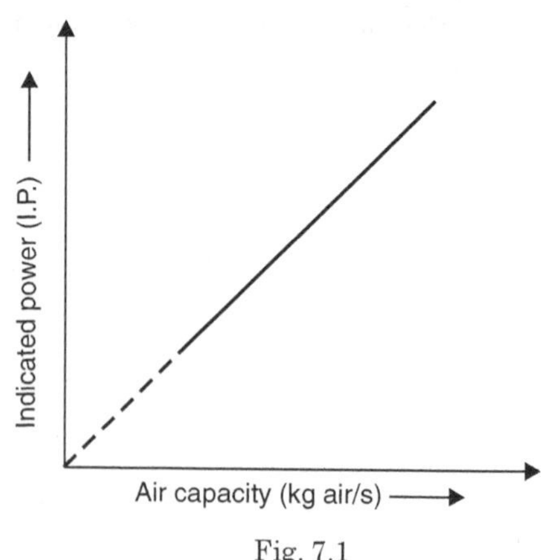

Fig. 7.1

- The *air capacity of an engine can be increased by :*
  (*i*) Better inlet-design ;
  (*ii*) Increasing r.p.m ;
  (*iii*) Supercharging (*i.e.* increasing inlet pressure) ;
  (*iv*) Cooling the inlet air.

7.2. IDEAL AIR CAPACITY

The **ideal air capacity** *corresponds to filling the displaced volume (i.e. piston swep volume) with fresh mixture at inlet conditions.*

Thus, for four stroke engines, the ideal air capacity per cylinder is given as :

$$(\dot{m}_a)_{\text{ideal}} = \frac{N}{2} V_s \rho_i \quad \text{...(7.3)}$$

$$= \frac{U_p}{4} A_p \rho_i \quad \text{...(7.4)}$$

$$\left[\because \; V_s = A_p \text{ (piston area)} \times L_p \text{ (piston length)} \text{ and } U_p = 2L_pN \text{ or } N = \frac{U_p}{2L_p} \right]$$

where $(\dot{m}_a)_{\text{ideal}}$ = Ideal flow rate of fresh mixture per unit time,

N = Engine revolution per unit time,

V_s = Engine piston swept volume,

ρ_i = Inlet gas density,

U_p = Mean piston velocity, and

A_p = Piston area.

Note. The fresh charge in C.I. engine consists of *air only* (dry air + water vapour) whereas in S.I. engine it consists of *air plus fuel*. The quantity of air taken in by an engine is practically unaffected by the presence of fuel in the air and hence it is assumed that only air is present.)

7.3. VOLUMETRIC EFFICIENCY

The **volumetric efficiency** *of an engine is defined as the ratio of actual air capacity to the ideal air capacity.* This is *equal to the ratio of mass of air which enters or is forced into the cylinder in suction stroke to the mass of free air equivalent to the piston displacement at intake temperature and pressure conditions.*

$$\eta_{\text{vol.}} = \frac{\text{Mass of charge actually induced}}{\text{Mass of charge represented by volume at intake temperature and pressure conditions}}$$

or

$$\eta_{\text{vol.}} = \frac{m_{\text{actual}}}{m_{\text{ideal}}} = \frac{m_{\text{actual}}}{\frac{N}{2} V_s \rho_i} = \frac{2\, m_{\text{actual}}}{N\, V_s\, \rho_i} = \frac{4\, m_{\text{actual}}}{U_p\, A_p\, \rho_i} \quad \text{...(7.5)}$$

where, m_{actual} = Measured valve of inlet gas or dry air mass flow rate, and

ρ_i = Measured value of inlet gas or dry air density.

Thus, indicated power (I.P.) may now be written as :

$$\text{I.P.} = m_{actual} (F/A \times C)\, \eta_{th\,(I)}$$

$$= \frac{N}{2}\, V_s\, \rho_i\, \eta_{vol.}\, (F/A \times C)\, \eta_{th\,(I)} \qquad ...(7.6)$$

Dividing by $\frac{N}{2} V_s$, we get indicated mean effective pressure (i.m.e.p.) given by

$$\text{i.m.e.p.} = \rho_i\, \eta_{vol.}\, (F/A \times C)\, \eta_{th\,(I)} \qquad ...(7.7)$$

From eqn. (7.7), we find that if the type of fuel, fuel-air ratio (F/A) and indicated thermal efficiency ($\eta_{th\,(I)}$) remain constant, then

$$\text{i.m.e.p.} \propto \rho_i\, \eta_{vol.}$$

Note. Power output of an engine is proportional to volumetric efficiency provided the combustion is complete.

7.4. EFFECT OF VARIOUS FACTORS ON VOLUMETRIC EFFICIENCY

It is desirable to have maximum volumetric efficiency in the intake of any engine ; it *varies with engine speed*. Fig. 7.2 shows a graph between volumetric efficiency and engine speed for a typical S.I. engine.

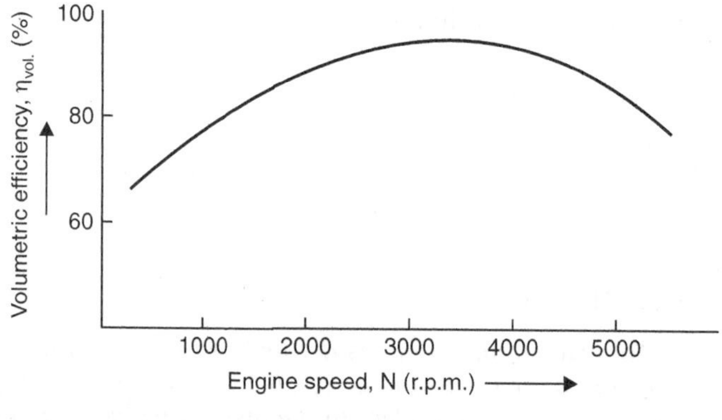

Fig. 7.2

— There is a certain speed at which the volumetric efficiency is maximum, decreasing at both higher and lower speeds. There are several physical and operating variables that shape this curve.

Effects of various factors which affect the volumetric efficiency are discussed below :

1. **Fuel :**

The volumetric efficiency of a naturally aspirated engine will always be less than 100% because fuel is also being added and the volume of the *fuel vapour will displace some incoming air. The type of fuel and how and when it is added will determine how much the volumetric efficiency is affected.*

- Systems with *carburettors or throttle body injection add fuel early* in the intake flow and generally have *lower overall volumetric efficiency*. This is because the fuel will immediately start to evaporate and fuel vapour will displace incoming air.

- *Multipoint injectors* which add fuel at the *intake valve ports* will have *better efficiency* because the air is displaced until after the intake manifold. Fuel evaporation does not occur until the flow is entering the cylinder at the intake valve.
- *Those engines that inject fuel directly into the cylinders after the intake valve is closed will experience no volumetric efficiency loss due to fuel evaporation.* Manifolds with late fuel addition may be designed to further increase volumetric efficiency by having large diameter runners. High velocity and turbulence to promote evaporation are *not needed.* They can also be operated cooler, which results in a *dense inlet air flow.*
- Fuels like *alcohol* which have a *smaller air-fuel ratio* will experience a *greater loss in volumetric efficiency.* Fuels with high heat of vaporisation will regain some of this lost efficiency due to the greater evaporation cooling that will occur with these fuels. This cooling will create a denser air-fuel flow for a given pressure, allowing for more air to enter the system. Alcohol has high heat of vaporisation, so some efficiency lost due to *air-fuel* is gained back again.
- Gaseous fuels like hydrogen and methane *displace more incoming* air than liquid fuels, which are only partially evaporated at the intake system. This must be considered when trying to modify engines made for gasoline fuel to operate on these gaseous fuels. It can be assumed that fuel vapour pressure in the intake system is between 1 to 10 percent of total pressure when gasoline-type liquid fuel is being used. When gaseous fuels or alcohol is being used, the fuel vapour pressure is often *greater* than 10 percent of the total. *Intake manifolds can be operated much cooler when gaseous fuel is used, as no vapourisation is required.* This will gain back some lost volumetric efficiency.
- *The later that fuel vaporises in the intake system, the better is the volumetric efficiency.* On the other hand, *the earlier that fuel vaporises, the better are the mixing process and cylinder-to-cylinder distribution consistency.*

2. **Heat transfer-High temperature :**

- All intake systems are hotter than the surrounding air temperature and will consequently heat the incoming air. This *lowers the density of the air, which reduces volumetric efficiency.*
- Intake manifolds of carburetted systems or throttle body injection systems are *purposely heated to enhance fuel evaporation.* At lower engine speeds, the air flow rate is slower and the air remains in the intake system for a longer time. It thus gets heated to higher temperatures at low speeds, which *lowers the volumetric efficiency curve in* Fig. 7.2 at the low-speed end.
- Some systems have been tried *which inject small amounts of water into the intake manifold.* This is *to improve the volumetric efficiency by increasing the resulting evaporative cooling that occurs.*

3. **Valve overlap :**

- At the top dead centre (T.D.C.) at the end of exhaust stroke and the beginning of the intake stroke, both intake and exhaust valves are open simultaneously for a brief moment. When this happens, *some exhaust gas can get pushed through the open intake valve back into the intake system.* The *exhaust then gets carried back into the cylinder with the intake air-fuel charge, displacing some of the incoming air and lowering volumetric efficiency.* This *problem is greatest at low engine speeds, when real time of valve overlap is greater.* This effects *lowers efficiency curve* in Fig. 7.2 at the low engine speed end.
- Other factors that affect the above problem are the *intake and exhaust valve location and compression ratio.*

4. **Fluid friction losses :**

- When air moves through any flow passage or past any flow restriction, it undergoes a *pressure drop.* For this reason, the pressure of air entering the cylinders is *less than the surrounding atmospheric air pressure,* and the amount of air entering the cylinder is subsequently *reduced.* The *viscous flow friction* that affects the air as it passes through the air filter, carburettor, throttle plate, intake manifold and intake valve *reduces the volumetric efficiency. Viscous drag which causes the pressure loss increases with square of flow velocity.* This results in decreasing the efficiency on the high-speed end of the curve in Fig. 7.2.
  - — A lot of development work has been carried out to reduce pressure losses in air intake systems. Smooth walls in the intake manifold, the avoidance of sharp corners and bends elimination of the carburettor, and close-fitting parts alignment with no gasket protrusions all contribute to decreasing intake pressure loss.
  - — One of the *greatest flow restriction is the flow through the intake valve.* To *reduce this restriction, the intake valve flow area has been increased by building multivalve engines having two or even three intake valves per cylinder.*
- The flow of air-fuel into the cylinders is usually diverted into a *rotational flow pattern* within the cylinder *to enhance evaporation, mixing and flame speed.* This flow pattern is *accomplished by shaping intake runners and contouring the surface of the valves and valve ports. This increases the inlet flow restriction and decreases volumetric efficiency.*
- In case the diameter of the intake manifold runners is increased, flow velocity will be decreased and pressure losses will be decreased. However, a decrease in velocity will result in poorer mixing of the air and fuel and less accurate cylinder-to-cylinder distribution. This *needs proper compromises in design.*
- In order to get better air-fuel mixing in some low performance, high fuel-efficient engines, the walls of the intake manifold are made *rough to enhance turbulence.* In these engines, high volumetric efficiency is *not as important.*

5. **Choked flow :**

- When choked flow occurs at some location in the intake system, it is the extreme case of flow restriction. When the air flow is increased to higher velocities, it eventually reaches sonic velocity at some point in the system. This *choked flow condition* is the *maximum flow rate that can be produced in the intake system regardless of how controlling conditions are changed.* This causes lowering of the efficiency curve on the high-speed end in Fig. 7.2.

The occurrence of choked flow takes place in the most restricted passage of the system, usually at the intake valve or in the carburettor throat on those engines with carburettors.

6. **Intake valve closure after B.D.C. :**

- The amount of air that ends up in the cylinder is affected by the timing of the closure of the intake valve. *The ideal time for the intake valve to close is when the pressure equalisation occurs between the air inside the cylinder and the air in the manifold.* If it closes *before this point, air that was still entering the cylinder is stopped and a loss of volumetric efficiency is experienced. If the valve is closed after the point, air being compressed by piston will force some air back out of the cylinder, again with a loss in volumetric efficiency.* This valve-closing point in the engine cycle, at which the pressure inside the cylinder is the same as the pressure in the intake manifold, is highly depended on *engine speed.*

- The position where the intake valve closes on most engine is *controlled by a crankshaft* and *cannot change with engine speed.* Thus the closing cycle position is designed for one engine speed, depending on the use for which the engine is designed.

7. **Exhaust residual :**

- All of the exhaust gases, during the exhaust stroke, do not get pushed out of the cylinder, by the piston, a *small residual being trapped in the clearance volume.* The amount of this residual depends on the *compression ratio,* and *somewhat on the location of the valve and valve overlap.*
- The exhaust gas residual, besides displacing some air, interacts with the air in two other ways. *When the very hot gas mixes with the incoming air it heats the air, lowers the gas density, and decreases volumetric efficiency.* This is *counteracted slightly, however, by the partial vacuum created in the clearance volume when the exhaust gas is in turn cooled by the incoming air.*

8. **Exhaust gas recycle (EGR) :**

- In several types of engines and all modern engines, some *exhaust gas is recycled* (EGR) *into the intake system to dilute the incoming air. This reduces combustion temperatures in the engine, which results in less nitrogen oxides in the exhaust.* Upto about *20 percent* of exhaust gases will be diverted back into the intake manifold, depending on how the engine is being operated. This *exhaust gas not only displace some incoming air, but it also heats the incoming air and lowers its density.* Due to both of these interactions the *volumetric efficiency of the engine is lowered.*
- In addition, engine crankcases are vented into the intake systems, displacing some of the incoming air and lowering the volumetric efficiency. *Gases forced through the crankcase can amount to about 1 percent of the total gas flow through the engine.*

9. **Piston speed and engine size :**

- It can be proved that inertia stress $\propto U_p^2$. This indicates that all geometrically similar engines reach maximum allowable inertia stresses at the same piston speed. Therefore, whether due to consideration of *maximum air capacity or limiting inertia dresses,* geometrically similar engines are generally *designed to run at the same speed.*
- *The power developed by an engine is not proportional to swept/displacement volume.*
- *Large engines develop less power per unit volume.*

10. **Design of inlet and exhaust systems :**

- The volumetric efficiency is affected by the inlet and exhaust pipe design (length and diameter) ; the effect of inlet pipe system being greater.
- By experimentation it has been found that, at certain speeds long inlet pipes give high volumetric efficiency.

7.5. INLET VALVE MACH INDEX

The flow of intake charge, in a reciprocating engine, takes place through the intake valve opening which varies during suction/induction operation. The maximum gas velocity (U_g) through this area is limited by the local sonic velocity (U_s). The following relation is used to choose the gas velocity :

$$U_g = \frac{A_p U_p}{K_i A_{iv}} \qquad ...(7.8)$$

where A_p = Piston area,

A_{iv} = Nominal intake valve opening area, and

K_i = Intake valve flow coefficient.

and
$$\frac{U_g}{U_s} = \frac{A_p\, U_p}{A_{iv}\, K_i\, U_s}$$
$$= \left(\frac{D_{cy}}{D_{iv}}\right)^2 \times \frac{U_p}{K_i\, U_s} = Z \quad \text{...(7.9)}$$

where D_{cy} = Cylinder diameter,

D_{iv} = Inlet valve diameter,

U_p = Mean piston speed,

U_s = Inlet sonic velocity, and

Z = Inlet valve mach index.

Fig. 7.3 shows a plot for $\eta_{vol.}$ *vs* Z (unaltered by varying inlet valve diameter, valve lift and valve design), from which it can be concluded that there is a particular value of Mach Index after which volumetric efficiency starts falling ; this is approximately $Z = 0.55$.

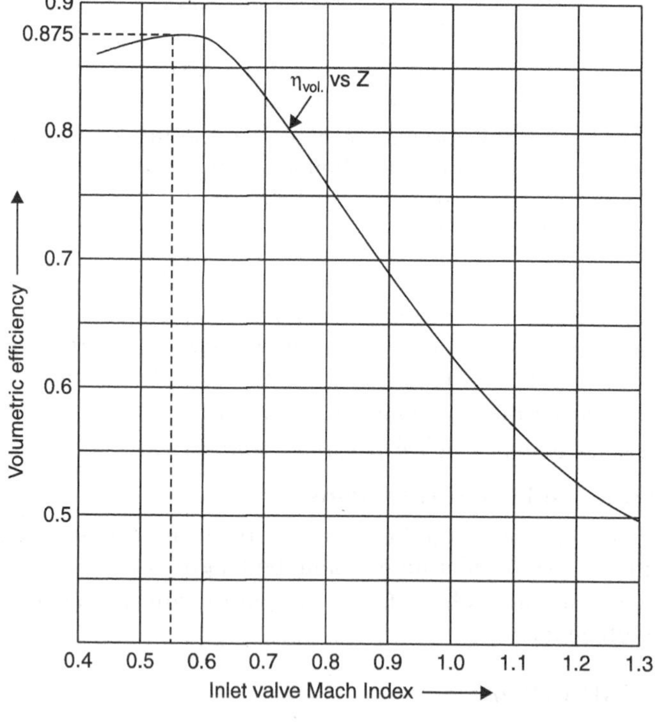

Fig. 7.3

WORKED EXAMPLES

Example 7.1. *A single-cylinder, 4-stroke cycle engine using CNG (compressed natural gas) as fuel has a cylinder, 20.3 cm bore × 30.5 cm stroke and runs at 300 rpm. If the volumetric efficiency of the engine based on conditions approaching the cylinder is 78% and the air/fuel ratio is 4 : 1, determine the volume of gas used per minute.*

Solution. *Given :* D = 20.3 cm = 0.203 m ; L = 30.5 cm = 0.305 m ; N = 300 r.p.m.

$\eta_{vol.}$ = 78%, A/F ratio = 4 : 1

Volume of gas used per minute :

$$\text{Stroke volume} = \frac{\pi}{4} \times 0.203^2 \times 0.305 = 0.009871 \text{ m}^3$$

$$\text{Volume inhaled} = \eta_{vol.} \times \text{stroke volume}$$
$$= 0.78 \times 0.009871 = 0.007699 \text{ m}^3$$

$$\text{Gas inhaled} = \frac{0.007699}{4+1} = 0.00154 \text{ m}^3$$

$$\text{Gas inhaled per minute} = 0.00154 \times \frac{300}{2} = \mathbf{0.231\ m^3/min.} \quad \textbf{(Ans.)}$$

Example 7.2. *A four-stroke, eight-cylinder engine is tested while running at 3600 r.p.m. The inlet air temperature is 15°C and the pressure is 760 mm of Hg. The total piston displacement volume is 4066 cm³. The air-fuel ratio of the engine is 14 : 1 and b.s.f.c. is 0.38 kg/kWh. Dynometer reading shows a power output of 86 kW. Find the volumetric efficiency of the engine.*

Solution. *Given :* N = 3600 r.p.m. ; Inlet temp. T = 15°C or 288 K ;

p = 760 mm Hg = 1.013 bar

V_s = 4066 cm³ or 4066 × 10^{-6} m³ ; A/F ratio = 14 : 1 ;

b.s.f.c. = 0.38 kg/kWh

B.P. = 86 kW

Volumetric efficiency, $\eta_{vol.}$:

Air consumption, $$m = \frac{86 \times 0.38 \times 14}{60} = 7.625 \text{ kg/min}$$

Also, $$pV = mRT$$

or $$V = \frac{mRT}{p} = \frac{7.625 \times 287 \times 288}{1.013 \times 10^5} = 6.222 \text{ m}^3/\text{min}$$

$$\text{Displacement or swept volume} = 4066 \times 10^{-6} \times \frac{3600}{2}$$
$$= 7.319 \text{ m}^3/\text{min}$$

$$\therefore \quad \eta_{vol.} = \frac{6.222}{7.319} = 0.85 \text{ or } \mathbf{85\%.} \quad \textbf{(Ans.)}$$

Example 7.3. *The airflow to a four-cylinder, four-stroke oil engine is measured by a 5 cm diameter orifice having a coefficient of discharge of 0.6. The engine having bore 10 cm and stroke 12 cm runs at 1200 r.p.m. Pressure drop across orifice is 4.6 cm of water and ambient temperature and pressure are 17°C and 1 bar respectively. Calculate the volumetric efficiency based on free air condition.*

Solution. *Given :* n = 4; d = 5 cm = 0.05 m; C_d = 0.6, D = 10 cm = 0.1 m ; L = 12 cm = 0.12 m ; N = 1200 r.p.m. ; h_w = 4.6 cm = 0.046 m ; T = 17 + 273 = 290 K ; p = 1 bar.

$\eta_{vol.}$:

$$\rho_a = \frac{p}{RT} = \frac{1 \times 10^5}{287 \times 290} = 1.2015 \text{ kg/m}^3$$

Head causing flow, metre of air,

$$h_a = \frac{h_w \rho_w}{\rho_a} = \frac{0.046 \times 1000}{1.2015} = 38.285 \text{ m}$$

Air velocity $= \sqrt{2gh_a} = \sqrt{2 \times 9.81 \times 38.285} = 27.407$ m/s

$$V_{actual} = C_d \times \text{area} \times \text{velocity}$$

$$= 0.6 \times \frac{\pi}{4} \times 0.05^2 \times 27.407 = 0.0323 \text{ m}^3/\text{s}$$

$$V_{swept} = n \times \frac{\pi}{4} \times D^2 \times L \times \text{no. of cycles/sec.}$$

$$= 4 \times \frac{\pi}{4} \times 0.1^2 \times 0.12 \times \left(\frac{1200}{60 \times 2}\right) = 0.0377 \text{ m}^3/\text{s}$$

$$\therefore \quad \eta_{vol.} = \frac{V_{actual}}{V_{swept}} = \frac{0.0323}{0.0377} = 0.8567 \text{ or } \mathbf{85.67\%.\ (Ans.)}$$

Example 7.4. *A single cylinder, 4-stroke cycle diesel engine with a displacement volume of 7000 cm^3 develops 14.7 kW at 450 r.p.m. with a specific fuel consumption of 0.272 kg/kWh. The fuel used is represented by the chemical formula, C_7H_{16}. If 30 percent excess air is used, determine at inlet conditions of 1.013 bar and 30°C,*

(i) The air-fuel ratio ; *(ii) The volumetric efficiency of the engine.*

For air, take R_{air} = 0.287 kJ/kg K. **(A.M.I.E., I.C. Engines)**

Solution. *Given :* $n = 1$, $k = 1/2$ (4-stroke cycle), $V_s = 7000$ cm^3 $= 7000 \times 10^{-6}$ m^3, $P = 14.7$ kW, $N = 450$ r.p.m., s.f.c. = 0.272 kg/kWh, Excess air = 30%,

Inlet conditions : $p = 1.013$ bar ; $T = 30 + 273 = 303$ K ; $R_{air} = 287$ J/kg K.

(*i*) **The air-fuel ratio :**

Chemical combustion equation for given fuel C_7H_{16} is given by the expression,

$$\underset{\substack{12 \times 7 + 1 \times 16 \\ = 100}}{C_7H_{16}} + \underset{\substack{11 \times 2 \times 16 \\ = 352}}{11O_2} = 7CO_2 + 8H_2O$$

$$\frac{\text{Mass of oxygen}}{\text{Mass of fuel}} = \frac{352}{100} = 3.52$$

Air-fuel ratio $= \dfrac{\text{Mass of air}}{\text{Mass of fuel}} = \dfrac{\text{Mass of oxygen} / 0.23}{\text{Mass of fuel}}$

$$= \frac{\text{Mass of oxygen}}{\text{Mass of fuel}} \times \frac{1}{0.23} = \frac{3.52}{0.23} = \mathbf{15.3.\ (Ans.)}$$

($\because$ Air contains oxygen 23% by weight)

(*ii*) **The volumetric efficiency of the engine, $\eta_{vol.}$:**

$$m_f = \text{s.f.c.} \times \text{power developed}$$

$$= 0.272 \times 14.7 = 3.999 \text{ kg/h}$$

$$m_a = 15.3 \times 3.999 = 61.185 \text{ kg/h}$$

Actual air supplied $= 61.185 + 61.185 \times \dfrac{30}{100} = 79.54$ kg/h

Mass of charge, m = Mass of actual air + mass of fuel

$$= 79.54 + 3.999 = 83.54 \text{ kg/h}$$

Volume of charge sucked $= \dfrac{mRT}{p}$

$$= \frac{(83.54/60) \times 287 \times 303}{1.013 \times 10^5} = 1.195 \text{ m}^3/\text{min}$$

Displacement volume/min. = Displacement volume/stroke × No. of cycles per min.

$$= 7000 \times 10^{-6} \times \frac{450}{2} = 1.575 \text{ m}^3/\text{min}$$

$$\therefore \quad \eta_{vol.} = \frac{\text{Volume of charge displaced}}{\text{Displacement volume}}$$

$$= \frac{1.195}{1.575} \times 100 = \mathbf{75.87\%.} \quad \textbf{(Ans.)}$$

Example 7.5. *A six-cylinder four-stroke S.I. engine having a piston displacement of 730 cm³ per cylinder developed 80 kW at 3100 r.p.m. and consumed 28 kg of petrol per hour. The calorific value of petrol is 44 MJ/kg. Determine :*

(*i*) *The volumetric efficiency of the engine if air-fuel ratio is 13 and the intake air is at 0.88 bar, 27°C ;*

(*ii*) *The brake thermal efficiency ;*

(*iii*) *The brake torque.*

Solution. *Given :* Number of cylinders = 6 ;

Piston displacement per cylinder $= 730 \text{ cm}^3 = 730 \times 10^{-6} \text{ m}^3$

Power produced per cylinder, B.P. = 80 kW at 3100 r.p.m.

Petrol consumed per hour = 28 kg

Calorific value of petrol, C = 44 MJ/kg

Air-fuel ratio = 13

Intake air conditions : 0.88 bar, 27°C

(*i*) **Volumetric efficiency, $\eta_{vol.}$:**

$$\eta_{vol.} = \frac{\text{Air consumed}}{\text{Swept volume}} = \frac{\dot{m}_a}{\rho_a \times V_s \times \frac{N}{2}}$$

where, $$\rho_a = \frac{p}{RT} = \frac{0.88 \times 10^5}{287 \times (27 + 273)} = 1.022 \text{ kg/m}^3$$

$$\therefore \quad \eta_{vol.} = \frac{(28 \times 13) / 60}{1.022 \times (730 \times 6 \times 10^{-6}) \times \frac{3100}{2}} = 0.874 \text{ or } \mathbf{87.4\%.} \quad \textbf{(Ans.)}$$

(*ii*) **The brake thermal efficiency, $\eta_{th\,(B)}$:**

$$\eta_{th\,(B)} = \frac{\text{B.P.}}{\dot{m}_f \times C}$$

$$= \frac{80 \times 10^3}{(28/3600) \times (44 \times 10^6)} = 0.234 \quad \text{or} \quad \mathbf{23.4\%.} \quad \textbf{(Ans.)}$$

(*iii*) **The brake torque, T :**

$$\text{B.P.} = \frac{2\pi NT}{60 \times 1000} \text{ kW, where } T \text{ is in Nm, and } N \text{ is in r.p.m.}$$

$$80 = \frac{2\pi \times 3100 \times T}{60 \times 1000}$$

$$T = \frac{80 \times 60 \times 1000}{2\pi \times 3100} = \mathbf{246.4 \text{ Nm.}} \quad \textbf{(Ans.)}$$

Example 7.6. *The volumetric efficiency of a petrol engine at full load is 80 percent, atmospheric conditions being 1.013 bar and 25°C, and inlet and exhaust pressures are equal to the atmospheric pressure. The compression ratio of the engine is 7.5. If the inlet temperature is raised to 45°C and exhaust pressure is raised to 1.15 bar, determine :*

(*i*) *The volumetric efficiency ;*

(*ii*) *The percentage change in indicated output of the engine.*

Solution. *Given :* η_{v_1} = 80% = 0.8 ; p_i = 1.013 bar ; p_e = 1.15 bar ; T_{i_1} = 25 + 273 = 298 *K* ;

T_{i_2} = 45 + 273 = 318 K ; r = 7.5.

(*i*) **The volumetric efficiency,** η_{v_2} **:**

For pressure change :

$$\frac{\eta_{v_2}}{\eta_{v_1}} = \frac{r - (p_e / p_i)_2^{1/\gamma}}{r - (p_e / p_i)_1^{1/\gamma}} = \frac{7.5 - (1.15 / 1.013)^{1/1.4}}{7.5 - (1.013 / 1.013)^{1/1.4}}$$

$$= \frac{7.5 - 1.0948}{7.5 - 1} = 0.9854$$

For inlet temperature change :

$$\frac{\eta_{v_2}}{\eta_{v_1}} = \sqrt{\frac{T_{i_2}}{T_{i_1}}} = \sqrt{\frac{318}{298}} = 1.033$$

Now volumetric efficiency, considering both pressure and temperature,

$$\eta_{v_2} = 0.8 \times 0.9854 \times 1.033 = 0.8143 \quad \text{or} \quad \mathbf{81.43\%.} \quad \textbf{(Ans.)}$$

(*ii*) **The percentage change in indicated output of the engine :**

Output $\propto (\rho_i \times \eta_v)$ or $\frac{\eta_v}{T_i}$ for constant inlet pressure

∴ *Percentage reduction in output*

$$= \frac{(0.80 / 298) - (0.814 / 318)}{(0.80 / 298)} = 1 - \frac{0.814 \times 298}{0.80 \times 318}$$

$$= 0.0465 \quad \text{or} \quad \mathbf{4.65\%.} \quad \textbf{(Ans.)}$$

Example 7.7. *A 4-stroke diesel engine has a compression ratio of 14 and works in ambient condition of 1.013 bar and 27°C. A supercharger is added to the engine which raises the inlet pressure to 1.3 bar and the inlet temperature to 60°C, other conditions remaining the same. Determine :*

(*i*) *The percentage change in charging efficiency ;*

(*ii*) *The percentage change in indicated output of the engine.*

Solution. *Given :* r = 14 ; p_1 = 1.013 bar ; p_2 = 1.3 bar ; T_1 = 27 + 273 = 300 *K* ;

T_2 = 60 + 273 = 333 *K*

(*i*) **The percentage change in charging efficiency :**

For change in pressure :

$$\frac{\eta_{v_2}}{\eta_{v_1}} = \frac{r - (p_e / p_i)_2^{1/\gamma}}{r - (p_e / p_i)_1^{11\gamma}} = \frac{14 - (1.013 / 1.3)^{1/1.4}}{14 - (1.013 / 1.013)^{1/1.4}} = \frac{14 - 0.837}{14 - 1} = 1.0125$$

For change in inlet temperature :

$$\frac{\eta_{v_2}}{\eta_{v_1}} = \sqrt{\frac{T_{i_2}}{T_{i_1}}} = \sqrt{\frac{333}{300}} = 1.054$$

∴ For both change in inlet pressure and temperature,

$$\frac{\eta_{v_2}}{\eta_{v_1}} = 1.0125 \times 1.054 = 1.067$$

∴ *The percentage increase in volumetric efficiency* = **6.7%. (Ans.)**

(*ii*) **The percentage change in indicated output of the engine :**

$$\text{Output} \propto \rho \times \eta_v$$

$$\rho_1 = \frac{p_1}{RT_1} = \frac{1.013 \times 10^5}{287 \times 300} = 1.176$$

$$\rho_2 = \frac{p_2}{RT_2} = \frac{1.3 \times 10^5}{287 \times 333} = 1.36 \text{ kg/m}^3$$

$$\therefore \quad \frac{P_2}{P_1} = \frac{\rho_2 \times \eta_{v_2}}{\rho_1 \times \eta_{v_1}} = \frac{1.36}{1.176} \times 1.067 = 1.234$$

or *Percentage increase in power* = **23.4%. (Ans.)**

Example 7.8. *A petrol engine operating at full throttle develops 32 kW with 80 percent mechanical efficiency at sea level where atmospheric conditions are 1.013 bar pressure and 35°C temperature. The engine is moved to a hill station whose altitude is 2000 m and temperature is 5°C. A drop of 10 mm of mercury barometer reading may be assumed for each 100 m of rise in altitude.*

Determine the percentage change in volumetric efficiency of the engine and the brake power of the engine if it runs at the same speed and full throttle.

Solution. *Given :* $(\text{I.P.})_1 = 32$ kW ; $\eta_{\text{mech.}} = 80\%$; $p_1 = 1.013$ bar ; $T_1 = 35 + 273 = 308\ K$; $T_2 = 5 + 273 = 278$ K

Percentage change in volumetric efficiency :

The ratio of p_e/p_i in both cases is 1. Therefore there is no change in volumetric efficiency due to inlet and exhaust pressure change.

For inlet temperature change,

$$\frac{\eta_{v_2}}{\eta_{v_1}} = \sqrt{\frac{T_{i_2}}{T_{i_1}}} = \sqrt{\frac{278}{308}} = 0.95$$

∴ *Percentage decrease in* $\eta_v = \dfrac{1 - 0.95}{1} \times 100 =$ **5%. (Ans.)**

Brake power of the engine, $(\text{B.P.})_2$:

Indicated power output (I.P.) $\propto \eta_v \times \rho_i$ or $\dfrac{\eta_v\, p_i}{T_i}$

$$(\text{I.P.})_1 \propto \frac{\eta_{v_1}\, p_{i1}}{T_{i_1}} \propto \frac{n_{v_1} \times 1.013}{308} \propto \eta_{v_1} \times 0.003289$$

Drop in pressure at hill station = $\rho g h$ N/m^2

$$= (13.6 \times 1000) \times 9.81 \times \left(\frac{10}{1000} \times \frac{2000}{100}\right) \times 10^{-5} \text{ bar} = 0.267 \text{ bar}$$

$$\therefore \quad p_2 = 1.013 - 0.267 = 0.746 \text{ bar}$$

$$(\text{I.P.})_2 \propto \frac{\eta_{v_2}\, p_{i_2}}{T_{i_2}} \propto \frac{0.95\, \eta_{v_1} \times 0.746}{278} \propto 0.002549\ \eta_{v_1}$$

$$\therefore \quad \frac{(\text{I.P.})_2}{(\text{I.P.})_1} = \frac{0.002549\, \eta_{v_1}}{0.003289\, \eta_{v_1}} = 0.775$$

$$\therefore \quad (\text{I.P.})_2 = (\text{I.P.})_1 \times 0.775 = \left(\frac{32}{0.8}\right) \times 0.775 = 31 \text{ kW}$$

Since the engine speed is same at two places, the friction power and hence mechanical efficiency remains unchanged.

$$\therefore \quad (\text{B.P.})_2 = 31 \times 0.8 = \mathbf{24.8\ kW.} \quad \textbf{(Ans.)}$$

Example 7.9. *A petrol engine, with a compression ratio of 7.5 develops 75 kW indicated power with ambient conditions of 1.01 bar and 27°C. A supercharger is added which increases the inlet pressure to 1.38 bar and inlet temperature to 48°C under otherwise identical conditions. If the volumetric efficiency of the engine is 81% without supercharging, calculate :*

(*i*) *The volumetric efficiency ;*

(*ii*) *Indicated power of the supercharged engine.*

Solution. *Given :* $r = 7.5$; I.P. = 75 kW ; $T_{i_1} = 27 + 273 = 300\ K$; $p_{i_1} = 1.01$ bar ; $p_{i_2} = 1.38$ bar, $T_{i_2} = 48 + 273 = 321$ K ; $\eta_{v_1} = 81\%$.

(*i*) **The volumetric efficiency of the supercharged engine η_{v_2} :**

For change in inlet pressure :

$$\frac{\eta_{v_2}}{\eta_{v_1}} = \frac{r - (p_e / p_i)_2^{1/\gamma}}{r - (p_e / p_i)_1^{1/\gamma}} = \frac{7.5 - (1.01/1.38)^{1/1.4}}{7.5 - (1.01/1.01)^{1/1.4}} = \frac{7.5 - 0.8}{7.5 - 1} = 1.03$$

For change in inlet temperature :

$$\frac{\eta_{v_2}}{\eta_{v_1}} = \sqrt{\frac{T_{i_2}}{T_{i_1}}} = \sqrt{\frac{321}{300}} = 1.0344$$

For both change in inlet pressure and temperature,

$$\frac{\eta_{v_2}}{\eta_{v_1}} = 1.03 \times 1.0344 = 1.0654$$

$$\therefore \quad \eta_{v_2} = \eta_{v_1} \times 1.0654 = 0.81 \times 1.0654 = 0.8629 \quad \text{or} \quad \mathbf{86.29\%.} \quad \textbf{(Ans.)}$$

(*ii*) **Indicated power of the supercharged engine, (I.P.)$_2$:**

$$\rho_{a_1} = \frac{p_1}{RT_1} = \frac{1.01 \times 10^5}{287 \times 300} = 1.173 \text{ kg/m}^3$$

$$\rho_{a_2} = \frac{p_2}{RT_2} = \frac{1.38 \times 10^5}{287 \times 321} = 1.498 \text{ kg/m}^3$$

$$\frac{(\text{I.P.})_2}{(\text{I.P.})_1} = \frac{\eta_{v_2} \times \rho_{a_2}}{\eta_{v_1} \times \rho_{a_1}}$$

or
$$(\text{I.P.})_2 = (\text{I.P.})_1 \times \frac{\eta_{v_2} \times \rho_{a_2}}{\eta_{v_1} \times \rho_{a_1}}$$

$$= 75 \times \frac{0.8629 \times 1.498}{0.81 \times 1.173} = \textbf{102 kW.} \quad \textbf{(Ans.)}$$

Example 7.10. *A 1-cylinder, 4-stroke cycle engine has a bore of 320 mm and stroke 380 mm. In a test the gas consumption was metered 0.25 m³/min. at 120 mm of water and 27°C. The air consumption was 3.36 kg/min at atmospheric pressure of 1.013 bar and temperature 27°C. The speed of the engine was 280 r.p.m. The calorific value of gas used was 18600 kJ/m³ at 25°C and 1.013 bar. Calculate :*

(i) The volumetric efficiency of the engine ;

(ii) The heating value of 1 m³ of the charge at 25°C and 1.013 bar.

Solution. *Given :* $n = 1$; $D = 320$ mm $= 0.32$ m ; $L = 380$ mm $= 0.38$ m ; $N = 280$ r.p.m. ;

C.V. = 18600 kJ/m³ at 25°C and 1.013 bar

(i) **Volumetric efficiency, $\eta_{vol.}$:**

Volume of air consumed at inlet condition,

$$V = \frac{mRT}{p} = \frac{3.36 \times 287 \times (27 + 273)}{1.013 \times 10^5} = 2.8558 \text{ m}^3\text{/min}$$

Gas supply pressure $= 1.013 + \dfrac{(120/1000)}{10.2} = 1.025$ bar (1 bar = 10.2 m)

Gas consumption at inlet condition

$$= 0.25 \times \frac{1.025}{1.013} = 0.253 \text{ m}^3\text{/min}$$

∴ Volume of mixture consumed in inlet condition

$$= 2.8558 + 0.253 = 3.109 \text{ m}^3\text{/min}$$

$$\eta_{vol.} = \frac{\text{Volume of mixture consumed at inlet condition}}{\text{Swept volume}}$$

$$= \frac{3.109}{\frac{\pi}{4} \times (0.32)^2 \times (0.38) \times (280/2)} = 0.727 \quad \text{or} \quad \textbf{72.7\%.} \quad \textbf{(Ans.)}$$

(ii) **The heating value of 1 m³ of the charge :**

The heating value of 1 m³ of the charge at 25°C and 1.013 bar

$$= \frac{\text{Volume of gas}}{\text{Volume of mixture}} \times 18600 = \frac{0.253}{3.109} \times 18600 = \textbf{1513.6 kJ/m}^3\textbf{.} \quad \textbf{(Ans.)}$$

☞**Example 7.11.** *A 6-cylinder, 4-stroke petrol engine with a bore of 125 mm and a stroke of 190 mm was supplied during a test with petrol of composition C = 82% and H_2 = 18% by mass. The dry exhaust composition by volume was CO_2 = 11.19%, O_2 = 3.61% and N_2 = 85.2%.*

(i) Determine the mass of air supplied per kg of petrol, the percentage of excess air and the volume of the mixture per kg of petrol at 17°C and 0.98 bar, which are the conditions for the mixture entering the cylinder during the test.

(ii) Also determine the volumetric efficiency of the engine based on intake conditions when mass of petrol used per hour during the test was 31 kg and the engine speed was 1600 r.p.m. The petrol is completely evaporated before entering the cylinder and the effect of its volume on the volumetric efficiency should be included.

Asume the following :

Density of petrol vapour as 3.35 times that of air at the same temperature and pressure. 1 kg of air at 0°C and 1.0132 bar occupies 0.7734 m³. Air contains 23% oxygen by mass.

Solution. *Given :* $n = 6$; $d = 125$ mm ; $l = 190$ mm ; $C = 82\%$; $H_2 = 18\%$; $CO_2 = 11.19\%$; $O_2 = 3.61\%$; $N_2 = 85.2\%$.

(*i*) Theoretical mass of air required per kg of fuel for complete combustion

$$= \frac{100}{23}\left(\frac{8}{3}C + 8H_2\right)$$

$$= \frac{100}{23}\left(\frac{8}{3} \times 0.82 + 8 \times 0.18\right) = 15.768 \text{ kg.}$$

Vol. of constituents per mole of dry flue gases (a)	*Molar mass* (b)	*Proportional mass of constituents* (c) = (a) × (b)	*Mass/kg of flue gas* d = c/Σ(c)	*Mass of carbon per kg of flue gas*
$CO_2 = 11.19$	44	492.36	0.16448	$\frac{0.16448 \times 12}{44} = 0.044858$
$O_2 = 3.61$	32	115.52	0.03859	
$N_2 = 85.20$	28	2385.60	0.79693	
Total 100.00		2993.48	1.00000	

$$\text{Mass of dry flue gases/kg of fuel} = \frac{\text{Mass of carbon per kg of fuel}}{\text{Mass of carbon per kg of flue gas}}$$

$$= \frac{0.82}{0.044858} = 18.28 \text{ kg.}$$

$$\text{Mass of unused air per kg of fuel} = \frac{100}{23}(18.28 \times 0.03859) = 3.067 \text{ kg.}$$

∴ **Air supplied/kg of fuel** = 15.768 + 3.067 = 18.835 kg. **(Ans.)**

∴ $$\textbf{Excess air} = \frac{3.067}{15.768} \times 100 = \textbf{19.45\%}. \quad \textbf{(Ans.)}$$

Now $$pV = mRT$$

For air $$R = \frac{pV}{mT} = \frac{(1.0132 \times 10^5) \times 0.7734}{1 \times 273}$$

$$= 287 \text{ kJ/kg K}$$

At the same temperature and pressure

$$\frac{\rho_{\text{petrol vapour}}}{\rho_{\text{air}}} = 3.35 = \frac{R_{\text{air}}}{R_{\text{petrol vapour}}}$$

∴ $$R_{\text{petrol vapour}} = \frac{287}{3.35} = 85.67$$

At 17°C and 0.98 bar,

Volume of 1 kg of petrol vapour $= \dfrac{mRT}{p}$

$$= \frac{1 \times 85.67 \times (273 + 17)}{0.98 \times 10^5} = 0.2535 \text{ m}^3$$

Volume of 18.835 kg of air $= \dfrac{mRT}{p}$

$$= \frac{18.835 \times 287 \times 290}{0.98 \times 10^5} = 15.996 \text{ m}^3$$

∴ Volume of mixture per kg of petrol

$= 0.2535 + 15.996 = 16.25 \text{ m}^3$. **(Ans.)**

(*ii*) Now, *volumetric efficiency*

$$= \frac{\text{Volume of air per min at intake condition}}{\text{Swept volume per min}}$$

$$= \frac{16.25 \times (31/60)}{6 \times \frac{\pi}{4}(0.125)^2 \times 0.19 \times \frac{1600}{2}}$$

$= 0.75$ or **75%. (Ans.)**

Example 7.12. *On testing a spark ignition engine it was observed that the volumetric efficiency is maximum when inlet valve Mach Index is 0.55 and the indicated torque, and indicated mean effective pressure occured at maximum volumetric efficiency.*

The engine having a bore of 110 mm and stroke 140 mm produces maximum indicated torque when running at 2400 r.p.m.

(*i*) *Determine the nominal diameter of the inlet valve.*

(*ii*) *If the same engine is required to develop maximum indicated power at 2800 r.p.m., how will the inlet valve size be modified ?*

(*iii*) *If the same engine runs at 2800 r.p.m. without any inlet valve modifications, how will volumetric efficiency get affected ?*

Pressure at intake valve = 0.88 bar ; Temperatue at intake valve = 340 K ; Inlet valve flow coefficient = 0.33.

Assume : Fuel-air mixture as perfect gas with $\gamma = 1.4$ *and* $R = 287$ *J/kg K.*

(*iv*) *What would be volumetric efficiency at maximum power speed of 4800 r.p.m., for unmodified engine.*

Solution. *Given :* $Z = 0.55$; $D_{cy} = 110$ mm $= 0.11$ m, $L = 140$ mm $= 0.14$ m ; $N = 2400$ r.p.m.; $p = 0.88$ bar, $T = 340$ K ; $K_i = 0.33$; $\gamma = 1.4$, $R = 287$ J/kg K.

(*i*) **Nominal diameter of inlet valve, D_{iv} :**

For the properties of mixture given in the data, the local sonic velocity of mixture of air-fuel at the inlet or suction valve is given by :

$$U_s = \sqrt{\gamma RT} = \sqrt{1.4 \times 287 \times 340} = 369.6 \text{ m/s}$$

Also,
$$\frac{U_g}{U_s} = \left(\frac{D_{cy}}{D_{iv}}\right)^2 \times \frac{U_p}{K_i U_s} = Z \qquad \text{...[Eqn. (7.9)]}$$

or
$$\left(\frac{D_{cy}}{D_{iv}}\right)^2 \times \frac{(2LN/60)}{K_i U_s} = Z \qquad \left[\because U_p = \text{piston speed} = 2LN/60\right]$$

$$\left(\frac{0.11}{D_{iv}}\right)^2 \times \left[\frac{(2 \times 0.14 \times 2400 / 60)}{0.33 \times 369.6}\right] = 0.55$$

or
$$D_{iv} = \left[\frac{(0.11)^2 \times (2 \times 0.14 \times 2400 / 60)}{0.33 \times 369.6 \times 0.55}\right]^{1/2} = 0.04495 \text{ m} \quad \text{or} \quad \mathbf{44.95\ mm.} \quad \mathbf{(Ans.)}$$

At this value of $Z = 0.55$, volumetric efficiency from Fig. 7.3 is 87.5% approximately.

(*ii*) Also for, the indicated torque, indicated mean effective pressure and volumetric efficiency to be maximum at 2800 r.p.m., $Z = 0.55$.

$$\therefore \quad D_{iv} = \left[\frac{(0.11)^2 \times (2 \times 0.14 \times 2800 / 60)}{0.33 \times 369.6 \times 0.55}\right]^{1/2} = 0.04855 \quad \text{or} \quad \mathbf{48.55\ mm.} \quad \mathbf{(Ans.)}$$

Thus $\eta_{vol.}$ does not get affected but remain constant as 87.5% at 2800 r.p.m., the inlet valve diameter is increased to 48.55 mm.

(*iii*) Again, without intake valve modification, $\frac{A_p}{A_{iv}}$ has to remain same *i.e.* the nominal diameter of the inlet valve = 44.95 mm, and the new value of U_p at 2800 r.p.m.

$$= \frac{2 \times 0.14 \times 2800}{60} = 13.067 \text{ m/s.}$$

Thus, the new value of Z is given by :

$$Z = \left(\frac{D_{cy}}{D_{iv}}\right)^2 \times \frac{U_p}{K_i\, U_s} = \left(\frac{0.11}{0.04495}\right)^2 \times \frac{13.067}{0.33 \times 369.6} = 0.64$$

Corresponding to $Z = 0.64$, from Fig. 7.3, $\eta_{vol.} = 84\%$ approximately. Indicated torque also drops in the same proportion as $\eta_{vol.}$

(*iv*) Again, U_p at 4800 r.p.m. $= \frac{2 \times 0.14 \times 4800}{60} = 22.4$ m/s

$$Z = \left(\frac{0.11}{0.04495}\right)^2 \times \frac{22.4}{0.33 \times 369.6} = 1.1$$

Corresponding to $Z = 1.1$, from Fig. 7.3, $\eta_{vol.} = 56\%$ app.

It may be noted that whereas speed has increased from 2400 r.p.m. to 4800 r.p.m. *i.e.* 100 percent ; the volumetric efficiency has fallen by $\frac{87.5 - 56}{87.5} = 0.36$ or **36%.**

— Thus, although power goes on *increasing* beyond 2400 r.p.m., the volumetric efficiency falls beyond 2400 r.p.m. *Finally the power reaches a maximum at 4800 r.p.m., and beyond this limit starts falling under the influence of very rapidly falling* $\eta_{vol.}$

— Hence, the *maximum torque-engine speed and maximum power-engine speed can be varied and fixed at the required values with these inlet valve modifications etc.*

HIGHLIGHTS

1. **Air capacity** is defined as the mass flow of fresh air through the engine per unit time.
2. The **ideal air capacity** corresponds to filling the displaced volume with fresh mixture at inlet conditions.

3. The **volumetric efficiency** of an engine is defined as the ratio of actual air capacity to the ideal air capacity. This is equal to the ratio of mass of air which enters or is forced into the cylinder in suction stroke to the mass of free air equivalent to piston displacement at intake temperature and pressure conditions.
4. Power output of an engine is proportional to volumetric efficiency provided the combustion is complete.
5. The volumetric efficiency of an engine is affected by many variables such as compression ratio, valve timing, induction and port design, mixture strength, latent heat of evaporation of fuel, heating of the induced charge, cylinder temperature and atmospheric conditions.

OBJECTIVE TYPE QUESTIONS

Fill in the Blanks or Say "Yes" or "No" :

1. At lower speeds volumetric efficiency is nearly constant, at high speeds it ....
2. The higher the altitude .... will be the volumetric efficiency.
3. The volumetric efficiency is defined as the ratio of ideal air capacity to actual air capacity of an engine.
4. The ideal air capacity is defined as the mass flow of fresh air through the engine per unit time.
5. Volumetric efficiency of 4-stroke cycle I.C. engines ... with increase in inlet temperature.
6. The maximum volumetric efficiency speed can be increased for 4-stroke I.C. engines by .... the inlet valve diameter.
7. The volumetric efficiency of 4-stroke cycle I.C. engine ... with increase in coolant temperature.
8. For 4-stroke cycle diesel engines exhaust gas temperature increases with load.
9. In a 4-stroke cycle I.C. engine, ideal air capacity varies directly as the piston speed.
10. Indicated mean effective pressure for a 4-storke I.C. engine varies inversely as inlet air density.
11. The volumetric efficiency of a 4-stroke I.C. engine varies directly as the diameter of the cylinder.
12. The volumetric efficiency of a 4-stroke I.C. engine varies ...... as inlet air density.
13. Inlet valve Mach Index for maximum volumetric efficiency for four stroke engine is approximately ......
14. The volumetric efficiency of a 4-stroke S.I. engine remains fairly constant with increase in F/A ratio.
15. The isentropic index of compression (γ) increases as the F/A ratio of octane-air mixture increases.

ANSWERS

1. falls rapidly	**2.** lower	**3.** No	**4.** No	**5.** decreases
6. increasing	**7.** decreases	**8.** Yes	**9.** Yes	**10.** No
11. No	**12.** inversely	**13.** 0.5	**14.** Yes	**15.** No.

THEORETICAL QUESTIONS

1. Define ideal air capacity of an engine. How does it differ from actual air capacity ?
2. How do you define volumetric efficiency of an I.C. engine ? How is it related to the power output of the engine ?
3. How is the volumetric efficiency affected by speed and altitude ?
4. Discuss briefly the effects of the following factors on the volumetric efficiency :
   (*i*) Fuel (*ii*) Heat transfer-high temperature
   (*iii*) Valve overlap (*iv*) Fluid friction losses
   (*v*) Choked flow (*vi*) Intake valve closure after B.D.C.
   (*vii*) Exhaust residual (*viii*) Exhaust gas recycle (EGR).
5. What is the effect of "Inlet Mach Number" on the volumetric efficiency of an engine ?

6. Why the inlet valve be kept open for a few degrees of crank angle even when the piston is on the compression stroke ?

   Assume that the engine under consideration is a high speed one.

7. "A 4-stroke I.C. engine is always economical and less pollutant than 2-stroke engine". Justify the statement.

UNSOLVED EXAMPLES

1. A six-cylinder four-stroke spark ignition engine having a piston displacement of 700 cm^3 per cylinder produced 78 kW at 3200 r.p.m. and consumed 27 kg of petrol per hour. The calorific value of petrol is 44 MJ/kg. Determine :

   (*i*) The volumetric efficiency of engine if air-fuel ratio is 12 and the intake air is at 0.9 bar, 32°C ;

   (*ii*) The brake thermal efficiency ;

   (*iii*) The brake torque. [**Ans.** (*i*) 78.16% ; (*ii*) 23.64% ; (*iii*) 233 Nm)]

2. On testing a S.I. engine it was observed that the volumetric efficiency is maximum when inlet valve Mach Index is 0.55 and the indicated torque, and indicated mean effective pressure occured at maximum volumetric efficiency.

   The engine having a bore of 120 mm and stroke 150 mm. produces maximum indicated torque when running at 2500 r.p.m.

   (*i*) Determine the nominal diameter of the inlet valve.

   (*ii*) If the same engine is required to develop maximum indicated power at 3000 r.p.m., how will the inlet valve size be modified ?

   (*iii*) If the same engine runs at 3000 r.p.m. without any inlet valve modifications, how will volumetric efficiency get affected ?

   Pressure at intake valve = 0.9 bar ; Temperature at intake valve = 350 K ; Inlet valve flow coefficient = 0.325.

   Assume : Fuel-air mixture as perfect gas with $\gamma = 1.4$ and R = 287 J/kg K.

   (*iv*) What would be volumetric efficiency at maximum power speed of 5000 r.p.m., for the un-modified engine ? [**Ans.** (*i*) 43 mm ; (*ii*) 47 mm ; (*iii*) 84% ; (*iv*) 55.25%]

8

Two Stroke Engines

8.1. General aspects—Construction and working—Comparison between two stroke cycle and four stroke cycle engines—Disadvantages of two stroke S.I. engine compared to two stroke C.I. engine—Reasons for use of two stroke engines for marine propulsion—Reasons for the use of two stroke S.I. engines for low horse power two wheelers 8.2. Intake for two stroke cycle engines. 8.3. Scavenging process. 8.4. Scavenging parameters. 8.5. Scavenging systems. 8.6. Crankcase scavenging. 8.7. Scavenging pumps and blowers—Highlights—Objective Type Questions—Theoretical Questions.

8.1. GENERAL ASPECTS

8.1.1. Construction and Working

- In 1878, Dugald-clerk, a British engineer introduced a cycle which could be completed in two *strokes of piston rather than four strokes* as is the case with the four stroke cycle engines. The engines using this cycle were called *two stroke cycle engines*. In this engine suction and exhaust strokes are eliminated. Here *instead of valves, ports are used. The exhaust gases are driven out from engine cylinder by the fresh charge of fuel entering the cylinder nearly at the end of the working stroke.*
- Fig. 8.1 shows a two-stroke petrol engine (used in scooters, motor cycle etc.) Refer Art. 2.12 also.

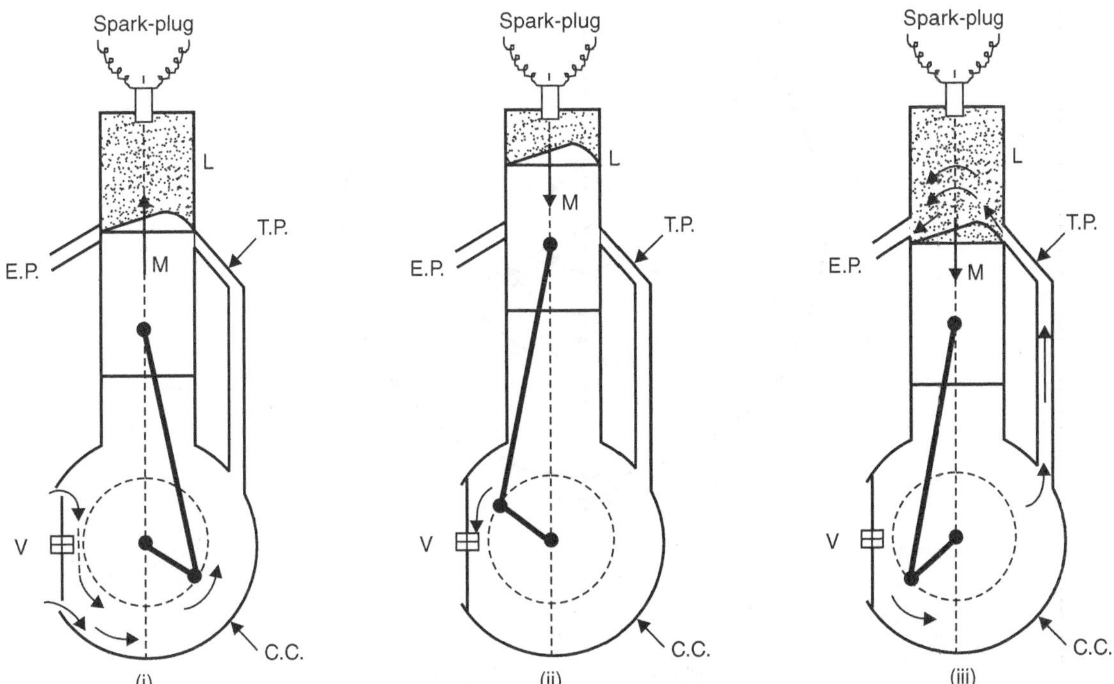

L = Cylinder ; *M* = Piston ; *C.C.* = Crankcase ; *V* = Valve ; *E.P.* = Exhaust port ; *T.P.* = Transfer port.

Fig. 8.1. Two stroke cycle engine (crankcase scavenged).

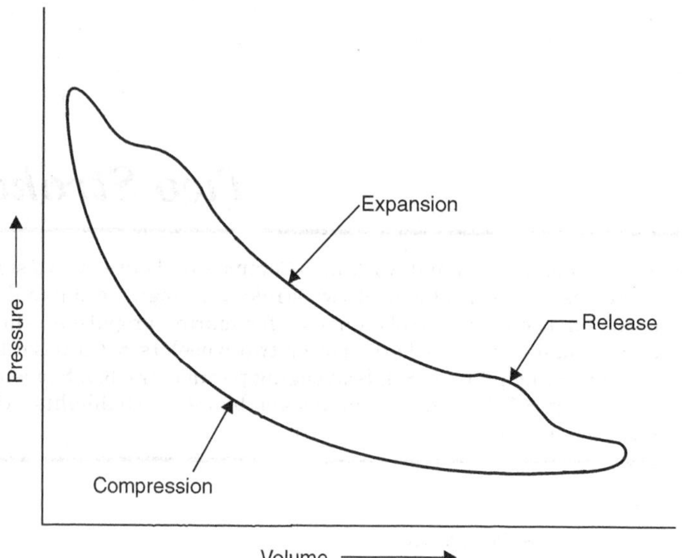

Fig. 8.2. p-V diagram for a two stroke cycle engine.

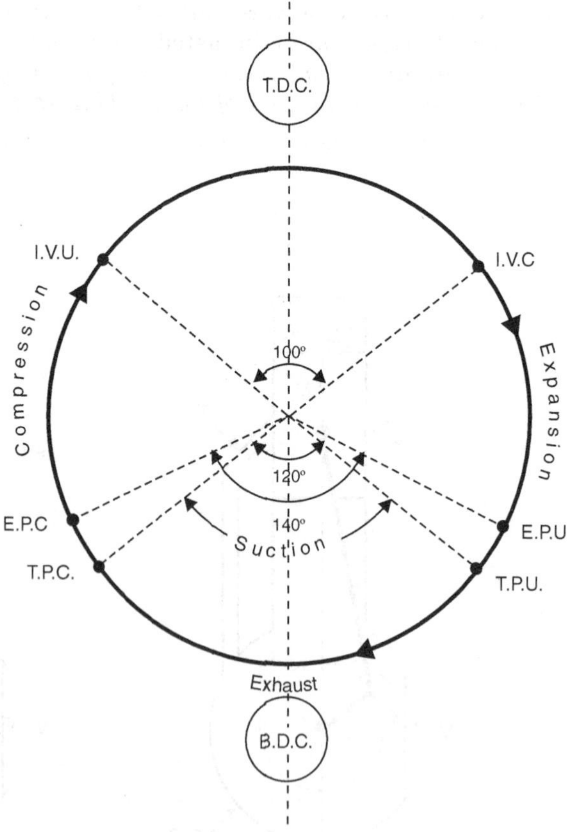

Fig. 8.3. Port timing diagram.

— The cyclinder L is connected to a closed crankcase C.C.
— During the upward stroke of the piston M, the gases in L are compressed and at the same time fresh air and fuel (petrol) mixture enters the crank chamber through the valve V.
— When the piston moves downwards, V closes and the mixture in the crank chamber is compressed.
— Refer Fig. 8.1 (*i*), the piston is moving upwards and is compressing an explosive charge which has previously been supplied to L. Ignition takes place at the end of the stroke. The piston then travels downwards due to expansion of the gases (Fig. 8.1 (*ii*)) and near the end of this stroke the piston uncovers the exhaust port (E.P.) and the burnt exhaust gases escape through this port (Fig. 8.1 (*iii*)).
— The transfer port (T.P.) then is uncovered immediately, and the compressed charge from the crank chamber flows into the cylinder and is deflected upwards by the hump provided on the head of the piston. It may be noted that the *incoming air petrol mixture helps the removal of gases from the engine-cylinder ; if, in case these exhaust gases do not leave the cylinder, the fresh charge gets diluted and efficiency of the engine will decrease.*
— The piston then again starts moving from B.D.C. to T.D.C. and the charge gets compressed when E.P. (exhaust port) and T.P. are covered by the piston ; thus the *cycle is repeated.*

- Fig. 8.2 show the *p-V* diagram for a two stroke cycle engine. It is *only for the main cylinder or the top side of the piston.*
- Fig. 8.3 shows self-explanatory port timing diagram for a two stroke cycle engine.

In a **two stroke Diesel cycle** *engine all the operations are the same as in the spark ignition (Otto cycle) engine with the differences ; firstly in this case,* ***only air*** *is admitted into cylinder instead of air fuel mixture and secondly* ***fuel injector is*** *fitted to supply the fuel instead of a sparking plug.*

Note. The top of the piston usually has a projection / hump to deflect the fresh air to sweep up to the top of the cylinder before flowing to the exhaust ports. This serves the following two purposes :

(*i*) To scavenge the upper part of the cylinder of combustion products.

(*ii*) To prevent the fresh charge from flowing directly to the exhaust ports.

The same objective can be achieved *without piston deflector by proper shaping of the transfer port.*

8.1.2. Comparison between Two-stroke Cycle and Four-stroke Cycle Engines

- For comparison between 2-stroke cycle and 4-stroke cycle engines refer Art. 2.14.
- For *all the petrol as well diesel two-stroke engines* a common ***disadvantage*** is *greater cooling and lubrication requirements due to one power stroke in each revolution of crank-shaft.* Due to *higher temperature the consumption of lubrication oil is also high in two-stroke engines.*

8.1.3. Disadvantages of Two-stroke S.I. Engine Compared to Two-stroke C.I. Engine

Following are the two main disadvantages from which the two-stroke S.I. engines suffer :

1. Loss of fuel
2. Idling difficulty.

- In case two cylinders are supplied the fuel after the closure of the exhaust ports, the fuel loss will be nil and the indicated thermal efficiency of the two-stroke engine will be comparable as the four-stroke engine. However, in S.I. engine using carburettor, the scavenging is done with fuel-air mixture and only the fuel mixed with the retained air is used for combustion.

— *In order to avoid the loss of fuel instead of carburettor fuel injection just before the exhaust port closure may be used.*

- At *low speeds* when m.e.p. (mean effective pressure) is reduced to about 2 bar, the two stroke S.I. engine *runs irregularly* and *may even stop.* This is owing to large amount of residual gas (more than in 4-stroke engine) mixing with small amount of charge. At low speeds there may be backfiring due to slow burning rate.

  — **Fuel injection** *improves idling and also eliminates backfiring as there is no fuel present in the inlet system.*

In case of C.I. engine there is neither fuel loss (as the charge is only air) nor difficulty in idling since there is no reduction in fresh charge (air).

8.1.4. Reasons for Use of Two-stroke C.I. Engines for Marine Propulsion

Two-stroke C.I. engines find wide use in marine propulsion for the following reasons :

1. More uniform torque, the ideal requirement for the propeller.

2. More cooling is required in two stroke engines, plenty of sea water is available for cooling.

3. In C.I. engines there in no loss of fuel in scavenging. Hence they have higher thermal efficiency.

4. *Propeller imposes the condition that maximum power must be developed at about 100 r.p.m. Two stroke engines may be made of slow speed, and with large displacement volume (over 60 cm bore) and of capacity 5000 kW and above. These slow speed engines can be coupled directly to the propeller of the ship, wihtout the necessity of gear reduction.*

- For marine propulsion, *two-stroke C.I. opposed engine (cross-head type) is mainly used.*

8.1.5. Reasons for the Use of Two-stroke S.I. Engines for Low Horse Power Two Wheelers

- When applied to S.I. engines, the Two-stroke cycle engine has certain disadvantages which have restricted its use to small low horse power engines.

  — In S.I. engines the charge consists of a mixture of air and fuel. *During scavenging both, inlet and exhaust ports are open simultaneously for sometime. Some part of the fresh charge escapes with exhaust which results in higher fuel consumption and lower thermal efficiency.*

- For *small two-wheeler engines the* ***fuel economy is not a vital factor.*** Here ***light-weight and low initial cost are the main considerations,*** *which are the main characteristics of two-stroke S.I. engines.*

8.2. INTAKE FOR TWO STROKE CYCLE ENGINES

- *In two stroke cycle engines inlet air must be input at a pressure greater than atmospheric.* At the start of the intake process, following blowdown, the cylinder is still filled with exhaust gas at atmospheric pressure. There is no exhaust stroke. Air under pressure enters the cylinder and pushes most of the remaining exhaust residual out of the still-open exhaust port. This is called **scavenging.** When most of the exhaust gas is out, the exhaust port closes and the cylinder is filled with air.

  — *At part throttle inlet pressure is low, and this results in poorer scavenging.*

- Generally following two methods are used for putting air into the cylinders :

  (*i*) Through normal intake valves ;

  (*ii*) Through intake slots in the cylinder walls.

  — *The intake air is pressurised using a supercharger, turbocharger, or crankcase compression.*

- There are open combustion chambers in the two stroke cycle engines. It would be extremely difficult to get proper scavenging in a cylinder with a divided chamber.
- In some automobile engines *standard-type superchargers are used and the air is input through intake valves with no fuel added.* The compressed air scavenges the cylinder and leaves it filled with air and a small amount of exhaust residual. After the intake valve is closed, fuel is injected directly into the combustion chamber by injectors mounted in the cylinder head. This is done to avoid HC pollution from fuel passing into the exhaust system, when both exhaust and intake valves are open. *In some automobile engines, air is injected with the fuel. This speeds evaporation and mixing, which is required because of the very short time of the compression stroke.*
  — Fuel injection pressure is of order of **500** to **600** kPa, while air injection pressure is slightly less at about 500 kPa.
  — For *"S.I. engine" fuel injection occurs early in the compression stroke, immediately after the exhaust valve closes.* In *"C.I. engines" the injection occurs late in the compression stroke, a short time before combustion starts.*
- In just about all two stroke cycle engines, *due to cost, crankcase compression is used to force air into and scavenge the cylinders.*
  — In these engines, air is introduced at atmospheric pressure into the cylinder below the piston through a one-way valve when the piston is near T.D.C. The power stroke pushes the piston down and compresses the air in the crankcase, which has been designed for this dual purpose. The compressed air then passes through an input channel into the combustion chambers. In *modern automobiles engines* the fuel is then added with injectors, as with supercharged engines the fuel is then added with injectors, as with supercharged engines. In *small engines the fuel is usually with a carburettor to the air as it enters the crankcase. This is done to keep the cost down on* small engines, simple carburettors being cheap to build. The fuel injectors will probably become more common as pollution laws become more stringent.
- In case of two stroke cycle engines *using crankcase compression, lubricating oil must be added to the inlet air.* The crankcase in these engines cannot be used as the oil reservoir as with most other engines. Instead, the surfaces of the engine components are lubricated by *oil vapour carried by the intake air.* In some engines, lubricating oil is mixed directly with the fuel and is vaporised in the carburettor along with the fuel. Other engines have a separate oil reservoir and feed lubricant directly into the intake air flow. *Two negative results* occur beacuse of this method of lubrications : (*i*) Some oil vapour gets into the exhaust flow during valve overlap and contributes directly to HC exhaust emissions ; (*ii*) Combustion is less efficient due to the poorer fuel quality of the oil.
  — *Engines which use superchargers or turbochargers generally use standard pressurised lubrication systems, with crankcase serving as the oil reservior.*
- *In order to avoid an excess of exhaust residual no pockets of stagnant flow or dead zones can be allowed in the scavenging process. This is controlled by :*
  (*i*) The size and position of the intake and exhaust slots or valves ;
  (*ii*) The geometry of the slots in the wall ;
  (*iii*) The contoured flow deflectors on the piston face.

8.3. SCAVENGING PROCESS

- In a two stroke engine because of non-availability of an exhaust stroke (unlike four-stroke engine) at the end of an expression stroke, its combustion chamber is left full of combustion products. *The process of clearing the cylinder after the expansion stroke is called* **scavenging process.** Scavenging process is the *replacement of combustion products in the cylinder from previous power stroke with fresh air charge to be burned in the next-cycle.* This process must be completed in a very short duration available between the end of the expansion stroke and start of the process of charging.

- The efficiency of a two stroke engine greatly depends on the effectiveness of the scavenging process.

  Inadequate/poor/bad scavenging leads to the following :

  (*i*) Low mean indicated pressure (m.e.p.) which results in high weight and high cost per kW (shaft output) for the engine.

  (*ii*) Low amount of oxygen availability which results in incomplete combustion leading to *higher specific fuel combustion.*

  (*iii*) Contamination of lubricating oil to a greater extent which *reduces the lubricating qualities* and eventually results in *increased wear of piston and cylinder liners.*

  (*iv*) *Higher mean temperature and greater heat stresses on the cylinder walls.*

- Scavenging process can be divided into the following *four distinct stages :*

  (*i*) **Pre-blowdown.** On the opening of inlet port, the gases expanding in the main cylinder tend to escape from it and to pre-discharge into the scavenge air mainfold. This pre-blowdown process ends with the opening of exhaust port.

  (*ii*) **Blowdown.** With the opening of exhaust ports, the gases existing in the cylinder at the end of expansion stroke discharge *simultaneously* into exhaust manifold ; consequently the pressure of main cylinder drops to a valve lower than existing in scavenge air manifold. This blowdown process terminates at the moment the gas pressure inside the cylinder attains a value slightly less than air pressure inside scavenge manifold.

  (*iii*) **Scavenging.** This phase of the scavenging process starts at the moment the spontaneous exhaust of gases from cylinder terminates and ends at the moment the exhaust ports are closed. The *scavenge air sweeps out all residual gases remaining in the main cylinder at the end of spontaneous exhaust and replaces them as completely as possible with fresh charge.*

  (*iv*) **Additional charging.** After the completion of scavenging phase, the fresh charge continues to flow till the scavenge ports are open and pressure in the cylinder rises. This last phase results in better filling of the cylinder.

- Fig. 8.4 shows a scavenging process on pressure-crank angle (p-θ) for a loop-scavenged cylinder of a typical two stroke engine.

  — It shows the adiabatic compression curve from B.D.C. when the exhaust pressure is 1.013 bar (atmospheric).

  — After the exhaust port opens at 70° before T.D.C., the pressure in the cylinder falls rapidly in the blowdown process (The ***blowdown*** is defined as *the crank angle from the point exhaust port opens to the point the cylinder pressure reaches exhaust pressure*). In practice due to inertia effect of the gases after the blowdown the pressure in the cylinder *normally falls below the exhaust pressure for a few degrees.* Immediately after the exhaust port opens, the inlet port also begins to open. The interval may be of the order of 20° of the crank angle. With the fall in cylinder pressure below scavenging pressure, the fresh charge gets introduced in the cylinder and continues as long as the inlet port is open, and the total inlet pressure becomes more than pessure in the cylinder.

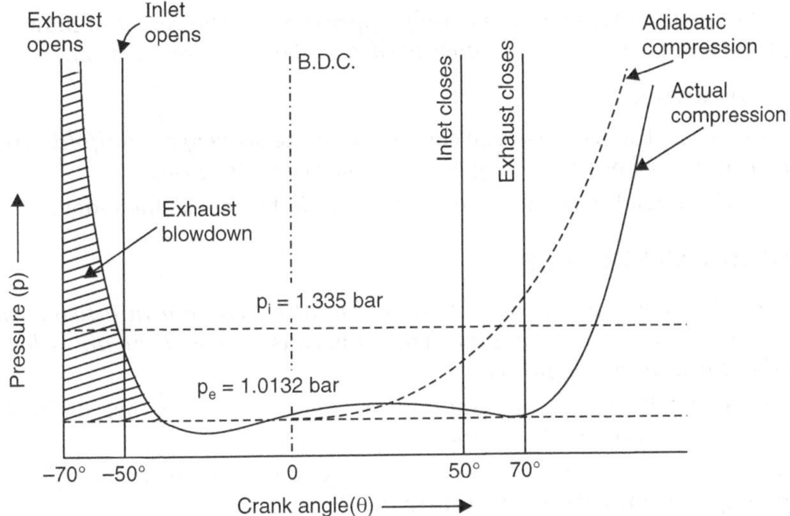

Fig. 8.4. Scavenging process on p-θ for a loop-scavenged cylinder of a typical two stroke engine.

— Whereas the gases flow into the inlet ports, the exhaust gases continue to flow out of exhaust port, due to the fact that these started in this direction at high velocity during blow-down. Also due to the fresh mixture entering through inlet port there is building up of pressure in the cylinder higher than the exhaust system pressure.

— **Scavenging angle.** It is defined as the *crank angle during which both inlet and exhaust ports are open.*

— **Scavenging period.** It is the *time period taken for scavenging angle.*

- After the closure of the two ports, cycle is completed by compression, combustion and expansion in the cylinder as in 4-stroke engine.

Theoretical Scavenging Processes :

Following are the three theoretical scavenging processes :

1. Perfect scavenging.
2. Perfect mixing.
3. Short circuiting.

1. Perfect scavenging

- In this type of scavenging, fresh air pumped into the cylinder by the blower through the inlet ports at the lower end of cylinder pushes the combustion products ahead of itself and of the cylinder through the exhaust valve at the other end.
- The air and combustion products *do not mix together.*
- So long as any products remain in the cylinder, the flow through the exhaust valves consists of products only.

2. Perfect mixing

- In this process, the *incoming fresh charge mixes completely and instantaneously with the cylinder contents and a portion of this mixture passes out of exhaust ports at a rate equal to that entering the cylinder.*
- The outgoing (homogeneous) mixture consists initially of combustion products only but then gradually changes to pure air.

- Since the result of this process closely approximates the result of many actual scavenging processes, therefore, it is *often used as a basis of comparison.*

3. **Short-circuiting**

- In this process, the *fresh charge coming from the scavenge manifold directly goes out of exhaust ports without removing combustion products/gases.*
- It results in a dead loss and its occurence must be checked/avoided.

8.4. SCAVENGING PARAMETERS

- *For the same power generation, more air input is required in a two stroke cycle engine than in a four stroke cycle engine.* This is because *some of the air is lost in the overlap period of the scavenging process.*
- A number of different intake and performance efficiencies are defined for the intake process of two stroke cycle engine.
- Volumetric efficiency of a four-stroke cycle engine can be replaced by either **delivery ratio ($R_{del.}$)** or **charging efficiency ($\eta_{ch.}$) :**

$$\textbf{Delivery ratio} = R_{del.} = \frac{m_{mi}}{V_s \rho_a} \quad ...(8.1)$$

$$\textbf{Charging efficiency} = \eta_{ch.} = \frac{m_{mt}}{V_s \rho_a} \quad ...(8.2)$$

where, m_{mi} = Mass of air-fuel mixture *ingested* into the cylinder,

m_{mt} = Mass of air-fuel mixture *trapped* in cylinder after all valves are closed,

V_s = Swept volume, and

ρ_a = Density of air at ambient conditions.

Typical values : $0.65 < R_{del.} < 0.95$

$0.50 < \eta_{ch.} < 0.75.$

— *Delivery ratio is greater than charging efficiency* because some of the air-fuel mixture ingested into the cylinder is lost out of the exhaust port before it is closed.

— In case of those engines that inject fuel after the valves are closed, the *mass of mixture in these equations* (8.1 and 8.2) *should be replaced with the mass of ingested air. Sometimes, ambient air density is replaced by the density of air in the inlet runner downstream of the supercharger.*

Other efficiencies :

$$\textbf{Trapping efficiency} = \eta_{trap.} = \frac{m_{mt}}{m_{mi}} = \frac{\eta_{ch.}}{R_{del.}} \quad ...(8.3)$$

$$\textbf{Scavenging efficiency} = \eta_{sc.} = \frac{m_{mt}}{m_{tc}} \quad ...(8.4)$$

$$\textbf{Relative charge} = C_{rel.} = \frac{m_{tc}}{V_s \rho_a} = \frac{\eta_{ch.}}{\eta_{sc.}} \quad ...(8.5)$$

where m_{tc} = Mass of total charge trapped in the cylinder, including exhaust residual.

Typical values : $0.65 < \eta_{trap.} < 0.80$

$0.75 < \eta_{sc.} < 0.90$

$0.60 < C_{rel.} < 0.90.$

Pressure loss co-efficient. It is defined as the *ratio between the main upstream and downstream pressures during the scavenging period and represents the pressure loss to which the scavenge air is subjected when it crosses the cylinder.*

Excess air factor (λ). The value ($R_{del.}$–1) is called the excess air factor. Thus if the R_{del} (delivery ratio) is 1.3 the excess air factor is 0.3.

Measurement of Scavenging Efficiency.

The following procedure is adopted in *diesel engines for measuring the scavenging efficiency :*

- A small sample of the combustion products is drawn just before the exhaust valve opens or during the earlier part of blowdown.
- The sample is analysed.
- The results obtained are compared with standard curves of exhaust products *vs.* F/A ratio. This determines the F/A ratio that must have existed in the cylinder before combustion.
- Knowing the quantity of fuel injected per cycle, the quantity of fresh air retained in the cylinder per cycle is determined. Air present in the residual gas is not considered as it represent a constant quantity which does not participate in combustion process.

8.5. SCAVENGING SYSTEMS

Different scavenging systems/arrangements *based on charge flow* are enumerated and described below :

1. Uniflow scavenging
2. Loop or reverse scavenging
3. Cross scavenging.

1. **Uniflow scavenging :**

It is the *most perfect method of scavenging.*

- The fresh charge is admitted at one end of the cylinder and the exhaust escapes at the other end. The air flow is from end to end, and little short-circuiting between the intake and exhaust openings is possible.
- The three available arrangements for uniflow scavenging are shown in Fig.8.5.

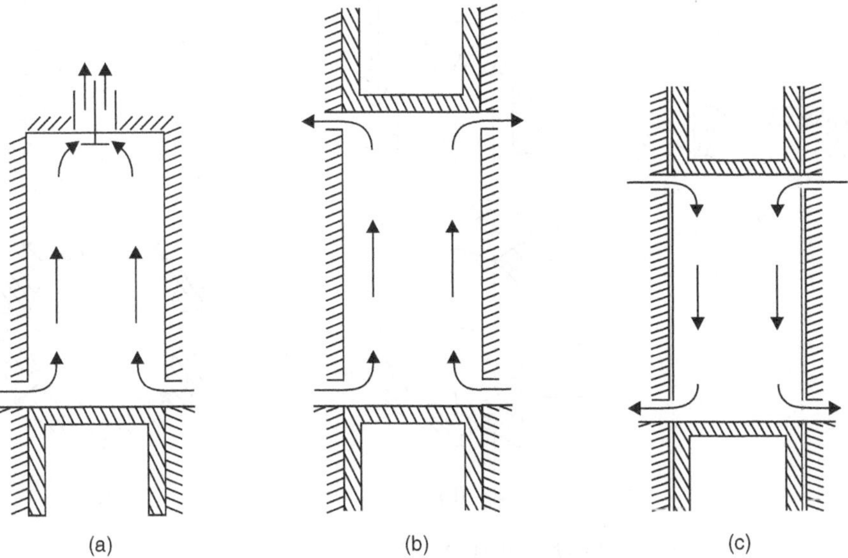

Fig. 8.5. Uniflow scavenging.

- — A *poppet valve* is used [Fig. 8.5 (*a*)] to admit the inlet charge or for the exhaust, as the case may be.
- — In Fig. 8.5 (*b*) the inlet and exhaust ports are both *controlled by separate pistons* that move in opposite directions (opposed piston engines).
- — In Fig. 8.5 (*c*) the inlet and exhaust ports are *controlled by the combined motion of piston and sleeve.*

- *All uniflow systems permit unsymmetrical port timings and supercharging.*
- Due to absence of any eddies or turbulence (at least theoretically) *it is easier in a uniflow scavenging system to push the combustion products out of the cylinder without mixing with it and short circuiting.* Thus this system ***has the highest scavenging efficiency.***
- Since this systems requires either opposed systems, poppet valves or sleeve valve (all of which increase the complication) its construction is not simple.

2. **Loop or reverse scavenging :**

In *loop or reverse scavenging, the fresh air first sweeps across the piston top, moves up and then down and finally out through the exhaust.* The system *avoids the short-circuiting of the cross-scavenged engine and thus improves upon its scavenging efficiency.*

- In the MAN type of loop scavenge, Fig. 8.6. (*a*), the exhaust and inlet ports are on the same side, the *exhaust above the inlet.*
- In the Schnuerle type, Fig. 8.6. (*b*), the ports are *side by side.*

  The Curtis type of scavenging, Fig. 8.6 (*c*), is similar to the Schnuerle type, except that *upwardly directed inlet ports are placed also opposite the exhaust ports.*

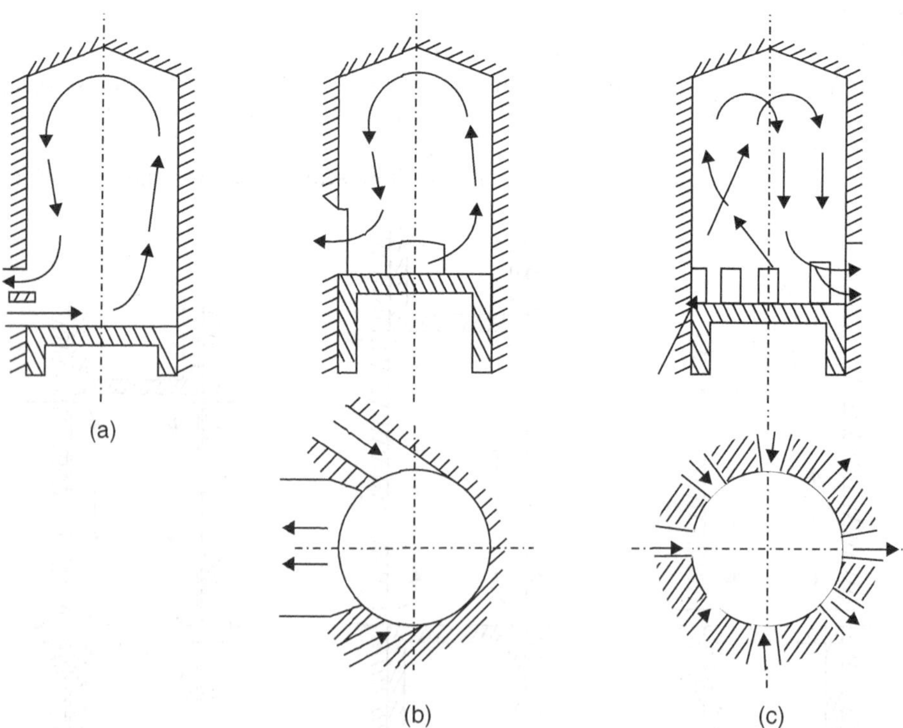

Fig. 8.6. Methods of loop scavenging.

- Owing to the absence of cams, valves and valve gear, loop or reverse scavenged *engines are simple and sturdy.* They have a *high resistance to thermal stresses and are thus much suited to higher supercharge.*
- In a loop scavenged two stroke engine, the *major mechanical problem* is that of obtaining an adequate oil supply to the cylinder wall consistent with reasonable lubricating oil consumption and cylinder wear.

3. **Cross-scavenging**

In this system the *inlet and exhaust ports are located on opposite sides of the cylinder (Fig. 8.7).*

The incoming flow is directed upwards by the deflector on the piston, and the cylinder head reverses the direction of flow, so that exhaust gases are forced through the exhaust port.

- In this type of arrangement the engine is structurally simpler than that with the uniflow scavenging (due to the absence of valves, distributors, and relative drive devices).
- The main demerit of this system is that *scavenging air is not able to get rid of the layer of exhaust gas near the wall resulting in poor scavenging.* A small portion of fresh charge goes directly into the exhaust port. These factors contribute towards poor b.m.e.p. of the cross-scavenged engines.

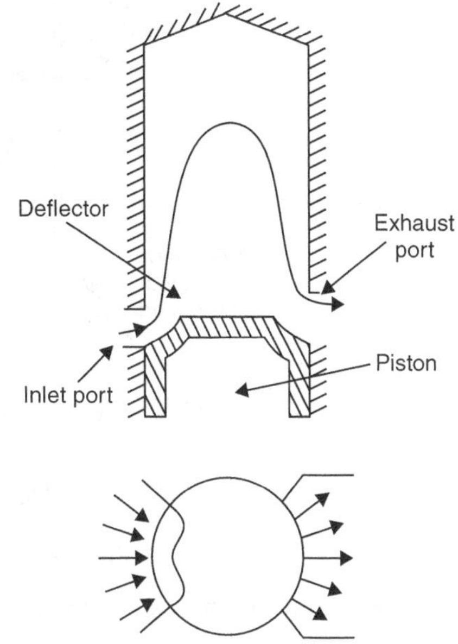

Fig. 8.7. Cross scavenging.

8.6. CRANKCASE SCAVENGING

This type of scavenging arrangement is employed in the simplest type of two stroke engine, and is shown in Fig. 8.8.

In this engine, the charge (fuel-air mixture in S.I. engine and air in C.I. engine) is compressed in the crankcase by the underside of the piston during the expansion stroke. There are three ports in this engine-*intake port at the crankcase, transfer port and the exhaust port.* The compressed charge passes through the transfer port into the engine cylinder flushling the products of combustion. This process is called *scavenging*, and this type of engine is called *crankcase scavenged engine.*

- As the piston moves down, it first uncovers the exhaust ports, and the cylinder pressure drops to atmospheric level as the combustion products escape through these ports.
- Further downward motion of the piston uncovers the transfer ports, permitting the slightly compressed mixture or air (depending upon the type of engine) in the crankcase to enter the engine cylinder. The top of the piston and the ports are usually *shaped in such a way that the fresh charge is directed towards the top of the cylinder before flowing towards the exhaust ports.* This is for the *purpose of scavenging the upper part of the cylinder of the combustion products and also to minimize the flow of fresh charge directly through the exhaust ports.* The projection on the piston is called the *deflector.*
- As the piston returns from B.D.C. the transfer ports and then the exhaust ports are closed and the compression of the charge begins. Motion of the piston during compression lowers the pressure in the crankcase so that the fresh charge is drawn into the crankcase through the inlet reed valve.

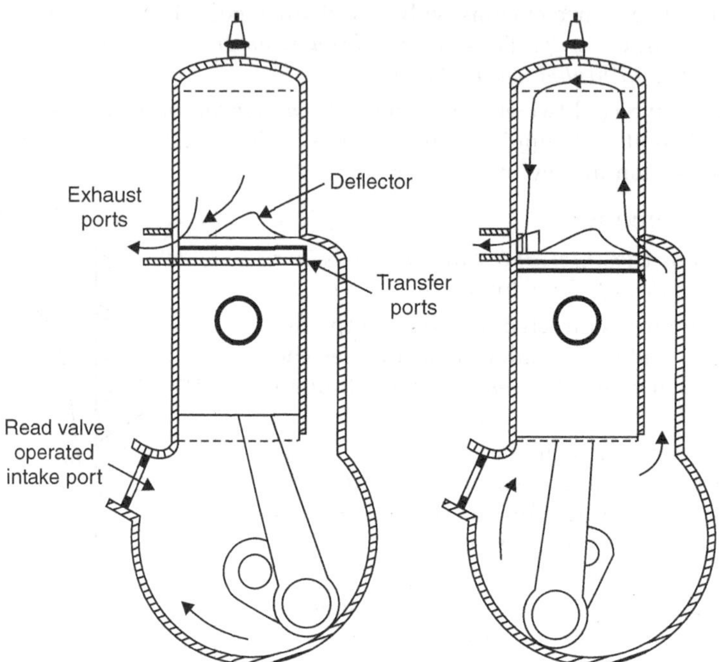

Fig. 8.8. Crankcase-scavenged two stroke engine.

- Ignition and expansion take place in the usual way, and the *cycle is repeated.*

Demerits :

1. This system is *very uneconomical and inefficient in operation.* This is owing to the fact that amount of air which can be used for scavenging is less than the swept volume of the cylinder due to *low volumetric efficiency of the crankcase which contains a large dead space.* Thus the delivery ratio ($R_{del.}$) is always *less than unity* and as such it is not possible to scavenge the cylinder completely of the combustion products and some residual gases always remain in the cylinder. Consequently the crankcase-scavenged engine has a lower m.e.p., typical valves being 3 to 4 bar. Since the charge transferred through the transfer port is only **40-50** percent of the cylinder volume, the engine output is strictly limited.
2. Due to mixing of the oil vapours from the crankcase with the scavenging air, oil consumption is increased.

In view of the above demerits the *crankcase scavenging is not preferred and a scavenging pump is essential for a high output two stroke engine.*

8.7. SCAVENGING PUMPS AND BLOWERS

Since the piston of a two stroke engine cannot carry out the pumping action, therefore, a separate pumping mechanism, called the *scavenging pump,* is needed to supply scavenging air to the cylinder.

Following types of pumps are used : Crankcase compression (Refer Art. 8.6), piston, roots and centrifugal blowers.

- **Piston type pump** shown in Fig. 8.9 is used for *low speed and single or two cylinder engines.*

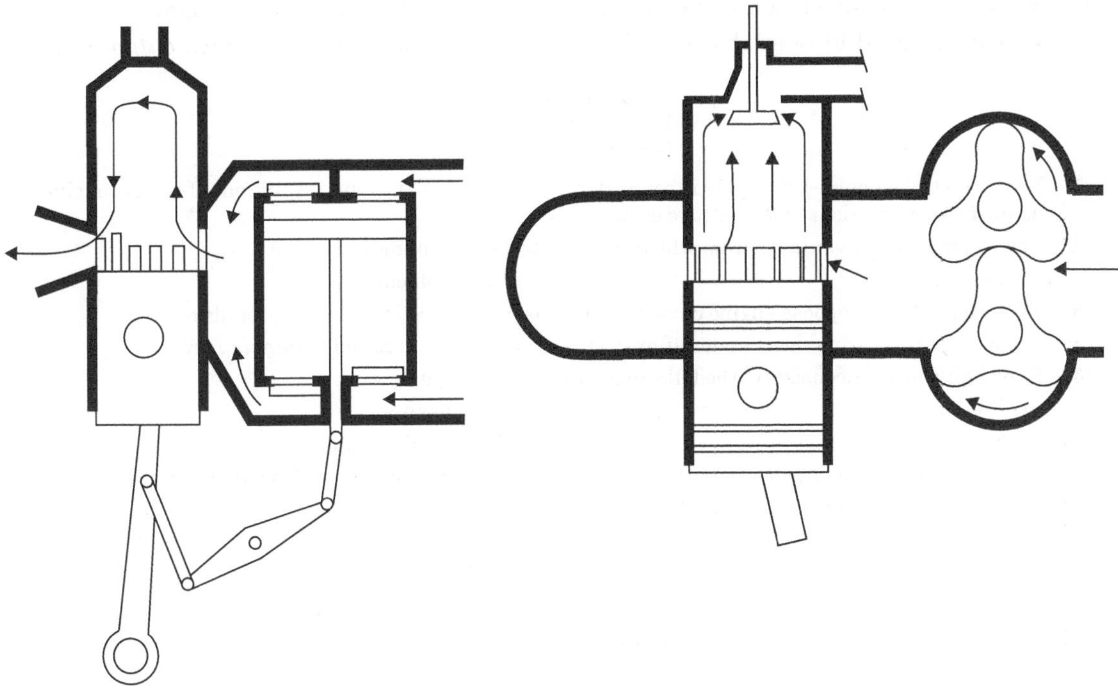

Fig. 8.9. Piston type pump.

Fig. 8.10. Roots blower.

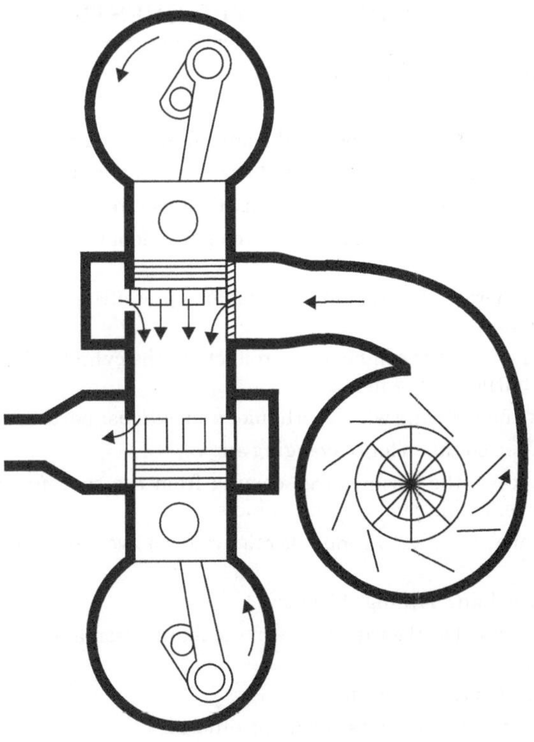

Fig. 8.11. Centrifugal blower.

- **Roots blower** shown in Fig. 8.10 is used for *small and medium engines.*
- **Centrifugal blower** shown in Fig. 8.11 is employed for *large and high output engines.*

HIGHLIGHTS

1. In a two stroke engines the cycle is completed in two strokes, *i.e.* one revolution of the crankshaft as against two revolutions of four-stroke cycle.
2. The two main disadvantages from which the two stroke S.I. engine suffer are :
   (*i*) Loss of fuel (*ii*) Idling difficulty.
3. The two stroke C.I. oppose engine cross-head type is mainly used for marine propulsion.
4. The process of clearing the cylinder after the expansion stroke is called *scavenging process.*
5. Scavenging process consists of the following four distinct stages :
   (*i*) Pre-blowdown (*ii*) Blowdown
   (*iii*) Scavenging (*iv*) Additional charging.
6. *Scavenging angle* is defined as the crank angle during which both inlet and exhaust ports are open.
7. *Scavenging period* is the time period taken for scavenging angle.
8. *Theoretical scavenging processes are :*
   (*i*) Perfect scavenging ; (*ii*) Perfect mixing ;
   (*iii*) Short-circuiting.
9. Scavenging systems, based on charge flow, are classified as follows :
   (*i*) Uniflow scavenging (*ii*) Loop or reverse scavenging
   (*iii*) Cross scavenging.

OBJECTIVE TYPE QUESTIONS

Fill in the Blanks or Say "Yes" or "No" :

1. In a two stroke engine instead of valves, ... are used.
2. For marine propulsion, two-stroke C.I. opposed engine (cross-head type) is mainly used.
3. For small two-wheeler engines light-weight and low initial cost are the main considerations.
4. In a two stroke cycle engine inlet air must be input at a pressure less than atmospheric.
5. In case of two stroke cycle engines using crankcase compression, lubricating oil must be added to the inlet air.
6. Engine which use ...... generally use standard supercharge pressurised lubrication systems, with crankcase serving as the oil reservour.
7. ...... process is the replacement of combustion products in the cylinder from previous power stroke with fresh air charge to be burned in the next cycle.
8. ...... angle is the crank angle during which both inlet and exhaust ports are open.
9. ...... period is the time period taken for scavenging angle.
10. In short-circuiting process, the fresh charge coming from the scavenge manifold directly goes out of exhaust ports without removing combustion products.
11. For the same power generation, ... air input is required in a two stroke cycle engine than in a four stroke cycle engine.
12. Delivery ratio is smaller than charging efficiency.
13. ...... loss co-efficient is defined as the ratio between the main upstream and downstream pressures during the scavenging period.
14. The valve ($R_{del.} - 1$) is called the ...... air factor.
15. ...... scavenging system has the highest scavenging efficiency.

16. In ...... scavenging, the fresh air first sweeps across the piston top, moves up and then down and finally out through the exhaust.

17. In ...... scavenging system the inlet and exhaust ports are located on opposite sides of the cylinder.

18. Crankcase scavenging arrangement is very uneconomical and inefficient in operation.

19. Piston type pump is used for high speed engines.

20. Roots blower is used for small and medium engines.

ANSWERS

1. ports	**2.** Yes	**3.** Yes	**4.** No	**5.** Yes
6. superchargers	**7.** Scavenging	**8.** Scavenging	**9.** Scavenging	**10.** Yes
11. more	**12.** No	**13.** Pressure	**14.** excess	**15.** Uniflow
16. loop	**17.** cross	**18.** Yes	**19.** No	**20.** Yes.

THEORETICAL QUESTIONS

1. "In a 2 stroke engine it is better to have deflector top type piston". Justify the statement.
2. Discuss the two main disadvantages of two-stroke cycle S.I. engine. How are these disadvantages avoided in the two stroke cycle C.I. engine ?
3. Discuss briefly 'Mist lubrication system'.
4. Why do the two stroke C.I. engine find wide use in marine propulsions ?
5. Why are two stroke S.I. engines more commonly used in low horse power two wheelers ?
6. Why is crankcase scavenging used only for low power engines ?
7. Why are two stroke diesel engines, for large power, more common than two-stroke S.I. engines ?
8. What is the reason that two-stroke engine is not used in car even though it develops theoretically twice power than that of four-stroke engine ?
9. Explain with suitable sketches the following scavenging processes :

   (*i*) Uniflow scavenging (*ii*) Loop scavenging.
10. Explain briefly crankcase scavenging.
11. Explain the scavenging process in two stroke engine. Disucss three scavenging processes used in two-stroke engine.
12. Define scavenging and scavenging efficiency. Explain with sketches different scavenging arrangements based on charge flow.
13. How the valve timings of a two stroke engine differ from that of four stroke cycle engine ?
14. What is the difference between the valve timing of a crankcase-scavenged and supercharged two stroke engine ?
15. Compare the relative merits and demerits of different scavenging systems.
16. How is the supercharging of two stroke engines done ?

9

Chemical Thermodynamics and Fuels (Conventional and Alternative)

9.1. Chemical thermodynamics—General aspects—Basic chemistry—Fuels—Combustion equations—Theoretical air and excess air—Stoichiometric air fuel (A/F) ratio—Air-fuel ratio from analysis of products—Analysis of exhaust and flue gas—Internal energy and enthalpy of reaction—Enthalpy of formation (ΔH_f)—Heating values of fuels— Adiabatic flame temperature—Chemical equilibrium—Actual combustion analysis. 9.2. Conventional fuels (for I.C. engines)—Introduction—Desirable properties of good I.C. engine fuels—Gaseous fuels—Liquid fuels—Structure of petroleum—Petroleum and composition of crude oil—Fuels for spark-ignition engines—Knock rating of S.I. engines fuels—Miscellaneous properties of S.I. engines fuels—Diesel fuel. 9.3. Alternative fuels for I.C. engines—General aspects—Advantages and disadvantages of using alternative fuels—Alcohol—Alcohol-gasoline fuel blends—Hydrogen—Natural gas (methane)—LPG and LNG—Biogas. Worked Examples—Highlights—Objective Type Questions—Theoretical Questions—Unsolved Examples.

9.1. CHEMICAL THERMODYNAMICS

9.1.1. General Aspects

In chemical thermodynamics the study of systems involving chemical reactions is an important topic. *A* ***chemical reaction*** *may be defined as the rearrangement of atoms due to redistribution of electrons.* In a chemical reaction the terms, **reactants** and the **products** are frequently used. '*Reactants*' comprise of initial constituents which start the reaction while '*products*' comprise of *final constituents* which are formed by the chemical reaction. Although the basic principles which will be discussed in this chapter apply to any chemical reaction, here main attention will be focused on an important type of chemical reaction—**"combustion"**.

9.1.2. Basic Chemistry

Before considering combustion problems it is necessary to understand the construction and use of chemical formulae. This involves elementary concepts which are discussed below briefly.

Atoms :

It is not possible to divide the chemical elements *indefinitely*, and the *smallest particle which can take part in a chemical change* is called an **'atom'.** If an atom is split as in nuclear reaction, the *divided atom does not retain the original chemical properties.*

Molecules :

It is rare to find elements to exist naturally as single atom. Some elements have atoms which exist in pairs, each pair forming a molecule (*e.g.* oxygen), and the atoms of each molecule are held together by stronger *inter-atomic forces*. The isolation of a molecule of oxygen would be tedious, but possible ; the isolation of an atom of oxygen would be a different prospect. The molecules of some substances are formed by the mating up of atoms of different elements. For example, water has a molecule which consists of two atoms of hydrogen and one atom of oxygen. The atoms of different elements have different masses and these values are important when a quantitative analysis is required. The actual masses are infinitesimally small, and the ratios of the masses of atoms are used. These ratios are indicated by **atomic weight** quoted on a scale which defines the atomic weight of oxygen as 16.

The symbols and molecular weight of some important elements, compounds and gases are given in the Table 9.1.

Table 9.1. Symbols and Molecular weights

Elements/Compounds/Gases	*Molecule*		*Atom*	
	Symbol	*Molecular weight*	*Symbol*	*Molecular weight*
Hydrogen	H_2	2	H	1
Oxygen	O_2	32	O	16
Nitrogen	N_2	28	N	14
Carbon	C	12	C	12
Sulphur	S	32	S	32
Water	H_2O	18	—	—
Carbon monoxide	CO	28	—	—
Carbondioxide	CO_2	44	—	—
Sulphurdioxide	SO_2	64	—	—
Marsh gas (Methane)	CH_4	16	—	—
Ethylene	C_2H_4	28	—	—
Ethane	C_2H_6	30	—	—

9.1.3. Fuels

Fuel may be *chemical* or *nuclear*. Here we shall consider briefly *chemical fuels* only.

A chemical fuel is a substance which releases heat energy on combustion. The principal combustible elements of each fuel are *carbon* and *hydrogen*. Though *sulphur* is a combustible element too but its presence in the fuel is considered to be *undesirable*.

Fuels can be classified according to whether :

(*i*) they occur in nature called **primary fuels** or are prepared called **secondary fuels** ;

(*ii*) they are in solid, liquid or gaseous state. The detailed classification of fuels can be given in a summary form as follows :

Type of fuel	*Natural (Primary)*	*Prepared (secondary)*
Solid	Wood	Coke
	Peat	Charcoal
	Lignite coal	Briquettes
Liquid	Petroleum	Gasoline
		Kerosene
		Fuel oil
		Alcohol
		Benzol
		Shale oil

Gaseous	Natural gas	Petroleum gas
		Producer gas
		Coal gas
		Coke-oven gas
		Blast furnace gas
		Carburetted gas
		Sewer gas

Solid fuels :

The most important *solid fuel* is **coal** and its various types are divided into groups according to their chemical and physical properties. *An accurate chemical analysis by mass of the important elements* in the fuel is called the **ultimate analysis**, the element usually included being *carbon, hydrogen, nitrogen* and *sulphur*. Another analysis of coal, called **proximate analysis**, gives the *percentages of moisture, volatile matter, combustible solid* (called fixed carbon), and *ash*. The fixed carbon is found as a remainder by deducting the percentages of the other quantities. The volatile matter includes the water derived from the chemical decomposition of the coal, the combustible gases (*e.g.* H_2, CH_4, C_2H_6 etc.), and tar.

Liquid fuels :

Most of the *liquid fuels* are *hydro-carbons* which exist in the liquid phase at atmospheric conditions. Petroleum oils are complex mixtures of sometimes hundreds of different fuels, but the necessary information to the engineer is the relative proportions of C, H_2, etc. as given by the ultimate analysis.

Gaseous fuels :

These fuels are chemically the simplest of the three groups. Some gaseous fuels exist naturally at atmospheric conditions ; example being CH_4 (methane) which is a paraffin. Other gaseous fuels are manufactured by the various treatments of coal. CO is an important gaseous fuel which is a constituent of other gas mixtures, and is also a product of the incomplete combustion of carbon.

9.1.4. Combustion Equations

In a combustion chamber proportionate masses of air and fuel enter where the chemical reaction takes place, and then the combustion products pass to the exhaust. By the conservation of mass the mass flow remains constant (*i.e.* total mass of *products* = total mass of *reactants*), but the reactants are chemically different from the products, and the products leave at a higher temperature. The *total number of atoms of each element concerned in the combustion remains constant, but the atoms are rearranged into groups having different chemical properties.* This information is expressed in the chemical equation which shows (*i*) the reactants and the products of combustion, (*ii*) the relative quantities of the reactants and products. The two sides of the equation must be **consistent**, each having the same number of atoms of each element involved.

The oxygen supplied for combustion is *usually* provided by *atmospheric air*, and it is necessary to use accurate and consistent analysis of air by *mass* and by *volume*. It is usual in combustion calculations to take air as 23.3% O_2, 76.7% N_2 by mass, and 21% O_2, 79% N_2 by volume. The small traces of other gases in dry air are included in nitrogen, which is sometimes called '*atmospheric nitrogen*'.

Some important combustion equations are given below :

1. **Combustion of hydrogen :**

$$2H_2 + O_2 = 2H_2O \quad ...(9.1)$$

The above equation of combustion of hydrogen tell us that :

(*i*) Hydrogen reacts with water to form steam or water.

(*ii*) Two molecules of hydrogen react with one molecule of oxygen to give two molecules of steam or water,

i.e., $$2 \text{ volumes } H_2 + 1 \text{ volume } O_2 \longrightarrow 2 \text{ volumes } H_2O$$

The H_2O may be liquid or a vapour depending on whether the product has been cooled sufficiently to cause condensation.

The *proportions by mass* are obtained by using *atomic weights* as follows :

$$2H_2 + O_2 \longrightarrow 2H_2O$$

$$\therefore \quad 2(2 \times 1) + 2 \times 16 \longrightarrow 2(2 \times 1 + 16)$$

i.e., $$4 \text{ kg } H_2 + 32 \text{ kg } O_2 \longrightarrow 36 \text{ kg } H_2O$$

or $$\mathbf{1 \text{ kg } H_2 + 8 \text{ kg } O_2 \longrightarrow 9 \text{ kg } H_2O}$$

[The same proportions are obtained by writing the equation (9.1) as : $H_2 + \frac{1}{2}O_2 \longrightarrow H_2O$, and this is sometimes done.]

It will be noted from equation (9.1) that the total volume of the *reactants* is

$$2 \text{ volumes } H_2 + 1 \text{ volume } O_2 = 3 \text{ volumes.}$$

The total volume of the *product* is only 2 volumes. There is therefore a *volumetric contraction on combustion.*

Since the oxygen is accompanied by nitrogen if air is supplied for the combustion, then this nitrogen should be included in the equation. As nitrogen is *inert* as far as chemical reaction is concerned, it will appear on both sides of the equation.

With one mole of oxygen there are 79/21 moles of nitrogen, hence equation (9.1) becomes,

$$2H_2 + O_2 + \frac{79}{21} N_2 \longrightarrow 2H_2O + \frac{79}{21} N_2 \quad ...(9.2)$$

2. **Combustion of carbon :**

(*i*) **Complete combustion of carbon to carbondioxide**

$$C + O_2 \longrightarrow CO_2 \quad ...(9.3)$$

and including the nitrogen,

$$C + O_2 + \frac{79}{21} N_2 \longrightarrow CO_2 + \frac{79}{21} N_2 \quad ...(9.4)$$

By volume :

$$0 \text{ volume C} + 1 \text{ volume } O_2 + \frac{79}{21} \text{ volumes } N_2 \longrightarrow 1 \text{ volume } CO_2 + \frac{79}{21} \text{ volumes } N_2$$

The volume of carbon is written as zero since the volume of solid is negligible in comparison with that of a gas.

By mass :

$$12 \text{ kg C} + (2 \times 16) \text{ kg } O_2 + \frac{79}{21} (2 \times 14) \text{ kg } N_2 \longrightarrow (12 + 2 \times 16) \text{ kg } CO_2 + \frac{79}{21} (2 \times 14) N_2$$

i.e., $$12 \text{ kg C} + 32 \text{ kg } O_2 + 105.3 \text{ kg } N_2 \longrightarrow 44 \text{ kg } CO_2 + 105.3 \text{ kg } N_2$$

or $$1 \text{ kg C} + \frac{8}{3} \text{ kg } O_2 + \frac{105.3}{12} \text{ kg } N_2 \longrightarrow \frac{11}{3} \text{ kg } CO_2 + \frac{105.3}{12} \text{ kg } N_2.$$

(*ii*) **The incomplete combustion of carbon.** The incomplete combustion of carbon occurs when there is an insufficient supply of oxygen to burn the carbon completely to carbondioxide.

$$2C + O_2 \longrightarrow 2CO \quad ...(9.5)$$

and including the nitrogen,

$$2C + O_2 + \frac{79}{21} N_2 \longrightarrow 2CO + \frac{79}{21} N_2 \quad ...(9.6)$$

By mass :

$$(2 \times 12) \text{ kg C} + (2 \times 16) \text{ kg } O_2 + \frac{79}{21}(2 \times 14) \text{ kg } N_2 \longrightarrow 2(12 + 16) \text{ kg CO} + \frac{79}{21}(2 \times 14) \text{ kg } N_2$$

i.e.,

$$24 \text{ kg C} + 32 \text{ kg } O_2 + 105.3 \text{ kg } N_2 \longrightarrow 56 \text{ kg CO} + 105.3 \text{ kg } N_2$$

or

$$1 \text{ kg C} + \frac{4}{3} \text{ kg } O_2 + \frac{105.3}{24} \text{ kg } N_2 \longrightarrow \frac{7}{3} \text{ kg CO} + \frac{105.3}{24} \text{ kg } N_2$$

If a further supply of oxygen is available then the combustion can continue to completion,

$$2CO + O_2 + \frac{79}{21} N_2 \longrightarrow 2CO_2 + \frac{79}{21} N_2 \quad ...(9.7)$$

By mass :

$$56 \text{ kg CO} + 32 \text{ kg } O_2 + \frac{79 \times 28}{21} \text{ kg } N_2 \longrightarrow 88 \text{ kg } CO_2 + \frac{79 \times 28}{21} \text{ kg } N_2$$

or

$$1 \text{ kg CO} + \frac{4}{7} \text{ kg } O_2 + \frac{105.3}{56} \text{ kg } N_2 \longrightarrow \frac{11}{7} \text{ kg } CO_2 + \frac{105.3}{56} \text{ kg } N_2.$$

9.1.5. Theoretical Air and Excess Air

The *minimum amount of air* that supplies sufficient oxygen for the complete combustion of all the carbon, hydrogen, and any other elements in the fuel that may oxidise is called the **"theoretical air"**. When complete combustion is achieved with theoretical air, the products contain no oxygen.

In practice, it is found that complete combustion is not likely to be achieved unless the amount of air supplied is somewhat greater than the theoretical amount. Thus 150 per cent theoretical air means that air actually supplied is 1.5 times the theoretical air.

The complete combustion of methane with minimum amount of theoretical air and 150 per cent theoretical air respectively is written as :

$$CH_4 + 2O_2 + 2\left(\frac{79}{21}\right) N_2 \longrightarrow CO_2 + 2H_2O + 2\left(\frac{79}{21}\right) N_2 \quad ...(9.8)$$

$$CH_4 + 2(1.5)\, O_2 + 2\left(\frac{79}{21}\right)(1.5)\, N_2 \longrightarrow CO_2 + 2H_2O + O_2 + 3\left(\frac{79}{21}\right) N_2 \quad ...(9.9)$$

(with 150 per cent theoretical air)

The amount of air actually supplied may also be expressed in terms of percent excess air. The excess air is the amount of air supplied over and above the theoretical air. Thus 150 per cent theoretical air is equivalent to 50 per cent excess air.

Note. For complete combustion of fuel we need air. As per theoretical basis there is a minimum amount of air which is required by the fuel to burn completely, *but always, air in excess is used because whole of air supplied for combustion purposes does not come in contact with the fuel completely and as such portion of fuel may be left unburnt.* But if a large quantity of excess air is used it exercises a *cooling effect on combustion process which however can be avoided by preheating the air. The weight of excess air supplied can be determined from the weight of oxygen which is left unused.* The amount of excess air supplied varies with the type of fuel and the firing conditions. It may approach a value of 100% but modern practice is to use 25% to 50% excess air.

9.1.6. Stoichiometric Air-Fuel (A/F) Ratio

Stoichiometric (or chemically correct) *mixture* of air and fuel is one that contains *just sufficient oxygen for complete combustion of the fuel.*

A *weak mixture* is one which has an *excess of air*.

A *rich* mixture is one which has a *deficiency* of air.

The percentage of excess air is given as under :

$$\%\text{age excess air} = \frac{\text{Actual A/F ratio} - \text{stoichiometric A/F ratio}}{\text{Stoichiometric A/F ratio}} \quad ...(9.10)$$

(where A and F denote *air* and *fuel* respectively).

The ratios are expressed as follows :

For *gaseous fuels* *By volume*

For solid and liquid fuels *By mass*

For *boiler* plant the mixture is usually greater than 20% weak ; for *gas turbines* it can be as much as 300% weak. *Petrol engines* have to meet various conditions of load and speed, and operate over a wide range of mixture strength. The following definition is used :

$$\text{Mixture strength} = \frac{\text{Stoichiometric A/F ratio}}{\text{Actual A/F ratio}} \quad ...(9.11)$$

The working value range between 80% (weak) and 120% (rich).

Note. The reciprocal of the air-fuel ratio is called the *fuel air F/A ratio.*

9.1.7. Air-Fuel Ratio From Analysis of Products

When analysis of combustion products is known air-fuel ratio can be calculated by the following methods :

1. **Fuel composition known**

(*i*) Carbon balance method (*ii*) Hydrogen balance method

(*iii*) Carbon hydrogen balance method.

2. **Fuel composition unknown**

Carbon-hydrogen balance method.

1. **Fuel composition known :**

(*i*) *Carbon balance method.* When the fuel composition is known, the carbon balance method is quite accurate if *combustion takes place with excess air and when free (solid) carbon is not present in the products.* It may be noted that the Orsat analysis will not determine the quantity of solid carbon in the products.

(*ii*) *Hydrogen balance method.* This method is used when solid carbon is suspected to present.

(*iii*) *Carbon hydrogen balance method.* This method may be employed *when there is some uncertainty about the nitrogen percentage reported by the Orsat analysis.*

2. **Fuel composition unknown :**

When the fuel composition is not known the carbon-hydrogen balance method has to be employed.

9.1.8. Analysis of Exhaust and Flue Gas

The combustion products are mainly gaseous. When a sample is taken for analysis it is usually cooled down to a temperature which is below the saturation temperature of the steam present. The *steam content* is therefore *not included* in the analysis, which is then quoted as the *analysis of the dry products.* Since the products are gaseous, it is usual to quote the analysis by *volume. An analysis which includes the steam in the exhaust is called a wet analysis.*

Practical analysis of combustion products

The most common means of analysis of the combustion products is the **Orsat apparatus** which is described on next page.

Construction. An Orsat's apparatus consists of the following :

(*i*) A burette

(*ii*) A gas cleaner

(*iii*) Four absorption pipettes 1, 2, 3, 4.

The pipettes are interconnected by means of a manifold fitted with cocks S_1, S_2, S_3 and S_4 and contain different chemicals to absorb carbondioxide (CO_2), carbonmonoxide (CO) and oxygen (O_2). Each pipette is also fitted with a number of small glass tubes which provide a greater amount of surface. These tubes are wetted by the absorbing agents and are exposed to the gas under analysis. The measuring burrette is surrounded by a *water jacket* to prevent, changes in temperature and density of the gas. The pipettes 1, 2, 3, 4 contain the following chemicals :

Pipette 1 : Contains *'KOH'* (caustic soda) *to absorb* CO_2 (carbondioxide)

Pipette 2 : Contains *an alkaline solution of 'pyrogallic acid' to absorb* O_2 (oxygen)

Pipette 3, 4 : Contain an *acid solution of 'cuprous chloride' to absorb CO* (carbonmonoxide)

Further-more the apparatus has a *levelling bottle* and a three way cock to connect the apparatus either to gases or to atmosphere.

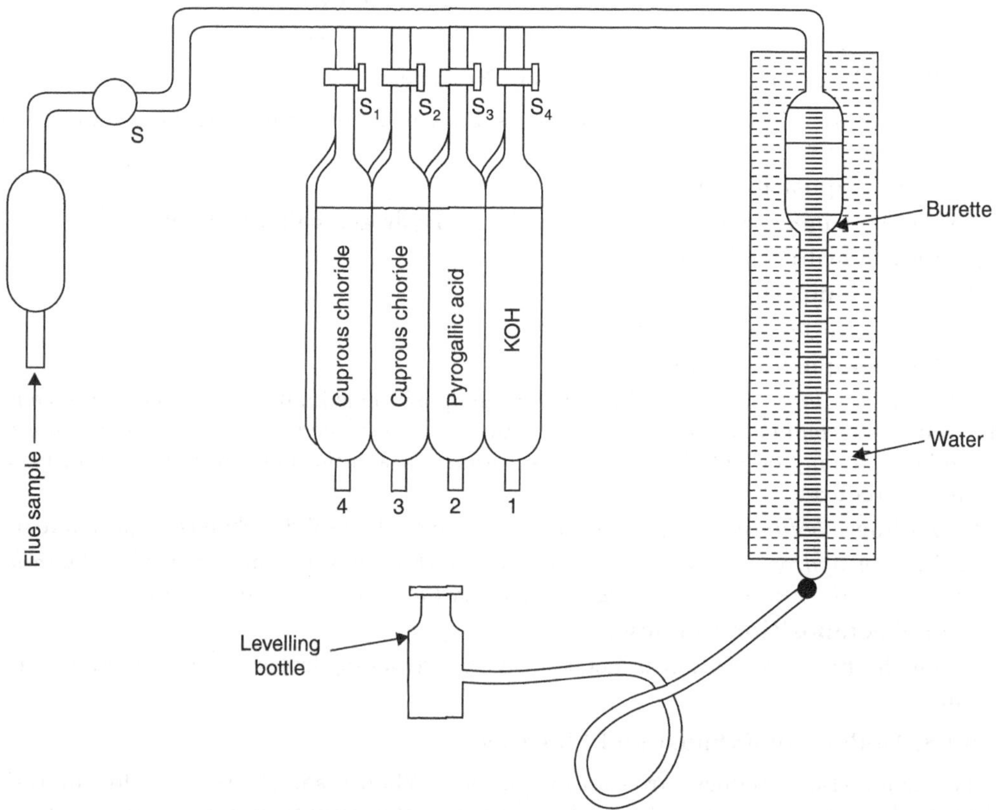

Fig. 9.1. Orsat apparatus.

Procedure. 100 cm^3 of gas whose analysis is to be made is drawn into the bottle by lowering the levelling bottle. The stop cock S_1 is then opened and the whole flue gas is forced to pipette 1. The gas remains in this pipette for sometime and most of the carbondioxide is absorbed. The levelling bottle is then lowered to allow the chemical to come to its original level. The volume of gas thus absorbed is read on the scale of the measuring bottle. The flue gas is then forced through the pipette 1 for a number of times to ensure that the whole of the CO_2 is absorbed. Further, the remaining flue gas is then forced to the pipette 2 which contains pyrogallic acid to absorb whole of O_2. The reading on the measuring burette will be the sum of volume of CO_2 and O_2. The oxygen content can then be found out by subtraction. Finally, as before the sample of gas is forced through the pipettes 3 and 4 to absorb carbonmonoxide completely.

The amount of nitrogen in the sample can be determined by subtracting from total volume of gas the sum of CO_2, CO and O_2 contents.

Orsat apparatus *gives an analysis of the dry products of combustion.* Steps may have been taken to remove the steam from the sample by condensing, but as the sample is collected over water it becomes saturated with water. The resulting analysis is nevertheless a true analysis of the dry products. This is because the volume readings are taken at a constant temperature and pressure, and the partial pressure of the vapour is constant. This means that the sum of the partial pressures of the remaining constituents is constant. The vapour then occupies the same proportion of the total volume at each measurement. Hence the vapour does not affect the result of the analysis.

Note. Quantitatively the dry product analysis can be used to calculate A/F ratio. This method of obtaining the A/F ratio is not so reliable as direct measurement of air consumption and fuel consumption of the engine. More caution is required when analysing the products of consumption of a *solid fuel* since some of the products do not appear in the flue gases (*e.g.* ash and unburnt carbon). The residual solid must be analysed as well in order to determine the carbon content, if any. With an engine using *petrol or diesel fuel* the exhaust may include unburnt particles of carbon and this quantity will not appear in the analysis. The exhaust from internal combustion engines may contain also some CH_4 and H_2 due to incomplete combustion. Another piece of equipment called the **Heldane apparatus** measures the CH_4 content as well as CO_2, O_2 and CO.

9.1.9. Internal Energy and Enthalpy of Reaction

The first law of thermodynamics can be applied to any system. Non-flow and steady-flow energy equations deducted from this law must be applicable to systems undergoing combustion processes.

It has been proved experimentally that the energy released, when a unit mass of a fuel undergoes complete combustion, depends on the *temperature at which the process is carried out.* Thus such quantities quoted are related to temperature. Now it will be shown that if the energy released by a fuel at one temperature is known then it can be calculated at other temperatures.

The process of combustion is defined as taking place from reactants at a state identified by the reference temperature T_0 and another property, either pressure or volume, to products at the same state.

Let U_{R_0} = Internal energy of the *reactants* (which is a mixture of fuel and air) at T_0,

U_{P_0} = Internal energy of *products* of combustion at T_0,

U_{R_1} = Internal energy of *reactants* at temperature T_1,

U_{P_1} = Internal energy of *products* at temperature T_1,

U_{R_2} = Internal energy of *reactants* at temperature T_2,

U_{P_2} = Internal energy of *products* at temperature T_2,

ΔU_0 = Constant volume heat of combustion,

Q = Heat transferred to the surroundings during the process, and

W = Work obtained during combustion process.

Analysis for a non-flow process involving combustion at 'constant volume' :

When the combustion process is carried out at *constant volume* then the non-flow energy equation, $Q = (U_2 - U_1) + W$, can be applied to give

$$Q = (U_{P_0} - U_{R_0}) \quad ...(9.12)$$

where, $W = 0$ for constant volume combustion

$U_1 = U_{R_0}$

$U_2 = U_{P_0}$

The internal energy change is independent of the path between the two states and depends only on the initial and final values and is given by the quantity Q. This is illustrated in Fig. 9.2 and *property diagram* of Fig. 9.3. The heat so transferred is called the *internal energy of combustion* at T_0 (or *constant volume heat of combustion*), and is denoted by ΔU_0. Thus,

$$\Delta U_0 = U_{P_0} - U_{R_0} \quad ...(9.13)$$

ΔU_0 is a *negative quantity* since the internal energy of the reactants includes the potential chemical energy and heat is transferred *from* the system.

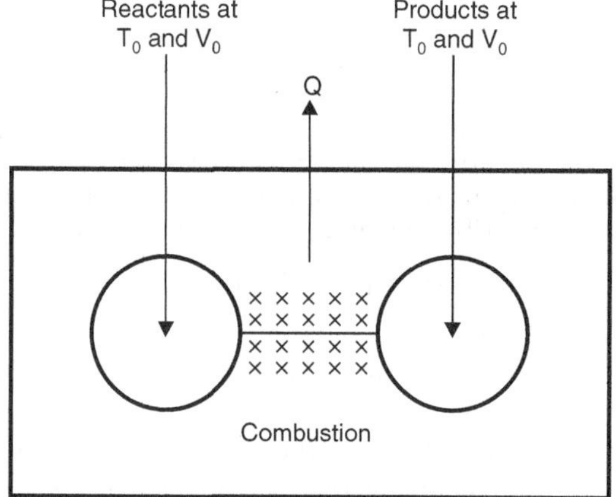

Fig. 9.2

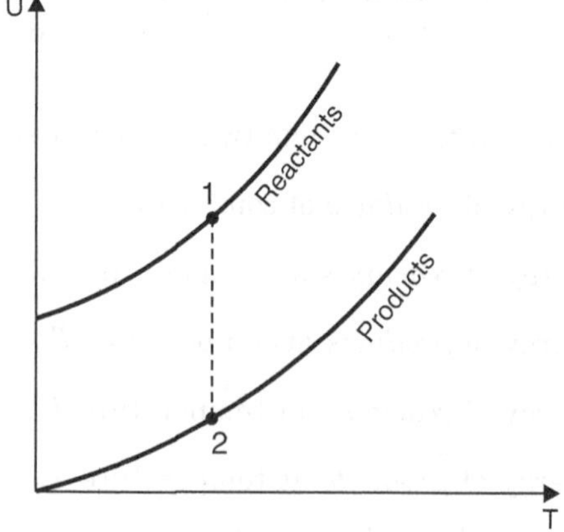

Fig. 9.3

It may be noted that in case of real constant volume combustion processes the initial and final temperatures will not be same as T_0 (reference temperature). The change in internal energy, for analytical purposes, between reactants at state 1 to products at state 2 can be considered in the following three steps (stages) :

(*i*) The change for the reactants from state 1 to T_0.

(*ii*) The constant volume combustion process from reactants to products at T_0.

(*iii*) The change for the products from T_0 to state 2.

The entire process can be thought of as taking place in piston-cylinder device as shown in Fig. 9.4.

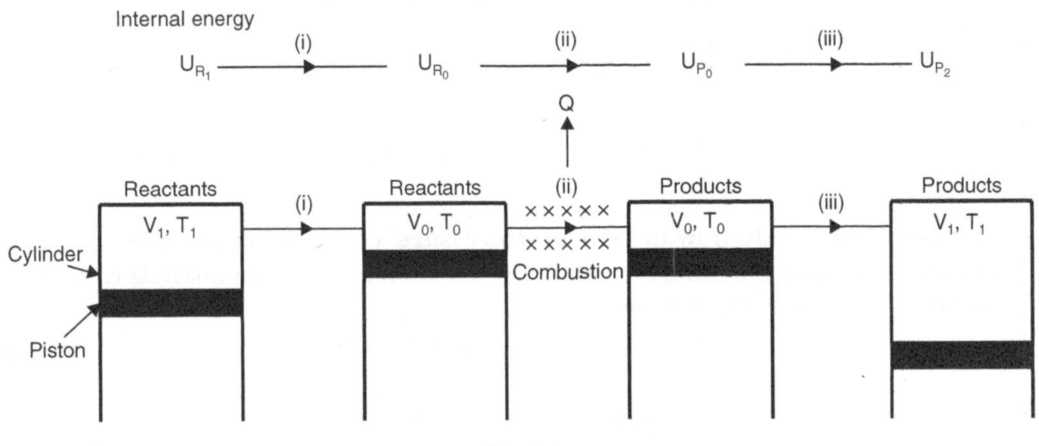

Fig. 9.4

Thus $U_2 - U_1$, the change in internal energy between states 1 and 2, can be written as $(U_{P_2} - U_{R_1})$ to show the chemical changes involved and this can be further expanded for analytical purposes as follows :

$$U_{P_2} - U_{R_1} = (U_{P_2} - U_{P_0}) + (U_{P_0} - U_{R_0}) + (U_{R_0} - U_{R_1})$$

i.e.

$$U_{P_2} - U_{R_1} = \underbrace{(U_{P_2} - U_{P_0})}_{\substack{\text{Products} \\ (iii)}} + \underbrace{\Delta U_0}_{(ii)} + \underbrace{(U_{R_0} - U_{R_1})}_{\substack{\text{Reactants} \\ (i)}} \qquad ...(9.14)$$

The values of $(U_{R_0} - U_{R_1})$ and $(U_{P_2} - U_{P_0})$ can be calculated from the following relations :

$$U_{R_0} - U_{R_1} = \sum_R n_i(u_{i_0} - u_{i_1}) \qquad ...(9.15)$$

where, u_i = Tabulated value of the internal energy for any constituent at the required temperature T_0 or T_1 in heat unit per mole,

n_i = Number of moles of the constituent, and

$\sum_R$ = Summation for all the constituents of the reactants denoted by i.

If *mass* base is used for tabulated values or calculation, then

$$U_{R_0} - U_{R_1} = \sum_R m_i(u_{i_0} - u_{i_1}) \qquad ...(9.16)$$

where, u_i = Internal energy per unit mass.

The above expression in terms of the *specific heats* (average values for the required temperature range) may be written as

$$U_{R_0} - U_{R_1} = \sum_R m_i c_{vi}(T_0 - T_1) = (T_0 - T_1) \sum_R m_i c_{vi} \quad ...(9.17)$$

For *products*, similar expressions may be written as :

$$U_{P_2} - U_{P_0} = \sum_P n_i(u_{i_2} - u_{i_0}) \quad ... \textit{on mole basis}$$

$$U_{P_2} - U_{P_0} = \sum_P m_i(u_{i_2} - u_{i_0}) \quad ... \textit{on mass basis}$$

$$U_{P_2} - U_{P_0} = \sum_P m_i c_{vi}(T_2 - T_0)$$

$$= (T_2 - T_0) \sum_P m_i c_{vi} \quad ... \text{in terms of } \textit{mean specific heats}$$

It may be noted that $n_i C_{vi} = m_i c_{vi}$.

Analysis for a steady flow or 'constant pressure' combustion process :

In such an analysis the *changes in enthalpy (H)* are important. An analysis carried out as above will give the following expressions :

$$H_{P_2} - H_{R_1} = \underbrace{(H_{P_2} - H_{P_0})}_{\text{Products}} + \Delta H_0 + \underbrace{(H_{R_0} - H_{R_1})}_{\text{Reactants}} \quad ...(9.18)$$

where, $\Delta H_0 = H_{P_0} - H_{R_0}$, and is always *negative* ...(9.19)

[ΔH_0 = *Enthalpy of combustion* at T_0 or the *constant pressure heat of combustion* at T_0]

Expressions for change of enthalpy of reactants and products :

Reactants :

$$H_{R_0} - H_{R_1} = \sum_R n_i(h_{i_0} - h_{i_1}) \quad ... \textit{on mole basis} \quad ...(9.20)$$

$$H_{R_0} - H_{R_1} = \sum_R m_i(h_{i_0} - h_{i_1}) \quad ... \textit{on mass basis} \quad ...(9.21)$$

$$H_{R_0} - H_{R_1} = \sum_R m_i c_{pi}(T_0 - T_1)$$

$$= (T_0 - T_1) \sum_R m_i c_{pi} \quad ... \text{in terms of } \textit{mean specific heats}$$

Products :

$$H_{P_2} - H_{P_0} = \sum_P n_i(h_{i_2} - h_{i_0}) \quad ... \textit{on mole basis} \quad ...(9.22)$$

$$H_{P_2} - H_{P_0} = \sum_P m_i(h_{i_2} - h_{i_0}) \quad ... \textit{on mass basis} \quad ...(9.23)$$

$$H_{P_2} - H_{P_0} = \sum_P m_i c_{pi}(T_2 - T_0)$$

$$= (T_2 - T_0) \sum_P m_i c_{pi} \quad ... \text{in terms of } \textit{mean specific heats}$$

...[9.23 (*a*)]

It may be noted that $n_i C_{pi} = m_i c_{pi}$

From the definition of the enthalpy of a perfect gas

$$H = U + pV = U + nR_0T$$

So if we are concerned only with gaseous mixtures in the reaction then for products and reactants

$$H_{P_0} = U_{P_0} + n_P R_0 T_0$$

and $$H_{R_0} = U_{R_0} + n_R R_0 T_0$$

where, n_P and n_R are the moles of products and reactants respectively and the temperature is the reference temperature T_0.

Thus, using equations (9.13) and (9.19), we have

$$\Delta H_0 = \Delta U_0 + (n_P - n_R)R_0T_0 \quad ...(9.24)$$

If there is no change in the number of moles during the reaction or if the reference temperature is absolute zero, then ΔH_0 and ΔU_0 will be equal.

9.1.10. Enthalpy of Formation (ΔH_f)

A combustion reaction is a particular kind of chemical reaction in which products are formed from reactants with the release or absorption of energy as heat is transferred to and from the surroundings. In some substances like hydrocarbon fuels which are many in number and complex in structure the *heat of reaction* or *combustion* may be calculated on the basis of known values of the enthalpy of formation, ΔH_f of the constituent of the reactants and products at the temperature T_0 (reference temperature). *The enthalpy of formation (ΔH_f) is the increase in enthalpy when a compound is formed from its constituent elements in their natural form and in a standard state.* The standard state is 25°C, and 1 atm. pressure, but it must be borne in mind that not all substances can exist in natural form, *e.g.* H_2O cannot be a vapour at 1 atm. and 25°C.

The expression of a particular reaction, for calculation purposes, may be given as :

$$\Delta H_0 = \sum_P n_i \Delta H_{f_i} - \sum_R n_i \Delta H_{f_i} \quad ...(9.25)$$

Typical values of ΔH_f for different substances at 25°C (298 K) in **kJ/mole** are given below :

S. No.	*Substance*	*Formula*	*State*	ΔH_f
1.	Oxygen	O	Gas	249143
		O_2	Gas	zero
2.	Water	H_2O	Liquid	– 285765
			Vapour	– 241783
3.	Carbon	C	Gas	714852
			Diamond	1900
			Graphite	zero
4.	Carbon monoxide	CO	Gas	– 111508
5.	Carbon dioxide	CO_2	Gas	– 393443
6.	Methane	CH_4	Gas	– 74855
7.	Methyl alcohol	CH_3OH	Vapour	– 240532
8.	Ethyl alcohol	C_2H_5OH	Vapour	– 281102
9.	Ethane	C_2H_6	Gas	– 83870
10.	Ethene	C_2H_4	Gas	51780
11.	Propane	C_3H_8	Gas	– 102900
12.	Butane	C_4H_{10}	Gas	– 125000
13.	Octane	C_8H_{18}	Liquid	– 247600

9.1.11. Heating Values of Fuels

If a fuel contains hydrogen water will be formed as one of the products of combustion. If this *water is condensed, a large amount of heat will be released* than if the water exists in the vapour phase. For this reason *two heating values* are defined ; the *higher or gross heating value* and the *lower* or *net heating value.*

The *higher heating value,* HHV, is obtained when the *water formed by combustion is completely condensed.*

The *lower heating value,* LHV, is obtained when the water formed by combustion exists *completely in the vapour phase.*

Thus :

$$(\text{HHV})_p = (\text{LHV})_p + m\, h_{fg} \quad \text{...(9.26)}$$

$$(\text{HHV})_v = (\text{LHV})_v + m(u_g - u_f) \quad \text{...(9.27)}$$

where, m = Mass of water formed by combustion,

h_{fg} = Enthalpy of vapourisation of water, kJ/kg,

u_g = Specific internal energy of vapour, kJ/kg, and

u_f = Specific internal energy of liquid, kJ/kg.

In almost all practical cases, the water vapour in the products is vapour, the lower value is the one *which usually applies.*

Fuel Calorimeters :

The heating value of a fuel is defined as the quantity of heat transferred from the calorimeter in order to reduce the temperature of the products to the initial reaction temperature. Heating values are reported as positive quantities and are used widely in the calculation of the thermal efficiency of power systems.

Two types of fuel calorimeters used for the determination of heating values are :

1. Bomb type calorimeter
2. Junkers-type calorimeter.

Bomb-type calorimeter. Refer Fig. 9.5. The bomb-type calorimeter, a *constant-volume* system, is initially charged with oxygen and a small sample of fuel. Subsequent to ignition and combustion, the heat is transferred from the products to a surrounding water bath. The heating value is calculated essentially from the measured temperature increase of the system mass. The calculated result is *usually reduced to a standard heating value at 25°C.* A heating value determined in a bomb-type calorimeter is designated as a *constant-volume higher heating value.* Water vapour formed during the reaction is completely condensed especially when a few drops of water are placed in the bomb prior to sealing in order to saturate the gaseous atmosphere.

Junkers-type calorimeter. Refer Fig. 9.6. Junkers-type calorimeter is designed to burn a *gaseous fuel under a steady flow conditions at atmospheric pressure.* Heat is transferred from the products to water flowing steadily through the outer jacket of the calorimeter. The operating conditions are adjusted to obtain a gas outlet temperature equal to the inlet temperature of the fuel and combustion air. From observed water temperatures and measured quantities of fuel and jacket water, the heating value is calculated and reduced to the corresponding value for 25°C operation. Some of the water vapour in the products condenses and drains from the calorimeter into a collecting vessel. This measured quantity of condensate is used in the subsequent conversion of the calorimetric heating value to the constant pressure higher and lower heating values that are based, respectively, upon complete and zero condensation of the water vapour formed during the combustion reaction.

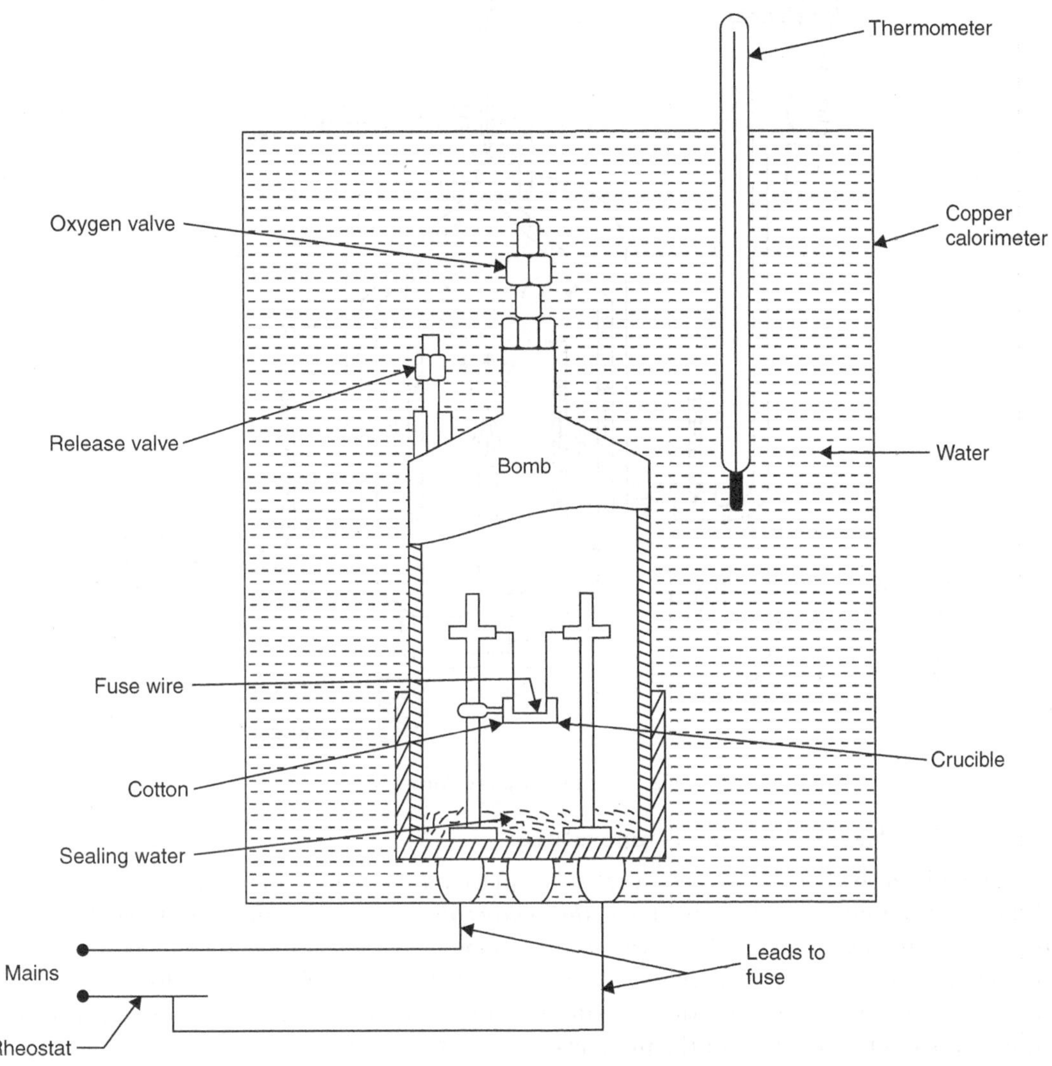

Fig. 9.5. Bomb calorimeter.

Note. Although the constant-pressure heating value and the enthalpy of combustion are developed from somewhat different concepts, there is a general similarity between these two terms. When the characteristics of the reaction are identical, the constant-pressure heating value and the corresponding enthalpy of combustion value will be numerically equal but of opposite sign. A corresponding similarity exists between the *constant-volume heating value and the internal energy of combustion.*

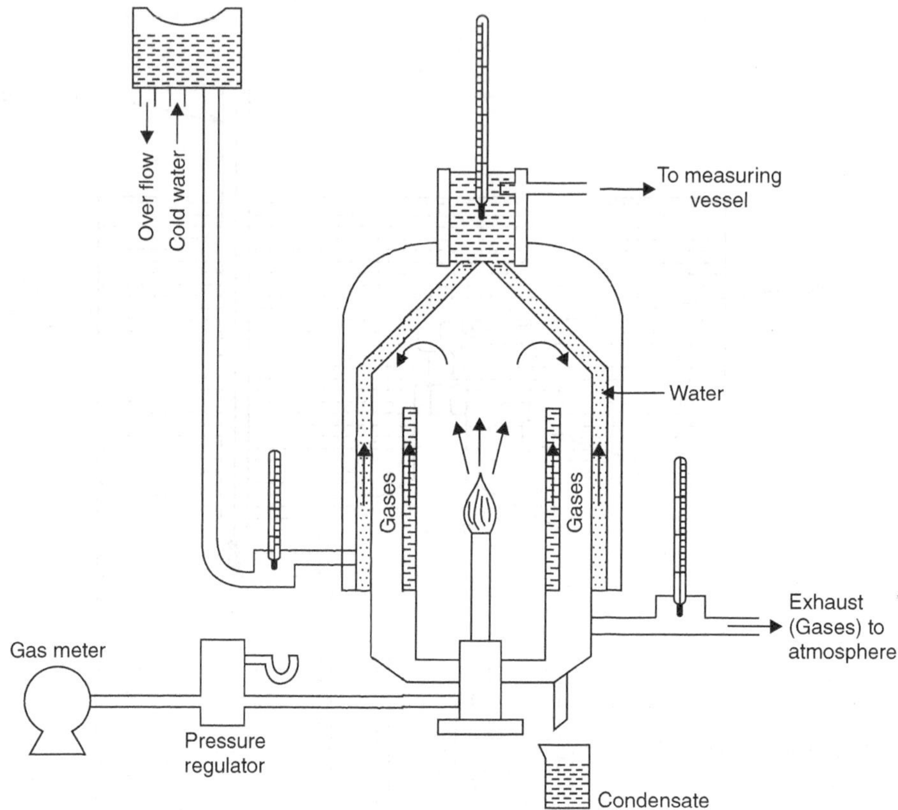

Fig. 9.6. Junkers gas calorimeter.

9.1.12. Adiabatic Flame Temperature

In a given combustion process, that takes place *adiabatically* and with no work or changes in kinetic or potential energy involved, *the temperature of the products is referred to as the 'adiabatic flame temperature'*. With the assumptions of no work and no changes in kinetic or potential energy, this is the *maximum temperature* that can be achieved for the given reactants because any heat transfer from the reacting substances and any incomplete combustion would tend to lower the temperature of the products.

The following points are worthnoting :

(*i*) *The maximum temperature achieved through adiabatic complete combustion varies with the type of reaction and percent of theoretical air supplied.*

An *increase in the air-fuel* ratio will effect a *decrease in the maximum temperature.*

(*ii*) For a given fuel and given pressure and temperature of the reactants, the *maximum adiabatic flame temperature that can be achieved is with a 'stoichiometric' mixture.*

(*iii*) The adiabatic flame temperature can be *controlled* by the *amount of excess air* that is used. This is important, for example, in *gas turbines*, where the maximum permissible temperature is determined by *metallurgical considerations* in the turbine, and *close control of the temperature of the products is essential.*

9.1.13. Chemical Equilibrium

The calculation of the *adiabatic flame temperature* is based, in part, on the assumption that the reaction goes to *completion*. Owing to *dissociation*, complete conversion of the reactants to the products is not accomplished. As a consequence of the failure to achieve complete conversion of the reactants, the maximum reaction temperature cannot attain the level of the theoretical adiabatic flame temperature.

The combination of CO and O_2 produces CO_2 together with a release of energy. In an adiabatic system no heat is transferred to the surroundings, hence the temperature of the mixture of the products and reacting substances rises rapidly. As the mixture temperature increases to higher levels the rate of dissociation of the CO_2 becomes increasingly more pronounced. Since the dissociation of CO_2 requires absorption of energy, *a condition is reached where the rate of evolution and the rate of absorption of energy are in balance. At this point no further increase in temperature can be observed* and the reaction is in *chemical equilibrium*. For this condition

$$C + O_2 \rightleftharpoons CO_2$$

At each temperature of the equilibrium mixture the substances participating in the reaction exist in unique proportions. For the combustion of CO the right-hand side of the equation

$$CO + \tfrac{1}{2}\, O_2 = (1 - x)\, CO_2 + x\, CO + \frac{x}{2}\, O_2 \qquad ...(9.28)$$

represents the distribution of the equilibrium products resulting from a reaction between CO and O_2. In this equation x denotes the fraction of dissociated CO_2. At low temperatures the fraction $(1 - x)$ approaches unity while at high temperatures $(1 - x)$ shows a substantial reduction in magnitude.

For the combustion of H_2 with O_2

$$H_2 + \tfrac{1}{2}\, O_2 \rightleftharpoons H_2O$$

and

$$H_2 + \tfrac{1}{2}\, O_2 = (1 - x)\, H_2O + x\, H_2 + \frac{x}{2}\, O_2 \qquad ...(9.29)$$

It is essential to distinguish between the effects of dissociation and the losses resulting from incomplete combustion of fuel. Incomplete combustion, which may be attributed to a number of factors, results in a discharge from the system of combustible substances. *Dissociation*, on the other hand, is of *transient nature*. Usually any appreciable degree of dissociation extends over a very short time interval at the highest level of temperature attained in the reaction. The gaseous products are likely to be discharged from the system at a temperature that is indicative of a low degree of dissociation. *For example, dissociation does not influence the heating value determined in a fuel calorimeter*. Although the maximum temperature attained in the calorimeter is limited by chemical equilibrium, the combustion process moves to completion with the decrease in the temperature of the products. The reduction in temperature is a result of heat transfer to the jacket water. Dissociation of the products is negligible at room temperature, which is essentially the calorimeter reaction temperature.

The temperature of the products discharged from the combustion chamber of the gas turbine power plant is limited to approximately 870°C by introduction of a *large quantity of excess air*. Absorption of energy in the water walls of a boiler furnace limits the outlet gas temperature to approximately 1100°C. The quantity of dissociated products at temperatures ranging upward to 1100°C is not appreciable. In the cylinder of I.C. engine, considerably higher maximum temperatures—that is, in excess of 1100°C are attained, hence in the analysis of this thermal power system consideration must be given to the effects of dissociation. Of particular significance is the effect of reduced maximum temperature on the system availability. As a result of heat transfer and work performed by the gaseous medium the products are discharged from the system at a temperature below the level at which an appreciable degree of dissociation is observed.

The proportions of the dissociated products in chemical equilibrium at temperature T are established from the *equilibrium constant.* The evaluation of the equilibrium constant is achieved in accordance with the analysis presented by Van't Hoff.

9.1.14. Actual Combustion Analysis

In evaluating the performance of an actual combustion process a number of different parameters can be defined depending on the nature of the process and the system considered. The combustion efficiency in a gas turbine for instance can be defined as

$$\eta_{combustion} = \frac{(F/A)_{ideal}}{(F/A)_{actual}} \quad ...(9.30)$$

where, $(F/A)_{ideal}$ = Fuel-air ratio required for adiabatic and complete combustion and in which the products would attain the adiabatic flame temperature.

In case of a *steam generator (boiler)*

$$\eta_{steam\ generator} = \frac{\text{Heat transferred to steam / kg fuel}}{\text{Higher heating value of the fuel}} \quad ...(9.31)$$

In case of an *internal combustion engine,*

$$\eta_{thermal} = \frac{W_{actual}}{\text{Heating value}} \quad ...(9.32)$$

9.2. CONVENTIONAL FUELS (FOR I.C. ENGINES)

9.2.1. Introduction

- I.C. engines can run on different kinds of fuel, including liquid, gaseous and even solid fuels. The properties and the character of the fuel exercise profound influence on the design, power output, efficiency, fuel consumption and the reliability and durability of the engine.
- The use of *solid fuels* present problems of complicated injection systems, as well as difficulties associated with solid residual ash, and hence are *not popular, gaseous fuels* present problems of storage and handling of large volumes. Hence for mobile *use its use gets restricted.* But gaseous fuels do find *use for stationary power plants particularly when gas is readily available at the location nearby. Thus liquid fuels find abundant use in I.C. engines.*

9.2.2. Desirable Properties of Good I.C. Engines Fuels

The fuels used in I.C. engines are designed to satisfy the performance requirements of the engine system in which they are used. Thus fuel should possess the following **properties :**

1. High energy density (kJ/kg).
2. Good combustion qualities.
3. High thermal stability.
4. Low toxicity.
5. Low pollution.
6. Easy transportation/transferability and storage.
7. Compatibility with the engine hardware.
8. Good fire safety.

9. Low deposit forming tendency.
10. Economically viable in very large quantities.
11. Easy mixing with air and low latent heat of evaporation (h_{fg}).
12. No chemical reaction with engine components through which it flows.

9.2.3. Gaseous Fuels

These fuels are used in S.I. engines. The different gaseous fuels are enumerated and discussed below :

1. Natural gas
2. Manufactured gases
3. By-product gases
4. Sewage sludge gas
5. Biogas.

1. **Natural gas :**

- It composition varies with source but mainly it contains CH_4 (75 to 95 percent) and remaining C_2H_6 and N_2. From some areas, the natural gas obtained contains H_2S which is much harmful to the engines.
- It is available with oil wells and is *colourless* and *odourless.*
- It is found in several parts of the world but particularly in U.S.A. It is also carried from the place of availability to the place of use through hundreds of kilometres pipeline.

2. **Manufactured gases :**

The gases are manufactured by various methods, discussed briefly below :

- ***Coal gas*** is manufactured by *heating soft coal in closed vessel.* The contents of the gas depend upon the type of coal and method of operation used in manufacturing.
  — A clean coal gas contains : 33% H_2 and 66% CH_4.
  — Its energy content is *50 percent of natural gas.*
- ***Water gas*** is *formed by using steam.* For its manufacture, the *water and air are passed alternately through a bed of hot carbon.*
  — It contains H_2, CO and N_2.

3. **By-product gases :**

- The gases produced during manufacture of other substances are known as *by-product gases.*
- ***Blast furnace gas*** is a by-product of steel plants. It contains CO and N_2. It contains large amount of dust particles ; therefore, it should be cleaned by an effective method before its use in the engine.

4. **Sewage sludge gas :**

- It contains CH_4 and CO_2 with very small percentage of H_2S.
- This gas is made available from present well developed sewage disposal plants.

5. **Biogas :**

- This gas is produced from the cow dung which is available in large quantities in India.
- It is easy to produce (with appropriate chemical reaction) and use locally.

Except natural gas, all other gases mentioned above are generally employed for running I.C. engines whose power is used locally to run different types of equipments like *small electric generators, pumps etc.*

Advantages of gaseous fuels :

(*i*) Easily compressed and stored.

(*ii*) Easily carried through pipes.

(*iii*) Easy starting of engines.

(*iv*) Easy to maintain A/F ratio in multi-cylinder engines, as compared to liquid fuels.

Disadvantages :

(*i*) High cost (on the basis of energy content)

(*ii*) High purifying cost.

(*iii*) Storage volume per unit energy very large.

(*iv*) As compared to engines using liquid fuels, the size and weight of the engine (kg/kW) is considerably large.

(*v*) The cost (capital and running) of the plants manufacturing gases is considerably high.

9.2.4. Liquid Fuels

Following are the three principal commercial types of liquid fuels :

1. Benzol ; 2. Alcohol ; 3. Refined products of petroleum.

1. **Benzol :**

- It consists of benzene (C_6H_6) and toulene (C_7H_8) and is obtained as a by-product of high temperature coal carbonization.
- It possesses anti-knock quality. As compared to gasoline, its heating value is low.

2. **Alcohol :**

- It has good anti-knock qualities.
- Its heating value is low as compared to gasoline.
- It is more expensive to produce.
- It is used as fuel blended with gasoline.
- It can be manufactured from grain, sugarcane and waste products.

3. **Refined products of petroleum :**

- It is the main source of liquid fuels for I.C. engines.
- It is used in the form of gasoline, kerosene, and diesel oil.

The liquid fuels are **classified** in two **groups :**

(*i*) *Liquid fuels which are vaporised easily : "Petrol"* and *"Alcohol"* . These are commonly used in S.I. engines.

(*ii*) *Liquid fuel which is directly injected in the combustion chamber :* "Diesel or fuel oil".

9.2.5. Structure of Petroleum

In I.C. engines the fuels which are usually used are *complex mixtures of hydrocarbons.* These fuels are obtained by *refining petroleum.*

Basically, petroleum is a mixture of hydrocarbons, compounds *made up exclusively of carbon and hydrogen atoms ;* it may also contain small quantities of other compounds having sulphur, oxygen and nitrogen (In some petroleum, metallic compounds such as derivatives of vanadium, iron, nickel, arsentic etc., are also found).

The constituents of petroleum are *classified* into the following four main groups :

	Constitute	***General formula***	
1.	*Paraffins*	C_nH_{2n+2}	(where *n* is the number of carbon atoms)
2.	*Olefins*	C_nH_{2n}	

3. *Naphthenes* C_nH_{2n}
4. *Aromatics* C_nH_{2n-6}

— Within each group also, the physical properties of individual compound differ according to the number of carbon and hydrogen atoms in the molecule.

— The physical differences between compounds, even in any group, influence the way fuel evaporates and hence the formation of combustible mixture.

— The difference in chemical properties of hydrocarbon from different groups affect the combustion process and hence the proportions of fuel and air requirements.

1. **Paraffins**

(*i*) ***Straight chain or normal paraffins :***

- It consists of a straight chain molecular structure as shown in Fig. 9.7.
- The names of hydrocarbon in this series end with *ane* as in methane, propane, hexane etc.
- The straight chain paraffins are saturated compounds as valency of carbon is fully utilised and therefore, they are very *stable*.

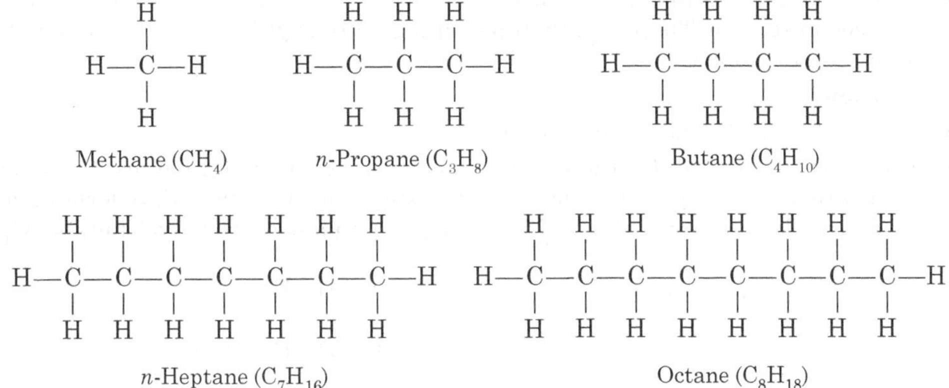

Fig. 9.7. Straight chain paraffin.

(*ii*) ***Branched chain or iso-paraffins :***

- The *carbon atoms are branched in these compounds.*
- Branched chain or iso-paraffins have an open structure which is branched as shown in Fig. 9.8. Iso-octane, triptane etc. are examples of this type.

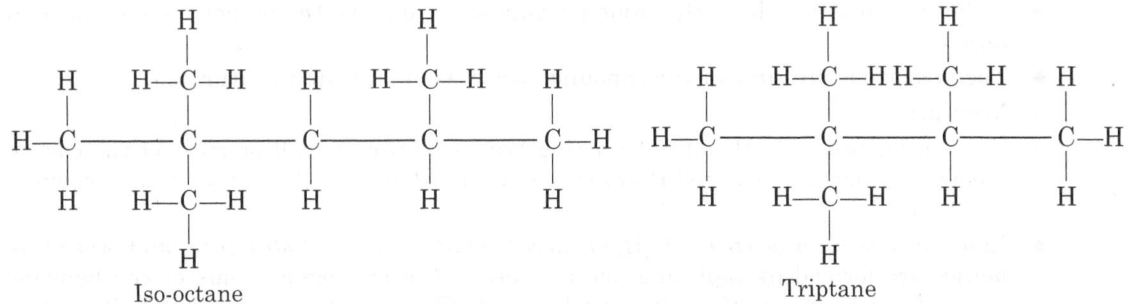

Fig. 9.8. Branched chain or iso-paraffins.

- Iso-paraffins are also *stable compounds* and *highly knock-resistant when used as S.I. engine fuels.*

2. **Olefins**

- These are *compounds with one or more double bonded carbon atoms in straight chain.* The names end with *ene* for one double bond and *adiene* for two double bonds. The examples are : Hexene and Butadiene (Fig. 9.9)

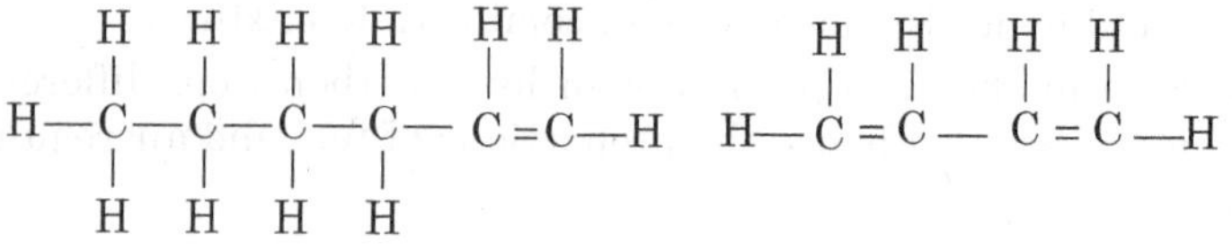

```
    H   H   H   H    H H              H   H     H   H
    |   |   |   |    | |              |   |     |   |
H—C—C—C—C— C=C—H       H— C = C — C = C —H
    |   |   |   |
    H   H   H   H
```

Hexene, C_6H_{12} (mono-olefin) Butadiene, C_4H_6 (diolefin)

Fig. 9.9. Olefins.

- The general formula for *mono-olefines* (single bond) is C_nH_{2n} and for *diolefins* (double bond) is C_nH_{2n-2}.
  — *Diolefins are more unstable than mono-olefins*
- Olefins are present in cracked gasoline. They form gummy deposits as they are readily oxidised in storage. Therefore, their percentages are kept low (less than 3%) in the fuels used in gas turbines.

3. **Napthenes**

- These are *ring structured compounds.*
- The chemical formula for these compounds is the same as for olefins, C_nH_{2n} but have each carbon atom joined by single bond to two other C atoms, thus forming a ring of structure. The examples, (Fig. 9.10) are : Cyclo-propane (C_3H_6), Cyclobutane (C_4H_8) etc.

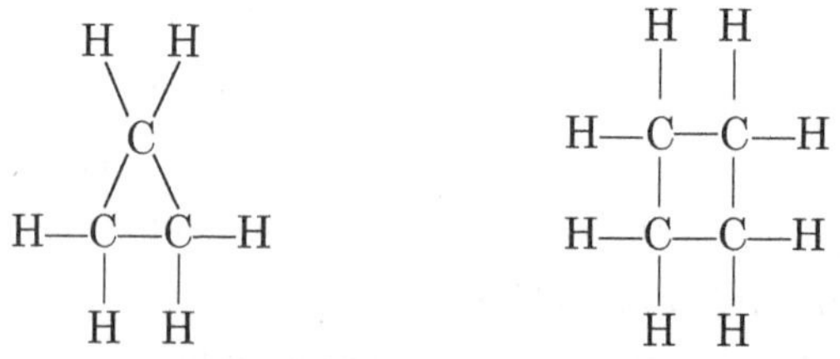

```
     H  H                 H  H
      \/                  |  |
      C               H—C—C—H
      /\                  |  |
  H—C—C—H            H—C—C—H
     |  |                 |  |
     H  H                 H  H
```

Cyclo-propane (C_3H_6) Cyclobutane (C_4H_8)

Fig. 9.10

- Although napthenes have the same formula as for olefins, the properties are *radically different.*
- The napthenes are *saturated* compounds whereas olefins are unsaturated.

4. **Aromatics**

- These compounds (C_nH_{2n-6}) have a ring type structure for all or most of the carbon atoms, to which are attached H or group of C and H atoms ; the examples are shown in Fig. 9.11.
- In all aromatics, a *benzene, (C_6H_6) molecule exists as central structure* and other aromatics are formed by replacing one or more of the hydrogen atoms of the benzene molecule with an organic radical such as paraffins, napthenes and olefins. By adding a methyl group (CH_3) ; benzene is converted to toluene ($C_6H_5CH_3$). The three double bonds make aromatics very active and therefore they are *highly unsaturated compounds.*

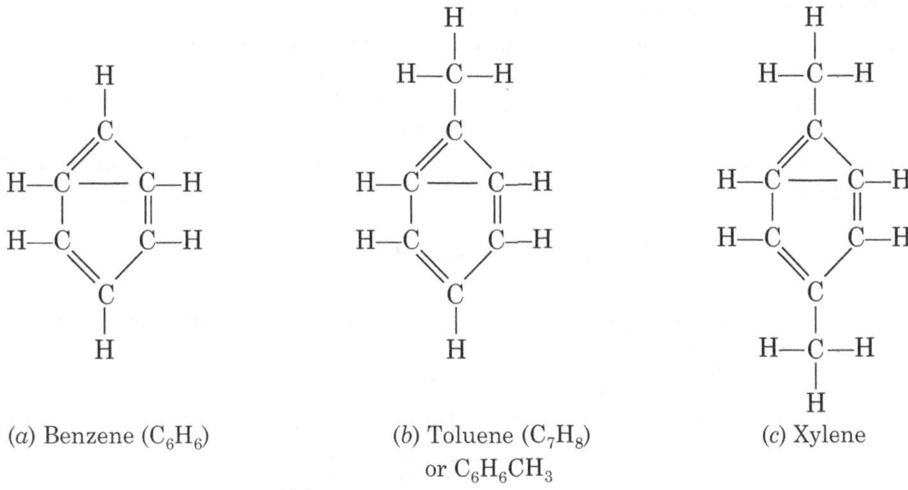

(*a*) Benzene (C_6H_6) (*b*) Toluene (C_7H_8) or $C_6H_6CH_3$ (*c*) Xylene

Fig. 9.11. Structures of aromatics.

— In most unaltered gasolines, both benzene and toluene are present to a modest extent.

Following are a few *special properties of aromatics :*

(*i*) Offer highest resistance to knocking in S.I. engines.

(*ii*) Suitability of these fuels for C.I. engines is just reverse of their suitability for S.I. engines.

— Therefore, *"paraffins" are most suitable fuels for C.I. engines* and *"aromatics" are most suitable fuels for S.I. engines.*

(*iii*) With the increase in the number of atoms in the molecular structure, the boiling temperature of fuel generally increases.

(*iv*) As the proportion of H_2-atoms to C-atoms in the molecule increases the calorific value of fuel increases. Thus, paraffins have lower calorific value whereas *aromatics have highest calorific value.*

9.2.6. Petroleum and Composition of Crude Oil

- Petroleum is a dark viscous oily liquid known as *rock oil* (In Greek, *petra*—rock, *oleum*—oil). It is formed from the bacterial decomposition of the remains of animals and plants which got buried under the sea millions of years ago. When these organisms died, they sank to the bottom and got covered by sand and clay. Over a period of millions of years, these remains got converted into hydrocarbons by heat, pressure and catalytic action. The hydrocarbons formed rose through porous rocks until they were trapped by impervious rocks forming an oil trap.
- Natural gas is found above the petroleum oil trapped under rocks. The crude petroleum is obtained by drilling a hole into the earth's crust and sinking pipes into it. When the pipe reaches the oil deposit, natural gas comes out with a great pressure. After the pressure has subsided, the crude oil is pumped out of the oil well. *This process of obtaining crude oil from its sources* is called **mining.**
- The *crude oil is a mixture of hydrocarbons* such as alkanes, cycloalkanes and aromatic hydrocarbons. It also contains a number of compounds having oxygen, nitrogen and sulphur. The actual composition of petroleum depends upon its place of origin.

Fractional distillation of crude oil :

- The crude petroleum obtained by mining is a dark coloured viscous liquid called **crude** oil. Before petroleum can be used for different purposes, it must be separated into various components. *The process of separating petroleum into useful fractions and removal of undesirable impurities is called* **refining**.
- The refining of petroleum is carried out by the process of **fractional distillation** *as* described below :

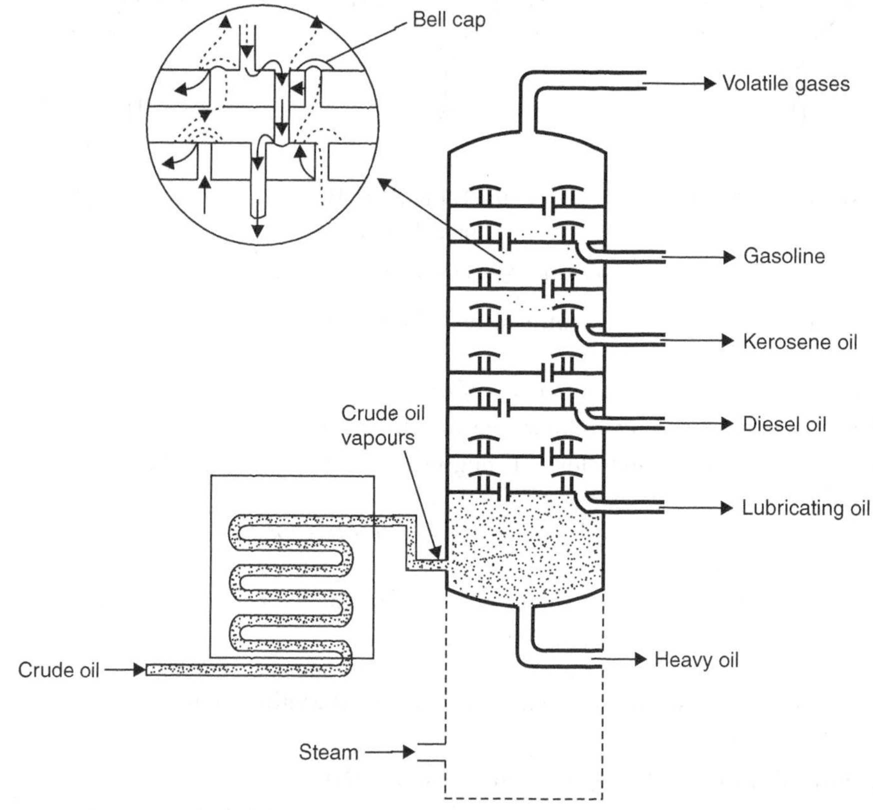

Fig. 9.12. Refining of petroleum.

— The refining of petroleum is done in big refineries. The first step in the refining process is neutralisation of crude oil by washing with acidic or basic solution as needed.

— Then the oil is heated in a furnace to about 675 K and the vapour thus obtained are introduced into a fractionating tower. The tower is divided into a number of compartments by means of shelves (trays) having holes (Fig. 9.12). The holes are covered with *bubble caps* which allow the lighter more volatile components to pass up the column while the heavier, less volatile components condense and flow into trays below. Each shelf is provided with an overflow pipe which keeps the liquid to a certain level and allows the rest to flow down to the lower shelf. Therefore, *during fractional distillation, the fractions with lower boiling points rise up the tower and condense at different levels depending upon the boiling points.* For example, the crude oil is fed at the base at about 675 K. At this temperature all the components except asphalt vapourise. As the mixture of hot vapour rises in the column, it cools. Therefore, the component with the highest boiling point liquefies first and is collected. Then a little, higher in the column, the component having slightly lower boiling point liquefies and so on. The residual gases which do not condense escape from the upper part of the tower. *The fractions are separated at different boiling points and are thus collected at different heights in the column.* The important fractions of petroleum refining are given in Table 9.2.

Table 9.2. Different fractions of petroleum refining.

Fraction (K)	*Boiling range (K)*	*Approximate composition*	*Uses*
1. Gaseous	113 to 303	C_1—C_5	As gaseous fuel, for producing carbon black and is also used for preparing ammonia, methyl alcohol and gasoline
2. Petroleum ether or ligroin	303 to 363	C_5—C_7	As a solvent for oils, fats, rubber and also in dry cleaning.
3. Gasoline or petrol (Sp. gravity = 0.7 to 0.8)	343 to 473	C_7—C_{12}	Mainly as a *motor fuel.*
4. Kerosene (Sp. gravity = 0.8 to 0.85)	448 to 548	C_{12}—C_{15}	As an illuminant fuel and for preparing *petrol gas.*
5. Gas oil, fuel oil, and diesel oil	523 to 673	C_{15}—C_{18}	In furnace oil, as a fuel for *diesel engines* and also in cracking.
6. Lubricating oils, greases and petroleum jelly	623 and higher	C_{16} and higher	Used mainly as *lubricants.*
7. Paraffin wax	melts between 325 to 330	C_{20} and higher	Used for manufacturing candles, waxed papers and for water proofing.
8. Petroleum coke	residue	C_{30} and higher	Used as artificial asphalt, fuel and also in making electrodes.

9.2.7. Fuels for Spark-Ignition Engines

Gasoline, a mixture of various hydrocarbons (such as paraffins, olefins, napthenes, and aromatics) is the major fuel used for S.I. engines. The composition depends upon the source of crude oil and the nature of refining process.

The following are the *requirements of an ideal gasoline* :

1. High calorific value
2. Knock-resistance
3. Easy to handle
4. Easy availability at reasonable cost
5. Quick evaporation (when injected by carburettor in the current of air)
6. Clean burning and no deposition of the residue.
7. No pre-ignition.
8. No tendency to decrease the volumetric efficiency of engine.

9.2.7.1. Volatility

- *"Volatility"* is commonly defined as *the evaporating tendency of a liquid fuel.*

- This quality of the fuel has great significance for carburetted engines. This will decide the fuel vapour to air ratio in the cylinder at the time of ignition. As $F_R = 0.6$ is the lowest limit for satisfactory ignition and flame propagation, therefore, volatility of fuel must ensure to give at least this fuel vapour to air ratio at the time of ignition under all conditions of operation including starting from cold.
- The volatility of gasoline is generally characterised by the following two laboratory tests.
  (*i*) ASTM distillation test (*ii*) Reid vapour test.

ASTM distillation test :

Fig. 9.13 shows the apparatus used for ASTM distillation test :

- 100 cubic centimetres of gasoline fuel is taken in the flask and heated. The flask is fitted with a thermometer to record the temperature of vapour being formed and collected in a graduated measuring cylinder.

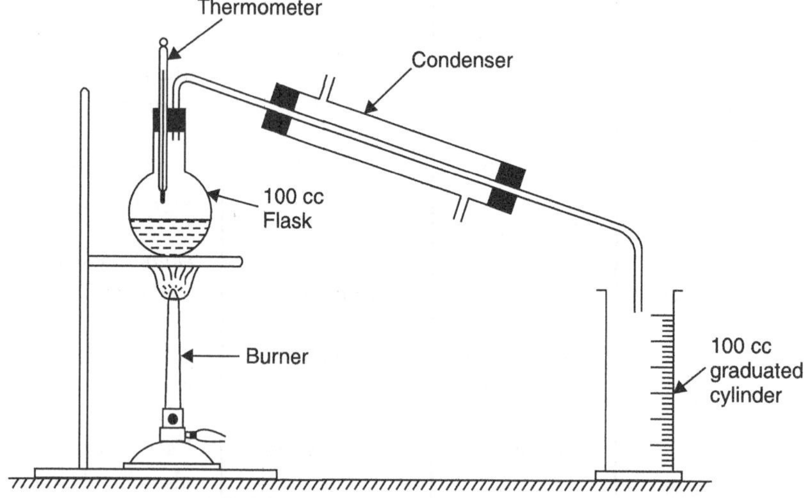

Fig. 9.13. ASTM distillation apparatus.

- When the first drop of condensed vapour drops from condenser, the temperature is recorded. This temperature is called *initial boiling point.*
- The vapour temperature is recorded at each successive 10 percent of condensed vapour collected. When 95 percent has been distilled the burner flame is increased and the maximum temperature is recorded as the *'end point'*. The mass of the residue in the flask is also recorded.

Fig. 9.14 shows distillation curves (ASTM) for various products of petroleum refining.

Reid vapour pressure

- The *volatility* of petrol is also defined in *terms of Reid vapour pressure.* This is a measure of the vapour pressure of oil at 38°C expressed as millimetres of mercury or in pounds per square inch pressure and indicates initial tendency of a fuel to vapour-lock.
- The apparatus used for determination of Reid vapour pressure is shown in Fig. 9.15.
  — A chilled fuel sample is placed in the Reid bomb and then immersed in water bath held at 38°C ; the *air chamber contains an air volume equal to four times the volume*

of fuel. The pressure indicated by the pressure gauge will indicate the *pressure rise due to vaporisation of fuel and increase in volume due to increase in temperature*. If the latter is subtracted from the total pressure rise we get the Reid vapour pressure (the increase in pressure due to vapourisation of a given quantity of fuel under a given quantity of fuel under given condition of temperature.

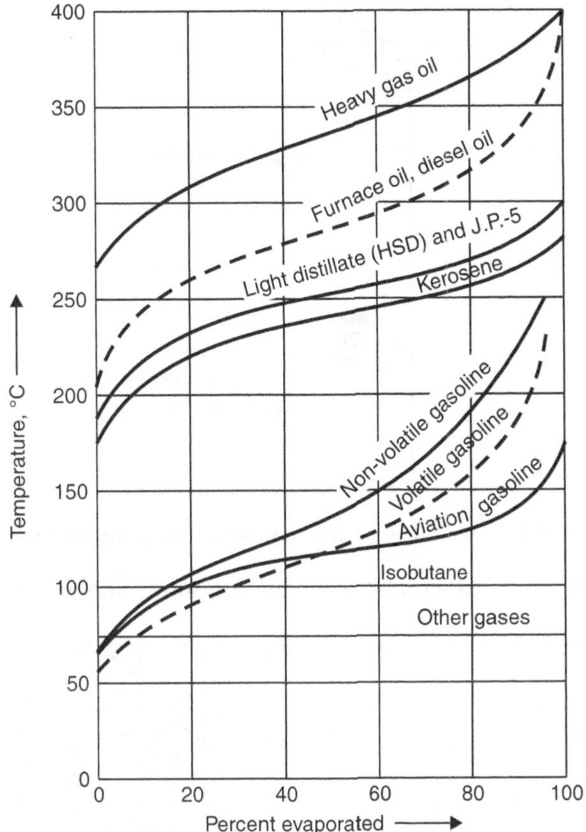

Fig. 9.14. Distillation curves (ASTM).

Equilibrium air distillation (EAD) :

- The ASTM distillation curve is not a true boiling point curve of the fuel. Therefore, it cannot directly relate to fuel performance in the engine. In this case, the fuel is allowed to evaporate into an air stream moving through a long tube with low velocity. The exit *vapour-air* ratio is measured as a function of *fuel-air* ratio. The tube should be sufficiently long to attain equilibrium. The tube represents the intake manifold of the engine and equilibrium of vapour-air is reached before entering the engine.
- In ASTM-test, the vapourisation of fuel is carried out in the presence of vapour of fuel so these curves can not be used directly, as actual evaporation of fuel takes place in the current of air. Therefore, to correlate the fuel performance in the engine, EAD test apparatus is used as described above.

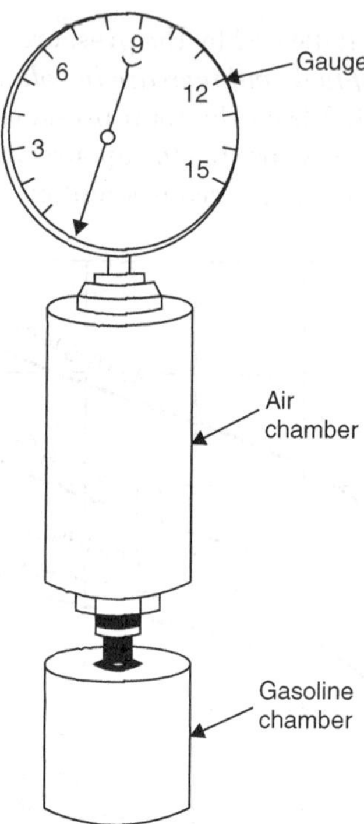

Fig. 9.15. Apparatus for determining Reid vapour pressure.

Fig. 9.16 shows the EAD test curves for a typical gasoline and also A/F ratio volatility curves.

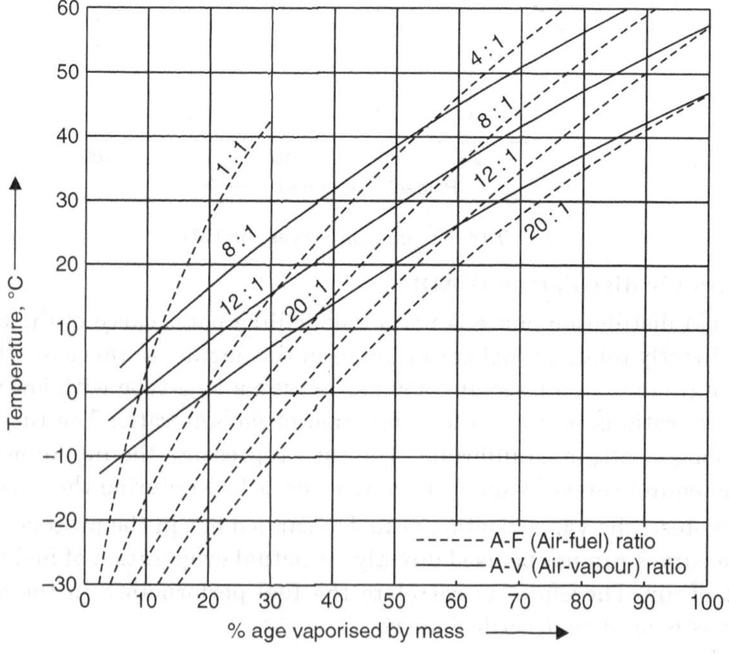

Fig. 9.16. EAD curves for a typical gasoline.

9.2.7.2. Effects of Volatility on S.I. Engine Performance

Volatility of a liquid is its tendency to evaporate under a given set of conditions. It is an extremely important characteristic of S.I. engine which affects engine performance and fuel economy characteristics.

— Cold starting of S.I. engine is improved if front end volatility is higher but it may lead to increased problems of hot starting and vapour lock.

— The *mid range* (20 to 80%) *portion* should be volatile enough to give satisfactory air-fuel ratios under a variety of operating conditions.

— Low *tail end volatility will help in good mixture distribution and hence good fuel economy.*

A few important effects of volatility on S.I. engine performance and enumerated and described below :

1. Starting and warm up
2. Vapour lock
3. Evaporation loss
4. Crankcase dilution
5. Operating range performance
6. Spark plug fouling
7. Formation of sludge deposits.

1. **Starting and warm up :**

- For easy starting of the engine a certain part of the gasoline should vaporise at room temperature. To fulfil this condition 1 to 10 per cent of the distillation curve must have a low boiling temperature. With the warming up of the engine, the temperature gradually increases to the operating temperature.
- For best warm-up, low distillation temperatures are desirable throughout the range of distillation curve.

2. **Vapour lock :**

- Vapour lock is a situation where *too lean a mixture is supplied to the engine.* The automotive fuel pump should handle both liquids and vapours. If the amount of fuel evaporated in the fuel system is very high the fuel pump is mainly pumping vapour and very little liquid will go to the engine. This results in very weak mixture which can not maintain engine output.

  Vapour lock causes the following :

  (*i*) Uneven running of an engine ;
  (*ii*) Stalling while idling ;
  (*iii*) Irregular acceleration ;
  (*iv*) Difficult starting when hot ;
  (*v*) Momentary stalling when running.

- The vapour lock tendency of the gasoline is related to *front end volatility.* The vapour-liquid ratio (V/L) of a gasoline directly correlates with the degree of vapour lock likely to be experienced in the fuel system. At V/L ratio of 24 vapour lock may *start,* and at V/L ratio of 36 vapour lock may be *very severe.* Therefore, the *volatility of the gasoline should be maintained as low as practicable to prevent this type of difficulty.*

3. **Evaporation loss :**

- The evaporation loss (from carburettors and storage tanks) *depends on the vapour pressure which is a function of fraction components and initial temperature.* These losses can be as high as 10 to 15 per cent.
- The evaporation loss *not only decreases the fuel economy, but also decreases its anti-knock quality as the lighter fraction have higher anti-knock properties.*

  The evaporation loss is a *reason for restricting the low end volatility of the fuels.*

4. **Crankcase dilution :**

- If very frequent starting of the engine with low engine temperature is necessary, very rich mixtures have to be supplied and some of un-evaporated fuel leaks past the piston rings and goes to crankcase. Consequently lubricating oil gets diluted. *This dilution decreases the viscosity of the lubricating oil and also washes away the lubricating oil film on engine cylinder walls.* It is found that the tendency of fuels to dilute the lubricating oils lies in the order of 90% ASTM temperature. Thus control of 90% ASTM temperature combined with proper ventilation of cranckcase reduces the dilution of crankcase oil.
- In the engines using *heavier fuels* like kerosene and other distillates, the problem of dilution and poor lubrication of pistons and rings may be *severe.*

5. **Operating range performance :**

- The acceleration of an engine depends upon its ability to deliver suddenly to intake an extra supply of fuel air mixture in a sufficiently vaporised form to burn quickly. Good acceleration occurs when air-fuel vapour ratio of 12 : 1 is supplied.
- The ability to accelerate falls off as available mixture becomes lean.

6. **Spark plug fouling :**

Spark plug fouling is caused due to deposition of some high boiling hydrocarbons. Lower the tail-end volatility less are the chances of spark plug fouling.

7. **Formation of sludge deposits :**

The sludge deposition inside an engine is caused by certain types of high boiling hydrocarbons. These deposits *can cause piston ring plugging and sticking and valve sticking resulting in poor operation and poor fuel economy.*

9.2.8. Knock Rating of S.I. Engines Fuels

9.2.8.1. Self ignition characteristics of fuels

- When the temperature of an air-fuel mixture is raised high enough, the mixture will *self-ignite* without the need of a spark plug or other external ignites. *The temperature above which this occurs is called the "self-ignition temperature (S.I.T.)".* This is the basic principle of ignition in a compression ignition engine. The compression ratio is high enough so that the temperature rises above S.I.T. during the compression stroke. Self-ignition then occurs when the fuel is injected into the combustion chamber. On the other hand, self-ignition (or pre-ignition, or auto-ignition) is not desirable in an S.I. engine, where a spark plug is used to ignite the air-fuel at the proper time in the cycle. *When self-ignition does occur in an S.I. engine higher than desirable, pressure pulses are generated. These high pressure pulses can cause damage to the engine and quite often are in the audible frequency range.* This phenomenon is often called **knock** or **ping.**

- Fig. 9.17. shows the basic process of what happens when self-ignition occurs.

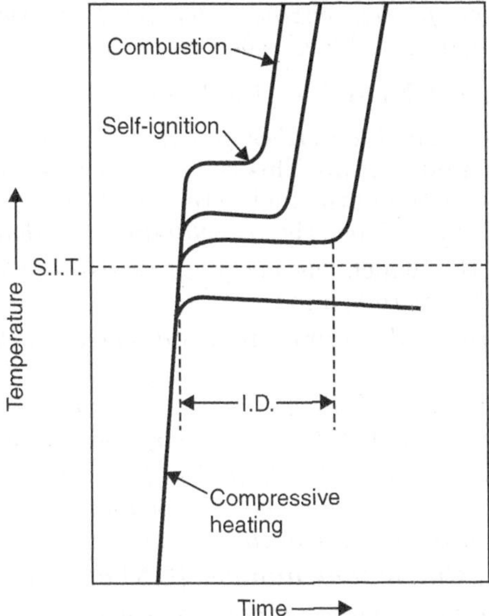

S.I.T. → Self-ignition temperature

I.D. → Ignition delay

Fig. 9.17. Self-ignition charactristics of fuels.

— If the temperature of a fuel is raised above the self-ignition temperature (S.I.T,), the fuel will spontaneously ignite after a short ignition delay (I.D.). *The higher above S.I.T. which the fuel is heated, the shorter will be I.D.*

— **Ignition delay** is generally a very small fraction of a second (generally of the order of *thousandths of a second*). During this time, pre-ignition reaction occurs, including oxidation of some fuel components and even cracking of some large hydrocarbon components into smaller hydro-carbon molecules. These pre-ignition reactions raise the temperature at local spots, which then promotes additional reactions until, finally, the actual combustion reaction occurs.

- Fig. 9.18. shows the cylinder pressure as a function of time in a typical S.I. engine combustion chamber showing (*i*) normal combustion, (*ii*) combustion with light knock and (*iii*) combustion with heavy knock.

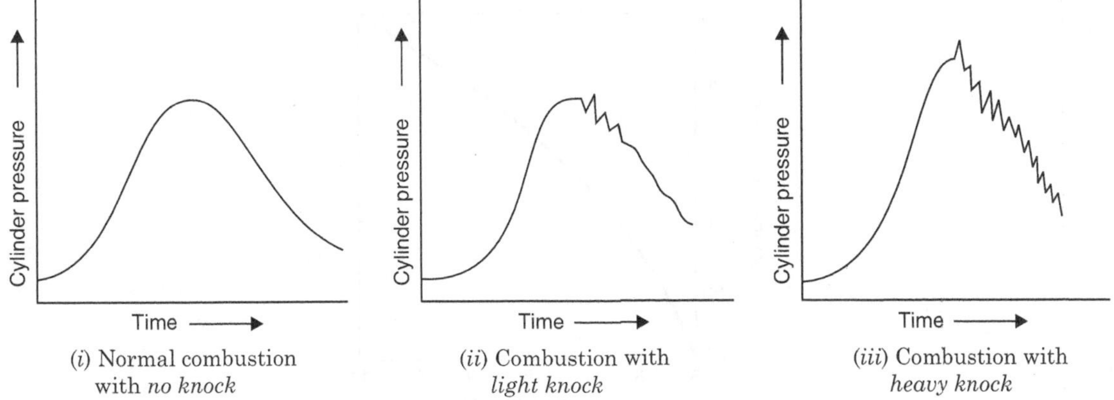

(*i*) Normal combustion with *no knock*

(*ii*) Combustion with *light knock*

(*iii*) Combustion with *heavy knock*

Fig. 9.18. Cylinder pressure as a function of time in a typical S.I. engine combustion chamber.

9.2.8.2. Highest Useful Compression Ratio (H.U.C.R.)

The highest useful compression ratio is the *highest compression ratio at which a fuel can be used without detonation in a specified test engine under specified operating conditions and the ignition and mixture strength being adjusted to give best efficiency.*

9.2.8.3. Octane number (ON) and engine knock

- *The property of a fuel which describes how fuel will or will not self-ignite is called the* **octane number or just octane**. This is a numerical scale generated by comparing the self-ignition characteristics of the fuel to that of standard fuels in a specific test engine at specific operating conditions. The *two standard reference fuels* are :
  - **Iso-octane** (C_8H_{18}) which has a *very high resistance to knock* and therefore is given an octane number of ***100*** ; and
  - **Normal heptane** (C_7H_{16}) which is *very prone to knock* and is therefore given a *zero* value.

Blends of these reference fuels define the knock resistance of intermediate octane numbers, and thus a blend of 10% *n*-heptane and 90% iso-octane by volume has an octane number of **90**.

- *The higher the octane number of fuel, the less likely it will self-ignite. Engines with low compression ratios can use fuels with lower octane numbers, but high-compression engines must use high-octane fuel to avoid self-ignition and knock.*

Test procedure for finding octane number (ON) of fuel :

To find ON of a fuel, the following *test procedure* is used :

- The test-engine is run at specified conditions using the fuel being tested. (The *specified/fixed conditions* to give maximum knock response are : Air inlet temperature, coolant temperature, engine speed, ignition advance setting, mixture strength etc.)
- *Compression ratio is adjusted until a standard level of knock is experienced.* (Intensity of knock is measured *with a magnetostriction knock detector*).
- The test fuel is then replaced with a mixture of the two standard fuels. The intake system of the engine is designed such that the blend of the two standard fuels can be varied to any percent from all iso-octane to all *n*-heptane.
- The blend of fuels is varied until the same knock characteristics are observed as with the test fuel.

  The percent of iso-octane in the fuel blend is the ON given to the test fuel. For instance, a fuel that has the same knock characteristics as a blend of 85% iso-octane and 15% *n*-heptane would have an ON of 85.

  The relationship between octane number and compression ratio is approximately shown in Fig. 9.19.

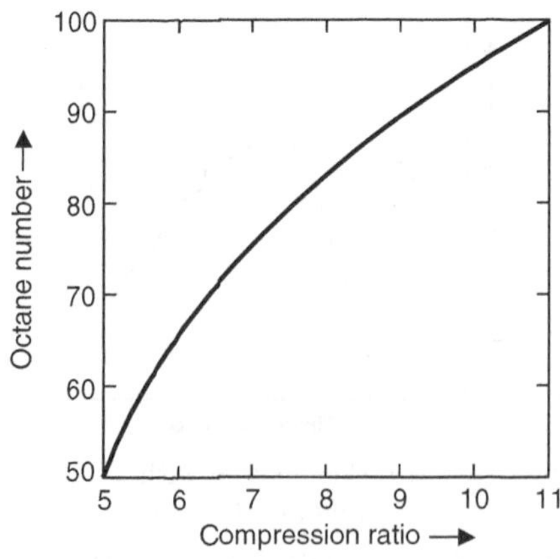

Fig. 9.19. Relationship between octane number and the highest useful compression ratio.

- There are several different tests used for rating octane numbers, each of which will give a slightly different ON value. The *two common methods* of rating gasoline and other automobile S.I. fuels are the **Motor method** and the **Research method.** These give the *motor octane number* (MON) and research octane number (RON). Another *less common method* is the **Aviation method** which is used for aircraft fuel and gives an Aviation Octane Number (AON).
  — The engine used to measure MON and RON was developed in the 1930s. It is a single cylinder, overhead valve engine that operates on the 4-stroke Otto cycle. It has a variable compression ratio which can be adjusted from 3 to 30.

Research Octave Number (RON) Method :

— This method measures anti-knock performance under *relatively mild operating condition.*

— It is considered to be similar to the detonation tendency of a fuel when the engine is *accelerating from 'low speed' in top gear with a wide open throttle under 'medium load'.* Inlet air temperature = 52°C ; coolant temperature = 100°C ; engine speed = 600 r.p.m. ; ignition advance setting = 13°C BTDC.

Motor octane number (MON) method :

— This method measures anti-knock performance under *relatively severe operating conditions.*

— It is considered to be similar to the detonation tendency of a fuel when the engine is driven at '*medium speed*' in top gear with a wide-open throttle under '*heavy load*'. Inlet air temperature = 150°C ; coolant temperature = 100°C ; engine speed = 900 r.p.m., ignition advance setting = 19–26° BTDC.

Fuel sensitivity (FS)

- The *difference in octane number between the Research method and the Motor method octane numbers is known as the fuel* ***sensitivity ;*** thus

$$\textit{Fuel sensitivity} = RON - MON$$

- Fuel sensitivity is a *good measure of how sensitive knock characteristics of a fuel will be to engine geometry.* A low FS number will usually mean that knock characteristics of that fuel are insensitive to engine geometry. FS numbers generally range from 0 to 10.

Antiknock index :

- The average of the two octane number rating methods, RON and MON, is very good antiknock quality indicator which is known as the **antiknock index** ; thus

$$\textbf{Antiknock index} = \frac{\text{RON} + \text{MON}}{2}$$

Advantages of high-octane fuel :

The *advantages of high-octane fuel* are as follows :

1. The engine can be operated at high compression ratio and therefore, with high efficiency without detonation.
2. The engine can be supercharged to high output without detonation.
3. Optimum spark advance may be employed raising both power and efficiency.

- High octane fuels (upto 100) can be produced by *refining techniques*, but it is done more cheaply, and more frequently, by the use of anti-knock *additives,* such *as tetraethyl lead.* (An addition of 1.1 cm^3 of tetraethyl lead to one litre of 80 octane petrol increases the octane number to 90). Fuels have been developed which have a higher anti-knock rating than iso-octane and this has lead to an extension of the octane scale.

9.2.9. Miscellaneous Properties of S.I. Engine Fuels

Miscellaneous properties of S.I. fuels are described below :

1. **Gum content**

- There is a tendency in some gasolines to deposit gum, a solid oxidation product, in fuel systems and on valve guides. Excessive gum formation often causes sticking of valves and plugging of fuel passages.
- The gum formation is reduced by mixing *inhibitors* (special chemicals) with gasoline.
- The *oxidised gasoline shows a loss of anti-knock quality.*

2. **Sulphur content**

The presence of sulphur content in gasolines is objectionable since it may lead to the formation of sulphuric acid in the presence of moisture. The sulphuric acid has corrosive effect on engine parts.

3. **Tetra-ethyl lead**

- It causes deposits on cylinder walls, spark plug and valves etc. which lead to the corrosion of spark plug and exhaust valves. These troubles are minimised by adding ethylene-dibromide ($C_2H_4Br_2$).
- It is a very dangerous poison acting on the skin and in vapour form, the lungs.

9.2.10. Diesel Fuel

- Diesel fuel (diesel oil, fuel oil) is obtainable over a large range of molecular weights and physical properties. It is classified by various methods, some using *numerical scales* and some designating it for *various uses.* Generally speaking, the *greater the refining done on a sample of fuel, the lower is its molecular weight, the lower is its viscosity, and the greater is its cost.*
  - — *"Numerical scales"* usually range from 1 to 5 or 6, with sub categories using alphabetical letters (*e.g.*, A1, 2D, etc.). *The lowest numbers have the lowest molecular weights and lowest viscosity. These are the fuels typically used in C.I. engines. Fuels with the largest numbers are very viscous and can only be used in large, massive heating units.* Each classification has acceptable limits set on various physical properties, such as viscosity, flask point, power point, cetane number, sulphur content etc.
  - — *Another method of classifying diesel fuel* to used in I.C. engines is to *designate it for its intended use.* These designations include, bus, truck, railroad, marine and stationary fuel, going from lowest molecular weight to highest.
- For convenience, diesel fuels for I.C. engines can be divided into two *extreme categories* :
  (*i*) Light diesel fuel (molecular weight 170 appr.)
  (*ii*) Heavy diesel fuel (molecular weight 200 appr.)

  Most diesel fuel used in engines will fit in this range.
  - — *Light diesel fuel* will be less viscous and easier to pump, will generally inject into smaller droplets, and will be more costly.
  - — *Heavy diesel* fuel can generally be used in larger engines with higher injection pressures and heated intake systems.

Often an automobile or light truck can use a less costly heavier fuel in the summer, but must change to a lighter, less viscous fuel in cold weather because of cold starting and fuel line pumping problems.

9.2.10.1. Cetane Number (CN)

- *The cetane number of a diesel fuel is a measure of its ignition quality.* When a fuel is injected into the hot compressed air in the cylinder, it must first be raised to a temperature high enough to ignite the air-fuel mixture. This requires a certain amount of time, known as *ignition delay.*

- Though ignition delay is affected by several engine design parameters such as compression ratio, injection rate, injection time inlet air temperature etc., it is also dependent on hydrocarbon composition of the fuel and to some extent on its volatility characteristic.
- *The cetane number is a numerical measure of the influence the diesel fuel has in determining the ignition delay.*
- *Higher the cetane rating of the fuel lesser is the propensity for diesel knock.*
- Ignition quality is usually determined by an **engine bench test** which *measures the ignition time delay under standard carefully controlled conditions.*
  - — In such a test, the unknown fuel is rated on a scale between 0 and 100 against a pair of pure hydrocarbon reference fuels. **Cetane** ($C_{16}H_{34}$) (*n*-hexadecane) a straight chain paraffin which has a *very high ignition quality (short delay)* and *does not readily knock*, is assigned to the top of the scale by a *cetane number of 100*, whereas **heptamethylnonane** (HMN) which has a very low *ignition quality (long delay)* and *readily knocks*, is represented at the bottom end of the scale by a cetane number of 15. *Originally,* the low ignition quality reference fuel was *alpha methyl napthalane* (C_{11} H_{10}) which was given a cetane number of zero. However, *heptamethylnonane, a, more stable compound but with a slightly better ignition quality (CN = 15), now replaces it.*

    Hence, the cetane number (CN) is shown by,

    Centane number (CN) = Percent centane

    0.15 × percent heptamethylnonane
  - — A standard single-cylinder pre-chamber variable compression ratio engine is used operating under fixed conditions : Inlet temperature = 65.5°C ; Jacket temperature = 100°C ; Speed = 900 r.p.m. ; Injection timing = 13° BTDC ; Injection pressure = 103.5 bar. The *engine is run on a supply of commercial fuel of unknown cetane number under standard operating conditions.* With the injection timing fixed to 13° BTDC, the compression ratio is varied until combustion commences at TDC (by observing the rapid rise in cylinder pressure) thereby producing a 13° delay period of 0.0024s at 900 r.p.m.. A *selection of reference fuel blends are then tested, where again the compression ratio is adjusted for each blend to obtain the standard 13° delay period. The percentage of cetane is one of the blends of reference fuels which gives exactly the same ignition delay (ignition quality) when subjected to the same compression ratio is called the* **cetane number** *of the fuel.* Thus, a commercial 40 cetane fuel would have an ignition delay performance equivalent to that of a blend of 40% cetane and 60% heptamethylnonane (HMN) by volume.
- *For higher speed engines the cetane number required is 50, for medium speed engine about 40, and for slow speed engines about 30.*
- Cetane number is the most important single fuel property which affects the exhaust emissions, noise and startability of a diesel engine. In general, *lower the cetane number higher are the hydrocarbon emissions and noise levels.* Low cetane fuels increase ignition delay so that start of combustion is near to top dead centre. This is similar to retarding of injection timing which is also known to result in higher hydrocarbon levels.
- In general, a *high octane value implies a low cetane value.*

- The relation between cetane number and delay period for a particular engine at a particular set of running conditions is illustrated in Fig 9.20.

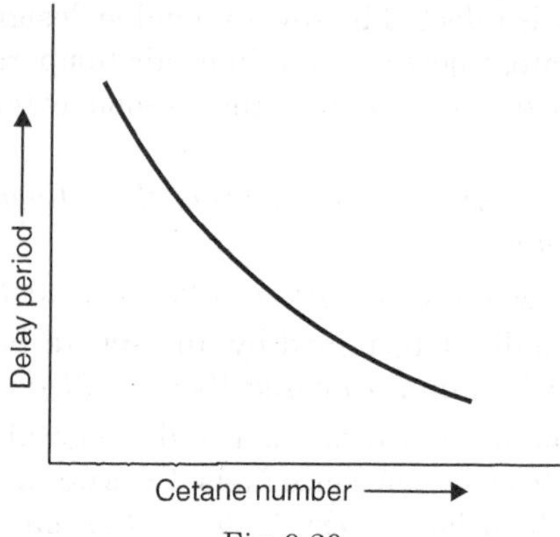

Fig. 9.20

- An approximate inverse relationship between cetane (CN) and octane (ON) number is given as :

$$CN = 60 - \frac{ON}{2}$$

(accurate within ± 5%)

9.2.10.2. Diesel Index (DI)

- The diesel index is a *cheap method of predicting ignition quality.*
- This scale is made possible because ignition quality is sensitive to hydrocarbon compositions ; that is, *paraffins have high ignition quality and aromatic and compounds have low ignition quality.*
- The *diesel index gives an indication of the ignition quality obtained from certain physical characteristics of the fuel as opposed to an actual determination in a test engine.* The index derived from the knowledge of aniline point American Petroleum Institute (API) gravity, which is put together as follows :

$$\text{Diesel index (DI)} = \text{Aniline point } (°F) \times \frac{\text{API gravity (deg)}}{100}$$

— The **aniline point of the fuel** is *the temperature at which equal parts of fuel and pure aniline dissolve in each other.* It therefore gives an indication of the chemical composition of the fuel since the more paraffinic the fuel the higher the solution temperature. Likewise, a higher API gravity reflects a low specific gravity and indicates a high paraffinic content, which corresponds to a good ignition quality.

Note. The correlation between the diesel-index and cetane number is *not exact* and *with certain fuel consumption it is not reliable but, nevertheless, it can be a useful indicator for estimating ignition quality.*

9.3. ALTERNATIVE FUELS FOR I.C. ENGINES

9.3.1. General Aspects

- The *crude oil and petroleum products,* sometimes during the 21st century will *become very scarce and costly to find and produce.* At the same time, there will likely be an increase in the number of automobiles and other I.C. engines. Although fuel economy of engines is greatly improved from the past and will probably continue to be improved, numbers only dictate that there will be a *great demand for fuel in the coming decades. Gasoline will become scarce and costly. Alternate fuel technology, availability, and use must and will become more common in the coming decades.*

- Although there have always been some I.C. engines fueled with non-gasoline or diesel oil fuels, there numbers have been relatively small. Because of the high cost of petroleum products, some third-world countries have for many years been using *manufactured alcohol as their main vehicle fuel.*
- Several pumping stations on natural gas pipelines use the pipeline gas to fuel the engines driving the pumps. This solves an otherwise complicated problem of delivering fuel to the pumping stations, many of which are in very isolated regions. Some large displacement engines have been manufactured especially for pipeline work. These consist of a bank of engine cylinders and a bank of compressor cylinders connected to the same crankshaft and contained in a single engine block similar to a V-style engine.
- *Another reason motivating the development of alternative fuels for the I.C. engine is concern over the emission problems of gasoline engines.* Combined with other air polluting systems, the large number of automobiles is a major contributor to the air quality problem of the world. Vast improvements have been made in reducing emissions given off by an automobile engine.
- Still another reason for alternative fuel development in India and other industrialised countries is the fact that a large percentage of crude oil must be imported from other countries which control the large oil fields.

Some **alternative fuels** *which can replace conventional fuels in I.C. engines are :*

(*i*) Alcohol (Methyl and ethyl)

(*ii*) Hydrogen

(*iii*) Natural gas

(*iv*) LPG (Liquified petroleum gas) and LNG (Liquified natural gas)

(*v*) Biogas.

9.3.2. Advantages and Disadvantages of using Alternative Fuels

Advantages of using alternative fuels :

1. Alcohols can be produced from highly reliable and long-lasting raw material sources like sugarcane, starchy materials, corns, potatoes etc. Thus they are renewable energy sources.
2. Biogas plants can be conveniently and economically installed in village and farms.
3. Manufacture of biogas from cow-dung will give, as a byproduct, manure of very high quality for use in the-farms.
4. Natural gas is available in plenty.
5. Coal gas is produced by coal gasification, and coal is abundant to last much longer than the liquid petroleums.
   — *Alcohols, biogas and natural gas have much higher octane number, and are suitable for use in S.I. engines with little modification. There is less problem of knocking or detonation. Higher compression ratios can be used to give more power and increased thermal efficiency.*
6. The exhaust from engines using gaseous fuels contains less pollutants. Use of hydrogen gives absolutely clean exhaust.

Disadvantages :

Use of alcohols as an alternative fuels have the following *disadvantages :*

1. Social problems due to prohibition, as it can be consumed as liquor by human beings.

2. The carburettor would *need modification*, as the stoichiometric *Air-fuel ratio with alcohols is quite low* (of the order of 10 : 1)
3. The calorific value of alcohol fuels is *low* as compared to that of diesel or petrol.
   — The *m.e.p.* and power output from a given size engine will be low with coal *gas and biogas.*
   — The handling and transportation of natural gas is costly.
   — High compressor power is required to compress it for storage otherwise it will need large storage space.
   — Hydrogen is highly explosive, and its handling is risky.

9.3.3. Alcohol

Alcohol is an attractive alternative fuel because it can be obtained from a number of sources, both natural and manufactured. *Methanol* (methyl alcohol) and *ethanol* (ethyl alcohol) are two kinds of alcohol that seem most promising and have had the most development as engine fuel.

Advantages :

The advantages of alcohol as a fuel are :

1. It is high octane fuel with anti-knock index numbers (octane number on fuel pump) of over 100. High octane numbers result, at least in part, from the high flame speed of alcohol.
   — Engines using high-octane fuel can run more efficiently by using higher compression ratios.
2. It can be obtained from a number of sources, both natural and manufactured.
3. It has high evaporative cooling (h_{fg}) which results in *cooler intake process and compression stroke.* This *raises the volumetric efficiency of the engine and reduces the required work input in the compression stroke.*
4. Generally *less overall emissions* when compared with gasoline.
5. Low sulphur content in the fuel.
6. When burned, it forms *more moles of exhaust which gives higher pressure and more power in the expansion stroke.*
7. The contamination of matter in alcohols is less dangerous than petrol or diesel because alcohols are less toxic to humans and has a recognizable taste.

Disadvantages :

1. *Low energy content* of the fuel (Almost twice as much alcohol as gasoline must be burned to give the same energy input to the engine).
2. The exhaust contains more aldehydes. If as much alcohol fuel was consumed as gasoline, *aldehyde emissions would be a serious exhaust pollutions problem.*
3. As compared to gasoline, alcohol is much more corrosive on copper, brass, aluminium, rubber and many plastics. This puts some restriction on the design and manufacturing of engines to be used with this fuel.
   — Methanol is very corrosive on metals.
4. In general, the ignition characteristic are poor.
5. Vapour lock in fuel delivery system.
6. Owing to low vapour pressure and evaporation, the cold weathering starting characteristics are poor.
7. Due to low vapour pressure, there is a danger of storage tank flammability. Air can leak into storage tanks and create a combustible mixture.

8. Alcohols have almost invisible flames, which is considered dangerous when handling fuel. Again, a small amount of gasoline remove this danger.
9. Low flame temperatures generate less NO_x, but the resulting lower exhaust temperatures take longer to heat the catalytic converter to an efficient operating temperature.
10. When refuelling an automobile, headaches and dizziness have been experienced (due to the strong odour of alcohol).

Note :

- Alcohols are considered as clean burning renewable alternative fuels which can come to our rescue to meet the a challenge of vehicular fuel oil scarcity and fouling of environment by exhaust emissions.
- Alcohols make very poor diesel engine fuels as their 'cetane number' is considerably lower.
  - Alcohols can be used in duel fuel engines or with assisted ignition in diesel engine. In a duel fuel mode, alcohol is inducted along with the air, compressed and then ignited by a pilot spray of diesel oil.

9.3.3.1. Methanol

- Of all the fuels being considered as an alternate to gasoline, methanol is one of the more promising and has experienced major research and development.
- Methanol can be obtained from many sources, both *fossil* and *renewable*. These include *coal, petroleum, natural gas, biomass, wood, landfills, and even the ocean.* However, any source that requires extensive manufacturing or processing raises the price of the fuel and requires an energy input back into the overall environmental picture, both unattractive.
- Methanol beleaves much like petroleum and so, it can be stored and shifted in the same manner.
  - It is more flexible fuel than hydrocarbon fuels permitting wider variation from ideal A/F ratios.
  - It has relatively good lean combustion characteristics compared to hydrocarbon fuels. *Its wider inflammability limits and higher flame speeds have showed higher thermal efficiency and lesser exhaust emissions compared with petrol engines.*
- Depending on gasoline-methanol mixture, some changes in fuel supply are essential. Simple modifications to the carburettor or fuel injection can allow methanol to replace petrol easily.

Some important features of methanol as fuel :

1. The specific heat consumption with methanol as fuel is 50 percent less than petrol engine.
2. Exhaust CO and HC are decreased continuously with blends containing higher and higher percentage of methanol. But exhaust aldehyde concentration shows a reversed trend.
3. Methanol can be used as supplementary fuel in heavy vehicles powered by C.I. engines with consequent saving in diesel oil and reduced exhaust pollution.

Advantages of methanol :

1. Owing to its *excellent anti-knock characteristics, it is much suitable for S.I. engines.*
2. Methanol use maintains good air quality (Methanol emits less amount of CO_2 and other polluting gases as compared to gasoline fueled vehicles).

3. Tertiary butial alcohol is used as an octane improving agent.
4. One percent methanol in petrol is used to prevent freezing of fuel in winter.
5. Iso-propyl alcohol is used as anti-icing agent in carburettor.
6. Addition of methanol causes methanol gasoline blend to evaporate at much faster rate than pure gasoline below its boiling point.

Performance of methanol as I.C. engine fuel :

The performance characteristics are considered as :

— Thermal efficiency verses A/F ratio ;

— Effect of speed on power output and specific heat consumption with petrol engine ;

— Effect of A/F ratio on exhaust emission.

- Fig. 9.21 (*a*) shows the effect of A/F ratio and speed on brake power.

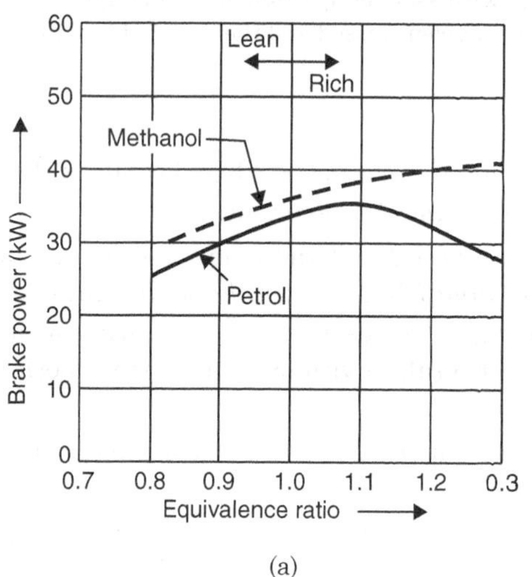

(a)

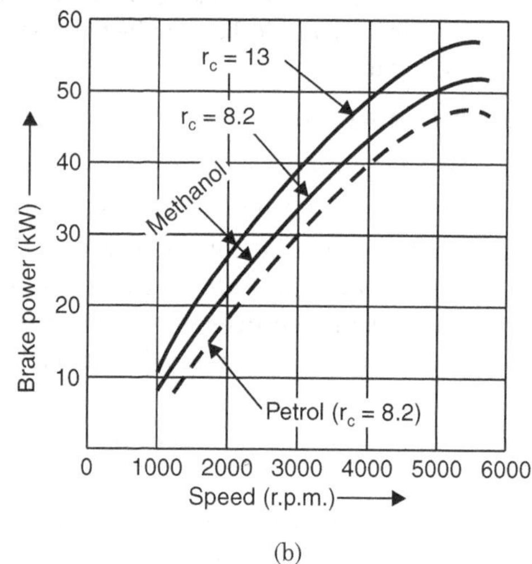

(b)

Fig. 9.21 Effect of A/F ratio and speed on brake power.

- Fig. 9.22 shows the effect of load on specific heat consumption.
- Fig. 9.23 shows the effect of speed on volumetric efficiency.

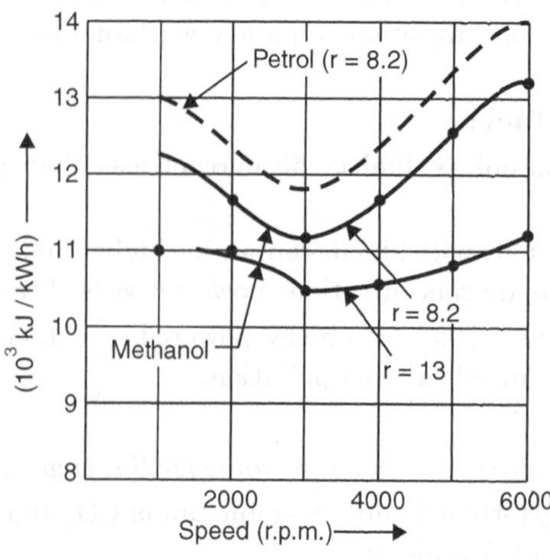

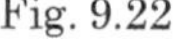
Fig. 9.22

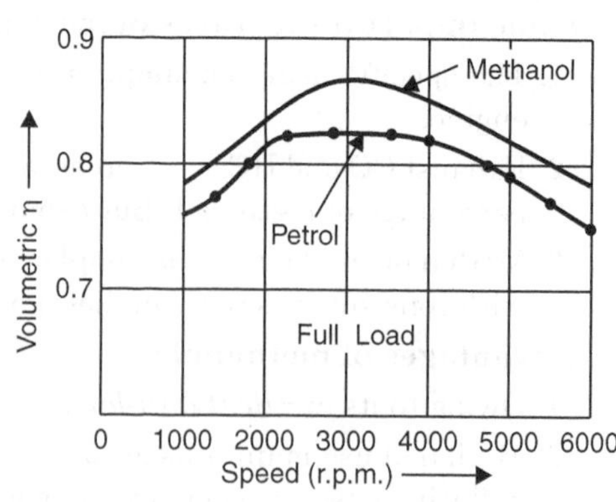

Fig. 9.23

- Fig. 9.24 to 9.27 show the effect of equivalence ratio on all important objectionable emissions, CO, HC, NO_x and aldehydes respectively, for Petrol and Methanol.

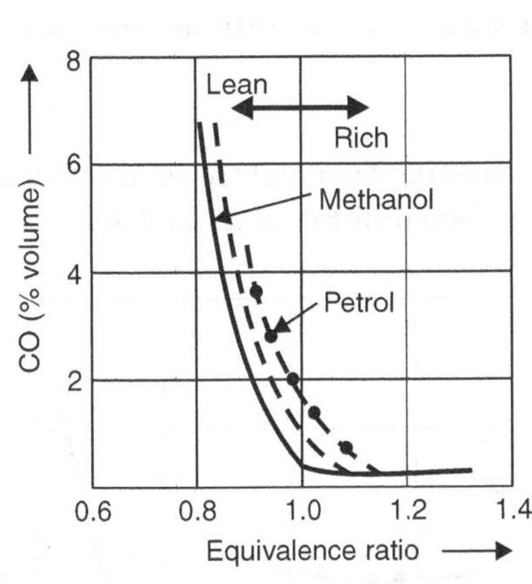

Fig. 9.24

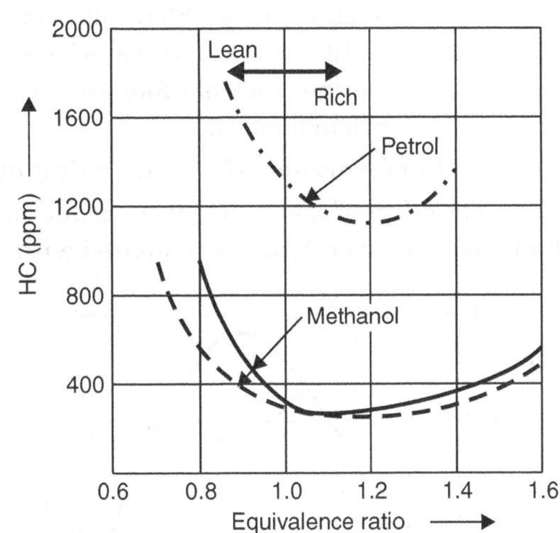

Fig. 9.25

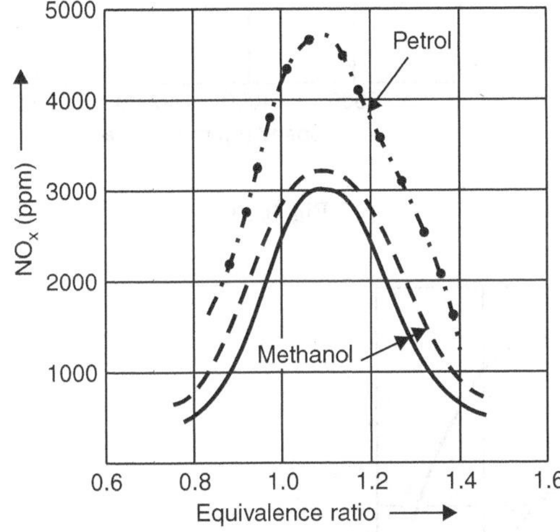

Fig. 9.26

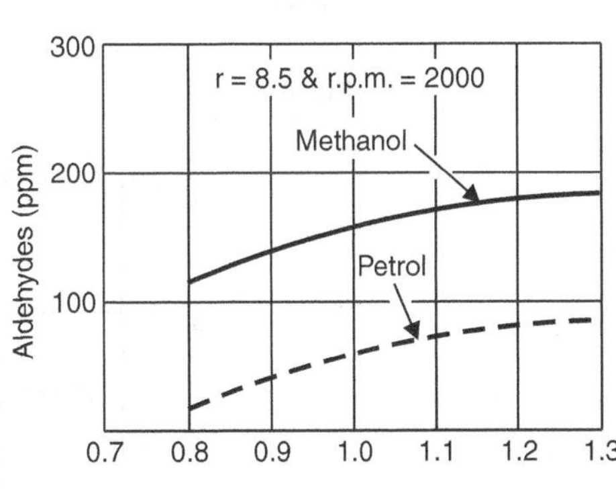

Fig. 9.27

9.3.3.2. Ethanol

- Ethanol can be *made from ethylene or from fermentation of grains and sugar. Much of it is made from corn, sugar beets, sugarcane, and even cellulose (wood and paper).* The present cost of ethanol is *high due to the manufacturing and processing required.* This would be reduced if large amounts of this fuel are used.
- Ethanol has less HC emissions than gasoline but more than methanol.
- Gasohol is a mixture of 90% gasoline and 10% ethanol. As with methanol, the development of systems using mixtures of gasoline and ethanol continues.

— Two mixture combinations that are important are E 85 (85% ethanol) and E10 (gasohol). E 85 is basically an alcohol fuel with 15% gasoline added to eliminate some of the problems of pure alcohol (*i.e.*, cold starting, tank flammability, etc.). E 10 reduces the use of gasoline with no modification needed to the automobile engine. Flexible-fuel engines are being tested which can operate on any ratio of ethanol-gasoline.

Performance of engine using ethanol :

The effect of speed on power output, brake specific heat consumption and thermal efficiency of an engine using ethanol is compared with gasoline engine as shown in Fig. 9.28 to 9.30.

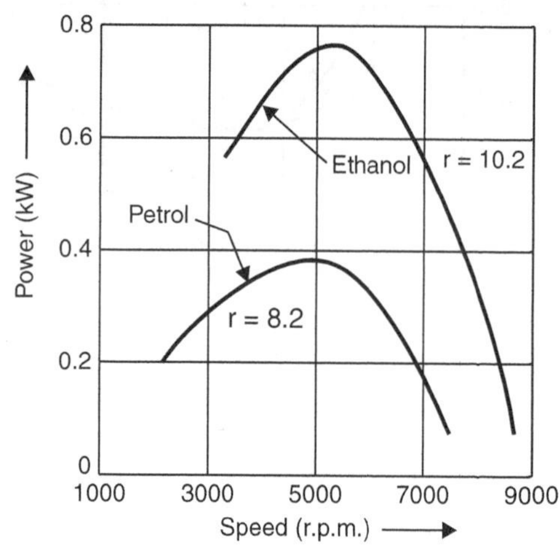

Fig. 9.28

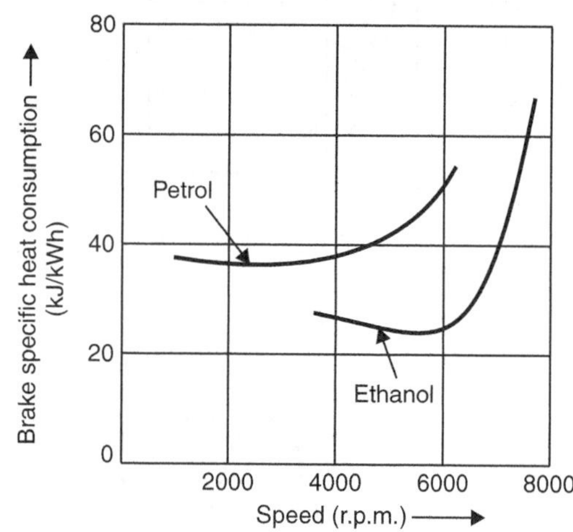

Fig. 9.29

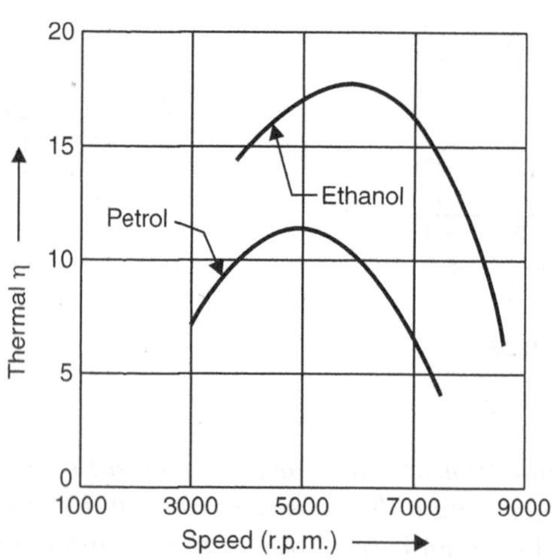

Fig. 9.30

- The *power output* of ethanol engine is *higher* compared to gasoline engine at all speeds.
- The *brake specific heat consumption* is *improved* with ethanol engine compared with petrol engine.

- The *maximum thermal efficiency* of ethanol engine is *higher* than petrol engine. The efficiency curve of ethanol engine is flat for a wide range of speed which indicates that the part load efficiency is much compared with petrol engine.
- The *engine torque* is *considerably higher* for ethanol as compared to petrol engine.

9.3.3.3. Properties of Methanol and Ethanol

The physical and chemical properties of gasoline, methanol and ethanol are listed in the Table 9.3.

Table 9.3. Important properties of gasoline, methanol, ethanol

S. No.	*Property*	*Gasoline nearly C_8H_{18} (Iso-Octane)*	*Methyl alcohol (CH_3OH)*	*Ethyl-alcohol (C_2H_5OH)*
1.	Molecular weight	114.2	32	46
2.	Boiling point at 1 bar °C	43 to 170	66	78
3.	Freezing point °C	– 107.4	– 161.8	–117.2
4.	Specific gravity (150°C)	0.72 to 0.75	0.79	0.79
5.	Latent heat (kJ/kg)	400	1110	900
6.	Viscosity (centipoise)	0.503	0.596	0.60
7.	Stoichiometric A : F (ratio)	14.6	6.45	9
8.	Mixture heating value (kJ/kg) (for stoic-mixture)	2930	3070	2970
9.	Ignition Limits (A/F)	8 to 19	2.15 to 12.8	3.5 to 17
10.	Self ignition Temp.	335	574	557
11.	Octane Number			
	(*a*) Research	80 to 90	112	111
	(*b*) Motor	85	91	92
12.	Cetane Number	15	3	8
13.	Lower C.V. (kJ/kg)	44100	19740	26880
14.	Vapour pressure at 38°C (bar)	0.48 to 1	0.313	0.17
15.	Flame speed (m/sec)	0.43	0.76	—
16.	Auto-ignition temperature (°C)	222	467	—

9.3.4. Alcohol-Gasoline Fuel Blends

- Normally straight alcohols are not used in automobile engines except methanol in racing cars. The alcohols can be used as blend with gasoline as this has the advantages that the *existing engines not be modified and TEL (Tetraethyl lead) can be eliminated from gasoline due to the Octane enhancing quality of alcohol.*
- If *the engine is to be operated using only pure alcohol, then some major modifications are required in the engine and fuel system as listed below :*
  - (*i*) The materials used with alcohols have to be changed since both alcohols are corrosive to many of the materials that are used with gasoline.
  - (*ii*) Adjustment of carburettor and fuel injection system to compensate for leaning effect.
  - (*iii*) Introduction of high energy ignition system with lean mixture.
  - (*iv*) Alteration in fuel pump and circulation system to avoid vapour lock as methanol vapourisation rate is very high.
  - (*v*) Increase in compression ratio to make better antiknock properties of the fuel.
  - (*vi*) Addition of detergent and volatile primers to reduce engine deposits and assit in cold starting.

(*vii*) Use of cooler running spark plugs for avoiding pre-ignition.

- The *blends* have the following important *advantages* over pure ethanol :
  (*i*) The engine can be started easily.
  (*ii*) No abnormal corrosion compared with pure ethanol.
  (*iii*) Lubrication in petrol and alcohol blend is more or less same.
  (*iv*) Some benzene is added to prevent the separation of layers of petrol and alcohol.

When blends are used, the following minor modifications in engine are required :

(*i*) The carburettor jet needs to be increased to increase the flow 1.56 times that of petrol.

(*ii*) The float is to be weighed down, to correct level due to higher specific gravity.

(*iii*) Modification of air inlet to get less air as blend requires less air for complete combustion than petrol.

(*iv*) Provision of a specific arrangement of heating the carburettor and intake manifold as lower vapour pressure of alcohol makes the starting difficult below 70°C.

- Fig. 9.31 shows the change in distillation by admixture of methanol and ethanol.
- Fig. 9.32 shows the leaning effect of alcohol admixture to petrol.
- Fig. 9.33 shows the antiknock quality of methanol-gasoline blends.

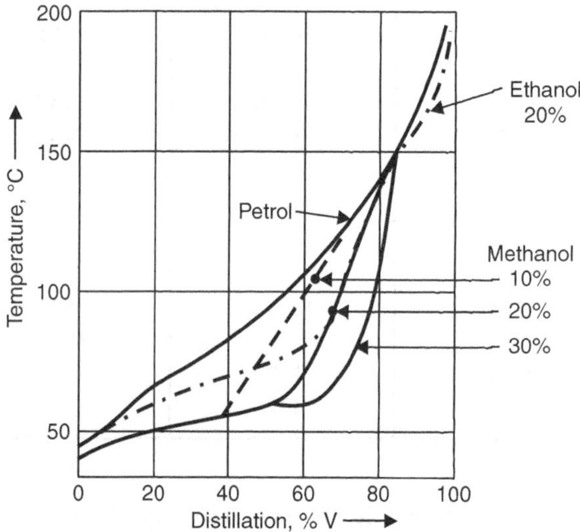

Fig. 9.31. Change in distillation by admixture of methanol and ethanol.

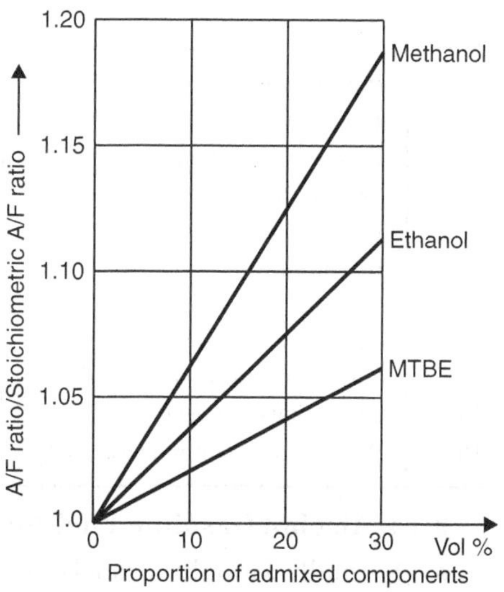

Fig. 9.32. Leaning effect of alcohol admixture to pertol.

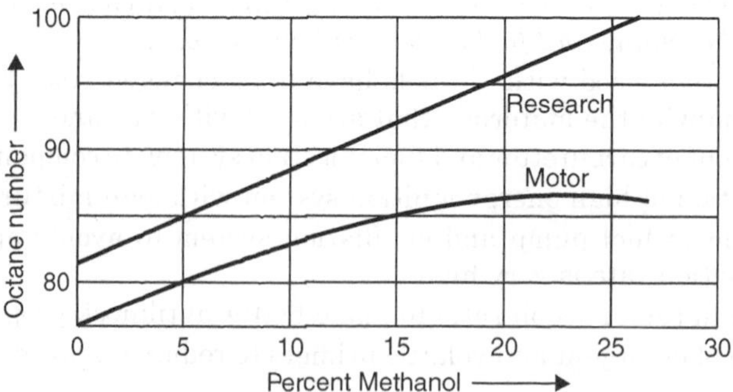

Fig. 9.33. Antiknock quality of methanol-gasoline blends.

9.3.5. Hydrogen

A number of companies have built automobiles with prototype or modified engines which operate on hydrogen.

— A H_2-powered car being developed in Melbourne University Department of Mechanical Engineering is achieving 40 percent energy saving over conventional petrol engine. The car is a converted Ford Cortina Wagon which carries enough fuel in 4-cylinders to travel upto 50 km at a speed of 130 km/h.

— German cars are so developed that they can be converted for driving either gasoline or H_2 by making a few simple adjustments. Presently a few vehicles are running on road as there are very few public filling stations for liquid H_2. Many more are planned for future.

Advantages of hydrogen as I.C. engine fuel :

1. *Low emissions.* Essentially no CO or HC in the exhaust as there is no carbon in the fuel. Most exhaust would be H_2O and N_2.
2. *Fuel availability.* There are a number of different ways of making hydrogen, including electrolysis of water.
3. Fuel leakage to environment is *not a pollutant.*
4. High energy content per volume when stored as liquid. This would give a large vehicle range for a given fuel tank.
5. Hydrogen-air mixture burns *ten-times faster* compared to gasoline-air mixture. Since the burning rate is considerably high, it is *more preferred in high speed engines.*
6. Hydrogen-ignition limits are much wider than gasolines. So it can burn easily and *give considerably higher efficiency.*
7. Hydrogen has high self-ignition temperature (S.I.T.) but very *little energy* (1/50 th of gasoline) is *required to ignite it.*
8. The exhaust heat can be used to *extract H_2 from the hybride reducing the load engine.*
9. Besides being a relatively clean burning renewable source, H_2 as I.C. engine fuel is *very efficient* as there are *no losses associated with throttling.*

Disadvantages :

1. The *handling of H_2 is more difficult* and *storage requires high capital and running cost particularly for liquid H_2.*
2. *Difficult to refuel.*
3. *Poor engine volumetric efficiency.* Any time a gaseous fuel is used in an engine, the fuel will displace some of the inlet air and poorer volumetric efficiency will result.
4. Fuel cost would be high at present day technology and availability.
5. Can detonate.
6. High NO_x emissions because of high flame temperature.
7. *In hydrogen engines there is a danger of back fire and induction ignition which can melt the carburettor. Therefore in H_2-fuel system, flame traps, flask back arresters are necessary. Additionally, crankcases must be vented to prevent accumulation of explosive mixtures.*

9.3.6. Natural Gas (Methane)

- *Natural gas* is a mixture of components, consisting mainly of **methane** (60–98%) with *small amounts of other hydrocarbon fuel components.* In *addition it contains various amounts of N_2, CO_2, He and traces of other gases.* Its sulphur content ranges from very little (sweet) to larger amounts (sour).

An *ideal composition of CNG* as an automative fuel is as follows :

Methane = 90% (minimum) ; *Ethane content* = 4% (maximum) ; *Propane content* = 1.7% (maximum) ; C_4 *and higher* = 0.7% (maximum) ; C_6 *and higher* = 0.2% (maximum) ; $(CO_2 + N_2)$ = 0.2% (maximum) ; *Hydrogen* = 0.1% (maximum) ; *Carbon monoxide* = 0.1% (maximum) ; *Oxygen* = 0.5% (maximum) ; *Sulphur* = 10% ppm (maximum).

- It is stored as **Compressed Natural Gas (CNG)** at pressures of 7 to 21 bar and a temperature around –160°C.
- *As a fuel it works best in an engine system with a single-throttle body fuel injector. This gives a longer mixing time, which is needed by this fuel.*
- Tests using CNG in various sized vehicles continue to be conducted by government agencies and private industry.

Properties of CNG :

The properties of CNG are almost similar to that of methane :

- Methane has very good antiknock qualities which means it *does not ignite readily. Antiknock Octane number of CNG is nearly 130, so it burns at much higher temperature compared with petrol unleaded (Octane No = 95) and diesel which have low octane number.*
- Owing to better antiknock quality of CNG it can be safely used in engines with a compression ratio as high as 12 : 1 compared with petrol (maximum 10 : 1).
  — *The CNG fuel used engines have higher thermal efficiencies* than *those fuelled by gasoline.* In addition to this, the reduction in the pollutants emitted by CNG engine is noticeable.
- CNG is *non-toxic* and *lighter than air so when leakage occurs it quickly disappears* unlike gasoline which paddles and evaporates.
- The presence of ethane and propane even in small percentages (5% and 2%, respectively) affect the burning properties of CNG. Both the gases try to lower the Octane characteristics and causes pre-ignition and reduced fuel efficiency.

Advantages of CNG :

(*i*) *High octane number* makes it a *very good S.I. engine fuel.*
(*ii*) *Low engine emissions.* Less aldehydes than with methanols.
(*iii*) It is cheap (It costs about 25 to 50% less than gasoline and more than 50% less than other alternative fuels, such as methanol and ethanol.
(*iv*) It is engine friendly.
(*v*) It is safe in operation.
(*vi*) Fuel fairly abundant world-wide. Natural gas is the second most abundant fuel available in India after coal.
(*vii*) Easy to tap.
(*viii*) It is odourless.
(*ix*) It is clean.

Disadvantages of CNG :

(*i*) Low energy density resulting in low engine performance.
(*ii*) Low engine volumetric efficiency because it is a gaseous fuel.
(*iii*) Need for large pressurised fuel storage tank.
(*iv*) Inconsistent fuel properties.
(*v*) Refueling is a slow process.
(*vi*) The storage cylinder takes a lot of space as the gas once filled has to travel at least of 400 km. But now a days there are byfuel and duel-fuel engines which can run on CNG and other fuel.

9.3.7. LPG and LNG

- **LPG** (*Liquified Petroleum Gas*) is mainly *propane* but may also contain a small proportion of butane and possibly, some ethane and a little pentane in heavier vapour form. The heavier fractions tend to occur in LPG produced by distillation of crude oil.
  - — Propane has a higher octane number, burns more clearly and saves on maintenance costs.
  - — Propane is gaining as a gasoline substitute because it costs 60% of petrol and gives 90% mileage of its fellow gasoline.
- **LNG** (*Liquified Natural Gas*) comes from dry natural reservoirs *mainly CH_4* with very small percentages of ethane and propane.
  - — The major difficulty encountered in the use of this gas is its boiling temperature − 161.5°C.

9.3.8. Biogas

9.3.8.1. Introduction

- The biogas is generally produced from by dung from different beasts as **cow**, *buffalo, goat, sheep, horse, donkey* and *elephant*. Some other sources are :

  (*i*) Sewage (*ii*) Crop residue
  (*iii*) Vegetable wastes (*iv*) Water hyacinth
  (*v*) Alga (*vi*) Poultry droppings
  (*vii*) Pig manure (*viii*) Ocean kelp.

- Biogas is produced by *digestion, pyrolysis* or *hydrogasification*. **Digestion** is a biological process that occurs in absence of oxygen and in the presence of anaerobic organism at ambient pressures and temperatures of 35–70°C. The container in which digestion takes place is known as the *digester*. Biogas plants have been built in various designs.

9.3.8.2. Composition and Properties of Biogas

- Its main combustible component is CH_4 and another major component is *CO_2 which reduces its octane number*. The components of biogas with composition are given below :

Component	*Composition (% volume)*
CH_4	50—60
CO_2	30—45
H_2 and N_2	5—10
H_2S and O_2	Traces

Octane rating : 110 with CO_2
130 without CO_2

- Biogas *possesses excellent antiknock properties* with an equivalent Octane number in excess of 120 compared with 87 for regular petrol.
- Its auto-ignition temperature is higher than petrol which makes it a *safer fuel.*
- Being a gas it mixes readily with air even at low temperature, therefore, there is no need to provide rich mixture during starting or idling.
- Although its calorific value is lesser than petrol, it is possible to use higher compression ratio for the same size engine thus making it possible to generate the same amount of power.

Use of biogas in S.I. engines :

S.I. engines can be operated on biogas after starting the engine by using petrol. Biogas can be used in these engines in *two forms* as :

(*i*) To run the engine entirely on biogas.

(*ii*) Duel fuel engine where engine can run on both fuels. (This arrangement is preferred these days).

9.3.8.3. Advantages of using biogas as fuel in C.I. engine

The biogas can be used in C.I. engine as a duel fuel and improves engine performance.

The following are the *advantages* of using biogas as fuel in C.I. engine :

(*i*) A uniform gas-air mixture is available in multi-cylinder engine at all times.

(*ii*) Due to clean operation of the engine there is virtually no CO-emission in exhaust.

(*iii*) When biogas is used as a fuel, NO_x emissions are reduced by about **60** per cent.

(*iv*) Soot is virtually eliminated and exhaust is found to have less pungent odour than that obtained while operating the engine with diesel oil.

WORKED EXAMPLES

Air-Fuel Ratio and Analysis of Products of Combustion

Example 9.1. *Calculate the amount of theoretical air required for the combustion of 1 kg of acetylene (C_2H_2) to CO_2 and H_2O.*

Solution. For combustion of acetylene (C_2H_2) the stoichiometric equation is written as

$$C_2H_2 + xO_2 \longrightarrow a\,CO_2 + b\,H_2O \qquad ...(i)$$

Balancing the *carbon atoms* on both sides of the combustion equation (*i*), we get

$$2C = a\,C \quad i.e. \quad a = 2$$

Now balancing *hydrogen atoms* on both sides, we get

$$2H = 2bH$$
$$b = 1$$

Thus, equation (*i*) becomes

$$C_2H_2 + x\,O_2 \longrightarrow 2CO_2 + H_2O$$

Now, balancing *oxygen atoms* in the above equation

$$2x = 2 \times 2 + 1 = 5 \quad i.e. \quad x = 2.5$$

Hence, the final combustion equation (*i*) is

$$C_2H_2 + 2.5O_2 \longrightarrow 2CO_2 + H_2O \qquad ...(ii)$$

Thus, for combustion of C_2H_2 in air, we get

$$C_2H_2 + 2.5O_2 + 2.5\left(\frac{79}{21}\right)N_2 \longrightarrow 2CO_2 + H_2O + 2.5\left(\frac{79}{21}\right)N_2$$

On a mass basis, this becomes

$$(2 \times 12 + 2 \times 1)\,C_2H_2 + 2.5\,(2 \times 16)\,O_2 + 2.5\left(\frac{79}{21}\right)(2 \times 14)\,N_2$$

$$\longrightarrow 2\,(12 + 2 \times 16)\,CO_2 + (1 \times 2 + 1 \times 16)\,H_2O + 2.5\left(\frac{79}{21}\right)(2 \times 14)\,N_2$$

$$26\text{ kg } C_2H_2 + 80\text{ kg } O_2 + 263.3\,N_2 \longrightarrow 88\text{ kg } CO_2 + 18\text{ kg } H_2O + 263.3\text{ kg } N_2$$

or $\quad 1\text{ kg } C_2H_2 + 3.076\text{ kg } O_2 + 10.12\text{ kg } N_2 \longrightarrow 3.38\text{ kg } CO_2 + 0.69\text{ kg } H_2O + 10.12\text{ kg } N_2$

i.e., Amount of air = 3.076 + 10.12 = 13.196 kg of air per kg of C_2H_2

Hence amount of theoretical air required for combustion of 1 kg acetylene = 13.196 kg. (Ans.)

Example 9.2. *Determine the gravimetric analysis of the products of complete combustion of acetylene with 200 per cent stoichiometric air.*

Solution. The Stoichiometric air equation (Example 9.1) is written as :

$$C_2H_2 + 2.5\ O_2 + 2.5\left(\frac{79}{21}\right) N_2 \longrightarrow 2CO_2 + H_2O + 2.5\left(\frac{79}{21}\right) N_2$$

If *200 per cent* stoichiometric air is used, the combustion equation for C_2H_2 becomes

$$C_2H_2 + (2)\ (2.5)\ O_2 + (2)\ (2.5)\left(\frac{79}{21}\right) N_2 \longrightarrow 2CO_2 + H_2O + (2)\ (2.5)\left(\frac{79}{21}\right) N_2 + (1)\ (2.5)\ O_2$$

$$(2 \times 12 + 2 \times 1) \text{ kg } C_2H_2 + (2)\ (2.5)\ (2 \times 16) \text{ kg } O_2 + (2)\ (2.5)\left(\frac{79}{21}\right)(2 \times 14) \text{ kg } N_2$$

$$\longrightarrow 2\ (12 + 2 \times 16) \text{ kg } CO_2 + (2 \times 1 + 16) \text{ kg } H_2O + (2)\ (2.5)\left(\frac{79}{21}\right)(2 \times 14) \text{ kg } N_2 + (1)\ (2.5)\ (2 \times 16) \text{ kg } O_2$$

or $\quad 26 \text{ kg } C_2H_2 + 160 \text{ kg } O_2 + 526.6 \text{ kg } N_2 \longrightarrow 88 \text{ kg } CO_2 + 18 \text{ kg } H_2O + 526.6 \text{ kg } N_2 + 80 \text{ kg } O_2$

or $\quad 1 \text{ kg } C_2H_2 + 6.15 \text{ kg } O_2 + 20.25 \text{ kg } N_2 \longrightarrow 3.38 \text{ kg } CO_2 + 0.69 \text{ kg } H_2O + 20.25 \text{ kg } N_2 + 3.07 \text{ kg } O_2$

Thus for 1 kg of fuel, the products will consist of

$$CO_2 = 3.38 \text{ kg}$$
$$H_2O = 0.69 \text{ kg}$$
$$O_2 = 3.07 \text{ kg}$$
$$N_2 = 20.25 \text{ kg}$$

Total mass of products $= 27.39$ kg

∴ Mass fractions are :

$$CO_2 = \frac{3.38}{27.39} = 0.123$$

$$H_2O = \frac{0.69}{27.39} = 0.025$$

$$O_2 = \frac{3.07}{27.39} = 0.112$$

$$N_2 = \frac{20.25}{27.39} = 0.739.$$

Hence the gravimetric analysis of the complete combustion is :

CO_2 = 12.3%, H_2O = 2.5%, O_2 = 11.2%, N_2 = 73.9%. (Ans.)

Example 9.3. *Calculate the theoretical air-fuel ratio for the combustion of octane, C_8H_{18}. The combustion equation is :*

$$C_8H_{18} + 12.5\ O_2 + 12.5\left(\frac{79}{21}\right) N_2 \longrightarrow 8CO_2 + 9H_2O + 12.5\left(\frac{79}{21}\right) N_2.$$

Solution. The air-fuel ratio on a mole basis is

$$A/F = \frac{12.5 + 12.5\left(\frac{79}{21}\right)}{1} = 59.5 \text{ mol air/mol fuel}$$

The theoretical air-fuel ratio on a mass basis is found by introducing the molecular weight of the air and fuel

$$A/F = \frac{59.5\,(28.97)}{(8 \times 12 + 1 \times 18)} = \textbf{15.08 kg air/kg fuel.} \quad \textbf{(Ans.)}$$

Example 9.4. *One kg of octane (C_8H_{18}) is burned with 200% theoretical air. Assuming complete combustion determine :*

(*i*) *Air-fuel ratio*

(*ii*) *Dew point of the products at a total pressure 100 kPa.*

Solution. The equation for the combustion of C_8H_{18} with theoretical air is

$$C_8H_{18} + 12.5\,O_2 + 12.5\left(\frac{79}{21}\right)N_2 \longrightarrow 8CO_2 + 9H_2O + 12.5\left(\frac{79}{21}\right)N_2$$

For *200% theoretical air* the combustion equation would be

$$C_8H_{18} + (2)\,(12.5)\,O_2 + (2)\,(12.5)\left(\frac{79}{21}\right)N_2$$

$$\longrightarrow 8CO_2 + 9H_2O + (1)\,(12.5)\,O_2 + (2)\,(12.5)\left(\frac{79}{21}\right)N_2$$

Mass of fuel = (1) (8 × 12 + 1 × 18) = 114 kg/mole

$$\text{Mass of air} = (2)\,(12.5)\left(1 + \frac{79}{21}\right) 28.97 = 3448.8 \text{ kg/mole of fuel}$$

(*i*) **Air-fuel ratio :**

$$\text{Air-fuel ratio, A/F} = \frac{\text{Mass of air}}{\text{Mass of fuel}} = \frac{3448.8}{114} = 30.25$$

i.e., **A/F = 30.25. (Ans.)**

(*ii*) **Dew point of the products, t_{dp} :**

Total number of moles of products

$$= 8 + 9 + 12.5 + (2)\,(12.5)\left(\frac{79}{21}\right) = 123.5 \text{ moles/mole fuel}$$

$$\text{Mole fraction of } H_2O = \frac{9}{123.5} = 0.0728$$

Partial pressure of H_2O = 100 × 0.0728 = 7.28 kPa

The saturation temperature corresponding to this pressure is 39.7°C which is also the dew-point temperature.

Hence **t_{dp} = 39.7°C. (Ans.)**

Note. The water condensed from the products of combustion usually contains some dissolved gases and therefore may be quite *corrosive*. For this reason the products of combustion are often *kept above the dew point until discharged to the atmosphere.*

Example 9.5. *One kg of ethane (C_2H_6) is burned with 90% of theoretical air. Assuming complete combustion of hydrogen in the fuel determine the volumetric analysis of the dry products of combustion.*

Solution. The complete combustion equation for C_2H_6 is written as :

$$C_2H_6 + 3.5\,O_2 \longrightarrow 2CO_2 + 3H_2O$$

The combustion equation for C_2H_6 for 90% theoretical air is written as :

$$C_2H_6 + (0.9)\,(3.5)\,O_2 + (0.9)\,(3.5)\left(\frac{79}{21}\right)N_2 \longrightarrow a\,CO_2 + b\,CO + 3H_2O + (0.9)\,(3.5)\left(\frac{79}{21}\right)N_2$$

By balancing *carbon atoms* on both the sides, we get

$$2 = a + b \quad ...(i)$$

By balancing *oxygen atoms* on both the sides, we get

$$(0.9)\,(3.5)\,(2) = 2a + b + 3 \quad ...(ii)$$

Substituting the value of b ($= 2 - a$) from eqn. (*i*) in eqn. (*ii*), we get

$$(0.9)\,(3.5)\,(2) = 2a + 2 - a + 3$$

$$6.3 = a + 5$$

$$\therefore \quad a = 1.3$$

and $$b = 2 - a = 2 - 1.3 = 0.7$$

Thus the combustion equation becomes :

$$C_2H_6 + (0.9)\,(3.5)\,O_2 + (0.9)\,(3.5)\left(\frac{79}{21}\right)N_2$$

$$\longrightarrow 1.3\,CO_2 + 0.7\,CO + 3H_2O + (0.9)\,(3.5)\left(\frac{79}{21}\right)N_2$$

Total number of moles of *dry products* of combustion

$$= 1.3 + 0.7 + (0.9)\,(3.5)\left(\frac{79}{21}\right)$$

$$= 1.3 + 0.7 + 11.85 = 13.85 \text{ moles/mole of fuel}$$

Volumetric analysis of dry products of combustion is as follows :

$$CO_2 = \frac{1.3}{13.85} \times 100 = \mathbf{9.38\%}. \quad \textbf{(Ans.)}$$

$$CO = \frac{0.7}{13.85} \times 100 = \mathbf{5.05\%}. \quad \textbf{(Ans.)}$$

$$N_2 = \frac{11.85}{13.85} \times 100 = \mathbf{85.56\%}. \quad \textbf{(Ans.)}$$

☞**Example 9.6.** *Methane (CH_4) is burned with atmospheric air. The analysis of the products on a 'dry' basis is as follows :*

$$CO_2 = 10.00\%,\ O_2 = 2.37\%,\ CO = 0.53\%,\ N_2 = 87.10\%.$$

(*i*) *Determine the combustion equation;* (*ii*) *Calculate the air-fuel ratio;*

(*iii*) *Percent theoretical air.*

Solution. (*i*) **Combustion equation :**

From the analysis of the products, the following equation can be written, keeping in mind that this analysis is on a *dry basis*.

$$x\,CH_4 + y\,O_2 + z\,N_2 \longrightarrow 10.0\,CO_2 + 0.53\,CO + 2.37\,O_2 + a\,H_2O + 87.1\,N_2$$

To determine all the unknown co-efficients let us find balance for each of the elements.

Nitrogen balance : $z = 87.1$

Since all the nitrogen comes from the air,

$$\frac{z}{y} = \frac{79}{21}\ ;\ y = \frac{87.1}{(79/21)} = 23.16$$

Carbon balance : $x = 10.00 + 0.53 = 10.53$

Hydrogen balance : $a = 2x = 2 \times 10.53 = 21.06$

Oxygen balance. All the unknown co-efficients have been solved for, and in this case the oxygen balance provides a check on the accuracy. Thus, y can also be determined by an oxygen balance

$$y = 10.00 + \frac{0.53}{2} + 2.37 + \frac{21.06}{2} = 23.16$$

Substituting these values for x, y, z and a, we have,

$$10.53\ CH_4 + 23.16\ O_2 + 87.1\ N_2 \longrightarrow 10.0\ CO_2 + 0.53\ CO + 2.37\ O_2 + 21.06\ H_2O + 87.1\ N_2$$

Dividing both sides by 10.53 we get the **combustion equation** per mole of fuel

$$\mathbf{CH_4 + 2.2\ O_2 + 8.27\ N_2 \longrightarrow 0.95\ CO_2 + 0.05\ CO + 2H_2O + 0.225\ O_2 + 8.27\ N_2.} \quad \textbf{(Ans.)}$$

(*ii*) **Air-fuel ratio A/F :**

The air-fuel ratio on a *mole basis* is

$2.2 + 8.27 =$ **10.47 moles air/mole fuel. (Ans.)**

The air-fuel ratio on a *mass basis* is found by introducing the molecular weights

$$\text{A/F} = \frac{10.47 \times 28.97}{(12 + 1 \times 4)} = \textbf{18.96 kg air/kg fuel.} \quad \textbf{(Ans.)}$$

The theoretical air-fuel ratio is found by writing the combustion equation for theoretical air.

$$CH_4 + 2O_2 + 2\left(\frac{79}{21}\right) N_2 \longrightarrow CO_2 + 2H_2O + (2)\left(\frac{79}{21}\right) N_2$$

$$\text{A/F}_{\text{theo.}} = \frac{\left[2 + (2)\left(\frac{79}{21}\right)\right] 28.97}{(12 + 1 \times 4)} = \textbf{17.24 kg air/kg fuel.} \quad \textbf{(Ans.)}$$

(*iii*) **Percent theoretical air :**

$$\text{Percent theoretical air} = \frac{18.96}{17.24} \times 100 = \mathbf{110\%.} \quad \textbf{(Ans.)}$$

Example 9.7. *The gravimetric analysis of a sample of coal is given as 82% C, 10% H_2 and 8% ash. Calculate :*

(*i*) *The stoichiometric A/F ratio;* (*ii*) *The analysis of the products by volume.*

Solution. (*i*) **The stoichiometric A/F ratio :**

1 kg of coal contains 0.82 kg C and 0.10 kg H_2

$\therefore$ 1 kg of coal contains $\frac{0.82}{12}$ moles C and $\frac{0.10}{2}$ moles H_2

Let the oxygen required for complete combustion $= x$ moles

Then the nitrogen supplied with the oxygen $= x \times \frac{79}{21} = 3.76x$ moles

For 1 kg of coal the combustion equation is therefore as follows :

$$\frac{0.82}{12} C + \frac{0.10}{2} H_2 + x\ CO_2 + 3.76x\ N_2 \longrightarrow a\ CO_2 + b\ H_2O + 3.76x\ N_2$$

Then, ***Carbon balance :*** $\frac{0.82}{12} = a \quad \therefore \quad a = 0.068$ moles

Hydrogen balance : $2 \times \frac{0.10}{2} = 2b \quad \therefore \quad b = 0.05$ moles

Oxygen balance : $2x = 2a + b \quad \therefore \quad x = \left(\frac{2 \times 0.068 + 0.05}{2}\right) = 0.093$ moles

The mass of 1 mole of oxygen is 32 kg, therefore, the mass of O_2 supplied per kg of coal
$= 32 \times 0.093 = 2.976$ kg

i.e., $$\text{Stoichiometric } A/F \text{ ratio} = \frac{2.976}{0.233}$$

(where air is assumed to contain 23.3% O_2 and 76.7% N_2 by mass)

Total moles of products $= a + b + 3.76x = 0.068 + 0.05 + 3.76 \times 0.093 = 0.467$ moles

(*ii*) **Analysis of the products by volume :**

The analysis of the products by volume is :

$$\mathbf{CO_2} = \frac{0.068}{0.467} \times 100 = \mathbf{14.56\%.} \quad \textbf{(Ans.)}$$

$$\mathbf{H_2} = \frac{0.05}{0.467} \times 100 = \mathbf{10.7\%.} \quad \textbf{(Ans.)}$$

$$\mathbf{N_2} = \frac{(3.76 \times 0.093)}{0.467} \times 100 = \mathbf{74.88\%.} \quad \textbf{(Ans.)}$$

☞**Example 9.8.** *Calculate the stoichiometric air-fuel ratio for the combustion of a sample of dry anthracite of the following composition by mass :*

Carbon (C) = 88 per cent — *Hydrogen (H_2) = 4 per cent*
Oxygen (O_2) = 3.5 per cent — *Nitrogen (N_2) = 1 per cent*
Sulphur (S) = 0.5 per cent — *Ash = 3 per cent*

If 30 per cent excess air is supplied determine :

(*i*) *Air-fuel ratio*

(*ii*) *Wet dry analysis of the products of combustion by volume.*

Solution. Stoichiometric air-fuel (A/F) ratio :

In case of a fuel with several constituents a *tubular method* is advisable, as shown below. Each constituent is taken separately and the amount of oxygen required for complete combustion is found from the chemical equation. The oxygen in the fuel is included in the column headed '*oxygen required*' as a *negative quantity*.

	Mass per kg coal	*Combustion equation*	*Oxygen required per kg of coal*	*Products per kg of coal*
C	0.88	$C + O_2 \longrightarrow CO_2$ 12 kg + 32 kg ⟶ 44 kg	$0.88 \times \frac{32}{12} = 2.346$ kg	$0.88 \times \frac{44}{12} = 3.23$ kg CO_2
H_2	0.04	$2H_2 + O_2 \longrightarrow 2H_2O$ 1 kg + 8 kg ⟶ 9 kg	$0.04 \times 8 = 0.32$ kg	$0.04 \times 9 = 0.36$ kg H_2O
O_2	0.035	—	– 0.035 kg	—
N_2	0.01	—	—	0.01 kg N_2
S	0.005	$S + O_2 \longrightarrow SO_2$ 32 kg + 32 kg ⟶ 64 kg	$0.005 \times \frac{32}{32} = 0.005$ kg	$0.005 \times \frac{64}{32} = 0.01$ kg SO_2
Ash	0.03	—	—	—
			Total O_2 = 2.636 kg	

From table :

O_2 required per kg of coal $= 2.636$ kg

$\therefore$ Air required per kg of coal $= \dfrac{2.636}{0.233} = 11.31$ kg

(where air is assumed to contain 23.3% O_2 *by mass*)

N_2 associated with this air $= 0.767 \times 11.31 = 8.67$ kg

$\therefore$ Total N_2 in products $= 8.67 + 0.01 = 8.68$ kg

The stoichiometric A/F ratio = 11.31/1. (Ans.)

When 30 per cent excess air is used :

(*i*) **Actual A/F ratio :**

Actual A/F ratio $= 11.31 + 11.31 \times \dfrac{30}{100} =$ **14.7/1. (Ans.)**

(*ii*) **Wet and dry analyses of products of combustion by volume :**

As per actual A/F ratio, N_2 supplied $= 0.767 \times 14.7 = 11.27$ kg

Also O_2 supplied $= 0.233 \times 14.7 = 3.42$ kg

(where air is assumed to contain $N_2 = 76.7\%$ and $O_2 = 23.3$)

In the products then, we have

$$N_2 = 11.27 + 0.01 = 11.28 \text{ kg}$$

and
$$\text{excess } O_2 = 3.42 - 2.636 = 0.784 \text{ kg}$$

The products are entered in the following table and the analysis by volume is obtained :

— In column 3 the percentage by mass is given by the mass of each product divided by the total mass of 15.66 kg.

— In column 5 the moles per kg of coal are given by equation $n = \dfrac{m}{M}$. The total of column 5 gives the total moles of wet products per kg of coal, and by subtracting the moles of H_2O from this total, the total moles of dry products is obtained as 0.5008.

— Column 6 gives the proportion of each constituent of column 5 expressed as a percentage of the total moles of the wet products.

— Similarly column 7 gives the percentage by volume of the dry products.

Product	*Mass/kg coal*	*% by mass*	*M*	*Moles/kg coal*	***% by vol. wet***	***% by vol. dry***
1	*2*	*3*	*4*	*5*	*6*	*7*
CO_2	3.23	20.62	44	0.0734	**14.10**	**14.66**
H_2O	0.36	2.29	18	0.0200	**3.84**	—
SO_2	0.01	0.06	64	0.0002 (say)	**0.04**	**0.04**
O_2	0.78	4.98	32	0.0244	**4.68**	**4.87**
N_2	11.28	72.03	28	0.4028	**77.34**	**80.43**
	15.66 kg			Total wet = 0.5208	100.00	100.00 **(Ans.)**
				$-H_2O$ = 0.0200		
				Total dry = 0.5008		

Example 9.9. *The following analysis relate to coal gas :*

$H_2 = 50.4$ *per cent* $CO = 17$ *per cent*

$CH_4 = 20$ *per cent* $C_4H_8 = 2$ *per cent*

$O_2 = 0.4$ *per cent* $N_2 = 6.2$ *per cent*

$CO_2 = 4$ *per cent.*

(*i*) *Calculate the stoichiometric A/F ratio.*

(*ii*) *Find also the wet and dry analyses of the products of combustion if the actual mixture is 30 per cent weak.*

Solution. The example is solved by a tabular method ; a specimen calculation is given below :

For CH_4 :

$$CH_4 + 2O_2 \longrightarrow CO_2 + 2H_2O$$

i.e., 1 mole CH_4 + 2 moles $O_2 \longrightarrow$ 1 mole CO_2 + 2 moles H_2O

There are 0.2 moles of CH_4 per mole of the coal gas, hence

0.2 moles CH_4 + 0.2 × 2 moles $O_2 \longrightarrow$ 0.2 moles CO_2 + 0.2 × 2 moles H_2O

∴ O_2 required for the CH_4 in the coal gas = 0.4 moles per mole of coal gas.

The oxygen in the fuel (0.004 moles) is included in column 4 as a *negative quantity*.

Product	*Moles/mole fuel*	*Combustion equation*	*O_2 moles/ mole fuel*	*Products CO_2*	*H_2O*
1	*2*	*3*	*4*	*5*	*6*
H_2O	0.504	$2H_2 + O_2 \longrightarrow 2H_2O$	0.252	—	0.504
CO	0.17	$2CO + O_2 \longrightarrow 2CO_2$	0.085	0.17	—
CH_4	0.20	$CH_4 + 2O_2 \longrightarrow CO_2 + 2H_2O$	0.400	0.20	0.40
C_4H_8	0.02	$C_4H_8 + 6O_2 \longrightarrow 4CO_2 + 4H_2O$	0.120	0.08	0.08
O_2	0.004	—	– 0.004	—	—
N_2	0.062	—	—	—	—
CO_2	0.04	—	—	0.04	—
			Total = 0.853	0.49	0.984

(*i*) **Stoichiometric A/F ratio :**

$$\text{Air required} = \frac{0.853}{0.21} = 4.06 \text{ moles/mole of fuel}$$

(where air is assumed to contain 21% O_2 by volume)

∴ *Stoichiometric A/F ratio* = **4.06/1 by volume. (Ans.)**

(*ii*) **Wet and dry analysis of the products of combustion if the actual mixture is 30% weak :**

Actual A/F ratio with 30% weak mixture

$$= 4.06 + \frac{30}{100} \times 4.06 = 1.3 \times 4.06 = 5.278/1$$

Associated N_2 = 0.79 × 5.278 = 4.17 moles/mole fuel

Excess oxygen = 0.21 × 5.278 – 0.853 = 0.255 moles

Total moles of N_2 in products = 4.17 + 0.062 = 4.232 moles/mole fuel.

Analysis by volume of wet and dry products

Product	*Moles/mole fuel*	*% by vol. (dry)*	*% by vol. (wet)*
CO_2	0.490	**9.97**	**8.31**
H_2O	0.984	—	**16.68**
O_2	0.255	**5.19**	**4.32**
N_2	4.170	**84.84**	**70.69**
	Total wet = 5.899 $- H_2O = 0.984$ Total dry = 4.915	100.00	100.00 **(Ans.)**

Example 9.10. *Find the stoichiometric air-fuel ratio for the combustion of ethyl alcohol (C_2H_6O), in a petrol engine. Calculate the air-fuel ratios for the extreme mixture strengths of 80 per cent and 130 per cent. Determine also the wet and dry analyses by volume of the exhaust gas for each mixture strength.*

Solution. The equation for combustion of ethyl alcohol is as follows :

$$C_2H_6O + 3O_2 \longrightarrow 2CO_2 + 3H_2O$$

— Since there are two atoms of carbon in each mole of C_2H_6O then there must be two moles of CO_2 in the products, giving two atoms of carbon on each side of the equation.

— Similarly, since there are six atoms of hydrogen in each mole of ethyl alcohol then there must be three moles of H_2O in the products, giving six atoms of hydrogen on each side of the equation.

— Then balancing the atoms of oxygen, it is seen that there are $(2 \times 2 + 3) = 7$ atoms on the right hand side of the equation, hence seven atoms must appear on the left hand side of the equation. There is one atom of oxygen in ethyl alcohol, therefore a further six atoms of oxygen must be supplied, and hence three moles of oxygen are required as shown.

Since the O_2 is supplied as air, the associated N_2 must appear in the equation,

i.e.,
$$C_2H_6O + 3O_2 + 3 \times \frac{79}{21} N_2 \longrightarrow 2CO_2 + 3H_2O + 3 \times \frac{79}{21} N_2$$

One mole of fuel has a mass of $(2 \times 12 + 1 \times 6 + 16) = 46$ kg. Three moles of oxygen have a mass of $(3 \times 32) = 96$ kg.

$$\therefore \ O_2 \text{ required per kg of fuel} = \frac{96}{46} = 2.09 \text{ kg}$$

$$\therefore \ \textbf{Stoichiometric A/F ratio} = \frac{2.09}{0.233} = \textbf{8.96/1.} \quad \textbf{(Ans.)}$$

Considering a mixture strength of 80% :

$$\text{Now, mixture strength} = \frac{\text{Stoichiometric A/F ratio}}{\text{Actual A/F ratio}}$$

i.e.,
$$0.8 = \frac{8.96/1}{\text{Actual A/F ratio}}$$

$$\therefore \ \textbf{Actual A/F ratio} = \frac{8.96}{0.8} = \textbf{11.2/1.} \quad \textbf{(Ans.)}$$

This means that 1/0.8 or 1.25 times as much air is supplied as is necessary for complete combustion. The exhaust will therefore contain 0.25 stoichiometric oxygen.

i.e., $$C_2H_6O + 1.25\left(3O_2 + 3 \times \frac{79}{21} N_2\right) \longrightarrow 2CO_2 + 3H_2O + 0.25 \times 3O_2 + 1.25 \times 3 \times \frac{79}{21} N_2$$

i.e., The products are :

$$2 \text{ moles } CO_2 + 3 \text{ moles } H_2O + 0.75 \text{ moles } O_2 + 14.1 \text{ moles } N_2$$

The total moles = 2 + 3 + 0.75 + 14.1 = 19.85

Hence wet analysis is :

$$CO_2 = \frac{2}{19.85} \times 100 = \mathbf{10.08\%}. \quad \textbf{(Ans.)}$$

$$H_2O = \frac{3}{19.85} \times 100 = \mathbf{15.11\%}. \quad \textbf{(Ans.)}$$

$$O_2 = \frac{0.75}{19.85} \times 100 = \mathbf{3.78\%}. \quad \textbf{(Ans.)}$$

$$N_2 = \frac{14.1}{19.85} \times 100 = \mathbf{71.03\%}. \quad \textbf{(Ans.)}$$

The total dry moles $= 2 + 0.75 + 14.1 = 16.85$

Hence dry analysis is :

$$CO_2 = \frac{2}{16.85} \times 100 = \mathbf{11.87\%}. \quad \textbf{(Ans.)}$$

$$O_2 = \frac{0.75}{16.85} \times 100 = \mathbf{4.45\%}. \quad \textbf{(Ans.)}$$

$$N_2 = \frac{14.1}{16.85} \times 100 = \mathbf{83.68\%}. \quad \textbf{(Ans.)}$$

Considering a mixture strength of 130% :

Now, $$1.3 = \frac{\text{Stoichiometric ratio}}{\text{Actual A/F ratio}}$$

∴ **Actual A/F ratio** $$= \frac{8.96}{1.3} = \mathbf{6.89/1}. \quad \textbf{(Ans.)}$$

This means that $\frac{1}{1.3}$ or 0.769 of the stoichiometric air is supplied. *The combustion cannot be complete, as the necessary oxygen is not available. It is usual to assume that all hydrogen is burned to H_2O, since hydrogen atoms have a greater affinity for oxygen than carbon atoms. The carbon in the fuel will burn to CO and CO_2, but the relative proportions have to be determined.*

Let, a = Number of moles of CO_2 in the products, and

b = Number of moles of CO in the products

Then the combustion equation is as follows :

$$C_2H_6O + 0.769\left(3O_2 + 3 \times \frac{79}{21} N_2\right) \longrightarrow a\, CO_2 + b\, CO + 3H_2O + 0.769 \times 3 \times \frac{79}{21} N_2$$

To find a and b a balance of carbon and oxygen atoms can be made,

i.e., ***Carbon balance :***

$$2 = a + b \quad ...(i)$$

and ***Oxygen balance :***

$$1 + 2 \times 0.769 \times 3 = 2a + b + 3$$

or
$$2.614 = 2a + b \quad ...(ii)$$

From eqn. (*i*) and (*ii*), we get $a = 0.614$, $b = 1.386$

i.e., The products are : 0.614 moles CO_2 + 1.386 moles CO + 3 moles H_2O + 8.678 moles N_2

The total moles = 0.614 + 1.386 + 3 + 8.678 = 13.678.

Hence wet analysis is :

$$CO_2 = \frac{0.614}{13.678} \times 100 = \mathbf{4.49\%}. \quad \textbf{(Ans.)}$$

$$CO = \frac{1.386}{13.678} \times 100 = \mathbf{10.13\%}. \quad \textbf{(Ans.)}$$

$$H_2O = \frac{3}{13.678} \times 100 = \mathbf{21.93\%}. \quad \textbf{(Ans.)}$$

$$N_2 = \frac{8.678}{13.678} \times 100 = \mathbf{63.45\%}. \quad \textbf{(Ans.)}$$

The total dry moles $= 0.614 + 1.386 + 8.678 = 10.678$

Hence dry analysis is :

$$CO_2 = \frac{0.614}{10.678} \times 100 = \mathbf{5.75\%}. \quad \textbf{(Ans.)}$$

$$CO = \frac{1.386}{10.678} \times 100 = \mathbf{12.98\%}. \quad \textbf{(Ans.)}$$

$$N_2 = \frac{8.678}{10.678} \times 100 = \mathbf{81.27\%}. \quad \textbf{(Ans.)}$$

Example 9.11. *For the stoichiometric mixture of example 9.10 calculate :*

(*i*) *The volume of the mixture per kg of fuel at a temperature of 50°C and a pressure of 1.013 bar.*

(*ii*) *The volume of the products of combustion per kg of fuel after cooling to a temperature of 130°C at a pressure of 1 bar.*

Solution. As before,

$$C_2H_6O + 3O_2 + 3 \times \frac{79}{21} N_2 \longrightarrow 2CO_2 + 3H_2O + 3 \times \frac{79}{21} N_2$$

$$\therefore \text{ Total moles reactants} = 1 + 3 + 3 \times \frac{79}{21} = 15.3$$

From equation, $pV = nR_0T$

$$V = \frac{nR_0T}{p} = \frac{15.3 \times 8.314 \times 10^3 \times (50 + 273)}{1.013 \times 10^5} = 405.6 \text{ m}^3\text{/mole of fuel}$$

In 1 mole of fuel there are (2 × 12 + 6 + 16) = 46 kg

(*i*) ∴ **Volume of reactants per kg of fuel** $= \dfrac{405.6}{46}$ **= 8.817 m³. (Ans.)**

When the products are cooled to 130°C the H_2O exists as steam, since the temperature is well above the saturation temperature corresponding to the partial pressure of the H_2O. (This must be so since the saturation temperature corresponding to the *total* pressure is 99.6°C, and the *saturation temperature decreases with pressure.* The total moles of the products is

$$= \left(2 + 3 + 3 \times \frac{79}{21}\right) = 16.3$$

From equation, $\quad pV = nR_0T$

$$V = \frac{nR_0T}{p} = \frac{16.3 \times 8.314 \times 10^3 \times (130 + 273)}{1 \times 10^5} = 546.14 \text{ m}^3\text{/mole of fuel.}$$

(*ii*) ∴ **Volume of products per kg of fuel** $= \dfrac{546.14}{46}$ **= 11.87 m³. (Ans.)**

☞**Example 9.12.** *The following is the composition of coal gas supplied to a gas engine :*

H_2 = 50.6 per cent ; CO = 10 per cent ; CH_4 = 26 per cent ; C_4H_8 = 4 per cent ; O_2 = 0.4 per cent ; CO_2 = 3 per cent ; N_2 = 6 per cent.

If the air-fuel ratio is 7/1 by volume, calculate the analysis of the dry products of combustion. It can be assumed that the stoichiometric A/F ratio is less than 7/1.

Solution. Since it is given that the actual *A/F ratio is greater than the stoichiometric,* therefore it follows that *excess air has been supplied.* The products will therefore consist of CO_2, H_2O, O_2 and N_2.

The combustion equation can be written as follows :

$$0.506H_2 + 0.1CO + 0.26CH_4 + 0.04C_4H_8 + 0.004O_2 + 0.03CO_2 + 0.06N_2 + 0.21 \times 7O_2 + 0.79 \times 7N_2$$
$$\longrightarrow a\, CO_2 + b\, H_2O + c\, O_2 + d\, N_2$$

Then,

Carbon balance : $\quad 0.1 + 0.26 + 4 \times 0.04 + 0.03 = a \qquad \therefore\ a = 0.55$

Hydrogen balance : $\quad 2 \times 0.506 + 4 \times 0.26 + 8 \times 0.04 = 2b \qquad \therefore\ b = 1.186$

Oxygen balance : $\quad 0.1 + 2 \times 0.004 + 2 \times 0.03 + 0.21 \times 7 \times 2 = 2a + b + 2c \quad \therefore\ c = 0.411$

Nitrogen balance : $\quad 2 \times 0.06 + 2 \times 0.79 \times 7 = 2d \qquad \therefore\ d = 5.59$

∴ Total moles of dry products = 0.55 + 0.411 + 5.59 = 6.65

Then analysis by volume is :

$$CO_2 = \frac{0.55}{6.55} \times 100 = 8.39\%. \quad \textbf{(Ans.)}$$

$$O_2 = \frac{0.411}{6.55} \times 100 = 6.27\%. \quad \textbf{(Ans.)}$$

$$N_2 = \frac{5.59}{6.55} \times 100 = 85.34\%. \quad \textbf{(Ans.)}$$

Example 9.13. *The following is the analysis (by weight) of a chemical fuel :*

Carbon = 60 per cent ; Hydrogen = 20 per cent ; Oxygen = 5 per cent ; Sulphur = 5 per cent and Nitrogen = 10 per cent.

Find the stoichiometric amount of air required for complete combustion of this fuel.

Solution. On the basis of 100 kg fuel let us assume an equivalent formula of the form :

$$C_aH_bO_cN_dS_e$$

From the given analysis by weight, we can write

$$12a = 60 \quad \text{or} \quad a = 5$$
$$1b = 20 \quad \text{or} \quad b = 20$$
$$16c = 5 \quad \text{or} \quad c = 0.3125$$
$$14d = 10 \quad \text{or} \quad d = 0.7143$$
$$32e = 5 \quad \text{or} \quad e = 0.1562$$

Then the formula of the fuel can be written as

$$C_5H_{20}O_{0.3125}N_{0.7143}S_{0.1562}$$

The combustion equation is

$$C_5H_{20}O_{0.3125}N_{0.7143}S_{0.1562} + x\,O_2 + x\left(\frac{79}{21}\right)N_2 \longrightarrow p\,CO_2 + q\,H_2O + r\,SO_2 + s\,N_2$$

Then,

Carbon balance : $5 = p \quad \therefore \quad p = 5$

Hydrogen balance : $20 = 2q \quad \therefore \quad q = 10$

Sulphur balance : $0.1562 = r \quad \therefore \quad r = 0.1562$

Oxygen balance : $0.3125 + 2x = (2p + q + 2r)$

i.e.,
$$x = p + \frac{q}{2} + r - \frac{0.3125}{2} = 5 + \frac{10}{2} + 0.1562 - \frac{0.3125}{2} = 9.99$$

Nitrogen balance : $0.7143 + 2x \times \frac{79}{21} = 2s$

$$\therefore \quad s = \frac{0.7143}{2} + x \times \frac{79}{21} = \frac{0.7143}{2} + 9.99 \times \frac{79}{21} = 37.94$$

Hence the combustion equation is written as follows :

$$C_5H_{20}O_{0.3125}N_{0.7143}S_{0.1562} + 9.99O_2 + 9.99\left(\frac{79}{21}\right)N_2 \longrightarrow 5CO_2 + 10H_2O + 0.1562SO_2 + 37.94N_2$$

$$\therefore \textbf{ Stoichiometric air required} = \frac{9.99 \times 32 + 9.99 \times \left(\frac{79}{21}\right) \times 28}{100} = \textbf{13.7 kg/kg of fuel. (Ans.)}$$

(**Note.** This example can also be solved by tabular method as explained in example 9.8.).

Example 9.14. *A sample of fuel has the following percentage composition by weight :*

Carbon = 84 per cent *Hydrogen = 10 per cent*

Oxygen = 3.5 per cent *Nitrogen = 1.5 per cent*

Ash = 1 per cent

(*i*) *Determine the stoichiometric air fuel ratio by mass.*

(*ii*) *If 20 per cent excess air is supplied, find the percentage composition of dry flue gases by volume.*

Solution. (*i*) **Stoichiometric air fuel ratio :**

On the basis of 100 kg of fuel let us assume an equivalent formula of the form :

$$C_aH_bO_cN_d$$

From the given analysis by weight, we can write

$$12a = 84 \quad i.e., \quad a = 7$$
$$1b = 10 \quad i.e., \quad b = 10$$
$$16c = 3.5 \quad i.e., \quad c = 0.218$$
$$14d = 1.5 \quad i.e., \quad d = 0.107$$

The formula of fuel is $C_7H_{10}O_{0.218}N_{0.107}$

The combustion equation is written as

$$C_7H_{10}O_{0.218}N_{0.107} + x\, O_2 + x\left(\frac{79}{21}\right) N_2 \longrightarrow p\, CO_2 + q\, H_2O + r\, N_2$$

Then,

Carbon balance : $7 = p \quad i.e., \quad p = 7$

Hydrogen balance : $10 = 2q \quad i.e., \quad q = 5$

Oxygen balance : $0.218 + 2x = (2p + q)$

or $0.218 + 2x = 2 \times 7 + 5 \quad i.e. \quad x = 9.39$

Nitrogen balance : $0.107 + 2x\left(\frac{79}{21}\right) = 2r$

or $0.107 + 2 \times 9.39 \times \frac{79}{21} = 2r \; i.e., \; r = 35.4$

Hence the combustion equation becomes

$$C_7H_{10}O_{0.218}N_{0.107} + 9.39O_2 + 9.39\left(\frac{79}{21}\right) N_2 \longrightarrow 7CO_2 + 5H_2O + 35.4N_2$$

$\therefore$ **Stoichiometric A/F ratio** $= \dfrac{9.39 \times 32 + 9.39 \times \frac{79}{21} \times 28}{100}$ $=$ **12.89. (Ans.)**

(*ii*) **Percentage composition of dry flue gases by volume with 20 per cent excess air :**

If 20 per cent excess air is used, the combustion equation becomes

$$C_7H_{10}O_{0.218}N_{0.107} + (1.2)(9.39)\, O_2 + (1.2)(9.39)\left(\frac{79}{21}\right) N_2$$
$$\longrightarrow 7CO_2 + 5H_2O + (0.2)(9.39)\, O_2 + (1.2)(35.4)\, N_2$$

Total number of moles of dry products of combustion

$$n = 7 + (0.2)(9.39) + (1.2)(35.4)$$
$$= 7 + 1.878 + 42.48 = 51.358$$

$\therefore$ **Percentage composition of dry flue gases by volume is as follows :**

$CO_2 = \dfrac{7}{51.358} \times 100 =$ **13.63%. (Ans.)**

$O_2 = \dfrac{1.878}{51.358} \times 100 =$ **3.66%. (Ans.)**

$N_2 = \dfrac{42.48}{51.358} \times 100 =$ **82.71%. (Ans.)**

Example 9.15. *Orsat analysis of the products of combustion of a hydrocarbon fuel of unknown composition is as follows :*

Carbon dioxide (CO_2) = *8%* *Carbon monoxide (CO) = 0.5%*

Oxygen (O_2) = *6.3%* *Nitrogen* (N_2) = *85.2%*

Determine the following :

(*i*) *Air-fuel ratio*

(*ii*) *Percent theoretical air required for combustion.*

Solution. From the given Orsat analysis the combustion equation is written as follows :

$$a\,C + b\,H + c\,O_2 + \left(\frac{79}{21}\right) c\,N_2 = 8CO_2 + 0.5CO + 6.3O_2 + x\,H_2O + 85.2N_2$$

Then,

Carbon balance : $a = 8 + 0.5 = 8.5$ *i.e.,* $a = 8.5$

Nitrogen balance : $\frac{79}{21}\,C = 85.2$ *i.e.,* $c = 22.65$

Oxygen balance : $C = 8 + \frac{0.5}{2} + 6.3 + \frac{x}{2}$

or $22.65 = 8 + 0.25 + 6.3 + \frac{x}{2}$ *i.e.,* $x = 16.2$

Hydrogen balance : $b = 2x = 2 \times 16.2 = 32.4$ *i.e.,* $b = 32.4$

(*i*) **Air-fuel ratio :**

The air supplied per 100 moles of dry products is

$$= 22.65 \times 32 + \left(\frac{79}{21}\right) \times 22.65 \times 28 = 3110.6 \text{ kg}$$

$\therefore$ **Air-fuel ratio** $= \frac{3110.6}{8.5 \times 12 + 32.4 \times 1}$ = **23.1 kg of air/kg of fuel. (Ans.)**

(*ii*) **Per cent theoretical air required for combustion :**

Mass fraction of carbon $= \frac{12 \times 8.5}{12 \times 8.5 + 32.4 \times 1} = 0.759$

Mass fraction of hydrogen $= \frac{32.4 \times 1}{12 \times 8.5 + 32.4} = 0.241$

Considering 1 kg of fuel, the *air required for complete combustion* is

$$= \left[0.759 \times \left(\frac{8}{3}\right) \times \frac{100}{23.3}\right] + \left[0.241 \times 8 \times \frac{100}{23.3}\right] = 16.96 \text{ kg}$$

$\therefore$ **Percent theoretical air required for combustion** $= \frac{23.1}{16.96} \times 100$ = **136.2%. (Ans.)**

Example 9.16. *The following is the volumetric analysis of the dry exhaust from an internal combustion engine :*

$$CO_2 = 8.9\% \text{ ; } CO = 8.2\% \text{ ; } H_2 = 4.3\% \text{ ; } CH_4 = 0.5\% \text{ and } N_2 = 78.1\%.$$

If the fuel used is octane (C_8H_{18}) *determine air-fuel ratio on mass basis :*

(*i*) *By a carbon balance.* (*ii*) *By a hydrogen oxygen balance.*

Solution. (*i*) As per analysis of dry products, the combustion equation is written as

$$a\ C_8H_{18} + 78.1N_2 + 78.1\left(\frac{21}{79}\right) O_2 \longrightarrow 8.9CO_2 + 8.2CO + 4.3H_2 + 0.5CH_4 + 78.1N_2 + x\ H_2O$$

Carbon balance : $8a = 8.9 + 8.2 + 0.5 = 17.6$ *i.e.* $a = 2.2$

$$\therefore \text{ Air-fuel (A/F) ratio } = \frac{78.1 \times 28 + 78.1 \times \frac{21}{79} \times 32}{2.2(8 \times 12 + 1 \times 18)}$$

$$= \frac{2186.8 + 664.3}{250.8} = \frac{2851.1}{250.8} = \mathbf{11.37.} \quad \textbf{(Ans.)}$$

(*ii*) In this case the combustion equation is written as

$$a\ C_8H_{18} + b\ O_2 + b\left(\frac{79}{21}\right) N_2 \longrightarrow 8.9CO_2 + 8.2CO + 4.3H_2 + 0.5CH_4 + b\left(\frac{79}{21}\right) N_2 + x\ H_2O$$

Carbon balance : $8a = 8.9 + 8.2 + 0.5 = 17.6$ *i.e.*, $a = 2.2$

Hydrogen balance : $18a = 4.3 \times 2 + 0.5 \times 4 + 2x$

or $18 \times 2.2 = 8.6 + 2 + 2x$ *i.e.*, $x = 14.5$

Oxygen balance : $2b = 8.9 \times 2 + 8.2 + x$

or $2b = 17.8 + 8.2 + 14.5$ *i.e.*, $b = 20.25$

$$\therefore \text{ Air-fuel (A/F) ratio} = \frac{(20.25 \times 32) + (20.25)\left(\frac{79}{21}\right) \times 28}{2.2(8 \times 12 + 1 \times 18)} = \frac{2781}{250.8} = \mathbf{11.09.} \quad \textbf{(Ans.)}$$

☞**Example 9.17.** *The exhaust from an engine running on benzole was measured with the help of Orsat apparatus. Orsat analysis showed a CO_2 content of 12%, but no CO. Assuming that the remainder of the exhaust contains only oxygen and nitrogen, calculate the air-fuel ratio of the engine.*

The ultimate analysis of benzole is C = 88% and H_2 = 12%.

Solution. 1 kg of fuel, consisting of 0.88 kg C and 0.12 kg H_2, can be written as 0.88/2 moles C and 0.12/2 moles H_2. Therefore, considering 1 mole of *dry exhaust gas* (D.E.G.) we can write the combustion equation as follows :

$$X\left(\frac{0.88}{12}C + \frac{0.12}{2}H_2\right) + Y\,O_2 + \frac{79}{21}\,Y\,N_2 \longrightarrow 0.12CO_2 + a\ O_2 + (0.88 - a)\ N_2 + b\ H_2O$$

[Let the D.E.G. contain a moles of O_2. The moles of CO_2 in 1 mole of D.E.G. are 0.12. Therefore the D.E.G. contains $(1 - a - 0.12) = (0.88 - a)$ moles of N_2.]

where, X = Mass of fuel per mole D.E.G.,

Y = Moles of O_2 per mole D.E.G.,

a = Moles of *excess* O_2 per mole D.E.G., and

b = Moles of H_2O per mole D.E.G.

Now,

Carbon balance : $\frac{0.88}{12}X = 0.12$ $\quad \therefore X = 1.636$

Hydrogen balance : $0.06X = b$ $\quad \therefore b = 0.06 \times 1.636 = 0.098$

Oxygen balance : $2Y = 2 \times 0.12 + 2a + b$ or $2Y = 0.24 + 2a + 0.098$

$\therefore Y = 0.169 + a$

Nitrogen balance : $\frac{79}{21} Y = (0.88 - a)$ $\therefore$ $Y = 0.234 - 0.266a$

Equating the expressions for Y gives

$$0.234 - 0.266a = 0.169 + a \qquad \therefore \quad a = 0.0513$$

i.e., $$Y = 0.169 + 0.0513 = 0.2203$$

$\therefore$ O_2 supplied = 0.2203 × 32 kg/mole D.E.G.

i.e., Air supplied $= \frac{0.2203 \times 32}{0.233} = 30.26$ kg/mole D.E.G.

Since $X = 1.636$, then, the fuel supplied per mole D.E.G. is 1.636 kg

$\therefore$ **A/F ratio** $= \frac{30.26}{1.636} =$ **18.5/1. (Ans.)**

Example 9.18. *The analysis of the dry exhaust from an internal combustion engine is as follows :*

Carbon dioxide (CO_2) = 15 per cent *Carbon monoxide (CO) = 3 per cent*
Methane (CH_4) = 3 per cent *Hydrogen (H_2) = 1 per cent*
Oxygen (O_2) = 2 per cent *Nitrogen (N_2) = 76 per cent*

Calculate the proportions by mass of carbon to hydrogen in the fuel, assuming it to be a pure hydrocarbon.

Solution. Let 1 kg of fuel contain x kg of carbon (C) and y kg hydrogen (H_2). Then considering 1 mole of D.E.G. and introducing X and Y, we can write

$$X\left(\frac{x\,C}{12} + \frac{y}{2}H_2\right) + Y\,O_2 + \frac{79}{21}\,Y\,N_2 \longrightarrow 0.15CO_2 + 0.03CO + 0.03CH_4$$

$$+ 0.01H_2 + 0.02O_2 + a\,H_2O + 0.76N_2$$

Then,

Nitrogen balance : $\frac{79}{21} Y = 0.76$ $\therefore$ $Y = 0.202$

Oxygen balance : $$Y = 0.15 + \frac{0.03}{2} + 0.02 + \frac{a}{2}$$

or $$0.202 = 0.15 + 0.015 + 0.02 + \frac{a}{2} \qquad \therefore \quad a = 0.034$$

Carbon balance : $\frac{Xx}{12} = 0.15 + 0.03 + 0.03$ $\therefore$ $Xx = 2.52$ ...(*i*)

Hydrogen balance : $\frac{Xy}{2} = 2 \times 0.03 + 0.01 + a = 0.06 + 0.01 + 0.034$

$\therefore$ $Xy = 0.208$ ...(*ii*)

Dividing equations (*i*) and (*ii*), we get

$$\frac{Xx}{Xy} = \frac{2.52}{0.208} \quad \text{or} \quad \frac{x}{y} = 12.1$$

i.e., **Ratio of C to H_2 in fuel** $= \frac{x}{y} = \frac{\mathbf{12.1}}{\mathbf{1}}$. **(Ans.)**

Internal Energy and Enthalpy of Combustion

Example 9.19. *ΔH_0 (enthalpy of combustion at reference temperature T_0) for benzene vapour (C_6H_6) at 25°C is – 3301000 kJ/mole with the H_2O in the liquid phase. Calculate ΔH_0 for the H_2O in the vapour phase.*

Solution. If H_2O remains as a vapour the heat transferred to the surroundings will be less than when the vapour condenses by the amount due to the change in enthalpy of the vapour during condensation at the reference temperature.

$$\Delta H_0 \text{ (vapour)} = \Delta H_0 \text{ (liquid)} + m_s\, h_{fgo}$$

where, m_s = Mass of H_2O formed, and

h_{fgo} = Change in enthalpy of steam between saturated liquid and saturated vapour at the reference temperature T_0

= 2441.8 kJ at 25°C

For the reaction :

$$C_6H_6 + 7.5O_2 \longrightarrow 6CO_2 + 3H_2O$$

3 moles of H_2O are formed on combustion of 1 mole of C_6H_6 ; 3 moles of H_2O

$$= 3 \times 18 = 54 \text{ kg } H_2O$$

$\therefore$ **ΔH_0 (vapour)** $= -\,3301000 + 54 \times 2441.8 = -$ **3169143 kJ/mole. (Ans.)**

Example 9.20. *Calculate ΔU_0 in kJ/kg for the combustion of benzene (C_6H_6) vapour at 25°C given that $\Delta H_0 = -\ 3169100$ kJ/mole and the H_2O is in the vapour phase.*

Solution. Given : $\Delta H_0 = -\ 3169100$ kJ

The combustion equation is written as

$$C_6H_6 + 7.5O_2 \longrightarrow 6CO_2 + 3H_2O \text{ (vapour)}$$

$$n_R = 1 + 7.5 = 8.5,\ n_P = 6 + 3 = 9$$

Using the relation, $\Delta U_0 = \Delta H_0 - (n_P - n_R)R_0T_0$

$$= -\,3169100 - (9 - 8.5) \times 8.314 \times (25 + 273)$$

$$= -\,3169100 - 1239 = -\,3170339 \text{ kJ/mole}$$

(It may be noted that ΔU_0 is negligibly different from ΔH_0)

1 mole of $C_6H_6 = 6 \times 12 + 1 \times 6 = 78$ kg

$$\therefore \qquad \mathbf{\Delta U_0} = \frac{-\,3170339}{78} = -\ \mathbf{40645\ kJ/kg.} \quad \textbf{(Ans.)}$$

Example 9.21. *ΔH_0 for CO at 60°C is given as – 285200 kJ/mole. Calculate ΔH_0 at 2500°C given that the enthalpies of gases concerned in kJ/mole are as follows :*

Gas	*60°C*	*2500°C*
CO	*9705*	*94080*
O	*9696*	*99790*
CO_2	*10760*	*149100*

Solution. The reaction equation is given by

$$CO + \tfrac{1}{2}\, O_2 \longrightarrow CO_2$$

Refer Fig. 9.34.

It can be seen from the property diagram of Fig. 9.34 that the enthalpy of combustion at temperature T, ΔH_T can be obtained from ΔH_0 and T_0 by the relationship

$$-\Delta H_T = -\Delta H_0 + (H_{R_T} - H_{R_O}) - (H_{P_T} - H_{P_O}) \quad ...(i)$$

where, $H_{R_T} - H_{R_O}$ = increase in enthalpy of the *reactants* from T_0 to T

and $H_{P_T} - H_{P_O}$ = increase in enthalpy of the *products* from T_0 to T.

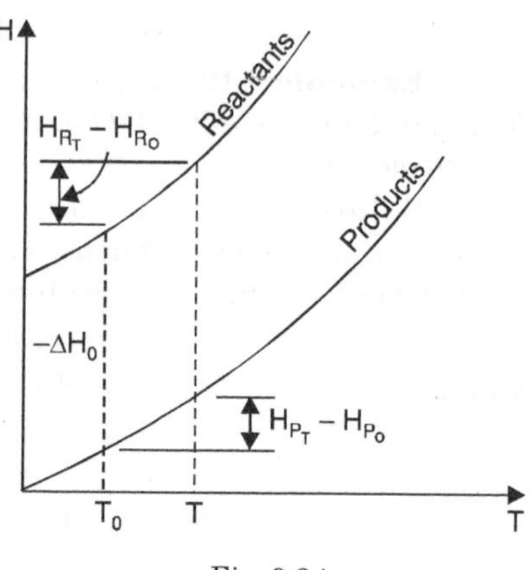

Fig. 9.34

Now, from the given data, we have

$$H_{R_O} = 1 \times 9705 + \tfrac{1}{2} \times 9696 = 14553 \text{ kJ}$$

$$H_{R_T} = 1 \times 94080 + \tfrac{1}{2} \times 99790 = 143975 \text{ kJ}$$

$$H_{P_O} = 1 \times 10760 = 10760 \text{ kJ}$$

$$H_{P_T} = 1 \times 149100 \text{ kJ} = 149100 \text{ kJ}$$

Using equation (*i*), we get

$$-\Delta H_T = +285200 + (143975 - 14553) - (149100 - 10760)$$

$$= 285200 + 129422 - 138340 = 276282$$

$\therefore$ $\mathbf{\Delta H_T = -276282}$ **kJ/mole. (Ans.)**

Heating Values of Fuels

Example 9.22. *The lower heating value of propane at constant pressure and 25°C is 2044009 kJ per kg mole. Find the higher heating value at constant pressure and at constant volume.*

Solution. (*i*) **Higher heating value at constant pressure, $(HHV)_p$:**

The combustion reaction for propane is written as

$$C_3H_8 + 5O_2 = 3CO_2 + 4H_2O$$

Now $$(HHV)_p = (LHV)_p + mh_{fg}$$

where, HHV = Higher heating value at constant pressure

LHV = Lower heating value

m = Mass of water formed by combustion

= 4 × 18 = 72 kg per kg mole

h_{fg} = Latent heat of vapourisation at given temperature per unit mass of water

= 2442 kJ/kg at 25°C

$\therefore$ $\mathbf{(HHV)_p}$ = 2044009 + 72 × 2442 = **2219833 kJ/kg. (Ans.)**

(*ii*) **Higher heating value at constant volume, $(HHV)_v$:**

Now $$(\Delta U) = \Delta H - \Delta n R_0 T$$

or $$-(HHV)_v = -(HHV)_p - \Delta n R_0 T$$

or $$(HHV)_v = (HHV)_p + \Delta n R_0 T$$

where, R_0 = Universal gas constant = 8.3143 kJ/kg mol K

$\Delta n = n_P - n_R$

[n_P = Number of moles of gaseous products
n_R = Number of moles of gaseous reactants]

Now, the reaction for higher heating value is

$$C_3H_8 + 5O_2 = 3CO_2 + 4H_2O \text{ (liquid)}$$

$$\Delta n = 3 - (1 + 5) = -3$$

$\therefore$ $\mathbf{(HHV)_v}$ = 2219833 − 3(8.3143)(25 + 273) = **2212400 kJ/kg. (Ans.)**

Example 9.23. *Calculate the lower heating value of gaseous octane at constant volume if $(\Delta U)_{25°C} = -5494977$ kJ for the reaction :*

$$[C_8H_{18} (gas) + 12.5O_2 (gas)] = [8CO_2 (gas) + 9H_2O (liquid)].$$

Solution. The given value of ΔU corresponds to the *higher heating value* at constant volume because the water in the products is in *liquid phase.*

$$\text{HHV} = 5494977 \text{ kJ/kg}$$

$$(\text{LHV})_v = (\text{HHV})_v - m(u_g - u_f)$$

$$m = 9 \times 18 = 162 \text{ kg/kg mole } C_8H_{18}$$

$$(u_g - u_f) = 2305 \text{ kJ/kg at } 25°\text{C}$$

$\therefore$ $\quad$ $(\textbf{LHV})_\textbf{v} = 5494977 - 162(2305) =$ **5121567 kJ/kg. (Ans.)**

Example 9.24. *Calculate the lower and higher heating values at constant pressure per kg of mixture at 25°C, for the stoichiometric mixtures of :*

(i) Air and benzene vapour (C_6H_6) and

(ii) Air and octane vapour (C_8H_{18}).

Given that the enthalpies of combustion at 25°C are :

$$C_6H_6 = -3169500 \text{ kJ/mole}$$

$$C_8H_{18} = -5116200 \text{ kJ/mole}$$

Both the above figures are for the case where the water in the products is in the vapour phase.

Solution. (*i*) **Air and benzene vapour :**

For benzene, the combustion equation is as follows :

$$C_6H_6(g) + 7.5O_2(g) + 7.5\left(\frac{79}{21}\right) N_2(g) = 6CO_2(g) + 3H_2O(g) + 7.5\left(\frac{79}{21}\right) N_2(g)$$

Since the water in the products is in *vapour phase*, therefore, the given value of enthalpy of combustion corresponds to the *lower heating* value at constant pressure.

i.e. $\quad$ $(\text{LHV})_p = 3169500$ kJ/mole

$$\textbf{(LHV)}_\textbf{v} \textbf{ per kg of mixture} = \frac{3169500}{(12 \times 6 + 6 \times 1) + (7.5 \times 32) + 7.5\left(\frac{79}{21}\right)(28)}$$

$$= \frac{3169500}{78 + 240 + 790} = \textbf{2861 kJ/kg.} \quad \textbf{(Ans.)}$$

Now, $\quad$ $(\text{HHV})_p = (\text{LHV})_p + mh_{fg}$

where, $(\text{HHV})_p$ = Higher heating value at constant pressure,

$(\text{LHV})_p$ = Lower heating value at constant pressure,

m = Mass of water formed by combustion,

= 3 × 18 = 54 kg/kg mole of fuel, and

h_{fg} = Latent heat of vapourisation at given temperature per unit mass of water

= 2442 kJ/kg at 25°C.

$\therefore$ $\quad$ $(\text{HHV})_p = 3169500 + 54 \times 2442 = 3301368$ kJ/mole

$$\text{Thus, } \textbf{(HHV)}_\textbf{p} \textbf{ per kg of mixture} = \frac{3301368}{78 + 240 + 790} = \textbf{2980 kJ/kg.} \quad \textbf{(Ans.)}$$

(*ii*) **Air and octane vapour :**

$$(\text{LHV})_p = 5116200 \text{ kJ/mole of } C_8H_{18}$$

For octane, the combustion equation is written as follows :

$$C_8H_{18}(g) + 12.5O_2(g) \longrightarrow 8CO_2(g) + 9H_2O(g) + 12.5\left(\frac{79}{21}\right) N_2(g)$$

(LHV)$_p$ per kg of mixture $= \dfrac{5116200}{(12 \times 8 + 18 \times 1) + 12.5 \times 32 + 12.5 \times \frac{79}{21} \times 28}$

$= \dfrac{5116200}{114 + 400 + 1317} =$ **2794 kJ/kg. (Ans.)**

$$(HHV)_p = (LHV)_p + mh_{fg}$$

$$m = 9 \times 18 = 162 \text{ kJ/kg mole of fuel}$$

$\therefore$ $$(HHV)_p = 5116200 + 162 \times 2442 = 5511804$$

Hence, **(HHV)$_p$ per kg of mixture** $= \dfrac{5511804}{114 + 400 + 1317} =$ **3010 kJ/kg. (Ans.)**

☞**Example 9.25.** *The higher heating value of kerosene at constant volume whose ultimate analysis is 88% and 12% hydrogen, was found to be 45670 kJ/kg. Calculate the other three heating values.*

Solution. Combustion of 1 kg of fuel produces the following products :

$$CO_2 = \frac{44}{12} \times 0.88 = 3.23 \text{ kg}$$

$$H_2O = \frac{18}{2} \times 0.12 = 1.08 \text{ kg}$$

At 25°C : $(u_g - u_f)$ *i.e.* $u_{fg} = 2304$ kJ/kg

$h_{fg} = 2442$ kJ/kg

(*i*) **(LHV)$_v$:**

$$(LHV)_v = (HHV)_v - m(u_g - u_f)$$

$$= 45670 - 1.08 \times 2304 = 43182 \text{ kJ/kg}$$

Hence **(LHV)$_v$ = 43182 kJ/kg. (Ans.)**

(*ii*) **(HHV)$_p$:, (LHV)$_p$:**

The combustion equation is written as follows :

$$1 \text{ mole fuel} + \frac{x}{32} O_2 \longrightarrow \frac{3.23}{44} CO_2 + \frac{1.08}{18} H_2O$$

i.e., $$\frac{x}{32} = \frac{3.23}{44} + \frac{1.08}{18 \times 2}$$

or $$x = 3.31 \text{ kg}$$

i.e., $$1 \text{ kg fuel} + 3.31 \text{ kg } O_2 = 3.23CO_2 + 1.08H_2O$$

Also, $$\Delta H = \Delta U + \Delta n R_0 T$$

i.e., $$-(HHV)_p = -(HHV)_v + \Delta n R_0 T$$

or $$(HHV)_p = (HHV)_v - \Delta n R_0 T$$

where, $\Delta n = n_p - n_R$

$$= \left(\frac{3.23}{44} - \frac{3.31}{32}\right)$$

$\left[\begin{array}{l} n_p = \text{Number of moles of gaseous products} \\ n_R = \text{Number of moles of gaseous reactants} \end{array}\right]$

Since in case of higher heating value, H_2O will appear in liquid phase

$$(\mathbf{HHV})_\mathbf{p} = 45670 - \left(\frac{3.23}{44} - \frac{3.31}{32}\right) \times 8.3143 \times (25 + 273)$$

$$= \mathbf{45744\ kJ/kg.} \quad \mathbf{(Ans.)}$$

$$(\mathbf{LHV})_\mathbf{p} = (\text{HHV})_p - 1.08 \times 2442 = 45774 - 1.08 \times 2442$$

$$= \mathbf{43107\ kJ/kg.} \quad \mathbf{(Ans.)}$$

Example 9.26. *The reaction equation of a fuel is represented by*

$$C_nH_m + aO_2 + bN_2 = 8CO_2 + 0.9CO + 8.8O_2 + dH_2O + 82.3\ N_2$$

Determine :

(i) The actual air-fuel ratio and the chemical formula of the fuel ;

(ii) The stoichiometric air-fuel ratio and the percent theoretical air used.

You may assume N_2/O_2 ratio in air = 3.76 : **(Bombay University)**

Solution. $C_nH_m + aO_2 + bN_2 = 8CO_2 + 0.9\ CO + 8.8O_2 + dH_2O + 82.3N_2$

Equating coefficients, we have

C : $n = 8 + 0.9 = 8.9$ ...(*i*)

H : $m = 2d$ ...(*ii*)

O_2 : $a = 8 + \frac{0.9}{2} + 8.8 + \frac{d}{2}$ ...(*iii*)

N_2 : $b = 82.3$ ...(*iv*)

Assume N_2/O_2 ratio in air = 3.76

It means $\frac{b}{a} = 3.76$

or $b = 3.76\ a$ ...(*v*)

On solving eqns. (*i*), (*ii*), (*iii*), (*iv*) and (*v*), we get

$$a = 21.89, \quad b = 82.3, \quad d = 9.28$$

$$m = 18.56, \quad n = 8.9$$

(*i*) **Chemical formula of fuel is : $C_{8.9}H_{18.56}$ (Ans.)**

$$\textbf{Actual air-fuel ratio} = \frac{m_a}{m_f} = \frac{21.89 \times 32 + 82.3 \times 28}{8.9 \times 12 + 18.56 \times 1} = \mathbf{23.97.} \quad \mathbf{(Ans.)}$$

(*ii*) **Stoichiometric air-fuel ratio :**

Stoichiometric air-fuel ratio can be found by finding the theoretical air required for complete combustion of 1 kg of fuel. Hence,

$$C_{8.9}\ H_{18.56} + 13.54\ O_2 = 8.9\ CO_2 + 9.28\ H_2O$$

Mass of oxygen required for combustion of 1 kg fuel

$$= \frac{13.54 \times 32}{8.9 \times 12 + 18.56 \times 1} = 3.456$$

Mass of air required for complete combustion of 1 kg fuel

$$= \frac{3.456}{0.23} = 15.03\ \text{kg}$$

Stoichiometric air-fuel ratio = 15.03 kg

Percentage theoretical air used $= \frac{23.97}{15.03} \times 100 =$ **159.48%. (Ans.)**

HIGHLIGHTS

Chemical Thermodynamics

1. A *chemical reaction* may be defined as the rearrangement of atoms due to redistribution of electrons. '*Reactants*' comprise of initial constituents which start the reaction while '*products*' comprise of final constituents which are formed by the chemical reaction.
2. A chemical fuel is a substance which releases heat energy on combustion.
3. The total number of atoms of each element concerned in the combustion remains constant, but the atoms are rearranged into groups having different chemical properties.
4. The amount of excess air supplied varies with the type of the fuel and the firing conditions. It may approach a value of 100 per cent but modern practice is to use 25% to 50% excess air.
5. *Stoichiometric* (or chemically correct) *mixture* of air and fuel is one that contains just sufficient oxygen for complete combustion of the fuel.
6. Mixture strength $= \frac{\text{Stoichiometric A/F ratio}}{\text{Actual A/F ratio}}$.
7. When analysis of combustion products is known air fuel ratio can be calculated by the following methods :

   (*a*) Fuel composition known

   (*i*) Carbon balance method (*ii*) Hydrogen balance method

   (*iii*) Carbon hydrogen balance method

   (*b*) Fuel composition unknown

   (*i*) Carbon hydrogen balance method.
8. The most common means of analysis of the combustion products is Orsat apparatus.
9. The *enthalpy of formation* (ΔH_f) is the increase in enthalpy when a compound is formed from its constituent elements in their natural form and in a standard state. The standard state is 25°C and 1 atm. pressure (but it must be borne in mind that not all substances can exist in natural form, *e.g.* H_2O cannot be a vapour at 1 atm. and 25°C).
10. $$(\text{HHV})_p = (\text{LHV})_p + m\, h_{fg}$$
    $$(\text{HHV})_v = (\text{LHV})_v + m(u_g - u_f)$$

    where, HHV = higher heating value

    LHV = lower heating value

    m = mass of water formed by combustion

    h_{fg} = enthalpy of vapourisation of water, kJ/kg

    u_g = specific internal energy of vapour, kJ/kg

    u_f = specific internal energy of liquid, kJ/kg.
11. In a given combustion process, that takes place adiabatically and with no work or changes in kinetic or potential energy involved, the temperature of the products is referred to as the '*adiabatic flame temperature*'.
12. For a given fuel and given pressure and temperature of the reactants, the maximum adiabatic flame temperature that can be achieved is with a 'stoichiometric' mixture.

Conventional Fuels

13. The constituents petroleum are classified into the following four groups :

    (*i*) Paraffins (*ii*) Olefins

    (*iii*) Naphthenes (*iv*) Aromatics.

14. The process of separating petroleum into useful fractions and removal of undesirable impurities is called *refining.*

15. *Volatility* is commonly defined as the *evaporating tendency of a liquid fuel.* The volatility of gasoline is generally characterised by the following two laboratory tests :

(*i*) ASTM distillation test (*ii*) Reid vapour test.

16. *Vapour lock* is a situation where too lean a mixture is supplied to the engine. The vapour lock tendency of the gasoline is related to front end volatility.

17. When self-ignition does occur in S.I. engine higher than desirable, pressure pulses are generated. The higher above self-ignition temperature which the fuel is heated, the shorter will be ignition delay.

18. The *highest useful compression ratio* (HUCR) is the highest compression ratio at which a fuel can be used without detonation in a specified test-engine under specified operating conditions and the ignition and mixture strength being adjusted to give best efficiency.

19. The property of a fuel which describes how fuel will or will not self-ignite is called the *Octane number* or just *Octane.* Engines with low compression ratios can use fuels with lower octane number, but high-compression engines must use high-octane fuel to avoid self-ignition and knock.

20. The difference in octane number between reseach method and motor method octane numbers is known as the *fuel sensitivity.*

21. Antiknock index $= \frac{\text{RON} + \text{MON}}{2}$.

22. High octane fuels (upto 100) can be produced by *refining techniques,* but it is done more cheaply, and more frequently, by the use of antiknock *additives* such as *tetraethyl lead.*

23. Cetane number of diesel fuel is a measure of its ignition quality.

In general, lower the cetane number higher are the hydrocarbon emissions and noise levels. In general, a high octane value implies a low cetane value.

24. Diesel index (D.I.) = Aniline point (°F) $\times \frac{\text{API gravity (deg)}}{100}$

Alternative Fuels

25. Some alternative fuels which can replace conventional fuels in I.C. engine are :

(*i*) Alcohol (methyl and ethyl) (*ii*) Hydrogen

(*iii*) Natural gas (*iv*) LPG and LNG

(*v*) Biogas.

26. The power output of ethanol engine is higher compared to gasoline engine at all speeds.

27. The CNG fuel used engines have higher thermal efficiencies than those fuelled by gasoline.

28. Biogas possesses excellent antiknock properties with an equivalent octane number in excess of 120 compared with 87 for regular petrol.

OBJECTIVE TYPE QUESTIONS

A. Choose the Correct Answer :

1. The smallest particle which can take part in a chemical change is called

(*a*) atom (*b*) molecule

(*c*) electron (*d*) compound.

2. A chemical fuel is a substance which releases ...... on combustion.

(*a*) chemical energy (*b*) heat energy

(*c*) sound energy (*d*) magnetic energy.

3. The most important solid fuel is

(*a*) wood (*b*) charcoal

(*c*) coal (*d*) all of the above.

4. For each mole of oxygen, number of moles of nitrogen required for complete combustion of carbon are
(*a*) 20/21 (*b*) 2/21
(*c*) 77/21 (*d*) 79/21.

5. Modern practice is to use ...... excess air.
(*a*) 5 to 10 per cent (*b*) 15 to 20 per cent
(*c*) 20 to 25 per cent (*d*) 25 to 50 per cent.

6. Stoichiometric air-fuel ratio by mass for combustion of petrol is
(*a*) 5 (*b*) 10
(*c*) 12 (*d*) 15.05.

7. An analysis which includes the steam in the exhaust is called
(*a*) dry analysis (*b*) wet analysis
(*c*) dry and wet analysis (*d*) none of the above.

8. The Orsat apparatus gives
(*a*) volumetric analysis of the dry products of combustion
(*b*) gravimetric analysis of the dry products of combustion
(*c*) gravimetric analysis of products of combustion including H_2O
(*d*) volumetric analysis of products of combustion including H_2O.

9. In the Orsat apparatus KOH solution is used to absorb
(*a*) carbon monoxide (*b*) carbon dioxide
(*c*) oxygen (*d*) none of the above.

10. Enthalpy of formation is defined as enthalpy of compounds at
(*a*) 25°C and 10 atmospheres (*b*) 25°C and 1 atmosphere
(*c*) 0°C and 1 atmosphere (*d*) 100°C and 1 atmosphere.

11. Bomb calorimeter is used to find the calorific value of ...... fuels.
(*a*) solid (*b*) gaseous
(*c*) solid and gaseous (*d*) none of the above.

12. When the fuel is burned and the water appears in the vapour phase, the heating value of fuel is called
(*a*) enthalpy of formation (*b*) lower heating value
(*c*) higher heating value (*d*) none of the above.

13. Heat released in a reaction at constant pressure is called
(*a*) entropy change (*b*) enthalpy of reaction
(*c*) internal energy of reaction (*d*) none of the above
(*e*) all of the above.

14. When the fuel is burned and water is released in the liquid phase, the heating value of fuel is called
(*a*) higher heating value (*b*) lower heating value
(*c*) enthalpy of formation (*d*) none of the above.

15. Choose the *correct* statement :
(*a*) Number of atoms of each constituent are not conserved in a chemical reaction.
(*b*) The mass of all the substances on one side of the equation may not be equal to the mass of all the substances on the other side.
(*c*) The number of atoms of each constituent are conserved in a chemical reaction.
(*d*) The number of moles of the reactants in a chemical equation are equal to the number of moles of the products.

ANSWERS

1. (*a*) **2.** (*b*) **3.** (*c*) **4.** (*d*) **5.** (*d*) **6.** (*d*) **7.** (*b*)
8. (*a*) **9.** (*b*) **10.** (*b*) **11.** (*a*) **12.** (*b*) **13.** (*b*) **14.** (*a*)
15. (*c*).

B. Fill in the Blanks or Say "Yes" or "No" :

1. The liquid fuels find ...... use in I.C. engines.
2. ...... gas is manufactured by heating soft coal in closed vessel.
3. Water gas is formed by using steam.
4. ...... gas is a by-product of steel plants.
5. Benzol is obtained as a by-product of high temperature coal carbonization.
6. Olefins are compounds with one or more double bonded carbon atoms in straight chain.
7. Diolefins are more ...... than mono-olefins.
8. Napthenes are ...... structured compounds.
9. Napthenes are unsaturated compounds.
10. Paraffins are most suitable fuels for ...... engines.
11. Aromatics are most suitable fuels for ...... engines.
12. The process of separating petroleum into useful fractions and removal of undesirable impurities is called ....
13. ...... is the evaporating tendency of a liquid fuel.
14. The ...... of petrol is also defined in terms of Reid vapour pressure.
15. ...... is a situation where too lean a mixture is supplied to the engine.
16. The vapour lock tendency of gasoline is related to ...... end volatility.
17. When self-ignition does occur in S.I. engine higher than desirable, ... pulses are generated.
18. The higher above S.I.T. which the fuel is heated, the longer will be ignition delay.
19. Ignition delay is generally a very small fraction of a second.
20. The property of fuel which describes how fuel will or will not self-ignite is called the ...... number.
21. The higher the octane number of fuel, the less likely it will self-ignite.
22. The difference in octane number between research method and the motor method octane numbers is known as the ....
23. Fuel sensitivity is a good measure of how sensitive knock characteristics of a fuel will be to engine geometry.
24. Anti-knock index = $\dfrac{\text{RON} - \text{MON}}{2}$.
25. The cetane number of a diesel fuel is a measure of its ignition quality.
26. Higher the cetane rating of the fuel lesser is the propensity for diesel knock.
27. In general, a high octane value implies a high cetane value.
28. Methanol is very corrosive on metals.
29. Ethanol has less HC emissions than gasoline but more than methanol.
30. The CNG used engines have lower thermal efficiencies than those fuelled by gasoline.

ANSWERS

1. abundant	**2.** coal	**3.** Yes	**4.** Blast furnace	**5.** Yes
6. Yes	**7.** unstable	**8.** ring	**9.** No.	**10.** C.I.
11. S.I.	**12.** refining	**13.** Volatility	**14.** Volatility	**15.** Vapour lock
16. front	**17.** pressure	**18.** No.	**19.** Yes	**20.** octane
21. Yes	**22.** fuel sensitivity	**23.** Yes	**24.** No	**25.** Yes
26. Yes	**27.** No	**28.** Yes	**29.** Yes	**30.** No.

THEORETICAL QUESTIONS

Chemical Thermodynamics

1. What is chemical thermodynamics ?
2. What is a chemical fuel ?
3. What are primary fuels ? List some important primary fuels.
4. What are secondary fuels ? List some important secondary fuels.
5. Write a short note on 'excess air'.
6. What do you mean by stoichiometric air fuel (A/F) ratio ?
7. Enumerate the methods by which air fuel ratio can be calculated when analysis of combustion products is known.
8. How is analysis of exhaust and flue gas carried out ?
9. Derive relations for internal energy and enthalpy of reaction.
10. What is enthalpy of formation (ΔH_f) = ?
11. Define heating value of fuel.
12. What is the difference between higher heating value (HHV) and lower heating value (LHV) of the fuel ?
13. Describe with the help of neat sketches the following calorimeters used for the determination of heating values :
    (*i*) Bomb calorimeter (*ii*) Junkers gas calorimeter.
14. What is 'adiabatic flame temperature' ?
15. Write a short note on chemical equilibrium.

Conventional Fuels

16. What are the desirable properties of good I.C. engines fuels ?
17. Enumerate and describe briefly the gaseous fuels.
18. How are constituents of petroleum classified ?
19. Explain briefly the chemical structure of petroleum.
20. What are five primary hydrocarbon families found in petroleum ? Which are chain types ? Which are ring types ? Which of primary families tends to be better S.I. engine fuel and C.I. engine fuel ?
21. What are different kinds of fuels used in an I.C. engine ?
22. What are the important properties which S.I. engine fuel possess ?
23. What are requirements of an ideal gasoline fuel ?
24. What is volatility ?
25. Discuss the significance of distillation curve.
26. Why volatility is an important quality of S.I. engine fuels ?
27. Explain briefly the following in regard to a fuel :
    (*i*) Vapour lock characteristics.
    (*ii*) Crankcase dilution.
28. "While volatility of the fuel is a determining factor in the selection of fuels for S.I. engines, ignition quality of the fuel is the primary deciding factor for C.I. engines". Discuss briefly the statements.
29. Distinguish clearly between 'Octane Numbers' and 'Cetane Number'. What is their significance in rating of fuels for S.I. and C.I. engines ?
30. What are the reference fuels for 'Octane Number' ?
31. What are the reference fuels for 'Cetane Number' ?
32. What is performance number (PN) ?
33. What is the significance of ASTM distillation curve ?
34. Explain the effect of fuel viscosity on diesel engine performance.
35. What qualities are desired in fuels to inhibit detonation ?

36. Give the advantages of using alternate fuels.

37. Discuss different properties of ethanol and methanol and compare them with gasoline.

38. Why blends of either ethanol or methanol are preferred over pure alcohol fuels ?

39. Give the advantages of alcohol as a fuel.

40. List the advantages of methanol as a fuel.

41. What modifications in engine are required when blends are used ?

42. State the advantages and disadvantages of hydrogen as I.C. engine fuel.

43. What is natural gas ?

44. What are the properties of CNG ?

45. What are the advantages and disadvantages of CNG ?

46. Explain briefly LPG and LNG.

47. What is Biogas ?

48. What are the properties of biogas ?

UNSOLVED EXAMPLES

1. Determine the gravimetric analysis of the products of complete combustion of acetylene (C_2H_2) with 125 per cent stoichiometric air. **[Ans.** CO_2 = 19.5%, H_2O = 3.9%, O_2 = 4.4%, N_2 = 72.2%]

2. One kg of ethane (C_2H_6) is burned with 80% of theoretical air. Assuming complete combustion of the hydrogen in the fuel determine the volumetric analysis of the dry products of combustion.

[Ans. CO_2 = 4.8%, CO = 11.2%, N_2 = 84%]

3. The gravimetric analysis of a sample of coal is given as 80% C, 12% H_2 and 8% ash. Calculate the stoichiometric A/F ratio and the analysis of the products by volume.

[Ans. CO_2 = 13.6%, H_2 = 12.2%, N_2 = 74.2%]

4. Calculate the stoichiometric air fuel ratio for the combustion of a sample of dry anthracite of the following composition by mass :

C = 90 per cent ; H_2 = 3 per cent ; N_2 = 1 per cent ; Sulphur = 0.5 per cent ; ash = 3 per cent.

If 20 per cent excess air is supplied determine :

(*i*) Air fuel ratio

(*ii*) Wet analysis of the products of combustion by volume.

[Ans. 11.25/1 (*i*) 13.5/1 ; (*ii*) CO_2 = 16.3%, H_2O = 0.03%, SO_2 = 3.51%, N_2 = 80.3%]

5. The following is the analysis of a supply of coal gas :

H_2 = 49.4 per cent ; CO = 18 per cent ; CH_4 = 20 per cent ; C_4H_8 = 2 per cent ; O_2 = 0.4 per cent ; N_2 = 6.2 per cent ; CO_2 = 4 per cent.

(*i*) Calculate the stoichiometric A/F ratio.

(*ii*) Find also the wet and dry analysis of the products of combustion if the actual mixture is 20 per cent weak. **[Ans.** (*i*) 4.06/1 by volume ; (*ii*) Wet analysis : CO_2 = 9.0%, H_2O = 17.5%, O_2 = 3.08%, N_2 = 70.4%. Dry analysis : CO_2 = 10.9%, O_2 = 3.72%, N_2 = 85.4%]

6. Find the stoichiometric air fuel ratio for the combustion of ethyl alcohol (C_2H_6O), in a petrol engine. Calculate the air fuel ratios for the extreme mixture strengths of 90% and 120%. Determine also the wet and dry analysis by volume of the exhaust gas for each mixture strength.

[Ans. 8.96/1 ; 9.95/1 ; 7.47/1, Wet analysis : CO_2 = 11.2%, H_2O = 16.8%, O_2 = 1.85%, N_2 = 70.2%

Dry analysis : CO_2 = 13.45%, O_2 = 2.22%, N_2 = 84.4%

Wet analysis : CO_2 = 6.94%, CO = 6.94%, H_2 = 20.8%, N_2 = 65.3%

Dry analysis : CO_2 = 8.7%, CO = 8.7%, N_2 = 82.5%]

7. For the stoichiometric mixture of Example 7.10 calculate :

(*i*) The volume of the mixture per kg of fuel at a temperature of 65°C and a pressure of 1.013 bar.

(*ii*) The volume of the products of combustion per kg of fuel after cooling to a temperature of 120°C at a pressure of 1 bar. **[Ans.** (*i*) 9.226 m^3 ; (*ii*) 11.58 m^3]

8. The chemical analysis of a fuel by weight is as follows :
Carbon = 50 per cent ; Hydrogen = 25 per cent ; Oxygen = 10 per cent ; Sulphur = 5 per cent and Nitrogen = 10 per cent.
Find the stoichiometric amount of air required for complete combustion of this fuel. **[Ans.** 14.26 kg]

9. The percentage composition of a fuel by weight is as follows :
Carbon = 89.3 per cent ; Hydrogen = 5 per cent ; Oxygen = 4.2 per cent ; Nitrogen = 1.5 per cent and the remainder ash. Determine the stoichiometric air fuel ratio by mass.
If 30 per cent excess air is supplied, find the percentage composition of dry flue gases by volume.
[Ans. 11.74 ; CO_2 = 14.3%, O_2 = 4.9%, N_2 = 80.8%]

10. Orsat analysis of the products of combustion of hydrocarbon fuel of unknown composition is as follows :
Carbon dioxide (CO_2) = 9% Carbon monoxide (CO) = 0.6%
Oxygen (O_2) = 7.3% Nitrogen (N_2) = 83.1%
Determine the following :
(*i*) Air-fuel ratio (*ii*) Per cent theoretical air required for combustion.
[Ans. (*i*) 22.1, (*ii*) 146.2%]

11. An Orsat analysis of the exhaust from an engine running on benzole showed a CO_2 content of 15 per cent, but no CO. Assuming that the remainder of the exhaust contains only oxygen and nitrogen, calculate the air-fuel ratio of the engine.
The ultimate analysis of benzole is C = 90 per cent and H_2 = 10%.
[Ans. 15.2/1]

12. The analysis of the dry exhaust from and internal-combustion engine gave :
CO_2 = 12 per cent ; CO = 2 per cent ; CH_4 = 4 per cent ; H_2 = 1 per cent ; O_2 = 4.5 per cent and the remainder nitrogen.
Calculate the proportions by mass of carbon to hydrogen in the fuel, assuming it to be a pure hydrocarbon.
[Ans. 7.35/1]

13. The following is the percentage analysis by mass of a fuel :
Hydrogen (H_2) = 10 per cent Oxygen (O_2) = 2 per cent
Sulphur (S) = 1 per cent Nitrogen (N_2) = 3 per cent
Determine the following :
(*i*) The amount of air required to completely burn 1 kg of this fuel
(*ii*) The products of combustion as a percentage by mass.
[Ans. 13.17 kg ; CO_2 = 21.7% ; H_2O = 6.35% ; SO_2 = 0.141% ; N_2 = 71.75%]

14. An Orsat analysis of the products of combustion resulting from the burning in air of a hydrocarbon fuel yielded the following :
Carbon dioxide (CO_2) = 12.2% Oxygen (O_2) = 1.1%
Carbon monoxide (CO) = 0.5% Nitrogen (N_2) = 86.2%
Determine : (*i*) The mass fraction of carbon in the fuel.
(*ii*) Air-fuel ratio.
(*iii*) Percent of air theoretically needed for complete combustion.
[Ans. (*i*) 80.4%, (*ii*) 16.57/1, (*iii*) 103.9%]

15. If the higher heating value at constant pressure (C_6H_6) at 25°C is 3298354 kJ/kg mole, determine its lower calorific value at constant pressure. **[Ans.** 316647 kJ/mole]

16. The lower heating value of propane at constant pressure and 25°C is 2042055 kJ/kg mole. Find the higher heating value at constant pressure and at constant volume.
[Ans. 2217816 kJ/kg mole, 2210333 kJ/kg mole]

17. The higher heating value of kerosene at constant volume whose ultimate analysis is 86% carbon and 14% hydrogen, was found to be 46890 kJ/kg. Calculate the other three heating values.
[Ans. $(LHV)_v$ = 43987 kJ/kg ; $(HHV)_p$ = 46977 kJ/kg ; $(LHV)_p$ = 43900 kJ/kg]

10

Fuel/Air Mixture Requirements

10.1. INTRODUCTION

- In general, we have already discussed in previous chapters about the profound influence of F/A ratio on S.I. engine power output and thermal efficiency. The discussion of adequate F/A ratio for each particular set of operating conditions is studied under the following two heads, namely
  (*i*) Steady running ;
  (*ii*) Transient operation.
- **Steady running** is *defined as mean continuous operation at a required speed and power output with normal temperatures.*
- **Transient operation** includes *starting, warming up*, and *changing from one speed or load* to another, specially for automotive vehicle engines during *acceleration* and *decelerations,* and also *idling.*

10.2. FUEL/AIR MIXTURE REQUIREMENTS FOR STEADY RUNNING

The specific torque or b.m.e.p. (brake mean effective pressure) is required to be developed at specific speed. Further it is desired that fuel consumption should be lowest, ensuring reliable and smooth operation. These requirements can be met with by using optimum F/A ratio.

- Fig. 10.1 shows the graphs of *i.m.e.p. (indicated mean effective pressure) and s.f.c. (specific fuel consumption at fixed engine speed at full throttle open.*
  - The left portion of the curve exhibits the *limit of lean mixture which causes explosions in the intake system, usually known as* **back-firing**.
  - Depending upon the type of engine, type of fuel and operating conditions, the limits on lean and rich side and also F/A ratios may vary.
  - The curve shown in Fig. 10.1 is a representative one since its shape is same for all S.I. engines. It may be noted that, for getting these curves optimum spark advance is used for each F/A ratio. If the spark advance is kept fixed for the best F/A ratio, this curve of lean and rich ends will get modified.

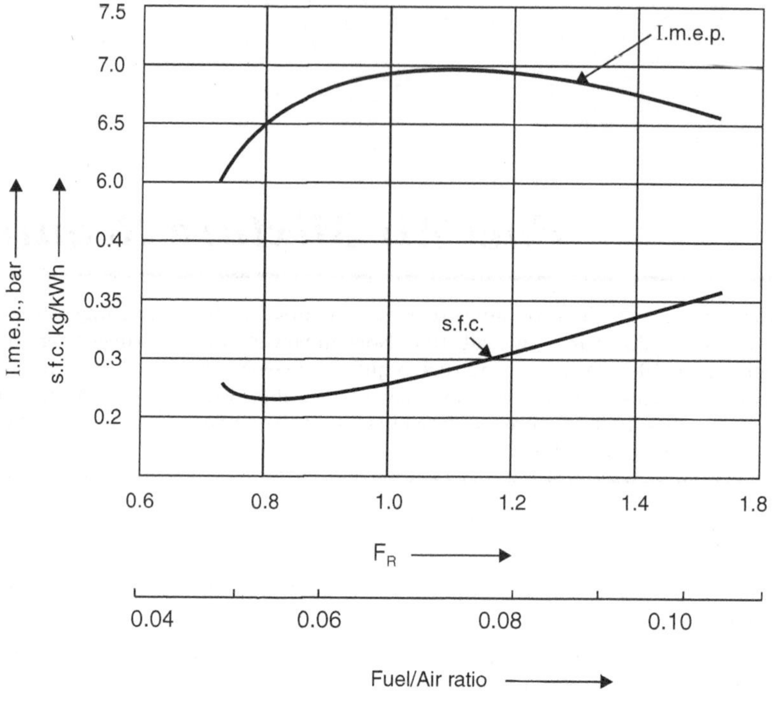

Fig. 10.1

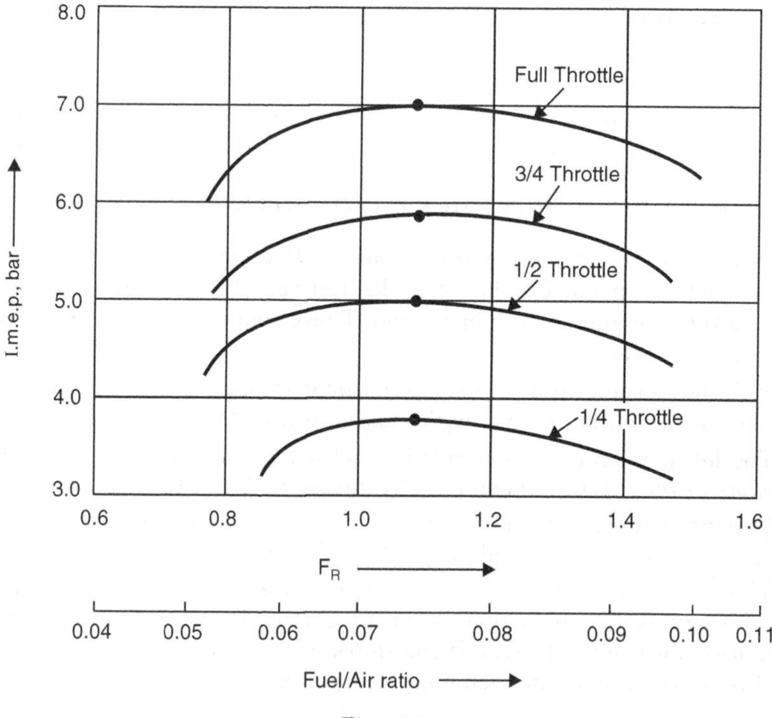

Fig. 10.2

- It has been observed from experiments that the F/A ratio which gives the highest i.m.e.p. (Refer Fig. 10.2) is more or less the same as the F/A ratio which gives the highest b.m.e.p. (Refer Fig. 10.3). This indicates that f.m.e.p. (frictional mean effective pressure) is *not* affected by F/A ratio. Hence for best power under all operation conditions the F/A ratio is same subject to the condition that distribution of fuel to various cylinders remains unchanged. In carburetted engines this requirement cannot be met with easily. At lean mixtures, the effect of lower flame speeds is small when, at each F/A ratio, spark timing is adjusted for highest m.e.p. (mean effective pressure).

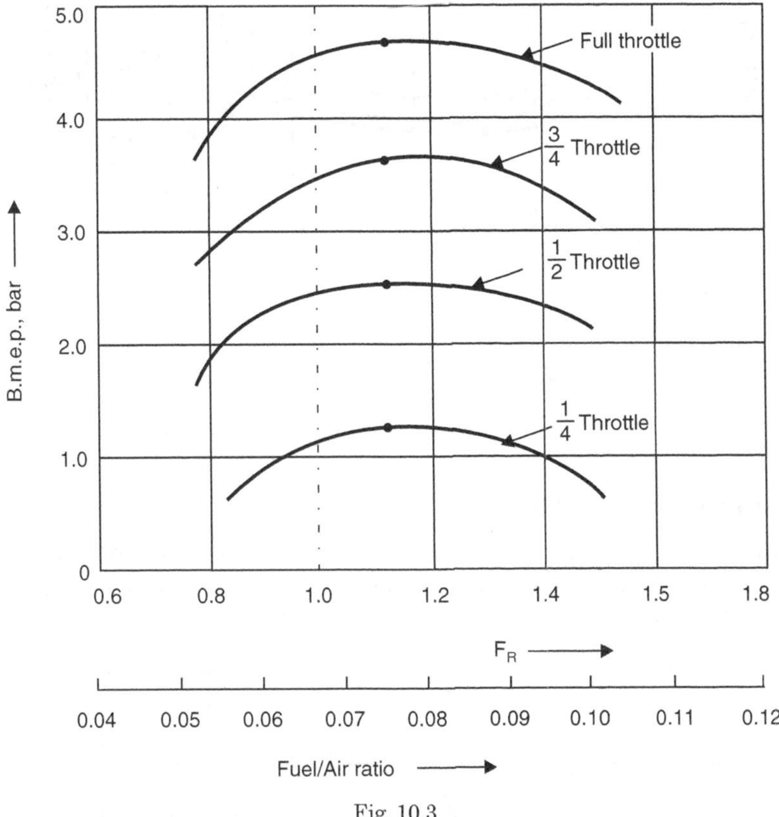

Fig. 10.3

- Fig. 10.4 shows graphs between specific fuel consumption versus F/A ratio at different throttle openings.
  - The minimum specific fuel consumption occurs at a point where, as the F/A ratio is reduced, the rate of increase in efficiency due to thermodynamic factor, is offset by rate of decrease in efficiency *due to increasing time losses.*
  - With optimum spark timing, the time losses become very large only when crank angle occupied by flame travel exceeds a certain value. As the mixture is made leaner, this value will be reached at high F/A ratio when the flame speed has already been slowed by throttling. Thus *any factor that tends to increase the crank angle occupied by the combustion (i.e., reduce the flame speeds), will tend to increase the F/A ratio for best economy.*

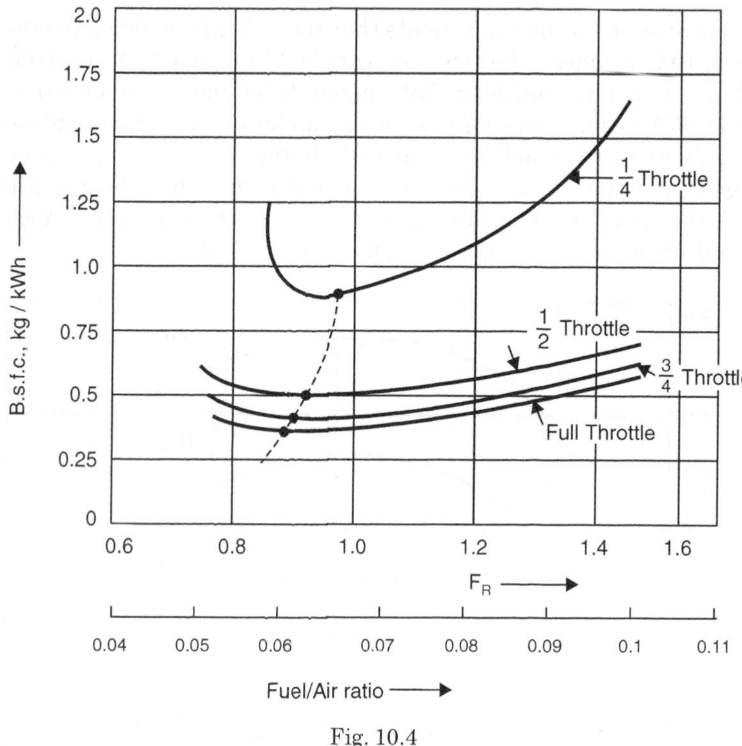

Fig. 10.4

10.3. OPTIMUM FUEL/AIR RATIOS

- Fig. 10.5 shows graphs between b.m.e.p. and F/A ratio for different speeds. It may be observed that all the curves are similar which leads to the conclusion that the *best economy F/A ratio is independent of speed.*

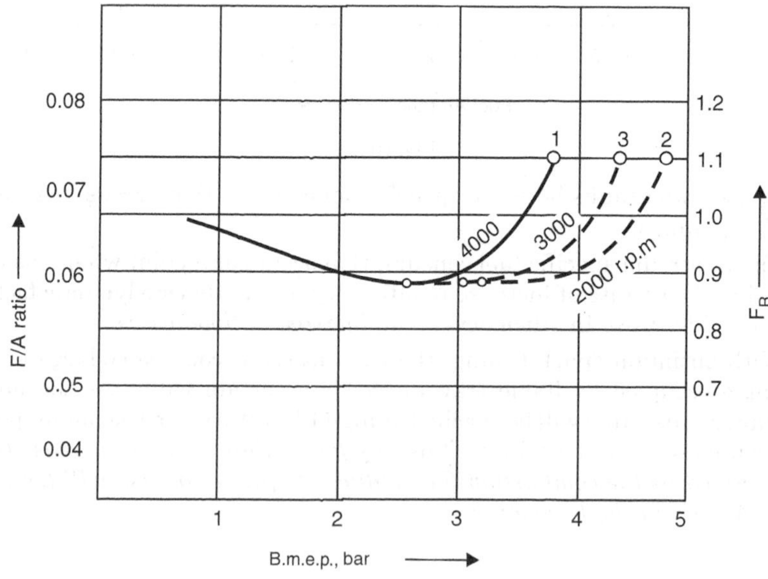

Fig. 10.5

- Fig. **10.6** shows a plot between load (ratio of actual b.m.e.p. to maximum b.m.e.p.) and F/A ratio.

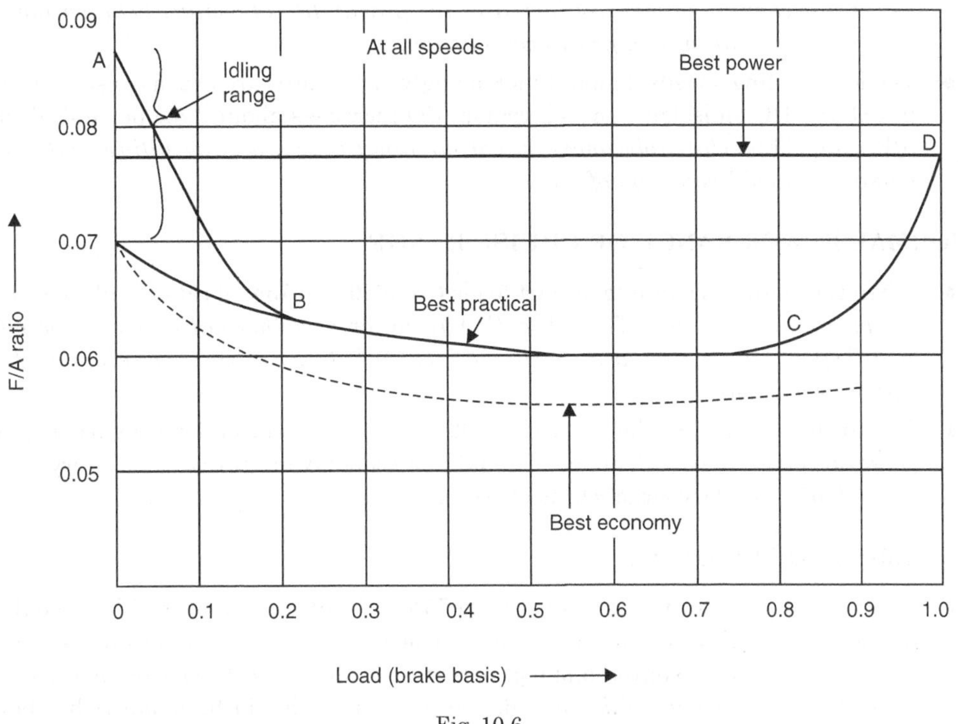

Fig. 10.6

— Starting with idling range, the curves show the F/A ratios in the idling range, best power, best economy etc. These curves are *practically independent of engine speed.*

10.4. IDLING AND LOW LOAD

- The *no load running mode of engine is called* **idling**. The air supply, during idling, is restricted by the nearly closed throttle and the suction pressure is very low. This condition of low pressure *gives rise to backflow of exhaust gases and air leakage from various parts of the engine intake system.*
- During idling and low load operation backflow during the valve overlap period occurs since the exhaust pressure is higher than the intake pressure, this *increases the amount of residual gases.* These gases expand during the suction stroke, thereby reducing the fresh mixture inhaled. The increased dilution causes the *combustion to be erratic* or *even impossible* which leads to *poor thermal efficiency and higher exhaust emissions.*
  — At idling and low loads the problem of dilution by residual gases becomes more pronounced because the exhaust temperature reduces with decreasing load, *i.e.* the density and hence *mass of residual gases increases.* Further, dilution of the charge occurs *due to air leakage past valves,* etc., at low intake manifold pressures obtained at low loads and idling.

- The *A/F ratio used for idling and low loads* (up to about 20% of full load) *should be* **rich** for *smooth operation* (F/A ratio 0.08 or A/F ratio $\simeq$ 12.5 : 1).
  — The *richening of mixture increases the probability of contact between fuel and air particles and thus improves combustion.*
- The lower idling speeds demand increasingly richer mixtures with consequent increase in CO and HC emissions. In rich region, CO increases about 2.8% per unit decrease in A/F ratio. Thus *close tolerances on carburettor jet sizes and on idling controls are demanded to avoid pollution of air.*

10.5. NORMAL POWER RANGE OR CRUISE RANGE

- When the engine is running at part loads (from 20% to 75% of rated load) the *objective is the maximum economy.* Thus, the *F/A ratio for best thermal efficiency or minimum specific fuel consumption is selected and incidently this ratio gives minimum HC* (hydrocarbon) *emissions.*
- As shown in Fig. 10.6, the F/A ratio 0.06 is the best compromise for various part loads operation of a modern S.I. engine (Actual best economy mixture at laboratory level works out to 0.055, but best practical is chosen).

10.6. MAXIMUM POWER RANGE

- The maximum power range lies between 75% to 100% rated power. When the throttle is full opened, the F/A ratio has to be made richer, because maximum torque is required at a given speed, or maximum b.m.e.p. is required. Moreover, the *richer mixture serves as coolant to prevent valve failure* and, therefore, F/A should be made richer before the throttle valve is wide open as shown by C-D portion of the curve in Fig. 10.6. *Incidentally, too rich mixtures inhibit NO_x and detonation.*
- The mixture requirement for maximum power is a rich mixture of A/F about 14 : 1 or F/A $\approx$ 0.07.
- In multi-cylinder engines the A/F ratios are slightly lower (*i.e.*, slightly richer mixture) to overcome maldistribution of air fuel mixture in different cylinders.
- In case of super-charged engines the best economy mixture is leaner than its naturally aspirated counterpart over the whole operational range.

Steady Running Mixture Requirements in Practical Operation

- Refer Fig. 10.6. The S.I. engines using gasoline show best economy F/A ratio very close to one shown in the figure, particularly, in the absence of severe detonation and with good distribution of fuel-air mixture from manifold to various cylinders at all speeds and loads.
- The following are the reasons due to which the departure from the best-brake economy curve shown in Fig. 10.6 may be practically needed :
  (*i*) To compensate for poor distribution.
  (*ii*) To allow for possible errors or variations in carburettor metering.
  (*iii*) To reduce the temperature of hot spots such as exhaust valve, spark plug points or piston crown, to help in cooling.
  (*iv*) To reduce or eliminate detonation.

10.7. TRANSIENT MIXTURE REQUIREMENTS

- *Transient conditions are those conditions at which, speed, load, temperatures or pressures are abnormal or changing rapidly,* like in :
  - — starting of an engine ;
  - — warming up of an engine ;
  - — acceleration of vehicle (*i.e.*, increased in load) ;
  - — deceleration of vehicle (*i.e.*, decrease in load).
- The transient mixture requirements are different from steady running mixture requirements because in the former case the *evaporation of the fuel may be incomplete, the quantity of liquid fuel in the inlet manifold may be increasing or decreasing, and distribution of fuel to various cylinders may be different.*

10.7.1. Starting and Warming Up Mixture Requirements

It has been observed that a very cold engine generally requires abnormally rich mixture at the carburettor in order to secure firing mixture in the cylinder. Thus the carburetion system must supply very rich mixtures for starting and the F/A ratio must be progressively reduced from this point during the warm up period until, the engine will run satisfactorily with the normal steady running F/A ratios.

— During starting a very rich mixture must be supplied, as much as 5 to 10 times the normal amount of petrol (A/F ratio 3 : 1 to 15 : 1, F/A ratio 0.3 to 0.07). With the warming up of the engine the amount of evaporated fuel increases and hence the mixture ratio should be progressively made leaner to avoid too much evaporated F/A ratio.

10.7.2. Mixture Requirements for Acceleration

- With regard to engines, the term 'acceleration' is generally used to refer to an increase in engine speed resulting from opening the throttle. The main purpose of opening the throttle, however, is to provide an increase in torque and whether or not an increase in speed follows depends on the nature of the load.
  - — With constant speed engines the throttle opening increases torque or b.m.e.p. at the governed speed.
- In order to impart acceleration when the throttle is opened, the manifold pressure increases, and fuel must be supplied to increase liquid content of the fuel of the manifold. If the carburettor supplies constant F/A ratio, the F/A ratio going to the manifold will becomes less during the time the liquid content of the manifold is being built up to a larger value when the throttle is suddenly opened. The reduction in the F/A ratio to the cylinders can be such as to *cause misfiring, backfiring,* or *even complete stopping of the engine*. In order to *avoid such a situation it is often found essential to increase the supply ratio by injecting into the manifold a quantity of fuel known as* **"Accelerating charge"**. The *optimum amount of accelerating charge is that which gives best power F/A ratio in the cylinder.*

In general, the required F/A ratios for various running conditions of the S.I. engine are listed below :

S. No.	Running condition	F/A ratio
1.	*Starting*	0.2 : 1
2.	*Warming*	0.15 : 1
3.	*Idling*	0.085 : 1
4.	*Running with maximum thermal efficiency (80% throttle)*	0.06 to 0.07 : 1
5.	*Running with developing maximum power (80% throttle)*	0.0775 to 0.08 : 1
6.	*Full throttle*	0.085 : 1
7.	*Acceleration*	0.1 : 1

10.8. EFFECTS OF OPERATING VARIABLES ON MIXTURE REQUIREMENTS

The effects of operating variables such as inlet and exhaust pressures, spark timing and friction are discussed below :

1. **Inlet and exhaust pressure :**

- *Reduction in inlet pressure due to throttling or operating at altitudes leads to reduction in flame speed, and increase in F/A ratio for best economy.*
- Increase in exhaust pressures result in reduced flame speeds and increase in F/A ratio for best economy.
- With constant throttling, simultaneous decrease of inlet and exhaust pressures at high altitudes also affects the F/A ratio depending upon the net effect.

2. **Spark timing**

Any *deviation from the optimum spark timing will lead to the increase of best economy F/A ratio, since it will increase the time losses.*

3. **Friction**

Keeping the i.m.e.p. constant, the increase in f.m.e.p. will result in the *increase of F/A ratio for best economy.*

10.9. MIXTURE REQUIREMENTS FOR DIESEL ENGINES

- An engine, under normal operating conditions and with injection timing fixed and one b.m.e.p., will require one F/A ratio. The F/A ratio controls the output of the engine. Thus with given $\eta_{vol.}$ and given intake air conditions (ρ_a), a given value of F/A ratio (F_R) will produce one i.m.e.p. and hence one b.m.e.p.
  — Fig. 10.7 shows graphs between i.m.e.p. and F_R for various factors of the products of $\rho_a \times \eta_{vol}$. The values of ρ_a have been chosen for below and above atmospheric conditions in order to cover high altitude un-supercharged and supercharged engines.
- The diesel engines do not show any definite low limit on F/A ratio and can burn fuel-air mixtures with very lean F/A ratio.

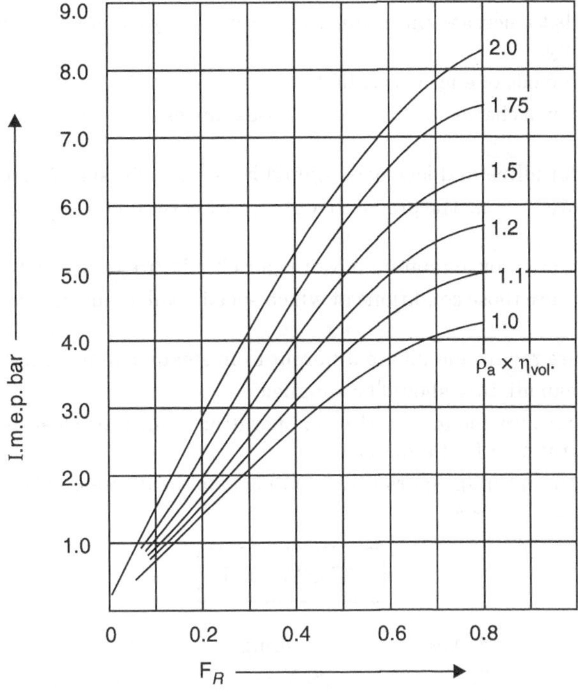

Fig. 10.7

- In diesel engines the fuel injection spray envelope contains in its evaporated portion, a heterogeneous mixture with local F/A ratios varying from $F_R = 0$ to $F_R = \infty$. Thus *compression ignition of most favourable local F/A ratio occurs initiating the flame which later serves as very hot source and assists in flame stabilisation and burning of fuel-air mixtures.* The practical high limit of overall F/A ratio, however, is set by smoke and deposits.
- The *smoke free combustion is seldom obtained above* $F_R = 0.8$ *and most diesel engines are never rated above this limit of F/A ratio for continuous operation.*

HIGHLIGHTS

1. *Steady running* is defined as mean continuous operation at a required speed and power output with normal temperatures.
2. *Transient operation* includes starting, warming up, and changing from one speed or load to another, specially for automotive vehicle engines during acceleration and decelerations, and also idling.
3. The richening of mixture increases the probability of contact between fuel and air particles and thus improves combustion.
4. The smoke free combustion is seldom obtained above $F_R = 0.8$ and most diesel engines are never rated above this limit of F/A ratio for continuous operation.

OBJECTIVE TYPE QUESTIONS

Fill in the Blanks or Say "Yes" or "No".

1. .... is defined as mean continuous operation at a required speed and power output with normal temperatures.

2. Any factor that tends to increase the crank angle occupied by combustion, will tend to increase the F/A ratio for best economy.
3. The no load running mode of engine is called ....
4. The increased dilution of charge due to exhaust gases causes the combustion to be erratic or even impossible.
5. The A/F ratio used for idling and low loads should be .... for smooth operation.
6. The richening mixture .... the probability of contact between fuel and air particles and thus .... combustion.
7. The mixture requirement for maximum power is a rich mixture of A/F of about 3 : 1.
8. Transient conditions are those conditions at which speed, load, temperatures or pressures are abnormal or changing rapidly.
9. The transient mixture requirements are different from steady running mixture requirements.
10. During starting, a lean mixture should be supplied.
11. Reduction in inlet pressure due to throttling or operating at altitudes leads to reduction in flame speeds and increase in F/A ratio for best economy.
12. Any deviation from the optimum spark timing will lead to the increase of best economy F/A ratio, since it will increase the .... losses.

ANSWERS

1. Steady running	**2.** Yes	**3.** idling	**4.** Yes	**5.** rich
6. increases, improves	**7.** No	**8.** Yes	**9.** Yes	**10.** No
11. Yes	**12.** time.			

THEORETICAL QUESTIONS

1. What do you mean by "Steady running" ?
2. What is a "Transient operation" ?
3. Explain briefly fuel/air mixture requirements for steady running.
4. Describe briefly optimum fuel/air ratios.
5. State the fuel/air mixture requirements for the following :
   (*i*) Idling and low load.
   (*ii*) Normal power range or cruise range.
   (*iii*) Maximum power range.
6. What are "Transient conditions" ?
7. Explain briefly the following :
   (*i*) Starting and warming up mixture requirements.
   (*ii*) Mixture requirements for acceleration.
8. What are the effects of following operating variables on mixture requirements ?
   (*i*) Inlet and exhaust pressure
   (*ii*) Spark timing
   (*iii*) Friction.
9. Write a short note on "Mixture requirements for diesel engines".

11

Carburetion and Carburettors

11.1. INTRODUCTION

Carburetion. *The process of preparing in the S.I. engine, a combustible fuel air mixture outside the engine cylinder is called* ***"carburetion".*** This complicated process is achieved in the *induction system.*

Carburettor. A *carburettor is a device which atomises the fuel and mixes it with air.* It is the *most* important part of the induction system.

— For several decades, carburettors were used on most S.I. engines as the means of adding fuel to the intake air. The basic principle on which the carburettor works is extremely simple, but by the 1980s, when fuel injectors finally replaced it as the main fuel input system, it had evolved into a complicated, sophisticated, expensive system. Carburettors are still found on a few automobiles, but vast majority of car engines use simpler, better controlled, more flexible fuel injector systems. Many small engines like those on lawn movers and model airplanes still use carburettors, although much simpler ones than those found on the automobile engines of the 1960s and 1970s. *This is to keep the cost of these engines down, simple carburettor being cheap to manufacture while fuel injectors require more costly control systems. Even on some of these small engines, carburettors are being replaced with fuel injectors as pollution laws become more stringent.*

11.2. INDUCTION SYSTEM

The schematic arrangement of induction system is shown in Fig. 11.1.

- The pipe that carries the prepared mixture to the engine cylinders is called the *intake manifold.*
- The carburettor is the focal point of the induction system.
- The fuel system, comprising the fuel supply tank and necessary fuel pumps, lines and filters supply liquid fuel to the carburettor.
- During the motion stroke vacuum is created in the cylinder which causes the air to flow through the carburettor and the fuel to be sprayed from the fuel jets. Due to the volatility

of the fuel, most of the fuel vaporises and forms a combustible fuel-air mixture. However, some of the larger drops may reach the cylinder in the liquid form and must be vaporised and mixed with air during the compression stroke before the electric spark ignites the mixture.

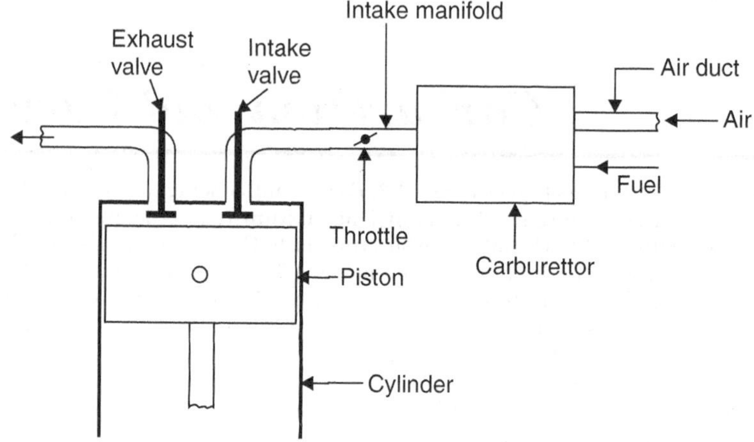

Fig. 11.1. Schematic arrangement of the 'Induction System'.

- The throttle located in the carburettor, regulates the quantity of the mixture.
- There is a limited range of A / F ratio in a homogenous mixture within which combustion in S.I. engines is sustainable. Outside this range, the ratio is too rich or too lean to sustain flame propagation. This range of useful A /F ratio is from approx. 20 : 1 (lean) to 8 : 1 (rich).

11.3. FACTORS INFLUENCING CARBURETION

The *various factors which influence the process of carburetion are as follows :*

1. The engine speed ; the time available for the preparation of the mixture.
2. The vaporisation characteristics of fuel.
3. The temperature of the incoming air.
4. The design of the carburettor.

- In case of modern high speed engines, the time duration available for the formation of mixture is very small and limited. The time duration for mixture formation and induction may be of the order of 10 to 5 milliseconds.
- Atomisation, mixing and vaporisation are the processes which require a finite time to occur. The time available for mixture formation is very small in high speed engines (For example, in an engine running at 3000 r.p.m., the induction process lasts for less than 0.02 second). For completion of these processes in such a small period a great ingenuity is required in designing the carburettor system.
- In order to achieve high quality carburetion within such a short time requires good vaporisation characteristics of the fuel which are ensured by presence of high volatile hydrocarbons in the fuel.
- The temperature is a factor which effectively controls vaporisation process of the fuel. If the temperature of the incoming air is high, it results in higher rates of vaporisation. The mixture temperature can be increased by heating the induction manifold but it will result in reducing power due to reduction in mass flow rate.

- For a S.I. engine, the design of carburetion system is very complicated owing to the fact that the air-fuel ratio required by it varies widely over its range of operation, particularly for an automotive engine. For idling as well as for maximum power rich mixture is required.

11.4. MIXTURE REQUIREMENTS

Fig. 11.2. shows the variation of mixture requirements from no-load to full-load in a S.I. engine.

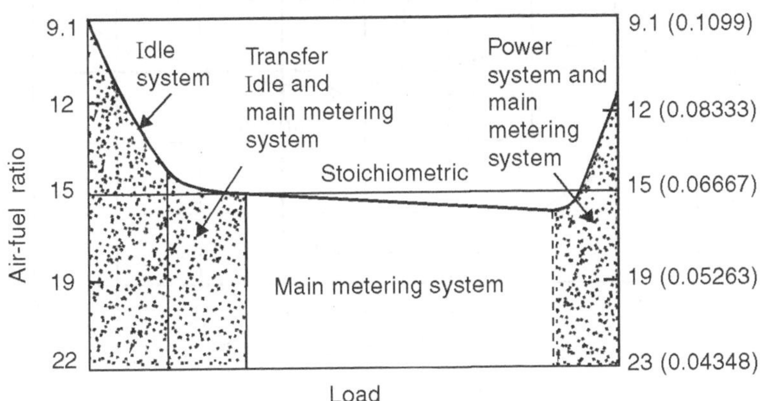

Fig. 11.2. Mixture requirements of automotive S.I. engine.

1. **Idling and low speed** (*From no-load to about 20% of rated power*) :

Idling refers to *no power demand.* During idling air supply is throttled and residual gases make up a large fraction of the charge at the end of the suction period. In addition, during valve overlap period some exhaust gases are drawn back into the cylinder. The result is that a chemically correct (stoichiometric) mixture of air and fuel (≈ 15 : 1) would be so diluted by residual gases that combustion would be erratic or impossible. *A rich mixture, therefore, must be supplied during idling (say A/F ratio* 11 : 1 or 12 : 1). *The richness should gradually change to slightly lean for the second range as* shown in Fig. 11.2.

2. **Cruising or normal power** (*from about 25% to about 75% of rated power*) :

In the *normal power range the main consideration is fuel economy.* Because mixture of fuel and air is never completely homogeneous the stoichiometric mixture of fuel and air will not burn completely and some fuel will be wasted. For this reason an *excess of air, say 10% above theoretically correct (≈ 16.5 : 1), is supplied in order to ensure complete burning of the fuel.*

3. **Maximum power** (*From 75% to 100% of rated power*):

Maximum power is obtained when all the air supplied is fully utilized. As the mixture is not completely homogeneous a rich mixture must be supplied to assure utilization of air (though this would mean wasting some fuel, which would pass in exhaust in unburned state). The *air-fuel ratio for maximum power is about 13 : 1.*

Running on the weakest mixture. *This results in high efficiency and there is fuel economy. On normal loads engines work on weak mixture.*

Running on richest mixture. *Engines run on rich mixture during idling and during the overload. The effect is lowering of efficiency and pollution problems.*

- Automobiles engines generally operate well below full power and a complicated system for making the mixture rich is neither called for nor economically advisable, although some means are employed to enrich the mixture. A more representative curve for an automobile engine is shown in Fig. 11.3.

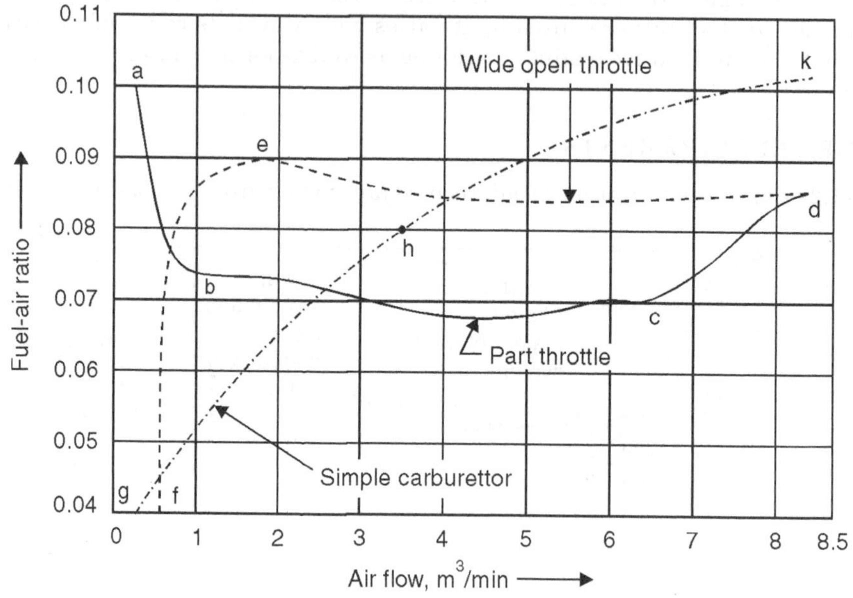

Fig. 11.3

— The portion of the curve from *d* to *e* shows the fuel-air requirements for wide open throttle (W.O.T.) and the load is further increased.

— The Fig. 11.3 exhibits that a simple elementary carburettor is incapable to provide the F/A ratio as desired at part throttle as shown by *a-b-c-d* or W.O.T. given by *d.e.f.* The simple carburettor gives the curve as shown by *g-h-k*.

11.5. DISTRIBUTION

Ideally, a carburettor should supply mixture of the same fuel-air ratio to each cylinder of a multi-cylinder engine ; this condition is very difficult to achieve practically.

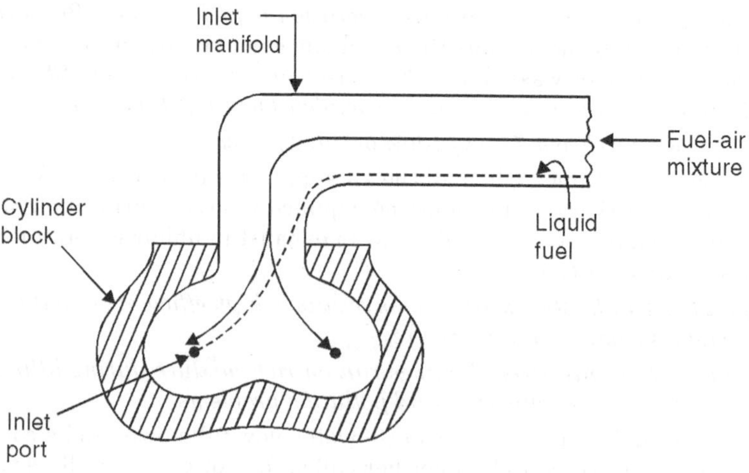

Fig. 11.4

- Since in a carburettor complete atomisation and vaporisation of the fuel is not achieved, therefore, the mixture passing through the intake manifold generally contains a certain amount of petrol in the droplet form. These droplets possess greater inertia than the gaseous mixture. Thus when there is an abrupt change in the direction of flow the droplets continue to move in their original direction (Refer Fig. 11.4) and consequently variation of A / F ratio between cylinders takes place, *the outer cylinders getting richer mixture than the inner cylinders.*
- *The uneven distribution is also caused owing to the existence of thin film of liquid film adhering to the inner walls of the inner manifold.*

The imbalance in air-fuel ratio in different cylinders can be *partially* corrected as follows :

(*i*) *By heating the mixture in the intake manifold to vaporize the droplets of liquid.* By doing so, however, the mass of the charge is reduced, resulting in reduction of power output.

(*ii*) *By supplying a rich overall air-fuel mixture so that the leanest cylinder receives the required air-fuel ratio.* This would, however, result in other cylinders getting a richer mixture than required.

11.6. TRANSIENT MIXTURE REQUIREMENTS

The function of a carburettor is not only to provide a suitable mixture for *steady-running* but also to supply mixture for transient conditions under which load, speed, temperature or pressure alter rapidly. The main transient conditions of operation are :

— Starting and warming up ;

— Acceleration, and deceleration.

The mixture requirements under transient conditions are different from those of steady running due to the following *reasons* :

(*i*) The evaporation may be incomplete ;

(*ii*) The quantity of liquid fuel in the inlet manifold may be increasing or decreasing ;

(*iii*) The distribution of fuel to various cylinders may be different.

1. **Starting and warming up requirements :**

When the engine is started from cold, its speed and temperature are low and as such much of 'heavy ends' (The hydrocarbons with high vapour pressures and low boiling points are called *'light ends'* and those which are less volatile are called 'heavy ends') supplied by the carburettor do not vaporise and remain in liquid form. Further vaporised fuel may recondense on coming in contact with cold cylinder walls and piston head. Thus, even when the F/A ratio at the carburettor is well within the normal combustion limits or petrol-air mixtures, the ratio of the *'evaporated fuel' to air in the cylinder may be too lean to ignite.* Consequently it is *necessary to supply a rich mixture during starting,* as much as 5 to 10 times the normal amount of petrol (A / F ratio 3 : 1 to 15 : 1 or F/A ratio 0.3 to 0.7), in order that 'light ends' are available for proper ignition. With the warming up of the engine there is an increase in the amount of evaporated fuel and hence the mixture ratio should be progressively made leaner, too rich *evaporated* F/A ratio is avoided.

2. **Acceleration requirements :**

- With regard to engines, the terms acceleration generally refers to an increase in engine speed resulting from opening the throttle. The main object of opening the throttle, however, is *to provide an increase in torque* and whether or not an increase in speed follows depends on the nature of load.
- Under steady running conditions, the fuel evaporated in the intake manifold moving much faster than the liquid film formed on the induction system walls, does not cause any problem. But when the throttle is *suddenly* opened *e.g.* during acceleration, the liquid

fuel lags behind and, the cylinder receives temporarily a lean mixture whilst actually, to produce more instantaneous power for acceleration, a rich mixture is needed. Hence, a suitable mechanism (acceleration pump) is required to provide rich mixture during the acceleration period.

Note. The petrol to be used should be carefully made to suit the engine and the climate of the place since too high volatility or too low volatility, both create difficulties in operation.

— *Too high volatility* may form bubbles in the carburettor and fuel lines particularly when the engine temperatures are high, which interfere with the supply and metering of the fuel and may disturb the F/A ratio so seriously that engine may stop working.

— *Too low volatility* may cause petrol to condense on the cylinder walls, diluting and removing the lubricating oil film ; ultimately the petrol may reach the cranckcase past the piston rings and dilute the engine oil. Condensation of petrol on cylinder walls also causes carbon deposits.

11.7. A SIMPLE OR ELEMENTARY CARBURETTOR

In order to understand a modern carburettor (a very complex device) it helps first to study a *simple carburettor* which supplies fuel-air mixture for cruising or normal range of speed and then to add other devices or attachments to take care of other function like *starting, idling, accelerating,* decelerating and other variable load and speed operations.

Fig. 11.5, shows a schematic diagram of a simple or elementary carburettor.

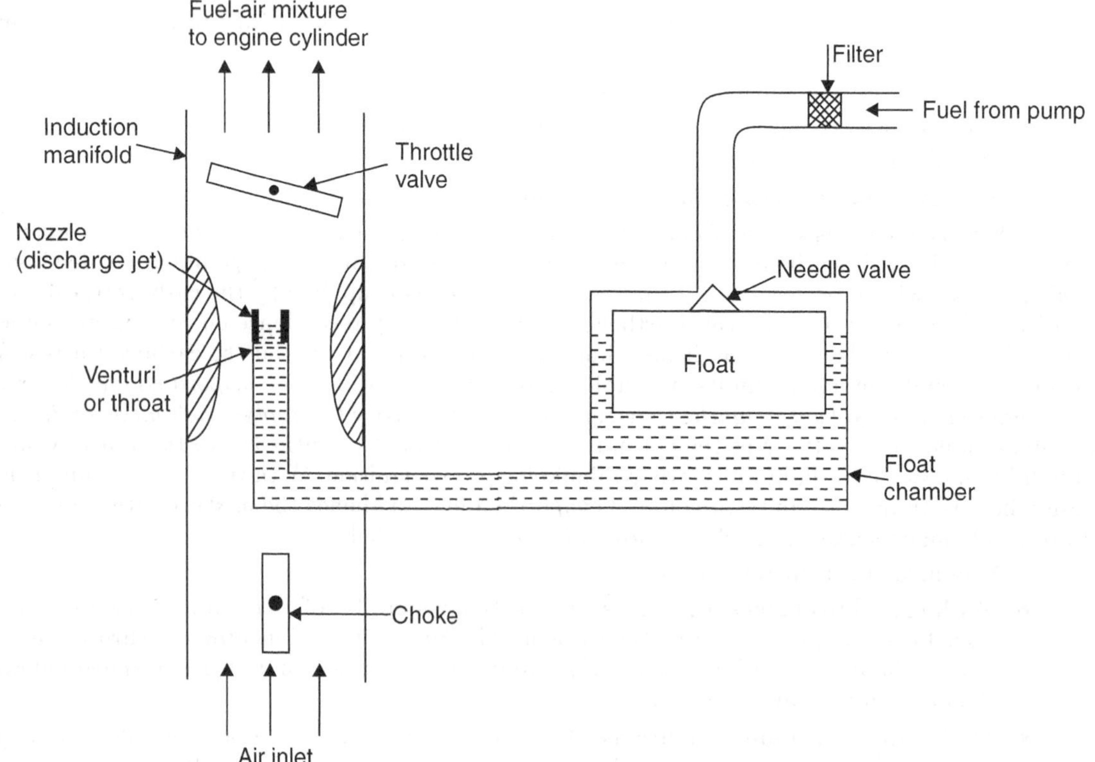

Fig. 11.5. A simple or elementary carburettor.

- It consists of a *float chamber, nozzle with metering orifice, venturi* and *throttle valve.*
  - — The **float chamber** is meant for storage of fuel. The fuel supplied under gravity action or by fuel pump enters the float chamber through a filter. The arrangement is such that when oil reaches a particular level the *needle/float valve* blocks the inlet passage and thus cuts off the fuel oil supply. On the fall of oil level, the *float* descends down, consequently intake passage open and again the chamber is filled with oil. Then the float and the needle/float valve maintains a constant fuel oil level in the float chamber. There is a *nozzle (discharge jet)* from which the fuel is sprayed into the air stream as it enters the inlet and passes through the venturi or throat. *The fuel level is slightly below the outlet of the jet when the carburettor is inoperative.*
- As the piston moves down in the engine cylinder, suction is produced in the cylinder as well as in the induction manifold as a result of which air flows through the carburettor. *The velocity of air increases as it passes through the constriction at the venturi and the pressure decreases due to conversion of a portion of pressure head into kinetic energy.* Due to decreased pressure at the venturi and hence by virtue of difference in pressure (between the float chamber and the venturi) the jet issues fuel oil into air stream. Since the jet has a fine bore, the oil issuing from the jet is in the form of *fine spray* ; it vaporises quickly and mixes with air. The air-fuel mixture enters the engine cylinder ; *it quantity being controlled by the position of the "throttle valve".*

Limitations :

(*i*) Although theoretically the air-fuel ratio supplied by a simple (single jet) carburettor should remain constant as the throttle goes on opening, actually *it provides increasingly richer mixture as the throttle is opened.* This is because of the reason that the density of air tends to decrease as the rate of flow increases.

— This fault is corrected by *using a number of compensating devices.*

(*ii*) During idling, however, the nearly closed throttle causes a reduction in the mass of air flowing through the venturi. At such low rates of air flow, the pressure difference between the float chamber and the fuel discharge nozzle becomes very small. It is not adequate enough to cause fuel to flow through the jet.

— This fault may be corrected by *using an idling jet which helps, in running the engine during idling.*

(*iii*) Carburettor does not have arrangement for providing rich mixture during starting and warm up.

— This limitation is taken of by using a *choke arrangement.*

11.8. COMPLETE CARBURETTOR

For meeting the demand of the engine under all conditions of operation, the following additional devices/systems are added to the simple carburettor :

1. Main metering system
2. Idling system
3. Power enrichment or economiser system
4. Acceleration pump system
5. Choke.

1. **Main metering system :**

The main metering system of a carburettor should be so designed as to supply a *nearly constant fuel-air ratio over a wide range of operation.* This F/A ratio is approximately equal to 0.064 (A/F ratio ≈ 15.6) for best economy at full throttle. In order to correct the tendency of the simple carburettor to give progressively richer mixtures with load speed, the following *automatic compensating devices* are incorporated in the main metering system :

(*i*) Compensating jet device.

(*ii*) Emulsion tube or air bleeding device.

(*iii*) Back suction control or pressure reduction method.

(*iv*) Auxiliary valve carburettor.

(*v*) Auxiliary port carburettor.

These devices are explained below :

(*i*) ***Compensating jet device :***

A schematic diagram of a compensating jet device is shown in Fig. 11.6.

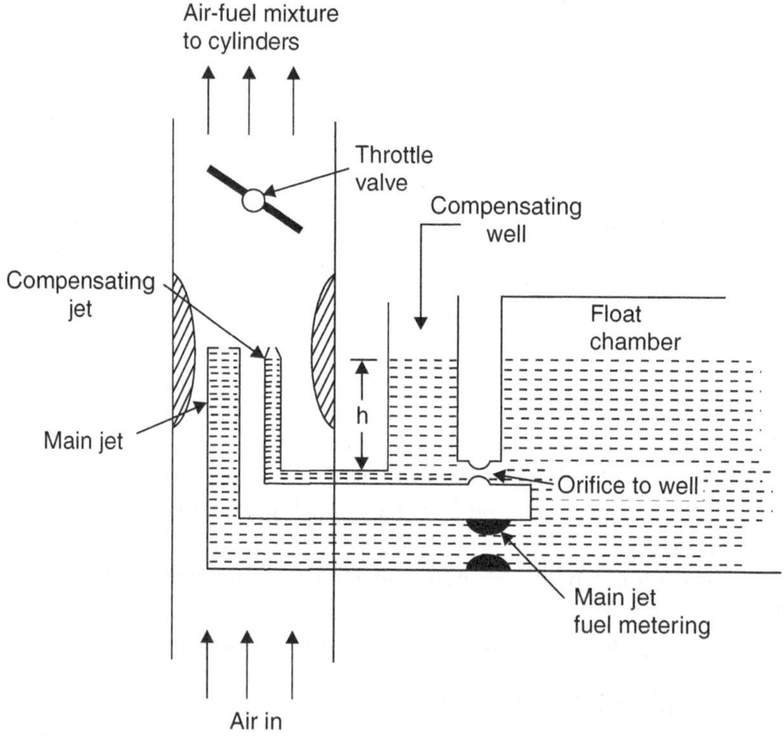

Fig. 11.6. A compensating jet device.

- In this device, in addition to the main jet, a compensating jet is provided which is in communication with a compensating well. The compensating well is also vented to atmosphere (like the main float chamber) ; it is supplied with fuel from the main float chamber through a restricting orifice.
- As the air flow increases, the level of fuel in the compensating well decreases, thus reducing the fuel supply through the compensating jet. The compensating jet thus progressively makes the mixture leaner as the main jet progressively makes the mixture richer, the sum of the two remaining constant as shown in Fig 11.7. The main jet and compensating curves are more less reciprocals of each other.

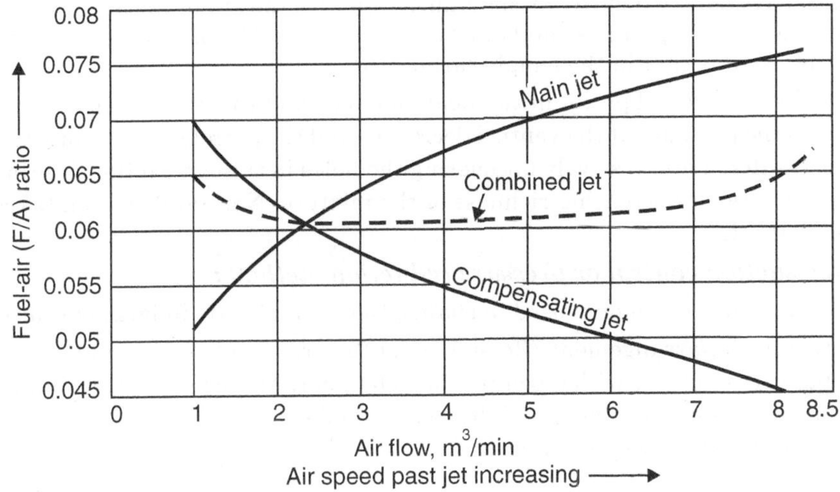

Fig. 11.7. Variation of F/A ratio vs. air flow with main and compensating jets.

— At even higher *rates* of air flow, when the compensating jet has been emptied, *air is bled through the compensating jet to continue the leaness effect*, and incidently to assist in *atomisation of fuel*.

(*ii*) ***Emulsion tube or air bleeding device* :**

The mixture correction in modern carburettor is done by air bleeding alone. Such an arrangement is shown in Fig. 11.8.

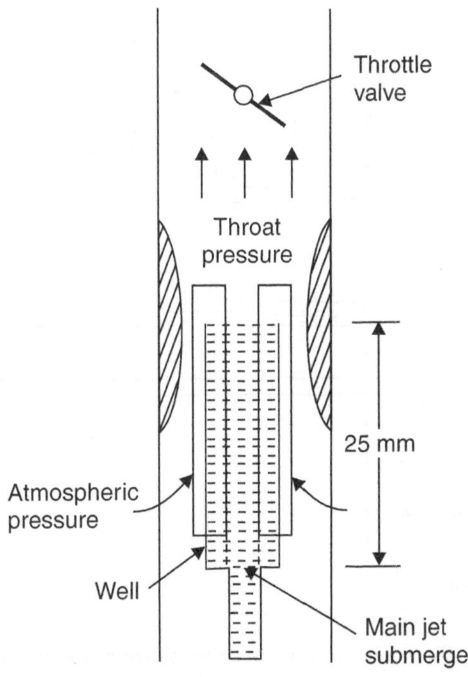

Fig. 11.8. Correction in modern carburettors by air bleeding.

- A main metering jet is fitted 25 mm below the petrol level in the float chamber and therefore it is called *submerged jet*. The jet is situated at the bottom of a well, the sides of which have holes which are in communication with the atmosphere. Air is drawn through the holes and the petrol is emulsified ; the pressure difference across the petrol column is not as great as that in the simple carburettor.

- Initially the level of petrol in the float chamber and the well is same. When throttle is opened the pressure at the venturi decreases and the petrol is drawn into the air stream. This results in progressively uncovering the holes in the central tube leading to decreasing F/A ratio or decreasing richness of the mixture. Normal flow then takes place from the main jet.

(*iii*) ***Back suction control or pressure reduction method :***

- This method is commonly used to change the air-fuel ratio in large carburettors.
- In this device/arrangement (Refer Fig. 11.9) a relatively large vent line connects the carburettor entrance with top of the float chamber. Another line, containing a very small orifice line, connects the top of the float chamber with the venturi / throat. A control valve is placed in the large vent line.

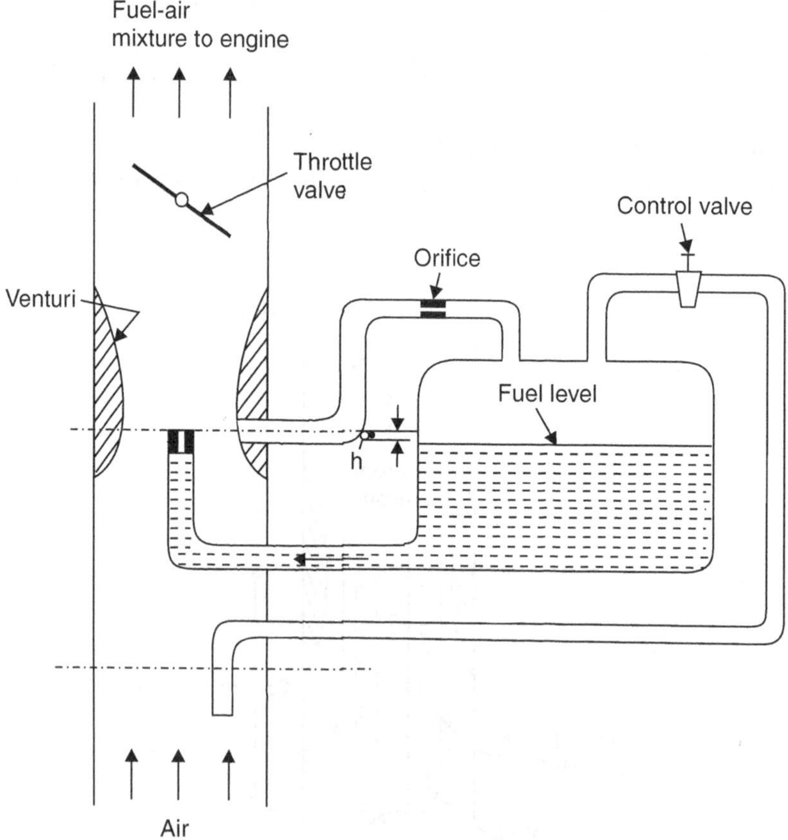

Fig. 11.9. Back-suction control or pressure reduction method.

— When the valve is wide open, the vent line is unrestricted and the pressure in the float chamber is atmospheric say p_1, and the pressure differential acting on the orifice is $(p_1 - p_2)$ when p_2 is the pressure at the throat. If the valve is closed, the float chamber communicates only with venturi throat and pressure on the fuel surface will be p_2. Then the carburettor depression Δp will be zero and no fuel can flow. By proper adjustment of control valve any pressure between p_1 and p_2 can be obtained in the float chamber, thus altering the quantity of fuel discharged by the nozzle.

(*iv*) ***Auxiliary valve carburettor :***

Fig. 11.10 shows an auxiliary valve carburettor. When load on the engine increases, the vacuum at the venturi throat also increases. This lifts the valve against the spring force and *consequently more air is admitted and the mixture is prevented from becoming over-rich.*

(*v*) ***Auxiliary port carburettor.***

- This method is used in *aircraft carburettors for altitude compensation.*
- Fig. 11.11 shows an auxiliary port carburettor. When the butterfly valve is opened, additional air is admitted and at the same time the depression at the venturi throat is reduced ; this results in decreasing the quantity of fuel drawn in.

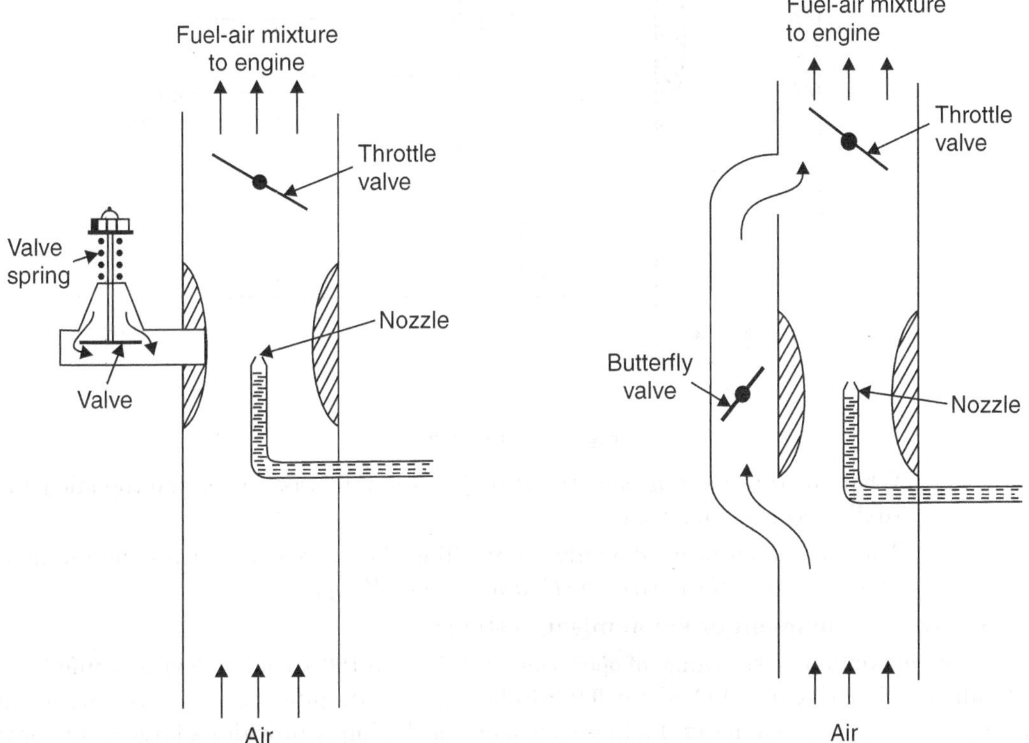

Fig. 11.10. An auxiliary valve carburettor. Fig. 11.11. An auxiliary port carburettor.

2. **Idling system :**

- As earlier discussed that at idling and low load an engine requires a rich mixture having about air-fuel ratio 12 : 1. The main metering system not only fails to supply enrich the mixture at low air flows but also cannot supply any fuel during idling operation. It is due to this reason that a separate idling jet must be incorporated in the basic carburettor.

- Fig. 11.12 shows an idling jet. It consists of a small fuel line from the float chamber to a point on the engine side of the throttle ; this line contains a fixed fuel orifice.
  - When *throttle is practically closed,* the full manifold suction operates on the outlet to this jet. Besides local suction is increased due to very high velocity past the throttle valve. Fuel therefore can be lifted by the additional height upto the discharge point, but this *occurs only at very low rates of air flow.*

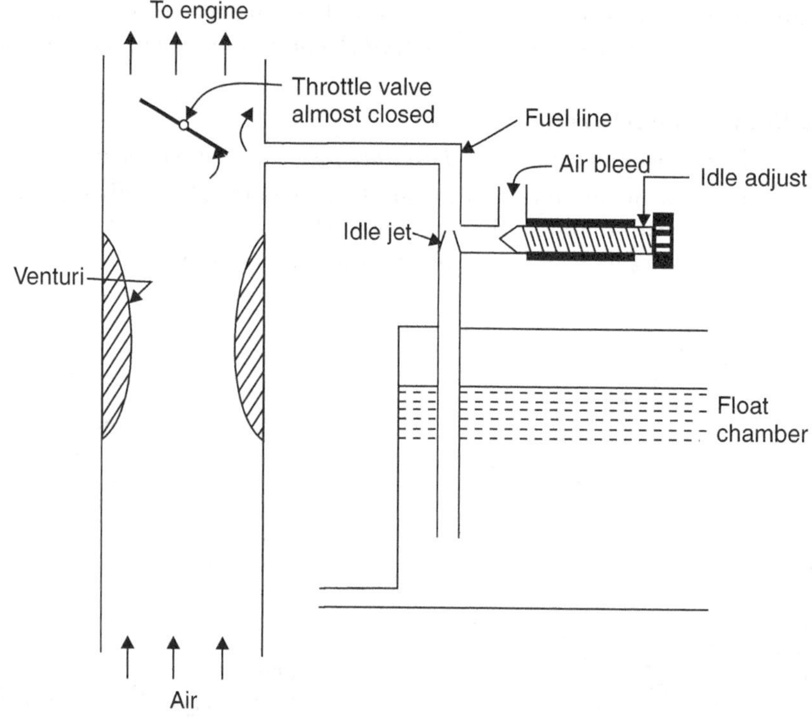

Fig. 11.12. Idling jet.

  - When the *throttle is opened,* the main jet gradually takes over and the idle jet eventually becomes ineffective.
  - The *idle adjust* (a needle valve controlling the air bleed, which is manually operated) *regulates the desired A/F ratio for the idling jet.*

3. **Power enrichment or economiser system :**

At the maximum power range of operation from 75% to 100% load, a device should be available to allow richer mixture (F/A about 0.08) to be supplied despite the compensating leanness. *Meter rod economiser* shown in Fig. 11.13 is such a device. It simply provides a large orifice opening to the main jet when the throttle is opened beyond specified limit. The rod may be tapered or stepped.

— An economiser is a valve which remains closed at normal cruise operation and gets opened to supply enriched mixture at full throttle operation. It regulates the additional fuel supply for the above operation. (The term economiser is rather *misleading.* It stems from the fact that such a device *provides a rich uneconomical mixture at high load demand without interfering with economical operation in the normal power range*).

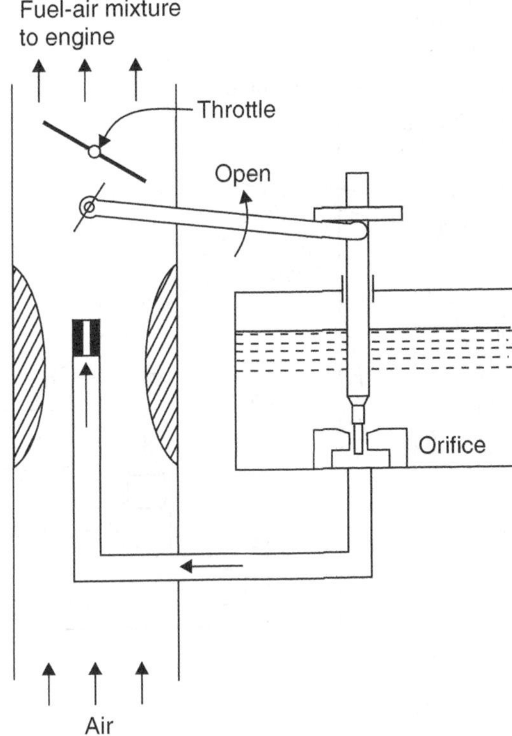

Fig. 11.13. Skelton outline of a meter rod economiser.

4. **Acceleration pump system :**

- Acceleration is a transient phenomenon. In order to accelerate the engine rapidly, a very rich mixture is required which a simple carburettor may not be able to supply. Rapid opening of throttle will be immediately followed by an increased air flow, but the inertia of liquid fuel (gasoline) will give momentarily lean mixture. Thus acceleration mixture required may not be met with in practice. To overcome this difficult situation an *acceleration pump is incorporated.*
- Fig. 11.14 shows an *acceleration pump.* It consists of a spring-load plunger. Also is provided a linkage mechanism so that when throttle is rapidly opened the plunger moves into the cylinder and forces an additional jet of fuel into the venturi. An arrangement is also provided which ensures that when throttle is opened slowly, the fuel in the pump cylinder is not forced into the venturi but leaks past plunger or some holes into the float chamber.
  - — In some carburettors, instead of providing mechanical linkages, an arrangement is made so *that the pump plunger is held up by manifold vacuum.* Whenever this vacuum is reduced by rapid opening of throttle a spring forces the plunger down pumping the fuel through the jet.

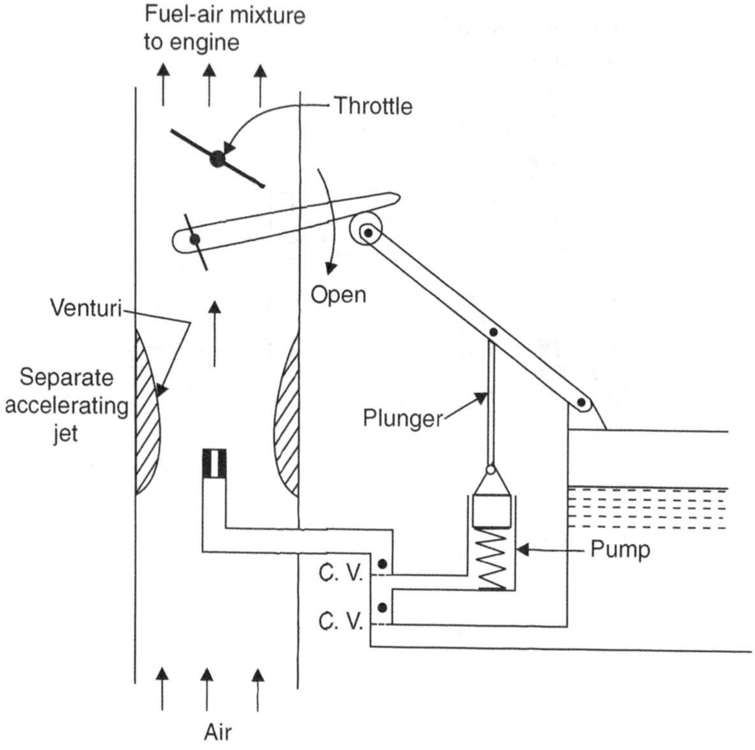

Fig. 11.14. Acceleration pump.

5. **Choke :**

- Starting of a vehicle which is kept stationary for a long period (may be overnight.) during cool winter seasons, is often more difficult. At low cranking speeds and intake temperatures a very rich mixture is required to initiate combustion. Sometimes as high as *five to ten times* more fuel (than usual mixtures) is required. The main reason is that very large fraction of fuel may remain liquid suspended in air even in the cylinder, and only the vapour fraction can provide a combustible mixture with air. The most popular method of providing such mixture is by the use of *choke.*
- A *choke* is simply butterfly valve located between the entrance to the carburettor and the venturi throat as shown in Fig. 11.15.
  - When the choke is partly closed, large pressure drop occurs at the venturi throat, would normally result from the amount of air passing through the venturi throat. The very large carburettor depression at the throat inducts large amount of fuel from the main nozzle and provides a very rich mixture so that the ratio of the evaporated fuel to air in the cylinder is within the combustible limits.
- Sometimes the choke valves are made with spring-loaded by-pass to ensure that large carburettor depression and excessive choking does not persist after the engine has started, and reached a desired speed. The choke can be made to operate automatically by means of a thermostat so that choke is closed when engine is cold and goes out of operation when the engine warms up after starting.

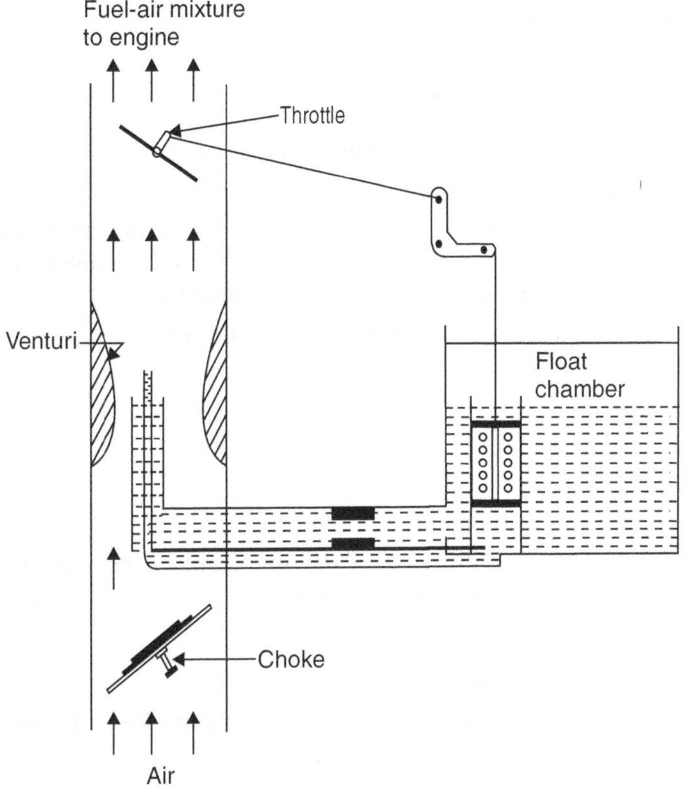

Fig. 11.15. Choke valve with spring-loaded by-pass.

- The provision of *auxiliary fuel jets* that are opened manually or automatically only as required, is an *alternative to the choke.*

11.9. CARBURETTORS

11.9.1. Essential Features of Good Commercial Carburettor for Automotive Engines

Carburettor is a *mixing device to supply the engine with air-fuel mixture. It atomizes the fuel and mixes it with air in varying proportions to meet the changing operating conditions of automotive engines.* It is required to provide the following *essential features* :

1. To meter and supply the proper quantity and proportion of air and fuel at correct strength under all conditions of load and speed of the engine of the car for

   (*i*) starting it easily from cold.

   (*ii*) providing a rich mixture for slow idling.

   (*iii*) providing a rich mixture for acceleration,

   (*iv*) providing a rich mixture for high speed, and

   (*v*) providing a rich mixture for low speed when moving up-gradient.

2. To operate satisfactorily when cold, or when hot

3. To operate satisfactorily both on level and hills

4. To overcome air-cleaner restrictions.
5. To withstand vibrations and road jerks.

11.9.2. Types of Carburettors

Carburettors, basically, are of the following *two types* :

1. Open choke type

- Here, the main orifice known as the *choke tube* or *venturi* is of *fixed dimensions,* and *metering is affected by varying the pressure drop across it.* Almost all carburettors, except S.U. carburettor, belong to this category of carburettors.
- The *important examples* of this type of carburettor are :
  (*i*) Zenith carburettor
  (*ii*) Solex carburettor
  (*iii*) Carter carburettor
  (*iv*) Stromberg carburettor.

2. Constant vacuum type

- In this type of carburettor the *area of the air passage is varied automatically while the pressure drop is kept approximately constant.*
- *Example* : S.U. carburettor.

Basic forms of carburettors :

Refer Fig. 11.16. Carburettor may be of the following *three basic forms.*

(*i*) Updraught (*ii*) Downdraught
(*iii*) Horizontal.

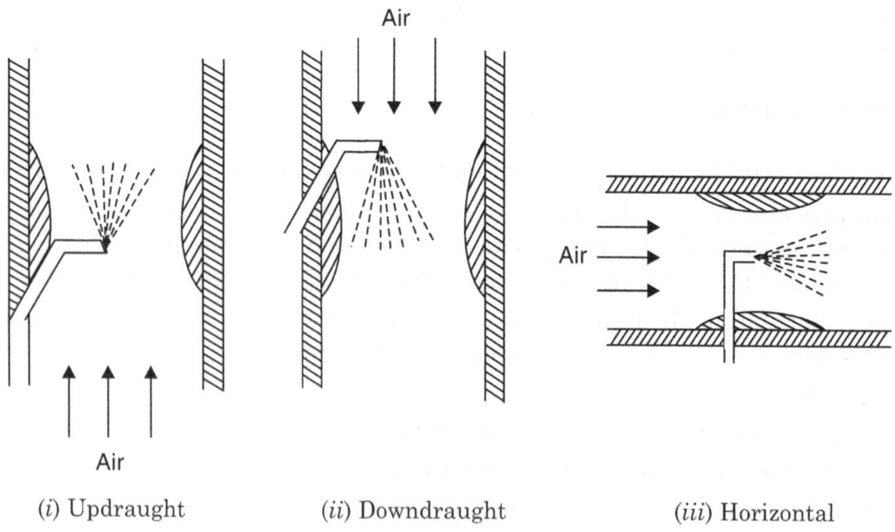

Fig. 11.16. Basic forms of carburettors.

- The *updraught variety* is now *obsolete* and is *only used where neither of the other types can be accommodated.*
- The *downdraught and horizontal types* of carburettors are *most widely used.*
  - The *advantage of downdraught variety is* that the *mixture is assisted by gravity in its passage into the engine induction tract*, and *at the same time the carburettor is usually reasonably accessible.*
  - The *horizontal type of carburettor has some advantage when under-bonnet space is limited.*

11.9.3. Description of Some Important Makes of Carburettors

Following carburettors will be described here :

1. Solex carburettors
2. Carter carburettors
3. S.U. Carburettor.

11.9.3.1. Solex Carburettor

This carburettor is made in various models and is used in Fiat, Standard and Willy's Jeep. It is famous for the following *characteristics* :

(*i*) *Easy starting ;*

(*ii*) *Good performance ;*

(*iii*) *Reliability.*

Fig. 11.17. shows the schematic arrangement of a solex carburettor. The unique feature of this carburettor is **Bi-starter** for *cold starting.*

The various components and the circuits for air and fuel for various ranges of operation are explained below :

1. **Normal running :**

- Solex carburettor comprises a conventional float (1) in a float chamber.
- The fuel is provided through the main metering jet (2) and the air by the choke tube or venturi (3).
- The fuel from the main jet passes into the well of air-bleed emulsion system ; (4) is the emulsion tube which has lateral holes.
- Air correction jet (5), calibrates the air entering through it and ensures automatically the correct balance of air and fuel.
- The metered emulsion of fuel and air is discharged through the spraying orifice or nozzle (6) drilled horizontally in the vertical stand-pipe in the middle of choke tube or venturi.
- (7) is the conventional butterfly valve.

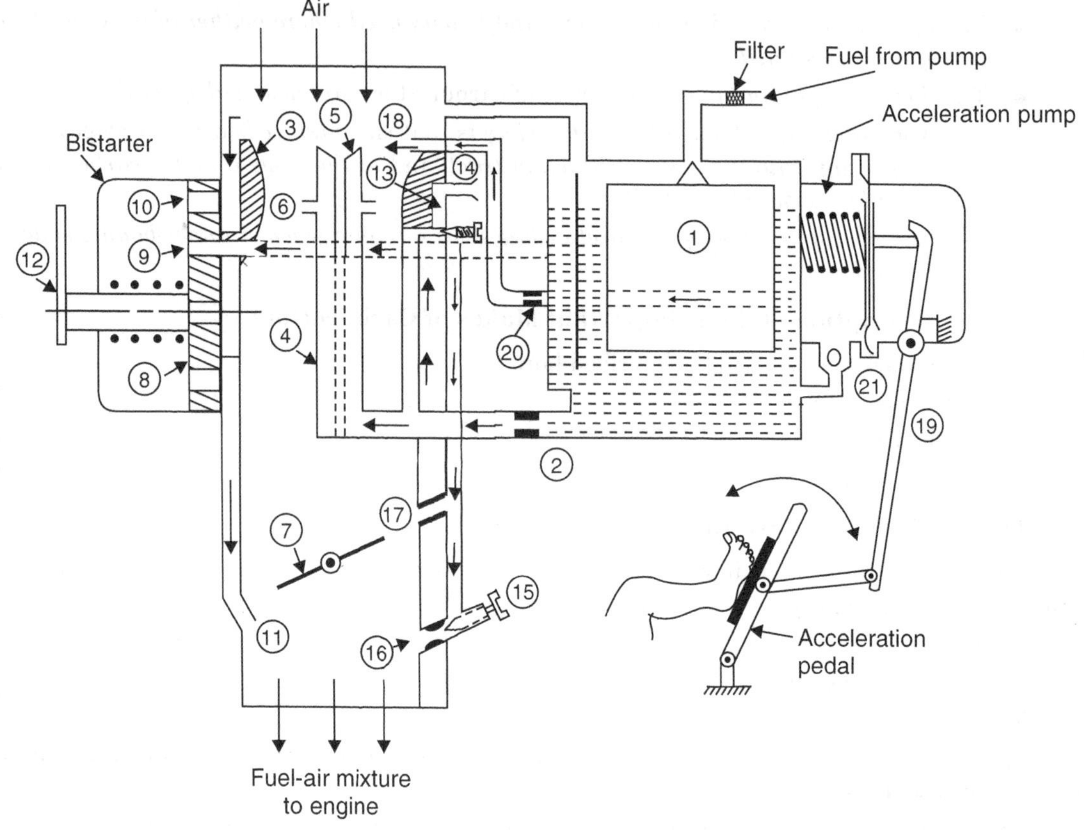

1. Conventional float
2. Main jet
3. Choke tube or venturi
4. Emulsion tube
5. Air correction jet
6. Spraying orifice or nozzles
7. Conventional butterfly valve
8. Flat disc with holes of different sizes
9. Starter petrol jet
10. Jet
11. Starting passage
12. Starter lever
13. Pilot jet
14. Small pilot air bleed orifice
15. Idling volume control screw
16. Idle port
17. By-pass orifice
18. Pump injector
19. Pump lever
20. Pump jet
21. Pump inlet valve.

Fig. 11.17. Schematic arrangement of a solex carburettor.

2. **Cold starting and warming :**

The *provision of a bi-starter or a progressive starter is the unique feature of solex carburettors.*

- The starter valve is in the form of a flat disc (8) with holes of different sizes. These holes connect the starter petrol jet (9) and starter air jet sides to the passage which opens into a hole just below the throttle valve at (11). Either bigger or smaller holes come opposite the passage, depending upon the position of the starter lever (12). The starter lever is operated by flexible cable from the dash board control. Initially, for starting richer mixture is required and after the engine starts, the richness required decreases. In the start position when the starter control is pulled out fully, bigger holes are the connecting holes. The throttle valve being in closed position the whole of the engine suction is applied to the starting passage (11), sucking petrol from jet (9) and air from jet (10). The jets and passages are so shaped that the mixture provided to the carburettor is rich enough for starting.

- After starting the engine, the starter lever is brought to the intermediate position, bringing the smaller holes in the starter valve (8) into the circuit, thus reducing the amount of petrol. Also in this position, the throttle valve is partly open, so that the petrol is also coming from the main jet. In this situation, the reduced mixture supply from the starter system, however, is sufficient to keep the engine running till it reaches the normal running temperature, when the starter is brought to *"off-position"*.

3. **Idling and slow running :**

- From the lower part of the well of the emulsion system a hole leads off to the pilot jet (13).
- At idling the throttle is practically closed and therefore the suction created by the engine on suction stroke gets communicated to the pilot jet. Fuel is inducted therefrom, and mixed with a small amount of air admitted through the small pilot air bleed orifice (14) and forms an emulsion which is conveyed down the vertical channel and discharged into the throttle body past the idling volume control screw (15). The *slow running adjustment screws allows the engine speed to be varied.*
- *By-pass orifice* (17) provided on the venturi side of the throttle valve *ensures the smooth transfer from idle and low speed circuit to the main jet circuit without occurrence of flat spot.*

4. **Acceleration :**

- In order to avoid flat spot during acceleration, a diaphragm type *acceleration* is incorporated (also known as economy system). This pump supplies spurts of extra fuel needed for acceleration through pump injector (18).
- Pump lever (19) is connected to the accelerator so that on pressing the pedal, the lever moves towards left, pressing the membrane towards left, thus forcing the petrol through pump jet (20) and injector (18). On making the pedal free, the lever moves the diaphragm back towards right creating vacuum towards left which opens the pump inlet valve (21) and thus admits the petrol from the chamber into pump.

11.9.3.2. Carter Carburettor

A *carter carburettor* is an American make carburettor and is used in *jeeps*. It is a *standard equipment on chevorolet and Pontiac series of cars.*

Fig. 11.18 shows the schematic arrangement of a downdraft type Carter carburettor. The brief description of the components and circuits is given below :

- The petrol (fuel) enters the conventional type float chamber (1).
- The air enters the carburettor from the top, a choke valve (2) in the passage remains open during normal working.
- This carburettor has a triple venturi diffusing type of choke, *i.e.* it has *three venturies*, the smallest (3) lies above the level in the float chamber, and the remaining two venturies (6) and (5) are below the fuel level (in the float chamber), one below the other.
- At very low speeds, suction in primary venturi (3) is sufficient to draw the fuel. The nozzle (4) enters the primary venturi at an angle, and throws the fuel up against the air stream evenly, thereby providing finely divided atomised fuel. The mixture from venturi (3) passes centrally through the secondary venturi (5) where it is surrounded by a blanket of air stream and finally this leads to the third main venturi (6), where again the fresh air supply insulates the stream from the secondary venturi. The fuel-air mixture

enters the engine in well mixed atomised state. *The multiple venturi gives more homogeneous and better mixture at very low speeds resulting in steady and smooth operation at low speeds. This arrangement also ensures adequately formed mixture at high speeds.*

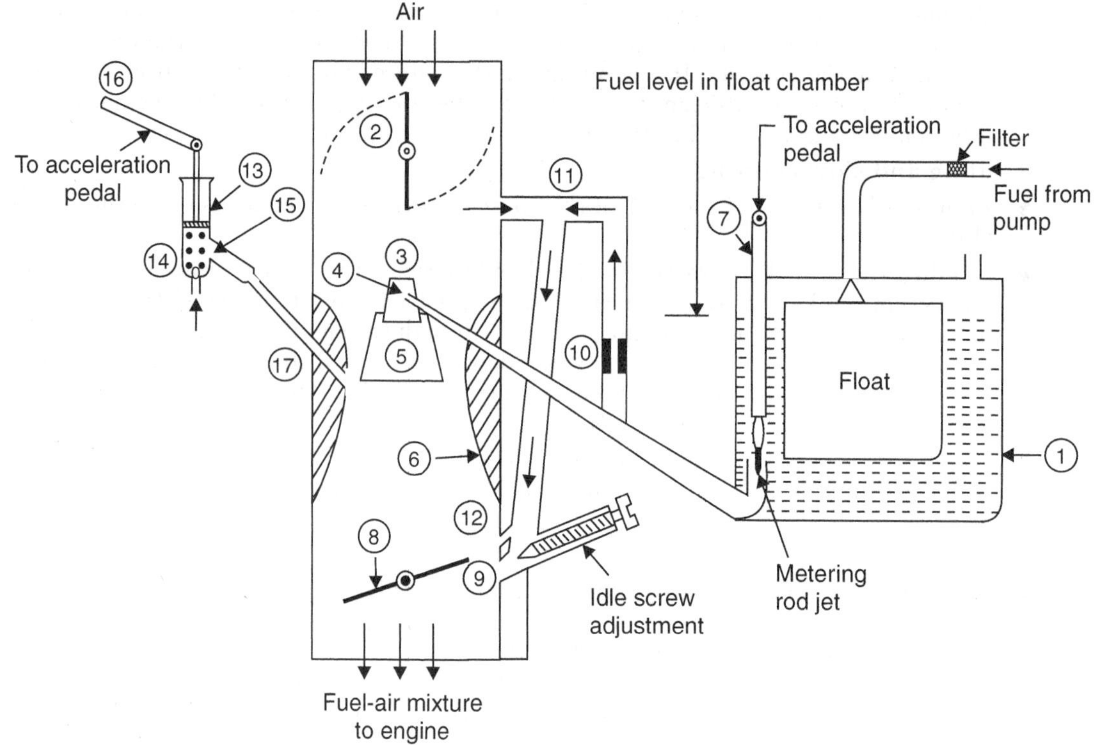

1. Float chamber
2. Choke valve
3. Primary venturi
4. Nozzle
5. Secondary venturi
6. Third (main) venturi
7. Metering rod
8. Throttle valve
9. Idle port
10. Idle feed jet
11. By pass
12. Low speed port
13. Plunger
14. Inlet check valve
15. Outlet check valve
16. Throttle control rod
17. Jet.

Fig. 11.18. Schematic arrangement of a downdraft Carter carburettor.

- In Carter carburettor mechanical metering method is used. In the fuel circuit there is a metering rod (7) (having two or more steps of diameter) which is actuated by a mechanism connected with the main throttle. *The amount of petrol drawn into the engine is governed by the area of opening between the metering rod jet and metering rod.*

1. **Starting circuit :**

- In order to start the engine, a *choke valve (2)* is incorporated in the air circuit. The choke valve is of butterfly type, one half of which is spring controlled. The valve is hinged at the centre.
  - When the engine is fully choked, the whole of the engine suction is applied at the main nozzle, which then delivers fuel. Since the airflow is quite small, very rich fuel-air mixture is supplied.

— When the engine starts/fires, the spring controlled half the choke valve is sucked open to provide correct quantity of air during the period of warm up.

2. **Idle and low speed circuit :**

- For *idling rich mixture in small quantity is required.*
- The throttle valve (8) is almost closed in idling condition.
- The entire suction pressure created by the piston in the engine, during suction stroke, is exerted at the idle port (9). Consequently, the petrol is drawn through the idle feed jet (10) through first by-pass (11) and a rich idle mixture is supplied. The throttle valve is opened further in low speed operation.

At this stage the fuel is delivered both by the main venturi and low speed port (12) through the idle passage.

3. **Acceleration pump circuit :**

- The *acceleration pump is employed to overcome flat spot in acceleration.*
- The pump consists of a plunger (13) working inside a cylinder consisting of inlet check valve (14) and outlet check valve (15). The pump plunger is connected to accelerator pedal by throttle control rod (16).
- On rapid opening the throttle by pressing the accelerator pedal, the pump is actuated and a small quantity of petrol is spurted into the choke tube by a jet (17). Releasing the accelerator pedal takes the plunger back by spring force and in the process sucks petrol from the float chamber for next operation.
- The acceleration pump *does not supply fuel continuously for heavy load but only provides an extra spurt of fuel during acceleration to avoid flat spot.*

11.9.3.3. S.U. Carburettor (constant vacuum variable choke)

— In general carburettors are of '*constant choke*' type ; examples being Zenith, Solex and Carter. S.U. carburettor differs from them being 'constant vacuum or depression' type with automatically variable choke.

— It is used in many British cars and was used in Hindustan Ambassador car.

Fig. 11.19 shows the S.U. carburettor schematically. The various components and circuits are described below :

1. **Normal operating conditions :**

- The full metering is accomplished by a **tapered needle** which is raised or lowered in a jet to alter the effective annular jet orifice, and hence fuel flow. The needle projects from underneath the flat face of the cylindrical air-valve, which alters the choke area as it is raised or lowered. The upper part of the air-valve is enlarged to form a piston which fits into the lower open end of the vacuum chamber. A spindle situated in the centre of the air-valve guides the assembly into the cylindrical vacuum chamber. To improve the accuracy and time response of the air-valve vertical movement with very small changes in engine demands, the friction between the air-valve spindle and guide is sometimes reduced by installing a ball-race between the two sliding surfaces.
- While the engine is running, the effect of the depression above the piston in the upper chamber, and the atmospheric pressure underneath, is to raise the air-valve and piston assembly against its own weight and the stiffness of the return spring. Since the doward load is almost constant, a constant depression is needed to keep the air-valve stationary in any raised position. *The amount the air-valve lifts depends on the flow rate of air which passes through the mixing chamber,* this being controlled by the engine speed and throttle opening position.

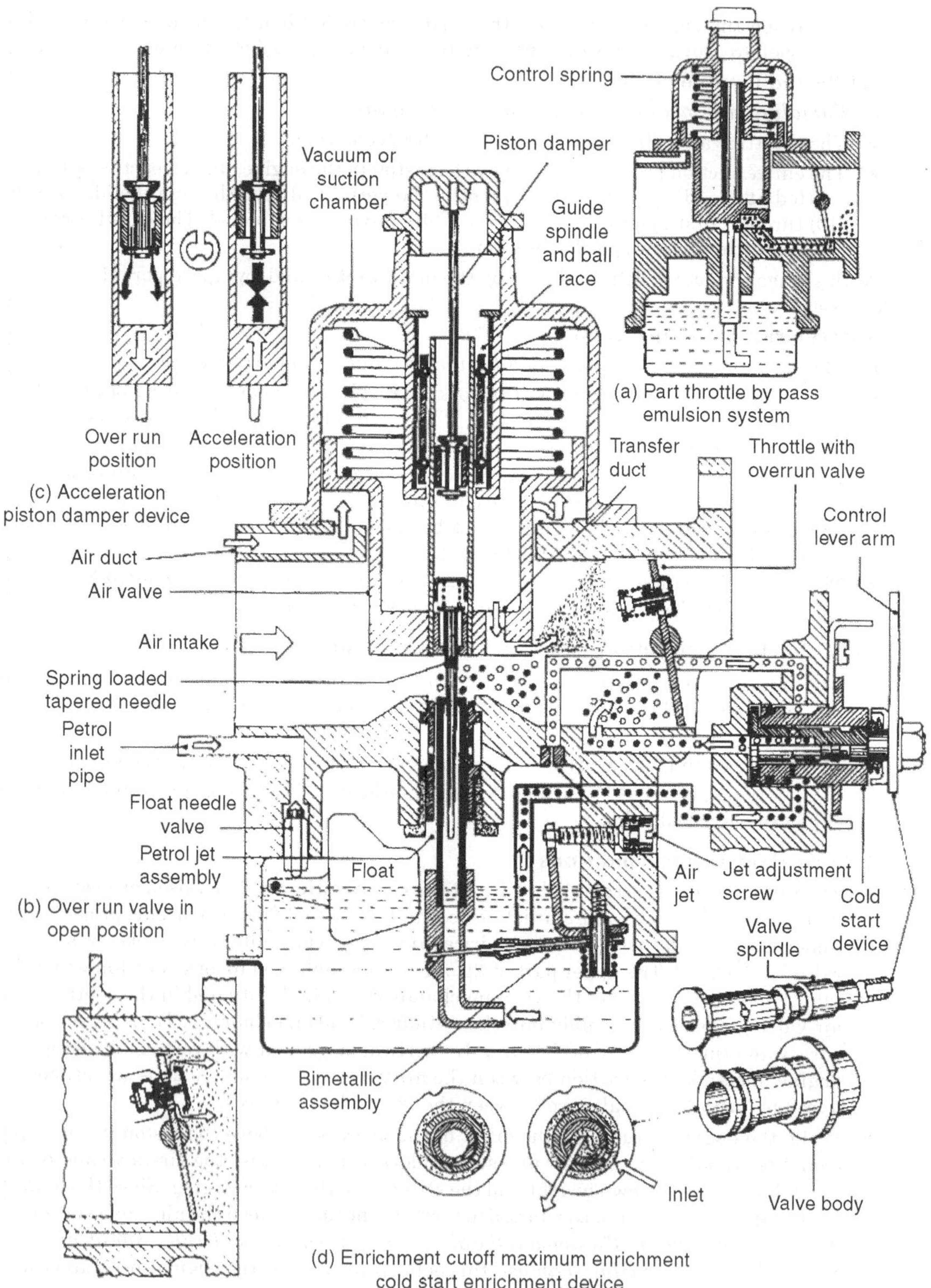

Fig. 11.19. S.U. carburettor (constant vacuum variable choke).

2. **Mixture adjustment and fuel temperature compensation :**

- The jet height initial adjustment and hence mixture strength can be made by altering the tilt of the *right angled lever* which is attached to a *spring-loaded retaining screw* and a bimetallic strip which extends to the petrol jet. To alter the jet height, the horizontal jet adjustment screw is screwed inwards to lower the jet and enrich the mixture, and outwards to raise the jet and weaken the mixture.
- In order to compensate for the variation in fuel viscosity within changing temperature and the reluctance of the fuel to flow through an orifice as its viscosity rises, a **bimetallic strip** *submerged in the fuel senses a temperature change and alters the effective jet size accordingly.* When the fuel temperature rises the bimetallic strip curls upwards and pushes the jet further into the tapered needle. Conversely, if the fuel becomes cooler, the strip bends downwards and lowers the jet to increase the annular jet orifice.

3. **Part throttle by pass emulsion system.** Refers Fig. 11.19 (*a*).

- This system is a passagaway which bypasses the mixing chamber, it spans the distance between the feed duct at the jet bridge and a discharge duct at the throttle butterfly edge.
- The bypass passage, with a small throttle opening, delivers a quantity of mixture in a well emulsified condition from the jet to a high depression point near the edge of the throttle. Since the bypass passagaway is much smaller than the mixing chamber bore, the mixture velocity through this passage will be much greater and therefore the air-fuel mixing will be that much more thorough.

4. **Overrun valve.** Refer Fig. 11.19 (*b*).

- Under *overrun working conditions,* the closed throttle will create a very high depression on the engine side of the throttle and in the induction manifold. Consequently, the *effective compression ratio will be low, burning will be slow and erratic, and the exhaust products will contain high valves of hydrocarbon. To improve the burning process so that more of the fuel is doing useful work* and less is passed through to the exhaust as incomplete combustion products, a **spring-loaded plate-valve** is *incorporated in the throttle butterfly disc.*
- When the engine is operating at overrun conditions, the manifold depression at some predetermined value will force open the spring loaded plate-valve to emit an additional quantity of correct air-fuel mixture. *The increased supply of air-fuel mixture will reduce the manifold depression with the result that the denser and better prepared mixture charge will improve combustion, and hence less unburnt products will be passed through to the exhaust.*

5. **Hydraulic damper (acceleration enrichment device).** Refers Fig. 11.19 (*c*).

- This device is incorporated *to enrich the mixture strength when the throttle is opened rapidly but it does not interfere with the normal air-valve lift or fall as the mixing chamber depression changes with respect to steady throttle opening.*
  - The damper valve is mounted on the lower end of a long stem inside the hollow guide spindle of the air-valve and is submerged in a light oil. The damper consists of a vertically positioned loose fitting sleeve, its underside resting on a spring clip attached to stem, while its upper end is chamfered so that it matches a conical seat formed on the central support stem.
- On rapid opening of the throttle, the sudden rise in depression in both mixing chamber and air valve upper chamber tends to jerk up the air valve assembly. Simultaneously, the viscous drag of oil in the hollow spindle will lift the sleeve and press it against its seat, and so the oil is thus temporarily trapped beneath the damper so that it prevents

any further upward movement of the air valve. For this brief period a temporary increase in the depression over the jet orifice is achieved, and more fuel will therefore be drawn to enrich the resultant mixture strength.

— When the change in engine speed steadies, the depression in the upper air-valve vaccum chamber will also stabilize and there will be a slight leakage of oil between the sleeve and its spindle bore. Consequently any oil pressure created underneath the sleeve damper will now be released enabling the sleeve to drop down onto spring clip-Oil will now move freely through the annular space made between the sleeve and its seat so that the *air-valve vertical movement can again react to small changes in demands of the engine.*

6. **Cold start device.** Refer Fig. 11.19 (*d*).

- A *cold start-device is in the form of a rotary-valve consisting of a cylindrical valve body, which has an annular groove in the middle region with a single radial hole drilled in its side.* Fitted inside the valve body is a spindle which has an axial hole bored half-way along from one end, while at the other end a control level is bolted. A double taper notched radial hole intersects the axial hole in the spindle. The whole assembly of the valve body and spindle is positioned in a larger hole made in the side of the floor chamber.
- In order to *cold start the engine* the *choke knob* situated on the instrument panel is *pulled out,* the interconnecting cable rotates the control lever and spindle to a position where the radial hole for both spindle and valve body are aligned. When the engine is cranked a high depression is created in the mixing chamber formed between the jet bridge and throttle valve, and this depression is conveyed to the axial hole in the control spindle where it then passes to the annular groove on the outside of the valve body. Here it divides and draws the fuel from the dip tube and atmospheric air from the float chamber by way of the air jet. The emulsified mixture is then drawn into the hollow spindle along the discharge passage duct and out into the mixing chamber.
- With warming up of the engine, the choke knob can be pushed back steadily, this rotates the control lever and spindle so that the *notched hole passageway becomes progressively smaller and thus restricts the quantity of air and petrol emulsion trying to enter the mixing chamber.*

11.9.3.4. Aircraft Carburettor

The *major difference* between an automobile carburettor and an aircraft carburettor is that whereas the former operates at ground level conditions, the latter operates at varying altitudes. With the increase in altitude the density of air decreases and A / F ratio which is proportional to air density/fuel density decreases, *i.e.* the fuel-air mixture goes on becoming richer with increase in altitude. The mixture will be about 40% richer at an altitude of about 7000 metres (since at this altitude the air density is nearly one-half that at ground level and hence A / F ratio about 0.7 times the valve at ground level).

In view of the above, it is imperative to provide in aircraft carburettors an **altitude mixture correction device** to reduce the quantity of fuel progressively with altitude. For this purpose, the following methods are employed :

(*i*) Air bleeding ;

(*ii*) Back suction control ;

(*iii*) Incorporating a metering pin.

The aircraft carburettors entail the following other *special features :*

- For fuel level system a *special float chamber is required.*
- For controlling/eliminating the formation of ice in the choke tube and on the throttle valve due to *low temperature* an automatic *de-icing unit* is fitted or *hot engine oil is arranged to flow around the carburettor barrel and through the hollow throttle valve.*

11.10. PETROL INJECTION

11.10.1. Drawbacks of Modern Carburettors

The modern carburettors have the following *drawbacks* :

1.The mixture supplied to various cylinders of a multi-cylinder engine *varies in quality and quantity*. Also, due to fuel condensation in induction manifold, the *mixture proportion is affected.*

2. Due to presence of several wearing parts, the carburettors operate at a lower efficiency.

3. Reduced volumetric efficiency due to non-availability of a free flow passage for the mixture owing to thc presence of choke tubes, throttle valves, jets, bends etc.

4. At low temperatures, freezing can occur (if special means to obviate this are not provided).

5. When the carburettor is tilted or during acrobatics in aircraft surging can occur (if means to avoid this are not provided).

6. In the absence of flame traps, backfiring may occur which may lead to ignition of fuel outside the carburettor.

11.10.2. Introduction to Fuel Injection

- The *function* of a fuel injection system is :
  (*i*) *To monitor the engine's operating variables,*
  (*ii*) *To transfer this information to a metering control, then*
  (*iii*) *To discharge and atomise the fuel into the incoming air stream.*
- The position where the fuel is injected into the air charge considerably influences the performance of the engine.

11.10.3. Direct Injection. Refer Fig. 11.20.

- In this type of layout the *fuel injectors are positioned in the cylinder-head so that fuel is directly discharged into each combustion chamber* (Fig. 11.20).
- With this arrangement it is essential that injection is timed to occur about 60° after T.D.C. on the induction stroke.
- Because of the shorter time period for fuel spray to mix with the incoming air charge, increased air turbulence is necessary. To compensate for the shorter permitted time for injection, atomising and mixing, the *injection pressure needs to be higher than for indirect injection.*
- *More overlap of exhaust and inlet valves can be utilized* compared with other carburetted or injected systems, so that incoming fresh air can assist in sweeping out any remaining exhaust gases from the combustion chambers.
- The injector nozzle and valve have to be designed to withstand the high operating pressures and temperatures of the combustion chamber, *this means that a more robust and costly injector unit is required.*
- *Generally, direct-injection air and fuel mixing is more thorough in large cylinders than in small ones because fuel droplet sizes do not scale down as the mixing space becomes smaller.*
- All condensation and wetting of the induction manifold and ports is *eliminated* but some spray may condense on the piston crown and cylinder walls.

11.10.4. Indirect Injection. Refer Fig. 11.20.

- In this arrangement the fuel is injected into the air *stream prior to entering the combustion chamber.* Fuel spray may be delivered from a single-point injection (S.P.I.) source, which is usually just upstream from the throttle (air intake side of the throttle), or it may be supplied from a multi-point injection (M.P.I) source, where the *injectors are positioned in each induction manifold branch pipe just in front of the inlet port.* (Fig. 11.20).

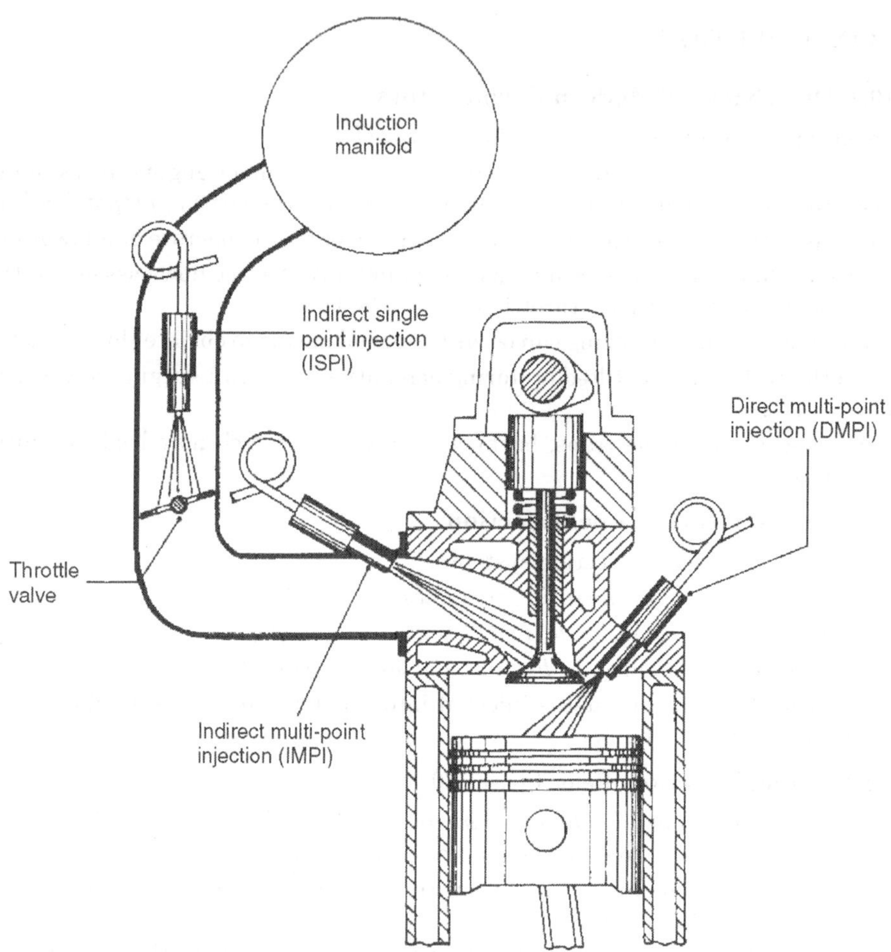

Fig. 11.20. Three-principal injector positions.

- Indirect injection can be discharged at relatively low pressure (2 to 6 bar) and need not be synchronized to the engine's induction cycle. Fuel can be discharged simultaneously to each induction pipe where it is mixed and stored until the inlet valve opens.
- Since indirect injection does not need to be timed, it *requires only low discharge pressures* and *the injectors are not exposed to combustion, the complexity of the operating mechanisms can be greatly reduced, which considerably lowers cost.*
- The single-point injection system has the same air and fuel mixing and distribution problems as a carburettor layout *but without venturi restriction* so that *higher engine volumetric efficiencies are obtained. High injection pressures,* compared with the carburettor discharge method of fuel delivery, *speed up and improve the atomization of the liquid spray.*
- The *multi-point injection* layout, in contrast to the single-point injection method *has no fuel distribution difficulties* since each injector discharges directly into its own induction port and the mixture then has only to move a short distance before it enters the cylinder. Since the induction manifold deals mainly with only induced air, the branch pipes can be enlarged and extended to maximize the ram effect of the incoming air charge.

A **major feature** *with petrol injection is that there is separate air and fuel metering and that fuel metering is precise under all engine operating condition's.*

11.10.5. Injection Considerations

The fuel can be discharged into the air stream, *using indirect injection arrangements,* by the following *two methods* :

1. Continuous injection.
2. Intermittent or pulsed injection.

1. **Continuous injection :**

In this arrangement, *the injector nozzle and valve are permanently open* while the engine is operating and the *amount of fuel discharged in the form of a spray is controlled by either varying the metering orifice or the fuel discharge pressure, or a combination of both of these possible variables.*

2. **Intermittent or pulsed injection :**

In this type of injection, *fuel is delivered from the injector in spray form at regular intervals with a constant fuel discharge pressure and the amount of fuel discharged is controlled by the time period the injector nozzle valve is open.*

- ***Timed injection.*** This where the start of delivery for each cylinder *occurs at the same angular point in the engine cycle,* this can be anything from 60° to 90° after T.D.C. on the induction stroke.
- ***Non-timed injection.*** In contrast to timed injection, this is where *all the injectors are programmed to discharge their spray at the same time, therefore each cylinder piston will be on a different part of the engine cycle.*

11.10.6. Comparison of Petrol Injection and Carburetted Fuel Supply Systems

Merits of petrol injection :

Following are the *merits of petrol engine system :*

1. In petrol injection system, *due to absence of venturi there is the minimum of air restriction so that higher engine volumetric efficiencies can be obtained with the corresponding improvement in power and torque.*

2. *The spots* for pre-heating the cold air and fuel mixture are *eliminated* so that *denser air enters the cylinder* when the engine has reached normal operating conditions.

3. As the manifold branch pipes are not greatly concerned with mixture preparation they *can be designed to utilize the inertia of the air charge to increase the engine's volumetric efficiency ;* (this does not apply for single point injection).

4. Because of *direct spray discharge* into each inlet port, *acceleration response is better.*

5. *Atomization of fuel droplets is generally improved* over normal speed and load driving conditons.

6. It is possible to *use greater inlet and exhaust valve overlap without poor idling, loss of fuel or increased exhaust pollution.*

7. The monitoring of engine operating parameters enables accurate matching of air and fuel requirements under normal speed and load conditions which *improves engine performance, fuel consumption and reduces exhaust pollution.*

8. Fuel injection equipment is *precise in metering injected fuel spray* into the intake ports over the complete engine speed, load and temperature operating range.

9. There is *precise fuel distribution* between engine cylinders even under full load conditions *with multi-point injection.*

10. *Multi-point injection does not require time for fuel transportation in the intake manifold and there is no manifold wall melting.*

11. With fuel injection, when cornering fast or due to heavy braking, fuel surge is eliminated.

12. The single point as well as multi-point injection systems are *particularly adaptable and suitable for supercharged engines.*

Demerits / Limitations of petrol injection :

Petrol injection system entails the following *demerits/limitations* :

1. Initial cost of equipment is high ; replacement parts are also expensive.
2. Increased care and attention required.
3. In order to diagnose fuel injection system faults and failures, special servicing equipment is necessary.
4. It is necessary to have considerably more mechanical and electrical knowledge to diagnose and rectify the faults of fuel equipment.
5. Injection equipment may be elaborately complicated, delicate to handle and impossible to service.
6. More electrical and mechanical components to go wrong.
7. Due to pumping and metering of the fuel there is increased mechanical and hydraulic noise.
8. Due to the fine working tolerances of the metering and discharging components, very careful filtration is needed.
9. To drive the fuel pressure pump or injection discharge devices, power (be it electrical or mechanical) is necessary.
10. More bulky and heavy (than that of a carburetted fuel supply system).

11.10.7. Electronic Fuel Injection

Fig. 11.21 shows the fuel injection system-L-Jetronic with air flow metering (developed by Robert Based Corp.) It consist of the following *units* :

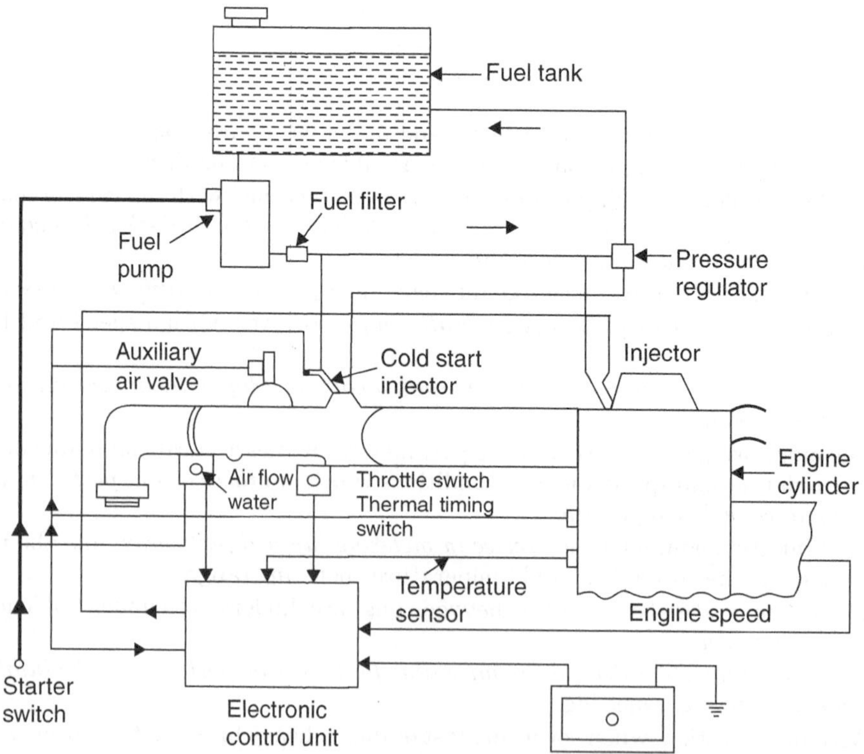

Fig. 11.21. Fuel injection system-L-Jetronic with air-flow metering.

1. Fuel delivery system
2. An induction system
3. Sensors and air flow control system
4. Electronic control unit.

1. **Fuel delivery system :**

- It consists of an electrically driven fuel pump which draws fuel from a *fuel tank.* The pump forces the oil through a *filter* into a line at the end of which is situated a *pressure regulator*, which in turn is connected to intake manifold.
- The pressure regulator keeps the pressure difference between the fuel pressure and the manifold pressure constant, so that the quantity of fuel injected is dependent on the injector open time only.

2. **Air induction system :**

- After passing the air filter, the incoming air flows through an air flow meter, which generates a voltage signal (depending on the quantity of air flow).
- Just behind the *throttle valve* is fitted a cold start *magnetic injection valve,* which injects additional fuel for cold start. This valve also supplies the extra fuel needed during warm-up period.
- An *auxiliary valve* (which by-passes the throttle valve) supplies the extra air required for idling (in addition to rich-air-fuel mixture). This extra air increases the engine speed after cold start to acceptable idling speed.
- To the throttle valve is attached a *throttle switch* equipped with a set of contacts which generate a sequence of voltage signals during the opening of throttle valve. The voltage *signals result in injection of additional fuel required for acceleration.*

3. **Electric control unit :**

- The *sensors* are incorporated to measure the operating data at different locations. The data measured by the sensors are transmitted to the *electronic control unit which computes the amount of fuel injected during each engine cycle. The amount of fuel injected is varied by varying the injector opening time only.*
- The *sensors* used are :
  - — Manifold pressure ;
  - — Engine speed ;
  - — Temperature at the intake manifold.

4. **Injection time :**

- For every revolution of the camshaft, the fuel is injected twice, each injection contributing half of a fuel quantity required for engine cycle.
- The injectors, at different phases of the operating cycle, are operating simultaneously.

11.11. THEORY OF SIMPLE CARBURETTOR

During the induction stroke, the air is sucked through the carburettor by the pressure difference across it created when the piston moves. As the air passes through the venturi, its velocity increases and reaches maximum (section 2-2, see Fig. 11.22) at venturi throat, this being the minimum area in the induction track (unless the throttle is sufficiently closed to provide a smaller area). As a result of suction created in the venturi fuel is sucked through the nozzle. The tip of the nozzle is z metres above the float chamber level ; this arrangement prevents spilling of petrol when vehicle is stationary. Let us find expressions for air flow neglecting and considering the compressibility of air.

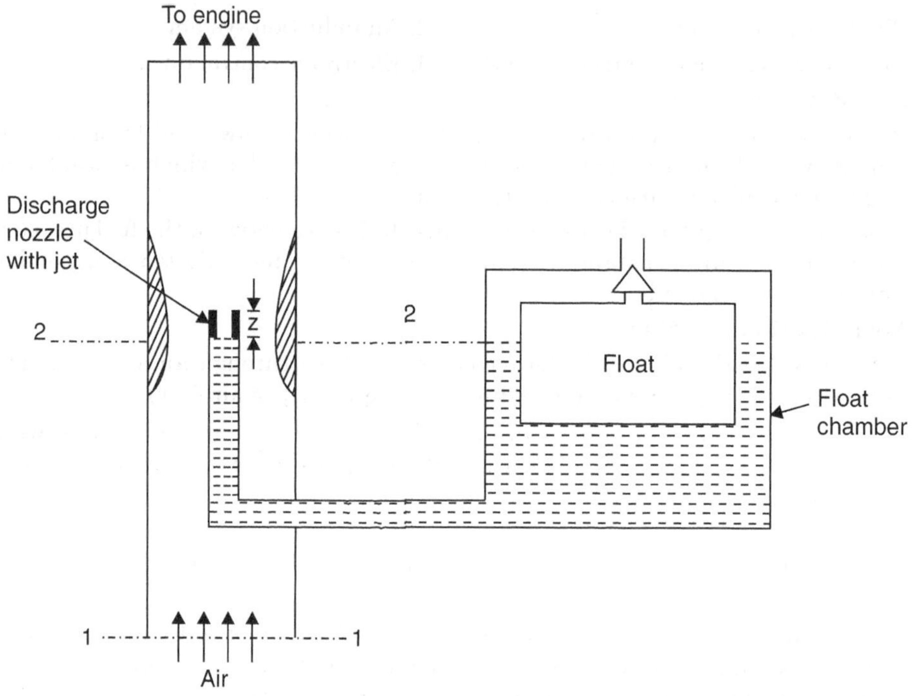

Fig. 11.22. Principle of a simple carburettor.

Case I. *Neglecting the compressibility of air*—**Approximate Analysis.**

Applying Bernoulli's equation at sections 1-1 and 2-2, the equation for air flow is given by

$$\frac{p_1}{\rho_a} + \frac{C_1^2}{2} = \frac{p_2}{\rho_a} + \frac{C_2^2}{2} \quad ...(11.1)$$

where, ρ_a = Density of air kg/m^3, and

p_1, p_2 = Pressure at sections 1-1 and 2-2 respectively.

$C_1 = C_2$ = Velocities at sections 1-1 and 2-2 respectively, m/s.

Assuming initial *velocity of air to be zero* (C_1 = **0**), *density of air* (ρ_a) to be *constant*, since air is *assumed incompressible*, we have

$$\frac{p_1}{\rho_a} = \frac{p_2}{\rho_a} + \frac{C_2^2}{2} \quad ...(11.2)$$

$$\therefore \quad C_2 = \sqrt{\frac{2(p_1 - p_2)}{\rho_a}} = \sqrt{\frac{2\Delta p_a}{\rho_a}} \quad ...(11.3)$$

where, $\Delta p_a = p_1 - p_2$

Mass of air per second $\dot{m}_a = C_2 A_2 \rho_a = A_2 \sqrt{2\rho_a \Delta p_a}$ ...(11.4)

where A_2 is area of venturi throat in m^2.

(The above equation gives *theoretical mass flow of air. The actual mass flow is obtained by multiplying the co-efficient of discharge of venturi*)

Similarly, for the flow of fuel, we have

$$\frac{p_1}{\rho_f} = \frac{p_2}{\rho_f} + \frac{C_f^2}{2} + gz \quad ...(11.5)$$

where, ρ_f = Constant density of fuel, and

C_f = Velocity of flow of fuel.

$$\therefore \quad C_f = \sqrt{\frac{2(p_1 - p_2 - gz\rho_f)}{\rho_f}} = \sqrt{\frac{2(\Delta p_a - gz\rho_f)}{\rho_f}} \quad ...(11.6)$$

[It may be noted that due to petrol surface being lower than the top of the jet by z metres the pressure difference becomes $(\Delta p_a - gz\rho_f)$ instead of Δp_a]

Mass of fuel per second, $\dot{m}_f$ (theoretical) $= A_f \rho_f = A_f \sqrt{2\rho_f(\Delta p_a - gz\rho_f)}$ ...(11.7)

where, Δ_f = Cross-sectional area of the fuel jet m^2.

$\therefore$ Air-fuel (A/F) ratio,

$$\frac{\dot{m}_a}{\dot{m}_f} = \frac{A_2\sqrt{2\rho_a \Delta p_a}}{A_f\sqrt{2\rho_f(\Delta p_a - gz\rho_f)}} = \frac{A_2}{A_f}\sqrt{\frac{\rho_a}{\rho_f}} \cdot \sqrt{\frac{\Delta p_a}{(\Delta p_a - gz\rho_f)}} \quad ...(11.8)$$

If C_{da} and C_{df} are the coefficients of discharge of venturi and fuel jet respectively, then

$$\frac{\dot{m}_a}{\dot{m}_f} = \frac{C_{da}}{C_{df}} \cdot \frac{A_2}{A_f} \cdot \sqrt{\frac{\rho_a}{\rho_f}} \cdot \sqrt{\frac{\Delta p_a}{\Delta p_a - gz\rho_f}} \quad ...(11.9)$$

If $z = 0$,

$$\frac{\dot{m}_a}{\dot{m}_f} = \frac{C_{da}}{C_{df}} \cdot \frac{A_2}{A_f}\sqrt{\frac{\rho_a}{\rho_f}} \quad ...(11.10)$$

Case II. *Taking into consideration the compressibility of air in account*—**Exact Analysis.**

When the compressibility of air is taken into account, the air flow will change but the fuel flow will remain unchanged. Applying steady flow energy equation (S.F.E.E.) at sections 1-1 and 2-2, we get,

$$h_1 + \frac{C_1^2}{2} + Q = h_2 + \frac{C_2^2}{2} + W$$

or

$$Q - W = (h_2 - h_1) + \frac{C_2^2 - C_1^2}{2}$$

where h_1, h_2 = Enthalpies at sections 1-1 and 2-2 respectively.

Since $Q = 0$, $W = 0$ and $C_1 = 0$

$$\therefore \quad C_2 = \sqrt{2(h_1 - h_2)}$$

Substituting $h_1 = c_p T_1$ and $h_2 = c_p T_2$, we get

$$C_2 = \sqrt{2c_pT_1\left(1 - \frac{T_2}{T_1}\right)} \quad ...(11.11)$$

Since the flow process between the atmosphere and the venturi throat is isentropic,

$$\therefore \quad \frac{T_2}{T_1} = \left(\frac{p_2}{p_1}\right)^{\frac{\gamma-1}{\gamma}} \quad ...(11.12)$$

Substituting eqn. (11.6) in eqn. (11.5), we get

$$C_2 = \sqrt{2c_pT_1\left[1 - \left(\frac{p_2}{p_1}\right)^{(\gamma-1)/\gamma}\right]} \quad ...(11.13)$$

Now, the mass of flow of air is constant from inlet to venturi throat, and is given by

$$\dot{m}_a = \frac{A_1 C_1}{v_1} = \frac{A_2 C_2}{v_2} \qquad ...(11.14)$$

where v_1, v_2 = Specific volumes at sections 1-1 and 2-2 respectively.

Since $p_1 v_1^{\gamma} = p_2 v_2^{\gamma}$, $\therefore$ $$v_2 = v_1 \left(\frac{p_1}{p_2}\right)^{1/\gamma} = \frac{RT_1}{p_1}\left(\frac{p_1}{p_2}\right)^{1/\gamma} \qquad ...(11.15)$$

Substituting the values of C_2 and v_2 in eqn. 11.8, we get

Theoretical mass flow of air, $$\dot{m}_a = \frac{A_2 p_1}{RT_1}\left(\frac{p_2}{p_1}\right)^{1/\gamma} \sqrt{2c_p T_1\left[1 - \left(\frac{p_2}{p_1}\right)^{(\gamma - 1)/\gamma)}\right]}$$

$$= \frac{A_2 p_1}{R\sqrt{T_1}} \sqrt{2c_p\left[\left(\frac{p_2}{p_1}\right)^{2/\gamma} - \left(\frac{p_2}{p_1}\right)^{(\gamma+1)/\gamma}\right]} \qquad ...(11.16)$$

$$\therefore \quad \frac{\dot{m}_a}{\dot{m}_f} = \frac{C_{da}}{C_{df}} \times \frac{A_2}{A_f} \times \frac{p_1}{R\sqrt{T_1}} \times \frac{\sqrt{2c_p\left[\left(\frac{p_2}{p_1}\right)^{2/\gamma} - \left(\frac{p_2}{p_1}\right)^{(\gamma+1)/\gamma}\right]}}{\sqrt{2\rho_f\,(\Delta p_a - gz\rho_f)}} \qquad ...(11.17)$$

Comments on Air-Fuel (A / F) ratio supplied by the carburettor :

1. From eqn (11.8), it may be observed that no fuel flow will take place when $\Delta p_a < gz\rho_f$. When $\Delta p_a > gz\rho_f$, the fuel flow will start and the mixture becomes progressively richer.

The minimum air velocity at throat to cause fuel flow, with given value of z, can be calculated as follows :

From eqn. 11.4, $$C_2 = \sqrt{\frac{2\Delta p_a}{\rho_a}} = \sqrt{\frac{2(gz\rho_f)}{\rho_a}} \qquad ...(11.18)$$

2. At higher air flows when $\Delta p_a >> gz\rho_f$ (*i.e.* fraction $gz\rho_f/\Delta p_a$ becomes negligible), the air-fuel ratio approaches

$$\frac{\dot{m}_a}{\dot{m}_f} = \frac{A_2}{A_f} \cdot \frac{C_{da}}{C_{df}}\sqrt{\frac{\rho_a}{\rho_f}} \qquad ...(11.19)$$

3. A reduction in the density of air reduces the value of A/F, *i.e.*, the mixture becomes richer. It happens at (*i*) high air flows when Δp_a is large, hence C_2 is large and ρ_2 becomes small, (*ii*) high altitudes where density of air is small.

Thus in a *simple carburettor the air-fuel mixture becomes progressively richer with increasing air flows and increasing altitudes* ; this is a *big drawback.* The other *drawbacks* are :

- It cannot supply rich mixture required during idling and low load operation.
- It cannot supply rich mixture for starting from cold.
- It cannot supply extra fuel needed during acceleration.

WORKED EXAMPLES

Example 11.1. *A four cylinder four-stroke engine having diameter and length of stroke as 100 m and 120 mm respectively is running at 1800 r.p.m. Its carburettor venturi has a 28 mm throat. Assuming co-efficient of air flow 0.8, density of air 1.2 kg/m³ and volumetric efficiency of the engine as 75 per cent, determine the suction at the throat.*

Solution. *Given* : $D = 100$ mm $= 0.1$ m ; $L = 120$ mm $= 0.12$ m ; $N = 1800$ r.p.m. ;

Throat diameter, $d_2 = 28$ mm $= 0.028$ m ; $C_{da} = 0.8$; $\rho_a = 1.2$ kg/m^3 ; $\eta_{vol.} = 75\%$.

Suction at the throat Δp_a :

Stroke volume $= \frac{\pi}{4} \times (0.1)^2 \times 0.12 \times 4 = 0.00377$ m^3

Actual volume per strokes $= \eta_{vol.} \times 0.00377 = 0.75 \times 0.00377 = 0.00283$ m^3

$\therefore$ Actual volume sucked per second

$$= 0.00283 \times \frac{1800}{2} \times \frac{1}{60} = 0.04245 \text{ m}^3/\text{s}$$

$$\dot{m}_a = 0.04245 \times 1.2 = 0.05094 \text{ kg/s}$$

As the initial temperature and pressure are not given, the problem is solved by approximate method *i.e.*, neglecting compressibility of the air.

$$\dot{m}_a = C_{da} \times A_2 = \sqrt{2\rho_a \, \Delta p_a} \qquad \text{...[Eqn. 11.4]}$$

$$0.05094 = 0.8 \times \frac{\pi}{4}(0.028)^2 \sqrt{2 \times 1.2 \times \Delta p_a}$$

$$= 7.63 \times 10^{-4} \sqrt{\Delta p_a}$$

$$\therefore \qquad \Delta p_a = \left(\frac{0.05094}{7.63 \times 10^{-4}}\right)^2 = 4457 \text{ N/m}^2 = \mathbf{0.04457 \text{ bar.}} \quad \textbf{(Ans.)}$$

Example 11.2. *A spark ignition engine on test consumes 5 kg/h of petrol when running on an air-fuel ratio of 16 : 1. The engine uses a single-jet carburettor having a fuel orifice area of 2 sq mm and the tip of the jet is 5 mm above the level of petrol in the float chamber, when the engine is not running. Calculate the depression in the venturi throat to maintain the required fuel flow rate through the carburettor. Assume specific gravity of petrol as 0.75 and the coefficient of discharge of the fuel orifice as 0.8. What area of venturi throat will be required to maintain the desired flow rate ? Density of air is 1.20 kg/m³ and the coefficient of discharge for venturi throat is 0.8. Neglect compressibility of air.* **(Roorkee University, AMIE, S-2000)**

Solution. *Given* : $\dot{m}_f = \frac{5}{3600} = 0.001389$ kg/s ; A/F ratio = 16 : 1 ;

Fuel orifice area, $A_f = 2$ mm$^2 = 2 \times 10^{-6}$ m^2 ; $z = 5$ mm $= 0.005$ m ;

Sp. gr. of petrol = 0.75 ; $C_{df} = 0.8$, $\rho_a = 1.2$ kg/m^3 ; $C_{da} = 0.8$.

Depression in venturi throat, Δp_a :

The actual fuel flow rate is given by,

$$\dot{m}_f = C_{df} \,.\, A_f \sqrt{2\rho_f(\Delta p_a - gz\rho_f)} \qquad \text{...[Eqn. (11.7)]}$$

where Δp_a is in N/m^2.

or $$0.001389 = 0.8 \times (2 \times 10^{-6}) \sqrt{2 \times (0.75 \times 1000)\,(\Delta p_a - 9.81 \times 0.005 \times 0.75 \times 1000)}$$

or $$\frac{0.001389}{0.8 \times 2 \times 10^{-6}} = 38.73 \sqrt{\Delta p_a - 36.79}$$

or $$\Delta p_a - 36.79 = \left(\frac{0.001389}{0.8 \times 2 \times 10^{-6} \times 38.73}\right)^2$$

or $$\mathbf{\Delta p_a = 539.2\ N/m^2.} \quad \textbf{(Ans.)}$$

Throat area, A_t :

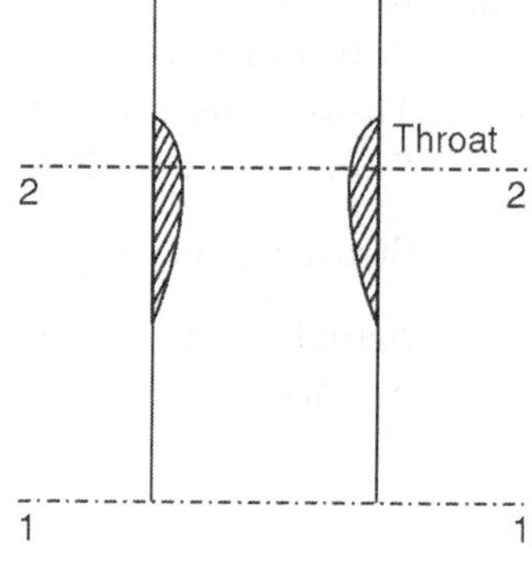

Fig. 11.23

Air flow rate, $\dot{m}_a = \frac{5}{3600} \times 16 = 0.02222$ kg/s

Also, $\dot{m}_a = C_{da} \times A_t \sqrt{2\rho_a \Delta p_a}$ ...(Eqn. (11.4)

(Here $A_2 = A_t$)

$$0.02222 = 0.8 \times A_t \sqrt{2 \times 1.2 \times 539.21}$$

$$\therefore \quad A_t = \frac{0.02222}{0.8\sqrt{2 \times 1.2 \times 539.21}} = 7.7209 \times 10^{-4}\ m^2$$

$$= \mathbf{7.7209\ cm^2.} \quad \textbf{(Ans.)}$$

Example 11.3. *The following data relate to a petrol engine :*

Petrol consumed per hour	*= 7.2 kg*
The specific gravity of the fuel	*= 0.75*
The temperature of air	*= 27°C*
The air fuel ratio	*= 1 : 15*
The diameter of the choke tube	*= 24 mm*

The height of top of the jet above the petrol level = 4.2 mm = 0.0042 m in the float chamber

The co-efficient of discharge for air	*= 0.8*
The co-efficient of discharge for fuel	*= 0.7*
Atmospheric pressure	*= 1.013 bar*

Calculate the diameter of the fuel jet of a simple carburettor.

Solution. *Given* : $\dot{m}_f = \frac{7.2}{3600}$ kg/s ; $\rho_f = 0.75 \times 1000 = 750$ kg/m^3 ; $T_1 = 27 + 273 = 300$ K ; A / F ratio = 1 : 15 ; $d_2 = 24$ mm = 0.024 m ; $z = 4.2$ mm = 0.0042 m ; $C_{da} = 0.8$; $C_{df} = 0.7$; $p_1 = 1.013$ bar.

Diameter of the fuel jet, d_f :

We know that, $$\rho_a = \frac{p_1}{RT_1} = \frac{1.013 \times 10^5}{(0.287 \times 1000) \times 300} = 1.176\ \text{kg/m}^3$$

Air flow rate, $$\dot{m}_a = A_2\, C_{da} \sqrt{2\rho_a \Delta p_a}$$

$$\frac{15 \times 7.2}{3600} = \frac{\pi}{4} \times (0.024)^2 \times 0.8 \sqrt{2 \times 1.176 \times \Delta p_a}$$

$$= 5.55 \times 10^{-4} \sqrt{\Delta p_a}$$

$$\therefore \quad \Delta p_a = \left(\frac{15 \times 7.2}{3600 \times 5.55 \times 10^{-4}}\right)^2 = 2922\ \text{N/m}^2$$

Fuel flow rate, $$\dot{m}_f = A_f\, C_{df} \sqrt{2\rho_f(\Delta p_a - gz\rho_f)}$$ ...[Eqn. 11.7]

$$\frac{7.2}{3600} = \frac{\pi}{4}(d_f)^2 \times 0.7 \ \sqrt{2 \times 750\,(2922 - 9.81 \times 0.0042 \times 750)}$$

$$= 1144.89\,(d_f)^2$$

$$\therefore \qquad \mathbf{d_f} = \left(\frac{7.2}{3600 \times 1144.89}\right)^{1/2} = 1.32 \times 10^{-3} \text{ m} \quad \text{or} \quad \mathbf{1.32\ mm.} \quad \mathbf{(Ans.)}$$

Example 11.4. *A simple carburettor under a certain condition delivers 5.45 kg/h of petrol with an air-fuel ratio of 15. The fuel jet area is 2 mm² with a coefficient of discharge of 0.75. If the tip of the fuel jet is 0.635 cm above the level of petrol in the float chamber and the venturi throat coefficient of discharge is assumed to be 0.80, calculate :*

(i) The venturi depression in cm of H_2O necessary to cause air and fuel flow at the desired rate.

(ii) The venturi throat diameter.

(iii) The velocity of air across the venturi throat.

You may take density of air = 1.29 kg/m³ and specific gravity of period = 0.72.

(Madras University)

Solution. *Given* : $\dot{m}_f = \frac{5.45}{3600} = 0.001514$ kg/s ; A/F ratio = 15 ; $A_f = 2$ mm² $= 2 \times 10^{-6}$ m² ; $C_{df} = 0.75$; $z = 0.635$ cm $= 0.00635$ m ; $C_{da} = 0.8$; $\rho_a = 1.29$ kg/m³ ; Sp. gr. of petrol = 0.72.

(*i*) **Venturi depression, Δp_a :**

$$\dot{m}_f = C_{df} \cdot A_f \ \sqrt{2\rho_f(\Delta p_a - gz\rho_f)} \qquad \text{...[Eqn. (11.7)]}$$

where Δp_a is in N/m².

$$0.001514 = 0.75 \times 2 \times 10^{-6} \ \sqrt{2 \times (0.72 \times 1000)\,(\Delta p_a - 9.81 \times 0.00635 \times 0.72 \times 1000)}$$

$$\frac{0.001514}{0.75 \times 2 \times 10^{-6}} = 37.95\sqrt{(\Delta p_a - 44.85)}$$

or
$$\Delta p_a - 44.85 = \left(\frac{0.001514}{0.75 \times 10^{-6} \times 37.95}\right)^2 = 707.37$$

$$\therefore \qquad \Delta p_a = 752.22 \text{ N/m}^2 = \frac{752.22}{9810} \text{ m of water} = \mathbf{7.67\ cm\ of\ H_2O.} \quad \mathbf{(Ans.)}$$

(*ii*) **Venturi throat diameter, D_t :**

$$\text{Air flow rate} = \frac{5.45}{3600} \times 15 = 0.02271 \text{ kg/s}$$

Also,
$$\dot{m}_a = C_{da}\,A_t\,\sqrt{2\rho_a \Delta p_a}$$

$$\therefore \qquad 0.02271 = 0.8 \times A_t \sqrt{2 \times 1.29 \times 752.22}$$

or
$$A_t = 6.444 \times 10^{-4} \text{ m}^2 = \frac{\pi}{4} D_t^2$$

$$\therefore \qquad \Delta_t = \left[\frac{6.444 \times 10^{-4} \times 4}{\pi}\right]^{1/2} = 0.0286 \text{ m} = \mathbf{2.86\ cm.} \quad \mathbf{(Ans.)}$$

(*iii*) **Velocity of air across the venturi throat C_t :**

$$C_t \text{ (or } C_2) = \sqrt{\frac{2(gz\rho_f)}{\rho_a}} \qquad \text{...(Eqn. 11.18)}$$

$$= \sqrt{\frac{2 \times 9.81 \times 0.00635 \times (0.72 \times 1000)}{1.29}} = \mathbf{8.34\ m/s.} \quad \textbf{(Ans.)}$$

Example 11.5. *A carburettor ; tested in the laboratory has its float chamber vented to atmosphere. The main metering system is adjusted to give an air-fuel ratio of 15 : 1 at sea level conditions. The pressure at the venturi throat is 0.8 bar. The atmospheric pressure is 1 bar. The same carburettor is tested again when an air cleaner is fitted at the inlet to the carburettor. The pressure drop to air cleaner is found to be 30 mm of Hg when air flow at sea level condition is 240 kg/h. Assuming zero tip and constant coefficient of flow, calculate (i) the throat pressure when the air cleaner is fitted and (ii) air-fuel ratio when the air cleaner is fitted.* **(Bombay University)**

Solution. *Given* : A / F ratio = 15 : 1 at sea level conditions ;

p_1 = 1 bar ; p_2 = 0.8 bar ;

(*i*) **The throat pressure when the air cleaner is fitted :**

Quantity of air flowing is same in both the cases.

$$(\dot{m}_a)_{\text{actual}} = C_{da} A_t \sqrt{2\rho_a(p_1 - p_2)}$$

When there is *no air cleaner,*

$$\Delta p_a = p_1 - p_2 = 1 - 0.8 = 0.2 \text{ bar}$$

When the *air cleaner is fitted,* let p_t be the throat pressure, then

$$\Delta p_a' = \left[1 - \left(1000 \times 13.6 \times 9.81 \times \frac{30}{1000} \times 10^{-5}\right) - p_t\right] \text{ bar}$$

$$= (0.96 - p_t) \text{ bar} \qquad (\because \ 1 \text{ bar} = 10^5 \text{ N/m}^2)$$

For the same air flow and constant coefficients,

$$\Delta p_a = \Delta p_a'$$

$$\therefore \quad 0.2 = 0.96 - p_t$$

or $\mathbf{p_t = 0.76}$ **bar. (Ans.)**

(*ii*) **Air-fuel ratio when the air cleaner is fitted :**

Without air cleaner, $\Delta p_f = \Delta p_a = 0.2$ bar

With air cleaner fitted (with float-chamber still vented to atmosphere),

$$\Delta p_f = 1 - 0.76 = 0.24 \text{ bar}$$

As Δp_f has increased more fuel will flow making the mixture richer.

$$\text{New A / F ratio} = \text{A / F ratio when air cleaner is not fitted} \times \sqrt{\frac{\Delta p_f \text{ without air cleaner}}{\Delta p_t \text{ with air cleaner}}}$$

$$= 15 \times \sqrt{\frac{0.2}{0.24}} = \mathbf{13.69.} \quad \textbf{(Ans.)}$$

Example 11.6. *A simple jet carburettor is required to supply 4.6 kg of air per minute. The pressure and temperature of air are 1.013 bar and 25°C respectively. Assuming flow to be isentropic and compressible and velocity coefficient as 0.8, calculate the throat diameter of the choke for air flow velocity of 80 m/s.*

Solution. *Given* : $\dot{m}_a = \frac{4.6}{60} = 0.0767$ kg/s ; p_1 = 1.013 bar, T_1 = 25 + 273 = 298 K ;

$C_1 = 0$; $C_2 = 80$ m/s ; $C_v = 0.8$.

Throat diameter d_2 :

Applying S.F.E.E. at sections 1-1 and 2-2, we have

$$h_1 + \frac{C_1^2}{2} + Q = h_2 + \frac{C_2^2}{2} + W$$

or $$h_1 = h_2 + \frac{C_2^2}{2} \qquad \left(\because \; Q = 0 \,;\, W = 0,\; C_1 = 0\right)$$

or $$C_2 = \sqrt{2(h_1 - h_2)}$$

$$= \sqrt{2c_p(T_1 - T_2)}$$

$$= \sqrt{2c_pT_1\left(1 - \frac{T_2}{T_1}\right)}$$

$$= \sqrt{2c_pT_1\left[1 - \left(\frac{p_2}{p_1}\right)^{\frac{\gamma-1}{\gamma}}\right]}$$

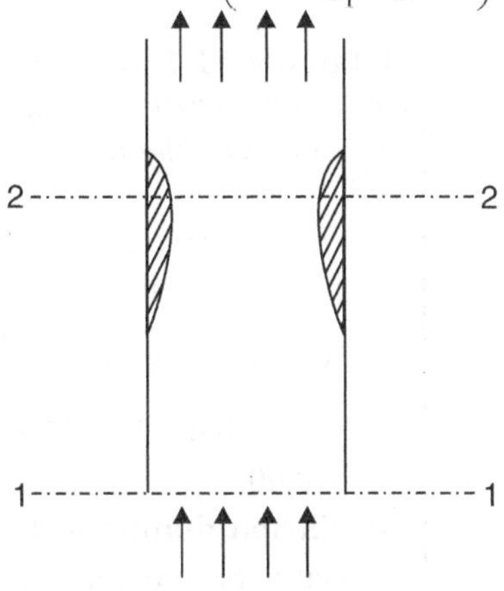

$\therefore$ $$(C_2)_{actual} = C_v\sqrt{2c_pT_1\left[1 - \left(\frac{p_2}{p_1}\right)^{\frac{\gamma-1}{\gamma}}\right]}$$

Fig. 11.24

$$c_p = R\left(\frac{\gamma}{\gamma - 1}\right) = 0.287 \times \frac{1.4}{4.1 - 1} = 1.005 \text{ kJ/kg K}$$

$$80 = 0.8\sqrt{2 \times 1.005 \times 1000 \times 298\left[1 - \left(\frac{p_2}{1.013}\right)^{\frac{1.4-1}{1.4}}\right]} = 619.15\sqrt{1 - \left(\frac{p_2}{1.013}\right)^{0.2857}}$$

or $$1 - \left(\frac{p_2}{1.013}\right)^{0.2857} = \left(\frac{80}{619.15}\right)^2 = 0.01669$$

or Throat pressure, $$p_2 = \left[(1 - 0.01669)^{\frac{1}{0.2857}}\right] \times 1.013 = 0.955 \text{ bar}$$

$$\rho_1 = \frac{p_1}{RT_1} = \frac{1.013 \times 10^5}{(0.287 \times 1000) \times 298} = 1.1844 \text{ kg/m}^3$$

Now $$pv^\gamma = \text{constant}$$

or $$\frac{p}{\rho^\gamma} = \text{constant}$$

or $$\frac{p_1}{\rho_1^\gamma} = \frac{p_2}{\rho_2^\gamma}$$

or $$\rho_2 = \rho_1\left(\frac{p_2}{p_1}\right)^{1/\gamma} = 1.1844\left(\frac{0.955}{1.013}\right)^{\frac{1}{1.4}} = 1.1356 \text{ kg/m}^3$$

We know that, $\dot{m}_a \, (= \rho AC) = \rho_2 A_2 C_2$

$$0.0767 = 1.1356 \times A_2 \times 80 \quad \text{(where } A_2 = \text{throat area)}$$

or
$$A_2 = 8.443 \times 10^{-4} \text{ m}^2$$

or
$$A_2 = 8.443 \times 10^{-4} = \frac{\pi}{4} d_2^2$$

$\therefore$ *Throat diameter,*
$$\mathbf{d_2} = \left(\frac{8.443 \times 10^{-4} \times 4}{\pi} \right)^{1/2} = 0.0328 \text{ m} \quad \text{or} \quad \mathbf{3.28 \text{ cm.}} \quad \textbf{(Ans.)}$$

Example 11.7. *A simple jet carburettor is required to supply 6 kg of air per minute and 0.45 kg of fuel of density 740 kg/m³. The air is initially at 1.013 bar and 27°C.*

(i) Calculate the throat diameter of the choke for a flow velocity of 92 m/s. Velocity coefficient = 0.8.

(ii) If the pressure drop across the fuel metering orifice is 0.75 of that at the choke, calculate the orifice diameter assuming $C_d = 0.60$. **(AMIE, S-2001 ; Nagpur University)**

Solution. *Given :* $\dot{m}_a = \frac{6}{60} = 0.1$ kg/s ; $\dot{m}_f = \frac{0.45}{60} = 0.0075$ kg/s ; $\rho_f = 740$ kg/m³ ;

$p_1 = 1.013$ bar ; $T_1 = 27 + 273 = 300$ K ; $C_2 = 92$ m/s ; $C_{da} = 0.8$;
$C_{df} = 0.60$.

(*i*) **Throat diameter, $\mathbf{D_2}$:**

Velocity of air at venturi throat,

$$C_2 = C_{da} \sqrt{2 c_p T_1 \left[1 - \left(\frac{p_2}{p_1} \right)^{\frac{\gamma - 1}{\gamma}} \right]}$$

$$92 = 0.8 \sqrt{2 \times 1.005 \times 1000 \times 300 \left[1 - \left(\frac{p_2}{p_1} \right)^{\frac{1.4 - 1}{1.4}} \right]}$$

or
$$1 - \left(\frac{p_2}{p_1} \right)^{\frac{0.4}{1.4}} = \left(\frac{92}{0.8} \right)^2 \times \frac{1}{2 \times 1.005 \times 1000 \times 300} = 0.021932$$

$\therefore$
$$\left(\frac{p_2}{p_1} \right)^{0.2857} = 0.97807 \; , \quad \therefore \quad \frac{p_2}{p_1} = 0.925$$

or
$$p_2 = 1.013 \times 0.925 = 0.937 \text{ bar}$$

Now,
$$p_1 v_1^{\gamma} = p_2 v_2^{\gamma}$$

$$v_2 = v_1 \left(\frac{p_1}{p_2} \right)^{1/\gamma}$$

$$= \frac{RT_1}{p_1} \left(\frac{1}{0.925} \right)^{\frac{1}{1.4}}$$

$$= \frac{287 \times 300}{1.013 \times 10^5} \left(\frac{1}{0.925} \right)^{0.7143} = 0.898 \text{ m}^3/\text{kg}$$

Now, $\dot{m}_a = \dfrac{A_2 C_2}{v_2}$

$\therefore$ Throat area, $A_2 = \dfrac{\dot{m}_a \times v_2}{C_2} = \dfrac{0.1 \times 0.898}{92} = 9.76 \times 10^{-4}\ \text{m}^2 = 9.76\ \text{cm}^2$

But $A_2 = \dfrac{\pi}{4} D_2^2 = 9.26$

$\therefore$ $\mathbf{D_2}$ (or D_t) $= \sqrt{\dfrac{9.76 \times 4}{\pi}}$ **= 3.525 cm. (Ans.)**

(*ii*) **Orifice diameter, $\mathbf{d_f}$:**

Pressure drop at venturi = 1.013 – 0.937 = 0.076 bar

Pressure drop at jet = 0.75 × 0.076 = 0.057 bar

Now, $\dot{m}_f = A_f C_{df} \sqrt{2\rho_f \Delta p}$

$0.0075 = A_f \times 0.6 \sqrt{2 \times 740 \times 0.057 \times 10^5} = 1742.68$

$\therefore$ $A_f = 4.304 \times 10^{-6}\ \text{m}^2$ or $4.304\ \text{mm}^2$

But, $A_f = \dfrac{\pi}{4} d_f^2 = 4.304$

$\therefore$ $\mathbf{d_f} = \sqrt{\dfrac{4 \times 4.304}{\pi}}$ **= 2.34 mm. (Ans.)**

Example 11.8. *The following data relate to a 4-stroke petrol engine of Hindustan Ambassador :*

Capacity of the petrol engine	*= 1489 c.c.*
Speed at which maximum power is developed	*= 4200 r.p.m.*
The volumetric efficiency (at the above speed)	*= 75 percent*
The air-fuel ratio	*= 13 : 1*
Theoretical air speed at choke (at peak power)	*= 85 m/s*
The co-efficient of discharge for venturi	*= 0.82*
The co-efficient of discharge of the main petrol jet	*= 0.65*
The specific gravity of petrol	*= 0.74*
Level of petrol surface below the choke	*= 6 mm*
Atmospheric pressure and temperature	*= 1.013 bar, 20°C respectively*

An allowance should be made for the emulsion tube, the diameter of which can be taken as 40 percent of the choke diameter.

Calculate the sizes of a suitable choke and main jet.

Solution. *Given* : V_s = 1489 c.c. = 1489×10^{-6} m³ = 0.001489 m³ ; N = 4200 r.p.m ; $\eta_{vol.}$ = 75% ; A/F ratio = 13 : 1 ; C_t (= C_2) = 85 m/s ; C_{da} = 0.82 ; C_{df} = 0.65 ; ρ_f = 0.74 × 1000 = 740 kg/m³ ; p_1 (= p_a) = 1.013 bar ; p_2 (= p_t) = ? ; T_1 (= T_a) = 20 + 273 = 293 K ; z = 6 mm = 0.006 m ; $d = 0.4\ D$.

Volume of air induced = $\eta_{vol.} \times V_s$

$$= \frac{0.75 \times 0.001489 \times 4200}{2 \times 60} = 0.03909\ \text{m}^3/\text{s}$$

∴ Mass flow of air, $\dot{m}_a = \frac{p_1 v}{RT_1} = \frac{1.013 \times 10^5 \times 0.03909}{0.287 \times 10^3 \times 293} = 0.04709$ kg/s

For compressible flow, velocity at throat,

$$C_t = \sqrt{\left[2T_a c_p \left\{1 - \left(\frac{p_t}{p_a}\right)^{\frac{\gamma - 1}{\gamma}}\right\}\right]} \quad \text{...[Eqn. (11.13)]}$$

$$85 = \sqrt{\left[2 \times 293 \times (1.005 \times 10^3) \left\{1 - \frac{p_t}{p_a}\right\}^{\frac{1.4-1}{1.4}}\right]}$$

$$85 = 767.4 \sqrt{\left\{1 - \left(\frac{p_t}{p_a}\right)^{0.2857}\right\}}$$

∴ $$\frac{p_t}{p_a} = \left[1 - \left(\frac{85}{767.4}\right)^2\right]^{\frac{1}{0.2857}} = 0.9576$$

or $$p_t = 1.013 \times 0.9576 = 0.97 \text{ bar}$$

Volume flow of air at choke,

$$v_t = 0.03909 \times \left(\frac{p_a}{p_t}\right)^{\frac{1}{\gamma}}$$

$$= 0.03909 \times \left(\frac{1.013}{0.97}\right)^{\frac{1}{1.4}} = 0.04032 \text{ m}^3\text{/s}$$

∴ $$A_t = \frac{v_t}{C_t \times C_{da}} = \frac{0.04032}{85 \times 0.82} = 0.0005785 \text{ m}^2 = 578.5 \text{ mm}^2$$

Now, $\frac{\pi}{4}(D^2 - d^2) = 578.5$

or $\frac{\pi}{4}[D^2 - (0.4\,D)^2] = 578.5$

or $\frac{\pi}{4} \times 0.84\,D^2 = 578.5$

or Choke dia., **D = 29.61 mm. (Ans.)**

Mass flow of fuel, $\dot{m}_f = \frac{\dot{m}_a}{13} = \frac{0.04709}{13} = 0.003622$ kg/s

$$\dot{m}_f = C_{df} \cdot A_j \sqrt{2\rho_f (\Delta p_a - gz\rho_f)} \quad \text{...[Eqn. (11.7)]}$$

$$0.003622 = 0.65 \times A_j \sqrt{2 \times 740 \{(1.013 - 0.97) \times 10^5 - 9.81 \times 0.006 \times 740\}}$$

$$= 1639.75\, A_j$$

$\therefore \quad A_j = 2.209 \times 10^{-6}\ m^2$ or $2.209\ mm^2$

$$A_j = \frac{\pi}{4} D_{jet}^2 = 2.209$$

$\therefore \quad$ **D_{jet} = 1.68 mm. (Ans.)**

Example 11.9. *The following data refer to a simple carburettor :*

Throat diameter	*= 18 mm*
Diameter of fuel orifice	*= 1.2 mm*
Co-efficient of air flow	*= 0.82*
Co-efficient of fuel flow	*= 0.65*
Level of petrol surface below the throat	*= 6 mm*
Density of air	*= 1.2 kg/m³*
Density of fuel	*= 750 kg/m³.*

Calculate :

(*i*) *The A/F ratio for a pressure drop of 0.065 bar when the nozzle lip is neglected ;*

(*ii*) *The A/F ratio when the nozzle lip is taken into account ;*

(*iii*) *The minimum velocity of air or critical air velocity required to start the fuel flow when nozzle lip is provided.*

Solution. *Given* : d_a = 18 mm = 0.018 m ;
d_f = 1.2 mm = 0.0012 m ; C_{da} = 0.82 ; C_{df} = 0.65 ; z = 6 mm = 0.006 m ;
ρ_a = 1.2 kg/m^3 ; ρ_f = 750 kg/m^3

(*i*) **A / F ratio when the nozzle lip is neglected :**

Air flow, $\quad \dot{m}_a = C_{da} A_a \sqrt{2\rho_a \Delta p_a}$ ...[Eqn. (11.4)]

Fuel flow, $\quad \dot{m}_f = C_{df} A_f \sqrt{2\rho_f \Delta p_a}$

$$\therefore \quad \text{A / F ratio} = \frac{C_{da}}{C_{df}} \cdot \frac{A_a}{A_f} \sqrt{\frac{\rho_a}{\rho_f}}$$

$$= \frac{0.82}{0.65} \times \left(\frac{0.018}{0.0012}\right)^2 \sqrt{\frac{1.2}{750}} = \textbf{11.35. (Ans.)}$$

$$\left[\because \frac{A_a}{A_f} = \frac{\frac{\pi}{4} d_a^2}{\frac{\pi}{4} d_f^2} = \left(\frac{d_a}{d_f}\right)^2\right]$$

(*ii*) **A / F ratio when the nozzle lip is taken into account :**

The *air flow will remain same*. The *fuel flow* will become,

$$\dot{m}_f = C_{df} \times A_f \sqrt{2\rho_f(\Delta p_a - gz\rho_f)} \quad \text{...[Eqn. (11.7)]}$$

$$\therefore \quad \text{A / F ratio} = \frac{C_{da}}{C_{df}} \cdot \frac{A_a}{A_f} \sqrt{\frac{\rho_a}{\rho_f}} \sqrt{\frac{\Delta p_a}{\Delta p_a - gz\rho_f}}$$

$$= \frac{0.82}{0.65} \times \left(\frac{0.018}{0.0012}\right)^2 \sqrt{\frac{1.2}{750}} \sqrt{\frac{0.065}{0.065 - (9.81 \times 0.006 \times 750/10^5)}}$$

$$= 11.35 \times \sqrt{\frac{0.065}{0.065 - 0.00044145}} = \textbf{11.39. (Ans.)}$$

(*iii*) **Minimum velocity of air, C_2 :**

The *flow of fuel when lip is provided will start only when the minimum velocity of air required to create requisite pressure difference for flow of fuel to overcome nozzle lip exists.*

∴ Pressure difference Δp_a must be *equal to* $gz\rho_f$. Assuming velocity at entrance of venturi, $C_1 = 0$,

$$\frac{p_1}{\rho_a} = \frac{p_2}{\rho_a} + \frac{C_2^2}{2}$$

or $$\frac{\Delta p_a}{\rho_a} = \frac{C_2^2}{2}$$

or $$\frac{C_2^2}{2} = \frac{gz\rho_f}{\rho_a}$$

or $$C_2 = \sqrt{\frac{2gz\rho_f}{\rho_a}} = \sqrt{\frac{2 \times 9.81 \times 0.006 \times 750}{1.2}} = \mathbf{8.58\ m/s.} \quad \textbf{(Ans.)}$$

Note. The **nozzle lip** is *the height of fuel nozzle in the throat above petrol surface in the carburettor. It is provided so that there is no spilling of fuel due to vibration or slight non-horizontal position of carburettor. This would avoid wastage of fuel and fire hazard.*

Example 11.10. *The following data refer to an eight-cylinder four-stroke petrol engine :*

Bore	= *110 mm*
Stroke	= *110 mm*
Composition of the fuel used	= *C = 84% ; H_2 = 16%*
Throat diameter of the choke tube	= *42 mm*
Volumetric efficiency at 300 r.p.m.	= *75% (referred to 0°C and 1.013 bar)*
The pressure depression	= *0.12 bar*
The temperature at the throat	= *15°C*
Characteristic gas constant : For air	= *287 J/kg K*
For fuel vapour	= *97 J/kg K*

If chemically correct air-fuel ratio is supplied for combustion, determine :

(*i*) *Fuel consumption in kg/h ;*

(*ii*) *The air velocity through the tube.*

Solution. *Given* : D = 110 mm = 0.11 m ; L = 110 mm = 0.11 m ; d_a = 42 mm = 0.042 m ; $\eta_{vol.}$ = 75%, N = 3000 r.p.m. ; R_a = 287 J / kg K ; R_v = 97 J/kg K ; Δp_a = 0.12 bar.

(*i*) **Fuel consumption in kg/h :**

The volume of mixture supplied at 0°C and 1.013 bar per minute

$$= \frac{\pi}{4} D^2 L \times 8 \times \frac{N}{2} \times \eta_{vol.}$$

$$= \frac{\pi}{4} \times (0.11)^2 \times 0.11 \times 8 \times \frac{3000}{2} \times 0.75 = 9.408 \text{ m}^3/\text{ min.}$$

Also
$$\begin{array}{ccccc} C & + & O_2 & = & CO_2 \\ 12 & & 32 & & 44 \end{array}$$

and
$$\begin{array}{ccccc} 2H_2 & + & O_2 & = & 2H_2O \\ 4 & & 32 & & 36 \\ 1 & & 8 & & 9 \end{array}$$

Thus, air required for combustion of 1 kg of fuel

$$= \left(0.84 \times \frac{32}{12} + 0.16 \times 8\right) \times \frac{100}{23} = 15.3 \text{ kg}$$

Thus A / F ratio $= \dfrac{m_a}{m_f} = 15.3$

The volume of one kg of air at 0°C and 1.013 bar,

$$v_a = \frac{R_a T}{p} = \frac{287 \times 273}{1.013 \times 10^5} = 0.773 \text{ m}^3/\text{kg}$$

Similarly volume of 1 kg of fuel vapour at 0°C and 1.013 bar,

$$v_f = \frac{R_f T}{p} = \frac{97 \times 273}{1.013 \times 10^5} = 0.2614 \text{ m}^3/\text{kg}$$

Thus $m_a / \text{min} \times v_a + m_f / \text{min} \times v_f = 9.408$

$15.3\, m_f / \text{min} \times v_a + m_f / \text{min} \times v_f = 9.408$

Thus, $$m_f/\text{min} = \frac{9.408}{15.3\, v_a + v_f} = \frac{9.408}{15.3 \times 0.773 + 0.2614} = 0.778 \text{ kg/min}$$

∴ Fuel consumption = 0.778 × 60 = **46.68 kg/h. (Ans.)**

(*ii*) **The air velocity through the tube,** C_2 $(= C_a)$

Density of air at the throat,

$$\rho_a = \frac{p_2}{R_a T_2}$$

$$= \frac{p_1 - \Delta p_a}{R_a T_2} \qquad (\because \; p_1 - p_2 = \Delta p_a)$$

$$= \frac{(1.013 - 0.12) \times 10^5}{287 \times (15 + 273)} = 1.08 \text{ kg/m}^3.$$

∴ Velocity at the throat in m/s,

$$\mathbf{C_a} = \frac{m_a}{A_a \rho_a} = \frac{15.3\, m_f}{\frac{\pi}{4} \times (0.42)^2 \times 1.08}$$

$$= \frac{15.3 \times (0.778 / 60)}{\frac{\pi}{4} \times (0.042)^2 \times 1.08} = \mathbf{132.59 \text{ m/s.}} \quad \textbf{(Ans.)}$$

Example 11.11. *Determine the air-fuel ratio supplied at 4500 m altitude by a carburettor which is adjusted to give an air-fuel ratio of 14 : 1 at sea level where air temperature is 25°C and pressure 1.013 bar.*

The temperature of air decreases with altitude as given by the expression,

$$t = t_s - 0.0064\, h$$

where h is the height in metres and t_s is sea level temperature in °C.

The pressure of air decreases with altitude as per relation :

$$h = 19300 \log_{10} \left(\frac{1.013}{p}\right)$$

where p is expressed in bar at altitude.

Solution.
$$t = t_s - 0.0064\,h$$
$$= 25 - 0.0064 \times 4500 = -3.8°\text{C}$$

Now
$$h = 19300 \log_{10}\left(\frac{1.013}{p}\right)$$
$$4500 = 19300 \log_{10}\left(\frac{1.013}{p}\right)$$

∴
$$\log_{10}\left(\frac{1.013}{p}\right) = \frac{4500}{19300} = 0.2332$$

or
$$\frac{1.013}{p} = 1.711$$

or
$$p = \frac{1.013}{1.711} = 0.592 \text{ bar}$$

Now,
$$\frac{\text{A / F ratio at altitude}}{\text{A / F ratio at sea level}} = \sqrt{\frac{\rho_{\text{altitude}}}{\rho_{\text{sea level}}}}$$

∴
$$\textbf{A/F ratio at altitude} = 14\sqrt{\frac{p_{\text{alt.}}/RT_{\text{alt.}}}{p_{\text{sea}}/RT_{\text{sea}}}}$$
$$= 14\sqrt{\frac{p_{\text{alt.}} \times T_{\text{sea}}}{p_{\text{sea}} \times T_{\text{alt.}}}}$$
$$= 14\sqrt{\frac{0.592 \times (25 + 273)}{1.013 \times (-3.8 + 273)}}$$
$$= \mathbf{11.26.} \quad \textbf{(Ans.)}$$

HIGHLIGHTS

1. The process of preparing, in the S.I. engine, a combustible fuel-air mixture outside the engine cylinder is called *carburetion*.
2. A *carburettor* is a device which atomises the fuel and mixes it with air.
3. The air-fuel ratio for maximum power is 13 : 1.
4. The following addition devices / systems are added to the simple carburettor :
   (*i*) Main metering system
   (*ii*) Idling system
   (*iii*) Power enrichment or economiser system
   (*iv*) Acceleration pump system
   (*v*) Choke.
5. Carburettors, basically, are of the following types :
   (*i*) Open choke type
   Examples : Zenith, Solex, Carter and Stromberg carburettors
   (*ii*) Constant vacuum type
   Example : S.U. carburettor.

OBJECTIVE TYPE QUESTIONS

Fill in the Blanks or Say "Yes" or "No" :

1. Carburettor is used for ....
2. For maximum power of S.I. engines the fuel air mixture ratio should be ....
3. Typical pressure in the induction manifold at the start of induction stroke of S.I. engine, under idling conditions is ....
4. Relative fuel-air ratio (F_R) for maximum power in S.I. engine may be ....
5. Relative fuel-air ratio (F_R) for maximum thermal efficiency of S.I. engine may be ....
6. For best thermal efficiency of S.I. engine, the fuel-air mixture ratio should be ....
7. In S.I. engine the process of preparing a combustible fuel-air mixture outside the engine cylinder is called ....
8. The pipe that carries the prepared mixture to the engine cylinders is called the intake manifold.
9. The A / F ratio for maximum power is not the same as the A / F ratio for maximum economy.
10. The maximum power is obtained at about .... A / F ratio.
11. At full throttle, maximum efficiency occurs at an A / F ratio of about ....
12. In stationary engines the designed air-fuel ratio is that which gives the maximum economy.
13. The richening of mixture increases the probability of contact between fuel and air particles and thus improves ....
14. In the carburettor, complete automization and vaporization of the fuel is achieved.
15. The volatility of fuel significantly affects the starting and warm up characteristics of the engine.
16. The term 'acceleration' with regard to engines, is generally used to refer to an increase in engine speed resulting from opening the throttle.
17. The petrol engine is .... governed.
18. In the modern carburettors the mixture correction is done by air bleeding alone.
19. A common method of changing the air-fuel ratio in large carburettors is the back suction control.
20. Almost all carburettors, except S.U. carburettor, are of open choke type.
21. The important difference between an aircraft and automobile carburettor is that the former operates at varying altitudes whereas the latter operates mostly at ground level conditions.
22. Engines fitted with petrol injection system can be used in tilt position which will cause surge trouble in carburettors.
23. Injection systems generate less noise.
24. Weight and bulk of petrol injection system is more than that of a carburettor.
25. A petrol injection system has increased volumetric efficiency.

ANSWERS

1. S.I. engine	**2.** rich	**3.** 0.3 bar	**4.** 1.2	**5.** 0.8
6. lean	**7.** carburetion	**8.** Yes	**9.** Yes	**10.** 12.5 : 1
11. 17 : 1	**12.** Yes	**13.** combustion	**14.** No	**15.** Yes
16. Yes	**17.** quantity	**18.** Yes	**19.** Yes	**20.** Yes
21. Yes	**22.** Yes	**23.** No	**24.** Yes	**25.** Yes.

THEORETICAL QUESTIONS

1. What do you mean by the term 'carburetion' ?
2. What is a carburettor ?
3. Draw a typical induction system of a petrol engine.
4. Enlist the factors which affect the process of carburetion.
5. Why is a choke used in a carburettor ?
6. Explain briefly the essential features of good commercial carburettor for automotive engines.
7. (*a*) Draw a curve representing variation of mixture requirements (fuel-air ratios) from no-load to full-load in a S.I. engine, mark the relative position of stoichiometric fuel-air ratio line and then explain why :
   (*i*) an idling engine requires a rich mixture ;
   (*ii*) a cruising engine requires an economy mixture ;
   (*iii*) maximum power demands a rich mixture ;
   (*b*) What will be the effects of prolonged running of such an engine on (*i*) the weakest mixture ; (*ii*) the richest mixture ?
8. With the help of a neat sketch explain the working principle of a simple carburettor.
9. A simple carburettor is inherently unsuitable to meet the varying mixture requirements of S.I. engine. What are the drawbacks of a simple carburettor ? How are they overcome by incorporating compensating devices ? Explain with the aid of suitable sketches wherever necessary.
10. Describe with suitable sketches the following systems of a modern carburettor :
    (*i*) Main metering system ;
    (*ii*) Idling system ;
    (*iii*) Economiser system ;
    (*iv*) Acceleration pump system.
11. With the help of a neat sketch describe the construction and working of a solex carburettor.
12. How the power and efficiency of the S.I. engine vary with (*a*) air-fuel ratio at full load ; (*b*) part load ?
13. Sketch and explain the fuel consumption loop in the S.I. engine.
14. Why a rich mixture is required for idling ?
15. Why a rich mixture is required for maximum power ?
16. What do you understand by transient mixture requirements ?
17. Why multi-cylinder engines require richer mixture than single cylinder engines ?
18. What are the basic types of carburettors ?
19. Explain with neat sketches the following types of carburettors :
    (*i*) Carter carburettor
    (*ii*) Solex carburettor
    (*iii*) S.U. carburettor.
20. Device an expression for A / F ratio
    (*a*) neglecting compressibility ;
    (*b*) taking compressibility into account.
21. State the special requirements of an aircraft carburettor ?
22. What is petrol injection ?
23. State the advantages and limitations of petrol injection.
24. Explain briefly 'continued' and 'timed' injection systems.
25. With the help of a neat sketch, explain briefly any petrol injection system.
26. What is the difference between 'Direct injection' and 'Indirect injection' ?
27. Explain the difference between 'continuous injection' and ' Intermittent or pulsed injection'.
28. Give the comparison between petrol injection and carburetted fuel supply systems.

UNSOLVED EXAMPLES

1. A four-cylinder four-stroke engine having diameter and length of stroke as 100 mm and 120 mm respectively is running at 2000 r.p.m. Its carburettor venturi has a 30 mm throat. Assuming coefficient of air flow 0.8, density of air 1.2 kg/m^3 and volumetric efficiency of the engine as 70 percent, determine the suction at the throat. **[Ans.** 0.0339 bar]

2. A simple jet carburettor is required to supply 6 kg of air per minute and 0.45 kg of fuel of density 740 kg/m^3. The air is initially at 1.013 bar and 27°C. Calculate the throat diameter of the choke for a flow velocity of 91 m/s. Velocity coefficient = 0.8.

   If the pressure drop across the fuel metering orifice is 0.75 of that at the choke, calculate orifice diameter assuming $C_d = 0.6$. **[Ans.** 35.25 mm ; 2.34 mm]

3. A 4-stroke petrol engine of Hindustan Ambassador has a capacity of 1489 c.c. It develops maximum power at 4200 r.p.m. The volumetric efficiency at this speed is 70 percent and the air/fuel ratio is 13 : 1. At peak power the theoretical air speed at choke is 90 m/s. The coefficient of discharge for venturi is 0.85 and that of the main petrol jet is 0.66. An allowance should be made for the emulsion tube, the diameter of which can be taken as 1/2.5 of the choke diameter. The petrol surface is 6 mm below the choke at this engine condition. The specific gravity of petrol is 0.74. Atmospheric pressure and temperature are 1.013 bar and 20°C respectively.

   Calculate the sizes of a suitable choke and main jet. **[Ans.** 27.35 mm ; 1.58 mm]

4. A single jet simple carburettor is to supply 6.11 kg/min. of air and 0.408 kg/min of petrol, density 768 kg/m^3. The air is initially at 1.027 bar and 15.5°C. Calculate the throat diameter of the venturi throat if the speed of air is 97.5 m/s, assuming a velocity coefficient of 0.84. Assume adiabatic expansion and γ for air as 1.4. If the drop across fuel metering orifice be 0.8 of the pressure at the throat ; calculate the orifice diameter assuming a coefficient as 0.66. **[Ans.** 2.05 mm]

5. An engine having a simple single jet carburettor consumes 6.5 kg of fuel/hour. The fuel density is 700 kg/m^3. The level of fuel in the float chamber is 3 mm below the top of the jet when the engine is not running. Ambient conditions are 1.01325 bar and 17°C. The jet diameter is 1.25 mm and its discharge coefficient is 0.6. The discharge coefficient of air is 0.85. Air-fuel ratio is 15. Determine the critical air velocity and the throat diameter (effective). Express the pressure depression in cm of water. Neglect compressibility of air. **[Ans.** 4.945 m/s ; 19.9 mm ; 43.99 cm]

6. An eight-cylinder 4-stroke petrol engine with bore and stroke of 100 mm each uses volatile fuel of composition C = 84%, H_2 = 16%. The throat diameter of choke tube is 40 mm. The volumetric efficiency at 3000 r.p.m. is 75 percent referred to 0°C and 1.01325 bar. The pressure depression is 0.116 bar and the temperature at throat is 16°C. If chemically correct A / F ratio is supplied for consumption, determine :

   (*i*) Fuel consumption in kg / h ;

   (*ii*) The air velocity through the tube.

   Take characteristic gas-constant R for air and fuel as 287 J/kg K and 97 J/kg K respectively.

   [Ans. 35.1 kg/h ; 116 m/s]

7. The venturi of a simple carburettor has a throat diameter of 20 mm and the coefficient of air flow is 0.85. The fuel orifice has a diameter of 1.25 mm and the coefficient of fuel flow is 0.66. The petrol surface is 5 mm below the throat. Assuming density of and fuel as 1.2 kg/m^3 and 750 kg/m^3 respectively, calculate :

   (*i*) The A/F ratio for a pressure drop of 0.07 bar when the nozzle lip is neglected ;

   (*ii*) The A/F ratio when the nozzle lip is taken into account ;

   (*iii*) The minimum velocity of air or critical air velocity required to start the fuel flow when nozzle lip is provided. **[Ans.** (*i*) 13.2 ; (*ii*) 13.235 ; (*iii*) 7.83 m/s]

8. A carburettor with float chamber vented to atmosphere is tested in a laboratory without the air cleaner. The A / F ratio as calculated is 15 at the atmospheric conditions of 1.008 bar. The pressure recorded at the throat is 0.812 bar.

   This carburettor is fitted with air cleaner and once again tested. The additional pressure drop due to air cleaner is 0.04 bar with the air flow at the atmospheric conditions to remain unchanged at 260 kg/h. Assuming negligible nozzle lip, same air flow in both cases and constant coefficient of flow determine :

   (*i*) The throat pressure with cleaner fitted ;

   (*ii*) The A / F ratio with cleaner fitted. **[Ans.** (*i*) 0.772 bar ; (*ii*) 13.67]

9. A 4-stroke petrol engine which when tested at sea level conditions of 30°C and 1.01325 bar gave the A / F ratio of 14. The same engine was once again tested at an altitude of 4000 m. Determine the A/F ratio at high altitude if the temperature varies with altitude as :

$t = t_s - 0.007\, h$;

The pressure varies as :

$$h = 8350 \ln (1.01325/p)$$

where, t_s = Temperature at sea level, °C

h = Height, m, and

p = Pressure, bar. [**Ans.** 11.05]

12

Fuel Injection Systems for C.I. Engines

12.1. INTRODUCTION

— In C.I. engines, the air is taken in during the suction stroke and compressed to a high pressure (28 to 70 bar) and high temperature (520° to 720°C) according to the compression ratio used (12 : 1 to 20 : 1). The high temperature of air at the end of stroke is sufficient to ignite the fuel.

— Fuel is injected into the cylinder at the end of the compression stroke ; the pressure of fuel injected lies between 100 to 200 bar. During the process of injection the fuel is broken into *very fine droplets*. The droplets vaporise taking the heat from the hot air and form a combustible mixture and start burning. As the burning starts, the vaporisation of fuel is accelerated as more heat is available. As the combustion progresses, the amount of oxygen available for burning reduces and therefore heat release is reduced.

— The period between the start of injection and start of ignition, called the *ignition delay*, is about **0.001** second for high speed engines and **0.002** second for low speed engines. The injection period covers about 25° of crank rotation. After the ignition the temperature and pressure rise rapidly. The *whole performance of engine is totally dependent on the delay period ; the lesser the delay period better is the engine performance.*

12.2. FUNCTIONAL REQUIREMENTS OF AN INJECTION SYSTEM

The *functional requirements of an injection system* are listed below :

1. Introduction of the fuel into the combustion chamber should take place within a precisely defined period of the cycle.
2. The metering of the amount of fuel injected per cycle should done very accurately.
3. The quantities of fuel metered should vary to meet the changing load and speed requirements.
4. The injection rate should be such that it results in the desired heat release pattern.
5. The injected fuel must be broken into very fine droplets.
6. The pattern of spray should be such as to ensure rapid mixing of fuel and air.
7. The beginning and end of the injection should be sharp.
8. The timing of injection, if desired, should change as per the requirements of load and speed.

9. The distribution of the metered fuel, in the case of multi-cylinder engines, should be uniform among various cylinders.
10. Besides above requirements, the weight and the size of the fuel injection system must be minimum. It should be cheaper to manufacture and least expensive to attend to ; adjust or repair.

For accomplishing these requirements the following **functional elements** are required in a fuel injection system :

1. *Pumping elements.* To move the fuel from the fuel tank to cylinder and piping etc.
2. *Metering elements.* To measure and supply the fuel at the rate demanded by the load and speed.
3. *Metering controls.* To adjust the rate of metering elements for changes in load and speed of the engine.
4. *Distributing elements.* To divide the metered fuel equally among the cylinders.
5. *Timing controls.* To adjust the start and the stop of injection.
6. *Mixing elements.* To atomise and distribute the fuel within the combustion chamber.

12.3. FUNCTIONS OF A FUEL INJECTION SYSTEM

The *main functions of a fuel injection system are :*

1. Filter the fuel.
2. Metre or measure the correct quantity of fuel to be injected.
3. Time the fuel injection.
4. Control the rate of fuel injection.
5. Automise or break up the fuel to fine particles.
6. Properly distribute the fuel in the combustion chamber.

- The injection systems are manufactured with *great accuracy,* especially the parts that actually meter and inject the fuel. Some of the tolerances between the moving parts are very small of the order of 1 micron. Such closely fitting parts require special attention during manufacture and hence the injection systems are costly.

12.4. FUEL INJECTION SYSTEMS

In compression ignition engines (diesel and semi-diesel) two methods of fuel injection are used. These are :

1. Air injection
2. Solid or airless injection.

12.4.1. Air Injection

In this method of fuel injection air is compressed in the compressor to a very high pressure (much higher than developed in the engine cylinder at the end of the compression stroke) and then injected through the fuel nozzle into the engine cylinder. *The rate of fuel admission can be controlled by varying the pressure of injection air.* Storage air bottles which are kept charged by an air compressor (driven by the engine) supply the high pressure air.

Advantages :

(*i*) It provides better automisation and distribution of fuel.

(*ii*) As the combustion is more complete, the b.m.e.p. is higher than with other types of injection systems.

(*iii*) Inferior fuels can be used.

Disadvantages :

This method is *not-used* now-a-days due to the following *reasons/disadvantages* :

(*i*) It requires a high pressure multi-stage compression. The large number of parts, the intercooler etc. make the system complicated and expensive.

(*ii*) A separate mechanical linkage is required to time the operation of fuel valve.

(*iii*) Due to the compression and the linkage the bulk of the engine increases. This also results in reduced B.P. due to power loss in operating the compression and linkage.

(*iv*) The fuel in the combustion chamber burns very near to injection nozzle which many times leads to overheating and burning of valve and its seat.

(*v*) The fuel valve sealing requires considerable skill.

(*vi*) In case of sticking of fuel valve the system becomes quite dangerous due to the presence of high pressure air.

12.4.2. Solid or Airless Injection

Injection of fuel directly into the combustion chamber without primary automisation is termed as **solid injection.** It is also termed as *mechanical injection.*

Main Components :

The *main components* of a fuel injection system are :

(*i*) *Fuel tank* ;

(*ii*) *Fuel feed pump* to supply the fuel from the main fuel tank to the injection pump ;

(*iii*) *Fuel filters* to prevent dust and abrasive particles from entering the pump and injectors ;

(*iv*) *Injection pump* to meter and pressurise the fuel for injection ;

(*v*) *Governor* to ensure that the amount of fuel is in accordance with variation in load ; and

(*vi*) *Fuel pipings and injectors* to take the fuel from the pump and distribute it in the combustion chamber by atomising it in fine droplets.

Main types of modern fuel injection systems :

1. Common-rail injection system.
2. Individual pump injection system.
3. Distributor system.

Atomisation of fuel oil has been secured by (*i*) *air blast* and (*ii*) *pressure spray.* Early diesel engines used air fuel injection at about 70 bar. This is sufficient not only to inject the oil, but also to atomise it for a rapid and thorough combustion. The expense of providing an air compressor and tank lead to the development of *"solid"* injection, using a liquid pressure of between 100 and 200 bar which is sufficiently high to atomise the oil it forces through spray nozzles. Great advances have been made in the field of solid injection of the fuel through research and progress in fuel pump, spray nozzles, and combustion chamber design.

1. **Common-rail injection system :**

Two types of common-rail injection systems are shown in Fig. 12.1 and 12.2 respectively.

- Refer Fig. 12.1. A single pump supplies high-pressure fuel to header, a relief valve holds pressure constant. The control wedge adjusts the lift of mechanical operated valve to set amount and time of injection.
- Refer Fig. 12.2. Controlled-pressure system has pump which maintains set head pressure. Pressure relief and timing valves regulate injection time and amount. Spring loaded spray valve acts merely as a check.

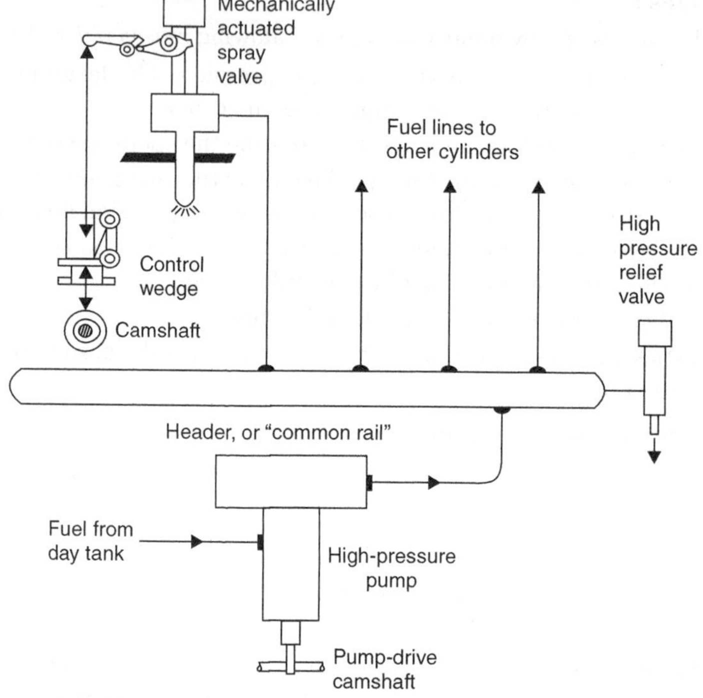

Fig. 12.1

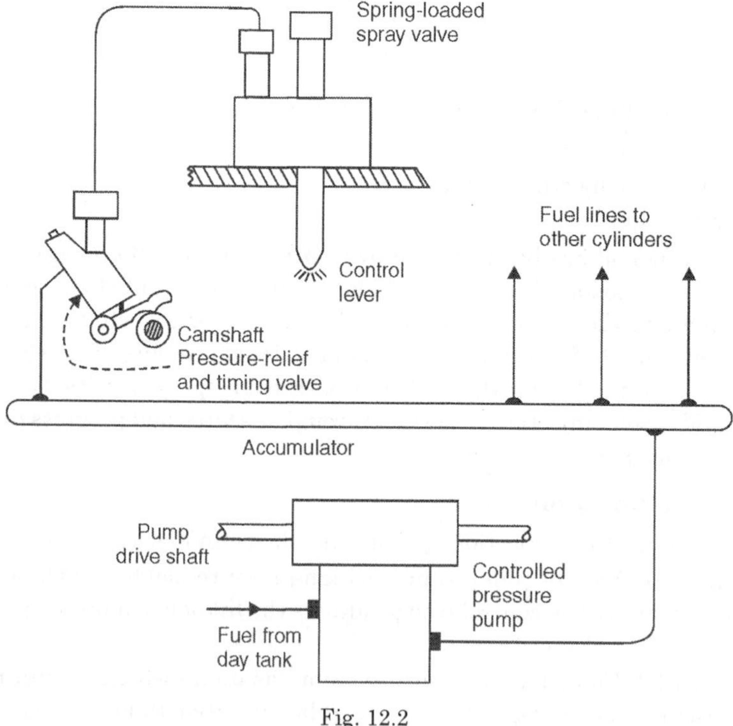

Fig. 12.2

Advantages :

(*i*) The system arrangement is simple and less maintenance cost.

(*ii*) Only one pump is sufficient for multi-cylinder engine.

(*iii*) It fulfills the requirements of either the constant load with variable speed or constant speed with variable load.

(*iv*) Variation in pump supply pressure will affect the cylinders uniformly.

Disadvantages :

(*i*) There is a tendency to develop leaks in the injection valve.

(*ii*) Very accurate design and workmanship are required.

2. Individual pump injection system :

- Refer Fig. 12.3. In this system an individual pump or pump cylinder connects directly to each fuel nozzle. Pump meters charge and control injection timing. Nozzles contain a delivery valve actuated by the fuel-oil pressure.

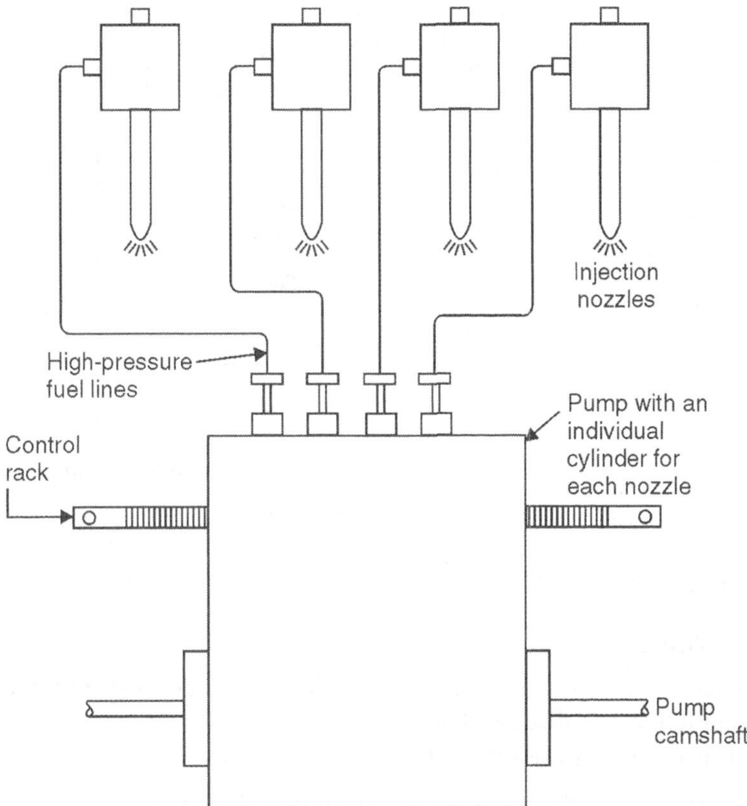

Fig. 12.3. Individual pump injection system.

- The design of this type of pump must be very accurate and precise as the volume of fuel injected per cycle is 1/20,000 of the engine displacement at full load and 1/100,000 of the engine displacement during idling. The time allowed for injecting such a small quantity of fuel is very limited (about 1/450 second at 1500 r.p.m. of the engine providing injection through 20° crank angle). The pressure requirements vary from 100 to 300 bar.

3. Distributor system :

Refer Fig. 12.4. In this system, the fuel is metered at a central point ; a pump pressurises, meters the fuel and times the injection. From here, the fuel is distributed to cylinders in correct firing order by cam operated poppet valves which open to admit fuel to the nozzles.

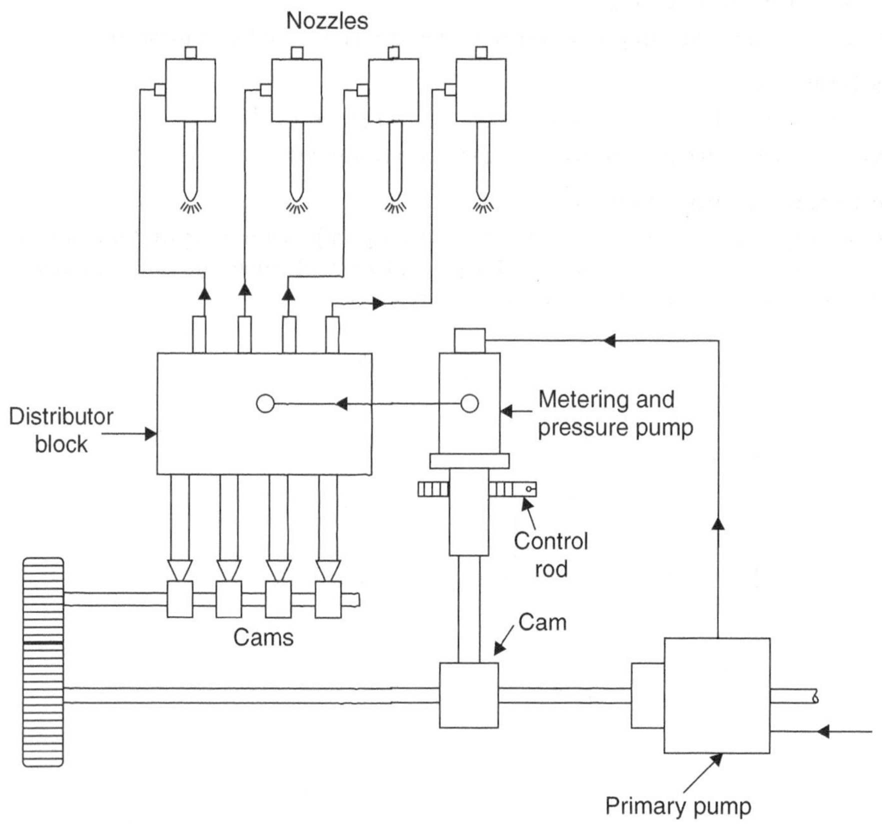

Fig. 12.4. Distributor system.

12.5. FUEL PUMP AND FUEL INJECTOR (ATOMISER)

12.5.1. Fuel Pump

A large number of igneous fuel pump designs have been developed by the manufacturers. Only one type fuel pump will be discussed here.

Bosch fuel injection pump : Refer Fig. 12.5.

- L is the **plunger** which is driven by a cam and tappet mechanism at the bottom (not shown), *B* is the **barrel** in which the plunger reciprocates. There is a rectangular vertical groove in the plunger which extends from top to another helical groove. *V* is the **delivery valve** which lifts off its seat under the liquid fuel pressure and the spring force(s). *The fuel pump is connected to fuel atomiser through the passage P. SP* and *Y* are the **spill** and **supply ports** respectively.

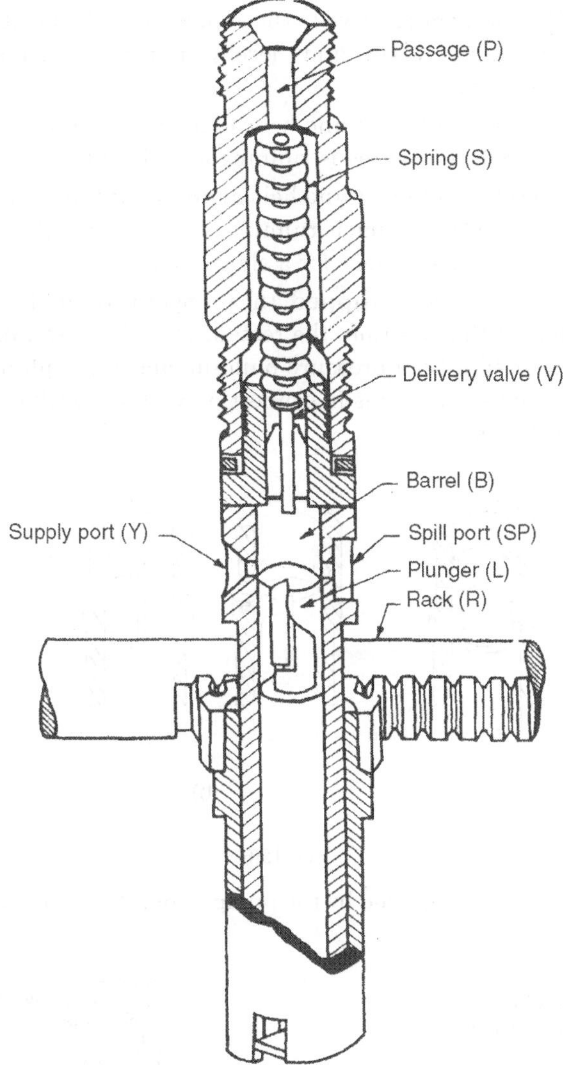

Fig. 12.5. Fuel pump.

- When the plunger is at its bottom stroke the ports *SP* and *Y* are uncovered (as shown in Fig. 12.5) oil from low pressure pump (not shown) after being filtered is forced into the barrel. When the plunger moves up due to cam and tappet mechanism, a stage reaches when both the ports *SP* and *Y* are closed and with the further upward movement of the plunger the fuel gets compressed. The high pressure thus developed lifts the delivery valve off its seats and fuel flows to atomiser through the passage *P*. With further rise of the plunger, at a certain moment, the port *SP* is connected to the fuel in the upper part of the plunger through the rectangular vertical groove by the helical groove, as a result of which a sudden drop in pressure occurs and the delivery valve falls back and occupies its seat against the spring force. The plunger is rotated by the rack *R* which is moved in or out by the governor. By changing the angular position of the helical groove (by rotating the plunger) of the plunger relative to the supply port, the length of stroke during which the oil is delivered can be varied and thereby quantity of fuel delivered to the engine is also varied accordingly.

— The positions of the plunger and helical groove at *the starting and end of the* **delivery stroke** when the engine is running at full load is shown in Fig. 12.6 (*a*).

— The positions of the plunger and helical groove at the *starting and end of the* **delivery stroke** when the engine is running at part load is shown in Fig. 12.6 (b). In this case, the delivery takes place for a shorter period.

— When the engine is to be **stopped,** the plunger is rotated to the position as shown in Fig. 12.6 (*c*). At this position, the rectangular slot is in line with the spill port and there is no possibility of pressure build-up above the plunger as the upper part of the plunger always remains in connection with the spill port. Therefore there is no delivery of the fuel.

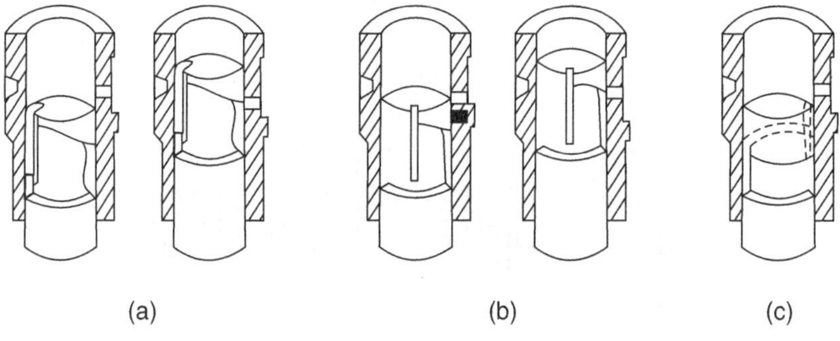

Fig. 12.6

— The amount of fuel supplied by the pump under different loads is shown in Fig. 12.7.

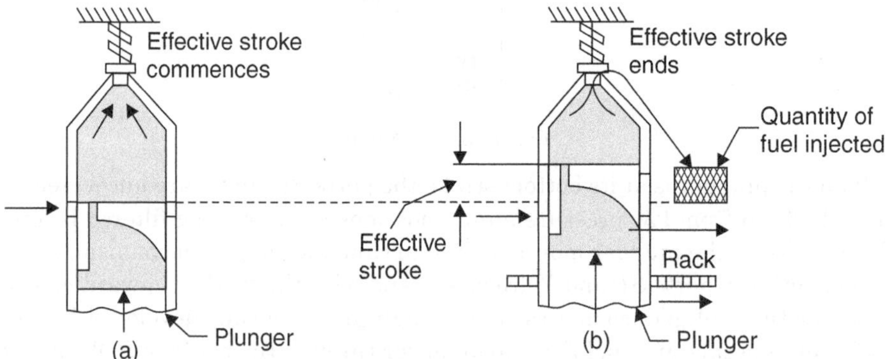

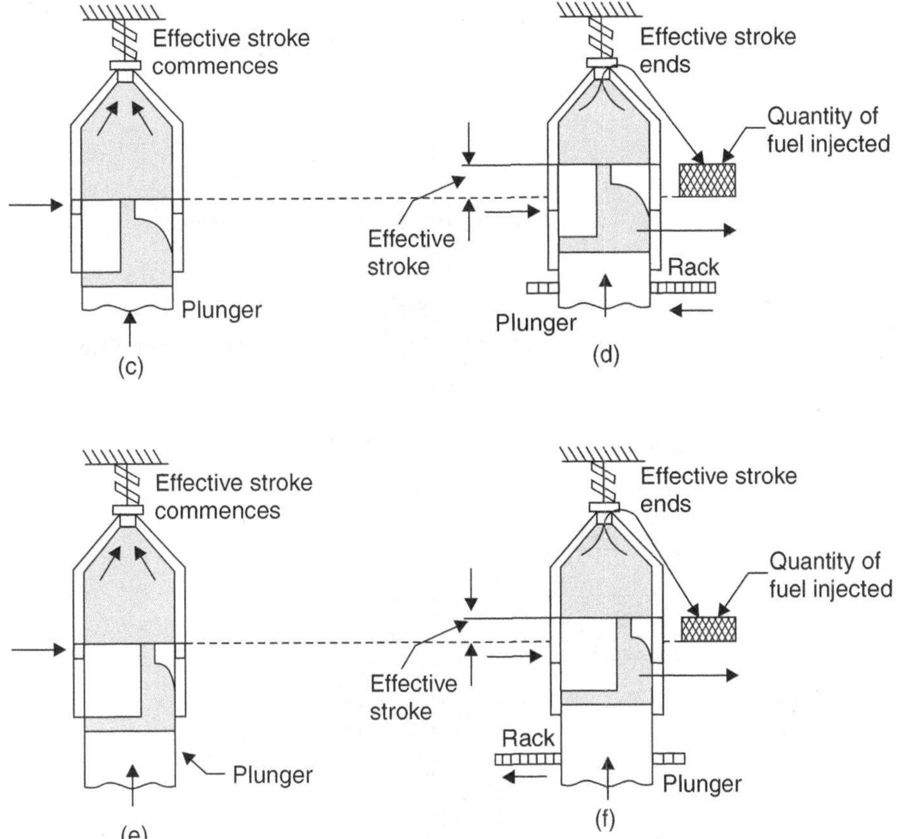

Fig. 12.7. Principle of helix bypass pump.

12.5.2. Fuel Atomiser or Injector : Refer Fig. 12.8.

- It consists of a nozzle valve (*NV*) fitted in the nozzle body (*NB*). The nozzle valve is held on its seat by a spring '*S*' which exerts pressure through the spindle *E*. '*AS*' is the adjusting screw by which the nozzle valve lift can be adjusted. Usually the nozzle valve is set to lift at 135 to 170 bar pressure. *FP* is the feeling pin which indicates whether valve is working properly or not.
- The fuel under pressure from the fuel pump enters the injector through the passages *B* and *C* and lifts the nozzle valve. The fuel travels down nozzle *N* and injected into the engine cylinder in the form of fine spray. Then the pressure of the oil falls, the nozzle valve occupies its seat under the spring force and fuel supply is cut off. Any leakage of fuel accumulated above the valve is led to the fuel tank through the passage *A*. The leakage occurs when the nozzle valve is *worn out*.

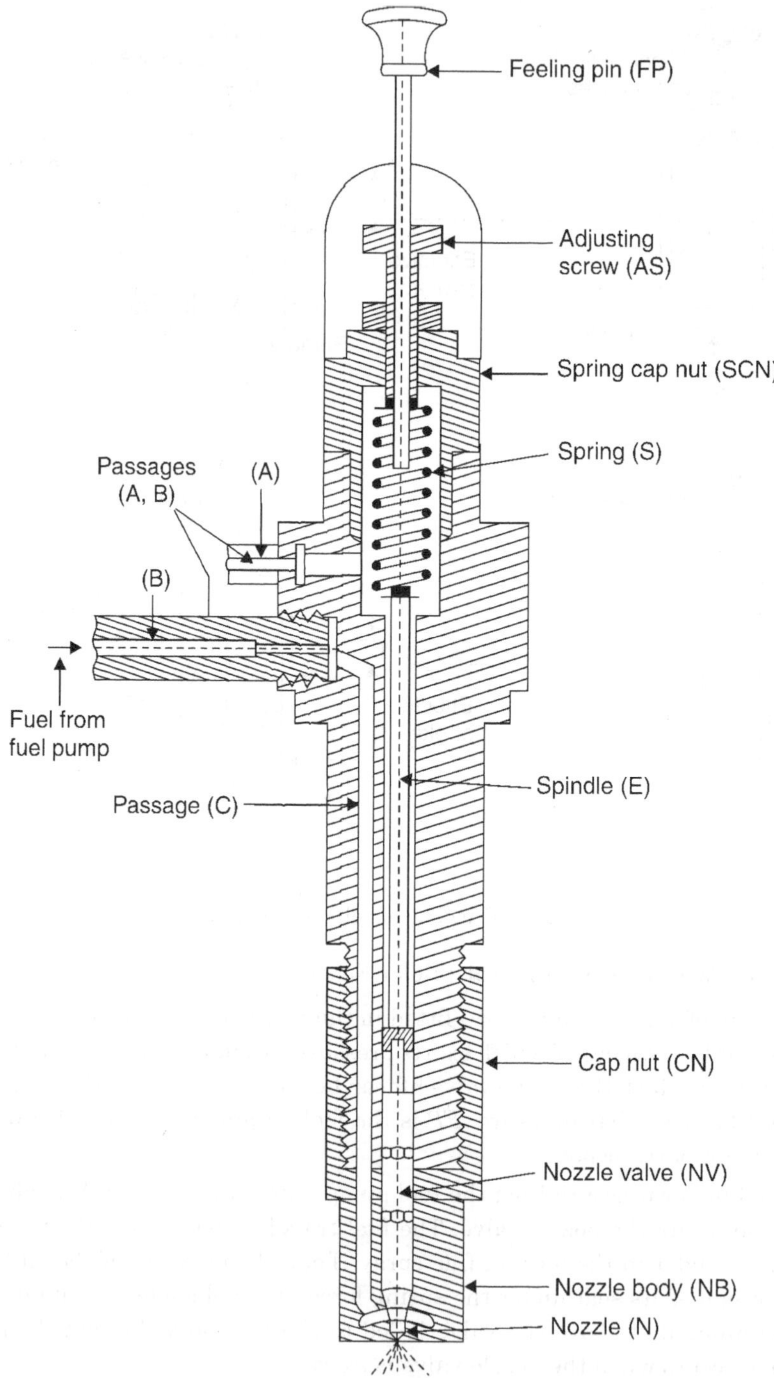

Fig. 12.8. Fuel atomiser or injector.

12.5.3. Faults, causes and remedies of injectors :

S. No.	Faults	Cause	Remedy
1.	*Pressure too high*	(*a*) Spring rate too high.	(*a*) Adjust if possible or replace spring with one having lower spring rate, or vary the number of shims.
		(*b*) Needle valve seized in almost closed position due to dirt etc.	(*b*) Soak in fuel oil. Then try to remove needle valve. If necessary replace with new needle valve and nozzle.
		(*c*) Carbon blocking injector holes.	(*c*) Clean with pricker or proke. Reverse flush using syringe and paraffin.
2.	*Pressure too low*	(*a*) Spring rate too low.	(*a*) Adjust if possible or replace spring with one having a higher spring rate.
		(*b*) Spring broken	(*b*) Replace with correct type.
		(*c*) Needle valve seized in position.	(*c*) Soak in fuel oil then try to remove the valve. If necessary replace with new needle nozzle.
3.	*Spray distorted*	(*a*) Damaged needle	(*a*) Replace.
		(*b*) Nozzle or holes carboned altering shape of orifice.	(*b*) Clean using pricker or proke. Reverse flush using syringe and paraffin.
4.	*Dribble from injector holes excessive*	Valve not seating correctly due to	
		(*a*) Dirt or carbon between needle seat and nozzle.	(*a*) Clean with paraffin ; do not wipe with fluffy cloth, but assemble wet from clean fuel. Hard carbon may be removed by using a solution of 60 grams caustic soda in 0.6 litres of water and add 15 grams of detergent. Boil for 1 to $1\frac{1}{2}$ hours.
		(*b*) Scoring or pitting on needle seat or nozzle seat :	(*b*) If possible lap needle and nozzle seat together.
		(*c*) Needle corroded, or binding nozzle due probably to faulty spindle or misalignment of nozzle and nozzle holder.	(*c*) Replace as necessary.
5.	*Excessive leakage off*	(*a*) Too much clearance between valve and hole in nozzle.	(*a*) Fit new nozzle and valve.
		(*b*) Nozzle cap loose.	(*b*) Tighten.
		(*c*) Injector body and nozzle separated by dirt or burns caused by lack of care in handling.	(*c*) Remove nozzle and examine pressure faces. Remove dirt or burns. Do not overtighten cap nut as a cure.

12.6. TYPES OF NOZZLES AND FUEL SPRAY PATTERNS

12.6.1. Main Requirements of an injector nozzle

The *main requirements of an injector nozzle are :*

1. To inject fuel at a sufficiently high pressure so that the fuel enters the cylinder with a *high velocity. Higher the velocity of the fuel smaller will be the droplet size. The momentum of smaller droplets is less, hence, penetration is also.*
2. Penetration should *not* be high so as to impinge on cylinder walls ; this may *result in poor starting.*
3. *Fuel supply and cut-off should be rapid ;* there should be *no dribbling.*

12.6.2. Classification and Description of Nozzles

The type of nozzle used is greatly dependent on the type of combustion chamber as open type or pre-combustion chamber. The nozzles are classified as per the type of orifice and its number used for injecting the fuel in the combustion chamber.

The nozzle are classified as :

1. Single hole nozzle.
2. Multi-hole nozzle
3. Circumferential nozzle
4. Pintle nozzle
5. Pintaux nozzle.

1. **Single hole nozzle.** Refer Fig. 12.9.

- This is the simplest type of nozzle and is *used in open combustion chambers.*
- It consists of a single hole bored centrally through the nozzle body and closed by the needle valve. The size of the hole is usually larger than 0.2 mm.
- Its spray cone angle varies from 5 to 15°. In some cases, a core is given a series of spiral grooves in order to impart a rotational motion to the fuel for bettering with air.

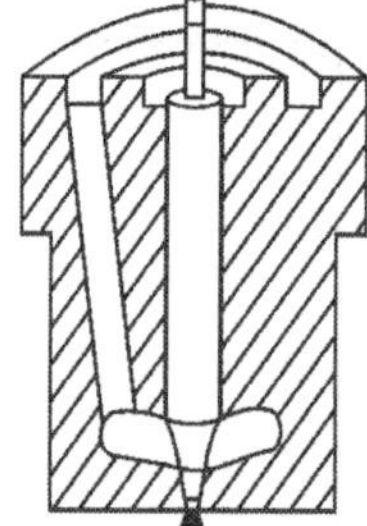

Fig. 12.9. Single hole nozzle.

Advantages :

Simple in construction and operation.

Disadvantages :

(*i*) *Very high injection pressure is required* because whole of the fuel passes through a single hole and, also, because the relative fuel velocity required is high.

(*ii*) This type of nozzle has a *tendency to dribble.*

(*iii*) As the spray angle is very narrow (usually about 15°), this does not facilitate good mixing unless higher air velocities are provided.

2. **Multi-hole nozzle.** Refer Fig. 12.10.

- This type of nozzle finds extensive use in automobile engines, particularly having open combustion chambers.

  It mixes the fuel with air properly even with slow air movement available with open combustion chambers.
- The number of holes varies from 4 to 18 ; the greater number provides better fuel distribution. The hole diameter lies between 0.25 to 0.35 mm and hole angle lies between 20° to 45°.

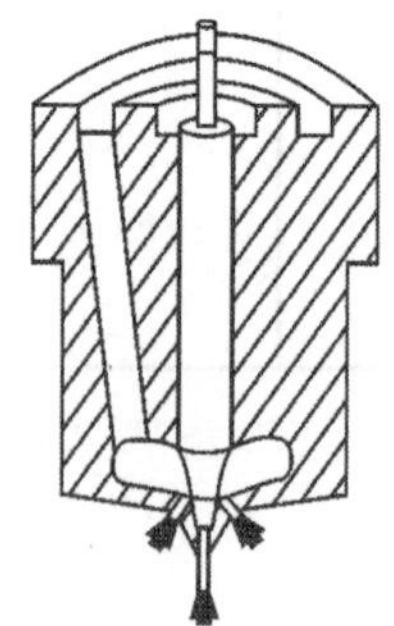

Fig. 12.10. Multi-hole nozzle.

- Usually the holes are drilled symmetrically but many times they are non-symmetrical to meet certain specific requirements of the combustion chamber.

Advantages :

(*i*) Gives *good automisation.*

(*ii*) *Distributes fuel properly* even with lower air motion available in open combustion chambers.

Disadvantages :

(*i*) Holes are small and *liable to clogging.*

(*ii*) *Dribbling* between injections.

(*iii*) *Very high injection pressures* (180 bar and above).

(*iv*) Close tolerance in manufacture (due to small holes) and hence *costly*).

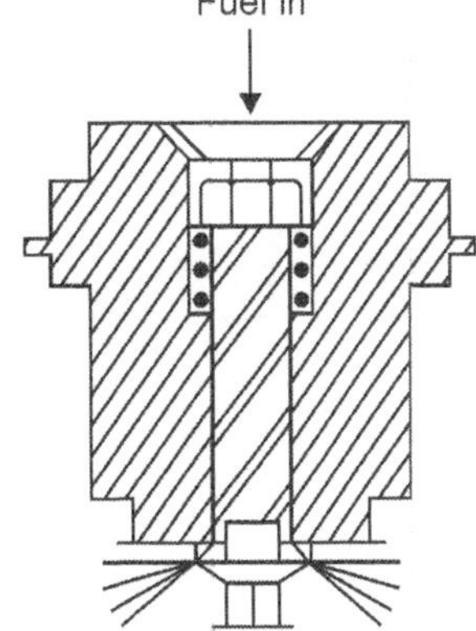

Fig. 12.11. Circumferential orifice.

3. **Circumferential nozzle.** Refer Fig. 12.11.

- Its spray characteristics are similar to a plate type opening.
- The injected fuel particles tend to be projected in the form of plane, with wide angle cone ; the purpose of which is to obtain as large an area of fuel spray as possible to come into contact with the air in the combustion chamber.

4. **Pintle nozzle :**

Fig. 12.12 shows a Pintle nozzle.

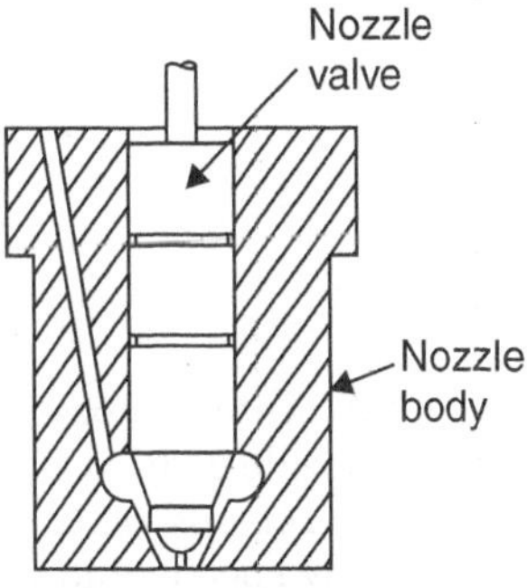

Fig. 12.12. Pintle nozzle.

- The stem of the nozzle valve is extended to form a pin or pintle which *protrudes through the mouth of the nozzle body.* It may be either cylindrical or conical in shape.
  — The size and shape of the pintle can be varied according to requirement. The spray core angle is generally 60°.
- When the valve lifts, the pintle partially blocks the orifice and thus does not allow the pressure drop to be greater. As the lift of the valve increases the entire orifice is uncovered and full area for flow is available. Thus *dribbling is avoided.*
  — *The spray obtained by the pintle nozzle is hollow conical spray.*

Advantages :

(*i*) It is self cleaning type and prevents the carbon deposition on the nozzle hole.

(*ii*) It avoids weak injection and dribbling.

(*iii*) It results in good atomisation.

(*iv*) Its injection characteristics are more near the required one.

Disadvantage :

Distribution and penetration poor, hence not suitable for open combustion chambers.

5. **Pintaux nozzle :**

- In case the fuel is injected in a direction *upstream the direction of air*, the delay period is reduced due to increased heat transfer between air and fuel. This results in good *cold starting performance.* However, if whole of the fuel is injected in this manner the efficiency of combustion is greatly reduced by flow of products of combustion back into the injection path. Thus in order to improve cold starting performance without any detrimental effect on efficiency a *pintaux nozzle* as shown in Fig. 12.13. is used.

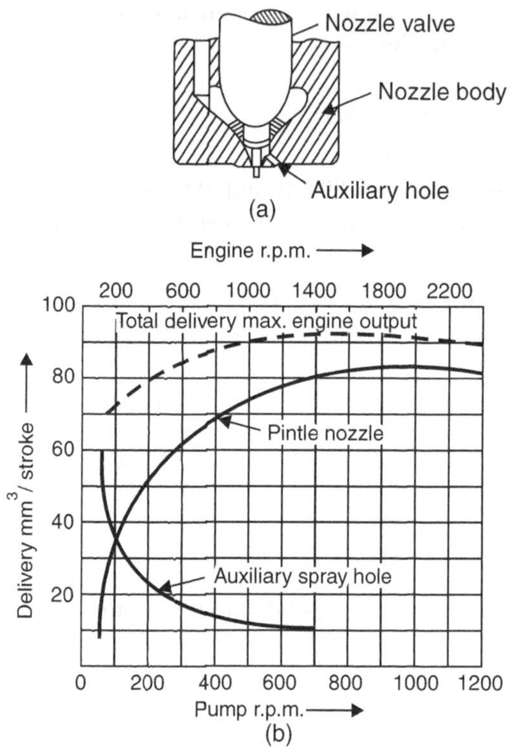

Fig. 12.13. Pintaux nozzle.

- A pintaux nozzle is a type of pintle nozzle which has an *auxiliary hole* drilled in the nozzle body. It *injects a small amount of fuel through this additional* hole (pilot injection) *in the upstream slightly before the main injection.* The needle valve does not lift fully at low speeds and most of the fuel is injected through the auxiliary hole, giving good cold starting performance.

Disadvantages / Drawbacks :

(*i*) The tendency of the auxiliary hole to choke.

(*ii*) The injections characteristics are even poorer than multi-hole nozzle.

Injection rate characteristics :

- *In order to avoid knocking in the engine it is always desirable to supply less quantity of fuel.*
- The characteristics of multi-hole, pintle and pintaux nozzles are shown respectively in Fig. 12.14 (*i*), (*ii*), (*iii*).
  - It is obvious from the figure the pintle nozzle gives desired characteristics as $(dm_f/d\theta)$ which is smaller at the beginning compared with multi-hole nozzle. The characteristics of pintaux nozzle is totally different because of fuel supply through auxiliary and main orifices.

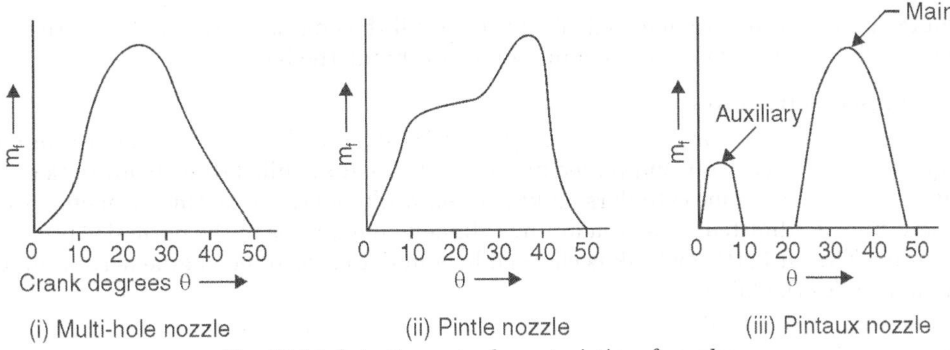

Fig. 12.14. Injection rate characteristics of nozzles.

Fuel spray pattern :

- For better and quick evaporation of fuel which is essential for better burning of fuel, the fuel pattern and relative direction of fuel particles with air are of significant importance.
- To give an idea, how the patterns help for better evaporation, two patterns are shown in Fig. 12.15.

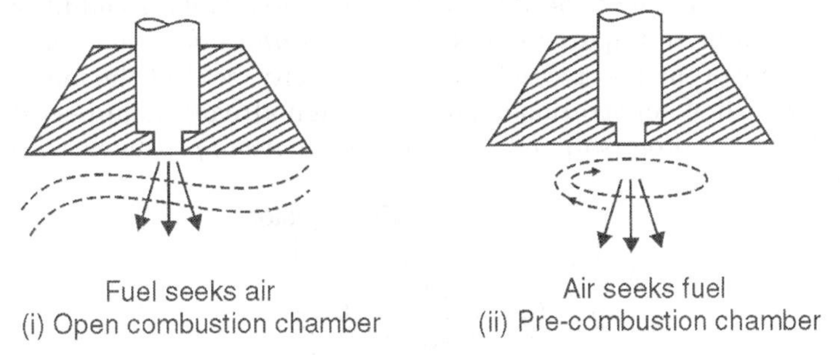

Fig. 12.15. Air and fuel movement for two types of combustion chambers.

12.7. ENGINE STARTING SYSTEMS

The following three are the commonly used starting systems in large and medium size engines :

1. Starting by an auxiliary engine
2. Use of electric motors or self starters
3. Compressed air system.

1. Starting by an auxiliary engine (generally petrol driven) :

In this system an auxiliary engine is mounted closed to the main engine and drives the latter through a clutch and gears. The clutch is first disengaged and the auxiliary engine started by hand or by a self starter motor. When it has warmed up and runs normally the drive gear is engaged through the clutch, and the main engine is cranked for starting. To avoid the danger of damage to drive gear it is desirable to have an over-running clutch or starter type drive.

2. Use of electric motors or self starters :

These are employed for small diesel and gasoline engines. A storage battery of 12 to 36 volts is used to supply power to an electric motor which is geared to the flywheel with arrangement for automatic disengagement after the engine has started. The motor draws a heavy current and is designed to be engaged continuously for about 30 seconds only, after which it is required to cool off, for a minute or so, and then re-engaged. This is done till the engine starts up. When the engine is running a small d.c. generator on the engine serves to charge the battery.

3. **Compressed air system :**

The compressed air system is commonly used for starting large diesel engines employed for stationary power plant service. Compressed air at about 17 bar supplied from an air tank or bottle is admitted to a few of the engine cylinders making them work like reciprocating air motors to run the engine shaft. Fuel is admitted to the remaining cylinders and ignites in the normal way causing the engine to start. The air bottle or tank is charged by a motor or gasoline engine driven compressor. The system includes the following :

(*i*) Storage tank/vessel

(*ii*) A safety valve

(*iii*) Interconnecting pipe work.

- *Small stationary engines of about 10 kW capacity are started by* ***hand cranking.***

12.8. FUEL INJECTION COMPUTATION IN C.I. ENGINES

- Refer Fig. 12.16. The fuel injector should develop a pressure which is *higher than the highest pressure desired to be obtained in the engine* and also additional pressure differential is available to impart *high velocity head and adequate atomisation.* There is an optimum value of the fuel particle size and velocity, so that the momentum is maximum and hence depth of penetration of spray through already compressed air in the cylinder is large. *Excessive atomisation is not very conducive to proper mixture formation.*

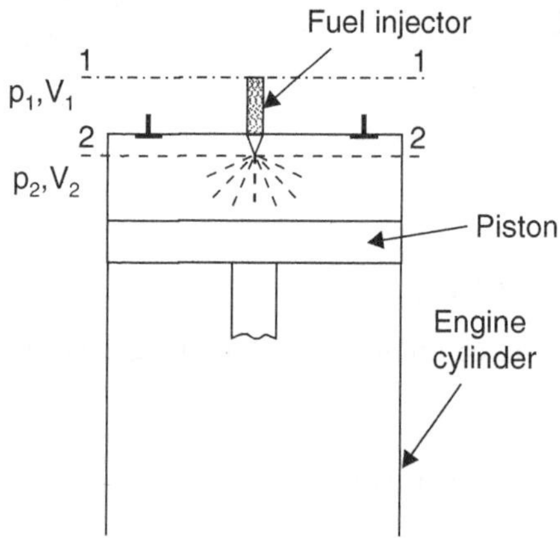

Fig. 12.16

Let p_1 = Injection pressure,

V_1 = Velocity at section 1-1,

p_2 = Pressure in the cylinder when injection of fuel takes place,

V_2 = Velocity at section 2-2,

ρ_f = Density of fuel, and

v_f = Specific volume of fuel (assumed in compressible).

Then,
$$\frac{V_1^2}{2} + p_1 v_f = \frac{V_2^2}{2} + p_2 v_f \qquad ...(12.1)$$

Neglecting V_1, being very small compared to V_2, we have

$$V_2 = \sqrt{2v_f (p_1 - p_2)} \quad ...(12.2)$$

$$= \sqrt{\frac{2(p_1 - p_2)}{\rho_f}} \quad ...(12.3)$$

or if V_f = Actual fuel velocity of injection (the velocity of fuel for good atomisation is of the order of 400 m/s), and

C_f = Flow coefficient of orifice,

Then,
$$V_f = C_f \sqrt{\frac{2(p_1 - p_2)}{\rho_f}} \quad ...(12.4)$$

The volume of the fuel injected per second Q_f is given by :

Q_f = Area of all orifices × fuel jet velocity × time of injection × number of injections per second for one orifice

or
$$Q_f = \left[\frac{\pi}{4} d_0^2 \times n_0\right] \times V_f \times \left[\frac{\theta}{360} \times \frac{60}{N}\right] \times \frac{N_i}{60} \quad ...(12.5)$$

where, d_0 = Diameter of fuel orifice, m^2,

n_0 = Number of orifices,

V_f = Velocity of flow of fuel through orifice,

θ = Duration of the injection in degrees of crank angle,

N = r.p.m., and

N_i = Number of injection per min.

$= \frac{\text{r.p.m.}}{2}$ ... for 4-stroke cycle engine

= r.p.m. ... for 2-stroke cycle engine

The rate of fuel injection is usually expressed in mm^3 / degree crank angle/litre cylinder volume to avoid the effect of engine size.

- Since the injection pressures employed are very high (to the tune of 100 to 150 bar), the assumption of incompressibility of fuel may lead to errors. In order to account for compressibility, a factor called ***"Coefficient of Compressibility"*** is introduced.

Coefficient of compressibility,

$$C_c = \frac{[v_{f(\text{sump})} - v_{f_1}]}{v_{f(\text{sump})}\,(p_1 - p_{\text{sump}})} \quad ...(12.6)$$

For pressures expressed in bar, accepted valve for $C_c \simeq 80 \times 10^{-8}$ per bar

Work of fuel compression per kg,

$$W_c = \frac{1}{2}(p_1 - p_{\text{sump}}) \times [v_{f(\text{sump})} - v_{f_1}] \quad ...(12.7)$$

Work of delivering fuel,

$$W_d = (p_1 - p_2) \times \text{volume of fuel injected} \quad ...(12.8)$$

Then, fuel pump work per kg,

$$W_p = W_c + W_d \quad ...(12.9)$$

WORKED EXAMPLES

Example 12.1. *A six-cylinder, four-stroke diesel engine develops 125 kW at 3000 r.p.m. Its brake specific fuel consumption is 200 g/kWh. Calculate the quantity of fuel to be injected per cycle per cylinder. Specific gravity of the fuel may be taken as 0.85.*

Solution. *Given :* $n = 6$; B.P. = 125 kW ; $N = 3000$ r.p.m. ; b.s.f.c. = 2.00 g/kWh ;
Sp. gr. of fuel = 0.85.

Fuel consumption per hour = b.s.f.c. × B.P.

$$= \frac{200}{1000} \times 125 = 25 \text{ kg}$$

∴ Fuel consumption per cylinder $= \frac{25}{n} = \frac{25}{6} = 4.167$ kg/h

$$\text{Fuel consumption per cycle} = \frac{\text{Fuel consumption per cylinder per min.}}{\text{No. of cycles per min.}}$$

$$= \frac{(4.167/60)}{(3000/2)} = 4.63 \times 10^{-5} \text{ kg} = 0.0463 \text{ g}$$

∴ Volume of fuel injected per cycle

$$\therefore \quad = \frac{\text{Fuel consumption per cycle}}{\text{Specific gravity of fuel}}$$

$$= \frac{0.0463}{0.85} = \mathbf{0.05447 \text{ c.c.}} \quad \textbf{(Ans.)}$$

Example 12.2. *A 6-cylinder 4-stroke C.I. engine develops 220 kW at 1500 r.pm. with brake specific fuel consumption of 0.273 kg/kWh. Determine the size of the single hole injector nozzle if the injection pressure is 160 bar and the pressure in the combustion chamber is 40 bar. The period of injection is 30° of crank angle. Specific gravity of fuel = 0.85 and orifice discharge coefficient = 0.9.*

Solution. *Given :* $n = n_0 = 6$; $N = 1500$ r.p.m. ; B.P. = 220 kW, b.s.f.c. = 0.273 kg/kWh
$\theta = 30°$, Sp. gr. of oil = 0.85, $C_f = 0.9$, $\Delta p = p_1 - p_2 = 160 - 40 = 120$ bar.

Diameter of the nozzle orifice, d_0 :

We know that, *actual* fuel velocity of injection,

$$V_f = C_f \sqrt{\frac{2(p_1 - p_2)}{\rho_f}} = C_f \sqrt{\frac{2\Delta p}{\rho_f}} \quad \text{...(Eqn. 12.4)}$$

$$= 0.9 \times \sqrt{\frac{2 \times 120 \times 10^5}{(0.85 \times 1000)}} = 151.23 \text{ m/s}$$

Volume of fuel injected per second,

$$Q_f = \frac{0.273 \times 220}{(0.85 \times 1000) \times 3600} = 1.963 \times 10^{-5} \text{ m}^3/\text{s}$$

Also, volume of fuel injected per second,

$$Q_f = \left[\frac{\pi}{4} d_0^2 \times n_0\right] \times V_f \times \left[\frac{\theta}{360} \times \frac{60}{N}\right] \times \frac{N_i}{60} \quad \text{...(Eqn. (12.5)}$$

(where N_i = No. of injection/min. $= \frac{1500}{2} = 750$)

$$1.963 \times 10^{-5} = \left[\frac{\pi}{4} d_0^2 \times 6\right] \times 151.23 \times \left[\frac{30}{360} \times \frac{60}{1500}\right] \times \frac{750}{60} = 29.694\ d_0^2$$

$$\therefore \quad \mathbf{d_0} = \left(\frac{1.963 \times 10^{-5}}{29.694}\right)^{1/2} = 8.13 \times 10^{-4} \text{m} \quad \text{or} \quad \mathbf{0.813\ mm.} \quad \textbf{(Ans.)}$$

Example 12.3. *Fuel injection in a single cylinder, 4-stroke cycle C.I. engine running at 650 r.p.m. takes place through a single orifice nozzle and occupies 28° of crank travel. The fuel consumption of the engine is 2.2 kg/hour and the fuel used has a specific gravity of 0.875. If injection pressure is 150 bar and the combustion chamber pressure is 32 bar estimate the volume of fuel injected per cycle and the diameter of the orifice. Take coefficient of discharge of orifice = 0.88.*

Solution. *Given* : $n = 1$, $N = 650$ r.p.m. ; $\theta = 28°$ of crank travel ;

Fuel consumption = 2.2 kg/h ; Sp. gr. = 0.875 ;

$$\Delta p = p_1 - p_2 = 150 - 32 = 118 \text{ bar} ; C_d = 0.88$$

Volume of fuel injected per cycle :

$$\text{Fuel to be injected per cycle} = \frac{\text{Fuel consumption per cylinder}}{\text{No. of cycles per min.}}$$

$$= \frac{(2.2 / 60)}{(650 / 2)} = 1.128 \times 10^{-4} \text{ kg}$$

$$\textit{Volume of fuel injected per cycle} = \frac{\text{Mass of fuel injected per cycle}}{\text{Density of fuel } (\rho_f)}$$

$$= \frac{1.128 \times 10^{-4}}{0.875 \times 1000} = 1.289 \times 10^{-7}$$

$$= \mathbf{0.1289\ cm^3.} \quad \textbf{(Ans.)}$$

Diameter of the orifice, $\mathbf{d_0}$:

$$\text{Time for fuel injection per cycle} = \left(\frac{\theta}{360} \times \frac{60}{N}\right) \text{sec.}$$

$$= \frac{28}{360} \times \frac{60}{650} = 0.00718 \text{s}$$

Mass of fuel injected per second,

$$m_f = \frac{\text{Fuel injected per cycle}}{\text{Time for fuel injection}} = \frac{1.128 \times 10^{-4}}{0.00718} = 0.0157 \text{ kg/s}$$

Actual velocity of per cycle injection,

$$V_f = C_f \sqrt{\frac{2\Delta p}{\rho_f}} = 0.88 \sqrt{\frac{2 \times 118 \times 10^5}{(0.875 \times 1000)}} = 144.5 \text{ m/s}$$

Now, $$m_f = A_0 \times V_f \times \rho_f$$

or $$0.0157 = \left(\frac{\pi}{4} d_0^2\right) \times 144.5 \times (0.875 \times 1000)$$

or $$d_0 = \left[\frac{0.0157 \times 4}{\pi \times 144.5 \times (0.875 \times 1000)}\right]^{1/2} = 3.976 \times 10^{-4} \text{ m} \simeq \mathbf{0.4\ mm.} \quad \textbf{(Ans.)}$$

Example 12.4. *A four-stroke engine using 0.272 kg/kWh fuel of 32° API develops 15 kW per cylinder at 2000 r.p.m. The fuel injection pressure is 120 bar and the combustion chamber pressure is 30 bar. If the duration of injection is 30° of crank travel and velocity coefficient is 0.9 determine the diameter of the fuel orifice.*

$$\textit{Table Sp. gr.} = \frac{141.5}{131.5 + {}^\circ API}.$$ **(Madras University)**

Solution. *Given* : s.f.c. = 0.272 kg/kWh, Power developed = 15 kW ;

$\Delta p = p_1 - p_2 = 120 - 30 = 90$ bar ; Duration of injection = 30° of crank angle ; $C_f = 0.9$

Diameter of the orifice, d_0 :

$$\text{Sp. gr.} = \frac{141.5}{131.5 + 32} = 0.8654$$

$$\text{Fuel consumption/cycle} = \frac{\text{s.f.c.} \times \text{kW}}{\text{cycle / hour}}$$

$$= \frac{0.272 \times 15}{\left(\frac{2000}{2}\right) \times 60} = 6.8 \times 10^{-5} \text{ kg}$$

$$\text{Duration of injection} = \frac{\theta}{360} \times \frac{60}{N} = \frac{30}{360} \times \frac{60}{2000} = 0.0025\text{s}$$

$$\therefore \quad m_f = \frac{6.8 \times 10^5}{0.0025} = 0.0272 \text{ kg/s}$$

Actual fuel velocity of injection,

$$V_f = C_f \sqrt{\frac{2\Delta p}{\rho_f}} = 0.9 \times \sqrt{\frac{2 \times 90 \times 10^5}{(0.8654 \times 1000)}} = 129.8 \text{ m/s}$$

Now,

$$m_f = A_f \times V_f \times \rho_f$$

$$= \frac{\pi}{4} d_0^2 \times V_f \times \rho_f$$

or

$$0.0272 = \frac{\pi}{4} d_0^2 \times 129.8 \times (0.8654 \times 1000)$$

$$\therefore \quad d_0 = \left[\frac{0.0272 \times 4}{\pi \times 129.8 \times (0.8654 \times 1000)}\right]^{1/2} = 5.55 \times 10^{-4} \text{ m}$$

$= 5.55 \times 10^{-4}$ m or **0.555 mm. (Ans.)**

Example 12.5. *A 4-stroke cycle C.I. engine develops 11 kW per cylinder while running at 1800 r.p.m. and using fuel oil of 32° API. Fuel injection occupies 32° of crank travel and takes place through a fuel injection orifice 0.47 mm diameter with flwo coefficient of 0.9. Fuel is injected at a pressure of 118.2 bar into combustion chamber where the pressure is 31.38 bar.*

Estimate the quantity of fuel injected in kg/kWh. Specific gravity of fuel oil is given by :

$$\frac{141.5}{131.5 + {}^\circ API}.$$

Solution. *Given* : Power developed = 11 kW ; N = 1800 r.p.m., θ = 32° of crank travel ; d_0 = 0.47 mm ; C_f = 0.9 ; $\Delta p = p_1 - p_2 = 118.2 - 31.38 = 86.82$ bar

$$\text{Sp. gr.} = \frac{141.5}{131.5 + {}^\circ\text{API}} = \frac{141.5}{131.5 + 32} = 0.8654$$

Actual fuel velocity of injection,

$$V_f = C_f \sqrt{\frac{2\Delta p}{\rho_f}} = 0.9 \sqrt{\frac{2 \times 86.82 \times 10^5}{(0.8654 \times 1000)}} = 127.48 \text{ m/s}$$

Now,

$$m_f = A_0 \times V_f \times \rho_f$$

$$= \frac{\pi}{4} d_0^{\,2} \times V_f \times \rho_f$$

$$= \frac{\pi}{4} \times \left(\frac{0.47}{1000}\right)^2 \times 127.48 \times (0.8654 \times 1000) = 0.01914 \text{ kg/s}$$

Time for fuel injection per cycle $= \dfrac{\theta}{360} \times \dfrac{60}{N} = \dfrac{32}{360} \times \dfrac{60}{1800} = 2.963 \times 10^{-3}$s

Mass of fuel injected per cycle $= 0.01914 \times 2.963 \times 10^{-3} = 5.671 \times 10^{-5}$ kg/cycle

Total number of cycles per hour $= \dfrac{1800}{2} \times 60 = 54000$

∴ **Fuel consumption in kg/kWh**

$$= 5.671 \times 10^{-5} \times 54000 \times \frac{1}{11} = \mathbf{0.278\ kg/kWh.} \quad \textbf{(Ans)}$$

Example 12.6. *An high-cylinder, four-stroke diesel engine has a power output of 386.4 kW at 800 r.p.m. The fuel consumption is 0.25 kg/kWh. The pressure in the cylinder at the beginning of injection is 32 bar and the maximum cylinder pressure is 55 bar. The injector is expected to be set at 207 bar and the maximum pressure at the injector is set to be about 595 bar. Calculate the orifice area required per injector if the injection takes place over 12° crank angle.*

Assume the following :

Specific gravity of fuel = 0.85 ; coefficient of discharge for the injector = 0.6 ; atmospheric pressure = 1.013 bar ; The effective pressure difference is the average pressure difference over the injection period. **(Madras University)**

Solution. *Given* : $n = 8$; Power output = 386.4 kW, N = 800 r.p.m.,

Fuel consumption = 0.25 kg / kWh, θ = 12° crank angle ; Sp. gr. = 0.85 ;

$C_f = 0.6$; $p_{atm.} = 1.013$ bar.

Orifice area reqd. per injection, A_0 :

$$\text{kW per cylinder} = \frac{386.4}{8} = 48.3$$

∴ Fuel consumption per cylinder = 48.3 × 0.25 = 12.075 kg / h or 0.2012 kg/ min.

Fuel to be injected per cycle $= \dfrac{0.2012}{(800/2)} = 5.03 \times 10^{-4}$ kg

Time for fuel injection per cycle $= \left(\dfrac{\theta}{360} \times \dfrac{60}{N}\right) = \dfrac{12}{360} \times \dfrac{60}{800} = 0.0025$ s

∴ Mass of fuel injected per second,

$$m_f = \frac{5.03 \times 10^{-4}}{0.0025} = 0.2012 \text{ kg/s}$$

Pressure difference at beginning = 207 – 32 = 175 bar

Pressure difference at end = 595 – 55 = 540 bar

Average pressure difference $= \dfrac{175 + 540}{2} = 357.5$ bar

Now, $$m_f = A_0 \times V_f \times \rho_f$$

$$= A_0 \times \left[C_f \sqrt{\frac{2\Delta p}{\rho_f}} \right] \times \rho_f = A_0 \times C_f \sqrt{2\Delta p.\rho_f}$$

or $$0.2012 = A_0 \times 0.6 \sqrt{2 \times (357.5 \times 10^5) \times (0.85 \times 1000)}$$
$$= 147915.5\, A_0$$

∴ $\mathbf{A_0} = 1.36 \times 10^{-6}\ \text{m}^2$ or **0.0136 cm²**. **(Ans.)**

Example 12.7. *A six-cylinder, four-stroke oil engine operates on A/F ratio = 20. The diameter and stroke of the cylinder are 100 mm and 140 mm respectively. The volumetric efficiency is 80 per-cent. The condition of air at the beginning of compression are 1 bar, 27° C.*

(*i*) *Determine the maximum amount of fuel that can be injected in each cylinder per second.*

(*ii*) *If the speed of the engine is 1500 r.p.m., injection pressure is 150 bar, air pressure during fuel injection is 40 bar and fuel injection is carried out for 20° crank angle, determine the diameter of the fuel orifice assuming only one orifice is used.*

Take, $\rho_f = 860\ kg/m^3$; $C_f = 0.67$. **(Roorkee University)**

Solution. *Given* : $n = n_0 = 6$; A / F ratio = 20 ; d = 100 mm = 0.1 m, l = 140 mm = 0.14 m ; $\eta_{vol.}$ = 80% ; p_a = 1 bar, T_a = 27 + 273 = 300 K, N = 1500 r.p.m. ; θ = 20° crank angle ; ρ_f = 960 kg/m³ ; C_f = 0.67, Δ_p = 150 – 40 = 110 bar.

(*i*) **Amount of fuel injected into each cylinder per cycle :**

Volume of air supplied per cylinder per cycle

$$= \text{Stroke volume} \times \eta_{vol.}$$

$$= \frac{\pi}{4} d^2 \times l \times \eta_{vol.} = \frac{\pi}{4} \times (0.1)^2 \times 0.14 \times 0.8 \simeq 8.8 \times 10^{-4} m^3$$

Mass of this air at suction conditions,

$$m_a = \frac{p_a V_a}{RT_a} = \frac{1 \times 10^5\, (8.8 \times 10^{-4})}{287 \times 300} = 1.022 \times 10^{-3}\ \text{kg/cycle}$$

$$\text{A / F ratio} = \frac{m_a}{m_f} = \frac{20}{1}$$

∴ $$m_f = \frac{m_a}{20} = \frac{1.022 \times 10^{-3}}{20} = 5.11 \times 10^{-5}\ \text{kg/cycle}$$

Time taken for fuel injection per cycle

$$= \left(\frac{\theta}{360} \times \frac{60}{N} \right) = \frac{20}{360} \times \frac{60}{1500} = 0.00222\text{s}$$

∴ *Amount/mass of fuel injected into each cylinder per second,*

$$\dot{m}_f = \frac{5.11 \times 10^{-5}}{0.00222} = \mathbf{0.023\ kg / s.}\quad \textbf{(Ans.)}$$

(*ii*) **Diameter of the fuel orifice, $\mathbf{d_0}$:**

The mass of fuel injected into each cylinder per second,

$$\dot{m}_f = A_0 \times V_f \times \rho_f$$

or $$0.023 = \frac{\pi}{4} d_0^{\,2} \times C_f \sqrt{\frac{2\Delta p}{\rho_f}} \times \rho_f$$

$$= \frac{\pi}{4} d_0^2 \times C_f \sqrt{2\Delta p \times \rho_f}$$

$$= \frac{\pi}{4} d_0^2 \times 0.67 \sqrt{2 \times 110 \times 10^5 \times 960} = 72381.11\, d_0^2$$

or $d_0 = 5.637 \times 10^{-4}$ m or **0.5637 mm. (Ans.)**

Example 12.8. *In a diesel fuel injection pump, the volume of fuel in the pump barrel before commencement of the effective stroke is 7 c.c. The diameter of the fuel line from pump to injector is 3 mm and is 700 mm long. The fuel in the injection valve is 2 c.c.*

(*i*) *To deliver 0.10 c.c. of fuel at a pressure of 150 bar, how much displacement the plunger undergoes ? Assume a pump inlet pressure of 1 bar ;*

(*ii*) *What is the effective stroke of the plunger if its diameter is 7 mm.*

Assume coefficient of compressibility of oil as 78.8 × 10⁻⁶ per bar at atmospheric pressure.

Solution. *Given* : The volume of fuel in the pump barrel before commencement of the effective stroke = 7 c.c.

The diameter and length of the fuel line from pump to injector = 3 mm, 700 mm,

Volume of fuel in the injection valve = 2 c.c.

Volume of fuel to be delivered = 0.10 c.c.

The pressure at which fuel to be delivered, p_1 = 150 bar

Atmospheric pressure, p_2 = 1 bar

Coefficient of compressibility, $C_c = 78.8 \times 10^{-6}$ per bar at atmospheric pressure

Diameter of plunger, d_p = 7 mm

(*i*) **Displacement of plunger :**

Coefficient of compressibility of oil,

$$C_c = \frac{\text{Change in volume per unit volume}}{\text{Difference in pressure causing compression}}$$

$$= \frac{(V_1 - V_2)}{V_1(p_1 - p_2)}$$

Total initial fuel volume,

V_1 = Volume of fuel in barrel + volume of fuel in the delivery line + volume of fuel in the injection valve

$$= 7 + \frac{\pi}{4}(0.3)^2 \times 70 + 2 = 13.95 \text{ c.c.}$$

No pressure is built up till the pump plunger closes the inlet port. Further advance of plunger will compress the fuel oil and raise the pressure to a required value. Once the delivery pressure is attained, further movement of plunger results in delivery of fuel oil at constant pressure.

Change in volume due to compression = $C_c(p_1 - p_2) \times V_1$

or $(V_1 - V_2) = 78.8 \times 10^{-6} \times (150 - 1) \times 13.95$

$= 0.16379$ c.c.

Total displacement of plunger

$= (V_1 - V_2) + 0.1 = 0.16379 + 0.1 =$ **0.26379 c.c. (Ans.)**

(*ii*) **Effective stroke of the plunger, l_p :**

$$\frac{\pi}{4} d_p^2 \times l_p = 0.26379 \quad \text{or} \quad \frac{\pi}{4} \times (0.7)^2 \times l_p = 0.26379$$

∴ $l_p = \dfrac{0.26379 \times 4}{\pi \times (0.7)^2} = 0.6854$ cm or **6.854 mm. (Ans.)**

Example 12.9. *At injection pressure of 145 bar, spray penetration of 22 cm in 16 milliseconds is obtained. Determine the time required for the spray to penetrate the same distance at an injection pressure of 235 bar. Assume the same orifice and the combustion chamber density. Combustion chamber pressure is 30 bar.*

Use the relation : $s = t\sqrt{\Delta p}$

where *s is the penetration in cm,*

t is the time in milliseconds, and

Δp is the pressure difference between the injection pressure and combustion chamber pressure.

Solution. *Given :* $p_1 = 145$ bar, $t_1 = 16$ milli-seconds, $p_2 = 230$ bar ;

$p_{cyl} = 30$ bar ; $s_1 = s_2 = 22$ cm

Here, $\Delta p_1 = p_1 - p_{cyl} = 145 - 30 = 115$ bar

$\Delta p_2 = p_2 - p_{cyl} = 235 - 30 = 205$ bar

From given relation $s = t\sqrt{\Delta p}$, we have

$$\frac{s_1}{s_2} = \frac{t_1}{t_2}\sqrt{\frac{\Delta p_1}{\Delta p_2}}$$

or

$$\mathbf{t_2} = \frac{s_2}{s_1} \times t_1\sqrt{\frac{\Delta p_1}{\Delta p_2}}$$

$$= 1 \times 16\sqrt{\frac{115}{205}} = \mathbf{11.98\ milli\text{-}seconds.} \quad \textbf{(Ans.)}$$

HIGHLIGHTS

1. Main types of modern fuel injection systems are :
   (*i*) Common rail injection system
   (*ii*) Individual pump injection system
   (*iii*) Distributor system.
2. The main components of a fuel injection system are :
   Fuel tank ; Fuel feed pump ; Fuel filters ; Injection pump ; Governor ; Fuel pipings and injectors.
3. The nozzle are classified as follows :
   (*i*) Single hole nozzle (*ii*) Multi-hole nozzle
   (*iii*) Circumferential nozzle (*iv*) Pintle nozzle
   (*v*) Pintaux nozzle.
4. Volume of fuel injected per min.,

$$Q_f = \left[\frac{\pi}{4} d_0^2 \times n_0\right] \times V_f \times \left[\frac{\theta}{360} \times N\right] \times N_i$$

where, d_0 = Diameter of fuel orifice, m[2],

n_0 = Number of orifices,

V_f = Velocity of flow of fuel through orifice,

θ = Duration of the injection in degrees of crank angle,

N = r.p.m., and

N_i = Number of injection per min.

= r.p.m./2 = for 4-stroke engine

= r.p.m. for 2-stroke engine.

OBJECTIVE TYPE QUESTIONS

Fill in the Blanks or Say "Yes" or "No" :

1. Fuel is injected into the cylinder at the end of ...... stroke.
2. The period between the start of injection and start of ignition is called ...... delay.
3. The beginning and end of the injection should be sharp.
4. The injected fuel may not be broken into very fine droplets.
5. Air injection system is largely used these days.
6. Inferior fuels can be used in air injection system.
7. Solid injection is also called ...... injection.
8. In ...... injection system only one pump is sufficient for multi-cylinder engine.
9. Very accurate design and workmanship are required in air injection system.
10. In ...... system, the fuel is metered at a central point ; a pump pressurises, meters the fuel and times the injection.
11. In ...... pump injection system an individual pump or pump cylinder connects directly to each fuel nozzle.
12. A multi-hole nozzle gives poor automisation.
13. A ...... nozzle is a type of pintle nozzle which has an auxiliary hole drilled in the nozzle body.

ANSWERS

1. compression	**2.** ignition	**3.** Yes	**4.** No	**5.** No
6. Yes	**7.** mechanical	**8.** common rail	**9.** Yes	**10.** distributor
11. individual	**12.** No	**13.** pintaux.		

THEORETICAL QUESTIONS

1. What are the functional requirements of an injection system ?
2. How are injection systems classified ? Describe them briefly. Why the air injection system is not used now-a-days ?
3. What are the essential requirements to be fulfilled by a fuel injection system for C.I. engines ? What is the most common injection system used in multi-cylinder diesel engines ?
4. What are the main functions of the nozzle ?
5. What are the most common types of nozzles used ?
6. What is the difference between air injection and solid injection ?
7. State the merits and demerits of multiple orifice nozzle and pintle type nozzle in C.I. engine injectors.
8. Bring out the differences in construction and working of printle and pintaux nozzles with the help of sketches and discuss their relative merits.
9. How are solid injection system classified ? Explain the working of common rail, individual pump and distributor systems with the help of neat sketches. Discuss their relative merits and demerits.

UNSOLVED EXAMPLES

1. A 6-cylinder 4-stroke diesel engine develops 89 kW at 2500 r.pm. Its brake specific fuel consumption is 245 g/kWh. Calculate the quantity of fuel to be injected per cycle per cylinder. Specific gravity of the fuel may be taken as 0.84. [**Ans.** 0.0576 c.c.]
2. A six-cylinder, four-stroke oil engine develops 200 kW at 1200 r.p.m. and consumes 0.3 kg/kWh. Determine the diameter of a single orifice injector if the injection pressure is 200 bar and combustion cham-

ber pressure is 40 bar. The injection is carried out for 30° rotation of a crank. Each nozzle on a cylinder is provided with single orifice.

Take $\rho_f = 900$ kg/m^3 and $C_f = 0.7$. **[Ans.** 1.005 mm]

3. A 4-stroke cycle C.I. engine develops 15 kW per cylinder at 2000 r.p.m. The specific fuel consumption is 0.275 kg/kWh fuel of 30° API. The injection pressure is 130 bar for 25° of crank angle. The pressure in the combustion chamber is 40 bar. The co-efficient of velocity = 0.875 and the specific gravity is given by :

   Sp. gr. = $\dfrac{141.5}{131.5 + {}^\circ API}$. Assuming fuel to be incompressible, determine the diameter of the fuel orifice.

   [Ans. 0.62 mm]

4. An eight-cylinder, four-stroke diesel engine has a power output of 368 kW at 800 r.p.m. The fuel consumption is 0.238 kg/kWh. The pressure in the cylinder at the beginning of injection is 35 bar and the maximum cylinder pressure is 60 bar. The injector is expected to be set at 210 bar and maximum pressure at the injector is set at about 600 bar. Calculate the orifice area required per injector if the injection takes place over 12° crank angle.

   Assume the following :

   Specific gravity of the fuel = 0.85 ; Coefficient of discharge for the injector = 0.6 ; Atmospheric pressure = 1.013 bar ; The effective pressure difference is the average pressure difference over the injection period.

   [Ans. 0.01234 cm^2]

5. In a diesel fuel injection pump the volume of the fuel in the pump barrel before the commencement of the effective stroke is 7 c.c. The fuel line from the pump to injector is 3 mm in diameter and 700 mm long and the fuel in the injection valve is 3 c.c.

   (*i*) Determine the pump displacement necessary to deliver 0.15 c.c. of fuel at a pressure of 200 bar. Assume a sump (atmospheric) pressure of 1 bar.

   (*ii*) Calculate the effective stroke of the plunger if the diameter of the plunger is 8 mm.

   Take the coefficient of compressibility of oil as 78.8×10^{-6} per bar at atmospheric pressure.

   [Ans. (*i*) 0.384 c.c. ; (*ii*) 7.65 mm]

6. A spray penetration of 10 cm is obtained in 15 milli-seconds when injection pressure was 150 bar. Find the time required to penetrate the same distance if the injection pressure is increased to 200 bar. The pressure in the combustion chamber under both conditions is 40 bar. Assume the density of fuel and air in cylinder remain same during injection under both conditions. **[Ans.** 12.4 milli-seconds]

13

Ignition Systems (S.I. Engines)

13.1. INTRODUCTION

- In S.I. engines the combustion process is initiated by a spark between the two electrodes of spark plug. This occurs just before the end of compression stroke. The ignition process must add necessary energy for *starting* and *sustaining* burning of the fuel till combustion takes place.
- Ignition is only *a pre-requisite of combustion. It does not influence the gross combustion process.* It is only a small scale phenomenon taking place within a specified small zone in the combustion chamber.
- Ignition only ensures initiation of combustion process and has no degree intensively or extensively.

Energy requirements for ignition :

— A spark energy below 10 millijoules is adequate to initiate combustion for A / F ratio 12-13 : 1 (Range of mixtures normally used) ; the duration of few micro-seconds is sufficient to start combustion.

— A spark can be struck between the gap in the two electrodes of the spark plug by sufficiently high voltage. There is a critical voltage called *breakdown voltage* below which no sparking would occur. In practice the pressure, temperature and density have a profound influence on the voltage required to cause the spark. Also, the striking voltage is increased due to the fouling factor of the electrodes owing to deposits and abrasion.

— For automotive engines, in normal practice, the spark energy to the tune of 40 millijoules and duration of about 0.5 millisecond is sufficient over entire range of operation.

13.2. REQUIREMENTS OF AN IGNITION SYSTEM

For an ignition system to be acceptable it must be *moderately priced, reliable and its performance must be adequate to meet all the demands imposed on it by various operating conditions.*

An ignition system should *fulfill the following requirements* :

1. It should have an adequate reserve of secondary voltage and ignition energy over the entire operating speed range of the engine.

2. It should consume the minimum of power and convert it efficiently to a high-energy spark across the spark-plug electrode gap.

3. It should have a spark duration which is sufficient to establish burning of the air-fuel mixture under all operating conditions.

4. It should have the ability to produce an ignition spark when a shunt (short) is established over the spark-plug electrode insulator surface, due possibly to carbon, oil or lead deposits, liquid fuel or water condensation.

5. Good performance at high speeds.

6. Longer life of breaker points and spark plug.

7. Good starting when the breaker points open slowly at cranking speed.

8. Good reproducibility of secondary voltage rise and maximum rise.

9. Adjustment of spark advance with speed and load.

- The basic source of electrical energy is either *battery*, a *generator* or a *magneto*.
  - The battery and generator normally provide 6 V or 12 V direct current, while the magneto provides an alternating current of higher voltage.
  - The low voltage (6 V or 12 V) is boosted to a very high potential of about 10 kV to 20 kV, in order to overcome the spark gap resistance and to release enough energy to initiate self propagating flame from within the combustible mixture.

13.3. BASIC IGNITION SYSTEMS

The following basic ignition systems are in use :

1. **Battery ignition system**—Conventional, transistor assisted.
2. **Magneto ignition system**—Low tension, high tension.
3. **Electronic ignition system.**

The difference between Battery and Magneto ignition systems lies only in the source of electrical energy. Whereas 'Battery ignition system' uses a battery, 'Magneto ignition system' uses a magneto to supply low voltage, all other system components being similar.

13.4. BATTERY (OR COIL) IGNITION SYSTEM

It is a commonly used system because of *its combined cheapness, convenience of maintenance, attention and general suitability*.

Construction. This system consists of the following components :

1. Battery (6 or 12 volts)
2. Ignition switch
3. Induction coil
4. Circuit/contact breaker
5. Condenser
6. Distributor.

Refer Fig. 13.1.

— One terminal of the battery is ground to the frame of the engine, and other is connected through the ignition switch to one primary terminal of the ignition coil (consisting of a *comparatively few turns of thick wire* wound round an iron core). The other primary terminal is connected to one end of the contact points of the circuit breaker and through closed points to ground. The primary circuit of the ignition coil thus gets completed when contact points of the circuit breaker are together and switch is closed. The secondary terminal of the coil is connected to the central contact of the distributor and hence to distributor rotor. The secondary circuit consists of secondary winding (consisting of a *large number of turns of fine wire*) of the coil, distributor and four spark plugs. The contact breaker is driven by a cam whose speed is *half* the engine speed (for *four stroke engines*) and breaks the primary circuit one for each cylinder during one complete cycle of the engine.

— The breaker points are held on contact by a spring except when forced apart by lobs of the cam.

— A *ballast resistor* is provided in series with the primary winding to regulate primary current. For starting purposes this resistor is by passed so that more current can flow in the primary circuit.

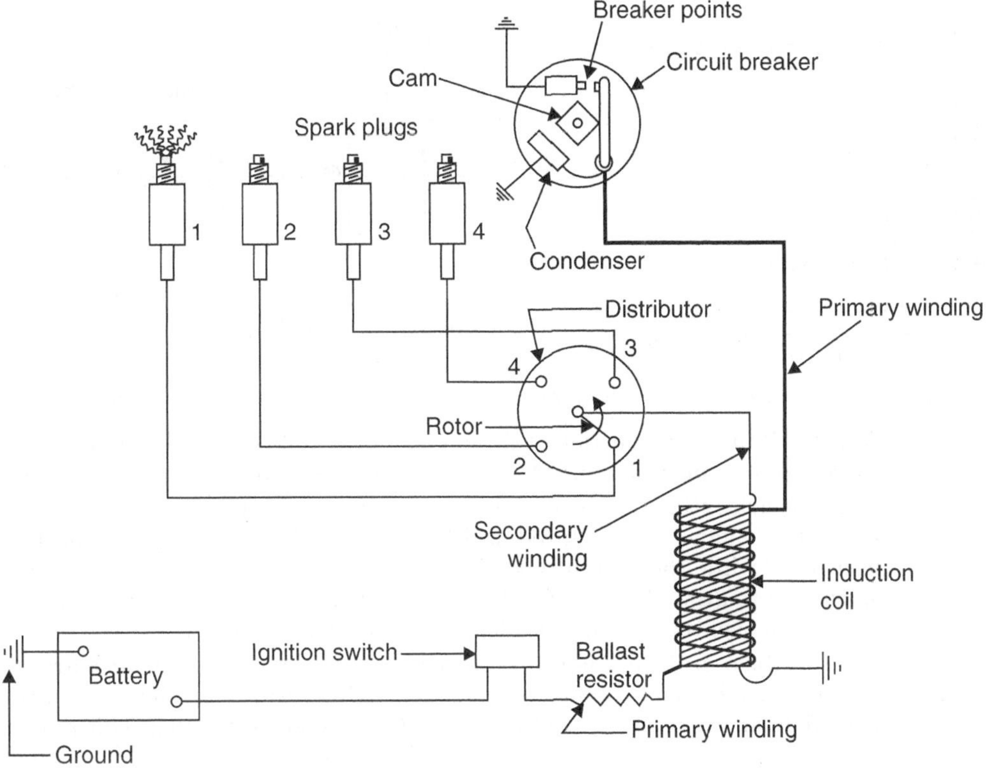

Fig. 13.1. Battery or coil ignition system.

Working :

— To start with, the ignition switch is made on and the engine is cranked *i.e.* turned by hand when the contacts touch, the current flows from battery through the switch, primary winding of the induction coil to circuit breaker points and the circuit is completed through the ground. A condenser connected across the terminals of the contact breaker points prevent the sparking at these points.

— The rotating cam breaks open the contacts immediately and breaking of this primary circuit brings about a change of magnetic field ; due to which a very high voltage to the tune of 8000 to 12000 V is produced across the secondary terminals. (The number of turns in the secondary winding may be 50 to 100 times than in primary winding). Due to high voltage the spark jumps across the gap in the spark plug and air fuel mixture is ignited in the cylinder.

- Fig. 13.2 shows the gradually building up of the primary current from the time the points close until they open.
- Fig. 13.3 shows a typical wave-form or pattern of the *normal ignition action*.
  — At point **L** the distributor opens and the magnetic field of the coil-primary winding *collapses* and consequently the secondary voltage, indicated by the firing line, rises to point **M**. The height of the firing line shows the voltage needed to jump the rotor gap and to ionize the gap between the spark plug electrodes. After the spark is

initiated the gap becomes ionized resulting in decreased gap resistance and a smaller voltage is then required to maintain the arc across the gap. The lower voltage and the spark duration is represented by the height and length of the spark line NP.

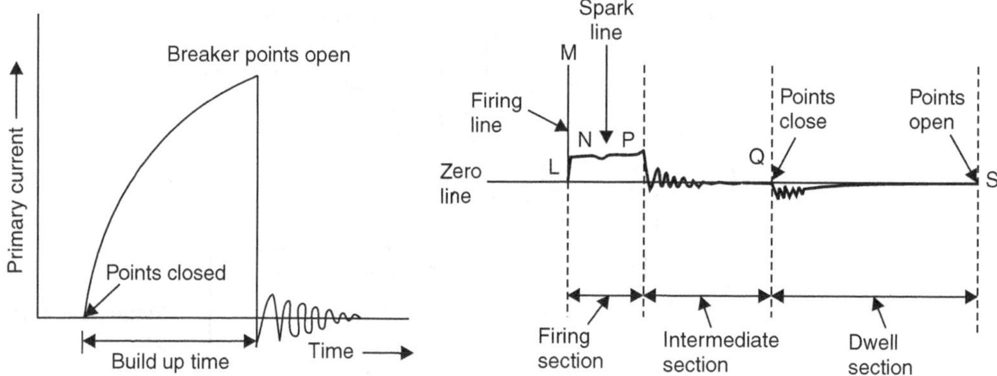

Fig. 13.2. Built up time for primary current. Fig. 13.3. Typical pattern of normal ignition action.

— At point **P** the major portion of the energy of the coil is expended and consequently there is a drop in the secondary voltage which result in extinguishing of the spark.

— Due to spark extinction the circuit becomes open, the current flow is stopped, and, hence the magnetic field (produced in the secondary winding, during the firing period NP while the current was flowing in the secondary winding and across the spark gap to ground) collapses, thereby, inducing a current in the primary winding, which eventually *flows into the condenser and charge it.*

— When voltage in the condenser becomes higher than that in the primary winding, it discharges back in the primary winding. This results in collapsing of the magnetic field and rebuilding up of voltage in the secondary winding. This pulsing back and forth, weaked each time, continues till whole of the energy is dissipated (Refer Fig. 13.3-intermediate section).

— At point **Q** the contact points close and remain so during *dwell period.* At the end of this period the points again open at S (there being no condenser action during the period, since it is shorted out across the closed points.)

Advantages :

1. It offers better sparks at low speeds, starting and for cranking purposes.
2. The initial cost of the system is low.
3. It is a reliable system and periodical maintenance required is negligible except for battery.
4. Items requiring attention can be easily located in more accessible position than those of magnetos.
5. The high speed engine drive is usually simpler than magneto drive.
6. Adjustment of spark timing has no deterimental effect over the complete ignition timing range.

Disadvantages :

1. With the increasing speed, sparking voltage drops.
2. Battery, the only unreliable component of the system needs regular attention. In case battery runs down, the engine cannot be started as induction coil fails to operate.
3. Because of battery, bulk of the system is high.

Description of Components of Battery Ignition System

1. The battery :

- The function of battery is to store electricity in the form of chemical energy, when required to convert the latter back into electrical energy.
- Motor vehicles use lead-acid batteries which have a series of positive and negative plates which are interpersed, the plates being immersed in a solution of dilute sulphuric acid, called the electrolyte. For compactness the plates are placed close together and separators are used to reduce the chance of shorting taking place.

Advantages of 12-V ignition system over 6-V system :

The following are the advantages of 12-V ignition system over 6-V system :

1. Considerably higher voltages are obtainable.
2. For transmitting equal power with excessive voltage drop, the cable in a 6-V system needs theoretically to be four times the thickness of 12-V system, cables.
3. Improved starting.
4. Adequate electric power to supply the increasing number of electrical accessories used.

2. The ignition coil :

- To create an adequate spark across the gap of sparking plug high electrical pressure is needed. Electrical pressure is measured in volts and the 12 volts supplied by the battery is totally inadequate.

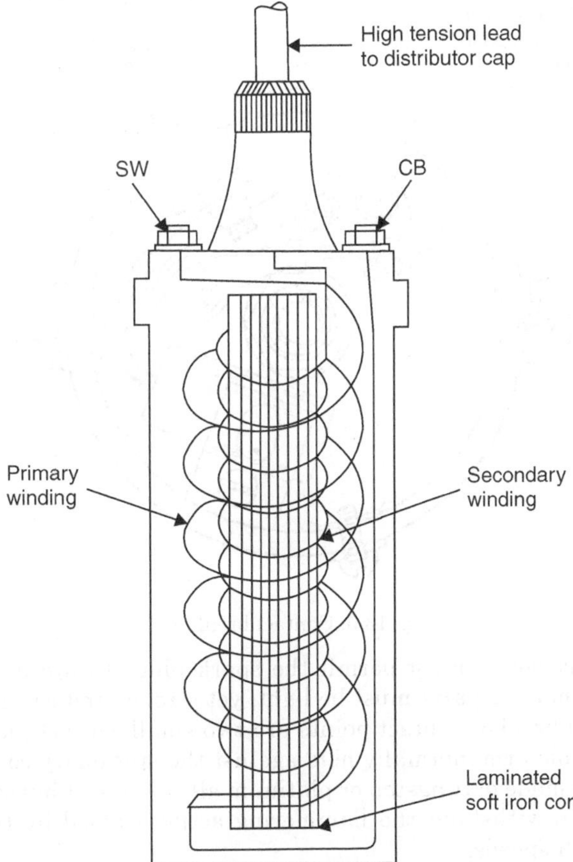

Fig. 13.4. The induction coil.

- *The function of ignition coil is to increase the voltage between 10,000 and 15,000 volts in some conditions*, although the voltage which occurs under normal running conditions is of the order of 4000–5000 volts.
- Two coils of insulated wire are wound on a laminated soft iron core. The *inner coil*, called the *secondary*, has more turns than the outer primary coil. There are about 20000 turns on the secondary and 400 turns on the primary.
- If a low voltage passing through the primary coil is switched off a higher voltage is induced in the secondary coil, the *increase being approximately in the same proportion to the number of turns of the two coils*. The core and windings are placed in an iron sheath. The entire assembly being housed in a sealed container (Fig. 13.4).
- A high tension lead from the centre of the coil carries the supply to the distributor. Two small terminals are situated either side of the high tension lead, one being connected to the contact breaker and marked CB and the other to the ignition switch identified by the letters SW.

3. **Contact breakers :**

- The distributor unit used on the modern motor vehicle contains the breaker contacts which make and break the primary current, and the distributor mechanism which supplies the secondary current to the plugs in their proper firing order. Fig. 13.5 shows how the distributor is connected into the typical battery ignition system.

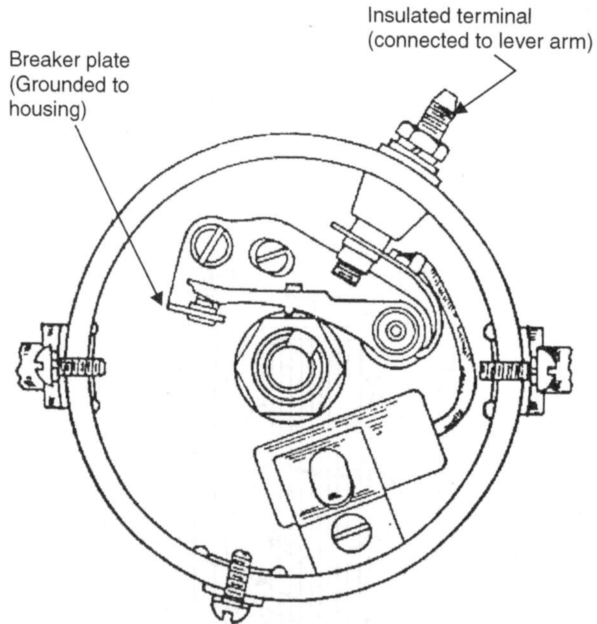

Fig. 13.5. Contact breakers.

- *Breaker contact points* must permit the spark plugs to fire accurately at high speed. Therefore the moving parts must be light, yet strong, and carefully built of high-grade materials. The breaker contact points are two small contact pieces one stationary and one on a movable arm, normally held against the stationary contact by spring tension. The points are made of tungsten or platinum alloy to resist burning and pitting, and are hard enough to withstand the hammering action caused by the rapid closing of the breaker at high speeds.

- A small cam, with a lobe for each time the breaker is required to open per revolution, revolves and pushes the movable breaker arm so that the contact points are separated. In this way, the current in the primary winding of the coil is interupted every time a spark is required in one of the cylinders.
- The usual circuit breaker used on modern vehicles is the closed-circuit type (Fig. 13.6). The contact points normally remain in the closed position, *being separated only when the breaker arm is lifted by a lobe of the breaker arm.* The closed circuit type is more adaptable to high speed engines because the points are in contact long enough to allow complete magnetisation of high tension coil. This results in especially good sparks at slow starting speeds, with less intense sparks at higher speeds when the time of the contact is shortened.

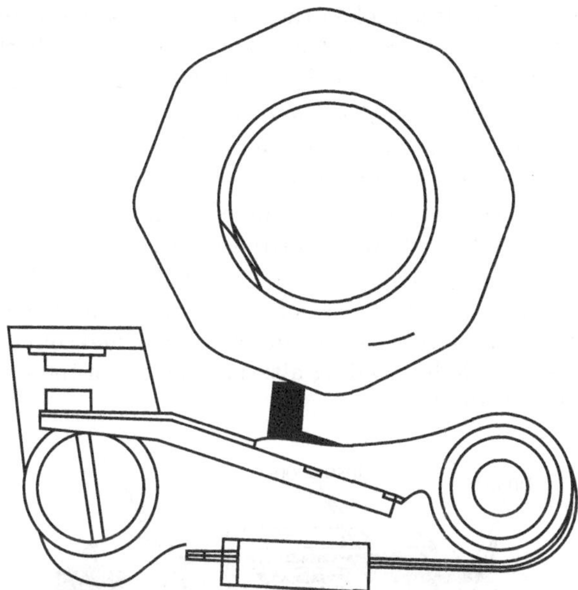

Fig. 13.6. Closed circuit type circuit breaker.

4. The distributor :

- In order for the coil to produce a high voltage from a low voltage supply the flow of electricity must be interrupted (switched off and on). The distributor contains the contact breaker and a cam, which is of rotary switch, that interrupts the supply. The *distributor cap* has a centre terminal which connects with the high tension terminal of the induction coil, and as many terminals equally spaced around it as there are spark plugs to fire. The cap is usually moulded of a highly resistant insulating material, such as bakelite or condensite which is moisture proof and possesses high insulating properties even under excessive heat. The terminals are of brass or metal alloy moulded in position, terminating on the underside either in the form of a button flush with the surface or in the form of a pin. The distributor head is usually held in place by two spring clips which snap on only when the head is in its proper position. Thus the head can be easily removed to inspect and adjust the rotor and breaker mechanism with no chance of replacing it incorrectly. The *rotor* or *distributor arm* is mounted on the upper end of the distributor shaft, on which the breaker cam is also located. The inside end of rotor makes contact with the centre terminal of the head, while the outside end in its rotation completes the circuit successively with the terminals leading to the spark plugs.

- The distributor may be considered as a revolving switch located in the secondary circuit, which connects the high-tension wire from coil to the proper spark plug at proper time.
- The distributor rotor must be well insulated to prevent grounding of the high-tension current, consequently it is usually moulded from an insulating material similar to that used in the distributor cap.
- It is designed to fit over the end of the liner shaft in one position only, to prevent its installation in a position which would throw it out of time with the breaker.
- The distributor may be of gap type or contact type ; classified according to the method used in completing the circuit between the rotor and the distributor head terminals.

5. **The condenser :**

- When the primary circuit is broken the current tends to continue flowing and a spark jumps across the separated contact breaker points. This effect is characteristic of an inductive circuit and is undesirable for the following *reasons* :
  (*i*) *It causes arcing, burning of the contact breaker points.*
  (*ii*) *The effectiveness of the coil is reduced since a sudden interruption of the circuit is essential to produce a high voltage.*
- The condenser absorbs and stores this inductive flow of current and causes the current in the coil to die away rapidly which increases the voltage in the secondary coil.
- A condenser is constructed strips of sheets of tin foil insulated by thin sheets of paraffined paper or mica. The alternate layers of tinfoil are connected in parallel, forming two groups, each group provided with a terminal for external connection. Rolled or cylindrical condensers are made by winding alternate layers of the tin foil and the insulation into a tight roll (Fig. 13.7).

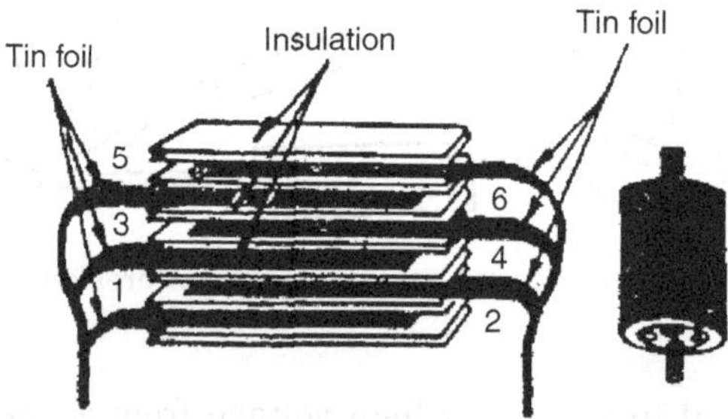

Fig. 13.7. The condenser.

Maintenance of ignition system :

Induction coil. (*i*) It is only necessary to keep the connections clean and tight.

(*ii*) Never leave the ignition switched on without the engine running. If the contact breaker points have come to rest in the closed position the coil will overheat and burn out.

Distributor. (*i*) Provide light grease or clean engine oil on the cam, but make sure it does not get into the contact faces of the contact breakers.

(*ii*) Wipe inside of distributor cap and the electrodes with a clean cloth.

(*iii*) Remove the rotor arm and inject light engine oil into the rotor arm spindle to lubricate the cam bearing.

(*iv*) Check that if the carbon brush is clean and moves freely in its holder.

(*v*) Burnt or blackened contact breaker points should be cleaned with a thin carborudum stone. Care must be taken that these surfaces are kept flat ; ensuring that the entire area of the points is in contact. Remove grease or metallic dust with a petrol moistened cloth.

(*vi*) Breaker gaps should be periodically checked, and if necessary reset.

(*vii*) If the engine does not fire disconnect the centre high tension lead and hold close to the cylinder head. Arrange the cam so that the points are closed, each time the points are separated, with the ignition switch on, a good spark should be obtained. This is only a rough check since the spark does not occur under the conditions preventing in the cylinder.

Condenser. The following symptoms may indicate a faulty condenser although they are usually checked by special equipment :

(*i*) Severe misfiring causing explosion in the silencer and carburettor.

(*ii*) Engine refuses to perform satisfactorily under load.

(*iii*) Arcing at contact breaker points causing blackened and pitted surfaces.

(*iv*) Weak spark.

Check connections and if fault still occurs replace the condenser.

High tension cables. Check that connections are clean and secure, and that the cable insulation is free from cracks.

Plugs. (*i*) Inspect and clean every 3000–4800 km and replace every 16000 km. When setting the gap adjust the side electrode and check with a wire gauge for preference.

(*ii*) It is not satisfactory to check the action of a plug by lying it on the cylinder head and rotating the engine. This is because it is not subjected to working conditions.

(*iii*) Plugs should be checked using a special machine which stimulates cylinder conditions and automatically gives the condition of the plugs.

13.5. MAGNETO IGNITION SYSTEM

- The magneto ignition system is similar in principle to the battery system except that the magnetic field in the core of the primary and secondary windings is produced by a *rotating permanent magnet* (Fig. 13.8). As the magnet turns, the field is produced from a positive maximum to a negative maximum and back again. As this magnetic field falls from a positive maximum value, a voltage and current are induced in the primary winding. The primary current produces a magnetic field of its own which keeps the total magnetic field surrounding the primary and secondary windings approximately constant. When the permanent magnet has turned for enough so that its contribution to the total field is strongly negative, the breaker points are opened and the magnetic field about the secondary winding suddenly goes from a high positive value to a high negative value. This induces a high voltage in the secondary winding which is led to the proper spark plug by the distributor.
- The magneto is an *efficient, reliable, self contained unit which is often preferred for aircraft engines because storage batteries are heavy and troublesome.* Special starting means are required, however, as the magneto will not furnish enough voltage for ignition at low speeds. Variation in ignition timing is more difficult with the magneto, since the breaker point must be opened when the rotating magnets are in the most favourable position. It is possible to change the engine crank angle at which the magneto points open without disturbing the relationship between point opening and magnet position by designing the attachment pad so that the entire magneto body may be rotated a few degrees about its own shaft. Obviously this method is *not as satisfactory as rotating a timer cam-plate.*

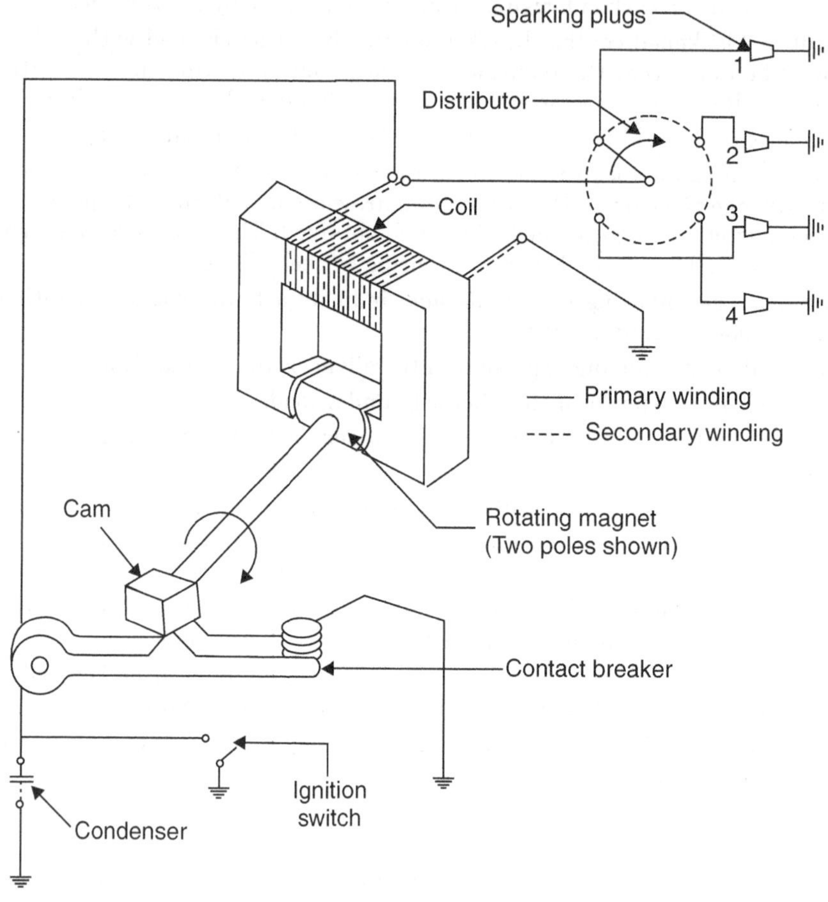

Fig. 13.8. Magneto ignition system.

Advantages :

1. The system is more reliable as there is no battery or connecting cable.
2. The system is more suitable for medium and very high speed engines.
3. With use of cobalt steel and nickel-aluminium magnet metals very light and compact units can be made which require very little room.
4. With recent development this system has become fairly reliable.

Disadvantages :

1. At low speeds and during cranking the voltage is very low. This has been overcome by suitable modifications in the circuit.
2. Adjustment of the spark timing *i.e.* advance or retard, has deterimental effect upon the spark voltage or energy.
3. The powerful sparks at high engine speeds cause burning of the electrodes.

Low tension magneto ignition system :

In a high tension magneto ignition system the main shortcoming is that the wirings carry a very high voltage current which may lead *to misfiring of the engine due to leakage*. This trouble can be avoided by using *low tension magneto system*. In this system the secondary winding is changed to limit the secondary voltage to *about 400 V and the distributor is replaced by a brush contact.* A step-up transformer is used to get high voltage.

Comparison Between Battery Ignition System and Magneto Ignition System

- In a **battery ignition system** a *6-12 V battery is used to provide primary voltage, and a separate ignition coil is required to boost up this voltage needed to operate the spark plug.*
- **Magneto** is a *special type of ignition system with its own electric generator to provide the necessary energy for the system.* It is mounted on the engine and replaces all the components of battery ignition system except the spark plug. The magneto when rotated by the engine is capable of producing a very high voltage and does not need a battery as a source of external energy.

The *differences between the battery ignition system and magneto ignition systems* are given in the table below :

S. No.	Aspect	Battery ignition system	Magneto ignition system
1.	*Current for primary circuit*	Obtained from battery.	Generated by the magneto.
2.	*Starting*	Difficult to start when battery is in discharged condition.	No problem of battery discharge.
3.	*Maintenance problems*	More, due to battery.	Less, due to absence of battery.
4.	*Intensity of spark at low speed*	Good	Poor
5.	*Efficiency*	Efficiency of the system *decreases* with the reduction in spark intensity as the *engine speed increases.*	Efficiency of the system *improves* as the engine speed increases due to high intensity spark.
6.	*Space occupied*	More	Less
7.	*Application*	In cars and light commercial vehicles.	In racing cars and two wheelers, aircrafts etc.

- Fig. 13.8(*a*) shows the plot between the breaker current vs. speed for battery and H.T. magneto ignition systems.

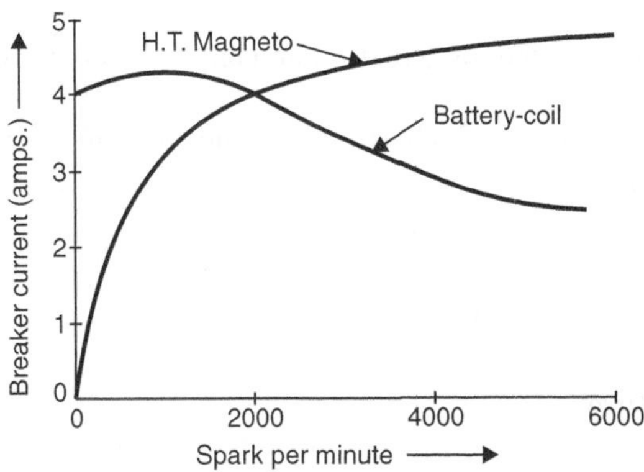

Fig. 13.8(*a*). Breaker current *vs* speed for battery-coil and magneto ignition systems.

— It can be observed that at start, in case of magneto ignition system, the current generated is very low since the starting cranking speed is low. The current increases with increase in the engine speed. Owing to the poor starting characteristics and effect of the spark timing changes on voltage generated, invariably battery ignition system is preferred to the magneto ignition system.

— However, the *battery ignition system is heavier and requires more maintenance than magneto ignition system.*

13.6. FIRING ORDER

Firing order *is the order in which various cylinders of a multi-cylinder engine fire. The firing order is arranged to have power impulses equally spaced, and from the point of view of balancing.*

Firing orders for various engines are given below :

No. of cylinders	Firing order
Two	1, 2
Three	1, 3, 2
Four	1, 4, 3, 2 or 1, 3, 4, 2
Six	1, 5, 3, 6, 2, 4 or 1, 4, 2, 6, 3, 5
Eight	1, 6, 2, 5, 8, 3, 7, 4 or 1, 8, 7, 3, 6, 5, 4, 2.

Eight (Vee) : Either of the following alternatives.

(A) 1*L*, 1*R*, 4*L*, 4*R*, 2*R*, 3*L*, 3*R*, 2*L*

(B) 1*L*, 4*R*, 2*R*, 2*L*, 3*R*, 3*L*, 4*L*, 1*R*

(C) 1*L*, 3*L*, 2*R*, 4*R*, 3*R*, 2*L*, 4*L*, 1*R*

(D) 1*L*, 4*R*, 4*L*, 2*L*, 3*R*, 3*L*, 2*R*, 1*R*

(E) 1*L*, 3*L*, 3*R*, 2*L*, 2*R*, 1*R*, 4*L*, 4*R*

L and *R* indicate cylinder on left and right hand side respectively. The firing order for a four-stroke engine with its cylinder numbered consecutively from 1 to n will be 1, 3, 5, 7, ... to n for one revolution of the crankshaft and 2, 4, 6, 8, ... to $(n - 1)$ for the next revolution.

13.7. IGNITION TIMING

- **Ignition timing** *is the correct instant for the introduction of spark near the end of compression stroke in the cycle. The ignition timing is fixed to obtain maximum power from the engine.*
- The correct instant for the introduction of a spark is mainly determined by the *ignition lag.*

Some of the important factors which affect ignition timings are :

(*i*) Compression ratio

(*ii*) Engine speed

(*iii*) Mixture strength

(*iv*) Combustion chamber design

(*v*) Throttle opening

(*vi*) Engine temperature

(*vii*) The type of fuel.

Spark Advance Mechanisms :

It is of significant importance that the point in the cycle where the spark occurs must be regulated *automatically* to ensure maximum power and economy at different loads and speeds.

Most of the engines are fitted with a mechanism which is integral with the distributor and automatically regulates the optimum spark advance to account for change of load and speed. The following *two mechanisms* are used :

1. Vacuum spark advance.
2. Centrifugal spark advance.

1. **Vacuum spark advance :**

- It is necessary to have vacuum advance of the spark timing since the lean mixtures require an earlier spark timing than the rich mixtures. Thus with the closure of the throttle, the spark must be advanced to get optimum performance.
- The ignition advance is obtained with the help of a spring loaded diaphragm connected to the venturi.
  - So long as the engine idles no advance is required. But when the throttle is opened vacuum is reduced, causing the diaphragm to move inwards. The diaphragm is so coupled with the contact breaker that when induction pressure is reduced, the angle of advance increases.
- This type of mechanism is strictly an *economy device* (when car is driven properly, increased fuel economy is obtained).

2. **Centrifugal spark advance :**

- The centrifugal spark advance is essential to compensate for the increase in speed of the engine.
- This mechanism is obtained by connecting the contact breaker cam to the driving shaft through a coupling which incorporates two centrifugal weights. With increase in engine speed, the weights fly out causing the cam to turn relative to the driving shaft in the direction of rotation and eventually affecting spark advance.

13.8. SPARK PLUGS

- *The main* **function** *of a spark plug is to conduct the high potential from the ignition system into the combustion chamber. It provides the proper gap across which spark is produced by applying high voltage, to ignite the combustion chamber.*
- A spark plug entails the following *requirements* :
  (*i*) It must withstand peak pressures up to atleast 55 bar.
  (*ii*) It must provide suitable insulation between two electrodes to prevent short circuiting.
  (*iii*) It must be capable of withstanding high temperatures to the tune of 2000°C to 2500°C over long periods of operation.
  (*iv*) It must offer maximum resistance to erosion burning away of the spark points irrespective of the nature of fuel used.
  (*v*) It must possess a high heat resistance so that the electrodes do not become sufficiently hot to cause the preignition of the charge within the engine cylinder.
  (*vi*) The insulating material must withstand satisfactorily the chemical reaction effects of the fuel and hot products of combustion.
  (*vii*) Gas tight joints between the insulator and metal parts are essential under all operating conditions.
  - The spark plug (Fig. 13.9) consists of a metal shell having two electrodes (oftenly made of nickel alloy with alloying elements of tungsten silicon or chromium) which are insulated from each other with an air gap. High tension current jumping from the supply electrode produces the necessary spark.

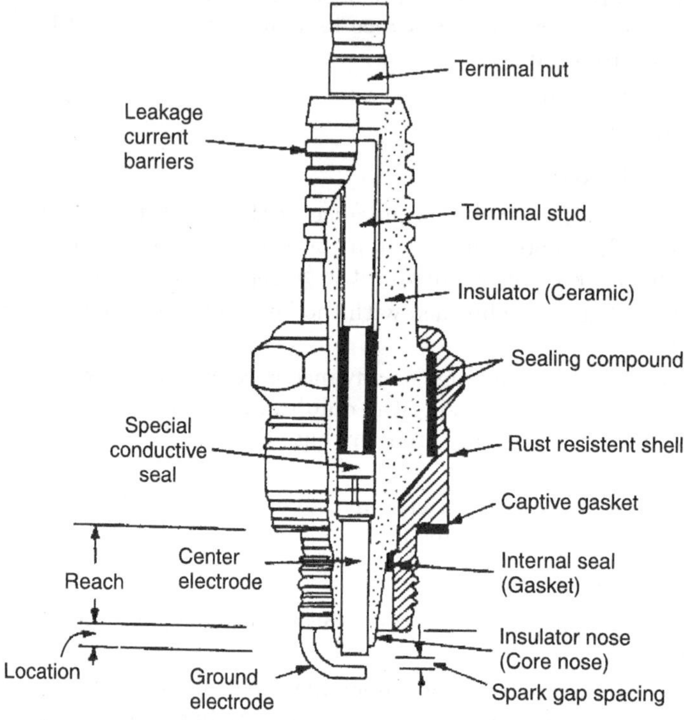

Fig. 13.9. Spark plug.

— Plugs are sometimes identified by the heat range or the relative temperature obtained during operation. The correct type of plug with correct width of gap between the electrodes are important factors.

— The spark plug gap can be easily checked by means of a feeler gauge and set as per manufacturer's specifications.

— It is most important that while adjusting the spark plug it is the outer earthed electrode *i.e.*, tip which is moved in or out gradually for proper setting of the gap. No bending force should be applied on the centre-electrode for adjusting the gap as this can cause crack and fracture of insulation and render the plug absolutely useless.

— Porcelain is commonly used as insulating material in spark plugs, as it is cheap and easy to manufacture. Mica can also be used as insulating material for spark plugs. Mica, however, cannot withstand high temperatures successfully.

Factors affecting establishment of arc across the air gap of the spark plug :

The problem of establishing an arc across the air gap of the spark plug is *affected by the following factors* :

(*i*) Air gap length. Greater the air gap, greater is the breakdown voltage.

(*ii*) Gap geometry. For pointed small electrodes less breakdown voltage is required.

(*iii*) Mixture density. High densities (*i.e.* high throttle openings) require higher breakdown voltages. It follows the **Paschen's law** ; $V = K\rho h$, where K, ρ and h are a constant (for given substance and electrode shape), density of the material and gap distance respectively.

(*iv*) Electrode temperature. High temperatures lead to low breakdown voltage.

(*v*) **Insulator's leakage resistance.** The carbon and metallic oxides from electrically-conductive coating on the insulator which, thus, shunt the secondary winding and reduce the maximum voltage that the secondary can impress across the spark plug.

(*vi*) **Voltage increase rate at the gap.** When the voltage is built up at a rapid rate by the ignition system, the effect of leakage gets minimized, and greater voltage is available.

(*vii*) **Fuel-air ratio.** The fuel-air ratio fixes the electrical properties of the mixture ; lean mixtures have higher breakdown voltage than those of slightly rich mixtures.

(*viii*) **Electrode material.**

(*ix*) **The presence of ionised gases in the gap.**

Difference between hot spark plug and cold spark plug :

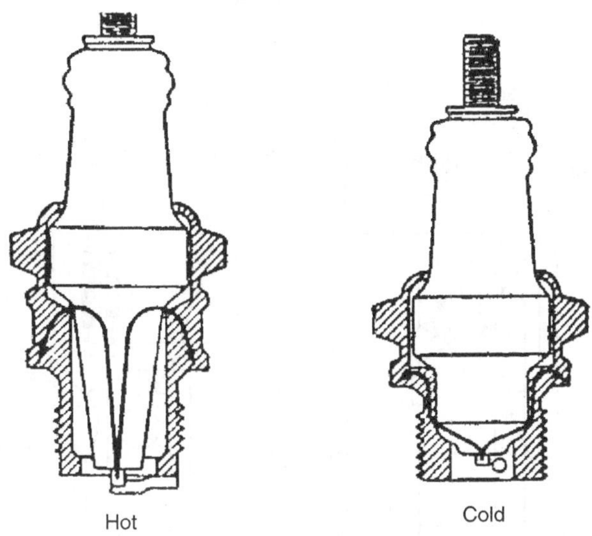

Fig. 13.10. Hot and cold spark plug.

The difference between hot and cold spark plugs is due to the *relative operating temperature range of the tip of the high tension electrodes.* The operating temperature is governed by the amount of heat transferred which depends upon the heat transfer path from the tip to the cylinder head and the amount of surface area exposed to the combustion gases. A **cold plug** *has a short heat transfer path and a small area exposed to the combustion gases, as compared to a* **hot plug**.

13.9. LIMITATIONS OF CONVENTIONAL IGNITION

The conventional ignition system entails the following *limitations* :

1. Frequent servicing and replacement of contact points.
2. Weaker sparks (or sometimes misfiring).
3. Inefficient at low speeds.
4. Poor starting ability.
5. Reduced ability to fire fouled plugs (since most of the spark energy is lost as current flows through low-resistance fouling deposits).
6. Poor reproducibility of the secondary voltage rise, peak voltage and firing time (owing to inherent erratic operation of the mechanical points).

13.10. ELECTRONIC IGNITION SYSTEMS

The limitations / disadvantages of conventional ignition system has led to the development of new ignition system concepts using *solid state devices*. In modern automobiles the following two types of electronic ignition systems are employed :

1. Transistorised coil ignition (TCI) system.
2. Capacitive discharge ignition (CDI) system.

These systems are described below :

1. **Transistorised coil ignition (TCI) system :**

- TCI system provides higher output voltage and uses electronic triggering to keep the required timing.
- It is categorised as high energy electronic ignition system.
- The circuit diagram of this system is shown in Fig. 13.11.

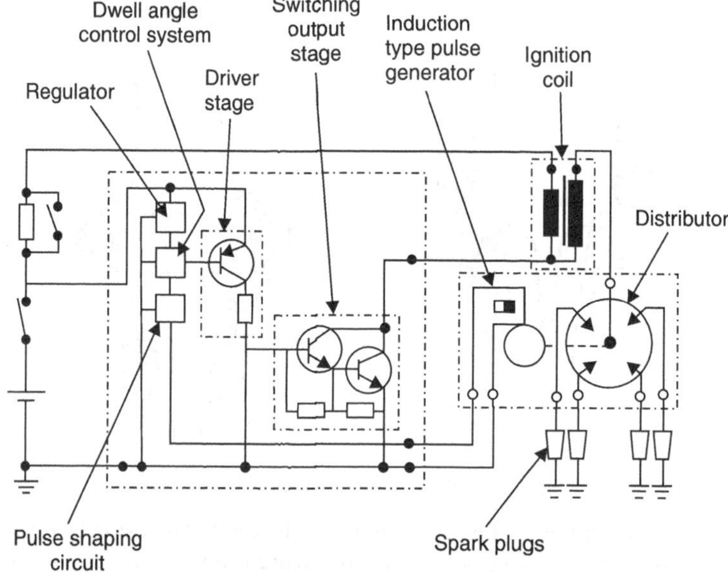

Fig. 13.11. Circuit diagram of TCI system.

— The function of *magnetic pulse generator* (*more oftenly used pulse generator*) *system is to detect the position of the distributor shaft and transmit an electrical pulse to an electronic control model.* This system replaces the contact breaker and the cam assembly of the conventional ignition system.

— The **module** switches off the flow of current inducing a high voltage in the secondary winding which is distributed to the spark plugs in the same fashion as in conventional breaker system.

— **Timing circuit** is contained in the **control model** that closes the primary circuit later so that the primary current build-up can occur for the next cycle.

Advantages :

The advantages of TCI system are :

1. Reduced wear of the components.
2. No problem faced in burning lean mixtures.

3. Reduced maintenance of the ignition system.
4. Increased reliability.
5. Increased spark plug life.

2. Capacitive discharge ignition (CDI) system :

Fig. 13.12 illustrates the capacitive discharge ignition system.

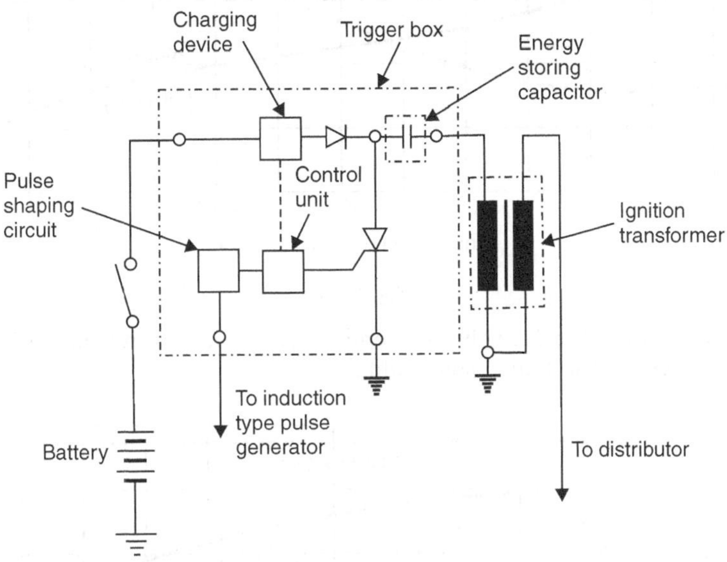

Fig. 13.12. Circuit diagram of CDI system.

- In this system a **capacitor** is used to store the energy, induction coil is *not* used. The magnitude of stored energy depends on the capacitance and the charging voltage of the capacitor.
- The primary voltage developed at the time of spark by the discharge of capacitor is boosted up by the **ignition transformer,** through the thyristor, to the high voltage required at the spark plug.
- The CDI **trigger box** comprises the *capacitor, thyristor power switch, charging device* to convert battery voltage to the charging voltage of 300 to 500 V, by means of pulses *via,* the voltage transformer, pulse shaping unit and control unit.

Advantages :

The *advantages of CDI system are* :

1. The capacitor in this system can store several thousands times more energy per unit of capacitance compared to that per unit inductance of the conventional TCI system. This permits a high output voltage even at high spark rates (speeds).

2. Due to the internal resistance being very small (about 50 Ω), faster voltage rise is obtainable and consequently the system is *relatively insensitive to side tracking.*

3. As in TCI system, the breaker points serve as a trigger only. This avoids frequent maintenance and increases life of the contact points.

4. As evident from Fig. 13.13, at low engine speeds in CDI systems low current is drawn from the battery and with increase in speed the current drawn also increases, unlike what happens in conventional or TCI system where high current is drawn at low speeds. This characteristic of CDI system *increases the low speed efficiency and allows easier cold starting.*

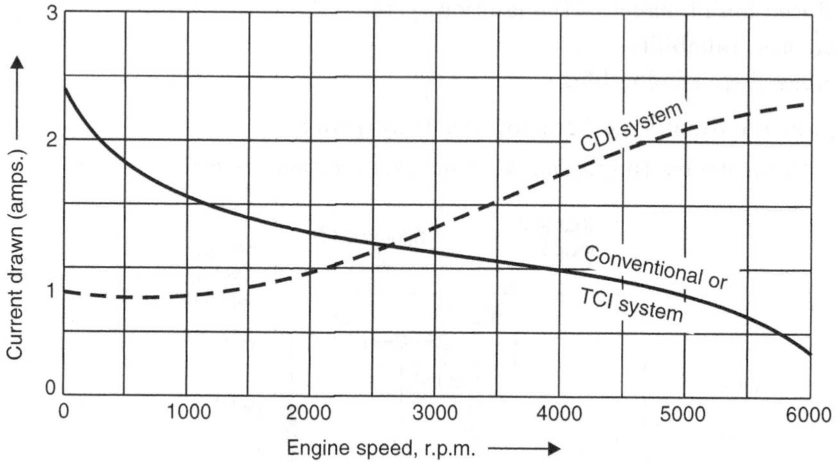

Fig. 13.13

5. The output voltage is relatively independent of engine speed (Fig. 13.14), this eliminates chances of misfiring of even the fouled spark plugs.

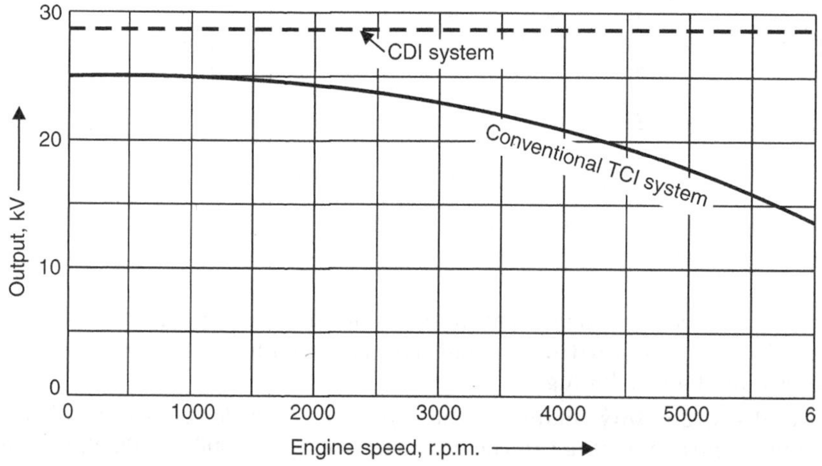

Fig. 13.14

Disadvantages :

The main *disadvantage* of CDI system is very rapid discharging of capacitor leading to strong spark but of *short duration* of about 0.1 ms. This may *cause ignition* problems when operating with rich mixtures.

HIGHLIGHTS

1. The *basic ignition systems* in uses are :
   (*i*) Battery ignition system ;
   (*ii*) Magneto-ignition system ;
   (*iii*) Electronic ignition system.

2. The *main components* of a battery ignition system are :
   (*i*) Battery (6 or 12 V)
   (*ii*) Ignition switch
   (*iii*) Induction coil
   (*iv*) Circuit/contact breaker
   (*v*) Condenser
   (*vi*) Distributor.
3. The *battery ignition system* is commonly used because of its combined cheapness, convenience of maintenance, attention and general suitability.
4. The *magneto* is an efficient, reliable, self contained unit, which is often preferred for aircraft engines because storage batteries are heavy and troublesome.
5. *Firing order* is the order in which various cylinders of a multi-cylinder engine fire. The firing order is arranged to have power impulses equally spaced, and from the point of view of balancing.
6. *Ignition timing* is the correct instant for the introduction of spark near the end of compression stroke in the cycle. The ignition timing is fixed to obtain maximum power from the engine.
7. The main function of a *spark plug* is to conduct the high potential from the ignition system into the combustion chamber. It provides the proper gap across which spark is produced by applying high voltage to ignite the combustion chamber.
8. A *cold spark plug* has a short heat transfer path and a small area exposed to the combustion gases, as compared to a hot spark plug.
9. In modern automobiles, the following two types of electronic ignition system are employed :
   (*i*) Transistorised coil ignition (TCI) system ;
   (*ii*) Capacitor discharge ignition (CDI) system.

OBJECTIVE TYPE QUESTIONS

Fill in the Blanks or Say "Yes" or "No" :

1. Ignition is only a pre-requisite of ....
2. A spark energy below 10 millijoules is adequate to initiate combustion A / F ratio 12-13 : 1.
3. Battery ignition system is rarely used.
4. A ballast resistor is provided in series with the primary winding to regulate .... current.
5. In a battery ignition system the sparking voltage .... with increase in engine speed.
6. The initial cost of battery ignition system is low.
7. The magneto-ignition system is more reliable than battery ignition system as there is no battery or connecting cable.
8. .... is a special type of ignition system with its own electric generator to provide the necessary energy for the system.
9. Battery ignition system occupies less space as compared to magneto-ignition system.
10. The battery ignition system is heavier and requires more maintenance that magneto-ignition system.
11. .... order is the order in which various cylinders of a multi-cylinder engine fire.
12. .... is the correct instant for the introduction of spark near the end of compression stroke in the cycle.
13. The main function of a .... is to conduct the high potential from the ignition system into the combustion chamber.
14. The spark plug must withstand pressures upto at least .... bar.
15. A .... plug has a short heat transfer path and a small area exposed to the combustion gases, or compared to .... plug.
16. TCI system provides lower output voltage.
17. TCI system is not much reliable.
18. In TCI system components are subjected to less wear.
19. In CDI system a .... is used to store the energy, induction coil is not used.
20. In CDI system, the output voltage is relatively independent of engine speed.

ANSWERS

1. combustion	**2.** Yes	**3.** No	**4.** primary	**5.** decreases
6. Yes	**7.** Yes	**8.** Magneto	**9.** No	**10.** Yes
11. Firing	**12.** Ignition timing	**13.** spark plug	**14.** 55	**15.** cold, hot
16. No	**17.** No	**18.** Yes	**19.** capacitor	**20.** Yes.

THEORETICAL QUESTIONS

1. What do you mean by the term "Ignition" ? How is it related with "combustion" ?
2. What are the requirements of an ignition system for an I.C. engine ?
3. Enumerate the basic ignition systems and describe any of them.
4. Describe with the help of a neat sketch a battery ignition system.
5. State the functions of an ignition coil and a condenser in the battery ignition system of a mutli-cylinder S.I. engine.
6. Name the various components of a battery ignition system and explain any three of them briefly.
7. State the advantages and disadvantages of a battery ignition system.
8. Describe a high tension magneto-ignition system. Also state its advantages and disadvantages.
9. What is the difference between 'ignition timing' and 'firing order' ?
10. What are the differences between battery and magneto-ignition systems ?
11. What do you understand by 'ignition timing' ? Enumerate the various factors which affect ignition timings.
12. What is the main difference between the battery and electronic systems ?
13. List various electronic ignition systems in use. Describe any one of them clearly stating its advantages over the conventional ignition system.
14. Differentiate between hot spark plug and cold spark plug.
15. Explain briefly spark advance mechanisms.
16. Explain briefly with a neat sketch the 'Transistorised coil ignition' (TCI) system.
17. How does 'capacitive discharge ignition system' differ from ' Transistorised coil ignition system' ?

14

Engine Friction and Lubrication

14.1. INTRODUCTION

- In an I.C. engine almost all machine parts have relative motion and rub against each other. To reduce this rubbing action lubrication is required, which increases the life of engine. The purpose of lubrication in I.C. engine is generally *two-fold* : (*i*) To reduce the rubbing action between different machine parts having relative motion with each other ; (*ii*) To remove the heat generated inside the cylinder.
- **Engine friction** *is defined as the difference between the indicated power (I.P.) (power developed inside the engine) and the brake power (B.P.) (power available at the crank-shaft) i.e.*

  Frictional power, F.P. = I.P. – B.P. ...(14.1)
- It is impossible to totally remove all the friction loss but it can be reduced by using lubrication between the parts which have relative motion with each other. *Increase in friction is ultimately dissipated as heat to the cooling water and it further increases the pump and fan power requirements also.*
- The frictional resistance between two moving parts having relative motion is mostly dependent on the following *factors* :
  - Lubricating oil properties
  - Surface condition
  - Materials of the surfaces
  - Rate of relative motion
  - Nature of relative motion
  - Quantity of lubricating oil.

14.2. TOTAL ENGINE FRICTION

The difference between I.P. and B.P. is known as ***total engine friction loss.*** This includes the following losses :

1. Direct frictional losses
2. Pumping loss
3. Blowby losses
4. Valve throttling losses
5. Combustion chamber pump loss
6. Power loss to drive the auxiliaries.

1. **Direct frictional losses :**

It includes bearing losses (main bearing, camshaft bearing), piston and cylinder friction loss etc. In reciprocating I.C. engines the frictional losses are comparatively higher.

2. **Pumping loss :**

- In *four-stroke* cycle engines an ample amount of power is used during intake and exhaust processes (Fig. 14.1).

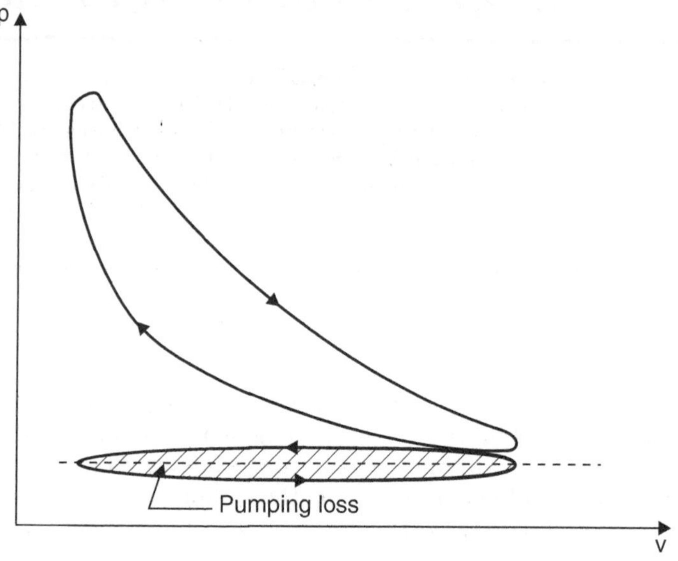

Fig. 14.1

- The pumping loss is *negligible* in *two-stroke* cycle engines since the incoming fresh mixture is used for scavenging the exhaust gases and charging the cylinder.

3. **Blowby losses :**

These losses are *caused due to the leakage of combustion products past the piston from the cylinder into the crankcase.*

— These losses *depend upon the inlet pressure and compression ratio.*

— These losses *increase directly with compression ratio but get reduced with an increase in engine speed.*

4. **Valve throttling losses :**

- The standard practice for sizing the *exhaust valve* is to make them a *certain percentage smaller than the inlet valves.* This usually results in an insufficiently sized exhaust valve and hence, results in exhaust pumping loss.
  — If due attention is not given to the valve size, valve timing and valve flow coefficients there may be a substantial loss with the increase in engine speed.
- The *inlet throttling loss* occurs due to the *restrictions* imposed by air cleaner, carburettor venturi, throttle valve, intake manifold and intake valve. All these restrictions lead to pressure loss. Similarly some pressure loss is necessary to exhaust the combustion products.

5. **Combustion chamber pump loss :**

This type of loss is caused due *to the pumping work required to pump gases into and out of the pre-combustion chamber*. Its exact value depends upon the orifice size (connecting the fore-combustion chamber and the main chamber), and the speed. *Higher the speed greater is the loss and smaller the orifice size greater is the loss.*

6. **Power loss to drive the auxiliaries :**

In order to drive auxiliaries such as water pump, oil pump, fuel pump, cooling fan and generator some power is needed. This is also considered as loss since a part of engine power developed is used for these purposes.

14.3. EFFECT OF ENGINE PARAMETERS ON ENGINE FRICTION

Following *parameters exercise their influence on engine friction* as described below :

1. Stroke-to-bore ratio
2. Cylinder size and number of cylinders
3. Piston rings
4. Compression ratio
5. Engine speed
6. Engine load
7. Cooling water temperature
8. Oil viscosity.

1. **Stroke-to-bore ratio**

- The effect of this parameter on engine friction and economy is not very significant.
- Lower stroke-to-bore ratio tends to decrease i.m.e.p. Its lower value reduces the friction losses as the surface area decreases with decreasing stroke-to-bore ratio with the same value of the stroke volume.

2. **Cylinder size and number of cylinders :**

- When a smaller number of large cylinders are used, the friction and economy improve. This is owing to the fact that the proportion between the working piston area and its friction producing area, *i.e.* circumference is reduced.
- Fig. 14.2 shows the effect of number of cylinders on the variation of friction for the same piston displacement.

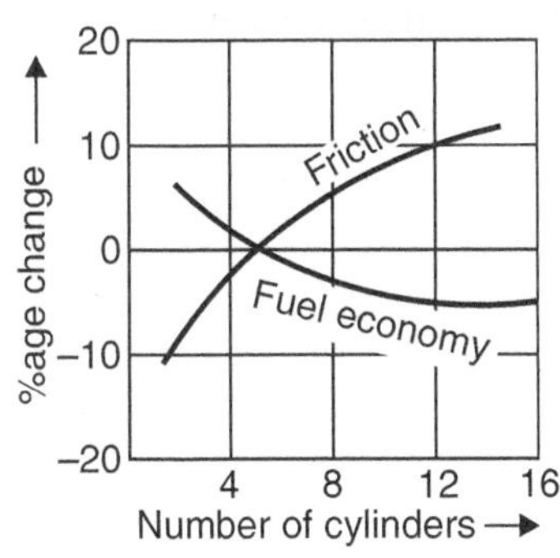

Fig. 14.2. Effect of number of cylinders on friction.

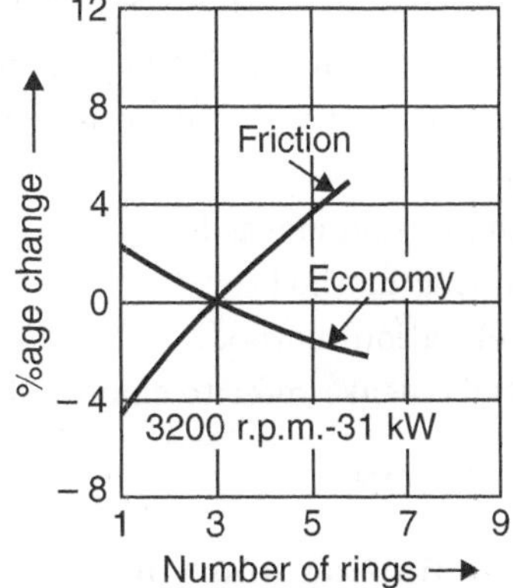

Fig. 14.3. Effect of number of rings on piston friction.

3. **Piston rings :**

The effect of number of piston rings on friction is not significant as the selection depends on the engine size, lightness required and material used for the rings. It is obvious from Fig. 14.3 that generally 3-rings provide best fuel economy.

4. **Compression ratio :**

The frictional mean effective pressure increases as the compression ratio is increased. But the *mechanical efficiency may even increase because of improvement in i.m.e.p.*

5. **Engine speed :**

The *mechanical friction increases with the increase in speed.* From Fig. 14.4. It is evident that there is nearly linear variation in f.m.e.p. with speed.

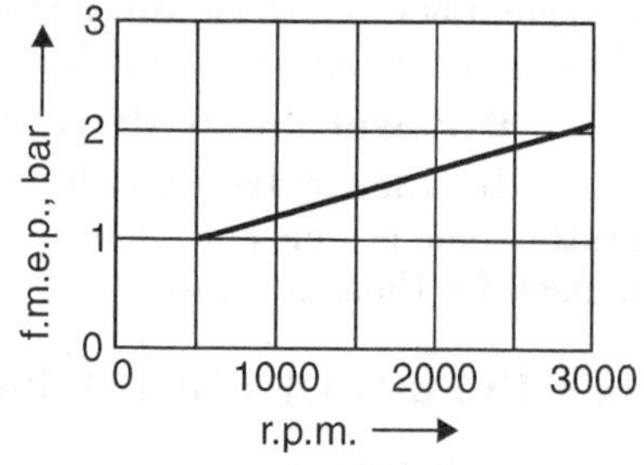

Fig. 14.4. Effect of speed on engine friction.

6. **Engine load :**

When the load on the engine increases, the i.m.e.p. also increases and friction loss also increases. However this increase in friction loss is compensated by decrease in viscosity of the lubricating oil due to higher temperature resulting from increased load.

7. **Cooling water temperature :**

- The *rise in cooling water temperature reduces the frictional loss as the viscosity of oil at higher temperature is lower which reduces the frictional loss.*

8. **Oil viscosity :**

Higher the viscosity of oil greater is the friction loss. As the oil temperature increases, the viscosity decreases and friction losses are reduced during a certain temperature range as shown in Fig. 14.5. If the temperature exceeds a certain value the local film is destroyed causing metal-to-metal contact.

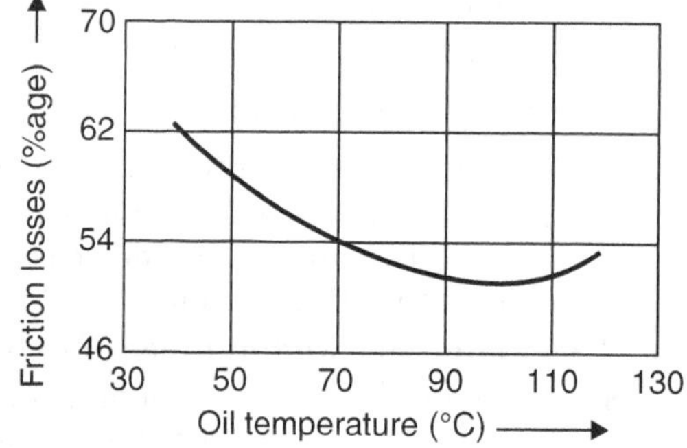

Fig. 14.5. Effect of oil temperature on friction.

14.4. DETERMINATION OF ENGINE FRICTION

The engine friction can be determined by the following five methods :

1. From I.P. and B.P. measurements
2. Morse test
3. Willian's line method
4. Motoring method
5. Deceleration method.

For details please refer to chapter 17.

14.5. LUBRICATION

14.5.1. Definition and Objects

Lubrication *is the admittance of oil between two surfaces having relative motion. The objects of lubrication may be one or more of the following :*

1. To reduce friction between the parts having relative motion.
2. To reduce wear of the moving part.
3. To cool the surfaces by carrying away heat generated due to friction.

4. To seal a space adjoining the surfaces.
5. To absorb shocks between bearings and other parts and consequently reduce noise.
6. To remove dirt and grit that might have crept between the rubbing parts.

14.5.2. Behaviour of a Journal in its Bearing

Fig. 14.6 shows the behaviour of a journal rotating in a bearing, the clearance between the two being shown very much exaggerated. The clearance space is supposed to be completely filled with oil at all times, which is possible by supplying the oil as fast as it runs out. If the shaft does not rotate it will sink to the bottom of the clearance space due to the load W, and the journal and bearing will touch as shown in Fig. 14.6 (*a*). The way in which the rotating shaft will build up pressure in the oil sufficient to separate the surfaces is shown in Fig. 4.16 (*b*) and (*c*). As the shaft starts to turn it will climb the bearing wall as shown in Fig. 14.6 (*b*). But as the speed increases, the moving journal tends to put oil into the wedge shaped area between the shaft and the bearing. As a result the oil pressure on right side becomes more than on the left side, and the journal is, therefore, forced away from the bearing wall. An equilibrium position is finally reached as shown in Fig. 14.6 (*c*) with the surfaces separated by a film of oil whose minimum thickness is r_o. The magnitude of r_o and position of the line of centres in Fig. 14.6 (*c*) will depend *on the load, the fluid properties of oil, the size and speed of shaft, the clearance and length of the bearing.*

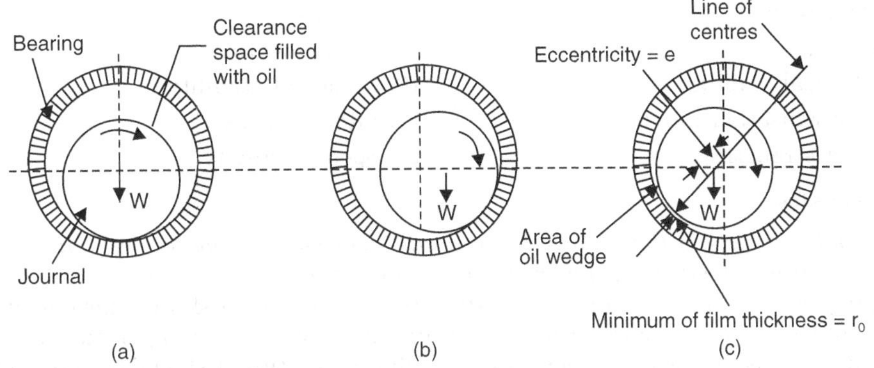

Fig. 14.6. Behaviour of a journal rotating in a bearing.

From above, the following points, about the items influencing bearing performance, can be concluded :

1. A slippery bearing material is desirable during starting and stopping, but once the oil film is established the bearing material is less important.
2. Higher is the speed of journal, more will be the oil pulled into the apex of the wedge of oil in the clearance space of Fig. 14.6 (*c*), and as a result, more supporting pressure will be developed.
3. An increase in supporting pressure will increase the oil film thickness r_o and decrease the eccentricity *e*.
4. If the eccentricity '*e*' is decreased, the supporting pressure must decrease because the sides of the wedge are more nearly parallel.
5. Oil film thickness r_o is also influenced by changing of clearance.

Film Lubrication. It is that type of lubrication in which *bearing surfaces are completely separated by a layer of film of lubricant and that the frictional resistance arises only due to relative movements of the lubricant layers.*

Boundary Lubrication :

- Under the hydro-dynamic condition the oil film supports the load. If the oil film becomes thin enough so as not to support the load without occasional metal to metal contact then journal friction developed is called *boundary friction* and the lubrication existing in this range is known as *boundary lubrication.* The word boundary friction is used because under this condition journal friction is neither completely dry and not completely fluid. In the boundary state, the kind of bearing material, the hardness and surface finish of the shaft, as well as type of lubricant all contribute to the amount of journal friction.
- When the load acting on the bearings is very high, the material itself deforms elastically against the pressure built up of the oil film. This type of lubrication, called **elasto-hydrodynamic** *lubrication*, occurs between *cams and followers, gear teeth, and rolling bearings when the contact pressures are extremely high.*

14.5.3. Properties of Lubricants

The chief qualities to be considered in selecting oil for lubrication are :

1. Viscosity
2. Flash point
3. Fire point
4. Cloud point
5. Pour point
6. Oiliness
7. Corrosion
8. Emulsification
9. Physical stability
10. Chemical stability
11. Neutralisation number
12. Adhesiveness
13. Film strength
14. Specific gravity.

1. **Viscosity**

- *It is the ability of the oil to resist internal deformation due to mechanical stresses and hence it is a measure of the ability of the oil film to carry a load.* A more viscous oil can carry a greater load, but it will offer greater friction to sliding movement of the one bearing surface over the other. Viscosity varies with the temperature and hence if a surface to be lubricated is normally at high temperature it should be supplied with oil of a higher viscosity.
- The viscosity is measured by *viscosimeter.* The important types of viscosimeters are :
  1. Saybolt universal viscosimeter
  2. Red wood viscosimeter
  3. Engler viscosimeter
  4. Barbey viscosimeter.

  The unit of viscosity is given as "*seconds saybolt*" or "*seconds redwood*" etc.
- The present-day method of expressing the rate at which the viscosity of an oil will change with temperature is by stating its *viscosity index* (V.I.). The oil is compared with *two reference oils having same viscosity at 99°C. One is paraffinic base oil (viscosity changes considerably with temperature), is arbitrarily assigned an index of* ***zero*** *and the other, a napthenic base oil (little change in viscosity with temperature) assigned an index of* ***100***.
  - *A high viscosity index indicates relatively smaller changes in viscosity of the oil with the temperature.*
  - The higher the viscosity index, the lower the rate at which its viscosity decreases with the increase in temperature. Although a high viscosity index is desirable in materials and much effort expended in improving the viscosity index of oils by

refining, it should be emphasized that viscosity-temperature characteristics are of little importance for oils that are to function at approximately constant temperature, such as *turbine oils.*

— *Viscosity index of oil is very important where extreme temperatures are encountered.*

- In order to improve the viscosity index of an oil certain compounds, called V.I. improvers, are added to it. These are long chain paraffinic compounds which enable to obtain an oil having easy starting characteristic of thin oils combined with good protection against high temperature.

2. **Flash point.** *It is defined as the lowest temperature at which the lubricating oil will flash when a small flame is passed across its surface.* The flash point of the oil should be *sufficiently high so as to avoid flashing of oil vapours at the temperatures occurring in common use.* High flash point oils are needed in air compressors.

3. **Fire point.** *It is the lowest temperature at which the oil burns continuously.* The fire point also *must be high in a lubricating oil, so that oil does not burn in service.*

4. **Cloud point.** When subject to low temperatures the oil changes from liquid state to a plastic or solid state. In some cases the oil starts solidifying which makes it to appear cloudy. The temperature at which this takes place is called the *cloud point.*

5. **Pour point.** *Pour point is the lowest temperature at which the lubricating oil will pour.* It is an indication of its ability to move at low temperatures. This property must be considered because of its effect on starting an engine in cold weather and on free circulation of oil through exterior feed pipes when pressure is not applied.

6. **Oiliness.** *This is the property which enables oil to spread over and adhere to the surface of the bearing.* It is most important in boundary lubrication.

7. **Corrosion.** A lubricant *should not corrode* the working parts and it must retain its properties even in the presence of foreign matter and additives.

8. **Emulsification.** A lubricating oil, when mixed with water is emulsified and loses its lubricating property. The *emulsification number is an index of the tendency of an oil to emulsify with water.*

9. **Physical stability.** A lubricating oil *must be stable physically at the lowest and highest temperatures between which the oil is to be used.* At the lowest temperature there should not be any separation of solids, and at the highest temperature it should not vapourise beyond a certain limit.

10. **Chemical stability.** A lubricating oil should also be stable chemically. There should not *be any tendency for oxide formation.*

11. **Neutralisation number.** An oil may contain certain impurities that are not removed during refining. The neutralisation number test is a simple procedure to determine acidity or alkalinity of an oil. It is the weight in milligrams of potassium hydroxide required to neutralise the acid content of one gram of oil.

12. **Adhesiveness.** *It is the property of lubricating oil due to which the oil particles stick with the metal surfaces.*

13. **Film strength.** *It is the property of a lubricating oil due to which the oil retains a thin film between the two surfaces even at high speed and load.* The film does not break and the two surfaces do not come in direct contact. Adhesiveness and film strength cause the lubricant to enter the metal pores and cling to the surfaces of the bearings and journals keeping them wet when the journals are at rest and presenting metal to metal contact until the film of lubricant is built up.

14. **Specific gravity.** It is a *measure of density of oil.* It is an indication regarding the grade of lubricant by comparing one lubricant with other. It is determined by a hydrometer which floats in the oil, and the gravity is read on the scale of the hydrometer at the surface of the oil.

14.5.3.1. Additives

Primarily, there are two types of oils : *Mineral* and *vegetable.* Both of them fulfill some basic requirements as lubricating oil. The vegetable oil is superior to mineral oil under some extreme conditions, but it is rarely used as it is too expensive oil. The *mineral oils are very commonly used for all lubricating purposes.*

— Simple mineral oils have most of the required characteristics as a good lubricant. However, varying operating conditions require some specific properties which it cannot meet, as high viscosity index and resistance to oxidation and corrosion. In order to achieve some required properties, different types of compounds, called *additives* are added.

1. **Detergents :**

- Control high-temperature deposits such as gums.
- If overbased, a detergent acts as effective acid neutraliser.

2. **Dispersants :**

- Control low-temperature deposits such as cold sludge and varnish deposits.

3. **Anti-wear additives :**

- These additives reduce wear and prevent scoring, galling and seizure.
- They provide an extra strength needed to ensure efficient lubrication under severe operating conditions.
- The additives used are : *Chlorine* and *phosphate compounds.*

4. **Rust inhibitors :**

- Reduce rusting by acid neutralisation of formation by protective film.

5. **Viscosity index (V.I.) improver :**

- *It prevents undue thinning of the oil as the temperature increases.*
- A high viscosity index is always desired for lubricating oils as it can be used in severe winter and summer conditions. *High molecule polymers* are used for the purpose as they are soluble in the oil.
- Recently *multigrade oils* have been produced which consist of low viscosity oils within thicking polymers. Their special ability is to adopt to both summer and winter conditions.

6. **Pour point additives :**

- The base oil at extremely low temperatures change into buttery solids, due to formation of wax crystals. If left unchecked, the wax crystals cogulate into honeycomb like structure which first inhibit and then totally suppress the oil flow. *These additives reduce pour point of oil by interfering with wax cystallization.*
- *Polymerized phenols or esters are used upto 1 percent of the oil* for the purpose.

7. **Anti-foam agents :**

- Reduce oil foaming by causing collapse of bubbles due to air entrainment.
- *Silicon esters are used as antifoam agents.*

8. **Anti-oxidants :**

- Reduce oil oxidation to protect alloy bearings against corrosive attack.

9. **Oiliness improvement :**

- Some substances such as colloidal graphite and zinc oxide when added to the oil are valuable in maintaining the oil film.

14.5.3.2. Oil contamination and sludge formation

- The lubricating oil after a certain operating period gets contaminated to the extent that it becomes unsuitable for further use. The *contamination* of the oil takes place due to *oxidation, dilution, water, formation of carbon, lead compounds, metals, dust and dirt.* When these contaminants mix with the oil, sludge is formed in an engine.
- **Sludge** is a black, brown or grey deposit having the consistency of mud. *Its formation takes place as a result of operation at low engine temperatures during starting, warming up, and idling periods.*

14.5.4. Types of Lubricants

Most lubricants are *oils* or *greases*. However, in special circumstances other fluids like *water, air etc. and solids* such as *graphite* may perform the function of lubrication. Synthetic lubricants are also used occasionally.

1. **Oils :**

The different lubricating oils are : *Mineral, Fatty* and *Synthetic.*

- The **mineral lubricating oils** are obtained from the residual mass left during crude petroleum distillation. In this treatment gasoline, kerosene oil, and gas oil fractions are recovered from the distillate. Paraffinic crudes generally give a higher yield of lubricating oils than napthenic crudes. The residue is subject to vacuum distillation. In several cases steam is also introduced into the system. Bubble towers are used for fractionalisation of oil into two or three fractions of different viscosities.
- **Fatty oils** from animal and vegetable origin are sometimes used alone but are frequently mixed with mineral oils. The fatty oils exhibit poor keeping quality and undergo decomposition. Fatty oils however, exhibit more oiliness than mineral oils of the same viscosity. Non drying oils such as olive oil, castor oil, rapeseed oil, lard and fish oil are mainly used for lubricating purposes. They are generally compounded with mineral oil to the extent of about 2 to 10 percent in order to increase oiliness of mineral oils. *Olive oil* is employed for lubricating textile machinery, as it can be easily washed out from the cloth without leaving any stain. ***Rapeseed oil*** is used for making lubricants for *railway engines*, while lard oil is used for producing lubricants for *internal combustion engines.* ***Caster oil*** is usually compounded with heavy mineral oils for making *extreme pressure lubricants.*
- **Synthetic lubricants** are named synthetic because they are not obtained directly from petroleum. Various applications require oils which will have, for example, a lower point, or better viscosity-temperature characteristics, or a higher degree of resistance to oxidation or isothermal decomposition than petroleum derived lubricants do. The synthetic lubricants presently in use are *classified* as :

  (*i*) Dibasic acid esters ;
  (*ii*) Organo-phosphate esters ;
  (*iii*) Silicate esters ;
  (*iv*) Silicon polymers ;

(*v*) Polyglycolethers and related compounds ; and

(*vi*) Fluorinated and chlorinated hydrocarbon compounds.

Classification of lubricating oils

- The lubricating oils are *normally classified according to their viscosity.* The SAE (Society of Automotive Engineers) method of assigning number to different oils is used universally.
  - — SAE has assigned a number to an oil whose viscosity at given temperatures falls in certain range. There are two temperatures used as reference for assigning the number to oils – 18°C and 99°C.
  - — SAE, 5 W, 10 W, 20 W grades are defined in terms of viscosity at – *18°C* and are the oils *which render starting of engine in cold weather easy.* SAE, 20, 30, 40 and 50 grades are defined in terms of viscosity at *99°C* ; these oils *work satisfactorily in normal and hot climates.* These numbers are merely used for classification of oil according to viscosity and *do not indicate the quality of* oil since factors like stability, oiliness etc. are not considered (by these numbers).

Table 14.1. SAE classification of Lubricanting Oils

SAE Viscosity Number	Viscosity Units	Viscosity Range			
		At – 18°C		At 99°C	
		Min.	Max.	Min.	Max.
5 W	(*a*) Centipoise	—	1200	—	—
	(*b*) SUS	—	6000	—	—
10 W	(*a*) Centipoise	1200	2400	—	—
	(*b*) SUS	6000	12000	—	—
20 W	(*a*) Centipoise	2400	9600	—	—
	(*b*) SUS	12000	48000	—	—
20	(*a*) Centistokes	—	—	5.7	9.6
	(*b*) SUS	—	—	45	58
30	(*a*) Centistokes	—	—	9.6	12.9
	(*b*) SUS	—	—	58	70
40	(*a*) Centistokes	—	—	12.9	16.8
	(*b*) SUS	—	—	70	85
50	(*a*) Centistokes	—	—	16.8	22.7
	(*b*) SUS	—	—	85	110

- **Multigrade oils.** It is possible to develop an oil with more than one viscosity at different temperatures. Thus an oil may be in SAE-30 grade at 99°C and in the SAE-10 W grade at – 18°C, oils of this type are known as *multigrade oils* ; other possible grades are : 5 W/20, 20 W/20, 20 W/40.

Following are the "*advantages*" *of multigrade oils* :

(*i*) No necessity to change the oil as per the ambient temperature.

(*ii*) Owing to ease of cranking at low ambient temperature the life of battery is extended.

(*iii*) Long engine life ; required viscosity is maintained under different operating temperatures.

(*iv*) Easy starting and short warming up period.

(*v*) Reduced oil consumption.

(*vi*) Prolonged mileage between the decarbonisation.

(*vii*) Due to execellent thermal and oxidation stability of multigrade oils the engine parts are protected against rust, corrosion and wear.

2. Greases :

- ***Lubricating grease*** *is a solid to semi-solid dispersion of a thickening agent in liquid lubricant. Other ingredients imparting special properties may be included.*
- Greases are normally used under conditions of lubrication for which oil is not as suitable or convenient. *Greases preform better than oils under conditions requiring* :

  (*i*) High bearing loads and shock loads.

  (*ii*) Slow journal speed.

  (*iii*) Temperature extremes.

  (*iv*) Cleanliness or avoidance of splash or drip.

  (*v*) Minimum attention.

  (*vi*) A seal against external contaminants.

  (*vii*) Large bearing clearances.

The various *types of greases used for lubrication are* :

(*i*) Calcium soap greases ;

(*ii*) Sodium soap greases ;

(*iii*) Aluminium soap greases ;

(*iv*) Mixed soap greases ;

(*v*) Barium soap greases ;

(*vi*) Lithium soap greases ;

(*vii*) Pure petroleum greases.

14.6. LUBRICATION SYSTEMS

14.6.1. Introduction

The *main parts of an engine which need lubrication* are as given under :

(*i*) Main crankshaft bearings. (*ii*) Big-end bearings.

(*iii*) Small end or gudgeon pin bearings. (*iv*) Piston rings and cylinder walls.

(*v*) Timing gears. (*vi*) Camshaft and camshaft bearings.

(*vii*) Valve mechanism. (*viii*) Valve guides, valve tappets and rocker arms.

Various *lubrication systems used for I.C. engines may be classified as* :

1. Wet sump lubrication system.
2. Dry sump lubrication system.
3. Mist lubrication system.

14.6.2. Wet Sump Lubrication System

These systems employ a large capacity oil sump at the base of crankchamber, from which the oil is drawn by a low pressure oil pump and delivered to various parts. Oil there gradually returns back to the sump after serving the purpose.

(*a*) **Splash system :**

- This system (Fig. 14.7) is used on some *small four stroke stationary engines*. In this case the caps on the big ends bearings of connecting rods are provided with scoops which, when the connecting rod is in the lowest position, just dip into oil troughs and thus direct the oil through holes in the caps to the big end bearings. Due to splash of oil it reaches the lower portion of the cylinder walls, crankshaft and other parts requiring lubrication. Surplus oil eventually flows back to the oil sump. Oil level in the troughs is maintained by means of a oil pump which takes oil from sump, through a filter.

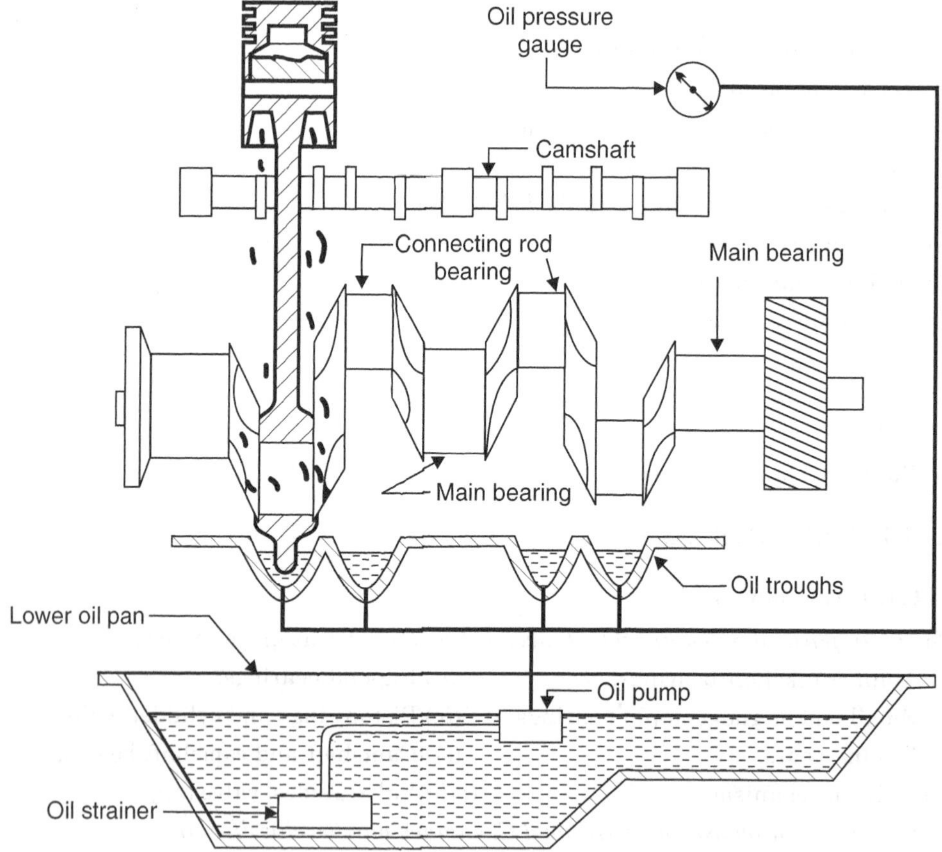

Fig. 14.7. Splash system.

- Splash system is suitable for *low and medium speed engines having moderate bearing load pressures*. For high performance engines, which normally operate at high bearing pressures and rubbing speeds this system does not serve the purpose.

(*b*) **Semi-pressure system :**

- This method is a combination of splash and pressure systems. It incorporates the advantages of both. In this case main supply of oil is located in the base of crank chamber. Oil is drawn from the lower portion of the sump through a filter and is delivered by means of a gear pump at pressure of about 1 bar to the main bearings. The big end bearings are lubricated by means of a spray through nozzles. Thus oil also lubricates the cams, crankshaft bearings, cylinder walls and timing gears. An oil pressure gauge is provided to indicate satisfactory oil supply.
- The system is *less costly to install as compared to pressure system. It enables higher bearing loads and engine speeds to be employed as compared to splash system.*

(*c*) **Full pressure system :**

- In this system, oil from oil sump is pumped under pressure to the various parts requiring lubrication. Refer Fig. 14.8. The oil is drawn from the sump through filter and pumped by means of a gear pump. Oil is delivered by the pressure pump at pressure ranging from 1.5 to 4 bar. The oil under pressure is supplied to main bearings of crankshaft and camshaft. Holes drilled through the main crankshafts bearing journals, communicate oil to the big end bearing and also small end bearings through hole drilled in connecting rods. A pressure gauge is provided to confirm the circulation of oil to the various parts. A pressure regulating valve is also provided on the delivery side of this pump to prevent excessive pressure.

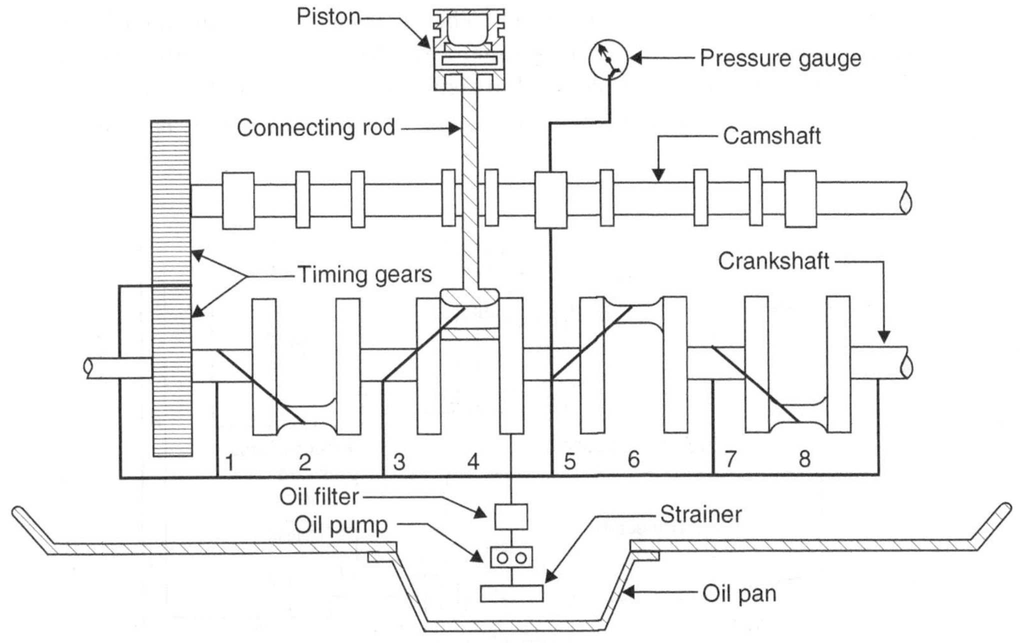

Fig. 14.8. Full pressure system.

- This system *finds favour from most of the engine manufacturers as it allows high bearing pressure and rubbing speeds.*
- The general arrangement of wet sump lubrication system is shown in Fig. 14.9. In this case oil is always contained in the sump which is drawn by the pump through a strainer.

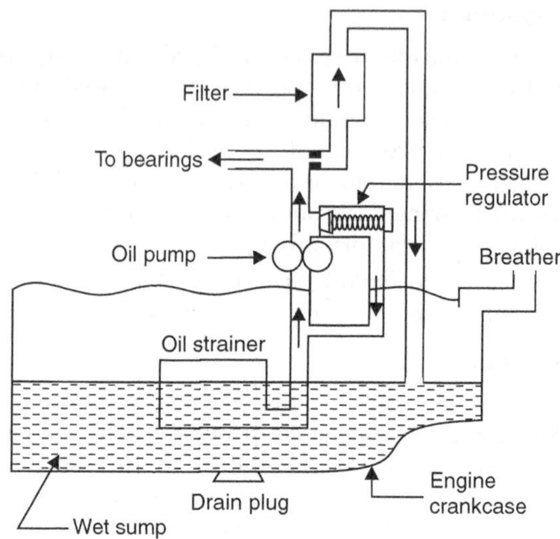

Fig. 14.9. Wet sump lubrication system.

14.6.3. Dry Sump Lubrication System

- Refer Fig. 14.10. In this system, the oil from the sump is carried to a separate storage tank outside the engine cylinder block. The oil from sump is pumped by means of a sump pump through filters to the storage tank. Oil from storage tank is pumped *to the engine cylinder* through oil cooler. Oil pressure may vary from 3 to 8 bar.
- *Dry sump lubrication system is generally adopted for high capacity engines.*

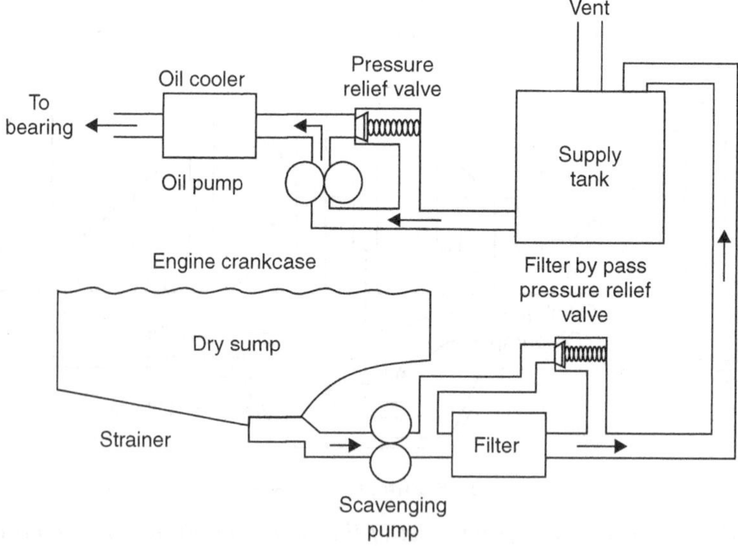

Fig. 14.10. Dry sump lubrication system.

14.6.4. Mist Lubrication System

- *This system is used for two stroke cycle engines.* Most of these engines are crankcharged, *i.e.*, they employ crankcase compression and thus, are not suitable for crankcase lubrication.
- These engines are lubricated by adding 2 to 3 per cent lubricating oil in the fuel tank. The oil and fuel mixture is induced through the carburettor. The gasoline is vaporised ; and the oil in the form of mist, goes via crankcase into the cylinder. The oil which impinges on the crankcase walls lubricates the main and connecting rod bearings, and rest of the oil which passes on the cylinder during charging and scavenging periods, lubricates the piston, piston rings and the cylinder.
- For good performance, F/A ratio used is also important. A F/A *ratio of 40 to 50 : 1 is optimum. Higher ratios increase the rate of wear and lower ratios result in spark plug fouling.*

Advantages :

1. System is simple.
2. Low cost (because no oil pump, filter etc. are required).

Disadvantages :

1. Some lubrication oil will burn and cause heavy exhaust emissions and deposits on piston crown, ring grooves and exhaust port and thus hamper the good performance of the engine.
2. Since the lubricating oil comes in contact with acidic vapours produced during the combustion process, it *rapidly loses its anti-corrosion properties* resulting in *corrosion damage of bearing.*
3. The oil and fuel must be thoroughly mixed for effective lubrication. This requires either separate mixing prior to use or use of some additive to give the oil good mixing characteristics.
4. Owing to higher exhaust temperature and less efficient scavening the crankcase oil is diluted. In addition some lubricating oil burns in combustion chamber. This results in *5 to 15 per cent higher lubricant consumption for two stroke engine of similar size.*
5. As there is no control over the lubricating oil, once introduced with fuel, most of the two stroke engines are *over-oiled most of the time.*

14.6.5. Lubrication of Different Engine Parts

1. **Lubrication of main bearings :**

- The *main bearings* are lubricated satisfactorily with the help of a ring (or chain) type feeder.
- The ring oiling system is shown in Fig. 14.11. It consists of a ring which is bigger in diameter than the shaft. There is one groove at the bearing cap and the ring is placed on the shaft. The lower part of the ring is dipped into an oil reservoir. Due to the rotation of the shaft, the ring rotates at a slower speed and carries the oil from the reservoir to the bearing. *This system works on the principle of adhesion.* In this system the lubricating oil can be fed only *when the shaft is rotating.* Instead of rings, sometimes *chains* are also used.

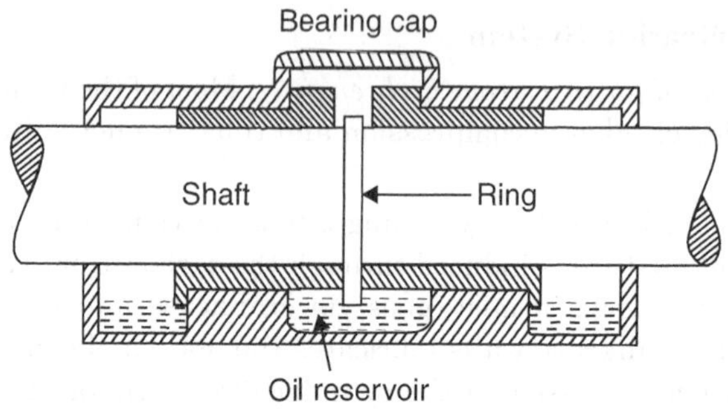

Fig. 14.11. Ring oiling system.

- This type of lubrication is *more useful for medium speed engines because at high speeds the oil will be thrown off due to centrifugal force and at low speeds, the amount of oil carried is inadequate.*

2. **Lubrication of cylinder and small end bearing of connecting rod :**

- The cylinder, small end bearing (gudgeon pin), valve gear pins, rocker shaft, crankpins etc. are lubricated by *drip system*.
- In drip system oil is fed to machine parts drop by drop, from an oil cup. Although it is not an efficient method yet it is often the most convenient way of lubricating the external parts of engines and machines. Valve gear pins, rocker shafts, main bearings of small engines, crankpins, cross head pins, line shaft bearings, and many other machine and engine parts are lubricated in this fashion.

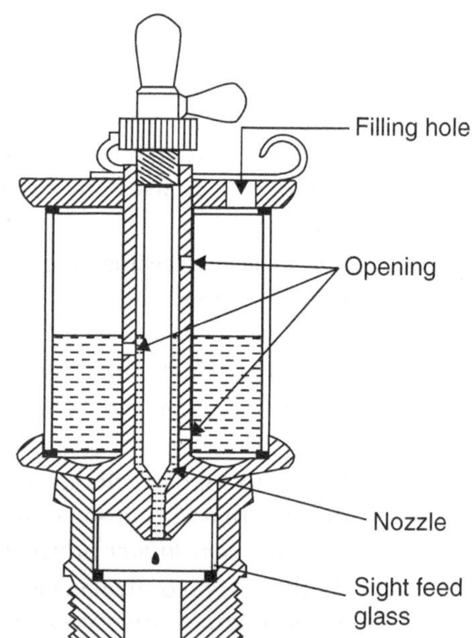

Fig. 14.12. Drop feed oil cup.

Fig. 14.12 shows a *drop feed oil cup*. It comprises of an oil reservoir or cup made of glass. At the top of the cup there is an opening called filling hole through which the lubrication oil is poured into the cup. The oil enters into the inner chamber through the openings and it falls drop by drop due to gravity by the nozzle. The passage of oil through the nozzle is controlled by a needle valve. A sight feed glass is provided to see the drops of oil.

3. **Lubrication of crank and gudgeon pin**

Fig. 14.13 shows *splash lubrication system*. The connecting rod is dipped into the oil of the crankcase. At the time of rotation the oil is splashed due to centrifugal force and it reaches the different parts requiring it. This type of lubrication is suitable for a crankshaft speed of **200** r.p.m. or more. The oil must be filtered or renewed periodically. Also the level of the oil should be maintained.

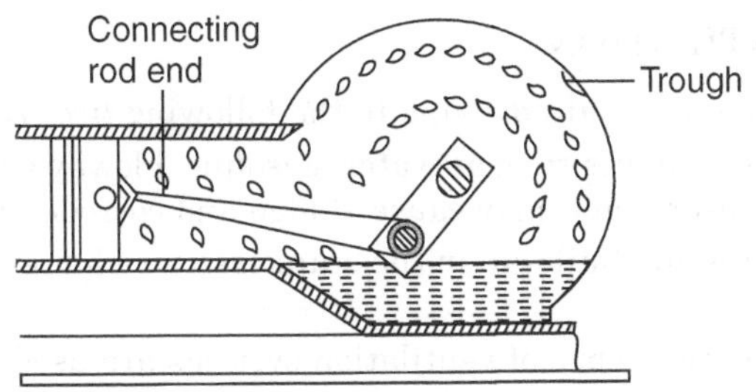

Fig. 14.13. Splash lubrication system.

Note. *Grease cup method* is used for lubricating rocker arms and reciprocating parts with a jerkey motion. It is also employed on parts that are not readily accessible and can be lubricated only at fairly long intervals.

14.6.6. Lubrication of Ball and Roller Bearings

The ball and roller bearings are lubricated for the following purposes :

1. To reduce friction and wear between the sliding parts of the bearing.
2. To prevent rusting or corrosion of bearing surfaces.
3. To protect the bearing surfaces from water, dirt etc.
4. To dissipate the heat.

For lubricating the ball and roller bearings, *generally oil or light grease is used.* Only pure mineral oil or a calcium-base grease should be used. If there is a possibility of moisture contact, then potassium or sodium base greases may be used. Another additional advantage of the grease is that it forms a seal to keep out dirt or any other foreign substance. The temperature should be kept below 90°C and in no case a bearing should operate above 150°C.

14.6.7. Oil Filters

All the lubricating oil (used for lubrication purpose) from the oil sump, must pass through an oil filter before it is supplied to the engine bearings. Bearings maintain very close tolerances and are likely to be damaged by any foreign abrasive material entering the lubrication line.

The filter arrangement may be of the following two types : (*i*) *By-pass type* ; (*ii*) *Full-flow type.*

1. **By-pass type filter arrangement :**

- In this arrangement only a small portion of the lubricating oil is passed through the filter and the remaining lubricating oil is directly supplied to the bearings by the oil pump at pre-set pressure, determined by the pressure regulating valve. Consequently a portion of the oil is continuously filtered.
- Since quantities of oil flowing through filter are *small,* a *very fine filter or a special filter impregnated with resin to avert disintegration due to moisture is used.* Such a fine paper/filter will remove all harmful contaminants.

2. **Full-flow type filter arrangement :**

- In this filter arrangement *whole of the oil is filtered* before it is supplied to various-bearings. Thus, the size of the filter is comparatively large.
- In this case, it is hardly possible to remove very fine particles because of high pressure required to pump oil through such filters.
- All the lubricating oil, in normal course, should be filtered approximately every half minute. A pressure relief valve is used to prevent excessive pressure build up after a cold start.

14.7. CRANKCASE VENTILATION

Crankcase ventilation is required owing to the following *two reasons* :

(*i*) The various contaminants such as water, gasoline, blowby gases etc. enter the crankcase due to several reasons, and may cause sludge and corrode metal parts.

(*ii*) To relieve any pressure build-up in the crankcase which may cause leakage of the crankshaft seal.

In practice, following two types of ventilation systems are used :

1. Open system 2. Closed system.

1. **Open system :**

- In this system, fresh air supply is inducted into the crankcase during the compression stroke (due to creation of small vacuum). The entering air picks up the contaminants (water vapour, gases and H_2SO_4 vapour) and discharge them to the atmosphere during expansion stroke.
- The main *disadvantage* of this system is that the *natural ventilation is quite inadequate during idling or running at low speeds.*

2. **Closed system :**

- In closed system the *fresh air supply is taken to the crankcase from the carburettor.*
- Air cleaner and the breather outlets are connected to the intake manifold through a PCV valve to ensure the burning of all the crankcase gases in combustion chamber.

HIGHLIGHTS

1. The difference between I.P. and B.P. is known as *total engine friction loss.*
2. *Lubrication* is the admittance of oil between two surfaces having relative motion.
3. *Film lubrication* is that type of lubrication in which bearing surfaces are completely separated by a layer of film of lubricant and that the frictional resistance arises only due to relative movement of the lubricant layers.
4. If the oil film becomes thin enough so as not to support the load without occassional metal-to-metal contact then journal friction developed is called *boundary friction* and the lubrication existing in this range is known as *boundary lubrication.*
5. *Viscosity* is the ability of the oil to resist internal deformation due to mechanical stresses and hence it is a measure of the ability of the oil film to carry a load.
6. *The mineral oils are very commonly used for all lubricating purposes.*
7. *Lubricating grease* is a solid to semi-solid dispersion of a thickening agent in liquid lubricant.
8. Lubrication systems are classified as follows :
   (*i*) Wet sump lubrication system
   (*ii*) Dry sump lubrication system
   (*iii*) Mist lubrication system used for *two stroke engines.*

OBJECTIVE TYPE QUESTIONS

Fill in the blanks or Say "Yes" or "No" :

1. Engine friction is defined as the difference between I.P. and ....
2. The pumping loss in two-stroke cycle engines is quite significant.
3. .... losses are caused due to the leakage of combustion products past the piston from the cylinder into the crankcase.

4. Blow by losses decrease directly with compression ratio.
5. Combustion chamber pump loss is caused due to the pumping work required to pump gases into and out of the pre-combustion chamber.
6. Lower stroke-to-bore ratio tends to .... i.m.e.p.
7. When a small number of large cylinders are used the friction and economy improve.
8. The effect of number of piston rings on friction is not significant.
9. The frictional mean effective pressure increases as the compression ratio is increased.
10. The mechanical friction .... with the increase in speed.
11. The rise in cooling water temperature .... the frictional loss.
12. When the load on the engine increases, the i.m.e.p. also increases.
13. Higher the viscosity of oil lower is the friction loss.
14. .... is the admittance of oil between two surfaces having relative motion.
15. .... lubrication is that type of lubrication in which bearing surfaces are completely separated by a layer of film of lubricant and that the frictional resistance arises only due to relative movement of the lubricant layers.
16. When the load on the bearings is very high, the material itself deforms elastically against the pressure built up of the oil film. This type of lubrication is called .... lubrication.
17. .... is the ability of the oil to resist internal deformation due to mechanical stresses and hence it is a measure of the ability of the oil film to carry a load.
18. The viscosity is measured by ....
19. A more viscous oil can carry a greater load.
20. .... is the method of expressing the rate at which the viscosity of an oil will change with temperature.
21. A low viscosity index indicates relatively smaller changes in viscosity of the oil with the temperature.
22. .... point is the lowest temperature at which the lubricating oil will flash when a small flame is passed across its surface.
23. .... point is the lowest temperature at which oil burns continuously.
24. .... point is the lowest temperature at which the lubricating oil will pour.
25. .... is the property of an oil which enables it to spread over and adhere to the surface of the bearing.
26. The .... number is an index of the tendency of an oil to emulsify with water.
27. .... is the property of lubricating oil due to which the oil particles stick with the metal surfaces.
28. Specific gravity is a measure of density of oil.
29. The .... oils are very commonly used for all lubricating purposes.
30. Viscosity index improver prevents undue thinning of the oil as the temperature increases.
31. Silicon easters are used as anti foam agents.
32. .... is a black, brown or grey deposit having the consistency of mud.
33. The mineral lubricating oils are obtained from the residue mass left during crude petroleum distillation.
34. Splash system is used on some small four-stroke stationary engines.
35. Semi-pressure system is a combination of splash and pressure systems.
36. .... sump lubrication system is generally adopted for high capacity engines.
37. .... lubrication system is used for two-stroke cycle engines.
38. In .... crankcase ventilation system fresh air supply is taken to the crankcase from the carburettor.
39. In .... crankcase ventilation system fresh air supply is inducted into the crankcase during the compression stroke.
40. Lubricating grease is a solid to semi-solid dispersion of a thickening agent in liquid lubricant.

ANSWERS

1. B.P.	**2.** No	**3.** Blowby	**4.** No	**5.** Yes
6. decrease	**7.** Yes	**8.** Yes	**9.** Yes	**10.** increases

11. reduces	**12.** Yes	**13.** No	**14.** lubrication	**15.** Film
16. elasto-hydrodynamic		**17.** viscosity	**18.** viscosimeter	**19.** Yes
20. viscosity index	**21.** No	**22.** Flash	**23.** Fire	**24.** Pour
25. Oiliness	**26.** emulsification	**27.** Adhesiveness	**28.** Yes	**29.** mineral
30. Yes	**31.** Yes	**32.** Sludge	**33.** Yes	**34.** Yes
35. Yes	**36.** Dry	**37.** Mist	**38.** closed	**39.** open
40. Yes.				

THEORETICAL QUESTIONS

1. How is 'engine friction' defined ?
2. State the importance of engine friction.
3. Enumerate the factors on which the frictional resistance between two moving parts having relative motion is dependent.
4. Discuss the components into which the total engine friction can be divided.
5. Explain briefly the following :
   (*i*) Direct frictional losses
   (*ii*) Valve throttling losses
   (*iii*) Combustion chamber pump loss
   (*iv*) Blowby losses.
6. State the effect of the following engine parameters on engine friction :
   (*i*) Stroke-to-bore ratio (*ii*) Cylinder size and number of cylinders
   (*iii*) Piston rings (*iv*) Compression ratio
   (*v*) Engine speed (*vi*) Engine load
   (*vii*) Cooling water temperature (*viii*) Oil viscosity.
7. Enumerate the methods by which engine friction can be determined.
8. Define the term 'lubrication'.
9. What are the objects of lubrication ?
10. Discuss the behaviour of a journal in its bearing.
11. Explain briefly the following :
    (*i*) Film lubrication
    (*ii*) Elasto-hydrodynamic lubrication
    (*iii*) Boundary lubrication.
12. Enumerate and discuss the chief qualities to be considered in selecting oil for lubrication.
13. Explain briefly the following properties of a lubricant :
    (*i*) Viscosity (*ii*) Flash point
    (*iii*) Oiliness (*iv*) Emulsification
    (*v*) Neutralisation number (*vi*) Adhesiveness.
14. What are additives ?
15. What do you understand by 'oil contamination and sludge formation' ?
16. How are lubricating oil classified ?
17. What are "Multigrade oils" ? What are their advantages ?
18. What is a grease ?
19. What is the importance of lubrication in I.C. engines ?
20. Enumerate lubrication systems and explain wet sump lubrication system with the help of a neat sketch.
21. Where is dry sump lubrication system preferred and why ?

22. What is the role of lubricating oil filters of automobile engines ?

23. How the lubrication of two wheeler is done ?

24. Explain briefly "Mist lubrication system".

25. What are the various desired properties of a lubricant ? Explain how additives help to achieve the desired properties ?

26. Explain briefly "Elasto-hydrodynamic lubrication" and "Boundary lubrication".

27. What do you understand by full-flow type and by-pass type oil filters ? When one is preferred over the other ?

28. What do you understand by "crankcase ventilation" ?

29. What is the difference between open crankcase and closed crankcase systems ?

15

Engine Cooling

15.1. NECESSITY OF ENGINE COOLING

In an I.C. engine, the temperature of the gases inside the engine cylinder may vary from 35°C or less to as high as 2750°C during the cycle. If an engine is allowed to run without external cooling, the cylinder walls, cylinder and pistons will tend to assume the average temperature of the gases to which they are exposed, which may be of the order of **1000** to 1500°C. Obviously at such high temperature ; the *metals will loose their characteristics and piston will expand considerably and sieze the liner. Of course theoretically thermal efficiency of the engine will improve without cooling but actually the engine will sieze to run.* If the cylinder wall temperature is allowed to rise above a certain limit, about 65°C, the lubricating oil will begin to evaporate rapidly and both cylinder and piston may be damaged. Also high temperature may cause excessive stress in some parts rendering them useless for further operation. In view of this, part of the heat generated inside the engine cylinder is allowed to be carried away by the cooling system. *Thus cooling system is provided on an engine for the following* **reasons :**

1. The even expansion of piston in the cylinder may result in seizure of the piston.
2. High temperatures reduce strength of piston and cylinder liner.
3. Overheated cylinder may lead to preignition of the charge, in case of spark ignition engines.
4. Physical and chemical changes may occur in lubricating oil which may cause sticking of piston rings and excessive wear of cylinder.
5. If the cylinder head temperature is high the volumetric efficiency and hence the power output of the engine is reduced.

Thus engine cooling is required to keep the temperature of the engine low in order to *avoid* :

(*i*) Loss of volumetric efficiency and hence power ;

(*ii*) Engine seizure ;

(*iii*) Danger of engine failure.

- *Almost 25 to 35 percent of total heat supplied in the fuel is removed by the* **cooling medium.**
- Heat carried away by *lubricating oil and heat lost by radiation* amounts to *3 to 5 per cent of the total heat supplied.*

  It must be noted that *heat carried away by the coolant is a dead loss because not only no useful work can be obtained from it but a part of the engine power is also used to remove this heat.* Hence, it is of paramount importance that this loss is *kept minimum* by the designer.

Table 15.1 gives the heat balance of prime movers.

Table 15.1. Heat Balance of Prime Movers

S. No.	Prime mover	% of Fuel energy				Total
		To power	To coolant	To exhaust	To radiation	
1.	*4-stroke S.I. engine*	26	30	32	12	100
2.	*Diesel engines :*					
	(*i*) 2-stroke	30	21	37	12	100
	(*ii*) 4-stroke					
	(*a*) Naturally aspirated	31	26	30	13	100
	(*b*) Turbo-charged	35	22	29	14	100
3.	*Gas turbine :*					
	(*i*) Simple cycle	15	—	70	15	100
	(*ii*) Regenerative cycle	15	—	65	20	100

Demerits of overcooling. Overcooling of the engine is *harmful* because of the following *reasons* :

1. At very low temperature, starting of engine becomes difficult.
2. Due to overcooling, engine life is reduced due to corrosion.
3. If the engine is overcooled some of the heat which could be used to expand the gases will be *lost.*
4. The fuel will not vaporise properly and some of the gases produced by combustion will condense on the cylinder walls. This *leads to dilution of the oil in the pump and the addition of harmful corrosive acids.* Removal of the oil film from the cylinder wall by unvaporised fuel *leads to increased cylinder bore wear.*
5. Inadeque lubrication of the engine, due to oil not being warm enough to flow freely, results in *greater frictional losses.*

In general, due to overcooling the **economy and life of the engine is reduced.**

Demerits of undercooling :

The following are the demerits of undercooling :

1. Undercooling can *cause engine seizure,* or *at least-shorten valve life and possible distortion of the cylinder block head or gasket.*
2. A hot-spot inside the combustion chamber may be sufficient to *cause pre-ignition, i.e.* to ignite the fuel before the spark plug does, thus *causing loss of power and possible damage to the engine components.*
3. Water in cooling system may boil and evaporate, and *should the oil film burn away additional friction and wear will occur between cylinder and piston.*

15.2. AREAS OF HEAT FLOW IN ENGINES

- The transfer of heat takes place due to difference in temperature and from higher temperature to lower temperature. Thus, there is a heat transfer *to the gases* during intake stroke and the first part of the compression stroke, but during the combustion and expansion processes the heat transfer takes place *from the gases to the walls.*
- The *hot combustion gases* give part of their heat to the following components :

(*i*) Cylinder liner

(*ii*) Piston and piston rings

(*iii*) Cylinder head

(*iv*) Exhaust valves and exhaust ports.

— Most of this heat is carried away by the cooling system while some is lost by direct radiation from the engine surfaces.

— The heat going to the surrounding air, to the structure of the engine and to the lubricating oil is usually small (less than 10%), whereas most *of the heat rejected goes either to the cooling system or as heat in exhaust gases.*

15.3. GAS TEMPERATURE VARIATION

During different processes of the cycle there is an appreciable temperature difference of gases inside the engine cylinder.

— At the beginning of induction stroke the temperature is that of clearance gases.

— As the cool mixture is inducted in the engine cylinder the temperature falls rapidly.

— During compression process the temperature increases and attains its maximum value at the end of combustion process.

— During the expansion process the temperature decreases and then drops very rapidly during the release process.

— In *actual engines,* there is some temperature drop during the exhaust process.

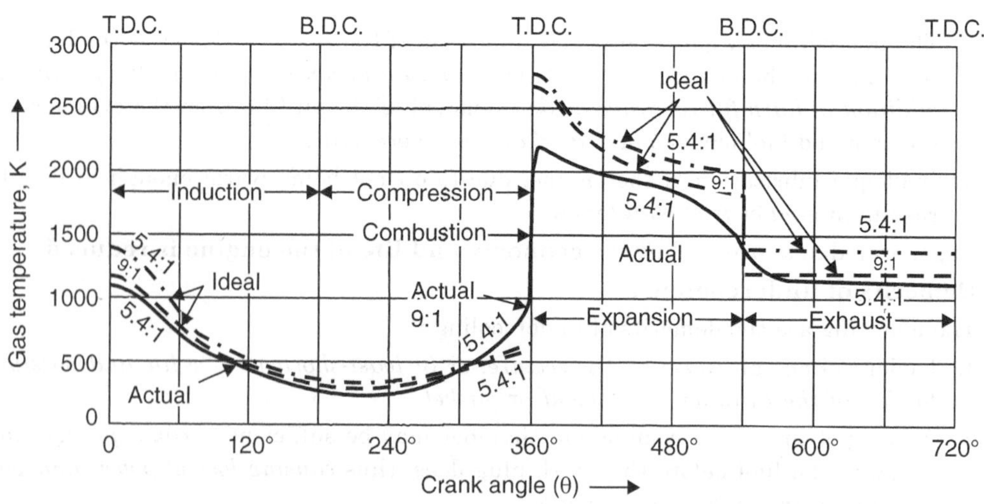

Fig. 15.1. Gas temperature variations for one cycle of 4-stroke engine.

Fig. 15.1 shows the gas temperature variations for one cycle (of a 4-stroke engine) and two compression ratios (*i.e.* 5.4 : 1 and 9 : 1) for chemically correct mixture of C_8H_{18} and air, plus clearance gas.

15.4. HEAT TRANSFER, TEMPERATURE DISTRIBUTION AND TEMPERATURE PROFILES

15.4.1. Heat Transfer

Following are the processes by which heat transfer takes places :

Conduction. '*Conduction*' is the transfer of heat from one part of a substance to another part of the same substance, or one substance to another in physical contact with it, without appreciable displacement of molecules forming the substance.

Convection. '*Convection*' is the transfer of heat within a fluid by mixing of one portion of the fluid with another.

Fuel or natural convection. It occurs when the fluid circulates by virtue of the natural differences in densities of hot and cold fluids. The denser portions of the fluid move downward because of the greater force of gravity, as compared with the force on the less dense.

Forced convection. When the work is done to blow or pump the fluid, it is said to be forced convection.

Radiation. *Radiation* is the transfer of heat through space or matter by means other than conduction or convection. This phenomenon is not very significant with regard to reciprocation I.C. engines.

- *At least 95 per cent of the heat transfer between the working fluid and engine components and the engine components and cooling fluid is effected by* ***"Forced convection"***.

The transmission of heat per unit time from a surface by convection is given by :

$$Q = h\,A\,(t_1 - t_2)$$

where, Q = Quantity of convective heat transferred,

h = Coefficient of convective heat transfer,

A = Area of surface, and

$(t_1 - t_2)$ = Temperature difference between the fluid and the surface.

The *units of coefficient of heat transfer* are :

$$h = \frac{Q}{A(t_1 - t_2)} = \frac{W}{m^2K} = \text{W/m}^2\text{K}$$

The coefficient of convective heat transfer '*h*' (also known as *film heat transfer coefficient*) may be defined as the *amount of heat transmitted* for a unit temperature difference between the fluid and unit area of surface in unit time. The value of '*h*' *depends on the types of fluids, their velocities and temperatures, dimensions of the pipe and the types of problems.* Since '*h*' depends upon several factors, it is difficult to frame a single equation to satisfy all the variations, however a dimensional analysis gives an equation for the purpose which is given as under :

$$\frac{hD}{k} = C\left(\frac{\rho VD}{\mu}\right)^a \left(\frac{c_p\mu}{k}\right)^b \left(\frac{D}{L}\right)^c \qquad \text{...(15.1)}$$

or

$$\text{Nu} = Z(\text{Re})^a\ (\text{Pr})^b \left(\frac{D}{L}\right)^c$$

where, Nu = Nusselt number $\left(\frac{hD}{k}\right)$,

$$\text{Re} = \text{Reynold's number}\left(\frac{\rho VD}{\mu}\right),$$

$$\text{Pr} = \text{Prandtl number}\left(\frac{c_p\mu}{k}\right),$$

$\frac{D}{L}$ = Diameter to length ratio,

C = A constant to be determined experimentally,

c_p = Specific heat at constant pressure,

k = Thermal conductivity,

ρ = Density,

μ = Dynamic viscosity, and

V = Velocity.

The overall heat transfer coefficient :

While dealing with the problems of fluid to fluid heat transfer across a metal boundary, it is usual to adopt an overall heat transfer coefficient U which *gives the heat transmitted per unit area-unit time per degree temperature difference between the bulk fluids on each side of the metal.*

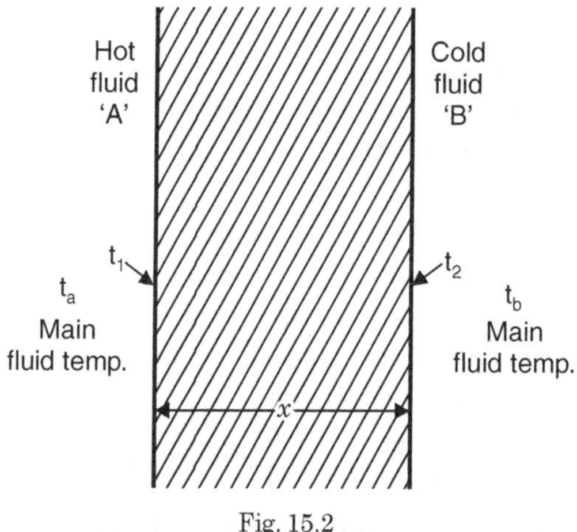

Fig. 15.2

Refer Fig. 15.2.

Let, h_a = Heat transfer coefficient from hot fluid to metal surface,

h_b = Heat transfer coefficient from metal surface to cold fluid, and

k = Thermal conductivity of metal wall.

The equations of heat flow through the fluids and the metal surface are as follows :

$$Q = h_a A\,(t_a - t_1) \quad ...(i)$$

$$Q = \frac{kA\,(t_1 - t_2)}{x} \quad ...(ii)$$

$$Q = h_b A\,(t_2 - t_b) \quad ...(iii)$$

By rearranging the equations (*i*), (*ii*) and (*iii*), we get

$$t_a - t_1 = \frac{Q}{h_a A} \quad ...(iv)$$

$$t_1 - t_2 = \frac{Qx}{kA} \qquad \text{...}(v)$$

$$t_2 - t_b = \frac{Q}{h_b \, . \, A} \qquad \text{...}(vi)$$

Adding equations (*iv*), (*v*) and (*vi*), we get

$$t_a - t_b = Q\left[\frac{1}{h_a \, . \, A} + \frac{x}{kA} + \frac{1}{h_b \, . \, A}\right]$$

$$Q = \frac{A(t_a - t_b)}{\frac{1}{h_a} + \frac{x}{k} + \frac{1}{h_b}} \qquad \text{...(15.2)}$$

If U is the overall coefficient of heat transfer then

$$Q = UA\,(t_a - t_b) = \frac{A\,(t_a - t_b)}{\frac{1}{h_a} + \frac{x}{k} + \frac{1}{h_b}}$$

or

$$U = \frac{1}{\frac{1}{h_a} + \frac{x}{k} + \frac{1}{h_b}} \qquad \text{...(15.3)}$$

It may be noticed from the above equation that if the individual coefficients differ greatly in magnitude only a change in the *least* will have any significant effect on the rate of heat transfer.

15.4.2. Temperature Distribution

Since piston cylinder liner and cylinder head come in direct contact with hot combustion gases, therefore, these components are subjected to very high temperatures. The temperature distribution

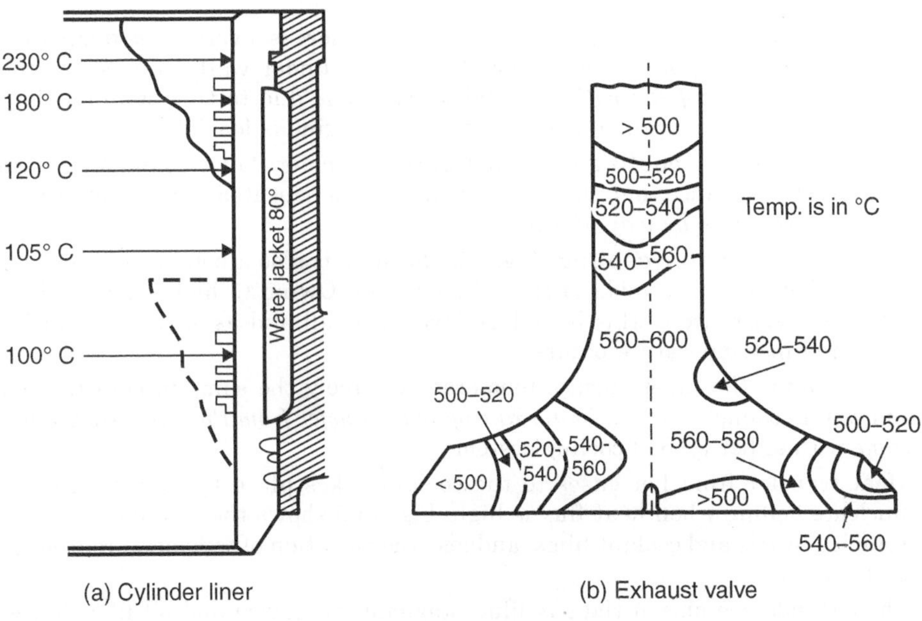

Fig. 15.3. Temperature distribution.

along a typical liner of a 4-stroke diesel engine is shown in Fig. 15.3 (*a*). The temperature isotherms for a typical engine exhaust valve is shown in Fig. 15.3 (*b*). The very wide variations of temperature at different exhaust valve sites lead to development of large thermal stresses.

15.4.3. Temperature Profiles

- Fig. 15.4 shows the representative temperature profiles across the cylinder barrel wall of S.I. engine.

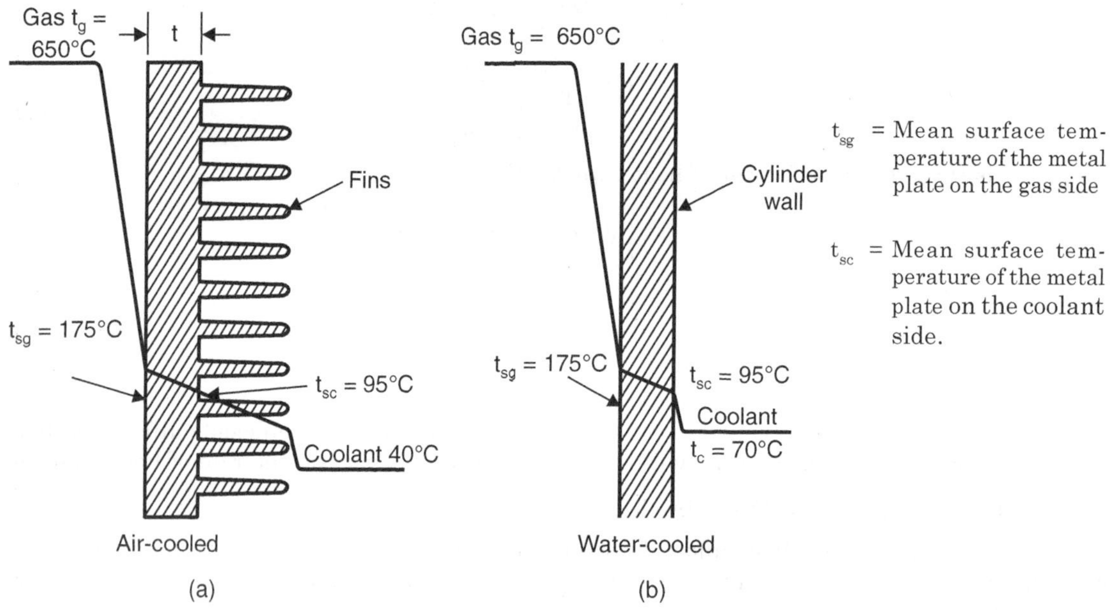

Fig. 15.4. Temperature profiles across cylinder barrel wall.

— It can be observed (from the Fig. 15.4) that there is a very large temperature fall in the 'boundary layer' of gas on the inner cylinder wall. This is due to *very low conductivity of the rather stagnant gas*. It is due to this stagnant layer of gas which acts as a protective layer that I.C. engine is feasible.

— In case of **air cooling** (Fig. 15.4 *a*) the boundary layer as on the gas side offers great resistance, but the effect of this is compensated largely by providing *more surface areas by way of cooling pins.*

— In case of **water cooling** (Fig. 15.4 *b*) there is also a boundary layer of the liquid coolant on the outside of the cylinder wall. Owing to high conductivity of water, the resistance of this boundary layer is small and as such comparatively small drop in temperature occurs.

For maintaining the inner surface temperature largely the same in both the cooling systems (air and water cooling), the *maximum rate of heat transfer from the gases* should be substantially the *same* with either type of cooling system.

- Heat transfer from hot gases to the coolant takes place by forced convection or by nucleate boiling when heat flux is high. Fig. 15.5 shows the cylinder (metal) wall surrounded by gas and coolant films, and also the variation of temperature from gas side to coolant side.

On the gas side are shown the gas film (stagnant gas layer) and oil film (lubricating oil layer). On the coolant side is the water film.

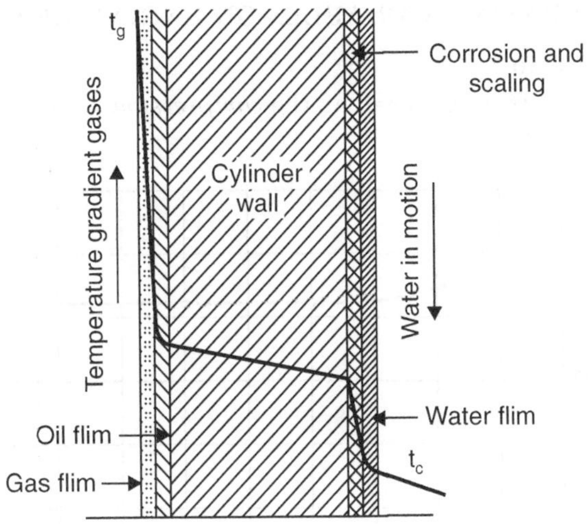

Fig. 15.5. Schematic diagram of heat transfer.

The general equation for heat transfer, using overall heat transfer coefficient, can be written as :

$$Q = UA\,(T_g - T_c)$$

where, U = Overall heat transfer coefficient, W/m^{2}K,

A = Surface area, m^2,

T_g = Gas side temperature, K, and

T_c = Coolant side temperature, K.

The overall heat transfer coefficient U is given by

$$U = \frac{1}{\dfrac{1}{h_g} + \dfrac{L}{k} + \dfrac{1}{h_c}} \qquad \text{...(15.4)}$$

where, h_g = Heat transfer coefficient on gas side, W/m^{2}K,

h_c = Heat transfer coefficient on coolant side, kJ/m^{2}K,

L = Thickness of the cylinder wall, m, and

k = Thermal conductivity, W/mK.

The maximum rate of heat transfer to external cooling medium is obtained when the following requirements are met with :

(*i*) The thickness of the gas film is minimum ;

(*ii*) The surface film of oil is minimum ;

(*iii*) Best conducting materials like aluminium and its alloys are used ;

(*iv*) The wall of the cylinder is of no greater thickness than strength and wear consideration require ;

(*v*) The external temperature is as low as possible relatively to the cycle mean temperature of the cylinder contents ;

(*vi*) Surface deposits such as those of carbon which deposit on piston crown, corrosion and scale due to hard water etc. are reduced to minimum.

15.5. EFFECTS OF OPERATING VARIABLES ON ENGINE HEAT TRANSFER

1. **Compression ratio :**

- Fig. 15.6 shows that, in general, as the *compression ratio increases, there is marginal reduction in heat rejection.*

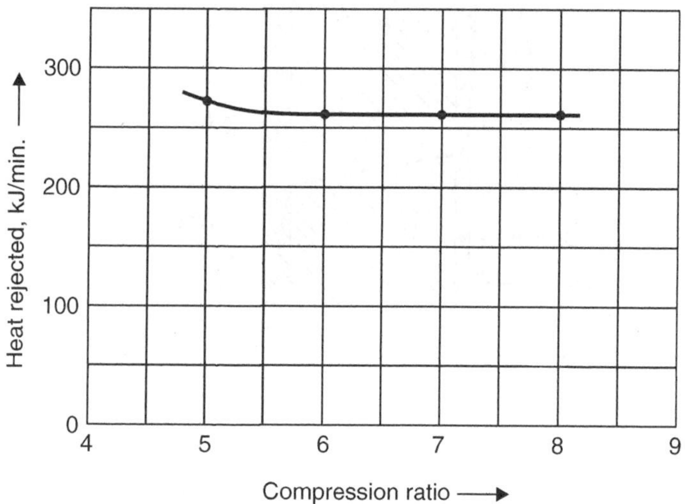

Fig. 15.6. Effect of compression ratio on heat rejected from the engine.

— As the compression ratio is increased the gas temperature near the T.D.C. increases slightly only. But due to greater expansion ratio, there is considerable decrease in gas temperature near the B.D.C. where large cylinder wall is exposed. Due to greater expansion the exhaust gas temperature is also reduced to a much lower value, and the heat rejected during blowdown is also less.

2. **Fuel-air ratio :**

- Air-fuel ratio, inlet temperature and exhaust pressure are the only variables which appear to influence gas temperature (t_g), it is illustrated in Fig. 15.7.

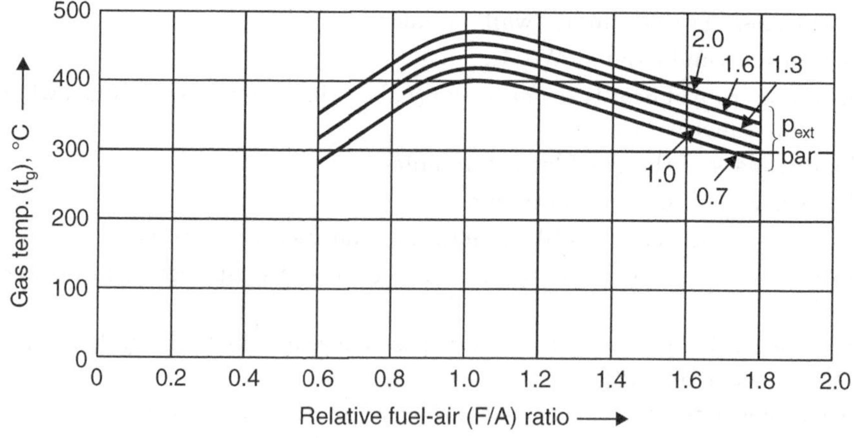

Fig. 15.7

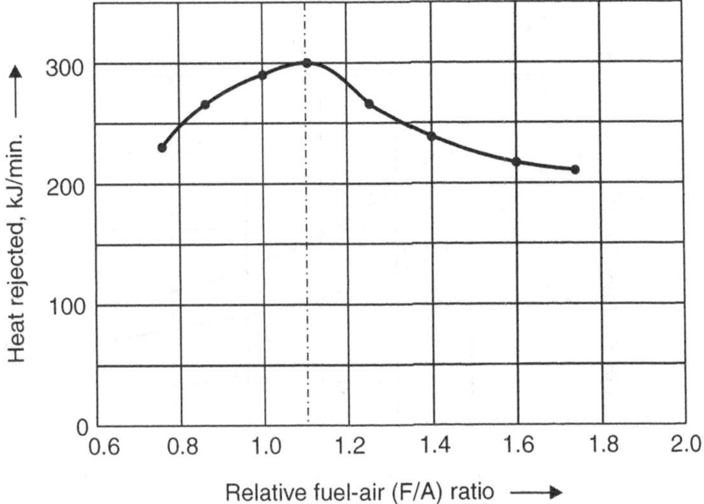

Fig. 15.8

— The temperature of cylinder gases and the flame speed are affected by F/A ratio changes.

— At *relative* F/A ratio of *1.12 maximum mean gas temperature occurs,* and *maximum heat rejection takes place.*

It is worthnoting that at leaner mixtures more heat is rejected and the engine has a tendency to overheat.

3. **Ignition timing :**

When the *spark advance is different from the optimum value the heat rejected to cooling system is increased.* It is due to the fact that any value of the spark advance other than the 'minimum spark advance' for best torque will reduce the power output in S.I. engines resulting in rejection of more quantity of heat.

4. **Load and speed :**

- In the case of S.I. engines the mass of air inhaled remains same, only fuel supplied is increased with the change of load. *This leads to excessive temperature of gases in the cylinder with increase in load. In case of S.I. engines the temperature variation is not much with changes in load.*
- Gas *temperatures remain at a higher average with increased speed.* If the load is constant, the heat input per cycle with the fuel increases with speed, atleast in the upper range, because of increased friction loss.

There is an increase in temperature of piston with speed.

15.6. COOLING AIR AND WATER REQUIREMENTS

- Fig. 15.9 shows the distribution of heat loss in for a four-stroke engine with respect to L / D (stroke / bore) ratio.
- Fig. 15.10 shows the amount of heat rejected to the coolant for small automotive diesel and petrol engines with cast iron and aluminium cylinder heads (as dependent on speed).

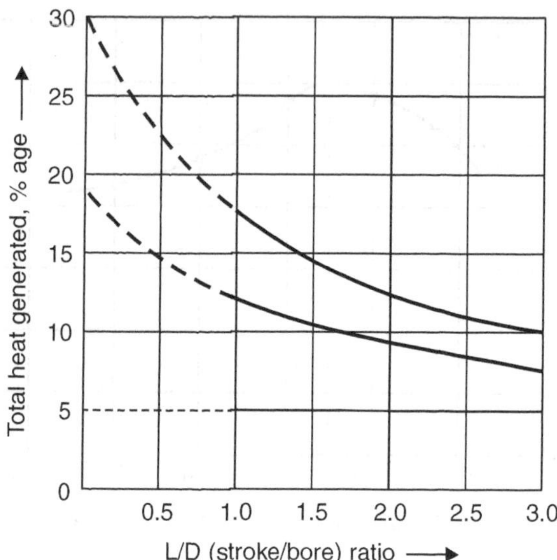

Fig. 15.9. Distribution of heat-loss in four-stroke engine.

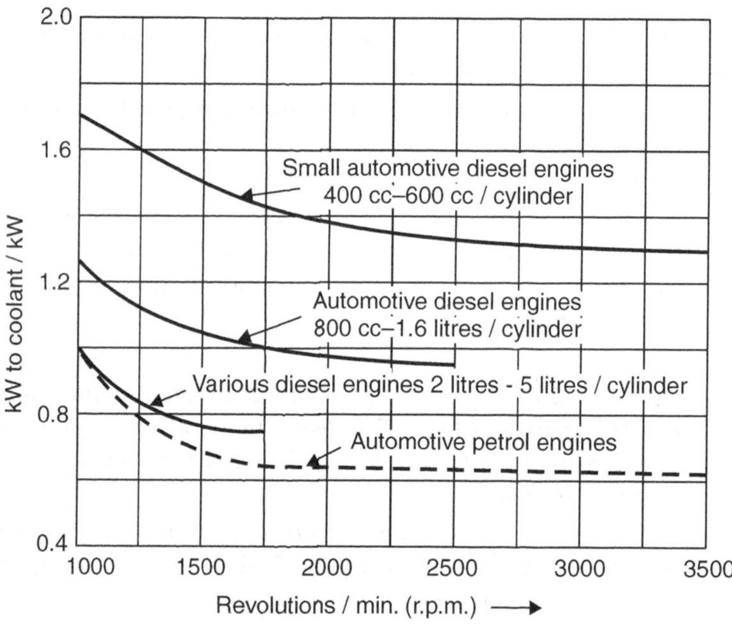

Fig. 15.10. Heat rejected by various I.C. engines.

Cooling air requirements :

- Fig. 15.11 shows the cooling air requirements for air-cooled radiator system of representative types of prime movers.
  — It is obvious from the figure that the gasoline engine requires much more air than a diesel engine and that the turbo-charged diesel engine requires less cooling air than naturally aspirated diesel engines.

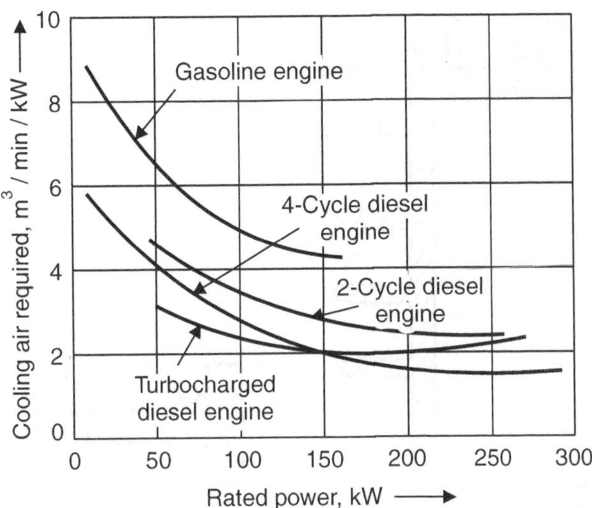

Fig. 15.11. Cooling air requirements.

Cooling water requirements. As is evident from Fig. 15.10, the heat rejected to the coolant greatly depends on the type of the engine. It can be seen that for a small high speed engine the heat rejected in the coolant can be as high as 1.3 times the B.P. developed, while for an open chamber engine it is only about 60% of the B.P. developed.

The quantity of water (Q_w) required for cooling is given by

$$Q_w = \frac{Z \times B.P.}{\Delta t_w} \quad ...(15.5)$$

where, Δt_w = *Permissible increase of temperature of cooling water, and*

Z = *Constant which depends upon fuel consumption and the compression ratio.*

- The heat flow to water jackets, on an average, is about 4200 kJ/kW-h for large engines and 500 to 5700 kJ/kW-h for small engines.
- Large heat stress is avoided if the temperature rise is limited to 10–12°C.
- The *outlet cooling water temperature* for various types of engines is as follows :

For large engines .... about 50°C

For medium engines .... 60 to 65°C

For automobile engines .... 80°C.

15.7. COOLING SYSTEMS

There are mainly following two methods/systems of cooling I.C. engines :

1. Air cooling
2. Water/liquid cooling.

15.7.1. Air-Cooling System

- In this system, heat is carried away by the air flowing over and around the cylinder.
- Here fins are cast on the cylinder head and cylinder barrel which provide additional conductive and radiating surface (See Fig. 15.12). The fins are arranged at right angles to cylinder axis. The number and dimensions should be adequate to take care of the surplus heat dissipation. From all points of view, the truncated conical fin with rounded edges, as shown in Fig. 15.13, accomplish the purpose.

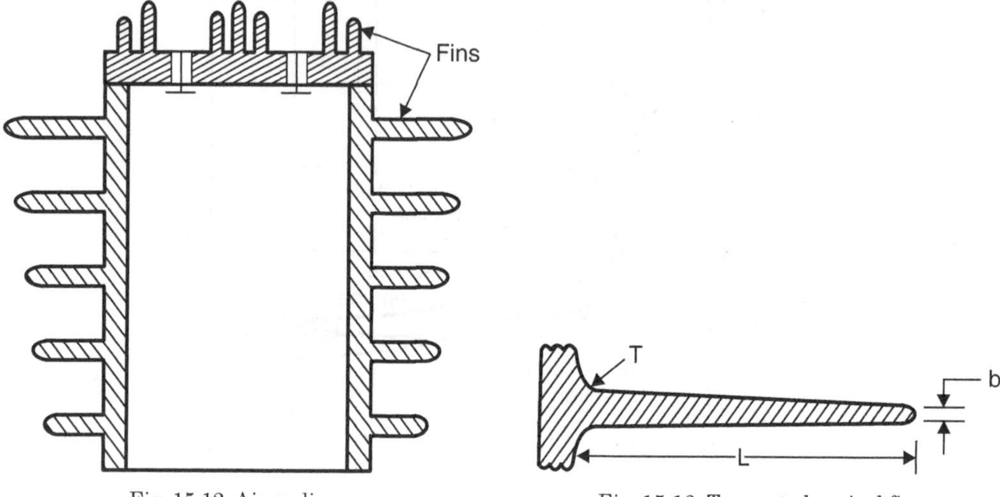

Fig. 15.12. Air cooling.

Fig. 15.13. Truncated conical fin.

A fairly good relationship for fin proportions is as follows :

$$L = 0.35\ D\ ;\ p = 0.10\ D\ ;\ r = 0.018\ D\ ;\ b = 2 \text{ to } 2.5 \text{ mm.}$$

where L, D, p, r and b are fin length, cylinder diameter, fin pitch, rounded edge radius and width at the fin tip respectively.

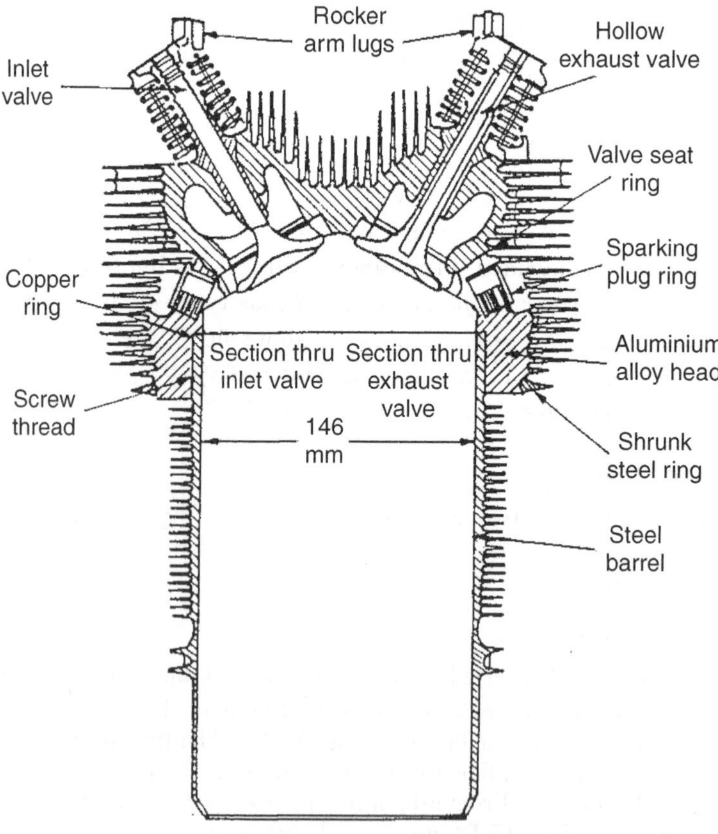

Fig. 15.14. Air-cooled cylinder (The Bristol "Mercury" S.I. engine).

- Fig. 15.14 shows a typical well designed air-cooled cylinder (the Bristol "Mercury" S.I. engine).
  - It is worth noting that the fin surface area which is function of fin height and spacing decreases down the cylinder compared to the fin surface area in the cylinder head comprising the exhaust valve space and spaces between the valves.
  - The *effective valve cooling requires methods of conducting the heat from the valve head to the cylinder, imposed guides and metal ; also to the lower part of the valve stem operating under cooler air conditions.* In order to enhance the conductivity of the exhaust valve, the practice of making the valve stem and even the heads *hollow* is becoming popular in large engines. *The hollow stem is filled with sodium, which melts at 97°C and boils at 880°C.* Thus at valve operating temperatures, the valve is filled with *conducting material.*

Applications. Air cooling system is used in the following engines :

1. "*Small engines*" and engines whose application gives extreme *importance to weight* such as "*aircraft engines*".
2. "*Industrial and agricultural engines*" where there can be a *strong objection to use of water as coolant.*

Advantages :

1. The design of the engine becomes simpler as no water jackets are required. The cylinder can be of identical dimensions and individually detachable and therefore cheaper to renew in case of accident etc.
2. Absence of cooling pipes, radiator etc. makes the cooling system simpler.
3. No danger of coolant leakage etc.
4. The engine is not subjected to freezing troubles etc. usually encountered in case of water-cooled engine.
5. The weight per B.P. of the air-cooled engine is less than that of water-cooled engine.
6. In this case engine is rather a self contained unit as it requires no external components *e.g.* radiator, headers, tank etc.
7. Installation of air-cooled engines is easier.
8. The control of cooling system is much easier than in water-cooled system.
9. An air-cooled engine can take up some degree of damage. A broken fin does not affect much while a hole in the radiator may stop a water-cooled engine.
10. High mean cylinder temperatures mean reduced carbon deposits on combustion chamber wall. This gives sustained engine performance.
11. The warm-up performance of air-cooled engine is better, this results in low wear of cylinders.

Disadvantages :

1. Their movement is noisy.
2. Non-uniform cooling.
3. The output of air-cooled engine is less than that of a liquid-cooled engine.
4. Maintenance is not easy.
5. Smaller useful compression ratio.
6. The volumetric efficiency of an air-cooled engine is lower due to high cylinder head temperatures.

15.7.2. Water/Liquid Cooling System

In this method of cooling engines, the cylinder walls and heads are provided with jackets through which the cooling liquid can circulate. The heat is transferred from cylinder walls to the liquid by convection and conduction. The liquid becomes heated in its passage through the jackets and is itself cooled by means of an air-cooled radiator system. The heat from liquid in turn is transferred to air.

- The coolant to be employed in liquid-cooling system should have the following *characteristics* :
  (*i*) Low freezing temperature.
  (*ii*) A high boiling point.
  (*iii*) A large latent heat of vaporisation.
  (*iv*) Non-corrosive.
  (*v*) Easily and cheaply available.

Water possesses these properties, and is mostly employed in liquid-cooling system.

Role of anti-freeze solution in water-cooling system. When the engine is kept in unheated areas and it is not in operation, and the temperatures are below freezing, the water in the cooling system is liable to *get frozen*. The *solid mass of water due to expansion (10 percent increase in volume) may cause fracture of the cylinder block, water jackets and pipes. The radiator may also be damaged. Anti-freeze mixtures or solution are added to water in extreme winter seasons to lower its freezing temperatures below the danger point.* Commonly used *anti-freeze materials* are :

(*i*) Denatured alcohol
(*ii*) Wood alcohol
(*iii*) Glycerine
(*iv*) Kerosene
(*v*) Sugar solution
(*vi*) Calcium or magnesium chloride
(*vii*) Ethylene glycol and propylene glycol.

In addition, *some oils are also used as anti-freeze solutions.*

Effect of warm-up period on S.I. engine fuel economy :

- When an engine is started from cold condition, it takes some time to reach its operating temperatures. Usual practice of warming up requires engine to run in idling condition for about 2 to 3 minutes. *This helps in an efficient and complete evaporation of fuel when it comes out of carburettor jet, and hence increases fuel economy.*
- Under cold conditions, a portion of fuel remains in the liquid state and get stuck to the cold walls of inlet manifold, cylinder and piston top, which may go in exhaust without or partial combustion. *Initial warm up period helps in complete evaporation of fuel, thereby improving fuel economy.*
- But *excessive warm up period is not desirable, as it decreases the efficiency.* Hence in *modern engines, heating of fresh charge during idling is provided. In water-cooled engines thermostat prevents water circulation during warm up period, which again reduces the warm up period.*

Methods used for circulating water around the engine cylinder :

Various methods are used for circulating water around the cylinder and cylinder head. These are :

1. Thermo-syphon cooling
2. Forced or pump cooling
3. Cooling with thermostatic regulator
4. Pressurised water cooling
5. Evaporative cooling.

1. **Thermo-syphon cooling**

- The *basic principle of thermo-syphon cooling* can be explained with reference to Fig. 15.15(*a*).

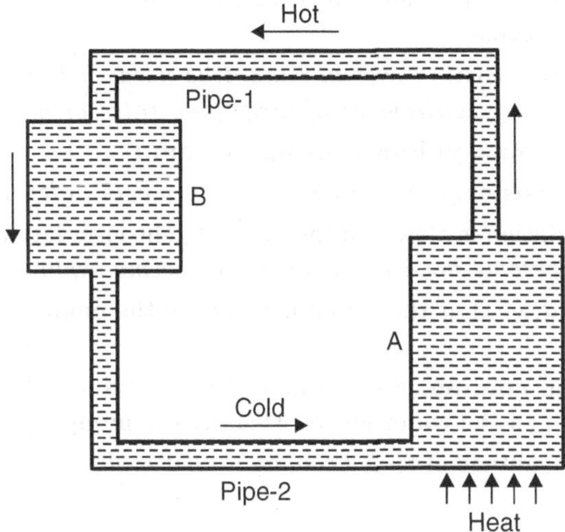

Fig. 15.15. (*a*) Principle of thermo-syphon.

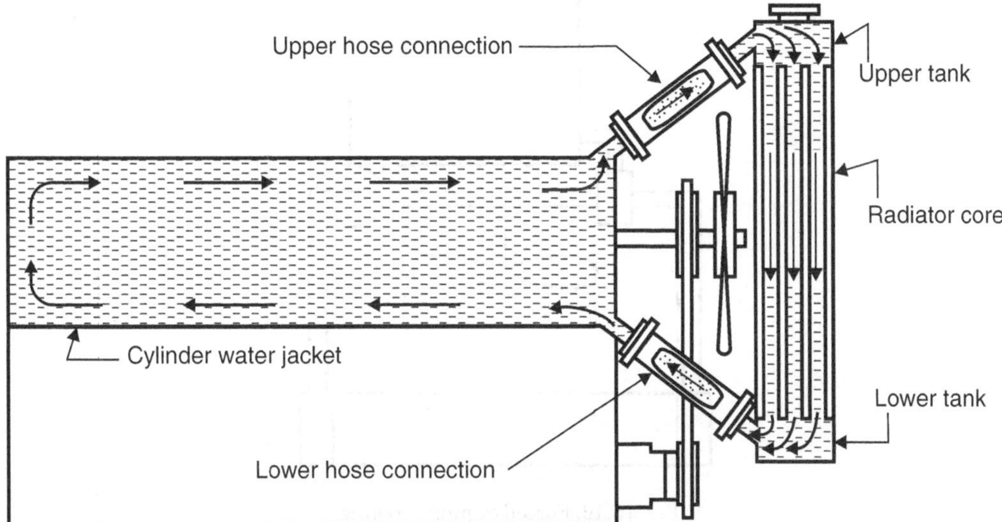

Fig. 15.15. (*b*) Thermo-syphon cooling system.

Heat is supplied to the fluid in the tank A. The hot fluid travels up, its place being taken up by comparatively cold fluid from the tank B through the pipe-2. The hot fluid flows through the pipe. Pipe-1 to the tank B where it gets cooled. Thus the fluid circulates through the system in the form of convection currents.

For engine application tank A represents *cylinder jackets* while tank B represents a *radiator*, and water acts as a *circulating fluid*. In order to ensure that coolest water is always made available to cylinder jackets, the water jackets are located *at a lower level than the radiator*.

- Fig. 15.15(*b*) shows the thermo-syphon cooling arrangement of an engine. The top of radiator is connected to the top of water jacket by a pipe, and the bottom of radiator to the bottom of the water jacket. Water travels down the radiator across which air is passed to cool it. The air flow can take place due to vehicle motion or a fan can be provided for the purpose.
- The *thermo-syphon cooling system cannot be used in high output engines due to its inability to meet the requirement of large flow rates of water.*

Limitations of thermo-syphon cooling system :

(*i*) The *cooling depends only on the temperature and is independent of the engine speed.*

(*ii*) The rate of circulation is slow and insufficient. The circulation of water starts only after the engine has *become hot enough to cause thermo-syphon action.*

(*iii*) This system requires that the radiator be above the engine for gravity flow of water to engine.

2. **Forced or pump cooling.** Refer Fig. 15.16. In this system, a pump is used to cause positive circulation of water in the water jacket. Usually the pump is belt driven from the engine.

Merit. The main advantage of forced cooling is that *cooling is ensured under all conditions of operation.*

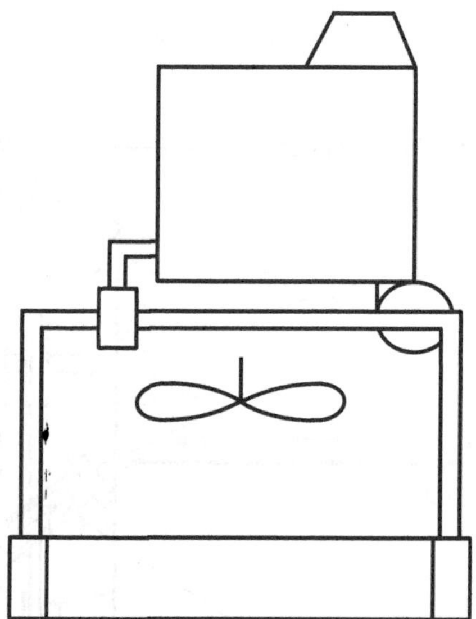

Fig. 15.16. Forced or pump cooling.

Demerits. (*i*) The cooling is not temperature dependent. Under certain circumstances, the engine may get overcooled.

(*ii*) The cooling requirement, while moving uphill, is increased because more fuel is burnt. The coolant circulation, however, is reduced which may lead to overheating of the engine.

(*iii*) The cooling ceases when the engine stops. This is undesirable since cooling must continue till the temperatures are reduced to normal values.

Comparison of thermo-syphon and forced cooling systems :

S. No.	Forced system	Thermo-syphon system
1.	Circulation of water by a centrifugal pump, belt driven from the engine.	Circulation of water by natural convection. No pump is used.
2.	Cooling is independent of temperature, but depends upon the engine speed.	Cooling depends only upon the temperature, and is independent of engine speed.
3.	Rate of circulation is fast.	Rate of cooling is slow and insufficient.
4.	The circulating water pump needs maintenance.	Simple, automatic, and no maintenance is required.
5.	Cooling is ensured under all conditions of operation.	The circulation of water starts only after the engine has become hot enough to cause thermosyphon action.
6.	Radiator position with respect to engine not restricted, can be placed anywhere.	The system requires that the radiator be placed above the engine for gravity flow of water to engine.
7.	It is costly.	It is cheaper.
8.	It is widely used.	It is not widely used.

3. **Thermostat Cooling :**

- Too lower cylinder barrel temperature, may result in severe corrosion damage due to condensation of acids on the barrel wall. To avoid such a situation it is customary to use a thermostat (a temperature controlling device) to stop flow of coolant below a pre-set cylinder barrel temperature. Most modern cooling system employ a thermo-static device which prevents the water in the engine jacket from circulating through the radiator for cooling until its temperature has reached to a value suitable for efficient engine operation.
- Fig. 15.17 shows a systematic diagram of a thermostatically controlled cooling system. Also shown is a typical thermostat (Fig. 15.18). It consists of bellows which are made of thin copper tubes, partially filled with a volatile liquid like ether or methyl alcohol. The volatile liquid changes into vapour at the correct working temperature, thus creating enough pressure to expand the bellows. The temperature at which the thermostat operates is set by the manufacturers and cannot be altered. The movement of the bellows opens the main valve in the ratio of temperature rise, increasing or restricting the flow of water from engine to the radiator. Hence when the normal temperature of the engine has been reached the valve opens and circulation of water commences. When the unit is closed the gas condenses and so the pressure falls. The bellows collapse and the thermostat seats on its seat and circulation around thermostat stops. When the thermostat valve is not open and the engine is running the water being pumped rises in pressure and causes the pressure relief valve to open. Thus the water completes its circulation through the by-pass as shown in Figs. 15.17 and 15.18. Now when the temperature of water around the engine-cylinder rises upto a certain limit, it causes the thermostat valve to open. The pressure of water being pumped falls and pressure relief valve closes. So the flow of cooling water in the normal circuit commences through the radiator. This accelerates the rise of temperature of the cylinder walls and water and more power is developed in a few moments of the starting of the engine.

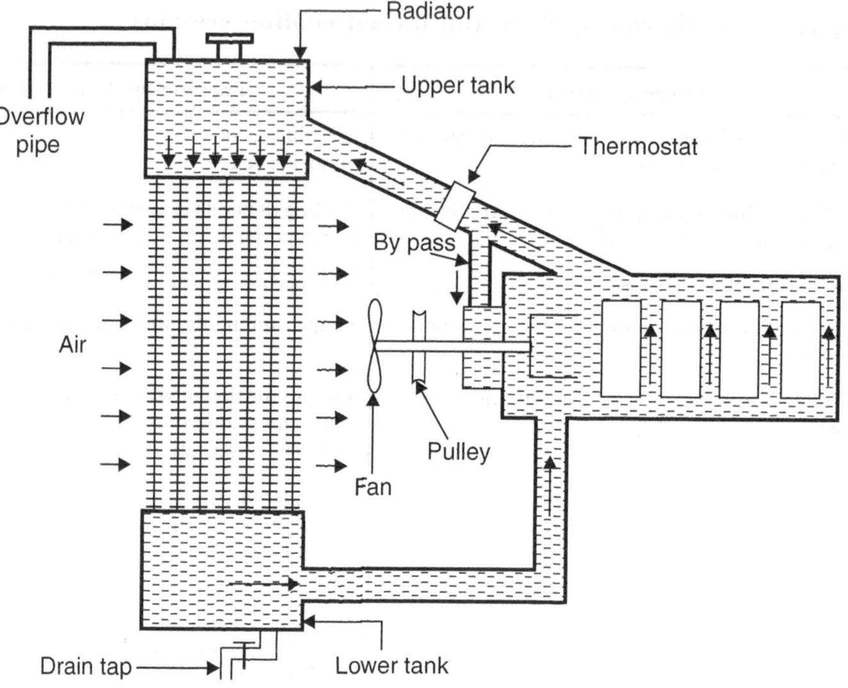

Fig. 15.17. Thermostatically controlled cooling system.

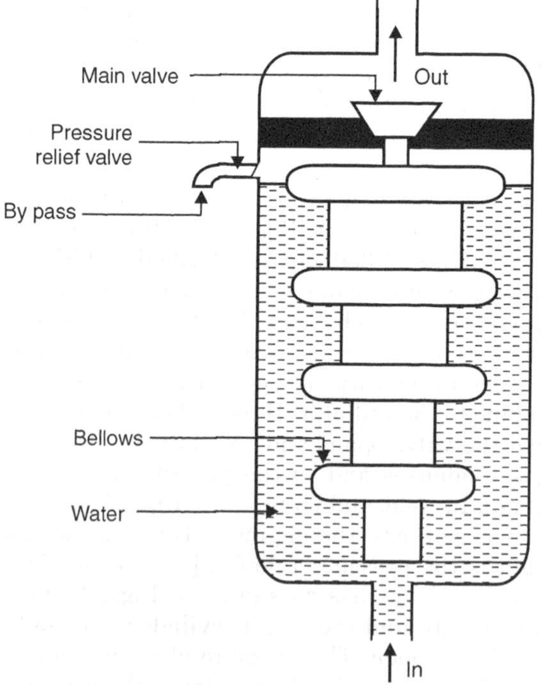

Fig. 15.18. Typical thermostat.

- Another method of warming up the radiator water upto the normal temperature is by utilising the shutter in the radiator in order to restrict the incoming air through the radiator till the engine warms up. Thereafter, the shutter is opened gradually so that the desired rate of cooling is achieved.

4. **Pressurised water cooling :**

Purpose of pressurising a liquid-cooled system

The rate of heat transfer depends upon the temperature difference between two medium, the area of exposed surface and conductivity of materials. In case of radiators, in order to reduce the size of radiators, the cooling system is sealed from the atmosphere, and the system is allowed to be under certain amount of pressure. Thus the advantage is taken of the fact that *temperature of the boiling water increases as the pressure on it is raised. This allows a greater heat transfer to occur in the radiator due to a larger temperature differential. In pressure cooling system moderate pressures, say upto 2 bar are commonly used.*

Method of pressurising

- As shown in Fig. 15.19, a cap is fitted on the radiator with two-valves, a safety valve which is loaded by a compression spring and a vacuum valve. When the coolant is cold both the valves are shut, but as the engine warms up the coolant temperature rises until it reaches a certain preset value corresponding to the desired pressure when the safety valve opens ; but if the coolant temperature falls during the engine operation the valve will close again until the temperature again rises to the equivalent pressure value.

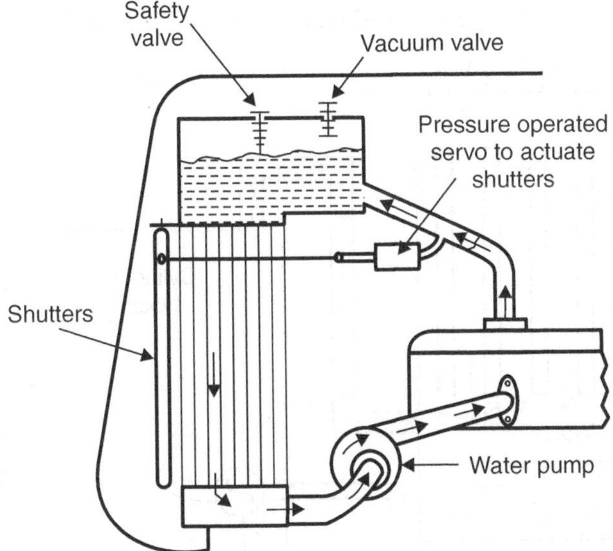

Fig. 15.19. Pressurised cooling.

- When the engine is switched off and the coolant cools down vacuum begins to form in the cooling system, but when the internal pressure falls below atmospheric the vacuum valve is opened by the higher outside pressure and the cooling system then attains atmospheric pressure.
- A safety valve is incorporated in the filler cap so that if an attempt is made to unscrew it while the system is under pressure, the first movement of the cap at once relieves the pressure and thus prevents the emission of steam or the blowing off the cap due to higher internal pressure.

Advantages of pressurised engine cooling over conventional thermo-syphon cooling system :

Following are the *advantages* of pressurised engine cooling over conventional thermo-syphon cooling system :

(*i*) Effective and positive cooling of all parts. Local overheating is avoided.

(*ii*) It can take overload easily because as the engine speed increases the water circulation also increases, and same effective cooling can be maintained at all the speeds.

(*iii*) In thermo-syphon system, the radiator should be kept well above the engine, to provide a height for natural circulation. There is no such requirement for pressurised forced pump system.

(*iv*) With pressurised system the coolant temperature is maintained higher. This reduces corrosion.

(*v*) Smaller coolant passages can be used. This reduces weight and bulk of the engine.

(*vi*) No loss of water by boiling and evaporation.

5. Evaporative cooling :

- In this system, also called steam or vapour cooling, the temperature of the cooling water is allowed to reach a temperature of 100°C. This method of cooling *utilises the high latent heat of vapourisation of water to obtain cooling with minimum of water.* Fig. 15.20 shows such a system. The cooling circuit is such that *coolant is always liquid but the steam formed is flashed off in the separate vessel.* The make up water so formed is sent back for cooling.
- This system is used for cooling of many types of *industrial engines.*

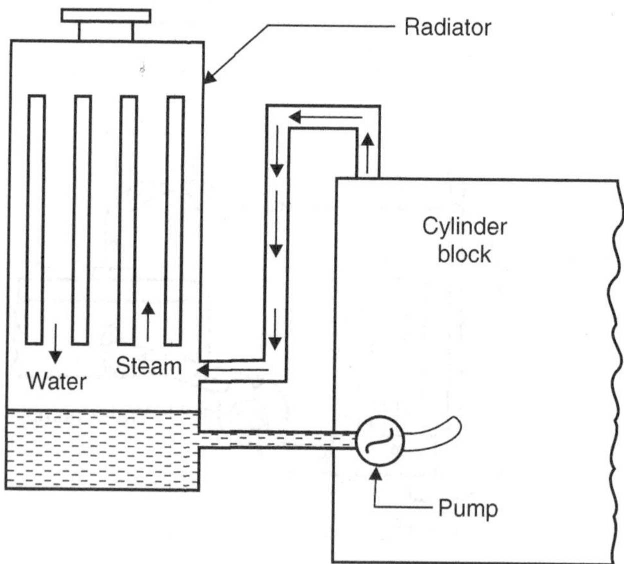

Fig. 15.20. Evaporating cooling.

Advantages and Disadvantages of liquid cooling :

Advantages :

1. Compact design of engine with appreciably smaller frontal area is possible.
2. The fuel consumption of high compression liquid-cooled engine is rather lower than for air-cooled one.

3. More even cooling of cylinder barrels and heads due to jacketing makes it easier to reduce the cylinder head and valve seating temperature.
4. In case of water-cooled engine installation is not necessary at the front of mobile vehicles, aircrafts etc. as the cooling system can be conveniently located wherever required. This is not possible in case of air-cooled engines.
5. The size of the engine does not involve serious problem as far as design of cooling system is concerned. In case of air-cooled engines particularly in high horse power range difficulty is encountered in circulation of required quantity of air for cooling purposes.
6. Volumetric efficiency of water-cooled engines is higher than that of air-cooled engines.

Disadvantages :

1. This is dependent system in which supply of water for circulation in the jacket is required.
2. Power absorbed by the pump for water circulation is considerably higher than that for cooling fans.
3. In the event of failure of cooling system serious damage may be caused to the engine.
4. Cost of system is considerably high.
5. System requires considerable attention for the maintenance of various parts of system.
6. The performance is weather sensitive.
7. The warm up performance is poor and has starting problems particularly in cold weather.

15.8. COMPONENTS OF WATER COOLING SYSTEM

The components of water-cooling system are enumerated briefly described below :

1. Water jacket
2. Water pump
3. Fan
4. Thermostat
5. Connecting loses
6. Radiators
7. Radiator cap (pressurised).

1. **Water jacket :**

- The cooling system starts with the water passages, parts and jackets which are usually cast into cylinder blocks and heads in the manufacturing process. Fig. 15.21. shows a section of a side-valve engine with arrows indicating heat from the combustion chamber being transferred (conducted) into the cooling water passages.
- Water is universally used as the coolant in the automobiles. *Only clean soft water should be used.* Hard water contains minerals which forms a scale on the inside surfaces of the cooling system, reducing its efficiency. *Incubators are available to reduce or prevent the formation of scale and rust and should always be used as the coolant.*
- Above or below a temperature of 4°C, water expands. In the liquid state this expansion can be taken up by the movement of the water into a vacant space, including the additional volume created by the expansion of the cylinder block. When frozen, the solid mass expands and exerts a force against the internal surfaces of the cooling system. The result could be a fractured cylinder block or damaged radiator, although the latter is less likely, due to the mass of water involved, and engine cracks will be revealed when turns back to a liquid. Antifreeze *lowers the freezing point of the water and*

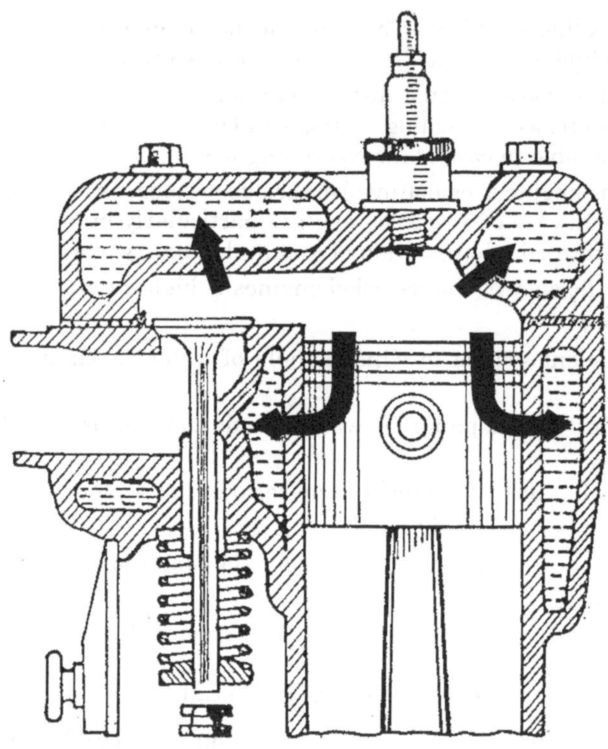

Fig. 15.21. Water jacket.

gives a margin of safety in cold condition. The proportion of antifreeze to water for various degree of protection is usually quoted by the manufacturers. An *ethylene-glycol based antifreeze combined with chemicals* to resist *corrosion* is quite popular.

2. **Water pump :**

- Water pumps entail the following *merits and functions :*

(*i*) Enables more efficient circulation and the radiator can be placed in a position which suits the body line of the vehicle, not necessarily at the fronts of the vehicle.

(*ii*) It is also not essential for the radiator header tank to be above the level of the engine cylinders, and the water passages can be reduced in size.

- Water pumps are usually of the *impeller* or *centrifugal type* and are driven from the crankshaft pulley, the degree of circulation achieved being in proportion to the engine *speed.*

— The impeller type has widely spaced vanes, so that should the impeller fail the vanes will offer little resistance to the natural thermo-syphon water flow.

Centrifugal-type pumps are more powerful and provide positive water circulation, the water entering the pump at the centre of the rotor and being thrown outwards by the curved vanes, so creating a pressure difference between the enter and discharge parts of the pump.

- Fig. 15.22 shows a typical pump assembly steps between the belt and pulley and broken or damaged vanes can reduce the effectiveness of the pump and cause overheating of the engine. If leaks are apparent, the pump should be inspected for defective gaskets and seals, and at the same time the bearings should be checked for wear and the pump casing scraped to remove scale.

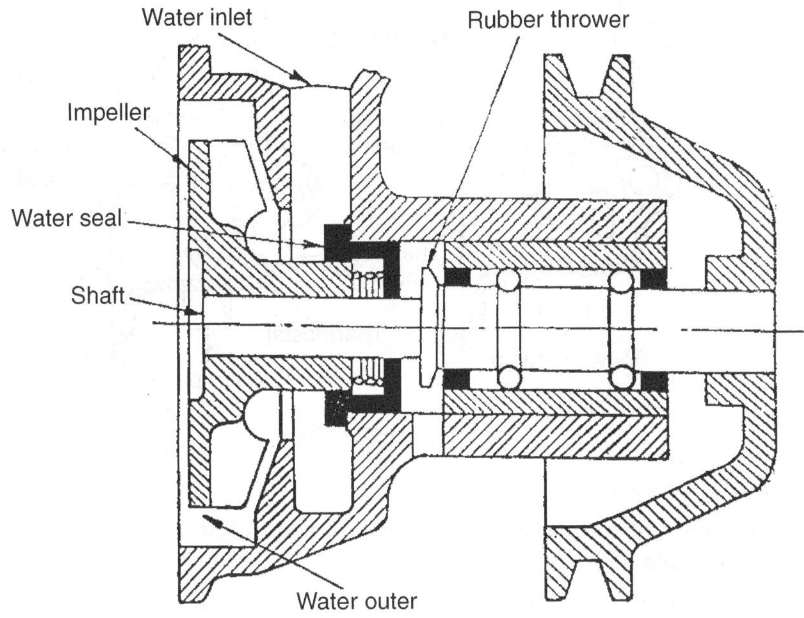

Fig. 15.22. Water pump.

3. **Fan :**

- The fan is generally fitted behind the radiator, usually being mounted on the water pump shaft and induces a flow of air through the radiator block. This function being especially important when vehicle is moving slowly, or when it is stationary and the engine is running. The fan is generally driven by a V-belt from the engine crankshaft and has a number of blades which are made from metal, nylon or some other plastic. As the number of blades increases, the fan size can be reduced.
- It is important that the fan belt is correctly adjusted, if it is too tight the dynamo and water pump bearing, may be damaged and excess power absorbed ; if it is too slack the efficiency of the cooling system and the dynamo output will be affected.
- On no account should the engine crankshaft be rotated by the fan blades as this will cause distortion to the blades and the assembly will run out of balance.

4. **Thermostat :**

- The thermostat is normally fitted either to the cylinder head water-outlet passage below the top hose, or in a special *hosing near the water pump.*
- The purpose of this device is primarily to restrict the circulation of water to cylinder head and block during the warm-up period. Then when a designed temperature is reached, the thermostat opens and the circulation of water starts in the radiator.
- The usual bronze bellow-type unit is shown in the Fig. 15.23. Expansion and contraction of vapour with the bellows gradually opens and closes the valve in proportion to the predetermined temperature range.

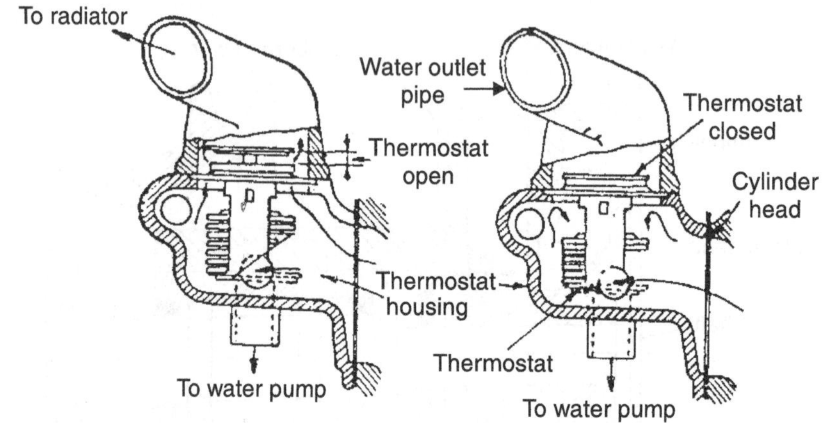

Fig. 15.23. Thermostat.

5. **Connecting hoses :**

- For convenience of connection and to isolate the radiator from engine movement and vibration, flexible hoses are used.
- They were originally of rubber, reinforced with canvas like large diameter garden hose-Neoprene and various plastics are now more suitable and are premoulded to the shapes and curves required for the installation.
- The ends of the hose are forced over the connecting spigots on the radiator and engine held in place by bore clamps or spring clips.

6. **Radiators :**

- The basic requirement of a radiator is to *provide a sufficiently large cooling area for transmission of heat from the coolant to the air.* The construction of the centre of the radiator or core varies, but in general the water passages terminate at a header tank at the bottom. In addition to an opening which enables the cooling system to be topped up the header tank allows for expansion and contraction of the coolant within the system.
- The principal types of radiator core are :
  (*i*) Film type
  (*ii*) Fin and tube
  (*iii*) Pack type.

Fig. 15.24 shows a radiator made of thin sheet brass and is typical of most cooling systems. As the water descends through hundreds of tubes or passage it is cooled by radiation to the air and returns to the engine via the bottom tank and the lower left connection. An overflow pipe from the filter neck can be clearly seen.

Maintenance of radiator. Periodically the radiator should be *reverse flushed.* This involves removing the radiator hoses and thermostat and flushing up through the radiator and *down* through the engine block. In other words flushing in the reverse direction to normal water flow. For reverse cases a solution of 0.45 kg. of washing soda to 4.5 litres of water can be used in the cooling system, but the system must be flushed after the solution has been drained off, and this method should not be used for aluminium blocks. Choked core leads to bad circulation and overheating, the latter also being caused by the air passages becoming blocked with dirt and dead insects.

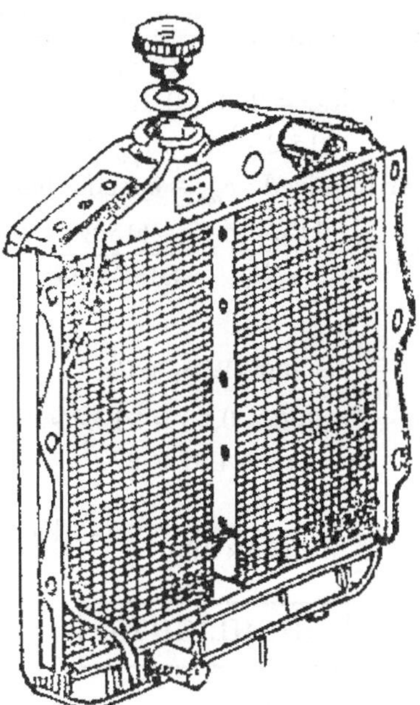

Fig. 15.24. Radiator.

7. **Radiator cap (pressurised) :**

- The greater the pressure acting on the surface of a liquid, the higher the temperature at which the liquid boils. If the cooling system is sealed off, the expanding water will increase the pressure of air in the header tank ; any steam formed will assist in raising the pressure, and in consequence will require a higher temperature to cause the water to boil, which means that any *arduous driving, will not result in water loss due to evaporation.*

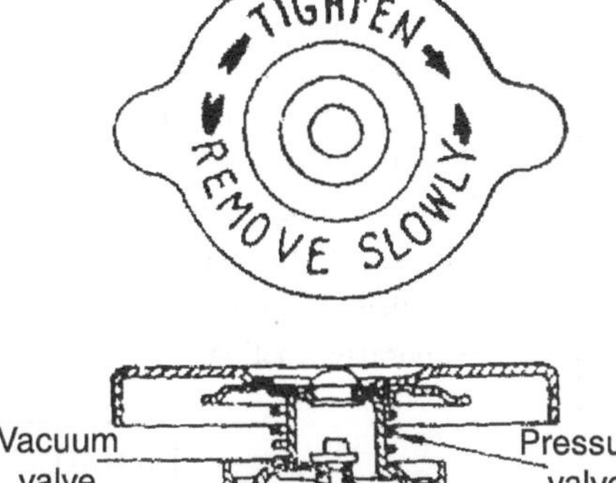

Fig. 15.25. Radiator pressrue cap.

- Fig. 15.24 shows a radiator pressure cap. A safety device is incorporated in the cap to reduce the risk of accident when the cap is removed due to hot water a steam being ejected under pressure from the radiator. In addition, a small lightly-loaded valve operating in reverse direction to the main valve prevents a vacuum being created in the system as it cools down, so preventing collapse or damage to the radiator or the connecting hoses.

Example 15.1. *Compare the quantity of water required for 90 kW petrol and diesel engines in which water is raised in temperature by 27°C in passing through the jackets. In petrol engine the percentage of energy going to coolant is 32 percent and in diesel engine 28 percent. The efficiencies of petrol and diesel engines are 25 percent and 30 per cent respectively.*

Solution. *Given :* B.P. = 90 kW ; Δt_w = 27°C ; η_{petrol} = 25%, η_{diesel} = 30%

Petrol engine :

Efficiency, $$\eta = \frac{\text{Work developed}}{\text{Heat supplied}}$$

$\therefore$ Heat supplied $= \dfrac{\text{Work developed}}{\eta_{petrol}} = \dfrac{90}{0.25} = 360$ kW or kJ/s

Energy/heat going to cooling water = 360 × 0.32 = 115.2 kJ/s

Also, $115.2 = m_w \times c_{pw} \times \Delta t_w$

(where, m_w = mass of cooling water required ; c_{pw} = specific heat of water at constant pressure)

$$\therefore \quad m_w = \frac{115.2}{c_{pw} \times \Delta t_w} = \frac{115.2}{4.187 \times 27} = 1.019 \text{ kg/s or } \mathbf{3668\ kg/h.} \quad \textbf{(Ans.)}$$

Diesel engine :

Heat supplied $\dfrac{90}{0.3} = 300$ kW or kJ/s

Heat going to cooling water = 300 × 0.28 = 84 kJ/s

Also, $84 = m_w \times 4.87 \times 27$

$$\therefore \quad m_w = \frac{84}{4.187 \times 27} = 0.743 \text{ kg/s or } \mathbf{2674.9\ kg/h.} \quad \textbf{(Ans.)}$$

HIGHLIGHTS

1. Almost 25 to 35% of total heat supplied in the fuel is removed by the cooling medium.
2. At least 95% of the total heat transfer between the working fluid and engine components and the engine components and cooling fluid is effected by "forced convection" process of heat transfer.
3. When the spark advance is different from the optimum value the heat rejected to cooling system is increased.
4. In *air-cooling* system, heat is carried away by the air flowing over and around the cylinder. Here fins are cast on the cylinder head and cylinder barrel which provide additional conductive and radiating surface.
5. In *water-cooling* system of cooling engines, the cylinder walls and heads are provided with jacket through which the cooling liquid can circulate.
6. The *methods* used for circulating water around the cylinder and cylinder head :
   (*i*) Thermo-system cooling
   (*ii*) Forced or pump cooling
   (*iii*) Cooling with thermostatic regulator
   (*iv*) Pressurised water cooling
   (*v*) Evaporative cooling.

OBJECTIVE TYPE QUESTIONS

Fill in the blanks or Say "Yes" or "No".

1. If the cylinder head temperature is high the volumetric efficiency and hence the power output of the engine is ....
2. Almost 25 to 35 percent of total heat supplied in the fuel is removed by the cooling medium.
3. Heat carried away by lubricating oil and heat lost by radiation amounts to 10 to 15 percent of the total heat supplied.
4. Heat carried away by the coolant is a dead loss.
5. At very low temperature, starting of engine becomes difficult.
6. Undercooling shortens valve life.
7. The overall heat transfer coefficient gives the heat transmitted per unit area-unit time per degree temperature difference between the bulk fluids on each side of the metal .

8. As the compression ratio is increased the gas temperature near the T.D.C. increases slightly only.
9. The temperature of cylinder gases and flame speed are not affected by F/A ratio changes.
10. At relative fuel-air ratio of 1.12 maximum mean gas temperature occurs, and maximum heat rejection takes place.
11. When the spark advance is different from the optimum value the heat rejected to cooling system is ....
12. In .... cooling system, heat is carried away by the air flowing over and around the cylinder.
13. The .... are arranged at right angles to cylinder axis.
14. Installation of air-cooled engines is easier.
15. Air-cooled engines are noisy.
16. The output of air-cooled engine is .... than that of water-cooled engine.
17. .... mixtures or solutions are added to water in extreme winter seasons to lower its freezing temperatures below the danger point.
18. The ...... cooling system cannot be used in high output engines due to its inability to meet the requirement of large flow rates of water.
19. In thermo-syphon cooling system the cooling depends only on the .... and is independent of the engine speed.
20. The .... cooling system requires that the radiator be above the engine for gravity flow of water to engine.
21. In .... cooling there is no loss of water by boiling and evaporation.
22. .... cooling utilises the high latent heat of vaporisation of water to obtain cooling with minimum of water.
23. Evaporative cooling system is used for cooling of many types of industrial engines.
24. In a water-cooling system power absorbed by the pump for water circulation is less than that for cooling fans.
25. Volumetric efficiency of water-cooled engine is .... than that of air-cooled engines.

ANSWERS

1. reduced	**2.** Yes	**3.** No	**4.** Yes	**5.** Yes
6. No	**7.** Yes	**8.** Yes	**9.** No	**10.** Yes
11. increased	**12.** air	**13.** Fins	**14.** Yes	**15.** Yes
16. less	**17.** Antifreeze	**18.** thermo-syphon	**19.** temperature	**20.** thermo-syphon
21. pressurised	**22.** Evaporative	**23.** Yes	**24.** No	**25.** higher.

THEORETICAL QUESTIONS

1. Why is cooling necessary for I.C. engines ?
2. Why overheating and overcooling of I.C. engines is harmful ?
3. State the demerits of overcooling and undercooling.
4. Discuss briefly the areas of heat flow in engines.
5. Explain briefly gas temperature variations for a 4-stroke cycle engine.
6. Explain briefly with a neat sketches the representative temperature profiles across the cylinder barrel wall of S.I. engine.
7. Explain the effects of operating variables on engine heat transfer :
   (*i*) Compression ratio
   (*ii*) Fuel-air ratio
   (*iii*) Ignition timing.
8. Describe briefly 'cooling air' and 'cooling water' requirements for I.C. engines.
9. What is the film coefficient ? On what factors the film coefficient generally depends ?

10. What are the two main types of cooling systems ? Where these systems are used ?

11. How engines are air-cooled ? What is the purpose of the fins in an air-cooled system ? What is the size and spacing of fins ?

12. State the applications, advantages and disadvantages of air-cooling system.

13. Name the various types of liquid cooling systems.

14. How is the circulation accomplished in a thermo-syphon system ? What is the draw back of this system ?

15. Explain with a sketch the forced circulation system. State its merits and demerits.

16. What are the advantages and disadvantages of liquid-cooling system ?

17. Explain with a neat sketch 'thermostat cooling' method of cooling I.C. engines.

18. Describe with a neat sketch the construction and working of a thermostat.

19. Explain briefly the following methods of water coolings

(*i*) Pressurised water cooling

(*ii*) Evaporative cooling.

20. Describe the operation of the thermostat ? What is the main advantage of using a thermostat in the cooling system ?

21. What is the advantage of pressurised cooling ? How pressurising is accomplished ?

22. What is evaporative cooling ?

23. Compare the merits and demerits of air and water cooling systems.

24. What is the function of a radiator ?

25. What are the various type of radiators ? Explain any one in detail.

26. Explain the role anti-freeze solutions in water-cooling system.

27. Discuss the advantages of pressurised engine cooling over conventional thermo-syphon cooling.

28. What is the purpose of pressurising a liquid-cooled system ? How is it accomplished ?

29. What is the effect of warm-up period on S.I. engine fuel economy ? Discuss the provisions made in recent S.I. engines to reduce warm-up period.

30. Why is cooling necessary for I.C. engines ? What kind of cooling system is employed for mobile units like automobiles ? What are the effects of load and speed on the heat loss through cylinder walls ?

16

Supercharging of I.C. Engines

16.1. PURPOSE OF SUPERCHARGING

The purpose of supercharging is *to raise the volumetric efficiency above that value which can be obtained by normal aspiration.*

- The engine is an air pump. Increasing the air consumption permits greater quantities of fuel to be added, and results in a greater potential output. *The indicated power produced is almost directly proportional to the engine air consumption.* While brake power is not so closely related to air consumption, it is nevertheless, dependent upon the mass of air consumed. It is desirable, then, that the *engine takes in greatest possible mass of air.*
- Three possible methods which might be utilized to increase the air consumption of an engine are :
  1. *Increasing the piston displacement,* but this increases the size and weight of the engine, and *introduces additional cooling problems.*
  2. *Running the engine at higher speeds,* which results in *increased fluid and mechanical friction losses, and imposes greater inertia stresses on engine parts.*
  3. ***Increasing the density of the charge,*** *such that a greater mass of charge is introduced into the same volume or same total piston displacement.*

The *last method* of increasing the air capacity of an engine is *widely used,* and is termed **supercharging.**

Supercharger :

- *The apparatus used to increase the air density is known as a* ***supercharger.*** It is merely a *compressor* which provides a denser charge to the engine, thereby enabling the consumption of a greater mass of charge with the same total piston displacement. During the process of compressing the charge, *the supercharger produces the following effects :*
  (*i*) *Provides better mixing of the air-fuel mixture.* The turbulent effect created by the supercharger assists in additional mixing of the fuel and air particles. The arrangement of certain types of superchargers, particularly the *centrifugal type,* also encourages more even distribution of the charge to the cylinders.

(*ii*) *The temperature of the charge is raised as it is compressed, resulting in a higher temperature within the cylinders.* This is partially beneficial in that it helps to produce better vapourisation of fuel (in case of S.I. engines) but deterimental in that it tends to lessen the density of the charge. The increase in temperature of the charge also effects the detonation of the fuel.

Supercharging tends to increase the possibility of detonation in a S.I. engine and lessen the possibility in a C.I. engine.

(*iii*) Power is required to drive the supercharger. This is usually taken from the engine and thereby removes, from over-all engine output, some of the gain in power obtained through supercharging.

Compressors used are of the following three types :

(*i*) ***Positive displacement type*** used with many reciprocating engines in stationary plants, vehicles and marine installations.

(*ii*) ***Axial flow type*** seldom used to supercharge reciprocating engines, it is widely used as the compressor unit of the *gas turbines.*

(*iii*) ***Centrifugal type*** *widely used* as the *supercharger for reciprocating engines, as well as compressor for gas turbines.* It is almost exclusively used as the supercharger with reciprocating power plants for *aircraft because it is relatively light and compact, and produces continuous flow rather than pulsating flow as in some positive displacement types.*

- A *correctly matched supercharger* will raise the cylinder's brake mean effective pressure (b.m.e.p.) *to well above that of a naturally aspirated engine without creating excessively high peak cylinder pressures ; the actual increase in the brake mean effective pressure is basically determined by the level of boost pressure the supercharged system is designed to deliver.*
- *Large commercial vehicle diesel engines are frequently turbocharged, with the objectives of raising b.m.e.p. (and therefore torque and power output) and at the same time reducing the engine's maximum speed.* The *other benefits of raising the cylinder mean pressure and decreasing the engine's limiting speed is that the engine mechanical losses and noise are reduced and there is an improvement in fuel consumption, normally, an added bonus of prolonged engine life expectancy.*

Object of supercharging. The objects of supercharging include one or more of the following :

1. *To increase the power output for a given weight and bulk of the engine* .... relates to *aircraft, marine and automotive engines.*
2. *To compensate for the loss of power due to altitute* .... Relates to *aircraft and other engines which are used at high altitudes.*
3. *To obtain more power from an existing engine.*

Effects of supercharging on performance of the engine :

1. The '*power output*' of a supercharged engine is *higher* than its naturally aspirated counterpart.
2. The '*mechanical efficiencies*' of supercharged engines are *slightly better* than the naturally aspirated engines.
3. Inspite of better mixing and combustion due to reduced delay a mechanically supercharged otto engine almost always have '*specific fuel consumption*' *higher than a naturally aspirated engine.*

16.2. SUPERCHARGING OF S.I. ENGINES

The schematic arrangement for supercharging S.I. engine is shown in Fig. 16.1.

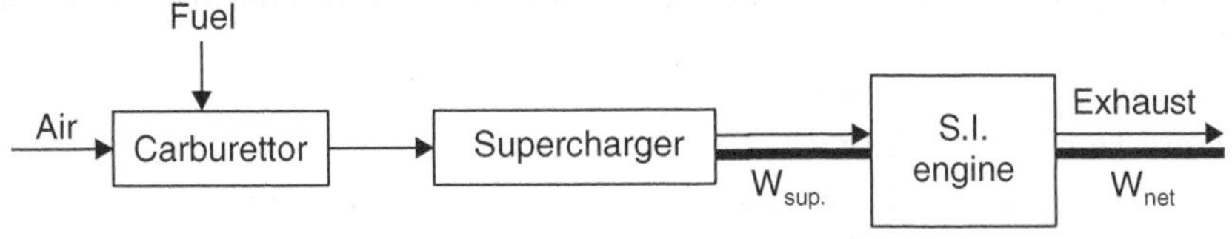

Fig. 16.1

The theoretical operating cycles for both the natural aspirated and supercharged petrol engines can be compared on pressure-volume (p-v) diagrams as shown in Fig. 16.2 and 16.3 respectively. *The larger upper loop is a measure of the positive power developed in the cylinder while the lower loop represents the negative power needed to fill the cylinder with fresh charge.*

16.2.1. Naturally Aspirated Cycle of Operation. Refer Fig. 16.2.

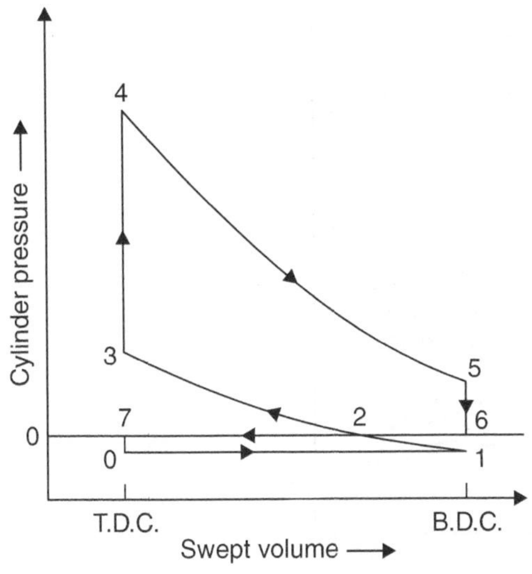

Fig. 16.2. Theoretical naturally aspirated petrol engine (constant volume) pressure-volume diagram.

Fig. 16.3. Theoretical supercharged petrol engine (constant volume) pressure-volume diagram.

- The *large area* enclosed in the upper loop (2-3-4-5-6-2) is *proportional to the useful work* performed by combustion on the piston whereas the *small loop area* (0-1-2-7-0), which is below the atmospheric line, is a *measure of the work done in inducing the fresh charge into the cylinder.*
- The *four phases* of the naturally aspirated engine's cycle are represented as follows :

Induction	:	0 to 1
Compression	:	1 to 3
Power	:	3, 4 and 5
Exhaust	:	5, 6 and 7

16.2.2. Supercharged Cycle of Operation. Refer Fig. 16.3.

- With this pressurised charged system the *large upper loop area* (1-2-3-4-1) represents a measure of the work done in moving the piston to and fro so that the crankshaft rotates.

Conversely, the *small loop area* (0-1-5-6-0) *which is above the atmospheric line represents the work done in pumping the fresh charge into the cylinder.*

- The *four phases* of the supercharged engine's cycle of operation are represented as follows :

Induction	:	0 to 1
Compression	:	1 to 2
Power	:	2, 3 and 4
Exhaust	:	4, 5 and 6.

16.2.3. Comparison of Actual Naturally Aspirated and Supercharged Engine Pressure-Volume Diagrams

Fig. 16.4 shows a direct comparison of how the actual cylinder pressure varies relative to the cylinder's swept volume for both a naturally aspirated and a *supercharging petrol engine with wide-open throttle.*

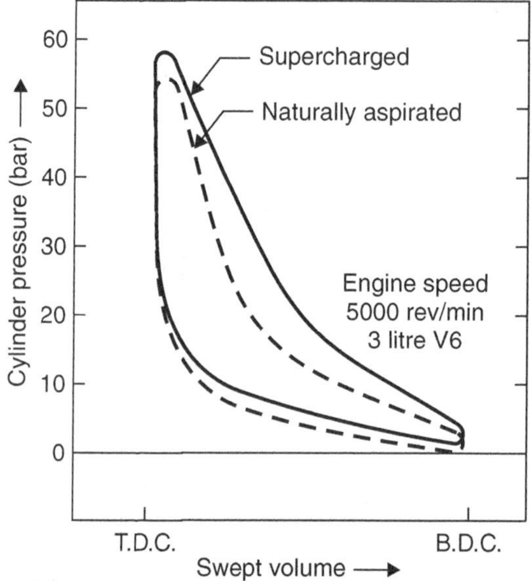

Fig. 16.4. Comparison of actual naturally aspirated and supercharged petrol engine pressure-volume diagrams.

— The two enclosed loops show that the lower compression stroke part of the curves for the supercharged engine is higher than for the naturally aspirated engine, whereas the peak cylinder pressures for the supercharged engine are only marginally higher.

— However, the *main advantage gained by supercharging the cylinders is that the vertical distance between the upper and lower curves for the supercharged engine is greater throughout the cylinder swept volume, which indicates that the mean effective pressure is greater.*

— Finally, the loop area enclosed by the supercharged engine is much larger, the *proportional difference being a measure of the increased power in the supercharged engine.*

16.2.4. Boost Pressure and Pressure Ratio

- **Boost pressure** *refers to the gauge pressure recorded when the air or mixture supply has passed through the supercharger.*

- **Pressure ratio** is the *ratio of absolute pressure to that of the atmospheric pressure.* (Here, absolute pressure = boost pressure + atmospheric pressure)

The *intensity of supercharging* can be broadly classified *as follows :*

S. No.	Particulars	Degree of charging			
		Naturally aspirated	**Low**	**Medium**	**High**
1.	*Boost pressure (bar)*	0.0 and below	0.0 — 0.5	0.5 — 1.0	1.0 and above
2.	*Pressure ratio*	1.0 and below	1.0 — 1.5 : 1	1.5 — 2.0 : 1	2.0 and above

16.2.5. The Effect of Pressure Ratio on Air Charge Temperature

The relationship between supercharger pressure ratio increase and air charge temperature is shown in Fig. 16.5, where it can be seen that as *the boost pressure increases, so does the discharge air temperature.* The lower curve shows the theoretical temperature rise with increased pressure ratio, whereas the band between the lower and upper curves is the *working, temperature variation likely to be encountered due to the turbulence and friction resistance generated by the compression process.*

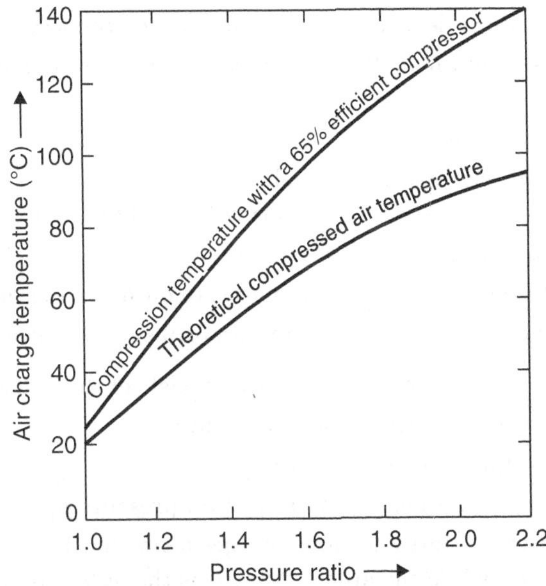

Fig. 16.5. Relationship between charge temperature and boost pressure ratio if heat is not dissipated.

Thus the minimum compression air temperatures with an intake temperature of 20°C for pressure ratios 1.2, 1.4, 1.6, 1.8, 2.0 and 2.2 are 35.7°C, 49.5°C, 62°C, 73.6°C, 84°C and 94°C respectively. In practice, with compressor efficiencies of the order of 60–75% and the churning, turbulence, and frictional factors, the actual output air charge temperature can be considerably higher, particularly at the higher boost levels.

16.2.6. Thermodynamic Cycle and Supercharging Power

Fig. 16.6 shows the thermodynamic cycle of a supercharged I.C. engine on the p-v diagram for an *ideal otto cycle.*

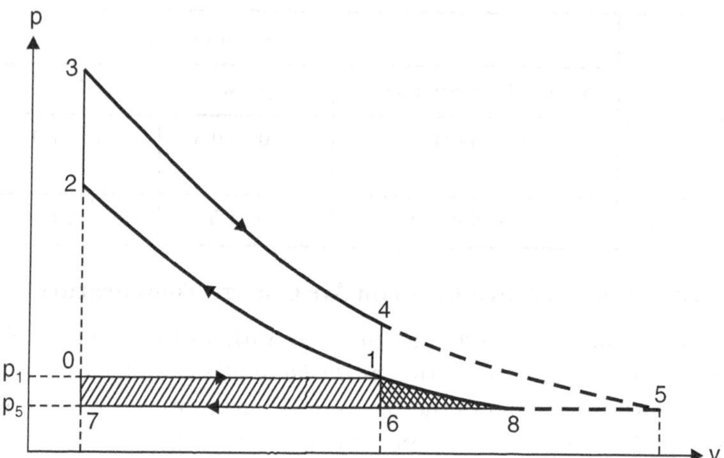

Fig. 16.6. Thermodynamic cycle of supercharged engine on p-v diagram for an ideal otto cycle.

- The pressure p_1 represents the *supercharging pressure* and p_5 is the *exhaust pressure.*
- The thermodynamic cycle, consists of the following *processes :*
  - (*i*) **0–1.** Admission of air *at the supercharging pressure* (which is *greater than* atmospheric pressure).
  - (*ii*) **1–2.** Isentropic compression.
  - (*iii*) **2–3.** Heat addition at constant volume (for diesel cycle, this will be replaced by a constant pressure process, representing heat addition at constant pressure.
  - (*iv*) **3–4.** Isentropic expansion.

    4–1–6. Heat rejection at constant volume (blow down to atmospheric pressure).

    6–7. Driving out exhaust at constant atmospheric pressure.

The thermodynamic cycle for the *supercharger* consists of the following *processes* :

(*i*) **7–6–8.** Admission of air at atmospheric pressure.

(*ii*) **8–1.** Isentropic compression to pressure p_1.

(*iii*) **1–0.** Delivery of supercharged air, at a constant pressure p_1.

- Area 8-6-7-0-1-8 represents the *supercharger work* (mechanically driven) in supplying air at a pressure p_1, while the area 1-2-3-4-1, is the *output of the engine.* Area 0-1-6-7-0 represents the *gain in work during the gas exchange process due to supercharging. Thus a part of the supercharger work is recovered.* However, the work represented by the area 1-6-8-1 *cannot be recovered and represents a loss of work.*

Supercharging power :

Refer Fig. 16.7.

p_1, v_1, T_1 = Initial conditions of air at entry to the supercharger ;

p_2, v_2, T_2 = Final conditions of air at exit from the supercharger.

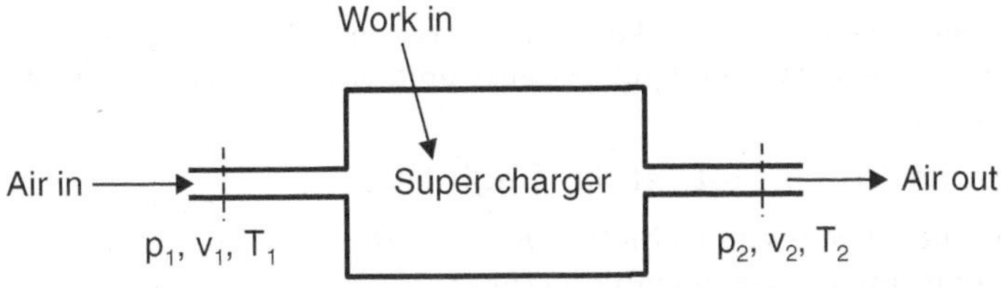

Fig. 16.7. Steady flow process.

Assuming adiabatic compression of air, the work done on the supercharger *per kg of air* is given by

$$W = -\int v\,dp = h_2 - h_1$$

$$= c_p (T_2 - T_1) = c_p T_1 \left(\frac{T_2}{T_1} - 1 \right)$$

$$= c_p T_1 \left[\left(\frac{p_2}{p_1} \right)^{\frac{\gamma-1}{\gamma}} - 1 \right] \times \frac{1}{\eta_{adi.}},$$

where, $\eta_{adi.}$ is the adiabatic efficiency of the supercharger/compressor.

∴ *Power required to drive the supercharger (P)* is then given by,

$$P = \frac{\dot{m}_a c_p T_1}{\eta_{adi.}} \left[\left(\frac{p_2}{p_1} \right)^{\frac{\gamma-1}{\gamma}} - 1 \right] \text{kW}$$

where, $\dot{m}_a$ = Amount of air in kg/s supplied by the supercharger, and

c_p = Specific heat of air in kJ/kg K.

This power may be supplied by :

— Gas turbine driven by the exhaust gas energy of the engine.

— Separate drive by motor or any other prime mover driving the supercharger.

— Connecting the supercharger to the engine output shaft.

In all the above cases the *gain in the power output of the engine would be many times the power required to drive the supercharger/compressor.*

Important points worth noting regarding supercharging of S.I. engines :

- Supercharging (of S.I. engines) is employed *only for aircraft and racing car engines,* due to the fact that the increase in supercharging pressure increases the *tendency to detonate and pre-ignite.*
- The supercharged petrol engines have a greater fuel consumption than naturally aspirated engines.
- Increased intake pressure and temperature *reduces ignition delay and increase flame speed.* The increased flame speeds make the petrol engine *more sensitive to fuel-air ratio and engine cannot run on weak mixtures without knock.* Rich mixtures are used to control detonation, *which further increases the specific fuel consumption of the engine.*
- Supercharging of petrol engines, because of poor fuel economy, is used in the following cases :
  — When a *large amount of power is needed,* or
  — When *more power is needed to compensate altitude loss.*

Note. In highly supercharged engine, knocking can be controlled by *injection of water in the combustion chamber* (however use of large amount of liquid becomes prohibitive). Alternatively the *charge may be cooled before it is fed to the engine.*

16.2.7. Supercharging Limits of S.I. Engines

It is primarily the '*knock*' which chiefly *limits the degree of supercharging in S.I. engines*. The knock limit is dependent upon the following factors :

(*i*) The type of fuel used ;

(*ii*) Mixture ratio ;

(*iii*) Spark advance ;

(*iv*) Design features of the engine (important being cooling systems and valve timing)

— It has been observed that for *volatile petroleum fuels* of high octane number the knocking and pre-ignition tendency is reduced at very rich and very lean mixture, and that the fuels of same octane value have different response to supercharging. In the case of alcoholic fuels the knock is reduced at rich mixtures (because of the cooling effect of high latent heat of these fuels).

— Very *lean and very rich mixtures give non-knocking operations. The strongest knocking,* however, occurs *near chemically correct mixtures.* The use of rich mixture results in a higher specific fuel consumption for a supercharged engine though knocking is controlled). A slight reduction in lean mixture makes the engine operation irregular and intermittent.

— The *ignition timings and thermal load on the engine affect the knock limit of S.I. engine.* The *ignition must be retarded at high intake pressure and temperature.*

- In general, supercharger pressure of 1.3 to 1.5 bar is used, which corresponds to about 30 to 50 percent supercharging.

16.3. SUPERCHARGING OF C.I. ENGINES

Fig. 16.8 shows the schematic arrangement and *p-v* diagram for a supercharged constant pressure (diesel) cycle.

- Unlike S.I. engines supercharging does not result in any combustion problem, *rather it improves combustion in diesel engine.* Increase in pressure and temperature of the intake air reduces significantly delay and hence the rate of pressure rise resulting in a *better,*

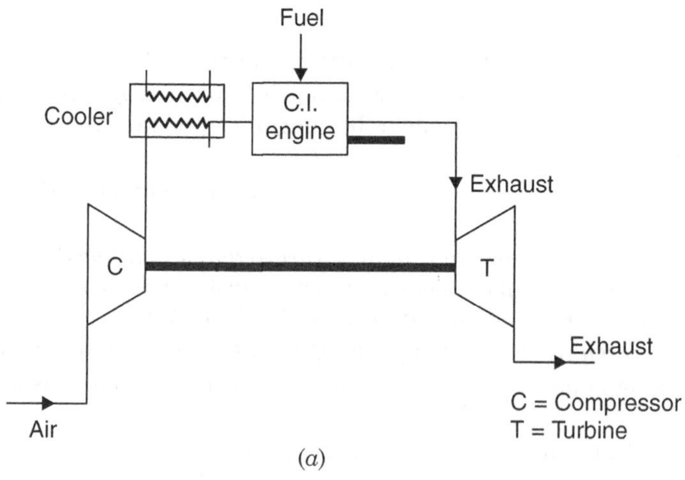

(*a*)

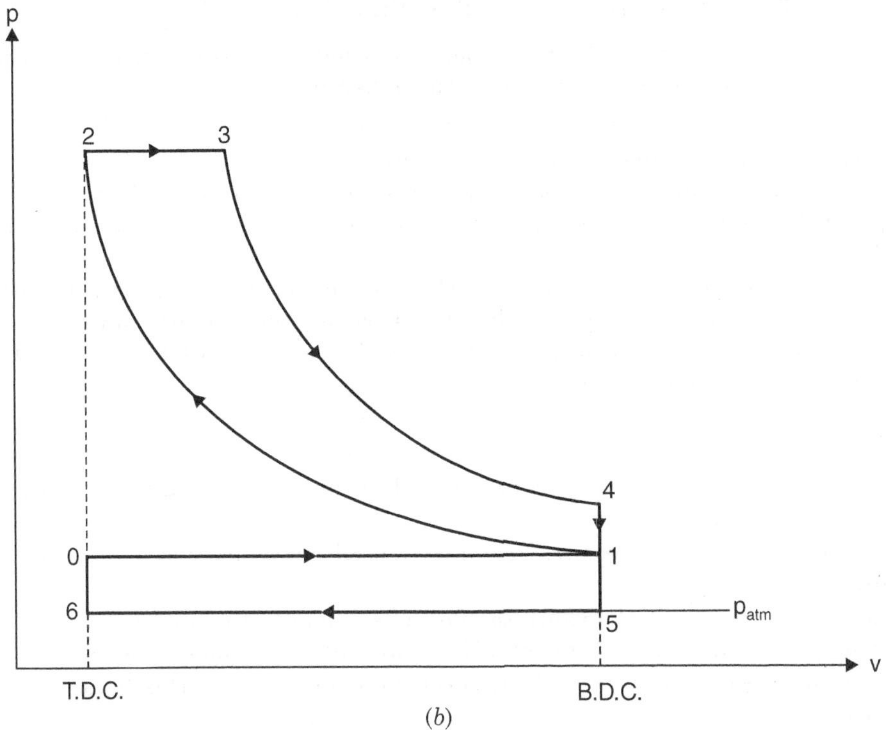

(*b*)

Fig. 16.8. Supercharging C.I. engine.

quieter and smoother combustion. This improvement in combustion allows a poor quality fuel to be used in a diesel engine and it is also not sensitive to the type of fuel used. *The increase in intake temperature reduces volumetric and thermal efficiency but increase in density due to pressure compensates for this and intercooling is not necessary except for highly supercharged engines.*

- *If an unsupercharged engine is supercharged it will increase the reliability and durability of the engine due to smoother combustion and lower exhaust temperatures. The degree of supercharging is limited by thermal and mechanical load on the engine and strongly depends on the type of supercharger used and design of the engine.*

16.3.1. Supercharging Limits of C.I. Engines

The supercharging limits for a C.I. engine (unlike S.I. engine where the limits of supercharging are due to combustion) is reached by ***thermal loading.*** The very high temperature of the piston and cylinder causes scuffing of piston rings and heavy liner wear.

It has been observed that load on bearing is increased due to increased pressure in the cylinder.

The *main considerations* in limiting the degree of supercharging of a C.I. engine are :

(*i*) Durability ;

(*ii*) Reliability ;

(*iii*) Fuel economy.

— The reliability of the engine decreases with the increase in maximum pressure in the cylinder. This also increases the thermal load on the engine due to the increase in the rate of heat release.

16.4. MODIFICATION OF AN ENGINE FOR SUPERCHARGING

Supercharging results in the increased output of a naturally-aspirated engine. The following *modifications* make the engine more suitable to supercharging :

- *Increase in the valve overlap period to permit complete scavenging of the clearance volume.*
- *Increase in the clearance volume by decreasing the compression ratio.*
- The injection system of a diesel engine must be modified to *supply increased amount of fuel* (this will require a nozzle of greater area than that required for the normally aspirated engine).
- In case of a turbocharged engine, in order to supply more energy to the turbocharger the exhaust value shall open a bit earlier. Furthermore, the exhaust manifold of such engine is insulated to reduce heat losses (whereas in case of a normally aspirated engine the exhaust manifold is water-cooled).

16.5. SUPERCHARGERS

The following types of superchargers are used to supercharge engines for various applications :

1. **Reciprocating compressor.** Very rarely used nowadays except for some stationary installations.

2. **Vane blower.** Due to the vanes the flow of air is pulsating and noisy and the speed is limited because of the radial motion of vanes. Nowadays these are almost obsolete.

3. **Lysholm compressor.** It produces a constant compression. Has limited use due to its mechanical complexity.

4. **Roots blower.** It is *suitable for low and medium speed engines for stationary and marine installations.*

5. **Centrifugal compressor.** It is simple, small, cheap and has a good efficiency in the range of pressure ratio of 1.5 to 3.0 and is *commonly used for supercharging.*

— The limited speed range of this compressor makes it suitable for constant speed type engines, such as aircraft engines. However, the exhaust driven centrifugal compressor is very popular because its operational speed matches with the exhaust turbine speed satisfactorily.

— *Disadvantage.* Occurrence of surge.

16.6. SUPERCHARGING ARRANGEMENTS

Fig. 16.9 shows the different supercharging arrangements.

- Fig. 16.9 (*i*) shows a compressor coupled to the engine with step-up gearing for increasing the speed of the compressor. Here, a portion of the engine output is used to drive the compressor. The net output increase due to supercharging is obtained by subtracting this power from the engine gross output. The purpose of aftercooler is to send cool air to the engine for further increasing the density of the intake air.
- Fig. 16.9 (*ii*) shows an arrangement in which compressor is driven by a turbine which is run by the exhaust of the engine. The compressor and the turbine are not mechanically coupled to the engine.
- Fig. 16.9 (*iii*) refers to an arrangement in which engine, turbine and compressor all are geared together. The wright Turbo-compound air plane engine is an example of this type of arrangement. Here if the turbine output is insufficient to run the compressor

particularly at part loads, then the remaining load of the compressor is taken care of by the engine.

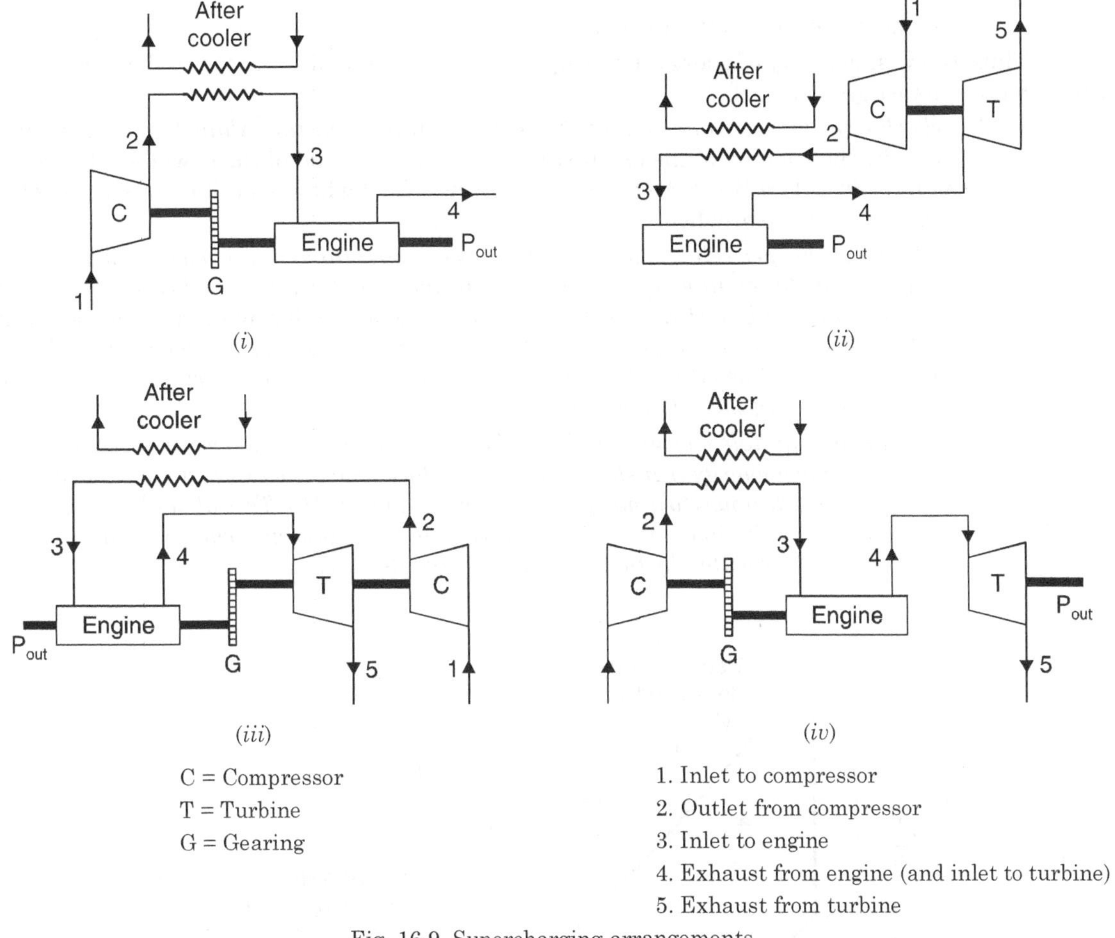

C = Compressor
T = Turbine
G = Gearing

1. Inlet to compressor
2. Outlet from compressor
3. Inlet to engine
4. Exhaust from engine (and inlet to turbine)
5. Exhaust from turbine

Fig. 16.9. Supercharging arrangements.

- Fig. 16.9 (*iv*) shows an arrangement of supercharging in which engine supplies its total power to the compressor and the exhaust gases from the engine run the turbine giving the power output. Such plants are called *fuel-piston engines.*

16.7. TURBOCHARGERS

16.7.1. Introduction

- ***Turbochargers*** *are centrifugal compressors driven by the exhaust gas turbines.* By utilising the exhaust energy of the engine it recovers a substantial part of energy which would otherwise go waste ; thus the *turbocharger will not draw upon the engine power.* These are *nowadays extensively used to supercharging almost all types of two stroke engines.*
- A *typical petrol engine may harness up to 30% of the energy contained in the fuel supplied to do useful work under optimum conditions but the remaining 70% of this energy is lost in the following way :*

— 7% heat energy to friction, pumping and dynamic movement ;

— 9% heat energy to surrounding air ;

— 16% heat energy to engine's coolant system ;

— 38% heat energy to outgoing exhaust gases.

Thus, the vast majority of energy, for design reasons, is allowed to escape to the atmosphere through the exhaust system.

- A *turbocharger utilizes a portion of the energy contained in the exhaust gas*—when it is released by the opening of the exhaust valve towards the end of the power stroke (something like 50° before B.D.C.)—to drive a turbine wheel which simultaneously propels a centrifugal compressor wheel.
  - The *turbocharger relies solely on extracting up to a third of the wasted energy passing out from the engine's cylinders to impart power to the turbine wheel and compressor wheel assembly*. However, this does produce a penalty by increasing the manifold's back-pressure and so making it more difficult for each successive burnt charge to be expelled from the cylinders. It therefore *impedes the clearing process in the cylinders during the exhaust strokes.*
  - The *ideal available energy which can be used to drive the turbocharger comes from the blow-down energy transfer which takes place when the exhaust valve opens and the gas expands down to atmospheric pressure* (Fig. 16.10). *This blow-down energy is represented* by the loop area 4-5-6 *whereas the boost pressure energy used to fill the cylinder is represented by the rectangular area 0-1-6-7.*

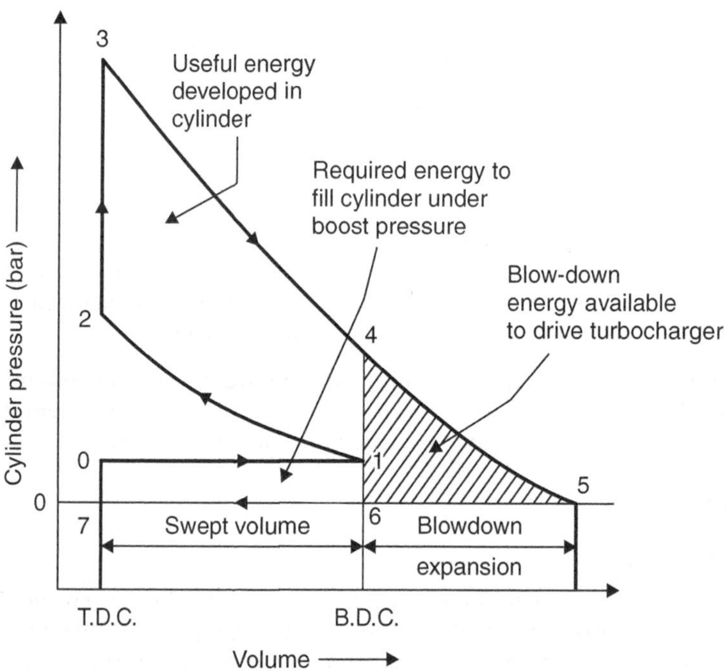

Fig. 16.10. Petrol engine cycle pressure-volume diagram showing available exhaust gas energy.

- *Turbocharged engines produce higher cylinder volumetric efficiencies compared with the normally aspirated induction systems. Therefore, there will be higher peak cylinder pressures which increase the mechanical loading of the engine components and could cause detonation in petrol engines. Therefore, it is usual to reduce the engine's compression ratio by a factor of one or two. Thus, a compression ratio of 10 : 1 for a normally aspirated engine would be derated to 9 : 1 for a low boost pressure or even reduced to 8 : 1 if a medium to high boost pressure is to be introduced. Similarly, for a direct injection diesel engine which might normally have a compression ratio of 16 : 1, when lightly turbocharged the compression ratio may be lowered to 15 : 1 and, if much higher supercharged inlet pressures are to be used, the compression ratio may have to be brought down to something like 14 : 1.*

- The *compression of the charge entering the cells of the impeller depends upon the centrifugal force effect which increases with the square of the rotational speed of the impellor wheel. Consequently, under light load and low engine speed conditions the energy released with the exhaust gases will be relatively small and is therefore insufficient to drive the turbine assembly at very high speeds.* Correspondingly, there will be very little extra boost pressure to make any marked improvement to the engine's torque and power output in the low-speed range of the engine. Thus, in effect, the turbocharged engine will operate with almost no boost pressure and with a reduced compression ratio compared with the equivalent naturally aspirated engine. Hence, *in the very low speed range, the turbocharged engine may have torque and power outputs and fuel consumption values which are inferior to the unsupercharged engine.*
- *Another inherent undesirable characteristic of turbochargers is that when the engine is suddenly accelerated there will be a small time delay before the extra energy discharged into the turbine housing volute can speed up the turbine wheel. Thus, during this transition period, there will be very little improvement in the cylinder filling process, and hence the rise in cylinder brake mean effective pressure will be rather sluggish.*

16.7.2. Altitude Compensation

Engine power outputs are tested and rated at sea-level where the atmospheric air is most dense ; however, as a vehicle climbs, its altitude is increased and the air becomes thinner, that is, less dense. The consequence is a decrease in volumetric efficiency as less air will be drawn into the cylinders per cycle, with a corresponding reduction in engine power since power is directly related to

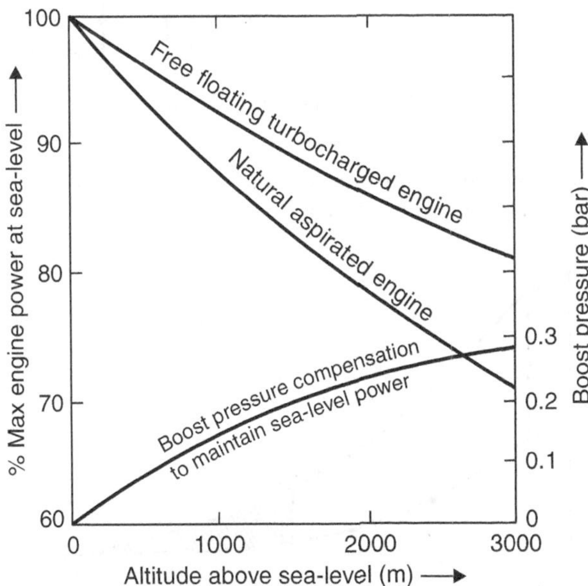

Fig. 16.11. Effect of altitude on rated engine power at sea level for both naturally-aspirated and turbocharged engines.

the actual mass of charge burnt in the cylinder's every power stroke. A *naturally aspirated engine will have its power output reduced by approximately 13% if it is operated approximately* ***1000*** *m above sea-level. Supercharging the cylinders enables the engine's rated power above sea-level to be maintained or even exceeded.*

With a turbocharged engine there will still be some power loss with the engine operating at high altitudes, but the *loss will be far less than if the engine breathing depended only on natural aspiration. As can be seen (Fig. 16.11) at 1000 m the power loss is only 8% compared with the naturally aspirated engine where the power decrease is roughly 13%.*

16.7.3. Turbocharging-Buchi System

- Exhaust gases coming out from the engine, when the exhaust valve opens, are at a pressure well above atmospheric pressure, and their temperature is also high. A part of the heat energy contained in the gases can be utilized by further expansion of the gases in an exhaust turbine down to atmospheric pressure. The extra power available from the turbine is used to drive the compressor. The compressor, in turn, will supply more air to the engine. Such utilization of the exhaust energy to drive the supercharger is called *Buchi system of turbocharging.*
  — Fig. 16.12 shows the arrangement of Buchi system.

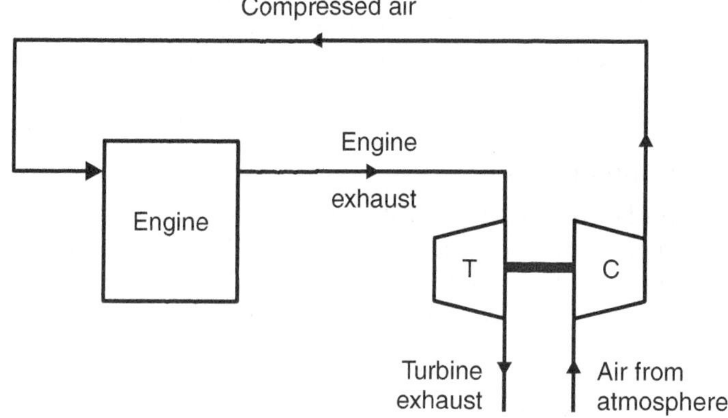

T = Turbine, C = Compressor

Fig. 16.12. Arrangement of Buchi system.

— Fig. 16.13 shows the additional energy equal to area 1-4-5 available when the expansion is carried down to atmosphere.

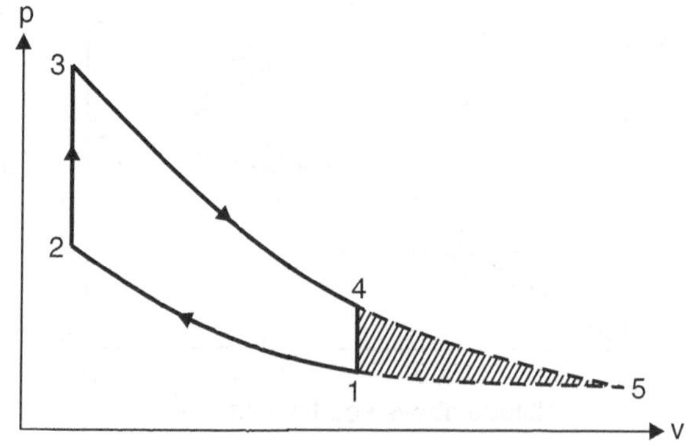

Fig. 16.13. Diagram shows additional energy available when expansion is carried down to atmosphere.

- The *power developed by the turbocharger is sufficient to drive the compressor, and overcome its mechanical friction.* The turbocharger is independent of the engine, and it is only connected to it by a simple exhaust pipe. The speed range of the turbocharger is from 20000 to 30000 r.p.m.
- For supplying adequate energy to the turbocharger, the *exhaust valve is opened much before the B.D.C.* in contrast to a naturally aspirated engine. *This permits the exhaust gases to escape at a higher pressure and temperature, giving turbocharger enough energy to drive the compressor. The loss in piston work due to early opening of the exhaust valve is more than offset by better charging and scavenging of the engine.*

16.7.4. METHODS OF TURBOCHARGING

The following are the main types of turbocharging methods :

1. **Constant pressure turbocharging :**

- The various cylinders discharge their exhaust into a common manifold at pressures higher than the atmospheric.

The exhaust gases (from all the cylinders) undergo expansion in the exhaust valves, without doing any work, to an approximately constant pressure in the common manifold and then enter the turbine. Thus the blow-down energy in the form of internal energy is converted into work in the turbine. The higher the pressure ratio of the turbine, the higher is the recovery of blow-down energy. During the whole of the cycle the exhaust gases are maintained at constant pressure to make use of a pure reaction turbine.

2. **Pulse turbocharging :**

- In this method of supercharging, as soon as the exhaust valve opens a considerable part of the blow-down energy is converted into *exhaust pulses.* These pulses enter the turbine (through narrow exhaust pipes by the shortest possible route) where a major proportion of the energy is recovered.
- In order that exhaust process of various cylinders do not interfere with one another, separate exhaust pipes are used.

3. **Pulse converter :**

This turbocharging method permits the advantages of the pulse and constant-pressure tubocharging methods simultaneously. The combination of these two systems is done by *connecting the different branches of exhaust manifolds together in a specially designed venturi junction, called "pulse converter", before the turbine.*

4. **Two-stage turbocharger :**

Two-stage turbocharging is defined as *the use of two turbochargers of different sizes in series, e.g.* a high pressure stage operating on pulse system and low-pressure stage on constant pressure operation. This type of arrangement is employed for diesel engines requiring very high degree of supercharging, b.m.e.p. ranging from 25 to 30 bar, which can not be obtained in a single-stage turbocharger.

5. **Miller turbocharging :**

- The system of turbocharging is based upon the idea of *increasing the expansion ratio relative to compression ratio by means of early closure of inlet valve as the boost pressure is increased.*
- The Miller turbocharging is not very attractive unless two-stage turbocharging is necessary because of other reasons such as need to reduce exhaust valve failures.

6. **Hyperbar turbocharging :**

Fig. 16.14 shows a hyperbar turbocharged engine. It consists of the following components :

1. A low compression (7 : 1) diesel engine ;
2. A high pressure ratio (upto 5.1) turbine ;
3. A by-pass control;
4. An auxiliary combustion chamber/ located between the direct exhaust valve and the turbocharger turbine.
   — Firstly, using an electrical starter, the turbocharger is started ; it is kept running by bypassing the air and injecting first into the auxiliary combustion chamber while the engine in inoperative.
   — Then after sometime when appropriate pressures and temperatures are reached the diesel engine is started.
   — The amount of air bypassed and the fuel injected into the auxiliary combustion chamber are controlled in accordance with operating conditions.

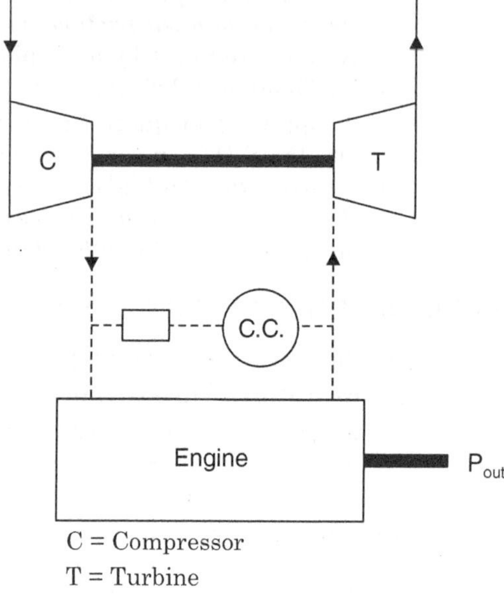

C = Compressor
T = Turbine
C.C. = Combustion chamber

Fig. 16.14. Hyperbar turbocharging.

Advantages :

1. High power-to-weight ratio.
2. High brake mean effective pressure to the tune of 30 bar can be obtained (peak pressure limited to 140 bar).
3. Thermal loading-moderate.
4. Surge pre-operation (due to by pass control).
5. Good torque and acceleration available.

Disadvantages :

1. Due to low compression ratio, there is high fuel consumption over the entire range of operation.
2. System is complex.

16.7.5. Limitations of Turbocharging

Turbocharging entails the following *limitations :*

1. Special exhaust manifolds are required for the turbocharging system.
2. In order to inject more fuel per unit time fuel injection needs modification.
3. In contrast to a naturally aspirated engine which can digest solid particles in the inlet air without undue stress, a turbocharged engine *can pass only the most minute material particles without damage.*
4. It is difficult to obtain good efficiency over a wide range of operations since the efficiency of the turbine blades is very sensitive to gas velocity.

WORKED EXAMPLES

Example 16.1. *An unsupercharged petrol engine develops 735 kW with air fuel ratio 12.8. The bsfc in 0.350 kg/kWh and mechanical efficiency 86%. The inlet pressure is 730 mm of mercury absolute and the mixture temperature is 325 K. The engine is supercharged to a pressure ratio of 1.6 by a supercharger of adiabatic efficiency 0.7 and mechanical efficiency 0.9. Assuming that air-fuel ratio remains unchanged and I.P. is proportional to inlet density, calculate the power required to run the supercharger. Assume that volumetric efficiency does not change due to supercharging.*

(Madras University)

Solution. *Given :* $\frac{p_2}{p_1} = 1.6$, $p_1 = 730$ mm of Hg abs., $T_1 = 325$ K

$\eta_{adi.} = 0.7$, $\eta_{mech.} = 0.9$

$$\frac{T_2}{T_1} = \left(\frac{p_2}{p_1}\right)^{\frac{\gamma-1}{\gamma}} = (1.6)^{\frac{1.4-1}{1.4}} = 1.1437$$

$\therefore$ $T_2 = 325 \times 1.1437 = 371.7$ K

Also, $$\eta_{adi.} = \frac{T_2 - T_1}{T_2' - T_1}$$

or $$T_2' = T_1 + \frac{T_2 - T_1}{\eta_{adi.}}$$

$\therefore$ $$T_2' = 325 + \frac{371.7 - 325}{0.7} = 391.7 \text{ K}$$

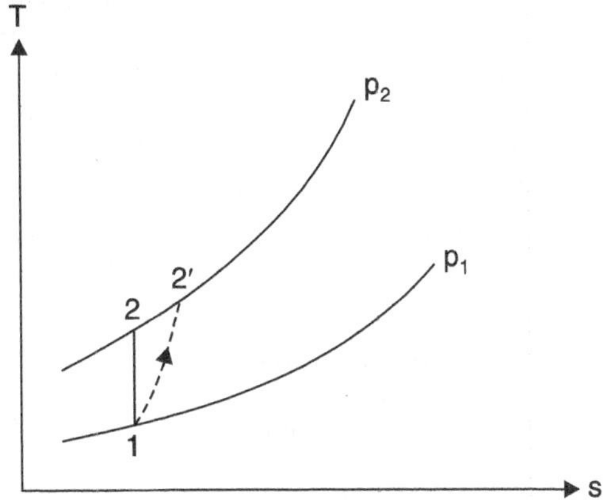

Fig. 16.15

Indicated work of supercharger $= c_p (T_2 - T_1)$ kJ/kg of air.

Work required to drive the supercharger

$$W_{sup} = \frac{c_p(T'_2 - T_1)}{\eta_{mech.}} = \frac{1.005(391.7 - 325)}{0.9} = 74.48 \text{ kJ/kg of air}$$

When unsupercharged :

Inlet pressure, $$p_1 = \frac{730}{1000} \times \frac{9.81 \times (13.6 \times 1000)}{1000} = 97.39 \text{ kN/m}^2$$

Density, $$\rho_{unsup} = \frac{p_1}{RT_1} = \frac{97.39}{0.287 \times 325} = 1.044 \text{ kg/m}^3$$

Air consumption, $$\dot{m}_{a(unsup).} = \frac{0.35 \times 735 \times 12.8}{3600} = 0.9047 \text{ kg/s}$$

When supercharged :

Density, $$\rho_{sup.} = \frac{p_2}{RT_2} = \frac{1.6 \times 97.39}{0.287 \times 391.7} = 1.386 \text{ kg/m}^3$$

Air consumption, $$\dot{m}_{a(sup.)} = m_{a(unsup.)} \times \frac{\rho_{sup}}{\rho_{unsup.}}$$

$$= 0.9147 \times \frac{1.386}{1.044} = 1.2143 \text{ kg/s}$$

∴ Power required to run the supercharger,

$$\mathbf{P_{sup}} = \dot{m}_{a(sup)} \times W_{sup}$$

$$= 1.2143 \times 74.48 = \mathbf{90.44\ kW.}\quad \textbf{(Ans.)}$$

Example 16.2. *A diesel engine operating on four-stroke cycle is to be designed to operate with following characteristics at sea level, where the mean conditions are 1.0132 bar and 10°C.*

B.P. = 260 kW, volumetric efficiency = 78% (at sea level free air conditions), specific fuel consumption = 0.247 kg/B.P.h. ; A/F ratio = 17 ; speed = 1500 r.p.m.

Calculate the required engine capacity and the anticipated brake mean effective pressure.

The engine is fitted with a supercharger so that it may be operated at an altitude of 2700 m where the atmospheric pressure is 0.72 bar. The power taken by a supercharger is 8 per cent of the total power produced by the engine and the temperature of the air leaving the supercharger is 32°C. The air-fuel ratio and thermal efficiency remain the same for the supercharged engine as when running unsupercharged at sea level, as does the volumetric efficiency. Calculate the increase of air pressure required at the supercharger to maintain the same net output of 260 kW. Take R = 0.287 kJ/kgK.

Solution. *Given :* p_1 = 1.0132 bar, T_1 = 10 + 273 = 283 K, B.P. = 260 kW,
$\eta_{vol.}$ = 78%, s.f.c. = 0.247 kg / B.P. *h*, A / F ratio = 1.7 : 1,
N = 1500 r.p.m.

Engine capacity :

Fuel consumption, $$m_f = \frac{s.f.c. \times \text{B.P.}}{60} \text{ kg/min.} = \frac{0.247 \times 260}{60} = 1.07 \text{ kg/min.}$$

Air consumption = Fuel consumption × A / F ratio
= 1.07 × 17 = 18.19 kg / min.

Air consumption per stroke $$= \frac{\text{Air consumption in kg / min.}}{\text{No. of cycles / min.}}$$

$$= \frac{\text{Air consumption in kg/min.}}{N/2} = \frac{18.19}{1500/2} = 0.0242 \text{ kg}$$

Let V_s be the swept volume, then mass of free air corresponding to swept volume $= \frac{p_1 V_s}{RT_1}$

$$= \frac{(1.0132 \times 10^5) \times V_s}{287 \times 283} = 1.247\ V_s \text{ kg}$$

Volumetric efficiency, $$\eta_{vol.} = \frac{\text{Mass of air taken in per stroke}}{\text{Mass of free air corresponding to swept volume}}$$

∴ $$0.78 = \frac{0.0242}{1.247\ V_s} \quad \text{or} \quad V_s = \frac{0.0242}{0.78 \times 1.247} = 0.02488 \text{ m}^3$$

i.e., *Engine capacity* = **0.02488 m³. (Ans.)**

Brake mean effective pressure, p_{mb} (bar) :

We know that $$\text{B.P.} = \frac{p_{mb}\ LANk \times 10}{6} \text{ kW}$$

$$260 = \frac{p_{mb} \times V_s \times Nk \times 10}{6} \text{ kW} \qquad (\because\ L \times A = V_s)$$

$$= \frac{p_{mb} \times 0.02488 \times 1500 \times \frac{1}{2} \times 10}{6}$$

$\therefore$ $\quad \mathbf{p_{mb} = 8.36\ bar.}$ **(Ans.)**

Supercharged engine :

Increase of pressure required :

Gross power produced by the engine $= 260 + 0.08 \times$ gross power

$\therefore$ Gross power produced by the engine $= \frac{260}{(1 - 0.08)} = 282.6$ kW

$\therefore$ Mass of air required $= \frac{18.19 \times 282.6}{260} = 19.77$ kg

$$\eta_{vol.} = \frac{\text{Mass of air taken in per cycle}}{\text{Mass of air corresponding to swept volume measured at outlet conditions of supercharger}}$$

$$0.78 = \frac{\frac{19.77}{1500/2}}{\frac{(p_2 \times 10^5) \times 0.02488}{287 \times (32 + 273)}} = \frac{0.02636}{0.0284 p_2}$$

or $\quad p_2 = \frac{0.02636}{0.78 \times 0.0284}$ bar $= 1.19$ bar

$\therefore$ **Increase of pressure required** $= 1.19 - 0.72 =$ **0.47 bar. (Ans.)**

Example 16.3. *A 4-stroke diesel engine having a capacity of 3600 cm³ develops 13 kW per m³ of free air induced per minute. It has a volumetric efficiency of 82 percent at 3000 r.p.m. referred to free air conditions of 1.0132 bar and 25°C. This engine is supercharged by a rotary compressor which develops a pressure ratio of 1.8 and has an isentropic efficiency of 75 per cent. This compressor is coupled to main shaft of the engine which supplies power to it. Estimate the increase in brake power due to surpercharging.*

Assume mechanical efficiency of 80 per cent and the air at intake to cylinder to be at the pressure equal to the delivery pressure from compressor and temperature equal to 4°C less than the delivery temperature from the compressor, and cylinder contains volume of charge equal to swept volume.

Solution. *Given :* Engine capacity $= 3600$ cm$^3 = 3600 \times 10^{-6}$ m^3 ;

Power developed $= 13$ kW/m^3 of free air induced per minute ;

Volumetric efficiency, $\eta_{vol} = 82\%$ at 3000 r.p.m. and

$p_1 = 1.0132$ bar, $T_1 = 25 + 2732 = 298$ K ;

Pressure ratio (rotary compressor) = 1.8

Isentropic efficiency of the compressor, $\eta_{isen.} = 75\%$

Mechanical efficiency, $\eta_{mech.} = 80\%$

Increase in B.P. due to supercharging :

Swept volume / min. $= (3600 \times 10^{-6}) \times \frac{3000}{2} = 5.4$ m^3 min.

$\therefore$ Unsupercharged volume induced per min.

$= $ Swept volume / min. $\times \eta_{vol}$

$= 5.4 \times 0.82 = 4.428$ m^3/min.

Rotary compressor delivery pressure = 1.8 × 1.0132 = 1.824 bar

Temperatures after isentropic compression,

$$T_2 = T_1 \left(\frac{p_2}{p_1}\right)^{\frac{\gamma-1}{\gamma}} = 298 \times (1.8)^{\frac{1.4-1}{1.4}} = 352.5 \text{ K}$$

Also, $\eta_{isen.} = \dfrac{T_2 - T_1}{T_2' - T_1}$

or $$0.75 = \frac{352.5 - 298}{T_2' - 298}$$

or $$T_2' = \left(\frac{352.5 - 298}{0.75}\right) + 298 = 370.7 \text{ K}$$

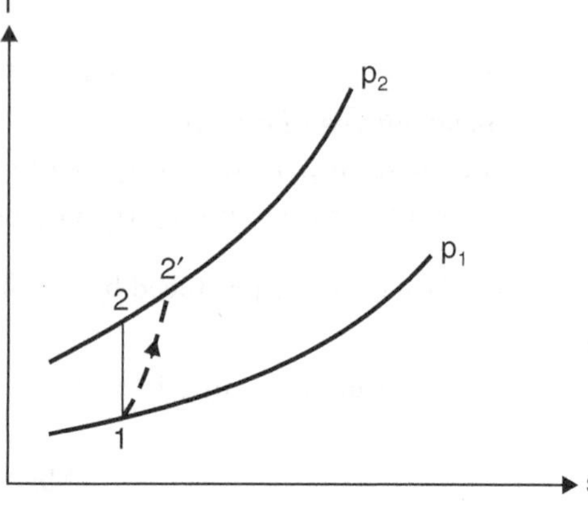

Fig. 16.16

The temperature of air at intake to the engine cylinder

$$= 370.7 - 4 = 366.7 \text{ K}$$

The compressor delivers 5.4 m^3/min. air at 1.824 bar and 370.7 K. Equivalent volume at 1.0132 bar and 298 K,

$$\frac{p_1 V_1}{T_1} = \frac{p_2 V_2}{T_2'}$$

$$V_1 = \frac{p_2 V_2 T_1}{p_1 T_2'} = \frac{1.824 \times 5.4 \times 298}{1.0132 \times 366.7} = 7.9 \text{ m}^3/\text{min.}$$

The increase in induced volume of air = 7.9 – 5.4 = 2.5 m^3/min.

Increase in I.P. from air induced = 13 × 2.5 = 32.5 kW

Increase in I.P. due to increased induction pressure

$$= \frac{\Delta p \times V_s/\text{min.}}{60 \times 1000} = \frac{(1.824 - 1.0132) \times 10^5 \times 5.4}{60 \times 1000} = 7.3 \text{ kW}$$

∴ Total increase in I.P. = 32.5 + 7.3 = 39.8 kW

∴ Increase in B.P. of the engine = 39.8 × $\eta_{mech.}$

= 39.8 × 0.8 = 31.84 kW

From this *must be deducted the power required to drive the rotary compressor,*

Mass of air delivered by the compressor,

$$\dot{m} = \frac{pV}{RT} = \frac{(1.824 \times 10^5) \times (5.4/60)}{287 \times 366.7} \text{ kg/s}$$

$$= 0.156 \text{ kg/s}$$

It may be noted that the delivery temperature from the compressor is 370.7 K and mass is calculated at 366.7 K because the volume occupied by this mass is known only at 366.7 K being the swept volume of the cylinder.

Power required by the compressor

$$= \frac{\dot{m}\, c_p\, \Delta T}{\eta_{mech.}} = \frac{0.156 \times 1.005 \times (370.7 - 298)}{0.8} = 14.25 \text{ kW}$$

∴ **Net increase in B.P.** = 31.84 – 14.25 = **17.59 kW. (Ans.)**

Example 16.4. *A 4-stroke diesel engine is designed to operate with the following characteristics at sea level, where the mean conditions are 1.013 bar and 10°C :*

B.P. = 250 kW ; Volumetric efficiency = 78 per cent (at sea level free air conditions) ; Specific fuel consumption = 0.245 kg/kWh ;

Air-fuel ratio = 17 : 1 ; Speed = 1500 r.p.m.

Determine the required engine capacity and the brake mean effective pressure.

If the engine is run at an altitude of 2700 m where the atmospheric pressure is 0.72 bar by fitting a supercharger directly and mechanically coupled to the engine ; the power consumed by the supercharger is 8 per cent of the total power produced by the engine and the temperature of air leaving the supercharger is 32°C. The air-fuel ratio and the thermal efficiency remain the same for the supercharged engine as when running unsupercharged at sea level, as does the volumetric efficiency.

Determine the increase of air pressure required at the supercharger to maintain the same net output of 250 kW. **(Kerala University)**

Solution. *Given :* $p_1 = 1.013$ bar ; $T_1 = 10 + 273 = 283$ K ; s.f.c. = 0.245 kg/kWh ;

Air-fuel ratio = 17 : 1 ; $N = 1500$ r.p.m. ; B.P. = 250 kW, $\eta_{vol.} = 78\%$; $p_2 = 0.72$ bar ; $T_2 = 32 + 273 = 305$ K.

(*a*) ***Unsupercharged engine at sea level :***

Engine capacity, V_s :

$$\text{Fuel consumption} = \frac{0.245 \times 250}{60} = 1.0208 \text{ kg/min.}$$

$$\therefore \quad \text{Air consumption} = 1.0208 \times 17 = 17.35 \text{ kg/min.}$$

$$\eta_{vol.} = \frac{\text{Mass of air taken in per cycle}}{\text{Mass of air corresponding to swept volume}}$$

$$0.78 = \frac{\dfrac{17.35}{(1500/2)}}{\dfrac{p_1 V_s}{RT_1}} = \frac{(17.35 / 750)}{(1.013 \times 10^5 \times V_s)/(287 \times 283)} = \frac{17.35 \times (287 \times 283)}{750 \times 1.013 \times 10^5 \times V_s}$$

or Swept volume,
$$V_s = \frac{17.35 \times (287 \times 283)}{0.78 \times 750 \times 1.013 \times 10^5} = \mathbf{0.0238\ m^3.} \quad \textbf{(Ans.)}$$

Brake mean effective pressure, p_{mb} :

$$\text{B.P.} = \frac{p_{mb}\ LANk \times 10}{6} \text{ kW}$$

$$250 = \frac{p_{mb} \times V_s \times 1500 \times \frac{1}{2} \times 10}{6} \qquad (\because \ V_s = L \times A)$$

or
$$250 = \frac{p_{mb} \times 0.0238 \times 750 \times 10}{6}$$

$$\therefore \quad \mathbf{p_{mb} = 8.4\ bar.} \quad \textbf{(Ans.)}$$

Supercharged engine :

Increase of pressure required :

$$\text{Gross power produced by the engine} = 250 + \frac{8}{100} \times \text{gross power}$$

$$\therefore \quad \text{Gross power} = \frac{250}{(1 - 0.08)} = 271.74 \text{ kW}$$

$$\therefore \quad \text{Mass of air required} = \frac{271.74 \times 17.35}{250} = 18.86 \text{ kg}$$

$$\eta_{vol.} = 0.78 = \frac{18.86/(1500/2)}{(p_2 V_s / RT_2)} = \frac{18.86 \times 2 \times RT_2}{1500 \times (p_2 \times 10^5)\, V_s}, \text{ where } p_2 \text{ is in bar}$$

$$= \frac{18.86 \times 2 \times 287 \times 305}{1500 \times p_2 \times 0.0238}$$

$$\therefore \qquad p_2 = \frac{18.86 \times 2 \times 287 \times 305}{0.78 \times 1500 \times 10^5 \times 0.0238} = 1.186 \text{ bar.}$$

$\therefore$ Increase of pressure required = 1.186 – 0.72 = **0.466 bar. (Ans.)**

Example 16.5. *An air compressor is being run by the entire output of a supercharged 4-stroke cycle oil engine. Air enters the compressor at 25°C and is passed on to a cooler where 1210 kJ/min. are rejected. The air leaves the cooler at 65°C and 1.75 bar. Part of this air flow is used to supercharge the engine which has a volumetric efficiency of 72 percent based on induction manifold condition of 65°C and 1.75 bar. The engine which has six cylinders of 100 mm bore and 110 mm stroke runs at 2000 r.p.m. and delivers an output torque of 150 Nm. The mechanical efficiency of engine is 80 per cent. Determine :*

(*i*) *The indicated mean effective pressure of the engine ;*

(*ii*) *The air consumption rate of the engine ;*

(*iii*) *The air flow into compressor in kg/min.*

Solution. *Given :* $T_1 = 25 + 273 = 298$ K ; $Q_{rej} = 1210$ kJ/min. ;
$p_2 = 1.75$ bar ; $\eta_{vol.} = 72\%$; $n = 6$; $D = 100$ mm = 0.1 m ;
$L = 110$ mm = 0.11 m ; $T_{out} = 150$ Nm ; $\eta_{mech.} = 80\%$; $N = 2000$ r.p.m.

(*i*) **The indicated mean effective pressure, p_{mi} :**

$$\text{B.P.} = \frac{2\pi NT}{60 \times 1000} \text{ kW} = \frac{2\pi \times 2000 \times 150}{60 \times 1000} = 31.41 \text{ kW}$$

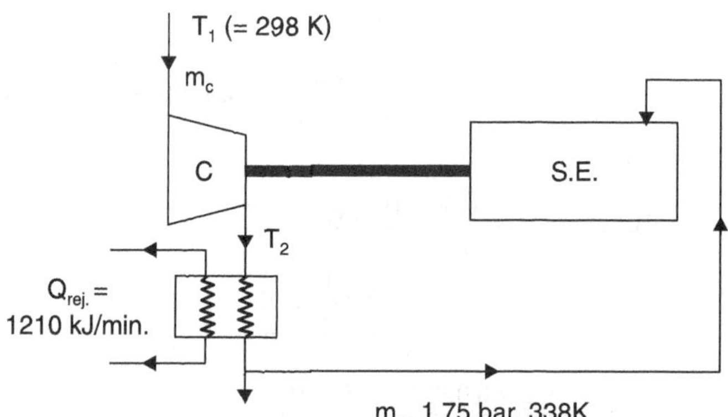

C = Compressor S.E. = Supercharged engine
m_e = Aspirated air mass flow into the engine.

Fig. 16.17

$$\text{I.P.} = \frac{\text{B.P.}}{\eta_{mech.}} = \frac{31.41}{0.8} = 39.26 \text{ kW}$$

But,
$$\text{I.P.} = \frac{n p_{mi}\, LANk \times 10}{6}$$

$$39.26 = \frac{6 \times p_{mi} \times 0.11 \times \frac{\pi}{4} \times (0.1)^2 \times 2000 \times \frac{1}{2} \times 10}{6} = 8.639\, p_{mi}$$

$\therefore$ $\mathbf{p_{mi} = 4.544}$ **bar. (Ans.)**

(*ii*) **Air consumption rate of the engine ; $\mathbf{m_c}$:**

Engine swept volume $= 6 \times \left(\frac{\pi}{4} D^2 L\right) \times \frac{N}{2}$ m³/min.

$$= 6 \times \frac{\pi}{4} \times 0.1^2 \times 0.11 \times \frac{2000}{2} = 5.184 \text{ m}^3/\text{min.}$$

Aspirated volume of air into engine

$$= \eta_{vol.} \times 5.184 = 0.72 \times 5.184 = 3.732 \text{ m}^3/\text{min.}$$

Aspirated air mass flow into the engine,

$$m_c = \frac{p_2 V_2}{RT_2} = \frac{(1.75 \times 10^5) \times 3.732}{287 \times 338} = \mathbf{6.732 \text{ kg/min.}} \quad \textbf{(Ans.)}$$

(*iii*) **The air flow into the compressor in kg/min, $\mathbf{m_c}$:**

Actual compressor work required from the engine to run the compressor = Gain in enthalpy of air in compressor

$$31.41 = \dot{m}_c \times c_p \times \Delta T$$
$$= \dot{m}_c \times 1.005 \times (T_2 - T_1)$$
$$= \dot{m}_c \times 1.005\, (T_2 - 298)$$

[where $\dot{m}_c$ = Mass of air handled by compressor in kg/s, and T_2 = Temperature of outlet air from compressor]

$\therefore$ $\dot{m}_c\, (T_2 - 298) = 31.25$ ...(*i*)

Also, in the cooler, $\dot{m}_c \times c_p \times [T_2 - (65 + 273)] = Q_{rej}/\text{sec.}$

or $$\dot{m}_c\, (T_2 - 338) = \frac{1210}{1.005 \times 60} = 20.07 \qquad ...(ii)$$

Dividing (*i*) by (*ii*), we get

$$\frac{T_2 - 298}{T_2 - 338} = \frac{31.25}{20.07} = 1.56$$

$$T_2 - 298 = 1.56\, (T_2 - 338) = 1.56\, T_2 - 527.28$$

$\therefore$ $T_2 = 409.4$ K or $136.4°$C

Thus, $$\dot{m}_c = \frac{31.25}{T_2 - 298} = \frac{31.25}{409.4 - 598} = 0.2805 \text{ kg/s}$$

or $m_c = 0.2805 \times 60 = \mathbf{16.83}$ **kg/min. (Ans.)**

HIGHLIGHTS

1. The purpose of *supercharging* is to raise the volumetric efficiency above that value which can be obtained by normal aspiration.
2. *Boost pressure* refers to the gauge pressure recorded when the air or mixture supply has passed through the supercharger.

3. *Pressure ratio* is the ratio of absolute pressure to that of the atmospheric pressure.
(Here, absolute-pressure = boost pressure + atmospheric pressure)
4. The supercharging limits for a C.I. engine (unlike S.I. engines where the limits of supercharging are due to combustion) is reached by *thermal loading.*
5. The following types of superchargers are used to supercharge engines for various applications :
Reciprocating compressors ; Vane blower ; lysholm compressor ;
Roots blower ; centrifugal compressor.
6. *Turbochargers* are centrifugal compressors driven by the exhaust gas turbines.
7. *Methods of turbocharging :*
(*i*) Constant pressure turbocharging.
(*ii*) Pulse turbocharging.
(*iii*) Pulse converter.
(*iv*) Two-stage turbocharger.
(*v*) Miller turbocharging.
(*vi*) Hyperbar turbocharging.

OBJECTIVE TYPE QUESTIONS

Fill in the Blanks or Say 'Yes' or 'No' :

1. The .... power produced is almost directly proportional to the engine air consumption.
2. Supercharging tends to lessen the possibility of detonation in a S.I. engine and increase the possibility in a C.I. engine.
3. The power output of a supercharged engine is .... than the naturally aspirated engines.
4. The mechanical efficiencies of supercharged engines are slightly better than naturally aspirated engines.
5. Inspite of better mixing and combustion due to reduced delay a mechanically supercharged otto engine almost always have specific fuel consumption higher than a naturally aspirated engine.
6. .... pressure refers to the gauge pressure recorded when the air or mixture supply has passed through the supercharger.
7. Supercharging of S.I. engines is employed only for aircraft and racing car engines, due to the fact that the increase in supercharging pressure increases the tendency to detonate and pre-ignite.
8. In highly supercharged engine, knocking can be controlled by injection of water in the combustion chamber.
9. Very lean and very rich mixtures give high knocking operations.
10. The ignition timings and thermal load on the engine do not affect the knock limit of S.I. engine.
11. Supercharging improves combustion in diesel engine.
12. The degree of supercharging in the case of a C.I. engine is limited by thermal and mechanical load on the engine and strongly depends on the type of supercharger used and design of the engine.
13. The reliability of the engine decreases with increase in maximum pressure in the cylinder.
14. Reciprocating compressor is commonly used for supercharging.
15. Roots blower is suitable for supercharging low and medium speed engines for stationary and marine installations.
16. .... are centrifugal compressors driven by the exhaust gas turbines.
17. Turbocharged engines produce higher cylinder volumetric efficiencies compared with the normally aspirated induction systems.
18. A naturally aspirated engine will have its power output reduced by approximately 13 per cent if it is operated approximately 1000 m above sea-level.
19. .... turbocharging is defined as the use of two turbochargers of different sizes in series.
20. .... turbocharging is based upon the idea of increasing the expansion ratio relative to compression ratio by means of early closure of inlet valve as the boost pressure is increased.

ANSWERS

1. indicated	**2.** No	**3.** higher	**4.** Yes	**5.** Yes
6. Boost	**7.** Yes	**8.** Yes	**9.** No	**10.** No
11. Yes	**12.** Yes	**13.** Yes	**14.** No	**15.** Yes
16. Turbocharging	**17.** Yes	**18.** Yes	**19.** Two-stage	**20.** Miller.

THEORETICAL QUESTIONS

1. What is supercharging ?
2. Enumerate the main objects of supercharging.
3. What is a supercharger ?
4. Explain briefly supercharging of S.I. engines.
5. Give the comparison of 'Actual naturally aspirated' and 'supercharged engine' pressure-volume diagrams.
6. Define the terms 'Boost pressure' and 'Pressure ratio'.
7. Explain briefly the effect of pressure ratio on air charge temperature.
8. Explain briefly the thermodynamic cycle of supercharged engine on p-v diagram for an ideal otto cycle.
9. Derive an expression for the power required for an I.C. engine supercharger.
10. What are supercharging limits of S.I. engines ?
11. Explain briefly supercharging of C.I. engines.
12. What are supercharging limits of C.I. engines ?
13. Enumerate the main considerations in limiting the degree of supercharging of a C.I. engine.
14. What modifications are necessary for a supercharged engine ?
15. List the types of superchargers used to supercharge engines for various applications.
16. Explain briefly with neat sketches various types of supercharging arrangements.
17. What are turbochargers ?
18. Describe the Buchi system of turbocharging.
19. Explain briefly the main types of supercharging methods.
20. What is Miller turbocharging ?
21. Explain briefly with a neat sketch Hyperbar turbocharging. State its advantages and disadvantages as well.
22. What is pulse converter ?
23. What are the limitations of supercharging ?
24. "S.I. engines are generally not supercharged". Justify this statement.
25. "Supercharging is preferred in diesel engine than petrol engine". Justify the statement.
26. "Supercharging is essential for an aircraft engine". Justify this statement.
27. Explain the factors that limit the extent of supercharging of S.I. and C.I. engines.
28. What do you mean by supercharging of I.C. engines ? Explain why supercharging is essential for the aircraft engines ?

UNSOLVED EXAMPLES

1. A 3000 cm^3 capacity 4-stroke diesel engine develops 14 kW/m^3 of free air induced per minute. It has a volumetric efficiency of 80% referred to free air conditions of 1.013 bar and 27°C, when running at 3500 r.p.m. It is proposed to boost the power of the engine by supercharging by a blower (driven mechanically from the engine) of pressure ratio 1.7 and isentropic efficiency of 75%. Taking overall mechanical efficiency as 80% determine the increase in B.P.

Assume that at the end of induction the cylinders contain a volume of charge equal to the swept volume, at the pressure and temperature of the delivery from the blower. [**Ans.** 27.7 kW]

2. It is required to design a 4-stroke diesel engine to operate with the following characteristics at sea level, where the mean conditions are 1.013 bar and 10°C.

B.P. = 250 kW ; volumetric efficiency = 78 percent (at sea level free air conditions) ; specific fuel consumption = 0.245 kg/kWh ; air-fuel ratio = 17 : 1 ; speed = 1500 r.p.m.

Determine the required engine capacity and anticipated brake mean effective pressure.

A supercharger is then fitted to the engine so that it may be operated at an altitude of 2700 m where the atmospheric pressure is 0.72 bar. The power consumed by the supercharger is 8 per cent of the total power produced by the engine and the temperature of the air leaving the supercharger is 32°C. Calculate the increase of air pressure required at the supercharger to maintain the same net output of 250 kW.

Assume that the air-fuel ratio and thermal efficiency remain the same for the supercharged engine as when running unsupercharged at sea level, as does the volumetric efficiency. [**Ans.** 0.466 bar]

3. Air enters an air compressor (run by a supercharged diesel engine) at 27°C and is passed on to a cooler where 1160 kJ/min. are rejected. The air leaves the cooler at 67°C and 1.7 bar. Some air from the compressor is bled after the cooler to supercharge the engine. The volumetric efficiency of the engine is 75 percent based on the intake manifold temperature of 67°C and pressure of 1.7 bar.

The engine has the following specifications :

B.P. = 30 kW ; Speed = 2000 r.p.m. ; Number of cylinders = 4 ; Bore = 100 mm ;

Stroke = 110 mm ; Mechanical efficiency = 80 percent.

Assuming the internal efficiency of the compressor as 90 percent, determine the following :

(*i*) The indicated mean effective pressure of the engine ;

(*ii*) The air consumption rate of the engine ;

(*iii*) The air handling capacity of the compressor. [**Ans.** (*i*) 6.5 bar ; (*ii*) 4.53 kg/min. ; 11.57 kg/min.]

4. The average indicated power developed in I.C. engine is 12.9 kW/m^3 of free air induced per minute. The engine is three-litre four-stroke engine running at 3500 r.p.m., and has a volumetric efficiency of 80%, referred to free conditions of 1.013 bar and 15°C. It is proposed to fit a blower, driven mechanically from the engine. The blower has an isentropic efficiency of 75% and works through a pressure ratio of 1.7, at the pressure and temperature of the delivery from the blower. Calculate the increase in B.P. to be expected from the engine. Take all the mechanical efficiencies as 80%. [**Ans.** 25.3 kW]

17

Testing and Performance of I.C. Engines

17.1. INTRODUCTION

- The primary task of the development engineer is to reduce the capital and running cost of the engine. This involves trial of various design concepts. The parameters are so enormous and different in nature that it is almost physically impossible to take care of all of them during the design of the engine. Therefore, it is necessary to conduct the test on the engine and determine the measures which should be taken to improve the engine performance. The nature and the type of the test to be conducted will depend upon a great number of factors such as, the degree of development of the particular design, the accuracy required, the funds available, the nature of the manufacturing company etc.
- The testing of the engine is necessary to verify the performance of the engine as per the specification of the manufacturer.

17.2. PERFORMANCE PARAMETERS

Engine performance is an indication of the degree of success with which it does its assigned job *i.e., conversion of chemical energy contained in the fuel into the useful mechanical work.*

In evaluation of engine performance certain basic parameters are chosen and the effect of various operating conditions, design concepts and modifications on these parameters are studied. The *basic performance parameters* are numerated and discussed below :

1. Power and mechanical efficiency
2. Mean effective pressure and torque
3. Specific output
4. Volumetric efficiency
5. Fuel-air ratio
6. Specific fuel consumption
7. Thermal efficiency and heat balance
8. Exhaust smoke and other emissions
9. Specific weight.

1. Power and mechanical efficiency :

(*i*) ***Indicated power.*** *The total power developed by combustion of fuel in the combustion chamber is called indicated power.*

$$\text{I.P.} = \frac{np_{mi}\, LANk \times 10}{6} \text{ kW} \qquad ...(17.1)$$

where , n = Number of cylinders,

p_{mi} = Indicated mean effective pressure, bar,

L = Length of stroke, m,

A = Area of piston, m^2, and

$k = \frac{1}{2}$ for 4-stroke engine

= 1 for 2-stroke engine.

(*ii*) ***Brake power (B.P.).*** *The power developed by an engine at the output shaft is called the brake power.*

$$\text{B.P.} = \frac{2\pi NT}{60 \times 1000} \text{ kW} \qquad ...(17.2)$$

where, N = Speed in r.p.m., and

T = Torque in N-m.

The difference between I.P. and B.P. is called *frictional power, F.P.*

i.e.,

$$\text{F.P.} = \text{I.P.} - \text{B.P.} \qquad ...(17.3)$$

The ratio of B.P. to I.P. is called *mechanical efficiency*

i.e., Mechanical efficiency,

$$\eta_{\text{mech.}} = \frac{\text{B.P.}}{\text{I.P.}} \qquad ...(17.4)$$

2. **Mean effective pressure and torque :**

"Mean effective pressure" is defined as hypothetical pressure which is thought to be acting on the piston throughout the power stroke. If it is based on I.P. it is called *indicated mean effective pressure* ($I_{\text{m.e.p.}}$ or p_{mi}) and if based on B.P. it is called brake mean effective pressure ($B_{\text{m.e.p.}}$ or p_{mb}). Similarly, frictional mean effective pressure ($F_{\text{m.e.p.}}$ or p_{mf}) can be defined as :

$$F_{\text{m.e.p.}} = I_{\text{m.e.p.}} - B_{\text{m.e.p.}} \qquad ...(17.5)$$

The torque and mean effective pressure are related by the engine size.

Since the power (P) of an engine is dependent on its size and speed, therefore it is not possible to compare engine on the basis of either power or torque. *Mean effective pressure is the true indication of the relative performance of different engines.*

3. **Specific output :**

It is defined as the *brake output per unit of piston displacement* and is given by :

$$\text{Specific output} = \frac{\text{B.P.}}{A \times L}$$

$$= \text{Constant} \times p_{mb} \times \text{r.p.m.} \qquad ...(17.6)$$

For the same piston displacement and brake mean effective pressure (p_{mb}) an engine running at higher speed will give more output.

4. **Volumetric efficiency :**

It is defined as the ratio of actual volume (reduced to N.T.P.) of the charge drawn in during the suction stroke to the swept volume of the piston.

The average value of this efficiency is from 70 to 80 per cent but in case of *supercharged engine* it may be *more* than 100 per cent, if air at about atmospheric pressure is forced into the cylinder at a pressure greater than that of air surrounding the engine.

5. **Fuel-air ratio :**

It is the ratio of the mass of fuel to the mass of air in the fuel-air mixture.

Relative fuel air ratio is defined as the ratio of the actual fuel-air ratio to that of stoichiometric fuel-air ratio required to burn the fuel supplied.

6. Specific fuel consumption (s.f.c.) :

It is the mass of fuel consumed per kW developed per hour, and is a criterion of economical power production.

i.e.,
$$\text{s.f.c.} = \frac{\dot{m}_f}{\text{B.P.}} \text{ kg/kWh.}$$

7. Thermal efficiency and heat balance :

Thermal efficiency. *It is the ratio of indicated work done to energy supplied by the fuel.*

If $\dot{m}_f$ = Mass of fuel used in kg/s, and

C = Calorific value of fuel (lower),

Then *indicated thermal efficiency* (based on I.P.),

$$\eta_{\text{th.}}\,(I) = \frac{\text{I.P.}}{\dot{m}_f \times C} \qquad ...(17.7)$$

and *brake thermal efficiency* (based on B.P.)

$$\eta_{\text{th.}}\,(B) = \frac{\text{B.P.}}{\dot{m}_f \times C} \qquad ...(17.8)$$

Heat balance sheet :

The performance of an engine is generally given by heat balance sheet.

To draw a heat balance sheet for I.C. engine, it is run at constant load. Indicator diagram is obtained with the help of an indicator. The quantity of fuel used in a given time and its calorific value, the amount, inlet and outlet temperature of cooling water and the weight of exhaust gases are recorded. After calculating I.P. and B.P. the heat in different items is found as follows :

Heat supplied by fuel

For *petrol and oil engines,* heat supplied = $m_f \times C$, where m_f and C are mass used per minute (kg) and lower calorific value (kJ or kcal) of the fuel respectively.

For *gas engines,* heat supplied = $V \times C$, where V and C is volume at N.T.P. (m^3/min.) and lower calorific value of gas respectively.

(*i*) ***Heat absorbed in I.P.***

Heat equivalent of I.P. (per minute) = I.P. × 60 kJ ...(17.9)

(*ii*) ***Heat taken away by cooling water***

If m_w = Mass of cooling water used per minute,

t_1 = Initial temperature of cooling water, and

t_2 = Final temperature of cooling water,

Then, heat taken away by water = $m_w \times c_w \times (t_2 - t_1)$ ...(17.10)

where, c_w = specific heat of water.

(*iii*) ***Heat taken away by exhaust gases***

If m_e = Mass of exhaust gases (kg/min),

c_{pg} = Mean specific heat at constant pressure,

t_e = Temperature of exhaust gases, and

t_r = Room (or boiler house) temperature,

Then heat carried away by exhaust gases = $m_e \times c_{pg}(t_e - t_r)$ ...(17.11)

Note. The mass of exhaust gases can be obtained by adding together mass of fuel supplied and mass of air supplied.

The heat balance sheet from the above data can be drawn as follows :

Item	*kJ*	*Per cent*
Heat supplied by fuel	...	...
(*i*) Heat absorbed in I.P.	...	...
(*ii*) Heat taken away by cooling water	...	...
(*iii*) Heat carried away by exhaust gases	...	...
(*iv*) Heat unaccounted for (by difference)	...	...
Total	...	...

8. **Exhaust smoke and other emissions :**

Smoke is an indication of incomplete combustion. It limits the output of an engine if air pollution control is the consideration. *Exhaust emissions* have of late become a matter of grave concern and with the enforcement of legislation on air pollution in many countries, it has become necessary to view them as performance parameters.

9. **Specific weight :**

It is defined as the weight of the engine in kg for each B.P. developed. It is an indication of the engine bulk.

17.3. BASIC MEASUREMENTS

To evaluate the performance of an engine following *basic measurements* are usually undertaken :

1. Speed
2. Fuel consumption
3. Air consumption
4. Smoke density
5. Exhaust gas analysis
6. Brake power
7. Indicated power and friction power
8. Heat going to cooling water
9. Heat going to exhaust.

1. **Measurement of speed :**

The speed may be measured by :

(*i*) Revolution counters
(*ii*) Mechanical tachometer
(*iii*) Electrical tachometer.

2. **Fuel measurement :**

The fuel consumed by an engine can be measured by the following methods :

(*i*) Fuel flow method
(*ii*) Gravemetric method
(*iii*) Continuous flow meters.

3. **Measurement of air consumption :**

The air consumption can be measured by the following methods :

(*i*) Air box method
(*ii*) Viscous-flow air meter.

(*i*) ***Air box method :***

Fig. 17.1 shows the arrangement of the system. It consists of air-tight chamber fitted with a sharp edged orifice of known coefficient of discharge. The orifice is located away from the suction connection to the engine. Due to the suction of engine, there is a pressure depression in the air box or chamber which causes the flow through the orifice. For obtaining a steady flow, the volume of chamber should be sufficiently large compared with the swept volume of the cylinder ; generally

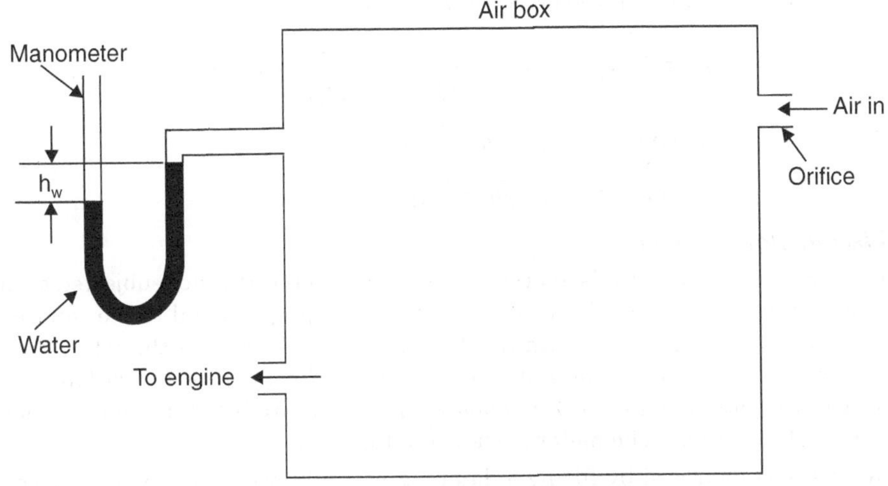

Fig. 17.1. Air-box method for measuring air.

500 to 600 times the swept volume. It is assumed that the intermittent suction of the engine will not effect the air pressure in the air box as volume of the box is sufficiently large, and pressure in the box remains same.

A water manometer is used to measure the pressure difference causing the flow through the orifice. The depression across the orifice *should not exceed 100 to 150 mm of water.*

Let A = Area of orifice, m^2,

d = Diameter of orifice, cm,

h_w = Head of water in cm causing the flow,

C_d = Coefficient of discharge for orifice,

ρ_a = Density of air in kg/m^3 under atmospheric conditions, and

ρ_w = Density of water in kg/m^3.

Head in metres of air (H) is given by :

$$H.\rho_a = \frac{h_w}{100}\,\rho_w$$

$$\therefore \qquad H = \frac{h_w}{100} \times \frac{\rho_w}{\rho_a} = \frac{h_w}{100} \times \frac{1000}{\rho_a} = \frac{10h_w}{\rho_a} \text{ m of air}$$

The velocity of air passing through the orifice is given by,

$$C_a = \sqrt{2gH} \text{ m/s} = \sqrt{2g\,\frac{10h_w}{\rho_a}} \text{ m/s}$$

The volume of air passing through the orifice,

$$V_a = C_d \times A \times C_a = C_d A\,\sqrt{2g\,\frac{10h_w}{\rho_a}} = 14\,AC_d\,\sqrt{\frac{h_w}{\rho_a}} \text{ m}^3\text{/s}$$

$$= 840\, AC_d \sqrt{\frac{h_w}{\rho_a}} \text{ m}^3/\text{min.}$$

Mass of air passing through the orifice is given by

$$m_a = V_a\rho_a = 14 \times \frac{\pi d^2}{4 \times 100^2} \times C_d \sqrt{\frac{h_w}{\rho_a}} \times \rho_a$$

$$= 0.0011\, C_d \times d^2 \sqrt{h_w \rho_a} \text{ kg/s}$$

$$= 0.066\, C_d \times d^2 \sqrt{h_w \rho_a} \text{ kg/min.} \qquad ...(17.12)$$

(*ii*) ***Viscous-flow air meter :***

Alcock viscous-flow air meter is another design of air meter. It is not subjected to the errors of the simple types of flow meters. With the air-box the flow is proportional to the square root of the pressure difference across the orifice. With the Alcock meter the air flows through a form of honey-comb so that flow is viscous. *The resistance of the element is directly proportional to the air velocity and is measured by means of an inclined manometer.* Felt pads are fitted in the manometer connections to damp out fluctuations. The meter is shown in Fig. 17.2.

The accuracy is improved by fitting a damping vessel between the meter and the engine to reduce the effect of pulsations.

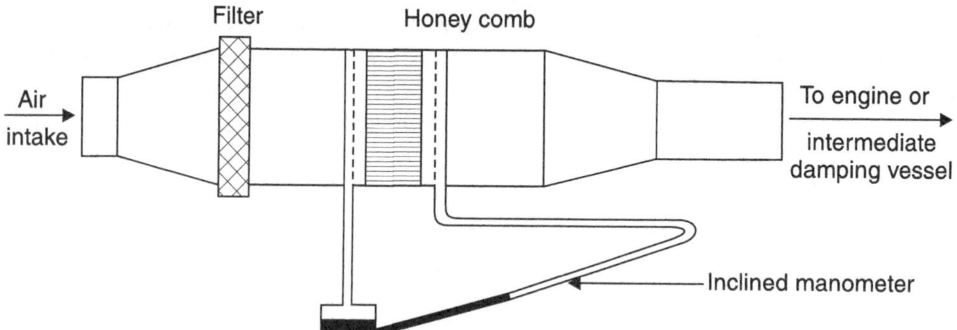

Fig. 17.2. Alcock viscous-flow air meter.

4. **Measurement of exhaust smoke :**

The following smoke meters are used :

(*i*) Bosch smoke meter
(*ii*) Hatridge smoke meter
(*iii*) PHS smoke meter.

5. **Measurement of exhaust emission :**

Substances which are emitted to the atmosphere from any opening down stream of the exhaust part of the engine are termed as "exhaust emissions". Some of the more commonly used instruments for measuring exhaust components are given below :

(*i*) Flame ionisation detector
(*ii*) Spectroscopic analysers
(*iii*) Gas chromatography.

6. **Measurement of B.P. :**

The B.P. of an engine can be determined by a brake of some kind applied to the brake pulley of the engine. The arrangement for determination of B.P. of the engine is known as *dynamometer*. The dynamometers are classified into following two classes :

(*i*) Absorption dynamometers
(*ii*) Transmission dynamometers.

(*i*) ***Absorption dynamometers.*** *Absorption dynamometers are those that absorb the power to be measured by friction.* The power absorbed in friction is finally dissipated in the form of heat energy.

Common forms of absorption dynamometers are :

—Prony brake —Rope brake

—Hydraulic brake —Fan brake

—Electrical brake dynamometers

(*a*) Eddy current dynamometer

(*b*) Swinging field d.c. dynamometer.

(*ii*) ***Transmission dynamometers.*** These are also called *torquemeters*. These are very accurate and are used where continuous transmission of load is necessary. These are used mainly in *automatic units.*

Here we shall discuss *Rope brake dynamometer* only :

Rope brake dynamometer :

Refer Fig. 17.3. A rope is wound round the circumference of the brake wheel. To prevent the rope from slipping small wooden blocks (not shown in the Fig. 17.3) are laced to rope. To one end of the rope is attached a spring balance (S) and the other end carries the load (W). The speed of the engine is noted from the tachometer (revolution counter).

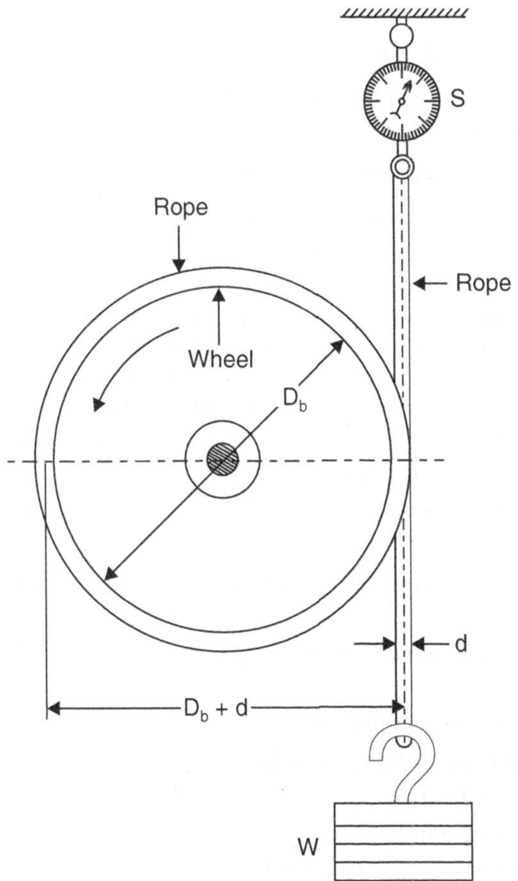

Fig. 17.3. Rope brake dynamometer.

If, W = Weight at the end of the rope, N,

S = Spring balance reading, N,

N = Engine speed, r.p.m.,

D_b = Diameter of the brake wheel, m,

d = Diameter of the rope, m, and

$(D_b + d)$ = Effective diameter of the brake wheel,

Then work/revolution = Torque × angle turned per revolution

$$= (W - S) \times \left(\frac{D_b + d}{2}\right) \times 2\pi = (W - S)(D_b + d) \times \pi$$

Work done/min. $= (W - S)\,\pi\,(D_b + d)\,N$

Work done/sec. $= \dfrac{(W - S)\,\pi\,(D_b + d)\,N}{60}$

$$\therefore \quad \text{B.P.} = \frac{(W - S)\,\pi\,(D_b + d)\,N}{60 \times 1000}\ \text{kW} \qquad ...(17.13)$$

$$= \frac{(W - S)\,\pi\,D_b N}{60 \times 1000} \quad ...... \text{ if } d \text{ is neglected}$$

$$\text{or} \quad \left(= \frac{T \times 2\pi N}{60 \times 1000}\ \text{kW}\right) \qquad ...(17.14)$$

Rope brake is cheap and easily constructed but not very accurate because of changes in friction coefficient of the rope with temperature.

Measurement of Indicated power (I.P.) :

The power developed in the engine cylinder or at the piston is necessarily greater than that at the crankshaft due to engine losses. Thus,

I.P. = B.P. + engine losses.

Indicated power is usually determined with the help of a p-V diagram taken with the help of an indicator. In case indicated power cannot be measured directly, it is made possible by measuring the brake power and also the engine losses. If the *indicator diagram* is available, the indicated power may be *computed by measuring the area of* diagram, either with a *planimeter* or by *ordinate method*, and dividing by the stroke measurement in order to obtain the mean effective pressure (m.e.p.).

i.e. $$p_{mi} = \frac{\text{Net area of diagram in mm}^2}{\text{Length of diagram in mm}} \times \text{Spring constant}$$

where p_{mi} is in bar. ...(17.15)

(The spring constant is given in *bar per mm of vertical movement of the indicator stylus.*)

Engine indicators :

The main types of engine indicators are :

1. Piston indicator

2. Balanced diaphragm type indicator :

(*i*) The Farnborough balanced engine indicator

(*ii*) Dickinson-Newell indicator

(*iii*) MIT balanced pressure indicator

(*iv*) Capacitance-type balance pressure indicator.

3. Electrical indicators

In addition to this, *optical indicators* are also used.

Calculation of indicated power (I.P.) :

If p_{mi} = Indicated mean effective pressure, bar,

A = Area of piston, m^2,

L = Length of stroke, m,

N = Speed of the engine, r.p.m., and

$k = \frac{1}{2}$ for 4-stroke engine

= 1 for 2-stroke engine,

Then, force on the piston $= p_{mi} \times A \times 10^5$ N

Work done per working stroke = Force × length of stroke

$= p_{mi} \times A \times 10^5 \times L$ N-m

Work done per second = Work done per stroke × Number of working stroke per second

$$= p_{mi} \times L \times A \times 10^5 \times \frac{N}{60} \times k \text{ N-m/s or J/s}$$

$$= \frac{p_{mi} \times LANk \times 10^5}{60 \times 1000} \text{ kW}$$

i.e., Indicated power, $\text{I.P.} = \frac{p_{mi}\ LANk \times 10}{6}$ kW

If n is the number of cylinders, then

$$\text{I.P.} = \frac{np_{mi}\ LANk \times 10}{6} \text{ kW} \quad ...(17.16)$$

Morse test :

This test is only applicable to *multi-cylinder engines.*

The engine is run at the required speed and the torque is measured. One cylinder is cut out, by shorting the plug if an S.I. engine is under test, or by disconnecting an injector if a C.I. engine is under test. *The speed falls because of the loss of power with one cylinder cut out, but is restored by reducing the load.* The torque is measured again when the speed has reached its original value. If the values of I.P. of the cylinders are denoted by I_1, I_2, I_3 and I_4 (considering a four-cylinder engine), and the power losses in each cylinder are denoted by L_1, L_2, L_3 and L_4, then the value of B.P., B at the test speed with all cylinders firing is given by

$$B = (I_1 - L_1) + (I_2 - L_2) + (I_3 - L_3) + (I_4 - L_4) \quad ...(i)$$

If *number 1 cylinder is cut out,* then the contribution I_1 is lost ; and if the losses due to that cylinder remain the same as when it is firing, then the B.P., B_1, now obtained at the same speed is

$$B_1 = (0 - L_1) + (I_2 - L_2) + (I_3 - L_3) + (I_4 - L_4) \quad ...(ii)$$

Subtracting equation (*ii*) from (*i*), we get

$$B - B_1 = I_1 \quad ...(17.17)$$

Similarly, $B - B_2 = I_2$ when cylinder number 2 is cut out

and $B - B_3 = I_3$ when cylinder number 3 is cut out

and $B - B_4 = I_4$ when cylinder number 4 is cut out.

Then, for the engine,

$$I = I_1 + I_2 + I_3 + I_4 \quad ...(17.18)$$

Assumptions :

1. Speed of engine and throttle opening or fuel injection setting has to remain same in both the cases of four cylinders working and three cylinders working with one cylinder cut out.
2. Frictional power and pumping power are functions of speed and must be kept constant.
3. Same throttle opening is expected to supply same quantity of fuel.
4. Cylinder individually will develop same power whether all four are working or only three cylinders are working with one cylinder consuming power.

Limitations :

1. This is applicable to multicylinder engines only (three or more cylinders).
2. Results are liable to errors due to change in mixture distribution and other conditions by cutting out one cylinder.

Cutting out of one cylinder may greatly affect the pulsations in exhaust system which may significantly change the engine performance by imposing different back pressures.

Engine Friction :

Frictional losses in an Engine

The total engine friction can be divided into *five main components.* These are :

1. Crankcase mechanical friction.
2. Blow-by losses (compression-expansion pumping loss).
3. Exhaust and inlet system throttling losses.
4. Combustion chamber pumping loop losses.
5. Piston mechanical friction.

1. ***Crankcase mechanical friction :***

It can be further sub-divided into :

(*i*) Bearing friction

(*ii*) Valve gear friction, and

(*iii*) Pump and miscellaneous friction.

The bearing friction includes the friction due to main bearing connecting rod bearing and other bearings. Bearing friction is viscous in nature and depends upon the oil viscosity, the speed, size and geometry of the journal.

The valve gear friction losses vary with the engine design variables and no general equation is available predicting them.

2. ***Blow-by losses :***

It is the phenomenon of *leakage of combustion products past the piston and piston rings from the cylinder to the crankcase.* These losses depend on the inlet pressure and compression ratio. *These losses vary as the square root of inlet pressure, and increase as the compression ratio is increased. Blow-by losses are reduced as the engine speed is increased.*

3. ***Exhaust and inlet throttling loss :***

The standard practice of sizing the exhaust valve is to make them a certain percentage *smaller* than the inlet valves. This usually results in an insufficiently sized exhaust valve and hence results in *exhaust pumping loss.*

4. ***Combustion chamber pumping loop losses :***

In the case of pre-combustion chamber engines an additional loss occurs. This is the loss occurring *due to the pumping work required to pump gasses into and out of the pre-combustion chamber*. The exact value of this would depend upon the orifice size connecting the pre-combustion chamber and the main chamber, and the speed. *Higher the speed greater is the loss and smaller the orifice size, greater is the loss.*

5. ***Piston mechanical friction :***

It can be sub-divided into :

(*i*) Viscous friction

(*ii*) Non-viscous-friction—(*a*) friction due to ring tension.

(*b*) friction due to gas pressure forces behind the ring.

The *viscous friction depends upon the viscosity of the oil and the temperature of the various parts of the piston.*

(*b*) Piston Mechanical friction can be sub-divided into :

(*i*) Viscous friction

(*ii*) Non-viscous friction

(*a*) friction due to ring tension.

(*b*) friction due to gas pressure forces behind the ring.

The viscous friction depends upon the viscosity of the oil and the temperature of the various parts of the piston. The degree to which the upper part of the piston can be lubricated also effects the viscous friction. The oil film thickness between piston and the cylinder is also affected by the piston side thrust and the resulting vibration.

Effect of engine variables on engine friction :

1. ***Effect of stroke/bore ratio.*** The effect of stroke / bore ratio on engine friction and economy is very small. High stroke / bore ratio engines have equally good friction mep values as that for low stroke / bore ratio engine. Indications are that at high speeds the higher stroke / bore ratio engine may be at some disadvantage.

2. ***Effect of cylinder size and number of cylinders.*** The friction and economy improves as a smaller number of larger cylinders are used. This is because the proportion between the working piston area and its friction producing area, *i.e.* circumference is reduced.

3. ***Effect of piston rings.*** The effect of number of piston ring is not very critical and this number is usually chosen on the basis of cost, size and other requirements rather than on the basis of their effect on friction.

4. ***Effect of compression ratio.*** Friction means effective pressure increases as the compression ratio is increased. But the mechanical efficiency either remains the constant or improves as the compression ratio is increased. If the displacement is varied to keep the maximum engine torque constant, this results in better part load friction characteristics.

5. ***Effect of engine speed.*** Engine friction *increases rapidly as the speed increases.* The best way to improve mechanical efficiency at high speed is to *increase the number of cylinders.*

6. ***Effect of oil viscosity.*** *Higher the oil viscosity greater is the friction loss.* The temperature of the oil in the crankcase significantly affects the friction losses, wear and service life of an engine. *As the oil temperature increases the viscosity decreases and friction losses are reduced during a certain temperature range.*

7. ***Effect of cooling water temperature.*** A rise in cooling water temperature *reduces* engine friction through its effect on oil viscosity.

During starting operation the temperature of both the oil and the water is low, hence, the viscosity is high. This results in high starting friction losses and rapid engine wear.

8. ***Effect of engine load.*** As the load increases the maximum pressure in the cylinder has a *tendency to increase slightly*. This results in *slightly higher friction values.*

Measurement of frictional power (F.P.) :

The frictional power of an engine can be determined by the following methods :

1. Willan's line method (used for C.I. engines only)
2. Morse test
3. Motoring test
4. Difference between I.P. and B.P.

1. ***Willan's line method :***

At a constant engine speed the load is reduced in increments and the corresponding B.P. and gross fuel consumption readings are taken. A graph is then drawn of fuel consumption against B.P. as in Fig. 17.4. The graph drawn in called the Willan's line (analogous to Willan's line for a steam engine), and is extrapolated back to cut the B.P. axis at the point *L*. The reading *OL* is taken as the power loss of the engine at that speed. The fuel consumption at zero B.P. is given by *OM* ; and if the relationship between fuel consumption and B.P. is assumed to be linear, then a fuel consumption *OM* is equivalent to a power loss of *OL*.

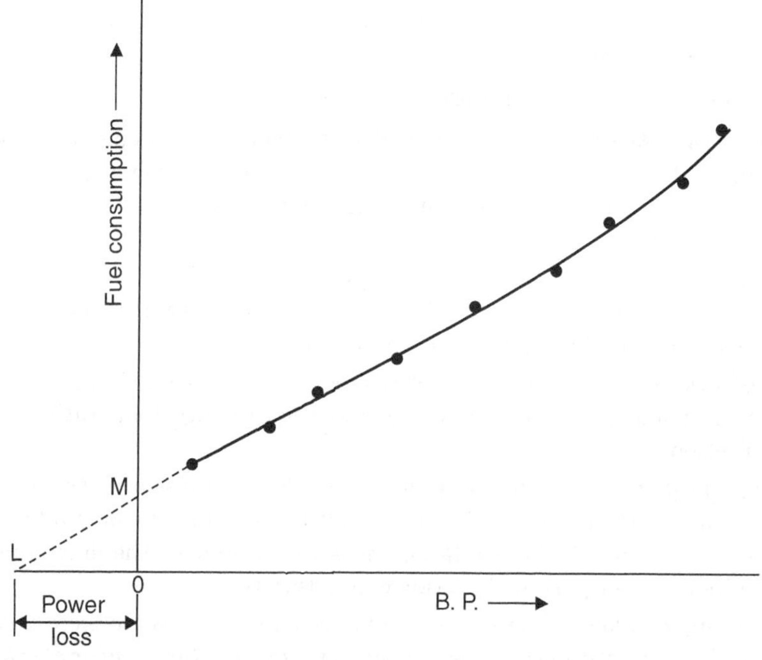

Fig. 17.4. Willan's line method.

- This method is used *only in case of unthrottled engines* as discussed below :
  — The Willians line is plotted for fuel consumption versus load (from no load to full load) *at constant speed.* The intersection of this line on the negative side of X-axis gives the friction power of the engine at that speed. The friction power (F.P.) is assumed constant from no load to full load at that constant speed. The F.P. includes not only the mechanical friction, but also pumping power.

For a throttled engine if such a trial is carried out, the throttle position has to be varied from almost closed at not load to full open at maximum load, to keep the engine speed constant. Therefore the pumping load will be bigger at no load, and reduce gradually as the load is increased. In other words, *the pumping power and therefore the F.P. will not remain constant, as is the assumption in Willians line method.*

Limitations :

The following are the *limitations* of William's line method.

1. Applicable mainly to C.I. engines only.

2. The fuel-consumption— brake power line is not straight line, but turns up slightly at low loads and considerably near full load. *Unless sufficient data are taken to accurately plot the straight line portion of the curve, the results will be significantly in error.*

2. ***Morse test :***

In 'Morse test' (already discussed), frictional power can be found by subtracting $(\text{B.P.})_n$ from $(\text{I.P.})_n$

i.e. $$\text{F.P.} = (\text{I.P.})_n - (\text{B.P.})_n$$

where n is the number of cylinders.

3. ***Motoring test :***

- In this test the engine is first run upto the desired speed by its own power and allowed to remain under the given speed and load conditions for sometime so that oil, water and engine component temperatures reach stable conditions. The power of the engine during this period is absorbed by a dynamometer (usually of electrical type). The fuel supply is then cut off and by suitable electric switching devices the dynamometer is converted to run as a motor to drive or 'motor' the engine at the same speed at which it was previously running. The power supply to the motor is measured which is a measure of F.P of the engine.
- Motoring test gives a very good insight into the various causes of losses and is much more powerful tool. *This test gives a higher value of F.P. as compared to that given by Willian's method.*

4. ***Difference between I.P. and B.P. :***

The method of finding the F.P. by finding the difference between I.P. as obtained from an indicator diagram, and B.P. as obtained by a dynamometer is the *ideal method.* However, due to difficulties in obtaining accurate indicator diagrams, especially at high engine speeds, this method is usually only used in research laboratories and its use at commercial level is very limited.

17.4. ENGINE PERFORMANCE CURVES

- The following parameters are of importance

  — Torque

  — Power

  — Specific fuel consumption and its inverse

  — I.C. engine efficiency

  — Brake mean effective pressure is defined as brake power/swept volume × cyclic frequency. In this relation swept volume is constant. Also brake power = torque × angular speed, thus at constant speed, the b.e.m.p. is directly proportional to torque and either may be used. The ratio of brake mean effective pressure to the indicated mean effective pressure (from indicator diagram) may seem to be equal to the ratio of brake power to the indicated power which is defined as mechanical efficiency. Thus indicator diagram and the output torque may be connected with a suitable allowance for mechanical efficiency.

- The torque-speed relations (Fig. 17.5) exhibit a curve even though it would seen reasonable to expect the mean effective pressure to be constant at all speeds. However, at low speeds leakage through valves etc. becomes of greater significance so that the m.e.p. falls and at high speeds the volumetic efficiency causes the induced mass to fall with a parallel fall in m.e.p. *The power curves are the product of the torque curves with speed.*

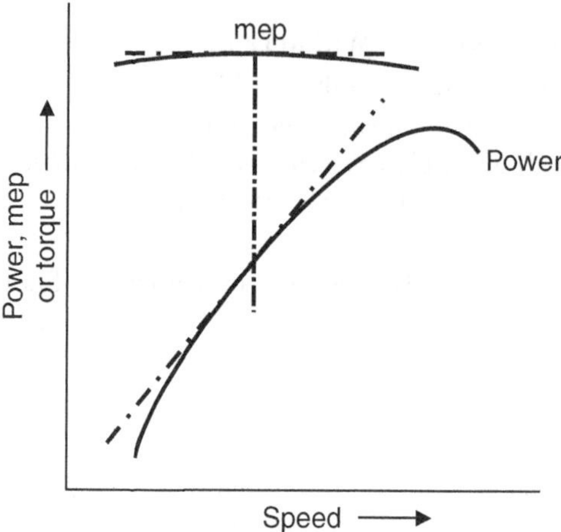

Fig.17.5. Power-speed and torque-speed curves for the I.C. engine.

Specific fuel consumption relations :

S.I. Engines : Refer Fig. 17.6. The curves are plotted for *constant throttle opening, constant speed and constant ignition setting.* The *only variable is the air-fuel ratio.* The effect of

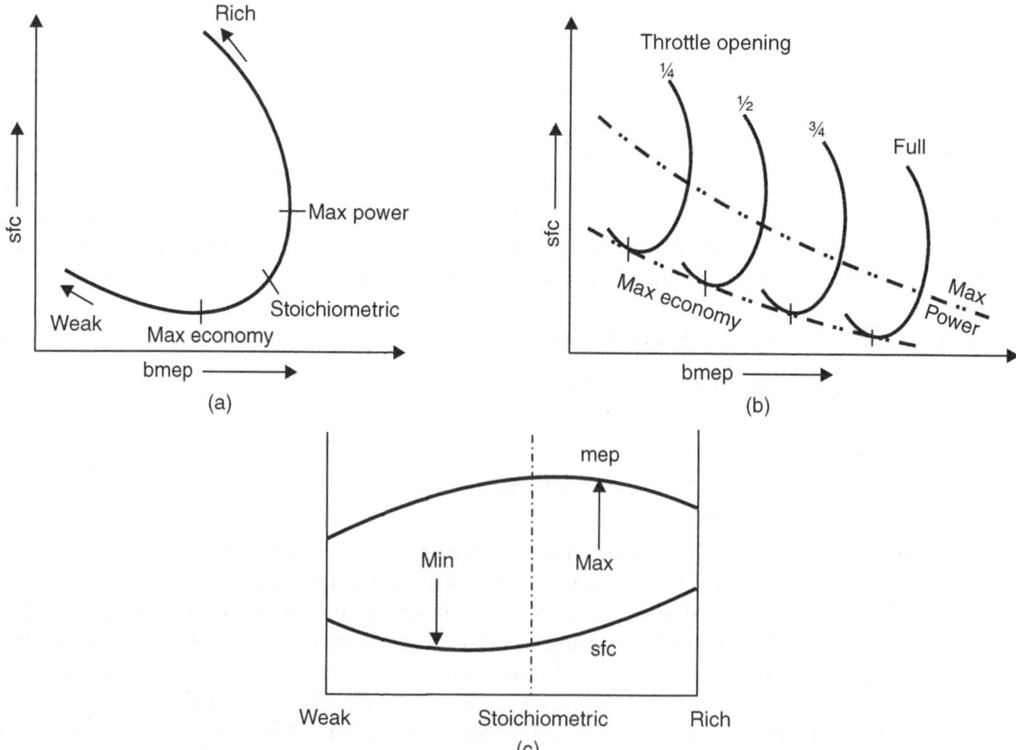

Fig. 17.6. Specific fuel consumption-brake mean effective pressure curves for the S.I. engine.

ignition setting or speed may by ascertained by producing a new family of curves. An alternative method of plotting these parameters is to use the air-fuel ratio as the abscissa. Here it can be seen that maximum economy occurs with a slightly weak mixture. This means that there is excess air and combustion is complete. Maximum power occurs with a slightly rich mixture when all the available oxygen is used. The I.C. engine efficiency is the inverse of the specific fuel consumption with the constant calorific value as a factor. Thus the curves of specific fuel consumption (s.f.c.) also represent efficiency. The maximum value of brake I.C. engine efficiency for S.I. and C.I. engines are of the order 35% and 40% respectively.

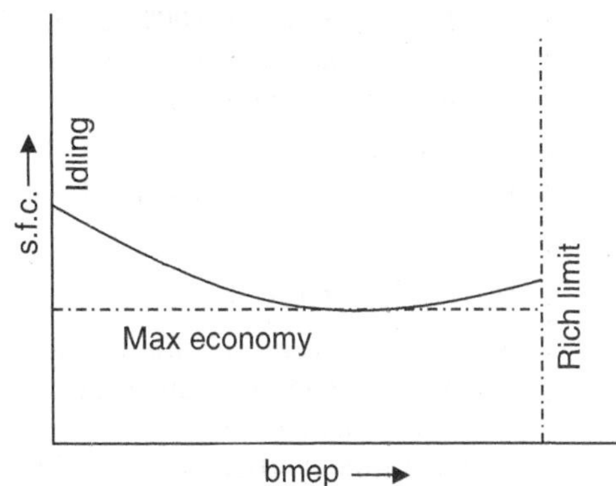

Fig. 17.7. Specific fuel consumption—brake mean effective pressure curve for the C.I. engine.

C.I. engine. The flat curve of Fig. 17.7 illustrates that at *part load the compression ignition engine is more economical than the spark ignition engine.* This is the benefit of quality control rather than quantity control of power.

17.5. COMPARISON OF PETROL AND DIESEL ENGINES—FUEL CONSUMPTION LOAD OUTPUTS AND EXHAUST COMPOSITION

I. Fuel Consumption :

Fig. 17.8 shows fuel consumption loops, for both petrol and diesel engines, plotted on a base of brake mean effective pressure (b.m.e.p).

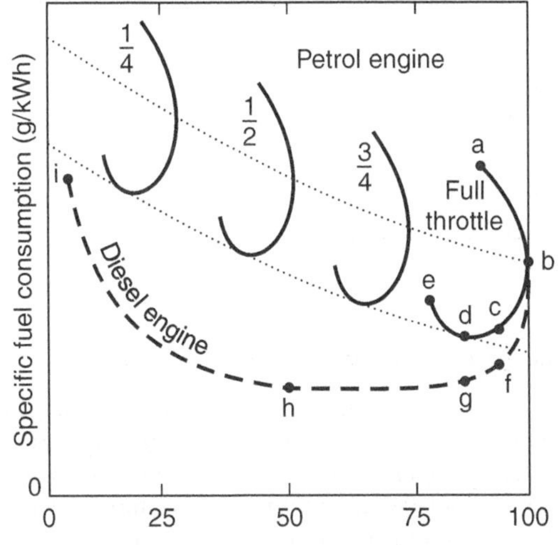

Fig. 17.8. Comparision of fuel consumption loops for petrol and diesel engines on a base of engine load (b.m.e.p.).

a = Excessively rich mixture gives slow and unstable combustion.

b = Maximum b.m.e.p. with something like 10–20% rich mixture.

c = Correct stoichiometric mixture of 14.7 : 1 by weight

d = Maximum thermal efficiency with something like 10–20% weak mixture (approaches ideal constant volume combustion).

e = Excessively weak mixture gives slow burning and popping back through air intake.

f = Maximum b.m.e.p. with satisfactory clear exhaust requires mixture strength of about 18 : 1 by weight.

g-h = Maximum thermal efficiency, minimum specific fuel consumption ranges between 50–85% of maximum b.m.e.p.

i = No-load (low speed idle) requires mixture strength 100–75 : 1 by weight.

- In case of a diesel *engine, load and speed output is controlled entirely by varying the quantity of fuel injected into the cylinder without misfiring occurring,* that is, from 0–100% of the maximum b.m.e.p. developed.
- With the petrol engines, however, if there was no throttle (full throttle position) the effects of varying the mixture strength from the richest position (a) to the weakest position

(*e*) produces a variations of b.m.e.p. (load) on only 25% of the maximum possible b.m.e.p. that is, from 75–100% of maximum b.m.e.p. (Fig. 17.8). Therefore the *petrol engine's output control can not be achieved alone by varying the mixture strength* and *therefore throttling the mixture coming into the cylinder becomes essential.*

The *diesel engine is thus* **"quality"** *controlled and the petrol engine is* **"quantity"** *controlled.*

- The different points on the consumption loops are listed on the right side of the Fig. 17.8.

II. Load Outputs :

- In case of a *diesel engine* the *pumping losses remain reasonably constant from full load to no load* whereas the pumping losses for a *petrol engine progressively rise as the b.m.e.p. is reduced due to the increased throttling.* This is reflected in the petrol engine by the mechanical efficiency reducing as b.m.e.p. decreases which causes the consumption loops to take up higher and higher levels of consumption as the intake becomes more throttled (Fig. 17.9). In contrast with the diesel engine, as the load is reduced the single consumption loop only begins to rise when the b.m.e.p. drops below about half the engine's maximum b.m.e.p. (Fig. 17.9).

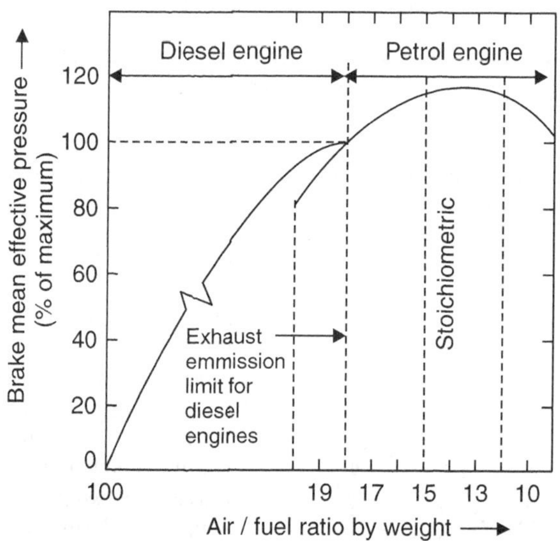

Fig. 17.9. Comparison of load (b.m.e.p.) for petrol and diesel engine on base of air-fuel ratio.

- A *petrol engine* can *operate effectively* under steady conditions over a *mixture strength ranging from 20 : 1 to 10 : 1,* whereas a *diesel engine* which can *normally only utilize 80% of its air charge with a reasonably clear exhaust, will operate from 18 : 1, that is 20% weak at full load to 100 : 1 at no load.* Consequently, the petrol engine will always have a 15 to 20% higher maximum b.m.e.p. than a similar diesel engine (Fig. 17.9).

III. Exhaust Composition

Fig. 17.10 shows the *comparison of composition of exhaust gases for petrol and diesel engines on a base of air-fuel ratio ;* this figure supplies the following information :

— The petrol engine has considerable amounts of carbon monoxide (CO) formed as the mixture moves from the stoichiometric (14.7 : 1 by weight) to the rich region of upto 10 : 1. This being within the petrol engine's mixture range when operating between idle and full power conditions. In contrast the diesel engine under full load never operates with mixture strengths greater than 20% rich, that is 18 : 1 by weight, so that no carbon monoxide (CO) is present in the exhaust (Fig. 17.10).

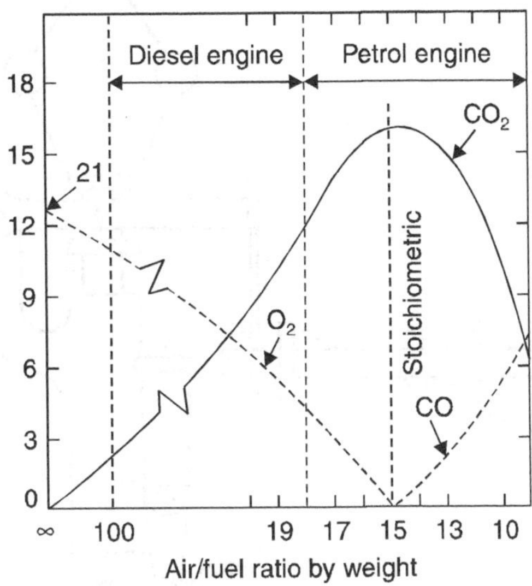

Fig. 17.10. Comparison of composition of exhaust gases for petrol and diesel engines on a base of air-fuel ratio.

— The carbon dioxide (CO_2) emission produced by the diesel engine relative to the petrol engine is always much lower, particularly as the engine load is reduced, whereas the petrol engine in the stoichimetric (14.7 : 1) band operates with the highest level of CO_2.

17.6. GOVERNING OF I.C. ENGINE

The function of a governor is to keep the speed of engine constant irrespective of the changes in load on the engine. The governor is usually of centrifugal type.

In petrol engine, the control is exercised by means of a throttle valve which is placed in intake manifold. The quantity of mixture entering the cylinder depends on the amount of opening of throttle valve. The position of throttle valve is controlled by the governor (centrifugal type). In diesel engines, the flow of fuel is controlled by centrifugal governor which actuates link rods which in turn operate some device on the fuel pump and consequently portion of the fuel by passes. The governor in plunger type injection pump alters the relative angular position of the plunger.

Following are the methods of governing I.C. engines :

(*i*) Hit and miss method

(*ii*) Quality governing

(*iii*) Quantity governing

(*i*) **Hit and miss method.** Refer Fig. 17.11. When the speed increases the permissible value the governor sleeve *S* gets lifted up, as a result of which the lever *A* lifts the distant piece *B*, so that the pecker *K* misses it. Thus the gas inlet valve *V* does not open and the usual charge does not enter the cylinder. This continues until the speed is reduced and *B* occupies its initial position. Explosions are thus missed intermittently but every charge is of normal strength. *This method is commonly used in gas engines.*

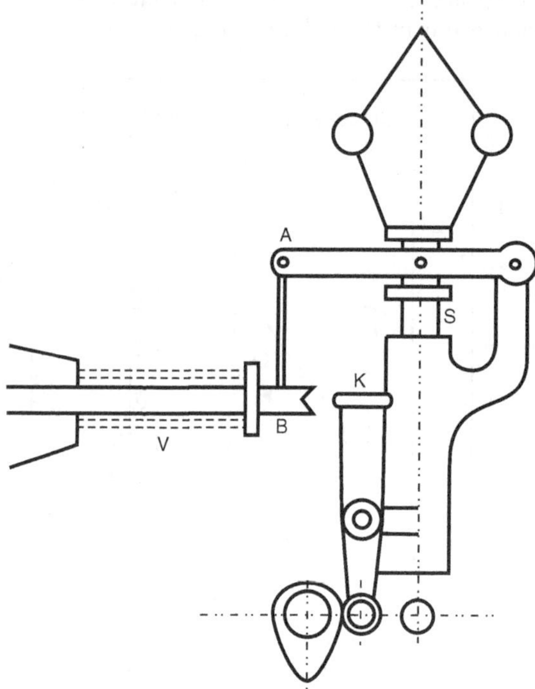

Fig. 17.11. Hit and miss governing.

(*ii*) **Quality governing.** In this method of governing, the *mixture strength is altered.* In gas engine it is effected by reducing the amount of gas supplied to the engine. This is accomplished by varying the lift of the gas valve. In oil engines, quality governing is carried out by varying the quality of fuel oil entering the cylinder per cycle, it is done by changing the angular position of the helical groove of the pump plunger.

In this type of governing, the ignition is not always satisfactory and thermal efficiency is reduced.

(*iii*) **Quantity governing.** Here *mixture strength remaining the same, the quantity of mixture entering the cylinder is altered.* When the speed is too high a lesser amount of charge is admitted into the cylinder. The compression ratio and air standard efficiency remain unchanged. The pressure after compression and during working stroke is lower, the less work thus obtained during the cycle reduces the speed. *This method is preferred for large engines.*

17.7. NOISE ABATEMENT

A lot of research and developement is in progress towards *reducing engine and exhause noise.* This can be accomplished by the following three ways :

1. Passive 2. Semi-active 3. Active

- Noise reduction is accomplished passively by *correct design and the use of proper materials.*
  — The use of *ribs and stiffners, composite materials, and sandwich construction is now routine.* This type of construction *reduces noise vibrations in the various engine components.*
- In **semi-active** *noise abatement systems,* **hydraulics** *are often used.*
  — Some engines are equipped with flywheels which have hydraulic passages through which fluid flows. At idle and other constant-speed operation, the system is designed

to give the flywheel the proper stiffness to absorb engine vibrations for frequencies at that condition. When acceleration occurs the flywheel fluid flows to other locations, changing the overall stiffness of the flywheel and making it more absorbent to the new vibration frequency.

— Some automobiles have *hydraulic engine mounts connecting the engine to the automobile body. Fluid in these mounts acts to absorb and dampen engine vibrations and isolate them from the passenger compartment.* Engine mounts using *electrorheological fluid are under development which will allow better vibration dampening at all frequencies.* The viscosity of these fluids can be changed by as much as a factor of 50 : 1 with the application of an external voltage. Engine noise (vibration) is sensed by *accelerometers* which feed this information into the engine management system (EMS). Here the frequency is analysed and proper voltage is applied to the engine mounts to best dampen that frequency. Response time is of the order of 0.005 second.

- **Active noise** abatemen is accomplished by *generating 'antinoise'* to cancel out engine exhaust noise. This is done by *sensing the noise with a receiver, analyzing the frequency of the noise, and then generating noise of equal frequency, but out of phase with the original noise.* If the noises are at the same frequency but *180° out of phase,* the wave fronts *cancel each other* and the noise is eliminated. This method works well with constant speed automobile engines. It requires additional electronic equipment (receiver, frequency analyzer, transmitter) than that used with normal EMS computers.

  — Some automobiles have receivers and transmitters mounted under seats in the passenger compartment as an active engine noise abatement system. Similar systems are used near the end of the tail pipe, a major source of engine-related noise.

- Noise reduction has been so successful that some automobiles are now equipped with a safety switch on the starter. At idle speed, the engine is so quiet that the safety switch is required to keep drivers from trying to start the engine when it is already running.

WORKED EXAMPLES

Example 17.1. *A two-stroke cycle internal combustion engine has a mean effective pressure of 6 bar. The speed of the engine is 1000 r.p.m. If the diameter of piston and stroke are 110 mm and 140 mm respectively, find the indicated power developed.*

Solution. Mean effective pressure (indicated), $p_{mi} = 6$ bar

Engine speed, $N = 1000$ r.p.m.

Diameter of the piston, $D = 110$ mm = 0.11 m

Stroke length, $L = 140$ mm = 0.14 m

Indicated power developed, P :

Indicated power, $$\text{I.P.} = \frac{np_{mi}\,LANk \times 10}{6} \text{ kW}$$

Here, n = No. of cylinders = 1,

and $k = 1$ ..... for 2-stroke cycle engine.

$$\therefore\ \text{I.P.} = \frac{1 \times 6 \times 0.14 \times \frac{\pi}{4} \times (0.11)^2 \times 1000 \times 1 \times 10}{6} = \textbf{13.3 kW.} \quad \textbf{(Ans.)}$$

Example 17.2. *A 4-cylinder four-stroke petrol engine develops 14.7 kW at 1000 r.p.m. The mean effective pressure is 5.5 bar. Calculate the bore and stroke of the engine, if the length of stroke is 1.5 times the bore.*

Solution. Number of cylinder, $n = 4$

Power developed, $P = 14.7$ kW

Engine speed, $N = 1000$ r.p.m.

Indicated mean effective pressure, $p_{mi} = 5.5$ bar

Length of stroke, $L = 1.5\ D$ (bore)

For four stroke cycle, $k = \frac{1}{2}$

L :, D :

Indicated power developed,

$$\text{I.P.} = \frac{np_{mi}\ LANk \times 10}{6}\ \text{kW}$$

$$14.7 = \frac{4 \times 5.5 \times 1.5D \times \left(\frac{\pi}{4} \times D^2\right) \times 1000 \times \frac{1}{2} \times 10}{6}$$

$$\therefore \quad D^3 = \frac{14.7 \times 6 \times 4 \times 2}{4 \times 5.5 \times 1.5 \times \pi \times 1000 \times 10} = 0.0006806$$

$\therefore$ $D =$ **0.0879 or 87.9 mm. (Ans.)**

and $L =$ **1.5 × 87.9 = 131.8 mm. (Ans.)**

Example 17.3. *A single-cylinder, four-stroke cycle oil engine is fitted with a rope brake. The diameter of the brake wheel is 600 mm and the rope diameter is 26 mm. The dead load on the brake is 200 N and the spring balance reads 30 N. If the engine runs at 450 r.p.m., what will be the brake power of the engine ?*

Solution. Diameter of the brake wheel, $D_b = 600$ mm $= 0.6$ m

Rope diameter, $d = 26$ mm $= 0.026$ m

Dead load on the brake, $W = 200$ N

Spring balance reading, $S = 30$ N

Engine speed, $N = 450$ r.p.m.

Brake power, B.P. :

Brake power is given by,

$$B.P. = \frac{(W - S)\,\pi\,(D_b + d)N}{60 \times 1000}\ \text{kW}$$

$$= \frac{(200 - 30)\pi(0.6 + 0.026) \times 450}{60 \times 1000} = \textbf{2.5 kW.} \quad \textbf{(Ans.)}$$

Example 17.4. *A four-cylinder, four-stroke, spark ignition engine develops a maximum brake torque of 160 Nm at 3000 r.p.m. Calculate the engine displacement, bore and stroke. The brake mean effective pressure at the maximum engine torque point is 960 kPa. Assume bore is equal to stroke.*

(GATE-1992)

Solution. *Given* : $n = 4$; $k = \frac{1}{2}$ (engine being 4-stroke cycle), $T_b = 160$ Nm,

N = 3000 r.p.m., $p_m = 960$ kPa $= 960 \times 10^3$ N/m^2 $= 9.6$ bar ; D = L

D, L, displacement :

Power developed $= \dfrac{2\pi NT_b}{60 \times 1000}$ kW $= \dfrac{p_m LANk \times 10}{6}$ kW

i.e., $$\frac{2\pi NT_b}{60 \times 1000} = \frac{p_m LANk \times 10}{6}$$

Substituting the values, we get

$$\frac{2\pi \times 3000 \times 160}{60 \times 1000} = \frac{9.6 \times D \times \frac{\pi}{4} \times D^2 \times 3000 \times \frac{1}{2} \times 10}{6}$$

$$50.265 = 18849.6\, D^3$$

$$\therefore \quad D = \left(\frac{50.265}{18849.6}\right)^{1/3} = 0.1387 \text{ m or } \mathbf{138.7 \text{ mm.}} \quad \textbf{(Ans.)}$$

$\therefore$ **L = D = 138.7 mm. (Ans.)**

Displacement $= \dfrac{\pi}{4} D^2 \times L = \dfrac{\pi}{4} \times 0.1387^2 \times 0.1387$

$= \mathbf{0.002095\ m^3}$. **(Ans.)**

Example 17.5. *A turbocharged six-cylinder diesel engine has the following performance details :*

(*i*) *Work done during compression and expansion = 820 kW*
(*ii*) *Work done during intake and exhaust = 50 kW*
(*iii*) *Rubbing friction in the engine = 150 kW*
(*iv*) *Network done by turbine = 40 kW*

If the brake mean effective pressure is 0.6 MPa, determine the bore and stroke of the engine taking the ratio of bore to stroke as 1 and engine speed as 1000 r.p.m. **(GATE-1998)**

Solution. *Given :* $p_{mb} = 0.6\ MP_a = 6$ bar ; $\dfrac{D}{L} = 1$; $N = 1000$ r.p.m.

D, L :

Net work available = 820 – (50 + 150 + 40) = 580 kW

$$\text{B.P} = \frac{n \times p_{mb}\, LANk \times 10}{6}$$

$$580 = \frac{6 \times 6 \times D \times \frac{\pi}{4} D^2 \times 1000 \times \frac{1}{2} \times 10}{6} = 23562\, D^3$$

$$D^3 = \left(\frac{580}{23562}\right)^{1/3} = 0.2908 \text{ m} \quad \text{or} \quad 290.8 \text{ mm}$$

Hence **D = L = 290.8 mm. (Ans.)**

Example 17.6. *A spark-ignition engine, designed to run on octane ($C_3 H_{18}$) fuel, is operated on methane (CH_4). Estimate the ratio of the power input of the engine with methane fuel to that with octane. In both cases the fuel ratio is stoichiometric, the mixture is supplied to the engine at the same conditions, the engine runs at the same speed, and has the same volumetric and thermal efficiencies. The heating value of methane is 50150 kJ/kg while that of octane in 44880 kJ/kg.* **(U.P.S.C.-1994)**

Solution. In spark ignition engine, the air standard efficiency is given as :

$$\eta_{\text{air-standard}} = 1 - \frac{1}{(r)^{\gamma - 1}},$$

where r is the compression ratio which depends on engine parameters and has no relevance to fuel *e.g.*, quantity, type, calorific value etc.

Indicated power of engine is given by :

$$\text{I.P.} = \frac{p_m\, LANk \times 10}{6} \text{ kW}$$

when L (length), A (area) and N (r.p.m.) depend on engine construction.

Only p_m (mean effective pressure) *depends on engine operation.*

For air-standard Otto cycle,

$$p_m = \frac{\text{Net work}}{\text{Displacement}} = \frac{Q_{\dot{m}} \times \text{efficiency}}{\text{Displacement}}$$

where $Q_{\dot{m}}$ is the energy supplied.

Since $Q_{\dot{m}}$ in proportional to the mass of the fuel supplied times calorific value of fuel, therefore

$$\frac{(\text{Power})_{\text{methane}}}{(\text{Power})_{\text{octane}}} = \frac{C_{\text{methane}}}{C_{\text{octane}}} = \frac{50150}{44880} = \mathbf{1.117}. \quad \textbf{(Ans.)}$$

Example 17.7. *A large diesel engine run on four-stroke cycle at 2000 r.p.m. The engine has a displacement of 25 litres and a brake mean effective pressure of 0.6 MN/m². It consumes 0.018 kg/s of fuel (calorific value = 42000 kJ/kg). Determine the brake power and brake thermal efficiency.*

(GATE-1999)

Solution. *Given* : $N = 2000$ r.p.m. ; $k = \frac{1}{2}$ (4-stroke cycle engine) ;

Displacement ($A \times L$) = 25 litres = $25 \times 10^{-3} = 0.025$ m³ ; $p_{mb} = 0.6$ MN/m² = 6 bar

$$\dot{m}_f = 0.018 \text{ kg/s} \; ; \; C = 42000 \text{ kJ/kg}$$

Brake power, B.P. :

$$\text{B.P.} = \frac{p_m LANk \times 10}{6}$$

$$= \frac{6 \times 0.025 \times 2000 \times \frac{1}{2} \times 10}{6} = \mathbf{250 \text{ kW}}. \quad \textbf{(Ans.)}$$

Brake thermal efficiency, $$\eta_{\text{th (B)}} = \frac{\text{B.P.}}{\dot{m}_f \times C}$$

$$= \frac{250}{0.018 \times 42000} = 0.3307 \text{ or } \mathbf{33.07\%}. \quad \textbf{(Ans.)}$$

Example 17.8. *A rope brake was used to measure the brake power of a single cylinder, four-stroke cycle petrol engine. It was found that the torque due to brake load is 175 N-m and the engine makes 500 r.p.m. Determine the brake power developed by the engine.*

Solution. Torque due to brake load, $T = 175$ N-m

Engine speed, $N = 500$ r.p.m.

Brake power, B.P. :

Brake power, $$\text{B.P.} = \frac{2\pi NT}{60 \times 1000} = \frac{2\pi \times 500 \times 175}{60 \times 1000} = \mathbf{9.16\ kW.}\quad \textbf{(Ans.)}$$

Example 17.9. *Following observations were recorded during a test on a single cylinder oil engine :*

Bore = 300 mm ; stroke = 450 mm ; speed = 300 r.p.m. ; i.m.e.p. = 6 bar ; net brake load = 1.5 kN ; brake drum diameter = 1.8 metres ; brake rope diameter = 2 cm.

Calculate : (i) Indicater power ; (ii) Brake power ; (iii) Mechanical efficiency.

(AMIE, Winter 1996)

Solution. Bore of engine cylinder, $D = 300$ mm $= 0.3$ m

Stroke length, $L = 450$ mm $= 0.45$ m

Engine speed, $N = 300$ r.p.m.

Indicated mean effective pressure, $p_{mi} = 6$ bar

Net brake load, $(W - S) = 1.5$ kN

Diameter of brake drum, $D_b = 1.8$ m

Brake rope diameter, $d = 2$ cm $= 0.02$ m

(*i*) **Indicated power, I.P. :**

$$\text{I.P.} = \frac{n\, p_{mi}\, LANk \times 10}{6} \quad \left[\text{where } k = \frac{1}{2} \text{ ...... four-stroke engine and } n = \text{no. of cylinders}\right]$$

$$= \frac{1 \times 6 \times 0.45 \times \frac{\pi}{4} \times 0.3^2 \times 300 \times \frac{1}{2} \times 10}{6} = \mathbf{47.71\ kW.}\quad \textbf{(Ans.)}$$

(*ii*) **Brake power, B.P. :**

$$\text{B.P.} = \frac{(W - S)\,\pi\,(D_b + d)N}{60} = \frac{1.5 \times \pi(1.8 + 0.02) \times 300}{60} = \mathbf{42.88\ kW.}\quad \textbf{(Ans.)}$$

(*iii*) **Mechanical efficiency, $\eta_{mech.}$:**

$$\eta_{mech.} = \frac{\text{B.P.}}{\text{I.P.}} = \frac{42.88}{47.71} = 0.8987 \text{ or } \mathbf{89.87\%.}\quad \textbf{(Ans.)}$$

Example 17.10. *The power output of an I.C. engine is measured by a roped brake dynamometer. The diameter of the brake pulley is 700 mm and the rope diameter is 25 mm. The load on the tight side of the rope is 50 kg mass and spring balance reads 50N. The engine running at 900 r.p.m. consumes fuel of calorific value of 44000 kJ/kg, at a rate of 4 kg/h.*

Assume g = 9.81 m/s². Calculate :

(i) Brake specific fuel consumption.

(ii) Brake thermal efficiency. **(GATE-1997)**

Solution. *Given* : $D_b = 700$ mm $= 0.7$ m , $d = 25$ mm, $= 0.025$ m,

$W = 50$ kg, $S = 50$ N ; $N = 900$ r.p.m. ; $C = 44000$ kJ/kg, $\dot{m}_f = 4$ kg/h

(*i*) **Brake specific fuel consumption, b.s.f.c :**

$$\text{Brake power, B.P} = \frac{(W - S)\,\pi\,(D_b + d) \times N}{60 \times 1000}\ \text{kW}$$

$$= \frac{(50 \times 9.81 - 50) \times \pi\,(0.7 + 0.025) \times 900}{60 \times 1000} = 15.05 \text{ kW}$$

Brake specific fuel consumption,

$$\text{b.s.f.c.} = \frac{m_f \text{ (kg/ h)}}{\text{B.P.(kW)}} = \frac{4}{15.05} = \mathbf{0.266 \text{ kJ/hW-h.}} \quad \textbf{(Ans.)}$$

(*ii*) **Brake thermal efficiency, $\eta_{th(B)}$:**

$$\eta_{th(B)} = \frac{\text{B.P.}}{\dot{m}_f \times \text{C}} \qquad \text{(where } \dot{m}_f = \text{Fuel used, in kg/s)}$$

$$= \frac{15.05}{(4 / 3600) \times 44000} = 0.3078 \quad \text{or} \quad \mathbf{30.78\%.} \quad \textbf{(Ans.)}$$

Example 17.11. *A four-cylinder four-stroke S.I. engine has a compression ratio of 8 and bore of 100 mm, with stroke equal to the bore. The volumetric efficiency of each cylinder is equal to 75%. The engine operates at a speed of 4800 r.p.m. with an air-fuel ratio 15.*

Given that the calorific value of fuel = 42 MJ/kg, atmospheric density = 1.12 kg/m³, mean effective pressure in the cylinder = 10 bar and mechanical efficiency of the engine = 80%, determine the indicated thermal efficiency and the brake power. **(GATE-1996)**

Solution. *Given* : Number of cylinders, $n = 4$, $k = \frac{1}{2}$ (engine being four-stroke),

$r = 8$, D = 100 mm = 0.1 m ; $L = D = 0.1$ *m* ; $\eta_{vol} =$ 75% ; $N = 4800$ r.p.m.,

Air-fuel ratio = 15 ; $C = 42$ MJ/kg ; $\rho = 1.12$ kg/m³ ; $p_{mi} = 10$ bar, $\eta_{mech} = 80\%$.

Indicated thermal efficiency, $\eta_{th.\,(I)}$:

Indicated power, $\quad \text{I.P.} = \dfrac{n\, p_{mi}\, LANk \times 10}{6}$

$$= \frac{4 \times 10 \times 0.1 \times \left(\frac{\pi}{4} \times 0.1^2\right) \times 4800 \times \frac{1}{2} \times 10}{6} \text{ kW} = 125.66 \text{ kW}$$

Air consumption $\quad = n \times \dfrac{\pi}{4} D^2 \times L \times \dfrac{N}{2} \times \eta_{vol.}$

$$= 4 \times \frac{\pi}{4} \times 0.1^2 \times 0.1 \times \frac{4800}{2} \times 0.75$$

$$= 5.655 \text{ m}^3\text{/min} = 0.09425 \text{ m}^3\text{/s}$$

Mass flow of air, $\quad \dot{m}_a = 0.09425 \times 1.12 = 0.1056$ kg/s

Fuel consumption, $\quad \dot{m}_f = \dfrac{0.1056}{\text{Air fuel ratio}} = \dfrac{0.1056}{15} = 0.00704$ kg/s

$\therefore \quad \eta_{th(I)} = \dfrac{\text{I.P.}}{\dot{m}_f \times C}$

$$= \frac{125.66 \times 10^3}{0.00704 \times 42 \times 10^6} = 0.425 \text{ or } \mathbf{42.5\%.} \quad \textbf{(Ans.)}$$

Brake power, B.P.

Brake power $\quad$ = Indicated power × η_{mech}

= 125.66 × 0.8 = **100.53 kW. (Ans.)**

Example 17.12. *A single-cylinder four-stroke diesel engine running at 1800 r.p.m. has a bore of 85 mm and a stroke of 110 mm. It takes 0.56 kg of air per minute and develops a brake power output of 6 kW while the air-fuel ratio is 20:1. The calorific value of the fuel used is 42550 kJ/kg, and the ambient air density is 1.18 kg/m³. Calculate.*

(i) The volumetric efficiency, and

(ii) Brake specific fuel consumption. **(GATE-1995)**

Solution. *Given* : $N = 1800$ r.p.m. ; $D = 85$ mm $= 0.085$ m ; $L = 110$ mm $= 0.11$ m ;

Air flow rate, $m = 0.56$ kg/min. ; B.P. = 6 kW ; Air-fuel ratio = 20 : 1 ;

$C = 42550$ kJ/kg ; $\rho = 1.18$ kg/m^3

(*i*) **The volumetric efficiency, $\eta_{vol.}$:**

$$\text{Volume displacement} = \frac{\pi}{4} D^2 \times L \times \frac{N}{2}$$

$$= \frac{\pi}{4} \times (0.085)^2 \times 0.11 \times \frac{1800}{2} = 0.5617 \text{ m}^3/\text{min}$$

Mass of air $= 0.5617 \times 1.18 = 0.663$ kg/min

$$\therefore \quad \text{Volumetric efficiency} = \frac{0.56}{0.663} = 0.845 \text{ or } \mathbf{84.5\%}. \quad \textbf{(Ans.)}$$

(*ii*) **Brake specific fuel consumption (b.s.f.c.) :**

$$\text{Fuel consumption} = \frac{0.56}{\text{Air-fuel ratio}} = \frac{0.56}{20} = 0.028 \text{ kg/min.}$$

$$\therefore \quad \text{Brake specific fuel consumption} = \frac{\text{Fuel used / hour}}{\text{B.P.}} \text{ kg/kW-h}$$

$$= \frac{0.028 \times 60}{6} = \mathbf{0.28 \text{ kg/ kW-h.}} \quad \textbf{(Ans.)}$$

Example 17.13. *Following data refer to a four-stroke double-acting diesel engine having cylinder diameter 200 mm and piston stroke 350 mm.*

m.e.p. on cover side	= *6.5 bar*
m.e.p. on crank side	= *7 bar*
Speed	= *420 r.p.m.*
Diameter of piston rod	= *20 mm*
Dead load on the brake	= *1370 N*
Spring balance reading	= *145 N*
Brake wheel diameter	= *1.2 m*
Brake rope diameter	= *20 mm*

Calculate the mechanical efficiency of the engine.

Solution. $p_{mi(\text{cover})} = 6.5$ bar, $p_{mi(\text{crank})} = 7$ bar, $D = 0.2$ m, $L = 0.35$ m,

$N = 420$ r.p.m., $d_{\text{rod}} = 20$ mm $= 0.02$ m, $W = 1370$ N, $S = 145$ N,

$D_b = 1.2$ m, $d = 0.02$ m, $k = \frac{1}{2}$ ... 4-stroke cycle engine

Mechanical efficiency ; $\eta_{\text{mech.}}$ **:**

Area of cylinder on cover end side,

$$A_{\text{cover}} = \pi/4 \, D^2 = (\pi/4) \times (0.2)^2 = 0.03141 \text{ m}^2$$

Effective area of cylinder on crank end side,

$$A_{crank} = \pi/4\ (D^2 - d_{rod}^2) = \pi/4\ (0.2^2 - 0.02^2) = 0.0311\ m^2$$

Indicated power on cover end side,

$$I.P._{(cover)} = \frac{p_{mi\,(cover)} \times LANk \times 10}{6}$$

$$= \frac{6.5 \times 0.35 \times 0.03141 \times 420 \times \frac{1}{2} \times 10}{6} = 25\ kW$$

Indicated power on crank end side,

$$I.P._{(crank)} = \frac{p_{mi\,(crank)} \times LANk \times 10}{6}$$

$$= \frac{7 \times 0.35 \times 0.0311 \times 420 \times \frac{1}{2} \times 10}{6} = 26.67\ kW$$

Total $\quad I.P. = 25 + 26.67 = 51.67\ kW$

Now, brake power, $\quad B.P. = \frac{(W - S)\,\pi\,(D_b + d)N}{60 \times 1000} = \frac{(1370 - 145)\pi(1.2 + 0.02) \times 420}{60 \times 1000}\ kW$

$$= 32.86\ kW$$

Mechanical efficiency, $\eta_{mech.} = \frac{B.P.}{I.P.} = \frac{32.86}{51.67} = 0.6359 =$ **63.59%. (Ans.)**

Example 17.14. *The following data refer to an oil engine working on Otto four-stroke cycle :*

Brake power $= 14.7$ *kW*
Suction pressure $= 0.9$ *bar*
Mechanical efficiency $= 80\%$
Ratio of compression $= 5$
Index of compression curve $= 1.35$
Index of expansion curve $= 1.3$
Maximum explosion pressure $= 24$ *bar*
Engine speed $= 1000$ *r.p.m.*
Ratio of stroke : bore $= 1.5$

Find the diameter and stroke of the piston.

Solution. Refer Fig. 17.12.

$$B.P. = 14.7\ kW,\ p_1 = 0.9\ bar,\ \eta_{mech.} = 80\%,\ r = 5,\ p_3 = 24\ bar$$

$$N = 1000\ r.p.m.,\ \frac{L}{D} = 1.5$$

$$\mathbf{D = ?,\ \ L = ?}$$

Compression ratio, $\quad r = \frac{V_1}{V_2} = \frac{V_4}{V_3}$

To find p_2, considering *compression process 1-2*, we have

$$p_1 V_1^{1.35} = p_2 V_2^{1.35}$$

or

$$\frac{p_2}{p_1} = \left(\frac{V_1}{V_2}\right)^{1.35} = (5)^{1.35} = 8.78$$

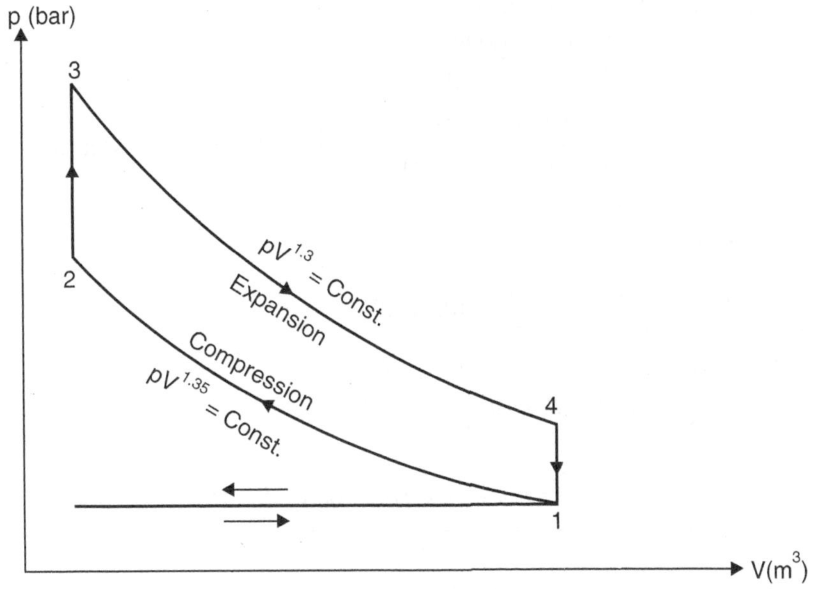

Fig. 17.12

$$\therefore \qquad p_2 = p_1 \times 8.78 = 0.9 \times 8.78 = 7.9 \text{ bar}$$

To find p_4, considering *expansion process 3-4*, we have

$$p_3 V_3^{\ 1.3} = p_4 V_4^{\ 1.3}$$

or
$$\frac{p_3}{p_4} = \left(\frac{V_4}{V_3}\right)^{1.3} = (5)^{1.3} = 8.1$$

$$\therefore \qquad p_4 = \frac{p_3}{8.1} = \frac{24}{8.1} = 2.96 \text{ bar}$$

Work done/cycle = Area 1-2-3-4

= Area under the curve 3-4 – area under the curve 1-2

$$= \frac{p_3V_3 - p_4V_4}{1.3 - 1} - \frac{p_2V_2 - p_1V_1}{1.35 - 1}$$

$$= \frac{p_3V_3 - p_4V_4}{0.3} - \frac{p_2V_3 - p_1V_4}{0.35} \qquad [\because \ V_1 = V_4 \text{ and } V_2 = V_3]$$

$$= \frac{10^5(24V_3 - 2.9V_4)}{0.3} - \frac{10^5(7.9V_3 - 0.9V_4)}{0.35}$$

$$= 10^5 \, [(80V_3 - 9.86V_4) - (22.57V_3 - 2.57V_4)]$$

$$= 10^5(80V_3 - 9.86V_4 - 22.57V_3 + 2.57V_4)$$

$$= 10^5(57.43V_3 - 7.29V_4)$$

$$= 10^5(57.43V_3 - 7.29 \times 5V_3) \qquad \left[\because \ \frac{V_4}{V_3} = 5\right]$$

$$= 10^5 \times 20.98 \ V_3 \text{ N-m.}$$

Mean effective pressure (theoretical),

$$p_m = \frac{\text{Work done / cycle}}{\text{Stroke volume } (V_s)}$$

$$= \frac{10^5 \times 20.98V_3}{(V_4 - V_3)} = \frac{10^5 \times 20.98V_3}{5V_3 - V_3} = 10^5 \times 5.245 \text{ N/m}^2 \text{ or } 5.245 \text{ bar.}$$

Now, $\eta_{mech.} = \frac{\text{B.P.}}{\text{I.P.}}$

$\therefore$ $\text{I.P.} = \frac{\text{B.P.}}{\eta_{mech.}} = \frac{14.7}{0.8} = 18.37 \text{ kW.}$

To find D and L :

$$\text{I.P.} = \frac{p_{mi} LANk \times 10}{6} \text{ kW}$$

$$14.7 = \frac{5.245 \times 1.5D \times \pi / 4 \times D^2 \times 1000 \times \frac{1}{2} \times 10}{6}$$

$\therefore$ $D^3 = \frac{14.7 \times 6 \times 4 \times 2}{5.245 \times 1.5 \times \pi \times 1000 \times 10} = 0.002855$

or $D = 0.1418 \text{ m} \simeq$ **0.142 m or 142 mm. (Ans.)**

and $L = 1.5D = 1.5 \times 142 =$ **213 mm. (Ans.)**

Example 17.15. *The following results refer to a test on a petrol engine :*

Indicated power = 30 kW ; brake power = 26 kW ;
Engine speed = 1000 r.p.m. ; fuel per brake-power hour = 0.35 kg ;
Calorific value of the fuel used = 43900 kJ/kg.

Calculate : (i) The indicated thermal efficiency,
(ii) The brake thermal efficiency, and
(iii) The mechanical efficiency.

Solution. Indicated power, I.P. = 30 kW
Brake power, B.P. = 26 kW
Engine speed, N = 1000 r.p.m.
Fuel per brake-power hour = 0.35 kg/B.P.h
Calorific value of the fuel used, C = 43900 kJ/kg
Now, fuel consumption per hour = 0.35 × 26 = 9.1 kg/h.

(*i*) **Indicated thermal efficiency,**

$$\eta_{th(I)} = \frac{\text{I.P.}}{\dot{m}_f \times C} = \frac{30}{\left(\frac{9.1}{3600}\right) \times 43900} = \mathbf{0.27 \text{ or } 27\%. \text{ (Ans.)}}$$

(*ii*) **Brake thermal efficiency,**

$$\eta_{th(B)} = \frac{\text{B.P.}}{\dot{m}_f \times C} = \frac{26}{\frac{9.1}{3600} \times 43900} \mathbf{0.234 \text{ or } 23.4\%. \text{ (Ans.)}}$$

(*iii*) **Mechanical efficiency,**

$$\eta_{mech.} = \frac{\text{B.P.}}{\text{I.P.}} = \frac{26}{30} = \mathbf{0.866 \text{ or } 86.6\%.} \quad \textbf{(Ans.)}$$

Example 17.16. *The output of an I.C. engine is measured by a rope brake dynamometer. The diameter of the brake pulley is 750 mm and rope diameter is 50 mm. The dead load on the tight side of the rope is 400 N and the spring balance reading is 50 N. The engine consumes 4.2 kg/h of fuel at rated speed of 1000 r.p.m. The calorific value of fuel is 43900 kJ/kg. Calculate :*

(*i*) *Brake specific fuel consumption, and*

(*ii*) *Brake thermal efficiency.*

Solution. Diameter of brake pulley, $D_b = 750 \text{ mm} = 0.75 \text{ m}$

Rope diameter, $d = 50 \text{ mm} = 0.05 \text{ m}$

Dead load, $W = 400\ N$

Spring balance reading, $S = 50\ N$

Consumption of fuel $= 4.2$ kg/h

Rated speed, $N = 1000$ r.p.m.

Calorific value of fuel, $C = 43900$ kJ/kg

Brake power,
$$\text{B.P.} = \frac{(W - S)\,\pi\,(D_b + d)N}{60 \times 1000} \text{ kW}$$

$$= \frac{(400 - 50)\pi(0.75 + 0.05) \times 1000}{60 \times 1000} = 14.66 \text{ kW.}$$

(*i*) **Brake specific fuel consumption**,

$$\text{s.f.c. (brake)} = \frac{4.2}{14.66} = \mathbf{0.286 \text{ kg/kWh.}} \quad \textbf{(Ans.)}$$

(*ii*) **Brake thermal efficiency,**

$$\eta_{th(B)} = \frac{\text{B.P.}}{\dot{m}_f \times C} = \frac{14.66}{\frac{4.2}{3600} \times 43900}$$

$(\dot{m}_f = \text{Fuel used in kg/s})$

$$= \mathbf{0.286 \text{ or } 28.6\%.} \quad \textbf{(Ans.)}$$

Example 17.17. *A six-cylinder, four-stroke 'Petrol engine' having a bore of 90 mm and stroke of 100 mm has a compression ratio of 7. The relative efficiency with reference to indicated thermal efficiency is 55% when the indicated specific fuel consumption is 0.3 kg/kWh. Estimate the calorific value of the fuel and fuel consumption (in kg/h), given that the imep is 8.5 bar and speed is 2500 r.p.m.* **(AMIE Summer, 1999)**

Solution. Number of cylinders, $n = 6$

Bore of each cylinder, $D = 90 \text{ mm} = 0.09 \text{ m}$

Stroke length, $L = 100 \text{ mm} = 0.1 \text{ m}$

Compression ratio, $r = 7$

Relative efficience, $\eta_{relative} = 55\%$

Indicates specific fuel consumption $= 0.3$ kg/kWh

Indicated mean effective pressure, imep = 8.6 bar

Engine speed, $N = 2500$ r.p.m.

Calorific value of fuel (C) and fuel consumption (in kg/h) :

$$\eta_{\text{air standard}} = 1 - \frac{1}{(r)^{\gamma-1}} = 1 - \frac{1}{(7)^{1.4-1}} = 0.5408$$

Now, $\eta_{\text{relative}} = \frac{\eta_{\text{th}(I)}}{\eta_{\text{air standard}}}$ or $\eta_{\text{th}(I)} = \eta_{\text{relative}} \times \eta_{\text{air standard}}$

or Indicated thermal efficiency, $\eta_{\text{th}(I)} = 0.55 \times 0.5408 = 0.297$

But $\eta_{\text{th}(I)} = \frac{\text{I.P.}}{\dot{m}_f \times C} = \frac{1}{(0.3/3600) \times C}$

or $0.297 = \frac{3600}{0.3 \times C}$ or $C = \frac{3600}{0.297 \times 0.3}$ = **40404 kJ/kg. (Ans.)**

Now, indicated power, $\text{I.P.} = \frac{np_{mi}LANk \times 10}{6}$ [where $k = \frac{1}{2}$ ...... four-stroke cycle engine]

$$= \frac{6 \times 8.6 \times 0.1 \times \frac{\pi}{4} \times 0.09^2 \times 2500 \times \frac{1}{2} \times 10}{6} = 68.39 \text{ kW}$$

∴ Fuel consumption $= 0.3 \times 68.39 =$ **20.52 kg/h. (Ans.)**

Example 17.18. *A 4-cylinder two-stroke cycle petrol engine develops 30 kW at 2500 r.p.m. The mean effective pressure on each piston is 8 bar and mechanical efficiency is 80%. Calculate the diameter and stroke of each cylinder of stroke-to-bore ratio 1.5. Also calculate the fuel consumption of the engine, if brake thermal efficiency is 28%. The calorific value of the fuel is 43900 kJ/kg.*

Solution. Number of cylinder, $n = 4$

Brake power, B.P. = 30 kW

Engine speed, $N = 2500$ r.p.m.

Mean effective pressure $p_{mi} = 8$ bar

Mechanical efficiency, $\eta_{\text{mech.}} = 80\%$

Length of stroke, $L = 1.5\ D$ (bore)

Brake thermal efficiency, $\eta_{\text{th(B)}} = 28\%$

Calorific value of the fuel, $C = 43900$ kJ/kg

$k = 1$ ...... for 2-stroke cycle engine

(*i*) **L = ?, D = ?**

$$\eta_{\text{mech.}} = \frac{\text{B.P.}}{\text{I.P.}}$$

$$0.8 = \frac{30}{\text{I.P.}}$$

∴ $$\text{I.P.} = \frac{30}{0.8} = 37.5 \text{ kW}$$

Also, $$\text{I.P.} = \frac{np_{mi}LANk \times 10}{6}$$

$$37.5 = \frac{4 \times 8 \times 1.5D \times \pi/4D^2 \times 2500 \times 1 \times 10}{6}$$

$\therefore \quad D^3 = \dfrac{37.5 \times 6 \times 4}{4 \times 8 \times 1.5 \times \pi \times 2500 \times 10} = 0.0002387$

or $\quad D = \mathbf{0.062\ m\ or\ 62\ mm.}$ **(Ans.)**

and $\quad L = 62 \times 1.5 = \mathbf{93\ mm.}$ **(Ans.)**

(*ii*) **Fuel consumption :**

Brake thermal efficiency, $\eta_{th(B)} = \dfrac{B.P.}{\dot{m}_j \times C}$ ($\dot{m}_f$ = Fuel used in kg/s)

$$0.28 = \frac{30}{\dot{m}_f \times 43900}$$

$$\therefore \quad \dot{m}_f = \frac{30}{0.28 \times 43900} = \mathbf{0.00244\ kg/s\ or\ 8.78\ kg/h.}\ \textbf{(Ans.)}$$

Example 17.19. *A six-cylinder, 4-stroke SI engine having a piston displacement of 700 cm*^{3} *per cylinder developed 78 kW at 3200 r.p.m. and consumed 27 kg of petrol per hour. The calorific value of petrol is 44 MJ/kg. Estimate :*

(*i*) *The volumetric efficiency of the engine if the air-fuel ratio is 12 and intake air is at 0.9 bar, 32°C.*

(*ii*) *The brake thermal efficiency, and*

(*iii*) *The brake torque.*

For air, R = 0.287 kJ/kgK. **(AMIE Summer, 1998)**

Solution. Number of cylinders = 6

Piston displacement per cylinder = 700 cm^3 or 700×10^{-6} m^3

Power developed, $P = 78$ kW

Speed of the engine, $N = 3200$ r.p.m.

Mass of fuel used, $m_f = 27$ kg/h

Calorific value of fuel, $C = 44$ MJ/kg

Air fuel ratio = 12

Intake air pressure, $p_1 (= p_a) = 0.9$ bar

Intake air temperature, $T_1 = (T_a) = 32 + 273 = 305$ K

For air, $R = 0.287$ kJ/kg K

(*i*) **Volumetric efficiency of the engine, η_{vol} :**

Mass of air, m_a = Air fuel ratio × mass of fuel

$= 12 \times 27 = 324$ kg/h

Also, $p_a V_a = m_a R_a T_a$ or $V_a = \dfrac{m_a R_a T_a}{P_a}$

$\therefore$ Volume of intake air, $V_a = \dfrac{324 \times 0.287 \times 305}{0.9 \times 10^2} = 315.126$ m^3/ h

Swept volume per hour = Piston displacement per cylinder × no. of cylinder × $\dfrac{N}{2} \times 60$

$= 700 \times 10^{-6} \times 6 \times \dfrac{3200}{2} \times 60 = 403.2$ m^3/ h

$\therefore$ Volumetric efficiency, $\eta_{vol.} = \dfrac{\text{Volume of intake air}}{\text{Swept volume}}$

$$= \frac{315.126}{403.2} = 0.781 \text{ or } \mathbf{78.1\%}. \quad \textbf{(Ans.)}$$

(*ii*) **Brake thermal efficiency η_{BT} :**

$$\eta_{BT} = \frac{\text{Brake work}}{\text{Heat supplied by fuel}} = \frac{78}{27 \times \frac{44 \times 10^3}{3600}} = \frac{78 \times 3600}{27 \times (44 \times 10^3)}$$

$$= \mathbf{0.2364 \text{ or } 23.64\%}. \quad \textbf{(Ans.)}$$

(*iii*) **The brake torque, T_B :**

$$P = \frac{2\pi N T_B}{60} \quad \text{or} \quad 78 = \frac{2\pi \times 3200 \times T_B}{60}$$

$$\therefore \quad T_B = \frac{78 \times 60}{2\pi \times 3200} = \mathbf{0.2328 \text{ kN/m}}. \quad \textbf{(Ans.)}$$

Example17.20. *A 6-cylinder, four-stroke gas engine with a stroke volume of 1.75 litres develops 26.3 kW at 504 r.p.m. The m.e.p. is 6 bar. Find the average number of times each cylinder misfires in one minute.*

Solution. Number of cylinders, $n = 6$

Stroke volume, $V_s = 1.75 \text{ litres} = 1.75 \times 10^{-3} \text{ m}^3$

Indicated power, I.P. = 26.3 kW

Engine speed, $N = 504$ r.p.m.

Mean effective pressure, $p_{mi} = 6$ bar

$k = 1/2$ ..... for 4-stroke engine.

Average number of times each cylinder misfires / min :

$$\text{I.P.} = \frac{np_{mi}LANk \times 10}{6} \text{ kW}$$

$$26.3 = \frac{6 \times 6 \times 1.75 \times 10^{-3} \times N \times 1/2 \times 10}{6} \qquad [\because \; LA = V_s = 1.75 \times 10^{-3} \text{ m}^3]$$

$$N = \frac{26.3 \times 6 \times 2 \times 1000}{6 \times 6 \times 1.75 \times 10} = 500 \text{ r.p.m.}$$

Actual number of fires in one minute $= \frac{500}{2} \times 6 = 1500$

Expected number of fires in one minute $= \frac{504}{2} \times 6 = 1512$

Number of misfires/min. $= 1512 - 1500 = 12.$

Average number of times each cylinder misfires in one minute $= \frac{12}{6} = \mathbf{2}.$ **(Ans.)**

Example 17.21. *The following data refer to a car engine having 4 cylinders.*

Bore = 75 mm ; stroke = 90 mm ; engine to rear axle ratio 39 : 8 ; wheel diameter with tyre fully inflated 650 mm. The petrol consumption for a distance of 3.2 km when car was moving at a speed of 48 km per hour was found to be 0.227 kg.

If the mean effective pressure is 5.625 bar, determine the indicated power and thermal efficiency. Calorific value of the petrol may be taken as 43470 kJ/kg.

Solution. Bore, $D = 75$ mm or 0.075 m

Stroke length, $L = 90$ mm or 0.09 m

Number of cylinders, $n = 4$

Engine to rear axle ratio $= 39 : 8$

Wheel diameter with tyre fully inflated = 650 mm or 0.65 m

Petrol consumption for a distance of 3.2 km at a speed of 48 km/h = 0.227 kg

Mean effective pressure, $p_{mi} = 5.625$ bar

Calorific value of petrol, $C = 43470$ kJ/kg

$k = 1/2$ ...... for 4-stroke engine.

Indicated power I.P. :

Speed of the car $= 48 \text{ km / h} = \dfrac{48 \times 1000}{60} = 800$ m/min.

In N_t are the revolutions made by the tyre per minute, then $\pi DN_t = 800$

$$\therefore \quad N_t = \frac{800}{\pi \times 0.650} = 392 \text{ r.p.m.}$$

As the rear axle ratio is 39 : 8.

$$\therefore N_e \text{ (speed of the engine shaft)} = \frac{392 \times 39}{8} = 1911 \text{ r.p.m.}$$

$$\text{I.P.} = \frac{np_{mi} LANk \times 10}{6} \text{ kW}$$

$$= \frac{4 \times 5.625 \times 0.09 \times \pi / 4 \times 0.075^2 \times 1911 \times \frac{1}{2} \times 10}{6}$$

= **14.25 kW. (Ans.)**

Indicated thermal efficiency :

To find indicated thermal efficiency, let us find $\dot{m}_f$ first :

Speed of the car $= \dfrac{48}{60} = 0.8$ km/min.

Time for covering 3.2 km $= \dfrac{3.2}{0.8} = 4$ min.

Amount of fuel consumed in 4 min. = 0.227 kg

$\therefore$ Fuel consumed/sec $= \dfrac{0.227}{4 \times 60} = 0.000946$ kg/s

Now, *Indicated thermal efficiency,*

$$\eta_{th(I)} = \frac{\text{I.P.}}{\dot{m}_f \times C} = \frac{14.25}{0.000946 \times 43470} = \mathbf{0.346 \text{ or } 34.6\%.} \quad \textbf{(Ans.)}$$

Example 17.22. *The following readings were taken during the test of a single-cylinder four-stroke oil engine :*

Cylinder diameter = *250 mm*

Stroke length = *400 mm*

Gross m.e.p. = *7 bar*

Pumping m.e.p. = *0.5 bar*

Engine speed = *250 r.p.m.*
Net load on the brake = *1080 N*
Effective diameter of the brake = *1.5 metres*
Fuel used per hour = *10 kg*
Calorific value of fuel = *44300 kJ/kg*

Calculate : (i) Indicated power ; (ii) Brake power ; (iii) Mechanical efficiency ; (iv) Indicated thermal efficiency.

Solution. $D = 250$ mm $= 0.25$ m, $L = 400$ mm $= 0.4$ m, $p_{mg} = 7$ bar,

$$p_{mp} = 0.5 \text{ bar}, N = 250 \text{ r.p.m.}, D_b = 1.5 \text{ m}, \dot{m}_f = \frac{10}{3600} = 0.00277 \text{ kg/s}$$

$$C = 44300 \text{ kJ/kg}, n = 1, (W - S) = 1080 \text{ N}$$

Net $\quad p_m = p_{mg} - p_{mp} = 7 - 0.5 = 6.5$ bar.

(*i*) **Indicated power I.P. :**

$$\text{I.P.} = \frac{np_m LANk \times 10}{6} = \frac{1 \times 6.5 \times 0.4 \times \pi/4 \times 0.25^2 \times 250 \times \frac{1}{2} \times 10}{6} \text{ kW} = 26.59 \text{ kW}.$$

(*ii*) **Brake power, B.P. :**

$$\text{B.P.} = \frac{(W - S)\pi D_b N}{60 \times 1000} \text{ kW} = \frac{1080 \times \pi \times 1.5 \times 250}{60 \times 1000} = 21.2 \text{ kW}.$$

(*iii*) **Mechanical efficiency, $\eta_{mech.}$:**

$$\eta_{\text{mech.}} = \frac{\text{B.P.}}{\text{I.P.}} = \frac{21.2}{26.59} = \mathbf{0.797 \text{ or } 79.7\%.} \quad \textbf{(Ans.)}$$

(*iv*) **Indicated thermal efficiency, $\eta_{th(I)}$:**

$$\eta_{\text{th(I)}} = \frac{\text{I.P.}}{\dot{m}_f \times C} = \frac{26.59}{0.00277 \times 44300} = \mathbf{0.216 \text{ or } 21.6\%.} \quad \textbf{(Ans.)}$$

Example 17.23. *The brake thermal efficiency of a diesel engine is 30 per cent. If the air-to-fuel ratio by weight is 20 and the calorific value of the fuel used is 41800 kJ/kg, what brake mean effective pressure may be expected at S.T.P. conditions ?*

Solution. Brake thermal efficiency, $\eta_{th(B)} = 30\%$
Air-fuel ratio by weight $= 20$
Calorific value of fuel used, $C = 41800$ kJ/kg

Brake mean effective pressure, p_{mb} :

$$\text{Brake thermal efficiency} = \frac{\text{Work produced}}{\text{Heat supplied}}$$

$$0.3 = \frac{\text{Work produced}}{41800}$$

$\therefore$ Work produced per kg of fuel $= 0.3 \times 41800 = 12540$ kJ

Mass of air used per *kg* of fuel $= 20$ kg

S.T.P. conditions refer to 1.0132 bar and 15°C

$$\text{Volume of air used} = \frac{mRT}{p} = \frac{20 \times 287 \times (273 + 15)}{1.0132 \times 10^5} = 16.31 \text{ m}^3$$

Brake mean effective pressure,

$$p_{mb} = \frac{\text{Work done}}{\text{Cylinder volume}} = \frac{12540 \times 1000}{16.31 \times 10^5} = \textbf{7.69 bar.} \quad \textbf{(Ans.)}$$

Example 17.24. *In a test on single-cylinder four-stroke cycle gas engine with explosion in every cycle, the gas consumption given by the meter was 0.216 m³ per minute ; the pressure and temperature of the gas being 75 mm of water and 17°C respectively. Air consumption was 2.84 kg/min., the temperature being 17°C and barometer reading 745 mm of mercury. The bore of the engine was 250 mm and stroke 475 mm and r.p.m. 240.*

Find volumetric efficiency of the engine referred to volume of charge at N.T.P. Assume R for air as 287 N m/kg K.

Solution. Gas consumption, $V_1 = 0.216$ m³/min.

Pressure of the gas $= 75$ mm of water

Temperature of gas, $T_1 = 17 + 273 = 290$ K

Air consumption $= 2.84$ kg/min

Temperature of air $= 17 + 273 = 290$ K

Barometer reading $= 745$ mm Hg

Bore of the engine, $D = 250$ mm $= 0.25$ m

Stroke of engine, $L = 475$ mm $= 0.475$ m

Engine speed, $N = 240$ r.p.m.

R for air $= 287$ N-m / kg K.

Volumetric efficiency, $\eta_{vol.}$:

Pressure of the gas, $p_1 = 745 + \dfrac{75}{13.6} = 750.5$ mm of mercury.

At N.T.P.

$$p_2 = 760 \text{ mm of mercury}$$
$$T_2 = 0 + 273 = 273 \text{ K}$$
$$V_2 = ?$$

To find volume of gas used at N.T.P. (V_2), using the relation :

$$\frac{p_1V_1}{T_1} = \frac{p_2V_2}{T_2}$$

$$\frac{750.5 \times 0.216}{290} = \frac{760 \times V_2}{273}$$

$$\therefore \quad V_2 = \frac{750.5 \times 0.216 \times 273}{760 \times 290} = 0.201 \text{ m}^3$$

Gas used per stroke $= \dfrac{0.201}{240/2} = 0.001675$ m³

Volume occupied by air at N.T.P. (V) :

$$pV = mRT$$

or

$$V = \frac{mRT}{p} = \frac{2.84 \times 287 \times 273}{1.0132 \times 10^5} = 2.196 \text{ m}^3\text{/min}$$

Air used per stroke $= \dfrac{2.196}{240/2} = 0.0183$ m³ at N.T.P.

Mixture of gas and air used per stroke

$$= 0.001675 + 0.0183 = 0.0199 \text{ m}^3$$

Volumetric efficiency, $\eta_{vol.} = \dfrac{\text{Actual volume of mixture drawn per stroke at N.T.P.}}{\text{Swept volume of system}}$

$$= \frac{0.0199}{\pi/4 \times 0.25^2 \times 0.475} \quad \textbf{0.853 or 85.3\%. (Ans.)}$$

Example 17.25. *The following particulars were obtained in a trial on a 4-stroke gas engine :*

Duration of trial	*= 1 hour*
Revolutions	*= 14000*
Number of missed cycle	*= 500*
Net brake load	*= 1470 N*
Mean effective pressure	*= 7.5 bar*
Gas consumption	*= 20000 litres*
L.C.V. of gas at supply condition	*= 21 kJ/litre*
Cylinder diameter	*= 250 mm*
Stroke	*= 400 mm*
Effective brake circumference	*= 4 m*
Compression ratio	*= 6.5 : 1*

Calculate : (i) Indicated power (ii) Brake power (iii) Mechanical efficiency (iv) Indicated thermal efficiency (v) Relative efficiency.

Solution. $N = \dfrac{14000}{60} = \dfrac{700}{3}$ r.p.m., $W - S = 1470$ N,

$p_{mi} = 7.5$ bar ; $V_g = \dfrac{20000}{3600} = 5.55$ litres/s,

$D = 250$ mm $= 0.25$ m, $L = 400$ mm $= 0.4$ m

$\pi D_b = 4$ m, $r = 6.5$, $n = 1$.

(*i*) **Indicated power, I.P. :**

$$\text{I.P.} = \frac{n p_{mi} LANk \times 10}{6}$$

$$Nk = \left(\frac{14000}{2} - 500\right)\Big/ 3600 = \frac{6500}{60} \text{ working cycles / min.}$$

$$\therefore \quad \text{I.P.} = \frac{1 \times 7.5 \times 0.4 \times \pi/4 \times 0.25^2 \times (6500/60) \times 10}{6}$$

$$= \textbf{26.59 kW. (Ans.)}$$

(*ii*) **Brake power, B.P. :**

$$\text{B.P.} = \frac{(W - S)\, \pi\, D_b N}{60 \times 1000} = \frac{1470 \times 4 \times (700/3)}{60 \times 1000} = \textbf{22.86 kW. (Ans.)}$$

(*iii*) **Mechanical efficiency, $\eta_{mech.}$:**

$$\eta_{mech.} = \frac{\text{B.P.}}{\text{I.P.}} = \frac{22.86}{26.59} = \textbf{0.859 or 85.9\%. (Ans.)}$$

(*iv*) **Indicated thermal efficiency, $\eta_{th.(I)}$:**

$$\eta_{th(I)} = \frac{\text{I.P.}}{V_g \times C} = \frac{26.59}{5.5 \times 21} = \mathbf{0.23 \text{ or } 23\%.} \quad \textbf{(Ans.)}$$

(*v*) **Relative efficiency, $\eta_{relative}$:**

$$\eta_{relative} = \frac{\eta_{thermal}}{\eta_{air\text{-}standard}}$$

$$\eta_{air\text{-}standard} = 1 - \frac{1}{(r)^{\gamma-1}} = 1 - \frac{1}{(6.5)^{1.4-1}} = 0.527 \text{ or } 52.7\%$$

$$\therefore \quad \eta_{relative} = \frac{0.23}{0.527} = \mathbf{0.436 \text{ or } 43.6\%.} \quad \textbf{(Ans.)}$$

Example 17.26. *The compression curve on the indicator diagram for a gas engine follows the law $pV^{1.3}$ = constant. At two points on the curve at $\frac{1}{4}$ stroke and $\frac{3}{4}$ stroke the pressures are 1.4 bar and 3.6 bar respectively. Determine the compression ratio of the engine. Calculate the thermal efficiency and the gas consumption per I.P. hour, if the relative efficiency is 0.4 and the gas has the calorific value of 18800 kJ/m³.*

Solution. Refer Fig. 17.13.

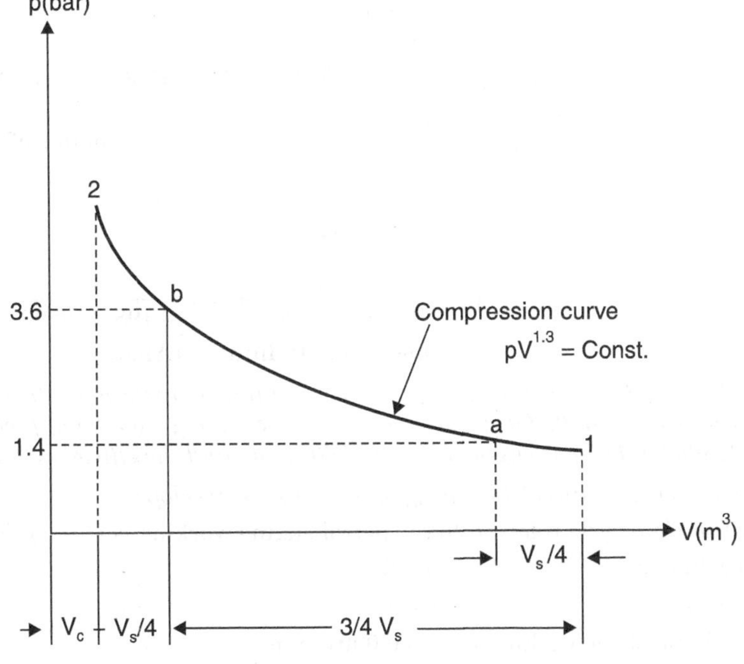

Fig. 17.13

Compression law, $pV^{1.3}$ = constant

Pressure at '*a*', p_a = 1.4 bar

Pressure at '*b*', p_b = 3.6 bar

Volume at '*a*', $V_a = V_c + 0.75V_s$

Volume at '*b*', $V_b = V_c + 0.25V_s$

Also $$p_a V_a^{1.3} = p_b V_b^{1.3}$$

or $$\frac{V_a}{V_b} = \left(\frac{p_b}{p_a}\right)^{1/1.3} = \left(\frac{3.6}{1.4}\right)^{1/1.3} = 2.067$$

Also $$\frac{V_a}{V_b} = \frac{V_c + 0.75V_s}{V_c + 0.25V_s} = 2.067$$

or $$(V_c + 0.75V_s) = 2.067\,(V_c + 0.25V_s)$$

or $$V_c + 0.75V_s = 2.067V_c + 0.516V_s$$

or $$0.234V_s = 1.067V_c \quad \text{or} \quad \frac{V_s}{V_c} = 4.56.$$

Compression ratio $$= \frac{V_s + V_c}{V_c} = \frac{V_s}{V_c} + 1 = 4.56 + 1 = \mathbf{5.56.} \quad \textbf{(Ans.)}$$

Air standard efficiency, $$\eta_{\text{air-standard}} = 1 - \frac{1}{(r)^{\gamma-1}} = 1 - \frac{1}{(5.56)^{1.4-1}} = 0.496 \text{ or } 49.6\%$$

$$\eta_{\text{relative}} = \frac{\eta_{\text{thermal}}}{\eta_{\text{air-standard}}}$$

$$0.4 = \frac{\eta_{\text{thermal}}}{0.496}$$

∴ $$\eta_{\text{thermal}} = 0.4 \times 0.496 = \mathbf{0.198 \text{ or } 19.8\%.} \quad \textbf{(Ans.)}$$

But $$\eta_{\text{thermal}} = \frac{\text{I.P.}}{V_g \times C}$$ (V_g = Volume of gas used in m³/s)

$$0.198 = \frac{1}{V_g \times 18800}$$

∴ $$V_g = \frac{1}{0.198 \times 18800} \text{ m}^3/\text{s} = \frac{1}{0.198 \times 18800} \times 3600$$
$$= \mathbf{0.967 \text{ m}^3/\text{I.P. hour.}} \quad \textbf{(Ans.)}$$

Example 17.27. *A 6-cylinder petrol engine has a volume compression ratio of 5 : 1. The clearance volume of each cylinder is 0.000115 m³. The engine consumes 10.5 kg of fuel per hour whose calorific value is 41800 kJ/kg. The engine runs at 2500 r.p.m. and the efficiency ratio is 0.65.*

Calculate the average indicated mean effective pressure developed.

Solution. The ideal cycle referred to the petrol engine working is Otto cycle.

Number of cylinder, $n = 6$

Compression ratio, $r = 5$

Clearance volume of each cylinder $= 0.000115$ m³

Fuel consumed $= 10.5$ kg/h

Calorific value of fuel, $C = 41800$ kJ/kg

Engine speed, $N = 2500$ r.p.m.

Efficiency ratio $= 0.65$.

Mean effective pressure developed, p_m :

Air-standard efficiency in case of Otto cycle is given by

$$\eta_{\text{air-standard}} = 1 - \frac{1}{(r)^{\gamma-1}} = 1 - \frac{1}{(5)^{1.4-1}} = 0.457 \text{ or } 47.5\%$$

Also,
$$\eta_{\text{ratio}} = \frac{\eta_{\text{thermal}}}{\eta_{\text{air-standard}}}$$

∴
$$\eta_{\text{thermal}} = \eta_{\text{ratio}} \times \eta_{\text{air-standard}} = 0.65 \times 0.475 = 0.308$$

But,
$$\eta_{\text{thermal(I)}} = \frac{\text{I.P.}}{\dot{m}_f \times C} \text{ or } 0.308 = \frac{\text{I.P.}}{\frac{10.5}{3600} \times 41800}$$

∴
$$\text{I.P.} = \frac{0.308 \times 41800 \times 10.5}{3600} = 37.55 \text{ kW} = 37.5 \times 10^3 \text{ N-m/s}$$

∴ Net work from one cycle per cylinder

$$= \frac{37.5 \times 10^3 \times 60}{6 \times (2500 / 2)} = 300 \text{ N-m}$$

Also,
$$r = \frac{V_s + V_c}{V_c} = 5$$

∴
$$V_c + V_s = 5V_c$$

or
$$V_s = 4V_c = 4 \times 0.000115 = 0.00046 \text{ m}^3$$

∴ *Mean effective pressure developed*

$$p_m = \frac{\text{W}_{\text{net}} \text{ per cycle}}{V_s} = \frac{300}{0.00046 \times 10^5} \text{ bar} = \mathbf{6.52\ bar.} \quad \textbf{(Ans.)}$$

☞**Example 17.28.** *A 2-cylinder C.I. engine with a compression ratio 13 : 1 and cylinder dimensions of 200 mm × 250 mm works on two stroke cycle and consumes 14 kg/h of fuel while running at 300 r.p.m. The relative and mechanical efficiencies of engine are 65% and 76% respectively. The fuel injection is effected upto 5% of stroke. If the calorific value of the fuel used is given as 41800 kJ/kg, calculate the mean effective pressure developed.*

Solution. Refer Fig. 17.14.

Diameter of cylinder, $D = 200 \text{ mm} = 0.2 \text{ m}$

Stroke length, $L = 250 \text{ mm} = 0.25 \text{ m}$

Number of cylinders, $n = 2$

Compression ratio, $r = 13$

Fuel consumption $= 14 \text{ kg / h}$

Engine speed, $N = 300 \text{ r.p.m.}$

Relative efficiency, $\eta_{\text{relative}} = 65\%$

Mechanical efficiency, $\eta_{\text{mech.}} = 76\%$

Cut-off $= 5\%$ of stroke

Calorific value of fuel, $C = 41800 \text{ kJ/kg}$

$k = 1$ ...... for two-stroke cycle engine

Cut-off ratio, $\rho = \frac{V_3}{V_2}$

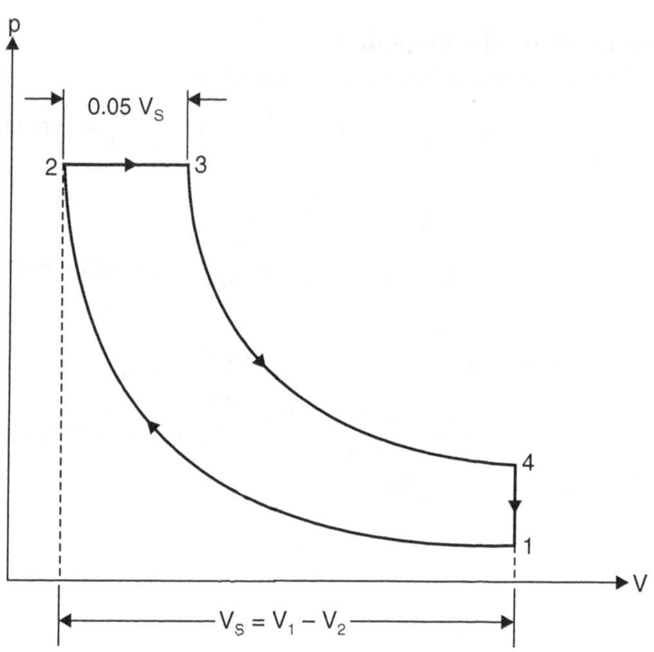

Fig. 17.14

Also $$V_3 - V_2 = 0.05V_s = 0.05(V_1 - V_2)$$

or $$V_3 - V_2 = 0.05\,(13V_2 - V_2) \qquad \left[\because \frac{V_1}{V_2} = r = 13\right]$$

or $$V_3 - V_2 = 0.05 \times 12V_2 = 0.6V_2.$$

$\therefore$ $$\frac{V_3}{V_2} = 1.6$$

$$\eta_{\text{air-standard}} = 1 - \frac{1}{\gamma(r)^{\gamma-1}}\left[\frac{\rho^\gamma - 1}{\rho - 1}\right]$$

$$= 1 - \frac{1}{1.4(14)^{1.4-1}}\left[\frac{1.6^{1.4} - 1}{1.6 - 1}\right]$$

$$= 1 - 0.248 \times 1.55 = 0.615\% \text{ or } 61.5\%$$

Also, $$\eta_{\text{relative}} = \frac{\eta_{\text{thermal}}}{\eta_{\text{air-standard}}}$$

$$0.65 = \frac{\eta_{\text{thermal}}}{0.615}$$

$\therefore$ $$\eta_{\text{thermal}} = 0.65 \times 0.615 = 0.4$$

But $$\eta_{\text{thermal (I)}} = \frac{\text{I.P.}}{\dot{m}_f \times C}$$

$$0.4 = \frac{\text{I.P.}}{\frac{14}{3600} \times 41800}$$

$$\therefore \qquad \text{I.P.} = \frac{0.4 \times 14 \times 41800}{3600} = 65 \text{ kW}$$

Now, $$\eta_{mech.} = \frac{\text{B.P.}}{\text{I.P.}}$$

$$0.76 = \frac{\text{B.P.}}{65}$$

$$\therefore \qquad \text{B.P.} = 0.76 \times 65 = 49.4 \text{ kW}$$

Mean effective pressure can be calculated based on I.P. or B.P. of the engine.

$$\text{I.P.} = \frac{np_{mi}LANk \times 10}{6}$$

where p_{mi} = indicated mean effective pressure

$$65 = \frac{2 \times p_{mi} \times 0.25 \times \pi / 4 \times 0.2^2 \times 300 \times 1 \times 10}{6}$$

$$\therefore \qquad p_{mi} = \frac{65 \times 6 \times 4}{2 \times 0.25 \times \pi \times 0.2^2 \times 300 \times 10} = \textbf{8.27 bar.} \quad \textbf{(Ans.)}$$

and brake mean effective pressure,

$$p_{mb} = 0.76 \times 8.27 = \textbf{6.28 bar.} \quad \textbf{(Ans.)}$$

Example 17.29. *Following data relate to 4-cylinder four-stroke petrol engine. Air-fuel ratio by weight = 16 : 1, calorific value of the fuel = 45200 kJ/kg, mechanical efficiency = 82%, air-standard efficiency = 52%, relative efficiency = 70%, volumetric efficiency = 78%, stroke/bore ratio = 1.25, suction conditions = 1 bar, 25°C, r.p.m. = 2400, power at brakes = 72 kW.*

Calculate : (i) Compression ratio *(ii) Indicated thermal efficiency*

(iii) Brake specific fuel consumption *(iv) Bore and stroke.*

Solution. Air fuel ratio by weight = 16 : 1

No. of cylinders, $n = 4$

Calorific value of fuel, $C = 45200$ kJ/kg

Mechanical efficiency, $\eta_{mech.} = 82\%$

Air-standard efficiency, $\eta_{air\text{-}standard} = 52\%$

Relative efficiency, $\eta_{relative} = 70\%$

Volumetric efficiency, $\eta_{vol.} = 78\%$

Stroke / bore ratio, = 1.25

Engine speed, $N = 2400$ r.p.m.

Suction conditions $p = 1$ bar, $T = 25 + 273 = 298$ K

Stroke / bore ratio = 1.25

Brake power, B.P. = 72 kW.

(*i*) **Compression ratio, r :**

For petrol engine, air standard efficiency is given by :

$$\eta_{air\text{-}standard} = 1 - \frac{1}{(r)^{\gamma-1}}$$

$$0.52 = 1 - \frac{1}{(r)^{1.4-1}} \text{ or } \frac{1}{(r)^{1.4-1}} = 0.48$$

or $$(r)^{0.4} = \frac{1}{0.48} = 2.08 \quad \text{or} \quad r = (2.08)^{1/0.4} = (2.08)^{2.5} = 6.2$$

i.e., *Compression ratio* = **6.2. (Ans.)**

(*ii*) **Indicated thermal efficiency, $\eta_{th(I)}$:**

$$\eta_{relative} = \frac{\eta_{thermal}}{\eta_{air\text{-}standard}} \text{ or } 0.7 = \frac{\eta_{thermal}}{0.52}$$

$$\therefore \quad \eta_{thermal\ (I)} = 0.7 \times 0.52 = 0.364 \quad \text{or} \quad 36.4\%$$

i.e., *indicated thermal efficiency* = **36.4%. (Ans.)**

(*iii*) **Brake specific fuel consumption (b.s.f.c.) :**

Indicated power, $$\text{I.P.} = \frac{\text{B.P.}}{\eta_{mech.}} = \frac{72}{0.82} = 87.8 \text{ kW}$$

Also, $$\eta_{th(I)} = \frac{\text{I.P.}}{\dot{m}_f \times C}$$

$$0.364 = \frac{87.8}{\dot{m}_f \times 45200} \quad \text{where, } \dot{m}_f = \text{Fuel used in kg/s}$$

$$\therefore \quad \dot{m}_f = \frac{87.8}{0.364 \times 45200} = 0.00533 \text{ kg/s}$$

Brake specific fuel consumption,

$$\text{b.s.f.c.} = \frac{\text{Fuel used / sec.}}{\text{B.P.}} = \frac{0.00533}{72} \text{ kg / kWs}$$

$$= \frac{0.00533}{72} \times 3600 \text{ kg/kWh} = \mathbf{0.2665\ kg\ /\ kWh.} \quad \textbf{(Ans.)}$$

(*iv*) **Bore and stroke :**

Mass of air-fuel mixture = 1 + 16 = 17 kg / kg of fuel

∴ For 0.00533 kg/s of fuel supplied to engine the mass of air-fuel mixture

$$= 17 \times 0.00533 = 0.0906 \text{ kg/s}$$

∴ Volume of air-fuel mixture supplied to the engine per sec.

$$\frac{mRT}{P} = \frac{0.0906 \times 287 \times (25 + 273)}{1 \times 10^5} = 0.07748 \text{ m}^3\text{/s}$$

$$\eta_{vol.} = \frac{\text{Mass of mixture supplied / sec.}}{\text{Swept volume}}$$

$$0.78 = \frac{0.07748}{\text{Swept volume}}$$

$$\therefore \quad \text{Swept volume} = \frac{0.07748}{0.78} = 0.0993 \text{ m}^3\text{/s}$$

But swept volume/sec. $$= \left(\frac{\pi}{4} D^2 \times L\right) \times \text{no. of cylinders} \times \frac{\text{r.p.m.}}{2 \times 60}$$

$$= \frac{\pi}{4} \times D^2 \times 1.25D \times 4 \times \frac{2400}{2 \times 60} = 0.0993$$

$$\therefore \quad D^3 = \frac{0.0993 \times 4 \times 2 \times 60}{\pi \times 1.25 \times 4 \times 2400} = 0.001264$$

$$\therefore \quad D = \mathbf{0.108\ m\ or\ 108\ mm.} \quad \textbf{(Ans.)}$$

and $\quad L = 108 \times 1.25 = \mathbf{135\ mm.} \quad \textbf{(Ans.)}$

Example 17.30. *A single-cylinder four-stroke gas engine has a bore of 180 mm and stroke of 340 mm and is governed on hit-and-miss principle. When running at 400 r.p.m. at full load, indicators cards are taken which give a working loop mean effective pressure of 6.4 bar, and a pumping loop mean effective pressure of 0.36 bar. Diagrams from the dead cycle give a mean effective pressure of 0.64 bar. The engine was run light at the same speed (i.e. with no load), and a mechanical counter recorded 46 firing strokes per minute.*

Calculate : (i) Full load brake power

(ii) Mechanical efficiency of the engine.

Solution. Number of cylinders, $n = 1$

Bore, $D = 180 \text{ mm} = 0.18 \text{ m}$

Stroke, $L = 340 \text{ mm} = 0.34 \text{ m}$

Engine speed, $N = 400$ r.p.m.

Working loop mean effective pressure $= 6.4$ bar

Pumping loop mean effective pressure $= 0.36$ bar

Mean effective pressure (dead cycle) $= 0.64$ bar

Firing strokes/min. $= 46$

Refer Fig. 17.15.

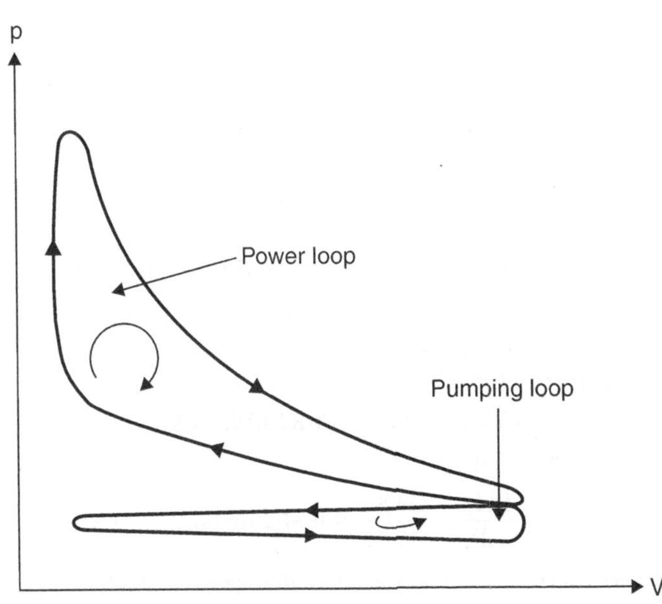

Fig. 17.15

(*i*) **Full load brake power. B.P. :**

Net indicated mean effective pressure,

$$p_{mi\,(net)} = \text{Working (or power) loop mean effective pressure} - \text{pumping loop mean effective pressure}$$

$$= 6.4 - 0.36 = 6.04 \text{ bar}$$

Also, working cycles/min $= 46$

and Dead cycles/min $= \left(\frac{400}{2} - 46\right) = 154$

Therefore, since there is no brake power output,

Frictional power, F.P. = (Net I.P.) – (pumping power of dead cycles) ...(*i*)

Now, $$\text{Net I.P.} = \frac{np_{mi}(\text{net}) \times LANk \times 10}{6}$$

$$= \frac{1 \times 6.04 \times 0.34 \times \pi / 4 \times 0.18^2 \times 46 \times 10}{6} \qquad [\because \; Nk = 46]$$

$$= 4 \text{ kW}$$

Pumping power of dead cycles

$$= \frac{np_{mi(d)}\, LANk \times 10}{6}$$

$$= \frac{1 \times 6.04 \times 0.34 \times \pi / 4 \times 0.18^2 \times 154 \times 10}{6} \qquad [\because \; Nk = 154]$$

$$= 1.42$$

Substituting the above values in eqn. (*i*), we get

$$\text{F.P.} = 4 - 1.42 = 2.58 \text{ kW}$$

At full load the engine fires regularly every two revolutions, and there are $\frac{400}{2} = 200$ firing strokes per minute.

$$\therefore \qquad \text{I.P.} = \frac{n\, p_{mi(net)}\, LANk \times 10}{6}$$

$$= \frac{1 \times 6.04 \times 0.34 \times \pi / 4 \times 0.18^2 \times 200 \times 10}{6} \qquad \left[\begin{aligned} Nk &= 400 \times \frac{1}{2} \\ &= 200 \end{aligned}\right]$$

$$= 17.42 \text{ kW}$$

Hence brake power, B.P.= (I.P. – F.P.)

= 17.42 – 2.58 = **14.84 kW. (Ans.)**

(*ii*) **Mechanical efficiency, $\eta_{mech.}$:**

$$\eta_{mech.} = \frac{\text{B.P.}}{\text{I.P.}} = \frac{14.84}{17.42} = \mathbf{0.852 \text{ or } 85.2\%.} \quad \textbf{(Ans.)}$$

Note. The F.P. is very nearly constant at a given engine speed ; and if the load is decreased giving lower values of B.P., then the variation in $\eta_{mech.}$ with B.P. is shown in Fig. 17.16. At zero B.P. at the same speed the engine is developing just sufficient power to overcome the frictional resistance.

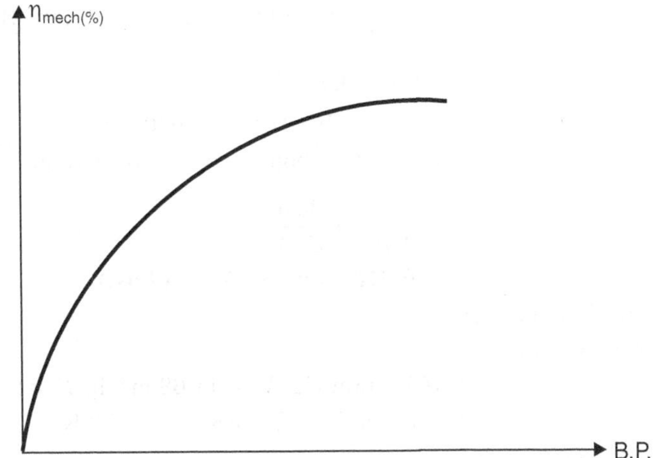

Fig. 17.16

Example 17.31. *During the trial of a gas engine following observations were recorded :*

Bore	*= 320 mm*
Stroke	*= 420 mm*
Speed	*= 200 r.p.m.*
Number of explosions/min.	*= 90*
Gas used	*= 11.68 m³/h*
Pressure of gas	*= 170 mm of water above atmospheric pressure*
Barometer	*= 755 mm (mercury)*
Mean effective pressure	*= 6.2 bar*
Calorific value of gas used	*= 21600 kJ/kg at N.T.P.*
Net load on brake	*= 2040 N*
Brake drum diameter	*= 1.2 m*
Ambient temperature	*= 25°C*

Calculate : (i) Mechanical efficiency, and (ii) Brake thermal efficiency.

Solution. $n = 1$, $D = 0.32$ m, $L = 0.42$ m, $N = 200$ r.p.m.,

$$Nk = 90,\ V_g = \frac{11.68}{3600} = 0.00324 \text{ m}^3/\text{s},$$

$$\text{Pressure of gas} = 755 + \frac{170}{13.6} = 767.5 \text{ mm Hg}$$

$$p_{mi} = 6.2 \text{ bar},\ C = 21600 \text{ kJ/kg at N.T.P.}$$

$$(W - S) = 1840 \text{ N},\ D_b = 1 \text{ m}.$$

(*i*) **Mechanical efficiency** :

As the number of explosions per minute is given as 90 per minute and r.p.m. of engine is 200 it shows that the engine is operating on four-stroke cycle.

Indicated power (I.P.) is given by the relation :

$$\text{I.P.} = \frac{np_{mi} LANk \times 10}{6}$$

$$= \frac{1 \times 6.2 \times 0.42 \times \pi / 4 \times 0.32^2 \times 90 \times 10}{6} \qquad [\because \; Nk = 90]$$

$$= 31.4 \text{ kW}$$

$$\text{B.P.} = \frac{(W - S)\, \pi\, D_b N}{60 \times 1000} = \frac{2040 \times \pi \times 1.2 \times 200}{60 \times 1000} = 25.6 \text{ kW}$$

$\therefore$ *Mechanical efficiency*, $\eta_{mech.} = \frac{\text{B.P.}}{\text{I.P.}} = \frac{25.6}{31.4}$

$$= \mathbf{0.815 \;\; or \;\; 81.5\%.} \quad \textbf{(Ans.)}$$

(*ii*) **Brake thermal efficiency :**

Volume of gas at N.T.P. :

$$p_1 = 767.5 \text{ mm Hg}, V_1 = 11.68 \text{ m}^3/\text{ h}, T_1 = 25 + 273 = 298 \text{ K}$$

$$p_2 = 760 \text{ mm Hg}, T_2 = 0 + 273 = 273 \text{ K}$$

Now, to find V_2, using the relation :

$$\frac{p_1 V_1}{T_1} = \frac{p_2 V_2}{T_2}$$

$$V_2 = \frac{p_1 V_1 T_2}{p_2 T_1} = \frac{767.5 \times 11.68 \times 273}{760 \times 298}$$

$$= 10.8 \text{ m}^3/\text{h}$$

Brake thermal efficiency,

$$\eta_{th.(B)} = \frac{\text{B.P.}}{V_g \times C} \qquad [V_g = \text{Volume of gas in m}^3/\text{s}]$$

$$= \frac{25.6}{\frac{10.8}{3600} \times 21600} = \mathbf{0.395 \text{ or } 39.5\%.} \quad \textbf{(Ans.)}$$

Example 17.32. *Air consumption for a four-stroke petrol engine is measured by means of a circular orifice of diameter 3.2 cm. The coefficient of discharge for the orifice is 0.62 and the pressure across the orifice is 150 mm of water. The barometer reads 760 mm of Hg. Temperature of air in the room is 20°C. The piston displacement volume is 0.00178 m³. The compression ratio is 6.5. The fuel consumption is 0.135 kg/min of calorific value 43900 kJ/kg. The brake power developed at 2500 r.p.m. is 28 kW. Determine :*

(*i*) *The volumetric efficiency on the basis of air alone.*

(*ii*) *The air-fuel ratio.*

(*iii*) *The brake mean effective pressure.*

(*iv*) *The relative efficiency on the brake thermal efficiency basis.*

Solution. Diameter of circular orifice, $d = 3.2 \text{ cm} = 0.032 \text{ m}$

Coefficient of discharge, $C_d = 0.62$

Pressure across orifice, $h_w = 150$ mm of water

Temperature of air in the room $= 20°\text{C}$

Piston displacement $= 0.00178 \text{ m}^3$

Compression ratio, $r = 6.5$

Fuel consumption $= 0.135$ kg/min

Calorific value of fuel, $C = 43900$ kJ/kg

Brake power, B.P. = 28 kW

Speed = 2500 r.p.m.

$k = \frac{1}{2}$ .... for 4-stroke cycle, engine.

(*i*) **Volumetric efficiency on the basis of air alone :**

Characteristic gas equation is written as

$$pV = mRT$$

or

$$\frac{m}{V} = \frac{p}{RT} = \frac{1.0132 \times 10^5}{287 \times (20 + 273)} = 1.2 \text{ kg/m}^3$$

Also $150 \text{ mm of } H_2O = \frac{150}{1000} \times 1000 = 150 \text{ kg/m}^2$

Thus head of air column causing flow,

$$H = \frac{150}{1.2} = 125 \text{ m}$$

Thus air flow through the orifice

$$= \text{Air consumption} = C_d \times A \times \sqrt{2gH}$$

$$= 0.62 \times \frac{\pi}{4} \times (0.032)^2 \times \sqrt{2 \times 9.81 \times 125} = 0.0247 \text{ m}^3/\text{s}$$

Therefore, air consumption per stroke

$$= \frac{0.0247 \times 60}{\left(\frac{2500}{2}\right)} = 0.001185 \text{ m}^3$$

$\therefore$ Volumetric efficiency, $\eta_{vol.} = \frac{\text{Air consumption of stroke}}{\text{Piston displacement}}$

$$= \frac{0.001185}{0.00178} = \mathbf{0.665 \quad or \quad 66.5\%.} \quad \textbf{(Ans.)}$$

(*ii*) **Air-fuel ratio :**

Mass of air drawn into the cylinder per min.

$$= 0.0247 \times 60 \times 1.2 = 1.778 \text{ kg/min}$$

$\therefore$ Air-fuel ratio $= \frac{1.778}{0.135} = \mathbf{13.67 : 1.}$ **(Ans.)**

(*iii*) **Brake mean effective pressure, p_{mb} :**

$$\text{B.P.} = \frac{n \times p_{mb} LANk \times 10}{6}$$

$$28 = \frac{1 \times p_{mb} \times 0.00178 \times 2500 \times \frac{1}{2} \times 10}{6} \qquad [\because \; LA = 0.00178 \text{ m}^3]$$

$\therefore$

$$p_{mb} = \frac{28 \times 6 \times 2}{0.00178 \times 2500 \times 10} = \mathbf{7.55 \text{ bar.}} \quad \textbf{(Ans.)}$$

(*iv*) **Relative efficiency :**

$$\eta_{\text{air-standard}} = 1 - \frac{1}{(r)^{\gamma-1}} = 1 - \frac{1}{(6.5)^{1.4-1}} = 0.527 \text{ or } 52.7\%$$

Brake thermal efficiency,

$$\eta_{\text{th.(B)}} = \frac{\text{B.P.}}{\dot{m}_f \times C} = \frac{28}{\frac{0.135}{60} \times 43900} = 0.2835 \text{ or } 28.35\%$$

$$\therefore \quad \eta_{\text{relative}} = \frac{\eta_{\text{thermal (B)}}}{\eta_{\text{air-standard}}} = \frac{0.2835}{0.527} = \mathbf{0.5379 \text{ or } 53.79\%.} \quad \textbf{(Ans.)}$$

Example 17.33. *A single-cylinder 4-stroke diesel engine gave the following results while running on full load :*

Area of indicator card	= *300 mm*^{2}
Length of diagram	= *40 mm*
Spring constant	= *1 bar/mm*
Speed of the engine	= *400 r.p.m.*
Load on the brake	= *370 N*
Spring balance reading	= *50 N*
Diameter of brake drum	= *1.2 m*
Fuel consumption	= *2.8 kg/h*
Calorific value of fuel	= *41800 kJ/kg*
Diameter of the cylinder	= *160 mm*
Stroke of the piston	= *200 mm*

Calculate : (i) Indicated mean effective pressure. (ii) Brake power and brake mean effective pressure. (iii) Brake specific fuel consumption, brake thermal and indicated thermal efficiencies.

Solution. $N = 400$ r.p.m., $W = 370\ N$, $S = 50\ N$, $D_b = 1.2$ m,

$m_f = 2.8$ kg / h, $C = 41800$ kJ / kg, $D = 0.16$ m, $L = 0.2$ m,

$k = \frac{1}{2}$ ...... for 4-stroke cycle engine.

(*i*) **Indicated mean effective pressure, p_{mi} :**

$$p_{mi} = \frac{\text{Area of indicator diagram or card} \times \text{spring costant}}{\text{Length of diagram}}$$

$$= \frac{300 \times 1}{40} = \mathbf{7.5 \text{ bar.}} \quad \textbf{(Ans.)}$$

Indicated power, $\text{I.P.} = \frac{np_{mi}LANk \times 10}{6} = \frac{1 \times 7.5 \times 0.2 \times \pi / 4 \times 0.16^2 \times 400 \times \frac{1}{2} \times 10}{6}$

$= 10.05$ kW.

(*ii*) **B.P. , p_{mb} :**

Brake power, $\text{B.P.} = \frac{(W - S)\,\pi\, D_b N}{60 \times 1000} = \frac{(370 - 50)\pi \times 1.2 \times 400}{60 \times 1000} = \mathbf{8.04 \text{ kW.}}$ **(Ans.)**

Also, $\text{B.P.} = \frac{np_{mb} \times LANk \times 10}{6}$

$$8.04 = \frac{1 \times p_{mb} \times 0.2 \times \pi / 4 \times 0.16^2 \times 400 \times \frac{1}{2} \times 10}{6}$$

$$\therefore \quad p_{mb} = \frac{8.04 \times 6 \times 4 \times 2}{0.2 \times \pi \times 0.16^2 \times 400 \times 10} = \textbf{6 bar.} \quad \textbf{(Ans.)}$$

(iii) **b.s.f.c., $\eta_{th(B)}$, $\eta_{th.(I)}$:**

Brake specific fuel consumption,

$$b.s.f.c = \text{Fuel consumption per B.P. hour}$$

$$= \frac{2.8}{8.04} = \textbf{0.348 kg / B.P. hour.} \quad \textbf{(Ans.)}$$

Brake thermal efficiency,

$$\eta_{th.(B)} = \frac{\text{B.P.}}{\dot{m}_f \times C} = \frac{8.04}{\frac{2.8}{3600} \times 41800} = \textbf{0.2473 or 24.73\%.} \quad \textbf{(Ans.)}$$

Indicated thermal efficiency,

$$\eta_{th.(I)} = \frac{\text{B.P.}}{\dot{m}_f \times C} = \frac{10.05}{\frac{2.8}{3600} \times 41800} = \textbf{0.3091 or 30.91\%.} \quad \textbf{(Ans.)}$$

Example 17.34. *A 4-cylinder, four-stroke cycle engine, 82.5 mm bore × 130 mm stroke develops 28 kW while running at 1500 r.p.m. and using a 20 per cent rich mixture. If the volume of the air in the cylinder when measured at 15.5°C and 762 mm of mercury is 70 per cent of the swept volume, the theoretical air-fuel ratio is 14.8, heating value of petrol used is 45980 kJ/kg and the mechanical efficiency of the engine is 90%, find :*

(i) The indicated thermal efficiency.

(ii) The brake mean effective pressure.

Take R = 287 N-m/kg K.

Solution. Number of cylinders, $n = 4$

Engine bore, $D = 0.0825$ m

Stroke length, $L = 0.13$ m

Brake power, B.P. = 28 kW

Engine speed, $N = 1500$ r.p.m.

Theoretical air-fuel ratio = 14.8

Calorific value of fuel, $C = 45980$ kJ/kg

Mechanical efficiency, $\eta_{mech.} = 90\%$.

(i) **Indicated thermal efficiency, $\eta_{th.(I)}$:**

Swept volume, $V_s = \pi/4\, D^2 L = \pi/4 \times 0.0825^2 \times 0.13 = 0.000695 \text{ m}^3$

$$\text{Volume of air drawn in} = \frac{70}{100} \times 0.000695 = 0.0004865 \text{ m}^3$$

Given : $p = \frac{762}{760} \times 1.0132 = 1.015$ bar

$V = 0.0004865 \text{ m}^3$ (calculated above)

$R = 287$ N-m / kg K

$$T = 15.5 + 273 = 288.5 \text{ K}$$

$\therefore$ m = Mass of air/stroke/cylinder

$$= \frac{pV}{RT} = \frac{1.015 \times 10^5 \times 0.0004865}{287 \times 288.5} = 0.000596 \text{ kg}$$

Theoretical mass of air used per minute

$$= 0.000596 \times \frac{1500}{2} \times 4 = 1.788 \text{ kg}$$

Theoretical air-fuel ratio = 14.8

$\therefore$ Theoretical mass of fuel used / min

$$= \frac{1.788}{14.8} = 0.1208 \text{ kg/min}$$

When using 20% rich mixture, then

$$\dot{m}_f = \text{Mass of fuel burnt / sec}$$

$$= \frac{0.1208}{60} \times \frac{120}{100} = 0.002416 \text{ kg/s}$$

Now, $$\eta_{mech.} = \frac{\text{B.P.}}{\text{I.P.}} = \frac{28}{\text{I.P.}}$$

$$0.9 = \frac{28}{\text{I.P.}}$$

$\therefore$ $$\text{I.P.} = \frac{28}{0.9} = 31.11 \text{ kW.}$$

Indicated thermal efficiency,

$$\eta_{th.(I)} = \frac{\text{I.P.}}{\dot{m}_f \times C} = \frac{31.11}{0.002416 \times 45980} = \mathbf{0.28 \text{ or } 28\%.} \quad \textbf{(Ans.)}$$

(*ii*) **Brake mean effective pressure, p_{mb} :**

$$\text{B.P.} = \frac{n \times p_{mb} LANk \times 10}{6}$$

$$28 = \frac{4 \times p_{mb} \times 0.13 \times \pi / 4 \times 0.0825^2 \times 1500 \times \frac{1}{2} \times 10}{6}$$

$\therefore$ $$p_{mb} = \frac{28 \times 6 \times 4 \times 2}{4 \times 0.13 \times \pi \times 0.0825^2 \times 1500 \times 10} = \mathbf{8.06 \text{ bar.}} \quad \textbf{(Ans.)}$$

Example 17.35. *During the test of 40 minutes on a single-cylinder gas engine of 200 mm cylinder bore and 400 mm stroke, working on the four-stroke cycle and governed by hit and miss method of governing, the following readings were taken :*

Total number of revolutions	*= 9400*
Total number of explosions	*= 4200*
Area of indicator diagram	*= 550 mm²*
Length of indicator diagram	*= 72 mm*
Spring number	*= 0.8 bar/mm*
Brake load	*= 540 N*

Brake wheel diameter = *1.6 m*
Brake rope diameter = *2 cm*
Gas used = *8.5 m*^{3}
Calorific value of gas = *15900 kJ/m*^{3}
Calculate : (i) Indicated power, (ii) Brake power, and
(iii) Indicated and brake thermal efficiencies.

Solution. $D = 0.2$ m, $L = 0.4$ m, $N_t = 9400$ r.p.m., $Nk = \frac{4200}{40} = 105$

$$(W - S) = 540\ N,\ D_b = 1.6 \text{ m},\ d = 0.02 \text{ m},\ V_g = \frac{8.5}{40 \times 60} = 0.00354 \text{ m}^3/\text{s}.$$

(*i*) **Indicated power, I.P. :**

Indicated mean effective pressure.

$$p_{mi} = \frac{\text{Area of indicator diagram} \times \text{spring number}}{\text{Length of the diagram}} = \frac{550 \times 0.8}{72}$$

$$= 6.11 \text{ bar}$$

$$\text{I.P.} = \frac{np_{mi}LANk \times 10}{6} = \frac{1 \times 6.11 \times 0.4 \times \pi / 4 \times 0.2^2 \times 105 \times 10}{6}$$

$$= 13.4 \text{ kW}.$$

(*ii*) **Brake power, B.P. :**

$$\text{B.P.} = \frac{(W - S)\,\pi\,(D_b + d)N}{60 \times 1000} = \frac{540 \times \pi\,(1.6 + 0.02) \times \left(\frac{9400}{40}\right)}{60 \times 1000}$$

$$= \mathbf{10.76\ kW.}\quad \textbf{(Ans.)}$$

(*iii*) **Indicated thermal efficiency :**

$$\eta_{\text{th.(I)}} = \frac{\text{I.P.}}{V_g \times C} = \frac{13.4}{0.00354 \times 15900} = \mathbf{0.238 \text{ or } 23.8\%.}\quad \textbf{(Ans.)}$$

Brake thermal efficiency,

$$\eta_{\text{th.(B)}} = \frac{\text{B.P.}}{V_g \times C} = \frac{10.76}{0.00354 \times 15900} = \mathbf{0.191 \text{ or } 19.1\%.}\quad \textbf{(Ans.)}$$

Example 17.36. *The following observations were recorded during the test on a 6-cylinder, 4-stroke Diesel engine :*

Bore = *125 mm*
Stroke = *125 mm*
Engine speed = *2400 r.p.m.*
Load on dynamometer = *490 N*
Dynamometer constant = *16100*
Air orifice diameter = *55 mm*
Coefficient of discharge = *0.66*
Head causing flow through orifice = *310 mm of water*
Barometer reading = *760 mm Hg*
Ambient temperature = *25°C*
Fuel consumption = *22.1 kg/h*

Calorific value of fuel = *45100 kJ/kg*
Per cent carbon in the fuel = *85%*
Per cent hydrogen in the fuel = *15%*
Pressure of air at the end of suction stroke = *1.013 bar*
Temperature at the end of suction stroke = *25°C*

Calculate : (i) Brake mean effective pressure, (ii) Specific fuel consumption, (iii) Brake thermal efficiency, (iv) Volumetric efficiency, and (v) Percentage of excess air supplied.

Solution. $n = 6$, $D = 0.125$ m, $L = 0.125$ m, $N = 2400$ r.p.m.

$W = 490$ N, C_D = dynamometer constant = 16100

d_0 = orifice diameter = 0.055 m, $C_d = 0.66$, $h_w = 310$ mm

$$\dot{m}_f = \frac{22.1}{3600} = 0.00614 \text{ kg/s}, \; C = 45100 \text{ kJ/kg},$$

$$k = \frac{1}{2} \text{ ..... for 4-stroke cycle engine.}$$

(*i*) **Brake mean effective pressure, p_{mb} :**

Brake power,
$$\text{B.P.} = \frac{W \times N}{C_D} = \frac{490 \times 2400}{16100} = 73 \text{ kW}$$

Also
$$\text{B.P.} = \frac{n p_{mb} LANk \times 10}{6}$$

$$73 = \frac{6 \times p_{mb} \times 0.125 \times \pi / 4 \times 0.125^2 \times 2400 \times \frac{1}{2} \times 10}{6}$$

$$\therefore \quad p_{mb} = \frac{73 \times 6 \times 4 \times 2}{6 \times 0.125 \times \pi \times 0.125^2 \times 2400 \times 10} = \mathbf{3.96 \text{ bar.}} \quad \textbf{(Ans.)}$$

(*ii*) **Specific fuel consumption, b.s.f.c. :**

$$\text{b.s.f.c.} = \frac{22.1}{73} = \mathbf{0.3027 \text{ kg / kWh.}} \quad \textbf{(Ans.)}$$

(*iii*) **Brake thermal efficiency, $\eta_{th.(B)}$:**

$$\eta_{th.(B)} = \frac{\text{B.P.}}{\dot{m}_f \times C} = \frac{73}{0.00614 \times 45100} = \mathbf{0.2636 \text{ or } 26.36\%.} \quad \textbf{(Ans.)}$$

(*iv*) **Volumetric efficiency, $\eta_{vol.}$:**

Stroke volume of cylinder = $\pi/4 \, D^2 \times L$

$= \pi/4 \times 0.125^2 \times 0.125 = 0.00153 \text{ m}^3$

The volume of air passing through the orifice of the air box per minute is given by,

$$V_a = 840 \, A_0 \, C_d \sqrt{\frac{h_w}{\rho_a}}$$

where, C_d = Discharge coefficient of orifice = 0.66

A_0 = Area of cross-section of orifice

$= \pi/4 \, d_0^2 = \pi/4 \times (0.055)^2 = 0.00237 \text{ m}^2$

h_w = Head causing flow through orifice in cm of water = $\frac{310}{10}$ = 31 cm

ρ_a = density of air at 1.013 bar and 25°C.

$$= \frac{P}{RT} = \frac{1.013 \times 10^5}{287 \times (25 + 273)} = 1.18 \text{ kg/m}^3$$

$\therefore$ Volume of air, $V_a = 840 \times 0.00237 \times 0.66 \sqrt{\frac{31}{1.18}} = 6.73 \text{ m}^3\text{/min}$

$\therefore$ Actual volume of air per cylinder

$$= \frac{6.73}{n} = \frac{6.73}{6} = 1.12 \text{ m}^3\text{/min}$$

$\therefore$ Air supplied per stroke per cylinder

$$= \frac{1.12}{(2400 / 2)} = 0.000933 \text{ m}^3$$

$$\therefore \quad \eta_{vol.} = \frac{\text{Volume of air actually supplied}}{\text{Volume of air theoretically required}}$$

$$= \frac{0.000933}{0.00153} = \mathbf{0.609 \text{ or } 60.9\%.} \quad \textbf{(Ans.)}$$

(*v*) **Percentage of excess air supplied :**

Quantity of air required per kg of fuel for complete combustion

$$= \frac{100}{23}\left[C \times \frac{8}{3} + H_2 \times \frac{8}{1}\right]$$

where C is the fraction of carbon and H_2 is the fraction of hydrogen present in the fuel respectively.

$$= \frac{100}{23}\left[0.85 \times \frac{8}{3} + 0.15 \times 8\right] = 15.07 \text{ kg / kg of fuel}$$

Actual quantity of air supplied per kg of fuel

$$= \frac{V_a \times \rho_a \times 60}{22.1} = \frac{6.73 \times 1.18 \times 60}{22.1} = 21.56 \text{ kg}$$

$\therefore$ *Percentage excess air* $= \frac{21.56 - 15.07}{15.07} \times 100 = \mathbf{43.06\%.}$ **(Ans.)**

HEAT BALANCE SHEET

Example 17.37. *The following observations were recorded in a test of one hour duration on a single-cylinder oil engine working on four-stroke cycle :*

Bore	*= 300 mm*
Stroke	*= 450 mm*
Fuel used	*= 8.8 kg*
Calorific value of fuel	*= 41800 kJ/kg*
Average speed	*= 200 r.p.m.*
m.e.p.	*= 5.8 bar*
Brake friction load	*= 1860 N*
Quantity of cooling water	*= 650 kg*
Temperature rise	*= 22°C*
Diameter of the brake wheel	*= 1.22 m*

Calculate : (i) Mechanical efficiency, and (ii) Brake thermal efficiency.
Draw the heat balance sheet.

Solution. $n = 1$, $D = 0.3$ m, $L = 0.45$ m, $m_f = 8.8$ kg / h, $C = 41800$ kJ/kg,
$N = 200$ r.p.m., $p_{mi} = 5.8$ bar, $(W - S) = 795$ N, $D_b = 1.22$ m,

$$k = \frac{1}{2} \text{ ..... for 4-stroke cycle engine}$$

$$m_w = 650 \text{ kg}, \; t_{w_2} - t_{w_1} = 22°\text{C}.$$

(*i*) **Mechanical efficient, $\eta_{mech.}$:**

Indicated power, $$\text{I.P.} = \frac{np_{mi}LANk \times 10}{6} = \frac{1 \times 5.8 \times 0.45 \times \pi / 4 \times 0.3^2 \times 200 \times \frac{1}{2} \times 10}{6}$$

$$= 30.7 \text{ kW}$$

Brake power, $$\text{B.P.} = \frac{(W - S)\, \pi\, DN}{60 \times 1000} = \frac{1860 \times \pi \times 1.22 \times 200}{60 \times 1000} = 23.76 \text{ kW}$$

$$\eta_{mech.} = \frac{\text{B.P.}}{\text{I.P.}} = \frac{23.76}{30.7} = \mathbf{0.773 \text{ or } 77.3\%.} \quad \textbf{(Ans.)}$$

(*ii*) **Brake thermal efficiency, $\eta_{th.(B)}$:**

$$\eta_{th.(B)} = \frac{\text{B.P.}}{\dot{m}_f \times C} = \frac{23.76}{\frac{8.8}{3600} \times 41800} = \mathbf{0.232 \quad or \quad 23.2\%.} \quad \textbf{(Ans.)}$$

Heat supplied $= 8.8 \times 41800 = 367840$ kJ/h.

(*i*) Heat equivalent of I.P.

$$= \text{I.P.} \times 3600 \text{ kJ/h}$$
$$= 30.7 \times 3600 = 110520 \text{ kJ/h.}$$

(*ii*) Heat carried away by cooling water

$$= m_w \times c_{pw} \times (t_{w_2} - t_{w_1})$$
$$= 650 \times 4.18 \times 22 = 59774 \text{ kJ/h.}$$

Heat balance sheet (hourly basis)

Item	*kJ*	*Per cent*
Heat supplied by fuel	*367840*	*100*
(*i*) Heat absorbed in I.P.	110520	30.05
(*ii*) Heat taken away by cooling water	59774	16.25
(*iii*) Heat carried away by exhaust gases, radiation etc. (by difference)	197546	53.70
Total	367840	100

Example 17.38. *In a trial of a single-cylinder oil engine working on dual cycle, the following observations were made :*

Compression ratio = *15*
Oil consumption = *10.2 kg/h*

Calorific value of fuel	*= 43890 kJ/kg*
Air consumption	*= 3.8 kg/min*
Speed	*= 1900 r.p.m.*
Torque on the brake drum	*= 186 N-m*
Quantity of cooling water used	*= 15.5 kg/min*
Temperature rise	*= 36°C*
Exhaust gas temperature	*= 410°C*
Room temperature	*= 20°C*
c_p for exhaust gases	*= 1.17 kJ/kg K*

Calculate : (i) Brake power,

(ii) Brake specific fuel consumption, and

(iii) Brake thermal efficiency.

Draw heat balance sheet on minute basis.

Solution. $n = 1$, $r = 15$, $m_f = 10.2$ kg/h, $C = 43890$ kJ/kg, $m_a = 3.8$ kg/min.,

$N = 1900$ r.p.m., $T = 186$ N-m, $m_w = 15.5$ kg/min, $t_{w_2} - t_{w_1} = 36°C$,

$t_g = 410°C$, $t_r = 20°C$, $c_p = 1.17$.

(*i*) **Brake power, B.P. :**

$$\text{B.P.} = \frac{2\pi NT}{60 \times 1000} = \frac{2\pi \times 1900 \times 186}{60 \times 1000} = \mathbf{37\ kW.} \quad \textbf{(Ans.)}$$

(*ii*) **Brake specific fuel consumption, b.s.f.c. :**

$$\text{b.s.f.c.} = \frac{10.2}{37} = \mathbf{0.2756\ kg/kWh.} \quad \textbf{(Ans.)}$$

(*iii*) **Brake thermal efficiency,**

$$\eta_{th(B)} = \frac{\text{B.P.}}{\dot{m}_f \times C} = \frac{37}{\frac{10.2}{3600} \times 43890} = \mathbf{0.2975\ or\ 29.75\%.} \quad \textbf{(Ans.)}$$

Heat supplied by the fuel per minute

$$= \frac{10.2}{60} \times 43890 = 7461 \text{ kJ/min}$$

(*i*) Heat equivalent of B.P.

$$= \text{B.P.} \times 60 = 37 \times 60 = 2220 \text{ kJ/min.}$$

(*ii*) Heat carried away by cooling water

$$= m_w \times c_{pw} (t_{w_2} - t_{w_1}) = 15.5 \times 4.18 \times 36 = 2332 \text{ kJ/min.}$$

(*iii*) Heat carried away by exhaust gases

$$= m_g \times c_{pg} \times (t_g - t_r)$$

$$= \left(\frac{10.2}{60} + 3.8\right) \times 1.17 \times (410 - 20) = 1811 \text{ kJ/min.}$$

Heat balance sheet (minute basis) :

Item	*kJ*	*Per cent*
Heat supplied by fuel	*7461*	*100*
(*i*) Heat absorbed in B.P.	2220	29.8
(*ii*) Heat taken away by cooling water	2332	31.2
(*iii*) Heat carried away by exhaust gases	1811	24.3
(*iv*) Heat unaccounted for (by difference)	1098	14.7
Total	7461	100

Example 17.39. *From the data given below, calculate indicated power, brake power and draw a heat balance sheet for a two-stroke diesel engine run for 20 minutes at full load :*

r.p.m. = *350*
m.e.p. = *3.1 bar*
Net brake load = *640 N*
Fuel consumption = *1.52 kg*
Cooling water = *162 kg*
Water inlet temperature = *30°C*
Water outlet temperature = *55°C*
Air used/kg of fuel = *32 kg*
Room temperature = *25°C*
Exhaust temperature = *305°C*
Cylinder bore = *200 mm*
Cylinder stroke = *280 mm*
Brake diameter = *1 metre*
Calorific value of fuel = *43900 kJ/kg*
Steam formed per kg of fuel in the exhaust = *1.4 kg*
Specific heat of steam in exhaust = *2.09 kJ/kg K*
Specific heat of dry exhaust gases = *1.0 kJ/kg K.*

Solution. N = 350 r.p.m., p_{mi} = 3.1 bar, $(W - S)$ = 640 N, m_f = 1.52 kg, m_w = 162 kg, t_{w_1} = 30°C, t_{w_2} = 55°C, m_a = 32 kg/kg of fuel, t_r = 25°C, t_g = 305°C, D = 0.2 m, L = 0.28 m, D_b = 1 m, C = 43900 kJ/kg, c_{ps} = 2.09, c_{pg} = 1.0 kJ/kg K, k = 1 ..... for two-stroke cycle engine.

(*i*) **Indicated power, I.P. :**

$$\text{I.P.} = \frac{np_{mi}\, LANk \times 10}{6}$$

$$= \frac{1 \times 3.1 \times 0.28 \times \pi / 4 \times 0.2^2 \times 350 \times 1 \times 10}{6} = \mathbf{15.9\ kW.}\quad \textbf{(Ans.)}$$

(*ii*) **Brake power, B.P. :**

$$\text{B.P.} = \frac{(W - S)\,\pi\, D_b N}{60 \times 1000} = \frac{640 \times \pi \times 1 \times 350}{60 \times 1000} = \mathbf{11.73\ kW.} \quad \textbf{(Ans.)}$$

Heat supplied in 20 minutes

$= 1.52 \times 43900 = 66728$ kJ

(*i*) Heat equivalent of I.P. in 20 minutes

$= \text{I.P.} \times 60 \times 20 = 15.9 \times 60 \times 20 = 19080$ kJ

(*ii*) Heat carried away by cooling water

$= m_w \times c_{pw} \times (t_{w_2} - t_{w_1})$

$= 162 \times 4.18 \times (55 - 30) = 16929$ kJ

Total mass of air $= 32 \times 1.52 = 48.64$ kg

Total mass of exhaust gases = Mass of fuel + mass of air

$= 1.52 + 48.64 = 50.16$ kg

Mass of steam formed $= 1.4 \times 1.52 = 2.13$ kg

$\therefore$ Mass of dry exhaust gases $= 50.16 - 2.13 = 48.03$ kg

(*iii*) Heat carried away by dry exhaust gases

$= m_g \times c_{pg} \times (t_g - t_r)$

$= 48.03 \times 1.0 \times (305 - 25) = 13448$ kJ

(*iv*) Heat carried away by steam $= 2.13[h_f + h_{fg} + c_{ps}\,(t_{sup} - t_s)]$

$$\left[\begin{array}{l}\text{At 1.013 bar pressure (atmospheric pressure assumed) :}\\ \qquad h_f = 417.5\ \text{kJ/kg},\ h_{fg} = 2257.9\ \text{kJ/kg}\end{array}\right]$$

$= 2.13\,[417.5 + 2257.9 + 2.09\,(305 - 99.6)]$

$= 6613$ kJ/kg ..... neglecting sensible heat of water at room temperature

Heat balance sheet (20 minute basis) :

Item	*kJ*	*Per cent*
Heat supplied by fuel	*66728*	*100*
(*i*) Heat equivalent of I.P.	19080	28.60
(*ii*) Heat carried away by cooling water	16929	25.40
(*iii*) Heat carried away by dry exhaust gases	13448	20.10
(*iv*) Heat carried away steam in exhaust gases	6613	9.90
(*v*) Heat unaccounted for (by difference)	10658	16.00
Total	66728	100.00

Example 17.40. *A six-cylinder, four-stroke CI engine is tested against a water brake dynamometer for which B.P. = WN/17 × 10^3 in kW where W is the brake load in newton and N is the speed of the engine in the r.p.m. The air consumption was measured by means of a sharp edged orifice. During the test following observations were taken :*

Bore = *10 cm*

Stroke = *14 cm*

Speed = *2500 r.p.m.*

Brake load = *480 N*
Barometer reading = *76 cm of Hg*
Orifice diameter = *3.3 cm*
Coefficient of discharge of orifice = *0.62*
Pressure drop across orifice = *14 cm of Hg*
Room temperature = *25°C*
Fuel consumption = *0.32 kg/min.*

Calculate the following :

(i) The volumetric efficiency ; (ii) The brake mean effective pressure (bmep) ; (iii) The engine torque ; (iv) The brake specific fuel consumption (bsfc). **(AMIE Summer, 2000)**

Solution. (*i*) **Volumetric efficiency,** η_{vol} :

V_s = Swept volume,

$$= \frac{\pi}{4} D^2 L \times \frac{N}{60 \times 2} \times \text{No. of cylinders, for 4-stroke engine. (where } N = \text{r.p.m.)}$$

$$= \frac{\pi}{4} (0.1)^2 \times 0.14 \times \frac{2500}{60 \times 2} \times 6 = 0.137 \text{ m}^3/\text{s}.$$

Barometer $= 76 \text{ cm Hg} = \left[\frac{76}{100} \times 13.6 \times 10^3 \times 9.81\right] \times 10^{-3} = 101.4 \text{ kN/m}^2$

$$\rho_a = \frac{p}{R_a T} = \frac{101.3}{0.287\,(273 + 25)} = 1.1844 \text{ kg/m}^3$$

$$\Delta p = 14 \text{ cm of Hg} = \frac{14}{100} \times 13.6 \times 1000 \times 9.81 = 18.678 \times 10^3 \text{ N/m}^2$$

$$\Delta p = \rho_a \times 9.81 \times h_a,$$

where, h_a = Head, m of air, causing flow

or $$h_a = \frac{18.678 \times 10^3}{1.1844 \times 9.81} = 1607.5 \text{ m of air}$$

V_a = Volume flow rate of air, at free air conditions

$$= C_d \ \frac{\pi}{4} (d_0)^2 \sqrt{2gh_a}$$

$$= 0.62 \times \frac{\pi}{4}\left(\frac{3.3}{100}\right)^2 \sqrt{2 \times 9.81 \times 1607.5} = 0.094 \text{ m}^3/\text{s}.$$

$$\%\eta_{vol} = \frac{V_a}{V_s} \times 100 = \frac{0.094}{0.137} \times 100 = \mathbf{68.6\%.} \quad \textbf{(Ans.)}$$

(*ii*) **The brake mean effective pressure,** $\mathbf{p_{mb}}$:

$$BP = \frac{WN}{17} \times 10^{-3} \text{ kW} = \frac{480 \times 2500}{17} \times 10^{-3} = 70.588 \text{ kW}$$

$$= p_{mb}\, LA \times \frac{N}{60} \times \frac{1}{2} \times 6, \text{ for six-cylinder, four-stroke engine}$$

or $$p_{mb} = \frac{70.588 \times 60 \times 2}{0.14 \times \frac{\pi}{4} (0.1)^2 \times 2500 \times 6} = \mathbf{513.57 \ kN/m^2.} \quad \textbf{(Ans.)}$$

(*iii*) **Engine torque, T :**

$$\text{B.P.} = 2\pi NT$$

or $$\text{Torque, } (T) = \frac{\text{B.P.}}{2\pi N} = \frac{70.588 \times 10^3}{2\pi \times \frac{2500}{60}} = \mathbf{269.63\ N\text{-}m.} \quad \textbf{(Ans.)}$$

(*iv*) **Brake specific fuel consumption, bsfc :**

$$bsfc = \frac{m_f \text{ (kg/h)}}{\text{B.P.}} = \frac{0.32 \times 60}{70.588} = \mathbf{0.272\ kg / kW\text{-}h.} \quad \textbf{(Ans.)}$$

Example 17.41. *During the trial of a single-acting oil engine, cylinder diameter 200 mm, stroke 280 mm, working on two-stroke cycle and firing every cycle, the following observations were made :*

Duration of trial	*= 1 hour*
Total fuel used	*= 4.22 kg*
Calorific value	*= 44670 kJ/kg*
Proportion of hydrogen in fuel	*= 15%*
Total number of revolutions	*= 21000*
Mean effective pressure	*= 2.74 bar*
Net brake load applied to a drum of 1 m diameter	*= 600 N*
Total mass of cooling water circulated	*= 495 kg*
Intel temperature of cooling water	*= 13°C*
Outlet temperature of cooling water	*= 38°C*
Air used	*= 135 kg*
Temperature of air in test room	*= 20°C*
Temperature of exhaust gases	*= 370°C*

Assume : c_p *(gases) = 1.005 kJ/kgK ;* c_p *(steam) at atmospheric pressure = 2.093 kJ/kgK.*

Calculate the thermal efficiency and draw up the heat balance. **(U.P.S.C., 1997)**

Solution. *Given* : $D = 200 \text{ mm} = 0.2 \text{ m}$; $L = 280 \text{ mm} = 0.28 \text{ m}$; $m_f = 4.22 \text{ kg/h}$;
$C = 44670 \text{ kJ/kg}$;

$$\text{r.p.m.} = \frac{21000}{60} = 350 \text{ ; } p_{mi} = 2.74 \text{ bar ; } D_b = 1 \text{ m ;}$$

$(W - S) = 600 \text{ N}$; $m_w = 495 \text{ kg/ h}$; $t_{w_1} = 13°\text{C}$, $t_{w_2} = 38°\text{C}$;
$m_a = 135 \text{ kg/ h}$, $t_r = 20° \text{ C}$, $t_g = 370°\text{C}$; $c_{pg} = 1.005 \text{ kJ/kg K}$;
$c_{ps} = 2.093 \text{ kJ/kg k}$

Thermal efficiency, $\eta_{th.}$:

Indicated power, $$\text{I.P.} = \frac{p_m LANk \times 10}{6}$$

$$= \frac{2.74 \times 0.28 \times \frac{\pi}{4} \times 0.2^2 \times 350 \times 1 \times 10}{6}$$

$$= 14.06 \text{ kW}$$ (k = 1, engine being 2-stroke cycle]

Thermal efficiency (indicated), $\eta_{th.(I)} = \dfrac{\text{I.P.}}{\dot{m}_f \times C}$

$$= \frac{14.06}{\dfrac{4.22}{3600} \times 44670} = 0.268 \quad \text{or} \quad \textbf{26.8\%. (Ans.)}$$

Brake power, $\quad \text{B.P.} = \dfrac{(W - S)\,\pi\, D_b N}{60 \times 1000}$ kW

$$= \frac{600 \times \pi \times 1 \times 350}{60 \times 1000} = 10.99 \text{ kW}$$

Heat balance sheet (minute basis) :

Heat input $= \dfrac{4.22}{60} \times 44670 = 3141.8$ kJ/ min.

(*i*) Heat equivalent of B.P. = 10.99 × 60 = 659.4 kJ/min

(*ii*) Heat lost to cooling water $= m_w \times c_{pw} \times (t_{w_2} - t_{w_1})$

$$= \frac{495}{60} \times 4.186 \times (38 - 13) = 863.4 \text{ kJ/min.}$$

Mass of exhaust gases (wet) = mass of air / min + mass of fuel /min

$$= \frac{135}{60} + \frac{4.22}{60} = 2.32 \text{ kg/min}$$

Steam in exhaust gases $= 9 \times H_2 \times$ mass of fuel used/min

$$= 9 \times \frac{15}{100} \times \frac{4.22}{60} = 0.095 \text{ kg/min}$$

Mass of dry exhaust gases / min,

m_g = Mass of exhaust gas (wet) – mass of H_2O produced/min

= 2.32 – 0.095 = 2.225 kg/min.

(*iii*) Heat carried away by dry exhaust gases

$$= m_g \times c_{pg} \times (t_g - t_r)$$

$$= 2.225 \times 1.005 \times (370 - 20) = 782.6 \text{ kJ/min}$$

(*iv*) Assuming that steam in exhaust gases exists as superheated steam at atmospheric pressure and exhaust gas temperature, the enthalpy of 1 kg of steam at atmospheric pressure

1.013 bar $\simeq$ 1 bar and 370°C

$$= h_{sup} - h$$

[where, h = sensible heat of water at room temperature].

$$= [h_f + h_{fg} + c_{ps}\,(t_{sup} - t_s) - h]$$

$$= 417.5 + 2257.9 + 2.093\,(370 - 99.6) - 1 \times 4.18 \times (20 - 0)]$$

$$= 3157.7 \text{ kJ/min}$$

Heat carried away by steam $= 0.095 \times 3157.7 \simeq 300$ kJ/min

Heat balance sheet (minute basis) :

Item	*kJ*	*Percent*
Heat supplied by fuel	*3141.8*	*100*
(*i*) Heat equivalent of B.P.	659.4	20.99
(*ii*) Heat carried away by cooling water	863.4	27.48
(*iii*) Heat carried away by dry exhaust gases	782.6	24.91
(*iv*) Heat carried away by steam	300	9.55
(*v*) Heat unaccounted for (by difference)	536.4	17.07
Total	3141.8	100

Example 17.42. *During a test on a two stroke oil engine on full load the following observations were recorded :*

Speed	*= 350 r.p.m.*
Net brake load	*= 590 N*
Mean effective pressure	*= 2.8 bar*
Oil consumption	*= 4.3 kg/h*
Jacket cooling water	*= 500 kg/h*
Temperature of jacket water at inlet and outlet	*= 25°C and 50°C respectively*
Air used per kg of oil	*= 33 kg*
Temperature of air in test room	*= 25°C*
Temperature of exhaust gases	*= 400°C*
Cylinder diameter	*= 220 mm*
Stroke length	*= 280 mm*
Effective brake diameter	*= 1 metre*
Calorific value of oil	*= 43900 kJ/kg*
Proportion of hydrogen in fuel oil	*= 15%*
Mean specific heat of dry exhaust gases	*= 1.0 kJ/kg K*
Specific heat of steam	*= 2.09 kJ/kg K*

Calculate : (i) Indicated power, and (ii) Brake power.

Also draw up heat balance sheet on minute basis.

Solution. $n = 1$, $N = 350$ r.p.m., $(W - S) = 590\ N$, $p_{mi} = 2.8$ bar

$m_f = 4.3$ kg/h, $m_w = 500$ kg/h, $t_{w_1} = 25°C$, $t_{w_2} = 50°C$

$m_a = 33$ kg/kg of oil, $t_r = 25°C$, $t_g = 400°C$, $D = 0.22$ m

$L = 0.28$ m, $D_b = 1$ m, $C = 43900$ kJ/kg, $c_{pg} = 1.0$, $c_{ps} = 2.09$

$k = 1$ ..... for two-stroke cycle engine.

(*i*) **Indicated power, I.P. :**

$$\text{I.P.} = \frac{np_{mi}LANk \times 10}{6}$$

$$= \frac{1 \times 2.8 \times 0.28 \times \pi / 4 \times 0.22^2 \times 350 \times 1 \times 10}{6} = \textbf{17.38 kW. (Ans.)}$$

(*ii*) **Brake power, B.P. :**

$$\text{B.P.} = \frac{(W - S)\,\pi\, D_b N}{60 \times 1000} = \frac{590 \times \pi \times 1 \times 350}{60 \times 1000} = \mathbf{10.81\ kW.} \quad \textbf{(Ans.)}$$

Heat supplied per minute $= \dfrac{4.3}{60} \times 43900 = 3146$ kJ/min.

(*i*) Heat equivalent of I.P. = 17.38 × 60 = 1042.8 kJ/min.

(*ii*) Heat lost to cooling water

$$= m_w \times c_{pw} \times (t_{w_2} - t_{w_1})$$

$$= \frac{500}{60} \times 4.18 \times (50 - 25) = 870.8 \text{ kJ/min}$$

Now,

$$\begin{array}{ccccc} 2H_2 & + & O_2 & = & 2H_2O \\ 4 & & 32 & & 36 \\ 1 & & 8 & & 9 \end{array}$$

i.e., 1 kg of H_2 produces 9 kg of H_2O

∴ Mass of H_2O produced per kg of fuel burnt

$$= 9 \times H_2 \times \text{mass of fuel used/min.}$$

$$= 9 \times 0.15 \times \frac{4.3}{60} = 0.0967 \text{ kg/min.}$$

Total mass of exhaust gases (wet)/min.

$$= \text{Mass of air/min.} + \text{mass of fuel/min.}$$

$$= \frac{(33 + 1) \times 4.3}{60} = 2.436 \text{ kg/min.}$$

Mass of dry exhaust gases/min.

$$= \text{Mass of wet exhaust gases/min} - \text{mass of } H_2O \text{ produced/min.}$$

$$= 2.436 - 0.0967 = 2.339 \text{ kg/min.}$$

(*iii*) Heat lost to dry exhaust gases

$$= m_g \times c_{pg} \times (t_g - t_r)$$

$$= 2.339 \times 1.0 \times (400 - 25) = 887 \text{ kJ/min.}$$

(*iv*) Assuming that steam in exhaust gases exists as superheated steam at atmospheric pressure and exhaust gas temperature, the enthalpy of 1 kg of steam at atmospheric pressure 1.013 ≃ 1 bar and 400°C

$$= h_{sup} - h \quad \text{(where } h \text{ is the sensible heat of water at room temperature)}$$

$$= [h_f + h_{fg} + c_{ps}\,(t_{sup} - t_s)] - 1 \times 4.18 \times (25 - 0)$$

$$= [417.5 + 2257.9 + 2.09\,(400 - 99.6)] - 104.5$$

$$= 3355 \text{ kJ/min.}$$

∴ Heat carried away by steam = 0.0967 × 3355 = 320.6 kJ/min.

Heat balance sheet (minute basis) :

Item	*kJ*	*Per cent*
Heat supplied by fuel	*3146*	*100*
(*i*) Heat equivalent of I.P.	1042.8	33.15
(*ii*) Heat carried away by cooling water	870.8	27.70
(*iii*) Heat carried away by dry gases	887	28.15
(*iv*) Heat carried away by steam	320.6	10.20
(*v*) Heat unaccounted for (by difference)	24.8	0.80
Total	314.6	100

Example 17.43. *During a test on a Diesel engine the following observations were made :*

The power developed by the engine is used for driving a D.C. generator. The output of the generator was 210 A at 200 V ; the efficiency of generator being 82%. The quantity of fuel supplied to the engine was 11.2 kg/h ; calorific value of fuel being 42600 kJ/kg. The air-fuel ratio was 18 : 1.

The exhaust gases were passed through a exhaust gas calorimeter for which the observations were as follows : Water circulated through exhaust gas calorimeter = 580 litres/h. Temperature rise of water through calorimeter = 36°C. Temperature of exhaust gases at exit from calorimeter = 98°C. Ambient temperature = 20°C.

Heat lost to jacket cooling water is 32% of the total heat supplied.

If the specific heat of exhaust gases be 1.05 kJ/kg K draw up the heat balance sheet on minute basis.

Solution. Output of generator = 210 A at 200 V

Generator efficiency = 82%

Fuel used = 11.2 kg/h

Calorific value of fuel = 42600 kJ/kg

Air-fuel ratio = 18 : 1

Mass of water circulated through calorimeter, m_c = 580 litres or 580 kg/h

Temperature rise of water, $t_{w_2} - t_{w_1}$ = 36°C

Temperature of exhaust gases at exit from calorimeter = 98°C

Ambient temperature = 20°C

Heat lost to jacket cooling water = 32% of the total heat supplied

Specific heat of exhaust gases = 1.05 kJ/kg K

Total power generated = $VI = 200 \times 210 = 42000$ W = 42 kW

Power available at the brakes of the engine, $\text{B.P.} = \dfrac{42}{0.82} = 51.22$ kW

Total heat supplied to the engine = Fuel supplied per min. × calorific value of fuel

$$= \frac{11.2}{60} \times 42600 = 7952 \text{ kJ/min.}$$

(*i*) Heat equivalent of B.P. = 51.22 × 60 = 3073 kJ/min

Mass of exhaust gases formed per minute

$$= \text{Fuel supplied/min.} \left(\frac{A}{F} \text{ ratio} + 1\right) ; \left[\frac{A}{F} \text{ ratio means air–Fuel ratio}\right]$$

$$= \frac{11.2}{60} (18 + 1) = 3.55 \text{ kg/min.}$$

(*ii*) Heat carried away by exhaust gases/min.

= Heat gained by water in exhaust gas calorimeter from exhaust gases + heat in exhaust gases at exit from exhaust gas calorimeter above room temperature.

$$= m_w \times c_{pw} \times (t_{w_2} - t_{w_1}) + m_g \times c_{pg}(t_g - t_r)$$

$$= \frac{580}{60} \times 4.18 \times 36 + 3.55 \times 1.05 \, (98 - 20)$$

$$= 1454.6 + 290.7 = 1745.3 \text{ kg/min.}$$

(*iii*) Heat lost to jacket cooling water

$$= 0.32 \times 7952 = 2544.6 \text{ kJ/min.}$$

Heat balance sheet (minute basis) :

Item	*kJ*	*Per cent*
Heat supplied	*7952*	*100*
(*i*) Heat equivalent of B.P.	3073	38.7
(*ii*) Heat carried away by exhaust gases	1745.3	21.9
(*iii*) Heat lost to jacket cooling water	2544.6	32.0
(*iv*) Heat unaccounted for (by difference)	589.1	7.4
Total	7952	100

Example 17.44. *During a trial of a single cylinder, 4-stroke diesel engine the following observations were recorded :*

Bore	= *340 mm*
Stroke	= *440 mm*
r.p.m.	= *400*
Area of indicator diagram	= *465 mm*^{2}
Length of diagram	= *60 mm*
Spring constant	= *0.6 bar/mm*
Load on hydraulic dynamometer	= *950 N*
Dynamometer constant	= *7460*
Fuel used	= *10.6 kg/h*
Calorific value of fuel	= *49500 kJ/kg*
Cooling water circulated	= *25 kg/min*
Rise in temperature of cooling water	= *25°C*
The mass analysis of fuel is :	
Carbon	= *84%*
Hydrogen	= *15%*
Incombustible	= *1%*

The volume analysis of exhaust gases is :

Carbon dioxide	= *9%*
Oxygen	= *10%*
Nitrogen	= *81%*
Temperature of exhaust gases	= *400°C*
Specific heat of exhaust gases	= *1.05 kJ/kg°C*
Ambient temperature	= *25°C*
Partial pressure of steam in exhaust gases	= *0.030 bar*
Specific heat of superheated steam	= *2.1 kJ/kg°C.*

Draw up heat balance sheet on minute basis.

Solution. $n = 1$, $D = 0.34$ m, $L = 0.44$ m, $N = 400$ r.p.m., $W = 950$ N,

C_d (dynamometer constant) = 7460, $m_f = 10.6$ kg/h,

$C = 49500$ kJ/kg, $m_w = 25$ kg/min, $(t_{w_2} - t_{w_1}) = 25°$C,

$t_g = 400°$C, $c_{pg} = 1.05$ kJ/kg°C, $c_{ps} = 2.1$ kJ/kg°C.

Mean effective pressure,

$$p_{mi} = \frac{\text{Area of indicator diagram} \times \text{Spring constant}}{\text{Length of indicator diagram}}$$

$$= \frac{465 \times 0.6}{60} = 4.65 \text{ bar}$$

Indicated power, $\text{I.P.} = \dfrac{np_{mi} \times LANk \times 10}{6}$

$$= \frac{1 \times 4.65 \times 0.44 \times \pi/4 \times 0.34^2 \times 400 \times \frac{1}{2} \times 10}{6} = 61.9 \text{ kW}$$

Brake power, $\text{B.P.} = \dfrac{W \times N}{C_d} = \dfrac{950 \times 400}{7460} = 50.9$ kW

Frictional power, F.P. = I.P. – B.P. = 61.9 – 50.9 = 11 kW

Heat supplied per minute

= Fuel used per min. × calorific value

$$= \frac{10.6}{60} \times 49500 = 8745 \text{ kJ/min.}$$

(*i*) Heat equivalent of B.P. = B.P. × 60 = 50.9 × 60 = 3054 kJ/min.

(*ii*) Heat lost in friction = F.P. × 60 = 11 × 60 = 660 kJ/min.

(*iii*) Heat carried away by cooling water

$$= m_w \times c_{pw} \times (t_{w_2} - t_{w_1})$$

$$= 25 \times 4.18 \times 25 = 2612.5 \text{ kJ/min.}$$

Mass of air supplied per kg of fuel

$$= \frac{N \times C}{33(CO + CO_2)} = \frac{81 \times 84}{33(0 + 9)} = 22.9 \text{ kg}$$

Mass of exhaust gases formed per kg of fuel

= 22.9 + 1 = 23.9 kg

Mass of exhaust gases formed/min.

$$= 23.9 \times \frac{10.6}{60} = 4.22 \text{ kg}$$

Mass of steam formed per kg of fuel

$$= 9 \times 0.15 = 1.35 \text{ kg}$$

∴ Mass of steam formed per min.

$$= 1.35 \times \frac{10.6}{60} = 0.238 \text{ kg/min.}$$

Mass of dry exhaust gases formed per min.

$$= 4.22 - 0.238 = 3.982 \text{ kg.}$$

(*iv*) Heat carried away by dry exhaust gases/min.

$$= m_g \times c_{pg} \times (t_g - t_r)$$
$$= 3.982 \times 1.05 \times (400 - 25) = 1568 \text{ kJ/min.}$$

Steam is carried away by exhaust gases. The temperature of steam is also the same as that of exhaust gases *e.g.* 400°C.

At partial pressure of steam 0.03 bar, the saturation temperature is 24.1°C. Therefore, steam is superheated.

Enthalpy of steam $= h_g + c_{ps}\,(t_{sup} - t_s)$

$$= 2545.5 + 2.1\,(400 - 24.1) = 3334.89 \text{ kJ/kg.}$$

(*v*) ∴ Heat carried by steam in exhaust gases

$$= 3334.89 \times 0.238 = 793.7 \text{ kJ/min.}$$

(*vi*) Heat unaccounted for

= Total heat supplied – heat equivalent of B.P.
– heat lost in friction – heat carried away by cooling water
– heat carried away by dry exhaust gases
– heat carried away by steam in exhaust gases

$$= 8745 - (3054 + 660 + 2612.5 + 1568 + 793.7)$$
$$= 56.8 \text{ kJ/min.}$$

Heat balance sheet on minute basis :

Item	*kJ*	*Per cent*
Heat supplied	*8745*	*100*
(*i*) Heat equivalent of B.P.	3054	34.92
(*ii*) Heat lost in friction	660	7.55
(*iii*) Heat carried away by cooling water	2612.5	29.87
(*iv*) Heat carried away by dry exhaust gases	1568	17.93
(*v*) Heat carried away by steam in exhaust gases	793.7	9.07
(*vi*) Heat unaccounted for	56.8	0.66
Total	8745	100

MORSE TEST

Example 17.45. *In a test of a 4-cylinder, 4-stroke engine 75 mm bore and 100 mm stroke, the following results were obtained at full throttle at a particular constant speed and with fixed setting of fuel supply of 6.0 kg/h.*

B.P. with all cylinder working = *15.6 kW*
B.P. with cylinder no. 1 cut-out = *11.1 kW*
B.P. with cylinder no. 2 cut-out = *11.03 kW*
B.P. with cylinder no. 3 cut-out = *10.88 kW*
B.P. with cylinder no. 4 cut-out = *10.66 kW*

If the calorific value of the fuel is 83600 kJ/kg and clearance volume is 0.0001 m³, calculate :

(i) Mechanical efficiency, *(ii) Indicated thermal efficiency, and*
(iii) Air standard efficiency.

Solution. B.P. = I.P. – F.P.

Assuming that the engine is running at constant speed the *frictional and pumping losses remain constant.* Now if one cylinder is cut-out it will not produce any power but the frictional loss and power lost in operating the valves will remain the same as the speed of the engine is constant. The B.P. reduction at the crankshaft due to one cylinder cut out will be *exactly equal to the I.P. produced by that cylinder.*

Therefore,

I.P. produced in cylinder 1, $IP_1 = BP - BP_1 = 15.6 - 11.1 = 4.5$ kW
I.P. produced in cylinder 2, $IP_2 = BP - BP_2 = 15.6 - 11.03 = 4.57$ kW
I.P. produced in cylinder 3, $IP_3 = BP - BP_3 = 15.6 - 10.88 = 4.72$ kW
I.P. produced in cylinder 4, $IP_4 = BP - BP_4 = 15.6 - 10.66 = 4.94$ kW
Total I.P. produced $= IP_1 + IP_2 + IP_3 + IP_4$

I.P. = 4.5 + 4.57 + 4.72 + 4.94 = 18.73 kW

(*i*) **Mechanical efficiency, $\eta_{mech.}$:**

$$\eta_{mech.} = \frac{\text{B.P.}}{\text{I.P.}} = \frac{15.6}{18.73} = \mathbf{0.833 \text{ or } 83.3\%.} \quad \textbf{(Ans.)}$$

(*ii*) **Indicated thermal efficiency, $\eta_{th(I)}$:**

$$\eta_{th.(I)} = \frac{\text{I.P.}}{\dot{m}_f \times C} = \frac{18.73}{\frac{6}{3600} \times 83600} = \mathbf{0.1344 \text{ or } 13.44\%.} \quad \textbf{(Ans.)}$$

(*iii*) **Air-standard efficiency, $\eta_{air\text{-}standard}$:**

Stroke volume, $V_s = \frac{\pi}{4} D^2 L = \frac{\pi}{4} \times 0.075^2 \times 0.1 = 0.0004417 \text{ m}^3$

Clearance volume, $V_c = 0.0001 \text{ m}^3$

Compression ratio, $r = \frac{V_s + V_c}{V_c} = \frac{0.0004417 + 0.0001}{0.0001} = 5.4$

$$\therefore \quad \eta_{air\text{-}standard} = 1 - \frac{1}{(r)^{\gamma - 1}} = 1 - \frac{1}{(5.4)^{1.4-1}} = \mathbf{0.49 \quad or \quad 49\%.} \quad \textbf{(Ans.)}$$

Example 17.46. *A 4-cylinder petrol engine has a bore of 60 mm and a stroke of 90 mm. Its rated speed is 2800 r.p.m. and it is tested at this speed against brake which has a torque arm of*

0.37 m. The net brake load is 160 N and the fuel consumption is 8.986 litres/h. The specific gravity of petrol used is 0.74 and it has a lower calorific value of 44100 kJ/kg. A Morse test is carried out and the cylinders are cut out in the order 1, 2, 3, 4 with corresponding brake loads of 110, 107, 104 and 110 N respectively. Calculate for this speed :

(*i*) *The engine torque,* (*ii*) *The brake mean effective pressure,*
(*iii*) *The brake thermal efficiency,* (*iv*) *The specific fuel consumption,*
(*v*) *Mechanical efficiency, and* (*vi*) *Indicated mean effective pressure.*

Solution. Number of cylinders, $n = 4$

Bore, $D = 60 \text{ mm} = 0.06 \text{ m}$

Stroke, $L = 90 \text{ mm} = 0.09 \text{ m}$

Speed, $N = 2800$ r.p.m.

Torque arm $= 0.37$ m

Net brake load $= 160$ N

Fuel consumption $= 8.986$ litres/h

$= 8.986 \times 1 \times 0.74$ kg/h

Specific gravity of petrol $= 0.74$

Calorific value $= 44100$ kJ/kg

(*i*) **Engine torque, T :**

Engine torque, T = Net brake load × torque arm

$= 160 \times 0.37 =$ **59.2 N-m. (Ans.)**

(*ii*) **Brake mean effective pressure, p_{mb} :**

Brake power, $$\text{B.P.} = \frac{2\pi NT}{60 \times 1000} = \frac{2\pi \times 2800 \times 59.2}{60 \times 1000} = 17.36 \text{ kW}$$

$$\text{B.P.} = \frac{np_{mb}LANk \times 10}{6}$$

$$17.36 = \frac{4 \times p_{mb} \times 0.09 \times \frac{\pi}{4} \times (0.06)^2 \times 2800 \times \frac{1}{2} \times 10}{6}$$

$$\therefore \quad p_{mb} = \frac{17.36 \times 6 \times 4 \times 2}{4 \times 0.09 \times \pi \times (0.06)^2 \times 2800 \times 10} = \mathbf{7.31 \text{ bar.}} \quad \textbf{(Ans.)}$$

(*iii*) **Brake thermal efficiency, $\eta_{th(B)}$:**

$$\eta_{th(B)} = \frac{\text{B.P.}}{\dot{m}_f \times C} = \frac{17.36}{\frac{(8.986 \times 1 \times 0.74)}{3600} \times 44100} = \mathbf{0.213 \text{ or } 21.3\%.} \quad \textbf{(Ans.)}$$

(*iv*) **Specific fuel consumption, s.f.c. :**

$$\text{s.f.c.} = \frac{m_f}{\text{B.P.}} = \frac{6.65}{17.36} = \mathbf{0.383 \text{ kg/kWh.}} \quad \textbf{(Ans.)}$$

(*v*) **Mechanical efficiency, $\eta_{mech.}$:**

Since the speed is constant, substituting the brake loads instead of the values of B.P. as follows :

$$IP_1 = BP - BP_1 = 160 - 110 = 50 \text{ N}$$
$$IP_2 = BP - BP_2 = 160 - 107 = 53 \text{ N}$$

$$IP_3 = BP - BP_3 = 160 - 104 = 56 \text{ N}$$
$$IP_4 = BP - BP_4 = 160 - 110 = 50 \text{ N}$$

Hence for the engine, the indicated load is given by

$$IP = IP_1 + IP_2 + IP_3 + IP_4 = 50 + 53 + 56 + 50 = 209 \text{ N}$$

$$\therefore \qquad \eta_{mech.} = \frac{BP}{IP} = \frac{160}{209} = \mathbf{0.765 \text{ or } 76.5\%.} \quad \textbf{(Ans.)}$$

(*vi*) **Indicated mean effective pressure, p_{mi}** :

$$\eta_{mech.} = \frac{p_{mb}}{p_{mi}}$$

$$\therefore \qquad p_{mi} = \frac{p_{mb}}{\eta_{mech.}} = \frac{7.31}{0.765} = \mathbf{9.55 \text{ bar.}} \quad \textbf{(Ans.)}$$

HIGHLIGHTS

Performance of I.C. engines. Some important relations :

(*i*) Indicated power (I.P.) $= \dfrac{np_{mi} \, LANk \times 10}{6}$ kW

(*ii*) Brake (B.P.) $= \dfrac{(W - S) \, \pi \, (D_b + d)N}{60 \times 1000}$ kW or $\left(= \dfrac{2\pi NT}{60 \times 1000} \text{kW} \right)$

(*iii*) Mechanical efficiency, $\eta_{mech.} = \dfrac{\text{B.P.}}{\text{I.P.}}$

(*iv*) Thermal efficiency (indicated), $\eta_{th.(I)} = \dfrac{\text{I.P.}}{\dot{m}_f \times C}$

and thermal efficiency (brake), $\eta_{th.(B)} = \dfrac{\text{B.P.}}{\dot{m}_f \times C}$

where $\dot{m}_f$ = mass of fuel used in kg/sec.

(*v*) $\eta_{relative} = \dfrac{\eta_{thermal}}{\eta_{air\text{-}standard}}$

(*vi*) Measurement of air consumption by **air box method** :

Volume of air passing through the orifice, $V_a = 840 \, AC_d \sqrt{\dfrac{h_w}{\rho_a}}$

and mass of air passing through the orifice,

$$m_a = 0.066 \, C_d \times d^2 \sqrt{h_w \rho_a} \text{ kg/min}$$

where, A = Area of orifice, m^2

d = Diameter of orifice, cm

h_w = Head of water of in 'cm' causing the flow

ρ_a = Density of air in kg/m^3 under atmospheric conditions.

OBJECTIVE TYPE QUESTIONS

Choose the Correct Answer :

1. The specific fuel consumption of a diesel engine as compared to that for petrol engines is
(*a*) lower (*b*) higher
(*c*) same for same output (*d*) none of the above.

2. The thermal efficiency of petrol engine as compared to diesel engine is
(*a*) lower (*b*) higher
(*c*) same for same power output (*d*) same for same speed.

3. Compression ratio of petrol engines is in the range of
(*a*) 2 to 3 (*b*) 7 to 10
(*c*) 16 to 20 (*d*) none of the above.

4. Compression ratio of diesel engines may have a range
(*a*) 8 to 10 (*b*) 10 to 15
(*c*) 16 to 20 (*d*) none of the above.

5. The thermal efficiency of good I.C. engine at the rated load is in the range of
(*a*) 80 to 90% (*b*) 60 to 70%
(*c*) 30 to 35% (*d*) 10 to 20%.

6. In case of S.I. engine, to have best thermal efficiency the fuel air mixture ratio should be
(*a*) lean (*b*) rich
(*c*) may be lean or rich (*d*) chemically correct.

7. The fuel air ratio, for maximum power of S.I. engines, should be
(*a*) lean (*b*) rich
(*c*) may be lean or rich (*d*) chemically correct.

8. In case of petrol engine, at starting
(*a*) rich fuel air ratio is needed (*b*) weak fuel air ratio is needed
(*c*) chemically correct fuel air ratio is needed (*d*) any fuel air ratio will do.

9. Carburettor is used for
(*a*) S.I. engines (*b*) Gas engines
(*c*) C.I. engines (*d*) none of the above.

10. Fuel injector is used in
(*a*) S.I. engines (*b*) Gas engines
(*c*) C.I. engines (*d*) none of the above.

11. Very high speed engines are generally
(*a*) Gas engines (*b*) S.I. engines
(*c*) C.I. engines (*d*) Steam engines.

12. In S.I. engine, to develop high voltage for spark plug
(*a*) battery is installed (*b*) distributor is installed
(*c*) carburettor is installed (*d*) ignition coil is installed.

13. In S.I. engine, to obtain required firing order
(*a*) battery is installed (*b*) distributor is installed
(*c*) carburettor is installed (*d*) ignition coil is installed.

14. For petrol engines, the method of governing employed is
(*a*) quantity governing (*b*) quality governing
(*c*) hit and miss governing (*d*) none of the above.

15. For diesel engines, the method of governing employed is
(*a*) quantity governing (*b*) quality governing
(*c*) hit and miss governing (*d*) none of the above.

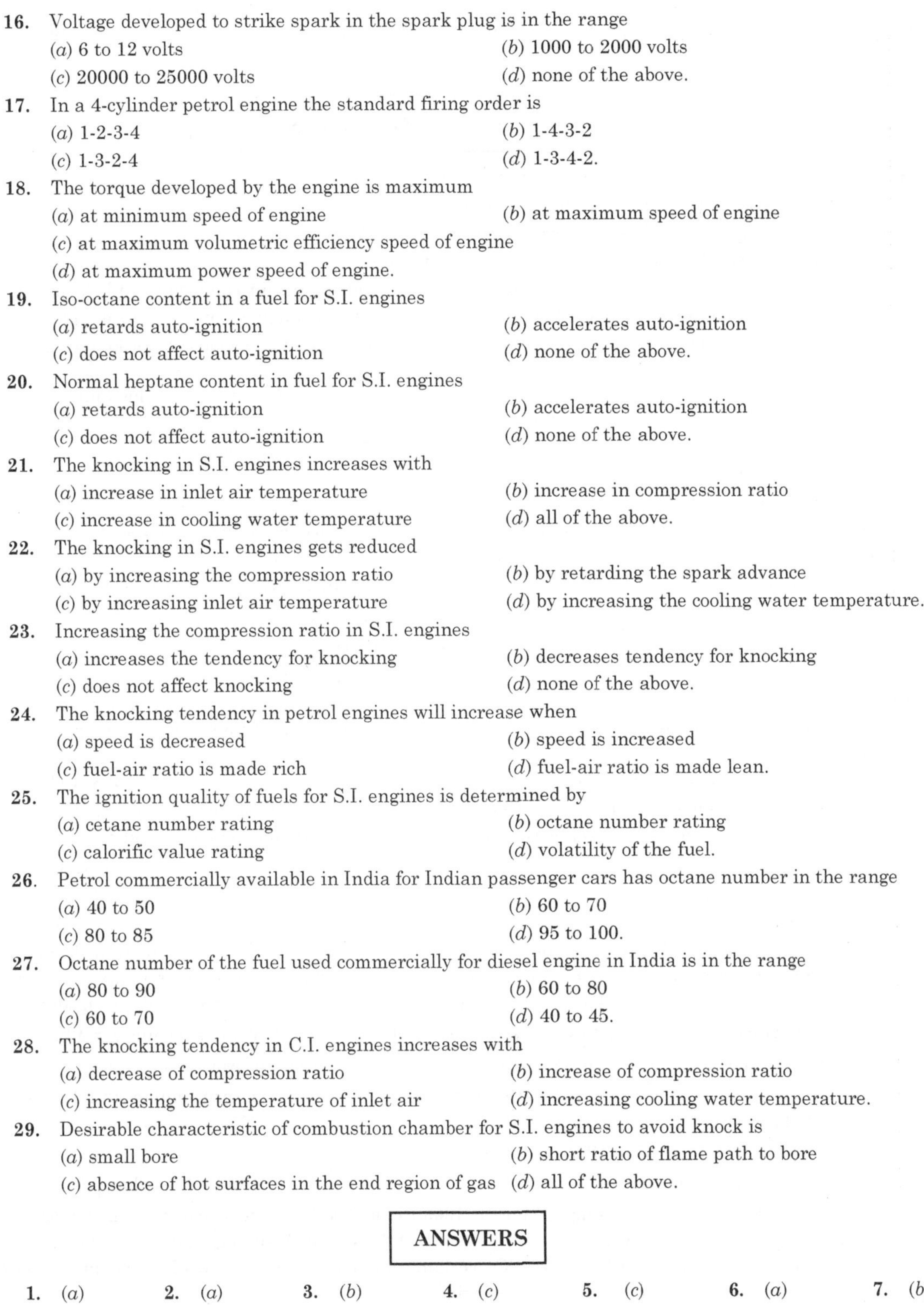

16. Voltage developed to strike spark in the spark plug is in the range
(*a*) 6 to 12 volts (*b*) 1000 to 2000 volts
(*c*) 20000 to 25000 volts (*d*) none of the above.

17. In a 4-cylinder petrol engine the standard firing order is
(*a*) 1-2-3-4 (*b*) 1-4-3-2
(*c*) 1-3-2-4 (*d*) 1-3-4-2.

18. The torque developed by the engine is maximum
(*a*) at minimum speed of engine (*b*) at maximum speed of engine
(*c*) at maximum volumetric efficiency speed of engine
(*d*) at maximum power speed of engine.

19. Iso-octane content in a fuel for S.I. engines
(*a*) retards auto-ignition (*b*) accelerates auto-ignition
(*c*) does not affect auto-ignition (*d*) none of the above.

20. Normal heptane content in fuel for S.I. engines
(*a*) retards auto-ignition (*b*) accelerates auto-ignition
(*c*) does not affect auto-ignition (*d*) none of the above.

21. The knocking in S.I. engines increases with
(*a*) increase in inlet air temperature (*b*) increase in compression ratio
(*c*) increase in cooling water temperature (*d*) all of the above.

22. The knocking in S.I. engines gets reduced
(*a*) by increasing the compression ratio (*b*) by retarding the spark advance
(*c*) by increasing inlet air temperature (*d*) by increasing the cooling water temperature.

23. Increasing the compression ratio in S.I. engines
(*a*) increases the tendency for knocking (*b*) decreases tendency for knocking
(*c*) does not affect knocking (*d*) none of the above.

24. The knocking tendency in petrol engines will increase when
(*a*) speed is decreased (*b*) speed is increased
(*c*) fuel-air ratio is made rich (*d*) fuel-air ratio is made lean.

25. The ignition quality of fuels for S.I. engines is determined by
(*a*) cetane number rating (*b*) octane number rating
(*c*) calorific value rating (*d*) volatility of the fuel.

26. Petrol commercially available in India for Indian passenger cars has octane number in the range
(*a*) 40 to 50 (*b*) 60 to 70
(*c*) 80 to 85 (*d*) 95 to 100.

27. Octane number of the fuel used commercially for diesel engine in India is in the range
(*a*) 80 to 90 (*b*) 60 to 80
(*c*) 60 to 70 (*d*) 40 to 45.

28. The knocking tendency in C.I. engines increases with
(*a*) decrease of compression ratio (*b*) increase of compression ratio
(*c*) increasing the temperature of inlet air (*d*) increasing cooling water temperature.

29. Desirable characteristic of combustion chamber for S.I. engines to avoid knock is
(*a*) small bore (*b*) short ratio of flame path to bore
(*c*) absence of hot surfaces in the end region of gas (*d*) all of the above.

ANSWERS

1. (*a*)	**2.** (*a*)	**3.** (*b*)	**4.** (*c*)	**5.** (*c*)	**6.** (*a*)	**7.** (*b*)
8. (*a*)	**9.** (*a*)	**10.** (*c*)	**11.** (*b*)	**12.** (*d*)	**13.** (*b*)	**14.** (*a*)

15. (*b*)	**16.** (*c*)	**17.** (*d*)	**18.** (*c*)	**19.** (*a*)	**20.** (*b*)	**21.** (*d*)
22. (*b*)	**23.** (*a*)	**24.** (*a*)	**25.** (*b*)	**26.** (*c*)	**27.** (*d*)	**28.** (*a*)
29. (*d*).						

THEORETICAL QUESTIONS

1. (*a*) What do you mean by performance of I.C. engine ?
   (*b*) Discuss briefly the basic performance parameters.
   (*c*) Discuss with suitable sketch the brake rope dynamometer.
2. Describe how the I.P. of a multicylinder engine is measured ?
3. Describe the method commonly used in laboratory for measuring the air supplied to an I.C. engine.
4. Derive the formula used for finding the mass of air supplied to an engine using an orifice tank.
5. Explain the phenomenon of auto-ignition. Explain how auto-ignition is responsible for knocking in S.I. engines.
6. Explain the phenomena of knocking in S.I. engine. What are the different factors which influence the knocking ? Describe the methods used to suppress it.
7. Explain the difference between (*i*) pre-ignition, (*ii*) auto-ignition and (*iii*) detonation.
8. What is meant by ignition delay ?
9. What are causes of knock in C.I. engines ?
10. What are the different methods used in C.I. engines to create turbulence in the mixture ?
    Explain its effect on power output and thermal efficiency of the engine.
11. What do you mean by 'octane number' and 'cetane number' of fuels ? How are they determined ?

UNSOLVED EXAMPLES

1. A single-cylinder petrol engine working on two-stroke cycle develops indicated power of 5 kW. If the mean effective pressure is 7.0 bar and the piston diameter is 100 mm, calculate the average speed of the piston.
   [**Hint.** Average piston speed = 2 *LN*.] [**Ans.** 109.1 m/s]
2. A 4-cylinder petrol engine works on a mean effective pressure of 5 bar and engine speed of 1250 r.p.m. Find the indicated power developed by the engine if the bore is 100 mm and stroke 150 mm.
   [**Ans.** 6.11 kW]
3. A 4-cylinder four-stroke S.I. engine is designed to develop 44 kW indicated power at a speed of 3000 r.p.m. The compression ratio used is 6. The law of compression and expansion is $pV^{1.3}$ = constant and heat addition and rejection takes place at constant volume. The pressure and temperature at the beginning of compression stroke are 1 bar and 50°C. The maximum pressure of the cycle is limited to 30 bar. Calculate the diameter and stroke of each cylinder assuming all cylinders have equal dimensions.
   Assume diagram factor = 0.8 and ratio of stroke/bore = 15. [**Ans.** D = 95 mm, L = 142.5 mm]
4. During the trial of a four-stroke diesel engine, the following observations were recorded :
   Area of indicator diagram = 475 mm², length of indicator diagram = 62 mm, spring number = 1.1 bar/mm, diameter of piston = 100 mm, length of stroke = 150 mm, engine r.p.m. = 375.
   Determine : (*i*) Indicated mean effective pressure
   (*ii*) Indicated power. [**Ans.** (*i*) 8.43 bar ; (*ii*) 3.1 kW]
5. A 4-cylinder, four-stroke diesel engine runs at 1000 r.p.m. The bore and stroke of each cylinder are 100 mm and 160 mm respectively. The cut off is 6.62% of the stroke. Assuming that the initial condition of air inside the cylinder are 1 bar and 20°C, mechanical efficiency of 75%, calculate the air-standard efficiency and brake power developed by the engine.

Also, calculate the brake specific fuel consumption if the air/fuel ratio is 20 : 1. Take R for air as 0.287 kJ/kg K and clearance volume as 0.000084 m^3. **[Ans.** 61.4%, 21.75 kW, 0.4396 kg/kWh]

6. During a trial of a two stroke diesel engine the following observations were recorded :

Engine speed = 1500 r.p.m., load on brakes = 120 kg, length of brake arm = 875 mm.

Determine :

(*i*) Brake torque (*ii*) Brake power. [**Ans.** (*i*) 1030 N-m ; (*ii*) 161.8 kW]

7. A four-stroke gas engine develops 4.2 kW at 180 r.p.m. and at full load. Assuming the following data, calculate the relative efficiency based on indicated power and air-fuel ratio used. Volumetric efficiency = 87%, mechanical efficiency = 74%, clearance volume = 2100 cm^3, swept volume = 9000 cm^3, fuel consumption = 5 m^3/h, calorific value of fuel = 16750 kJ/m^3. [**Ans.** 50.2%, 7.456 : 1]

8. During the trial of a four-stroke cycle gas engine the following data were recorded :

Area of indicator diagram = 565.8 mm^2
Length of indicator diagram = 74.8 mm
Spring index = 0.9 bar/mm
Cylinder diameter = 220 mm
Stroke length = 430 mm
Number of explosions/min = 100

Determine : (*i*) indicated mean effective pressure
(*ii*) Indicated power. [**Ans.** (*i*) 6.8 bar ; (*ii*) 18.5 kW]

9. The following observations were recorded during a trial of a four stroke engine with rope brake dynamometer :

Engine speed = 650 r.p.m., diameter of brake drum = 600 mm, diameter of rope = 50 mm, dead load on the brake drum = 32 kg, spring balance reading = 4.75 kg.

Calculate the brake power. [**Ans.** 5.9 kW]

10. The following data refer to a four stroke petrol engine :

Engine speed = 2000 r.p.m., ideal thermal efficiency = 35%, relative efficiency = 80%, mechanical efficiency = 85%, volumetric efficiency = 70%.

If the engine develops 29.42 kW brake power calculate the cylinder swept volume. [**Ans.** 0.00185 m^3]

11. A single cylinder four-stroke gas engine has a bore of 178 mm and a stroke of 330 mm and is governed by hit and miss principle. When running at 400 r.p.m. at full load, indicator cards are taken which give a working loop mean effective pressure of 6.2 bar, and a pumping loop mean effective pressure of 0.35 bar. Diagrams from the dead cycle give a mean effective pressure of 0.62 bar. The engine was run light at the same speed (*i.e.* with no load), and a mechanical counter recorded 47 firing strokes per minute. Calculate :

(*i*) Full load brake power (*ii*) Mechanical efficiency of the engine.

[**Ans.** (*i*) 13.54 kW ; (*ii*) 84.7%]

12. During a 60 minutes trial of a single cylinder four stroke engine the following observations were recorded :

Bore = 0.3 m, stroke = 0.45 m, fuel consumption = 11.4 kg, calorific value of fuel = 42000 kJ/kg, brake mean effective pressure = 6.0 bar, net load on brakes = 1500 N, r.p.m. = 300, brake drum diameter = 1.8 m, brake rope diameter = 20 mm, quantity of jacket cooling water = 600 kg, temperature rise of jacket water = 55°C, quantity of air as measured = 250 kg, exhaust gas temperature = 420°C, c_p for exhaust gases = 1 kJ/kg K, ambient temperature = 20°C.

Calculate : (*i*) Indicated power ; (*ii*) Brake power ;
(*iii*) Mechanical efficiency (*iv*) Indicated thermal efficiency.

Draw up a heat balance sheet on minute basis. [**Ans.** (*i*) 47.7 kW, (*ii*) 42.9 kW, (*iii*) 89.9%, (*iv*) 35.86%]

13. A quality governed four-stroke, single-cylinder gas engine has a bore of 146 mm and a stroke of 280 mm. At 475 r.p.m. and full load the net load on the friction brake is 433 N, and the torque arm is 0.45 m. The indicator diagram gives a net area of 578 mm^2 and a length of 70 mm with a spring rating of 0.815 bar/mm.

Calculate : (*i*) The indicated power (*ii*) Brake power

(*iii*) Mechanical efficiency. [**Ans.** (*i*) 12.5 kW (*ii*) 9.69 kW (*iii*) 77.5%]

14. A two-cylinder four stroke gas engine has a bore of 380 mm and a stroke of 585 mm. At 240 r.p.m. the torque developed is 5.16 kN-m.

Calculate : (*i*) Brake power (*ii*) Mean piston speed in m/s

(*iii*) Brake mean effective pressure. [**Ans.** (*i*) 129.8 kW (*ii*) 4.68 m/s ; (*iii*) ; 4.89 bar]

15. The engine of Problem 14 is supplied with a mixture of coal gas and air in the proportion of 1 to 7 by volume. The estimated volumetric efficiency is 85% and the calorific value of the coal gas is 16800 kJ/m^3. Calculate the brake thermal efficiency of the engine. [**Ans.** 27.4%]

16. A 4-cylinder, four-stroke diesel engine has a bore of 212 mm and a stroke of 292 mm. At full load at 720 r.p.m., the b.m.e.p. is 5.93 bar and the specific fuel consumption is 0.226 kg/kWh. The air/fuel ratio as determined by exhaust gas analysis is 25 : 1. Calculate the brake thermal efficiency and volumetric efficiency of the engine.

Atmospheric conditions are 1.01 bar and 15°C and calorific value for the fuel may be taken as 44200 kJ/kg. [**Ans.** 36% ; 76.5%]

17. A 4-cylinder petrol engine has an output of 52 kW at 2000 r.p.m. A Morse test is carried out and the brake torque readings are 177, 170, 168 and 174 N-m respectively. For normal running at this speed the specific fuel consumption is 0.364 kg/kWh. The calorific value of fuel is 44200 kJ/kg.

Calculate : (*i*) Mechanical efficiency

(*ii*) Brake thermal efficiency of the engine. [**Ans.** (*i*) 82% ; (*ii*) 22.4%]

18. A V-8 four-stroke petrol engine is required to give 186.5 kW at 440 r.p.m. The brake thermal efficiency can be assumed to be 32% at the compression ratio of 9 : 1. The air/fuel ratio is 12 : 1 and the volumetric efficiency at this speed is 69%. If the stroke to bore ratio is 0.8, determine the engine displacement required and the dimensions of the bore and stroke. The calorific value of the fuel is 44200 kJ/kg, and the free air conditions are 1.013 bar and 15°C. [**Ans.** 5.12 litres ; 100.6 mm ; 80.5 mm]

19. During the trial (60 minutes) on a single cylinder oil engine having cylinder diameter 300 mm, stroke 450 mm and working on the four stroke cycle, the following observations were made :

Total fuel used = 9.6 litres, calorific value of fuel = 45000 kJ/kg, total number of revolutions = 12624, gross indicated mean effective pressure = 7.24 bar, pumping i.m.e.p. = 0.34 bar, net load on the brake = 3150 N, diameter of brake wheel drum = 1.78 m, diameter of the rope = 40 mm, cooling water circulated = 545 litres, cooling water temperature rise = 25°C, specific gravity of oil = 0.8.

Determine : (*i*) Indicated power. (*ii*) Brake power.

(*iii*) Mechanical efficiency.

Draw up the heat balance sheet on minute basis. [**Ans.** (*i*) 77 kW ; (*ii*) 61.77 kW ; (*iii*) 80.22%]

20. The following results were obtained on full load during a trial on a two stroke oil engine :

Engine speed	= 350 r.p.m.
Net brake load	= 600 N
m.e.p.	= 2.75 bar
Oil consumption	= 4.25 kg/h
Temperature rise of jacket cooling water	= 25°C
Air used per kg of oil	= 31.5 kg
Temperature of air in test room	= 20°C
Temperature of exhaust gases	= 390°C

Following data also apply to the above test :

Cylinder diameter	= 220 mm
Stroke	= 280 mm
Effective brake diameter	= 1 metre
Calorific value of oil	= 45000 kJ/kg
Proportion of hydrogen in fuel oil	= 15%
Partial pressure of steam in exhaust gases	= 0.04 bar

Mean specific heat of exhaust gases = 1.0 kJ/kg K
Specific heat of superheated steam = 2.1 kJ/kg K
Specific heat of water = 4.186 kJ/kg K

Determine : (*i*) Indicated power. (*ii*) Brake power.
(*iii*) Mechanical efficiency.

Draw up heat balance sheet for the test. [**Ans.** (*i*) 17.1 kW ; (*ii*) 11 kW ; (*iii*) 64.33%]

21. A 4-cylinder, four-stroke diesel engine developes 83.5 kW at 1800 r.p.m. with specific fuel consumption of 0.231 kg/kWh, and air/fuel ratio of 23 : 1. The analysis of fuel is 87% carbon and 13% hydrogen, and the calorific value of the fuel is 43500 kJ/kg. The jacket cooling water flows at 0.246 kg/s and its temperature rise is 50 K. The exhaust temperature is 316°C. Draw up an energy balance for the engine. Take R = 0.302 kJ/kg K and c_p = 1.09 kJ/kg K for the dry exhaust gases and c_p = 1.86 kJ/kg K for superheated steam. The temperature in the test house is 17.8°C, and the exhaust gas pressure is 1.013 bar.

[**Ans.** B.P. = 35.8%, cooling water = 22.1%, exhaust = 24%, radiation and unaccounted = 16.7%]

22. During the trial of a single cylinder, 4-stroke, diesel engine the following observations were recorded : Bore = 350 mm, stroke = 450 mm, r.p.m. = 400, area of indicator diagram = 472 mm^2, length of indicator diagram = 62 mm, spring constant = 0.59 bar/mm, load on hydraulic dynamometer = 970 N, dynamometer constant = 7500, fuel used = 10.78 kg/h, calorific value of fuel = 50000 kJ/kg, cooling water circulated = 24 litres/min, rise in temperature of cooling water = 24°C. The main analysis of fuel is : carbon = 85%, hydrogen = 14%, incombustibles = 1%. The volume analysis of exhaust gases is : carbon dioxide = 8%, oxygen = 11%, nitrogen = 81%. Temperature of exhaust gases = 380°C, specific heat of exhaust gases = 1.05 kJ/kg°C, ambient temperature = 20°C, partial pressure of steam in exhaust gases = 0.03 bar, specific heat of superheated steam = 2.1 kJ/kg°C.

Draw up the heat balance sheet on minute basis.

[**Ans.** (*i*) Heat equivalent of B.P. = 34.55%, (*ii*) heat lost in friction = 8.73%,
(*iii*) heat carried away by cooling water = 26.84%,
(*iv*) heat in dry exhaust gases = 19.54%,
(*v*) heat carried away by steam in exhaust gases = 7.24%,
(*vi*) heat unaccounted for = 3.10%]

23. A 4-cylinder petrol engine has a bore of 57 mm and a stroke of 90 mm. Its rated speed is 2800 r.p.m. and it is tested at this speed against a brake, which has a torque arm of 0.356 m. The net brake load is 155 N and the fuel consumption is 6.74 litres/h. The specific gravity of petrol used is 0.735 and it has a lower calorific value of 44200 kJ/kg. A Morse test is carried out and the cylinders are cut out in order 1, 2, 3, 4 with corresponding brake loads 111, 106.5, 104.2 and 111 N, respectively. Calculate for this speed :

(*i*) The engine torque. (*ii*) The brake mean effective pressure.
(*iii*) The brake thermal efficiency. (*iv*) The specific fuel consumption.
(*v*) The mechanical efficiency. (*vi*) The indicated mean effective pressure.

[**Ans.** (*i*) 55.2 N-m ; (*ii*) 7.55 bar ; (*iii*) 26.6% ; (*iv*) 0.306 kg/kWh ; (*v*) 82.8% ; (*vi*) 9.12 bar]

18

Air Pollution from I.C. Engines and Its Control

18.1. INTRODUCTION

- **Air pollution** *can be defined as an addition to our atmosphere of any material which will have a deleterious effect on life upon our planet.* Besides I.C. engines other sources such as electric power stations, industrial and domestic fuel consumers also add pollution.
- There has been a great concern, in recent years, that the *internal combustion engines* is responsible for too much atmospheric pollution, which is detrimental to human health and the environment. Thus concerted efforts are being made to *reduce the responsible pollutants emitted from the exhaust system without sacrificing power and fuel consumption.*

18.2. POLLUTANTS

18.2.1. Pollution Derived from Combustion Products

Pollutants are produced by the *incomplete burning* of the air-fuel mixture in the combustion chamber. The *major pollutants* emitted from the exhaust due to *incomplete combustion* are :

1. Carbon monoxide (CO)
2. Hydrocarbons (HC)
3. Oxides of nitrogen (NO_x).

Other products produced are *acetylene, aldehydes* etc. If, however, combustion is complete the only products being expelled from the exhaust would be *water vapour* which is harmless, and *carbon dioxide,* which is an inert gas and, as such it is not directly harmful to humans.

1. **Carbon monoxide (CO) :**

- It is a *colourless gas* of about the same density as air.
- It is a *poisonous gas* which, when inhaled, replaces the oxygen in the blood stream so that the body's metabolism can not function correctly.
- Small amounts of CO concentrations, when breathed in, slow down physical and mental activity and produces headaches, while large concentration will kill.

Mechanism of formation of CO :

CO is generally formed when the mixture is rich in fuel. The amount of CO formed increases as the mixture becomes more and more rich in fuel. A small amount of CO will come out of the exhaust even when the mixture is slightly lean in fuel. This is due to the fact that equilibrium is not established when the products pass to the exhaust. At the high temperature developed during the combustion, the products formed are unstable, and the following reactions take place before the equilibrium is established.

$$2H_2O + O_2 \longrightarrow 2(1-y)\,H_2O + 2yH_2 + yO_2$$

where, y is the fraction of H_2O dissociated.

$$C + O_2 \longrightarrow CO_2 \longrightarrow (1-x)CO_2 + xCO + \frac{x}{2}O_2$$

As the products cool down to exhaust temperature, major part of CO reacts with oxygen to form CO_2. However, a relatively small amount of CO will remain in exhaust, its concentration increasing with rich mixtures.

2. **Hydrocarbons (HC) :**

- Hydrocarbons, derived from unburnt fuel emitted by exhausts, engine crankcase fumes and vapour escaping from the carburettor are also harmful to health.

Mechanism of formation of HC

— Due to existence of local very rich mixture pockets at much lower temperatures than the combustion chambers, unburnt hydrocarbons may appear in the exhaust.

— The hydrocarbons also appear due to flame quenching near the metallic walls.

A significant portion of this unburnt hydrocarbon may burn during expansion and exhaust strokes if the oxygen concentration and exhaust temperature is suitable for complete oxidation. Otherwise a large amount of hydrocarbon will go out with the exhaust gases.

3. **Oxides of nitrogen (NO_x) :**

- Oxides of nitrogen and other obnoxious substances are produced in very small quantities and, in certain environments, can cause pollution ; while *prolonged exposure is dangerous to health.*

Mechanism of formation of nitric oxide (NO)

— At high combustion temperatures, the following chemical reactions take place behind the flame :

$$N_2 + O_2 \longrightarrow 2NO$$
$$N_2 + 2H_2O \longrightarrow 2NO + 2H_2$$

Chemical equilibrium calculations show that a significant amount of NO will be formed at the *end of combustion*. The majority of NO formed will however decompose at the low temperatures of exhaust. But due to very low reaction rate at the exhaust temperature a part of NO formed remains in exhaust. It is far in excess of the equilibrium composition at that temperature as the formation of NO freezes at low exhaust temperatures.

— *The NO formation will be less in rich mixtures than in lean mixtures.*

Smoke or particulate

- Solid particles are usually formed by *dehydrogenation, polymerisation* and *agglomeration.*
- In the combustion process of different hydrocarbons, acetylene (C_2H_2) is formed as an intermediate product. These acetylene molecules after simultaneous polymerisation and dehydration produce carbon particles, which are the main constituent of the particulate.

Aldehydes

- Due to very slow chemical reaction during delay period in the diesel engines, *aldehydes are formed as intermediate products.* In some parts of the spray the aldehydes will be left after the initial reactions. These aldehydes may be oxidised in the later part of the cycle, if the mixture temperature is high, and if there is sufficient oxygen.
- *At heavy loads, due to lack of oxygen, an increase in aldehyde emission in the exhaust is observed.*

Following points are **worth noting :**

1. If the *air-fuel mixture is* ***too rich*** there is insufficient air for complete combustion and some of the fuel will not be burnt or at least only partly burnt. Since hydrogen has a greater affinity for oxygen, hydrogen will take all the oxygen it needs leaving the carbon with a deficiency of oxygen. As *a result of the shortage of oxygen a percentage of the carbon will be converted to carbon monoxide as well as carbon dioxide, and, with very rich mixtures, particles of pure unburnt carbon may be expelled from the exhaust as "black smoke".*

Incomplete combustion due to partial oxidation of the hydrocarbon fuel also produces other products such as *acetylene* and *aldehyde.* These products, when expelled from the exhaust, leave an unpleasant smell and are particularly noticeable during engine warm-up when a rich mixture is provided.

2. If the *mixture is made* ***too weak*** it is unlikely that the atomised liquid fuel will be thoroughly mixed throughout the combustion chamber so that *slow burning, incomplete combustion and misfiring may result.*

A further characteristic of weak mixtures is that the excess oxygen (which has not taken part in the combustion process) at very high temperature is able to combine with some of the nitrogen that constitutes about three-quarters of air, to form oxides of nitrogen such as *nitrogen peroxide* (NO_2). *The amount of nitrogen peroxide produced will increase as the mixture weakens until it peaks at just over an air-fuel ratio of 15.5 : 1, beyond this point the temperature of combustion begins to fall below that necessary for formation of nitrogen peroxide so that with further reduction in mixture strength the amount of nitrogen peroxide progressively decreases.*

18.2.2. Mixture Strength and Combustion Product Characteristics

- The *chemically correct air-fuel ratio by mass for complete combustion is known as* **stoichiometric ratio.**
- Refer Fig. 18.1. It shows how the three main exhaust pollutants products (CO, HC, NO_x) vary from different air-fuel ratio operating on either side of the stoichiometric ratio for a very rich mixture (11 : 1) to a very lean mixture (18 : 1).
  - — The amount of CO produced in the exhaust is about 8% for an 11 : 1 air-fuel, ratio, but this percentage steadily decreases to zero as the mixture is reduced to just beyond the stoichiometric ratio (on the lean side).
  - — HC produced in the exhaust gases amounts to about **1100** parts per million (ppm) with a rich 11 : 1 air-fuel ratio and, as the mixture strength approaches the stoichiometric ratio, it progressively falls to around **500** ppm. A further weakening of the mixture to 18 : 1 air-fuel ratio only reduces HC content to approximately 350 ppm.
  - — *Oxides of nitrogen* products formed during combustion are very low at **100** ppm with a rich air-fuel ratio of 11 : 1. As the mixture strength approaches the stoichiometric ratio it rises fairly rapidly to **2000** ppm, and a further reduction of the mixture strength to 15 : 1 peaks the oxides of nitrogen to something like **2,300** ppm, weakening the mixture beyond this point rapidly reduces it until, at an 18 : 1 air-fuel ratio, it is **1000** ppm.

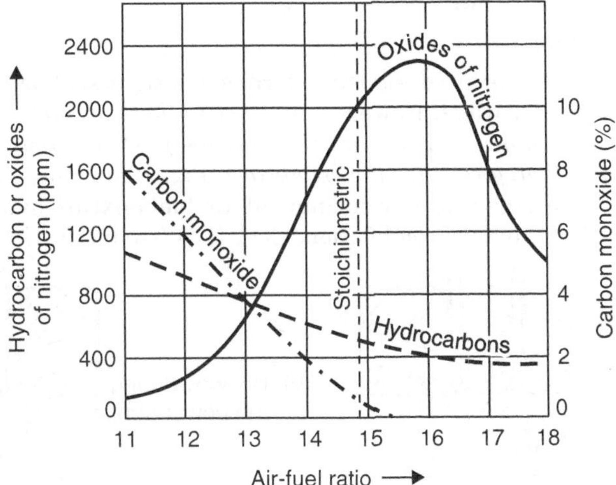

Fig. 18.1. Effects of mixture strength on exhaust composition of a petrol engine.

18.3. SPARK IGNITION (S.I.) ENGINE EMISSIONS

The following are the three main sources from which pollutants are emitted from the S.I. engine :

1. **The crankcase.** *Where piston blow-by fumes and oil mist are vented to the atmosphere.*

2. **The fuel system.** *Where evaporative emissions from the carburettor or petrol injection air intake and fuel tank are vented to the atmosphere.*

3. **The exhaust system.** *Where the products of incomplete combustion are expelled from the tail pipe into the atmosphere.*

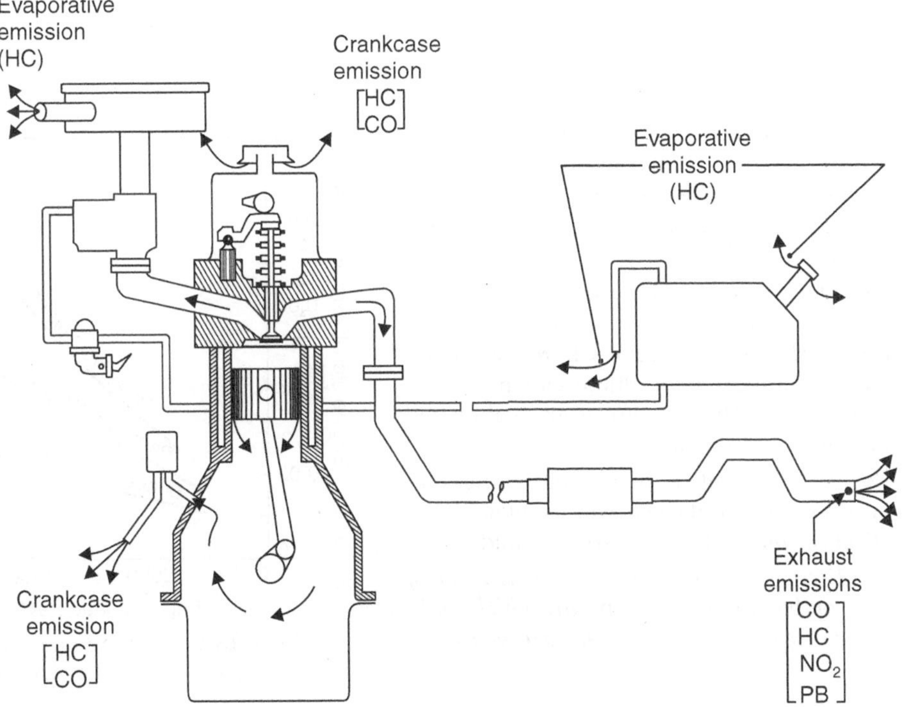

Fig. 18.2. Spark ignition engine emissions.

18.3.1. Crankcase Emission

Piston ring blow-by :

- The piston and its rings are designed to form a gas-tight seal between the sliding piston spirt and the cylinder walls. However, in *practice there will always be some compressed charge and burnt fumes which manage to escape past the compression and oil control piston rings and therefore enter the crankcase* (Fig. 18.3). These gases which find their way past the piston ring belt may be unburnt air-fuel mixture hydrocarbons, or burnt (or partially burnt) products of combustion, CO_2, H_2O (steam) or CO.

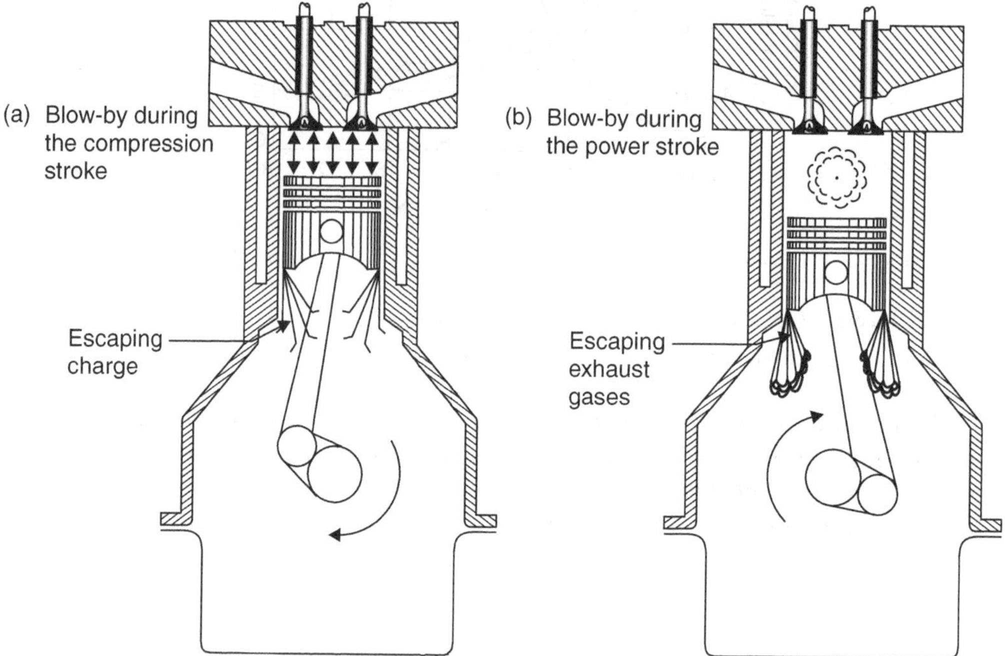

Fig. 18.3. Piston ring blow-by.

- Piston blow-by increases with engine speed and, in particular, as the piston rings and cylinder bore wears (Fig. 18.4), the blow-by becomes more noticeable in the upper speed range.
- Blow-by takes place between the piston ring gap, piston-ring to piston-groove clearance and in the T.D.C. region where the piston ring circumferential shape can not accurately follow the contour of an oval or bell mounter cylinder wall.
- Besides effectively reducing the engine compression ratio and the power developed, the *effects of piston blow-by* are two-fold :

  (*i*) *It can lead to a high concentration of combustible air-fuel mixture which could cause an explosion in the crankcase.*

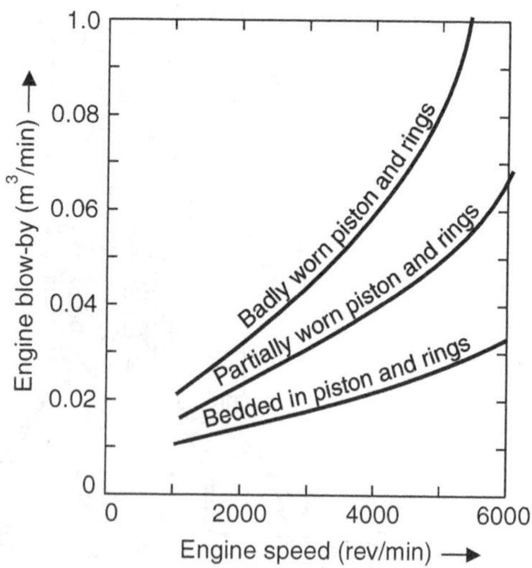

Fig. 18.4. Effect of engine speed on piston and ring blow-by.

(*ii*) The air-fuel mixture, partially burnt and fully burnt vapour fumes, will *condense and contaminate the engine's lubricating oil.*

Since it is impossible to eliminate piston blow-by completely, *an organised connection is deliberately created which circulates the crankcase and rocker or camshaft cover spaces and consequently carries the unwanted fumes out with it.* The removal of blow-by gases and vapour fumes from the crankcase is obtained by *creating a partial depression at the outlet location so* that blow-by gases under pressure (escaping between the piston and cylinder wall) are attracted towards the lower pressure region of the crankcase, at which point they are expelled.

Following are the *two methods of creating the extraction depression :*

(*i*) *The road draught crankcase ventilation system.*

The unacceptable limitation of this system of crankcase ventilation, due to expulsion of gas and fume vapour (HC and CO) into the atmosphere thereby contributing to pollution, has made this method of internal purging of the engine *obsolete.*

(*ii*) *The induction manifold vacuum positive crankcase ventilation system.*

- *The blow-by HC emissions are about 20% of the total HC emission from the engine ; this is increased to about 30% if the rings are worn.*

18.3.2. Evaporative Emission

Evaporative emissions account for 15 to 25% of total hydrocarbon emission from a gasoline engine. The following are two main sources of evaporative emissions :

(*i*) The fuel tank

(*ii*) The carburettor.

(*i*) **Fuel tank losses.** The main factors governing the tank emissions are *fuel volatility* and the *ambient temperature* but the *tank design and location can also influence the emissions as location* affects *the temperature.* Insulation of tank and vapour collection systems have all been explored with a view to reduce the tank emission.

(*ii*) **Carburettor losses.** Carburettor emission may be divided with following two *categories* :

(*i*) Running losses ; (*ii*) Parking losses.

— Although most internally vented carburettors have an external vent which open at idle throttle position, the existing pressure forces prevent outflow of vapours to the atmosphere. Internally vented carburettor may enrich the mixture which in turn increases exhaust emission.

— Carburettor losses are significant only during hot condition when the vehicle is in operation. The fuel volatility also affects the carburettor emissions.

18.3.3. Exhaust Emission

The different constituents which are exhausted from S.I. engine and different factors which will affect percentages of different constituents are discussed below :

1. **Hydrocarbons (HC) :**

The emission amount of HC (due to incomplete combustion) is closely related to :

— *design variables* (such as induction system and combustion chamber design) ;

— *operating variable* (such as A/F ratio, speed, load) ;

— mode of operation (such as idling, running or accelerating).

The following *factors affect HC emission :*

(*i*) *Surface/Volume (S/V) ratio.* Fig. 18.5 shows the effect of S/V ratio on HC emission.

(*ii*) *Wall quenching.*

(*iii*) *Incomplete combustion.* When the mixture supplied is rich or lean ; the flame propagation becomes weak which causes incomplete combustion and results in HC emission. The incomplete flame propagation is caused by the following factors :

(*a*) Law charge temperature ;

(*b*) Too rich or too lean mixture ;

(*c*) Poor condition of the ignition system ;

(*d*) Non-uniform fuel in the mixture supplied to the engine ;

(*e*) Large exhaust residual gases left in the cylinder.

(*iv*) *Spark plug timing.* It is observed that HC emission is reduced by retarding the spark plug timing during low speed (lower than 40 km/h) but has no effect when running at 40 km/h.

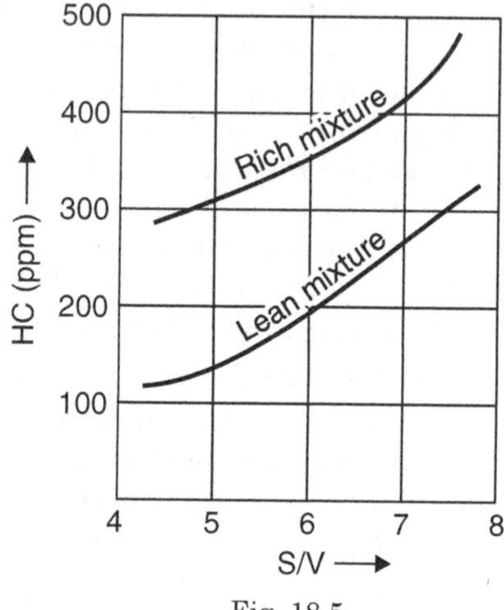

Fig. 18.5

(*v*) *Compression ratio.* It has been observed through experiments that *emission of HC in exhaust is decreased with an increase in compression ratio.*

2. **Carbon monoxide (CO) :**

- If the oxidation of CO to CO_2 is not complete, CO remains in the exhaust.
- It can be said theoretically that, the petrol engine exhaust can be made free from CO by operating it at A/F ratio = 15. However, some CO is always present in the exhaust even at lean mixture and can be as high as 1 per cent.
- The *percentage of CO increases during engine idling but decreases with speed.*
  — Whatever may be condition of running at any load or speed, and A/F ratio, it *is not possible to completely eliminate CO and 0.5 percent is considered a reasonable goal.*
- CO emissions are *lowest during acceleration and at steady* speeds. They are, however, *high during idling and reach maximum during deceleration.*

3. **Oxides of nitrogen (NO_x) :**

- Oxides of nitrogen occur mainly in the form of NO and NO_2 and are generally formed at high temperature.
- The maximum NO_x levels are observed with A/F ratios of about 10 percent above stoichiometric. More air than this reduces peak temperature and therefore NO_x concentration falls, even free O_2 is available.
- The following factors affect the formation of NO_x :

  (*i*) *A/F ratio ;*

  (*ii*) *r.p.m. ;*

  (*iii*) *Angle of advance.* The decreasing angle of advance decreases appreciably the formation of NO_x.

  — It has also been observed that *NO_x increases with increasing manifold pressure, engine load and compression ratio.* This characteristic is different from HC and CO emission which are nearly independent of engine load except for idling and deceleration.

Lead emission

- *Lead emissions come only from S.I. engines.*
- In the fuel, lead is present as lead tetroethyl or tetramethyl, to control the self ignition tendency of fuel-air mixtures that is responsible for knock (to improve the octane rating of the fuel).

- Major portion of the lead that enters the engine is emitted from the exhaust which forms very small particles of oxides and oxyhalides in the atmosphere. A portion of the lead particles falls to the ground very quickly, others are small enough to remain suspended in the atmosphere sometime, before they fall out, usually after coagulation with other dusty material in air.
- It may not be possible to eliminate lead completely from all petrols immediately because a large number of existing engines rely upon the lubrication provided by a lead film to prevent rapid wear of exhaust valve seats. However, a very small lead content would be adequate for the purpose.

Following points are worth noting :

- Both the *flow rate* and *pollutant concentration*, for exhaust emissions, can change with the mode operation. Both must be considered in determining emissions.
  - *Under constant high speed conditions, exhaust* HC *concentrations are low while the flow rates are high. During accelerations the flow rate is low but* HC *concentration is high.*
  - The *concentration* of HC in the crankcase and evaporative losses is virtually *independent* of operating conditions, but the *flow rates* from each of these sources *change* during various operations.

    Thus, on km basis CO and HC emissions decrease with increasing driving speed while NO_x emissions are relatively not affected.
- In *a poorly maintained engine the exhaust pollution is more.*
  - An automatic choke sticking in the closed position or a very dirty air cleaner element can reduce air-fuel ratio, generally increasing HC and CO emissions.
  - A misfire allows an entire air-fuel charge to be exhausted without combustion.

18.4. S.I. ENGINE EMISSION CONTROL

The main methods, among various methods, for S.I. engine emission control are :

1. Modification in the engine design and operating parameters.
2. Treatment of exhaust products of combustion.
3. Modification of the fuels.

18.4.1. Modification in the Engine Design and Operating Parameters

Engine design modification improves upon the emission quality. A few parameters which improve an emission are discussed below :

1. **Combustion chamber configuration :**

Modification of combustion chamber involves avoiding flame quenching zones where combustion might otherwise be incomplete and resulting in high HC emission. This includes :

- Reduced surface to volume (S/V) ratio ;
- Reduced squish area ;
- Reduced space around piston ring ;
- Reduced distance of the top piston ring from the top of the piston.

2. **Lower compression ratio :**

- Lower compression ratio reduces the quenching effect by reducing the quenching area, thus *reducing HC.*
- Lower compression ratio also *reduces NO_x* emissions due to lower maximum temperature.
- Lower compression, however, *reduces thermal efficiency and increases fuel consumption.*

The decrease in compression ratio is becoming very important design parameter, because it will result in decrease in pollutants from the new engine. Besides, the octane *number requirements* will *decrease* and *high leaded gasolines will not be necessary, or the lead compound will be phased out to be replaced by un-leaded gasoline.*

3. **Modified induction system :**

In a multi-cylinder engine it is always difficult to supply designed A/F ratio under all conditions of load and power. This can be achieved by proper design of induction system or using high velocity or multi-choke carburettors.

4. **Ignition timing :**

- The ignition timing control is so adjusted as to provide normal required spark advance during cruising and retard the same for idle running. *NO_x emissions are reduced due to lowering of maximum combustion temperatures. Also HC emission gets reduced due to high exhaust temperatures. However, cooling requirements increase. The fuel economy also suffers to some extent accompanied by some power loss.* Thus a judicial balance needs to be struck between fuel economy, power loss and pollutants.

5. **Reduced value overlap :**

- Increased overlap allows some fresh charge to escape directly and increase emission level. This can be controlled by *reducing value overlap.*
- A new variable valve timing (VVT) allows for controlled scheduling of valve timing events ; improves engine performance. It is also claimed VVT system will work best with petrol injection. This system is also applicable to petrol as well as to diesel engines.

18.4.2. Exhaust Gas Oxidation

The *exhaust gas coming out of exhaust manifold is treated to reduce HC and CO emissions.*
The devices used to accomplish it are discussed below :

1. **After-burner :**

- An *"after-burner"* (Fig. 18.6) is *a burner where air is supplied to the exhaust gases and mixture is burnt with the help of ignition system.* The HC and CO which are formed in the engine combustion because of inadequate O_2 and inadequate time to burn are further brunt by providing air in a separate box, known as *after-burner.* The after-burner is located very near to the exhaust manifold with an intention that there is no fall in the temperature of exhaust. The oxidation of HC in the after-burner depends upon the temperature of exhaust and mixing provided in the after-burner. NO_x emission is *not* affected by the air injection.

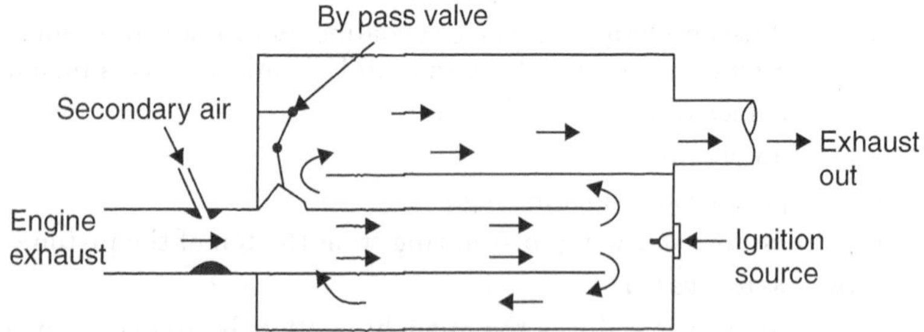

Fig. 18.6. Direct flame after-burner.

- This arrangement/system is not successful in reducing the emission owing to the difficulty in sustaining the combustion during low HC emissions due to high heat losses over a large area.

2. **Exhaust manifold reactor :**

- The exhaust manifold reactor is a further development of after-burner where the design is changed so as to minimise the heat loss and to provide sufficient time for mixing of exhaust and secondary air.
- Fig. 18.7 shows a General Motors, air injection system.

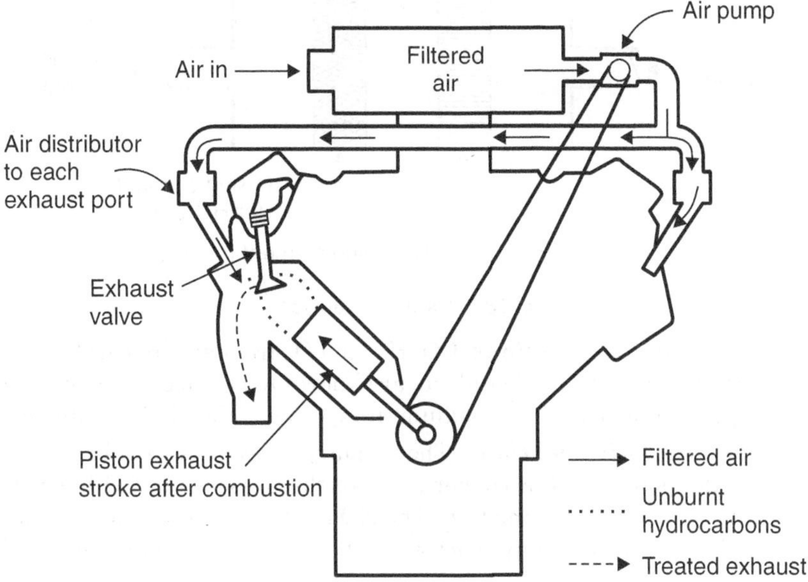

Fig. 18.7. General motors air injection system.

— Here a positive displacement vane pump, driven by the engine, inducts air from the air cleaner or from separate air filter.

— The air passes into an internal or external distributing manifold, with tubes feeding a metered amount into the exhaust port of each cylinder and close to the exhaust valve.

— Since the exhaust gases are at high temperature, the injected air reacts with HC, CO and aldehydes to reduce greatly the concentration of such emissions.

The injected air is closely metered otherwise it can decrease the temperature of the exhaust gas.

- In earlier type of reactor developed by Du Pont the entry of exhaust gases was radial and the air flow peripheral.

3. **Catalytic converter :**

- A *catalytic converter is a device which is placed in the vehicle exhaust system to reduce HC and CO by oxidising catalyst and NO by reducing catalyst.*
- The *basic requirements of a catalytic converter are :*

  (*i*) High surface area of the catalyst for better reactions.

  (*ii*) Good chemical stability to prevent any deterioration in performance.

  (*iii*) Low volume heat capacity to reach the operating temperatures.

(*iv*) Physical durability with attrition resistance.

(*v*) Minimum pressure drop during the flow of exhaust gases through the catalyst bed ; this will not increase back pressure of the engine.

Fig. 18.8 shows a catalytic converter, developed by the Ford Company. It consists of two separate elements, one for NO_x and the other for HC/CO emissions. The secondary air is injected ahead of the first element. The flow in the converter is axial.

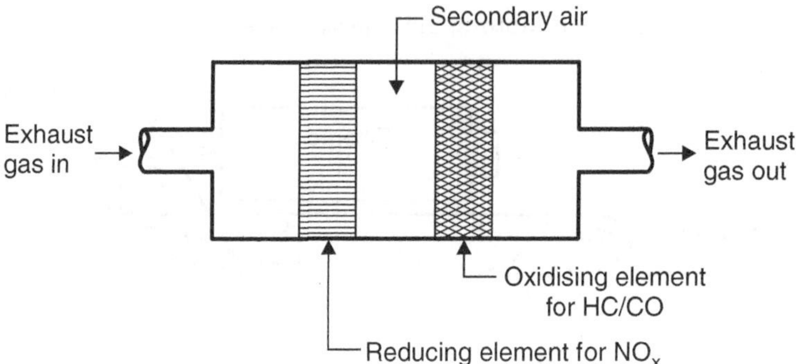

Fig. 18.8. Catalytic converter.

— ***Oxidation catalytic reactions.*** CO, HC and O_2 from air are catalytically converted to CO_2 and H_2O and number of catalysts are known to be effective noble metals like platinum and plutonium, copper, vanadium, iron, cobalt, nickel, chromium etc.

— ***Reduction catalytic reactions.*** The primary concept is to offer the NO molecule an activation site, say nickel or copper grids in the presence of CO but not O_2 which will cause oxidation, to from N_2 and CO_2. The NO may react with a metal molecule to form an oxide which then in turn, may react with CO to restore the metal molecule.

Rhodium is best catalyst to control NO_x *but A/F ratio must be within a narrow range of 14.6 : 1 to 14.7 : 1.*

Major drawbacks of catalytic converter are as under :

(*i*) Owing to the exothermic reactions in the catalyst bed the exhaust systems are *hotter than normal.*

(*ii*) Cars equipped with such converter *should not use leaded fuel as lead destroys complete catalytic activity.*

(*iii*) If the fuel contains sulphur (as diesel oil) emission of SO_3 is increased.

Three-way, Two-way and noble metal catalytic converters :

1. **Three-way catalytic converter :**

If an engine is operated at all times with an air-fuel ratio close to stoichiometric, then both NO reduction and CO and HC oxidation can be done in a single catalyst bed. The catalyst effectively brings the exhaust gas composition to a near equilibrium state at these exhaust conditions, *i.e.*, a composition of CO_2, H_2O and N_2. Enough reducing gas will be present to reduce NO, and enough O_2 to oxidize the CO and hydrocarbons (HC). Such a converter is called **three-way catalytic converter,** *since it removes all the three pollutants.* There is a narrow band of air-fuel ratios near stoichiometric in which high conversion efficiencies for all three pollutants are available. *Commercial three-way catalysts contain platinum, rhodium with some* A_2O_3, *NiO and* CeO_2. Alumina is the preferred support material.

2. Two-way catalytic converter :

In two-way catalytic converter, the exhaust line has two catalytic converters in it. A reduction catalyst is required to reduce NO_2. An oxidation catalyst is placed downstream of the reduction catalyst to convert the excess HC and CO. The advantage of two-way converter is that it *allows a partial decoupling of emission control from engine operations.* Therefore the conversion efficiencies for HC and CO are very high at normal exhaust temperatures.

3. Noble metal catalytic converter :

They use noble metals as catalyst materials. Platinum or Platinum and Palladium are applied to a ceramic support which has been treated with an aluminium oxide wash coat. This results in an extremely porous structure providing a large surface area to stimulate the combination of O_2 with HC and CO. This oxidation process converts most of these compounds to water vapour and CO_2.

Failures of catalytic converter :

The failures of catalytic converter may be due to the following *factors/reasons* :

1. Converter melt down
2. Carbon deposit
3. Catalyst fracture
4. Poisoning.

The reason for catalytic converter being so popular :

The most common range of exhaust gas temperatures for S.I. engines is 400–600°C, and the exhaust gas may contain modest amount of O_2 (when lean) and more substantial amount of CO (when rich). In contrast diesel engines always operate lean. The exhaust therefore contains substantial amount of oxygen and is at low temperatures (200–500°C) removal of gaseous pollutants from the exhaust gases after they leave the engine may be either thermal or catalytic.

In order to oxidize the hydrocarbons in the gas phase without catalyst (thermal), a resident time of the order of 50 minutes, and a temperature in excess of 700°C are required. To oxidize CO temperatures in excess of 700°C are required. Temperatures high enough for this purpose can be obtained by spark retard (with loss in efficiency) and insulation of exhaust parts and manifold. The residence time can be increased by increasing the exhaust manifold volume to form a thermal reactor. However this approach has limited application.

Catalytic oxidation of CO and HC in the exhaust can be achieved at temperatures as low as 250°C. Thus effective removal of the pollutants occur over a much wider range of exhaust temperatures than can be achieved with thermal oxidation.

The only satisfactory method known for the removal of NO from exhaust gases involve catalytic processes. Removal of NO by catalytic oxidation to NO_2 requires *temperatures less than 400°C* (from equilibrium conditions), and subsequent removal of NO_2 produced. Catalytic reaction of NO with added NH_3 is not practicable because of the transient variation in NO produced in the engine.

Reduction of NO by CO, HC or H_2 in the exhaust to produce N_2 is the preferred catalytic process. *It is only feasible in S.I. engine exhaust.*

Therefore, use of catalytic convertor for CO, HC and NO_2 removal in IC engines has become wide-spread.

18.4.3. Exhaust Emission Control by Fuel Variation

- The ability of a fuel to burn in mixtures leaner than stoichiometric ratio is a rough indication of its potential emission reducing characteristics and reduced fuel consumption.
- If gasoline is changed to propane as engine fuel CO emission can substantially be reduced with reduced HC and NO_x ; and in changing from propane to methane the CO as well HC emission touch zero level and only the NO_x remains as a significant factor.
- From pollution point of view, both methane and steam reformed hexane are very attractive fuels but we are unable to use at present for want of technological progress.

18.4.4. Blow-by Control

- *The basic principle of crankcase blow-by control system is recirculation of vapours back to the inlet manifold.*
- Fig. 18.9 shows a positive crankcase ventilation (PCV) system.
  — The filtered air is drawn from the air cleaner and passes on to the crankcase. The *air and the blow-by gases passes through flow calibrated PCV valve before being drawn into the intake manifold in order to restrict the air flow while idling when the vacuum is high at the manifold and the blowby is less.* PCV valve is spring loaded. At wide open throttle, the air flow gets unrestricted but flow rate is metered by the valve opening.

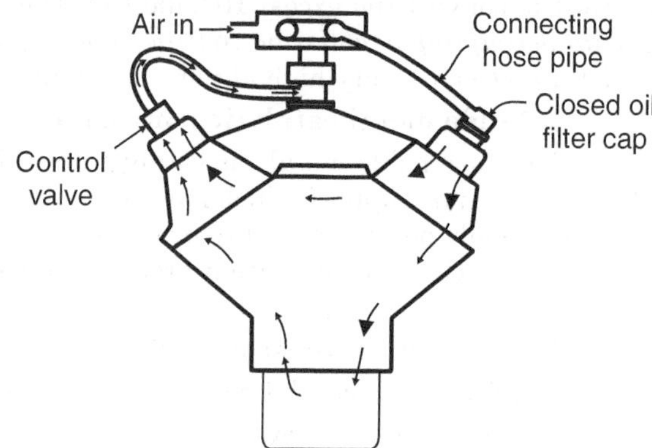

Fig. 18.9. PCV system.

  — In view of the added blowby flow, the carburettor has to be calibrated. Since blow-by gas and air by-pass the carburettor in entering the manifold, the failure of PCV will give improper fuel ratios. Due to improper seating of the PCV valve at high vacuum, large quantity of air and also lubricating oil mist will enter the manifold. This will lead to lean mixtures. Also, the valve getting stuck to due to contaminants, will lead to rich mixtures.
  — In case of valve failure, in this system, blow-by is unable to go to atmosphere.

18.4.5. Evaporation Emission Control Device (EECD)

In one of general motors system the filter pipe is sealed with a pressure cap. It comprises a built in vacuum relief to allow air to enter as the fuel gets consumed. An inverted bowl is placed inside the tank. It traps air as tank is filled to allow for liquid fuel volume expansion. The air gets released from the bowl by two small orifices in the top. The tank is ventilated to a cannister which has activated carbon particles. This can hold 0.2 kg of fuel vapour upon shut down. But when the engine is running, filtered air is drawn through the bottom of cannister, removing the absorbed vapour. This mixture is sent to air cleaner or to the intake manifold, in proportion to air flow rate.

18.4.6. Control of Oxides of Nitrogen (NO_x)

The concentration of oxides of nitrogen in the exhaust is closely related to the peak cycle temperature. The following are the three methods (investigated so far) for reducing peak cycle temperature and thereby reducing NO_x emission.

1. Exhaust gas recirculation (EGR)
2. Catalyst
3. Water injection.

1. **Exhaust gas recirculation (EGR) :**

- This method is *commonly used to reduce NO_x* in petrol as well as diesel engines. In S.I engines, about 10 percent recirculation reduces NO_x emission by 50 percent. Unfortunately, the consequently poorer combustion directly increases HC emission and calls for mixture enrichment to restore combustion regularity which gives a further indirect increase of both HC and CO.

- Fig. 18.10 shows the arrangement of exhaust gas recirculation (EGR) system. A portion (about 10 to 15%) of the exhaust gases is recirculated to cylinder intake charge, and this reduces the quantity of O_2 available for combustion. The exhaust gas for recirculation is taken (as shown in Fig. 18.10) through an orifice and passed through control valve for regulation of the quantity of recirculation.

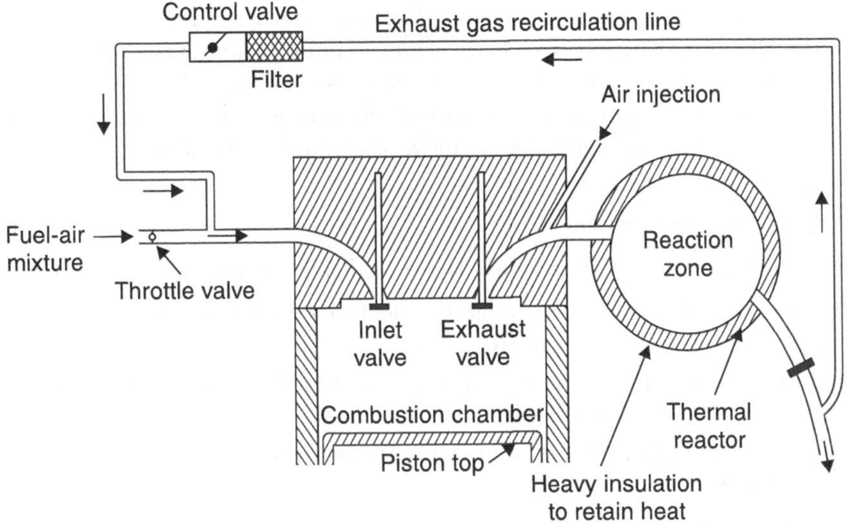

Fig. 18.10. EGR system.

- The effect of A/F ratio of NO_x emission taking EGR parameter is shown in Fig. 18.11.
  - — It may be observed from the figure that, maximum emission of NO_x takes place during lean mixture limits when gas recirculation is least effective. Whereas, for emission of hydrocarbon (HC) and carbon monoxide (CO) lean mixture is preferred, 15 percent recycling reduces NO_x by 80 percent but increases HC and CO by 50 to 80%. These are two conflicting requirements of this emission control system and this problem has been solved by adopting package system which have both NO_x and HC/CO control devices.

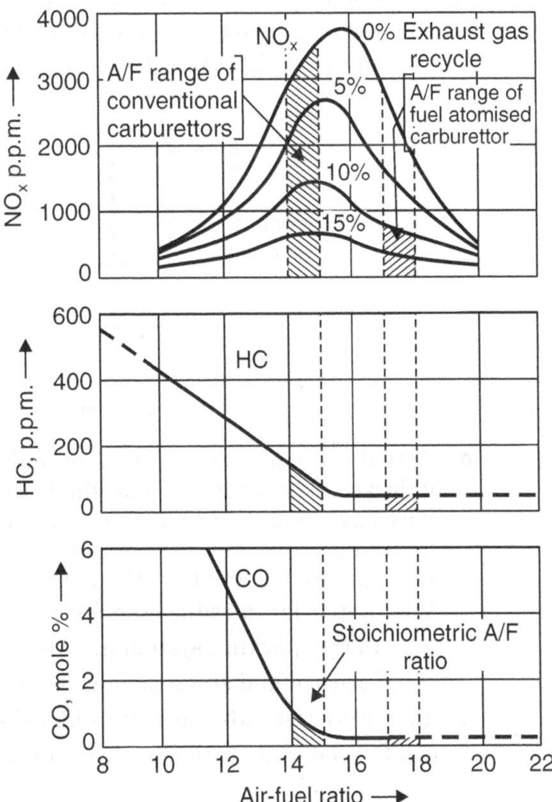

Fig. 18.11. Effect of recycling of gas on NO_x concentration.

2. Catalyst. A few types of catalysts have been tested to reduce the emission of NO_x; a copper catalyst has been used in the presence of CO for this purpose. The research is going on to develop a good catalyst.

3. **Water injection.** It has been observed that the specific fuel consumption decreases a few percent at medium water injection rate. Attempts have been made to use water as a device for controlling the NO_x. This method, because of its complexity, is rarely used.

18.4.7. Total Emission Control Packages

Earlier (previous articles) we have seen that any method which is used to decrease NO_x tries to increase HC and CO and vice-versa. Thus it is of paramount importance to develop a method/system which should reduce emissions of NO_x, HC, CO to a desired level simultaneously. After a long and detailed experimental study of various possible systems, the following two systems/packages have been developed to achieve the required results :

1. Thermal reactor package
2. Catalytic converter package.

Using this approach, the following are the *three basic methods of emission control :*

(*i*) Thermal reactors, which rely on homogeneous oxidation to control CO and HC ;

(*ii*) Oxidation catalyst for CO and HC ;

(*iii*) Dual catalyst system (here a reduction catalyst for NO_x and an oxidation catalyst for CO and HC are connected in series).

— Where control of NO_x is required with the first two methods, EGR is added to the system.

1. **Thermal reactor package :**

- A *thermal reactor* is a chamber which is designed to provide adequate residence time for allowing appreciable oxidation of CO and HC to take place. For enhancing the conversion of CO to CO_2 the exhaust temperature is increased by retarding the spark.

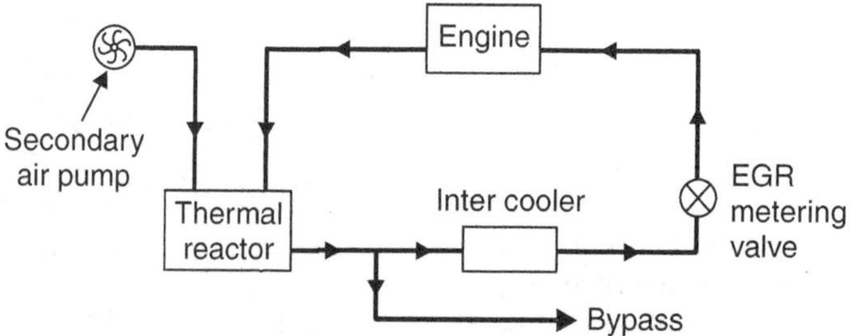

Fig. 18.12. Thermal reactor package (Ford).

- Actual thermal reactor (made of high nickel steel) that is used on a car consists of two enlarged exhaust manifolds which allow greater residence time for burning HC and CO with oxygen in the pumped in air. For keeping a flame constantly burning (and there by assuming complete combustion) a secondary air pump injects fresh air into the reactor ; this reduces HC and CO. About 10 to 75 percent of the gas is recirculated after cooling in the intercooler to reduce the formation of NO_x.

— In this packing system are also included the following :

(*i*) Enriched and stage carburettor temperature controls ;

(*ii*) Crankcase valve to control blow-by gases ;

(*iii*) Special evaporation control valves.

— In this package emission of NO_x, HC and CO are reduced to a required level but at the cost of 20 per cent less power and 10 per cent more fuel consumption.

— This converter can be employed for a run of 1.5 lacs km.

2. Catalytic converter package :

The *working principle of this package is to control the emission levels of various pollutants by changing the chemical characteristics of the exhaust gases.* The catalytic converter package as compared to thermal reactor package requires non-leaded fuel as lead reduces the catalytic action.

The *major advantage* of this converter (as compared to thermal reactor) is that it allows a partial decoupling of emission control from engine operation in that the conversion efficiencies for HC and CO are very high at normal exhaust temperatures.

- Fig. 18.13 shows the arrangement of *catalytic converter package.*

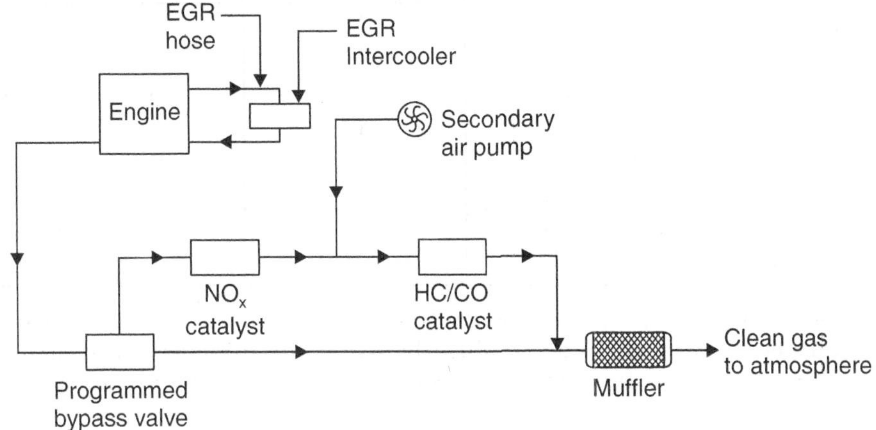

Fig. 18.13. Catalytic converter package.

— Converters for HC and CO and NO_x are arranged as shown in the figure. The NO_x catalyst is the first element in the gas flow path, *does not cause release of any heat.* The next is HC/CO catalyst, which releases heat to such a great extent that may cause overheating and burning of the element. This is taken care of by injecting air through secondary air pump.

— A bypass valve ahead of converter is used to increase the converter life (to about 0.8 lcas km).

— For better control of NO_x, exhaust gas is circulated via an intercooler back to air cleaner.

- For this system, the power loss is about 30% and the fuel consumption is about 10% more than normal.

18.5. DIESEL ENGINE EMISSIONS

- Emissions from diesel engines can be classified in the same categories as those for the gasoline engines but the level of emission in these categories vary considerably. Typical level, of the constituents of the exhaust products of combustion in 4-stroke cycle and 2-stroke cycle are given in Table 18.1 at idling, accelerating, partial load and full load.

Table 18.1

Engine exhaust constituents	Concentration as measured in exhaust products			
	Idling	Accelerating	Partial load operation	Fuel load operation
Two-stroke cycle engine				
1. CO%	0.01	0.25	0.01	0.35
2. CO_2%	0.85	5.5	3.8	5.30
3. HC ppm	250	500	350	550
4. NO_x ppm	200	1200	1100	1250
5. RCH ppm	17.0	9.5	1.0	5.5
6. Smoke Hartrige Unit	4	44	4	10
Four-stroke cycle engine				
(Un-supercharged)				
1. CO%	0.02	0.08	0.04	0.25
2. CO_2%	2.5	3.5	5.5	6.7
3. HC ppm	180	330	210	150
4. NO_x ppm	330	920	590	780
5. RCHO ppm	8.0	7.5	5.0	1.5
6. Smoke, Hartrige Unit	4	44	4	10

- As shown in Fig. 18.14, it is worth noting that NO_x concentration in unsupercharged 4-stroke-cycle diesel engines vary linearly with power (output). The design and operating parameters to control NO_2 are the same as for S.I. engines already discussed.

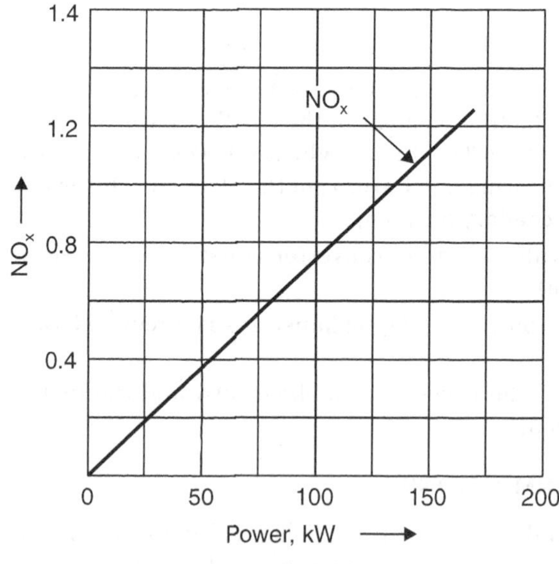

Fig. 18.14

18.6. DIESEL SMOKE AND CONTROL

18.6.1. Exhaust Smoke

In C.I. engines, any volume of the combustion chamber in which the fuel is burnt at a relatively F/A ratio greater than 1.5 ($F_R > 1.5$) at pressures developed in these engines produces **soot.** The quantity of soot formed depends on the following *factors* :

(*i*) The local F/A ratios ;

(*ii*) The type of fuel ;

(*iii*) The pressure.

If this soot is able to find *adequate air* (O_2) which on the whole is much in excess of the requirements of perfect combustion, it will *burn completely*. If it is unable to find air (O_2) in the combustion cycle, it will *pass* as *exhaust*. And if the *quantity is sufficient, it will be visible and that is called* **smoke.** The *colour of the smoke depends on the size of soot particles.*

— Formation of smoke is basically a process of conversion of molecules of hydrocarbon fuels into particles of soot. It should be noted that soot is not *carbon* but simply an aggomeration of very large polybenzenoid free radicals. It is also observed that soot formation during the early part of the actual combustion process is common to all diesel engines but it is consumed during later part of combustion.

— Pyrolysis of fuel molecules themselves is thought to be responsible for soot formation. Fuel heated with insufficient O_2 will give carbonaceous deposits. It is believed that the "heavy ends" of diesel fuel may pyrolyze to yield the type of smoke that is observed from the diesel engine. This is believed to be the path of formation of polycyclic aromatic hydrocarbons (benzo-pyrene) found in soot.

— Many theories have been put forward for the formation of smoke but the basic reactions leading to the formation of smoke are not fully known.

- The smoke of a diesel engine is, in general, of two basic types :

  (*i*) ***Blue-white smoke.*** *It is caused by liquid droplets of lubricating oil or fuel oil while starting from cold.*

  Owing to low lower surrounding temperatures the combustion products are at a relatively low temperature and intermediate products of combustion *do not burn.* This results in *bluish white smoke* when exhausted. This type of smoke is also formed when lubricating oil flows past piston rings.

  (*ii*) ***Black smoke :***

- It consists of carbon particles suspended in the exhaust gas and depends largely upon A/F ratio.
- It increases rapidly with the increase in load and available air is depleted.

18.6.2. Causes of Smoke

It is known that the *cause of smoke is incomplete burning of fuel inside the combustion chamber.* The two major reasons for incomplete combustion are :

(*i*) Incorrect A/F ratio ;

(*ii*) Improper mixing.

These might result due to the design factors discussed below :

1. **Injection system :**

The following injection characteristics substantially *increase the smoke levels.*

- Unsuitable droplet size.
- Inadequate or excess penetration.

- Excessive duration of injection.
- Secondary injection.
- Improper dispersion atomisation.

2. **Rating :**

- It has been observed that smoke limited power is reached much before the thermal load limited power ; and the inevitability of smoke in exhaust is inherent in less loads. Thus it seems, smoke is unavoidable in CI engine, only its level can be kept as low as possible.

3. **Load :**

- Smoke may be defined as visible products of combustion. Rich mixtures (higher loads) results in higher smoke because of *non-availability of oxygen for combustion.*
- The smoke level rises from no-load to full-load (Refer Fig. 18.15). During the first part (no-load, say, about half load), the smoke level is more or less constant, as there is always excess air present. However in higher load range there is an *abrupt* rise in smoke level due to less available oxygen.

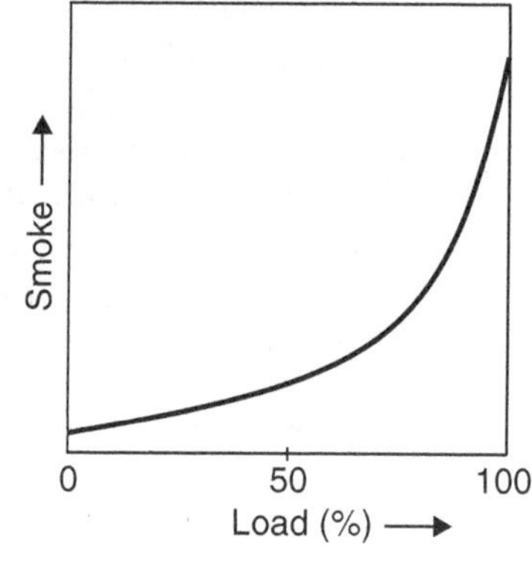

Fig. 18.15

4. **Fuel :**

- The white smoke produced in an engine depends upon the quality of fuel.
- Generally, more volatile fuels give less smoke than heavier fuels of similar cetane number.
- The cetane number exercises no effect on production of black smoke.

5. **Fuel-air ratio :**

- The smoke increases with the increasing fuel-air ratio. This increase in smoke occurs with as much as 25% excess air in the cylinder, clearly indicates that, even in the presence of excess oxygen, diesel engine has a mixing problem.

6. **Engine type and speed :**

- The smoke levels at higher loads are higher in naturally aspirated engines than turbocharged engines because the latter have adequate oxygen even at full-load.
- At low as well as high speed the smoke is worse.

7. **Maintenance :**

- The smoke levels greatly depend upon the condition of the engine. Good maintenance is a must for lower smoke levels (the maintenance affects the injection characteristics and the quantity of lubricating oil which passes across the piston rings and thus exercises a significant effect on engine tendency to generate smoke).

18.6.3. Measurement of Smoke

Two basic types of smoke meters for measuring smoke density are :

(*i*) *Filter darkening type.* (Examples : Bosch smoke meter ; Vem Brand smoke meter).

(*ii*) *Light extinction type.* (Examples : Hartridge smoke meter ; UTAC smoke meter).

The *'light extinction meter'* can be used for *continuous measurements* while the *'filter type'* can be *used only under steady state conditions.*

1. **Bosch smoke meter :**

- It is a filter darkening type of smoke meter.

- In this meter, a measured volume of exhaust gas is drawn through a filter paper which is blackened to various degrees depending upon the carbon present in the exhaust gas. The *density of "soot"* is *measured by timing out amount of light reflected from the sooted paper.*
- The specifications of sample volume, filter paper size etc. are well defined.

2. **Hatridge smoke meter :**

- It works on the principle of high extinction.
- In this meter exhaust sample is passed through tubes of about 0.46 m length which has light source at one end and photocell or solar cell at the other end. The amount of light passed through this smoke column is used as indication of level of smoke.
- This type of meter is useful for continuous testing and can be employed in vehicles.

18.6.4. Control of Smoke

There is hardly any successful method to control the soot except the *engine has to run at lower load i.e.* ***derating*** *and* ***maintain*** *the engine at best possible condition.*

The other methods which may be used for the control of smoke are :

1. **Smoke suppress out additives :**

- It has been observed that some barium compounds when added in fuel reduce the temperature of combustion and avoid the soot formation. It is further observed that if the soot is formed, the barium compounds break them in very fine particles and reduce the smoke.
- The use *of barium sals,* however, *enhances the deposit formation tendencies of engine and reduces the life of the fuel filter.*

2. **Fumigation :**

- It is a method of *introducing small amount of fuel with the intake manifold.* This *initiates pre-combustion reactions* before and during the compression stroke resulting in reduced chemical delay (because the intermediate products such as peroxides and aldehydes react more rapidly with oxygen than original hydrocarbons. The shortening *in chemical delay period curbs thermal cracking which is responsible for soot formation.*
- Fumigation rate of about 11 to 15% gives the best smoke improvement.

3. **Catalytic mufflers :**

- The use of catalytic mufflers, unlike petrol engine, are *not very effective.*
- Much development is needed in such-devices before they can be put to use.

18.6.5. Diesel Odour and Control

It has been observed through some experiments that *the products of partial oxidation are the main cause of odour in diesel exhaust.* This partial oxidation may be because of either very lean mixtures such as during idling or due to quenching effect.

The *following factors affect odour production :*

1. Fuel-air ratio
2. Engine operation mode
3. Engine type
4. Fuel consumption
5. Odour suppressant additives.

Control of Odour :

- Several manufacturers chaim that odour additive compounds can reduce the intensity of odour, but it has been found in practice that these additives hardly have any effect on odour formation etc.

- The control of odours by using catalyst is under development. It has been found experimentally that a few oxidation catalysts reduce intensity of odour.

18.7. COMPARISON OF GASOLINE AND DIESEL EMISSIONS

Fig. 18.16 shows the comparison of emissions and odour from gasoline and diesel engines.

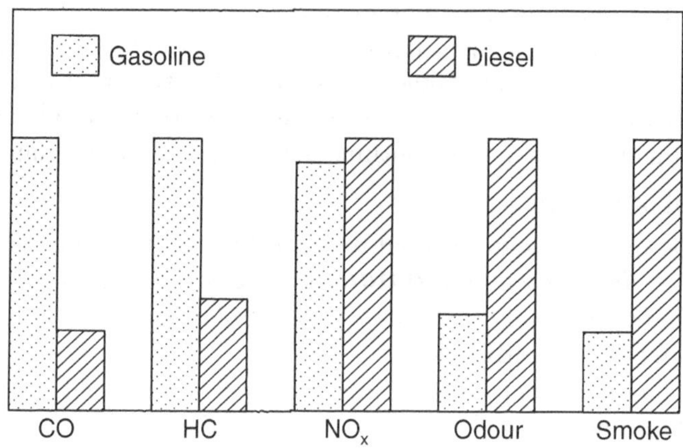

Fig. 18.16. Comparison of emissions and odour from gasoline and diesel engines.

- It is evident from the figure that there is a marked difference between the products of combustion of gasoline and diesel engines.
- Whereas gasoline petrol engines have a somewhat similar emission pattern, all diesel engines have different emission characteristics.

The combustion in a diesel engine occurs over a wide range of fuel-air ratios ranging from very lean mixture to very rich mixture while the gasoline combustion is of relatively homogeneous nature.

In order to account for various differences in the design and operation principles of gasoline and diesel engine the following *correction factor* is applied for comparing these two types of engines :

Correction factor = 15/(%CO_2 + %CO + %C)

- If the diesel engines are maintained properly, they have very little CO in their exhaust and a small quantity of smoke. On the other hand gasoline engines exhaust significant amount of CO and unburnt hydrocarbon (UBHC). Thus diesel engine is cleaner as compared to petrol engine.

18.8. ZERO EMISSION

- For *"zero emission"* the Electronic-Catalytic Converter (ECC) assumes significance. In the recent development India-based Hydrodrive Systems and Controls has launched common Electronic-Catalytic Converter for automobile fuels *viz* ; the diesel, leaded and unleaded gasoline.
- *ECC excites the fuel molecules with microwave electronics and does molecular engineering of fuel for complete combustion.*
- ECC can be installed as a pre-engine device.
- ECC is the *first catalytic converter to offer fuel savings due to improved combustion.*

The comparison between ECC and conventional CC is given in the chart below :

S. No.	Electronic Catalytic Converter	Conventional Exhaust Catalytic Converter
1.	Used as a pre-engine fuel converter.	Used as an exhaust converter.
2.	Cleans the process of combustion.	Cleans the exhaust gases.
3.	Improves fuel properties for better combustion.	Does not improve fuel properties.
4.	Works with all fuels-diesel, leaded, unleaded gasoline, CNG and LPG.	Does not work with leaded petrol.
5.	Decarbonizes the engine combustion chamber with use and keeps it clean.	Does not decarbonize the combustion chamber.
6.	Reduces engine noise and improves engine smoothness.	Does not reduce engine noise or engine smoothness.
7.	Enhances fuel economy.	Does not enhance fuel economy.
8.	Works on microwave electronics, plasma chemistry, particle accelerator principle.	Works on chemical oxidation and reduction principle.
9.	No Nobel metal or chemical is used. Fuel flows through open-ended pipe, also acting as wave-guide and particle accelerator.	Noble metal substrates or chemicals are used.
10.	Engine pick up is improved.	No change in Engine pick up.
11.	Same size for all engines, all fuels and cubic capacity.	Different sizes for different engines cubic capacity and different fuels.
12.	No warming up time required for cold start emission control.	Warm up time required for cold start emission control unless provided with electrical heating for the catalyst.

18.9. AIR POLLUTION FROM GAS TURBINES AND ITS CONTROL

- The pollution from gas turbines is *low* as compared to pollution from convention piston engines, due to the fact that *former operate at much leaner mixture* used in the latter, and still further, the *exhaust is diluted due to the presence of excess air.*
- As compared to S.I. engines, a gas turbine exhaust about 1 to 20% HC, 1 to 10% CO and 30 to 80% NO_x. Besides this exhaust also contains oxides of sulphur, aerosol , smoke particles and odourants.

Control of pollutants emanated from gas turbines :

- **Reduction of CO concentration** can be accomplished by using following methods :

(*i*) By improving atomisation of fuel.

(*ii*) By redistributing the air flow to bring the primary zone equivalence ratio to an optimum value of around 0.85.

(*iii*) By increasing the residence time.

(*iv*) By reducing the film cooling air.

(*v*) By fuel staging (In this technique supply of fuel is cut-off to some nozzles and diverted to the remainder ; it reduces emissions at low power conditions by improving the quality of automisation).

- **Reduction of UBHC (unburnt hydrocarbon)** requires the same treatment as employed for reducing CO.
- **Reduction of NO_x.** The NO_x emission increases exponentially as : $NO_x \, \alpha (e)^{0.09T}$, where T is reaction temperature in K.

Thus, for reducing NO_x concentration, it will be required to lower the reaction temperature T. The reduction in both flame temperature and residence time are readily accomplished by increasing the air flow in the primary zone but increases the production of HC and CO. Thus any attempt made to reduce HC and CO increases NO_x and vice-versa.

— The NO_x level as well as the smoke levels can be reduced by *providing lean combustion zone.*

18.10. EFFECTS OF ENGINE EMISSIONS ON HUMAN HEALTH

The *effects of different engine emissions* on *human health are discussed below :*

1. **Sulphur dioxide (SO_2) :**

- It is an irritant gas and affects the mucous membrane when inhaled. In the presence of water vapour it forms sulphurous and sulphuric acids. These acids cause severe bronchospasma at very low levels of concentration.
- Diseases like bronchitis and asthama are aggrevated by a high concentration of SO_2.

2. **Carbon-monoxide (CO) :**

- It has a strong affinity (200 times) for combining with the haemoglobin of the blood to form carboxyhaemoglobin. This reduces the ability of the haemoglobin to carry oxygen to the blood tissues.
- *CO affects the central nervous system.*
- It is also responsible for heart attacks and a high mortality rate.

3. **Oxides of nitrogen (NO_x) :**

- These are known to cause occupational diseases. It is estimated that eye and nasal irritation will be observed after exposure to about 15 p.p.m. of nitrogen oxide, and pulmonary discomfort after brief exposure to 25 p.p.m. of nitrogen oxide.
- It also aggrevates diseases like bronchitis and asthama.

4. **Hydrocarbon vapours :**

- They are primarily irritating.
- They are major contributors to eye and respiratory irritation caused by photochemical smog.

5. **Compounds of Incomplete combustion :**

Exhaust discharge from IC engines carry compounds of incomplete combustion (polycyclic organic compounds and aliphatic hydrocarbons), which act as carcinogenic agents and are responsible for lungs cancer.

6. **Lead :**

- Inorganic lead compounds (discharged from vehicles using leaded petrol) cause a variety of human health disorders.
- The effects include gastrointestinal damage, liver and kidney damage, abnormality in fertility and pregnancy etc.

7. **Smoke :**

It is visible carbon particles.

- It causes irritation in eyes and lungs, and visibility reduction. It also causes other respiratory diseases.

Generally speaking, susceptibility to the effects of exhaust emissions is greatest amongst infants and the elderly. Those with chronic diseases of lungs or heart are thought to be at great risk.

HIGHLIGHTS

1. *Air pollution* can be defined as an addition to our atmosphere of any material which will have a deleterious effect on life upon our planet.
2. The major pollutants emitted from the exhaust due to incomplete combustion are :
   CO, HC, NO_x and other products *e.g.* acetylene, aldehydes etc.
3. CO is generally formed when mixture is rich in fuel.
4. NO formation will be less in rich mixtures than in lean mixtures.
5. The chemical correct air-fuel ratio by mass for complete combustion is known as *stoichiometric ratio.*
6. The blowby HC emissions are about 20% of the total HC emission from the engine ; this is increased to about 30% if the rings are worn.
7. The following factors affect the formation of NO_x :
   (*i*) A/F ratio ; (*ii*) r.p.m. ; (*iii*) Angle of advance.
8. In a poorely maintained engine the exhaust pollution is more.
9. A *catalytic converter* is a device which is placed in the vehicle exhaust system to reduce HC and CO by oxidising catalyst and NO by reducing catalyst.
10. Rhodium is the best catalyst to control NO_x but A/F ratio must be within a narrow range of 14.6 : 1 to 14.7 : 1.
11. The basic principle of crankcase blowby control system is recirculation of vapours back to the intake manifold.
12. Oxides of nitrogen can be controlled by :
    (*i*) Exhaust gas recirculation ; (*ii*) Catalyst ; (*iii*) Water injection.
13. The working principle of *catalytic converter package* is to control the emission levels of various pollutants by changing the chemical characteristics of the exhaust gases.

OBJECTIVE TYPE QUESTIONS

Fill in the Blanks or Say "Yes" or "NO" :

1. ...... are produced by incomplete burning of the air-fuel mixture in the combustion chamber.
2. The major pollutants emitted from the exhaust due to incomplete combustion are : ......, ..., ...
3. If the combustion is complete, the only products being expelled from the exhaust would be water vapour and carbondioxide.
4. CO is a colourless and ...... gas.
5. CO has about the same density as air.
6. CO is generally formed when the mixture is ...... in fuel.
7. Solid particles are usually formed by dehydrogenation, polymerisation and agglomeration.
8. At heavy loads, due to lack of oxygen, a decrease in aldehyde emission is observed.
9. The NO formation is more in rich mixture than in lean mixtures.
10. The chemically correct air-fuel ratio by mass for complete combustion is known as ...... ratio.
11. Oxides of nitrogen products formed during combustion are very low at 100 p.p.m. with a rich air-fuel ratio of 11 : 1.
12. The three main sources from which pollutants are emitted from the S.I. engines are : The crankcase, the fuel system and the ...... system.
13. Piston blow-by ...... with engine speed.
14. Evaporative emissions account for 15 to 25% of total hydrocarbon emission from a gasoline engine.
15. The two main sources of evaporative emissions are : The tank and the ....
16. The emission of HC in exhaust is decreased with an increase in compression ratio.
17. The percentage of CO_2 decreases during engine idling but increases with speed.

18. CO emissions are lowest during acceleration and at steady speeds.
19. Oxides of nitrogen occur mainly in the form of NO and NO_2 and are generally formed at higher temperature.
20. The formation of NO_x is affected by the factor : A/F ratio, R.p.m. ; ....
21. NO_x increases with increasing manifold pressure, engine load and compression ratio.
22. Lead emissions come only from ...... engines.
23. Under high speed conditions, exhaust HC concentrations are low while the flow rates are high.
24. In a poorly maintained engine the exhaust pollution is more.
25. Lower compression increases thermal efficiency and reduces fuel consumption.
26. NO_x emissions are reduced due to lowering of maximum combustion temperatures.
27. An after-burner is a burner where air is supplied to the exhaust gases and mixture is burnt with the help of ignition system.
28. ...... is the best catalyst to control NO_x but A / F ratio must be within a narrow range of 14.6 : 1 to 14.7 : 1.
29. The ability of a fuel to burn in mixtures leaner than stoichiometric ratio is a rough indication of its potential emission reducing characteristics and reduced fuel consumption.
30. The basic principle of crankcase blowby control system is recirculation of vapours back to the intake manifold.
31. NO_x emission can be reduced by : Exhaust gas recirculation ; catalyst ; ....
32. A thermal reactor is a chamber which is designed to provide adequate residence time for allowing appreciable oxidation of CO and HC.
33. The working principle of ...... package is to control the emission levels of various pollutants by changing the chemical characteristics of the exhaust gases.
34. The quantity of soot formed depends upon the local F/A ratios, the type of fuel and the pressure.
35. ...... smoke is caused by liquid droplets of lubricating oil or fuel oil while starting from cold.
36. ...... smoke consists of carbon particles suspended in the exhaust gas and depends largely upon A/F ratio.
37. The cause of smoke is incomplete burning of fuel inside the combustion chamber.
38. ...... may be defined as visible product of combustion.
39. Rich mixtures (higher loads) result in high smoke because of non-availability of oxygen for combustion.
40. The smoke ...... with the increasing F/A ratio.

ANSWERS

1. Pollutants	**2.** CO, HC, NO_x	**3.** Yes	**4.** poisonous	**5.** Yes
6. rich	**7.** Yes	**8.** No	**9.** No	**10.** stoichiometric
11. Yes	**12.** exhaust	**13.** increases	**14.** Yes	**15.** carburettor
16. Yes	**17.** No	**18.** Yes	**19.** Yes	
20. angle of advance	**21.** Yes	**22.** S.I.	**23.** Yes	**24.** Yes
25. No	**26.** Yes	**27.** Yes	**28.** Rhodium	**29.** Yes
30. Yes	**31.** Water injection	**32.** Yes	**33.** Catalytic converter	**34.** Yes
35. Black-white	**36.** Black	**37.** Yes	**38.** Smoke	**39.** Yes
40. increases.				

THEORETICAL QUESTIONS

1. What do you mean by "Air pollution" ?
2. What are the main sources of pollutants from gasoline/petrol engine ?
3. What are the main pollutants emitted by petrol engine ?

4. State the mechanism of formation of CO.
5. How are hydrocarbons (HC) formed ?
6. Explain briefly the mechanism of formation of nitric oxide (NO).
7. What are the effects of the following factors on the exhaust emission ?
   (*i*) Air fuel ratio ; (*ii*) Surface volume ratio ;
   (*iii*) Engine speed.
8. Explain briefly the various sources from which pollutants are emitted from S.I. engine.
9. Define the crankcase blowby and explain how it can be controlled.
10. Discuss briefly the following with regard to S.I. Engines :
   (*i*) Crankcase emission (*ii*) Evaporative emission
   (*iii*) Exhaust emission.
11. What are the sources of HC formation in petrol engine ? Explain different factors which affect HC formation.
12. Explain briefly the factors which affect the formation of NO_x.
13. What are the sources of HC formation in petrol engine ? Explain various factors which affect the HC formation.
14. What are the sources of evaporative emission in petrol engines ? How can it be controlled ?
15. What do you mean by crankcase blowby ? How can it be controlled.
16. Explain briefly various methods by which S.I. engine emission can be controlled.
17. Explain in detail any two of the following methods of S.I. engine emission control :
   (*i*) Modification in the engine design and operating parameters.
   (*ii*) Treatment of exhaust products of combustion.
   (*iii*) Modification of the fuels.
18. Describe with sketches the following methods of petrol exhaust emission control :
   (*i*) After-burner (*ii*) Exhaust manifold reactor
   (*iii*) Catalytic converter system.
19. Explain briefly the exhaust gas recirculation (EGR) device for the control of NO_x.
20. What do you mean by "Total emission control packages" ? Describe with neat sketches two types of total emission control packages.
21. Explain briefly the following :
   (*i*) Three-way catalytic converter (*ii*) Two-way catalytic converter
   (*iii*) Noble metal catalytic converter.
22. Explain briefly the "evaporation emission control device (EECD)".
23. Discuss the emissions from diesel engines. On what factors these emissions depend ?
24. Define 'smoke' and discuss different factors which affect smoke formation in C.I. engines.
25. What is the cause of diesel smoke ? Discuss the ways by which diesel smoke can be controlled.
26. What is the mechanism of smoke formation ?
27. How can the smoke intensity be measured ?
28. Describe two important types of smoke meters.
29. Explain the effect of engine load on diesel engine smoke.
30. What is diesel odour ? Explain the effect of different factors on the formation of odour in C.I. engines.
31. How odour can be controlled ?
32. Compare the diesel engine and gasoline engine emissions.
33. Discuss the air pollution from gas turbines and compare its with emissions from petrol engines.
34. Give the comparison between "electronic catalytic converter" and "conventional exhaust catalytic converter".
35. Discuss the effects of emissions on human health.

19

Miscellaneous Engines

19.1. DUAL-FUEL AND MULTI-FUEL ENGINES

19.1.1. Dual-Fuel Engines

19.1.1.1. Introduction

- Owing to various technical and financial reasons, some engines are designed to operate, *using a combination of two fuels.* For instance, in some third-world countries *dual-fuel engines are used because of high cost of diesel fuel.*
- Large C.I. engines are run on a *combination of methane and diesel oil.* Methane is the main fuel because it is more cheaply available. However, methane is not a good C.I. fuel by itself because it *does not readily self-ignite* (due to its *high octane number*). *A small amount of diesel oil is injected at the proper cycle time. This ignites in a normal manner and initiates combustion in the methane-air mixture filling the cylinder.*
- On these type of engines, *combination of fuel input system are needed.*
- *Duel-fuel operation combines in a simple manner the possibility of operating a diesel engine on liquid fuels such as diesel oil or gas oil and on gaseous fuels such as natural gas, sewage gas and cook oven gas etc.*
  - The engine can be *switched from dual-fuel operation almost instantaneously in case of emergency.*

19.1.1.2. Working of Dual-fuel Engine

Fig. 19.1 shows a diesel engine with necessary modifications, which it requires when it *works on duel-fuel.*

The working of a dual-fuel engine, which works on a diesel cycle, is given below :

- The mixture of air and gaseous fuel is compressed in the cylinder similar to air compression in diesel engine.
- A small amount of liquid fuel, (diesel fuel), called *pilot fuel* or the *secondary fuel* (5 to 7% of the fuel at full load) is injected near T.D.C., irrespective of load on the engine. This pilot injection acts as a *source of ignition.* The gas-air mixture in the vicinity of injected spray ignites at number of places establishing a number of flame-front and combustion continues rapidly and smoothly. (In a 4-stroke cycle engine the gas is supplied in the

intake manifold where it mixes with the incoming air to form a homogeneous mixture. In 2-stroke cycle engines the gas may be injected at low pressures, from 1.5 to 3.5 bar, directly into the combustion chamber after the closure of the exhaust ports). It is interesting to note that in the duel fuel engine, the combustion starts similar to C.I. engine but it propagates by flame similar to S.I. engine.

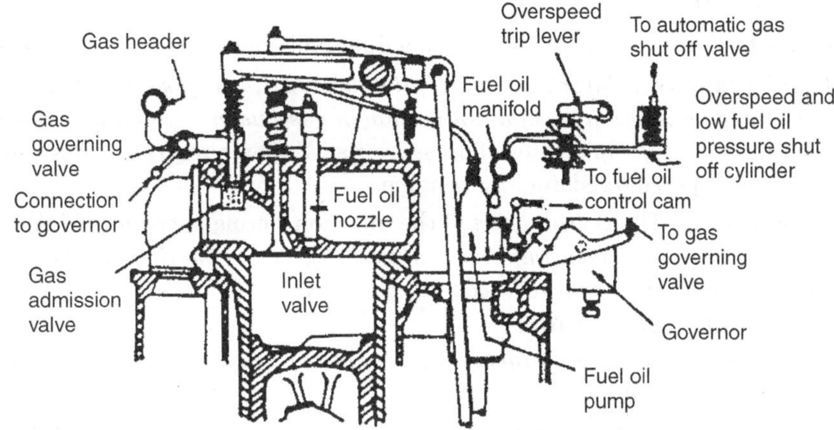

Fig. 19.1. Dual-fuel engine.

- The engine output (power) is generally controlled by *altering the amount* of primary gaseous fuel added to inlet manifold (the pilot oil quantity is usually kept constant).
- This type of engine is *capable of running on either gas or diesel oil or a combination of these two over a wide range of temperature ratios.*
- Fig. 19.2 shows the p-θ diagram for a duel-fuel engine working on different fuels (change from one fuel to another being quite smooth).

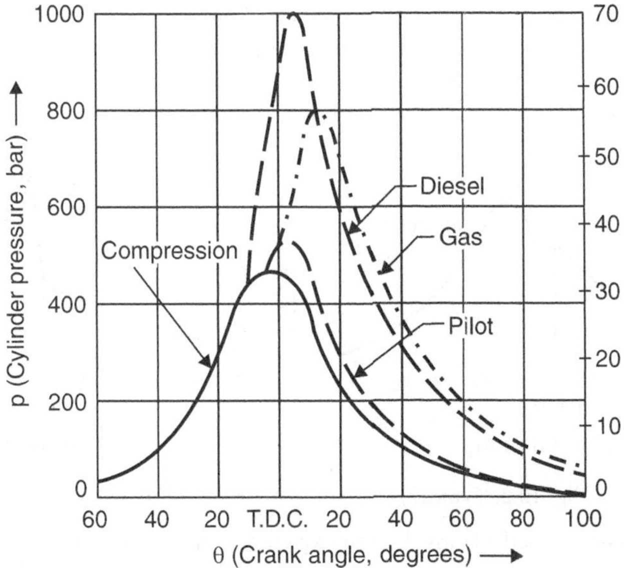

Fig. 19.2. Typical p-θ diagram for a dual-fuel engine.

19.1.1.3. Factors affecting combustion in a dual-fuel engine :

The following factors affect combustion in duel-fuel engine :

1. Pilot fuel quantity
2. Injection timing
3. Cetane number of pilot-fuel
4. Inlet temperature
5. Type of gaseous fuels
6. Throttling
7. Mixture strength.

The *effects of these factors on combustion* in a duel-fuel engine are briefly discussed below :

1. **Pilot fuel quantity.** A *large quantity of pilot fuel results in knocking.* (The *phenomenon of knock in a duel-fuel engine is of nature of auto-ignition of gaseous mixture in the neighbourhood of injected spray*), because of very rapid rates of pressure rise. Superchargers engines use about same pilot fuel as used by unsupercharged duel-fuel engine.

2. **Injection timing.** The *overall effect* of the injection timing is *not very high* except that at a *slightly retarded timing there is some improvement in the efficiency but the rate of pressure rise also increases making the engine more near to knocking condition.*

3. **Centane number of pilot fuel :**

- The *use of low cetane number fuels results in poor performance of the engine* and *greatly affects the combustion.*
- The ignition quality of fuel, in general, exercises little effect on combustion in duel-fuel engines as compared to the ignition quality of the primary fuel.

4. **Inlet temperature.** The *temperature of inlet charge greatly affects power output, ignition limits, the rate of pressure rise* and hence the *knocking limits of a particular fuel-air mixture.*

5. **Type of gaseous fuels.** The type of gaseous fuel used decides amply the knock-limited output of the duel-fuel engine. Methane (main constituent of natural gas) does not undergo decomposition during the compressions in the engine and is *more resistant to knock, pre-ignition and back-firing from cylinder into gas/air inlet than other gases (e.g. town gas etc.).*

6. **Throttling.** When throttling is used in duel-fuel unit *maximum cylinder pressure reduces greatly.*

7. **Mixture strength.** With the increase in mixture strength the ignition delay reduces rapidly and then again rises sharply. A further increase in the mixture strength leads to failure of ignition.

- Fig. 19.3 shows the p-θ diagrams for normal burning and knocking for duel-fuel engines.

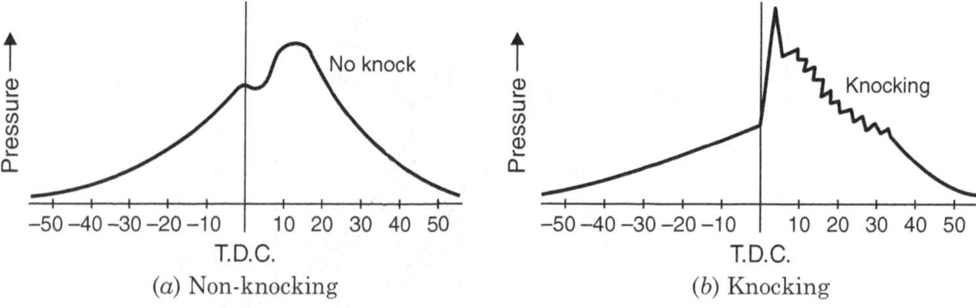

(*a*) Non-knocking (*b*) Knocking

Fig. 19.3. Typical p-θ diagrams for a dual-fuel engine.

— The pressure oscillations during expansion are high and similar to S.I. engine knocking condition. *A small increase in gaseous fuel beyond a limit can cause a very severe knocking which may damage the engine.*

— The *knock tendency of the engine can be reduced by supplying excess air with low initial temperature and by using the antiknock additive* (*e.g.* tetramethyl lead).

19.1.1.4. Performance of dual-fuel engines

- Fig. 19.4 shows the relative performance of dual-fuel engines.

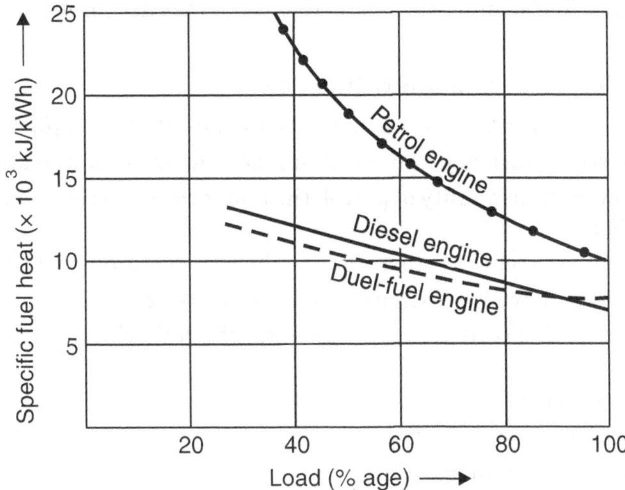

Fig. 19.4. Comparison of performance of dual-fuel engine with other engines.

— At *part load conditions,* in case of *duel-fuel engine, the thermal efficiency is low (38.5% against 40% of diesel engine) and specific fuel consumption high* because of increased delay periods at weak mixture of air and gas. This requires *either adjustment of mixture strength or injecting more pilot fuel at part loads.*

— *At full load, duel-fuel engine is superior to diesel engine since the gaseous fuel fills all parts of the combustion chamber and allows more air to take part in the combustion.*

19.1.1.5. Advantages and applications of dual-fuel engines

Advantages :

1. A duel-fuel engine can run on either of the fuel and diesel requirement is hardly 5 percent if it runs on gas.
2. Reduced pollution.
3. It is preferred when cheap gas is easily available.
4. Owing to clean combustion in the engine the wear and tear of the engine as well as consumption of lubricating oil are reduced.
5. The engine's utility is considerably increased due to the possibility of instantaneous change over from gas to diesel and vice-versa.
6. Best suited for low pressure liquified gas (LPG) which evaporates very easily.
7. Very attractive power generation system because of its greater flexibility of operation compared with conventional diesel engines.

Applications :

1. A typical use of duel-fuel engine is to produce synthetic gas (mixture of CO and H_2) by burning CH_4 and simultaneously developing power.
2. A lot of conventional fuel can be saved by using duel-fuel engines particularly for irrigation purposes.

19.1.2. Multi-fuel Engines

19.1.2.1. Introduction

A **multi-fuel engine** is *one which can operate satisfactorily (with substantially unchanged performance and efficiency) on a wide variety of fuels ranging from diesel oil, crude oil, IP-4 to lighter fuel like gasoline, and even normal lubricating oil.* The use of low ignition quality fuels require higher compression ratio for burning, therefore, diesel engines are prefered to use multi-fuels in the engine.

19.1.2.2. Requirements of a multi-fuel engine

The following *requirements* must be met with by a *multi-fuel engine* :

1. Good combustion efficiency and minimum heat losses from the engine.
2. In view of low ignition quality of petrol, the temperature of the combustion chamber should be comparatively higher.
3. The engine must be able to start under sub-zero conditions without any external aid.
4. The engine must have low exhaust smoke and low noise levels.

In order to meet with the above requirements, the following design features of the engine need be considered :

(*i*) High compression ratio

(*ii*) Large stroke / bore ratio

(*iii*) Open combustion chamber

(*iv*) Injection pump.

19.1.2.3. Difficulties associated with multi-fuel operation

The multi-fuel operation entails the following *difficulties :*

1. Tendency of vapour lock in the fuel pump while using lighter fuels.
2. Tendency of increased wear in the fuel pump due to lower lubricity of gasoline.
3. In view of differences in heating values and compressibility of fuels, different volumes (of fuels) need be injected.

19.1.2.4. Performance of multi-fuel engine

- Fig. 19.5 shows the performance of multi-fuel engine.

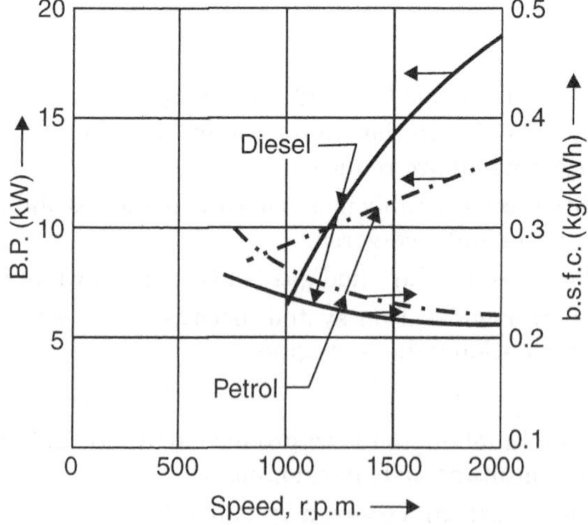

Fig. 19.5

— The engine output is 15 to 20 percent less with petrol as fuel compared with diesel oil at *higher speed.* The difference is less in lower speeds.

— The b.s.f.c. (brake specific fuel consumption) for petrol engine is also higher at lower speed compared with diesel fuel. The difference narrows down at higher speeds.

19.2. STRATIFIED CHARGE ENGINE

19.2.1. Introduction

- The **stratified charge engine** is *usually defined as a S.I. engine* (stratified diesel engine has also been developed) *in which the mixture in the zone of spark plug is very much richer than that in the rest of the combustion chamber i.e one which burns leaner overall fuel-air mixtures.*
  - **Charge stratification** *means providing different fuel-air mixture strengths at various places in the combustion chamber.*
- Whereas several S.I. engines are designed to have a homogeneous air-fuel mixture throughout the combustion chamber, some modern stratified charge engines are designed to have a *different air-fuel ratio at different locations within the combustion chamber. A rich mixture that ignites readily is desired around the spark plug, while the major volume of the combustion chamber is filled with a very lean mixture that gives good fuel economy. Special intake systems are necessary to supply this non-homogeneous mixture combination of multiple valves and multiple fuel injectors, alongwith flexible valve and injection timing are used to accomplish the desired results.*
- Some stratified charge S.I. engines are operated with no throttle, which raises the volumetric efficiency. Speed is controlled by proper timing and quantity of fuel input.
- The *stratified charge engine combines the advantages of both petrol engines* (very good full load power characteristics *e.g.* high degree of air utilisation, high speed and flexibility) *and diesel engines* (good part-load characteristics) and *at the same time avoids as far as possible their disadvantages.*

The following are the *advantages of burning overall fuel-air mixtures :*

(*i*) Higher thermodynamic efficiency.

(*ii*) Reduced air pollution.

19.2.2. Classification

According to the method of formation of the heterogeneous mixture in the combustion chamber, the stratified charge engines can be classified as follows :

I. ***Those using fuel injection and positive ignition (including swirl stratified charge engines)***

(*a*) *Stratification by fuel injection and positive ignition :*

1. Ricardo system
2. Pre-chamber stratified charge engine
3. Volkswagon PCI stratified charge engine
4. Broderson method of stratification.

(*b*) *Swirl stratified charge engine :*

1. Witzby swirl stratification process
2. Texaco combustion process (TCP)
3. Ford combustion process (FCP)
4. Ford PROCO
5. Deutz combustion process (AD-process).

II. ***Stratification by carburetion alone***

1. Russian stratified charge concept.
2. Institute Francias Du Petrols (IFP) process.
3. Honda CVCC (Compound Vertex Controlled Combustion) engine.

Here we shall discuss in detail only Texaco combustion process (TCP).

Texaco combustion process :

The carburetted S.I. engine has several problems and areas of potential weaknesses, these are given below :

(*i*) The end charge has a long residence time. It may become highly reactive, and it may cause knock. Therefore a fuel of high octane rating is required.

(*ii*) Homogeneous mixtures with the fuel-air limits of propagating a flame yield relatively low enthalpy efficiencies and also relatively high HC and NO_x emissions.

(*iii*) Throttling as a means of controlling output induces a pumping loss.

(*iv*) Flame quenching at the walls adds to air pollution.

To overcome these weaknesses (and hopefully without adding others) stratified charged engine has been developed.

- The Texaco combustion process (TCP) is illustrated in Fig. 19.6. This system was patented by Texas company (USA) in 1949.

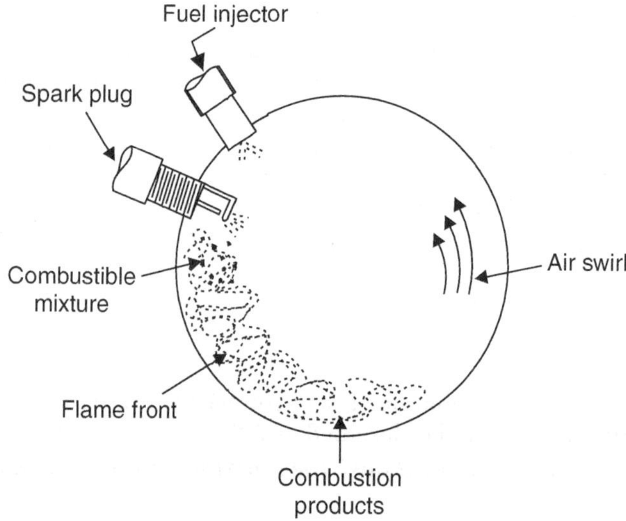

Fig. 19.6. Barber's Texaco combustion process.

— In this system, *the fuel spray is injected at the end of the compression stroke toward spark plug which is entrained by swirl air and forms almost stoichiometric mixture near spark plug.*

— The *charge from the injected region starts burning successively but the flame front located between the spark plug and the injector remains almost stationary.*

- This system entails the following *advantages :*

(*i*) Gives good performance over whole range of load and speed with a wide range of fuels ranging from premium gasoline to high cetane diesel fuel.

(*ii*) Besides giving better part load efficiency the starting and warm-up characteristics of TCP are very good due to employment of cup combustion chamber.

(*iii*) The inherent knock resistance of TCP allows the use of higher compression ratio or turbocharging without imposing any octane requirement.

(*iv*) The lean overall mixture can be used.

(*v*) Low exhaust smoke.

- TCP system has proved very versatile as regards multi-fuel requirements and are *used in military vehicle engines.*

19.2.3. Advantages and Disadvantages of Stratified Charge Engines

The advantages and disadvantages of stratified charged engines are listed below :

Advantages :

1. Compact, lightweight design and good fuel economy.
2. Good part-load efficiency.
3. Exhibit multi-fuel capability.
4. The rich mixture near the spark plug and lean mixture near the piston surface *provides cushioning to the explosive combustion.*
5. *Resist the knocking* and provide smooth combustion resulting in smooth and quiet engine operation over the entire speed and load range.
6. Low level of exhaust emissions ; NO_x is *reduced* considerably.
7. Usually *no starting problem.*
8. Can *tolerate wide quality of fuels.*
9. Can be manufactured by the existing technology.

Disadvantages :

1. For a given engine size, charge stratification results in *reduced power.*
2. These engines create *high noise level at low load conditions.*
3. *More complex design* to supply rich and lean mixture, and quantity is varied with load on the engine.
4. *Higher weight* than that of a conventional engine.
5. Unthrottled stratified charge engines *emit high percentage of HC* (due to either incomplete combustion of lean charge or occasional misfire of the charge at part load conditions.
6. Reliability is yet to be well established.
7. Higher manufacturing cost.

19.3. STIRLING ENGINE

19.3.1. Stirling Cycle. Please refer Art 3.11.

The Stirling cycle is ***superior*** *to the Carnot cycle because of the following reasons :*

1. The Stirling cycle is *practicable,* whereas the *Carnot cycle cannot be realised in practice* due to wide variation in speed during a cycle (alternately very high for adiabatic and very slow for the isothermal part of the cycle).

2. The workoutput per cycle and *m.e.p. of Stirling cycle are high. The Carnot cycle needs a very long stroke and produces a very narrow strip of work giving a low m.e.p.*

19.3.2. Working Principle of Stirling Engine

The basic principle of working of stirling engine is the same as that of conventional engine. *The alternate compression at low temperature and expansion at high temperature of a working fluid is the basis for the stirling engine.* However, the working fluid is heated in a radically different manner. **It burns fuel outside the engine itself, and continuously.**

The following features *distinguish* the stirling engine from other heat engines :

1. The air and fuel are *externally burned* and the heat generated is transferred to the working fluid (may be air or any other suitable gas) of stirling engine where the *working fluid works in a closed cycle.*

2. The cyclic flow of working fluid within the engine is achieved solely through geometric volume changes, and without the use of intermittently closed valves or ports.

3. An intermittent flow heat exchanger stores a large portion of the heat of the working fluid after expansion and subsequently returns it to the working fluid after compression, thereby accomplishing thermal regeneration.

Fuel used for stirling engine :

Since the stirling engine is an *external combustion engine,* it *possesses multi-fuel capacity.* It can use any petroleum fraction such as *gasoline, diesel, methanol-gasoline blends, etc., with no octane or cetane requirements.* Thus the stirling engine has the *desirable characteristic of adaptability to changing fuel availability.*

19.3.3. Differences between Carnot and Stirling Engines

The differences between Carnot and Stirling engines are given below :

S. No.	Carnot Engine	Stirling Engine
1.	Its cycle consists of *two isentropic and two isothermal processes.*	Its cycle consists of *two isothermal and two constant volume processes.*
2.	No heat is supplied during isentropic process.	Regenerative heat exchange takes place during constant volume processes. This is an ideal cycle and the net heat flow to the working fluid is zero in the constant volume processes, or the net effect of these processes is adiabatic.
3.	For its operation only one piston-cylinder is required.	For its operation two cylinder piston are required (the displacer piston and the power piston).
4.	The cycle is *not practicable* as alternately the isothermal and isentropic processes are to be slow and fast respectively. The cycle is of *academic importance only.*	The cycle is *practicable,* as all the processes are performed at the same speed. Engines working on Stirling cycle have been in existence.
5.	It requires a very long stroke, and produces, *a very low m.e.p.,* due to a narrow indicator diagram.	It does not require a long stroke, and hence the *m.e.p. is relatively quite high.*

19.3.4. Engine Geometry and Driving mechanism

The following are the three basic configurations/arrangements of engine layout :

Type 1 **:** *Two separate cylinders, cold and hot sides of the engine, each with working piston*—**Two piston mechanism.**

Type 2 **:** A *Single cylinder with two pistons, one is the working piston and other is a displacer*—**Two piston engine.**

Type 3 **:** Two *cylinders with a working piston in the cooled cylinder and displacer piston with the heated cylinder*—**Piston displacer engine.**

19.3.4.1. Two piston mechanism

Fig. 19.7 shows the arrangement of the system. The whole mechanism can be divided into two systems :

(*i*) The displacer system ; (*ii*) The power system.

— *The function of the* ***displacer system*** *is to heat and cool the working fluid.*

— *The function of the* ***power system*** *is to obtain the power during expansion and supply power during compression.*

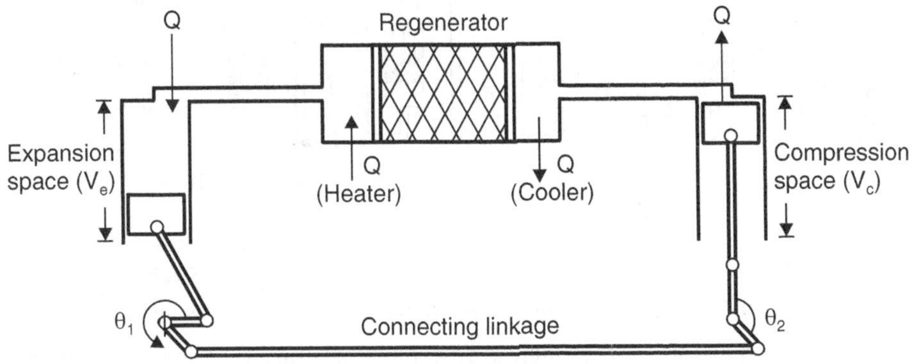

Fig. 19.7. Two piston mechanism.

The total working volume (V_t) is given by :

$$V_t = V_e + V_c + V_d$$

where, V_e = Volume of expansion space,

V_c = Volume of compression space,

V_d = Volume of regenerator, heater, cooler and connecting tubes.

This total volume undergoes changes due to the motion of the two piston.

When the crank rotates through one revolution, the following processes take place :

1. The working fluid is passed from hot space to the cold space via *heater, regenerator* and *cooler*. This is accomplished by phasing the two pistons such that there will be small change in the total working volume. Thus, the *heat associated with the working fluid is passed on to the regenerator at almost constant volume.*

2. The working fluid is compressed when it is mainly within the cold space and the cooler.

3. After the working fluid undergoes compression, it is passed from the cold space to hot space through the *cooler, regenerator* and *heater*. During this operation the heat of the regenerator is transferred back to the working fluid at *constant volume.*

4. The working fluid being located mainly within hot space and the heater, an expansion takes place thereby producing work output.

— In this system, these four processes are *not distinct and separate* but *overlap* each other, whereas in an ideal cycle the heat is rejected and received at constant volume and the compression and expansion occur isothermally.

— So long as the compression and expansion are isothermal and regeneration is effective, the thermal efficiency of the engine is least affected with the variation of total volume (V_t).

19.3.4.2. Two piston engine

- Fig. 19.8 shows a new arrangement (the arrangement discussed earlier in Art. 19.3.4.1 is unsuitable for higher capacity engines) of a double acting suitable for *high capacity engine*. Here, compression of the working fluid occurs on one side and expansion on the other side of the engine.

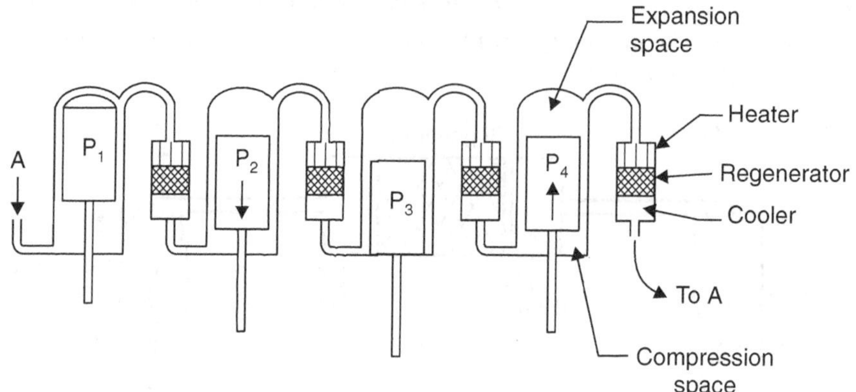

Fig. 19.8. Two-piston double-acting in-line 4-cylinder stirling engine.

— The top space of each cylinder is an expansion space (hot space) and bottom space is compression space (cold space).

— The hot space of the cylinder is connected to the cold space of the other cylinder through heater, regenerator and cooler.

— The pistons move with a suitable phase shift (90° for 4-cylinder engine).

- The *major drawback* of such an engine *is greater difficulty in obtaining a uniform flow through the cooler, regenerator and heater.* In view of greater aerodynamic and thermodynamic complexities involved this engine is more *difficult to be optimised for high efficiency.*

19.3.4.3. Piston displacer engine

- The piston displace engine consists of a cylinder into which two pistons–one is known as *displacer piston* and other *power piston.* The cylinder of displacer piston is divided into hot space (expansion) and cold space (compression). The displacer piston reciprocates within the hot portion of the cylinder (open one) while the power piston reciprocates within the cold portion of the cylinder (lower one).
- *The main function of the displacer piston is to heat and cool the working fluid while the main function of the power-piston is to compress and expand the gas.*

Thus, the working of stirling engine is completely governed by relative motion of these two pistons.

19.3.4.4. Advantages and disadvantages

The following are the advantages and disadvantages of a stirling engines vis-a-vis conventional I.C. engine.

Advantages :

1. Since it is an external combustion engine, it possesses *multi-fuel capability.*
2. Owing to rhombic drive (this mechanism consists of two crankshafts rotating in opposite directions, each having a crank on it ; two synchronising gears are used to time these crankshafts which in turn drive the output shaft), it is a *perfectly balanced engine.*

3. *Reduced exhaust emission* (due to overall lean mixtures and exhaust gas circulation).
4. *Part load efficiency considerably higher.*
5. The engine can be *safely overloaded for a brief period.*
6. *Higher thermal efficiency* (Typical value is 45% at 100 bar and 300°C).
7. *Noise-free engine* (because combustion takes place outside the engine).
8. The *starting of the engine is reliable* (because it depends only on the ignition of fuel in the burner).
9. *Suitable for a variety of applications* (*e.g.* automobiles, ship propulsion or power generation) since a stirling engine has almost *constant torque characteristics over the entire range of speed.*
10. *No lubricating oil required* (the use of rollsock seal has eliminated the use of lubricating oil).

Disadvantages :

1. *Complex engine design* (due to rhombic drive, regenerator, heater and cooler).
2. *Requires large quantity of cooling water and bigger radiator.*
3. The engines *requires a blower* to force the air through the preheater and the combustion chamber. This reduced the engine efficiency, and increases the noise.
4. The *major disadvantage* of a stirling engine is its **high cost.**

19.3.4.5. Comparison between Stirling Engine and I.C. engine

The comparison between stirling engine and I.C. engine is given (in tabular form) below :

S. No.	Aspects	I.C. engine	Stirling Engine
1.	*Fuel addition*	Atomised fuel is added in air, either before or after compression.	Continuous combustion takes place in combustion chamber and energy is transferred into or out of engine through an heat exchanger.
2.	*Valve or ports*	Valves or ports required.	Valves or ports (for induction or exhaust) not required since the gas charge remains permanently inside the engine only.
3.	*Speed variation*	Achieved by controlling the amount of fuel injected or the quantity of mixture.	Achieved by changing the mass of working gas within the engine.
4.	*Heat rejection*	Achieved by high temperature exhaust.	Achieved through heat exchanger and hence it requires large cooling system.
5.	*Noise*	Noisy (due to presence of valves and periodic explosion).	Noiseless, smooth working.
6.	*Cooling system*	Sealing of cooling of system not absolutely essential.	Should be perfectly sealed.
7.	*Sealing*	The sealing can be done by the use of oil.	This engine does not use oils since it will cause contamination of the working fluid.

19.4. THE WANKEL ROTARY COMBUSTION (RC) ENGINE

19.4.1. Introduction

- Scheffel in 1952, to get a patent for *rotary engine,* utilised the *principle that oval or elliptical rotors can be designed to maintain contact, while turning about fixed centres, and that three or more rotors can be run to enclose between them a continuously varying volume.* The four volumes between the rotors, with the suitable arrangement of ports, ignition system and adequate compression ratio, could be made to execute a *four-phase otto cycle.* However, this design failed due to its complexity and great difficulties/problems involved in its manufacture.
- Felia Wankel (German inventor), in 1954, got a patent for design of four-phase rotary engine working on the otto cycle principle.
- Later Dr. Froede made certain modifications and an engine was developed, called as KKM (Kreiskolben motor) ; now popularly known as Wankel Rotating Combustion (RC) engine.

19.4.2. Construction and Working

Construction. Refer Fig. 19.9. It consists of the following parts :

(*i*) **Rotor** (Three lobed).

(*ii*) **Eccentric or output shaft** with its integral eccentric. No connecting rod is required as the rotor rotates directly on the eccentric shaft. The *output torque is transmitted to the shaft through eccentric.*

(*iii*) **Internal and external timing gears.** They maintain the phase relationship between the rotation of the rotor and the eccentric shaft and eventually *control the orbital motion of the rotor.*

Working :

- The Wankel engine works on the *four-phase principle.*

  (The word *phase* corresponds to *stroke* of the reciprocating engine)

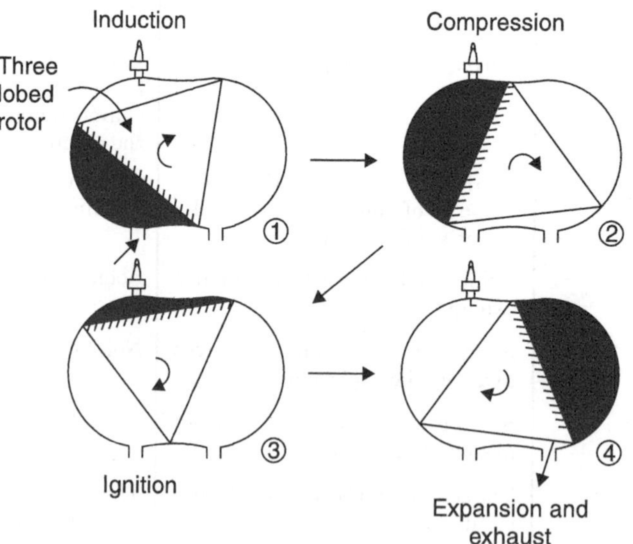

Fig. 19.9. The Wankel engine.

- The engine having three lobed rotor is driven eccentrically in a casing in such a way that there are *three separate volumes trapped between the rotor and the casing. These three volumes perform "induction", "compression", "combustion", "expansion" and "exhaust" processes in sequences.* There are three power impulses for each revolution of the rotor, and since the eccentric or output shaft rotates at three times the speed of the rotor, *there is only one power impulse for each revolution of the output shaft of a single bank rotary engine.*

One complete *thermodynamic cycle* is completed over 360° rotation of the rotor ; the suction phase takes 90° of rotor movement and so the also other three phases. One *thermodynamic phase* is completed every 270° rotation of the output shaft, since the output shaft makes three revolution for every single rotation of the rotor.

19.4.3. Features

1. *Simple construction, less mechanical loss, smooth motion and does not require a cranking mechanism.*
2. *Good power volume ratio.*
3. *No reciprocating parts* and hence no balancing problem and complicated engine vibrations eliminated.
4. Due to the absence of intake-exhaust valve mechanism, the *correct timings* for opening and closing (the ports) *can be maintained even at high speeds.*
5. *Low torque fluctuation.*

There are problems in the design, notabty of sealing and of heat transfer but these have been overcome sufficiently well for spark ignition engine to be marketed.

19.4.4. Constructional and Other Details of Wankel Engine

1. **Rotor housing and housing materials :**

- **Rotor** are generally made from *high-grade malleable spheroidal graphite iron.*
- The **rotor housing** is an *aluminium silicon alloy,* bonded to the cylinder-bore walls in thin sheet metal, the outer surfaces of which have a saw-tooth finish to improve adhesion and thermal conductivity. This lining is then given a hard chromium vanadium plating, which in turn is plated with more chrome but in a thin, porous and oil retaining layer. End and intermediate rotor housings are made from implated high silicon aluminium alloy.
- **Apex and side seal blades** can be made from *cast-iron* but the more popular types are made from *hard carbon material.*
- Both the **leaf and washer springs** can be made from *beryllium* copper, which has the ability to retain its elasticity when operating under working temperatures.

2. **Rotor seals :**

- The *planetary motion of the rotor within the epitrochoid bore of the rotor housing is designed to maintain a contact between the triangular corner of the rotor and the cylinder walls.* Peripheral radial corner blades, known as the *apex seals,* are necessary to prevent gas leakage between the three cylinder spaces created by the three-sided rotor. Similarly, *side seals* between the flat rotor sides and the end and intermediate housing side walls are essential to stop engine oil reaching the cylinders, and gas from combustion escaping into the eccentric output shaft region.
- The gas-tight sealing between the rotor and housing may be considered in terms of primary and secondary sealing areas.
  - The *primary sealing areas* are those between the sealing elements (apex and side blades) and the cylinder housing bore and side walls.

— The *secondary sealing areas* are those between the sealing elements and grooves, plots and trunnion counter bores.

— The total sealed system around the rotor is known as a *sealed grid.*

- In order to ensure *minimum leakage at each rotor apex,* slotted cylindrical trunnion blocks are counter-sunk on both sides of the rotor at each rotor corner, the object being to provide a much longer leakage path at different meeting points where the apex and side blades overlap.

3. **Combustion and spark-plug location :**

- The improvement in combustion process can generally be made by installing *two spark-plugs :* one in the *lead combustion space and the other in the trailing region of the combustion-chamber.*

A tilting or rocking action takes place when the rotor is near the T.D.C. position and combustion commences. This tends to expand the gases in the leading portion of the combustion chamber but to compress them in the trailing region.

— *Positioning a plug in the high-pressure trailing region of the combustion chamber tends to produce slightly more power and use less fuel,* whereas if the plug *is located in the leading portion of the chamber, better starting and idle running is obtained.*

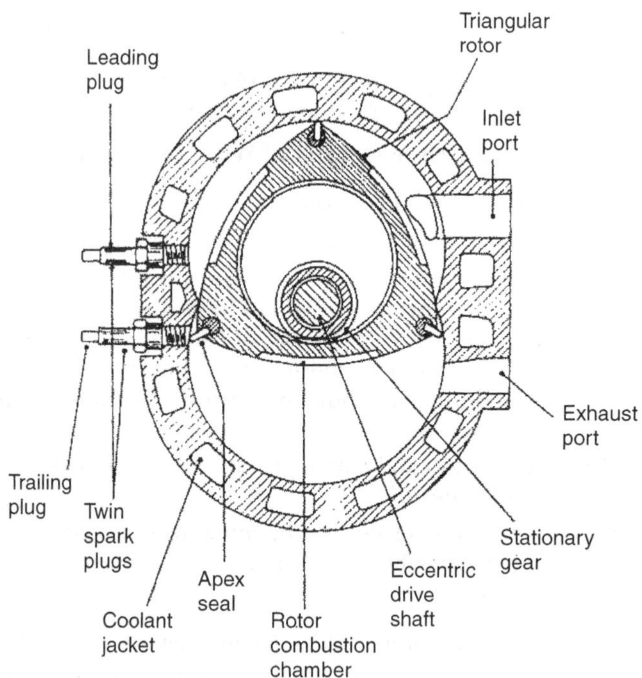

Fig. 19.10. Wankel engine-sectioned view.

— For achieving high efficiency it is necessary to have rapid combustion and pressure rise, but the backward flame propagation of the burning gases in the chamber is opposed by the forward movement of the gases as they are swept around with the rotor, which, in effect, slow down the speed of combustion and therefore does not give optimum performance.

- *Efficient combustion is produced by having a very advanced timing,* this *can be partially achieved by positioning a spark-plug in the trailing region of the bore wall as far as possible from the waist or minor axis of the epitrochoid wall profile.*

4. **Cooling system :**

- The engine block is made up of the following five separate section housings :
  - — The front and rear end housings ;
  - — Two rotor bore housings ;
  - — Intermediate housing.

All of the above housings are clamped together by bolts.

- ***Coolant circulation*** is *provided around the bore walls by axial passages* cast *into each housing section, while the side walls of the twin cylinders are cooled by extended coolant passages adjacent to the side walls cast in both the end covers and the intermediate housing.*

Clearance and Swept Volume, Compression Ratio, Engine Displacement Capacity

1. **Clearance and swept volume :**

The **clearance volume** (V_c) is *defined as the volume created between the curved chamber bore, flat side walls and one of the rotor flanks when in the nearest-approach position (T.D.C.)* adjacent to the waisted portion of the epitrochoid bore.

The **swept volume** (V_s) is the difference *of the maximum volume to which the chamber expands (B.D.C.) as the rotor revolves and the minimum volume of the combustion chamber at the point of greatest compression (T.D.C.)* It may be given by the following :

$$V_s = 3\sqrt{3}\ RWe$$

where, R = Radius of rotor from its centre to one of the apex tips (cm),

W = Width of the rotor (cm), and

e = Distance between the axes of the output shaft and eccentric lobe, known as the *eccentricity* (cm).

2. **Compression ratio :**

The Compression ratio (r), will be the conventional formula, as follows :

$$r = \frac{V_s + V_c}{V_c}$$

(where V_s and V_c are the swept volume and clearance volume respectively, in cm)

$$\therefore \qquad r = \frac{(3\sqrt{3}RWe) + V_c}{V_c}.$$

3. **Engine displacement capacity :**

In a Wankel engine there are *three power strokes per rotor revolution,* and the output shaft completes three revolutions for every one revolution of the rotor. Therefore, there is *one power stroke per revolution of the output shaft.* This, therefore, can be seen to be equivalent to a two-cylinder four-stroke engine which also has a power stroke every crankshaft revolution. Because of this reasoning a *two-chamber swept volume equivalence is generally recognized.*

The displacement capacity (V_d) is therefore *twice the swept volume of a single chamber multiplied by the number of rotors ;* that is

$$V_d = 2\ V_s\ n$$

where, V_s = Swept volume, and

n = Number of rotors.

19.4.5. Performance of Wankel Engine

1. Power and specific fuel combustion :

Fig. 19.11 shows the effect of the speed on the power output and specific fuel consumption for I.C. engine (conventional) and Wankel engine.

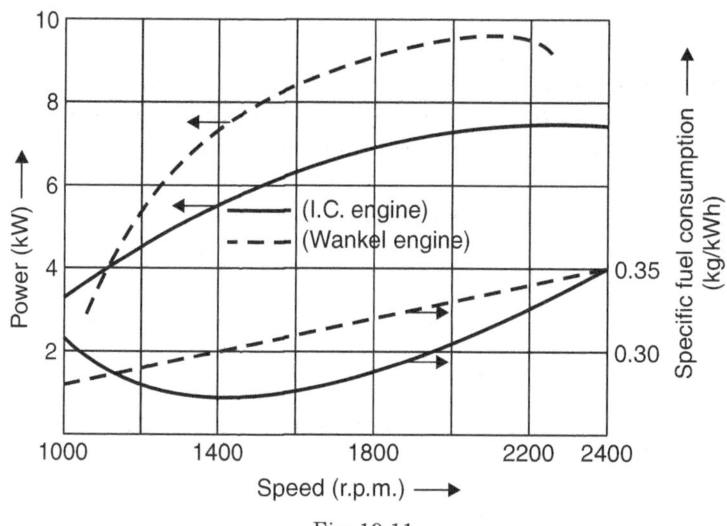

Fig. 19.11

— The *power of Wankel engine,* as compared with conventional I.C. engine, *increases rapidly.*

— The *frictional losses of Wankel engine,* with increase in speed, are *less* than I.C. engines (conventional).

— The *specific fuel consumption of wankel engine* is *higher* than I.C. engines (conventional).

2. Brake thermal efficiency :

- The thermal efficiency of Wankel engine is *less* than the conventional I.C. engine since the *cooling loss is more.*
- The thermal efficiency *can be increased by creating squish flow by providing recess in the rotor.*

3. **Exhaust emission.** Fig. 19.12 shows the emission of HC, CO and NO_x from Wankel engine (engine speed being 1450 r.p.m.) taking A/F ratio as variable.

— The presence of unburnt HC in the exhaust gases is due to quenching of the flame by the surface of the combustion chamber. The HC presence in this layer depends upon A/F ratio, temperature and turbulence in the layer.

— The variation of CO (percentage) is also shown in the Fig. 19.12. The emission of CO is higher at rich A/F ratio and decreases with increasing A/F ratio as sufficient air is available to burn CO as mixture is lean.

— The formation of NO_x depends upon O_2 content and peak cycle temperature. Since the maximum temperature in Wankel engine is *lower* than conventional I.C. engine, therefore *NO_x formation is low.*

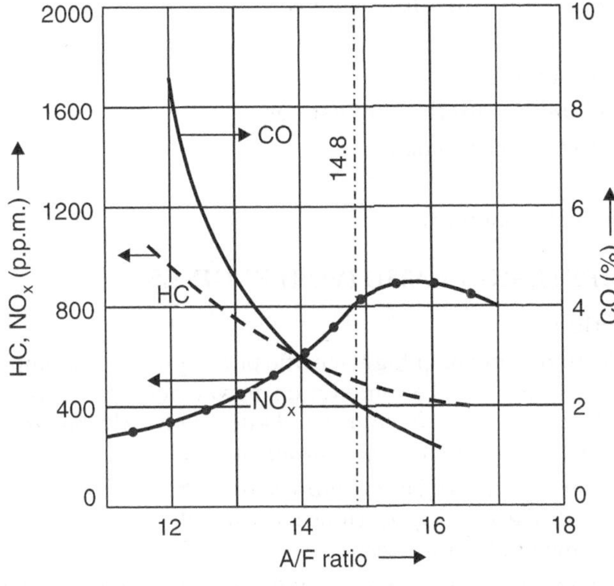

Fig. 19.12

19.4.6. Advantages and Applications of Rotary Combustion Engines

Advantages :

1. Lighter in weight, High power to weight ratio.
2. Compact, occupies less space.
3. Engine completely balanced, hence free from vibration and inertia forces.
4. Less noise.
5. Low grade petrol can be used, as it is less knock prone.
6. Low maintenance requirements.
7. The size of the engine is appreciably smaller than conventional I.C. engines for the same output.
8. It generates power after every rotation whereas 4-stroke engine develops power after every two revolutions.

Applications :

1. High performance motor cycle (Comotor, Luxemberg)
2. Snow mobile (Outboard marine in USA)
3. Outboard engine (Yanmar)
4. Helicopter engine (Curtis Wright)
5. General purpose engines
6. Industrial engines (Ingersoll-Rand, USA)
7. Passenger Cars (Several models in Japan and Germany).

19.4.7. Why Wankel Rotary Engine could not become Successful ?

Wankel rotary engine could not become successful mainly *due to the failure of apex seal on the straight line contact between the rotor apex and the casing.* Also the side *gas seals and corner seals have not been successful in preventing, mixture and gas leakage.* Further **problems / limitations** *with the Wankel engine are :*

1. High compression ratio difficult to achieve.
2. Inefficient combustion.
3. High surface to volume ratio.
4. Scraping of wall layers into the exhaust ports.
5. Lubrication and cooling problem.
6. Shorter life.
7. Higher specific fuel consumption.

19.5. VARIABLE COMPRESSION RATIO (VCR) ENGINES

19.5.1. Introduction

- It is always desired to develop high specific power output, accompanied by good reliability and longer engine life. The maximum power output can be obtained if maximum amount of fuel can be burnt efficiently and effectively. All the *methods which are used to increase power out entail a number of problems.*
- It has been observed that *a fixed compression ratio engine cannot meet the various requirements of high specific output.* Hence, the development of variable compression ratio (VCR) engine seems to be a necessity.
  - In *the VCR engine a "high compression ratio" is employed for good stability and low load operation and a "low compression ratio" is used at full-load to allow the turbocharger to boost the intake pressures without increasing the peak cycle pressure.* It is worth noting that the greater volume of clearance volume at lower compression ratio would result in increased air intake for the same peak compression pressure and consequently more output will be obtained.
- A S.I. engine can be used as a VCR engine. The concept of variable compression ratio, however, can be more suitably used with the turbocharged diesel engine because of the following *reasons :*
  (*i*) The *VCR concept is beneficial only at part load* and the part-load efficiency of the diesel engine is higher than that of the gasoline engine.
  (*ii*) The diesel engine has better multi-fuel capability.

19.5.2. Methods to Obtain Variable Compression Ratio

Variable compression ratio can be obtained :

(*i*) ***By changing the clearance volume.*** It *requires raising or lowering of either entire crankshaft or piston crown.*

(*ii*) ***By changing both the clearance and swept volume of the engine.*** This method requires a variable throw crankshaft for changing the stroke.

The following mechanisms have been proposed for the VCR engine.

- Fig. 19.13, shows the mechanism by which the displacement can be varied by altering the stroke length. It is not favoured since it mechanism is quite complex.
- Fig. 19.14, shows the mechanism of Varimax engine developed by Tecquiment Ltd. (U.K.) which uses the movement of crankshaft for altering the compression ratio. This mechanism changes the compression ratio from 4.5 : 1 to 20 : 1.
- The most promising VCR mechanism is the use of special piston to lower or raise the piston skirt as developed by British International Combustion Engine, Research Institute (BICERI).

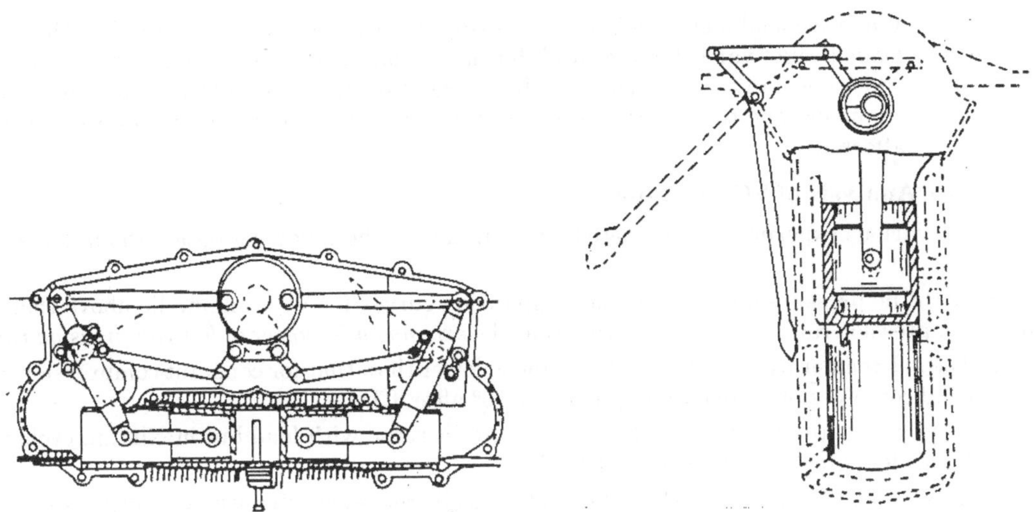

Fig. 19.13. VCR engine by changing the stroke length.

Fig. 19.14. Varimax VCR engine.

Fig. 19.15 shows the VCR system, as developed by BICERI.

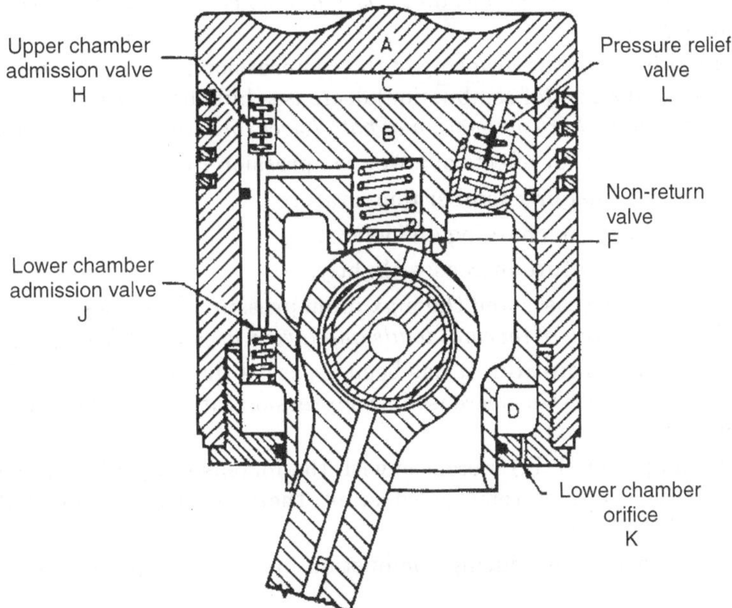

Fig. 19.15. BICERI variable compression ratio engine piston.

— It consists of two main parts A and B as shown in the figure. The part B, known as *carrier* is mounted on the gudgeon pin and part A, known as *shell,* slides on carrier B to vary the clearance volume. These two parts of the piston are so arranged that two chambers C and D are formed between them. The chambers are kept full of lubricating oil via a hole in the connecting rod and non-return valve F from the lubricating system.

— The gas load is carried by the oil in the upper chamber C.

— When the load increases the gas pressure is increased to a pre-set value, the spring loaded relief valve L opens and discharges oil to the main sump. The piston shell slides down to a position decided by the relationship between the oil pressure in two chambers and the cylinder gas pressure, and thus a charge in compression ratio is affected.

19.5.3. Analysis of VCR Engine

From various experiments conducted on VCR engine the following *observations* have been made :

1. Both the pressure and temperature expansion curves for VCR-engine lie above those for constant compression ratio engine, indicating that the *expansion is slower at low compression ratios.*

2. The gas temperature for the VCR-engine is lower than that for constant compression ratio engine for full compression stroke and upto about 50° after T.D.C.

3. Boost pressure and mean cycle temperature increase with load but maximum cycle temperature for VCR engine is lower at higher loads.

4. In case of VCR engines, both b.s.f.c. and i.s.f.c. increase significantly with increase in load.

19.5.4. Performance of VCR Engine

The performance aspects of VCR-engine are discussed below :

1. **Power output.** It has been observed that for the same engine dimension a *VCR-engine develops more power* (*i.e.* it is *very compact and has a high power-to-weight ratio without any penalty on specific fuel consumption*).

2. **Thermal loads.** With the lowering of the compression ratio the first-stage of combustion increases while the second stage (of combustion) decreases but the overall heat release duration is shortened. This leads to *smoother combustion of both lower and higher compression ratios.*

The following are the other effects of lower compression ratio :

(*i*) Reduction in combustion chamber temperature.

(*ii*) There is an increase in charging efficiency.

(*iii*) The exhaust temperature increases slightly.

(*iv*) The ignition lag increases and the maximum pressure decreases. The use of VCR principle *results in lowering of thermal loads and a very high specific output.*

3. **Specific fuel consumption.** It has been observed that the thermal efficiency of the engine decreases considerably with the decrease in compression ratio. This effect, however, is counterbalanced by the following factors :

(*i*) The frictional power (F.P.) increases with the increase in peak cylinder pressure. Since the peaks pressure in VCR engine remains constant, therefore, the F.P. of VCR-engine remains constant irrespective of the load.

(*ii*) Lower rate of expansion during combustion provides adequate time for combustion to complete.

Owing to above mentioned factors it is expected that the b.s.f.c. of VCR-engine should not be much higher than the conventional engine.

4. **Engine noise.** *The noise emanating from the engine depends upon the peak pressure in the cylinder and rate of pressure rise.* The peak pressure affects the lower frequency noise while the rate of pressure rise affects the high frequency noise. *As the peak pressure in VCR-engine remains same irrespective of load on engine, the low frequency noise is reduced. However, as the compression ratio is reduced (i.e. high rates of pressure rise) the noise from the VCR-engine is mainly high frequency noise.*

5. **Starting.** As the VCR-engine uses higher compression ratio at low loads, it has a *good cold starting and idling performance at low ambient temperatures.*

6. **Multi-fuel capability.** The VCR-engine has a good multi-fuel capability due to higher compression ratio at starting and part-load operation.

— The opposed piston type engine is especially suitable for multi-fuel operation.

19.6. FREE-PISTON ENGINE PLANT

Free-piston engine plants are the conventional gas turbine plants *with the difference that the air compressor and combustion chamber are replaced by a free piston engine.* The working of such a plant is explained in Fig. 19.16.

Refer Fig. 19.16. A free-piston engine unit comprises of five cylinders with two assemblies of pistons that move opposite to each other. The pistons are powered by a diesel cylinder located in the centre. Each assembly has a diesel piston at the centre of the unit. These are rigidly connected to large-diameter pistons reciprocating in the large air-compressor cylinders. Each assembly has air-cushion cylinders at the end.

Refer Fig. 19.16 (*a*) :

— The pistons are at their innermost position.

— The diesel cylinder has charge of compressed air at about 100 bar ready for firing.

— The air-compression cylinders are filled with air at atmospheric pressure ; the inlet valves IV_1 and IV_2 have just closed. The air trapped in the *air-cushion cylinders,* also called *bounce cylinders,* is at its lowest pressure.

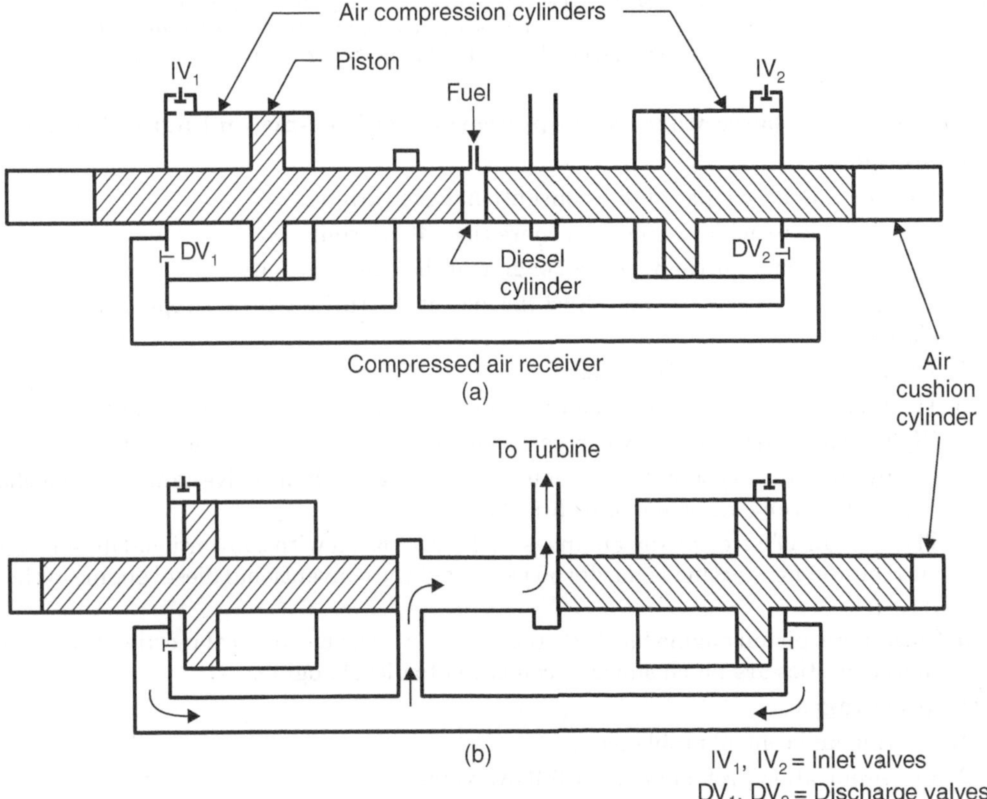

Fig. 19.16. Free-piston engine plant.

— The discharge valves DV_1 and DV_2 in the *air compression cylinders* are held shut by the high-pressure air in the compressed-air receiver that connects the air compression cylinders and the diesel cylinders.

— The compressed air-charge in the diesel cylinder is at high temperature from work of compression, so that a charge of fuel injected immediately ignites and burns. The resulting sudden rise of pressure in the diesel cylinder forces the pistons apart. As they move apart, the exhaust port *leading to the turbine* uncovers first. The combustion products at a pressure of 3 to 6 bar and temperature of about 550°C rush out *through the exhaust to the turbine.* As the piston continue to move to their outermost positions, the inlet port from the compressed-air receiver is uncovered and compressed air from the receiver enters the diesel cylinder to scavenge out the combustion products and fill the cylinder with a fresh charge of air [Fig. 19.16 (*b*)]. During this outward stroke, the air in the cushion cylinders has been compressed. This air now expands and pushes on the end pistons to return the assemblies back to the innermost position, in turn compressing the fresh air charge trapped in the diesel cylinder by the opposed piston closing on each other.

— *Thermal efficiencies of 35%* and more can be developed by these plants. This can be achieved *because the 1900°C combustion gas in the diesel cylinder does work directly on the piston in compressing air. The higher the temperature of gases doing work, the more efficient the process.*

Note. The temperature to the tune of 1900°C can be tolerated because *it appears only instantaneously at ignition and cyclically.* During most of the cycle the cylinder is being cooled by relatively cooler air charges and expanding combustion products. In addition, the diesel cylinder walls are cooled by a water jacket. The materials problem of the gas turbine is alleviated considerably by being required to handle temperatures of 550°C.

Advantages, Disadvantages and Applications of Free-piston Engine Arrangement

Advantages :

1. Less air rate as compared to a conventional gas turbine.
2. It is possible to achieve efficiency more than 40 percent.
3. Lighter and smaller than a diesel engine of the same capacity.
4. The gas turbine is about one third the size of the turbine for a simple open gas turbine plant.
5. The free-piston is vibrationless.
6. The thermal efficiency is considerably higher than simple gas turbine unit.
7. Higher volumetric efficiency can be obtained than conventional diesel engine.
8. Highly simplified system (since connecting rod, crankshaft, valves and valve mechanism required for conventional engine are absent).
9. The starting of a free piston engine is easier compared with conventional diesel engine as air required for starting is hardly 50 percent of the conventional diesel engine of the same capacity.
10. Since a free piston engine has little thermal as well as mechanical inertia, its acceleration characteristics are much superior compared to diesel engine.

Disadvantages :

1. Starting and control problems.
2. Synchronisation problem not yet full overcome.
3. The specific fuel consumption is higher than conventional diesel engine, particularly at part load conditions.
4. The fuel supply can be varied only in a limited range. Any fluctuation in the fuel supply will make the operation unstable.

Applications :

The following are the *applications of free-piston engines* :

1. Widely used as a submarine air compressor units.
2. Suitable for power generation in medium power range. Below 300 kW diesel engines are indispensable as free piston engine of comparable sizes are not being built on commercial scale.
3. Free piston engines are specially suitable for pumping oil ; also the same oil can be used as fuel.

HIGHLIGHTS

1. *Duel-fuel operation,* combines in a simple manner the possibility of operating a diesel engine on liquid fuels such as diesel oil or gas oil and on gaseous fuels such as natural gas, sewage gas and cook oven gas etc.
2. At full-load, duel-fuel engine is superior to diesel engine.
3. A *multi-fuel engine* is one which can operate satisfactorily on a wide variety of fuels ranging from diesel oil, crude oil, JP-4 to lighter fuels like gasoline, and even normal lubricating oil.
4. The *stratified charge engine* is usually defined as a S.I. engine (stratified diesel engine has also been developed) in which the mixture in the zone of spark plug is very much richer than that in the rest of the combustion chamber *i.e.* one which burns leaner overall fuel-air mixtures. *Charge stratification* means providing different fuel-air mixture strengths at various places in the combustion chamber.
5. The basic principle of working of *stirling engine* is the same as that of conventional engine. *The alternate compression at low temperature and expansion at high temperature of a working fluid in the basis for the stirling engine.*
6. In the *VCR-engine* a high compression is used for good stability and low load operation and a low compression is used at full-load to allow the turbocharger to boost the intake pressures without increasing the peak cycle pressure.
7. *Free-piston engine plants* are the conventional gas turbine plants with the *difference that the air compressor and combustion chamber are replaced by free-piston engine.*

OBJECTIVE TYPE QUESTIONS

Fill in the Blanks or Say "Yes" or "No" :

1. A ...... engine is capable of running on either gas or diesel oil or a combination of these two over a wide range of temperature ratio.
2. The use of low octane number fuels in duel-fuel engine results in poor performance of the engine and greatly affects the combustion.
3. In a duel-fuel engine the temperature of inlet charge has no effect on the knocking limits of a particular fuel-air mixture.
4. At full-load, duel-fuel engine is superior to diesel engine.
5. A duel-fuel engine is preferred when cheap gas is easily available.
6. A ...... engine is one which can operate satisfactorily on a wide variety of fuels.
7. ...... means providing different fuel-air mixture strengths at various places in the combustion chamber.
8. The stratified charge engine combines the advantages of both petrol and diesel engines.
9. A stratified charge engine exhibits multi-fuel capability.
10. The stirling engine is an external combustion engine.
11. The part-load efficiency of a stirling engine is very low.
12. In case of stirling engine no lubricating oil is required.

13. The Wankel rotary combustion engine has a poor power volume ratio.
14. Rotors of the Wankel engine are made from high-grade malleable spheroidal graphite iron.
15. The specific fuel consumption of Wankel engine is lower than conventional I.C. engine.
16. Rotary combustion engines are less noisy.
17. The VCR concept is beneficial only at par-load.
18. The use of VCR principle results in lowering of thermal loads and a very high specific output.
19. In a free-piston engine arrangement it is possible to achieve efficiency more than 40%.
20. The free piston is vibrationless.

ANSWERS

1. duel-fuel	**2.** Yes	**3.** No	**4.** Yes	**5.** Yes
6. multi-fuel	**7.** charge stratification		**8.** Yes	**9.** Yes
10. Yes	**11.** No.	**12.** Yes	**13.** No.	**14.** Yes
15. No	**16.** Yes	**17.** Yes	**18.** Yes	**19.** Yes
20. Yes.				

THEORETICAL QUESTIONS

1. What is a duel-fuel engine ?
2. Explain with a neat diagram the working of a duel-fuel engine.
3. Enumerate and explain briefly the various factors which affects combustion in a duel-fuel engine.
4. Discuss briefly the performance characteristics of duel-fuel engines.
5. State the advantages and applications of duel-fuel engines.
6. What is a multi-fuel engine ?
7. What are the requirements of a multi-fuel engine ?
8. List the difficulties associated with multi-fuel operation.
9. Discuss briefly the performance characteristics of multi-fuel engine.
10. What is a stratified charge engine ?
11. Give the classification of stratified charge engines.
12. Explain with a neat sketch Texaco combustion process.
13. State the advantages and disadvantages of stratified charge engines.
14. What is the working principle of stirling engine ?
15. List the features which distinguish the stirling engine from other heat engines.
16. The stirling cycle is superior to the carnot cycle, how ?
17. List the fuels which are used in a stirling engine.
18. Give the differences (in a tabular form) between carnot and stirling engines.
19. Explain briefly the following with regard to a stirling engine :
(*i*) Two piston mechanism
(*ii*) Two piston engine.
20. State the advantages and disadvantages of a stirling engine vis-a-vis conventional I.C. engine.
21. Give the comparison between stirling engine and I.C. engine.
22. Describe with sketches the working principle of Wankel rotary combustion engine.
23. List the features of the Wankel engine.

24. Explain briefly with regard to Wankel engine :
(*i*) Clearance and swept volume ; (*ii*) Compression ratio ;
(*iii*) Engine displacement capacity.

25. Describe briefly the performance characteristics of Wankel engine.

26. State the advantages and applications of rotary combustion engines.

27. Wankel rotary engine could not become successful, why ?

28. What are the advantages of variable compression ratio engine ?

29. Discuss the important designs of variable compression ratio engines and comment on their salient points.

30. Discuss the performance aspects of VCR-engine.

31. Describe with a neat sketch a free-piston engine plant.

32. State the advantages, disadvantages and applications of free-piston engines.

20
Air Compressors

20.1. GENERAL ASPECTS

The compressed air finds application in the following *fields* :

1. It is widely employed for powering small engines, generally those of portable nature. Compressed air is used in such diversified fields as :

(*i*) operating tools in factories ;

(*ii*) operating drills and hammers in road building ;

(*iii*) excavating ;

(*iv*) tunneling and mining ;

(*v*) starting diesel engines ; and

(*vi*) operating brakes on buses, trucks and trains.

2. A large quantity of air at moderate pressure is used in smelting of various metals such as melting iron, in blowing converters, and cupola work.

3. Large quantities of air are used in the air-conditioning, drying, and ventilation fields. In many of these cases, there is little resistance to the flow of air ; and hence it does not have to be compressed (*i.e.*, measurably decreased in volume). For such cases fans serve the purpose of moving the air to the desired location. In other cases, particularly in drying work, there is appreciable resistance to the flow of air and a compressor of some sort is required to build up sufficient pressure to overcome the resistance to flow.

The function of a compressor is to take a definite quantity of fluid (usually gas, and most often air) *and deliver it at a required pressure.*

An air compressor takes in atmospheric air, compresses it and delivers the high-pressure air to a storage vessel from which it may be conveyed by the pipeline to wherever the supply of compressed air is required. Since the process of compressing the gas requires that work should be done upon it, it will be clear that a compressor must be driven by some form of prime-mover. Of the energy received by the compressor from the prime-mover, some will be absorbed in work done against friction, some will be lost to radiation and any coolant which might be employed to cool the machine, and the rest will be maintained within the air itself. The prime-mover converts only a fraction of the heat it receives from the source into work, and so far as the compressor alone is concerned, the energy which it receives is that which is available at the shaft of the prime-mover.

The general arrangement of a compressor set is shown diagrammatically in Fig. 20.1.

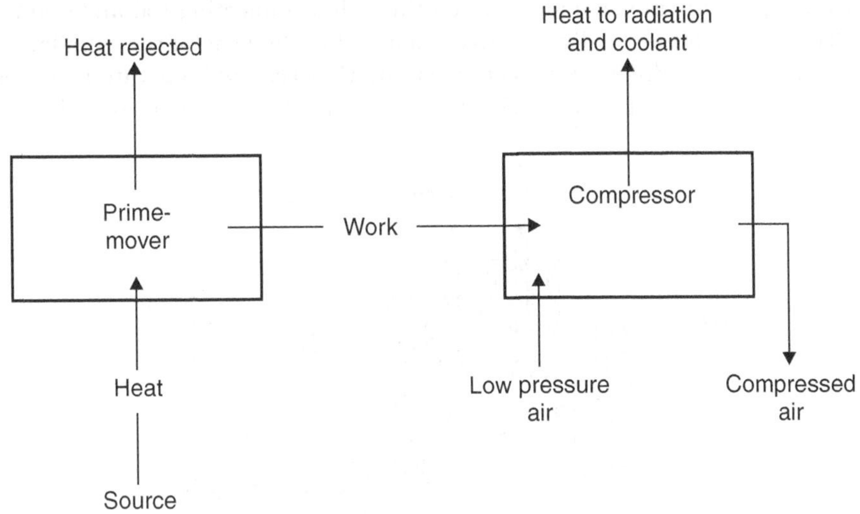

Fig. 20.1. General arrangement of a compressor set.

20.2. CLASSIFICATION OF AIR COMPRESSORS

Air and gas compressors are *classified* into *two main types* :

1. Reciprocating compressors ; and
2. Rotary compressors.

— According to whether or not the process of compressing is carried out in one unit or in several similar units in the one machine, a compressor may be *single-stage*, or *multi-stage*.

— Again, in case of reciprocating compressors, the air may be compressed in the cylinder on one side of the piston only, or use may be made of both piston faces. Such compressors are *single-acting* and *double-acting,* respectively.

— ***Centrifugal compressors,*** which are of the rotary type, may be single or double entry, which means that the compressor is filled with either one or two air intakes according to whether it is of the former or latter type when compression takes place in one or two units, respectively.

Air compressors may be classified in another manner, this time from an aspect of the *use to which they are put.*

— For example, air pumps and exhausters are used to produce vacua, their job being to remove air from a particular system to create a low pressure therein.

— *Blowers* and *superchargers* are essentially air compressors, but the *increase in pressure* which they produce is only small, and upto, say *0.7 to 1.05 bar.*
— A *booster is an air or gas compressor which is employed to raise the pressure of air/gas which has already been compressed.* It is where a slightly higher pressure is required, or where a loss of pressure has occurred in a long delivery line.

20.3. RECIPROCATING COMPRESSORS

20.3.1. Construction and Working of a Reciprocating Compressor (Single-stage)

Fig. 20.2 (*a*) shows a sectional view of a single-stage reciprocating compressor. It consists of a piston which reciprocates in a cylinder, driven through a connecting rod and crank mounted in a crankcase. There are inlet and delivery valves mounted in the head of the cylinder. These valves are usually of the pressure differential type, meaning that they will operate as the result of the difference of pressures across the valve. The working of this type of compressor is as follows :

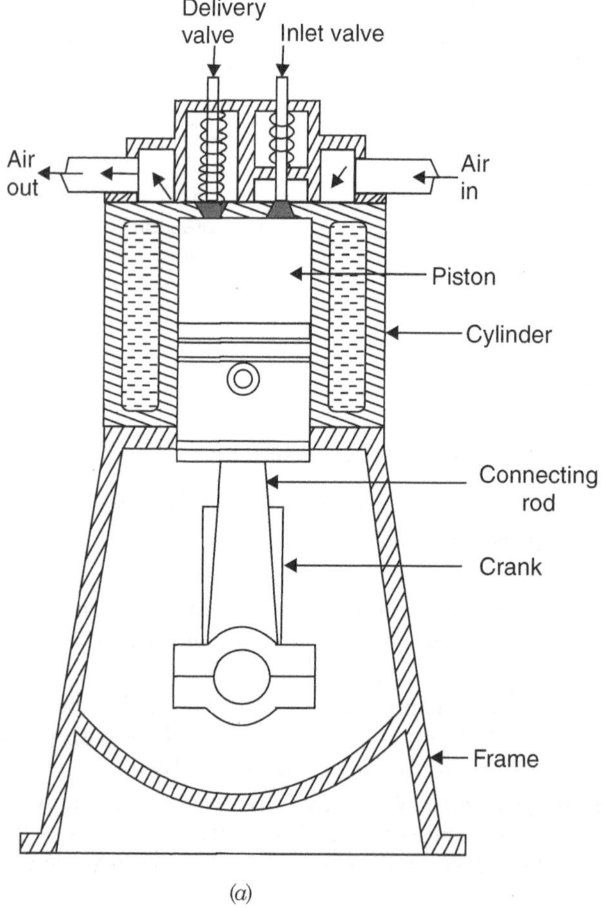

Fig. 20.2. (*a*) Sectional view of a single-stage reciprocating compressor.

As shown in Fig. 20.2 (*b*), the piston is moving down the cylinder and any residual compressed air left in the cylinder after the previous compression will expand and will eventually reach a pressure slightly below intake pressure early or in the stroke. This means that the pressure outside the inlet valve is now higher than on the inside and hence the inlet valve will lift off its seat. A stop is provided to limit its lift and to retain it in its valve seating. Thus a fresh charge of air will be aspirated into the cylinder for the remainder of the *induction stroke,* as it is called. During this stroke the delivery valve will remain closed, since the compressed air on the outside of this valve is at a much higher pressure than the induction stroke.

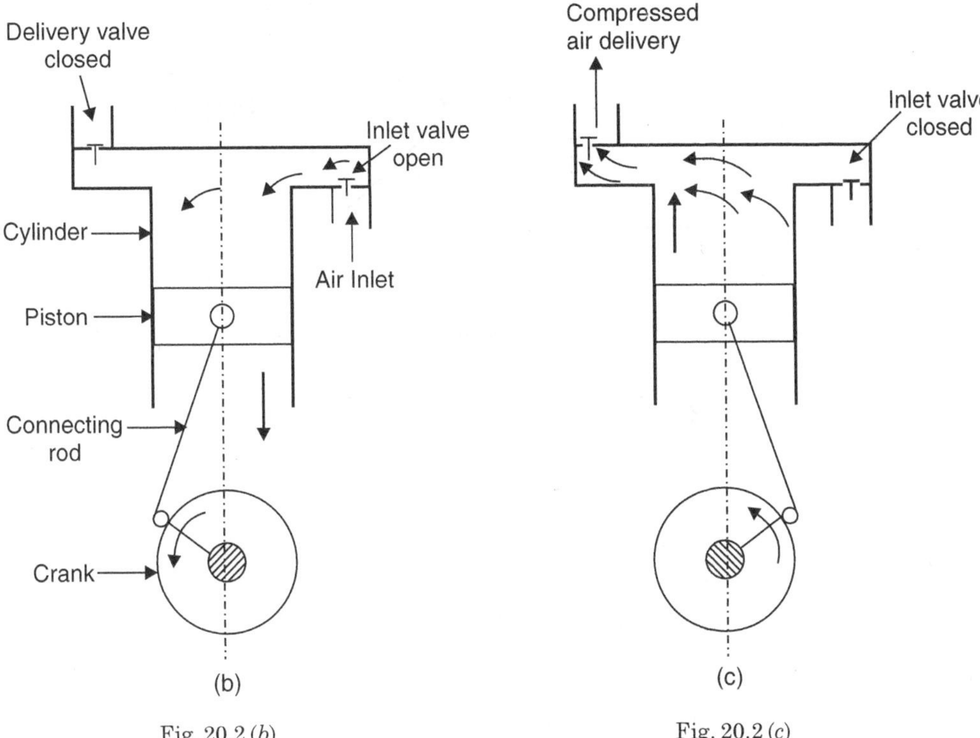

Fig. 20.2 (*b*) Fig. 20.2 (*c*)

As shown in Fig. 20.2 (*c*) the piston is now moving upwards. At the beginning of this upward stroke, a slight increase in cylinder pressure will have closed the inlet valve. Since both the inlet and delivery valves are now closed, the pressure of air will rapidly rise because it is now locked up in the cylinder. Eventually a pressure will be reached which is slightly in excess of the compressed air pressure on the outside of the delivery valve and hence the delivery valve will lift. The compressed air is now delivered from the cylinder in the remainder of the stroke. Once again there is a stop on the delivery valve to limit its lift and to retain it in its seating. At the end of compression stroke piston once again begins to move down the cylinder, the delivery valve closes ; the inlet valve eventually opens and the cycle is repeated.

As air is locked up in the cylinder of a reciprocating compressor then the *compression pressure* for this type of compressor *can be very high. It is limited by the strength of the various parts of the compressor and the power of the driving motor.*

It may be noted that there is *intermittent flow of air in a reciprocating air compressor.*

20.3.2. Single-stage Compressor : Equation for Work (*neglecting clearance*)

In Fig. 20.3 is shown a theoretical p-V diagram for a single-stage reciprocating air compressor, neglecting clearance.

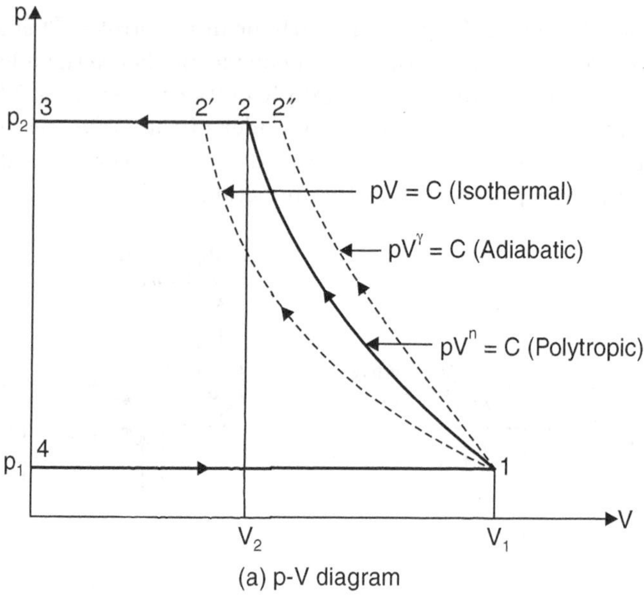

(a) p-V diagram

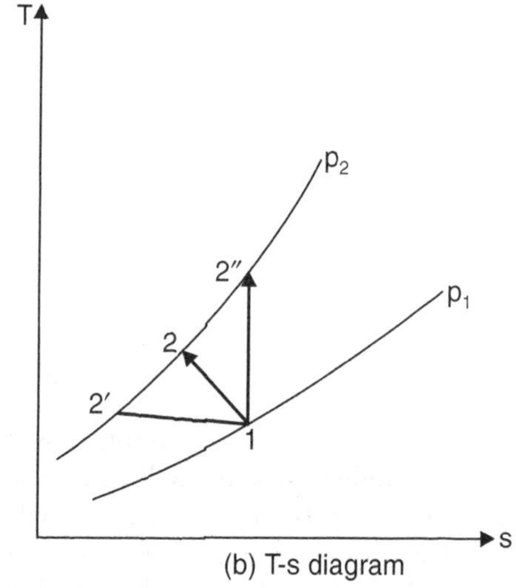

(b) T-s diagram

Fig. 20.3. Theoretical p-V and T-s diagrams for a single-stage reciprocating air compressor.

The sequence of operations as represented on the diagram, are as follows :

(*i*) *Operation 4-1* : Volume of air V_1 aspirated into the compressor at pressure p_1 and temperature T_1.

(*ii*) *Operation 1-2* : Air compressed according to the law $pV^n = C$ from p_1 to pressure p_2. Volume decreases from V_1 to V_2. Temperature increases from T_1 to T_2.

(*iii*) *Operation 2-3* : Compressed air of volume V_2 and at pressure p_2 with temperature T_2 delivered from the compressor.

During compression, due to its excess temperature above the compressor surroundings, the air will lose some heat. Thus, neglecting the internal effect of friction which is small in the case of the reciprocating compressor, the index n *is less than* γ, the adiabatic index. Since work must be put into an air compressor to run it, every effort is made to reduce this amount of work input. Inspection of p-V diagram shows the frictionless adiabatic as 1-2″ and that if compression were along the isothermal 1-2′ instead of polytropic 1-2 then the work done, given by the area of the diagram, would be reduced and, infact, would then be *'minimum'. Isothermal compression cannot be achieved in practice but an attempt is made to approach the isothermal case by cooling the compressor either by addition of cooling fins or a water jacket to the compressor cylinder.* For a reciprocating compressor, a comparison between the actual work done during compression and the ideal isothermal work done is made by means of the ***isothermal efficiency.***

This is defined as,

$$\text{Isothermal efficiency} = \frac{\text{Isothermal work done}}{\text{Actual work done}}$$

Thus, the higher the isothermal efficiency, the more nearly has the actual compression approached the ideal isothermal compression.

Total shaft work done/cycle, W = Area 41234

or W = Area under 4-1 – Area under 1-2 – Area under 2-3

$$= p_1V_1 - \frac{p_2V_2 - p_1V_1}{n-1} - p_2V_2$$

$$= (p_1V_1 - p_2V_2) - \left(\frac{p_2V_2 - p_1V_1}{n-1}\right) = (p_1V_1 - p_2V_2) + \left(\frac{p_1V_1 - p_2V_2}{n-1}\right)$$

$$= \left(1 + \frac{1}{n-1}\right)(p_1V_1 - p_2V_2)$$

$$\therefore \quad W = \left(\frac{n}{n-1}\right)(p_1V_1 - p_2V_2) \quad \text{...(20.1)}$$

This equation can be modified as follows :

$$W = \frac{n}{n-1}(p_1V_1 - p_2V_2) = \frac{n}{n-1} \cdot p_1V_1\left(1 - \frac{p_2V_2}{p_1V_1}\right) \quad \text{...(20.2)}$$

Now $\quad p_1V_1^n = p_2V_2^n$

$$\therefore \quad \frac{V_2}{V_1} = \left(\frac{p_1}{p_2}\right)^{1/n}$$

and substituting this into eqn. (20.2), we have

$$W = \frac{n}{n-1}\,p_1V_1\left\{1 - \frac{p_2}{p_1}\left(\frac{p_1}{p_2}\right)^{1/n}\right\} = \frac{n}{n-1}\,p_1V_1\left\{1 - \frac{p_2}{p_1}\left(\frac{p_2}{p_1}\right)^{-\frac{1}{n}}\right\}$$

$$= \frac{n}{n-1}\,p_1V_1\left\{1 - \left(\frac{p_2}{p_1}\right)^{1-\frac{1}{n}}\right\} = \frac{n}{n-1}\,p_1V_1\left\{1 - \left(\frac{p_2}{p_1}\right)^{\frac{n-1}{n}}\right\} \quad \text{...(20.3)}$$

The solution to this equation will always come out *negative* showing that work must be done *on* the compressor. Since only the *magnitude* of the work done is required from the expression then it is often written,

$$W = \frac{n}{n-1} p_1V_1 \left\{\left(\frac{p_2}{p_1}\right)^{\frac{n-1}{n}} - 1\right\} \qquad ...(20.4)$$

$$= \frac{n}{n-1} mRT_1 \left\{\left(\frac{p_2}{p_1}\right)^{\frac{n-1}{n}} - 1\right\} \qquad ...(20.5)$$

If the air delivery temperature T_2 is required then this can be obtained by using this equation :

$$\frac{T_2}{T_1} = \left(\frac{p_2}{p_1}\right)^{\frac{n-1}{n}} \quad \text{or} \quad T_2 = T_1 \left(\frac{p_2}{p_1}\right)^{\frac{n-1}{n}} \qquad ...(20.6)$$

20.3.3. Equation for Work (with clearance volume)

In practice, all reciprocating compressors will have a *clearance volume.* The clearance volume is *that volume which remains in the cylinder after the piston has reached the end of its inward stroke.*

Refer Fig. 20.4. At point 1, the cylinder is full of intake air, volume V_1 and the piston is about to commence its compression stroke. The air is compressed polytropically according to some law $pV^n = C$ to delivery pressure p_2 and volume V_2. At 2 the delivery valve theoretically opens and for the remainder of the stroke, 2 to 3, the compressed air is delivered from the cylinder. At 3 the piston has reached the end of its inward stroke and so on, delivery of compressed air ceases at 3. V_3

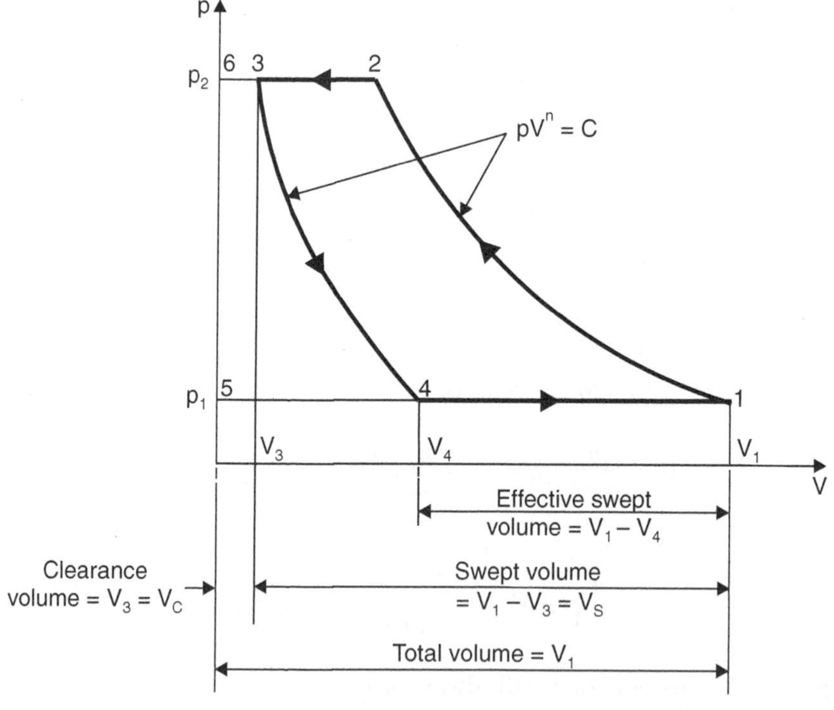

Fig. 20.4

is the clearance volume and is filled at this stage with compressed air. As the piston begins the intake stroke *this residual air will expand,* according to some polytropic law $pV^n = C$, and it is not *until the pressure has reduced to intake pressure* at 4 that the inlet valve will begin to open thus permitting the intake of a fresh charge of air. For the remainder of the intake stroke a fresh charge is taken into the cylinder. This volume $(V_1 - V_4)$ is *effective swept volume.*

Work done/cycle, W = Net area 12341 = Area 51265 – Area 54365

Assuming the polytropic index to be same for both compression and clearance expansion, then,

$$W = \frac{n}{n-1} p_1V_1 \left\{\left(\frac{p_2}{p_1}\right)^{\frac{n-1}{n}} - 1\right\} - \frac{n}{n-1} p_4V_4 \left\{\left(\frac{p_3}{p_4}\right)^{\frac{n-1}{n}} - 1\right\} \quad ...(20.7)$$

But $p_4 = p_1$ and $p_3 = p_2$, then eqn. (20.7) becomes,

$$W = \frac{n}{n-1} p_1V_1 \left\{\left(\frac{p_2}{p_1}\right)^{\frac{n-1}{n}} - 1\right\} - \frac{n}{n-1} p_1V_4 \left\{\left(\frac{p_2}{p_1}\right)^{\frac{n-1}{n}} - 1\right\}$$

$$= \frac{n}{n-1} p_1(V_1 - V_4) \left\{\left(\frac{p_2}{p_1}\right)^{\frac{n-1}{n}} - 1\right\} \quad ...(20.8)$$

20.3.4. Volumetric Efficiency

Refer Fig. 20.4. *The* ***volumetric efficiency*** *of a compressor is the ratio of free air delivered to the displacement of the compressor.* It is also the *ratio of effective swept volume to the swept volume.*

i.e., $$\text{Volumetric efficiency} = \frac{\text{Effective swept volume}}{\text{Swept volume}} = \frac{V_1 - V_4}{V_1 - V_3} \quad ...(20.9)$$

Because of presence of clearance volume, volumetric efficiency is always *less than unity.* As a percentage, it usually varies from *60% to 85%.*

The ratio, $$\frac{\text{Clearance volume}}{\text{Swept volume}} = \frac{V_3}{V_1 - V_3} = \frac{V_c}{V_s} = k \quad ...(20.10)$$

is the *clearance ratio.*

As a percentage, this ratio will have a value, in general, of between *4% and 10%.* The greater the pressure ratio through a reciprocating compressor, then the greater will be the effect of the clearance volume since the clearance air will now expand through a greater volume before intake conditions are reached. The cylinder size and stroke being fixed, however will mean that $(V_1 - V_4)$, the effective swept volume, will reduce as the pressure ratio increases and thus the volumetric efficiency reduces.

$$\text{Volumetric efficiency, } \eta_{vol.} = \frac{V_1 - V_4}{V_1 - V_3}$$

$$= \frac{(V_1 - V_3) + (V_3 - V_4)}{(V_1 - V_3)} = 1 + \frac{V_3}{V_1 - V_3} - \frac{V_4}{V_1 - V_3}$$

$$= 1 + \frac{V_3}{V_1 - V_3} - \frac{V_4}{V_1 - V_3} \cdot \frac{V_3}{V_3} = 1 + \frac{V_3}{V_1 - V_3} - \frac{V_3}{V_1 - V_3} \cdot \frac{V_4}{V_3}$$

$$= 1 + k - k \,.\, \frac{V_4}{V_3}$$

$$= 1 + k - k \left(\frac{p_3}{p_4}\right)^{1/n} \qquad \left| \begin{array}{l} p_3 V_3^{\,n} = p_4 V_4^{\,n} \\ \dfrac{V_4}{V_3} = \left(\dfrac{p_3}{p_4}\right)^{1/n} \end{array} \right|$$

or $$\eta_{vol.} = 1 + k - k \left(\frac{p_2}{p_1}\right)^{1/n} \qquad (\because \; p_3 = p_2, p_4 = p_1) \qquad \text{...(20.11)}$$

or $$\eta_{vol.} = 1 + k - k \left(\frac{V_1}{V_2}\right) \qquad \text{...(20.12)}$$

The above equations are valid if the index of expansion and compression is same. However, it may be noted that the *clearance volumetric efficiency is dependent on only the index of expansion of the clearance volume from V_3 to V_4*. Thus, if the index of compression = n_c and index of expansion = n_e, the volumetric efficiency is given by

$$\eta_{vol.} = 1 + k - k \left(\frac{p_3}{p_4}\right)^{1/n_e} \qquad \text{...(20.13)}$$

$$= 1 + k - k \left(\frac{p_2}{p_1}\right)^{1/n_e} \qquad \text{...(20.14)}$$

$$= 1 + k - k \left(\frac{V_4}{V_3}\right) \qquad \text{...(20.15)}$$

In this case volumetric efficiency = $1 + k - k \left(\dfrac{V_1}{V_2}\right)$.

In practice the air that is sucked in during the induction (suction) stroke gets heated up while passing through the hot valves and coming in contact with hot cylinder walls. There is wire drawing effect through the valves resulting in drop in pressure. Thus the ambient conditions are different from conditions obtained at state 1 in Fig. 20.4.

Let $p_{amb.}$ = Pressure of ambient air, and

$T_{amb.}$ = Temperature of ambient air

$$\therefore \qquad \frac{p_{amb.} V_{amb.}}{T_{amb.}} = \frac{p_1 (V_1 - V_4)}{T_1}$$

Thus, $$V_{amb.} = \frac{p_1 \times T_{amb.}}{T_1 \times p_{amb.}} \times (V_1 - V_4)$$

Thus volumetric efficiency referred to ambient conditions may be written as

$$\eta_{vol.\,(amb.)} = \frac{V_{amb.}}{V_1 - V_3} = \frac{p_1 \times T_{amb.}}{T_1 \times p_{amb.}} \times \frac{V_1 - V_4}{V_1 - V_3}$$

But from eqn. (20.11)

$$\frac{V_1 - V_4}{V_1 - V_3} = 1 + k - k \left(\frac{p_2}{p_1}\right)^{1/n}$$

$$\therefore \qquad \eta_{vol.\,(amb.)} = \frac{p_1 \times T_{amb.}}{T_1 \times p_{amb.}} \left[1 + k - k \left(\frac{p_2}{p_1} \right)^{1/n} \right] \qquad ...(20.16)$$

$$= \frac{p_1 \times T_{amb.}}{T_1 \times p_{amb.}} \left[1 + k - k \left(\frac{V_2}{V_1} \right) \right] \qquad ...(20.17)$$

(*This efficiency should not be used for finding out the dimensions of the cylinder. For finding out the dimensions of the cylinder, the volumetric efficiency based on suction condition only should be used*).

Fig. 20.5 shows the manner in which the volumetric efficiency varies with delivery pressure. Theoretically, the volumetric efficiency is 100% when the delivery pressure equals that of the surroundings, and in fact no compression takes place at all. It decreases rapidly with increase in delivery pressure at first, and then more slowly for increase in delivery pressure at higher pressure.

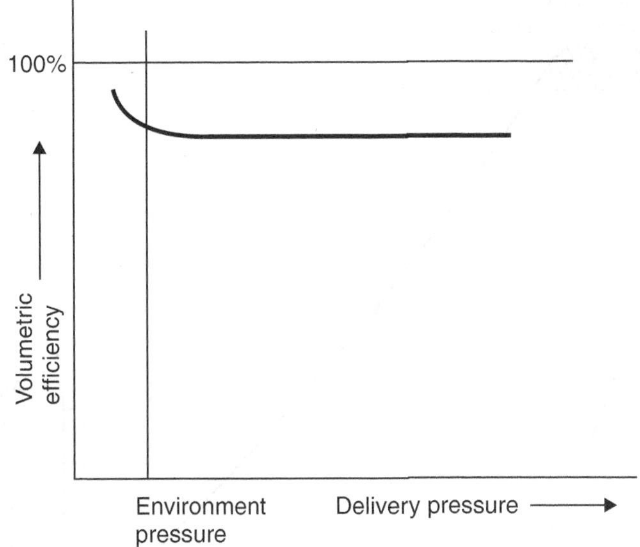

Fig. 20.5. Variation of volumetric efficiency with delivery pressure.

The *volumetric efficiency is lowered* by any of the following conditions :

(*i*) Very high speed

(*ii*) Leakage past the piston

(*iii*) Too large a clearance volume

(*iv*) Obstruction at inlet valves

(*v*) Overheating of air by contact with hot cylinder walls.

(*vi*) Inertia effect of air in suction pipe.

By paying careful attention in the design of the compressor to these causes of loss, an improvement in volumetric efficiency can be obtained.

20.3.5. Actual p-V (indicator) Diagram for Single-stage Compressor

Fig. 20.6 shows an actual compressor diagram. 1234 is the theoretical p-V diagram (already discussed). At point 4, when the clearance air has reduced to atmospheric pressure. The inlet valve in practice will not open. There are two main reasons for this : (*i*) there must be a pressure difference across the inlet valve in order to move it and (*ii*) inlet valve inertia. Thus, the pressure drops away until the valve is forced off its seat. Some *valve bounce* will then set in, as shown by the wavy line, and eventually intake will become near enough-steady at some pressure below atmospheric pressure. This negative pressure difference, called the *intake depression* settles naturally, showing that what is called suction is really the atmospheric air forcing its way into the cylinder against reduced pressure. A similar situation occurs at 2, at the beginning of compressed air delivery. There is a constant pressure rise, followed by valve bounce and the pressure then settles at some pressure above external delivery pressure. Compressed air is usually delivered into a tank called the receiver and hence external delivery pressure is sometimes called the receiver pressure.

Other small effects at inlet and delivery would be *gas inertia* and *turbulence.*

The practical effects (discussed above) are responsible for the addition of the two small shaded *negative work areas* shown in Fig. 20.6.

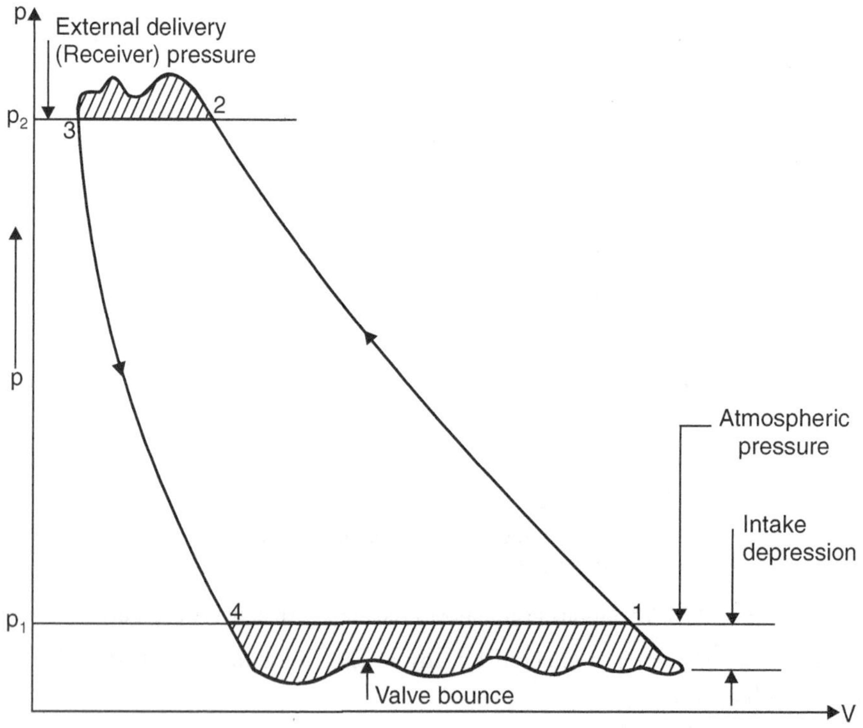

Fig. 20.6. Actual compressor p-V diagram.

20.3.6. Multistage Compression

In a single-stage reciprocating compressor if the delivery is restricted the delivery pressure will increase. If the delivery pressure is increased too far, however, certain disadvantages appear. Referring to Fig. 20.7 assume that the single-stage compressor is compressing to pressure p_2, the complete cycle to 1234. Clearance air expansion will be 3-4 and mass flow through the compressor will be controlled by the effective swept volume $(V_1 - V_4)$. Assume now that a restriction is now placed on delivery. The delivery pressure becomes p_5 say, and the cycle is then 1567, clearance expansion being 6-7. The mass flow through the compressor is now controlled by effective swept volume ($V_1 - V_7$), which is less than ($V_1 - V_4$). In the limit, assuming the compressor to be strong enough, the compression would take place 1-8, where V_s = clearance volume, in which case there would be no delivery. It is seen, therefore, that as the *delivery pressure for a single-stage, reciprocating compressor is increased so the mass flow through the compressor decreases.* Note, also that as *the delivery pressure is increased, so also will the delivery temperature increase.* Referring to Fig. 20.7, $T_8 > T_5 > T_2$. *If high temperature air is not a requirement of the compressed air delivered, then, any increase in temperature represents an energy loss.*

If high pressure is to be delivered by a single-stage machine then it will require heavy working parts in order to accommodate the high pressure ratio through the machine. This will increase the balancing problem and the high torque fluctuation will require a heavier flywheel installation. Such disadvantages can be overcome by multi-stage compression.

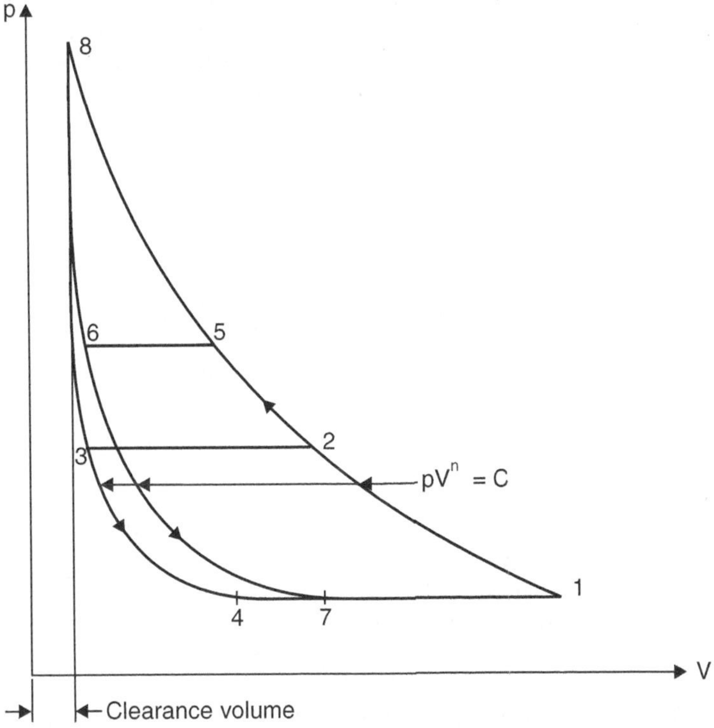

Fig. 20.7

Multi-stage compression is *very efficient* and is now a days almost universally adopted except for compressors where the overall pressure rise required is small. The method is not only advantageous from a thermodynamic point of view, but also has mechanical advantages over single-stage compression.

Advantages :

The important *advantages of multi-stage compression* can be summed up as follows :

1. The *air can be cooled* at pressures intermediate between intake and delivery pressures.
2. The *power* required to drive a multi-stage machine is *less* than would be required by a single-stage machine delivering the same quantity of air at the same delivery pressure.
3. Multi-stage machines have *better mechanical balance.*

4. The pressure range (and hence also the temperature range) may be kept within desirable limits. This results in (*i*) *reduced losses due to air leakage* (*ii*) *improved lubrication,* due to lower temperatures and (*iii*) *improved volumetric efficiency.*
5. The cylinder, in a single-stage machine, must be robust enough to withstand the delivery pressure. The down pressure cylinders of a multi-stage machine may be *lighter in construction* since the maximum pressure there in is low.

Disadvantages :

Inspite of all these advantages, a multi-stage compressor with intercoolers is likely to be more expensive in initial cost than a single-stage compressor of the same capacity.

Multi-stage reciprocating compressors

Multistage compression is a series arrangement of cylinders in which compressed air from the cylinder before, becomes the intake air for the cylinder which follows. This is illustrated in Fig. 20.8. The low pressure ratio in the low-pressure cylinder means that the clearance air expansion is reduced and the effective swept volume of this cylinder is increased. Since in this cylinder which controls the mass flow through the machine, because it is this cylinder which introduces the air into the machine, then there is greater mass flow through the multi-stage arrangement than the single-stage machine.

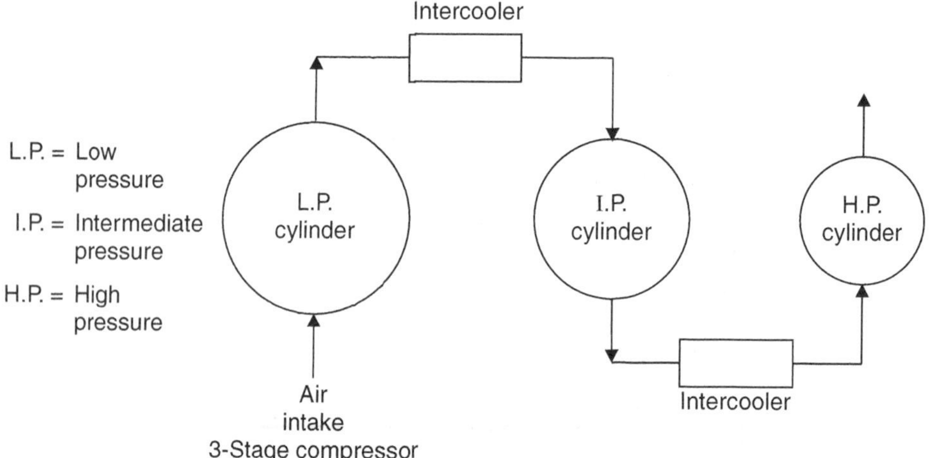

Fig. 20.8

If an *intercooler* is installed between cylinders, in which the compressed air is cooled between cylinders, then the *final delivery temperature is reduced.* This reduction in temperature means a reduction in internal energy of the delivered air, and since this energy must have come from the input energy required to drive the machine, this results in a *decrease in input work requirement for a given mass of delivered air.*

It is common to find machines with either two or three stages of compression. *The complexity of the machinery limits the number of stages.*

Refer Fig. 20.8. The cylinders are shown with diameters which decrease as the pressure increases. This is because, as the pressure increases, so the volume of a given mass of gas decreases. There is continuity of mass flow through a compressor and hence each following cylinder will require a smaller volume due to its increased pressure range. This reduction in volume is usually accomplished by reducing the cylinder diameter.

Fig. 20.9 shows cycle arrangements in the development of the *ideal conditions* required for multi-stage compression. For simplicity, clearance is neglected.

Referring to Fig. 20.9, the overall pressure range is p_1 to p_3. Cycle 8156 is that of single-stage compressor. Cycles 8147 and 7456 are that of a two-stage compressor without intercooling between cylinders. Cycles 8147 and 7236 are that of a two-stage compressor with *perfect intercooling* between cylinders.

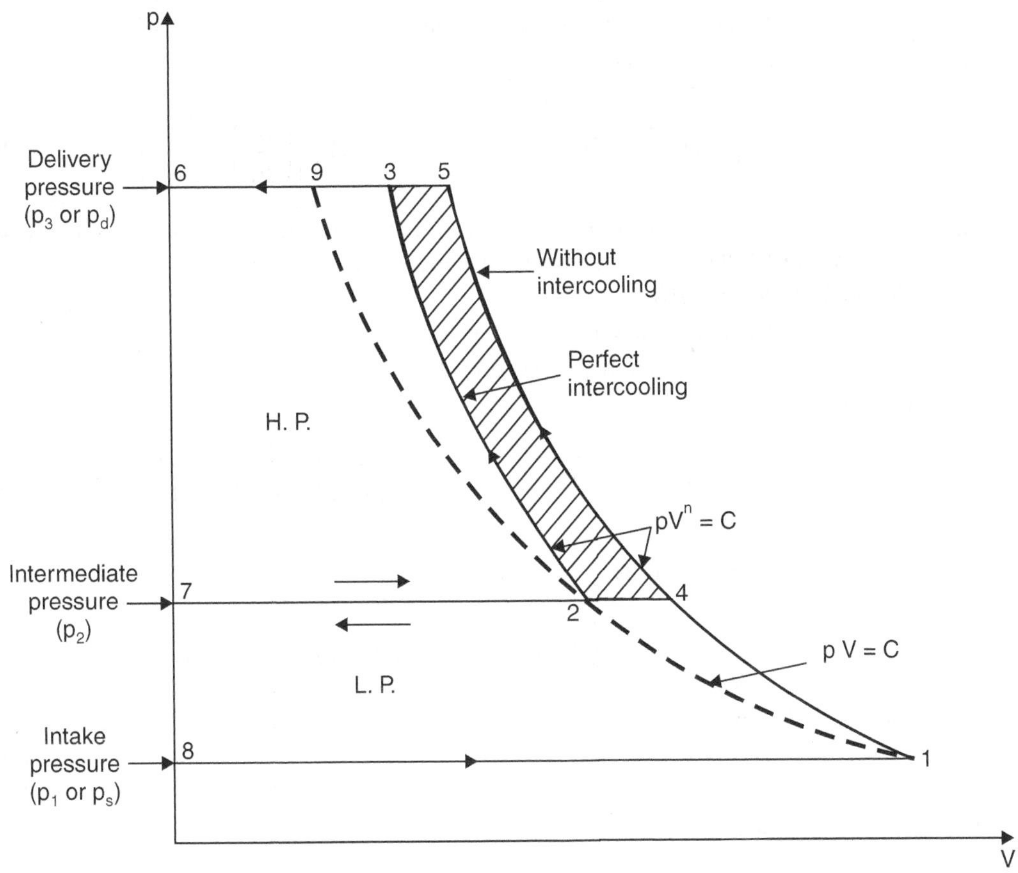

Fig. 20.9

'Perfect intercooling' means that after the initial compression in the L.P. cylinder, with its consequent temperature rise, the air is cooled in an intercooler back to its original temperature. This means, referring to Fig. 21.9, $T_2 = T_1$, in which case point 2 lies on isothermal through point 1. This shows that multi-stage compression, with perfect intercooling, approaches more closely the ideal isothermal compression than in the case with single-stage compression.

Ideal conditions for multi-stage compressors :

Case 1. Single-stage compressor :

As earlier stated cycle 8156 is that of a single-stage compressor, neglecting clearance. For this cycle,

$$W = \frac{n}{n-1} p_1 V_1 \left[\left(\frac{p_5}{p_1} \right)^{\frac{n-1}{n}} - 1 \right] \quad ...(20.18)$$

Delivery temperature, $T_5 = T_1 \left(\frac{p_5}{p_1}\right)^{\frac{n-1}{n}}$ ...(20.19)

Case 2. Two-stage compressor :

(*i*) ***Without intercooling***

8147 ...... Low pressure cycle

7456 ...... High pressure cycle.

For this arrangement work done,

$$W = \frac{n}{n-1} p_1V_1 \left[\left(\frac{p_4}{p_1}\right)^{\frac{n-1}{n}} - 1\right] + \frac{n}{n-1} \cdot p_4V_4 \left[\left(\frac{p_5}{p_4}\right)^{\frac{n-1}{n}} - 1\right] \quad ...(20.20)$$

This will give the same result as that of eqn. (20.18). The final delivery temperature will also be given by eqn. (20.19), because there is no intercooling.

(*ii*) ***With perfect intercooling***

8147 ...... Low pressure cycle

7236 ...... High pressure cycle.

For this arrangement,

$$W = \frac{n}{n-1} p_1V_1 \left[\left(\frac{p_4}{p_1}\right)^{\frac{n-1}{n}} - 1\right] + \frac{n}{n-1} p_2V_2 \left[\left(\frac{p_3}{p_2}\right)^{\frac{n-1}{n}} - 1\right] \quad ...(20.21)$$

Delivery temperature is given by

$$T_3 = T_2 \left(\frac{p_3}{p_2}\right)^{\frac{n-1}{n}} = T_1 \left(\frac{p_3}{p_2}\right)^{\frac{n-1}{n}}, \text{ Since } T_2 = T_1 \quad ...(20.22)$$

Now, since $T_2 = T_1$, then

$$p_2V_2 = p_1V_1 \quad ...(20.23)$$

Also $$p_4 = p_2 \quad ...(20.24)$$

Inserting eqns. (20.23) and (20.24) in eqn. (20.21)

$$W = \frac{n}{n-1} p_1V_1 \left[\left(\frac{p_2}{p_1}\right)^{\frac{n-1}{n}} + \left(\frac{p_3}{p_2}\right)^{\frac{n-1}{n}} - 2\right] \quad ...(20.25)$$

Now, inspection of Fig. 20.9 *shows the shaded area 2453 which is the work saving which occurs as a result of using an intercooler.*

Conditions for minimum work

It will be observed from the Fig. 20.9 that as intermediate pressure $p_2 \to p_1$, then ares 2453 $\to$ 0. Also as $p_2 \to p_3$, then area 2453 $\to$ 0. This means, therefore, that an intermediate pressure p_2 exists which makes area 2453 a maximum. This is the condition when W is a minimum.

Inspection of eqn. 20.25 shows that for minimum W, $[(p_2/p_1)^{n-1/n} + (p_3/p_2)^{n-1/n}]$ must be minimum, all other parts of the equation being constant in this consideration and p_2 is the variable.

$$\text{Hence, for } minimum, W, \quad \frac{dW}{dp_2} = \frac{d\left[\left(\frac{p_2}{p_1}\right)^{\frac{n-1}{n}} + \left(\frac{p_3}{p_2}\right)^{\frac{n-1}{n}}\right]}{dp_2} = 0$$

Differentiating, with respect to p_2,

$$\frac{1}{(p_1)^{\frac{n-1}{n}}} \times \left(\frac{n-1}{n}\right) p_2^{\left(\frac{n-1}{n}\right)-1} + (p_3)^{\frac{n-1}{n}} \times -\left(\frac{n-1}{n}\right)(p_2)^{-\left(\frac{n-1}{n}\right)-1} = 0$$

or
$$\frac{1}{p_1^{\frac{n-1}{n}}} \times \left(\frac{n-1}{n}\right)(p_2)^{-1/n} = (p_3)^{\frac{n-1}{n}} \times \left(\frac{n-1}{n}\right) p_2^{\frac{-2n+1}{n}}$$

or
$$\frac{p_2^{-1/n}}{p_2^{\left(\frac{-2n+1}{n}\right)}} = (p_1 p_3)^{\frac{n-1}{n}}$$

or
$$p_2^{-1/n} \times p_2^{-\left(\frac{-2n+1}{n}\right)} = (p_1 p_3)^{\frac{n-1}{n}}$$

or
$$p_2^{-1/n} \; p_2^{\frac{2n-1}{n}} = (p_1 p_3)^{\frac{n-1}{n}}$$

or
$$p_2^{\frac{2n-2}{n}} = (p_1 p_3)^{\frac{n-1}{n}}$$

$\therefore$
$$p_2^2 = p_1 p_3 \qquad \text{...(20.26)}$$

or
$$p_2 = \sqrt{p_1 p_3} \qquad \text{...(20.27)}$$

and
$$\frac{p_2}{p_1} = \frac{p_3}{p_2} \qquad \text{...(20.28)}$$

or *pressure ratio per stage is equal.*

p_2 obtained from eqn. (20.27) will give ideal intermediate pressure which, with perfect intercooling, will give the minimum *W*.

With these ideal conditions, inserting equations (20.23), (20.24) and (20.28) into eqn. (20.21) shows that *there is equal work per cylinder.*

Hence,
$$W = \frac{2n}{n-1} p_1 V_1 \left[\left(\frac{p_2}{p_1}\right)^{\frac{n-1}{n}} - 1\right] \qquad \text{...(20.29)}$$

Inserting eqn. (20.27) in eqn. (20.29), we get

$$W = \frac{2n}{n-1} p_1 V_1 \left[\left\{\frac{(p_1 p_3)^{1/2}}{p_1}\right\}^{\frac{n-1}{n}} - 1\right]$$

$$= \frac{2n}{n-1} p_1 V_1 \left[\left\{\left(\frac{p_3}{p_1}\right)^{1/2}\right\}^{\frac{n-1}{n}} - 1\right]$$

$$\therefore \qquad W = \frac{2n}{n-1} p_1 V_1 \left[\left(\frac{p_3}{p_1} \right)^{\frac{n-1}{2n}} - 1 \right] \qquad ...(20.30)$$

Note that p_3 / p_1 is the pressure ratio through the compressors.

Case 3. *Multistage compressor*

From the analysis of compressor so far, for a *single-stage* compressor,

$$W = \frac{n}{n-1} p_1 V_1 \left[\left(\frac{p_2}{p_1} \right)^{\frac{n-1}{n}} - 1 \right]$$

For a *two-stage* compressor,

$$W = \frac{2n}{n-1} p_1 V_1 \left[\left(\frac{p_3}{p_1} \right)^{\frac{n-1}{2n}} - 1 \right]$$

It seems reasonable to assume, therefore, that for a *three-stage* machine,

$$W = \frac{3n}{n-1} p_1 V_1 \left[\left(\frac{p_4}{p_1} \right)^{\frac{n-1}{3n}} - 1 \right]$$

and for *x-stage* compressor,

$$W = \frac{xn}{n-1} p_1 V_1 \left[\left(\frac{p_{(x+1)}}{p_1} \right)^{\frac{n-1}{xn}} - 1 \right] \qquad ...(20.31)$$

This equation is very important, since it *applies to any type of compressor or motor, and even to vapour engines, provided n = or < γ.*

Note that $\frac{p_{(x+1)}}{p_1}$ is the pressure ratio through the compressor, in each case.

Note, also, that since for an ideal compressor there is equal work per cylinder, for an x-stage compressor

$$W = \frac{xn}{n-1} p_1 V_1 \left[\left(\frac{p_2}{p_1} \right)^{\frac{n-1}{n}} - 1 \right] \qquad ...(20.32)$$

To determine the intermediate pressures for an *x-stage* machine running under ideal conditions, use is made of eqn. (20.28).

Here it is shown that the pressure ratio per stage is equal.

Hence, for an *x-stage* machine,

$$\frac{p_2}{p_1} = \frac{p_3}{p_2} = \frac{p_4}{p_3} \; ..... \; \frac{p_{(x+1)}}{p_x} = Z, \textit{ say} \qquad ...(20.33)$$

From this,

$$p_2 = Zp_1$$

$$p_3 = Zp_2 = Z^2 p_1$$

$$p_4 = Zp_3 = Z^3 p_1$$

$$\vdots$$

$$p_{x+1} = Zp_x = Z^x p_1$$

$$\therefore \quad Z^x = \frac{p_{x+1}}{p_1}$$

or
$$Z = \sqrt[x]{\frac{p_{(x+1)}}{p_1}} = x\sqrt{\text{(Pressure ratio through compressor)}} \qquad ...(20.34)$$

Inserting the value of Z in eqn. (20.33) will determine the intermediate pressures.

In the event of *intercooling being imperfect we must treat each stage as a separate compressor*, in which case 'x' in eqn. (20.31) will be unity. With this special value of 'x' the power per stage can be calculated, and finally the total power is the sum of the powers per stage :

$$W = \frac{n_1}{n_1 - 1} p_1 V_1 \left[\left(\frac{p_2}{p_1}\right)^{\frac{n_1 - 1}{n_1}} - 1\right] + \frac{n_2}{n_2 - 1} p_2 V_2 \left[\left(\frac{p_3}{p_2}\right)^{\frac{n_2 - 1}{n_2}} - 1\right] + ...$$

Heat rejection per stage per kg of air :

If the air is cooled to its initial temperature the whole of the work done in compression must be rejected to the cooling medium.

Hence for a *single-stage* the heat rejected is given by,

Heat rejected
$$W = \frac{n}{n - 1} p_1 V_1 \left[\left(\frac{p_2}{p_1}\right)^{\frac{n - 1}{n}} - 1\right] \qquad ...(20.35)$$

and since for 1 kg of air, $p_1 V_1 = RT_1$ and $\frac{T_2}{T_1} = \left(\frac{p_2}{p_1}\right)^{\frac{n-1}{n}}$, then eqn. (20.35) may be written as

$$W = \frac{n}{n - 1} RT_1 \left[\frac{T_2}{T_1} - 1\right] \text{ per kg of air}$$

$$= \frac{n}{n - 1} \frac{R}{J} (T_2 - T_1) \text{ heat units} \qquad \left[\begin{array}{l} J = 1......\text{S.I. units} \\ J = 427...... \text{M.K.S. units} \end{array}\right]$$

But
$$\frac{R}{J} = c_p - c_v$$

$$W = \frac{n}{n - 1} (c_p - c_v)(T_2 - T_1) \qquad ...(20.36)$$

$\therefore$ ***Heat rejected with perfect intercooling***

$$= \left[c_p + c_v \left(\frac{\gamma - n}{n - 1}\right)\right] (T_2 - T_1) \text{ per kg of air} \qquad ...(20.37)$$

$\left[\textbf{Note.} \ \frac{n}{n-1}(c_p - c_v) = c_p + \frac{c_v(\gamma - n)}{n - 1}\right]$

The first term in eqn. (20.37) represents the *heat rejected at constant pressure in the intercooler* ; whilst the second term represents the *heat rejected during compression alone ;* and writing $c_v = R/J\,(\gamma - 1)$ it may be reduced to the form

$$\frac{\gamma - n}{\gamma - 1} \times \text{work done in heat units}$$

To find the change in entropy (s) during the first stage of compression, we have, from the definition of entropy,

$$ds = \frac{dW}{T} = \frac{dQ}{T} \qquad (\because \text{ Work done} = \text{Heat rejected})$$

Differentiating eqn. (20.37) and dividing by T,

$$ds = \frac{dW}{T} = \left[c_p + c_v\left(\frac{\gamma - n}{n - 1}\right)\right]\frac{dT}{T}$$

Integration gives the change in s as

$$(s_2 - s_1) = \left[c_p + c_v\left(\frac{\gamma - n}{n - 1}\right)\right]\log_e T_2/T_1$$

$$= \frac{n}{n - 1}\,(c_p - c_v)\log_e T_2/T_1 \qquad \left(\text{Substituting, } \gamma = \frac{c_p}{c_v}\right) \qquad ...(20.38)$$

For the complete isothermal *two-state* compression the change in entropy,

$$(s_2 - s_1) = \frac{R}{J}\log_e p_3/p_1 = (c_p - c_v)\log_e p_3/p_1 \qquad ...(20.39)$$

But if the work done in *stage compression is* to be *minimum*

$$\frac{T_2}{T_1} = \left(\frac{p_2}{p_1}\right)^{\frac{n-1}{n}} = \left(\frac{p_3}{p_1}\right)^{\frac{n-1}{2n}} \qquad ...(20.40)$$

By inserting eqn. (20.40) in eqn. (20.38)

$$(s_2 - s_1) = \frac{n}{n - 1}\,(c_p - c_v)\log_e\left(\frac{p_3}{p_1}\right)^{\frac{n-1}{2n}}$$

$$= \left(\frac{c_p - c_v}{2}\right)\log_e p_3/p_1 \qquad ...(20.41)$$

Comparing eqn. (20.39) and eqn. (20.41) it will be seen that the one is half the value of the other ; hence the work done per stage is minimum when increase in entropy per stage

$$= \frac{\text{Isothermal increase in entropy for whole compression}}{\text{Number of stages}}$$

and the *maximum temperature per stage is constant* and equal to T_2.

Actual p-V (indicator) diagram for two-stage compressor

The actual indicator diagram for a two-stage compressor is shown in Fig. 20.10. The wavy lines during induction and delivery strokes are due to "Flatter" of the disc valves. The L.P. and H.P. diagrams *overlap due to pressure drop in the intercooler,* and, of course, clearance effects are plainly visible in the actual indicator diagram. *The inertia and friction effects which result in valve flatter increase the area of the diagram slightly, and hence their effect is to increase the total work of compression.*

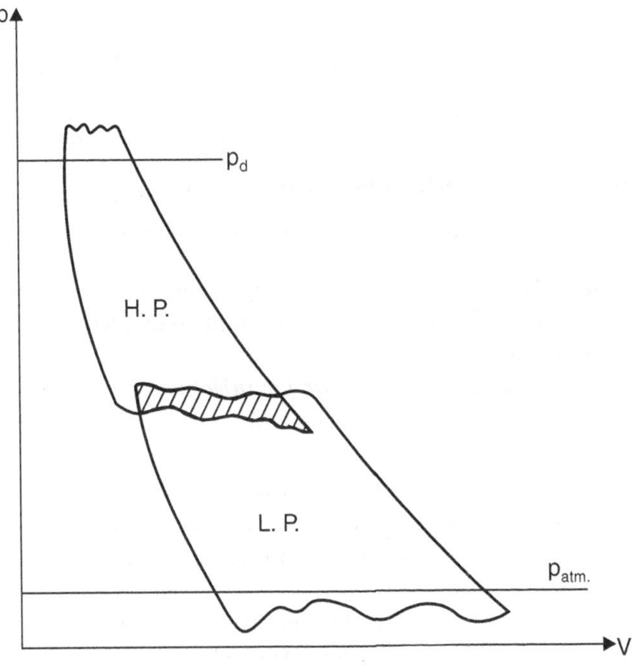

Fig. 20.10

20.3.7. Efficiency of Compressor

The theoretical horse power of a compressor is calculated on the assumption that the compression curve of p-V diagram is an isothermal. Then,

Isothermal work done/cycle $= \text{Area of } p\text{-}V \text{ diagram}$

$$= p_1 V_1 \log_e r$$

Isothermal power $$= \frac{p_1 V_1 \log_e r \times N}{60 \times 1000} \text{ kW} \qquad ...(20.42)$$

[**In M.K.S. units**

Isothermal horse power $$= \frac{p_1 V_1 \log_e r \times N}{4500} \qquad ...(20.42\ (a))$$

where, N = number of cycles/min.]

The indicated power of a compressor is the power obtained from the actual indicator card taken during a test on the compressor,

Compressor efficiency $$= \frac{\text{isothermal horse power}}{\text{indicated horse power}}$$

Isothermal efficiency $$= \frac{\text{isothermal horse power}}{\text{shaft horse power}}$$

where the shaft horse power is the brake horse power required to drive the compressor. A usual value of isothermal efficiency is about 70 per cent.

The '*adiabatic efficiency*' of an air compressor is the ratio of the horse power required to drive the compressor compared with the area of the hypothetical indicator diagram assuming adiabatic compression.

or Adiabatic efficiency, $$\eta_{adiabatic} = \frac{\left(\frac{\gamma}{\gamma-1}\right)p_1V_1\left[\left(\frac{p_2}{p_1}\right)^{\frac{\gamma-1}{\gamma}}-1\right]\times\frac{N}{4500}}{\text{B.H.P. required to drive the compressor}} \quad ...(20.43)$$

20.3.8. How to Increase Isothermal Efficiency ?

The following methods are employed to achieve nearly isothermal compression for high speed compressors :

(1) **Spray injection.** This method (used some years ago) assimilates the practice of injecting water into the compressor cylinder towards the compression stroke with the object of cooling the air. It entails the following *demerits* :

(*i*) It necessitates the use of a *special gear for injection* ;

(*ii*) The injected water *interferes with the cylinder lubrication and attacks cylinder walls and valves*, and

(*iii*) The water mixed with air should be *separated before using the air.*

(2) **Water jacketing.** It consists in circulating water around the cylinder through the water jacket which helps to cool the air during compression. *This method is commonly used in all types of reciprocating air compressors.*

(3) **Inter-cooling.** When the speed of the compressor is high and pressure ratio required is also high with single-stage compression water jacketing proves to be less effective. The use of inter-cooling is restored to in addition to the water jacketing by dividing the compression into two or more stages. The air compressed in the first stage is cooled in an intercooler (heat exchanger) to its original temperature before passing it on to the following (second) stage.

(4) **External fins.** The small capacity air compressors can be effectively cooled by using fins on their external surfaces.

(5) **By a suitable choice of cylinder proportions.** By providing a short stroke and a large bore in conjunction with sleeve valves, a much greater surface is available for cooling, and the surface of the cylinder head is far more effective in this respect than the surface of the barrel, because the periodic motion of the piston does not allow the barrel to be exposed to the air for a sufficient time for heat to flow away. Moreover the air is compressed against the cylinder cover. Unfortunately clearance increases as the square of the bore, but in the Broom-Wade compressor this increase is compensated for by the mechanically operated valve.

Mechanical efficiency. In general, the mechanical efficiency is the ratio of the mechanical output to the mechanical input. For an air compressor,

Mechanical efficiency, $$\eta_{mech.} = \frac{\text{I.H.P. of compressor}}{\text{Shaft horse power}}.$$

20.3.9. Clearance in Compressors

The *clearance volume* consists of the following *two spaces* :

(*i*) The space between the cylinder end and the piston to allow for wear and to give mechanical freedom and

(*ii*) The space for the reception of valves.

In high-class H.P. compressors the clearance volume may be as little as 3 per cent of the swept volume, lead fuse wire being used to determine the actual width of the gap between the cylinder end and the piston, whilst in cheap L.P. compressors the clearance may be 6 per cent of the swept volume, in which the thickness of a flattened ball of putty is a measure of gap.

The direct effect of clearance is to make the volume taken in per stroke less than the swept volume, and because of the necessary increase in size of the compressor (to maintain the output) the power to drive the compressor is slightly increased. *The maximum compression pressure is also controlled by the clearance volume.*

The *value of clearance* may be expressed as follows :

(*i*) Since less precision is required in machining and erection, a large clearance cheapens a compressor and tends to increase its reliability.

(*ii*) A variable clearance is a convenient and safe way of controlling the output of a constant speed compressor.

(*iii*) Increasing the clearance in one stage throws more work on the stage below. In this way the temperature rise in the higher stages, consequent on controlling the output by throttling the L.P. suction, may be limited.

20.3.10. Effect of Clearance Volume

The ***clearance volume*** is *the volume within the cylinder when the piston it at the end of its inward travel plus the volume within the passages leading to the valves. The effect of clearance volume is to reduce the volume actually aspirated.* Therefore clearance volume should be as small as possible, but it cannot be reduced to zero since, for mechanical reasons, the piston face cannot be allowed to come into contact with cylinder head. The clearance volume of the compressor is given as *a percentage of stroke volume.*

The *p-V* diagram for a single-stage and a single-acting air-compressor with clearance volume is shown in Fig. 20.11. At the end of delivery stroke, the high pressure air is left in the

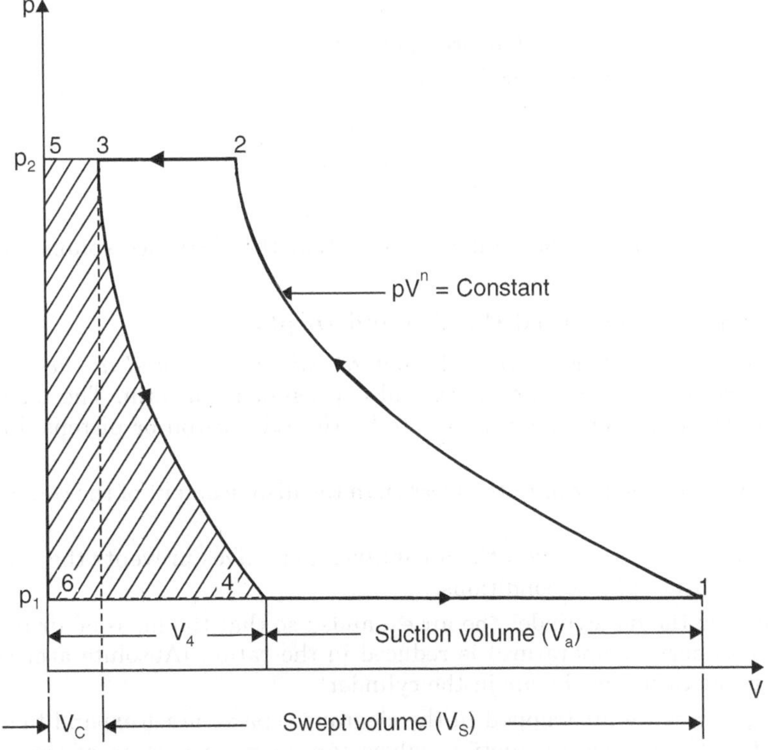

Fig. 20.11

clearance volume and *suction for the second cycle starts only when the air pressure falls to the atmospheric pressure.* This is represented by the expansion curve 3-4. Assuming the compression and expansion of the air follow the same law, the work done per cycle is given by the area 1-2-3-4-1 on p-V diagram.

$$\therefore \quad W(\text{area 1-2-3-4-1}) = W_{comp.} (\text{area 1-2-5-6-1}) - W_{exp.} (\text{area 3-4-6-5-3})$$

$$\therefore \quad W = \frac{n}{n-1} p_1 V_1 \left[\left(\frac{p_2}{p_1} \right)^{\frac{n-1}{n}} - 1 \right] - \frac{n}{n-1} p_4 V_4 \left[\left(\frac{p_3}{p_4} \right)^{\frac{n-1}{n}} - 1 \right]$$

As $\quad p_3 = p_2$ and $p_4 = p_1$

$$\therefore \quad W = \frac{n}{n-1} p_1 (V_1 - V_4) \left[\left(\frac{p_2}{p_4} \right)^{\frac{n-1}{n}} - 1 \right] \quad \text{...(20.44)}$$

$$= \frac{n}{n-1} p_1 V_a \left[\left(\frac{p_2}{p_1} \right)^{\frac{n-1}{n}} - 1 \right] \quad \text{...(20.45)}$$

where $V_a = V_1 - V_4$ is the actual volume of free-air delivered per cycle.

$$\therefore \quad W = \frac{n}{n-1} m_1 R T_1 \left[\left(\frac{p_2}{p_1} \right)^{\frac{n-1}{n}} - 1 \right] \quad \text{...(20.46)}$$

where m_1 is the actual mass of air delivered per cycle.

$\therefore$ Work delivered per kg of air delivered

$$= \frac{n}{n-1} R T_1 \left[\left(\frac{p_2}{p_1} \right)^{\frac{n-1}{n}} - 1 \right] \quad \text{...(20.47)}$$

[From eqns. (20.35) and (20.47) it is obvious that the clearance volume *does not affect the work of compression per kg of air.*]

20.3.11. Free Air Delivered (F.A.D.) and Displacement

The *free air delivered* (F.A.D.) is the *actual volume delivered at the stated pressure reduced to intake temperature and pressure,* and expressed in m^3/min. The displacement is the actual volume in m^3/min swept out per minute by the L.P. piston or pistons during the suction strokes.

The free air delivered per minute is *less* than the displacement of the compressor because of the following *reasons* :

1. The fluid resistance through the air intake, and valves prevents the cylinder being fully charged with air at atmospheric conditions.

2. On entering the hot cylinder the air expands ; so that the mass of air present (compared with that at atmospheric temperature) is reduced in the ratio : (Absolute atmospheric temperature)/(Absolute temperature of the air in the cylinder).

3. The high-pressure air trapped in the clearance space, must expand to a pressure below atmospheric before the automatic suction valves can open ; a portion of the suction stroke is therefore wasted in effecting this expansion.

4. A certain loss is caused by the leakage.

20.3.12. Compressor Performance

By compressor performance, we generally mean the mass of air delivered per minute per B.P. (or B.H.P.) on the machine.

For a machine of given capacity and numerial pressure the *performance of a compressor* is influenced by the following *factors* :

(*i*) The pressure range per cylinder.

(*ii*) The number of stages employed.

(*iii*) The clearance volume.

(*iv*) The speed of the machine.

(*v*) The cooling efficiency.

(*vi*) The air intake piping.

(*vii*) The type and disposition of the valves.

20.3.13. Effect of Atmospheric Conditions on the Output of a Compressor

A low barometer and a high temperature (as encountered at considerable elevations during day time in tropical countries) is responsible for an appreciable diminution in the mass output of compressors which have to operate under these conditions. The volumetric efficiency (when referred to a standard atmosphere) falls by about 3% per 300 mm increase in elevation, and 1% per 5°C increase in temperature. As a result of the considerable reduction in temperature after sundown, and accompanying humidity, power plant in tropical climates runs considerably better at night.

20.3.14. Control of Compressors

Compressor control may be carried out in many different ways, depending on the circumstances in which they are used ; *e.g.*

1. A compressor, directly driven by a steam engine, may be controlled by a combined centrifugal governor on the steam engine and an air-pressure regulator, the control consisting in an adjustment of the speed to suit the load. The mechanism operates either the steam throttle or varies the cut-off. This is suitable where the prime-mover may be run at reduced speeds without too great drop in efficiency.

2. Where the drive is by means of electrical motors it will usually be necessary to keep the speed constant (it may be inevitable with synchronous motors), and then some unloading device may be used to blow low-pressure air off to the atmosphere. By artificially obtructing the low-pressure intake and thus lowering the intake pressure in addition to the mass aspired, the temperature of delivery may be raised to a dangerous value due to the higher pressure ratio, and this method is therefore *not to be recommended.*

3. A method, commendable because it affords some control over the volumetric efficiency, *is to provide* ***variable clearance control.*** This is achieved by having *air pockets adjacent to the cylinder*, which are brought into communication with the cylinder by automatically operated valves.

4. With mechanically operated valves it is usual to hold the suction valve open for part of the compression stroke.

In many cases a combination of these, and other, controls may actually be used.

20.3.15. Arrangement of Reciprocating Compressors

As earlier discussed the reciprocating air compressors may be classified into single and multi-stage machines, and they may be single or double-acting. In the latter case, air is compressed alternately on either side of the piston, and consequently there are two compression strokes per revolution per cylinder. In some machines compression in the various stages take place in separate cylinders, the pistons in which are independently actuated from separate cranks upon the crankshaft. In others a compound cylinder, which forms either two or even three stages, fitted with a single piston, is employed and again, two pistons may be connected together each reciprocating in its own cylinder.

Fig. 20.12 shows three arrangements in diagram form :

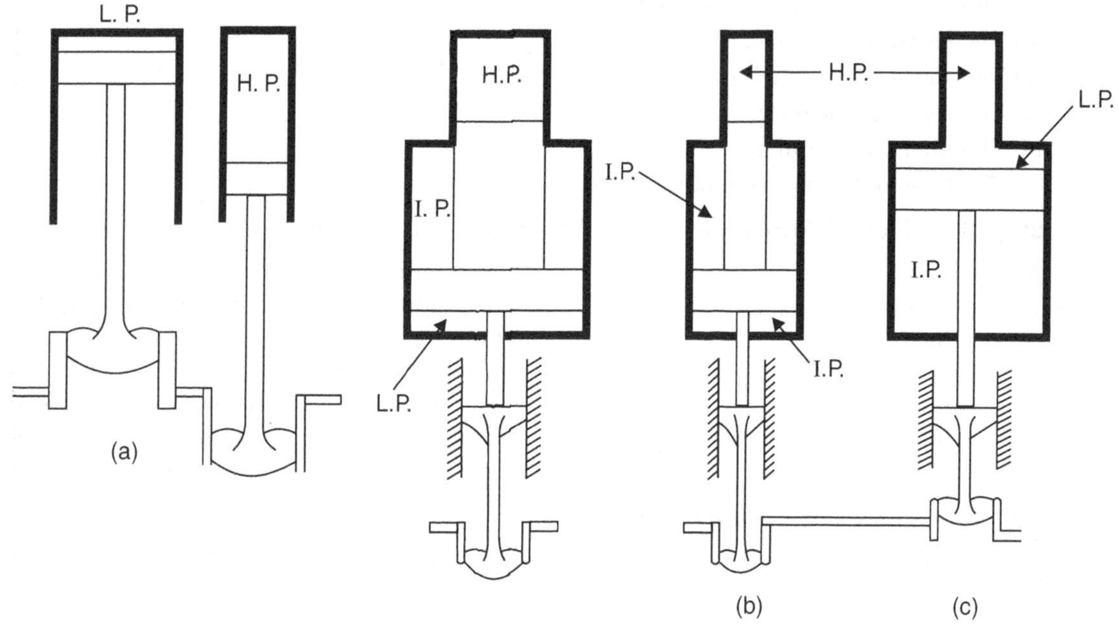

Fig. 20.12. Arrangement of reciprocating compressors.

Fig. 20.12 (*a*) shows a two-crank, two-stage single-acting air compressor. In this arrangement each piston is actuated independently from a separate crank. The cranks are kept 180° out of phase.

Fig. 20.12 (*b*) shows a single-crank, three-stage single-acting air compressor. Here compression of the air in the low pressure cylinder takes place on the down-stroke of the piston, whilst on the up-stroke compression in the intermediate and high pressure stages occurs simultaneously.

Fig. 20.12 (*c*) shows a two-crank, three-stage air compressor having double-acting low and intermediate pressure cylinders and two single-acting high pressure cylinders, compression in these high pressure cylinders taking place only during the up-stroke of the pistons.

Many such arrangements are possible but their discussion is beyond the scope of this book.

20.3.16. Intercooler

The cooler which is placed in between stages is called **Intercooler.** With the object of removing moisture, coolers are sometimes fitted after the last stage, and for this reason are called '***Aftercoolers***', but it should be understood that *aftercoolers cannot influence the work done in compression.*

Intercoolers and aftercoolers are simple heat exchangers in which heat is removed from air which has been compressed and its temperature has risen as a result of compression. A simplified

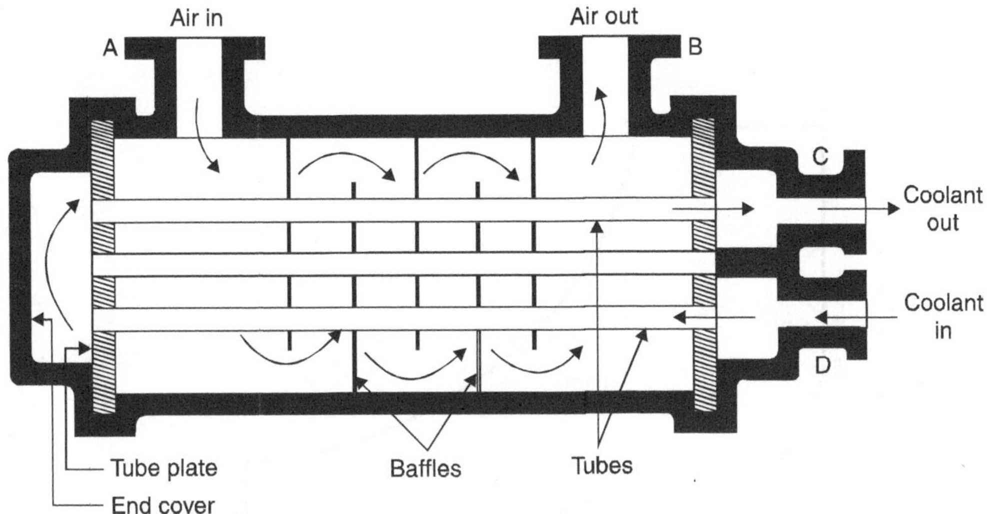

Fig. 20.13. Intercooler.

section through an intercooler is shown in Fig. 20.13. The cooling water passes through the tubes which are secured between two tube plates, and the air circulates over the tubes through a system of baffles.

20.3.17. Compressed Air Motors

Compressed air is employed in a wide variety of applications in industry. For some purposes air-operated motors are the most suitable forms of power, especially where there are safety requirements to be met as in mining applications. Pneumatic breakers, picks, spades, rammers, vibrators, riveters, etc. form a range of hand tools which have wide applications in constructional work. They are light in construction and suitable for operation in remote situations for which other forms of power tools may not be suitable. The action required of such tools, with the associated simplicity and robustness of construction, is obtained with an air-operated design.

The most general types of motors are :

(1) Piston type or Reciprocating type.

(2) Rotary type.

20.3.18. Reciprocating Air Motor

The cycle in the reciprocating type air motor is reverse of that in the reciprocating compressor. Air is supplied to the air motor from an air receiver in which the air is at approximately ambient temperature. There is a pressure drop in the air line between the receiver and the motor. The air expands in a motor cylinder to atmospheric pressure in a manner which is polytropic (*i.e.*, the expansion is internally reversible and the law of expansion is pV^n = constant, where $n < \gamma$, and is usually about 1.3). If the air is initially at ambient temperature, then this form of expansion will bring about a reduction in air temperature as lower pressures are reached. The temperatures reached may be sufficiently low to be below the dew point of the moisture in the air, this may be condensed, and the water formed may even be cooled to its freezing point. This may lead to the formation of ice in the cylinder with the consequence of blocked valves. To prevent this condition it may be necessary to *pre-heat the air* to an initial temperature which is high enough *to prevent the ice formation.* This heating of air causes an increase in volume at supply pressure and reduces the demand from the compressor. Further, the temperature at which the heat transfer is required is low, and a low grade supply of heat or 'waste heat' may be utilized for the purpose.

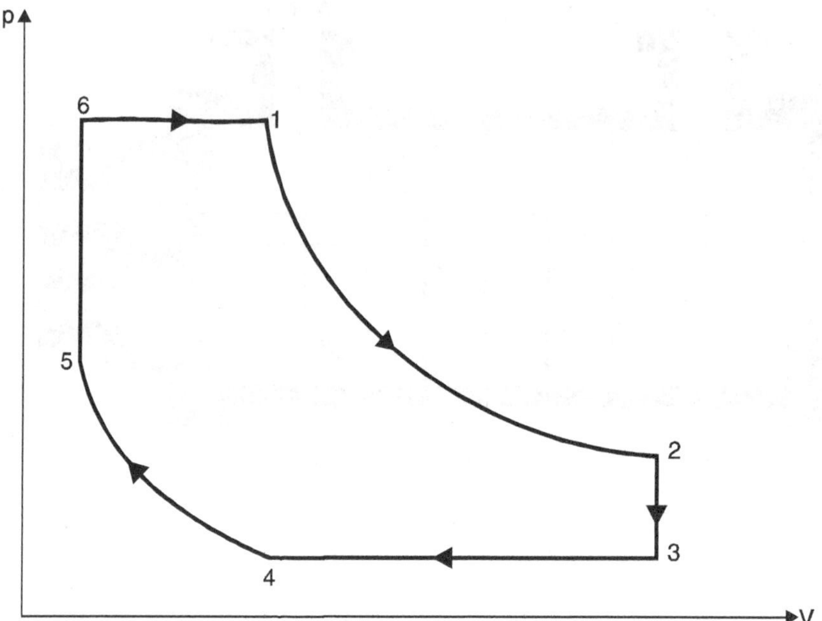

Fig. 20.14. Hypothetical diagram for a reciprocating air motor.

Fig. 20.14 shows a hypothetical diagram for a reciprocating air motor. The sequence of operations is as follows :

(*i*) **Operation 1-2** : The air expands from 1 (p_1) to 2 (p_2) at the end of the stroke (according to the law $pV^n = C$).

(*ii*) **Operation 2-3.** : Blow down (release) of air from 2 to 3 (at constant volume).

(*iii*) **Operation 3-4** : Air is exhausted from 3 to 4, and at 4 compression of the trapped or cushion air begins.

(*iv*) **Operation 4-5-6** : Air at supply pressure p_6, is admitted to the cylinder at the point 5 where it mixes irreversibly with the cushion air. The pressure in the cylinder is rapidly brought upto the inlet valve, p_6.

(*v*) **Operation 6-1** : The supply of air is made at constant pressure behind the moving piston to the point of cut-off at 1. The cut-off ratio is given by

$$\text{Cut-off ratio} = \frac{V_1 - V_6}{V_3 - V_6}$$

The effect of cushion air is to give a smoother running motor. The position of the point 5 depends on the point of initial compression 4, and on the law of compression pV^n = constant. The conditions may be such that the points 5 and 6 coincide. The analysis of such a diagram is best carried out from basic principles.

Note. For a given power, a reciprocating air compressor consumes less air than the rotary form, because of the reduced leakage and of the greater expansion that is possible. This however is only secured at the expense of a heavier, and more costly mechanism.

20.3.19. Rotary Type Air Motor

The *air turbine* is valveless, small in size, light in weight, and requires no internal lubrication, but air friction is high, and any dampness in the air causes rapid deterioration of the blading at low temperatures.

The sliding blade eccentric drum type requires internal lubrication, and even so the slots, in which the blades move, wear rapidly.

The toothed wheel type has a smaller friction and can expand damp air without internal deterioration. In the "herringbone" type expansion is possible together with a high starting torque and extreme mechanical simplicity. This commends the turbine for colliery work inspite of its extravagance on air.

Example 20.1. *A single-stage reciprocating compressor takes 1 m^3 of air per minute at 1.013 bar and 15°C and delivers it at 7 bar. Assuming that the law of compression is $pV^{1.35}$ = constant, and the clearance is negligible, calculate the indicated power.*

Solution. Volume of air taken in, $V_1 = 1\ m^3/min$

Intake pressure, $p_1 = 1.013$ bar

Initial temperature, $T_1 = 15 + 273 = 288$ K

Delivery pressure, $p_2 = 7$ bar

Law of compression : $pV^{1.35}$ = constant

Indicated power I.P. :

Mass of air delivered per min.,

$$m = \frac{p_1V_1}{RT_1} = \frac{1.013 \times 10^5 \times 1}{287 \times 288} = 1.226 \text{ kg/min}$$

Delivery temperature, $$T_2 = T_1 \left(\frac{p_2}{p_1}\right)^{(n-1)/n}$$

$$= 288 \left(\frac{7}{1.013}\right)^{(1.35-1)/1.35} = 475.2 \text{ K}$$

Indicated work $$= \frac{n}{n-1} mR(T_2 - T_1) \text{ kJ/min}$$

$$= \frac{1.35}{1.35-1} \times 1.226 \times 0.287\ (475.2 - 288) = 254 \text{ kJ/min}$$

i.e., Indicated power I.P $= \frac{254}{60}$ = **4.23 kW. (Ans.)**

Example 20.2. *If the compressor of example 21.1 is driven at 300 r.p.m. and is a single-acting, single-cylinder machine, calculate the cylinder bore required, assuming a stroke to bore ratio of 1.5 : 1. Calculate the power of the motor required to drive the compressor if the mechanical efficiency of the compressor is 85% and that of the motor transmission is 90%.*

Solution. Volume dealt with per minute at inlet = 1 m^3/min.

∴ Volume drawn in per cycle $= \frac{1}{300} = 0.00333\ m^3$/cycle

i.e., Cylinder volume $= 0.00333\ m^3$

$$\therefore \quad \frac{\pi}{4} D^2L = 0.00333$$

(where D = bore, L = stroke)

i.e., $$\frac{\pi}{4} D^2 (1.5 \times D) = 0.00333 \quad \text{or} \quad D^2 = \frac{0.00333 \times 4}{\pi \times 1.5}$$

i.e., *Cylinder bore,* $D = 0.1414$ m or **141.4 mm. (Ans.)**

$$\text{Power input to the compressor} = \frac{4.23}{0.85} = 4.98 \text{ kW}$$

$$\therefore \quad \textit{Motor power} = \frac{4.98}{0.9} = \mathbf{5.53\ kW.} \quad \textbf{(Ans.)}$$

Example 20.3. *An air compressor takes in air at 1 bar and 20°C and compresses it according to law* $pv^{1.2}$ = *constant. It is then delivered to a receiver at a constant pressure of 10 bar.* $R = 0.287$ *kJ/kg K. Determine :*

(*i*) *Temperature at the end of compression ;*

(*ii*) *Workdone and heat transferred during compression per kg of air.*

Solution. Refer Fig. 20.15.

$$T_1 = 20 + 273 = 293 \text{ K} ; p_1 = 1 \text{ bar} ; p_2 = 10 \text{ bar}$$

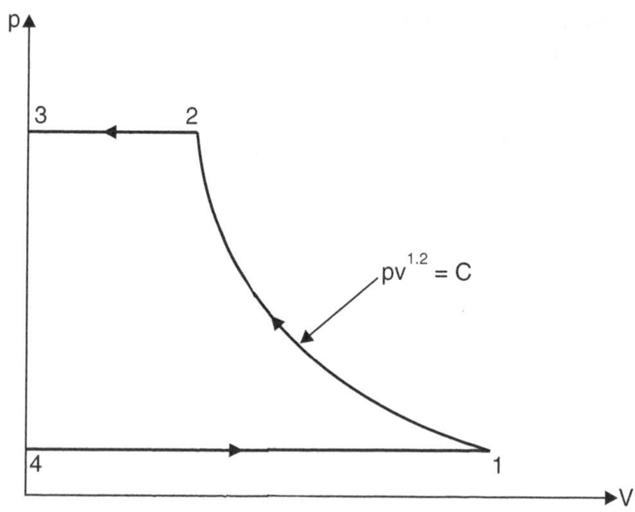

Fig. 20.15

Law of compression : $pv^{1.2} = C$; R = **0.287** J kJ/kg K

(*i*) **Temperature at the end of compression, T_2 :**

For *compression process 1-2,* we have

$$\frac{T_2}{T_1} = \left(\frac{p_2}{p_1}\right)^{\frac{n-1}{n}} = \left(\frac{10}{1}\right)^{\frac{1.2-1}{1.2}} = 1.468$$

or $$T_2 = T_1 \times 1.468 = 293 \times 1.468 = 430 \text{ K or } \mathbf{157°C.} \ \textbf{(Ans.)}$$

(*ii*) **Work done and heat transferred during compression per kg of air :**

$$\textit{Work done, } W = mRT_1 \frac{n}{n-1}\left[\left(\frac{p_2}{p_1}\right)^{\frac{n-1}{n}} - 1\right] \quad \text{...[Eqn. (20.5)]}$$

$$= 1 \times 0.287 \times 293 \times \left(\frac{1.2}{1.2-1}\right)\left[\left(\frac{10}{1}\right)^{\frac{1.2-1}{1.2}} - 1\right] = \mathbf{236.13\ kJ/kg\ of\ air.}\ \textbf{(Ans.)}$$

Heat transferred during compression,

$$Q = W + \Delta U$$

$$= \frac{p_1v_1 - p_2v_2}{n-1} + c_v(T_2 - T_1)$$

$$= \frac{R(T_1 - T_2)}{n-1} + c_v(T_2 - T_1) = (T_2 - T_1)\left[c_v - \frac{R}{n-1}\right]$$

$$= (430 - 293)\left[0.718 - \frac{0.287}{1.2 - 1}\right] = -\ \mathbf{98.23\ kJ/kg.}\quad \textbf{(Ans.)}$$

Negative sign indicates heat rejection.

Example 20.4. *Following data relate to a performance test of a single-acting 14 cm × 10 cm reciprocating compressor :*

Suction pressure	*= 1 bar*
Suction temperature	*= 20°C*
Discharge pressure	*= 6 bar*
Discharge temperature	*= 180°C*
Speed of compressor	*= 1200 r.p.m.*
Shaft power	*= 6.25 kW*
Mass of air delivered	*= 1.7 kg/min*

Calculate the following :

(*i*) *The actual volumetric efficiency ;*

(*ii*) *The indicated power ;*

(*iii*) *The isothermal efficiency ;*

(*iv*) *The mechanical efficiency ;*

(*v*) *The overall isothermal efficiency.* **(AMIE Summer, 2000)**

Solution. *Given :* p_1 = 1 bar ; T_1 = 20 + 273 = 293 K ; p_2 = 6 bar ; T_2 = 180 + 273 = 453 K ; N = 1200 r.p.m., P_{shaft} = 6.25 kW ; m_a = 1.7 kg/min.

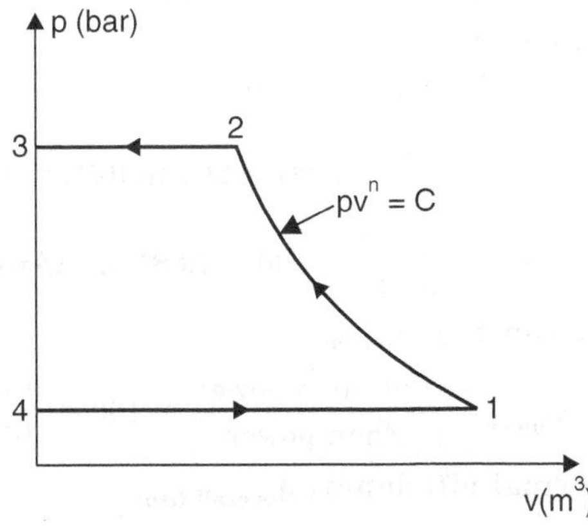

Fig. 20.16

(*i*) **The actual volumetric efficiency, $\eta_{vol.}$:**

Displacement volume (m³/min)

$$V_d = \frac{\pi}{4} D^2 L \times N \qquad \text{(for single-acting compressor)}$$

$$= \frac{\pi}{4} \times \left(\frac{14}{100}\right)^2 \times \left(\frac{10}{100}\right) \times 1200 = 1.8473 \text{ m}^3\text{/min}$$

$$F.A.D. = \frac{mRT_1}{p_1} = \frac{1.7 \times (0.287 \times 1000) \times 293}{1 \times 10^5} = 1.4295 \text{ m}^3\text{/min.}$$

$$\therefore \quad \eta_{vol.} = \frac{F.A.D.}{V_d} \times 100 = \frac{1.4295}{1.8473} \times 100 = \mathbf{77.38\%.} \quad \textbf{(Ans.)}$$

(*ii*) **The indicated power, I.P. :**

$$\frac{T_2}{T_1} = \left(\frac{p_2}{p_1}\right)^{\frac{n-1}{n}} \quad \text{or} \quad \frac{n-1}{n} = \frac{\ln (T_2/T_1)}{\ln (p_2 p_1)}$$

or $$\frac{1}{n} = 1 - \frac{\ln (453/293)}{\ln (6/1)}$$

or $$n = 1.32$$

Hence, index of compression, $n = 1.32$

$$\therefore \quad \text{Indicated power,} \quad \text{I.P.} = \frac{n}{n-1} mRT_1 \left\{\left(\frac{p_2}{p_1}\right)^{\frac{n-1}{n}} - 1\right\}$$

$$= \frac{1.32}{1.32-1} \times \frac{1.7}{60} \times 0.287 \times 293 \left\{\left(\frac{6}{1}\right)^{\frac{1.32-1}{1.32}} - 1\right\}$$

$$= 5.346 \text{ kJ/s or kW}$$

i.e., $$\textbf{I.P. = 5.346 kW.} \quad \textbf{(Ans.)}$$

(*iii*) **Isothermal efficiency, η_{iso} :**

Isothermal power $= mRT_1 \ln (p_2/p_1)$

$$= \frac{1.7}{60} \times 0.287 \times 293 \ln (6/1) = 4.269 \text{ kJ/s or kW}$$

$$\therefore \quad \eta_{iso} = \frac{4.269}{5.346} \times 100 = \mathbf{79.85\%.} \quad \textbf{(Ans.)}$$

(*iv*) **The mechanical efficiency, η_{mech} :**

$$\eta_{mech} = \frac{\text{Indicated power}}{\text{Shaft power}} \times 100 = \frac{5.346}{6.25} \times 100 = \mathbf{85.5\%.} \quad \textbf{(Ans.)}$$

(*v*) **The overall isothermal efficiency, $\eta_{overall\ (iso)}$:**

$$\eta_{overall\ (iso)} = \frac{\text{Isothermal power}}{\text{Shaft power}} \times 100 = \frac{4.269}{6.25} \times 100 = \mathbf{68.3\%.} \quad \textbf{(Ans.)}$$

Example 20.5. *(a) Show that the compressor work obtained from the analysis of a conventional card with clearance and polytropic processes for the reciprocating compressor is identical to that obtained from the analysis of a reversible steady flow rotary compressor where in certain mass of a gas is compressed from the initial condition of pressure p_1 and t_1 respectively to the final pressure p_2 in accordance to $pv^n = C$.*

(b) A low pressure, water jacketed steady flow rotary compressor compresses polytropically 6.75 kg/min of air from 1 atm. and 21°C to 0.35 bar gauge and 43°C. Neglecting the change in kinetic energy find the work and mass of water circulated if the temperature rise of the cooling water is 3.3°C. Take c_p (for air) = 1.003 kJ/kg K. **(P.U.)**

Solution. (*a*) In reciprocating compressors the work required is

$$W = \frac{n}{n-1} p_1 V_1 \left[\left(\frac{p_2}{p_1} \right)^{\frac{n-1}{n}} - 1 \right]$$

$$= \frac{n}{n-1} mR\,(T_2 - T_1) \quad \text{...(1)}$$

When compression is adiabatic, $n = \gamma$

Work transfer in rotary compressor is determined by *applying steady flow energy equation.*

(*i*) **With isentropic flow.** Applying steady flow energy equation, we get

$$h_1 + \frac{C_1^2}{2g} + W = h_2' + \frac{C_2^2}{2g}$$

For $C_1 = C_2$, $W = h_2' - h_1 = c_p(T_2' - T_1)$

where T_2' is the temperature after isentropic compression

Since $$c_p = \frac{\gamma R}{\gamma - 1}$$

$$W = \frac{\gamma}{\gamma - 1} R(T_2' - T_1) \quad \text{...(2)}$$

Eqn. (1) is similar to eqn. (2) for unit mass.

(*ii*) **With compression polytropic.** In actual practice due to internal heating there is increase of work done above isentropic work, and work done is

$$W = c_p(T_2 - T_1) = \frac{\gamma}{\gamma - 1} R(T_2 - T_1)$$

where T_2 is the actual temperature, *i.e.*, obtained by using the relationship $pv^n = C$.

(*iii*) **With cooled compression.** Some heat is being taken away by cooling of compressor and so

$$W = c_p(T_2 - T_1) + Q$$

(*b*) **Work,** $$W = m\,c_p\,(T_2 - T_1)$$

$$= 6.75 \times 1.003\,(43 - 21) = \textbf{148.94 min.} \quad \textbf{(Ans.)}$$

If the compression would have been isentropic

$$T_2' = T_1\,(r_p)^{\frac{\gamma - 1}{\gamma}} = (21 + 273)\left(\frac{1.35}{1}\right)^{\frac{1.4-1}{1.4}} = 320.3°\text{K} \quad \text{or} \quad 47.3°\text{C}$$

Heat rejected to cooling water

$$= m\,c_p(T_2' - T_2)$$

$$= 6.75 \times 1.003\,(47.3 - 43) = 29.11 \text{ kJ}$$

Mass of cooling water, m_w $= \dfrac{29.11}{c_{pw} \times (t_{w2} - t_{w1})} = \dfrac{29.11}{4.18 \times 3.3} =$ **2.11 kg/min. (Ans.)**

Example 20.6. *A single-stage double-acting air compressor is required to deliver 14 m^3 of air per minute measured at 1.013 bar and 15°C. The delivery pressure is 7 bar and the speed 300 r.p.m. Take the clearance volume as 5% of the swept volume with the compression and expansion index of $n = 1.3$. Calculate :*

(*i*) *Swept volume of the cylinder ;* (*ii*) *The delivery temperature ;*

(*iii*) *Indicated power.*

Solution. Quantity of air to be delivered = 14 m^3/min

Intake pressure and temperature $p_1 = 1.013$ bar,

$T_1 = 15 + 273 = 288$ K

Delivery pressure, $p_2 = 7$ bar

Compressor speed, $N = 300$ r.p.m.

Clearance volume, $V_c = 0.05\ V_s$

Compression and expansion index, $n = 1.3$

(*i*) **Swept volume of the cylinder, V_s :**

Swept volume, $V_s = V_1 - V_3 = V_1 - V_c = V_1 - 0.05\ V_s$

$\therefore$ $V_1 = 1.05\ V_s$

Volume induced per cycle $= (V_1 - V_4)$

and $$V_1 - V_4 = \frac{14}{300 \times 2} = 0.0233 \text{ m}^3$$

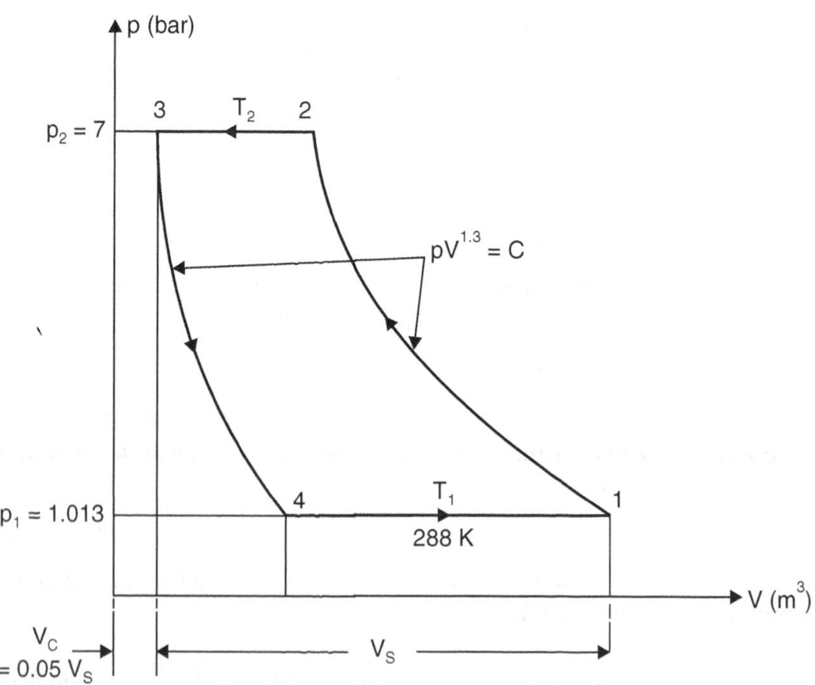

Fig. 20.17

Now, $$V_1 = 1.05\ V_s \text{ and } \frac{V_4}{V_3} = \left(\frac{p_2}{p_1}\right)^{1/n} = \left(\frac{7}{1.013}\right)^{1/1.3} = 4.423$$

i.e., $V_4 = 4.423\ V_3 = 4.423 \times 0.05\ V_s = 0.221\ V_s$

$\therefore$ $(V_1 - V_4) = 1.05\ V_s - 0.221\ V_s = 0.0233$

$$\therefore \quad V_s = \frac{0.0233}{(1.05 - 0.221)} = 0.0281 \text{ m}^3$$

i.e., *Swept volume of the cylinder* $= \mathbf{0.0281\ m^3}$. **(Ans.)**

(*ii*) **The delivery temperature, T_2 :**

Using the relation,
$$\frac{T_2}{T_1} = \left(\frac{p_2}{p_1}\right)^{\frac{n-1}{n}}$$

$$\therefore \quad T_2 = T_1 \times \left(\frac{p_2}{p_1}\right)^{\frac{n-1}{n}} = 288 \times \left(\frac{7}{1.013}\right)^{\frac{1.3-1}{1.3}} = 450 \text{ K}$$

$\therefore$ *Delivery temperature* $= 450 - 273 = \mathbf{177°C}$. **(Ans.)**

(*iii*) **Indicated power :**

$$\text{Indicated power} = \frac{n}{n-1}\, p_1(V_1 - V_4) \left\{\left(\frac{p_2}{p_1}\right)^{\frac{n-1}{n}} - 1\right\}$$

$$= \frac{1.3-1}{1.3} \times \frac{1.013 \times 10^5 \times 14}{10^3 \times 60} \left\{\left(\frac{7}{1.013}\right)^{\frac{1.3-1}{1.3}} - 1\right\} \text{ kW}$$

$= 57.56$ kW.

i.e., *Indicated power* $= \mathbf{57.56\ kW}$. **(Ans.)**

☞**Example 20.7.** *A single-stage, double-acting compressor has a free air delivery (F.A.D.) of 14 m³/min. measured at 1.013 bar and 15°C. The pressure and temperature in the cylinder during induction are 0.95 bar 32°C. The delivery pressure is 7 bar and index of compression and expansion, n = 1.3. The clearance volume is 5% of the swept volume. Calculate :*

(*i*) *Indicated power required ;* (*ii*) *Volumetric efficiency.*

Solution.

Free air delivery, F.A.D. = 14 m³/min. (measured at 1.013 bar and 15°C)

Induction pressure, $p_1 = 0.95$ bar

Induction temperature, $T_1 = 32 + 273 = 305$ K

Delivery pressure, $p_2 = 7$ bar

Index of compression and expansion, $n = 1.3$

Clearance volume, $V_3 = V_c = 0.05\ V_s$

(*i*) **Indicated power :**

Mass delivered per minute,
$$m = \frac{pV}{RT} = \frac{1.013 \times 10^3 \times 14}{0.287 \times 288 \times 10^3} = 17.16 \text{ kg/min.}$$

[where F.A.D. per minute is V at p(= 1.013 bar) and T(15 + 273 = 288 K)]

To find T_2, using the equation,

$$\frac{T_2}{T_1} = \left(\frac{p_2}{p_1}\right)^{\frac{n-1}{n}}$$

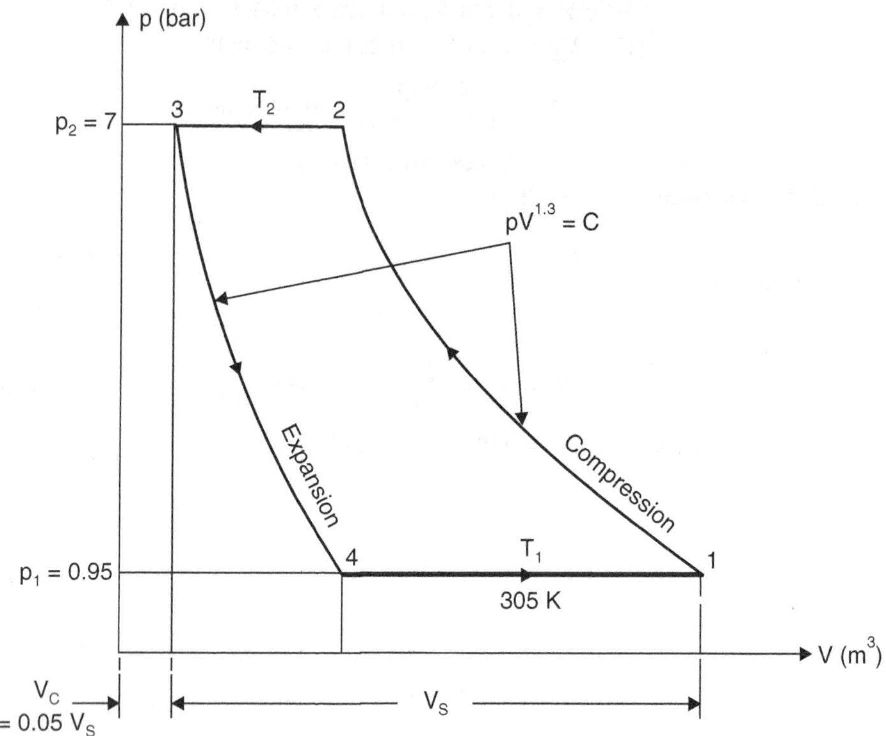

Fig. 20.18

$$\therefore \qquad T_2 = T_1 \left(\frac{p_2}{p_1}\right)^{\frac{n-1}{n}} = 305 \times \left(\frac{7}{0.95}\right)^{\frac{1.3-1}{1.3}} = 483.5 \text{ K.}$$

Indicated power $= \dfrac{n}{n-1} \; mR(T_2 - T_1)$

$$= \frac{1.3}{1.3-1} \times 17.16 \times 0.287 \; (483.5 - 305) = 3809.4 \text{ kJ/min.}$$

$\therefore$ **Indicated power** $= \dfrac{3809.4}{60} = \mathbf{63.49 \text{ kW.}}$ **(Ans.)**

(*ii*) **Volumetric efficiency :**

Using the relation,

$$\frac{V_4}{V_3} = \left(\frac{p_3}{p_4}\right)^{1/n} = \left(\frac{p_2}{p_1}\right)^{1/1.3} = \left(\frac{7}{0.95}\right)^{1/1.3} = 4.65$$

$$\therefore \qquad V_4 = 4.65 \times V_3 = 4.65 \times 0.05 \; V_s = 0.233 \; V_s$$

$$\therefore \qquad V_1 - V_4 = V_1 - 0.233 V_s = 1.05 \; V_s - 0.233 \; V_s = 0.817 \; V_s$$

Now $\qquad m = \dfrac{pV}{RT} = \dfrac{p_1 (V_1 - V_4)}{RT_1}$

i.e., F.A.D./cycle, $V = (V_1 - V_4) \dfrac{T}{T_1} \cdot \dfrac{p_1}{p}$

(where p_1 and T_1 are the suction conditions)

$$\therefore \quad V = 0.817\ V_s \times \frac{288}{305} \times \frac{0.95}{1.013} = 0.723\ V_s$$

$\therefore$ Volumetric efficiency, $\eta_{vol.} = \dfrac{V}{V_s} = \dfrac{0.723\ V_s}{V_s} = 0.723$ or **72.3%. (Ans.)**

Example 20.8. (*a*) *What is meant by volumetric efficiency of a reciprocating compressor ? How is it affected by (i) speed of the compressor ; (ii) delivery pressure, and (iii) throttling across the valves.*

(*b*) *State at least six uses of compressed air.*

(*c*) *The free air delivered by a single-stage double-acting reciprocating compressor, measured at 1 bar and 15°C of free air, is 16 m³/min. The pressure and temperature of air inside the cylinder during suction are 0.96 bar and 30°C respectively and delivery pressure is 6 bar. The compressor has a clearance of 4% of the swept volume and the mean piston speed is limited to 300 m/min. Determine :*

(*i*) *Power input to the compressor if mechanical efficiency is 90% and compression efficiency 85% ;*

(*ii*) *Stroke and bore if the compressor runs at 500 r.p.m.*

Take index of compression and expansion = 1.3. **(AMIE Summer, 2001)**

Solution. (*a*) ***Volumetric efficiency*** of *a compressor is defined as the ratio of Free Air Delivered (FAD) to the swept volume.* ***FAD*** is the *volume of air delivered by the compressor measured at some reference condition* (which may be the ambient condition or the standard sea level condition). FAD is *less* than the swept volume due to the following *reasons* :

(*i*) Throttling and pressure drop at inlet valve and passages ;

(*ii*) Heating of inlet air by coming in contact with hot cylinder walls ; and

(*iii*) Re-expansion of compressed air retained in the clearance volume.

Effects of parameters on volumetric efficiency :

(*i*) ***Speed of compressor.*** As the speed is increased the pressure drop in the inlet passage and the inlet valve increases. Further the air temperature during intake also increases due to less available for cooling. Both of these factors *reduce* volumetric efficiency of compressor *with increase of its speed.*

(*ii*) ***Delivery pressure.*** Refer Fig. 20.19. With increase of delivery pressure the pressure ratio increases and hence during inward stroke *a-b*, the effective swept volume is reduced. The volume of air delivered (FAD) is reduced from V_d to V_d' when the delivery pressure in increased from p_d to p_d'. Thus the volumetric efficiency is *decreased* when the *delivery pressure is increased.*

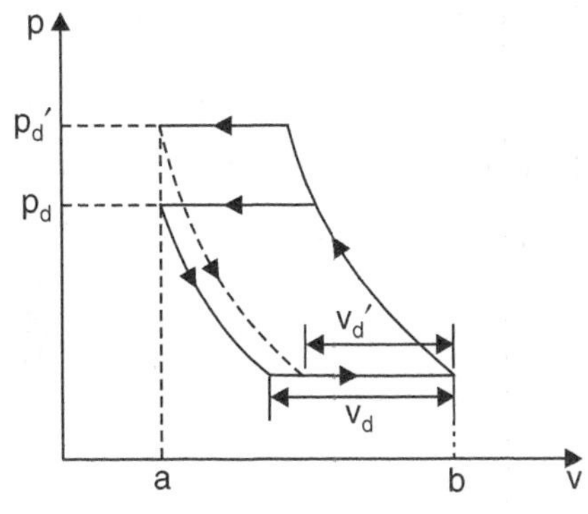

Fig. 20.19

(*iii*) ***Throttling across the valves.*** Throttling across the inlet valve reduces the pressure in the cylinder at the end of the inlet stroke. Further, throttling at the inlet and delivery valves increases the pressure ratio. Both of these effects would *reduce the FAD, and hence the volumetric efficiency of the compressor.*

(*b*) **Uses of compressed air :**

(*i*) Driving a compressed air engine (air motor).

(*ii*) Driving pneumatic tools.

(*iii*) Spray painting.

(*iv*) Cleaning surfaces by air blast.

(*v*) Conveying solid and powdered materials in pipelines.

(*vi*) Blast furnaces and boiler furnaces.

(*vii*) Pneumatic controls.

(*viii*) Inflating tyres of automobiles and tractors.

(*c*) Refer Fig. 20.20.

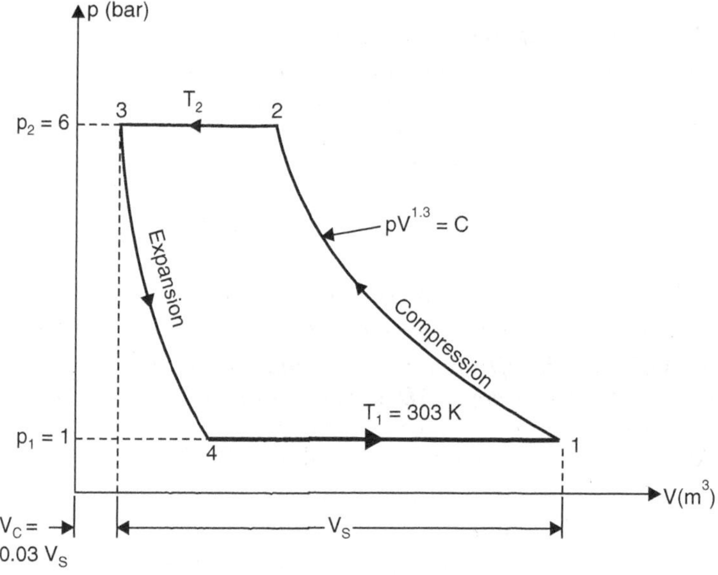

Fig. 20.20

(*d*) F.A.D. = 16 m^3/min (measured at 1 bar and 15°C) ; p_1 = 0.96 bar, T_1 = 30 + 273 = 303 K, n = 1.3 ; $V_3 = V_c = 0.04\ V_s$; p_2 = 6 bar ; $\eta_{mech.}$ = 90% ; $\eta_{comp.}$ = 85% ; Piston speed = 300 m/min ; N = 500 r.p.m.

(*i*) **Power input to compressor :**

Mass flow rate of compressor

$$m = \frac{pV}{RT} = \frac{(1 \times 10^5) \times 16}{287 \times 288} = 19.36 \text{ kg/min.}$$

[where F.A.D. per minute is V at p (= 1 bar) and T (15 + 273 = 288 K)]

To find T_2, using the equation,

$$\frac{T_2}{T_1} = \left(\frac{p_2}{p_1}\right)^{\frac{n-1}{n}}$$

$$\therefore \qquad T_2 = T_1 \times \left(\frac{p_2}{p_1}\right)^{\frac{n-1}{n}} = 303 \times \left(\frac{6}{0.96}\right)^{\frac{1.3-1}{1.3}} = 462.4 \text{ K}$$

$$\textbf{Power input to compressor} = \left[\frac{n}{n-1} mR\,(T_2 - T_1)\right] \times \frac{1}{\eta_{mech.} \times \eta_{comp.}}$$

$$= \left[\left(\frac{1.3}{1.3-1}\right) \times 19.36 \times 0.287\,(462.4 - 303)\right] \times \frac{1}{0.9 \times 0.85}$$

$$= 5016.9 \text{ kJ/min} = \mathbf{83.6 \text{ kJ/s (kW)}}. \quad \textbf{(Ans.)}$$

(*ii*) **Stroke (L) and bore (D) :**

Piston speed = $2LN$

$\therefore \quad 300 = 2 \times L \times 500$

or $\quad L = 0.3$ m or **300 mm. (Ans.)**

$$\text{F.A.D.} = \frac{\pi}{4} D^2 L \times 2N \times \eta_{vol.} \text{ ...... for double-acting air compressor} \quad ...(1)$$

To find $\eta_{vol.}$ proceed as follows :

$$\frac{V_4}{V_3} = \left(\frac{p_3}{p_4}\right)^{1/n} = \left(\frac{p_2}{p_1}\right)^{1/1.3} = \left(\frac{6}{0.96}\right)^{1/1.3} = 4.094$$

$\therefore \quad V_4 = 4.094 \times V_3 = 4.094 \times 0.04\,V_s = 0.1637\,V_s$

$\therefore \quad V_1 - V_4 = V_1 - 0.1637\,V_s = 1.04\,V_s - 0.1637\,V_s = 0.8763\,V_s$

Now $\quad m = \dfrac{pV}{RT} = \dfrac{p_1(V_1 - V_4)}{RT_1}$

i.e., F.A.D./cycle, $\quad V = (V_1 - V_4)\,\dfrac{T}{T_1} \cdot \dfrac{p_1}{p}$

(where p_1 and T_1 are suction conditions)

$\therefore \quad V = 0.8763\,V_s \times \dfrac{288}{303} \times \dfrac{0.96}{1} = 0.799\,V_s$

$\therefore \quad \eta_{vol.} = \dfrac{V}{V_s} = \dfrac{0.799 V_s}{V_s} = 0.799$

Substituting the values in eqn. (1), we get

$$16 = \frac{\pi}{4} D^2 \times 0.3 \times 2 \times 500 \times 0.799$$

$\therefore \quad \mathrm{D} = \left(\dfrac{16 \times 4}{\pi \times 0.3 \times 2 \times 500 \times 0.799}\right)^{1/2} = 0.29$ m or **290 mm. (Ans.)**

Example 20.9. *A single-stage single-acting air compressor delivers 0.6 kg of air per minute at 6 bar. The temperature and pressure at the end of suction stroke are 30°C and 1 bar. The bore and stroke of the compressor are 100 mm and 150 mm respectively. The clearance is 3% of the swept volume. Assuming the index of compression and expansion to be 1.3, find :*

(*i*) *Volumetric efficiency of the compressor,*

(*ii*) *Power required if the mechanical efficiency is 85%, and*

(*iii*) *Speed of the compressor (r.p.m).* **(N.U.)**

Solution. Refer Fig. 20.21.

Mass of air delivered, $\quad m = 0.6$ kg/min.

Delivery pressure, $\quad p_2 = 6$ bar

Induction pressure, $\quad p_1 = 1$ bar

Induction temperature, $\quad T_1 = 30 + 273 = 303$ K

Bore, $D = 100$ mm $= 0.1$ m

Stroke length, $L = 150$ mm $= 0.15$ m

Clearance volume, $V_c = 0.03\ V_s$

Mechanical efficiency, $\eta_{mech.} = 85\%$

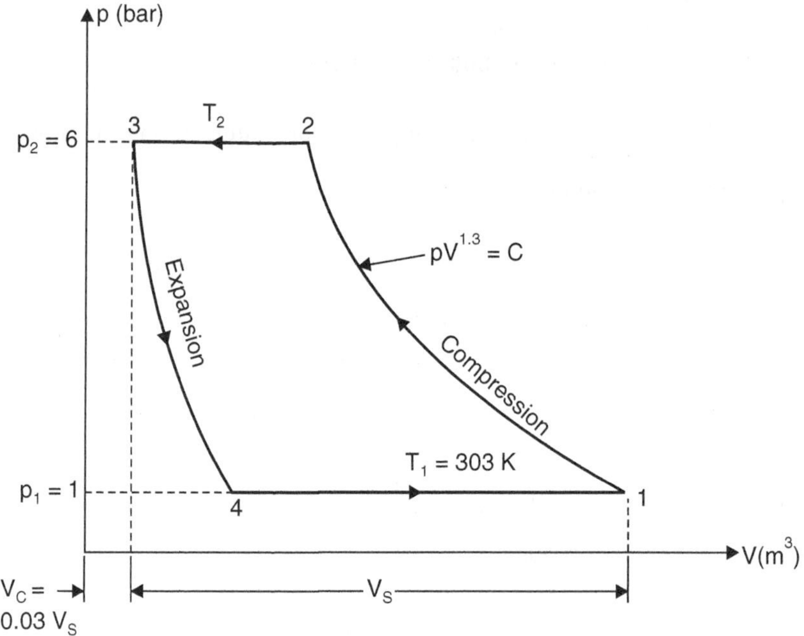

Fig. 20.21

(*i*) **Volumetric efficiency of compressor, $\eta_{vol.}$:**

$$\eta_{vol.} = 1 + k - k \left(\frac{p_2}{p_1}\right)^{1/n} \quad \text{...(Eqn. 20.12)}$$

where $k = \dfrac{V_c}{V_s} = 0.03$

$$\therefore \quad \eta_{vol.} = 1 + 0.03 - 0.03 \left(\frac{6}{1}\right)^{\frac{1}{1.3}} = 0.91096 \text{ or } \mathbf{91.096\%}. \quad \textbf{(Ans.)}$$

(*ii*) **Power required to drive the compressor :**

$$\text{Indicated power} = \frac{n}{n-1}\ mRT_1 \left[\left(\frac{p_2}{p_1}\right)^{1/n} - 1\right]$$

$$= \frac{1.3}{1.3-1} \times \frac{0.6}{60} \times 0.287 \times 303 \left[\left(\frac{6}{1}\right)^{\frac{1.3-1}{1.3}} - 1\right] = 1.93 \text{ kW}$$

$$\therefore \quad \text{Power required to drive the compressor} = \frac{1.93}{\eta_{mech.}} = \frac{1.93}{0.85} = \mathbf{2.27\ kW.} \quad \textbf{(Ans.)}$$

(*iii*) **Speed of the compressor, N (r.p.m.) :**

Free air delivery, F.A.D. $= \frac{mRT_1}{p_1} = \frac{0.6 \times 0.287 \times 1000 \times 303}{1 \times 10^5} = 0.5218$ m³/min.

Displacement volume $= \frac{\text{F.A.D.}}{\eta_{vol.}} = \frac{0.5218}{0.91096} = 0.5728$ m³/min.

Also, $0.5728 = \frac{\pi}{4} D^2 L \times N$ (for single-acting compressor)

or $0.5728 = \frac{\pi}{4} \times 0.1^2 \times 0.15 \times N$

∴ Speed of compressor $N = \frac{0.5728 \times 4}{\pi \times 0.1^2 \times 0.15} =$ **486.2 r.p.m. (Ans.)**

☞ **Example 20.10.** *An air compressor having stroke length of 88 cm and clearance volume of 2 per cent of the swept volume delivers air at a pressure of 8.2 bar. In order to study the effect of clearance on free air delivery and work expended, the compressor was overhauled and a distance piece of 0.55 cm was fitted between the cylinder head and the cylinder. The compressor was then commissioned under the changed clearance. Calculate :*

(*i*) *Percentage change in the volume of free air delivered, and*

(*ii*) *Percentage change in power expended.*

Before and after overhauling the piston had a suction pressure 1.025 bar and the index of compression and expansion was 1.3. **(M.U.)**

Solution. Stroke length of the air compressor, $L = 88$ cm

Clearance volume, $V_c = \frac{2}{100} \times V_s = 0.02 V_s$, where V_s is the swept volume

The pressure at which air is delivered, $p_2 = 8.2$ bar

Index of compression and expression, $n = 1.3$

(*i*) **Percentage change in the volume of free air delivered :**

Before overhauling :

Stroke volume = Area of the piston × Stroke length

$= A$ (cm²) × 88 (cm) = 88 A cm³

Clearance volume, $V_c = 0.02 \times 88 A = 1.76 A \text{ cm}^3 = V_3$

Cylinder volume, $V_1 = V_s + V_c = 88 A + 1.76 A = 89.76 A \text{ cm}^3$

Considering *expansion process 3-4*, we have

$$p_3 V_3^n = p_4 V_4^n$$

or
$$V_4 = V_3 \left(\frac{p_3}{p_4}\right)^{1/n}$$

or
$$V_4 = 1.76 A \left(\frac{8.2}{1.025}\right)^{\frac{1}{1.3}} = 8.712 A \text{ cm}^3$$

∴ Volume of free air sucked in $= V_1 - V_4 = (89.76 A - 8.712 A) = 81.048 A \text{ cm}^3$

After overhauling :

When the distance piece is inserted the clearance space is increased, consequently for the same stroke volume, the cylinder volume would increase.

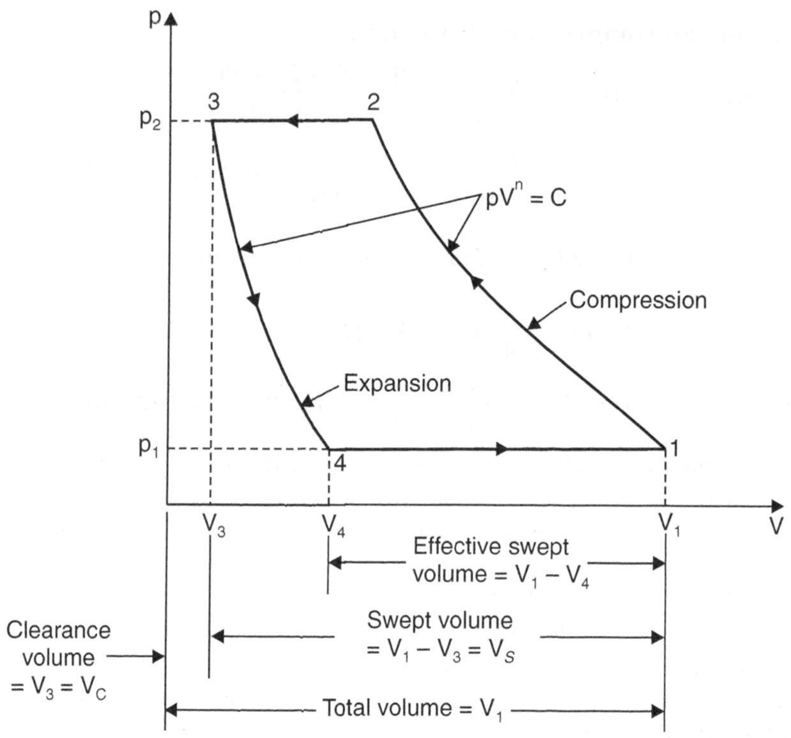

Fig. 20.22

Clearance volume, $V_3 = (1.76 + 0.55)A = 2.31\ A\ \text{cm}^3$

Cylinder volume, $V_1 = 88\ A + 2.31\ A = 90.31\ A\ \text{cm}^3$

From *expansion curve 3-4*, we have

$$p_3 V_3^n = p_4 V_4^n$$

or
$$\frac{V_4}{V_3} = \left(\frac{p_3}{p_4}\right)^{1/n} = \left(\frac{p_2}{p_1}\right)^{\frac{1}{n}} = \left(\frac{8.2}{1.028}\right)^{\frac{1}{1.3}} = 4.95$$

or
$$V_4 = V_3 \times 4.95 = 2.31\ A \times 4.95 = 11.43\ A$$

∴ Volume of free air sucked in $= V_1 - V_4$

$$= 90.31\ A - 11.43\ A = 78.88\ A\ \text{cm}^3$$

Percentage change in free air delivery

$$= \frac{81.048A - 78.88A}{81.048A} \times 100 = \mathbf{2.67\%}.\ \textbf{(Ans.)}$$

(*ii*) **Percentage change in power expended :**

$$p = \frac{n}{n-1} \times p_1 \times \text{volume of free air sucked in} \left[\left(\frac{p_2}{p_1}\right)^{\frac{n-1}{n}} - 1\right]$$

From the above expression it is evident that with index '*n*' and pressures p_1 and p_2 remaining unchanged, the power required to run the compressor is *directly proportional to free air sucked in*.

∴ *Percentage change in power expended* = **2.67% (decrease). (Ans.)**

Example 20.11. *A 4-cylinder double-acting compressor is required to compress 30 m³/min. of air at 1 bar and 27°C to a pressure of 16 bar. Determine the size of motor required and cylinder dimensions if the following data is given :*

Speed of the compressor, $N = 320$ *r.p.m.*
Clearance volume, $V_c = 4\%$
Stroke to bore ratio, $L/D = 1.2$
Mechanical efficiency, $\eta_{mech.} = 82\%$
Value of index, $n = 1.32.$

Assume no pressure change in suction valves and the air gets heated by 12°C during suction stroke.

Solution. Refer Fig. 20.11.

$$\text{Net work done} = \frac{n}{n-1} \times p_1 \times (V_1 - V_4)\left[\left(\frac{p_2}{p_1}\right)^{\frac{n-1}{n}} - 1\right]$$

where $V_1 - V_4$ = suction volume = 30 m³/min (given) = $\frac{30}{60}$ = 0.5 m³/s

$$\therefore \quad \text{Work done} = \frac{1.32}{1.32-1} \times 1 \times 10^5 \times 0.5\left[\left(\frac{16}{1}\right)^{\frac{1.32-1}{1.32}} - 1\right] = 197648.9 \text{ Nm/s}$$

$$\text{Theoretical power} = \frac{197648.9}{1000} = 197.64 \text{ kW}$$

$$\therefore \quad \textbf{Motor power} = \frac{197.64}{\eta_{mech.}} = \frac{197.64}{0.82} = \mathbf{241\ kW} \quad \textbf{(Ans.)}$$

$$\text{Volumetric efficiency, } \eta_{vol.} = \left[1 + k - k\left(\frac{p_2}{p_1}\right)^{1/n}\right] \times \frac{p_i T_a}{p_a T_i}$$

(Suffix 'i' and 'a' stand for inside and atmospheric conditions)

i.e.,
$$\eta_{vol.} = \left[1 + 0.04 - 0.04\left(\frac{16}{1}\right)^{1/1.32}\right] \times \frac{1 \times (273 + 27)}{1 \times (273 + 39)} = 0.686 \text{ or } 68.6\%$$

Now swept volume of one cylinder

$$= \frac{30}{4} \times \frac{1}{2 \times 320} \times \frac{1}{0.686} = 0.01708 \text{ m}^3$$

$$\therefore \quad \frac{\pi}{4} D^2 L = 0.01708 \quad \text{or} \quad \frac{\pi}{4} D^2 \times 1.2\, D = 0.01708$$

$$\therefore \quad D^3 = \frac{0.01708 \times 4}{\pi \times 1.2}$$

$$\therefore \quad \mathbf{D} = 0.263 \text{ m} \quad \text{or} \quad \mathbf{263\ mm.} \quad \textbf{(Ans.)}$$

and
$$\mathbf{L} = 1.2 \times 263 = \mathbf{315.6\ mm.} \quad \textbf{(Ans.)}$$

Example 20.12. *A two-cylinder single-acting air compressor is to deliver 16 kg of air per minute at 7 bar from suction conditions 1 bar and 15°C. Clearance may be taken as 4% of stroke volume and the index for both compression and re-expansion as 1.3. Compressor is directly* *coupled to a four-cylinder four-stroke petrol engine which runs at 2000 r.p.m. with a brake mean effective pressure of 5.5 bar. Assuming a stroke-bore ratio of 1.2 for both engine and compressor and a mechanical efficiency of 82% for compressor, calculate the required cylinder dimensions.*

Solution. Refer Fig. 20.11.

Amount of air delivered per cylinder $= \frac{16}{2} = 8$ kg/min.

Suction conditions : $p_1 = 1$ bar, $T_1 = 15 + 273 = 288$ K

From the gas equation, $p_1(V_1 - V_4) = mRT_1$

or $$V_1 - V_4 = \frac{mRT_1}{p_1} = \frac{8 \times 287 \times 288}{1 \times 10^5} = 6.61 \text{ m}^3/\text{min.}$$

$$= \frac{6.61}{2000} = 0.003305 \text{ m}^3/\text{stroke}$$ ..... compressor being single-acting

From expansion curve, $$\frac{V_4}{V_3} = \left(\frac{p_3}{p_4}\right)^{1/n} = \left(\frac{7}{1}\right)^{1/1.3} = 4.467$$

$\therefore$ $$V_4 = 4.467\, V_3 = 0.04\, V_s \times 4.467 = 0.1787\, V_s$$

(where V_s is the swept volume)

Since $$V_1 = V_3 + V_s = 0.04\, V_s + V_s = 1.04\, V_s$$

$\therefore$ $$V_1 - V_4 = 1.04\, V_s - 0.1787\, V_s = 0.8613\, V_s$$

But $V_1 - V_4$ is also equal to 0.003305

$\therefore$ $$0.8613\, V_s = 0.003305$$

i.e., $$V_s = \frac{0.003305}{0.8613} = 0.003837 \text{ m}^3$$

If L_c = length of stroke of compressor, and

D_c = diameter of the cylinder of the compressor, then

$L_c = 1.2\, D_c$ ....(given)

$\therefore$ $$\frac{\pi}{4} D_c^2 \times L_c = V_s$$

or $$\frac{\pi}{4} D_c^2 \times 1.2\, D_c = 0.003837 \quad \text{or} \quad D_c^3 = \frac{0.003837 \times 4}{\pi \times 1.2}$$

i.e., $\mathbf{D_c}$ = 0.1596 m or **159.6 mm. (Ans.)**

$\mathbf{L_c}$ = 159.6 × 1.2 = **191.5 mm. (Ans.)**

Now indicated power of the compressor

$$= \frac{n}{n-1} \times mRT_1 \left[\left(\frac{p_2}{p_1}\right)^{\frac{n-1}{n}} - 1\right]$$

$$= \frac{1.3}{1.3-1} \times \frac{16}{60} \times 0.287 \times 288 \left[\left(\frac{7}{1}\right)^{\frac{1.3-1}{1.3}} - 1\right] = 54.1 \text{ kW}$$

Brake power of the engine $$= \frac{54.1}{\eta_{mech.}} = \frac{54.1}{0.82} = 65.97 \text{ kW}$$

Now, $$65.97 = \frac{n_e\, p_{mb}\, L_e\, AN}{60}$$

If D_e = diameter of the engine cylinder,

L_e = length of the stroke of engine = 1.2 D_e,

n_e = number of engine cylinders,

Then, $$65.97 = \frac{4 \times (5.5 \times 10^5) \times 1.2\, D_e \times \frac{\pi}{4} \times D_e^2 \times 2000}{60 \times 10^3}$$

$$\therefore \quad D_e^3 = \frac{65.97 \times 60 \times 10^3 \times 4}{4 \times 10^5 \times 5.5 \times 1.2 \times \pi \times 2000} = 0.0009545 \text{ m}^3$$

i.e., $\mathbf{D_e}$ = 0.0984 m or **98.4 mm. (Ans.)**

$\mathbf{L_e}$ = 1.2 × 98.4 = **118.1 mm. (Ans.)**

Example 20.13. *A single-stage double-acting air compressor delivers air at 7.5 bar. The pressure and temperature at the end of suction stroke are 1 bar and 25°C. It delivers 2.2 m³ of free air per minute when the compressor is running at 310 r.p.m. The clearance volume is 5% of stroke volume. The pressure and temperature of ambient air are 1.03 and 20°C.*

Determine : (i) Volumetric efficiency of the compressor ;

(ii) Diameter and stroke of the cylinder if both are equal, and

(iii) I.P. of the compressor and B.P. if the mechanical efficiency is 85%.

Take : Index of compression = 1.25, and Index of expansion = 1.3.

Solution. Refer Fig. 20.23.

Given : p_1 = 1 bar, p_2 = 7.5 bar, T_1 = 25 + 273 = 298 K, $V_{amb.}$ = 2.2 m³

N = 310 r.p.m., $V_c = 0.05\, V_s$, $p_{amb.}$ = 1.03 bar, $T_{amb.}$ = 20 + 273 = 293 K, $\eta_{mech.}$ = 85%.

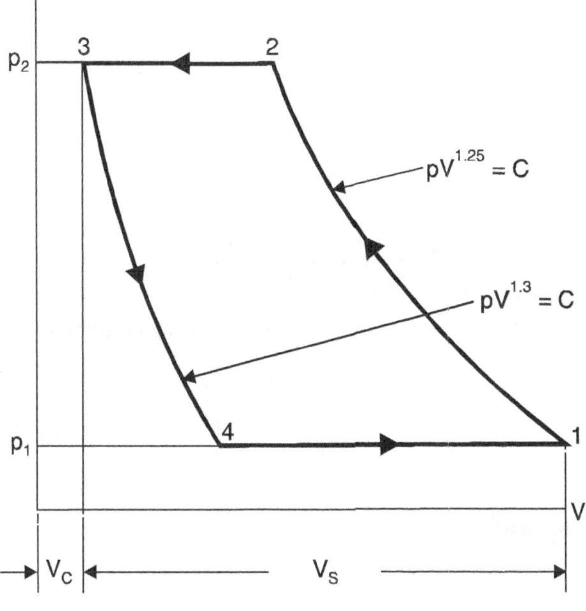

Fig. 20.23

(*i*) **Volumetric efficiency of the compressor $\eta_{vol.}$:**

Volumetric efficiency $= \dfrac{V_1 - V_4}{V_s}$

From the *expansion curve 3-4,*

$$p_2 V_3^{1.3} = p_4 V_4^{1.3}$$

or $$p_2 V_c^{1.3} = p_1 V_4^{1.3} \qquad (\because \; V_3 = V_c \text{ and } p_4 = p_1)$$

$$V_4 = V_c \left(\frac{p_2}{p_1}\right)^{\frac{1}{1.3}} = V_c \left(\frac{7.5}{1}\right)^{1/1.3} = V_c\,(7.5)^{0.769} = 4.71\ V_c$$

But $$V_c = 0.05\ V_s$$

$\therefore$ $$V_4 = 4.71 \times 0.05\ V_s = 0.2355\ V_s$$

$\therefore$ $$\eta_{vol.} = \frac{(V_s + V_c) - 0.2355\ V_s}{V_s}$$

$$= \left(\frac{V_s + 0.05\ V_s - 0.2355\ V_s}{V_s}\right) = 0.814 \text{ or } \mathbf{81.4\%.} \quad \textbf{(Ans.)}$$

(*ii*) **Diameter and stroke of the cylinder D and L :**

The volume of air delivered at suction condition is given by

$$V_1 = \frac{p_{amb}\, V_{amb}\, T_1}{p_1\, T_{amb}} = \frac{1.03 \times 2.2 \times 298}{1 \times 293} = 2.305 \text{ m}^3/\text{min.}$$

The volume delivered per minute is given by

$$V_1 = (V_s \times \eta_{vol.} \times \text{r.p.m.}) \times 2$$

$\therefore$ $$2.305 = V_s \times 0.814 \times 310 \times 2$$

$\therefore$ $$V_s = \frac{2.305}{0.814 \times 310 \times 2} = 0.00456 \text{ m}^3 \text{ or } 4560 \text{ cm}^3$$

$\therefore$ $$\pi/4\ D^2 L = 4560$$

$$\pi/4\ D^3 = 4560 \qquad (\because \; D = L)$$

$\therefore$ $$D = \left(\frac{4560 \times 4}{\pi}\right)^{1/3} = 17.97 \text{ cm} \simeq \mathbf{18\ cm.} \quad \textbf{(Ans.)}$$

$\therefore$ **L = 18 cm. (Ans.)**

(*iii*) **I.P. of compressor and B.P :**

The work done per cycle of operation for *double-acting* compressor is given b

$$W = 2 \times \left[\frac{n_1}{n_1 - 1} \cdot p_1 V_1 \left\{\left(\frac{p_2}{p_1}\right)^{\frac{n_1 - 1}{n_1}} - 1\right\} - \frac{n_2}{n_2 - 1} \cdot p_4 V_4 \left\{\left(\frac{p_3}{p_4}\right)^{\frac{n_2 - 1}{n_2}} - 1\right\}\right]$$

as $$p_3 = p_2 \text{ and } p_4 = p_1$$

$$= 2\left[\frac{n_1}{n_1 - 1} \cdot p_1 V_1 \left\{\left(\frac{p_2}{p_1}\right)^{\frac{n_1 - 1}{n_1}} - 1\right\} - \frac{n_2}{n_2 - 1} \cdot p_1 V_4 \left\{\left(\frac{p_2}{p_1}\right)^{\frac{n_2 - 1}{n_2}} - 1\right\}\right] \qquad ...(i)$$

where, n_1 = Index of compression,

n_2 = Index of expansion,

$V_1 = V_c + V_s = 0.05\ V_s + V_s = 1.05\ V_s$, and

$V_4 = 0.235\ V_s$.

Inserting these values in the eqn. (*i*), we get

$$\text{W/cycle} = 2 \times p_1 V_s \left[\frac{n_1}{n_1 - 1} \times 1.05 \left\{ \left(\frac{p_2}{p_1} \right)^{\frac{n_1 - 1}{n_1}} - 1 \right\} - \frac{n_2}{n_2 - 1} \times 0.2355 \left\{ \left(\frac{p_2}{p_1} \right)^{\frac{n_2 - 1}{n_2}} - 1 \right\} \right]$$

$$= 2 \times 1 \times 10^5 \times 0.00456 \left[\frac{1.25}{1.25 - 1} \times 1.05 \left\{ (7.5)^{0.25/1.25} - 1 \right\} \right.$$

$$\left. - \frac{1.3}{1.3 - 1} \times 0.2355 \left\{ (7.5)^{0.3/1.3} - 1 \right\} \right]$$

$$= 912\ [(5 \times 1.05 \times 0.496) - 1.0205 \times 0.59]$$

$$= 1825.7 \text{ Nm/cycle}$$

$$\therefore \quad \text{I.P.} = \frac{\text{Work done / cycle} \times \text{r.p.m.}}{60 \times 1000}$$

$$\frac{1825.7 \times 310}{60 \times 1000} = 9.43 \text{ kW}$$

$$\therefore \quad \text{B.P.} = \frac{\text{I.P.}}{\eta_{mech.}} = \frac{9.43}{0.85} = \mathbf{11.09\ kW.\ (Ans.)}$$

Example 20.14. *A single-stage single-acting compressor delivers 14 m³ of free air per minute from 1 bar to 7 bar. The speed of compressor is 310 r.p.m. Assuming that compression and expansion follow the law $pV^{1.35}$ = constant and clearance is 5% of the swept volume, find the diameter and stroke of the compressor. Take L = 1.5 D. The temperature and pressure of air at the suction are same as atmospheric air.*

Solution. We know that,

$$n_{\text{vol.}} = 1 + k - k \left(\frac{p_2}{p_1} \right)^{1/n} = 1 + 0.05 - 0.05 \left(\frac{7}{1} \right)^{1/1.35} \quad \left(\because \ k = \frac{V_c}{V_s} = \frac{0.05\ V_s}{V_s} = 0.05 \right)$$

$$= 0.839 \text{ or } 83.9\%$$

The free air delivered per minute is given by,

$$V_s \times \eta_{\text{vol.}} \times 310 = 14$$

$$\therefore \quad V_s = \frac{14}{\eta_{\text{vol.}} \times 310} = \frac{14}{0.839 \times 310} = 0.0538 \text{ m}^3$$

But,
$$V_s = \frac{\pi}{4} D^2 L$$

or
$$0.0538 = \frac{\pi}{4} D^2 \times 1.5D$$

or
$$D = \left(\frac{0.538 \times 4}{\pi \times 1.5} \right)^{1/3} = 0.77 \text{ m or } \mathbf{77\ cm.} \quad \mathbf{(Ans.)}$$

$$L = 1.5\ D = 1.5 \times 77 = \mathbf{115.5\ cm.\ (Ans.)}$$

Example 20.15. (*a*) *Define the term 'overall volumetric efficiency' with reference to a reciprocating compressor. Discuss the parameters in brief which affect it.*

(*b*) *Show the effect of increase in compression ratio in a single-stage reciprocating compressor on p-v diagram and give its physical explanation.*

(*c*) *A double-acting air compressor having size (D × L) 33 × 35 cm and clearance 5 per cent runs at 300 r.p.m. It takes in air at 0.95 bar and 25°C. The delivery pressure is 4.5 bar and the index of compression, n = 1.25. The free air conditions are 1.013 bar and 20°C. Work out the following :*

(*i*) *Show the process on p-v diagram for cover and crank ends ;* (*ii*) *The free air delivered as reckoned from the apparent volumetric efficiency ;* (*iii*) *The heat rejected during compression ;* (*iv*) *The power needed to drive the compressor if mechanical efficiency is 80%.*

(AMIE Summer, 1999)

Ans. (*a*) ***Overall volumetric efficiency*** *is defined as the ratio of Free Air Delivered (FAD) referred to the ambient conditions to the swept volume or displacement of the compressor.* The volume of free air delivered is less than the displacement volume due to clearance. At the end of the delivery stroke the clearance space is filled with compressed air. On the inward stroke, the air will be admitted only after the clearance air is expanded to the inlet conditions. This would reduce the effective swept volume. Further, in practice the air that is sucked in during the induction stroke gets heated up while passing through the hot valves and coming in contact with hot cylinder walls. There is also wire drawing effect through the valves resulting in drop in pressure. Thus the conditions obtained at the end of induction stroke (p_1, T_1) are different from the ambient conditions (p_{amb}, T_{amb}). Therefore the overall volumetric efficiency is

$$\eta_{v,\,\text{overall}} = \frac{p_1}{T_1} \times \frac{T_{amb}}{p_{amb}} \left[1 - C\left\{ \left(\frac{p_2}{p_1} \right)^{1/n} - 1 \right\} \right]$$

where, C = Clearance ratio (ratio of clearance volume to swept volume), and

n = Index of re-expansion.

The parameters which affect the overall volumetric efficiency are as given below :

(*i*) *Engine speed.* Higher the engine speed, greater is the wire drawing effect, and also higher the temperature of cylinder walls, lower the efficiency.

(*ii*) Leakage past the piston lowers the volumetric efficiency.

(*iii*) Too large a clearance volume will lower the clearance volumetric efficiency, and hence the overall volumetric efficiency.

(*iv*) Obstructions at inlet valves will increase wire drawing, and hence lower the volumetric efficiency.

(*v*) Inertia effects of air in suction pipe may lower the volumetric efficiency.

(*b*) The ***effect of increase in compression ratio on p-V diagram*** is as shown in Fig. 20.24.

By increasing the delivery from p_2 to p_2',

(*i*) The pressure ratio is increased from $\left(\frac{p_2}{p_1}\right)$ to $\left(\frac{p_2'}{p_1}\right)$

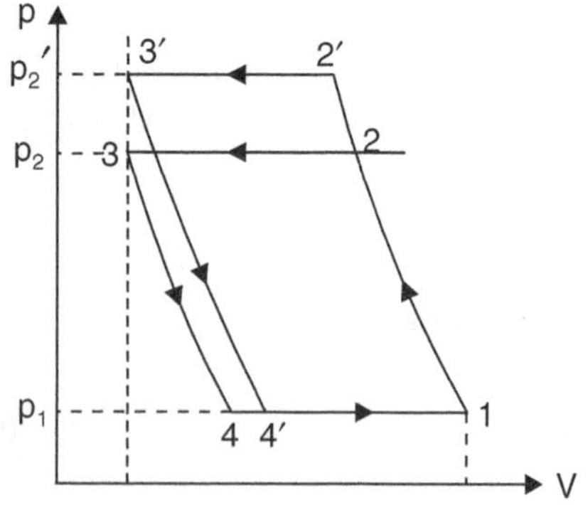

Fig. 20.24

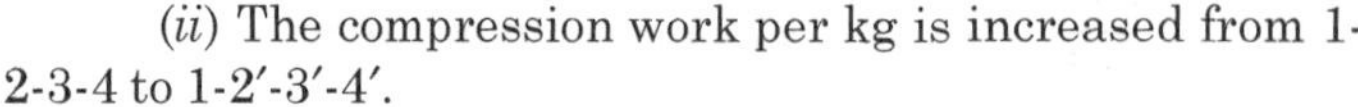

(*ii*) The compression work per kg is increased from 1-2-3-4 to 1-2′-3′-4′.

(*iii*) The effective swept volume is decreased from $(V_1 - V_4)$ to $(V_1 - V_4')$. This is due to the fact that by increasing the pressure ratio the pressure of air in the clearance volume, at the end of delivery stroke, is increased. The re-expansion 3′-4′ occupies a large fraction of swept volume (as compared to 3-4). This reduces the effective suction stroke with increase in pressure ratio.

(*c*) (*i*) **Process on p-V diagram :**

Neglecting the effect of piston rod the processes for cover and crank ends are shown on the *p-V* diagram as indicated in Fig. 20.25.

(*ii*) **The free air delivered :**

$$V_S = \frac{\pi}{4} D^2 L \times 2N \text{ for double-acting}$$

$$\frac{\pi}{4}\left(\frac{33}{100}\right)^2 \times \frac{35}{100} \times 2 \times 300 = 17.96 \text{ m}^3/\text{min.}$$

$$\eta_{vol,\,C} = \text{clearance volumetric efficiency}$$

$$= 1 - C\left[\left(\frac{p_2}{p_1}\right)^{1/n} - 1\right] = 1 - 0.05\left[\left(\frac{4.5}{0.95}\right)^{1/1.25} - 1\right]$$

$$= 0.8765$$

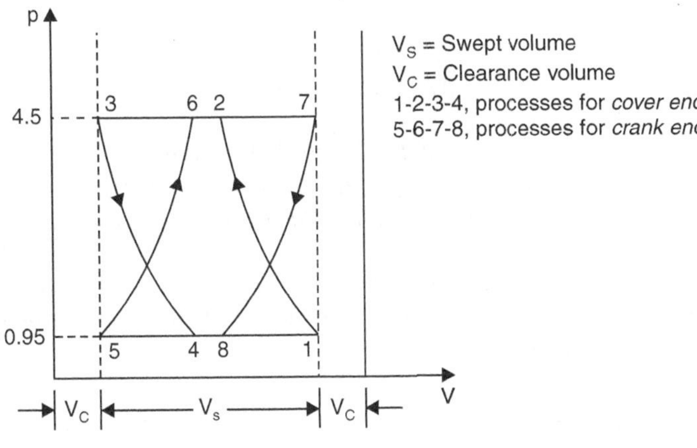

Fig. 20.25

The actual air drawn in per min.

$$= (V_1 - V_4) = V_a = \eta_{vol,\,C} \times V_s$$

$$= 0.8765 \times 17.96 = 15.74 \text{ m}^3/\text{min.}$$

This volume of air drawn is measured at 0.95 bar and 25°C. The free air conditions are 1.013 bar and 20°C.

Therefore the free air delivered, in m^3/min is

$$= \frac{p_{suc}}{p_{amb}} \times \frac{T_{amb}}{T_{suc}} \times 15.74$$

$$= \frac{0.95}{1.013} \times \frac{273 + 20}{273 + 25} \times 15.74 = \mathbf{14.51\ m^3/min.}\ \textbf{(Ans.)}$$

(*iii*) **Heat rejected during compression :**

Mass of air delivered per min.

$$\dot{m}_a = \frac{(0.95 \times 10^5) \times 15.74}{287 \times 298} = 17.48 \text{ kg/min.}$$

$$T_2 = T_1 \left(\frac{p_2}{p_1}\right)^{n-1/n} = 298 \left(\frac{4.5}{0.95}\right)^{0.25/1.25} = 406.7 \text{ K}$$

Heat rejected during compression,

$$Q_2 = mc_v \frac{\gamma - n}{n - 1} (T_2 - T_1)$$

$$= 17.48 \times 0.717 \times \frac{1.4 - 1.25}{1.25 - 1} (406.7 - 298)$$

$$= 817.4 \text{ kJ/min.} \quad \textbf{(Ans.)}$$

(*iv*) **Power needed to drive the compressor P :**

$$P = \frac{1}{\eta_{mech.}} \left[\frac{n}{n-1} mRT_1 \left\{\left(\frac{p_2}{p_1}\right)^{\frac{n-1}{n}} - 1\right\}\right]$$

$$= \frac{1}{0.8} \left[\frac{1.25}{0.25} \times \frac{17.48}{60} \times 0.287 \times 298 \left\{\left(\frac{4.5}{0.95}\right)^{\frac{0.25}{1.25}} - 1\right\}\right] \text{kJ/s}$$

$$= \textbf{56.82 kJ/s or kW.} \quad \textbf{(Ans.)}$$

Example 20.16. *Air at 103 kPa and 27°C is drawn in L.P. cylinder of a two-stage air compressor and is isentropically compressed to 700 kPa. The air is then cooled at constant pressure to 37°C in an intercooler and is then again compressed isentropically to 4 MPa in the H.P. cylinder, and is delivered at this pressure. Determine the power required to run the compressor if it has to deliver 30 m³ of air per hour measured at inlet conditions.* **(M.U.)**

Solution. Refer Fig. 20.26.

Pressure of intake air (L.P. cylinder), $p_1 = 103$ kPa

Temperature of intake air, $T_1 = 27 + 273 = 300$ K

Pressure of air entering H.P. cylinder, $p_2 = 700$ kPa

Temperature of air entering H.P. cylinder, $T_2 = 37 + 273 = 310$ K

Pressure of air after compression in H.P. cylinder, $p_3 = 4$ MPa or 4000 kPa

Volume of air delivered $= 30$ m^3/h

Power required to run the compressor, P :

$$\text{Mass of air compressed, } m = \frac{(103 \times 10^3) \times 30}{(0.287 \times 1000) \times 300} = 35.89 \text{ kg/h}$$

For the *compression process 1-2′*, we have

$$\frac{T_2'}{T_1} = \left(\frac{p_2}{p_1}\right)^{\frac{\gamma - 1}{\gamma}} = \left(\frac{700}{103}\right)^{\frac{1.4-1}{1.4}} = 1.7289 \text{ or } T_2' = 300 \times 1.7289 = 518.7 \text{ K}$$

Similarly for the *compression process 2-3*, we have

$$\frac{T_3}{T_2} = \left(\frac{p_3}{p_2}\right)^{\frac{\gamma-1}{\gamma}} = \left(\frac{4000}{700}\right)^{\frac{1.4-1}{1.4}} = 1.6454 \text{ or } T_3 = 310 \times 1.645 = 510.1 \text{ K}$$

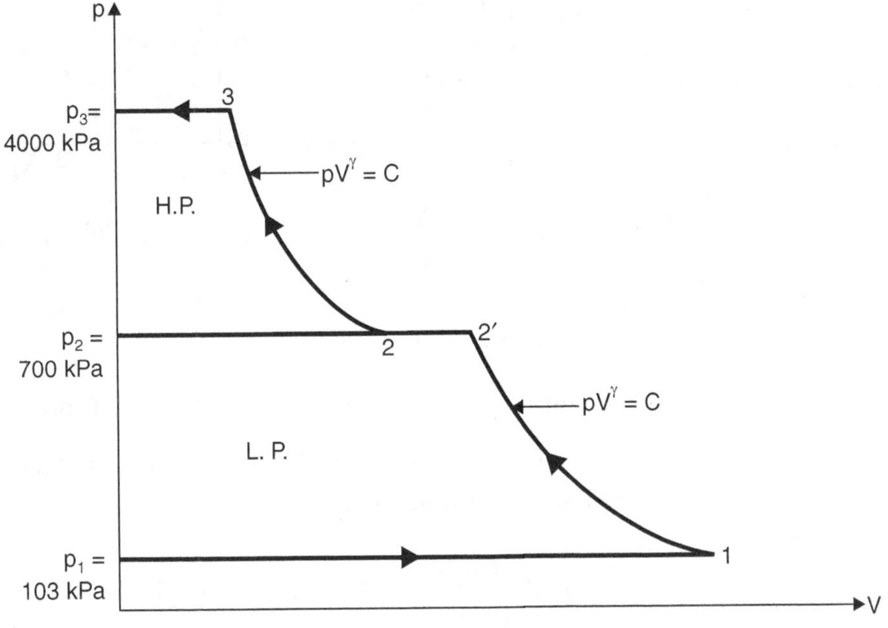

Fig. 20.26

∴ Work required to run the compressor,

$$W = \frac{\gamma}{\gamma - 1} [mR (T_2' - T_1) + mR (T_3 - T_2)]$$

$$= \frac{\gamma}{\gamma - 1} \times mR [(T_2' - T_1) + (T_3 - T_2)]$$

$$= \frac{1.4}{1.4 - 1} \times \frac{35.89}{3600} \times 0.287 [(518.7 - 300) + (510.1 - 310)] = 4.194 \text{ kN m/s}$$

Hence power required to run the compressor = **4.194 kW. (Ans.)**

Example 20.17. *A trial on a two-stage single-acting reciprocating air compressor gave the following data :*

Free air delivered	*= 6 m^3/min*
Atmospheric pressure and temperature	*= 1 bar and 27°C*
Delivery pressure	*= 40 bar*
Speed	*= 400 r.p.m.*
Intermediate pressure	*= 6 bar*
Temperature at the inlet to the second stage	*= 27°C*
Law of compression	*= $pV^{1.3}$ = constant*
Mechanical efficiency	*= 80%*
Stroke of L.P.	*= diameter of L.P. = stroke of H.P.*

Calculate :

(i) Cylinder-diameters ;

(ii) Power required, neglect clearance. **(AMIE Winter, 1998)**

Solution. Let, $D_{L.P.}$ = dia of L.P. cylinder, and

$D_{H.P.}$ = dia of H.P. cylinder.

L = Stroke = $D_{L.P.}$

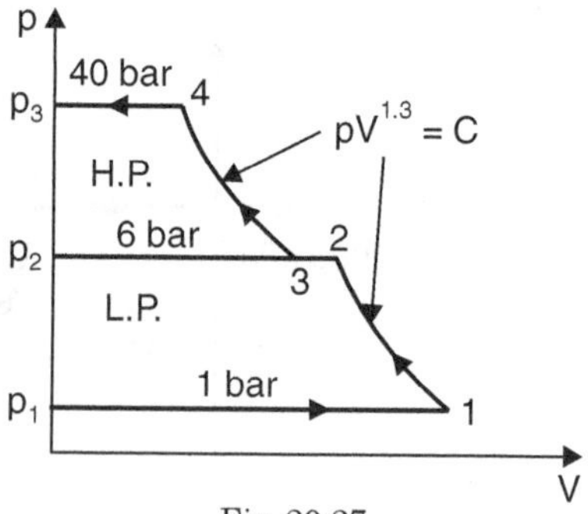

Assuming a volumetric efficiency of 100%,

$$\text{F.A.D.} = \frac{\pi}{4} D^2_{L.P.} \times \text{r.p.m. (for single-acting)}$$

or
$$6 = \frac{\pi}{4} D^2_{L.P.} \times D_{L.P.} \times 400 \text{ m}^3/\text{min}$$

$\therefore$ $D_{L.P.} = 0.2673$ m = **267.3 mm = L. (Ans.)**

Fig. 20.27

Swept volume of H.P. cylinder

= Volume of air at 27°C and 6 bar

$$= 1 \text{ m}^3/\text{min} = \frac{\pi}{4} D^2_{H.P.} \times L \times \text{r.p.m.} = \frac{\pi}{4} D^2_{H.P.} \times 0.2673 \times 400$$

$$[\because \; p_1 V_{L.P.} = p_2 V_{H.P.} \text{ or } 1 \times 6 = 6 \times V_{H.P.} \text{ or } V_{H.P.} = 1 \text{ m}^3/\text{min}]$$

$\therefore$ $D_{H.P.} = 0.109$ m or **109 mm. (Ans.)**

$$\text{Indicated work} = \frac{n}{n-1} p_1 V_1 \left[\left(\frac{p_2}{p_1} \right)^{\frac{n-1}{n}} - 1 \right] + \frac{n}{n-1} p_2 V_3 \left[\left(\frac{p_3}{p_2} \right)^{\frac{n-1}{n}} - 1 \right]$$

$$= \frac{n}{n-1} mRT_1 \left[\left(\frac{p_2}{p_1} \right)^{\frac{n-1}{n}} - 1 \right] + \frac{n}{n-1} mRT_3 \left[\left(\frac{p_3}{p_2} \right)^{\frac{n-1}{n}} - 1 \right]$$

$$= \frac{n}{n-1} mRT_1 \left[\left(\frac{p_2}{p_1} \right)^{\frac{n-1}{n}} + \left(\frac{p_3}{p_2} \right)^{\frac{n-1}{n}} - 2 \right], \text{ as } T_1 = T_3.$$

Also,
$$m = \frac{p_1 V_1}{RT_1}$$

$$= 100 \times \frac{6}{60} \times \frac{1}{0.287\,(273 + 27)} = 0.116 \text{ kg/s}$$

$$\therefore \text{ Indicated work} = \frac{1.3}{0.3} \times 0.116 \times 0.287 \times 300 \left[\left(\frac{6}{1} \right)^{\frac{0.3}{1.3}} + \left(\frac{40}{6} \right)^{\frac{0.3}{1.3}} - 2 \right]$$

$$= 45.94 \text{ kJ/s (kW)}$$

$$\text{Power required} = \frac{45.99}{0.8} = \textbf{58.42 kW. (Ans.)}$$

Example 20.18. *A two-stage single-acting reciprocating compressor takes in air at the rate f 0.2 m³/s. The intake pressure and temperature of air are 0.1 MPa and 16°C. The air is com-*

pressed to a final pressure of 0.7 MPa. The intermediate pressure is ideal and intercooling is perfect. The compression index in both the stages is 1.25 and the compressor runs at 600 r.p.m. Neglecting clearance, determine :

(*i*) *The intermediate pressure,*

(*ii*) *The total volume of each cylinder,*

(*iii*) *The power required to drive the compressor, and*

(*iv*) *The rate of heat rejection in the intercooler.*

Take c_p = 1.005 kJ/kg K and R = 0.287 kJ/kg K. **(AMIE)**

Solution. Intake volume, $V_1 = 0.2\ \text{m}^3/\text{s}$

Intake pressure, $p_1 = 0.1$ MPa

Intake temperature, $T_1 = 16 + 273 = 289$ K

Final pressure, $p_3 = 0.7$ MPa

Compression index in both stages, $n_1 = n_2 = n = 1.25$

Speed of the compressor, $N = 600$ r.p.m.

$c_p = 1.005$ kJ/kg K ; $R = 0.287$ kJ/kg K

(*i*) **The intermediate pressure, $\mathbf{p_2}$:**

$$p_2 = \sqrt{p_1 p_3} = \sqrt{0.1 \times 0.7} = 0.2646 \text{ MPa} \qquad \text{......(Perfect intercooling)}$$

(*ii*) **The total volume of each cylinder, $\mathbf{V_{s_1}}$, $\mathbf{V_{s_2}}$:**

We know that $$V_{s_1} \times \frac{N}{60} = V_1 \quad \text{or} \quad V_{s_1} \times \frac{600}{60} = 0.2$$

$$\therefore \quad V_{s_1} \text{ (Volume of L.P. cylinder)} = \frac{60 \times 0.2}{600} = \mathbf{0.02\ m^3}. \quad \textbf{(Ans.)}$$

Also $$p_1 V_{s_1} = p_2 V_{s_2} \quad \text{or} \quad V_{s_2} = \frac{p_1 V_{s_1}}{p_2}$$

$$\therefore \quad V_{s_2} \text{ (Volume of H.P. cylinder)} = \frac{0.1 \times 0.02}{0.2646} = \mathbf{0.00756\ m^3}. \quad \textbf{(Ans.)}$$

(*iii*) **The power required to drive the compressor, P :**

$$P = \frac{2n}{n-1}\, p_1 V_1 \left[\left(\frac{p_3}{p_1} \right)^{\frac{n-1}{2n}} - 1 \right] \qquad \text{......(Perfect intercooling)}$$

$$= \frac{2 \times 1.25}{1.25 - 1} \times (0.1 \times 10^3) \times 0.2 \left[\left(\frac{0.7}{0.1} \right)^{\frac{1.25-1}{2 \times 1.25}} - 1 \right] = \mathbf{42.96\ kW}. \quad \textbf{(Ans.)}$$

(*iv*) **The rate of heat rejection in the intercooler :**

Mass of air handled, $$m = \frac{p_1 V_1}{R T_1} = \frac{(0.1 \times 10^3) \times 0.2}{0.287 \times 289} = 0.241 \text{ kg/s}$$

Also, $$\frac{T_2}{T_1} = \left(\frac{p_2}{p_1} \right)^{\frac{n-1}{n}} \quad \text{or} \quad \frac{T_2}{289} = \left(\frac{0.2646}{0.1} \right)^{\frac{1.25-1}{1.25}} \quad \text{or} \quad T_2 = 351.1 \text{ K}$$

$\therefore$ Heat rejected in the intercooler $= m \times c_p \times (T_2 - T_1)$

$= 0.241 \times 1.005 \times (351.1 - 289) = 15.04$ kJ/s or **15.04 kW. (Ans.)**

Example 20.19. *A two-stage air compressor with complete intercooling delivers air to the mains at a pressure of 30 bar, the suction conditions being 1 bar and 15°C. If both cylinders have the same stroke, find the ratio of cylinders diameters, for the efficiency of compression to be a maximum. Assume the index of compression to be 1.3.*

Solution. Refer Fig. 20.28.

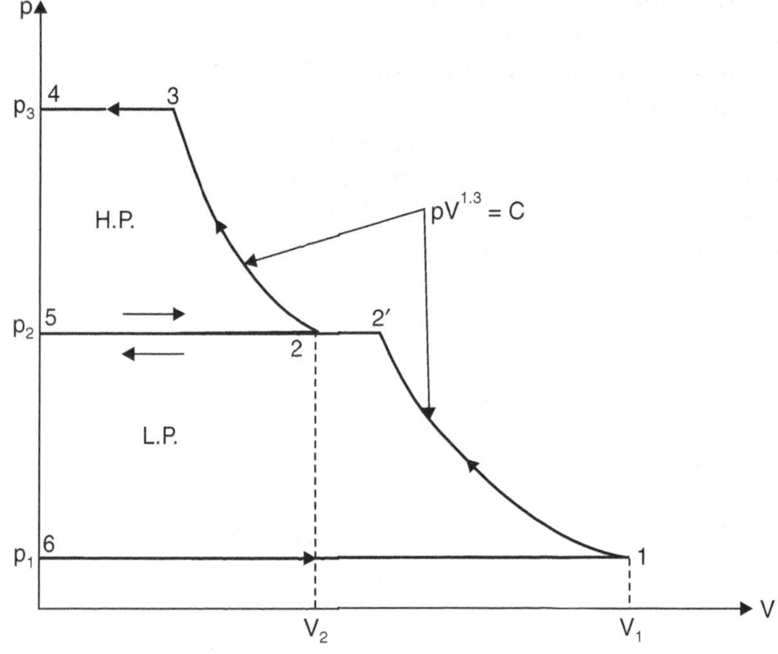

Fig. 20.28

Volume of L.P. cylinder $= V_1$

Volume of H.P. cylinder $= V_2$

If $D_{L.P.}$ and $D_{H.P.}$ are the diameters of the low pressure cylinder and high pressure cylinder respectively, then

$$\frac{V_1}{V_2} = \frac{\frac{\pi}{4} D^2_{L.P.}}{\frac{\pi}{4} D^2_{H.P.}} \quad \text{or} \quad \frac{D_{L.P.}}{D_{H.P.}} = \sqrt{\frac{V_1}{V_2}}$$

From the *curve 1-2′* following the law ;

$pV^{1.3} = C$, we have

$$p_1 V_1^{1.3} = p_2 V_2'^{1.3} \quad \text{or} \quad \frac{V_1}{V_2'} = \left(\frac{p_2}{p_1}\right)^{1/1.3}$$

But for *maximum efficiency*

$$p_2 = \sqrt{p_1 p_3} = \sqrt{1 \times 30} = 5.48 \text{ bar}$$

$$\therefore \quad \frac{V_1}{V_2'} = (5.48)^{1/1.3} = 3.7 \qquad ...(i)$$

Now temperature T_2' at the end of compression in the low pressure cylinder

$$T_2' = T_2 \times \left(\frac{p_2}{p_1}\right)^{\frac{n-1}{n}} = (15 + 273)\left(\frac{5.48}{1.0}\right)^{\frac{1.3-1}{1.3}} = 426.4 \text{ K.}$$

From *constant pressure process of 2′-2*

$$\frac{V_2}{T_2} = \frac{V_2'}{T_2'}$$

i.e.,

$$\frac{V_2'}{V_2} = \frac{T_2'}{T_2} = \frac{426.4}{(15 + 273)} = 1.48 \qquad \ldots(ii)$$

From (*i*) and (*ii*),

$$\frac{V_1}{V_2} = \frac{V_1}{V_2'} \times \frac{V_2'}{V_2} = 3.7 \times 1.48 = 5.476$$

$$\therefore \qquad \frac{D_{L.P.}}{D_{H.P.}} = \sqrt{\frac{V_1}{V_2}} = \sqrt{5.476} = 2.34.$$

i.e., **Ratio of cylinder diameters = 2.34. (Ans.)**

Example 20.20. *In a two-stage air compressor the pressures are atmospheric 1.0 bar ; intercooler 7.4 bar ; delivery 42.6 bar. Assuming complete intercooling to the original temperature of 15°C and compression index n = 1.3, find the work done in compressing 1 kg of air.*

If both cylinders have the same stroke and the piston diameters are 9 cm and 3 cm and the volumetric efficiency of the compressor is 90 per cent, will the intercooler pressure be steady or will it rise or fall as the compressor continues working ? Justify your answer. **(P.U.)**

Solution. With *complete intercooling, work done,*

$$W = \frac{n}{n-1} RT_1 \left[\left(\frac{p_2}{p_1}\right)^{\frac{n-1}{n}} + \left(\frac{p_3}{p_2}\right)^{\frac{n-1}{n}} - 2\right]$$

$$= \frac{1.3}{1.3-1} \times 0.287 \times 288 \left[\left(\frac{7.4}{1.0}\right)^{\frac{1.3-1}{1.3}} + \left(\frac{42.6}{7.4}\right)^{\frac{1.3-1}{1.3}} - 2\right]$$

$$= 358.17\,(1.587 + 1.497 - 2) = \mathbf{388.25\ kJ.}\quad \textbf{(Ans.)}$$

Since *intercooling is perfect,*

$$p_1 V_1 = p_2' V_2' \quad \text{or} \quad \frac{V_1}{V_2'} = \frac{7.4}{1.0} = 7.4$$

Ratio of effective cylinder volumes

$$= \frac{\text{Effective volume of L.P. cylinder}}{\text{Volume of H.P. cylinder}} = \frac{\pi/4 \times (0.09)^2 \times l \times 0.9}{\pi/4 \times (0.03)^2 \times l} = 8.1$$

As the ratio of effective cylinder volumes is more than the ratio of the volumes obtained from *p-V* diagram, *more air is supplied to high pressure cylinder than can hold, therefore H.P. cylinder would suck less air from intercooler than received from L.P. cylinder. Thus pressure in the intercooler will rise.*

It may be noted that the effective volume of L.P. cylinder neglecting clearance is actual volume × volumetric efficiency. It applies to the *L.P. cylinder only as the air is sucked in this cylinder.*

Example 20.21. *A single-acting two-stage air compressor deals with 4 m³/min of air under atmospheric conditions of 1.016 bar and 15°C with a speed of 250 r.p.m. The delivery pressure is 78.65 bar. Assuming complete intercooling find the minimum power required by the compressor and the bore and stroke of the compressor. Assume a piston speed of 3 m/s, mechanical efficiency of 75% and volumetric efficiency of 80% per stage. Assume the polytropic index of compression in both the stages to be n = 1.25 and neglect clearance.* **(N.U.)**

Solution. In the Fig. 20.29. 1-2 shows compression in L.P. cylinder, 2-3 shows cooling before compression in H.P. cylinder and 3-4 compression in H.P. cylinder.

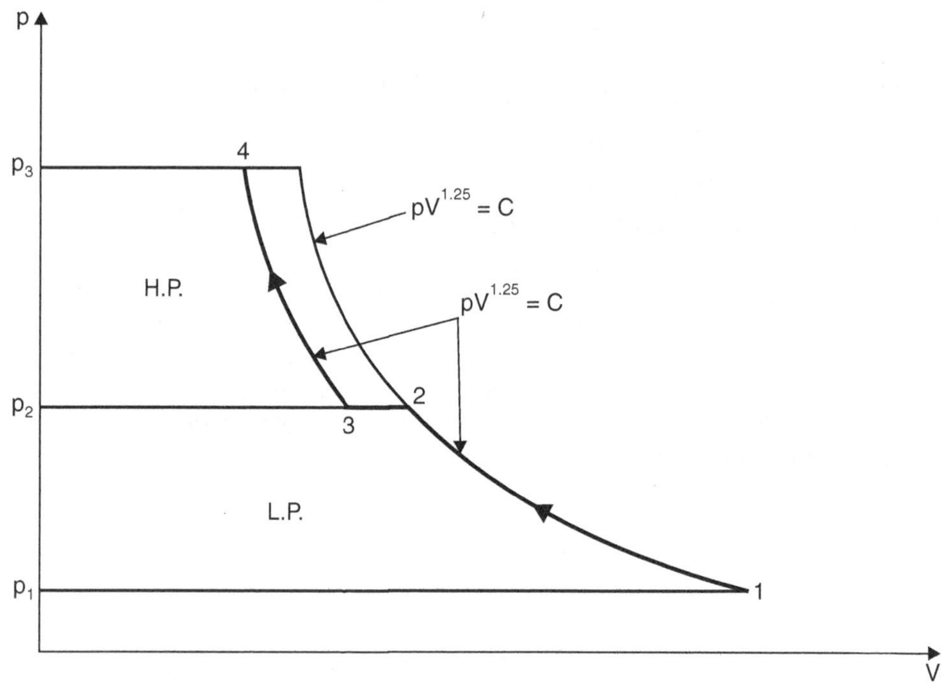

Fig. 20.29

Piston speed $= 2lN$

$\therefore$ **Stroke length,** $l = \dfrac{\text{Piston speed}}{2N} = \dfrac{3 \times 60}{2 \times 250}$ = **0.36 m. (Ans.)**

Volume handled $= \pi/4\ d^2 l \times N \times \eta_{vol.}$

$$4 = \pi/4\ d^2 \times 0.36 \times 250 \times 0.8$$

$\therefore$ **Diameter,** $d = \left(\dfrac{4 \times 4}{\pi \times 0.36 \times 250 \times 0.8}\right)^{1/2}$ = **0.266 m. (Ans.)**

Mass handled by compressor,

$$m = \frac{pV}{RT} = \frac{1.016 \times 10^5 \times 4}{287 \times 288} = 4.916 \text{ kg/min.} = 0.0819 \text{ kg/s.}$$

Intermediate pressure, $p_2 = \sqrt{p_3 p_1} = \sqrt{78.65 \times 1.016} = 8.94$ bar

Temperature at the end of first stage compression,

$$T_2 = T_1 \times (r_p)^{\frac{n-1}{n}} = 288 \times \left(\frac{8.94}{1.016}\right)^{\frac{1.25-1}{1.25}} = 444.9 \text{ K} \qquad \left[\text{where } r_p = \frac{p_2}{p_1}\right]$$

Work required

$$= 2 \times \frac{n}{n-1} \times mR\,(T_2 - T_1) \times \frac{1}{\eta_{mech.}}$$

$$= 2 \times \frac{1.25}{1.25-1} \times 0.0819 \times 287\,(444.9 - 288) \times \frac{1}{0.75} \times \frac{1}{1000}$$

$$= \mathbf{49.17 \text{ kW.}} \quad \textbf{(Ans.)}$$

Example 20.22. *In a single-acting two-stage reciprocating air compressor 4.5 kg of air per min. are compressed from 1.013 bar and 15°C through a pressure ratio of 9 to 1. Both stages have the same pressure ratio, and the law of compression and expansion in both stages is $pV^{1.3}$ = constant. If the intercooling is complete, calculate :*

(i) The indicated power

(ii) The cylinder swept volumes required.

Assume that the clearance volumes of both stages are 5% of their respective swept volumes and that the compressor runs at 300 r.p.m.

Solution. Refer Fig. 20.30.

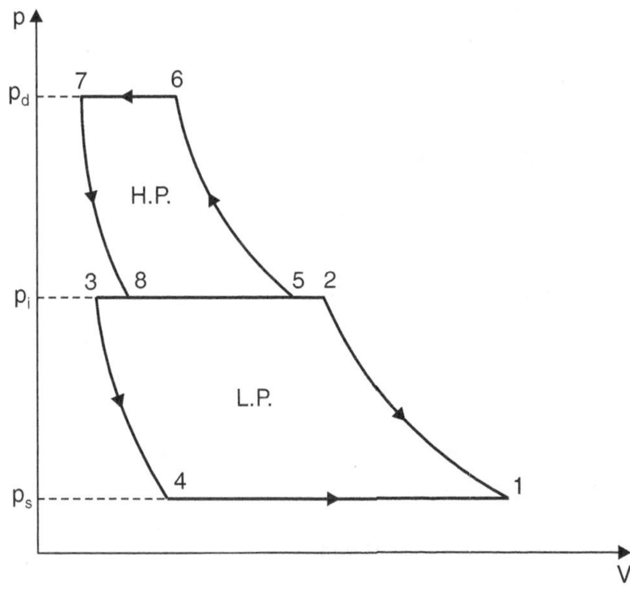

Fig. 20.30

Amount of air compressed, $m = 4.5$ kg/min.

Suction conditions, $p_s = 1.013$ bar, $T_s = 15 + 273 = 288$ K

Pressure ratio, $\dfrac{p_d}{p_s} = 9$

Also $\dfrac{p_i}{p_s} = \dfrac{p_d}{p_i}$ ... (Given)

Compression, expansion index, $n = 1.3$

Clearance volume in each stage = 5% of swept volume

Speed of the compressor, $N = 300$ r.p.m.

(*i*) **Indicated power :**

$$\therefore \quad \frac{p_i}{p_s} = \frac{p_d}{p_i}$$

$$\therefore \quad p_i^2 = p_s \times p_d = p_s \times 9p_s = 9p_s^2$$

$$\therefore \quad p_i = 3\,p_s \quad i.e., \quad \frac{p_i}{p_s} = 3$$

Now using the equation, $$\frac{T_i}{T_s} = \left(\frac{p_i}{p_s}\right)^{\frac{n-1}{n}} = (3)^{\frac{1.3-1}{1.3}}$$

$$\therefore \quad T_i = T_s \times (3)^{0.3/1.3} = 288 \times (3)^{0.3/1.3} = 371 \text{ K}$$

Now as *n, m* and temperature difference are the same for both stages, then the *work done in each stage is the same.*

Total work required per min. $$= 2 \times \frac{n}{n-1}\, mR\,(T_i - T_s)$$

$$= 2 \times \frac{1.3}{1.3-1} \times 4.5 \times 0.287\,(371 - 288) = 929 \text{ kJ/min.}$$

$\therefore$ **Indicated power** $= \frac{929}{60} = \mathbf{15.48\ kW.}$ **(Ans.)**

(*ii*) **The cylinder swept volumes required :**

The mass induced per cycle, $m = \frac{4.5}{300} = 0.015$ kg/cycle.

This mass is passed through each stage in turn

For the L.P. pressure cylinder (Fig. 20.30)

$$V_1 - V_4 = \frac{mRT_s}{p_s} = \frac{0.015 \times 287 \times 288}{1.013 \times 10^5} = 0.0122 \text{ m}^3\text{/cycle}$$

$$\eta_{vol.} = \frac{V_1 - V_4}{V_s} = 1 + k - k\left(\frac{p_i}{p_s}\right)^{1/n} = 1 + 0.05 - 0.05\,(3)^{1/1.3}$$

$$= 0.934 \qquad \left(k = \frac{V_c}{V_s} = 0.05\right)$$

i.e., $$\eta_{vol.} = 0.934$$

$$\therefore \quad V_s = \frac{V_1 - V_4}{\eta_{vol.}} = \frac{V_1 - V_4}{0.934} = \frac{0.0122}{0.934} = 0.0131 \text{ m}^3\text{/cycle}$$

i.e., **Swept volume of L.P. cylinder $V_{s(L.P.)} = 0.0131$ m³. (Ans.)**

For the high pressure stage, a mass of 0.015 kg/cycle is drawn in at 15°C and a pressure of $p_i = 3 \times 1.013 = 3.039$ bar

i.e., Volume drawn in $$= \frac{0.015 \times 287 \times 288}{3.039 \times 10^5} = 0.00408 \text{ m}^3\text{/cycle}$$

$$\eta_{vol.} = 1 + k - k\left(\frac{p_d}{p_i}\right)^{1/n}$$

and since $\frac{V_c}{V_s}$ (= k) is the same as for the low pressure stage and also $\frac{p_d}{p_i} = \frac{p_i}{p_s}$ then $\eta_{vol.}$ is 0.934 as above.

∴ **Swept Volume of H.P. stage,**

$$V_{s(H.P.)} = \frac{0.00408}{0.934} = \mathbf{0.004367\ m^3.}\ \textbf{(Ans.)}$$

It may be noted that the clearance ratio $\left(k = \frac{V_c}{V_s}\right)$ is the same in each cylinder, and the suction temperatures are the same since *intercooling is complete*, therefore the swept volumes are in the ratio of the suction pressures,

i.e.,
$$V_{s(H.P.)} = \frac{V_{L.P.}}{3} = \frac{0.0131}{3} = 0.00437\ m^3.$$

Example 20.23. *A two-stage compressor delivers 2.2 m³ free air per minute. The pressure and temperature of air at the suction are 1 bar and 25°C respectively. The pressure at the delivery is 55 bar. The clearance in L.P. cylinder is 5% and also in H.P. cylinder is 5% of the stroke. Assuming perfect intercooling between the two stages, find the minimum power required to run the compressor at 210 r.p.m.*

If the strokes of both the cylinders are equal to the diameter of the L.P. cylinder, find the diameters and strokes. What is the ratio of the cylinder volumes ; Law of compression and re-expansion in both the cylinders is $pV^{1.3}$ = *constant.*

Solution. Refer Fig. 20.31.

Free air delivered, $V_1 = 2.2\ m^3/min$, $p_1\ (= p_s) = 1$ bar, $T_1 = 25 + 273 = 298$ K ;
$p_d = 55$ bar ; $N = 210$ r.p.m. ; Law of compression and expansion ;
$pV^{1.3}$ = constant.

Clearance in each of L.P. and H.P. cylinders = 5% of the stroke

Minimum power required to run the compression, P :

For *two-stage compressor* (with *perfect intercooling*) work done is given by,

$$W = \frac{2n}{n-1}\, p_1V_1\left[\left(\frac{p_d}{p_s}\right)^{\frac{n-1}{2n}} - 1\right]$$

$$= \frac{2n}{n-1}\, mRT_1\left[\left(\frac{p_d}{p_s}\right)^{\frac{n-1}{2n}} - 1\right] \qquad (\because\ p_1V_1 = mRT_1)$$

But
$$m = \frac{p_1V_1}{RT_1} = \frac{1\times10^5\times2.2}{287\times298} = 2.57\ \text{kg/min}$$

∴
$$W = \frac{2\times1.3}{(1.3-1)}\times 2.57\times 287\times 298\left[\left(\frac{55}{1}\right)^{\frac{(1.3-1)}{2\times1.3}} - 1\right]$$

$$= 1119919\ \text{Nm/min.}$$

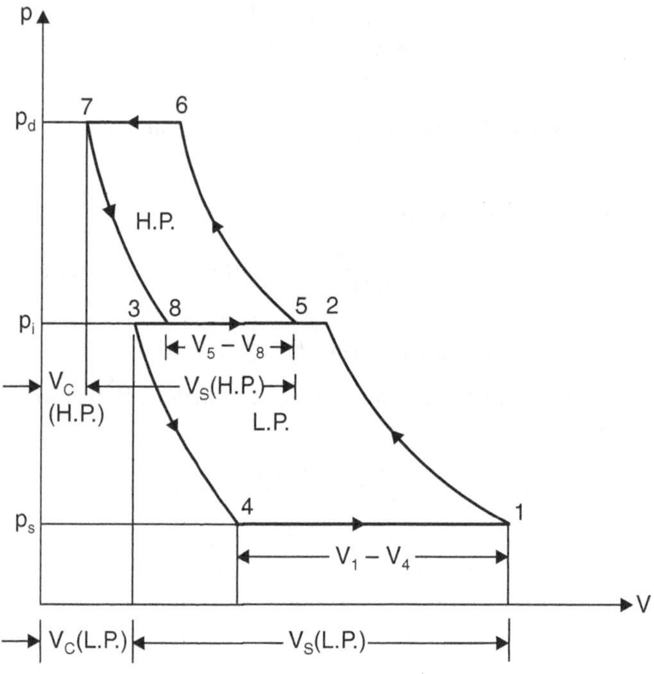

Fig. 20.31

$$\therefore \quad P = \frac{1119919}{60 \times 1000} = \mathbf{18.66\ kW.}\quad \textbf{(Ans.)}$$

Diameters and strokes :

We know that $\quad p_i = \sqrt{p_s p_d} \quad$ (p_i = Intermediate pressure)

$$= \sqrt{1 \times 55} = 7.4 \text{ bar}$$

The volumetric efficiency of L.P. cylinder is given by

$$n_{vol.(1)} = 1 + k - k \left(\frac{p_2}{p_1}\right)^{1/n} \quad (p_2 = p_i)$$

$$= 1 + 0.05 - 0.05 \left(\frac{7.4}{1}\right)^{1/1.3} = 0.817 \text{ or } 81.7\%$$

The volumetric efficiency of the H.P. cylinder is same, because

$$\frac{p_d}{p_i} = \frac{p_i}{p_s} \quad \text{and} \quad \frac{V_{c(\text{H.P.})}}{V_{s(\text{H.P.})}} = \frac{V_{c(\text{L.P.})}}{V_{s(\text{L.P.})}}$$

Free air or air at suction condition delivered per minute is given by,

Free air $\quad = (V_1 - V_4) \times \text{r.p.m.}$

$$= V_{s(\text{L.P.})}\, \eta_{vol.(\text{L.P.})} \times \text{r.p.m.} = V_{s(\text{L.P.})} \times 0.817 \times 210 = 2.2$$

$$\therefore \quad V_{s(\text{L.P.})} = \frac{2.2}{0.817 \times 210} \text{ m}^3 = \frac{2.2 \times 10^6}{0.817 \times 210} = 12822.7 \text{ cm}^3$$

$$\pi/4\ D^2_{\text{L.P.}}\ L_{\text{L.P.}} = 12822.7$$

$$\pi/4\ D^3_{L.P.} = 12822.7 \qquad [\because\ L_{L.P.} = L_{H.P.} = D_{L.P.}] \quad \text{(Given)}$$

$$\therefore \quad D_{L.P.} = \left(\frac{12822.7 \times 4}{\pi}\right)^{1/3} = \mathbf{25.37\ cm.}\quad \textbf{(Ans.)}$$

$$\therefore \quad L_{L.P.} = L_{H.P.} = \mathbf{25.37\ cm.}\quad \textbf{(Ans.)}$$

Since $\eta_{vol.(L.P.)} = \eta_{vol.(H.P.)}$

$$\therefore \quad D^2_{L.P.}\ p_1 = D^2_{H.P.}\ p_i$$

$$\therefore \quad D^2_{H.P.} = \frac{D^2_{L.P.}\ p_s}{p_i} \qquad (\because\ p_1 = p_s)$$

$$D_{H.P.} = \left(\frac{25.37^2 \times 1}{7.4}\right) = \mathbf{9.33\ cm.}\quad \textbf{(Ans.)}$$

Ratio of cylinder volumes, $\frac{V_1}{V_s}$:

As the points 1 and 5 are on isothermal line we have

$$p_1V_1 = p_5V_5$$

(where V_1 and V_5 are the cylinder volumes of L.P. and H.P.)

$$\frac{V_1}{V_5} = \frac{p_5}{p_1} = \frac{p_i}{p_s} = \frac{7.4}{1} = \mathbf{7.4.}\quad \textbf{(Ans.)}$$

Example 20.24. *A two-stage double-acting air compressor, operating at 220 r.p.m. takes in air at 1.0 bar and 27°C. The size of the L.P. cylinder is 360 × 400 mm ; the stroke of H.P. cylinder is the same as that of the L.P. cylinder and the clearance of both the cylinders is 4%. The L.P. cylinder discharges the air at a pressure of 4.0 bar. The air passes through the intercooler so that it enters the H.P. cylinder at 27°C and 3.80 bar, finally it is discharged from the compressor at 15.2 bar. The value of n in both the cylinders is 1.3, c_p = 1.0035 kJ/kg K and R = 0.287 kJ/kg K.*

Calculate : (i) The heat rejected in the intercooler ;

(ii) Diameter of H.P. cylinder ;

(iii) The power required to drive H.P. cylinder.

Solution. Refer Fig. 20.32.

Speed of compressor, N = 220 r.p.m.

p_1 = 1.0 bar, p_5 (= p_8) = 3.8 bar, p_2 (= p_3) = 4.0 bar, p_6 (= p_7) = 15.2 bar

T_1 = 27 + 273 = 300 K, T_5 = 27 + 273 = 300 K,

Clearance of L.P. and H.P. cylinders = 4%

Value of 'n' for both the cylinders = 1.3

c_p = 1.0035 kJ/kg K ; R = 0.287 kJ/kg K

Diameter of L.P. cylinder, $D_{L.P.}$ = 360 mm = 0.36 m

.Stroke of L.P. cylinder, $L_{L.P.}$ = 400 mm = 0.4 m

Swept volume of L.P. cylinder/min

$$= \pi/4\ D^2_{L.P.} \times L_{L.P.} \times (220 \times 2) = \pi/4 \times 0.36^2 \times 0.4 \times (220 \times 2)$$

= 17.91 m³/min ...... since compressor is double-acting

Volumetric efficiency referred to condition at 1

$$= 1 + k - k\left(\frac{p_2}{p_1}\right)^{1/n} = 1 + 0.04 - 0.04\left(\frac{4.0}{1.0}\right)^{1/1.3}$$

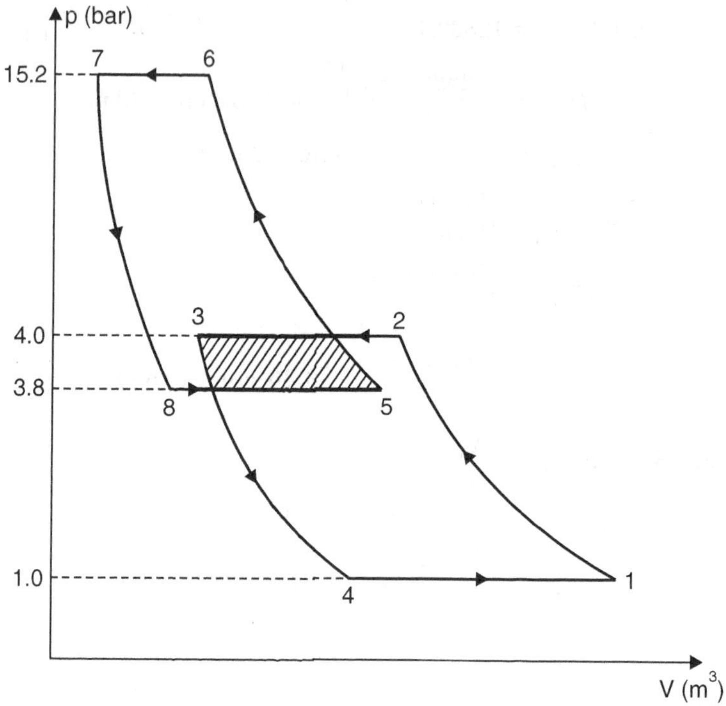

Fig. 20.32

$$= 0.9238 \text{ (or } 92.38\%) \qquad \left[\because\ k = \frac{V_c}{V_s} = \frac{4}{100} = 0.04\right]$$

∴ Volume of air drawn, referred to condition at 1

$$= 0.9238 \times 17.91 = 16.54 \text{ m}^3\text{/min}$$

Thus, mass of air/min, $m = \dfrac{p_1 V_1}{RT_1} = \dfrac{1.0 \times 10^5 \times 16.54}{0.287 \times 300 \times 10^3} = 19.21$ kg/min

Also,
$$T_2 = T_1 \times \left(\frac{p_2}{p_1}\right)^{\frac{n-1}{p_n}} = 300 \left(\frac{4}{1}\right)^{\frac{1.3-1}{1.3}} = 413 \text{ K}$$

(*i*) **Heat rejected in the intercooler :**

Heat rejected in the intercooler

$$= mc_p(T_2 - T_5) = 19.21 \times 1.0035\ (413 - 300)$$
$$= \mathbf{2178.3\ kJ/min.} \quad \textbf{(Ans.)}$$

(*ii*) **Diameter of H.P. cylinder :**

Volume of air drawn in H.P. cylinder per minute

$$V_s = \frac{mRT_5}{p_5} = \frac{19.21 \times 0.287 \times 300 \times 10^3}{3.8 \times 10^5} = 4.352 \text{ m}^3\text{/min.}$$

Since the pressure ratio and clearance percentage in H.P. and L.P. cylinders are the same, therefore, the volumetric efficiency of both cylinders is same referred to condition at the start of compression.

$\therefore$ Swept volume of H.P. cylinder $= \dfrac{4.352}{0.9238} = 4.71 \text{ m}^3/\text{min}$

i.e., $\pi/4\, D^2_{\text{H.P.}} \times L_{\text{H.P.}} \times (2 \times 220) = 4.71$

$\pi/4\, D^2_{\text{H.P.}} \times 0.4 \times (2 \times 220) = 4.71$ $\quad [\because\ L_{\text{H.P.}} = L_{\text{L.P.}} = 0.4 \text{ m}]$

$$\therefore \quad D_{\text{H.P.}} = \left(\frac{4.71 \times 4}{\pi \times 0.4 \times 2 \times 220}\right)^{1/2} = 0.1846 \text{ m or } 184.6 \text{ mm} \quad \text{(given)}$$

i.e., *Diameter of H.P. cylinder* = **184.6 mm. (Ans.)**

(*iii*) **Power required to drive H.P. cylinder :**

Since the initial temperature and pressure ratio in the L.P. and the H.P. cylinders are the same, $T_6 = T_2$

$\therefore$ Power required for H.P. cylinder

$$= \frac{n}{n-1} mR(T_2 - T_1) = \frac{1.3}{1.3-1} \times \frac{19.12}{60} \times 0.287\,(413 - 300)$$

= **45 kW. (Ans.)**

Example 20.25. *A single-acting two-stage air compressor delivers air at 18 bar. The temperature and pressure of the air before the compression in L.P. cylinder are 25°C and 1 bar. The discharge pressure of L.P. cylinder is 4.2 bar. The pressure of air leaving the intercooler is 4 bar and the air is cooled to 25°C. The diameter and stroke of L.P. cylinder are 40 cm and 50 cm respectively. The clearance volume is 5% stroke in both cylinders. The speed of the compressor is 200 r.p.m. Assuming the index of compression and re-expansion in both cylinders as 1.25, c_p for air = 1.004 kJ/kg K, find :*

(*i*) *Power required to run the compressor, and*

(*ii*) *Heat rejected in intercooler/min.*

Solution. Fig. 20.33 (*a* and *b*) shows the process on *p-V* and *T-s* diagrams

Given : $p_s = 1$ bar, $p_i = 4.2$ bar, $p_i' = 4.0$ bar,

$p_d = 18$ bar, $T_5 = T_1 = 25 + 273 = 298$ K

$D_{\text{L.P.}} = 0.4$ m, $L_{\text{L.P.}} = 0.5$ m

$N = 200$ r.p.m. , $V_c = 5\%$ of stroke in both the cylinders, and

$c_p = 1.004$ kJ/kg K

$$V_{s(\text{L.P.})} = \frac{\pi}{4} D^2_{\text{L.P.}} L_{\text{L.P.}} = \frac{\pi}{4} \times 0.4^2 \times 0.5 = 0.0628 \text{ m}^3$$

The volumetric efficiency of L.P. cylinder is given by,

$$\eta_{vol.\,(\text{L.P.})} = 1 + k - k\left(\frac{p_i}{p_s}\right)^{1/n}$$

$$= 1 + 0.05 - 0.05\left(\frac{4.2}{1}\right)^{1/1.25} \qquad \left(k = \frac{V_c}{V_s} = 0.05\right)$$

$$= 0.892 \text{ or } 89.2\%$$

$$\therefore \quad (V_1 - V_4) = \eta_{vol.\,(\text{L.P.})} \times V_{s_1} = 0.892 \times 0.0628 = 0.056 \text{ m}^3$$

Considering the *compression curve 1-2*

$$\frac{T_2}{T_1} = \left(\frac{p_i}{p_s}\right)^{\frac{n-1}{n}}$$

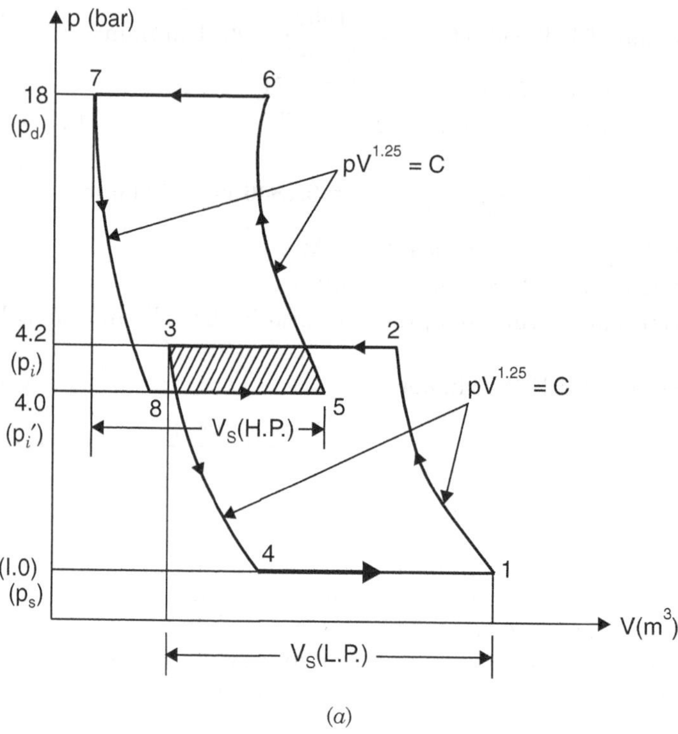

p (bar)
18
(p_d)
7
6
pV^1.25 = C
4.2
(p_i)
4.0
(p_i′)
3
2
8
5
pV^1.25 = C
V_S(H.P.)
(l.0)
(p_s)
4
1
V(m^3)
V_S(L.P.)

(a)

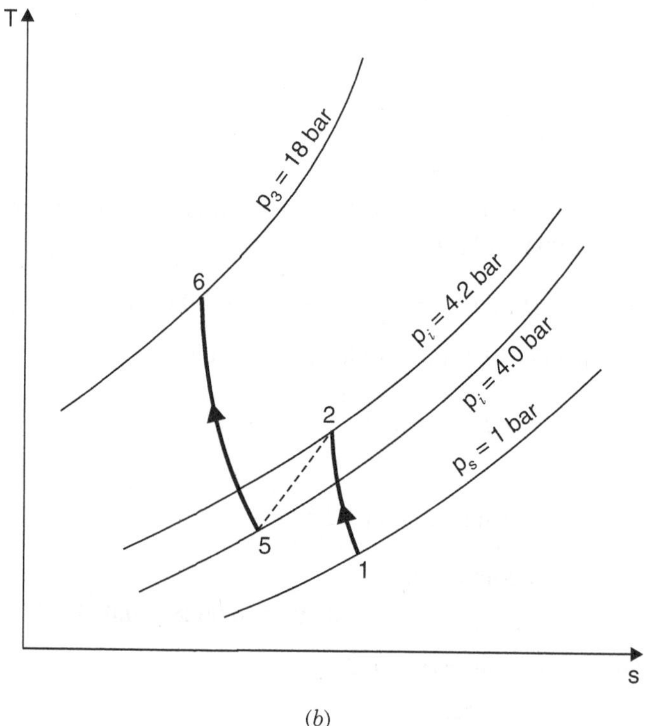

T
p_3 = 18 bar
6
p_i = 4.2 bar
p_i = 4.0 bar
p_s = 1 bar
2
5
1
s

(b)

Fig. 20.33

$$T_2 = T_1 \left(\frac{p_i}{p_s}\right)^{\frac{n-1}{n}} = 298 \left(\frac{4.2}{1}\right)^{\frac{1.25-1}{1.25}} = 397 \text{ K}$$

$$W_{(L.P.)}/\text{cycle} = \frac{n}{n-1} \, mRT_1 \left[\left(\frac{p_i}{p_s}\right)^{\frac{n-1}{n}} - 1\right]$$

$$W_{(H.P.)}/\text{cycle} = \frac{n}{n-1} \, mRT_5 \left[\left(\frac{p_d}{p_i}\right)^{\frac{n-1}{n}} - 1\right]$$

As $T_5 = T_1$ (given) $\left[\begin{array}{l} m = \text{Mass of air of volume } (V_1 - V_4) \text{ at suction condition of L.P. cylinder} \\ \therefore \quad m = \dfrac{1 \times 10^5 \times 0.056}{287 \times 298} = 0.0655 \text{ kg / stroke} \end{array}\right]$

$\therefore$ Total work done/cycle, $W = W_{(L.P.)} + W_{(H.P.)}$

$$= \frac{n}{n-1} \, mRT_1 \left[\left(\frac{p_i}{p_s}\right)^{\frac{n-1}{n}} + \left(\frac{p_d}{p_i'}\right)^{\frac{n-1}{n}} - 2\right]$$

$$= \frac{1.25}{1.25-1} \times 0.0655 \times 287 \times 298 \left[\left(\frac{4.2}{1}\right)^{\frac{1.25-1}{1.25}} + \left(\frac{18}{4.0}\right)^{\frac{1.25-1}{1.25}} - 2\right]$$

$$= 28009.8 \times (1.332 + 1.351 - 2) = 19130.7 \text{ Nm}$$

(*i*) **Power required to run the compressor**

$$\text{I.P.} = \frac{W \times N \text{ (r.p.m)}}{60 \times 1000} \text{ kW} = \frac{19130.7 \times 200}{60 \times 1000} = \mathbf{63.77 \text{ kW.}} \quad \textbf{(Ans.)}$$

(*ii*) **Heat rejected in intercooler/min**

$$= (m \times \text{r.p.m.}) \times c_p \times (T_2 - T_1)$$
$$= (0.0655 \times 200) \times 1.004 \, (397 - 298)$$
$$= \mathbf{1302.08 \text{ kJ/min.}} \quad \textbf{(Ans.)}$$

☞**Example 20.26.** *A single-acting two-stage compressor with complete intercooling delivers 10.5 kg/min of air at 16 bar. The suction occurs at 1 bar and 27°C. The compression and expansion processes are reversible, polytropic index n = 1.3. Calculate :*

(i) The power required to drive the compressor

(ii) The isothermal efficiency

(iii) The free air delivery

(iv) The heat transferred in intercooler

The compressor runs at 440 r.p.m.

(v) If the clearance ratios for L.P. and H.P. cylinders are 0.04 and 0.06 respectively, calculate the swept and clearance volumes for each cylinder.

Solution. *Given :* $p_1 = 1.0$ bar, $p_2 = 4.0$ bar, $p_3 = 16.0$ bar

$T_1 = 27 + 273 = 300$ K, $n = 1.3$,

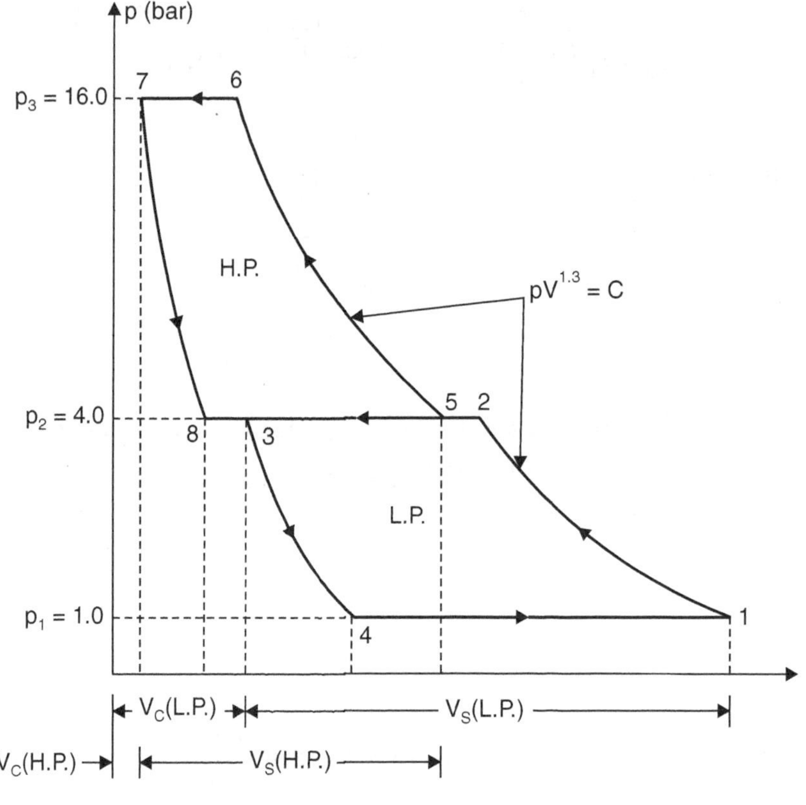

Fig. 20.34

Amount of air delivered = **10.5 kg/min**

Clearance ratio for H.P. cylinder, k = **0.04**

Clearance ratio for H.P. cylinder, k = **0.06**

The pressure ratio per stage = $\sqrt{p_1 p_3} = \sqrt{1 \times 16} = 4$

(*i*) **Power required :**

Work done in two stages with *perfect intercooling*

$$= \frac{2n}{n-1}\, mRT_1 \left[\left(\frac{p_3}{p_1}\right)^{\frac{n-1}{2n}} - 1\right]$$

$$= \frac{2 \times 1.3}{1.3 - 1} \times \frac{10.5}{60} \times 0.287 \times 300 \left[(16)^{\frac{1.3-1}{2 \times 1.3}} - 1\right] = \mathbf{49.21\ kW.} \quad \textbf{(Ans.)}$$

(*ii*) **Isothermal efficiency :**

Isothermal work $= mRT_1 \log_e \left(\frac{p_3}{p_1}\right)$

$$= \frac{10.5}{60} \times 0.287 \times 300 \times \log_e \left(\frac{16}{1}\right) = \mathbf{41.77\ kW.} \quad \textbf{(Ans.)}$$

Isothermal efficiency $= \dfrac{41.77}{49.21} = 0.8488$ or **84.88%. (Ans.)**

(*iii*) **Free air delivery (F.A.D.) :**

Free air delivery, $V = \dfrac{mRT_1}{p_1} = \dfrac{10.5 \times 0.287 \times 300 \times 10^3}{1 \times 10^5} = \mathbf{9.04\ m^3/min.}$ **(Ans.)**

(*iv*) **Heat transferred in intercooler :**

Temperature at the end of compression

$$T_2 = T_1 \times \left(\frac{p_2}{p_1}\right)^{\frac{n-1}{n}} = 300\left(\frac{4}{1}\right)^{\frac{1.3-1}{1.3}} = 413 \text{ K}$$

Heat transferred in intercooler

$$= m \times c_p \times (T_2 - T_5) = m \times c_p (T_2 - T_1) \qquad [\because \; T_5 = T_1]$$

$$= \frac{10.5}{60} \times 1.005\,(413 - 300) = \mathbf{19.87\ kW.}\ \textbf{(Ans.)}$$

(*v*) **Swept and clearance volumes :**

Volumetric efficiency for L.P. stage,

$$\eta_{vol.(L.P.)} = 1 + k - k\left(\frac{p_2}{p_1}\right)^{\frac{1}{n}} = 1 + 0.04 - 0.04(4)^{1/1.3} = 0.9238 \text{ or } 92.38\%$$

Similarly volumetric efficiency of H.P. stage,

$$\eta_{vol.\,(H.P.)} = 1 + 0.06 - 0.06(4)^{1/1.3} = 0.8857 \text{ or } 88.57\%$$

L.P. Stage :

(*i*) **Swept volume,** $\mathbf{V}_{\mathbf{s(L.P.)}} = (V_1 - V_3) = \dfrac{\text{Free air delivery}}{\text{Speed} \times \eta_{vol.\,(L.P.)}}$.

$$= \frac{9.04}{440 \times 0.9238} = \mathbf{0.0222\ m^3.}\ \textbf{(Ans.)}$$

(*ii*) **Clearance volume,** $\mathbf{V}_{\mathbf{c\ (L.P.)}} = 0.04\ (V_1 - V_3)$ or $0.04\ V_{s(L.P.)}$

$$= 0.04 \times 0.0222 = \mathbf{0.000888\ m^3.}\ \textbf{(Ans.)}$$

H.P. Stage :

(*i*) **Swept volume** $\mathbf{V}_{\mathbf{s\ (H.P.)}} = (V_5 - V_7)$

$$= \frac{\text{Free air delivery}}{\text{Stage pressure ratio} \times \text{speed} \times \eta_{vol.\,(H.P.)}}$$

$$= \frac{9.04}{4 \times 440 \times 0.8857} = \mathbf{0.0058\ m^3.}\ \textbf{(Ans.)}$$

(*ii*) **Clearance volume,** $\mathbf{V}_{\mathbf{c\ (H.P.)}} = 0.06 \times (V_5 - V_7)$ or $0.06\ V_{s(H.P.)}$

$$= 0.06 \times 0.0058 = \mathbf{0.000348\ m^3.}\ \textbf{(Ans.)}$$

Example 20.27. *The pressure limits of a 3-stage compressor are 1.05 bar and 40 bar. The compressor supplies 3 m³ of air per minute. The law of compression is* $pV^{1.25}$ *= Constant. Calculate on one minute basis :*

(*i*) *Indicated work done assuming conditions to be those for maximum efficiency ;*

(*ii*) *Isothermal work between the same pressure limits ;*

(*iii*) *Isothermal efficiency ;*

(*iv*) *Indicated work if the machine were of one-stage only ;*

(*v*) *Percentage saving in work done to using three stages instead of one.*

Solution.

(*i*) **Work done/min.**

$$= \frac{3n}{n-1} p_1V_1 \left[\left(\frac{p_d}{p_s}\right)^{\frac{n-1}{3n}} - 1\right]$$

$$= \frac{3 \times 1.25}{1.25 - 1} \times 1.05 \times 10^5 \times 3 \left[\left(\frac{40}{1.05}\right)^{\frac{1.25-1}{3 \times 1.25}} - 1\right]$$

$$= \mathbf{1297725.7\ Nm/min.}\quad \textbf{(Ans.)}$$

(*ii*) **Isothermal work done/min** $= 10^5 \, p_1V_1 \log_e \dfrac{p_d}{p_s}$

$$= 10^5 \times 1.05 \times 3 \times \log_e \frac{40}{1.05} = \mathbf{1146628.1\ Nm.}\quad \textbf{(Ans.)}$$

(*iii*) **Isothermal efficiency** $= \dfrac{1146628.1}{1297725.7} = \mathbf{0.883\ or\ 88.3\%.}$ **(Ans.)**

(*iv*) **Single-stage, work done/min.**

$$= \frac{n}{n-1} p_1V_1 \left[\left(\frac{p_d}{p_s}\right)^{\frac{n-1}{n}} - 1\right]$$

$$= \frac{1.25}{1.25 - 1} \times 1.05 \times 10^5 \times 3 \left[\left(\frac{40}{1.05}\right)^{\frac{1.25-1}{1.25}} - 1\right]$$

$$= \mathbf{1686780.2\ Nm.}\quad \textbf{(Ans.)}$$

(*v*) **% Work saved** $= \dfrac{1686780.2 - 1297725.7}{1686780.2} = 0.23$ or **23%.** **(Ans.)**

Example 20.28. *A 3-stage compressor is used to compress air from 1.0 bar to 36 bar. The compression in all stages follows the law $pV^{1.25} = C$. The temperature of air at the inlet of compressor is 300 K. Neglecting the clearance and assuming perfect intercooling, find out the indicated power required in kW to deliver 15 m^3 of air per minute measured at inlet conditions and intermediate pressures also. Take R = 0.287 kJ/kg K.*

Solution. *Given :* $p_2 = 1.0$ bar, $p_4 = 36$ bar, $n = 1.25$, $R = 0.287$ kJ/kg K

$T_1 = 300$ K

As there is *perfect intercooling,*

$$\frac{p_2}{p_1} = \frac{p_3}{p_2} = \frac{p_4}{p_3} = \left(\frac{p_4}{p_1}\right)^{1/3} = \left(\frac{36}{1}\right)^{1/3} = 3.302$$

∴ **Intermediate pressures,**

$$\mathbf{p_2} = 3.302\, p_1 = 3.302 \times 1 = \mathbf{3.302\ bar.}\quad \textbf{(Ans.)}$$

$$\mathbf{p_3} = 3.302\, p_2 = 3.302 \times 3.302 = \mathbf{10.9\ bar.}\quad \textbf{(Ans.)}$$

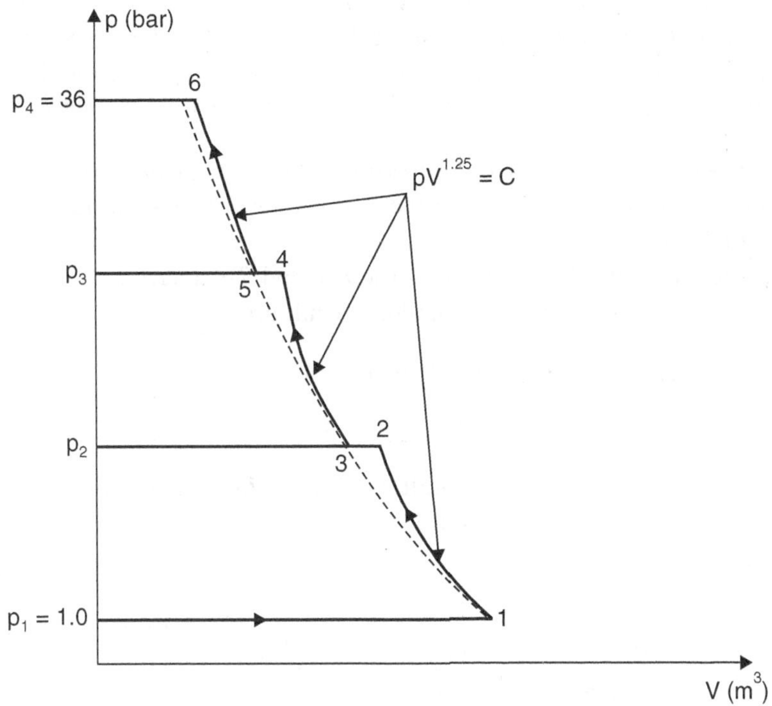

Fig. 20.35

Now $$T_2 = T_1 \left(\frac{p_4}{p_1}\right)^{\frac{n-1}{n} \times \frac{1}{x}}$$ [where x = no. of stages = 3]

$$\therefore \quad T_2 = 300 \left(\frac{36}{1}\right)^{\left(\frac{1.25-1}{1.25}\right) \times \frac{1}{3}} = 380.9 \text{ K}$$

Mass of air handled per minute

$$= \frac{p_1 V_1}{RT_1} = \frac{1.0 \times 10^5 \times 15}{0.287 \times 1000 \times 300} = 17.42 \text{ kg/min.}$$

Total work done in three stages (in kJ/s)

$$= \left[\frac{n}{n-1} . mR\,(T_2 - T_1)\right] \times x$$

$$= \left[\frac{1.25}{(1.25-1)} \times \frac{17.42}{60} \times 0.287 \times (380.9 - 300)\right] \times 3$$

$$= 101.11 \text{ kJ/s or } 101.11 \text{ kW}$$

i.e., **Indicated power required = 101.11 kW. (Ans.)**

☞**Example 20.29.** *A 3-stage double-acting compressor, operating at 200 r.p.m. takes in air at 1.0 bar and 20°C. The low pressure cylinder size is 350 mm × 400 mm. The intermediate pressure cylinder and the high pressure cylinder have the same stroke as the low pressure cylinder. The discharge pressures from the first stage and second stage are 4.0 bar and 16.0 bar and the air is finally delivered at 64.0 bar. The air is cooled to initial temperature in the intercooler after each stage and there is a drop of pressure of 0.2 bar in each of the intercoolers. The clearance volume in each cylinder is 4% of the stroke volume, but the compression indices are 1.2, 1.25 and*

1.3 for compression and expansion in each of the 1st, 2nd and 3rd stages respectively. Neglect the effect of piston rods and assume $R = 0.287$ kJ/kg K, and $c_p = 1.00$ kJ/kg K. Determine :

(i) Heat rejected in each of the intercoolers and also during compression process in each stage. Also find the heat rejected in the after-cooler if the delivered air is cooled to initial temperature.

(ii) The diameter of the intermediate pressure and the high pressure stage cylinders.

(iii) The shaft power required to drive the compressor with mechanical efficiency of 80%. Take $c_p = 1.005$ kJ/kg K.

Solution. Fig. 20.36 [(*a*) and (*b*)] shows the *p-V* and *T-s* diagrams.

The swept volume of low pressure cylinder per minute,

$$V_{S(L.P.)} = \frac{\pi}{4} D^2_{L.P.} L \times (\text{r.p.m.}) \times 2$$

$$= \frac{\pi \times 0.35^2 \times 0.4 \times 200 \times 2}{4} = 15.394 \text{ m}^3/\text{min.}$$

[r.p.m. is multiplied by 2 since the compressor is *double-acting*]

Volumetric efficiency referred to conditions at 1,

i.e., 1 bar and 20°C are :

$$\eta_{vol.\ (1st\ stage)} = 1 + k - k\left(\frac{p_2}{p_1}\right)^{\frac{1}{n}} = 1 + 0.04 - 0.04\left(\frac{4}{1}\right)^{\frac{1}{1.2}}$$

$$= 0.913 \text{ or } 91.3\%$$

$$\eta_{vol.\ (2nd\ stage)} = 1 + 0.04 - 0.04\left(\frac{p_6}{p_5}\right)^{\frac{1}{1.25}} = 1.04 - 0.04\left(\frac{16}{3.8}\right)^{\frac{1}{1.25}}$$

$$= 0.9136 \text{ or } 91.36\%$$

$$\eta_{vol.\ (3rd\ stage)} = 1 + 0.04 - 0.04\left(\frac{p_{10}}{p_9}\right)^{\frac{1}{1.3}} = 1.04 - 0.04\left(\frac{64}{15.8}\right)^{\frac{1}{1.3}}$$

$$= 0.9227 \text{ or } 92.27\%$$

Volume of air taken in at 1 bar 20°C

$$(V_1 - V_4)/\text{min} = V_{S(L.P.)}/\text{min.} \times \eta_{vol.\ (1st\ stage)}$$

$$= 15.394 \times 0.913 = 14.05 \text{ m}^3/\text{min.}$$

And mass of air/min.,

$$m = \frac{p_1(V_1 - V_4)}{RT_1} = \frac{1 \times 10^5 \times 14.05}{0.287 \times 293 \times 10^3} = 16.708 \text{ kg/min.}$$

Also $$T_2 = T_1\left(\frac{p_2}{p_1}\right)^{\frac{n-1}{n}} = 293\left(\frac{4}{1}\right)^{\frac{1.2-1}{1.2}} = 369 \text{ K}$$

And $$T_6 = T_5\left(\frac{p_6}{p_5}\right)^{\frac{n-1}{n}} = 293\left(\frac{16}{3.8}\right)^{\frac{1.25-1}{1.25}} = 390.6 \text{ K.}$$

And $$T_{10} = T_9\left(\frac{p_{10}}{p_9}\right)^{\frac{n-1}{n}} = 293\left(\frac{64}{15.8}\right)^{\frac{1.3-1}{1.3}} = 404.6 \text{ K}$$

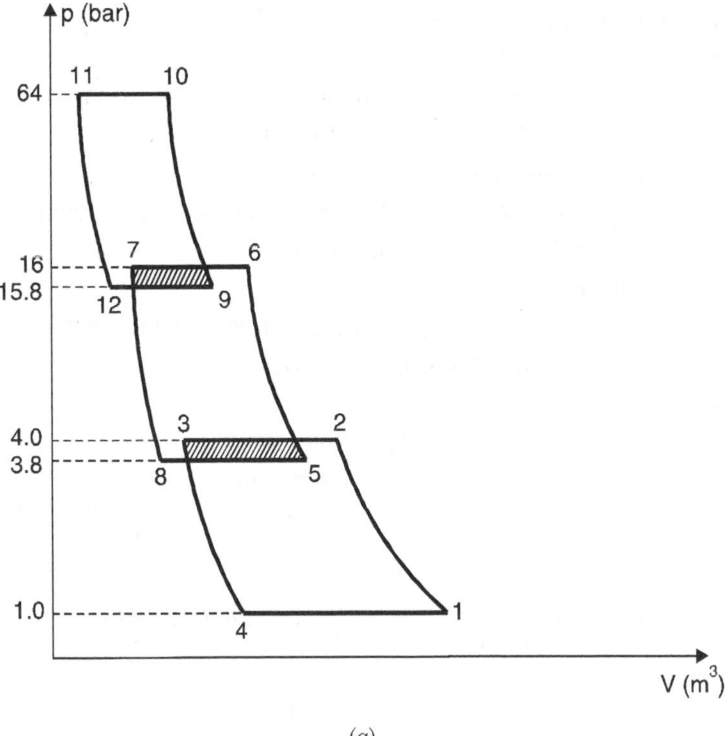
p (bar)
64
11
10
16
15.8
7
6
12
9
4.0
3.8
3
2
8
5
1.0
1
4
V (m³)

(a)

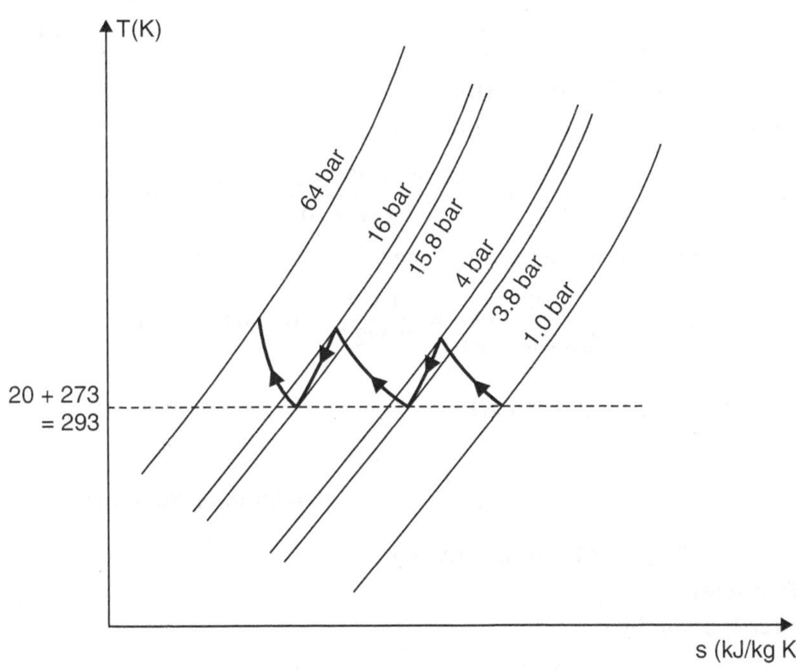
T(K)
64 bar
16 bar
15.8 bar
4 bar
3.8 bar
1.0 bar
20 + 273
= 293
s (kJ/kg K)

(b)

Fig. 20.36

(*i*) **Heat rejected in each cooler :**

Heat rejected in intercooler *after 1st stage*

$$= mc_p\,(T_2 - T_5) = 16.708 \times 1.005\,(369 - 293)$$
$$= \mathbf{1276.16\ kJ/min.}\quad \textbf{(Ans.)}$$

And heat rejected in intercooler *after 2nd stage*

$$= mc_p(T_6 - T_9) = 16.708 \times 1.005 \times (390.6 - 293)$$
$$= \mathbf{1638.85\ kJ/min.}\quad \textbf{(Ans.)}$$

And heat rejected in the *after-cooler*

$$= mc_p(T_{10} - T_1) = 16.708 \times 1.005\,(404.6 - 293)$$
$$= \mathbf{1873.93\ kJ/min.}\quad \textbf{(Ans.)}$$

(*ii*) $\mathbf{D_{I.P.}}$; $\mathbf{D_{H.P.}}$:

Volume drawn in intermediate pressure cylinder/min.

$$(V_5 - V_8)/\text{min.} = \frac{mRT_5}{p_5} = \frac{16.708 \times 0.287 \times 293 \times 10^3}{3.8 \times 10^5} = 3.69\ \text{m}^3/\text{min.}$$

$\therefore$ Swept volume of intermediate cylinder/min

$$V_{s(\text{intermediate})}/\text{min} = \frac{3.69}{\eta_{vol.(2\text{nd stage})}} = \frac{3.69}{0.9136} = 4.039\ \text{m}^3/\text{min.}$$

or
$$\frac{\pi}{4}\,D^2_{\text{I.P.}}\,L \times (r.p.m.) \times 2 = 4.039$$

$\therefore$
$$D_{\text{I.P.}} = \left[\frac{4.039 \times 4}{\pi \times L \times (r.p.m.) \times 2}\right]^{1/2} = \left[\frac{4.039 \times 4}{\pi \times 0.4 \times 200 \times 2}\right]^{1/2} = 0.179\ \text{m or } 179\ \text{mm}$$

i.e.,
$$D_{\text{I.P.}} = \mathbf{179\ mm.}\quad \textbf{(Ans.)}$$

Volume drawn in high pressure cylinder/min

$$(V_9 - V_{12})/\text{min} = \frac{mRT_9}{p_9} = \frac{16.708 \times 0.287 \times 293 \times 10^3}{15.8 \times 10^5} = 0.889\ \text{m}^3/\text{min.}$$

$\therefore$ Swept volume of high pressure cylinder,

$$V_{s(\text{H.P.})}/\text{min} = \frac{0.889}{\eta_{vol.\,(2\text{nd stage})}} = \frac{0.889}{0.9227} = 0.9364\ \text{m}^3/\text{min.}$$

or
$$\frac{\pi}{4}\,\text{D}^2_{\text{H.P.}}\,L \times (r.p.m.) \times 2 = 0.9364$$

$\therefore$
$$D_{\text{H.P.}} = \left(\frac{0.9364 \times 4}{\pi \times 0.4 \times 200 \times 2}\right)^{1/2} = 0.0863\ \text{m or } 86.3\ \text{mm}$$

i.e.,
$$D_{\text{H.P.}} = \mathbf{86.3\ mm.}\quad \textbf{(Ans.)}$$

(*iii*) **Shaft power :**

Shaft power is given by :

$$\left[\frac{1.2}{1.2-1}\,mR\,(T_2 - T_1) + \frac{1.25}{1.25-1}\,mR\,(T_6 - T_5) + \frac{1.3}{1.3-1}\,mR\,(T_{10} - T_9)\right] \times \frac{1}{60 \times \eta_{mech.}}$$

$$= \frac{1}{60 \times 0.80} mR \left[\frac{1.2}{0.2} \times (369 - 293) + \frac{1.25}{0.25} (390.6 - 293) + \frac{1.3}{0.3} \times (404.6 - 293) \right]$$

$$= \frac{16.708 \times 0.287}{60 \times 0.8} (456 + 488 + 483.6) = 142.6 \text{ kW}$$

i.e., **Shaft power = 142.6 kW. (Ans.)**

Heat rejected during compression process during a stage above may not be confused with heat rejected per stage, which includes heat transferred during suction and delivery, if any. Thus heat rejected during compression process 1-2.

$$= \left[c_v \left(\frac{\gamma - n}{\gamma - 1} \right) (T_2 - T_1) \right] \times m$$

Also $c_p = c_v + R, \quad \therefore \quad c_v = c_p - R = 1.005 - 0.287 = 0.718$

and $\gamma = \frac{c_p}{c_v} = \frac{1.005}{0.718} = 1.4$

Therefore, heat transfer

$$= \left[0.718 \left(\frac{1.4 - 1.2}{1.4 - 1} \right) (369 - 293) \right] \times 16.708 = \mathbf{455.86 \text{ kJ/min.}} \quad \textbf{(Ans.)}$$

Similarly for **2nd stage** compression process 5-6

Heat transfer $= \left[0.718 \left(\frac{1.4 - 1.2.5}{1.4 - 1} \right) (390.6 - 293) \right] \times 16.708 = \mathbf{439.07 \text{ kJ/min.}}$ **(Ans.)**

Similarly for **3rd stage** compression process 9-10

Heat transfer $= \left[0.718 \left(\frac{1.4 - 1.3}{1.4 - 1} \right) (404.6 - 293) \right] \times 16.708 = \mathbf{334.69 \text{ kJ/min.}}$ **(Ans.)**

Example 20.30. *A multi-stage air compressor is to be designed to elevate the pressure from 1 bar to 125 bar such that stage pressure ratio will not exceed 4. Determine :*

(i) Number of stages *(ii) Exact stage-pressure ratios*

(iii) Intermediate pressures.

Solution. *(i)* **Number of stages, x :**

Assuming *perfect intercooling,* the condition for *minimum* work of compression in multi-stage compression is

$$\frac{(p)_{x+1}}{(p)_x} = \left[\frac{(p)_{x+1}}{(p)_1} \right]^{1/x}$$

In this case $\frac{(p)_{x+1}}{(p)_x}$ is restricted to 4 and $\frac{(p)_{x+1}}{(p)_1} = \frac{125}{1} = 125$

Hence $4 = (125)^{1/x}$

$$\log_e 4 = \frac{1}{x} \log_e 125$$

$$1.386 = \frac{1}{x} \times 4.828$$

$\therefore \qquad x = \dfrac{4.828}{1.386} = 3.48$ say 4

Hence the number of stages, x = 4. (Ans.)

(*ii*) **Exact stage pressure ratios :**

Again using the relation

$$\frac{(p)_{x+1}}{(p)_x} = \left[\frac{(p)_{x+1}}{(p)_1}\right]^{1/x} = \left(\frac{125}{1}\right)^{1/4} = \mathbf{3.343.} \quad \textbf{(Ans.)}$$

(*iii*) **Intermediate pressures :**

Extreme pressures are already fixed

i.e., $\qquad p_1 = 1$ bar, $p_{4+1} = p_5 = 125$ bar

Also $\qquad \dfrac{(p)_{4+1}}{(p)_4} = \dfrac{p_5}{p_4} = 3.343$

$\therefore \qquad p_4 = \dfrac{p_5}{3.343} = \dfrac{1.25}{3.343} =$ **37.39 bar. (Ans.)**

Similarly $\qquad \dfrac{p_4}{p_3} = 3.343$

$\therefore \qquad p_3 = \dfrac{p_4}{3.343} = \dfrac{37.39}{3.343} =$ **11.18 bar. (Ans.)**

and $\qquad p_2 = \dfrac{p_3}{3.343} = \dfrac{11.18}{3.343} =$ **3.344 bar. (Ans.)**

Example 20.31. *The requirement is to compress air at 1 bar and 25°C and deliver it at 160 bar using multi-stage compression and intercoolers. The maximum temperature during compression must not exceed 125°C and also cooling in the intercooler is done so as not to drop the temperature below 40°C. The law of compression followed is $pV^{1.25}$ = constant for all stages.*

Calculate : (i) Number of stages required,

(ii) Work input per kg of air, and

(iii) Heat rejected in the intercoolers.

Take R = 0.287 kJ/kg K, c_v = 0.71 kJ/kg K.

Solution. Let, p_s = Suction pressure = 1 bar,

T_s = Suction temperature = 25 + 273 = 298 K,

p_1 = Delivery pressure from 1st stage = entry pressure to 2nd stage,

T_1 = Delivery temperature after every stage,

p_d = Delivery pressure from (*N* + 1)th stage,

x = Number of stages after the 1st stage, and

T_{ratio} = Temperature ratio for 2nd stage, 3rd stage ..... *n*th stage.

(*i*) **Number of stages :**

Thus, $$\frac{p_1}{p_s} = \left(\frac{T_1}{T_s}\right)^{\frac{n}{n-1}}$$

$$= \left(\frac{125+273}{25+273}\right)^{\frac{1.25}{1.25-1}} = \left(\frac{398}{298}\right)^5 = 4.25$$

$\therefore$ $$p_1 = p_s \times 4.25 = 1.0 \times 4.25 = 4.25 \text{ bar}$$

Also $$\frac{T_2}{T_1} = \frac{T_3}{T_2} = \frac{T_4}{T_3} = ... \frac{T_x}{T_{x-1}}$$

or $$\frac{p_x}{p_1} = (T_{ratio})^{\frac{xn}{n-1}}$$

and $$T_{ratio} = \frac{(125 + 273)}{(40 + 273)} = \frac{398}{313} = 1.2715$$

or $$\frac{160}{4.25} = (1.2175)^{\frac{x \times 1.25}{(1.25-1)}} \text{ or } 37.65 = (1.2715)^{5x}$$

$$\log_e 37.65 = 5x \times \log_e 1.2715$$

$$3.628 = 5x \times 0.24$$

$\therefore$ $$x = \frac{3.628}{5 \times 0.24} = 3.02 \text{ or say } 3$$

Hence, number of stages $= 3 + 1 = 4$. **(Ans.)**

(*ii*) **Work done per kg of air :**

Pressure ratio in 1st stage = 4.25

Pressure ratio in the following stage = $\left(\frac{160}{4.25}\right)^{1/3}$ = 3.351 bar

Temperature leaving 2nd, 3rd and 4th stages = 125 + 273 = 398 K (Given)

Work done/kg in 1st stage = $\frac{n}{n-1} RT_s \left[\left(\frac{p_1}{p_s}\right)^{\frac{n-1}{n}} - 1\right]$

$$= \frac{1.25}{1.25-1} \times 0.287 \times 298 \left[(4.25)^{\frac{1.25-1}{1.25}} - 1\right] = 143.5 \text{ kJ}$$

Work done in the following 3 stages

$$= 3 \times \frac{1.25}{1.25-1} \times 0.287 \times 313 \left[(3.351)^{\frac{1.25-1}{1.25}} - 1\right] = 368.6 \text{ kJ}$$

Total work done/kg = 143.5 + 368.6 = **512.1 kJ/kg. (Ans.)**

(*iii*) **Heat rejected in the intercoolers :**

Heat rejected in the intercoolers

$= 3c_p (398 - 313)$

$= 3 \times 0.997 \times 85$ $[\because c_p = c_v + R = 0.71 + 0.287 = 0.997 \text{ kJ/kg}]$

= **254.23 kJ/kg. (Ans.)**

☞**Example 20.32.** *A 3-stage air compressor supplying air blast for an oil engine has a rated capacity of 10.5 m³ of free air/min, and is driven by the main engine at 100 r.p.m. The pressure at the suction to L.P. cylinder is 1 bar and at delivery from H.P. cylinder is 95 bar. The fractional clearances are 0.04 and 0.07 for I.P. and H.P. cylinders. Assuming the temperature at the end of suction in all cylinders is 25°C i.e., perfect intercooling, stage pressures in geometric* *progression and the law of compression $pV^{1.25}$ = constant, calculate the swept volume of each cylinder. The free air conditions are 1.013 bar and 15°C.*

Solution. Refer Fig. 20.37. Since the pressures are in geometric progression (given),

$$\frac{p_d}{p_{i2}} = \frac{p_{i2}}{p_{i1}} = \frac{p_{i1}}{p_s} = z \text{ (Pressure ratio).}$$

$$p_{i1} = zp_s$$
$$p_{i2} = p_{i1}z = z^2 p_s$$
$$p_d = p_{i2}z = z^3 p_s$$

$\left[\begin{array}{l} p_d = \text{Delivery pressure (H.P. cylinder)} \\ p_{i2} = \text{Intermediate pressure-2} \\ p_{i1} = \text{Intermediate pressure-1} \\ p_s = \text{Suction pressure (L.P. cylinder)} \end{array}\right]$

$$\therefore \quad z = \left(\frac{p_d}{p_s}\right)^{1/3}$$

$$= \left(\frac{95}{1}\right)^{1/3} = 4.56$$

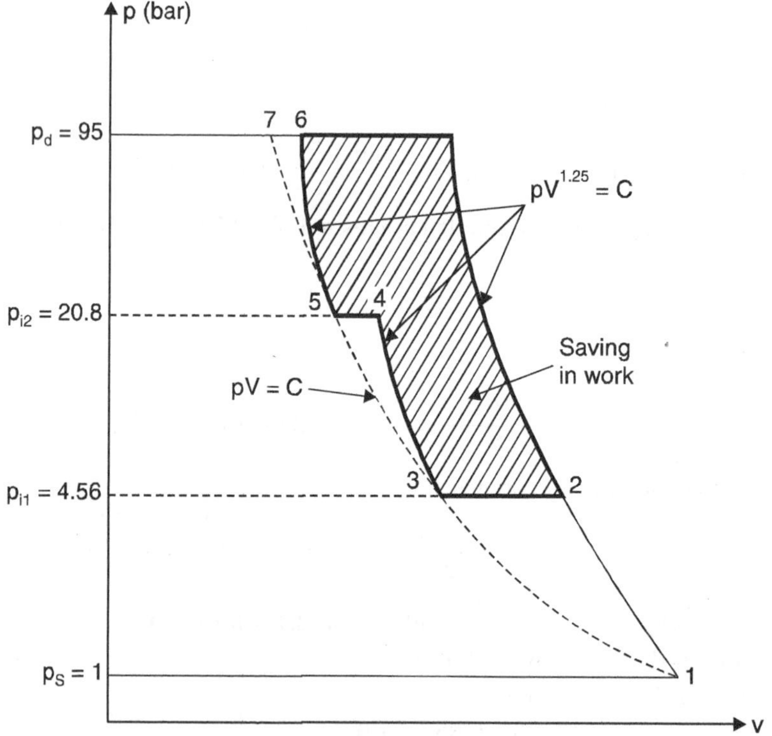

Fig. 20.37

$$\therefore \quad p_{i1} = zp_s = 4.56 \times 1 = 4.56 \text{ bar}$$

$$p_{i2} = zp_{i1} = 4.56 \times 4.56 \times 20.8 \text{ bar}$$

$$\eta_{vol.(L.P.)} = 1 + k - k\,(z)^{1/n}$$

$$= 1 + 0.04 - 0.04\,(4.56)^{1/1.25} = 0.905 \text{ or } 90.5\%$$

Similarly, $\eta_{vol.\,(L.P.)} = \eta_{vol.\,(H.P.)} = 1 + 0.07 - 0.07\,(4.56)^{1/1.25} = 0.834$ or 83.4%

Swept volume of each cylinder

Volume of free air reduced to suction condition of L.P. cylinder is given by

$$\frac{p_1 V_1}{T_1} = \frac{p_{amb.} V_{amb.}}{T_{amb.}}$$

$$V_1 = \frac{p_{amb.} V_{amb.} T_1}{T_{amb.} p_1}$$

$$= \frac{1.013 \times 10^5 \times 10.5 \times (273 + 25)}{(273 + 15) \times 1 \times 10^5} = 11.0 \text{ m}^3/\text{min.}$$

$\therefore$ Swept capacity of L.P. cylinder $= \dfrac{11.0}{\eta_{vol.(L.P)} \times \text{r.p.m.}}$

$$= \frac{11.0}{0.905 \times 100} = \mathbf{0.1215 \text{ m}^3.} \quad \textbf{(Ans.)}$$

Again volume of free air reduced to suction conditions of I.P. cylinder

$$= \frac{1.013 \times 10^5 \times 10.5 \times 298}{288 \times 4.56 \times 10^5} = 2.41 \text{ m}^3/\text{min}$$

Swept capacity of I.P. cylinder

$$= \frac{2.41}{0.834 \times 100} = \mathbf{0.089 \text{ m}^3/\text{min.}} \quad \textbf{(Ans.)}$$

Again volume of free air reduced to suction conditions of H.P. cylinder

$$= \frac{1.013 \times 10^5 \times 10.5 \times 298}{288 \times 20.8 \times 10^5} = 0.529 \text{ m}^3/\text{min}$$

Swept capacity of H.P. cylinder

$$= \frac{0.529}{0.834 \times 100} = \mathbf{0.00634 \text{ m}^3.} \quad \textbf{(Ans.)}$$

Example 20.33. *Show that the heat rejected per stage per kg of air in a reciprocating compressor with perfect intercooling is given by*

$$\left[c_p + c_v\left(\frac{\gamma - n}{n - 1}\right)\right](T_2 - T_1)$$

where $(T_2 - T_1)$ = *Temperature rise during compression,*

n = *Polytropic index,*

γ = *Adiabatic index, and*

c_p, c_v = *Two specific heats of air.*

Solution. In a compressor heat is rejected in two stages :

(*i*) *During compression* when heat is rejected to the cylinder walls ;

(*ii*) *During intercooling.*

Heat rejected during *compression*

$$= \frac{\gamma - n}{\gamma - 1} \times \text{compression work} = \frac{\gamma - n}{\gamma - 1}\left(\frac{p_2 v_2 - p_1 v_1}{n - 1}\right)$$

$$= \frac{\gamma - n}{\gamma - 1}\left(\frac{T_2 - T_1}{n - 1}\right) R \text{ ...... per kg of air}$$

For perfect intercooling the temperature of air after compression T_2 must be reduced to initial temperature T_1.

Heat rejected in *intercooling*

$$= c_p (T_2 - T_1) \text{ ..... per kg of air}$$

$\therefore$ Total heat rejected per kg of air

$$= \frac{\gamma - n}{\gamma - 1} \times \frac{R}{n-1} (T_2 - T_1) + c_p (T_2 - T_1) = \left[c_p + \frac{R}{\gamma - 1}\left(\frac{\gamma - n}{n-1}\right)\right] (T_2 - T_1)$$

$$= \left[c_p + c_v \left(\frac{\gamma - n}{n - 1}\right)\right] (T_2 - T_1). \textbf{ Proved.}$$

Example 20.34. *The cylinder of an "Air motor" has a bore of 6.35 cm and a stroke of 11.4 cm. The supply pressure is 6.3 bar, the supply temperature 24°C, and exhaust pressure is 1.013 bar. The clearance volume is 5% of the swept volume and the cut-off ratio is 0.5. The air is compressed by the returning piston after it has travelled through 0.95 of its stroke. The law of compression and expansion is $pV^{1.3}$ = constant. Calculate the temperature at the end of expansion and the indicated power of the motor which runs at 300 r.p.m.*

Also calculate the air supplied per minute. Take R = 0.287 kJ/kg K.

Solution. Refer Fig. 20.14.

$$\text{Swept volume} = \frac{\pi \times (0.0635)^2 \times 0.114}{4} = 0.000361 \text{ m}^3$$

$$\text{Clearance volume} = V_6 = V_5 = 0.05 \times 0.000361 = 0.000018 \text{ m}^3$$

$$V_1 = \frac{0.000361}{2} + 0.000018 = 0.000198 \text{ m}^3$$

$$V_2 = 0.000361 + 0.000018 = 0.000379 \text{ m}^3$$

$$V_4 - V_5 = 0.05 \times 0.000361 \simeq 0.000018 \text{ m}^3$$

$$\therefore \quad V_4 = 0.000018 + 0.000018 = 0.000036 \text{ m}^3$$

$$p_1 V_1^n = p_2 V_2^n$$

$$\therefore \quad p_2 = p_1 \left(\frac{V_1}{V_2}\right)^n = 6.3 \left(\frac{0.000198}{0.000379}\right)^{1.3} = 2.71 \text{ bar}$$

Temperature at the end of expansion :

$$T_2 = T_1 \left(\frac{V_1}{V_2}\right)^{n-1} = 297 \left(\frac{0.000198}{0.000379}\right)^{0.3} = 244.4 \text{ K}$$

i.e., Temperature after expansion = 244.4 – 273 = **– 28.6°C. (Ans.)**

Indicated power of the motor :

$$p_5 = p_4 \left(\frac{V_4}{V_5}\right)^n = 1.013 \left(\frac{0.000036}{0.000018}\right)^{1.3} = 2.494 \text{ bar}$$

Work done per cycle = Area 1234561

i.e., Work done

$$= p_1(V_1 - V_6) + \frac{(p_1V_1 - p_2V_2)}{n-1} - p_3(V_3 - V_4) - \left(\frac{p_5V_5 - p_4V_4}{n-1}\right)$$

$\therefore$ Work done/cycle $= 10^5 \times 6.3\ (0.000198 - 0.000018)$

$$+ \frac{10^5\ (6.3 \times 0.000198 - 2.71 \times 0.000379)}{(1.3 - 1)}$$

$$- 10^5 \times 1.013\ (0.000379 - 0.000036)$$

$$- \frac{10^5\ (2.494 \times 0.000018 - 1.013 \times 0.000036)}{(1.3 - 1)}$$

$$= 113.4 + 73.44 - 34.74 - 2.808 = 149.29 \text{ Nm}$$

$\therefore$ $$\text{I.P.} = \frac{149.29 \times 300}{60 \times 1000} = \mathbf{0.746\ kW.}\quad \textbf{(Ans.)}$$

Air supplied per minute :

The mass induced per cycle is given by $(m_1 - m_4)$. It is necessary to determine the *temperature of air at 4,* which *can be taken as equal to that at 3.* It is assumed that the air in the cylinder at the point 2 expands isentropically to the exhaust pressure.

$\therefore$ $$T_3 = T_2 \left(\frac{p_3}{p_2}\right)^{\frac{\gamma - 1}{\gamma}} = 244.4 \left(\frac{1.013}{2.71}\right)^{0.4/1.4} = 184.4 \text{ K}$$

i.e., $$m_4 = \frac{p_4 \times V_4}{RT_4} = \frac{1.013 \times 10^5 \times 0.000036}{287 \times 184.4} = 0.0000689 \text{ kg}$$

Also, $$m_1 = \frac{p_1 V_1}{RT_1} = \frac{6.3 \times 10^5 \times 0.000198}{287 \times 297} = 0.001463 \text{ kg}$$

$\therefore$ Induced mass/cycle $= (0.001463 - 0.0000689)$ kg

$\therefore$ Mass of air supplied/min $= (0.001463 - 0.0000689) \times 300 =$ **0.418 kg/min. (Ans.)**

20.4. ROTARY COMPRESSORS

- **Rotary compressors** *are the machines which develop pressure and have a rotor as their primary element when compared with the piston sliding mechanism of the reciprocating compressor.*
- *Whenever large quantities of air or gas are required at relatively low pressure* **rotary compressors** *are employed.*

20.4.1. Classification

The rotary compressors are *classified* as follows :

1. *Displacement (positive) compressors :*
   (*i*) Roots blower
   (*ii*) Sliding vane compressor
   (*iii*) Lysholm compressor
   (*iv*) Screw compressor.
2. *Steady-flow (or Non-positive displacement) compressors :*
   (*i*) Centrifugal (or radial) compressor
   (*ii*) Axial flow compressor.

20.4.2. Displacement Compressors

'Displacement Compressors' are those compressors in which air is compressed by being trapped in the reduced space formed by two sets of engaging surfaces.

20.4.2.1. Roots Blower

The two lobe type is shown in Fig. 20.38, but three and four lobe versions are in use for higher pressure ratios. One of the rotors is connected to the drive and the second rotor is gear driven from the first. In this way the rotors rotate in phase and the profile of the lobes is of cycloidal or involute form giving correct making of the lobes to seal the delivery side from the inlet side. This sealing continues untill delivery commences. There must be some clearance between the lobes, and between the casing and the lobes, to reduce wear, this clearance forms a leakage path which has an increasingly adverse effect on efficiency as the pressure ratio increases.

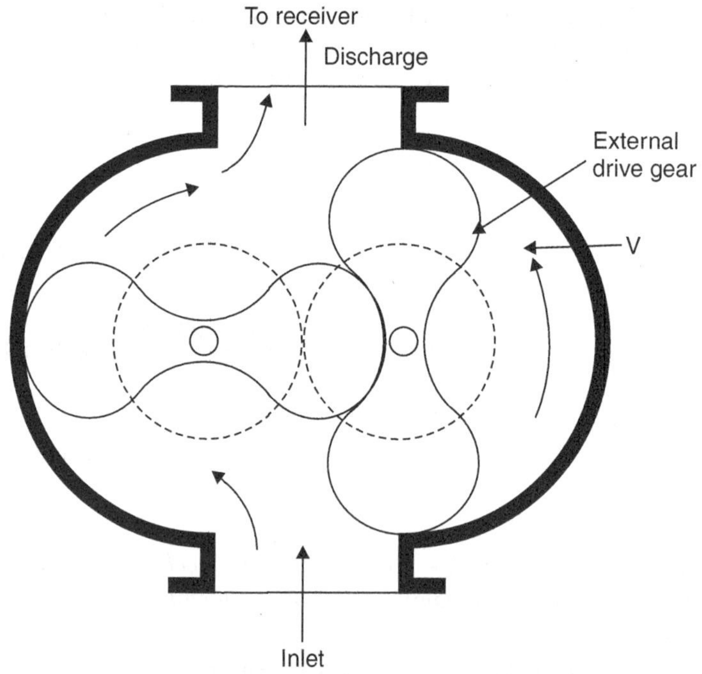

Fig. 20.38. Roots blower, two lobe rotors.

As each side of the each lobe faces its side of the casing a volume of gas V, at pressure p_1, is displaced towards the delivery side at constant pressure. A further rotation of the rotor opens this volume to the receiver, and the gas flows back from the receiver, since this gas is at a-higher pressure. The gas induced is compressed irreversibly by that from the receiver, to the pressure p_2 and then delivery begins. This process is carried out *four times* per revolution of the driving shaft.

For this machine the p-V diagram is shown in Fig. 20.39, in which the pressure rise from p_1 to p_2 is shown as an irreversible process at constant volume.

Work done per cycle $= (p_2 - p_1)V$

$\therefore$ Work done per revolution $= 4(p_2 - p_1)V$ ...(20.48)

If V_s is the volume dealt with per minute at p_1 and T_1 then

Work done/min. $= (p_2 - p_1)V_s$ ...[20.48 (*a*)]

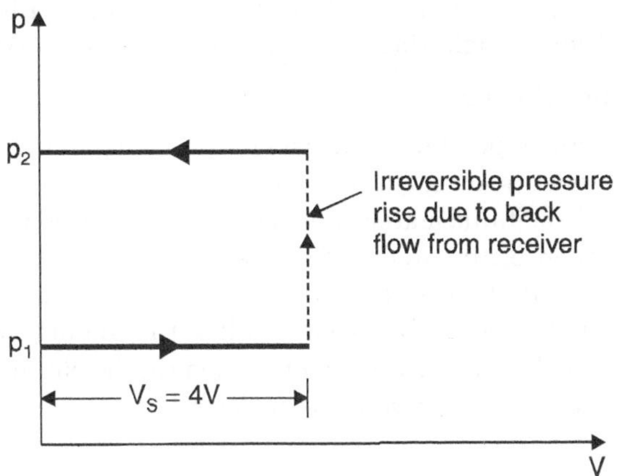

Fig. 20.39. *p-V* diagram for roots blower.

The ideal compression process from p_1 to p_2 is a reversible adiabatic (*i.e.*, isentropic) process. The work done per minute ideally is given by,

$$\text{Work done/min.} = \frac{\gamma}{\gamma - 1} p_1 V_s \left\{ \left(\frac{p_2}{p_1} \right)^{\frac{\gamma - 1}{\gamma}} - 1 \right\}$$

Then a comparison may be made on the basis of a Roots efficiency,

i.e., Roots efficiency

$$= \frac{\text{Work done isentropically}}{\text{Actual work done}}$$

$$= \frac{\frac{\gamma}{\gamma - 1} p_1 V_s \left\{ \left(\frac{p_2}{p_1} \right)^{\frac{\gamma - 1}{\gamma}} - 1 \right\}}{V_s (p_2 - p_1)}$$

$$= \frac{\frac{\gamma}{\gamma - 1} p_1 V_s \left\{ (r)^{\frac{\gamma - 1}{\gamma}} - 1 \right\}}{p_1 V_s (r - 1)} \quad \left(\text{where } r = \frac{p_2}{p_1} = \text{pressure ratio} \right)$$

Also,

$$\frac{\gamma}{\gamma - 1} = \frac{c_p}{R}$$

$$\therefore \quad \text{Roots efficiency} = \frac{c_p}{R} \left[\frac{(r)^{\frac{\gamma - 1}{\gamma}} - 1}{(r - 1)} \right] \quad \text{...(20.49)}$$

In case of a Roots air blower values of pressure ratio, *r* of 1.2, 1.6, and 2 give values for the Roots efficiency of 0.945, 0.84 and 0.765 respectively. These values show that the *efficiency decreases as the pressure ratio increases.*

This machine has a number of *imperfections but is well suited to such tasks as the scavenging and supercharging of I.C. engines.*

Roots blowers are built for capacities from 0.14 m^3/min. to 1400 m^3/min, and pressure ratios of the order of 2 to 1 for a single-stage machine and 3 to 1 for a two-stage machine.

20.4.2.2. Vane Type Blower

Refer Fig. 20.40. A vane type blower consists of a rotor mounted eccentrically in the body, and supported by ball and roller bearings in the end covers of the body. The rotor is slotted to take the blades which are of a non-metallic material, usually fibre or carbon. As each blade moves past the inlet passage, compression begins due to decreasing volume between the rotor and casing. Delivery begins with the arrival of each blade at the delivery passage. This type of compression differs from that of the Roots blower in that some or all of the compression is obtained before the trapped volume is opened to delivery. Further compression can be obtained by the back-flow of air from the receiver which occurs in an irreversible manner.

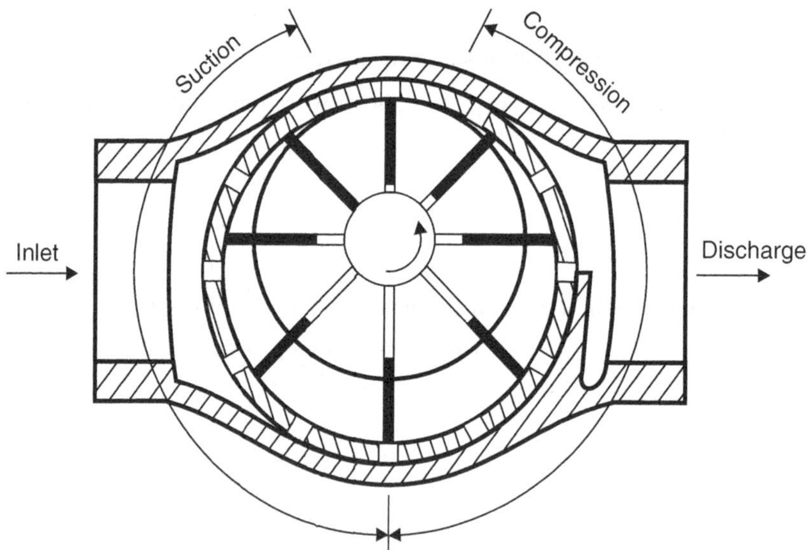

Fig. 20.40. Vane type blower.

Fig. 20.41 shows the p-V diagram. V_s is the induced volume at pressure p_1 and temperature T_1. Compression occurs to the pressure p_i, the ideal form for an uncooled machine being isentropic. At this pressure the displaced gas is opened to the receiver and gas flowing back from the receiver raises the pressure irreversibly to p_2. The work done per revolution with N vanes is given by the following expression :

$$W = N \frac{\gamma}{\gamma - 1} p_1 V_1 \left[\left(\frac{p_i}{p_1} \right)^{\frac{\gamma - 1}{\gamma}} - 1 \right] + N(p_2 - p_i)V_i \qquad \text{...(20.50)}$$

- The vane blowers *require less work* compared to roots blower for the same capacity and pressure rise.
- They are commonly used to deliver upto 150 m^3 of air per minute at pressure ratio upto 8.5.
- The speed limit of a vane blower is 3000 r.p.m.

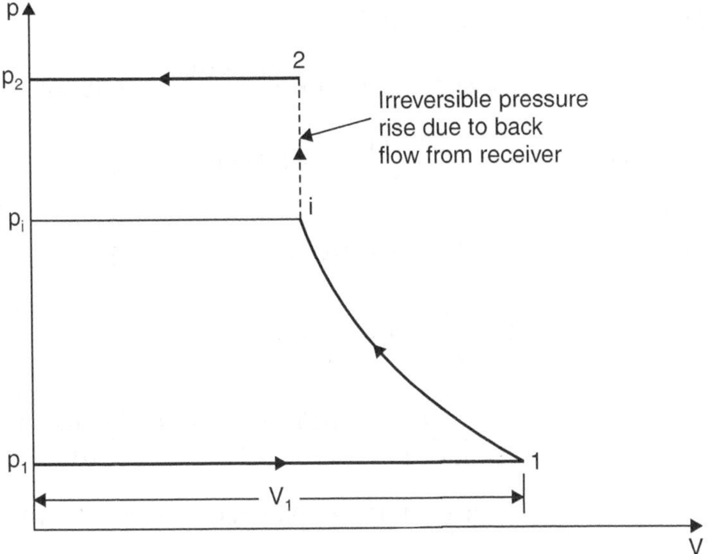

Fig. 20.41. *p-V* diagram for vane blower.

Example 20.35. *Compare the work inputs required for a Roots blower and a Vane type compressor having the same induced volume of 0.03 m³/rev., the inlet pressure being 1.013 bar and the pressure ratio 1.5 to 1. For the Vane type assume that internal compression takes place through half the pressure range.*

Solution. Inlet pressure, $p_1 = 1.013$ bar

Pressure ratio, $\dfrac{p_2}{p_1} = 1.5$

$\therefore$ $p_2 = 1.5\, p_1 = 1.5 \times 1.013 = 1.52$ bar.

For the **Roots blower**, refer Fig. 20.42.

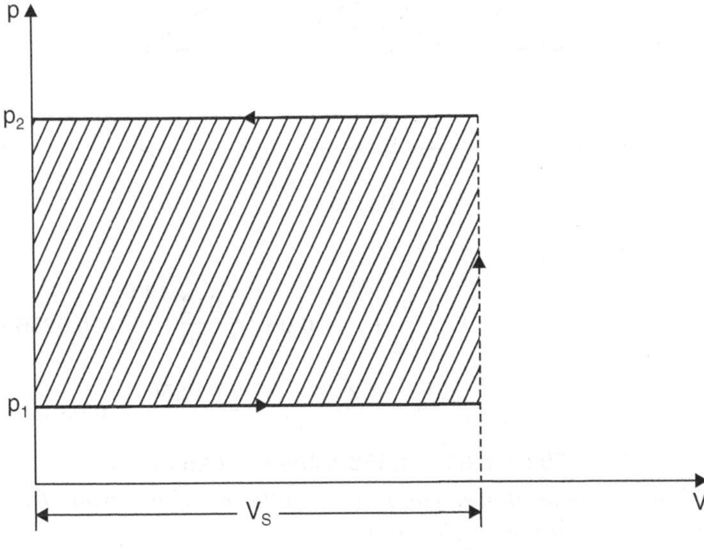

Fig. 20.42

Work done/rev. $= (p_2 - p_1)V_s$

$$= (1.52 - 1.013) \times \frac{10^5 \times 0.03}{10^3} = \mathbf{1.52\ kJ.}\quad \textbf{(Ans.)}$$

For the **Vane type,** $p_i = \dfrac{1.52 + 1.013}{2} = 1.266$ bar

Refer Fig. 20.43.

Work required $= (\text{Area } A + \text{Area } B)$

Now, $$\text{Area } A = \frac{\gamma}{\gamma - 1}\, p_1 V_s \left[\left(\frac{p_i}{p_1}\right)^{(\gamma - 1)/\gamma} - 1\right]$$

$$= \frac{1.4}{1.4 - 1} \times \frac{1.013 \times 10^5 \times 0.03}{10^3}\left[\left(\frac{1.266}{1.013}\right)^{\frac{1.4-1}{1.4}} - 1\right] \text{ kJ/rev.}$$

$$= 3.5 \times 1.013 \times 100 \times 0.03 \times 0.066 = 0.702 \text{ kJ/rev.}$$

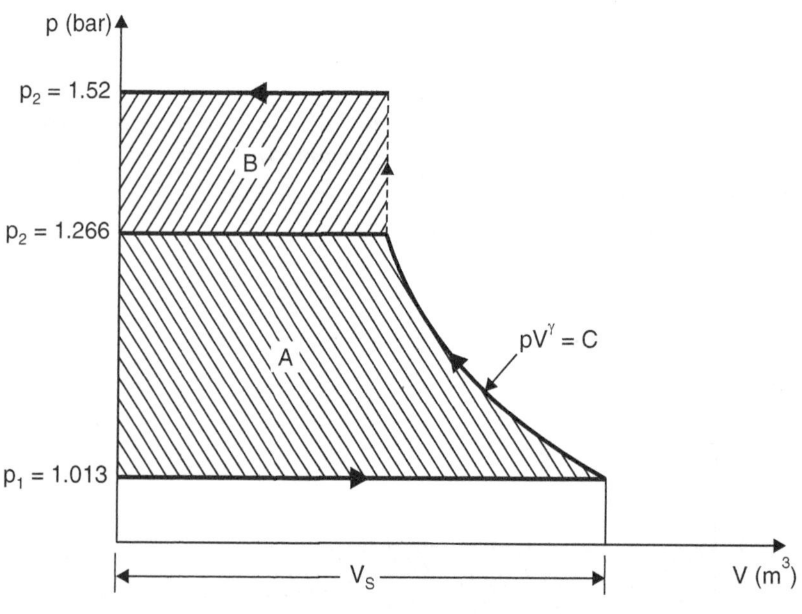

Fig. 20.43

Area $B = (p_2 - p_1)V_b$

Now, $$V_b = V_s\left(\frac{p_1}{p_2}\right)^{1/\gamma} = 0.03\left(\frac{1.013}{1.266}\right)^{1/1.4} = 0.0256 \text{ m}^3$$

Area $$B = \frac{(1.52 - 1.266) \times 10^5 \times 0.0256}{10^3} = 0.65 \text{ kJ/rev.}$$

∴ Work required = 0.702 + 0.65 = **1.352 kJ/rev. (Ans.)**

Example 20.36. *A roots blower compresses 0.08 m^3 of air from 1.0 bar to 1.5 bar per revolution. Calculate the compressor efficiency.*

Solution. Volume of air to be compressed, $V = 0.08 \text{ m}^3$

Intake pressure, $p_1 = 1.0$ bar

Pressure after compression, $p_2 = 1.5$ bar

Actual work done, $W_{actual} = (p_2 - p_1)\, V = 10^5(1.5 - 1.0) \times 0.08 = 4000$ Nm.

Also ideal work done per revolution is given by,

$$W_{ideal} = \frac{\gamma}{\gamma - 1} p_1 V_1 \left[\left(\frac{p_2}{p_1} \right)^{(\gamma - 1)/\gamma} - 1 \right]$$

$$= \frac{1.4}{1.4 - 1} \times 1.0 \times 10^5 \times 0.08 \left[\left(\frac{1.5}{1.0} \right)^{\frac{1.4 - 1}{1.4}} - 1 \right] = 3438.89 \text{ Nm.}$$

$$\therefore \quad \eta_{compressor} = \frac{W_{ideal}}{W_{actual}} = \frac{3438.89}{4000} = 0.8597 \text{ or } \mathbf{85.97\%}. \quad \textbf{(Ans.)}$$

20.4.3. Steady-flow Compressors

The compressors in which compression occurs by transfer of kinetic energy from a rotor are called ***Steady-flow compressors.***

The centrifugal type of compressor was used in the earliest gas turbine units for aircraft.

- For low pressure ratios (no greater than about 4 : 1) the centrifugal compressor is *lighter* and is able to *operate effectively over a wide range of mass flows at any one speed,* than its axial-flow counterpart.
- For *larger units with higher pressure ratios the axial flow compressor is more efficient and is usually preferred.* For industrial and large marine gas turbine plants axial flow compressors are usually used, although some units *may employ two or more centrifugal compressors.*

— For *aircraft* the trend has been to higher pressure ratios, and the compressor is usually of the *axial flow.* In aircraft units the advantage of the smaller diameter axial flow compressor can offset the disadvantage of the increased length and weight compared with an equivalent centrifugal compressor.

Advantages *of centrifugal compressors over axial flow compressors :*

1. Smaller length.
2. Contaminated atmosphere does not deteriorate the performance.
3. Can perform efficiently over wide range of mass flows at any speed.
4. Cheaper to produce.
5. More robust.
6. Less prone to icing troubles at high altitudes.

Disadvantages :

1. Large frontal area.
2. Lower maximum efficiency.

Uses : The centrifugal compressors are used in :

(*i*) Superchargers (*ii*) Turbo-prop.

- **Centrifugal compressors** *are preferred where simplicity, light weight, ruggedness are more important than maximum efficiency and smaller diameter.*

20.4.3.1. Static and Total Head Values

As compared to reciprocating compressors the *velocities encountered in centrifugal compressors are very large and therefore total head quantities should be considered* while analysing centrifugal compressors. The *total head quantities take into account the kinetic energy of the air passing through the compressor.*

Let us consider, a horizontal passage of varying area through which air is flowing (Fig. 20.48). Applying steady flow energy equation to the system for 1 kg of air flow (assuming no external heat transfer and work transfer to the system), we get

$$u_1 + p_1 v_1 + \frac{C_1^2}{2} = u_2 + p_2 v_2 + \frac{C_2^2}{2}$$

$$\left[u_1 + \frac{p_1 v_1}{J} + \frac{C_1^2}{2gJ} = u_1 + \frac{p_2 v_2}{J} + \frac{C_2^2}{2gJ} \quad \text{...... in } \textit{MKS units} \right]$$

$$h_1 + \frac{C_1^2}{2} = h_2 + \frac{C_2^2}{2}$$

$$\therefore \quad c_p T_1 + \frac{C_1^2}{2} = c_p T_2 + \frac{C_2^2}{2}$$

$$\therefore \quad c_p T + \frac{C_2^2}{2} = \text{constant} \quad \text{...(20.51)}$$

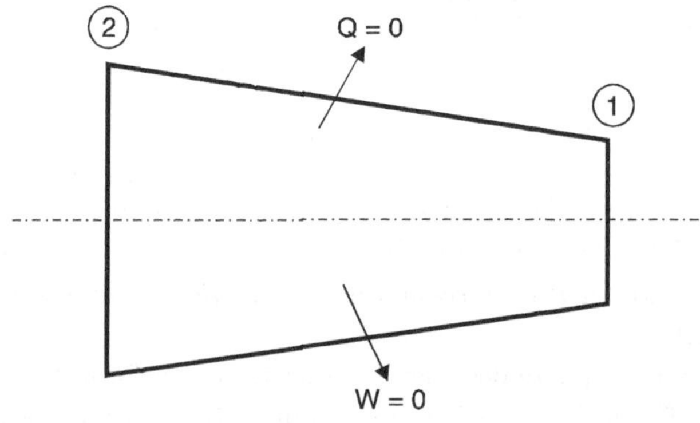

Fig. 20.44

Temperature '*T*' is called the ***static temperature*** or the *temperature* of the *air measured by the thermometer when the thermometer is moving at the air velocity.* If the moving air is brought to rest under reversible conditions, the total kinetic energy of the air is converted into heat energy, increasing the temperature and pressure of the air. This temperature and pressure of the air is known as ***"stagnation"*** or ***"total head"*** *temperature and pressure.* The total head temperature and pressure are denoted by a suffix notation '0'.

$$\therefore \quad c_p T + \frac{C^2}{2} = c_p T_0 \quad \text{...(20.52)}$$

where T_0 is known as *total head or stagnation temperature*

$$\therefore \qquad T_0 - T = \frac{C^2}{2c_p} \qquad \text{...(20.53)} \qquad \text{or} \qquad h_0 - h = \frac{C^2}{2} \qquad \text{...(20.54)}$$

For finding the total head pressures, use the equation,

$$\frac{p_0}{p} = \left(\frac{T_0}{T}\right)^{\frac{\gamma}{\gamma - 1}} \qquad \text{...(20.55)}$$

where, p = Static pressure,

T = Static temperature,

p_0 = Stagnation pressure, and

T_0 = Stagnation temperature.

Adiabatic process and isentropic process :

- In **adiabatic process** the system does not exchange any heat with the surroundings *i.e.,* no heat enters or leaves the working fluid externally. The *ideal reversible adiabatic process* is called **isentropic process** *and in this process entropy remains constant.*

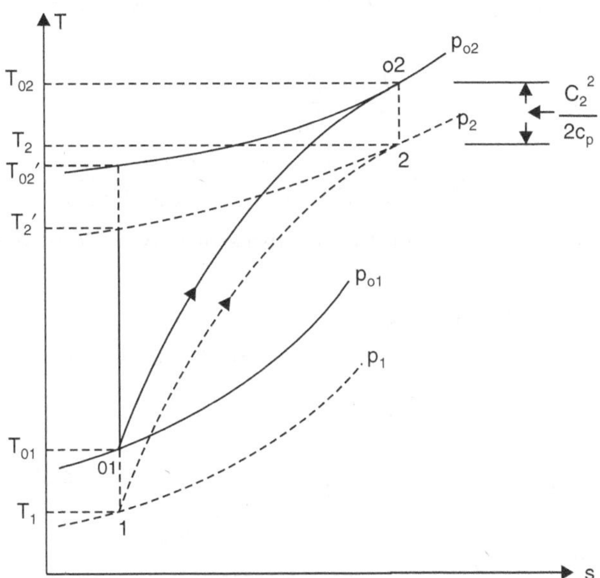

Fig. 20.45. Total head and static quantities on T-s diagram.

- During adiabatic compression in a rotary compressor there is friction between molecules of air and between air and blade passages, eddies formation and shocks at entry and exit. These factors *cause internal generation of heat and consequently the maximum temperature reached would be more than that for adiabatic compression.* This *results in a progressive increase in entropy.* Such a process *though adiabatic is not isentropic.* The heat generated by friction etc. may be removed continuously with the result that the process might not involve any entropy change. The process *would then be isentropic but not adiabatic as heat has been transferred.*

Isentropic efficiency. *"Isentropic efficiency" of rotary compressor* may be defined as *the ratio of isentropic temperature rise to actual temperature rise.*

With reference to Fig. 20.45, which represents a combined diagram for static and stagnation (total head) values, we have

$$\text{Isentropic efficiency} = \frac{\text{Isentropic temperature rise}}{\text{Actual temperature rise}}$$

$$\eta_{isen} = \frac{T_{02}' - T_{01}}{T_{02} - T_{01}} \quad \text{...based on total values} \quad \text{...[20.56 (a)]}$$

$$= \frac{T_2' - T_1}{T_2 - T_1} \quad \text{...based on static values} \quad \text{...[20.56 (b)]}$$

During compression process work has to be imparted to the impeller. The energy balance equation would then yield :

$$c_pT_1 + \frac{C_1^2}{2} = c_pT_2 + \frac{C_2^2}{2} - W$$

or $$c_pT_{01} = c_pT_{02} - W \quad \text{or} \quad W = c_p(T_{02} - T_{01}) \quad \text{...(20.57)}$$

Thus the work input is the product of specific heat at constant pressure and temperature rise. *This relation is true both for adiabatic and isentropic processes.*

From eqn. 20.56,

Isentropic efficiency, $$\eta_{isen} = \frac{T_{02}' - T_{01}}{T_{02} - T_{01}}$$

or $$\eta_{isen} = \frac{c_p(T_{02}' - T_{01})}{c_p(T_{02} - T_{01})} = \frac{\text{Isentropic work}}{\text{Actual work}} \quad \text{...(20.58)}$$

Thus the isentropic efficiency of a rotary compressor may be defined *as the ratio of isentropic compression work to actual compression work.*

20.4.3.2. Centrifugal Compressor

Fig. 20.46 shows a centrifugal compressor (with double sided impeller). It consists of the following *parts* :

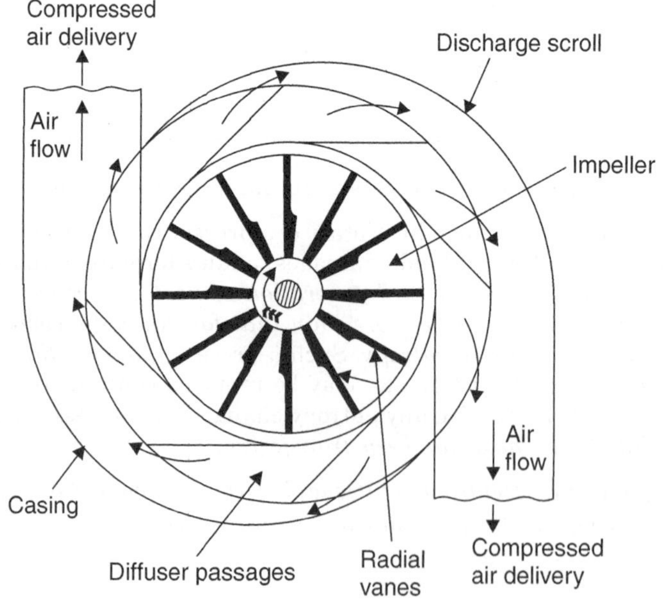

Fig. 20.46. Centrifugal compressor.

1. *Curved radial vanes.* A series of curved radial vanes are attached to and rotate with the shaft.

2. *Impeller.* The impeller is a disc fitted with radial vanes. The impeller is *generally forged or die-casted of low silicon aluminium alloy.*

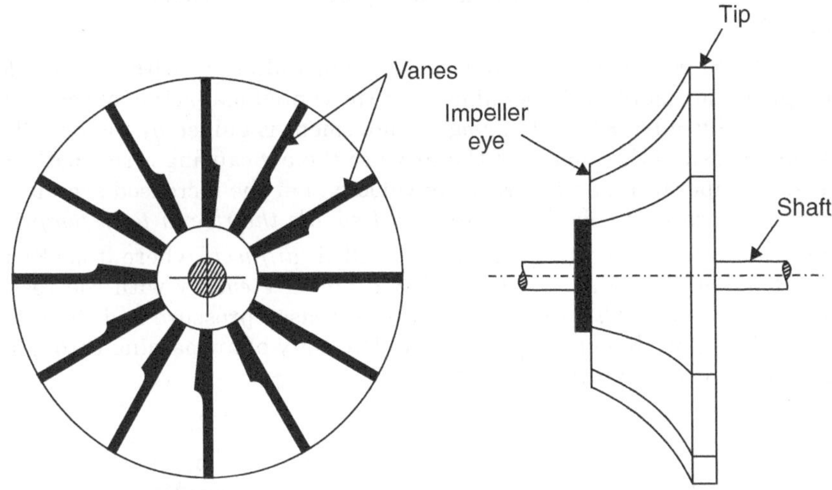

Fig. 20.47. Impeller (single-eyed) and radial vanes of centrifugal compressor.

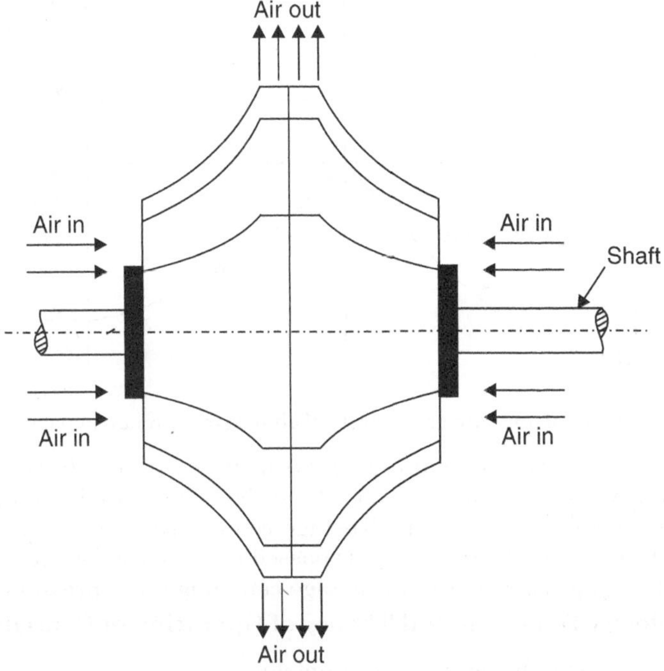

Fig. 20.48. Double eyed impeller.

The impeller may be single-eyed or double sided (having eye on either side of the compressor, so that air is drawn in on both sides) as shown in Fig. 20.47 and 20.48 respectively. The advantage of double sided impeller is that the *impeller is subjected to approximately equal forces in an axial direction.*

3. *Casing.* The casing surrounds the rotating impeller.

4. *Diffuser.* The diffuser is housed in a radial portion of the casing.

Working :

- Air *enters the eye of the "impeller"* at a mean radius r_m with a low velocity C_1 and atmospheric pressure p_1. Depending upon the centrifugal action of the impeller, the air moves radially outwards and during its movement is guided by the impeller vanes. The impeller transfers the energy of the drive to the air causing a rise both in static pressure and temperature, and increase in velocity. Let the increased pressure and velocity be p_2 and C_2 respectively. The *work input equals the rise in total temperature.*
- The air now *enters the diverging passage* called "*diffuser*" where it is efficiently slowed down. The *kinetic energy is converted into pressure energy* with the result that *there is a further rise in static pressure.* Let the increased pressure and the reduced velocity be p_3 and C_3. The changes of pressure and velocity of air passing through the impeller and diffuser are shown in Fig. 20.49.

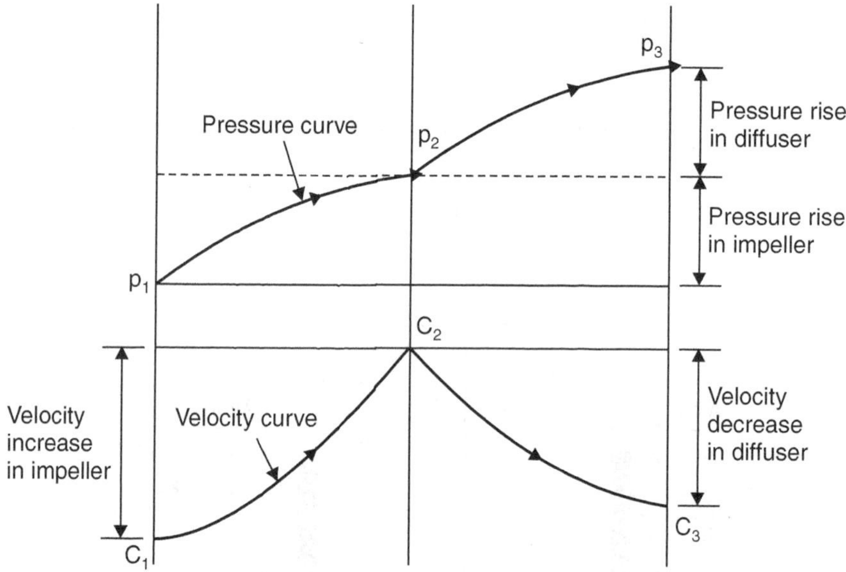

Fig. 20.49. Variations of pressure and velocity of air passing through impeller and diffuser.

- In practice *nearly half the total pressure is achieved in impeller and remaining half in the diffuser.* A pressure ratio of 4.5 : 1 can be achieved with *single-stage* centrifugal compressor. For *higher ratios,* multi-stage compressors are used. In multi-stage compressors, the outlet of the first stage is passed to the second stage and so on. A pressure ratio of 12 : 1 is possible with multi-stage centrifugal compressors.

20.4.3.2.1. Velocity Diagrams and Theory of Operation of Centrifugal Compressors

Let, C_{bl_1} = Mean blade velocity at entrance,

C_{bl_2} = Mean blade velocity at exit,

C_1 = Absolute velocity at inlet to the rotor,

C_2 = Absolute velocity at outlet to the rotor,

C_{r_1} = Relative velocity of air at entry of rotor,

C_{r_2} = Relative velocity of air at exit of rotor,

C_{w_1} = Velocity of whirl at inlet,

C_{w_2} = Velocity of whirl at outlet,

C_{f_1} = Velocity of flow at inlet,

C_{f_2} = Velocity of flow at outlet,

α_1 = Exit angle from the guide vane or inlet angle of the guide vane,

β_1 = Inlet angle to the rotor or impeller,

β_2 = Outlet angle from the rotor or impeller, and

α_2 = Inlet angle to the diffuser.

Fig. 20.50 shows the velocity diagrams for the inlet and outlet of the impeller. It is *assumed that the entry of the air is 'axial'*, therefore the *whirl component at the inlet* (C_{w_1}) is *zero* and therefore $C_1 = C_{f_1}$. The enlarged views of inlet and outlet velocity diagrams are shown in Fig. 20.51 (*a*) and 20.51 (*b*). To *avoid shock at entry and exit the blade must be parallel to the relative velocity* of *air at inlet or outlet* and therefore β_1 and β_2 are the *impeller* blade angles at the inlet and outlet. The *diffuser* blade angle *must be parallel to the absolute velocity of air from the impeller* (C_2), therefore α_2 is the diffuser blade angle at the inlet and α_3 is the diffuser blade angle at the outlet. If the *discharge from the diffuser is circumferential,* then its blade angle at outlet (α_3) should be *as small as possible.*

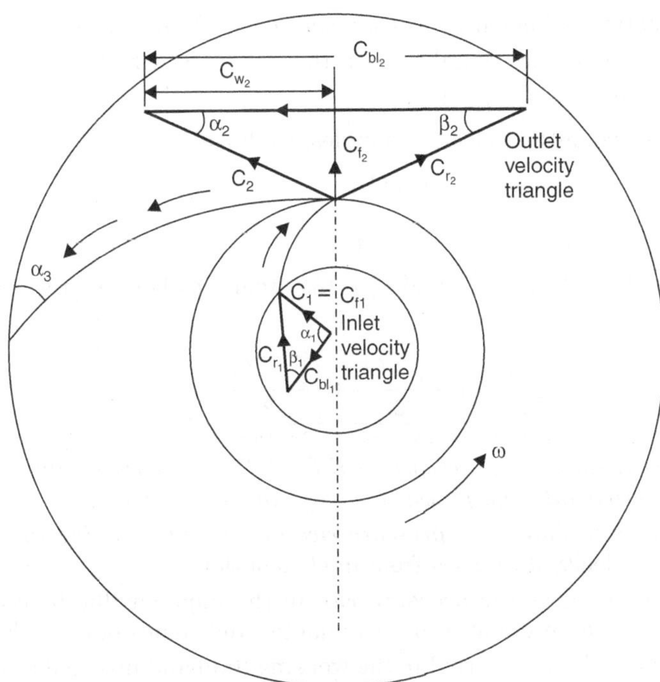

Fig. 20.50

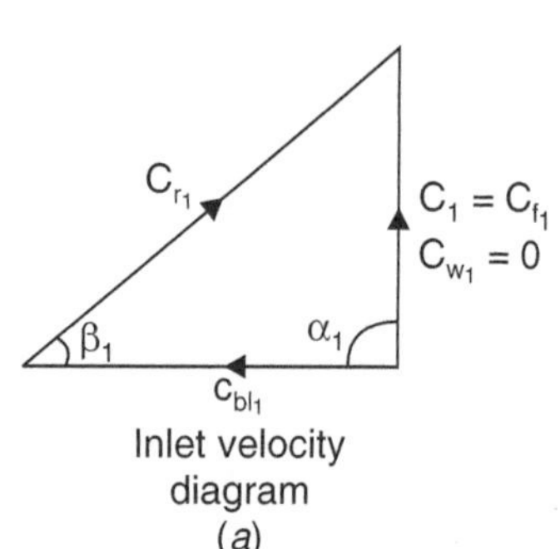

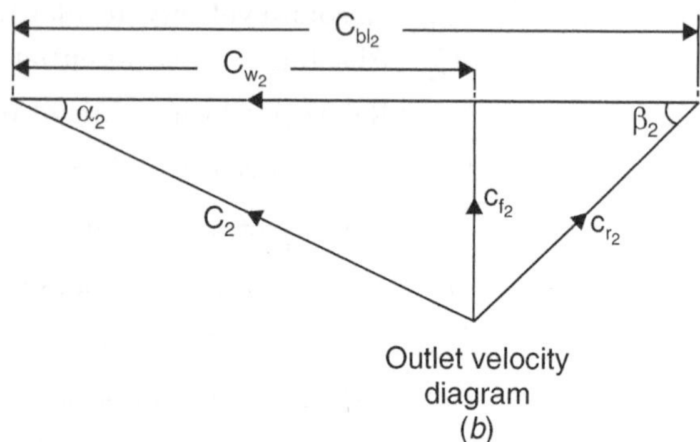

Fig. 20.51. Velocity diagrams.

Work done by impeller (Euler's work) :

The work supplied to a fluid in a stage of compressor may be found by applying the moment of momentum theorem. Consider 1 kg of working fluid passing through the impeller. The theoretical torque which must be supplied to the impeller will be equal to the rate of change of moment of momentum experienced by the working fluid.

The theoretical torque = $(C_{w2} \cdot r_2 - C_{w1} r_1)$, where r_1 and r_2 are the radii at the inlet and outlet of the impeller respectively.

If w is the angular velocity in rad/s, the work done on 1 kg of fluid will be,

$$W = \text{Theoretical torque} \times \text{angular velocity}$$

$$= (C_{w2} \cdot r_2 - C_{w1} \cdot r_1)\ \omega$$

$$= (C_{w2} \cdot r_2 \cdot \omega - C_{w1} \cdot r_1 \cdot \omega)$$

or

$$W = C_{w2} C_{bl_2} - C_{w1} C_{bl_1} = h_{o2} - h_{o1} = c_p(T_{o2} - T_{o1}) \quad ...(20.59)$$

The above equation is known as *Euler's equation* or *Euler's work*.

If the working fluid enters radially *i.e.*, if there is no prewhirl, $C_{w1} = 0$, then

$$W = C_{w2} \cdot C_{bl_2} \text{ J/kg} \quad ...(20.60)$$

Using the inlet and outlet velocity triangles, we have

$$C_{r1}^{\ 2} = C_{bl_1}^2 + C_1^{\ 2} - 2C_{bl_1} C_{w1} \quad ...(i)$$

$$C_{r_2}^2 = C_{bl_2}^2 + C_2^{\ 2} - 2C_{bl_2} C_{w2} \quad ...(ii)$$

Inserting the values of $C_{w2} \cdot C_{bl_2}$ and $C_{w1} \cdot C_{bl_1}$ from the above expressions (*i*) and (*ii*) in eqn. (20.59), we get

$$W = \underbrace{\frac{C_2^{\ 2} - C_1^{\ 2}}{2}}_{\text{First term}} + \underbrace{\frac{C_{r_1}^2 - C_{r_2}^2}{2}}_{\text{Second term}} + \underbrace{\frac{C_{bl_2}^2 - C_{bl_1}^2}{2}}_{\text{Third term}} \quad ...(20.61)$$

— The *first term* shows the *increase in K.E. of 1 kg of working fluid in the impeller that has to converted into the pressure energy in the 'diffuser'.*

— The *second term* shows the *pressure rise in the impeller due* to *'diffusion action'* (as the relative velocity decreases from inlet to outlet).

— The *third term* shows the pressure rise in the impeller due to *'centrifugal action'* (as the working fluid enters at a lower diameter and comes out at a higher diameter).

Thus the fraction of K.E. imparted to the working fluid and inverted into pressure energy in impeller is given by

$$\frac{C_{r_1}^2 - C_{r_2}^2}{2} + \frac{C_{bl_2}^2 - C_{bl_1}^2}{2} = \int_1^2 \frac{dp}{\rho} \qquad ...(20.62)$$

where ρ is the density.

If the diffuser outlet velocity is C_4, then

$$\frac{C_2^2 - C_4^2}{2} = \int_2^4 \frac{dp}{\rho} + \Delta h_{\text{dif. loss}}$$

i.e., the difference of K.E. at the impeller outlet and diffuser outlet introduces the work partly utilised for pressure increase, and partly irreversibly inverted into heat due to losses in diffuser.

- **Power required per impeller** for $\dot{m}$ kg of air flow in one second,

$$P = \frac{\dot{m}\, C_{w_2} C_{bl_2}}{1000} \text{ kW, using eqn. (20.60)} \qquad ...(20.63)$$

— If the blade is radial (ideal case), then the velocity diagram at the outlet of the impeller is as shown in Fig. 20.51. As $C_{w_2} = C_{bl_2}$, the *work done per kg of air flow per second* is given by

$$W = C_2^2 \qquad ...(20.64)$$

Since the air cannot leave the impeller at a velocity greater than the impeller tip velocity, the maximum work supplied per kg of air flow per second is given by eqn. (20.64).

— Now consider the steady flow at the inlet and outlet of the impeller, assuming the heat transfer during the flow of air through the impeller is zero.

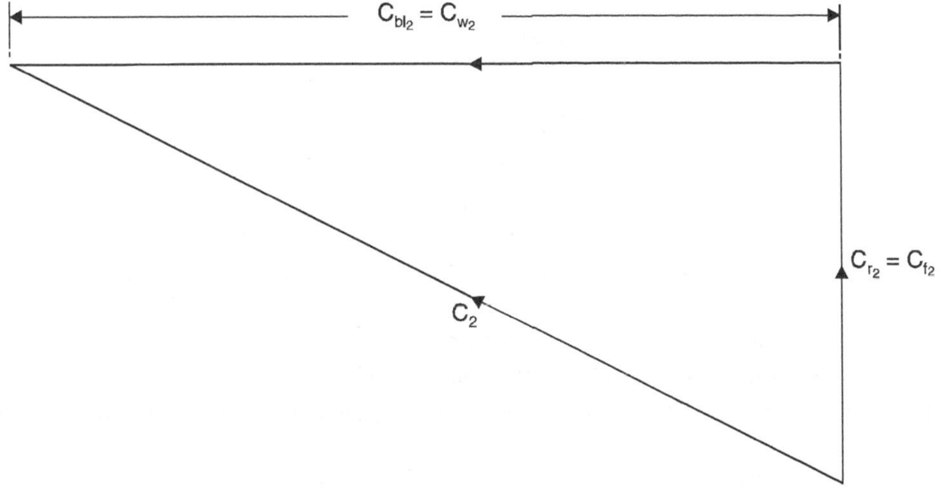

Fig. 20.52

$$h_1 + \frac{C_1^2}{2} + W = h_2 + \frac{C_2^2}{2}$$

$$W = \left(h_2 + \frac{C_2^2}{2}\right) - \left(h_1 + \frac{C_1^2}{2}\right)$$

$$= c_p\left(T_2 + \frac{C_2^2}{2c_p}\right) - c_p\left(T_1 + \frac{C_1^2}{2c_p}\right)$$

$$= c_p\, T_{02} - c_p T_{01} = c_p(T_{02} - T_{01}) \qquad ...(20.65)$$

$$\therefore \qquad W = c_p T_{01}\left[\frac{T_{02}}{T_{01}} - 1\right] = c_p T_{01}\left[\frac{T_2\left(\dfrac{p_{02}}{p_2}\right)^{\frac{\gamma-1}{\gamma}}}{T_1\left(\dfrac{p_{01}}{p_1}\right)^{\frac{\gamma-1}{\gamma}}} - 1\right]$$

$$= c_p T_{01}\left[\left(\frac{p_{02}}{p_{01}}\right)^{\frac{\gamma-1}{\gamma}} - 1\right] = c_p T_{01}\left[(r_{p0})^{\frac{\gamma-1}{\gamma}} - 1\right] \qquad ...(20.66)$$

where r_{p0} is the *pressure ratio based on stagnation pressures.*

In most practical problems, $C_1 = C_2$, then equations (20.65) and (20.66) are reduced to

$$W = c_p\,(T_2 - T_1) \qquad ...(20.67)$$

$$= c_p T_1\left[(r_p)^{\frac{\gamma-1}{\gamma}} - 1\right] \qquad ...(20.68)$$

Now from eqns. (20.64) and (20.66)

$$C_2^{\,2} = c_p T_{01}\left[\left(\frac{p_{02}}{p_{01}}\right)^{\frac{\gamma-1}{\gamma}} - 1\right]$$

$$\therefore \qquad \frac{p_{02}}{p_{01}} = \left[\frac{C_2^{\,2}}{c_p T_{01}} + 1\right]^{\frac{\gamma}{\gamma-1}} \qquad ...(20.69)$$

If $\qquad C_1 = C_2$

$$\frac{p_2}{p_1} = \left[\frac{C_2^{\,2}}{c_p T_1} + 1\right]^{\frac{\gamma}{\gamma-1}} \qquad ...(20.70)$$

Thus as per eqn. (20.66) the *power input to the compressor* depends upon the following *factors* :

(*i*) Mass flow of air through the compressor.

(*ii*) Total temperature at the inlet of the compressor.

(*iii*) Total pressure ratio of the compressor which depends upon the square of the impeller tip velocity.

In the above analysis :

p_{01} = Stagnation pressure at inlet of the compressor ;
T_{01} = Stagnation temperature at inlet ;
p_{02} = Stagnation pressure at outlet ;
T_{02} = Stagnation temperature at outlet ;
p_1 = Static pressure at inlet ;
T_1 = Static temperature at inlet ;
p_2 = Static pressure at outlet ;
T_2 = Static temperature at outlet.

20.4.3.2.2. Width of Blades of Impeller and Diffuser

If the mass of the air flowing per second is constant and is known, then the width of blades of impeller and diffuser can be calculated as follows :

Let, $\dot{m}$ = Mass of air flowing per second,

b_1 = Width (or height) of impeller at inlet,

C_{f_1} = Velocity of flow at inlet of the impeller,

v_1 = Volume of 1 kg of air at the inlet,

r_1 = Radius of impeller at the inlet,

Then, $$\dot{m} = \frac{\text{Volume of air flowing per second}}{\text{Volume of 1 kg of air}} = \frac{2\pi r_1\, b_1 \times C_{f_1}}{v_1}$$

But as the air is trapped radially,

$$C_{f_1} = C_1$$

$$\therefore \quad \dot{m} = \frac{2\pi r_1\, b_1 \times C_1}{v_1} \quad \text{...(20.71)}$$

i.e., $$b_1 = \frac{\dot{m}\, v_1}{2\pi_1\, r_1\, C_1} \quad \text{...[20.71 (a)]}$$

Similarly the width of impeller blade at the outlet can be found by using suffix 2 in eqn. (20.71)

$$\dot{m} = \frac{2\pi r_2\, b_2 \times C_{f_2}}{v_2} \quad \text{...(20.72)}$$

The width or height of the impeller blades at the outlet and height of diffuser blade at the inlet should be same theoretically.

The width or height of the diffuser blades at the outlet, is given by

$$\dot{m} = \frac{2\pi r_d\, b_d \times C_{fd}}{v_d} \quad \text{...(20.73)}$$

where suffix '*d*' represents the quantities at the *outlet of the diffuser*.

If, n = Number of blades on the impeller, and

t = Thickness of the blade,

then eqns. (20.71), (20.72) and (20.73) are expressed as follows :

$$\dot{m} = \frac{(2\pi r_1 - nt) b_1 C_{f_1}}{v_1} \quad \text{...(20.74)}$$

$$\dot{m} = \frac{(2\pi r_2 - nt) b_2 C_{f_2}}{v_2} \quad \text{...(20.75)}$$

$$\dot{m} = \frac{(2\pi r_d - nt) b_d C_{f_d}}{v_d} \quad \text{...(20.76)}$$

20.4.3.2.3. Isentropic Efficiency of the Compressor

The following *losses* occur when air flows through the impeller :

(*i*) Friction between the air layers moving with relative velocities and friction between the air and flow passages (*ii*) Shock at entry

(*iii*) Turbulence caused in air.

The losses mentioned above cause an increase in enthalpy of the air without increase of pressure therefore the *actual temperature* of air coming out from the compressor is *more* than the temperature of air if it is compressed isentropically. The *actual work* required for the same increase in pressure ratio is *more* due to *irreversibilities*. The actual and isentropic compression for the same pressure ratio is shown in Fig. 20.53.

The isentropic efficiency is given by the relation,

$$\eta_{isen} = \frac{\text{Isentropic work}}{\text{Actual work}} = \frac{h_{02}' - h_{01}}{h_{02} - h_{01}}$$

$$= \frac{T_{02}' - T_{01}}{T_{02} - T_{01}} \quad ...(20.77)$$

provided specific heat at constant pressure (c_p) remains constant.

If $C_1 = C_2$, then $$\eta_{isen} = \frac{T_2' - T_1}{T_2 - T_1} \quad ...(20.78)$$

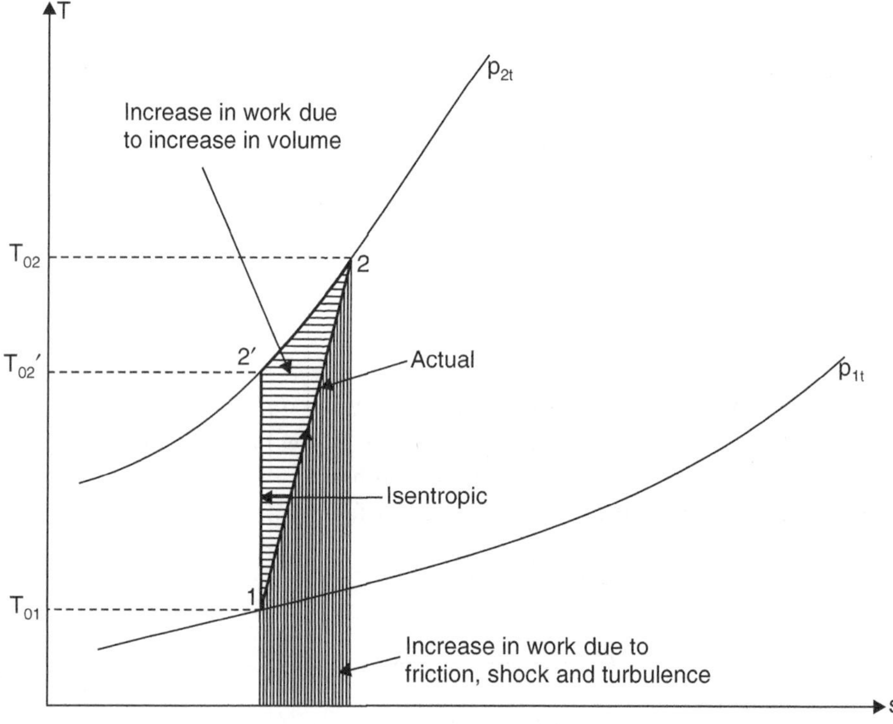

Fig. 20.53. Actual and isentropic compression on *T-s* diagram.

$$= \frac{\dfrac{T_2'}{T_1} - 1}{\dfrac{T_2}{T_1} - 1} = \frac{\left(\dfrac{p_2}{p_1}\right)^{\frac{\gamma - 1}{\gamma}} - 1}{\left(\dfrac{p_2}{p_1}\right)^{\frac{n - 1}{n}} - 1}, \quad n > \gamma.$$

20.4.3.2.4. Slip Factor and Pressure Co-efficient

In the earlier analysis it was assumed that $C_{w_2} = C_{bl_2}$ but this condition is not satisfied in actual practice *due to secondary flow effects* and therefore in actual compressors $C_{w_2} < C_{bl_2}$.

The difference between $(C_{bl_2} - C_{w_2})$ is known as **slip.**

Slip factor (ϕ_s). It is defined as the *ratio of actual whirl component* (C_{w_2}) *and the ideal whirl component* (C_{bl_2})

$$\therefore \qquad \phi_s = \frac{C_{w_2}}{C_{bl_2}} = 1 \text{ if } C_{w_2} = C_{bl_2} \qquad ...(20.79)$$

The stagnation (total head) pressure ratio is given by

$$\frac{p_{02}}{p_{01}} = \left(\frac{T_{02}'}{T_{01}}\right)^{\frac{\gamma-1}{\gamma}} = \left(1 + \frac{T_{02}' - T_{01}}{T_{01}}\right)^{\frac{\gamma-1}{\gamma}}$$

Substituting the value of $(T_{02}' - T_{01})$ from eqn. (20.77), we get

$$\frac{p_{02}}{p_{01}} = \left[1 + \frac{\eta_{isen}\,(T_{02} - T_{01})}{T_{01}}\right]^{\frac{\gamma-1}{\gamma}} \qquad ...(20.80)$$

As per eqn. [20.60 (*a*)], the actual work done per kg of air is given by

$$c_p(T_{02} - T_{01}) = C_{bl_2} - C_{w_2}$$

The actual work done per kg of air by the compressor is always greater than $C_{bl_2} C_{w_2}$ due to fluid friction and windage losses, therefore the actual work is obtained by multiplying $C_{bl_2} C_{w_2}$ by a factor ϕ_w known as **work factor** or **power input factor.**

$$\therefore \qquad c_p(T_{02} - T_{01}) = \phi_w C_{bl_2} C_{w_2}$$

$$\therefore \qquad T_{02} - T_{01} = \frac{\phi_w C_{bl_2} C_{w_2}}{c_p} \qquad ...(20.81)$$

Now substituting the value of eqn. (20.81) into eqn. (20.80), we have

$$\frac{p_{02}}{p_{01}} = \left[1 + \frac{\eta_{isen}\,\phi_w C_{bl_2} C_{w_2}}{c_p\, T_{01}}\right]^{\frac{\gamma-1}{\gamma}}$$

Now substituting the value of C_{w_2} from eqn. (20.79), we have

$$\frac{p_{02}}{p_{01}} = \left[1 + \frac{\eta_{isen}\,\phi_w\,\phi_s\, C_{bl}^{\,2}}{c_p\, T_{01}}\right]^{\frac{\gamma-1}{\gamma}} \qquad ...(20.82)$$

Pressure Co-efficient (ϕ_p). It is defined as the *ratio of isentropic work to Euler work.*

$$\therefore \qquad \phi_p = \frac{\text{Isentropic work}}{\text{Euler work}} = \frac{c_p\,(T_{02}' - T_{01})}{C_{bl_2}\, C_{w_2}}$$

Using eqn. (20.77) and assuming the vanes of the impeller are radial and counter flow of fluid is neglected ($C_{w_2} = C_{bl_2}$)

$$\phi_p = \frac{c_p\,\eta_{isen}\,(T_{02} - T_{01})}{C_{bl_2}^{\,2}}$$

Now using eqn. (20.81) which is

$$c_p(T_{02} - T_{01}) = \phi_w \; C_{bl_2} \; C_{w_2} \; = \phi_w \phi_s C_{bl_2}^{\;2}$$

$$\phi_p = \frac{\phi_w \; \phi_s \; C_{w_2}^{\;2} \; \eta_{isen.}}{C_{bl_2}^{\;2}} = \phi_w \phi_s \; \eta_{isen.} \qquad ...(20.83)$$

20.4.3.2.5. The Effect of Impeller Blade Shape on Performance. The following shapes of blades are utilized in the impellers of centrifugal compressors :

1. *Backward-curved blades* ($\beta_2 < 90°$)

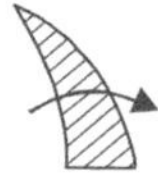

2. *Radial-curved blades* ($\beta_2 = 90°$)

3. *Forward-curved blades* ($\beta_2 > 90°$)

Fig. 20.54, shows the relative performance of these blades. Centrifugal effects on the curved blades *create a bending moment and produce increased stresses which reduce the maximum speed at which the impeller can run.*

- Normally backward blades/vanes with β_2 between 20-25° are employed except in cases where high head is the major consideration.
- Sometimes compromise is made between the low energy transfer (backward-curved vanes) and high outlet velocity (forward-curved vanes) by using *radial vanes.*

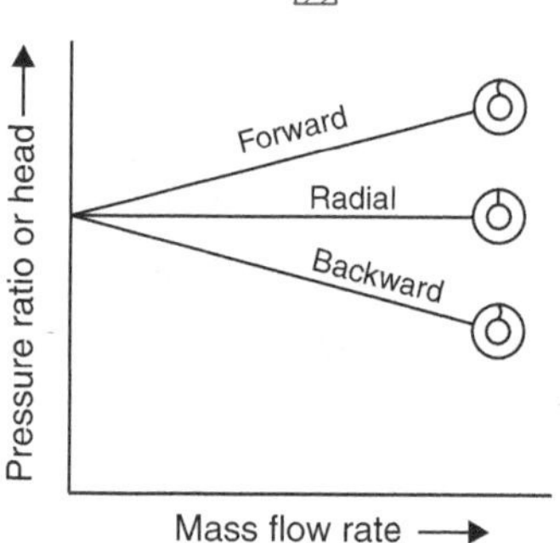

Fig. 20.54. Characteristics of backward-curved, radial-curved, and forward-curved vanes.

Advantages of radial-blade impellers :

1. Can be manufactured *easily.*
2. Lowest unit blade stress for a given diameter and rotational speed, hence *highest weight.*
3. Free from complex bending stresses.
4. Equal energy conversion in impeller and diffuser, *giving high pressure ratios with good efficiency.*

In view of the above reasons, the *impeller with radial blades has been the logic choice of the designers of aircraft centrifugal compressors.*

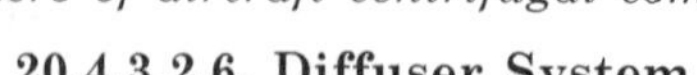

20.4.3.2.6. Diffuser System

In a centrifugal compressor, the diffuser plays a significant role in overall compression process. Whereas the *impeller* is designed to impart energy to the air by increasing its velocity as efficiently as possible, the '**diffuser**' *converts this imparted kinetic energy to pressure-rise.* For a radial-bladed impeller, the diffuser contributes about one-half of the overall static pressure-rise.

In the *vaned diffuser* the vanes are used to remove the whirl of the fluid at a higher rate than is possible by a simple increase in radius, thereby reducing the length of flow path and diameter. The *vaned diffuser is advantageous where small size is important as in the case of aircraft engines.*

Fig. 20.55 shows a typical vaned diffuser. There is a clearance between the impeller and vane leading edges amounting to about 10 to 20% of the diameter for compressors. This *space* constitutes a vaneless diffuser and its functions are :

(*i*) To smooth out velocity variation between the impeller tip and vanes.

(*ii*) To reduce the circumferential pressure gradient at impeller tip.

(*iii*) To reduce the Mach number at entry to the vanes.

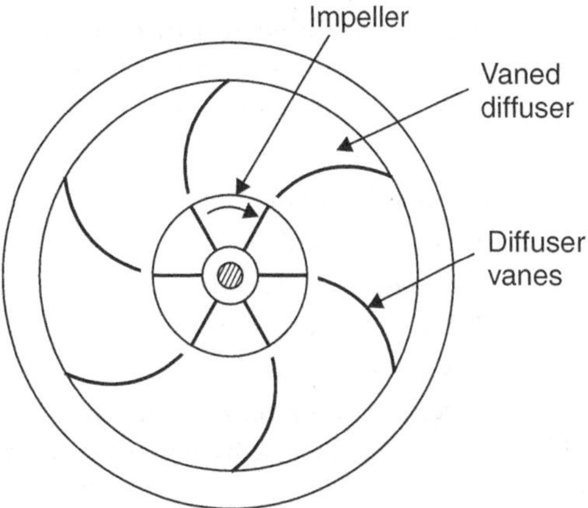

Fig. 20.55. Typical vaned diffuser.

The flow follows an approximately *logarithmic spiral path* to the vanes after which it is constrained by the diffuser channels. For rapid diffusion the axis of the channel is straight and tangential to the spiral. The design of the passages is usually based on the simple channel theory with an equivalent degree of divergence ranging between 8 to 12° to control separation.

- *The number of diffuser vanes has a direct bearing on the size and the efficiency of the diffuser.*
- *When the number of diffuser passages is less than the number of impeller passages a more uniform total flow results.s*

Diffuser efficiency is *defined as,*

$$\eta_d = \frac{(p_2 - p_1)}{\frac{w}{2g}(C_1^2 - C_2^2)} \qquad ...(20.84)$$

Suffices 1 and 2 denote upstream and downstream conditions of diffuser and w is the weight density.

In order to achieve higher diffuser efficiency, the following points need be considered at the time of design :

1. The entrance blade angle of the diffuser must be such that *air impinges on it with a small angle of attack.*

2. Sudden changes in flow are to be avoided. After the air leaves the diffuser, it must be turned through 90° to flow in an axial direction to the combustion chamber. This turning may be achieved by vanes installed in the diffuser below.

3. The area of the flow passage must be large enough to handle the air and it must expand within certain maximum and reasonable limits.

20.4.3.2.7. Losses in Centrifugal Compressors. The losses in a centrifugal compressor may be categorized as follows :

1. **Friction losses.** These losses are proportional to C^2 and hence proportional to m^2.

2. **Incidence loss.** These losses in terms of drag coefficient C_D is proportional to $C_D C^2$.

Fig. 20.56 shows the variation of losses with respect to the mass flow rate.

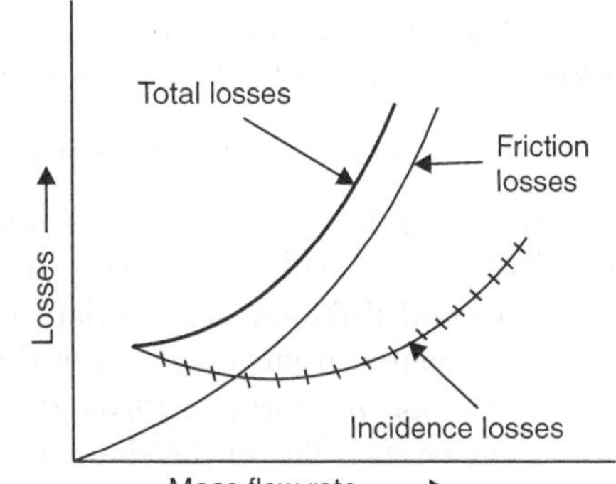

Fig. 20.56. Variation of losses with respect to mass flow rate.

20.4.3.2.8. Selection of Compressors Geometrics. The various compressors' geometrics are selected on the basis of the following :

1. Number of blades in impeller
2. Blade angles
3. Impeller diameters
4. Impeller widths
5. Impeller material
6. Vaneless diffuser
7. Vaned diffuser.

1. **Number of blades in impeller.** The optimum number of blades which gives the best efficiency can be chosen by experience for a particular requirement. A number of empirical relations are available for calculating the optimum number of blades.

- *Vincent* suggests that the optimum blade number varies from 18 to 22 for radial bladed impeller having diameter from 25 cm to 36 cm.

2. **Blade angles :**

- Among the outlet and inlet blade angles, the former influence the latter to a great extent.
- The inlet blade angle, within the reasonable limit, does not affect the performance and, its optimum value is about 30° to 35°.

3. **Impeller diameters.** For *radial blades,* the tip impeller diameter is calculated by the following equations :

$$\text{Actual work input} = \phi_w \phi_s \, C_{bl}^2 \text{ J/kg} \qquad ...(i)$$

and
$$C_{bl_2} = \frac{\pi D_2 N}{60} \qquad ...(ii)$$

After knowing the tip diameter, the inlet diameter is calculated from the value of diameter ratio. Generally $\frac{d_2}{d_1}$ varies from 1.6 to 2.

4. **Impeller width.** If b_1 and b_2 are the blade width at inlet and outlet of impeller, then neglecting the thickness of the blades it is calculated by the equation

$$m = \pi d_1 b_1 \, C_{f_1} \rho_1 = \pi d_2 b_2 \, C_{f_2} \rho_2$$

Generally $C_{f_1} = C_{f_2}$.

5. **Impeller material.** The impeller of the centrifugal compressor is generally forged or die casted of low silicon aluminium alloy.

6. **Vaneless diffuser.** The function of vaneless diffuser or space is to stabilize the flow for shockless entry into the bladed diffuser and to invert some portions of K.E. into pressure energy. The diameter ratio of vaneless to impeller tip diameter varies from $\frac{D_3}{D_2}$ = 1/0.06 to 1.12.

Since the flow in the vaneless diffuser is assumed to be logarithmic spiral, hence $\alpha_2 = \alpha_3$. Generally $b_2 = b_3$ = width of vaneless diffuser. In some cases $b_3 > b_2$.

7. **Vaned diffuser.** The outlet diameter of the vaned diffuser depends upon the choice for the velocity desired from the outlet of the compressor.

By the *use of bladed diffuser the dimensions of the machine can be reduced due to reduction in diffuser size.* The outlet diameter is calculated by the equation,

$$m = \pi d_4 b_4 \, C_4 \rho p_4 \sin \alpha_4$$

where $b_4 = b_2 = b_3$.

— The number of vanes may vary from 10 to 30 but *should not coincide with number of vanes in impeller* to avoid resistance.

— $\frac{D_4}{D_3}$ may vary from 1.25 to 1.6 and the maximum diffusion angle is around 10°.

- *Isentropic efficiency of vaned diffuser at design condition is higher than that of vaneless but its off-design performance is poor than that of vaneless.*

20.4.3.2.9. Compressor Characteristics. When flow is taking place in an impeller channel, there are certain inlet losses, friction and separation losses and discharge losses in the diffuser. If these losses and the effect of slip and non-uniform distribution of radial velocity around the periphery of the impeller are taken into account, the head-capacity characteristic for the backward-curved vanes would take the form *LM* as shown in Fig. 20.57.

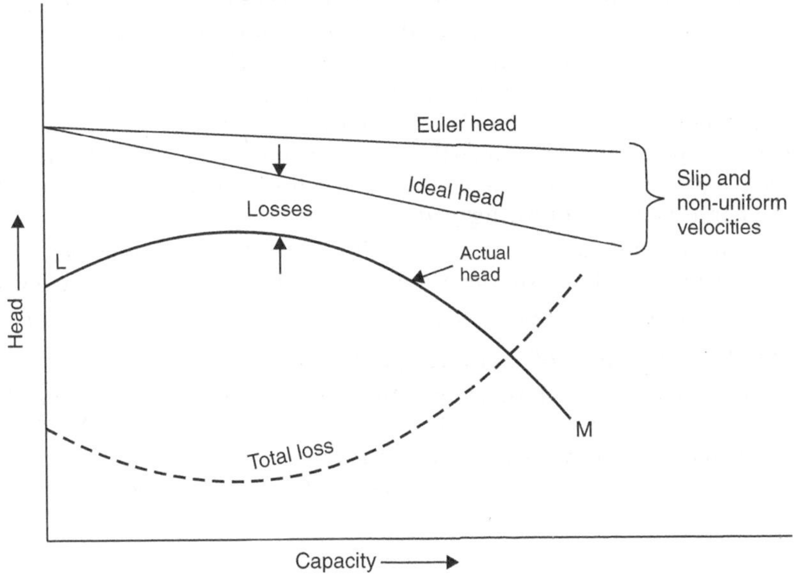

Fig. 20.57. Actual characteristics of a centrifugal compressor.

20.4.3.2.10. Surging and Choking. Fig. 20.58 shows a typical characteristics of a centrifugal compressor at one particular sped. *Consider that the compressor is running at point N* :

- If now the *resistance to flow is increased* (say, by, closing the valve provided at the delivery line of the compressor), the equilibrium point moves to *M*.
- Any further restriction to the flow will cause the operating point to shift to the left, ultimately arriving at point *L*. At this point *maximum pressure ratio is obtained.* If the flow is still reduced from this point then the pressure ratio will *reduce.*

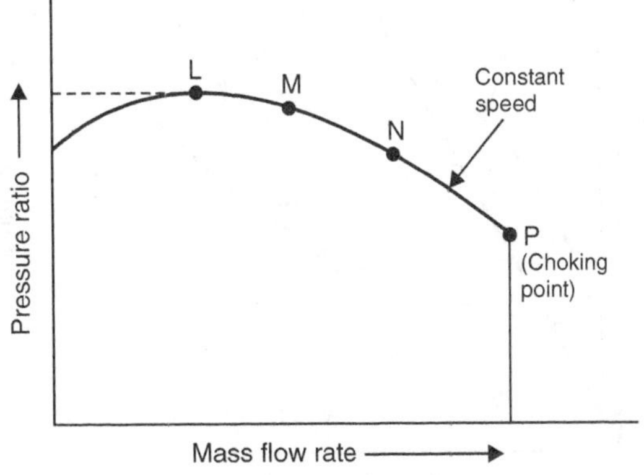

Fig. 20.58. A typical characteristic at one particular speed.

At this moment, there is a higher pressure in the downstream system than the compressor delivery and the *flow stops* and *may even reverse in direction.*

- *After a short interval of time compressor again starts to deliver fluid.* The *pressure starts to increase from a very low value and operating point moves from right to left.* If the downflow conditions are unchanged then once again the flow will break down after point *L* and the cycle will be repeated with *a high frequency*. This phenomenon is known as **surging** or **pumping**. This *instability will be severe in compressors producing high pressure ratios*, which may ultimately *lead to physical damage due to impact loads and high-frequency vibration.*
- Owing to this particular phenomenon of surging or pumping at *low mass flow rates the compressor cannot be operated at any point left of the maximum pressure ratio point.* That is, it cannot be operated on the positive slope of the characteristic.

On the characteristic, the following *situation occurs at higher mass flow rate points* :

At a constant rotor speed the tangential velocity component at the impeller tip *remains constant.* With the *increase in mass flow the pressure ratio decreases* and hence the density is *decreased.* Consequently, the radial velocity is increased considerably, which increases the absolute velocity and incidence angle at the diffuser vane tip. Thus, there is rapid progression towards a *choking state.* The slope of the characteristic therefore steepens and finally after point '*P*' mass flow *cannot be increased any further.* The characteristic finally becomes *vertical.* The point '*P*' on the characteristic curve is called **choking** *point.*

When the compressor is used with gas turbine then the characteristics must be matched properly otherwise troubles will be experienced either due to surging or due to low efficiency.

Note. The *choking mass flow can be varied by changing the impeller rotational speed.*

20.4.3.2.11. Performance of Centrifugal Compressors

- Fig. 20.59 (*a*) shows the relationship between pressure ratio, power and efficiency curves *versus* flow rate for various values of speeds such as N_1, N_2 etc. At *a certain speed, efficiency increases as the flow rate increases and reaches a maximum value after which it decreases.* Accordingly as *the flow rate increases the power consumed also increases.*

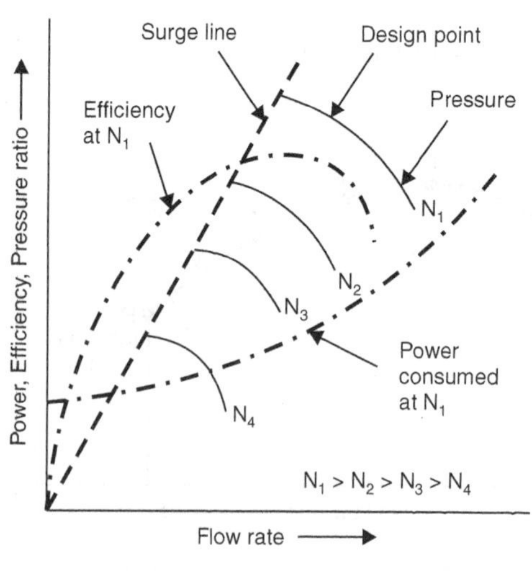

(*a*)

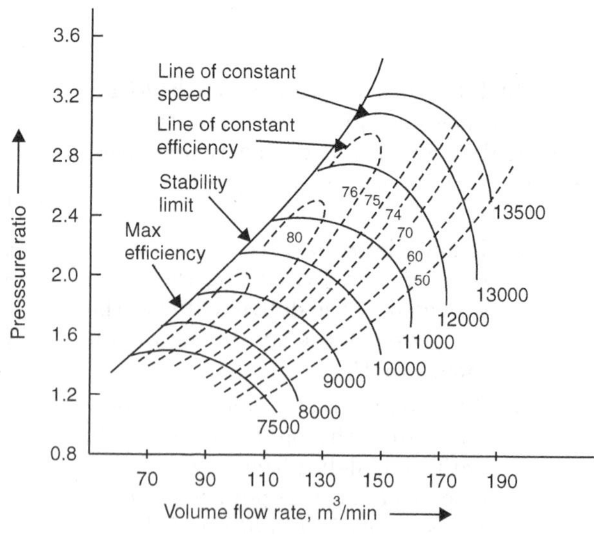

(*b*)

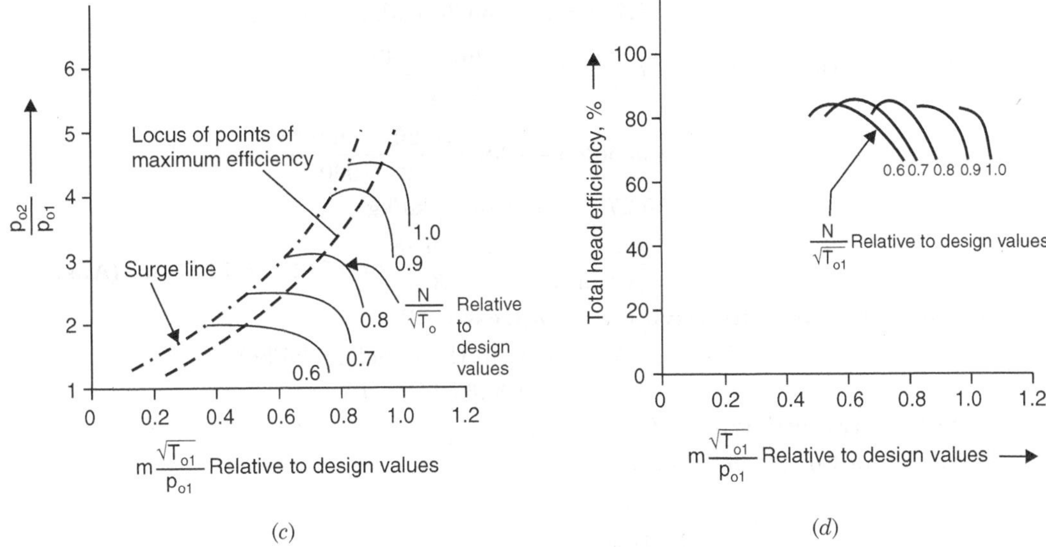

Fig. 20.59. Performance curves of centrifugal compressor.

- Fig. 20.59 (*b*) shows the performance and constant efficiency curves.

  Such a plot does not take into account the varying inlet temperature and pressure. In addition to this, these plots cannot show the comparison of performance for similar compressors of different sizes. To account for all these, the *performance curves are plotted with 'dimensionless parameters'*. These dimensionless parameters are : Pressure ratio, $\frac{p_2}{p_1}$; speed parameter, $\frac{N_1}{\sqrt{T_1}}$ and flow parameter $\frac{m\sqrt{T_1}}{p_1}$ [Fig. 20.59 (*c*) and (*d*)].

Example 20.37. *A centrifugal compressor used as a supercharger for aero-engines handles 150 kg/min. of air. The suction pressure and temperature are 1 bar and 290 K. The suction velocity is 80 m/s. After compression in the impeller the conditions are 1.5 bar 345 K and 220 m/s. Calculate :*

(*i*) *Isentropic efficiency.*

(*ii*) *Power required to drive the compressor.*

(*iii*) *The overall efficiency of the unit.*

It may be assumed that K.E. of air gained in the impeller is entirely converted into pressure in the diffuser.

Solution. *Given :* $\dot{m} = \frac{150}{60} = 2.5$ kg/s ; $p_1 = 1$ bar ; $T_1 = 290$ K ; $C_1 = 80$ m/s ;

$p_2 = 1.5$ bar ; $T_2 = 345$ K ; $C_2 = 220$ m/s.

(*i*) **Isentropic efficiency, $\eta_{isen.}$:**

$$\frac{T_2'}{T_1} = \left(\frac{p_2}{p_1}\right)^{\frac{\gamma-1}{\gamma}} = \left(\frac{1.5}{1}\right)^{\frac{1.4-1}{1.4}} = 1.1228$$

or $$T_2' = 290 \times 1.1228 = 325.6 \text{ K}$$

$$\therefore \quad \text{Isentropic work done} = c_p(T_2' - T_1) + \frac{C_2^2 - C_1^2}{2 \times 1000}$$

$$= 1.005(325.6 - 290) + \frac{(220)^2 - (80)^2}{2 \times 1000}$$

$$= 35.778 + 21 = 56.78 \text{ kJ/kg.}$$

$$\text{Work done in the impeller} = c_p(T_2 - T_1) + \frac{(220)^2 - (80)^2}{2 \times 1000}$$

$$= 1.005(345 - 290) + \frac{(220)^2 - (80)^2}{2 \times 1000}$$

$$= 55.275 + 21 = 76.27 \text{ kJ/kg}$$

$$\therefore \quad \eta_{isen} = \frac{\text{Isentropic work}}{\text{Actual work}} = \frac{56.78}{76.27} = 0.7445 \text{ or } \mathbf{74.45\%}. \quad \textbf{(Ans.)}$$

(*ii*) **Power required to drive the compressor, P :**

$$P = \dot{m} \times \text{Work done in the impeller (kJ/kg)}$$

$$= 2.5 \times 76.27 = \mathbf{190.67 \text{ kW}}. \quad \textbf{(Ans.)}$$

(*iii*) **The overall efficiency of the unit, $\eta_{overall}$:**

As K.E. gained in the impeller is converted into pressure, hence

$$c_p(T_3 - T_2) = \frac{C_2^2 - C_1^2}{2 \times 1000}$$

$$1.005(T_3 - 345) = \frac{(220)^2 - (80)^2}{2 \times 1000}$$

or $$T_3 = 365.9 \text{ K.}$$

The *pressure of air after leaving the diffuser, p_3* :

$$\frac{T_3}{T_2} = \left(\frac{p_3}{p_2}\right)^{\frac{\gamma-1}{\gamma}}$$

or $$\frac{p_3}{p_2} = \left(\frac{T_3}{T_2}\right)^{\frac{\gamma}{\gamma-1}} = \left(\frac{365.9}{345}\right)^{\frac{1.4}{1.4-1}} = 1.2286$$

$$\therefore \quad p_3 = 1.5 \times 1.2286 = 1.843 \text{ bar.}$$

After isentropic compression, the *delivery temperature from diffuser, T_3'* :

$$\frac{T_3'}{T_1} = \left(\frac{p_3}{p_1}\right)^{\frac{\gamma-1}{\gamma}} = \left(\frac{1.843}{1}\right)^{\frac{1.4-1}{1.4}} = 1.191$$

or $$T_3' = 290 \times 1.191 = 345.39 \text{ K}$$

$$\therefore \quad \eta_{overall} = \frac{T_3' - T_1}{T_3 - T_1} = \frac{345.39 - 290}{365.9 - 290} = 0.7298 \text{ or } \mathbf{72.98\%}. \quad \textbf{(Ans.)}$$

Example 20.38. *A single inlet-type centrifugal compressor handles 528 kg/min. of air. The ambient air conditions are 1 bar and 20°C. The compressor runs at 20000 r.p.m. with isentropic efficiency of 80 percent. The air is compressed in the compressor from 1 bar static pressure to 4.0 bar total pressure. The air enters the impeller eye with a velocity of 145 m/s with no prewhirl. Assuming that the ratio of whirl speed to tip speed is 0.9, calculate :*

(*i*) *Rise in total temperature during compression if the change in K.E. is negligible.*

(*ii*) *The tip diameter of the impeller.*

(*iii*) *Power required.*

(*iv*) *Eye diameter if the hub diameter is 12 cm.*

Solution. *Given* : $\dot{m} = \frac{528}{60} = 8.8$ kg/s ; $p_1 = 1$ bar, $T_1 = 20 + 273 = 293$ K ;

$N = 20000$ r.p.m. ; $\eta_{isen} = 80\%$; $p_{01} = 1$ bar ; $p_{02} = 4.0$ bar ;

$C_1 = 145$ m/s ; $\frac{C_{w2}}{C_{bl_2}} = 0.9$; $d_h = 10$ cm $= 0.1$ m.

(*i*) **Rise in total temperature during compression if the change in K.E. is negligible :**

Refer Fig. 20.60. The suffix '0' indicates the total values.

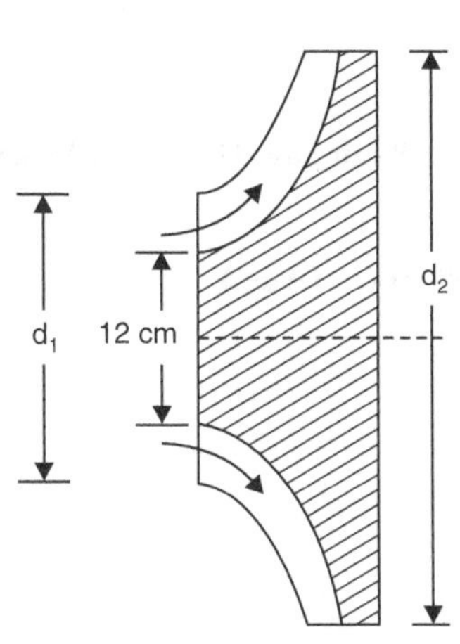

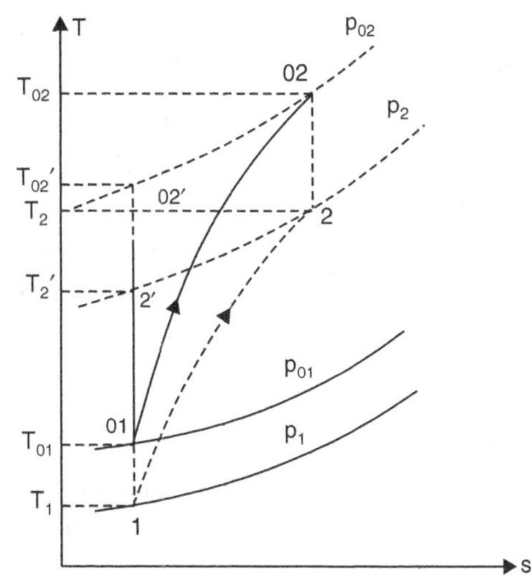

Fig. 20.60

The stagnation temperature at inlet to the machine,

$$T_{o1} = T_1 + \frac{C_1^2}{2c_p} = 293 + \frac{(145)^2}{2 \times 1.005 \times 1000} = 303.5 \text{ K}$$

Now,
$$\frac{T_{o1}}{T_1} = \left(\frac{p_{o1}}{p_1}\right)^{\frac{\gamma-1}{\gamma}} \quad \text{or} \quad p_{01} = p_1 \times \left(\frac{T_{o1}}{T_1}\right)^{\frac{\gamma}{\gamma-1}}$$

or
$$p_{o1} = 1 \times \left(\frac{303.5}{293}\right)^{\frac{1.4}{1.4-1}} = 1.131 \text{ bar}$$

$$\frac{T_{o2}'}{T_{o1}} = \left(\frac{p_{o2}}{p_{o1}}\right)^{\frac{\gamma-1}{\gamma}} = \left(\frac{4.0}{1.131}\right)^{\frac{1.4-1}{1.4}} = 1.435$$

$\therefore \quad T_{o2}' = 303.5 \times 1.435 = 435.5$ K.

$\therefore$ ***Isentropic rise in total temperature*** $= 435.5 - 303.5 = 132$°C

Hence, *Actual rise in total temperature* $= \frac{132}{\eta_{isen}} = \frac{132}{0.8} =$ **165°C. (Ans.)**

(*ii*) **The tip diameter of the impeller, d_2 :**

Work consumed by the compressor $= c_p \times (\Delta t)_{actual}$

$= 1.005 \times 165 = 165.8$ kJ/kg

Work consumed by the compressor is also given by Euler's equation without prewhirl as :

$$W = \frac{C_{w_2} \times C_{bl_2}}{1000} \text{ kJ/kg} = 165.8 \text{ kJ/kg}$$

But $$\frac{C_{w_2}}{C_{bl_2}} = 0.9 \quad \therefore \quad C_{w_2} = 0.9 C_{bl_2}$$

$\therefore$ $$165.8 = \frac{C_{bl_2}^2 \times 0.9}{1000}$$

or $$C_{bl_2} = \left(\frac{165.8 \times 1000}{0.9}\right)^{1/2} = 429.2 \text{ m/s}$$

But $$C_{bl_2} = 429.2 = \frac{\pi d_2 N}{60} = \frac{\pi d_2 \times 20000}{60}$$

$\therefore$ $$d_2 = \frac{429.2 \times 60}{\pi \times 20000} = 0.4098 \text{ m} \quad \text{or} \quad 40.98 \text{ cm say } \mathbf{41 \text{ cm.}} \quad \textbf{(Ans.)}$$

(*iii*) **Power required, P :**

$$P = \dot{m} \times 165.8 = 8.8 \times 165.8 = \mathbf{1459 \text{ kW.}} \quad \textbf{(Ans.)}$$

(*iv*) **Eye diameter if hub diameter is 12 cm, d_1 :**

From continuity equation, we have

$$\dot{m} = \frac{\pi}{4} (d_1^2 - d_h^2) \times C_1 \times \rho_1$$

But density at entry is given by,

$$\rho_1 = \frac{p_1}{RT_1} = \frac{1 \times 10^5}{287 \times 293} = 1.189 \text{ kg/m}^3$$

$\therefore$ $$8.8 = \frac{\pi}{4} (d_1^2 - 0.12^2) \times 145 \times 1.189$$

or $$d_1^2 = \frac{8.8 \times 4}{\pi \times 145 \times 1.189} + 0.12^2 = 0.07939 \text{ m}^2$$

$\therefore$ $$d_1 = 0.2818 \text{ m or } \mathbf{28.2 \text{ cm.}} \quad \textbf{(Ans.)}$$

Example 20.39. *A centrifugal compressor running at 10000 r.p.m. delivers 660 m³/min. of free air. The air is compressed from 1 bar and 20°C to a pressure ratio of 4 with an isentropic efficiency of 82%. Blades are radial at outlet of impeller and flow velocity of 62 m/s may be assumed throughout constant. The outer radius of impeller is twice the inner and the slip factor may be assumed as 0.9. The blade area coefficient may be assumed 0.9 at inlet. Calculate :*

(*i*) *Final temperature of air.* (*ii*) *Theoretical power.*

(*iii*) *Impeller diameters at inlet and outlet.*

(*iv*) *Breadth of impeller at inlet.*

(*v*) *Impeller blade angle at inlet.*

(*vi*) *Diffuser blade angle at inlet.*

Solution. *Given* : $N = 10000$ r.p.m. ; Volume of air delivered, $V = 660$ m³/min. ;

$p_1 = 1$ bar, $T_1 = 20 + 273 = 293$ K ; $r_p = 4$, $\eta_{isen} = 0.82$; $C_{f_2} = 62$ m/s ;

$r_2 = 2r_1$; $\phi_s = 0.9$; Blade area coefficient, $k_a = 0.9$.

(*i*) **Final temperature of air, T_2 :**

$$\frac{T_2'}{T_1} = (r_p)^{\gamma-1/\gamma} = (4)^{\frac{1.4-1}{1.4}} = 1.486$$

$\therefore \quad T_2' = 293 \times 1.486 = 435.4 \text{ K}$

Now, $\quad \eta_{isen} = \dfrac{T_2' - T_1}{T_2 - T_1}$

or $\quad T_2 = T_1 + \dfrac{T_2' - T_1}{\eta_{isen}} = 293 + \dfrac{435.4 - 293}{0.82} = $ **466.7 K. (Ans.)**

(*ii*) **Theoretical power, P :**

Mass flow rate, $\dot{m} = \dfrac{PV}{RT} = \dfrac{1 \times 10^5 \times (660/60)}{287 \times 293} = 13.08 \text{ m}^3/\text{s}$

$\therefore \quad P = \dot{m} c_p (T_2 - T_1)$

$= 13.08 \times 1.005(466.7 - 293) =$ **2283.3 kW. (Ans.)**

(*iii*) **Impeller diameters at inlet and outlet, d_1, d_2 :**

For radial blades, work input to the compressor is given by,

$$\text{Work done} = \frac{\phi_s C_{bl_2}^2}{1000} = c_p(T_2 - T_1)$$

Here T_2 is the final temperature of air from the exit of compressor.

or $$C_{bl_2} = \left[\frac{1000 \times c_p(T_2 - T_1)}{\phi_s}\right]^{1/2} = \left[\frac{1000 \times 1.005(466.7 - 293)}{0.9}\right]^{1/2} = 440.4 \text{ m/s}$$

Also, $$C_{bl_2} = \frac{\pi d_2 N}{60} = 440.4$$

or $$d_2 = \frac{60 \times 440.4}{\pi \times 10000} = 0.8411 \text{ m} \quad \text{or} \quad \mathbf{84.11 \text{ cm.}} \quad \textbf{(Ans.)}$$

$$\therefore \quad d_1 = \frac{d_2}{2} = \frac{84.11}{2} = \mathbf{42.06 \text{ cm.}} \quad \textbf{(Ans.)}$$

(*iv*) **Breadth of impeller at inlet, b_1 :**

Volume flow rate $= 2\pi r_1 b_1 C_{f_1} k_a$, where k_a is the blade area coefficient

$$\therefore \quad b_1 = \frac{\text{Volume flow rate}}{2\pi r_1 \,.\, C_{f_1} \,.\, k_a}$$

$$= \frac{(660/60)}{2\pi \times (0.4206/2) \times 62 \times 0.9} = 0.1492 \text{ m} \quad \text{or} \quad \mathbf{14.92 \text{ cm.}} \quad \textbf{(Ans.)}$$

(*v*) **Impeller blade angle at inlet β_1 :**

$$\tan \beta_1 = \frac{C_{f_1}}{C_{bl_1}} = \frac{62}{(440.4/2)} = 0.2816$$

$$\therefore \quad \beta_1 = \tan^{-1}(0.2816) = \mathbf{15.73°.} \quad \textbf{(Ans.)}$$

(*vi*) **Diffuser blade angle at inlet, α_2 :**

$$\tan \alpha_2 = \frac{C_{f_2}}{\phi_s \,.\, C_{bl_2}} = \frac{62}{0.9 \times 440.4} = 0.1564$$

$$\therefore \quad \alpha_2 = \tan^{-1}(0.1564) = \mathbf{8.9°.} \quad \textbf{(Ans.)}$$

Example 20.40. *A centrifugal blower compresses 4.8 m³/s of air from 1 bar and 20°C to 1.5 bar. The index of compression n is 1.5. The flow velocity at inlet and outlet of the machine is the same and equal to 65 m/s. The inlet and outlet impeller diameters are 0.32 m and 0.62 m respectively. The blower rotates at 8000 r.p.m. Calculate :*

(*i*) *The blade angles at inlet and outlet of the impeller.*
(*ii*) *The absolute angle at the tip of the impeller.*
(*iii*) *The breadth of blade at inlet and outlet.*

It may be assumed that no diffuser is employed and the whole pressure increase takes place in the impeller and the blades have negligible thickness.

Solution. *Given* : $V_1 = 4.8$ m³/s ; $p_1 = 1$ bar ; $T_1 = 20 + 273 = 293$ K ; $n = 1.5$

$C_{f1} = C_{f2} = 65$ m/s ; $d_1 = 0.32$ m ; $d_2 = 0.62$ m ; $N = 8000$ r.p.m.

The temperature at the outlet of the compressor,

$$\frac{T_2'}{T_1} = \left(\frac{p_2}{p_1}\right)^{\frac{\gamma-1}{\gamma}} = \left(\frac{1.5}{1}\right)^{\frac{1.5-1}{1.5}} = 1.1447$$

$\therefore \quad T_2' = 293 \times 1.1447 = 335.4$ K

The peripheral velocity at inlet,

$$C_{bl_1} = \frac{\pi d_1 N}{60} = \frac{\pi \times 0.32 \times 8000}{60} = 134 \text{ m/s}$$

The tip peripheral velocity at outlet,

$$C_{bl_2} = \frac{\pi d_2 N}{60} = \frac{\pi \times 0.62 \times 8000}{60} = 259.7 \text{ m/s}$$

Work done $= c_p(T_2' - T_1) = \dfrac{C_{bl_2} C_{w_2}}{1000}$

$$\therefore \quad C_{w_2} = \frac{1.005(335.4 - 293) \times 1000}{259.7} = 164.1 \text{ m/s}$$

(*i*) **The blade angles at inlet and outlet of the impeller, β_1, β_2 :**

$$\tan \beta_1 = \frac{C_{f1}}{C_{bl_1}} = \frac{65}{134} \quad \therefore \quad \beta_1 = \mathbf{25.88°}. \quad \textbf{(Ans.)}$$

$$\tan \beta_2 = \frac{C_{f2}}{C_{bl_2} - C_{w2}} = \frac{65}{259.7 - 164.1} = 0.6799 \quad \therefore \quad \beta_2 = \mathbf{34.2°}. \quad \textbf{(Ans.)}$$

(*ii*) **The absolute angle at the tip of the impeller, α_2 :**

$$\tan \alpha_2 = \frac{C_{f2}}{C_{w2}} = \frac{65}{164.1} = 0.3961 \quad \therefore \quad \alpha_2 = \mathbf{21.6°}. \quad \textbf{(Ans.)}$$

(*iii*) **The breadth of blade at inlet and outlet, b_1, b_2 :**

Discharge at the inlet, $V_1 = 2\pi r_1 b_1 C_{f1}$

$4.8 = 2\pi \times 0.16 \times b_1 \times 65$

$$\therefore \quad b_1 = \frac{4.8}{2\pi \times 0.16 \times 65} = 0.0734 \text{ m} \quad \text{or} \quad \mathbf{7.34 \ cm.} \quad \textbf{(Ans.)}$$

Let V_2 be the discharge at the outlet, then

$$V_2 = \frac{p_1 V_1}{T_1} \times \frac{T_2}{p_2} \qquad \left[\because \frac{p_1 V_1}{T_1} = \frac{p_2 V_2}{T_2} \quad \text{Taking } T_2' = T_2\right]$$

$$= \frac{1 \times 10^5 \times 4.8}{293} \times \frac{335.4}{1.5 \times 10^5} = 3.66 \text{ m}^3\text{/s}$$

$\therefore \quad V_2 = 3.66 = 2\pi r_2 \,.\, b_2 \,.\, C_{f2} = 2\pi \times 0.31 \times b_2 \times 65$

or

$$b_2 = \frac{3.66}{2\pi \times 0.31 \times 65} = 0.0289 \text{ m} \quad \text{or} \quad \mathbf{2.89 \ cm.} \quad \textbf{(Ans.)}$$

Example 20.41. *A centrifugal compressor delivers 16.5 kg/s of air with a total head pressure ratio of 4 : 1. The speed of the compressor is 15000 r.p.m. Inlet total head temperature is 20°C, slip factor 0.9, power input factor 1.04 and 80% isentropic efficiency. Calculate :*

(*i*) *Overall diameter of the impeller.*

(*ii*) *Power input.*

Solution. *Given :* $\dot{m}$ = 16.5 kg/s ; Pressure ratio, r_{op} = 4 ; N = 15000 r.p.m. ; T_{01} = 20 + 273 = 293 K

$$\phi_s = 0.9 \; ; \; \phi_w = 1.04 \; ; \; \eta_{isen} = 80\%.$$

(*i*) **The overall diameter of the impeller, D :**

$$\frac{T_{02}'}{T_{01}} = \left(\frac{p_{02}}{p_{01}}\right)^{\frac{\gamma-1}{\gamma}} = (4)^{\frac{1.4-1}{1.4}} = 1.486$$

$$\therefore \quad T_{02}' = T_{01} \times 1.486 = 293 \times 1.486 = 435.4 \text{ K}$$

$$\eta_{isen} = \frac{T_{02}' - T_{01}}{T_{02} - T_{01}} \quad \text{or} \quad 0.8 = \frac{435.4 - 293}{T_{02} - T_{01}}$$

or $$T_{02} - T_{01} = \frac{435.4 - 293}{0.8} = 178 \text{ K}$$

$$\therefore \quad \text{Work done} = \frac{\phi_w \phi_s C_{bl_2}^2}{1000} = c_p(T_{02} - T_{01})$$

or $$C_{bl_2} = \left[\frac{1000 \times c_p(T_{02} - T_{01})}{\phi_w \phi_s}\right]^{1/2} = \left[\frac{1000 \times 1.005 \times 178}{1.04 \times 0.9}\right]^{1/2} = 437.2 \text{ m/s}$$

Also, $$C_{bl_2} = \frac{\pi D N}{60} = 437.2$$

$$\therefore \quad D_2 = \frac{437.2 \times 60}{\pi \times 15000} = 0.5567 \text{ m} \quad \text{or} \quad \mathbf{55.67\ cm.} \quad \textbf{(Ans.)}$$

(*ii*) **Power input, P :**

$$P = \dot{m} c_p (T_{02} - T_{01})$$
$$= 16.5 \times 1.005 \times 178 = \mathbf{2951.7\ kW.} \quad \textbf{(Ans.)}$$

☞**Example 20.42.** *The following data pertain to a centrifugal compressor :*

Total pressure ratio	*= 3.6 : 1*
Diameter of inlet eye of compressor impeller	*= 35 cm*
Axial velocity at inlet	*= 140 m/s*
Mass flow	*= 12 kg/s*
The velocity in the delivery duct	*= 120 m/s*
The tip speed of impeller	*= 460 m/s*
Speed of impeller	*= 16000 r.p.m.*
Total head isentropic efficiency	*= 80%*
Pressure co-efficient	*= 0.73*
Ambient conditions	*= 1.013 bar and 15°C.*

Calculate :

(*i*) *The static pressure and temperature at inlet and outlet of compressor.*

(*ii*) *The static pressure ratio.*

(*iii*) *Work of compressor per kg of air.*

(*iv*) *The theoretical power required.*

Solution. *Given :* r_p (pressure ratio) = 3.6

$$T_{01} = 15 + 273 = 288 \text{ K} \; ; \; \eta_{isen} = 0.8$$

$$\dot{m} = 12 \text{ kg/s} \; ; \; \phi_p \text{ (Pressure co-efficient)} = 0.73 \; ;$$

$$C_{bl_2} = 120 \text{ m/s}$$

(*i*) **The static pressure and temperature at inlet and outlet of compressor :**

Total head temperature rise

$$\Delta T_0 = \frac{T_{01}\left(r_p^{\frac{\gamma-1}{\gamma}} - 1\right)}{\eta_{isen}} = \frac{288\left[(3.6)^{\frac{1.4-1}{1.4}} - 1\right]}{0.8} = 159.1°\text{C}$$

$\therefore$ $\mathbf{T_{02}} = T_{01} + \Delta T_0 = 288 + 159.1 = \mathbf{447.1\ K.}$ **(Ans.)**

The static temperature at exist is

$$\mathbf{T_2} = T_{02} - \frac{C_{bl_2}^2}{2c_p} = 447.1 - \frac{120^2}{2 \times 1.005 \times 10^3}$$

$$= 447.1 - 7.1 = \mathbf{440\ K.} \textbf{ (Ans.)}$$

The *static pressure at exit is*

$$p_2 = p_{02} - \frac{\rho_2 C_{bl_2}^2}{2} \qquad \left(\text{where } \rho_2 = \frac{p_2}{RT_2}\right)$$

i.e., $$p_2 = p_{02} - \frac{p_2 \times 120^2}{2 \times 0.287 \times 440 \times 10^3} = p_{02} - 0.057\, p_2$$

But $p_{02} = 1.013 \times 3.6 = 3.65$ bar

$\therefore$ $p_2 = 3.65 - 0.057\, p_2$ or $1.057\, p_2 = 3.65$

$\therefore$ $\mathbf{p_2 = 3.45\ bar.}$ **(Ans.)**

To calculate *static conditions at inlet :*

$$T_{01} = T_1 + \frac{140^2}{2c_p} = T_1 + \frac{140^2}{2 \times 1.005 \times 10^3} = T_1 + 9.75$$

$\therefore$ $\mathbf{T_1} = 288 - 9.75 = \mathbf{278.25\ K.}$ **(Ans.)**

$$p_1 = p_{01} - \frac{p_1 \times 140^2}{2 \times 0.287 \times 278.25 \times 10^3} = 1.013 - 0.123\, p_1$$

$\therefore$ $\mathbf{p_1 = 0.9\ bar.}$ **(Ans.)**

(*ii*) **Static pressure ratio** $= \dfrac{p_2}{p_1} = \dfrac{3.45}{0.9} = \mathbf{3.83.}$ **(Ans.)**

(*iii*) **Work done on air** $= c_p \times \Delta T_0$

$= 1.005 \times 159.1 =$ **159.89 kJ/kg of air. (Ans.)**

(*iv*) **Theoretical power required to drive compressor**

$= \dot{m}\, c_p\, \Delta T_0 = 12 \times 1.005 \times 159.1 =$ **1918.7 kW. (Ans.)**

Example 20.43. *Air at a temperature of 300 K flows in a centrifugal compressor running at 18000 r.p.m. The other data given is as follows :*

Isentropic total head efficiency = 0.76

Outer diameter of blade tip = 550 mm

Slip factor = 0.82

Calculate :

(*i*) *The temperature rise of air passing through the compressor.*

(*ii*) *The static pressure ratio.*

Assume that the absolute velocities of air at inlet and exit of the compressor are same. Take c_p *= 1.005 kJ/kg K.*

Solution. $$C_{bl_2} = \frac{\pi DN}{60} = \frac{\pi \times (550/1000) \times 18000}{60} = 518.36 \text{ m/s.}$$

Work done per kg of air $= (C_{w_2} . C_{bl_2} - C_{w_1} . C_{bl_1})$

But $C_{w_1} = 0$ and $\phi_s = \dfrac{C_{w_2}}{C_{bl_2}}$

$\therefore$ Work done per kg of air $= C_{bl_2}^2 \times \phi_s = (518.36)^2 \times 0.82 = 220331$ W

$= 220.331$ kW ...(*i*)

(*i*) **Temperature rise of air, $\mathbf{T_1 - T_2}$:**

Also work done $= c_p(T_2 - T_1)$ ...(*ii*)

Equating (*i*) and (*ii*), we get

$$\mathbf{T_2 - T_1} = \frac{220.331}{c_p} = \frac{220.331}{1.005} = \mathbf{219.23°C.} \quad \textbf{(Ans.)}$$

$$\eta_{isen} = \frac{T_2' - T_1}{T_2 - T_1}$$

i.e., $$0.76 = \frac{T_2' - 300}{219.23} \quad \text{or} \quad T_2' = 466.6 \text{ K.}$$

(*ii*) **The static pressure ratio :**

The static pressure ratio is given by

$$\frac{p_2}{p_1} = \left(\frac{T_2'}{T_1}\right)^{\frac{\gamma}{\gamma - 1}} = \left(\frac{466.6}{300}\right)^{\frac{1.4}{1.4 - 1}} = \mathbf{4.69.} \quad \textbf{(Ans.)}$$

20.4.3.3. Axial Flow Compressor

20.4.3.3.1. Construction and Working

In an axial flow compressor, the *flow proceeds* throughout the compressor in a direction essentially *parallel to the axis of the machine.*

Construction. Refer. 20.61.

- An axial flow compressor consists of adjacent rows of rotor (moving) blades and stator (fixed) blades. The rotor blades are mounted on the rotating drum and stator blades are fixed to the casing stator. *One stage* of the machine comprises a row of rotor blades followed by a row of stator blades.
- For efficient operation the blades are of air foil section based on aerodynamic theory. The blades are so designed that wasteful losses due to shock and turbulence are prevented and the blades are free from *stalling troubles.* (The blades are said to *be stalled when the air stream fails to follow the blade contour*). Whereas the compressor blades have aerofoil section, the turbine blades have profiles formed by a number of circular arcs. This is so because the acceleration process being carried out in the converging blade passages of a reaction turbine is much more efficient and stable process as compared with the diffusing or decelerating process being carried out in the diverging passage between the blades of an axial flow compressor.

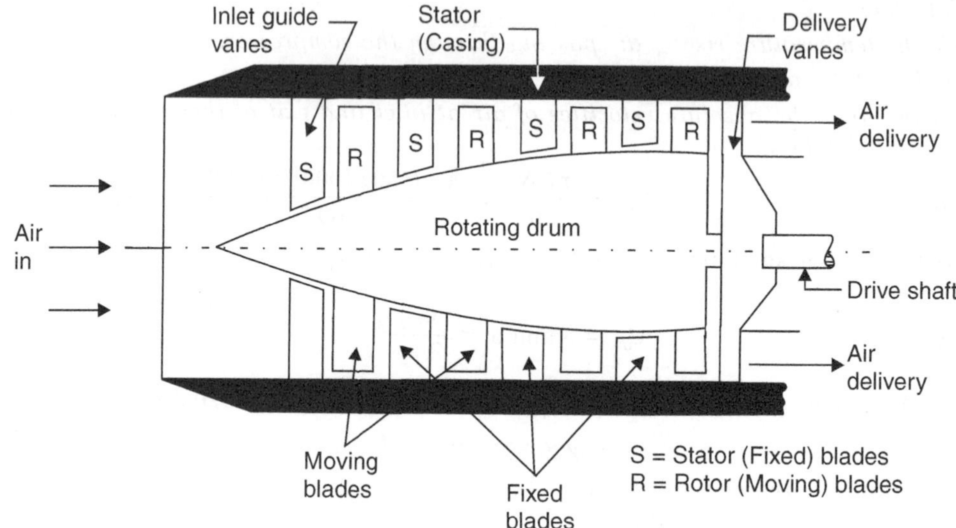

Fig. 20.61. Axial flow compressor.

- The annular area is usually reduced from inlet to outlet of the compressor. This is to *keep the flow velocity constant throughout the compressor length.* In the diverging passages of the moving blades, there is rise in temperature due to diffusion. The absolute velocity is also increased due to work input.

The *"fixed blades"* serve the following two *purposes* :

(*i*) Convert a *part of the K.E. of the fluid into pressure energy. This conversion is achieved by diffusion process carried out in the diverge blade passages.*

(*ii*) *Guide and redirect the fluid flow so that entry to the next stage is without shock.*

Working

Basically, the compression is performed in a similar manner to that of the centrifugal type. The work input to the rotor shaft is transferred by the moving blades to the air, thus accelerating it. The blades are so arranged that the *spaces between the blades form diffuser passages, and hence the velocity of the air relative to the blades is decreased as the air passes through them, and there is a rise in pressure. The air is then further diffused in the stator blades, which are also arranged to form diffuser passages.* In the fixed stator blades the *air is turned through an angle so that its direction is such that it can be allowed to pass to a second row of moving rotor blades. It is usual to have a relatively large number of stages and to maintain a constant work input per stage* (*e.g.*, from 5 to 14 stages have been used).

— The necessary reduction in volume may be allowed by flaring the stator or by flaring the rotor. It is more common to use a flared rotor, and this type is shown diagrammatically in Fig. 20.61.

— It is usually arranged to have an *equal temperature rise in the moving and the fixed blades, and to keep the axial velocity of air constant throughout the compressor.* Thus *each stage of the compression is exactly similar with regard to air velocity and blade inlet and outlet angles.*

— A diffusing flow is less stable than a converging flow, and for this reason the blade shape and profile is much more important for a compressor than for a reaction turbine. The *design of compressor blades is based on aerodynamic theory and an aerofoil shape* is used.

Note. Two forms of rotors have been used namely the *drum* and *disc types*. The **disc type** is used where consideration of *low weight* is more important than cost as in *aircraft applications*. The drum type is more suitable for *static industrial applications*. In some applications, combination of both types has been used.

Materials. The following materials are used for the various components of an axial flow compressor :

1. ***Rotor bladings.*** The materials listed below are in the *increasing* order of weight and their ability to withstand high temperature :

(*i*) Fibrous composites (*ii*) Aluminium

(*iii*) Titanium (*iv*) Steel

(*v*) Nickel alloy.

2. ***Rotor :***

- For *rotor shafts* and disc .... *"steel."*
- *Aircraft engines may use titanium at the front stages and "nickel alloy" in the rest.*

3. ***Stator bladings :***

- Same materials as that of rotor but *steel is the most common.*

4. ***Castings.*** These may be of *cast magnesium, aluminium, steel or iron or fabricated from titanium or steel.*

— NC (Numerically controlled) machines make dies and the blades are manufactured by precision forging. Blades are also machined by CNC copying machines.

20.4.3.3.2. Velocity Diagrams and Work Done of a stage of Axial Flow Compressors. Fig. 20.62 shows the velocity triangles for one stage of an axial flow compressor. All angles are measured from the axial direction and the blade velocity C_{bl} is taken to be *same* at blade entry and exist. This is because the air enters and leaves the blades at *almost equal radii.*

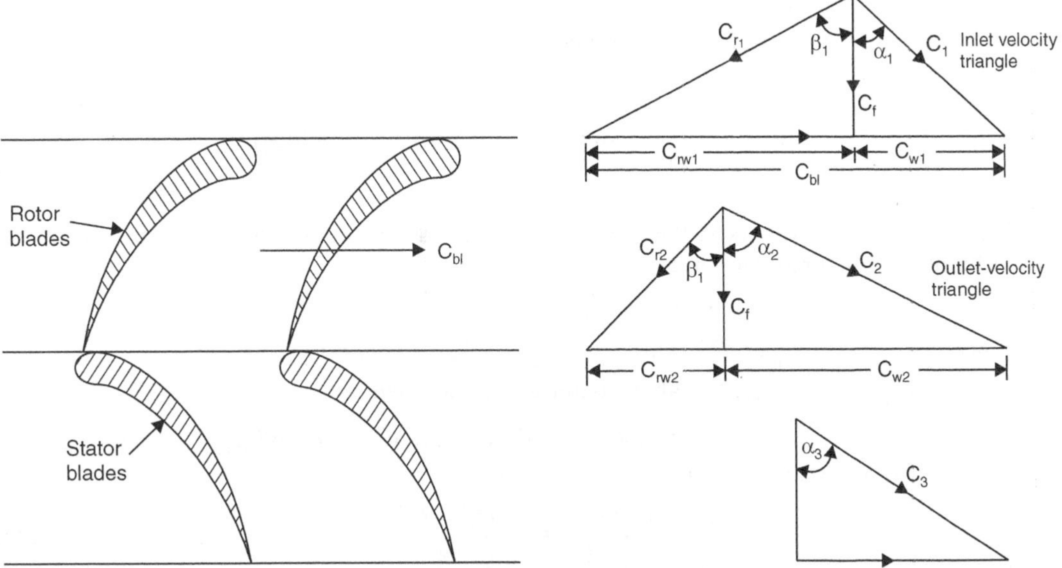

Fig. 20.62. Velocity diagrams for axial flow compressor.

- Air approaches the rotor blade with absolute velocity C_1 and at an angle α_1. The relative velocity C_{r1}, obtained by the vectorial addition of absolute velocity C_1 and blade velocity C_{bl}, has the inclination β_1 with the axial direction.

- Due to diffusion in the diverging passages formed by rotor blades, there is some pressure rise. This is at the expense of relative velocity and so the relative velocity decreases from C_{r1} to C_{r2}. Since work is being done on the air by rotor blades, the air would ultimately leave the rotor with increased absolute velocity C_2.
- The air then enters the stator blades and the diffusion and deceleration takes place in the diverging passage of stator blades. Finally the air leaves the stator blades with velocity C_3 at an angle α_3 and is redirected to the *next stage*. Generally it is *assumed that absolute velocity C_3 leaving the compressor stage equals the approach velocity C_1.*

From the velocity triangles, we have

$$\frac{C_{bl}}{C_f} = \tan \alpha_1 + \tan \beta_1 \qquad ...(20.85)$$

and,

$$\frac{C_{bl}}{C_f} = \tan \beta_2 + \tan \alpha_2 \qquad ...(20.86)$$

Assume *1 kg of flow of air* through the *compressor stage.*

From Newton's second law of motion,

Tangential force per kg = $C_{w2} - C_{w1}$

Work absorbed by the stage per kg of air,

$$W_{st} = C_{bl}(C_{w2} - C_{w1}) = c_p(T_{02} - T_{01})_{act} \qquad ...(20.87)$$

where C_{w1} and C_{w2} are the whirl components of absolute velocities at inlet and exit of rotating blades.

(It may be noted that here whirl component at the entrance of the compressor is not *zero* because of the fact that air flows *axially and not radially.*)

The expression for work done may be put in terms of flow/axial velocity and air angles.

$$W_{st} = C_{bl}\, C_f \;\; (\tan \alpha_2 - \tan \alpha_1) \qquad ...(20.88)$$

From eqns. (20.85) and (20.86), we have

$$W_{st} = C_{bl}\, C_f \;\; (\tan \beta_1 - \tan \beta_2) \qquad ...(20.89) \qquad \left[\because \; \frac{C_{w1}}{C_f} = \tan \alpha_1, \text{ and } \; \frac{C_{w2}}{C_f} = \tan \alpha_2\right]$$

The work input to the axial flow compressor may also be obtained by the Euler's equation. For each kg of air delivered, we have

$$W_{st} = C_{bl_2}\, C_{w2} - C_{bl_1}\, C_{w1}$$

Since $\quad C_{w2} - C_{w1} = [(C_{bl} - C_{rw2}) - (C_{bl} - C_{rw1})] = C_{rw1} - C_{rw2}$

Further $C_{bl_1} = C_{bl_2} = C_{bl}$, the above equation is modified as,

$$W_{st} = C_{bl}(C_{rw1} - C_{rw2}) = C_{bl}\, \Delta C_{rw} = c_p\, (\Delta T_0)_{act}$$

By use of velocity triangle and cosine theorem,

$$W_{st} = \underset{\text{(First term)}}{\frac{C_{r1}^2 - C_{r2}^2}{2}} + \underset{\text{(Second term)}}{\frac{C_2^2 - C_3^2}{2}} \qquad ...(20.90)$$

Here $\quad C_3 = C_1$

— The *first term* on the right side of the above equation introduces the part of the work supplied by a rotating cascade, which is converted into pressure due to *diffusion action* in rotating cascade itself.

— The second term represents the increment of K.E. in rotating cascade that has to be converted into pressure energy in stationary cascade. Comparing this equation to the work input to centrifugal compressor, we find that the term centrifugal action $\left[\frac{C_{bl_2}{}^2 - C_{bl_1}{}^2}{2}\right]$ is missing in axial flow compressors. Due to this reason the *pressure ratio per stage in axial flow compressor is much less than that of centrifugal compressor.*

The stage temperature rise, regardless of efficiency of compression, will be given by the equation

$$(\Delta T)_{act} = \frac{C_{bl} C_f}{c_p} (\tan \beta_1 - \tan \beta_2) \qquad ...(20.91)$$

Pressure rise in isentropic flow through a cascade :

Consider the incompressible isentropic and steady flow through a cascade from uniform condition 1 to uniform condition 2. From Bernoulli's equation, we have

$$\frac{p_1}{\rho} + \frac{C_1^2}{2} = \frac{p_2}{\rho} + \frac{C_2^2}{2}$$

Thus the static isentropic pressure rise may be expressed in terms of the inlet dynamic head.

$$(\Delta p)_{isen} = (p_2)_{isen} - \rho \left(\frac{C_1^2 - C_2^2}{2} \right)$$

$$= (p_2)_{isen} - \frac{\rho}{2} [(C_{f1}{}^2 + C_{w1}{}^2) - (C_{f2}{}^2 + C_{w2}{}^2)]$$

Since $\qquad C_{f1} = C_{f2} = C_f$

$$\therefore \qquad (\Delta p)_{isen} = \frac{\rho}{2} (C_{w1}{}^2 - C_{w2}{}^2) = \frac{\rho}{2} C_f^2 (\tan^2 \alpha_1 - \tan^2 \alpha_2) \qquad ...(20.92)$$

$[\because \; C_{w1} = C_f \tan \alpha_1$, and $C_{w2} = C_f \tan \alpha_2]$

20.4.3.3.3. Degree of Reaction

Degree of reaction (R_d) is defined as *the ratio of pressure rise in the compressor stage.*

i.e.,
$$R_d = \frac{\text{Pressure rise in the rotor blades}}{\text{Pressure rise in the stage}}$$

Pressure rise in the compressor stage equals work input per stage and is

$$= C_{bl} \, (C_{w_2} - C_{w_1})$$

Pressure rise in the rotor blades is at the expense of K.E. and is

$$= \frac{C_{r_1}{}^2 - C_{r_2}{}^2}{2}$$

$$\therefore \qquad R_d = \frac{C_{r_1}{}^2 - C_{r_2}{}^2}{2 C_{bl} \, (C_{w_2} - C_{w_1})} \qquad ...(20.93)$$

Refer inlet and outlet velocity triangles :

$$C_{w_2} = C_{bl} - C_f \tan \beta_2$$

$$C_{w_1} = C_{bl} - C_f \tan \beta_1$$

$$\therefore \quad C_{w_2} - C_{w_1} = C_f(\tan\beta_1 - \tan\beta_2) \quad ...(20.94)$$

Similarly from velocity triangles,

$$C_{r_1}^{\ 2} = (C_f)^2 + (C_f \tan\beta_1)^2$$

$$C_{r_2}^{\ 2} = (C_f)^2 + (C_f \tan\beta_2)^2$$

$$\therefore \quad C_{r_1}^{\ 2} - C_{r_2}^{\ 2} = C_f^{\ 2}(\tan^2\beta_1 - \tan^2\beta_2) \quad ...(20.95)$$

So $$R_d = \frac{C_f^{\ 2}(\tan^2\beta_1 - \tan^2\beta_2)}{2C_{bl}.C_f(\tan\beta_1 - \tan\beta_2)} = \frac{1}{2}\frac{C_f}{C_{bl}}(\tan\beta_1 + \tan\beta_2) \quad ...(20.96)$$

Degree of reaction is usually kept as **0.5**,

$$\therefore \quad 0.5 = \frac{1}{2}\cdot\frac{C_f}{C_{bl}}(\tan\beta_1 + \tan\beta_2)$$

or $$\frac{C_{bl}}{C_f} = \tan\beta_1 + \tan\beta_2$$

But $$\frac{C_{bl}}{C_f} = \tan\alpha_1 + \tan\beta_1 = \tan\alpha_2 + \tan\beta_2 \quad \text{(from velocity triangles)}$$

$$\therefore \quad \frac{C_{bl}}{C_f} = \tan\beta_1 + \tan\beta_2 = \tan\alpha_1 + \tan\beta_1 = \tan\alpha_2 + \tan\beta_2$$

From this $\alpha_1 = \beta_2$; $\alpha_2 = \beta_1$

So with 50% reaction blading, the compressors have symmetrical blades and with this type of set up losses in flow path are amply reduced.

In *symmetrical blades, the tip clearance and fluid friction losses are minimum.*

20.4.3.3.4. Polytropic Efficiency. The work input to a compressor, with usual notations,

$$W = c_p(T_{02} - T_{01})$$

$$= c_p(T_{02}' - T_{01}) \times \left(\frac{T_{02} - T_{01}}{T_{02}' - T_{01}}\right) = c_p\left(\frac{T_{02} - T_{01}}{T_{02}' - T_{01}}\right) \times (T_{02'} - T_{01})$$

$$= c_p\,\eta_{isen}\,T_{01}\left(\frac{T_{02}'}{T_{01}} - 1\right) = c_p\,\eta_{isen}\,T_{01}\left[\left(\frac{p_{02}}{p_{01}}\right) - 1\right] \quad ...(20.97)$$

Eqn. (20.97) indicates that for the same isentropic efficiency η_{isen} and pressure ratio $\frac{p_{02}}{p_{01}}$, the *work input is proportional to the initial temperature.* Thus in a compressor consisting of several stages of equal isentropic efficiency, each succeeding stage will have to perform more work because it has to deal with a fluid of increased temperature delivered to it from the preceding stage. Thus the overall isentropic efficiency which is a useful measure of the overall performance of the machine cannot be used to compare two compressors having different pressure ratios. *Overall efficiency will be less for a compressor operating at a higher pressure ratio.*

For comparing the performance of compressors with different stages, the concept of polytropic efficiency is introduced.

Polytropic efficiency *is the isentropic efficiency of one stage of a multistage compressor.* This small stage efficiency is *constant for all stages of a compressor with infinite number of stages.*

Let us consider the compression process of a multistage compressor on *T-s* plot of Fig. 20.63.

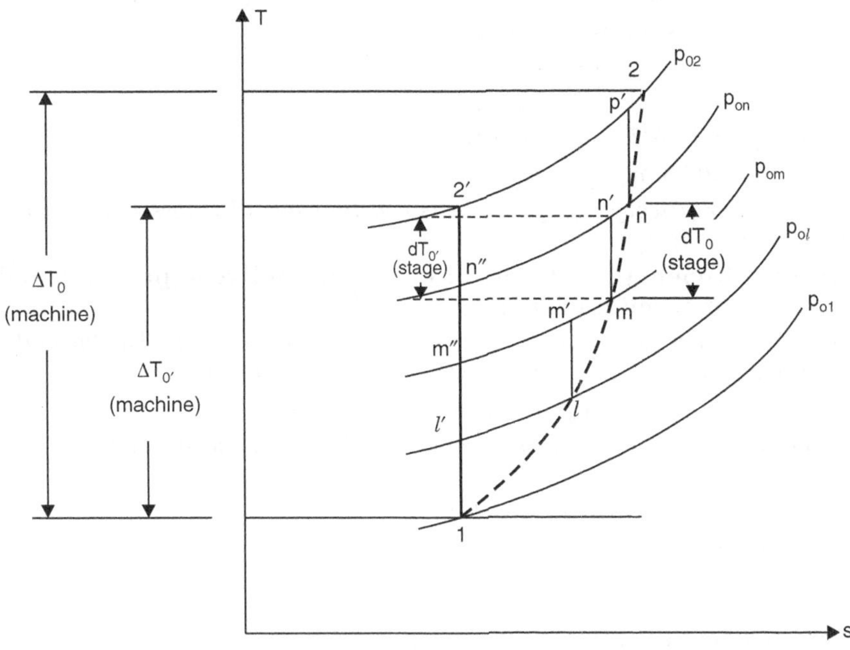

Fig. 20.63. Concept of polytropic efficiency.

The gas is being compressed from pressure p_{01} to p_{02} in *four stages of equal pressure ratio.* p_{ol}, p_{om} and p_{on} are the intermediate pressures.

Now by definition :

Overall isentropic efficiency (stagnation) for the *machine (m/c),*

$$\eta_{isen}(m/c) = \frac{\Delta T_0'}{\Delta T_0}$$

Stagnation isentropic efficiency for a *stage, (st)*

$$\eta_{isen(st)} = \frac{dT_0'}{dT_0}$$

The total actual temperature rise ΔT_0 can be represented as :

$$\Delta T_o = \frac{(\Delta T_o')_{machine}}{\eta_{isen\,(m/c)}} \quad ...(i)$$

or

$$\Delta T_o = \frac{\Sigma\,(dT_0')_{stage}}{\eta_{isen\,(st)}} \quad ...(ii)$$

Equating expressions (*i*) and (*ii*), we get

$$\frac{(\Delta T_o')_{machine}}{\eta_{isen\,(m/c)}} = \frac{\Sigma\,(dT_o')_{stage}}{\eta_{isen\,(st)}}$$

or

$$\frac{\eta_{isen\,(m/c)}}{\eta_{isen\,(st)}} = \frac{(\Delta T_0')_{machine}}{\Sigma\,(dT_0')_{stage}} \quad ...(20.98)$$

From Fig. 20.63, we have :

$$(\Delta T_0')_{machine} = (1\text{-}l') + (l'\text{-}m'') + (m''\text{-}n'') + (n''\text{-}2')$$

$$\Sigma(dT_0')_{stage} = (1\text{-}l') + (l\text{-}m') + (m\text{-}n') + (n\text{-}p')$$

On *T-s* plot, the constant pressure lines *'diverge' towards right*

i.e., $l\text{-}m' > l'\text{-}m''$

and $m\text{-}n' > m''\text{-}n''$

So we can say that $\Sigma(dT_o')_{stage} > (\Delta T_o')_{machine}$

Thus from eqn. 20.98, $\eta_{isen(st)} > \eta_{isen\ (m/c)}$

The *small stage efficiency* $\eta_{isen(st)}$ *which is constant for all stages is called polytropic efficiency and is designated by* η_p.

"Polytropic efficiency" in terms of entry and delivery pressure and temperatures and the ratio of specific heats :

Refer Fig. 20.63. The actual compression path 1-2 is *irreversible, but the end states 1 and 2 are in equilibrium and lie on the same polytropic path characterised by :*

$$p_o v_o^{\,n} = \text{constant } z_1$$

Let the corresponding reversible isentropic path 1-2′ with end states 1 and 2′ characterised by :

$$p_o v^{\gamma} = \text{constant } z_2$$

For irreversible path, we may write

$$p_o = z_1 \rho_o^{\,n}$$

On differentiation, we get $dp_o = z_1 n \rho_o^{\,n-1}.d\rho_o = n\dfrac{p_o}{\rho_o}\,d\rho_o$ ...(20.99)

Now, the characteristic gas equation may be written as :

$$\frac{p_o}{\rho_o} = RT_o \qquad \text{or} \qquad \rho_o = \frac{p_o}{RT_o}$$

On differentiation, we get $d\rho_o = \dfrac{1}{R}\left[\dfrac{T_o.dp_o - p_o.dT_o}{T_o^2}\right]$ ...(20.100)

From eqns. 20.99 and 20.100, we have

$$dp_o = n\frac{p_o}{\rho_o}\frac{1}{R}\left[\frac{T_o.dp_o - p_o.dT_o}{T_o^2}\right] = n\,.\,RT_o.\frac{1}{R}\left[\frac{T_o.dp_o - p_o.dT_o}{T_o^2}\right]$$

$$= n\left[\frac{T_o.dp_o - p_o.dT_o}{T_o}\right]$$

$$dp_o = n.dp_o - \frac{n.\,p_o.dT_o}{T_o}$$

or $$\frac{n.\,p_o.dT_o}{T_o} = n.dp_o - dp_o = dp_o(n-1)$$

or Actual stage temperature, $dT_o = dp_o\left(\dfrac{n-1}{n}\right)\dfrac{T_o}{p_o}$ ...(20.101)

A similar treatment to the ideal compression process following the law $p_o v_o^{\gamma} = z_2$ would give the stage isentropic temperature dT_o' expressed as

$$dT_o' = dp_o\left(\frac{\gamma-1}{\gamma}\right)\frac{T_o}{p_o} \qquad ...(20.102)$$

Thus, by the definition of polytropic or small stage efficiency, we get :

$$\eta_p = \frac{dT_o'}{dT_o} = \frac{dp_o\left(\frac{\gamma-1}{\gamma}\right)\frac{T_o}{p_o}}{dp_o\left(\frac{n-1}{n}\right)\frac{T_o}{p_o}}$$

or
$$\eta_p = \left(\frac{\gamma-1}{\gamma}\right)\left(\frac{n}{n-1}\right) \qquad ...(20.103)$$

Eqn. 20.103 gives the value of polytropic efficiency in terms of exponent n and the adiabatic exponent γ.

Substituting the value of $\frac{n}{n-1}$ from eqn. (20.101) into eqn. (20.103) we have :

$$\eta_p = \frac{\gamma-1}{\gamma}.\frac{dp_o}{p_o}.\frac{T_o}{dT_o}$$

or
$$\eta_p \frac{dT_o}{T_o} = \frac{\gamma-1}{\gamma}.\frac{dp_o}{p_o}$$

Integrating between the two end states 1 and 2, we get

$$\eta_p \ln\left(\frac{T_{02}}{T_{01}}\right) = \frac{\gamma-1}{\gamma}\ln\left(\frac{p_{02}}{p_{01}}\right)$$

or
$$\eta_p = \frac{\frac{\gamma-1}{\gamma}\ln\left(\frac{p_{02}}{p_{01}}\right)}{\ln\left(\frac{T_{02}}{T_{01}}\right)} = \frac{\ln\left(\frac{p_{02}}{p_{01}}\right)^{\frac{\gamma-1}{\gamma}}}{\ln\left(\frac{T_{02}}{T_{01}}\right)} \qquad ...(20.104)$$

The eqn. (20.104) gives the polytropic efficiency in terms of *pressure ratio* $\frac{p_{02}}{p_{01}}$, *temperature ratio* $\frac{T_{02}}{T_{01}}$ and the *ratio of specific heat* γ.

20.4.3.3.5. Flow Coefficient, Head or Work Coefficient, Deflection coefficient and Pressure Coefficient

1. Flow coefficient (ϕ_f). The flow coefficient of axial flow compressor stage is defined as,

$$\phi_f = \frac{C_{f1}}{C_{bl}}$$

Since
$$C_{bl} = C_{rw1} + C_{w1} = C_{f1}(\tan\beta_1 + \tan\alpha_1)$$

$$\phi_f = \frac{C_{f1}}{C_{f1}(\tan\beta_1 + \tan\alpha_1)} = \frac{1}{\tan\beta_1 + \tan\alpha_1} \qquad ...(20.105)$$

Also,
$$C_{f1} = C_{f2} = C_f$$

$\therefore$
$$\phi_f = \frac{C_{f2}}{C_{f2}(\tan\beta_2 + \tan\alpha_2)} = \frac{1}{\tan\beta_2 + \tan\alpha_2} \qquad ...(20.106)$$

2. Head or work coefficient (ϕ_h). *It is defined as the ratio of actual work done to the kinetic energy corresponding to the mean peripheral velocity*. Thus,

$$\phi_h = \frac{c_p\,\Delta T}{C_{bl}^2/2} = \frac{2\,(C_{w2} - C_{w1})}{C_{bl}} = 2\left(\frac{\tan\alpha_2 - \tan\alpha_1}{\tan\beta_2 + \tan\alpha_2}\right) \qquad ...(20.107)$$

3. Deflection coefficient (ϕ_{def}). It is defined as,

$$\phi_{def} = \frac{C_{bl}(C_{w2} - C_{w1})}{C_{bl}^2} = \frac{C_{w2} - C_{w1}}{C_{bl}} \quad \text{or} \quad \phi_h = 2\,\phi_{def} \quad ...(20.108)$$

4. Pressure coefficient (ϕ_p). *It is defined as the ratio of isentropic work done to kinetic energy corresponding to the peripheral velocity.* Thus,

$$\phi_p = \frac{c_p\,\Delta T_{isen}}{C_{bl}^2/2} = \eta_{isen}\,\phi_h \quad ...(20.109)$$

20.4.3.3.6. Pressure Increase in a Stage of an Axial Flow Compressor and Number of Stages

The pressure ratio is expressed as

$$\frac{p_2}{p_1} = \left[1 + \eta_{isen}\frac{T_2 - T_1}{T_1}\right]^{\gamma/(\gamma-1)} \quad ...(20.110)$$

Let T_1 and T_2 denote the temperature of the working fluid at inlet and outlet of *rotating blades*. Hence the temperature increase is

$$\Delta T_R = T_2 - T_1 = \frac{C_{r1}^2 - C_{r2}^2}{2 \times c_p}$$

$$\therefore \quad \frac{p_2}{p_1} = \left[1 + \eta_R\frac{\Delta T_R}{T_1}\right]^{\gamma/(\gamma-1)} \quad ...(20.111)$$

The temperature rise in *stationary blades* is given by,

$$\Delta T_s = T_3 - T_2 = \frac{C_2^2 - C_3^2}{2 \times c_p}$$

$$\therefore \quad \frac{p_3}{p_2} = \left[1 + \eta_s\frac{\Delta T_s}{T_2}\right]^{\gamma/(\gamma-1)}$$

Hence pressure rise in the stationary blades is

$$\Delta p_s = p_3 - p_2 = p_2\left[\left\{1 + \eta_s\frac{\Delta T_s}{T_2}\right\}^{\gamma/\gamma - 1} - 1\right]$$

The pressure increase in a stage is

$$\Delta p_{st} = \Delta p_R + \Delta p_s$$

and

$$\Delta T_{st} = \Delta T_R + \Delta T_s$$

The stagnation pressure ratio is given by

$$\frac{p_{o2}}{p_{o1}} = \left[1 + \eta_{o\,isen}\frac{\Delta T_o}{T_{o1}}\right]^{\gamma/(\gamma-1)}$$

If the work done per stage is assumed to be the same, then the number of stages (N) is given by,

$$N = \frac{\Delta T_o'}{(\Delta T_o')_{stage}} \quad ...(20.112)$$

If the pressure ratio per stage be the same, then

$$(r_p)_{stage} = \frac{p_{02}}{p_{01}} = \frac{p_{03}}{p_{02}} = \frac{p_{0(N+1)}}{p_{0N}}$$

The overall pressure ratio is given by

$$r_p = \left[(r_p)_{stage}\right]^N \quad ...(20.113)$$

or

$$N = \frac{\ln (r_p)}{\ln\left[(r_p)_{stage}\right]} \quad ...(20.114)$$

where $(r_p)_{stage}$ varies from 1.12 to 1.2.

20.4.3.3.7. Losses in Axial Flow Compressor Stage

In actual practice, various losses occur while the fluid flows through a compressor stage. The total pressure loss arises in three ways :

1. Profile losses on the surface of the blades.
2. Skin friction on the annulus walls.
3. Secondary flow losses.

The various losses represented on graph between stage efficiency and flow coefficient is shown in Fig. 20.64.

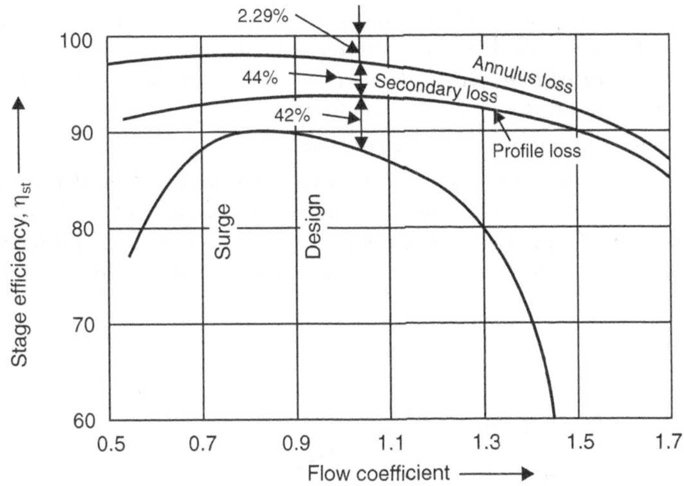

Fig. 20.64. Losses in compressor stage.

1. Profile losses on the surface of the blades :

- By profile losses, we mean the total pressure loss of two dimensional rectilinear cascade arising from the skin friction on the *surface* and due to the mixing of flow particles after the blades.
- These losses are usually determined experimentally.

2. Skin friction loss on the annulus walls :

- The wall friction total pressure losses arising from the skin friction on the annulus walls and the secondary losses are *difficult to analyse as boundary layer growth on these walls is a complex three-dimensional phenomena.*
- Empirical relations (by Howell, Haller) are available for calculating drag coefficient.

3. Secondary flow losses :

- In an axial flow compressor blade channels, certain secondary flows are *produced by combined effects of curvature and boundary layer.*
- *Secondary flow is produced when a streamwise components of velocity is developed*

from the deflection of an initially sheared flow. Such secondary flow occurs when a developed pipe flow enters a bend, when a sheared flow passes over an aerofoil of finite thickness or an aerofoil of finite lift or when a boundary level meets an obstacle normal to the surface over which it is flowing (*e.g.*, a wind blowing past a telegraph pole).

- One of most important engineering aspects of secondary flow occurs in *axial turbo-machinery aerodynamics where boundary layers growing on the casing and hub walls of the machines are deflected by rows of blades-stationary and rotating.*

20.4.3.3.8. Surging, Choking and Stalling—Compressor Characteristics

Surging. In axial flow and centrifugal compressors *"surging" is an unstable limit of operation.* ***Surging*** *is caused due to unsteady, periodic and reversal of flow through the compressor when the compressor has to operate at less mass flow rate than a predetermined value (a value corresponding to maximum pressure).* As the flow is drastically reduced than this predetermined value, this surge can reach such a magnitude as to endanger the compressor and in many cases *mechanical failures may result.* The alternating stresses to which the rotor of the machine is subjected during this irregular working condition, *may damage compressor bearings, rotor blading and scales.* Severe surge have been known to *bend the rotor shaft.*

Choking. When the *pressure ratio is unity* (*i.e.*, there is no compression), *theoretically mass flow rate becomes maximum.* This generally occurs when the Mach number corresponding to relative velocity at inlet becomes *sonic. The maximum mass flow rate possible in compressor is known as* ***choking flow.*** *"Choking" means fixed mass flow rate regardless of pressure ratio (i.e., characteristic becomes vertical).*

Fig. 20.65, shows the *compressor characteristics.*

In the compressor where the flow is against the pressure gradient the incidence loss due to incorrect fluid angles relative to the blades becomes more important and pressure ratio (r_p) falls sharply at conditions away from the design point. This loss, added to the friction loss which will increase with increase of mass flow rate, gives a pressure ratio-mass flow rate relation as shown in Fig. 20.65.

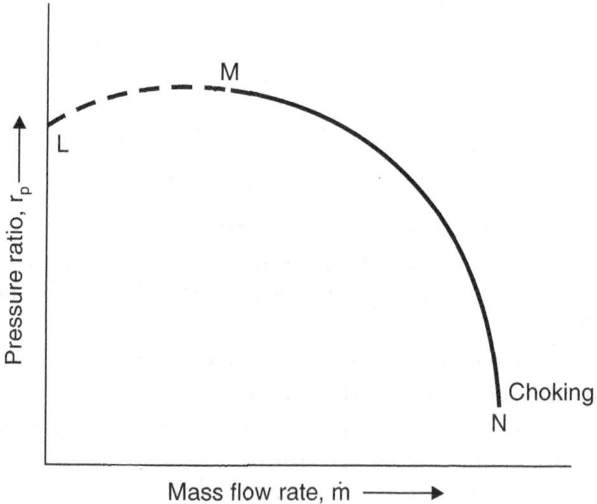

Fig. 20.65. The compressor characteristics.

- At point *N*, the compressor is *choked* and is *passing the maximum mass flow rate.*
- On the section *MN* of the curve the *flow is stable.* A fall in mass flow rate will result in a rise in pressure ratio which will tend to restore the fall.

- On the section *LM* of the curve, the flow is *not stable*. A fall in mass flow rate will be accompanied by a fall in pressure ratio. In this situation any small disturbance causing a check in mass flow will cause a fall in pressure ratio and the *flow may reverse at some point*. When the temporary disturbance is removed, the flow will pick up and it is found that small disturbances cause the flow to *oscillate rapidly*. The oscillations is noisy and can, if allowed to continue, cause structural damage in the compressor. It is called '*surge*' and the point *M* on the curve marks the limit of useful operation of the compressor. If a compressor is running normally at the point where surge usually commences it is possible to induce surge merely by passing the hand across the inlet. It is found that *compressor efficiency is highest at point adjacent to M* and it is therefore advisable to able to *operate as close to M as possible*.

Stalling. "*Stalling*" *of a stage of axial flow compressor is defined as the aerodynamic stall or the breakway of the flow from suction side of the blade aerofoil.* It may be due to *lesser flow rate* than designed value or due to *non-uniformity in the blade profile*. Thus stalling is *ahead phenomenon of surging*.

A multi-stage compressor may operate stable in the unsurged region with one or more of the stages stalled and rest of the stages unstalled. In other words, *stalling is a local phenomenon whereas surging is a complete system phenomena*.

20.4.3.3.9. Performance of Axial Flow Compressor

- Fig. 20.66 (*a*) shows the relationship between pressure ratio, power and efficiency *versus* flow rate for various values of speeds such as N_1, N_2, etc. At a *certain speed, efficiency increases as the flow rate increases and reaches a maximum value after which it decreases*. Accordingly as *the flow rate increases the power consumed also increases*.

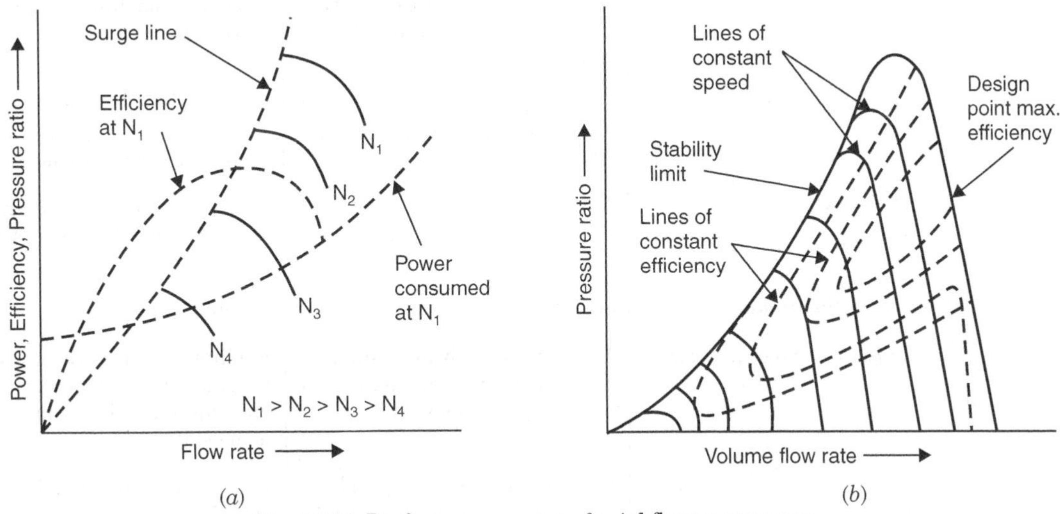

Fig. 20.66. Performance curves of axial flow compressor.

- Fig. 20.66 (*b*) shows the performance and constant efficiency curves.

— Such a plot does not take into account the varying inlet temperature and pressure. In addition to this, these plots cannot show the comparison of performance for similar compressors of different sizes. To, account for all these, the *performance curves are plotted with 'dimensionless parameters'*. These dimensionless parameters are : Pressure ratio $\frac{p_2}{p_1}$; speed parameter, $\frac{N_1}{\sqrt{T_1}}$ and flow parameter $\frac{m\sqrt{T_1}}{p_1}$, Refer Fig. 20.67 (*a* and *b*).

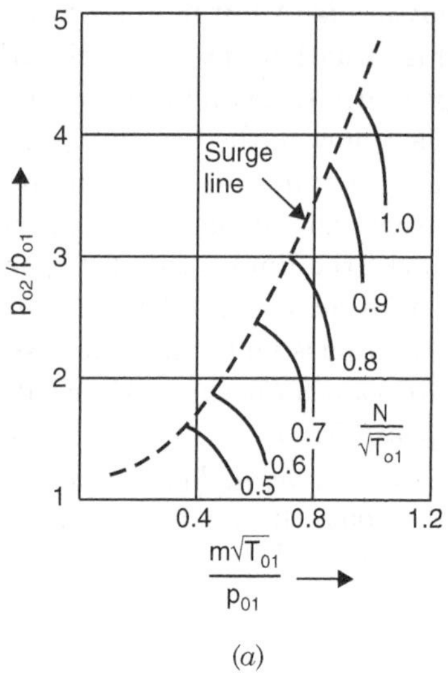

(a)

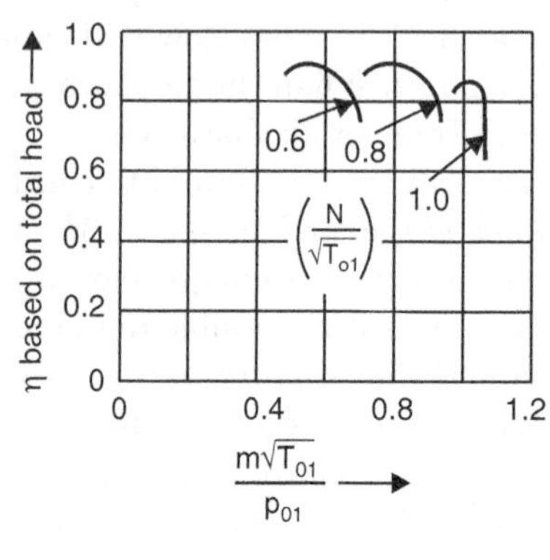

(b)

Fig. 20.67

20.5. COMPARISON BETWEEN RECIPROCATING AND CENTRIFUGAL COMPRESSORS

S. No.	Aspects	Reciprocating compressors	Centrifugal compressors
1.	*Vibration problems*	Greater vibration problems (due to the presence of reciprocating parts the machine is poorly balanced)	Less vibrational problems since the machine does not have reciprocating parts.
2.	*Mechanical efficiency*	Lower (due to the presence of several sliding or bearing members)	Higher comparatively (due to the absence of numerous sliding or bearing members)
3.	*Installed first-cost*	Higher	Lower (where pressure and volume conditions are favourable.
4.	*Pressure ratio per stage*	About 5 to 8	About 3 to 4.5.
5.	*Capability to deliver pressure*	High pressure (By multistaging, high delivery pressure upto 5000 atm. may be achieved).	Medium pressure (By multistaging, the delivery pressure upto 400 atm. may be achieved).
6.	*Capability of delivering volume of air/gas*	Small (By using multicylinders, the volume may be increased).	Greater (per unit of building space).
7.	*Flexibility in capacity and pressure range*	Greater	No flexibility in capacity and pressure range.
8.	*Maintenance expenses*	Higher	Lower
9.	*Continuity of service*	Lesser	Greater
10.	*Compression efficiency*	Higher, at compression ratio above 2.	Higher, at compression ratio less than 2.

11.	*Adaptability*	Adaptability to low speed drive	Adaptability to high speed, low maintenance cost drivers such as turbines
12.	*Operating attention*	More	Less
13.	*Mixing of working fluid with lubricating oil*	Always a chance	No chance
14.	*Suitability*	For low, medium and high pressures and low and medium gas volumes.	For low and medium pressures and large gas volumes.

20.6. COMPARISON BETWEEN RECIPROCATING AND ROTARY AIR COMPRESSORS

S. No.	Aspects	Reciprocating air compressors	Rotary air compressors
1.	*Suitability*	Suitable for low discharge of air at high pressure	Suitable for handling large volumes of air at low pressures.
2.	*Operational speed*	Low	Usually high
3.	*Air supply*	Pulsating	Continuous
4.	*Balancing*	Cyclic vibrations occur	Less vibrations
5.	*Lubricating system*	Generally complicated	Generally simple lubrication systems are required
6.	*Quality of air delivered*	Generally contaminated with oil	Air delivered is relatively more clean.
7.	*Air compressor size*	Large for the given discharge	Small for same discharge
8.	*Free air handled*	250—300 m^3/min	2000—3000 m^3/min
9.	*Delivery pressure*	High	Low
10.	*Usual standard of compression*	Isothermal compression	Isentropic compression

20.7. COMPARISON BETWEEN CENTRIFUGAL AND AXIAL FLOW COMPRESSORS

S. No.	Aspects	Centrifugal Compressors	Axial Flow Compressors
1.	*Type of flow*	Axial (Parallel to the direction of axis of the machine)	Radial
2.	*Pressure ratio per stage*	High, about 4.5 : 1. Thus unit is *compact*—In supersonic compressors, the pressure ratio is about 10 but at the cost of efficiency. Operation is *not* so difficult and risky.	Low, about 1.2 : 1. This is due to absence of centrifugal action. To achieve the pressure ratio equal to that per stage in centrifugal compressor 10 to 20 stages are required. Thus the unit is *less compact* and *less rugged*.
3.	*Isothermal efficiency*	About 80 to 82%	About 86 to 88% (with modern aerofoil blades)
4.	*Frontal area*	Larger	Smaller (This makes the axial flow compressors *more suitable for jet engines due to less drag*).

5.	*Flexibility of operation*	More (due to adjustable pre-whirl and diffuser vanes)	Less
6.	*Part load performance*	Better	Poor
7.	*Effect of deposit formation on the surface of impeller rotor*	*Performance not adversely affected*	*Performance adversely affected*
8.	*Starting torque required*	Low	High
9.	*Suitability for multistaging*	Slightly difficult	More suitable for multistaging
10.	*Delivery pressure possible*	Upto 400 bar	upto 20 bar
11.	*Applications*	Used in blowing engines in steel mills, low pressure refrigeration, big central air conditioning plants, fertiliser and industry, supercharging I.C. engines, gas pumping in long distance pipe lines etc. — Previously it was used in jet engines	Mostly used in jet engines (due to higher efficiency and smaller frontal area). Also preferred in power plant gas turbines and steel mills.
12.	*Efficiency vs. speed curve*	More flat (Fig. 24.68)	Less flat comparatively Fig. 24. 68

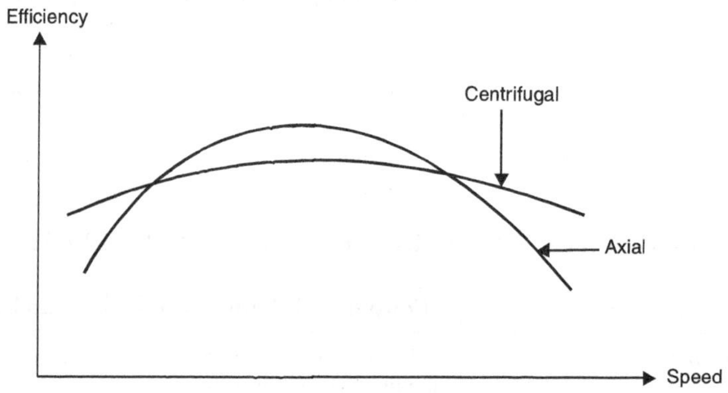

Fig. 20.68

Example 20.44. *The following data relate to an axial flow compressor :*

$$C_{bl} = 240 \text{ m/s}, C_f = 190 \text{ m/s}, \alpha_1 = 45° ; \alpha = 14° ; \rho = 1 \text{ kg/m}^3$$

Calculate :

(*i*) *The pressure rise* (*ii*) *The work done per kg of air.*

Solution. *Given* : C_{bl} = 240 m/s ; C_f = 190 m/s ; α_1 = 45° ; α_2 = 14°, ρ = 1 kg/m^3

(*i*) **The pressure rise, Δp :**

The pressure rise through a ring of rotating blades,

$$\Delta p = \frac{\rho}{2} C_f^2(\tan^2 \alpha_1 - \tan^2 \alpha_2) \quad \text{...[Eqn. (20.92)]}$$

$$= \frac{1}{2} \times \frac{(190)^2}{10^5} [(\tan 45°)^2 - (\tan 14°)^2] = \mathbf{0.169 \text{ bar.}} \quad \textbf{(Ans.)}$$

(*ii*) **The work done per kg of air, W :**

$$W = C_{bl}\, C_f\,(\tan \alpha_1 - \tan \alpha_2)$$

$$= \frac{240 \times 190}{10^3}(\tan 45° - \tan 14°) = \mathbf{34.23\ kW.} \quad \textbf{(Ans.)}$$

Example 20.45. *An axial flow compressor having eight stages and with 50% reaction design compresses air in the pressure ratio of 4 : 1. The air enters the compressor at 20°C and flows through it with a constant speed of 90 m/s. The rotating blades of compressor rotate with a mean speed of 180 m/s. Isentropic efficiency of the compressor may be taken as 82%. Calculate :*

(*i*) *Work done by the machine* (*ii*) *Blades angles.*

Assume $\gamma = 1.4$ *and* $c_p = 1.005$ *kJ/kg K.*

Solution. $$\eta_{isen} = \frac{T_2' - T_1}{T_2 - T_1}$$

Also $$\frac{T_2'}{T_1'} = \left(\frac{p_2}{p_1}\right)^{\frac{\gamma - 1}{\gamma}} = (4)^{\frac{1.4-1}{1.4}} = (4)^{\frac{0.4}{1.4}} = 1.486$$

∴ $$T_2' = (20 + 273) \times 1.486 = 435.4 \text{ K}$$

∴ $$\eta_{isen} = \frac{435.4 - 293}{T_2 - 293} \quad \text{or} \quad 0.82 = \frac{142.4}{T_2 - 293}$$

∴ $$T_2 = \frac{142.4}{0.82} + 293 = 466.6 \text{ K}$$

Work required/kg $= c_p(T_2 - T_1) = 1.005(466.6 - 293) = \mathbf{174.47\ kJ/kg.}$ **(Ans.)**

Now, work done/kg $= \text{Number of stages} \times C_{bl}(C_{w_2} - C_{w_1})$

$$174.47 = 8 \times C_{bl}\, C_f\,(\tan \alpha_2 - \tan \alpha_1) \quad \text{[Refer Fig. 20.62]}$$

∴ $$\tan \alpha_2 - \tan \alpha_1 = \frac{174.47 \times 1000}{8 \times 180 \times 90} = 1.346$$

For 50% reaction blading, $\alpha_2 = \beta_1$ and $\alpha_1 = \beta_2$

∴ $$1.346 = \tan \beta_1 - \tan \alpha_1$$

Now, $$\tan \alpha_1 + \tan \beta_1 = \frac{C_{bl}}{C_f} = \frac{180}{90} = 2$$

i.e., $$\tan \beta_1 - \tan \alpha_1 = 1.346 \quad ...(i)$$

$$\tan \beta_1 + \tan \alpha_1 = 2 \quad ...(ii)$$

From (*i*) and (*ii*), we get

$$2 \tan \beta_1 = 3.346$$

∴ $$\beta_1 = \mathbf{59.1°} = \alpha_2. \quad \textbf{(Ans.)}$$

and $$\alpha_1 = \mathbf{18.1°} = \beta_2. \quad \textbf{(Ans.)}$$

Example 20.46. *An axial flow compressor with an overall isentropic efficiency of 85% draws air at 20°C and compresses it in the pressure ratio of 4 : 1. The mean blade speed and flow velocity are constant throughout the compressor. Assuming 50% reaction blading and taking blade velocity as 180 m/s and work input factor as 0.82, calculate :*

(*i*) *Flow velocity*

(*ii*) *Number of stages*

Take $\alpha_1 = 12°$, $\beta_1 = 42°$.

Solution. Given : $\eta_{isen} = 85\%$, $T_1 = 20 + 273 = 293$ K

Pressure ratio, $\frac{p_2}{p_1} = 4$, $C_{bl} = 180$ m/s

Work input factor $= 0.82$

$$\frac{T_2'}{T_1} = \left(\frac{p_2}{p_1}\right)^{\frac{\gamma-1}{\gamma}} = (4)^{\frac{1.4-1}{1.4}} = 1.486$$

$\therefore$ $T_2' = 293 \times 1.486 = 435.4$ K

Now $\eta_{isen} = \frac{T_2' - T_1}{T_2 - T_1}$

$$0.85 = \frac{435.4 - 293}{T_2 - 293}$$

$\therefore$ $T_2 = 460.5$ K

Theoretical work required per kg

$$= c_p(T_2 - T_1) = 1.005(460.5 - 293) = 168.33 \text{ kJ/kg}$$

From velocity Δs (Fig. 20.62)

$$\frac{C_{bl}}{C_f} = \tan \alpha_1 + \tan \beta_1 = \tan 12° + \tan 42° = 0.212 + 0.9 = 1.112$$

$\therefore$ $C_f = \frac{C_{bl}}{1.112} = \frac{180}{1.112} =$ **161.8 m/s. (Ans.)**

Work done per stage $= C_{bl}(C_{w_2} - C_{w_1}) \times$ work input factor

Now, $C_{w_2} = C_f \tan \alpha_2 = 161.8 \tan 42° = 145.7$ m/s $(\because \alpha_2 = \beta_1)$

and $C_{w_1} = C_f \tan \alpha_1 = 161.8 \tan 12° = 34.4$ m/s

$\therefore$ Work done per stage $= 180(145.7 - 34.4) \times 0.82 \times 10^{-3}$ kJ/kg $= 16.4$ kJ/kg

$\therefore$ Number of stages $= \frac{168.33}{16.43} \simeq 10$

i.e., **Number of stages = 10. (Ans.)**

Example 20.47. *In an eight stage axial flow compressor, the overall stagnation pressure ratio achieved is 5 : 1 with an overall isentropic efficiency of 92 per cent. The inlet stagnation temperature and pressure at inlet are 290 K and 1 bar. The work is divided equally between the stages. The mean blade speed is 160 m/s and 50% reaction design is used. The axial velocity through the compressor is constant and is equal to 90 m/s. Calculate :*

(*i*) *The blade angles.* (*ii*) *The power required.*

Solution. *Given* : $N = 8$; $r_p = 5 : 1$; $\eta_{isen} = 92\%$; $T_{01} = 290$ K ; $p_{01} = 1$ bar ; $C_{bl} = 160$ m/s ; Degree of reaction = 50% ; $C_f = 90$ m/s.

(*i*) **The blade angles, $\alpha_1, \beta_1, \alpha_2, \beta_2$:**

Refer to Fig. 20.69 for velocity diagrams. Since the degree of reaction is 50% reaction, the *blades are symmetrical* and hence the velocity diagrams are *identical.* Thus

$$\alpha_1 = \beta_2 \text{ and } \alpha_2 = \beta_1$$

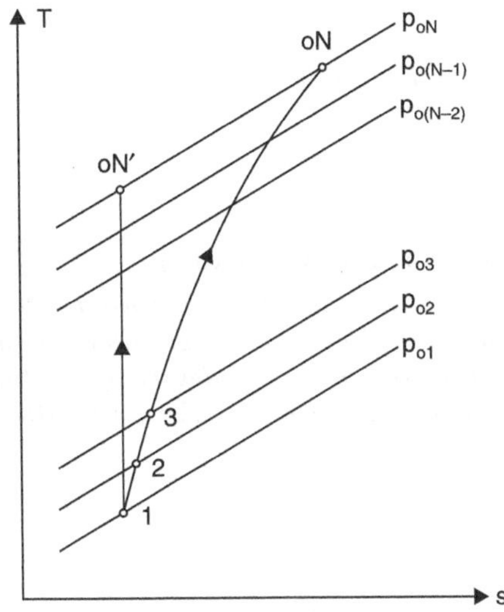

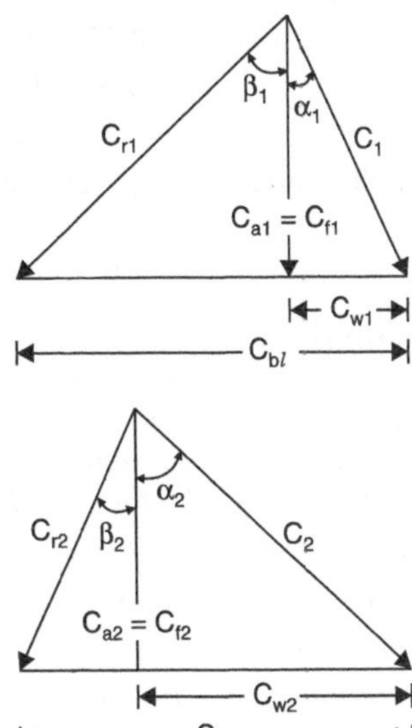

Fig. 20.69

Let suffix N denotes the *number of stages.*

With isentropic compression the temperature of air leaving the compressor stage is

$$T_{0N}' = T_{01}\left(\frac{p_{0N}}{p_{01}}\right)^{(\gamma-1/\gamma)} = 290 \times (5)^{\frac{1.4-1}{1.4}} = 459.3 \text{ K}$$

But $$\eta_{isen} = \frac{T_{0N}' - T_{01}}{T_{0N} - T_{01}}$$

$$0.92 = \frac{459.3 - 290}{T_{oN} - 290}$$

$\therefore$ $$T_{oN} = \frac{459.3 - 290}{0.92} + 290 = 474 \text{ K}$$

The work consumed by the compressor

$$= c_p(T_{oN} - T_{o1}) = (C_{w2} - C_{w1})\, C_{bl} \times \text{N}$$

or $$c_p(T_{oN} - T_{o1}) = C_f\,(\tan \alpha_2 - \tan \alpha_1)\, C_{bl}\,.\, N$$

$\therefore$ $$\tan \alpha_2 - \tan \alpha_1 = \frac{c_p(T_{oN} - T_{o1})}{C_f . C_{bl} . N} = \frac{1.005\,(474 - 290) \times 10^3}{90 \times 160 \times 8} = 1.605 \quad ...(i)$$

From velocity triangles, we have

$$\frac{C_{bl}}{C_f} = \tan \alpha_1 + \tan \beta_1 = \frac{160}{90} = 1.778 \quad ...(ii)$$

Adding (*i*) and (*ii*), we get

$$\tan \beta_1 = \frac{1.605 + 1.778}{2} = 1.6915 \quad (\because \ \alpha_1 = \beta_1)$$

or $$\beta_1 = \tan^{-1}(1.6915) = 59.4°$$

$\therefore$ $$\boldsymbol{\beta_1 = \alpha_2 = 59.4°.} \quad \textbf{(Ans.)}$$

Putting the value of $\tan \beta_1$ in (*ii*), we have

$$\tan \alpha_1 + 1.6915 = 1.778$$

or $$\tan \alpha_1 = 0.0865 \quad \text{or} \quad \alpha_1 = \tan^{-1}(0.0865) = 4.94°$$

$\therefore$ $$\alpha_1 = \beta_2 = \mathbf{4.94°.} \quad \textbf{(Ans.)}$$

(*ii*) **The power required by compressor P :**

$$P = \dot{m}\, c_p(T_{0N} - T_{01})$$
$$= 1 \times 1.005(474 - 290) = \mathbf{184.9\ kW.} \quad \textbf{(Ans.)}$$

Example 20.48. *In an axial flow compressor, the overall stagnation pressure ratio achieved is 4 with overall stagnation isentropic efficiency 85 per cent. The inlet stagnation pressure and temperature are 1 bar and 300 K. The mean blade speed is 180 m/s. The degree of reaction is 0.5 at the mean radius with relative air angles of 12° and 32° at the rotor inlet and outlet respectively. The work done factor is 0.9. Calculate :*

(*i*) *Stagnation polytropic efficiency.*

(*ii*) *Number of stages.*

(*iii*) *Inlet temperature and pressure.*

(*iv*) *Blade height in the first stage if the hub-tip ratio is 0.42, mass flow rate 19.5 kg/s.*

Solution. *Given* : $r_p = \dfrac{p_{0N}}{p_{01}} = 4$; $\eta_{isen} = 85\%$; $p_{01} = 1$ bar ; $T_{01} = 300$ K

$C_u = 180$ m/s ; Degree of reaction, $R_d = 0.5$. Work done factor = 0.9 ; Hub-tip ratio = 0.42 ; $\dot{m} = 19.5$ kg/s.

Refer Fig. 20.70. For 50% reaction, the inlet and outlet velocity diagrams are identical. Hence $\alpha_1 = \beta_2 = 12°$; $\alpha_2 = \beta_1 = 32°$.

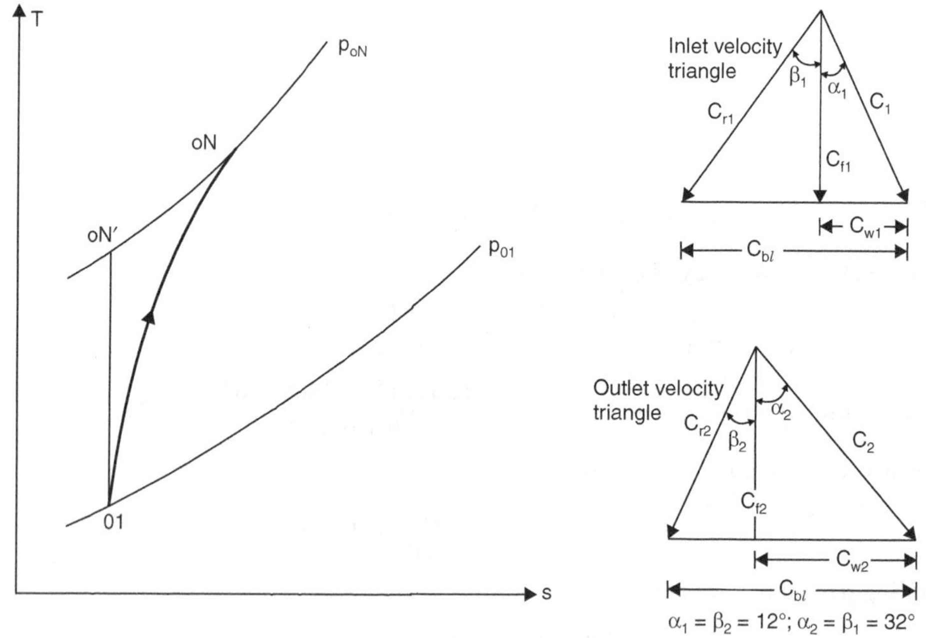

Fig. 20.70

(*i*) **Stagnation polytropic efficiency, η_p :**

The temperature at the end of the compression stage due to isentropic expansion is

$$T_{0N}' = T_{01}\left(\frac{p_{0N}'}{p_{01}}\right)^{\frac{\gamma-1}{\gamma}} = 300\ (4)^{\frac{1.4-1}{1.4}} = 445.8\ \text{K}$$

$$\eta_{0\ isen} = 0.85 = \frac{T_{0N}' - T_{01}}{T_{0N} - T_{01}} = \frac{445.8 - 300}{T_{0N} - 300}$$

$$\therefore \quad T_{0N} = \frac{445.8 - 300}{0.85} + 300 = 471.5\ \text{K}$$

Now
$$\eta_p = \frac{\ln\left(\frac{p_{0N}}{p_{01}}\right)^{\frac{\gamma-1}{\gamma}}}{\ln\left(\frac{T_{0N}}{T_{01}}\right)} \qquad \text{...[Eqn. (20.104)]}$$

or
$$\eta_p = \frac{\ln\ (4)^{\frac{1.4-1}{1.4}}}{\ln\left(\frac{471.5}{300}\right)} = \frac{0.396}{0.452} = 0.8761 \quad \text{or} \quad \mathbf{87.61\%.\ \ (Ans.)}$$

(*ii*) **Number of stages, N :**

From the configuration of velocity triangles,

$$\frac{C_{bl}}{C_f} = \tan\alpha_1 + \tan\beta_1 = \tan 12° + \tan 32° = 0.8374$$

$$\therefore \quad C_f = \frac{C_{bl}}{0.8374} = \frac{180}{0.8374} = 214.9\ \text{m/s}$$

Now,
$$C_{w1} = C_f \tan\alpha_1 = 214.9 \times \tan 12° = 45.68\ \text{m/s}$$
$$C_{w2} = C_f \tan\alpha_2 = 214.9 \times \tan 32° = 134.28\ \text{m/s}$$

Work consumed per stage

$$= C_{bl}(C_{w2} - C_{w1}) \times \text{work done factor} \qquad \text{[Eqn. 20.87)]}$$
$$= \frac{180\ (134.28 - 45.68) \times 0.9}{1000} = 14.35\ \text{kJ/kg}$$

Total work consumed by the compressor

$$= c_p(T_{oN} - T_{o1}) = 1.005(471.5 - 300) = 172.36\ \text{kJ/kg}$$

$$\therefore \quad \text{Number of stages, N} = \frac{172.36}{14.35} = \mathbf{12\ stages.\ \ (Ans.)}$$

(*iii*) **Inlet temperature and pressure, T_1, p_1 :**

The absolute velocity C_1 at exit from the guide vanes and approaching to moving blades of first stage,

$$C_1 = \frac{C_f}{\cos\alpha_1} = \frac{214.9}{\cos 12°} = 219.7\ \text{m/s}$$

Temperature,
$$T_1 = T_{o1} - \frac{C_1^2}{2c_p} = 300 - \frac{(219.7)^2}{2 \times 1.005 \times 1000} = \mathbf{276\ K.\ \ (Ans.)}$$

Assuming reversible flow through the guide vanes put ahead of the first stage,

$$\frac{p_1}{p_{o1}} = \left(\frac{T_1}{T_{o1}}\right)^{\frac{\gamma}{\gamma-1}}$$

or
$$p_1 = 1 \times \left(\frac{276}{300}\right)^{\frac{1.4}{1.4-1}} = \mathbf{0.7469\ bar.\ \ (Ans.)}$$

(*iv*) **Blade height in the first stage, l :**

The density of air approaching to first stage,

$$\rho_1 = \frac{p_1}{RT_1} = \frac{0.7469 \times 10^5}{287 \times 276} = 0.9429 \text{ kg/m}^3$$

From the continuity equation,

$$\rho_1 A_1 C_f = \dot{m} = 19.5$$

$$0.9429 \times \pi r_1^2[1 - (0.42)^2] \times 241.9 = 19.5$$

or $$r_1 = \left[\frac{19.5}{0.9429 \times \pi \times [1 - (0.42)^2] \times 241.9}\right]^{1/2} = 0.1818 \text{ m or } 18.18 \text{ cm}$$

But $$\frac{r_h}{r_1} = 0.42 \quad \therefore \quad r_h = 18.18 \times 0.42 = 7.636 \text{ cm}$$

Hence height of the blade in the first stage,

$$l = r_1 - r_h = 18.18 - 7.636 = \mathbf{10.544 \text{ cm.}} \quad \textbf{(Ans.)}$$

Example 20.49. *A multistage axial flow compressor delivers 20 kg/s of air. The inlet stagnation condition is 1 bar and 17°C. The power consumed by the compressor is 4350 kW.*

Calculate :

(*i*) *Delivery pressure.*

(*ii*) *Number of stages.*

(*iii*) *Overall isentropic efficiency of the compressor.*

Assume temperature rise in the first stage is 15°C, the polytropic efficiency of compression is 88% and the stage stagnation pressure ratio is constant.

Solution. *Given :* $\dot{m}$ = 20 kg/s ; p_{o1} = 1 bar ; T_{o1} = 17 + 273 = 290 K ; P = 4350 kW ; T_{o2} = 15 + 290 = 305 K ; η_p = 88%.

(*i*) **Delivery pressure, P_{oN} :**

Now, $$\eta_p = \frac{\ln\left(\frac{p_{o2}}{p_{o1}}\right)^{\frac{\gamma-1}{\gamma}}}{\ln\left(\frac{T_{o2}}{T_{o1}}\right)} \quad \text{or} \quad 0.88 = \frac{\ln\left(\frac{p_{o2}}{p_{o1}}\right)^{0.2857}}{\ln\left(\frac{305}{290}\right)}$$

or $$\ln\left(\frac{p_{o2}}{p_{o1}}\right)^{0.2857} = 0.88 \times \ln\left(\frac{305}{290}\right) = 0.04438$$

or $$\left(\frac{p_{o2}}{p_{o1}}\right)^{0.2857} = e^{0.04438} = 1.0454 \quad \text{or} \quad \frac{p_{o2}}{p_{o1}} = (1.0454)^{\frac{1}{0.2857}} = 1.168$$

The power required by the compressor,

$$P = \dot{m} \times c_p \times (T_{oN} - T_{o1})$$

$$4350 = 20 \times 1.005 \times (T_{oN} - 290)$$

or $$T_{oN} = \frac{4350}{20 \times 1.005} + 290 = 506.4 \text{ K}$$

Also, the polytropic efficiency of the compressor is given as

$$\eta_p = \left(\frac{\gamma - 1}{\gamma}\right)\left(\frac{n}{n-1}\right) \quad ...(20.103)$$

or $$0.88 = \left(\frac{1.4-1}{1.4}\right)\left(\frac{n}{n-1}\right)$$ or $$\frac{n}{n-1} = \frac{0.88 \times 1.4}{(1.4-1)} = 3.08$$

During polytropic compression from first stage to last, we have

$$\frac{p_{oN}}{p_{o1}} = \left(\frac{T_{oN}}{T_{o1}}\right)^{\frac{n}{n-1}} = \left(\frac{506.4}{290}\right)^{3.08} = 5.567$$

$\therefore$ $p_{oN} = 1 \times 5.567 =$ **5.567 bar. (Ans.)**

(*ii*) **Number of stages, N :**

As the pressure ratio for each stage is the same hence

$$\frac{p_{02}}{p_{01}} = \frac{p_{03}}{p_{02}} = \frac{p_{04}}{p_{03}} = \ldots\ldots \frac{p_{0N}}{p_{0(N-1)}}$$

where suffix N indicates the number of stages.

$$\therefore \quad \left(\frac{p_{02}}{p_{01}}\right)^N = \frac{p_{0N}}{p_{01}}$$

Taking log on both the sides, we get

$$N = \ln\left(\frac{p_{o2}}{p_{o1}}\right) = \ln\left(\frac{p_{oN}}{p_{o1}}\right)$$

or $$N = \frac{\ln\left(\frac{p_{oN}}{p_{o1}}\right)}{\ln\left(\frac{p_{o2}}{p_{o1}}\right)} = \frac{\ln(5.567)}{\ln(1.168)} = 11. \quad \textbf{(Ans.)}$$

(*iii*) **Overall isentropic efficiency of the compressor, $(\eta_{overall})_{isen}$:**

With isentropic compression the temperature of air leaving the compressor is,

$$T_{oN}' = T_{o1}\left(\frac{p_{oN}}{p_{o1}}\right)^{\frac{\gamma-1}{\gamma}} = 290\,(5.567)^{\frac{1.4-1}{1.4}} = 473.6 \text{ K}$$

$$\therefore \quad (\eta_{overall})_{isen} = \frac{T_{oN}' - T_{o1}}{T_{oN} - T_{o1}} = \frac{473.6-290}{506.4-290} = 0.8484 \text{ or } \mathbf{84.84\%.} \quad \textbf{(Ans.)}$$

HIGHLIGHTS

1. An air compressor takes in atmospheric air, compresses it and delivers the high pressure air to a storage vessel from which it may be conveyed by a pipeline to wherever the supply of compressed air is required.
2. Air and gas compressors are classified into two main types :
   (*i*) Reciprocating compressor (*ii*) Rotary compressors.
3. **Single stage compressor**

   Equation for work (neglecting clearance),

$$W = \frac{n}{n-1} mRT_1 \left\{\left(\frac{p_2}{p_1}\right)^{\frac{n-1}{n}} - 1\right\}$$

Equation for work (with clearance volume),

$$W = \frac{n}{n-1} p_1 (V_1 - V_4) \left\{ \left(\frac{p_2}{p_1} \right)^{\frac{n-1}{n}} - 1 \right\}.$$

4. **'Volumetric efficiency'** of a compressor is the ratio of free air delivered to the displacement of the compressor.

$$\eta_{vol.} = 1 + k - k \left(\frac{p_3}{p_4} \right)^{\frac{1}{n_e}} = 1 + k - k \left(\frac{p_2}{p_1} \right)^{\frac{1}{n_e}}$$

$$= 1 + k - k \left(\frac{p_4}{p_3} \right) = 1 + k - k\, (V_1/V_2)$$

where, $k = \frac{V_c \text{ (clearance volume)}}{V_s \text{ (swept volume)}}$ = Clearance ratio.

5. Multi-stage compression is very efficient and is now a days almost universally adopted except for compressors where the overall pressure rise required is small.
6. In a two-stage compressor efficiency will be maximum when

$$\frac{p_2}{p_1} = \frac{p_3}{p_2} \quad \text{or} \quad p_2 = \sqrt{p_1 p_3}$$

For x-stage compressor the work to be supplied,

$$W = \frac{xn}{n-1} p_1 V_1 \left[\left(\frac{p_{x+1}}{p_1} \right)^{\frac{n-1}{xn}} - 1 \right]$$

This equation is very important, since it applies to any type of compressor or motor, and even to vapour engines, provided n = or < γ.

7. An **air-turbine** is valveless, small in size, light in weight, and requires no internal lubrication, but air friction is high, and any dampness in the air causes rapid deterioration of the blading at low temperatures.
8. **Displacement compressors** are those compressors in which air is compressed by being trapped in the reduced space formed by two sets of engaging surfaces.
9. **Steady-flow compressors** are those compressors in which compression occurs by transfer of kinetic energy from a rotor.
10. **Slip factor** is defined as the ratio of actual whirl component and ideal whirl component.
11. **Pressure co-efficient** is defined as the ratio of isentropic work to Euler work.
12. **Degree of reaction (R_d)** is defined as the ratio of pressure rise in the compressor stage

i.e.,
$$R_d = \frac{\text{Pressure rise in the rotor blades}}{\text{Pressure rise in the stage}}.$$

OBJECTIVE TYPE QUESTIONS

Choose the Correct Answer :

1. The work input to air compressor is minimum if the compression law followed is

(a) $pV^{1.35} = C$ (b) Isothermal $pV = C$

(c) Isentropic $pV^{\gamma} = C$ (d) $pV^{1.2} = C$.

2. For reciprocating air compressor the law of compression desired is isothermal and that may be possible by
   (a) very low speeds
   (b) very high speeds
   (c) any speed as speed does not affect the compression law
   (d) none of the above.
3. Work input to the air compressor with 'n' as index of compression
   (a) increases with increase in value of n
   (b) decreases with increase in value of n
   (c) remains same whatever the value of n
   (d) first increases and then decreases with increase of value of n.
4. Work done in a single-stage, single-acting air compressor without clearance per kg of air delivered when the compression process is isothermal is given by
   (a) $\frac{n}{n-1} p_1 v_1 \left[\left(\frac{p_2}{p_1}\right)^{\frac{n-1}{n}} - 1\right]$
   (b) $\frac{\gamma}{\gamma-1} p_1 v_1 \left[\left(\frac{p_2}{p_1}\right)^{\frac{\gamma-1}{\gamma}} - 1\right]$
   (c) $p_1 v_1 \log_e \frac{p_2}{p_1}$
   (d) $\frac{n}{n-1} p_1 v_1 \log_e \frac{p_2}{p_1}$.
5. The clearance volume in reciprocating air compressors is provided to
   (a) to reduce the work done per kg of air delivered
   (b) to increase the volumetric efficiency of the compressor
   (c) to accommodate valves in the head of the compressor
   (d) to create turbulence in the air to be delivered.
6. With increase in clearance volume, the ideal work of compressing 1 kg of air
   (a) increases
   (b) decreases
   (c) remains same
   (d) first increases and then decreases.
7. In reciprocating air compressors the clearance ratio is given by
   (a) $\frac{\text{Total volume of cylinder}}{\text{Clearance volume}}$
   (b) $\frac{\text{Swept volume of cylinder}}{\text{Clearance volume}}$
   (c) $\frac{\text{Clearance volume}}{\text{Swept volume of cylinder}}$
   (d) $\frac{\text{Clearance volume}}{\text{Total volume of cylinder}}$.
8. With suction pressure being atmospheric, increase in delivery pressure with fixed clearance volume
   (a) increase volumetric efficiency
   (b) decreases volumetric efficiency
   (c) does not change volumetric efficiency
   (d) first increases volumetric efficiency and then decreases it.
9. Mechanical efficiency of reciprocating air compressor is expressed as
   (a) $\frac{\text{B.P.}}{\text{I.P.}}$
   (b) $\frac{\text{I.P.}}{\text{B.P.}}$
   (c) $\frac{\text{F.P.}}{\text{B.P.}}$
   (d) $\frac{\text{F.P.}}{\text{I.P.}}$.
10. For the same overall pressure ratio, the leakage of air past the piston for multi-stage compression as compared to single-stage compression, is
   (a) more
   (b) less
   (c) constant
   (d) may be more or less.

11. Work done in a two stage reciprocating air compressor with imperfect cooling is given by

(a) $\frac{n}{n-1} p_1V_1\left[\left(\frac{p_2}{p_1}\right)^{\frac{n-1}{n}} - 1\right] + \frac{n}{n-1} p_2V_2\left[\left(\frac{p_3}{p_2}\right)^{\frac{n-1}{n}} - 1\right]$

(b) $\frac{2n}{n-1} p_1V_1\left[\left(\frac{p_2}{p_1}\right)^{\frac{n-1}{n}} - 1\right]$

(c) $\frac{2n}{n-1} p_1V_1\left[\left(\frac{p_3}{p_1}\right)^{\frac{n-1}{2n}} - 1\right]$

(d) $\frac{n-1}{2n} p_1V_1\left[\left(\frac{p_3}{p_1}\right)^{\frac{n-1}{2n}} - 1\right]$.

12. Work done in a two-stage reciprocating air compressor with perfect intercooling is given by

(a) $\frac{n}{n-1} p_1V_1\left[\left(\frac{p_2}{p_1}\right)^{\frac{n-1}{n}} + \left(\frac{p_3}{p_1}\right)^{\frac{n-1}{n}} - 2\right]$

(b) $\frac{2n}{n-1} p_1V_1\left[\left(\frac{p_3}{p_1}\right)^{\frac{n-1}{n}} - 1\right]$

(c) $\frac{n}{n-1} p_1V_1\left[\left(\frac{p_3}{p_1}\right)^{\frac{n-1}{n}} - 1\right]$

(d) $\frac{n}{n-1} p_1V_1\left[\left(\frac{p_2}{p_1}\right)^{\frac{n-1}{n}} \left(\frac{p_3}{p_2}\right)^{\frac{n-1}{n}} - 2\right]$.

13. In reciprocating air compressor the method of controlling the quantity of air delivered is done by

(a) throttle control (b) blow-off control
(c) clearance control (d) all of the above.

14. The efficiency of Vane type air compressor as compared to Roots air compressor for the same pressure ratio is

(a) more (b) less
(c) same (d) may be more or less.

15. In centrifugal air compressor the pressure developed depends on

(a) impeller tip velocity (b) inlet-temperature
(c) compression index (d) all of the above.

16. In a centrifugal air compressor the pressure ratio is increased by

(a) increasing the speed of impeller keeping its diameter fixed
(b) increasing the diameter of the impeller keeping its speed constant
(c) reducing inlet temperature, keeping impeller diameter and speed fixed
(d) all of the above.

Answers

1. (b)	**2.** (a)	**3.** (a)	**4.** (c)	**5.** (c)	**6.** (c)	**7.** (c)
8. (b)	**9.** (b)	**10.** (b)	**11.** (a)	**12.** (b)	**13.** (d)	**14.** (a)
15. (d)	**16.** (d).					

THEORETICAL QUESTIONS

1. Enumerate the applications of compressed air.
2. State how are the air compressors classified ?
3. Describe with a neat sketch the construction and working of a single-stage single-acting reciprocating air compressor.

4. Prove that the work done/kg of air in a compressor is given by

$$W = RT_1 \frac{n}{n-1}\left[(r_p)^{\frac{n-1}{n}} - 1\right], \text{ where } r_p = \text{pressure ratio.}$$

5. Prove that the volumetric efficiency of a single stage compressor is given by

$$\eta_{vol.} = 1 + k - k\left(\frac{p_2}{p_1}\right)^{\frac{1}{n}}, \text{ where } k = \frac{V_c}{V_s}.$$

6. State the conditions which lower the volumetric efficiency.
7. Explain with a neat sketch actual p-V diagram for a single-stage compressor.
8. What do you mean by multistage compression ? State its advantages.
9. What is the effect of atmospheric conditions on the output of a compressor ?
10. Explain briefly with a neat sketch a reciprocating air motor.
11. Write a short note on 'rotary type air motor.'
12. What is a rotary compressor ? How are rotary compressors classified ?
13. What is a centrifugal compressor ? How does it differ from an axial flow compressor ?
14. What is Euler's work ?
15. What is a slip factor and a pressure co-efficient ?
16. Describe briefly an axial-flow compressor.
17. Draw the velocity diagrams of an axial-flow compressor.
18. What do you mean by 'surging' and 'choking' ?
19. Prove that the work done in two-stage compressor per kg of air delivered with perfect intercooling is given by

$$W/\text{kg} = \frac{2n}{n-1} RT_1\left[\left(\frac{p_d}{p_s}\right)^{\frac{n-1}{2n}} - 1\right]$$

Using the above equation, prove that the work done/kg of air in 'x' stages with perfect intercooling is given by

$$W/\text{kg} = \frac{xn}{n-1} RT_1\left[\left(\frac{p_d}{p_s}\right)^{\frac{n-1}{xn}} - 1\right].$$

20. Prove that the heat rejected (per kg of air) with perfect intercooling

$$= \left[c_p + c_v\left(\frac{\gamma - n}{n-1}\right)\right](T_2 - T_1).$$

21. Explain with a neat sketch the actual p-V diagram for a two-stage compressor.
22. Define the following efficiencies as applied to reciprocating air compressor :
    (*i*) Compressor efficiency.
    (*ii*) Isothermal efficiency.
    (*iii*) Adiabatic efficiency.
    (*iv*) Mechanical efficiency.
23. Write short notes on any three of the following :
    (*i*) Clearance in compressors.
    (*ii*) Free air delivered (F.A.D.) and displacement.
    (*iii*) Compressor performance.
    (*iv*) Control of compressors.
    (*v*) Arrangement of reciprocating compressors.
    (*vi*) Intercooler.
    (*vii*) Compressed air motors.

UNSOLVED EXAMPLES

1. Air is to be compressed in a single-stage reciprocating compressor from 1.013 bar and 15°C to 7 bar. Calculate the indicated power required for a free air delivery of 0.3 m^3/min, when the compression process is :

   (*i*) Isentropic (*ii*) Reversible isothermal

   (*iii*) Polytropic, with $n = 1.25$. **[Ans.** 1.31 kW ; 0.98 kW ; 1.19 kW]

2. The compressor of the above example is to run at 1000 r.p.m. If the compressor is single-acting and has a stroke/bore ratio of 1.2/1, calculate the bore size required. **[Ans.** 68.3 mm]

3. A single-stage, single-acting air compressor running at 1000 r.p.m. delivers air at 25 bar. For this purpose the induction and free air conditions can be taken as 1.013 bar and 15°C, and the free air delivery as 0.25 m^3/min. The clearance volume is 3% of the swept volume and the stroke/bore ratio is 1.2 : 1. Calculate the bore and stroke and the volumetric efficiency of this machine. Take the index of compression and expansion as 1.3. Calculate also the indicated power and the isothermal efficiency.

   [Ans. 73.2 mm ; 87.84 mm ; 67.6% ; 2 kW ; 67.5%]

4. A single-acting compressor is required to deliver air at 70 bar from an induction pressure of 1 bar, at the rate of 2.4 m^3/min measured at free-air conditions of 1.013 bar and 15°C. The temperature at the end of the induction stroke is 32°C. Calculate the indicated power required if the compression is carried out in two stages with an ideal intermediate pressure and complete intercooling. The index of compression and expansion for both stages is 1.25. What is the saving in power over single-stage compression ?

   If the clearance volume is 3% of the swept volume in each cylinder, calculate the swept volumes of the cylinders. The speed of the compressor is 750 r.p.m.

   If the mechanical efficiency of the compressor is 85%, calculate the power output in kilowatts of the motor required. **[Ans.** 22.7 kW ; 6 kW ; 0.00396 m^3, 0.000474 m^3, 26.75 kW]

5. A single-cylinder, single-acting air compressor running at 300 r.p.m. is driven by a 23 kW electric motor. The mechanical efficiency of the drive between motor and compressor is 87%. The air inlet conditions are 1.013 bar and 15°C and the delivery pressure is 8 bar. Calculate the free-air delivery in m^3/min, the volumetric efficiency, and the bore and stroke of the compressor. Assume that the index of compression and expansion is $n = 1.3$, that the clearance volume is 7% of the swept volume and that the bore is equal to the stroke. **[Ans.** 4.47 m^3/min ; 73% ; 296 mm]

6. A two-stage air compressor consists of three cylinders having the same bore and stroke. The delivery pressure is 7 bar and the free air delivery is 4.2 m^3/min. Air is drawn in at 1.013 bar, 15°C and an intercooler cools the air to 38°C. The index of compression is 1.3 for all the three cylinders. Neglecting clearance calculate :

   (*i*) The intermediate pressure (*ii*) The power required to drive the compressor

   (*iii*) The isothermal efficiency. **[Ans.** (*i*) 2.19 bar (*ii*) 16.3 kW (*iii*) 84.5%]

7. A two-stage double-acting air compressor, operating at 200 r.p.m., takes in air at 1.013 bar and 27°C. The size of the L.P. cylinder is 350 × 380 mm ; the stroke of H.P. cylinder is the same as that of the L.P. cylinder and the clearance of both the cylinders is 4%. The L.P. cylinder discharges the air at a pressure of 4.052 bar. The air passes through the intercooler so that it enters the H.P. cylinder at 27°C and 3.85 bar, finally it is discharged from the compressor at 15.4 bar. The value of '*n*' in both cylinders is 1.3. $c_p = 1.0035$ kJ/kg K and $R = 0.287$ kJ/kg K.

   Calculate : (*i*) The heat rejected in the intercooler

   (*ii*) The diameter of H.P. cylinder (*iii*) The power required to drive the H.P. cylinder.

   [Ans. (*i*) 1805.68 kg/min. (*ii*) 179.5 mm (*iii*) 37.3 kW]

8. A single-acting two-stage compressor with complete intercooling delivers 10 kg/min of air at 16 bar. The suction occurs at 1 bar and 15°C. The expansion and compression processes are reversible polytropic with polytropic index $n = 1.25$. Calculate :

   (*i*) The power required. (*ii*) The thermal efficiency.

   (*iii*) The free air delivery. (*iv*) Heat transferred in intercooler.

(*v*) If the clearance ratios for L.P. and H.P. cylinders are 0.04 and 0.06 respectively, calculate the swept and clearance volumes for each cylinder.

The speed of the compressor is 400 r.p.m. [**Ans.** (*i*) 44 kW, (*ii*) 86.8% (*iii*) 8.266 m^3/min (*iv*) 15.41 kW (*v*) 0.0225 m^3, 0.0009 m^3 ; 0.00588 m^3, 0.000353 m^3]

9. A two-stage air compressor with complete intercooling delivers air to the mains at a pressure of 30 bar, the suction conditions being 1 bar and 27°C. If both cylinders have the same stroke, find the ratio of the cylinders diameters, for the efficiency of compression to be a maximum. Assume the index of compression to be 1.3. [**Ans.** 2.337]

10. A 4-cylinder double-acting compressor is required to compress 25 m^3/min of air at 1 bar and 25°C to a pressure of 15 bar. Determine the size of the motor required and the cylinder dimensions if the following additional data is given :

Clearance volume = 5%, $L/D = 1.2$, r.p.m. = 300, Mechanical efficiency = 80%, Value of index, $n = 1.35$.

Assume no pressure change in suction valves and the air gets heated by 10°C during suction stroke.

[**Ans.** D = 550 mm, L = 680 mm]

11. A three-stage compressor is used to compress hydrogen from 1.04 bar to 35 bar. The compression in all stages follows the law $pV^{1.25} = C$. The temperature of hydrogen at inlet of compressor is 288 K. Neglecting clearance and assuming perfect intercooling, find :

(*i*) Indicated power required to deliver 14 m^3 of hydrogen per minute measured at inlet conditions.

(*ii*) Intermediate pressures.

Take R = 4125 J/kg K. [**Ans.** (*i*) 96.2 kW (*ii*) 3.354 bar, 10.815 bar]

12. A three-stage reciprocating air compressor compresses air from 1 bar and 17°C to 35 bar. The law of compression is $pV^{1.25}$ = C and is same for all the stages of compression.

Assuming perfect intercooling and neglecting clearance and valve resistance, find the minimum power required to compress 15 m^3/min of free air. Also find the intermediate pressures.

[**Ans.** 100.3, kW, 3.271 bar, 10.7 bar]

13. A three-stage single-acting air compressor running in an atmosphere at 1.013 bar and 15°C has a free air delivery of 2.83 m^3/min. The suction pressure and temperature are 0.98 bar and 32°C respectively. Calculate the indicated power required, assuming complete intercooling, n = 1.3, and that the machine is designed for minimum work. The delivery pressure is to be 70 bar. [**Ans.** 24.2 kW]

14. Using the data of example 13 determine heat loss to the cylinder jacket cooling water and heat loss to the intercooler circulating water, per minute. [**Ans.** 90 kJ/min, 375 kJ/min.]

15. A four-stage compressor works between limits of 1 bar and 112 bar. The index of compression in each stage is 1.28, the temperature at the start of compression in each stage is 32°C and the intermediate pressures are so chosen that the work is divided equally among the stages. Neglecting clearance find :

(*i*) The volume of free air delivered per kWh at 1.013 bar and 15°C.

(*ii*) The temperature at delivery from each stage.

(*iii*) The isothermal efficiency. [**Ans.** (*i*) 6.24 m^3/kWh, (*ii*) 122°C, (*iii*) 87.87%]

16. A multi-stage air compressor is to be designed to elevate the pressure from 1 bar to 120 bar such that stage pressure ratio will not exceed 4. Determine :

(*i*) Number of stages. (*ii*) Exact stage pressure ratio.

(*iii*) Intermediate pressures. [**Ans.** (*i*) 4, (*ii*) 3.31 (*iii*) 36.25 bar, 10.95 bar, 3.31 bar]

17. In an ideal four-stage reciprocating air compressor, the inlet pressure is 96 kN/m^2 and inlet temperature is 300 K. The air is delivered at a pressure of 27.6 MN/m^2. The compressor is designed for the minimum power requirement and has perfect intercooling. The reversible compression and expansion processes both conform to the relation $pV^{1.2} = C$. Determine :

(*i*) Intermediate pressures. (*ii*) The air delivery temperature.

(*iii*) The ideal isothermal efficiency.

For air, which may be assumed to be a perfect gas, the specific gas constant is 0.28702 kJ/kg K.

[**Ans.** (*i*) 395 kN/m^2, 1628 kN/m^2 and 7 is N/m^2 (*ii*) 380 K (*iii*) 88.6%]

Rotary compressors

18. Air at 1.013 bar and 15°C is to be compressed at the rate of 5.6 m^3/min to 11.75 bar. Two machines are considered : (*i*) the roots blower ; and (*ii*) a sliding vane rotary compressor. Compare the powers required, assuming for the vane type that internal compression takes place through 75% of the pressure rise before delivery takes place, and that the compressor is an ideal uncooled machine. [**Ans.** 6.88 kW, 5.75 kW]

19. Air is compressed in a two-stage vane type compressor from 1.013 bar to 8.75 bar. Assuming equal pressure ratio in each stage, calculate the power required. Assume that in each compression is complete and that intercooling between stages is 75% complete. Calculate also the capacity of the high pressure stage in cubic metres per minute for a free air delivery of 42 m^3/min measured at 1.013 bar and 15°C. The machine is uncooled except for the intercooler and operates in an ideal manner. [**Ans.** 187 kW ; 15.6 m^3/min]

20. A roots blower compresses 0.06 m^3 of air from 1.0 bar to 1.45 bar per revolution. Calculate the compressor efficiency. [**Ans.** 87.11%]

21. Free air of 30 m^3/min is compressed from 1.013 bar to 2.23 bar. Calculate the power required (*i*) if the compression is carried out in roots blower, (*ii*) if the compression is carried out in vane blower. Assume that there is 25% reduction in volume before the back-flow occurs, and (*iii*) the isentropic efficiency in each case.

[**Ans.** (*i*) 60.85 kW (*ii*) 48.46 kW (*iii*) 73.69%, 92.53%]

22. A centrifugal compressor is designed to have a pressure ratio of 3.5 : 1. The inlet eye of the compressor impeller is 30 cm in diameter. The axial velocity at inlet is 130 m/s and the mass flow is 10 kg/s. The velocity in the delivery duct is 115 m/s. The tip speed of the impeller is 450 m/s and runs at 16000 r.p.m. with total head isentropic efficiency of 78% and pressure co-efficient of 0.72. The ambient conditions are 1.013 bar and 15°C.

Calculate :

(*i*) The static pressure ratio

(*ii*) The static pressure and temperature at inlet and outlet of compressor

(*iii*) Work of compressor per kg of air, and

(*iv*) The theoretical power required. [**Ans.** (*i*) 4.21 (*ii*) 0.917 bar, 279.6 K, 3.86 bar, 461.07 K (*iii*) 180.29 kJ/kg of air (*iv*) 1802.9 kW]

23. Air at a temperature of 290 K flows in a centrifugal compressor running at 20000 r.p.m. The other data given is as follows :

Slip factor = 0.80 ; Isentropic total head efficiency = 0.75 ; Outer diameter of blade tip = 500 mm. Determine :

(*i*) The temperature rise of air passing through the compressor.

(*ii*) The static pressure ratio.

Assume that the velocities of air at inlet and exit of the compressor are same.

[**Ans.** (*i*) 218.62°C (*ii*) 4.8]

24. An axial flow air compressor of 50% reaction design has blades with inlet and outlet angles of 45° and 10° respectively. The compressor is to produce a pressure ratio of 6 : 1 with an overall isentropic efficiency of 0.85 when the air inlet temperature is 40°C. The blade speed and axial velocity are constant throughout the compressor. Assuming a value of 200 m/s for the blade speed, find the number of stages required when the work factor is (*i*) unity (*ii*) 0.89 for all stages. [**Ans.** (*i*) 9 (*ii*) 10]

25. A centrifugal compressor running at 9000 r-p.m. delivers 600 m^3/min of free air. The air is compressed from 1 bar and 20°C to a pressure ratio of 4 with an isentropic efficiency of 0.82. Blades are radial at outlet of impeller and the flow velocity of 62 m/s may be assumed throughout constant. The outer radius of the impeller is twice the inner and the slip factor may be assumed as 0.9. The blade area coefficient may be assumed as 0.9 at the inlet. Calculate :

(*i*) Final temperature of air.

(*ii*) Theoretical power.

(*iii*) Impeller diameters at inlet and outlet.

(*iv*) Breadth of the impeller at inlet.

(*v*) Impeller blade angle at inlet.

(*vi*) Diffuse blade angle at inlet. [**Ans.** (*i*) 466.85 K ; (*ii*) 2077.7 kW ; (*iii*) 46.745 cm, 94.9 cm ; (*iv*) 12.2 cm, (*v*) 15.7°, (*vi*) 8.9°]

26. A single inlet type centrifugal compressor handles 8 kg/s of air. The ambient air conditions are 1 bar and 20°C. The compressor runs at 22000 r.p.m. with isentropic efficiency of 82%. The air is compressed in the compressor from 1 bar static pressure to 4.2 bar total pressure. The air enters the impeller eye with a velocity of 150 m/s with no prewhirl. Assuming that the ratio of whirl speed to tip speed is 0.9, calculate :

(*i*) Rise in total temperature during compression if the change in K.E. is negligible.

(*ii*) The tip diameter of the impeller.

(*iii*) Power required.

(*iv*) Eye diameter if the hub diameter is 10 cm.

[**Ans.** (*i*) 167.67°C ; (*ii*) 37.56 cm ; (*iii*) 1348 kW ; (*iv*) 25.9 cm]

27. In an axial flow compressor, the overall stagnation pressure ratio achieved is 4 with overall stagnation isentropic efficiency 86 percent. The inlet stagnation pressure and temperature are 1 bar and 320 K. The mean blade speed is 190 m/s. The degree of reaction is 0.5 at the mean radius with relative air angles of 10° and 30° at rotor inlet and outlet respectively. The work done factor is 0.9. Calculate :

(*i*) Stagnation polytropic efficiency.

(*ii*) Number of stages.

(*iii*) Inlet temperature and pressure.

(*iv*) Blade height in the first stage if the hub-tip ratio is 0.4, mass flow rate is 20 kg/s.

[**Ans.** (*i*) 88.4% ; (*ii*) 11 ; (*iii*) 287.39 K, 0.6864 bar, (*iv*) 11.4 cm]

28. A multistage axial flow compressor delivers 18 kg/s of air. The inlet stagnation condition is 1 bar and 20°C. The power consumed by the compressor is 4260 kW. Calculate :

(*i*) Delivery pressure.

(*ii*) Number of stages.

(*iii*) Overall isentropic efficiency of the compressor.

Assume temperature rise in the first stage is 18°C. The polytropic efficiency of compression is 0.9 and the stage stagnation pressure ratio is constant. [**Ans.** (*i*) 6.41 bar ; (*ii*) 10 ; (*iii*) 87.24%]

21

Gas Turbines and Jet Propulsion

21.1. GAS TURBINES—GENERAL ASPECTS

Probably a wind-mill was the first turbine to produce useful work, wherein the there is no pre-compression and no combustion. The characteristic features of a gas turbine as we think of the name today include a *compression process* and a *heat addition* (or combustion) *process*. The gas turbine represents perhaps the most satisfactory way of producing very large quantities of power in a self-contained and compact unit. The gas turbine may have a future use in conjunction with the oil engine. For smaller gas turbine units, the inefficiencies in compression and expansion processes become greater and to improve the thermal efficiency it is necessary to use a heat exchanger. In order that a small gas turbine may compete for economy with the small oil engine or petrol engine it is necessary that a compact effective heat exchanger be used in the gas turbine cycle. The thermal efficiency of the gas turbine alone is still quite modest 20 to 30% compared with that of a modern steam turbine plant 38 to 40%. It is possible to construct combined plants whose efficiencies are of order of 45% or more. Higher efficiencies might be attained in future.

The following *are* the major fields of application of gas turbines :

1. Aviation
2. Power generation
3. Oil and gas industry
4. Marine propulsion.

The efficiency of a gas turbine is not the criteria for the choice of this plant. A gas turbine is used in aviation and marine fields because it is *self contained, light weight not requiring cooling water and generally fit into the overall shape of the structure*. It is selected for power generation because of its *simplicity, lack of cooling water, needs quick installation and quick starting*. It is used in oil and gas industry because of *cheaper supply of fuel and low installation cost.*

The gas turbines have the following limitations : *(i) They are not self starting ; (ii) low efficiencies at part loads ; (iii) non-reversibility ; (iv) higher rotor speeds and (v) overall efficiency of the plant low.*

21.2. CLASSIFICATION OF GAS TURBINES

The gas turbines are mainly divided into two groups :

1. **Constant pressure combustion gas turbine**

(*a*) Open cycle constant pressure gas turbine

(*b*) Closed cycle constant pressure gas turbine.

2. Constant volume combustion gas turbine

In almost *all the fields open cycle gas turbine plants are used. Closed cycle plants were introduced at one stage because of their ability to burn cheap fuel.* In between their progress remained slow because of availability of cheap oil and natural gas. Because of rising oil prices, now again, the attention is being paid to closed cycle plants.

21.3. MERITS OF GAS TURBINES

(*i*) **Merits over I.C. engines :**

1. The mechanical efficiency of a gas turbine (95%) is quite high as compared with I.C. engine (85%) since the I.C. engine has a large number of sliding parts.
2. A gas turbine does not require a flywheel as the torque on the shaft is continuous and uniform. Whereas a flywheel is a must in case of an I.C. engine.
3. The weight of gas turbine per H.P. developed is less than that of an I.C. engine.
4. The gas turbine can be driven at a very high speeds (40000 r.p.m.) whereas this is not possible with I.C. engines.
5. The work developed by a gas turbine per kg of air is more as compared to an I.C. engine. This is due to the fact that gases can be expanded upto atmospheric pressure in case of a gas turbine whereas in an I.C. engine expansion upto atmospheric pressure is not possible.
6. The components of the gas turbine can be made lighter since the pressures used in it are very low, say 5 bar compared with I.C. engine, say 60 bar.
7. In the gas turbine the ignition and lubrication systems are much simpler as compared with I.C. engines.
8. Cheaper fuels such as paraffine type, residue oils or powdered coal can be used whereas special grade fuels are employed in petrol engine to check knocking or pinking.
9. The exhaust from gas turbine is less polluting comparatively since excess air is used for combustion.
10. Because of low specific weight the gas turbines are particularly suitable for use in aircrafts.

Demerits of gas turbines

1. The thermal efficiency of a simple turbine cycle is low (15 to 20%) as compared with I.C. engines (25 to 30%).
2. With wide operating speeds the fuel control is comparatively difficult.
3. Due to higher operating speeds of the turbine, it is imperative to have a speed reduction device.
4. It is difficult to start a gas turbine as compared to an I.C. engine.
5. The gas turbine blades need a special cooling system.
6. One of the main demerits of a gas turbine is its *very poor thermal efficiency at part loads,* as the quantity of air remains same irrespective of load, and output is reduced by reducing the quantity of fuel supplied.
7. Owing to the use of nickel-chromium alloy, the manufacture of the blades is difficult and costly.
8. For the same output the gas turbine produces five times exhaust gases than I.C. engine.
9. Because of prevalence of high temperature (1000 K for blades and 2500 K for combustion chamber) and centrifugal force the life of the combustion chamber and blades is short/small.

(*ii*) **Merits over steam turbines :**

The gas turbine entails the following *advantages over steam turbines* :

1. Capital and running cost less.
2. For the same output the space required is far less.
3. Starting is more easy and quick.
4. Weight per H.P. is far less.
5. Can be installed anywhere.
6. Control of gas turbine is much easier.
7. Boiler along with accessories not required.

21.4. CONSTANT PRESSURE COMBUSTION GAS TURBINES

21.4.1. Open Cycle Gas Turbines

Refer Fig. 21.1. The fundamental gas turbine unit is one operating on the open cycle in which a rotary compressor and a turbine are mounted on a common shaft. Air is drawn into the compressor and after compression passes to a combustion chamber. Energy is supplied in the combustion chamber by spraying fuel into the air stream, and the resulting hot gases expand through the turbine to the atmosphere. In order to achieve net work output from the unit, the turbine must develop more gross work output than is required to drive the compressor and to overcome mechanical losses in the drive. The products of combustion coming out from the turbine are exhausted to the atmosphere as they cannot be used any more. The working fluids (air and fuel) must be replaced continuously as they are exhausted into the atmosphere.

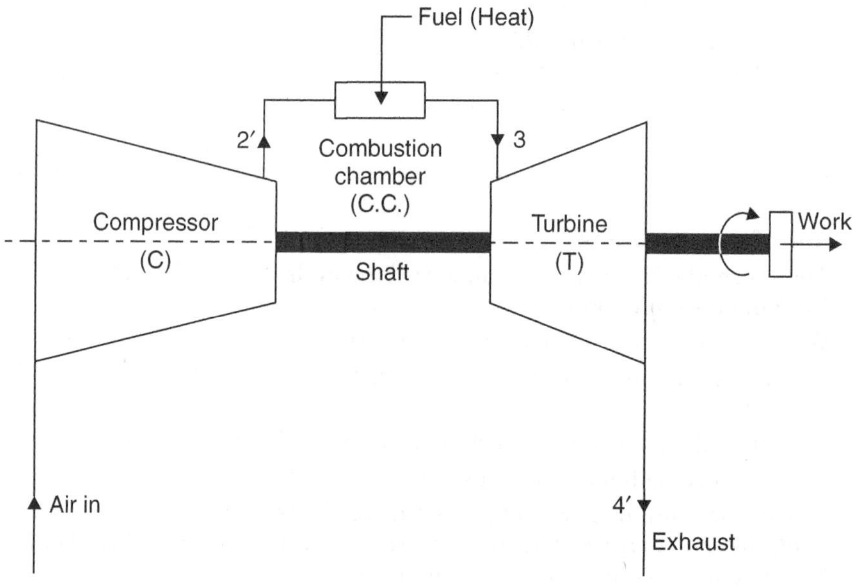

Fig. 21.1. Open cycle gas turbine.

If pressure loss in the combustion chamber is neglected, this cycle may be drawn on a T-s diagram as shown in Fig. 21.2.

- 1-2′ represents : *irreversible adiabatic compression.*
- 2′-3 represents : *constant pressure heat supply in the combustion chamber.*
- 3-4′ represents : *irreversible adiabatic expansion.*

- 1-2 represents : *ideal isentropic compression.*
- 3-4 represents : *ideal isentropic expansion.*

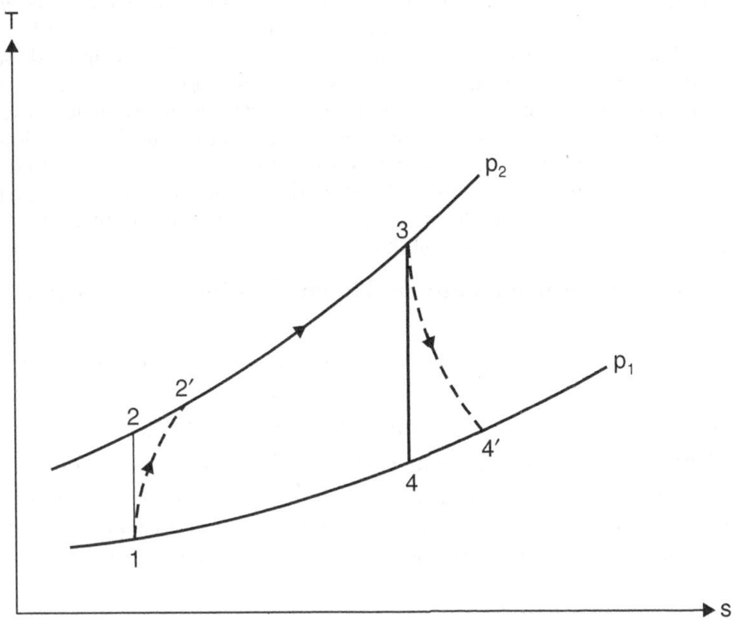

Fig. 21.2

Assuming change in kinetic energy between the various points in the cycle to be negligibly small compared with enthalpy changes and then applying the flow equation to each part of cycle, for unit mass, we have

Work input (compressor) $= c_p\,(T_2' - T_1)$

Heat supplied (combustion chamber) $= c_p\,(T_3 - T_2')$

Work output (turbine) $= c_p\,(T_3 - T_4')$

$\therefore$ Net work output = Work output – Work input

$= c_p\,(T_3 - T_4') - c_p(T_2' - T_1)$

and
$$\eta_{thermal} = \frac{\text{Net work output}}{\text{Heat supplied}}$$

$$= \frac{c_p\,(T_3 - T_4') - c_p\,(T_2' - T_1)}{c_p(T_3 - T_2')}$$

Compressor isentropic efficiency, η_{comp}

$$= \frac{\text{Work input required in isentropic compression}}{\text{Actual work required}}$$

$$= \frac{c_p(T_2 - T_1)}{c_p(T_2' - T_1)} = \frac{T_2 - T_1}{T_2' - T_1} \qquad ...(21.1)$$

Turbine isentropic efficiency, $\eta_{turbine}$

$$= \frac{\text{Actual work output}}{\text{Isentropic work output}}$$

$$= \frac{c_p(T_3 - T_4')}{c_p(T_3 - T_4)} = \frac{T_3 - T_4'}{T_3 - T_4} \qquad ...(21.2)$$

Note. With the variation in temperature, the value of the specific heat of a real gas varies, and also in the open cycle, the specific heat of the gases in the combustion chamber and in turbine is different from that in the compressor because fuel has been added and a chemical change has taken place. Curves showing the variation of c_p with temperature and air/fuel ratio can be used, and a suitable mean value of c_p and hence γ can be found out. It is usual in gas turbine practice to assume fixed mean value of c_p and γ for the expansion process, and fixed mean values of c_p and γ for the compression process. In an open cycle gas turbine unit the mass flow of gases in turbine is greater than that in compressor due to mass of fuel burned, but it is possible to neglect mass of fuel, since the air/fuel ratios used are large. Also, in many cases, air is bled from the compressor for cooling purposes, or in the case of air-craft at high altitudes, bled air is used for de-icing and cabin air-conditioning. This amount of air bled is approximately the same as the mass of fuel injected therein.

21.4.2. Methods for Improvement of Thermal Efficiency of Open Cycle Gas Turbine Plant

The following methods are employed *to increase the specific output and thermal efficiency of the plant :*

1. Intercooling 2. Reheating 3. Regeneration.

1. **Intercooling.** A compressor in a gas turbine cycle utilises the major percentage of power developed by the gas turbine. The work required by the compressor can be reduced by compressing the air in two stages and *incorporating an intercooler* between the two as shown in Fig. 21.3. The corresponding *T-s* diagram for the unit is shown in Fig. 21.4. The actual processes take place as follows :

1-2′ ... L.P. (Low pressure) compression
2′-3 ... Intercooling
3-4′ ... H.P. (High pressure) compression
4′-5 ... C.C. (Combustion chamber)-heating
5-6′ ... T (Turbine)-expansion

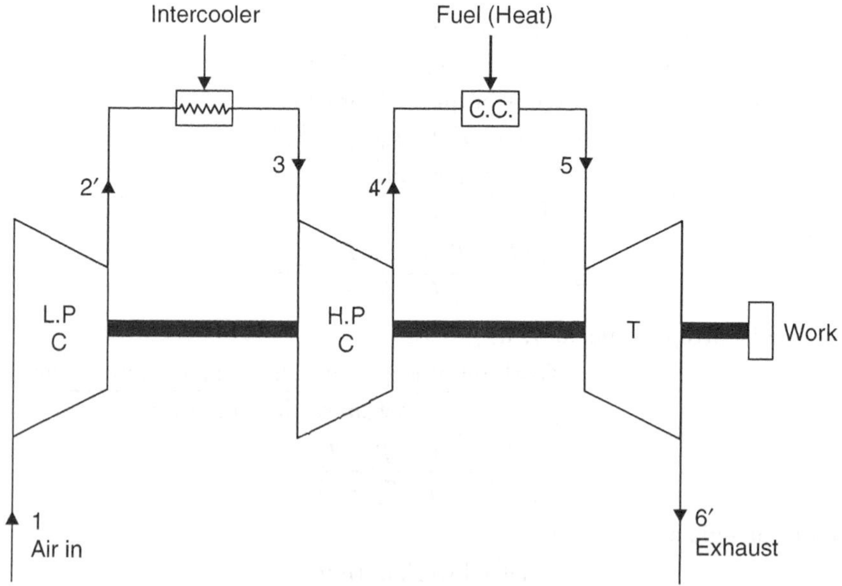

Fig. 21.3. Turbine plant with intercooler.

The ideal cycle for this arrangement is 1-2-3-4-5-6 ; the compression process without intercooling is shown as 1-L' in the actual case, and 1-L in the ideal isentropic case.

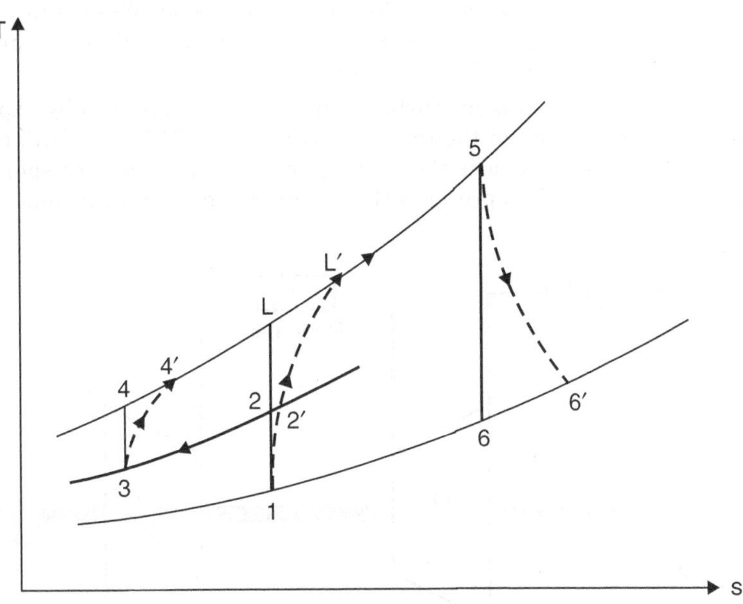

Fig. 21.4. *T-s* diagram for the unit.

Now,

Work input (*with intercooling*)

$$= c_p(T_2' - T_1) + c_p(T_4' - T_3) \quad ...(21.3)$$

Work input (*without intercooling*)

$$= c_p(T_L' - T_1) = c_p(T_2' - T_1) + c_p(T_L' - T_2') \quad ...(21.4)$$

By comparing equation (21.4) with equation (21.3) it can be observed that the *work input with intercooling is less than the work input with no intercooling*, when $c_p\ (T_4' - T_3)$ is *less* than $c_p(T_L' - T_2')$. This is so if it is assumed that isentropic efficiencies of the two compressors, operating separately, are each equal to the isentropic efficiency of the single compressor which would be required if no intercooling were used. Then $(T_4' - T_3) < (T_L' - T_2')$ since the pressure lines diverge on the *T-s* diagram from left to the right.

$$\text{Again, work ratio} = \frac{\text{Net work output}}{\text{Gross work output}}$$

$$= \frac{\text{Work of expansion} - \text{Work of compression}}{\text{Work of expansion}}$$

From this we may conclude that *when the compressor work input is reduced then the work ratio is increased.*

However the heat supplied in the combustion chamber when intercooling is used in the cycle, is given by,

Heat supplied *with intercooling* $= c_p(T_5 - T_4')$

Also the heat supplied when intercooling is not used, with the same maximum cycle temperature T_5, is given by

Heat supplied *without intercooling* $= c_p\ (T_5 - T_L')$

Thus, the *heat supplied when intercooling is used is greater than with no intercooling. Although the net work output is increased by intercooling it is found in general that the increase in heat to be supplied causes the thermal efficiency to decrease.* When intercooling is used a supply of cooling water must be readily available. The additional bulk of the unit may offset the advantage to be gained by increasing the work ratio.

2. **Reheating.** The output of a gas turbine can be amply improved by expanding the gases in two stages with a *reheater* between the two as shown in Fig. 21.5. The H.P. turbine drives the compressor and the L.P. turbine provides the useful power output. The corresponding *T-s* diagram is shown in Fig. 21.6. The line *4′-L′* represents the expansion in the L.P. turbine if reheating is *not* employed.

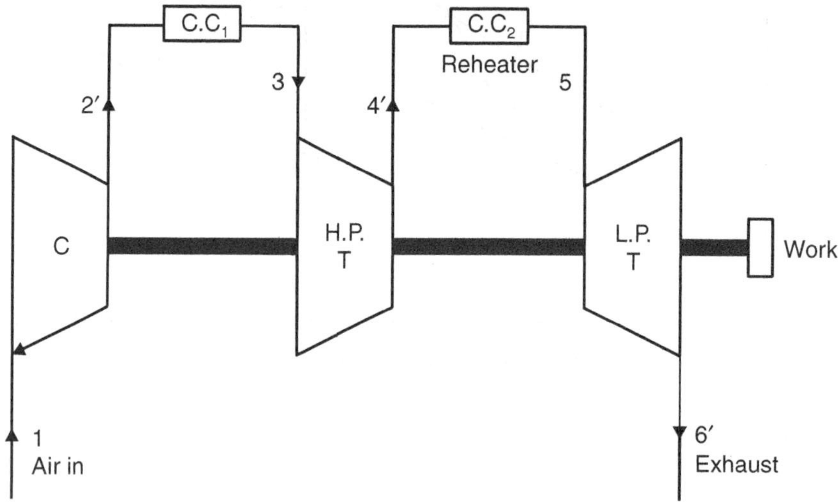

Fig. 21.5. Gas turbine with reheater.

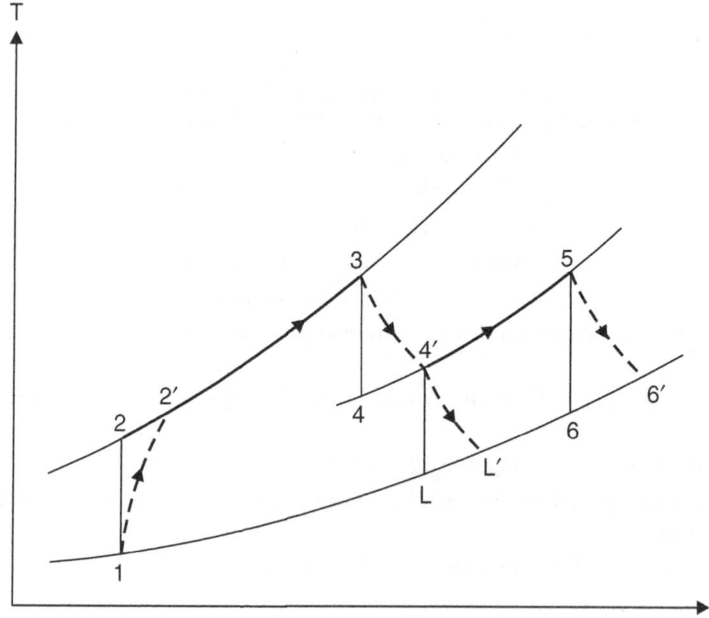

Fig. 21.6. *T-s* diagram for the unit.

Neglecting mechanical losses the work output of the H.P. turbine must be exactly equal to the work input required for the compressor i.e., $c_{pa}(T_2' - T_1) = c_{pg}(T_3 - T_4')$

The work output (net output) of L.P. turbine is given by,

Net work output (with reheating) $= c_{pg}(T_5 - T_6')$

and Net work output (without reheating) $= c_{pg}(T_4' - T_L')$

Since the pressure lines diverge to the right on T-s diagram it can be seen that the temperature difference $(T_5 - T_6')$ is always *greater* than $(T_4' - T_L')$, so that *reheating increases the net work output.*

Although net work is increased by reheating the heat to be supplied is also increased, and the net effect *can be to reduce the thermal efficiency*

Heat supplied $= c_{pg}(T_3 - T_2') + c_{pg}(T_5 - T_4')$.

Note. c_{pa} and c_{pg} stand for specific heats of air and gas respectively at constant pressure.

3. **Regeneration.** The exhaust gases from a gas turbine carry a large quantity of heat with them since their temperature is far above the ambient temperature. They can be used to heat the air coming from the compressor thereby reducing the mass of fuel supplied in the combustion chamber. Fig. 21.7 shows a gas turbine plant with a regenerator. The corresponding T-s diagram is shown in Fig. 21.8. 2′-3 represents the heat flow into the compressed air during its passage through the heat exchanger and 3-4 represents the heat taken in from the combustion of fuel. Point 6 represents the temperature of exhaust gases at discharge from the heat exchanger. The maximum temperature to which the air could be heated in the heat exchanger is ideally that of exhaust gases, but less than this is obtained in practice because a temperature gradient must exist for an unassisted transfer of energy. The *effectiveness* of the heat exchanger is given by :

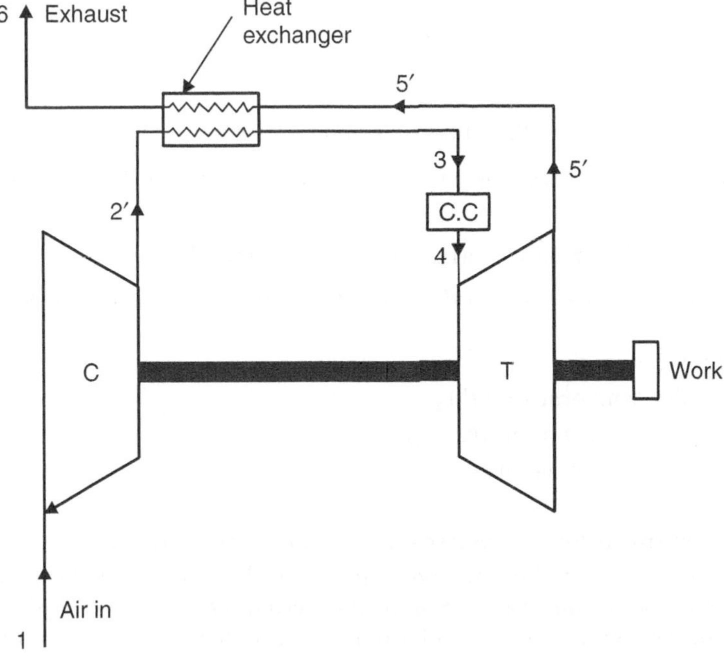

Fig. 21.7. Gas turbine with regenerator.

Effectiveness, $\varepsilon = \dfrac{\text{Increase in enthalpy per kg of air}}{\text{Available increase in enthalpy per kg of air}}$

$$= \frac{(T_3 - T_2')}{(T_5' - T_2')} \quad \ldots(21.5)$$

(assuming c_{pa} and c_{pg} to be equal)

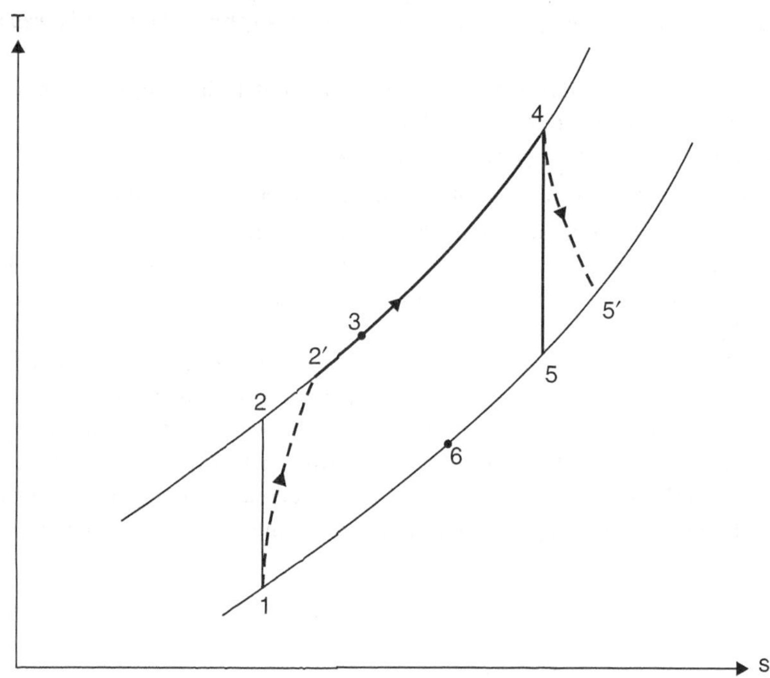

Fig. 21.8. *T-s* diagram for the unit.

A heat exchanger is usually used in large gas turbine units for marine propulsion or industrial power.

21.4.3. Effect of Operating Variables on Thermal Efficiency

The thermal efficiency of *actual open cycle* depends on the following thermodynamic variables :

(*i*) Pressure ratio

(*ii*) Turbine inlet temperature (T_3)

(*iii*) Compressor inlet temperature (T_1)

(*iv*) Efficiency of the turbine ($\eta_{turbine}$)

(*v*) Efficiency of the compressor (η_{comp}).

Effect of turbine inlet temperature and pressure ratio :

If the permissible turbine inlet-temperature (with the other variables being constant) of an *open cycle gas turbine power plant* is increased its *thermal efficiency* is *amply improved*. A practical limitation to increasing the turbine inlet temperature, however, is the ability of the material available for the turbine blading to *withstand the high rotative and thermal stresses.*

Refer Fig. 21.9. For *a given turbine inlet temperature, as the pressure ratio increases, the heat supplied as well as the heat rejected are reduced. But the ratio of change of heat supplied is not the same as the ratio of change heat rejected.* As a consequence, there *exists an optimum pressure ratio producing maximum thermal efficiency for a given turbine inlet temperature.*

As the pressure ratio increases, the thermal efficiency also increases until it becomes maximum and then it drops off with a further increase in pressure ratio (Fig. 21.10). Further, as the turbine inlet temperature increases, the peaks of the curves flatten out giving a greater range of ratios of pressure optimum efficiency.

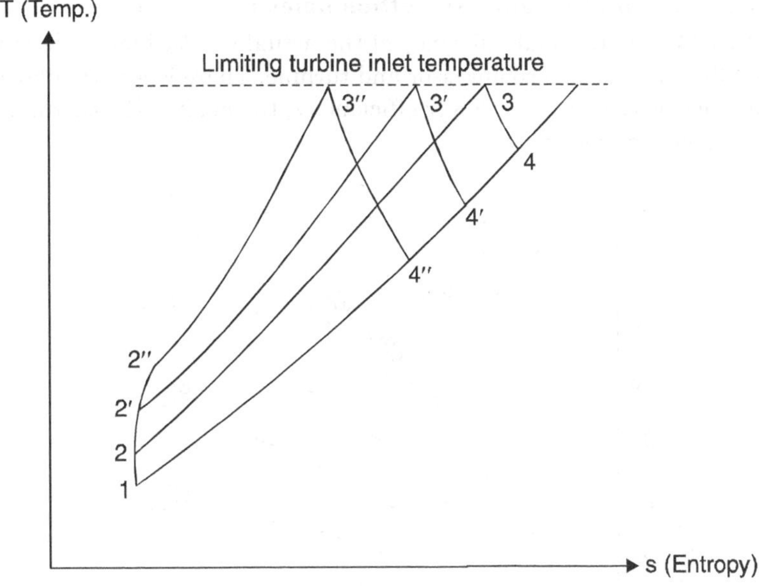

Fig. 21.9

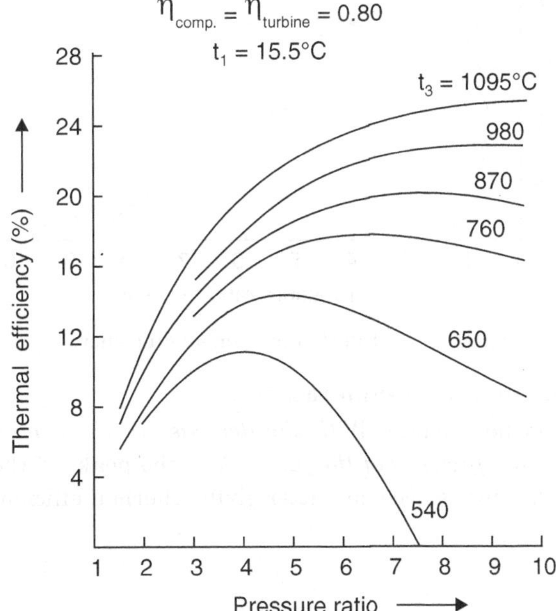

Fig. 21.10. Effect of pressure ratio and turbine inlet temperature.

Following particulars are worthnoting :

Gas temperatures	*Efficiency (gas turbine)*
550 to 600°C	20 to 22%
900 to 1000°C	32 to 35%
Above 1300°C	more than 50%

Effect of turbine and compressor efficiencies :

Refer Fig. 21.11. The thermal efficiency of the actual gas turbine cycle is very sensitive to variations in the efficiencies of the compressor and turbine. There is a particular pressure ratio at which maximum efficiencies occur. For lower efficiencies, the peak of the thermal efficiency occurs at lower pressure ratios and vice versa.

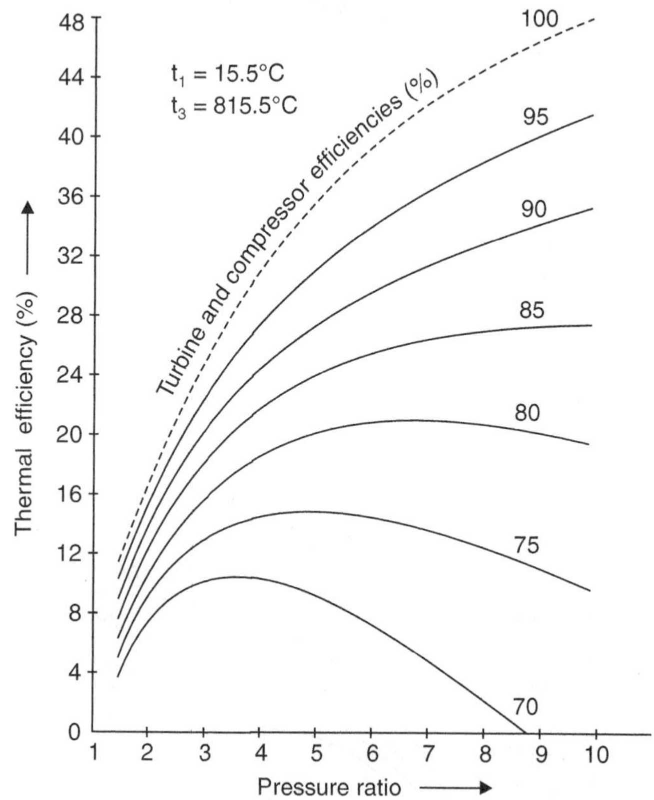

Fig. 21.11. Effect of component efficiency.

Effect of compressor inlet temperature :

Refer Fig. 21.12 (on next page). *With the decrease in the compressor inlet temperature there is increase in thermal efficiency of the plant.* Also the peaks of thermal efficiency occur at high pressure ratios and the curves become flatter giving thermal efficiency over a wider pressure ratio range.

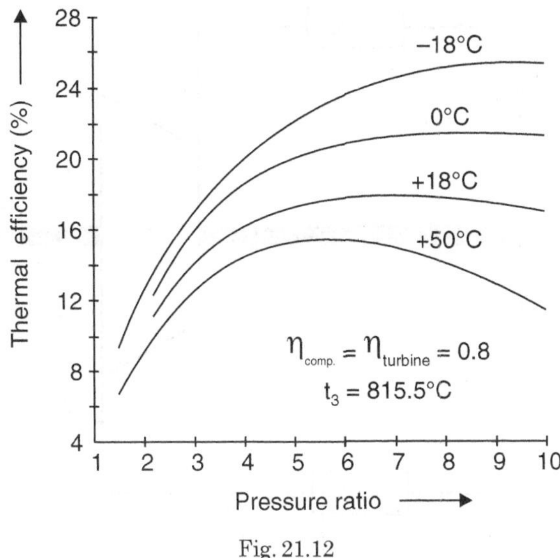

Fig. 21.12

21.4.4. Closed Cycle Gas Turbine (Constant pressure or joule cycle).

Fig. 21.13 shows a gas turbine operating on a constant pressure cycle in which the closed system consists of air behaving as an ideal gas. The various operations are as follows : Refer Figs. 21.14 and 21.15.

Operation 1-2 : The air is compressed isentropically from the lower pressure p_1 to the upper pressure p_2, the temperature rising from T_1 to T_2. No heat flow occurs.

Operation 2-3 : Heat flow into the system increasing the volume from V_2 to V_3 and temperature from T_2 to T_3 whilst the pressure remains constant at p_2. Heat received $= mc_p\ (T_3 - T_2)$.

Operation 3-4 : The air is expanded isentropically from p_2 to p_1, the temperature falling from T_3 to T_4. No heat flow occurs.

Operation 4-1 : Heat is rejected from the system as the volume decreases from V_4 to V_1 and the temperature from T_4 to T_1 whilst the pressure remains constant at p_1. Heat rejected $= mc_p\ (T_4 - T_1)$

$$\eta_{air\text{-}standard} = \frac{\text{Work done}}{\text{Heat received}}$$

$$= \frac{\text{Heat received/cycle} - \text{Heat rejected/cycle}}{\text{Heat received/cycle}}$$

$$= \frac{mc_p(T_3 - T_2) - mc_p(T_4 - T_1)}{mc_p\,(T_3 - T_2)} = 1 - \frac{T_4 - T_1}{T_3 - T_2}$$

Now, from isentropic expansion

$$\frac{T_2}{T_1} = \left(\frac{p_2}{p_1}\right)^{\frac{\gamma - 1}{\gamma}}$$

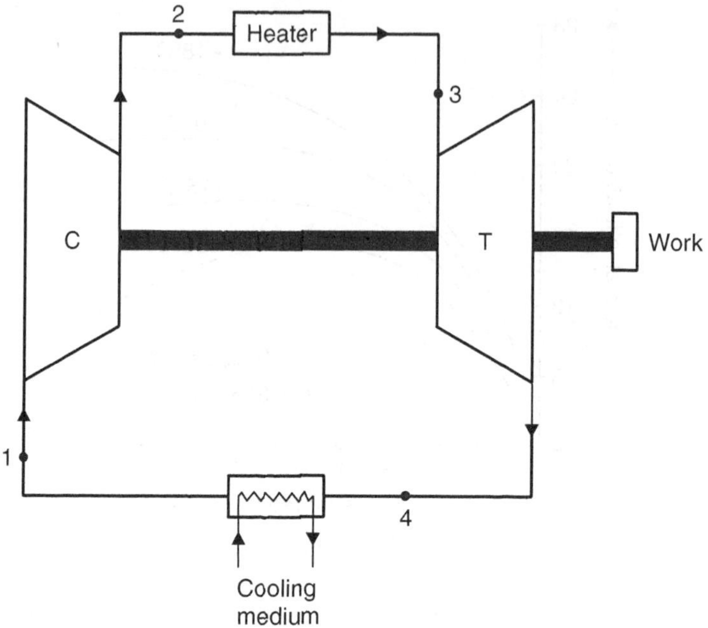

Fig. 21.13. Closed cycle gas turbine.

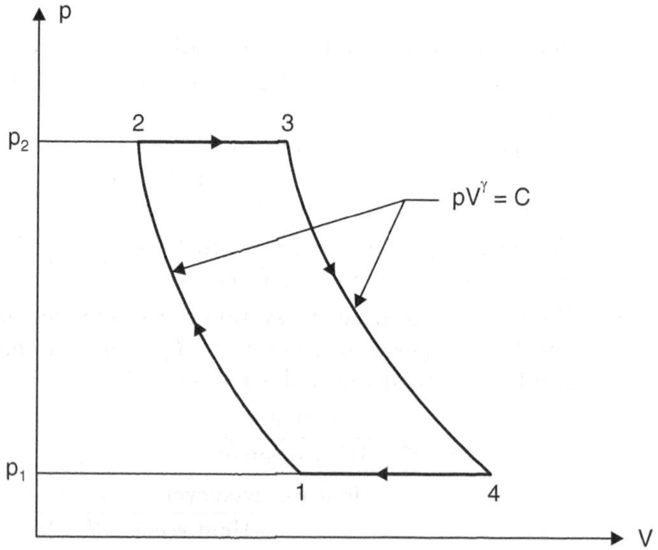

Fig. 21.14. *p-V* diagram.

$$T_2 = T_1 \, (r_p)^{\frac{\gamma - 1}{\gamma}}, \text{ where } r_p = \text{Pressure ratio}$$

Similarly $$\frac{T_3}{T_4} = \left(\frac{p_2}{p_1}\right)^{\frac{\gamma - 1}{\gamma}} \text{ or } T_3 = T_4 \, (r_p)^{\frac{\gamma - 1}{\gamma}}$$

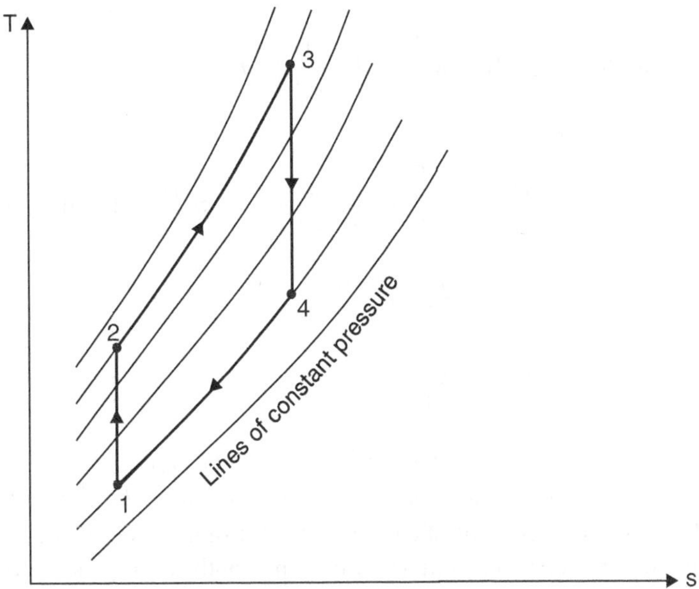

Fig. 21.15. *T-s* diagram.

$$\therefore \quad \eta_{air\text{-}standard} = 1 - \frac{T_4 - T_1}{T_4(r_p)^{\frac{\gamma-1}{\gamma}} - T_1(r_p)^{\frac{\gamma-1}{\gamma}}} = 1 - \frac{1}{(r_p)^{\frac{\gamma-1}{\gamma}}} \quad ...(21.6)$$

The expression shows that the *efficiency of the ideal joule cycle increases with the pressure ratio. The absolute limit of pressure is determined by the limiting temperature of the material of the turbine at the point at which this temperature is reached by the compression process alone, no further heating of the gas in the combustion chamber would be permissible and the work of expansion would ideally just balance the work of compression so that no excess work would be available for external use.*

Now we shall prove that the *pressure ratio for maximum work is a function of the limiting temperature ratio.*

Work output during the cycle

$$= \text{Heat received/cycle} - \text{Heat rejected/cycle}$$

$$= mc_p (T_3 - T_2) - mc_p (T_4 - T_1) = mc_p (T_3 - T_4) - mc_p (T_2 - T_1)$$

$$= mc_p T_3 \left(1 - \frac{T_4}{T_3}\right) - T_1 \left(\frac{T_2}{T_1} - 1\right)$$

In case of a given turbine the minimum temperature T_1 and the maximum temperature T_3 are prescribed, T_1 being the temperature of the atmosphere and T_3 the maximum temperature which the metals of turbine would withstand. Consider the specific heat at constant pressure c_p to be constant. Then,

Since, $$\frac{T_3}{T_4} = (r_p)^{\frac{\gamma-1}{\gamma}} = \frac{T_2}{T_1}$$

Using the constant $$\text{'}z\text{'} = \frac{\gamma - 1}{\gamma},$$

we have, work output/cycle
$$W = K\left[T_3\left(1 - \frac{1}{r_p^{\,z}}\right) - T_1(r_p^{\,z} - 1)\right]$$

Differentiating with respect to r_p

$$\frac{dW}{dr_p} = K\left[T_3 \times \frac{z}{r_p(z+1)} - T_1 z r_p^{(z-1)}\right] = 0 \text{ for a maximum}$$

$$\therefore \quad \frac{zT_3}{r_p^{(z+1)}} = T_1 z (r_p)^{(z-1)}$$

$$\therefore \quad r_p^{\,2z} = \frac{T_3}{T_1}$$

$$\therefore \quad r_p = (T_3/T_1)^{1/2z} \text{ i.e., } r_p = (T_3/T_1)^{\frac{\gamma}{2(\gamma-1)}}$$

Thus the *pressure ratio for maximum work is a function of the limiting temperature ratio.*

Fig. 21.16 shows an arrangement of closed cycle stationary gas turbine plant in which air is continuously circulated. This ensures that the air is not polluted by the addition of combustion waste product, since the heating of air is carried out in the form of heat exchanger shown in the diagram as air heater. The air exhausted from the power turbine is cooled before readmission to L.P. compressor. The various operations as indicated on *T-s* diagram (Fig. 21.17) are as follows :

Operation 1-2′ : Air is compressed from p_1 to p_x in the L.P. compressor.

Operation 2′-3 : Air is cooled in the intercooler at constant pressure p_x.

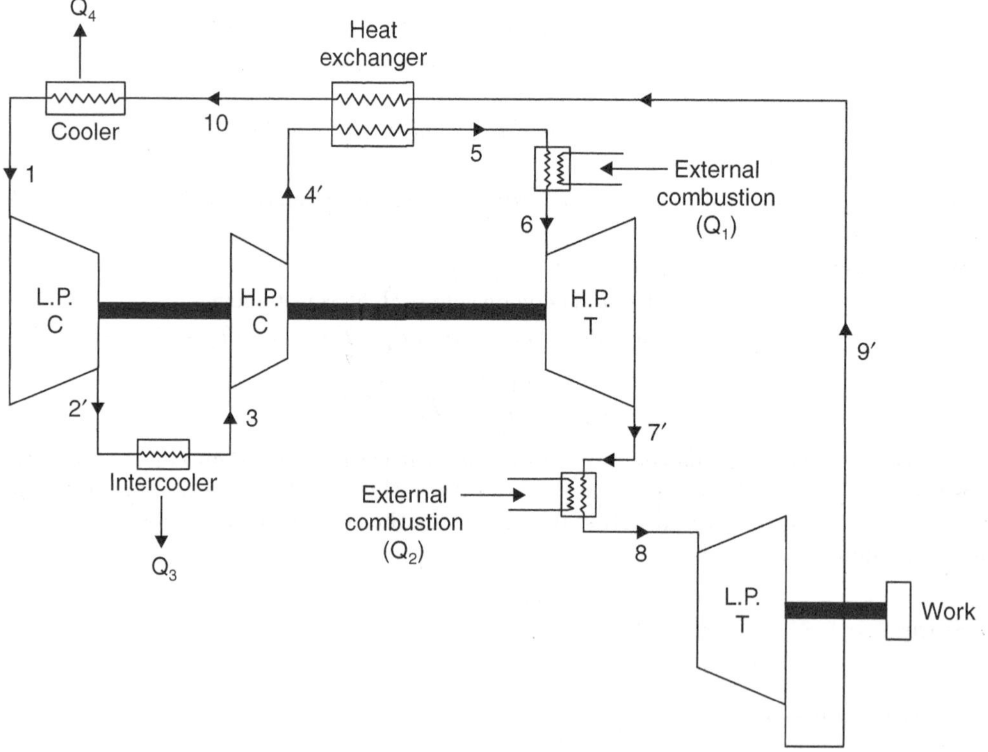

Fig. 21.16. Closed cycle gas turbine plant.

Operation 3-4′ : Air is compressed in the H.P. compressor from p_x to p_2.

Operation 4′-5 : High pressure air is heated at constant pressure by exhaust gases from power turbine in the heat exchanger to T_5.

Operation 5-6 : High pressure air further heated at constant pressure to the maximum temperature T_6 by an air heater (through external combustion).

Operation 6-7′ : The air is expanded in the H.P. turbine from p_2 to p_x producing work to drive the compressor.

Operation 7′—8 : Exhaust air from the H.P. turbine is heated at constant pressure in the air heater (through external combustion) to the maximum temperature $T_8 (= T_6)$.

Operation 8-9′ : The air is expanded in the L.P. turbine from p_x to p_1, producing energy for a flow of work externally.

Operation 9′-10 : Air from L.P. turbine is passed to the heat exchanger where energy is transferred to the air delivered from the H.P. compressor. The temperature of air leaving the heat exchanger and entering the cooler is T_{10}.

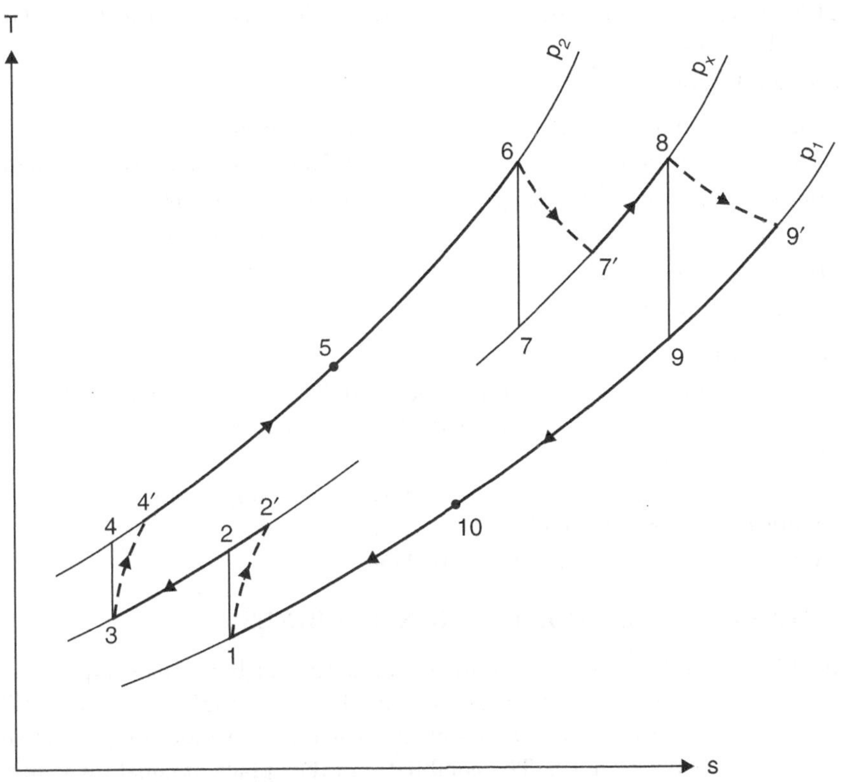

Fig. 21.17. *T-s* diagram for the plant.

Operation 10-11 : Air cooled to T_1 by the cooler before entering the L.P. compressor.

The energy balance for the whole plant is as follows :

$$Q_1 + Q_2 - Q_3 - Q_4 = W$$

In a closed cycle plant, in practice, the control of power output is achieved by varying the mass flow by the use of a reservoir in the circuit. The *reservoir maintains the design pressure and temperature and therefore achieves an approximately constant level of efficiency for varying* loads. In this cycle since it is closed, *gases other than air with favourable properties can be used* ; furthermore it is possible to burn solid fuels in the combustion heaters. *The major factor responsible for inefficiency in this cycle is the large irreversible temperature drop which occurs in the air heaters between the furnace and circulating gas.*

Note 1. In a closed cycle gas turbines, although air has been extensively used, the use of 'helium' which though of a lower density, has been inviting the attention of manufacturers for its use, for large output gas turbine units. *The specific heat of helium at constant pressure is about 'five times' that of air, therefore for each kg mass flow the heat drop and hence energy dealt with in helium machines is nearly five times of those in case of air.* The *surface area of the heat exchanger for helium can be kept as low as 1/3* of that *required for gas turbine plant using air as working medium.* For the same temperature ratio and for the plants of the same output the *cross-sectional area required for helium is much less than that for air.* It may therefore be concluded that the *size of helium unit is considerably small comparatively.*

2. Some gas turbine plants work on a combination of two cycles the open cycle and the closed cycle. Such a combination is called the *semi-closed cycle. Here a part of the working fluid is confined within the plant and another part flows from and to atmosphere.*

21.4.5. Merits and Demerits of Closed Cycle Gas Turbine Over Open Cycle Gas Turbine

Merits of closed cycle :

1. Higher thermal efficiency
2. Reduced size
3. No contamination
4. Improved heat transmission
5. Improved part load efficiency
6. Lesser fluid friction
7. No loss of working medium
8. Greater output
9. Inexpensive fuel.

Demerits of closed cycle :

1. Complexity
2. Large amount of cooling water is required. This limits its use to stationary installation or marine use where water is available in abundance.
3. Dependent system.
4. The weight of the system per H.P. developed is high comparatively, therefore not economical for moving vehicles.
5. Requires the use of a very large air heater.

21.5. CONSTANT VOLUME COMBUSTION TURBINES

Refer Fig. 21.18. In a constant volume combustion turbine, the compressed air from an air compressor C is admitted into the combustion chamber D through the valve A. When the valve A is closed, the fuel is admitted into the chamber by means of a fuel pump P. Then the mixture is ignited by means of a spark plug S. The combustion takes place at constant volume with increase of pressure. The valve B opens and the hot gases flow to the turbine T, and finally, they are discharged, into atmosphere. The energy of the hot gases is thereby converted into mechanical energy. For continuous running of the turbine these operations are repeated.

The main demerit associated with this type of turbine is that the *pressure difference and velocities of hot gases are not constant ; so the turbine speed fluctuates.*

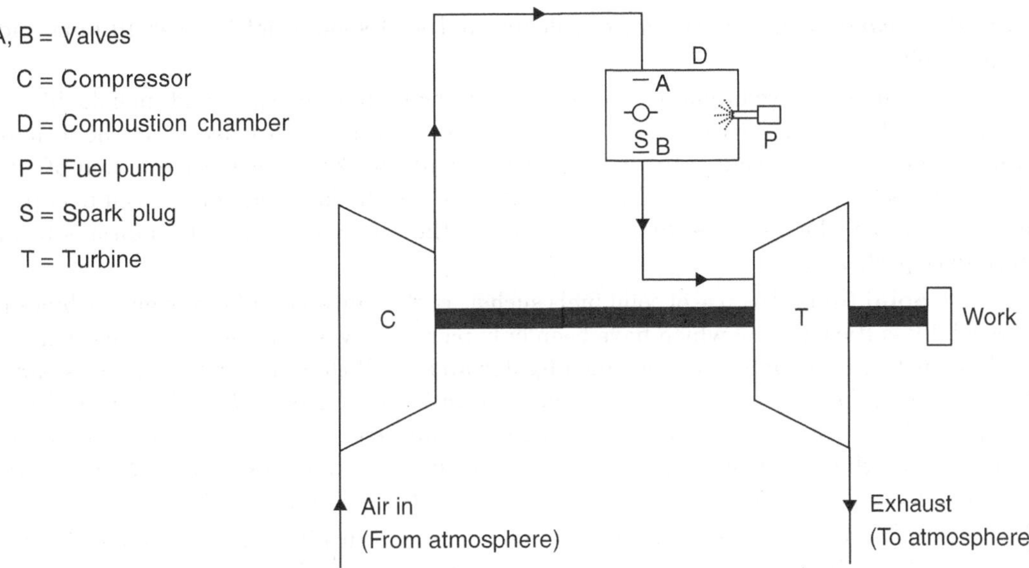

Fig. 21.18. Constant volume combustion gas turbine.

21.6. USES OF GAS TURBINES

Gas turbines find wide *applications* in the following *fields* :

1. Supercharging
2. Turbo-jet and turbo-propeller engines
3. Marine field
4. Railway
5. Road transport
6. Electric power generation
7. Industry.

21.7. GAS TURBINE FUELS

The various fuels used in gas turbines are enumerated and discussed below :

1. Gaseous fuels
2. Liquid fuels
3. Solid fuels

1. **Gaseous fuels.** *Natural gas is the ideal fuel for gas turbines*, but this is not available everywhere.

Blast furnace and *producer gases* may also be used for gas turbine power plants.

2. **Liquid fuels.** Liquid fuels of petroleum origin such as distillate oils or residual oils are most commonly used for gas turbine plant. The essential qualities of these fuels include *proper volatility, viscosity and calorific value*. At the same time it *should be* free *from any contents of moisture* and *suspended impurities that would log the small passages of the nozzles and damage valves and plungers of the fuel pumps.*

Minerals like *sodium, vanadium* and *calcium* prove *very harmful* for the *turbine blading* as these build deposits or corrode the blades. The sodium in ash should be less than 30% of the vanadium content as otherwise the ratio tends to be critical. The actual sodium content may be between 5 ppm to 10 ppm (part per million). If the vanadium is over 2 ppm, the magnesium in ash tends to become critical. *It is necessary that the magnesium in ash is at least three times the quantity of vanadium.* The content of calcium and lead should not be over 10 ppm and 5 ppm respectively.

Sodium is removed from residual oils by mixing with 5% of water and then double centrifuging when sodium leaves with water. Magnesium is added to the washed oil in the form of epsom salts, before the oil is sent into the combustor. This checks the corrosive action of vanadium. Residual oils burn with less ease than distillate oils and the latter are often used to start the unit from cold, after which the residual oils are fed in the combustor. In cold conditions residual oils need to be preheated.

3. **Solid fuels.** The use of solid fuels such as coal in pulverised form in gas turbines presents several difficulties most of which have been only partially overcome yet. The pulverising plant for coal in gas turbines applications is much lighter and small than its counterpart in steam generators. *Introduction of fuel in the combustion chamber of a gas turbine is required to be done against a high pressure whereas the pressure in the furnace of a steam plant is atmospheric.* Furthermore, the *degree of completeness of combustion in gas turbine applications has to be very high as otherwise soot and dust in gas would deposit on the turbine blading.*

Some practical applications of solid fuel burning in turbine combustors have been commercially, made available in recent years. In one such design finely crushed coal is used instead of pulverised fuel. This fuel is carried in stream of air tangentially into one end of a cylindrical furnace while gas comes out at the centre of opposite end. As the fuel particles roll around the circumference of the furnace they are burnt and a high temperature of about 1650°C is maintained which causes the mineral matter of fuel to be converted into a liquid slag. The slag covers the walls of the furnace and runs out through a top hole in the bottom. The result is that fly-ash is reduced to a very small content in the gases. In *another design* a regenerator is used to transfer the heat to air, the combustion chamber being located on the outlet of the turbine, and the combustion is carried out in the turbine exhaust stream. The advantage is that only clean air is handled by the turbine.

Example 21.1. *The air enters the compressor of an open cycle constant pressure gas turbine at a pressure of 1 bar and temperature of 20°C. The pressure of the air after compression is 4 bar. The isentropic efficiencies of compressor and turbine are 80% and 85% respectively. The air-fuel ratio used is 90 : 1. If flow rate of air is 3.0 kg/s, find :*

(*i*) *Power developed.*

(*ii*) *Thermal efficiency of the cycle.*

Assume $c_p = 1.0$ *kJ/kg K and* $\gamma = 1.4$ *of air and gases*

Calorific value of fuel = 41800 kJ/kg.

Solution. $p_1 = 1$ bar ; $T_1 = 20 + 273 = 293$ K

$p_2 = 4$ bar ; $\eta_{compressor} = 80\%$; $\eta_{turbine} = 85\%$

Air-fuel ratio = 90 : 1 ; Air flow rate, $m_a = 3.0$ kg/s

(*i*) **Power developed, P :**

Refer Fig. 21.19 (*b*)

$$\frac{T_2}{T_1} = \left(\frac{p_2}{p_1}\right)^{\frac{\gamma-1}{\gamma}} = \left(\frac{4}{1}\right)^{\frac{1.4-1}{1.4}} = 1.486$$

$$\therefore \quad T_2 = (20 + 273) \times 1.486 = 435.4 \text{ K}$$

$$\eta_{compressor} = \frac{T_2 - T_1}{T_2' - T_1}$$

$$0.8 = \frac{435.4 - 293}{T_2' - 293}$$

$$\therefore \qquad T_2' = \frac{435.4 - 293}{0.8} + 293 = 471 \text{ K}$$

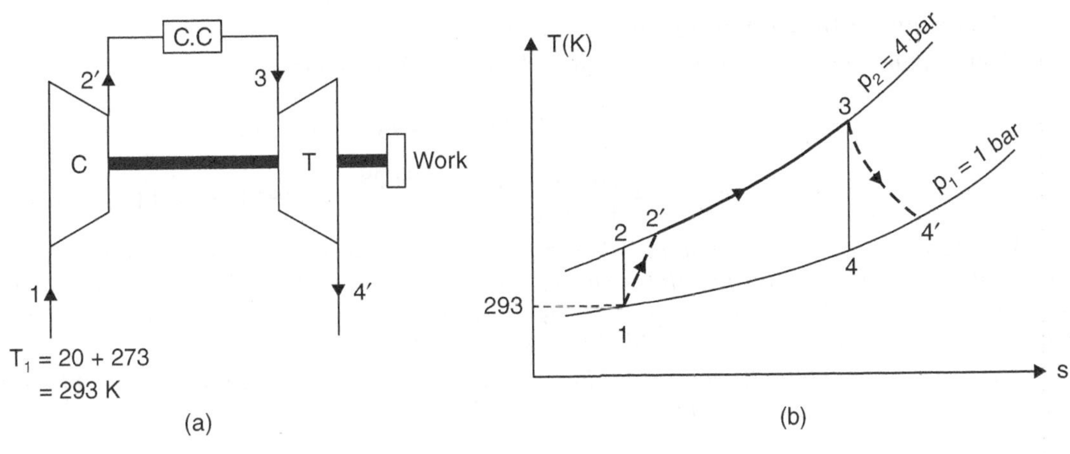

Fig. 21.19

Heat supplied by fuel = Heat taken by burning gases

$$m_f \times C = (m_a + m_f)\, c_p(T_3 - T_2')$$

(where m_a = mass of air, m_f = mass of fuel)

$$\therefore \qquad C = \left(\frac{m_a}{m_f} + 1\right) c_p\,(T_3 - T_2')$$

$$\therefore \qquad 41800 = (90 + 1) \times 1.0 \times (T_3 - 471)$$

i.e.,

$$T_3 = \frac{41800}{91} + 471 = 930 \text{ K}$$

Again,

$$\frac{T_4}{T_3} = \left(\frac{p_4}{p_3}\right)^{\frac{\gamma - 1}{\gamma}} = \left(\frac{1}{4}\right)^{0.4/1.4} = 0.672$$

$$\therefore \qquad T_4 = 930 \times 0.672 = 624.9 \text{ K}$$

$$\eta_{turbine} = \frac{T_3 - T_4'}{T_3 - T_4}$$

$$0.85 = \frac{930 - T_4'}{930 - 824.9}$$

$$\therefore \qquad T_4' = 930 - 0.85\,(930 - 624.9) = 670.6 \text{ K}$$

$$W_{turbine} = m_g \times c_p \times (T_3 - T_4')$$

(where m_g is the mass of hot gases formed per kg of air)

$$\therefore \qquad W_{turbine} = \left(\frac{90 + 1}{90}\right) \times 1.0 \times (930 - 670.6)$$

$$= 262.28 \text{ kJ/kg of air.}$$

$$W_{compressor} = m_a \times c_p \times (T_2' - T_1) = 1 \times 1.0 \times (471 - 293)$$
$$= 178 \text{ kJ/kg of air}$$
$$W_{net} = W_{turbine} - W_{compressor}$$
$$= 262.28 - 178 = 84.28 \text{ kJ/kg of air.}$$

Hence *power developed*, $P = 84.28 \times 3 =$ **252.84 kW/kg of air. (Ans.)**

(*ii*) **Thermal efficiency of cycle, $\eta_{thermal}$:**

Heat supplied per kg of air passing through combustion chamber

$$= \frac{1}{90} \times 41800 = 464.44 \text{ kJ/kg of air}$$

$$\therefore \quad \eta_{thermal} = \frac{\text{Work output}}{\text{Heat supplied}} = \frac{84.28}{464.44} = 0.1814 \text{ or } \mathbf{18.14\%}. \quad \textbf{(Ans.)}$$

Example 21.2. *A gas turbine unit has a pressure ratio of 6 : 1 and maximum cycle temperature of 610°C. The isentropic efficiencies of the compressor and turbine are 0.80 and 0.82 respectively. Calculate the power output in kilowatts of an electric generator geared to the turbine when the air enters the compressor at 15°C at the rate of 16 kg/s.*

Take $c_p = 1.005$ kJ/kg K and $\gamma = 1.4$ for the compression process, and take $c_p = 1.11$ kJ/kg K and $\gamma = 1.333$ for the expansion process.

Solution. $T_1 = 15 + 273 = 288$ K ; $T_3 = 610 + 273 = 883$ K ; $\frac{p_2}{p_1} = 6$,

$\eta_{compressor} = 0.80$; $\eta_{turbine} = 0.82$; Air flow rate = 16 kg/s

For compression process : $c_p = 1.005$ kJ/kg K, $\gamma = 1.4$

For expansion process : $c_p = 1.11$ kJ/kg K, $\gamma = 1.333$

In order to evaluate the net work output it is necessary to calculate temperatures T_2' and T_4'. To calculate T_2' we must first calculate T_2 and then use the isentropic efficiency.

For an isentropic process, $$\frac{T_2}{T_1} = \left(\frac{p_2}{p_1}\right)^{\frac{\gamma - 1}{\gamma}} = (6)^{\frac{1.4 - 1}{1.4}} = 1.67$$

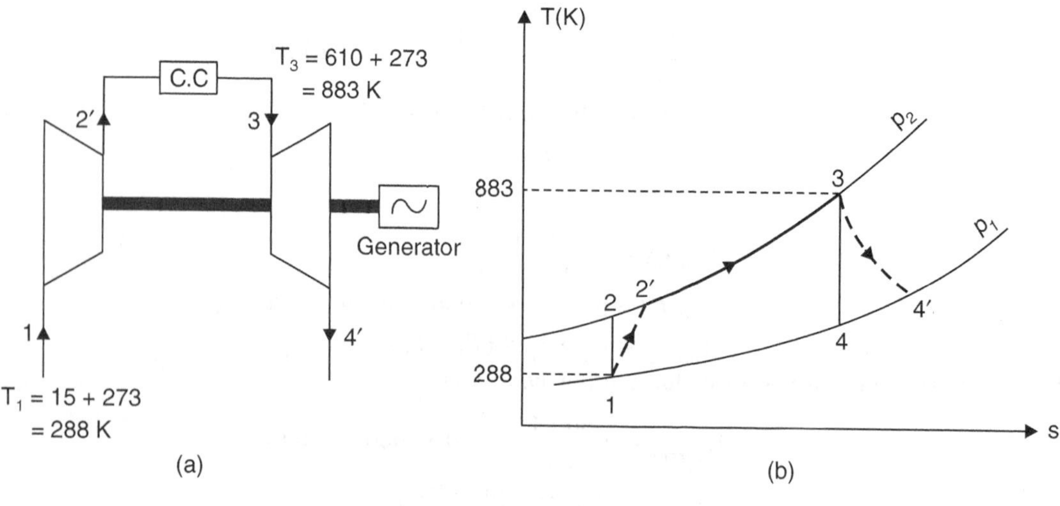

Fig. 21.20

$\therefore \qquad T_2 = 288 \times 1.67 = 481 \text{ K}$

Also, $\eta_{compressor} = \dfrac{T_2 - T_1}{T_2' - T_1}$

$$0.8 = \frac{481 - 288}{T_2' - T_1}$$

$\therefore \qquad T_2' = \dfrac{481 - 288}{0.8} + 288 = 529 \text{ K}$

Similarly for the turbine, $\dfrac{T_3}{T_4} = \left(\dfrac{p_3}{p_4}\right)^{\frac{\gamma-1}{\gamma}} = \left(\dfrac{p_2}{p_1}\right)^{\frac{\gamma-1}{\gamma}} = (6)^{\frac{1.333-1}{1.333}} = 1.565$

$\therefore \qquad T_4 = \dfrac{T_3}{1.565} = \dfrac{883}{1.565} = 564 \text{ K}$

Also, $\eta_{turbine} = \dfrac{T_3 - T_4'}{T_3 - T_4} = \dfrac{883 - T_4'}{883 - 564}$

$\therefore \qquad 0.82 = \dfrac{883 - T_4'}{883 - 564}$

$\therefore \qquad T_4' = 883 - 0.82\,(883 - 564) = 621.4 \text{ K}$

Hence,

Compressor work input, $W_{compressor} = c_p\,(T_2' - T_1)$
$= 1.005\,(529 - 288) = 242.2 \text{ kJ/kg}$

Turbine work output, $W_{turbine} = c_p\,(T_3 - T_4')$
$= 1.11\,(883 - 621.4) = 290.4 \text{ kJ/kg}$

$\therefore$ Net work output, $W_{net} = W_{turbine} - W_{compressor}$
$= 290.4 - 242.2 = 48.2 \text{ kJ/kg}$

Power in kilowatts $= 48.2 \times 16 =$ **771.2 kW. (Ans.)**

Example 21.3. *A gas terbine unit receives air at 1 bar and 300 K and compresses it adiabatically to 6.2 bar. The compressor efficiency is 88%. The fuel has a heating valve of 44186 kJ/kg and the fuel-air ratio is 0.017 kJ/kg of air.*

The turbine internal efficiency is 90%. Calculate the work of turbine and compressor per kg of air compressed and thermal efficiency.

For products of combustion, c_p = 1.147 kJ/kg K and γ = 1.333. **(U.P.S.C. 1997)**

Solution. *Given :* $p_1\ (= p_4) = 1$ bar, $T_1 = 300$ K ; $p_2 (= p_3) = 6.2$ bar ; $\eta_{compressor} = 88\%$;
$C = 44186$ kJ/kg ; Fuel-air ratio = 0.017 kJ/kg of air, $\eta_{turbine} = 90\%$;
$c_p = 1.147$ kJ/kg K ; $\gamma = 1.333$.

For isentropic compression process 1-2 :

$$\frac{T_2}{T_1} = \left(\frac{p_2}{p_1}\right)^{\frac{\gamma-1}{\gamma}} = \left(\frac{6.2}{1}\right)^{\frac{1.4-1}{1.4}} = 1.684$$

$\therefore \qquad T_2 = 300 \times 1.684 = 505.2 \text{ K}$

Now, $\eta_{compressor} = \dfrac{T_2 - T_1}{T_2' - T_1}$

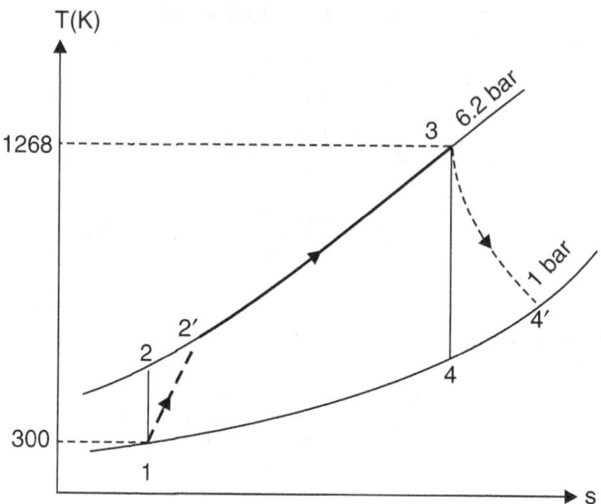

Fig. 21.21

$$0.88 = \frac{505.2 - 300}{T_2' - 300}$$

$$T_2' = \left(\frac{505.2 - 300}{0.88} + 300\right) = 533.2 \text{ K}$$

Heat supplied $= (m_a + m_f) \times c_p(T_3 - T_2') = m_f \times C$

or $$\left(1 + \frac{m_f}{m_a}\right) \times c_p(T_3 - T_2') = \frac{m_f}{m_a} \times C$$

or $$(1 + 0.017) \times 1.005(T_3 - 533.2) = 0.017 \times 44186$$

$\therefore$ $$T_3 = \frac{0.017 \times 44186}{(1 + 0.017) \times 1.005} + 533.2 = 1268 \text{ K}$$

For isentropic expression process 3-4 :

$$\frac{T_4}{T_3} = \left(\frac{p_4}{p_3}\right)^{\frac{\gamma - 1}{\gamma}} = \left(\frac{1}{6.2}\right)^{\frac{1.333 - 1}{1.333}} = 0.634$$

$\therefore$ $$T_4 = 1268 \times 0.634 = 803.9 \text{ K} \qquad (\because \gamma_g = 1.333 \text{ ......Given})$$

Now, $$\eta_{\text{turbine}} = \frac{T_3 - T_4'}{T_3 - T_4}$$

$$0.9 = \frac{1268 - T_4'}{1268 - 803.9}$$

$\therefore$ $$T_4' = 1268 - 0.9(1268 - 803.9) = 850.3 \text{ K}$$

$$W_{\text{compresssor}} = c_p(T_2' - T_1) = 1.005(533.2 - 300) = 234.4 \text{ kJ/kg}$$

$$W_{\text{turbine}} = c_{pg}(T_3 - T_4') = 1.147(1268 - 850.3) = 479.1 \text{ kJ/kg}$$

$$\text{Net work} = W_{\text{turbine}} - W_{\text{compressor}}$$
$$= 479.1 - 234.4 = 244.7 \text{ kJ/kg}$$

Heat supplied per kg of air

$$= 0.017 \times 44186 = 751.2 \text{ kJ/kg}$$

$\therefore$ Thermal efficiency, $\eta_{th.} = \dfrac{\text{Net work}}{\text{Heat supplied}}$

$$= \frac{244.7}{751.2} = 0.3257 \text{ or } \mathbf{32.57\%.} \quad \textbf{(Ans.)}$$

Example 21.4. *Find the required air-fuel ratio in a gas turbine whose turbine and compressor efficiencies are 85% and 80%, respectively. Maximum cycle temperature is 875°C. The working fluid can be taken as air (c_p = 1.0 kJ/kg K, γ = 1.4) which enters the compressor at 1 bar and 27°C. The pressure ratio is 4. The fuel used has calorific value of 42000 kJ/kg. There is a loss of 10% of calorific value in the combustion chamber.* **(GATE 1998)**

Solution. *Given :* $\eta_{turbine}$ = 85% ; $\eta_{compressor}$ = 80% ; T_3 = 273 + 875 = 1148 K, T_1 = 27 + 273 = 300 K ; c_p = 1.0 kJ/kg K ; γ = 1.4, p_1 = 1 bar, p_2 = 4 bar (Since pressure ratio is 4) ; C = 42000 kJ/kg K, η_{cc} = 90% (since loss in the combustion chamber is 10%)

For isentropic compression 1-2 :

$$\frac{T_2}{T_1} = \left(\frac{p_2}{p_1}\right)^{\frac{\gamma-1}{\gamma}} = (4)^{\frac{1.4-1}{1.4}} = 1.486$$

$$\therefore \quad T_2 = 300 \times 1.486 = 445.8 \text{ K}$$

$$\eta_{compressor} = \frac{T_2 - T_1}{T_2' - T_1}$$

or
$$0.8 = \frac{445.8 - 300}{T_2' - 300}$$

or
$$T_2' = \frac{445.8 - 300}{0.8} + 300 = 482.2 \text{ K}$$

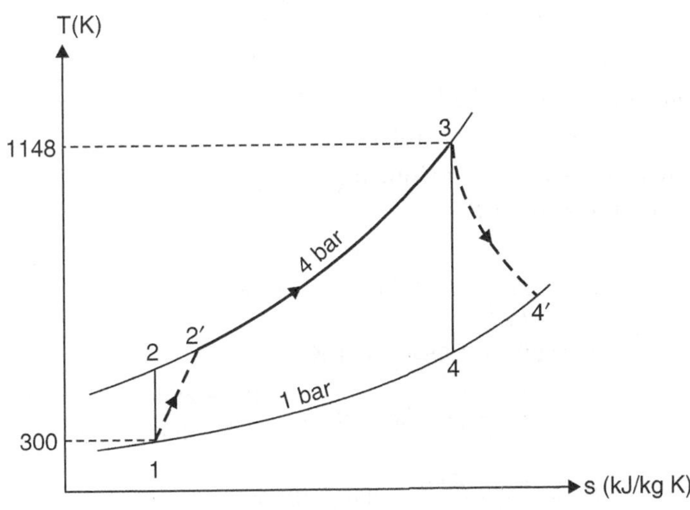

Fig. 21.22

Now, heat supplied by the fuel = heat taken by the burning gases

$$0.9 \times m_f \times C = (m_a + m_f) \times c_p \times (T_3 - T_2')$$

$$\therefore \quad C = \left(\frac{m_a + m_f}{m_f}\right) \times \frac{c_p(T_3 - T_2')}{0.9} = \left(\frac{m_a}{m_f} + 1\right) \times \frac{c_p(T_3 - T_2')}{0.9}$$

or

$$42000 = \left(\frac{m_a}{m_f} + 1\right) \times \frac{1.00\,(1148 - 482.2)}{0.9} = 739.78\left(\frac{m_a}{m_f} + 1\right)$$

$$\therefore \quad \frac{m_a}{m_f} = \frac{42000}{739.78} - 1 = 55.77 \text{ say } 56$$

$\therefore$ **A/F ratio = 56 : 1. (Ans.)**

Example 21.5. *Calculate the thermal efficiency and work ratio of the plant is example 21.4, assuming that c_p for the combustion process is 1.11 kJ/kg K.*

Solution. Heat supplied $= c_p(T_3 - T_2')$

$= 1.11\,(883 - 529) = 392.9$ kJ/kg

$$\eta_{thermal} = \frac{\text{Net work output}}{\text{Heat supplied}} = \frac{48.2}{392.9} = 0.1226 \text{ or } \mathbf{12.26\%}. \quad \textbf{(Ans.)}$$

Now,
$$Work\ ratio = \frac{\text{Net work output}}{\text{Gross work output}} = \frac{48.2}{W_{turbine}} = \frac{48.2}{290.4} = \mathbf{0.166}. \quad \textbf{(Ans.)}$$

Example 21.6. *In a constant pressure open cycle gas turbine air enters at 1 bar and 20°C and leaves the compressor at 5 bar. Using the following data : Temperature of gases entering the turbine = 680°C, pressure loss in the combustion chamber = 0.1 bar, $\eta_{compressor}$ = 85%, $\eta_{turbine}$ = 80%, $\eta_{combustion}$ = 85%, γ = 1.4 and c_p = 1.024 kJ/kg K for air and gas, find*

(i) The quantity of air circulation if the plant develops 1065 kW.

(ii) Heat supplied per kg of air circulation.

(iii) The thermal efficiency of the cycle.

Mass of the fuel may be neglected. **(AMIE Winter, 2000)**

Solution. *Given :* p_1 = 1 bar, p_2 = 5 bar, p_3 = 5 – 0.1 = 4.9 bar, p_4 = 1 bar,

T_1 = 20 + 273 = 293 K, T_3 = 680 + 273 = 953 K,

$\eta_{compressor}$ = 85%, $\eta_{turbine}$ = 80%, $\eta_{combustion}$ = 85%,

For air and gases : c_p = 1.024 kJ/kg K, γ = 1.4

Power developed by the plant, P = 1065 kW.

(i) **The quantity of air circulation, m_a :**

For *isentropic compression 1-2,*

$$\frac{T_2}{T_1} = \left(\frac{p_2}{p_1}\right)^{\frac{\gamma-1}{\gamma}} = \left(\frac{5}{1}\right)^{\frac{1.4-1}{1.4}} = 1.584$$

$$\therefore \quad T_2 = 293 \times 1.584 = 464 \text{ K}$$

Now,
$$\eta_{compressor} = \frac{T_2 - T_1}{T_2' - T_1} \quad i.e., \quad 0.85 = \frac{464 - 293}{T_2' - 293}$$

$$\therefore \quad T_2' = \frac{464 - 293}{0.85} + 293 = 494 \text{ K}$$

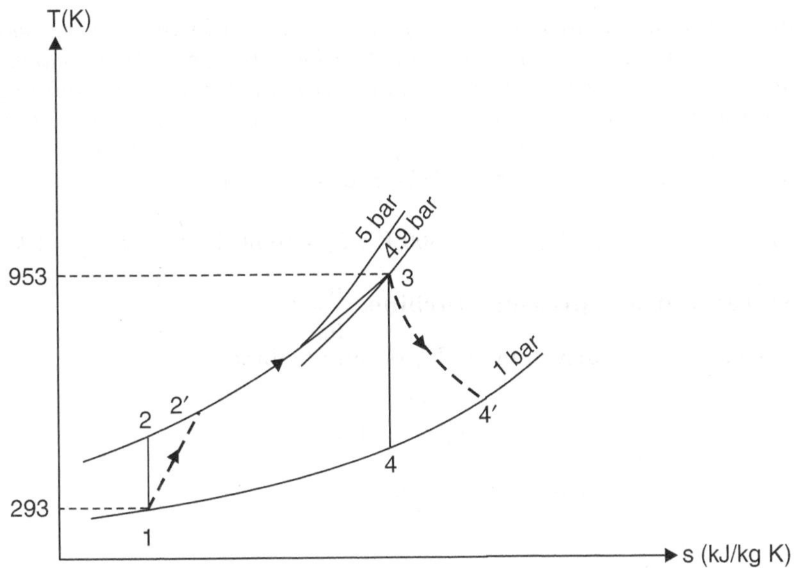

Fig. 21.23

For *isentropic expansion process 3-4,*

$$\frac{T_4}{T_3} = \left(\frac{p_4}{p_3}\right)^{\frac{\gamma-1}{\gamma}} = \left(\frac{1}{4.9}\right)^{\frac{1.4-1}{1.4}} = 0.635$$

$\therefore$ $$T_4 = 953 \times 0.635 = 605 \text{ K}$$

Now, $$\eta_{turbine} = \frac{T_3 - T_4'}{T_3 - T_4}$$

$$0.8 = \frac{953 - T_4'}{953 - 605}$$

$\therefore$ $$T_4' = 953 - 0.8(953 - 605) = 674.6 \text{ K}$$

$$W_{compressor} = c_p\,(T_2' - T_1) = 1.024\,(494 - 293) = 205.8 \text{ kJ/kg}$$

$$W_{turbine} = c_p\,(T_3 - T_4') = 1.024\,(953 - 674.6) = 285.1 \text{ kJ/kg.}$$

$\therefore$ $$W_{net} = W_{turbine} - W_{compressor} = 285.1 - 205.8 = 79.3 \text{ kJ/kg of air}$$

If the mass of air is flowing is m_a kg/s, the power developed by the plant is given by

$$P = m_a \times W_{net} \text{ kW}$$

$$1065 = m_a \times 79.3$$

$\therefore$ $$m_a = \frac{1065}{79.3} = 13.43 \text{ kg.}$$

i.e., **Quantity of air circulation = 13.43 kg. (Ans.)**

(*ii*) **Heat supplied per kg of air circulation :**

Actual heat supplied per kg of air circulation

$$= \frac{c_p\,(T_3 - T_2')}{\eta_{combustion}} = \frac{1.024\,(953 - 494)}{0.85} = 552.9 \text{ kJ/kg.}$$

(*iii*) **Thermal efficiency of the cycle,** $\eta_{thermal}$ **:**

$$\eta_{thermal} = \frac{\text{Work output}}{\text{Heat supplied}} = \frac{79.3}{552.9} = \mathbf{0.1434} \text{ or } \mathbf{14.34\%}. \quad \textbf{(Ans.)}$$

Example 21.7. *In a gas turbine the compressor is driven by the high pressure turbine. The exhaust from the high pressure turbine goes to a free low pressure turbine which runs the load. The air flow rate is 20 kg/s and the minimum and maximum temperatures are respectively 300 K and 1000 K. The compressor pressure ratio is 4. Calculate the pressure ratio of the low pressure turbine and the temperature of exhaust gases from the unit. The compressor and turbine are isentropic.* c_p *of air and exhaust gases = 1 kJ/kg K and* $\gamma = 1.4$. **(GATE 1995)**

Solution. *Given :* $\dot{m}_a = 20$ kg/s ; $T_1 = 300$ K ; $T_3 = 1000$ K, $\frac{p_2}{p_1} = 4$; $c_p = 1$ kJ/kg K ; $\gamma = 1.4$,

Pressure ratio of low pressure turbine, $\frac{p_4}{p_5}$ **:**

Since the compressor is driven by high pressure turbine,

$$\therefore \quad \frac{T_2}{T_1} = \left(\frac{p_2}{p_1}\right)^{\frac{\gamma-1}{\gamma}} = (4)^{\frac{0.4}{1.4}} = 1.486$$

or $$T_2 = 300 \times 1.486 = 445.8 \text{ K}$$

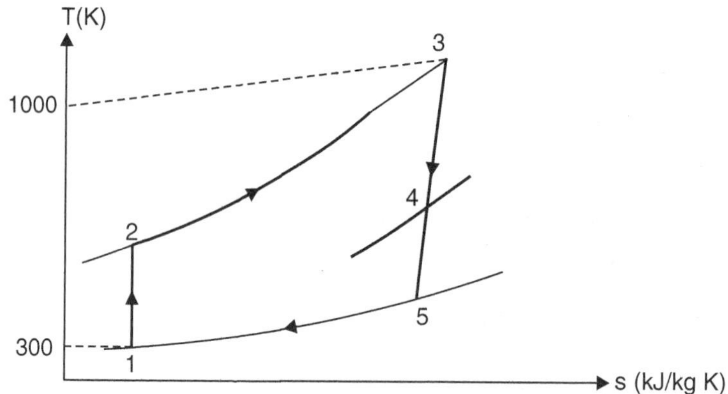

Fig. 21.24

Also, $$\dot{m}_a c_p (T_2 - T_1) = \dot{m}_a c_p (T_3 - T_4) \text{ (neglecting mass of fuel)}$$

or $$T_2 - T_1 = T_3 - T_4$$

$$445.8 - 300 = 1000 - T_4, \quad \text{or} \quad T_4 = 854.2 \text{ K}$$

For process 3-4 :

$$\frac{T_3}{T_4} = \left(\frac{p_3}{p_4}\right)^{\frac{\gamma-1}{\gamma}}, \quad \text{or} \quad \frac{p_3}{p_4} = \left(\frac{T_3}{T_4}\right)^{\frac{1.4}{0.4}}$$

or $$\frac{p_3}{p_4} = \left(\frac{1000}{854.2}\right)^{3.5} = 1.736$$

Now, $$\frac{p_3}{p_4} = \frac{p_3}{p_5} \times \frac{p_5}{p_4} = 4 \times \frac{p_5}{p_4} \qquad \left(\because \frac{p_3}{p_5} = \frac{p_2}{p_1} = 4\right)$$

$$\therefore \quad \frac{p_5}{p_4} = \frac{1}{4}\left(\frac{p_3}{p_4}\right) = \frac{1}{4} \times 1.736 = 0.434$$

Hence pressure ratio of low pressure turbine $= \frac{p_4}{p_5} = \frac{1}{0.434} = $ **2.3. (Ans.)**

Temperature of the exhaust from the unit T_5 :

$$\frac{T_4}{T_5} = \left(\frac{p_4}{p_5}\right)^{\frac{\gamma-1}{\gamma}} = (2.3)^{\frac{1.4-1}{1.4}} = 1.269$$

$$\therefore \quad \mathbf{T_5} = \frac{T_4}{1.269} = \frac{854.2}{1.269} = \mathbf{673\ K.}$$

Example 21.8. *In an air-standard regenerative gas turbine cycle the pressure ratio is 5. Air enters the compressor at 1 bar, 300 K and leaves at 490 K. The maximum temperature in the cycle is 1000 K. Calculate the cycle efficiency, given that the efficiency of the regenerator and the adiabatic efficiency of the turbine are each 80%. Assume for air, the ratio of specific heats is 1.4. Also, show the cycle on a T-s diagram.* **(GATE 1997)**

Solution. *Given :* $p_1 = 1$ bar ; $T_1 = 300$ K, $T_2' = 490$ K ; $T_3 = 1000$ K

$$\frac{p_2}{p_1} = 5, \ \eta_{turbine} = 80\%, \ \varepsilon = 80\% = 0.8 \ ; \ \gamma = 1.4$$

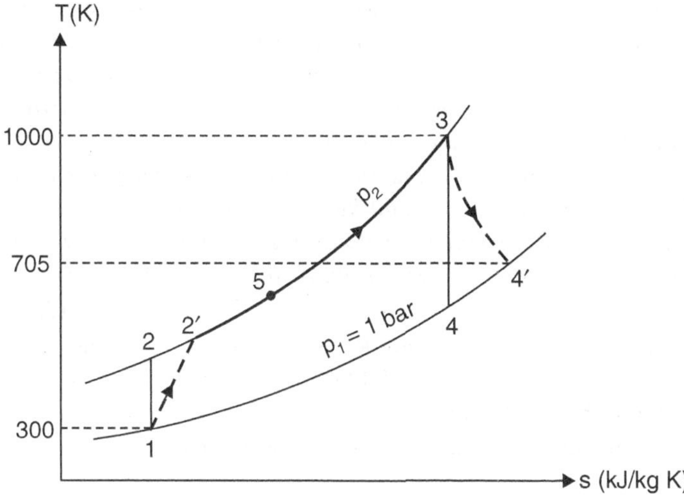

Fig. 21.25

Now, $$\frac{T_3}{T_4} = \left(\frac{p_2}{p_1}\right)^{\frac{\gamma-1}{\gamma}} = (5)^{\frac{1.4-1}{1.4}} = 1.5838$$

$$\therefore \quad T_4 = \frac{T_3}{1.5838} = \frac{1000}{1.5838} = 631.4 \text{ K}$$

Also, $$\eta_{turbine} = \frac{T_3 - T_4'}{T_3 - T_4}$$

or $$0.8 = \frac{1000 - T_4'}{1000 - 631.4}$$

$$\therefore \quad T_4' = 1000 - 0.8\ (1000 - 631.4) = 705 \text{ K}$$

Effectiveness of heat exchanger, $\varepsilon = \dfrac{T_5 - T_2'}{T_4' - T_2'}$

or $$0.8 = \frac{T_5 - 490}{705 - 490}$$

$\therefore \qquad T_5 = 0.8\ (705 - 490) + 490 = 662\ \text{K}$

Work consumed by compressor $= c_p\ (T_2' - T_1)$

$= 1.005\ (490 - 300) = 190.9\ \text{kJ/kg}$

Work done by turbine $= c_p\ (T_3 - T_4')$

$= 1.005\ (1000 - 705) = 296.5\ \text{kJ/kg}.$

Heat supplied $= c_p\ (T_3 - T_5)$

$= 1.005\ (1000 - 662) = 339.7\ \text{kJ/kg}$

$$\therefore \quad \text{Cycle efficiency, } \eta_{\text{cycle}} = \frac{\text{Network}}{\text{Heat supplied}}$$

$$= \frac{\text{Turbine work} - \text{Compressor work}}{\text{Heat supplied}}$$

$$= \frac{296.5 - 190.9}{339.7} = 0.31 \quad \text{or} \quad \mathbf{31\%.} \quad \textbf{(Ans.)}$$

Example 21.9. *A gas turbine plant consists of two turbines. One compressor turbine to drive compressor and other power turbine to develop power output and both are having their own combustion chambers which are served by air directly from the compressor. Air enters the compressor at 1 bar and 288 K and is compressed to 8 bar with an isentropic efficiency of 76%. Due to heat added in the combustion chamber, the inlet temperature of gas to both turbines is 900°C. The isentropic efficiency of turbines is 86% and the mass flow rate of air at the compressor is 23 kg/s. The calorific value of fuel is 4200 kJ/kg. Calculate the output of the plant and the thermal efficiency if mechanical efficiency is 95% and generator efficiency is 96%. Take c_p = 1.005 kJ/kg K and γ = 1.4 for air and c_{pg} = 1.128 kJ/kg K and γ = 1.34 for gases.*

(AMIE Summer, 2001)

Solution. *Given :* $p_1 = 1$ bar ; $T_1 = 288$ K ; $p_2 = 8$ bar, $\eta_{(\text{isen})} = 76\%$; $T_3 = 900°\text{C}$ or 1173 K, $\eta_{\text{T(isen.)}} = 86\%$, $m_a = 23$ kg/s ; C.V. = 4200 kJ/kg ; $\eta_{\text{mech.}} = 95\%$; $\eta_{\text{gen.}} = 96\%$; $c_p = 1.005$ kJ/kg ; $\gamma_a = 1.4$; $c_{pg} = 1.128$ kJ/kg K ; $\gamma_g = 1.34$.

The arrangement of the plant and the corresponding *T-s* diagram are shown in Fig. 21.26 (*a*), (*b*) respectively.

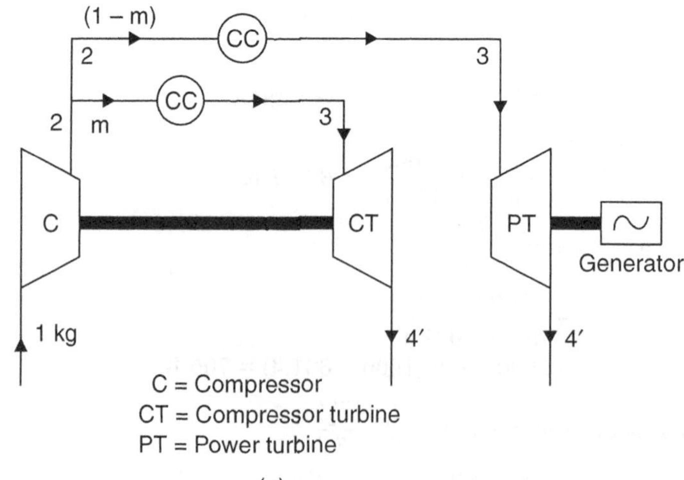

(a)

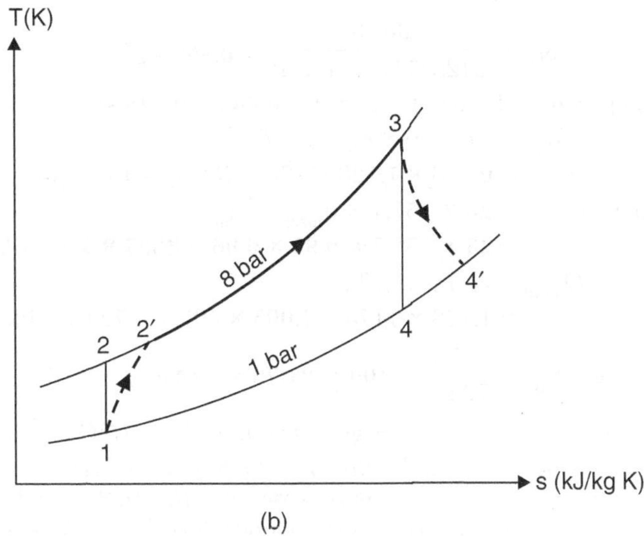

Fig. 21.26

Considering *isentropic compression process 1-2,* we have

$$\frac{T_2}{T_1} = \left(\frac{p_2}{p_1}\right)^{\frac{\gamma - 1}{\gamma}} = \left(\frac{8}{1}\right)^{\frac{1.4-1}{1.4}} = 1.811$$

$\therefore$ $$T_2 = 288 \times 1.811 = 521.6 \text{ K}$$

Also, $$\eta_{C(\text{isen.})} = \frac{T_2 - T_1}{T_2' - T_1}$$

or $$0.76 = \frac{521.6 - 288}{T_2' - 288}$$

or $$T_2' = \frac{521.6 - 288}{0.76} + 288 = 595.4 \text{ K}$$

Considering *isentropic expansion process 3-4,* we have

$$\frac{T_4}{T_3} = \left(\frac{p_4}{p_3}\right)^{\frac{\gamma - 1}{\gamma}} = \left(\frac{1}{8}\right)^{\frac{1.34-1}{1.34}} = 0.59$$

$\therefore$ $$T_4 = 1173 \times 0.59 = 692.1 \text{ K}$$

Also, $$\eta_{T(\text{isen.})} = \frac{T_3 - T_4'}{T_3 - T_4}$$

or $$0.86 = \frac{1173 - T_4'}{1173 - 692.1}$$

$\therefore$ $$T_4' = 1173 - 0.86(1173 - 692.1) = 759.4 \text{ K}$$

Consider 1 kg of air flow through compressor

$$W_{compressor} = c_p\,(T_2' - T_1') = 1.005(595.4 - 288) = 308.9 \text{ kJ}$$

This is equal to work of compressor turbine.

$\therefore$ $$308.9 = m_1 \times c_{pg}\,(T_3 - T_4'),\ \text{neglecting fuel mass}$$

or $$m_1 = \frac{308.9}{1.128\,(1173 - 759.4)} = 0.662 \text{ kg}$$

and flow through the power turbine = 1 – m = 1 – 0.662 = 0.338 kg

$$\therefore \quad W_{PT} = (1 - m) \times c_{pg}(T_3 - T_4')$$
$$= 0.338 \times 1.128\,(1173 - 759.4) = 157.7 \text{ kJ}$$

$\therefore$ **Power output** $= 23 \times 157.7 \times \eta_{mech.} \times \eta_{gen.}$
$$= 23 \times 157.7 \times 0.95 \times 0.96 = \mathbf{3307.9\ kJ.} \quad \textbf{(Ans.)}$$

$$Q_{input} = c_{pg}T_3 - c_{pa}T_2'$$
$$= 1.128 \times 1173 - 1.005 \times 595.4 = 724.7 \text{ kJ/kg of air}$$

Thermal efficiency, $\eta_{th} = \frac{157.7}{724.7} \times 100 = \mathbf{21.76\%.}$ **(Ans.)**

☞**Example 21.10.** *Air is drawn in a gas turbine unit at 15°C and 1.01 bar and pressure ratio is 7 : 1. The compressor is driven by the H.P. turbine and L.P. turbine drives a separate power shaft. The isentropic efficiencies of compressor, and the H.P. and L.P. turbines are 0.82, 0.85 and 0.85 respectively. If the maximum cycle temperature is 610°C, calculate :*

(i) The pressure and temperature of the gases entering the power turbine.

(ii) The net power developed by the unit per kg/s mass flow.

(iii) The work ratio.

(iv) The thermal efficiency of the unit.

Neglect the mass of fuel and assume the following :

For compression process c_{pa} *= 1.005 kJ/kg K and* γ *= 1.4*

For combustion and expansion processes ; c_{pg} *= 1.15 kJ/kg and* γ *= 1.333.*

Solution. *Given :* $T_1 = 15 + 273 = 288$ K, $p_1 = 1.01$ bar, Pressure ratio $= \frac{p_2}{p_1} = 7$,

$$\eta_{compressor} = 0.82,\ \eta_{turbine\,(H.P.)} = 0.85,\ \eta_{turbine\,(L.P.)} = 0.85,$$

Maximum cycle temperature, $T_3 = 610 + 273 = 883$ K

(*i*) **Pressure and temperature of the gases entering the power turbine, p_4' and T_4' :**

Considering *isentropic compression 1-2*, we have

$$\frac{T_2}{T_1} = \left(\frac{p_2}{p_1}\right)^{\frac{\gamma - 1}{\gamma}} = (7)^{\frac{1.4-1}{1.4}} = 1.745$$

$$\therefore \quad T_2 = 288 \times 1.745 = 502.5 \text{ K}$$

Also $$\eta_{compressor} = \frac{T_2 - T_1}{T_2' - T_1}$$

$$0.82 = \frac{502.5 - 288}{T_2' - 288}$$

$$\therefore \quad T_2' = \frac{502.5 - 288}{0.82} + 288 = 549.6 \text{ K}$$

$$W_{compressor} = c_{pa}(T_2' - T_1) = 1.005 \times (549.6 - 288) = 262.9 \text{ kJ/kg}$$

Now, the *work output of H.P. turbine = Work input to compressor*

$$\therefore \quad c_{pg}(T_3 - T_4') = 262.9$$

i.e., $$1.15(883 - T_4') = 262.9$$

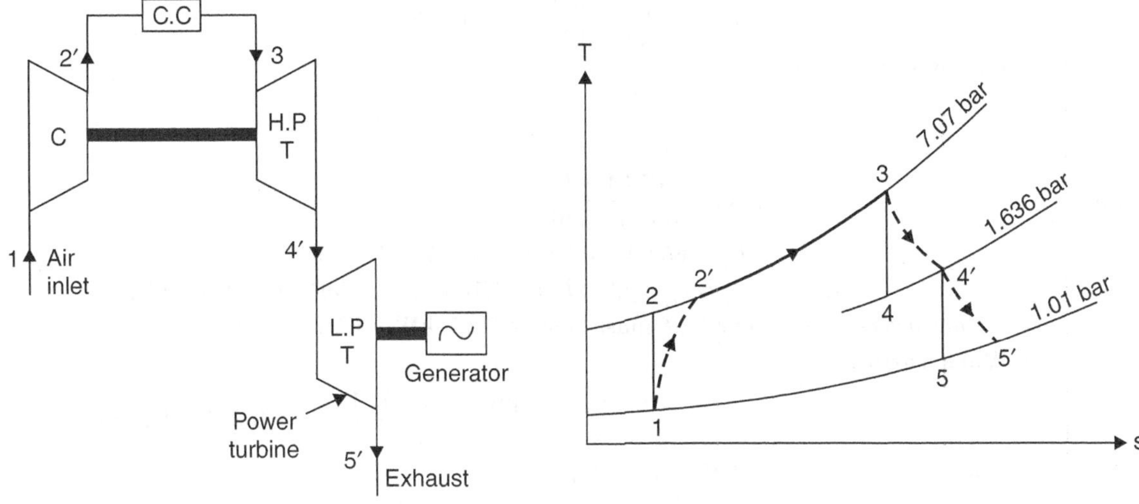

Fig. 21.27

$$\therefore \quad T_4' = 883 - \frac{262.9}{1.15} = 654.4 \text{ K}$$

i.e., *Temperature of gases entering the power turbine* = **654.4 K. (Ans.)**

Again, for H.P. turbine :

$$\eta_{turbine} = \frac{T_3 - T_4'}{T_3 - T_4} \quad i.e., \quad 0.85 = \frac{883 - 654.4}{883 - T_4}$$

$$\therefore \quad T_4 = 883 - \left(\frac{883 - 654.4}{0.85}\right) = 614 \text{ K}$$

Now, considering *isentropic expansion process 3-4,* we have

$$\frac{T_3}{T_4} = \left(\frac{p_3}{p_4}\right)^{\frac{\gamma - 1}{\gamma}}$$

or

$$\frac{p_3}{p_4} = \left(\frac{T_3}{T_4}\right)^{\frac{\gamma}{\gamma - 1}} = \left(\frac{883}{614}\right)^{\frac{1.33}{0.33}} = 4.32$$

i.e.,

$$p_4 = \frac{p_3}{4.32} = \frac{7.07}{4.32} = 1.636 \text{ bar}$$

i.e., *Pressure of gases entering the power turbine* = **1.636 bar. (Ans.)**

(*ii*) **Net power developed per kg/s mass flow, P :**

To find the power output it is now necessary to calculate T_5'.

The pressure ratio, $\frac{p_4}{p_5}$, is given by $\frac{p_4}{p_3} \times \frac{p_3}{p_5}$

i.e.,

$$\frac{p_4}{p_5} = \frac{p_4}{p_3} \times \frac{p_2}{p_1} = \frac{7}{4.32} = 1.62 \qquad (\because \quad p_2 = p_3 \text{ and } p_5 = p_1)$$

Then,

$$\frac{T_4'}{T_5} = \left(\frac{p_4}{p_5}\right)^{\frac{\gamma - 1}{\gamma}} = (1.62)^{\frac{0.33}{1.33}} = 1.127$$

$$\therefore \qquad T_5 = \frac{T_4'}{1.127} = \frac{654.4}{1.127} = 580.6 \text{ K.}$$

Again, for L.P. turbine

$$\eta_{turbine} = \frac{T_4' - T_5'}{T_4' - T_5}$$

i.e.,

$$0.85 = \frac{654.4 - T_5'}{654.4 - 580.6}$$

$$\therefore \qquad T_5' = 654.4 - 0.85\,(654.4 - 580.6) = 591.7 \text{ K}$$

$$W_{\text{L.P. } turbine} = c_{pg}\,(T_4' - T_5') = 1.15\,(654.4 - 591.7) = 72.1 \text{ kJ/kg}$$

Hence *net power output* (per kg/s mass flow) = **72.1 kW. (Ans.)**

(*iii*) **Work ratio :**

$$Work\ ratio = \frac{\text{Net work output}}{\text{Gross work output}} = \frac{72.1}{72.1 + 262.9} = \mathbf{0.215.} \quad \textbf{(Ans.)}$$

(*iv*) **Thermal efficiency of the unit, $\eta_{thermal}$:**

Heat supplied $= c_{pg}(T_3 - T_2') = 1.15\,(883 - 549.6) = 383.4$ kJ/kg

$$\therefore \qquad \eta_{thermal} = \frac{\text{Net work output}}{\text{Heat supplied}} = \frac{72.1}{383.4} = 0.188 \text{ or } \mathbf{18.8\%.} \quad \textbf{(Ans.)}$$

Example 21.11. *The pressure ratio of an open-cycle gas turbine power plant is 5.6. Air is taken at 30°C and 1 bar. The compression is carried out in two stages with perfect intercooling in between. The maximum temperature of the cycle is limited to 700°C. Assuming the isentropic efficiency of each compressor stage as 85% and that of turbine as 90%, determine the power developed and efficiency of the power plant, if the air flow is 1.2 kg/s. The mass of fuel may be neglected, and it may be assumed that c_p = 1.02 kJ/kg K and γ = 1.41.* **(P.U.)**

Solution. Refer Fig. 21.28.

Pressure ratio of the open-cycle gas turbine = 5.6

Temperature of intake air, $T_1 = 30 + 273 = 303$ K

Pressure of intake air, $p_1 = 1$ bar

Maximum temperature of the cycle, $T_5 = 700 + 273 = 973$ K

Isentropic efficiency of each compressor, $\eta_{comp.} = 85\%$

Isentropic efficiency of turbine, $\eta_{turbine} = 90\%$

Rate of air flow, $\dot{m}_a = 1.2$ kg/s

$c_p = 1.02$ kJ/kg K and $\gamma = 1.41$.

Power developed and efficiency of the power plant :

Assuming that the pressure ratio in each stage is same, we have

$$\frac{p_2}{p_1} = \frac{p_4}{p_3} = \sqrt{\frac{p_4}{p_1}} = \sqrt{5.6} = 2.366$$

Since the pressure ratio and the isentropic efficiency of each compressor is the same then the *work input required for each compressor is the same* since both the compressors have the same inlet temperature (*perfect intercooling*) *i.e.,* $T_1 = T_3$ and $T_2' = T_4'$:

Now,
$$\frac{T_2}{T_1} = \left(\frac{p_2}{p_1}\right)^{\frac{\gamma - 1}{\gamma}} = (2.366)^{\frac{1.41 - 1}{1.41}} = 1.2846 \quad \text{or} \quad T_2 = 303 \times 1.2846 = 389.23 \text{ K}$$

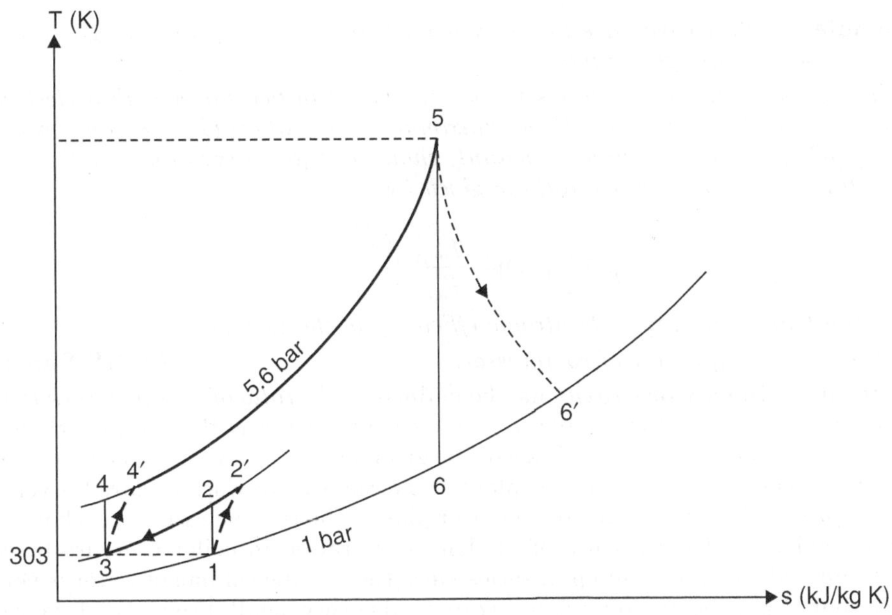

Fig. 21.28

Also, $$\eta_{comp.} = \frac{T_2 - T_1}{T_2' - T_1} \quad \text{or} \quad 0.85 = \frac{389.23 - 303}{T_2' - 303}$$

or $$T_2' = \frac{389.23 - 303}{0.85} + 303 = 404.44 \text{ K}$$

Work input to 2-stage compressor, $W_{comp.} = 2 \times m \times c_p(T_2' - T_1)$

$$= 2 \times 1.2 \times 1.02(404.44 - 303) = 248.32 \text{ kJ/s}$$

For turbine, we have

$$\frac{T_5}{T_6} = \left(\frac{p_5}{p_6}\right)^{\frac{\gamma - 1}{\gamma}} = (5.6)^{\frac{1.41 - 1}{1.41}} = 1.65 \quad \text{or} \quad T_6 = \frac{T_5}{1.65} = \frac{973}{1.65} = 589.7 \text{ K}$$

Also, $$\eta_{turbine} = \frac{T_5 - T_6'}{T_5 - T_6}$$

or $$0.9 = \frac{973 - T_6'}{973 - 589.7} \quad \text{or} \quad T_6' = 973 - 0.9(973 - 589.7) = 628 \text{ K}$$

∴ Work output of turbine, $W_{turbine} = m \times c_p(T_5 - T_6')$

$$= 1.2 \times 1.02(973 - 628) = 422.28 \text{ kJ/s}$$

Net work output, $W_{net} = W_{turbine} - W_{comp.}$

$$= 422.28 - 248.32 = 173.96 \text{ kJ/s or kW}$$

Hence *power developed* = **173.96 kW. (Ans.)**

Heat supplied, $Q_s = m \times c_p \times (T_5 - T_4')$

$$= 1.2 \times 1.02 \times (973 - 404.44) = 695.92 \text{ kJ/s}$$

∴ Power plant efficiency, $$\eta_{th} = \frac{W_{net}}{Q_s} = \frac{173.96}{695.92} = 0.25$$ or **25%. (Ans.)**

Example 21.12. *(a) Why are the back work ratios relatively high in gas turbine plants compared to those of steam power plants ?*

(b) In a gas turbine plant compression is carried out in two stages with perfect intercooling and expansion in one stage turbine. If the maximum temperature (T_{max} K) and minimum temperature (T_{min}K) in the cycle remain constant, show that for maximum specific output of the plant, the optimum overall pressure ratio is given by

$$r_{opt} = \left(\eta_T \cdot \eta_C \cdot \frac{T_{max}}{T_{min}} \right)^{\frac{2\gamma}{3(\gamma - 1)}}$$

where γ = *Adiabatic index ;* η_T = *Isentropic efficiency of the turbine.*

η_C = *Isentropic efficiency of compressor.* **(AMIE Summer, 1998)**

Solution. (*a*) **Back work ratio** may be defined as *the ratio of negative work to the turbine work in a power plant.* In gas turbine plants, air is compressed from the turbine exhaust pressure to the combustion chamber pressure. This work is given by $- \int vdp$. As the specific volume of air is very high (even in closed cycle gas turbine plants), the compressor work required is very high, and also *bulky compressor is required.* In steam power plants, the turbine exhaust is changed to liquid phase in the condenser. The pressure of condensate is raised to boiler pressure by condensate extraction pump and boiler feed pump in series since the specific volume of water is very small as compared to that of air, the pump work ($- \int vdp$), is also very small. From the above reasons, the back work ratio

$$= \frac{- \int vdp}{\text{Turbine work}}$$

for gas *turbine plants is relatively high compared to that for steam power plants.*

(*b*) Refer Fig. 21.29.

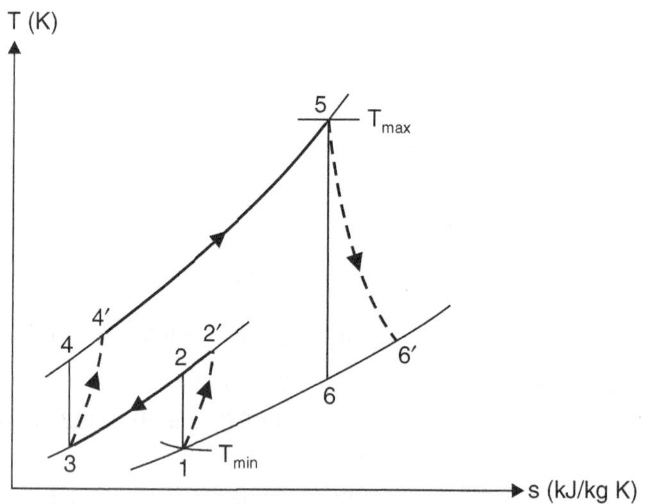

Fig. 21.29

Assuming optimum pressure ratio in each stage of the compressors $\sqrt{r}$,

$$\frac{T_2}{T_1} = \left(\frac{p_2}{p_1} \right)^{\frac{\gamma - 1}{\gamma}}$$

or $$T_2 = T_{min} \times (r)^{\frac{\gamma - 1}{2\gamma}}$$

$$W_{compressor} = 2[c_p (T_2' - T_1)] \text{ for both compressors}$$

$$= 2c_p \frac{T_2 - T_1}{\eta_C} = \frac{2c_p}{\eta_C} T_{min} \left[(r)^{\frac{\gamma - 1}{2\gamma}} - 1 \right], \text{ as } T_1 = T_{min}$$

Also, $$\frac{T_5}{T_6} = \left(\frac{p_2}{p_1} \right)^{\frac{\gamma - 1}{\gamma}} = (r)^{\frac{\gamma - 1}{\gamma}}$$

$\therefore$ $$T_6 = \frac{T_5}{(r)^{\gamma - 1/\gamma}} = \frac{T_{max}}{(r)^{\gamma - 1/\gamma}}, \text{ as } T_5 = T_{max}$$

$$W_{turbine} = c_p (T_5 - T_6') = c_p \left[T_{max} - \frac{T_{max}}{(r)^{\gamma - 1/\gamma}} \right] \eta_T, \text{ as } \eta_T = \frac{T_5 - T_6'}{T_5 - T_6}$$

$$= c_p T_{max} \left[1 - \frac{1}{(r)^{\gamma - 1/\gamma}} \right] \eta_T$$

$$W_{net} = W_{turbine} - W_{compressor}$$

$$= c_p \eta_T T_{max} \left[1 - \frac{1}{(r)^{\gamma - 1/\gamma}} \right] - \frac{2c_p}{\eta_C} T_{min} [(r)^{\frac{\gamma - 1}{2\gamma}} - 1]$$

For maximum work output,

$$\frac{dW_{net}}{dr} = 0$$

or $$- c_p \eta_T T_{max} \left(- \frac{\gamma - 1}{\gamma} \right) (r)^{-\left(\frac{\gamma - 1}{\gamma}\right) - 1} - \frac{2c_p}{\eta_C} T_{min} \left(\frac{\gamma - 1}{2\gamma} \right) (r)^{\frac{\gamma - 1}{2\gamma} - 1} = 0$$

or $$\eta_T \eta_C \frac{T_{max}}{T_{min}} = (r)^{3(\gamma - 1)/2\gamma}, \text{ on simplification.}$$

Hence, the optimum pressure ratio is

$$r_{opt} = \left[\eta_T \cdot \eta_C \cdot \frac{T_{max}}{T_{min}} \right]^{\frac{2\gamma}{3(\gamma - 1)}} \text{...... } \textbf{Proved.}$$

Example 21.13. *In a gas turbine the compressor takes in air at a temperature of 15°C and compresses it to four times the initial pressure with an isentropic efficiency of 82%. The air is then passed through a heat exchanger heated by the turbine exhaust before reaching the combustion chamber. In the heat exchanger 78% of the available heat is given to the air. The maximum temperature after constant pressure combustion is 600°C, and the efficiency of the turbine is 70%. Neglecting all losses except those mentioned, and assuming the working fluid throughout the cycle to have the characteristic of air find the efficiency of the cycle.*

Assume R = 0.287 kJ/kg K and γ = 1.4 for air and constant specific heats throughout.

Solution. *Given :* $T_1 = 15 + 273 = 288$ K, Pressure ratio, $\frac{p_2}{p_1} = \frac{p_3}{p_4} = 4$, $\eta_{compressor} = 82\%$.

Effectiveness of the heat exchanger, $\varepsilon = 0.78$,

$\eta_{turbine} = 70\%$, Maximum temperature, $T_3 = 600 + 273 = 873$ K.

Efficiency of the cycle η_{cycle} :

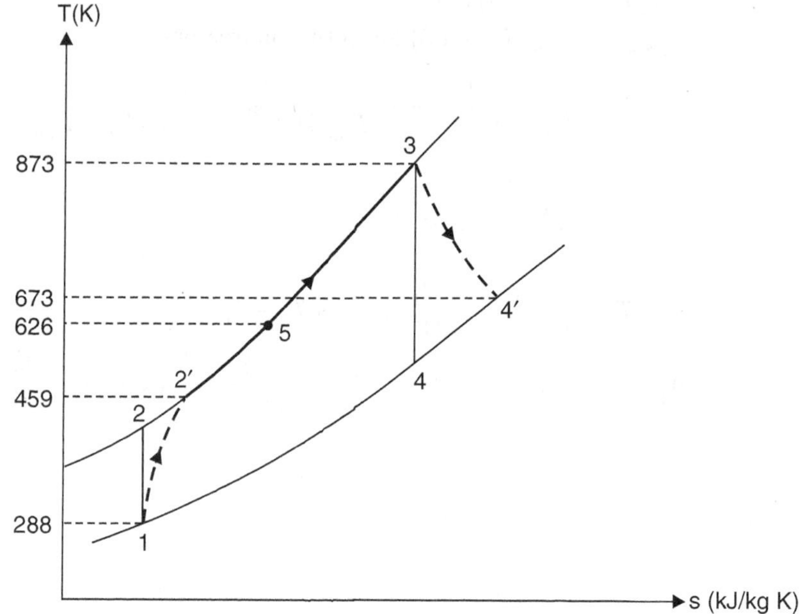

Fig. 21.30

Considering the *isentropic compression 1-2*, we have

$$\frac{T_2}{T_1} = \left(\frac{p_2}{p_1}\right)^{\frac{\gamma-1}{\gamma}} = (4)^{\frac{1.4-1}{1.4}} = 1.486$$

$\therefore$ $$T_2 = 288 \times 1.486 = 428 \text{ K}$$

Now, $$\eta_{compressor} = \frac{T_2 - T_1}{T_2' - T_1}$$

i.e., $$0.82 = \frac{428 - 288}{T_2' - 288}$$

$\therefore$ $$T_2' = \frac{428 - 288}{0.82} + 288 = 459 \text{ K}$$

Considering the *isentropic expansion process 3-4*, we have

$$\frac{T_3}{T_4} = \left(\frac{p_3}{p_4}\right)^{\frac{\gamma-1}{\gamma}} = (4)^{\frac{1.4-1}{1.4}} = 1.486$$

$\therefore$ $$T_4 = \frac{T_3}{1.486} = \frac{873}{1.486} = 587.5 \text{ K.}$$

Again, $$\eta_{turbine} = \frac{T_3 - T_4'}{T_3 - T_4} = \frac{873 - T_4'}{873 - 587.5}$$

i.e., $$0.70 = \frac{873 - T_4'}{873 - 587.5}$$

$\therefore \quad T_4' = 873 - 0.7(873 - 587.5) = 673 \text{ K}$

$$W_{compressor} = c_p(T_2' - T_1)$$

But $\quad c_p = R \times \frac{\gamma}{\gamma - 1} = 0.287 \times \frac{1.4}{1.4 - 1} = 1.0045 \text{ kJ/kg K}$

$\therefore \quad W_{compressor} = 1.0045(459 - 288) = 171.7 \text{ kJ/kg}$

$$W_{turbine} = c_p(T_3 - T_4') = 1.0045(873 - 673) = 200.9 \text{ kJ/kg}$$

$\therefore \quad \text{Net work} = W_{turbine} - W_{compressor} = 200.9 - 171.7 = 29.2 \text{ kJ/kg.}$

Effectiveness for heat exchanger, $\varepsilon = \frac{T_5 - T_2'}{T_4' - T_2'}$

i.e., $\quad 0.78 = \frac{T_5 - 459}{673 - 459}$

$\therefore \quad T_5 = (673 - 459) \times 0.78 + 459 = 626 \text{ K}$

$\therefore$ Heat supplied by fuel per kg

$$= c_p(T_3 - T_5) = 1.0045(873 - 626) = 248.1 \text{ kJ/kg}$$

$\therefore \quad \eta_{cycle} = \frac{\text{Net work done}}{\text{Heat supplied by the fuel}} = \frac{29.2}{248.1} = 0.117$ or **11.7%. (Ans.)**

Example 21.14. *A gas turbine employs a heat exchanger with a thermal ratio of 72%. The turbine operates between the pressures of 1.01 bar and 4.04 bar and ambient temperature is 20°C. Isentropic efficiencies of compressor and turbine are 80% and 85% respectively. The pressure drop on each side of the heat exchanger is 0.05 bar and in the combustion chamber 0.14 bar. Assume combustion efficiency to be unity and calorific value of the fuel to be 41800 kJ/kg.*

Calculate the increase in efficiency due to heat exchanger over that for simple cycle.

Assume c_p is constant throughout and is equal to 1.024 kJ/kg K, and assume $\gamma = 1.4$.

For simple cycle the air-fuel ratio is 90 : 1, and for the heat exchange cycle the turbine entry temperature is the same as for a simple cycle.

Solution. Simple Cycle. Refer Fig. 21.31.

$$\frac{T_2}{T_1} = \left(\frac{p_2}{p_1}\right)^{\frac{\gamma - 1}{\gamma}} = \left(\frac{4.40}{1.01}\right)^{\frac{1.4 - 1}{1.4}} = 1.486$$

$\therefore \quad T_2 = 293 \times 1.486 = 435.4$

Also, $\quad \eta_{compressor} = \frac{T_2 - T_1}{T_2' - T_1}$

$$0.8 = \frac{435.4 - 293}{T_2' - 293}$$

$\therefore \quad T_2' = \frac{435.4 - 293}{0.8} + 293 = 471 \text{ K}$

Now, $\quad m_f \times C = (m_a + m_f) \times c_p \times (T_3 - T_2')$

[m_a = mass of air, m_f = mass of fuel]

$\therefore \quad T_3 = \frac{m_f \times C}{c_p\,(m_a + m_f)} + T_2' = \frac{1 \times 41800}{1.024\,(90 + 1)} + 471 = 919.5 \text{ K}$

Also, $\quad \frac{T_4}{T_3} = \left(\frac{p_4}{p_3}\right)^{\frac{\gamma - 1}{\gamma}}$

or $$T_4 = T_3 \times \left(\frac{p_4}{p_3}\right)^{\frac{\gamma - 1}{\gamma}} = 919.5 \times \left(\frac{1.01}{3.9}\right)^{\frac{1.4-1}{1.4}} = 625 \text{ K}$$

Again, $$\eta_{turbine} = \frac{T_3 - T_4'}{T_3 - T_4}$$

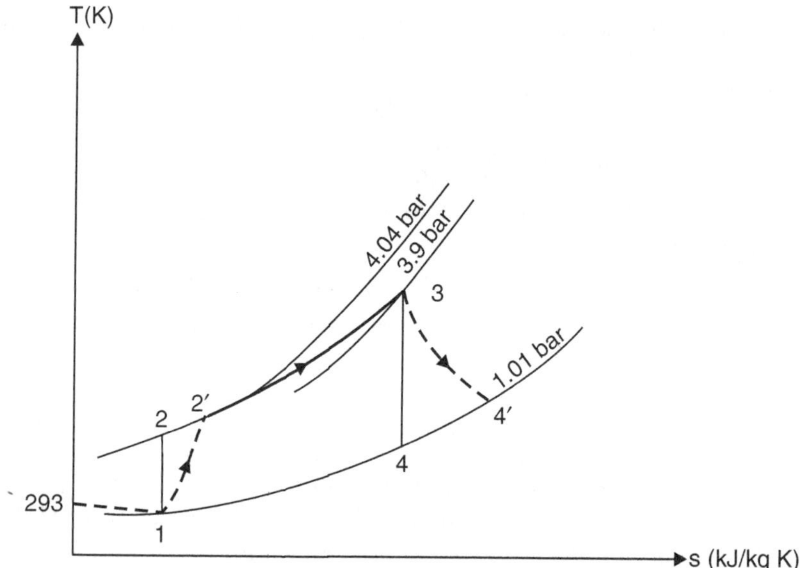

Fig. 21.31

$\therefore$ $$0.85 = \frac{919.5 - T_4'}{919.5 - 625}$$

$\therefore$ $$T_4' = 919.5 - 0.85(919.5 - 625) = 669 \text{ K}$$

$$\eta_{thermal} = \frac{(T_3 - T_4') - (T_2' - T_1)}{(T_3 - T_2')}$$

$$= \frac{(919.5 - 669) - (471 - 293)}{(919.5 - 471)} = \frac{72.5}{448.5} = 0.1616 \text{ or } \mathbf{16.16\%}. \quad \textbf{(Ans.)}$$

Heat Exchanger Cycle. Refer Fig. 21.32 (*a, b*)

$T_2' = 471$ K (as for simple cycle) ; $T_3 = 919.5$ K (as for simple cycle)

To find T_4' :

$$p_3 = 4.04 - 0.14 - 0.05 = 3.85 \text{ bar} ; \; p_4 = 1.01 + 0.05 = 1.06 \text{ bar}$$

$\therefore$ $$\frac{T_4}{T_3} = \left(\frac{p_4}{p_3}\right)^{\frac{\gamma-1}{\gamma}} = \left(\frac{1.06}{3.85}\right)^{\frac{1.4-1}{1.4}} = 0.69$$

i.e., $$T_4 = 919.5 \times 0.69 = 634 \text{ K}$$

$$\eta_{turbine} = \frac{T_3 - T_4'}{T_3 - T_4} \; ; \quad 0.85 = \frac{919.5 - T_4'}{919.5 - 634}$$

$\therefore$ $$T_4' = 919.5 - 0.85 \,(919.5 - 634) = 677 \text{ K}$$

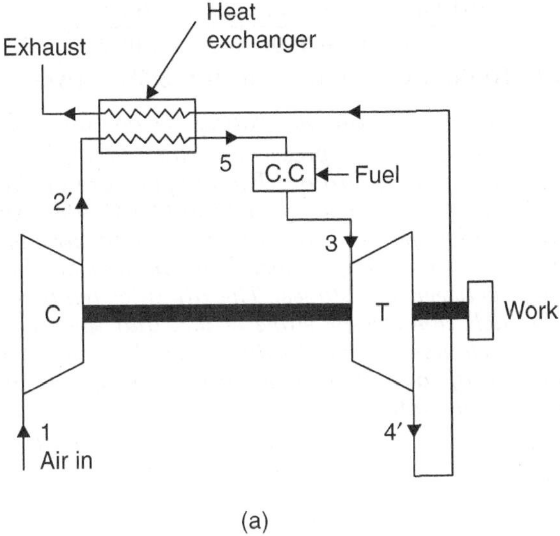

(a)

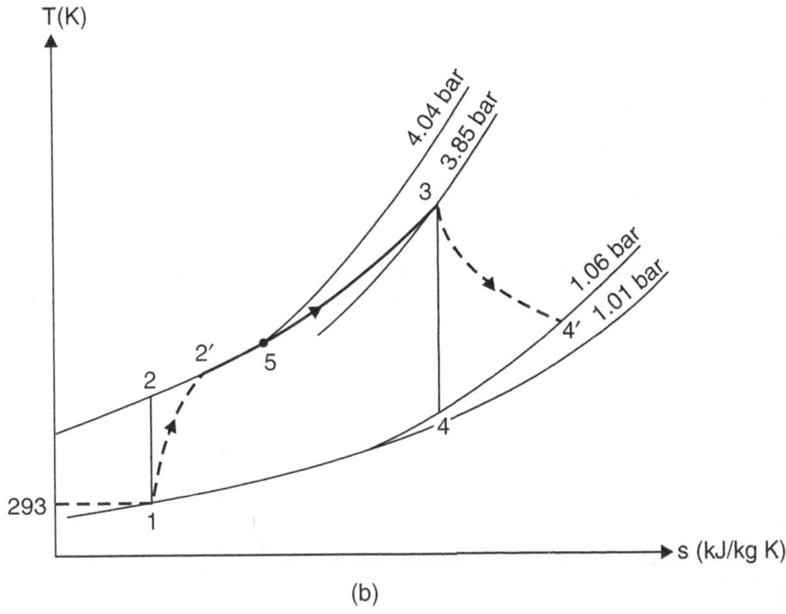

(b)

Fig. 21.32

To find T_5 :

Thermal ratio (or effectiveness),

$$\varepsilon = \frac{T_5 - T_2'}{T_4' - T_2'} \qquad \therefore \quad 0.72 = \frac{T_5 - 471}{677 - 471}$$

$$\therefore \qquad T_5 = 0.72(677 - 471) + 471 = 619 \text{ K}$$

$$\eta_{thermal} = \frac{(T_3 - T_4') - (T_2' - T_1)}{(T_3 - T_5)}$$

$$= \frac{(919.5 - 677) - (471 - 293)}{(919.5 - 619)} = \frac{64.5}{300.5} = 0.2146 \text{ or } \mathbf{21.46\%}$$

$\therefore$ **Increase in thermal efficiency** = 21.46 – 16.16 = **5.3%. (Ans.)**

☞**Example 21.15.** *A 4500 kW gas turbine generating set operates with two compressor stages ; the overall pressure ratio is 9 : 1. A high pressure turbine is used to drive the compressors, and a low-pressure turbine drives the generator. The temperature of the gases at entry to the high pressure turbine is 625°C and the gases are reheated to 625°C after expansion in the first turbine. The exhaust gases leaving the low-pressure turbine are passed through a heat exchanger to heat air leaving the high pressure stage compressor. The compressors have equal pressure ratios and intercooling is complete between the stages. The air inlet temperature to the unit is 20°C. The isentropic efficiency of each compressor stage is 0.8, and the isentropic efficiency of each turbine stage is 0.85, the heat exchanger thermal ratio is 0.8. A mechanical efficiency of 95% can be assumed for both the power shaft and compressor turbine shaft. Neglecting all pressure losses and changes in kinetic energy calculate :*

(i) The thermal efficiency ; *(ii) Work ratio of the plant ;*

(iii) The mass flow in kg/s.

Neglect the mass of the fuel and assume the following :

For air : $c_{pa} = 1.005$ *kJ/kg K and* $\gamma = 1.4$

For gases in the combustion chamber and in turbines and heat exchanger, $c_{pg} = 1.15$ *kJ/kg K and* $\gamma = 1.333$.

Solution. Refer Fig. 21.33

Given : $T_1 = 20 + 273 + 293$ K, $T_6 = T_8 = 625 + 273 = 898$ K

Efficiency of each compressor stage = 0.8

Efficiency of each turbine stage = 0.85, $\eta_{mech.} = 0.95$, $\varepsilon = 0.8$

(i) **Thermal efficiency,** η_{thermal} **:**

Since the pressure ratio and the isentropic efficiency of each compressor is the same then the work input required for each compressor is the same since both compressors have the same air inlet temperature *i.e.,* $T_1 = T_3$ and $T_2' = T_4'$.

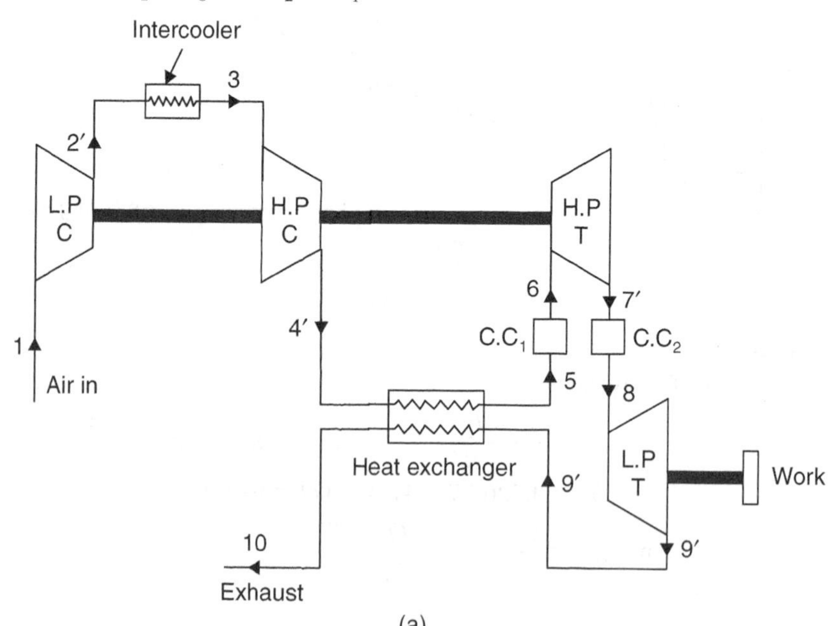

(a)

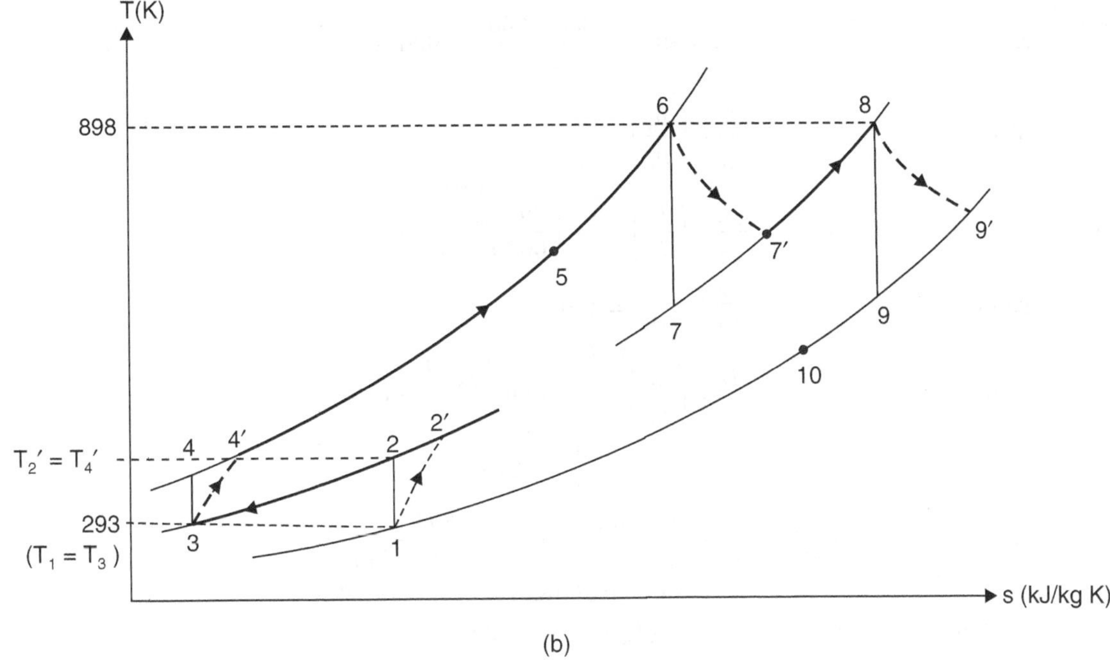

(b)

Fig. 21.33

Also, $\dfrac{T_2}{T_1} = \left(\dfrac{p_2}{p_1}\right)^{\frac{\gamma - 1}{\gamma}}$ and $\dfrac{p_2}{p_1} = \sqrt{9} = 3$

$$\therefore \qquad T_2 = (20 + 273) \times (3)^{\frac{1.4 - 1}{1.4}} = 401 \text{ K}$$

Now, $$\eta_{compressor} \text{ (L.P.)} = \frac{T_2 - T_1}{T_2' - T_1}$$

$$0.8 = \frac{401 - 293}{T_2' - 293}$$

i.e., $$T_2' = \frac{401 - 298}{0.8} + 293 = 428 \text{ K}$$

Work input per compressor stage

$$= c_{pa}(T_2' - T_1) = 1.005\ (428 - 293) = 135.6 \text{ kJ/kg}$$

The H.P. turbine is required to drive both compressors and to overcome mechanical friction.

i.e., Work output of H.P. turbine $= \dfrac{2 \times 135.6}{0.95} = 285.5$ kJ/kg

$$\therefore \qquad c_{pg}\ (T_6 - T_7') = 285.5$$

i.e., $$1.15\ (898 - T_7') = 285.5$$

$$\therefore \qquad T_7' = 898 - \frac{285.5}{1.15} = 650 \text{ K}$$

Now, $$\eta_{turbine \text{ (H.P.)}} = \frac{T_6 - T_7'}{T_6 - T_7} \ ; \quad 0.85 = \frac{898 - 650}{898 - T_7}$$

$\therefore \qquad T_7 = 898 - \left(\dfrac{898 - 650}{0.85}\right) = 606 \text{ K}$

Also, $\qquad \dfrac{T_6}{T_7} = \left(\dfrac{p_6}{p_7}\right)^{\frac{\gamma - 1}{\gamma}}$

or $\qquad \dfrac{p_6}{p_7} = \left(\dfrac{T_6}{T_7}\right)^{\frac{\gamma}{\gamma - 1}} = \left(\dfrac{898}{606}\right)^{\frac{1.333}{0.333}} = 4.82$

Then, $\qquad \dfrac{p_8}{p_9} = \dfrac{9}{4.82} = 1.86$

Again, $\qquad \dfrac{T_8}{T_9} = \left(\dfrac{p_8}{p_9}\right)^{\frac{\gamma - 1}{\gamma}} = (1.86)^{\frac{1.333 - 1}{1.333}} = 1.16$

$\therefore \qquad T_9 = \dfrac{T_8}{1.16} = \dfrac{898}{1.16} = 774 \text{ K}$

Also, $\qquad \eta_{turbine\,(L.P.)} = \dfrac{T_8 - T_9'}{T_8 - T_9}\ ; \quad 0.85 = \dfrac{898 - T_9'}{898 - 774}$

$\therefore \qquad T_9' = 898 - 0.85\,(898 - 774) = 792.6 \text{ K}$

$\therefore$ Net work output $= c_{pg}(T_8 - T_9') \times 0.95$

$= 1.15\,(898 - 792.6) \times 0.95 = 115.15$ kJ/kg

Thermal ratio or effectiveness of heat exchanger,

$$\varepsilon = \frac{T_5 - T_4'}{T_9' - T_4'} = \frac{T_5 - 428}{792.6 - 428}$$

i.e., $\qquad 0.8 = \dfrac{T_5 - 428}{792.6 - 428}$

$\therefore \qquad T_5 = 0.8\,(792.6 - 428) + 428 = 719.7 \text{ K}$

Now, $\qquad$ Heat supplied $= c_{pg}(T_6 - T_5) + c_{pg}(T_8 - T_7')$

$= 1.15\,(898 - 719.7) + 1.15\,(898 - 650) = 490.2$ kJ/kg

$\therefore \qquad \eta_{thermal} = \dfrac{\text{Net work output}}{\text{Heat supplied}} = \dfrac{115.15}{490.2}$

$= 0.235$ or **23.5%. (Ans.)**

(*ii*) **Work ratio :**

Gross work of the plant $= W_{turbine\,(H.P.)} + W_{turbine\,(L.P.)}$

$= 285.5 + \dfrac{115.15}{0.95} = 406.7$ kJ/kg

$\therefore$ **Work ratio** $= \dfrac{\text{Net work output}}{\text{Gross work output}} = \dfrac{115.15}{406.7} =$ **0.283. (Ans.)**

(*iii*) **Mass flow in ṁ :**

Let the mass flow be $\dot{m}$, then

$\dot{m} \times 115.15 = 4500$

$\therefore \qquad \dot{m} = \dfrac{4500}{115.15} = 39.08$ kg/s

i.e., **Mass flow = 39.08 kg/s. (Ans.)**

Example 21.16. *In a closed cycle gas turbine there is two-stage compressor and a two-stage turbine. All the components are mounted on the same shaft. The pressure and temperature at the inlet of the first-stage compressor are 1.5 bar and 20°C. The maximum cycle temperature and pressure are limited to 750°C and 6 bar. A perfect intercooler is used between the two-stage compressors and a reheater is used between the two turbines. Gases are heated in the reheater to 750°C before entering into the L.P. turbine. Assuming the compressor and turbine efficiencies as 0.82, calculate :*

(*i*) *The efficiency of the cycle without regenerator.*

(*ii*) *The efficiency of the cycle with a regenerator whose effectiveness is 0.70.*

(*iii*) *The mass of the fluid circulated if the power developed by the plant is 350 kW,*

The working fluid used in the cycle is air. For air : $\gamma = 1.4$ *and* $c_p = 1.005$ *kJ/kg K.*

Solution. *Given :* $T_1 = 20 + 273 = 293$ K, $T_5 = T_7 = 750 + 273 = 1023$ K, $p_1 = 1.5$ bar, $p_2 = 6$ bar, $\eta_{compressor} = \eta_{turbine} = 0.82$.

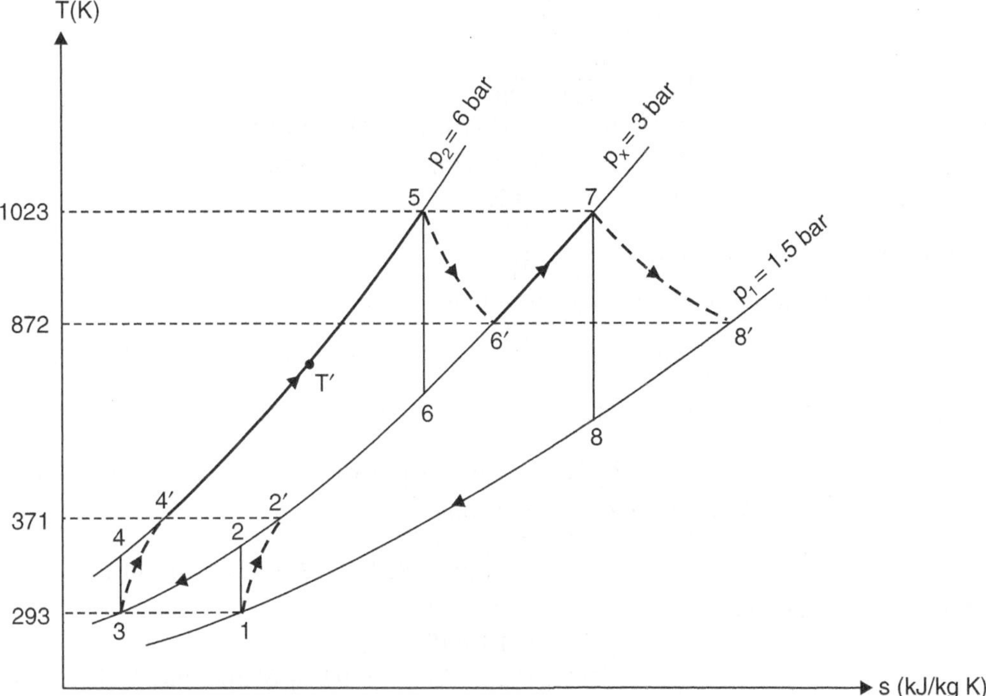

Fig. 21.34

Effectiveness of regenerator, $\varepsilon = 0.70$, Power developed, $P = 350$ kW.

For air : $c_p = 1.005$ kJ/kg, K, $\gamma = 1.4$

As per given conditions : $T_1 = T_3$, $T_2' = T_4'$

$$\frac{T_2}{T_1} = \left(\frac{p_2}{p_1}\right)^{\frac{\gamma-1}{\gamma}} \quad \text{and} \quad p_x = \sqrt{p_1 p_2} = \sqrt{1.5 \times 6} = 3 \text{ bar}$$

Now
$$T_2 = T_1 \times \left(\frac{p_2}{T_1}\right)^{\frac{\gamma-1}{\gamma}} = 293 \times \left(\frac{3}{1.5}\right)^{\frac{1.4-1}{1.4}} = 357 \text{ K}$$

$$\eta_{compressor\ (L.P.)} = \frac{T_2 - T_1}{T_2' - T_1}$$

$$0.82 = \frac{357 - 293}{T_2' - 293}$$

$$\therefore \quad T_2' = \frac{357 - 293}{0.82} + 293 = 371 \text{ K } i.e., T_2' = T_4' = 371 \text{ K}$$

Now,
$$\frac{T_5}{T_6} = \left(\frac{p_5}{p_6}\right)^{\frac{\gamma-1}{\gamma}} = \left(\frac{p_2}{p_x}\right)^{\frac{1.4-1}{1.4}} \quad \left[\because \begin{array}{l} p_5 = p_2 \\ p_6 = p_x \end{array}\right]$$

$$\frac{1023}{T_6} = \left(\frac{6}{3}\right)^{0.286} = 1.219$$

$$\therefore \quad T_6 = \frac{1023}{1.219} = 839 \text{ K}$$

$$\eta_{turbine\ (H.P.)} = \frac{T_5 - T_6'}{T_5 - T_6}$$

$$0.82 = \frac{1023 - T_6'}{1023 - 839}$$

$$\therefore \quad T_6' = 1023 - 0.82\,(1023 - 839) = 872 \text{ K}$$

$$T_8' = T_6' = 872 \text{ K as } \eta_{turbine\ (H.P.)} = \eta_{turbine\ (L.P.)}$$

and
$$T_7 = T_5 = 1023 \text{ K}$$

Effectiveness of regenerator, $\varepsilon = \dfrac{T' - T_4'}{T_8' - T_4'}$

where T' is the temperature of air coming out of regenerator

$$\therefore \quad 0.70 = \frac{T' - 371}{872 - 371} \quad i.e., \quad T' = 0.70\,(872 - 371) + 371 = 722 \text{ K}$$

Net work available, $W_{net} = \left[W_{T(L.P.)} + W_{T(L.P.)}\right] - \left[W_{C(H.P.)} + W_{C(L.P.)}\right]$

$= 2\left[W_{T(L.P.)} - W_{C(L.P.)}\right]$ as the *work developed by each turbine is same and work absorbed by each compressor is same.*

$$\therefore \quad W_{net} = 2c_p\,[(T_5 - T_6') - (T_2' - T_1)]$$
$$= 2 \times 1.005\,[(1023 - 872) - (371 - 293)] = 146.73 \text{ kJ/kg of air}$$

Heat supplied per kg of air *without regenerator*

$$= c_p(T_5 - T_4') + c_p(T_7 - T_6')$$
$$= 1.005\,[(1023 - 371) + (1023 - 872)] = 807 \text{ kJ/kg of air}$$

Heat supplied per kg of air *with regenerator*

$$= c_p(T_5 - T') + c_p(T_7 - T_6')$$
$$= 1.005\,[(1023 - 722) + (1023 - 872)]$$
$$= 454.3 \text{ kJ/kg}$$

(*i*) $\eta_{\text{thermal (without regenerator)}} = \dfrac{146.73}{807} = 0.182$ or **18.2%. (Ans.)**

(*ii*) $\eta_{\text{thermal (with regenerator)}} = \dfrac{146.73}{454.3} = 0.323$ or **32.3%. (Ans.)**

(*iii*) **Mass of fluid circulated, $\dot{m}$:**

Power developed, $P = 146.73 \times \dot{m}$ kW

$\therefore \quad 350 = 146.73 \times \dot{m}$

i.e., $\quad \dot{m} = \dfrac{350}{146.73} = 2.38 \text{ kg/s}$

i.e., *Mass of fluid circulated* **= 2.38 kg/s. (Ans.)**

☞**Example 21.17.** *The air in a gas turbine plant is taken in L.P. compressor at 293 K and 1.05 bar and after compression it is passed through intercooler where its temperature is reduced to 300 K. The cooled air is further compressed in H.P. unit and then passed in the combustion chamber where its temperature is increased to 750°C by burning the fuel. The combustion products expand in H.P. turbine which runs the compressors and further expansion is continued in L.P. turbine which runs the alternator. The gases coming out from L.P. turbine are used for heating the incoming air from H.P. compressor and then expanded to atmosphere.*

Pressure ratio of each compressor = 2, isentropic efficiency of each compressor stage = 82%, isentropic efficiency of each turbine stage = 82%, effectiveness of heat exchanger = 0.72, air flow = 16 kg/s, calorific value of fuel = 42000 kJ/kg, c_v (for gas) = 1.0 kJ/kg K, c_p (gas) = 1.15 kJ/kg K, γ (for air) = 1.4, γ (for gas) = 1.33.

Neglecting the mechanical, pressure and heat losses of the system and fuel mass also determine the following :

(i) The power output. *(ii) Thermal efficiency.*

(iii) Specific fuel consumption.

Solution. *Given :* $T_1 = 293$ K, $T_3 = 300$ K, $\dfrac{p_2}{p_1} = \dfrac{p_4}{p_3} = 2$, $T_6 = 750 + 273 = 1023$ K,

$\eta_{compressor} = 82\%$, $\eta_{turbine} = 82\%$, $\varepsilon = 0.72$, $\dot{m}_a = 16$ kg/s, $C = 42000$ kJ/kg,

$c_{pa} = 1.0$ kJ/kg K, $c_{pg} = 1.15$ kJ/kg K, γ (for air) = 1.4, γ (for gas) 1.33.

$$\frac{T_2}{T_1} = \left(\frac{p_2}{p_1}\right)^{\frac{\gamma-1}{\gamma}} = (2)^{\frac{1.4-1}{1.4}} = 1.219$$

$\therefore \quad T_2 = 293 \times 1.219 = 357 \text{ K}$

Also, $\quad \eta_{compressor} = \dfrac{T_2 - T_1}{T_2' - T_1}$

$\therefore \quad 0.82 = \dfrac{357 - 293}{T_2' - 293}$

$\therefore \quad T_2' = \left(\dfrac{357 - 293}{0.82}\right) + 293 = 371 \text{ K}$

Similarly, $\quad \dfrac{T_4}{T_3} = \left(\dfrac{p_4}{p_3}\right)^{\frac{\gamma-1}{\gamma}} = (2)^{\frac{1.4-1}{1.4}} = 1.219$

$\therefore \quad T_4 = 300 \times 1.219 = 365.7 \text{ K and } \eta_{compressor} = \dfrac{T_4 - T_3}{T_4' - T_3}$

$\therefore \quad 0.82 = \dfrac{365.7 - 300}{T_4' - 300}$

i.e., $\quad T_4' = \left(\dfrac{365.7 - 300}{0.82}\right) + 300 = 380 \text{ K}$

Work output of H.P. turbine = Work input to compressor.

Neglecting mass of fuel we can write

$$c_{pg}(T_6 - T_7') = c_{pa}[(T_2' - T_1) + (T_4' - T_3)]$$

$$1.15(1023 - T_7') = 1.0[(371 - 293) + (380 - 300)]$$

or $\quad 1.15(1023 - T_7') = 158$

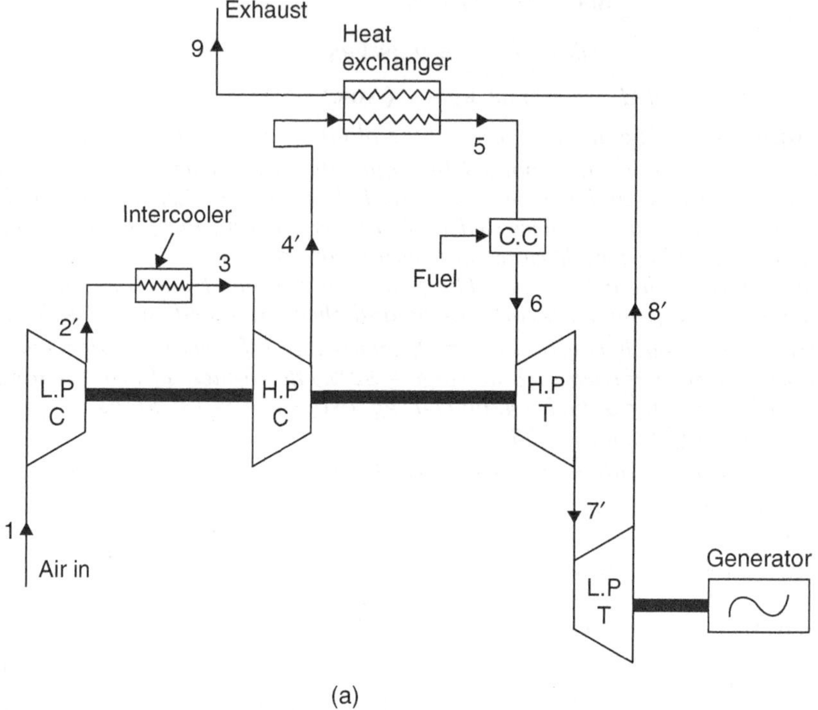
Exhaust
9
Heat
exchanger
5
Intercooler
3
4′
C.C
Fuel
6
8′
2′
L.P
C
H.P
C
H.P
T
7′
1
Air in
Generator
L.P
T

(a)

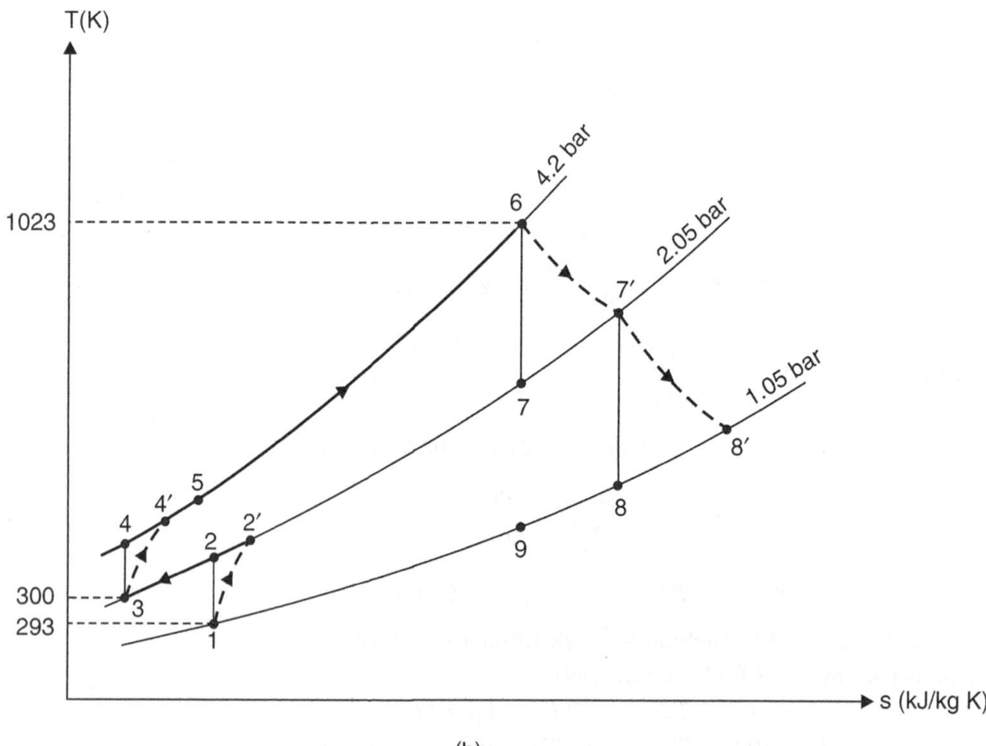
T(K)
4.2 bar
6
1023
2.05 bar
7′
7
1.05 bar
8′
5
4
4′
2
2′
8
9
300
3
293
1
s (kJ/kg K)

(b)

Fig. 21.35

$\therefore \qquad T_7' = 1023 - \dfrac{15.8}{1.15} = 886 \text{ K}$

Also, $\qquad \eta_{turbine\ (\text{H.P.})} = \dfrac{T_6 - T_7'}{T_6 - T_7}$

i.e., $\qquad 0.82 = \dfrac{1023 - 886}{1023 - T_7}$

$\therefore \qquad T_7 = 1023 - \left(\dfrac{1023 - 886}{0.82}\right) = 856 \text{ K}$

Now $\qquad \dfrac{T_6}{T_7} = \left(\dfrac{p_6}{p_7}\right)^{\frac{\gamma - 1}{\gamma}}$

$\therefore \qquad \dfrac{p_6}{p_7} = \left(\dfrac{T_6}{T_7}\right)^{\frac{\gamma}{\gamma - 1}} = \left(\dfrac{1023}{856}\right)^{\frac{1.33}{1.33 - 1}} = 2.05$

i.e., $\qquad p_7 = \dfrac{p_6}{2.05} = \dfrac{4.2}{2.05} = 2.05 \text{ bar} \qquad [\because \ p_6 = 1.05 \times 4 = 4.2 \text{ bar}]$

$$\dfrac{T_7'}{T_8} = \left(\dfrac{p_7}{p_8}\right)^{\frac{\gamma - 1}{\gamma}} = \left(\dfrac{2.05}{1.05}\right)^{\frac{1.33 - 1}{1.33}} = 1.18$$

$$T_8 = \dfrac{T_7'}{1.18} = \dfrac{886}{1.18} = 751 \text{ K}$$

Again, $\qquad \eta_{turbine\ (\text{L.P.})} = \dfrac{T_7' - T_8'}{T_7' - T_8}$

$$0.82 = \dfrac{886 - T_8'}{886 - 751}$$

$\therefore \qquad T_8' = 886 - 0.82\ (886 - 751) = 775 \text{ K}$

(*i*) **Power output :**

Net power output $= c_{pg}\ (T_7' - T_8')$

$= 1.15\ (886 - 775) = 127.6$ kJ/kg

$\therefore$ Net output per second $= \dot{m} \times 127.6$

$= 16 \times 127.6 = 2041.6$ kJ/s = **2041.6 kW. (Ans.)**

(*ii*) **Thermal efficiency :**

Effectiveness of heat exchanger, $\varepsilon = \dfrac{T_5 - T_4'}{T_8' - T_4'}$

i.e., $\qquad 0.72 = \dfrac{T_5 - 380}{778 - 380}$

$\therefore \qquad T_5 = 0.72(775 - 380) + 380 = 664 \text{ K}$

Heat supplied in combustion chamber per second

$= \dot{m}_a\ c_{pg}\ (T_6 - T_5)$

$= 16 \times 1.15(1023 - 664) = 6605.6$ kJ/s

$\therefore \qquad \eta_{thermal} = \dfrac{2041.6}{6605.6} = 0.309$ or **30.9%. (Ans.)**

(*iii*) **Specific fuel consumption** :

If m_f is the mass of fuel supplied per kg of air, then

$$m_f \times 42000 = 1.15(1023 - 664)$$

$$\therefore \quad \frac{1}{m_f} = \frac{42000}{1.15(1023 - 664)} = \frac{101.7}{1}$$

∴ Air-fuel ratio $= 101.7 : 1$

∴ Fuel supplied per hour $= \dfrac{16 \times 3600}{101.7} = 566.37$ kg/h

∴ *Specific fuel consumption* $= \dfrac{566.37}{2041.6} =$ **0.277 kg/kWh. (Ans.)**

☞**Example 21.18.** *Air is taken in a gas turbine plant at 1.1 bar 20°C. The plant comprises of L.P. and H.P. compressors and L.P. and H.P. turbines. The compression in L.P. stage is upto 3.3 bar followed by intercooling to 27°C. The pressure of air after H.P. compressor is 9.45 bar. Loss in pressure during intercooling is 0.15 bar. Air from H.P. compressor is transferred to heat exchanger of effectiveness 0.65 where it is heated by the gases from L.P. turbine. After heat exchanger the air passes through combustion chamber. The temperature of gases supplied to H.P. turbine is 700°C. The gases expand in H.P. turbine to 3.62 bar and air then reheated to 670°C before expanding in L.P. turbine. The loss of pressure in reheater is 0.12 bar. Determine :*

(*i*) *The overall efficiency* (*ii*) *The work ratio*

(*iii*) *Mass flow rate when the power generated is 6000 kW.*

Assume : Isentropic efficiency of compression in both stages = 0.82.

Isentropic efficiency of expansion in turbines = 0.85.

For air : $c_p = 1.005$ *kJ/kg K,* $\gamma = 1.4$.

For gases : $c_p = 1.15$ *kJ/kg K,* $\gamma = 1.33$.

Neglect the mass of fuel.

Solution. *Given :* $T_1 = 20 + 273 = 293$ K, $p_1 = 1.1$ bar, $p_2 = 3.3$ bar, $T_3 = 27 + 273 = 300$ K, $p_3 = 3.3 - 0.15 = 3.15$ bar, $p_4 = p_6 = 9.45$ bar, $T_6 = 973$ K, $T_8 = 670 + 273 = 943$ K, $p_8 = 3.5$ bar,

$\eta_{compressors} = 82\%$, $\eta_{turbines} = 85\%$, Power generated = 6000 kW,

Effectiveness, $\varepsilon = 0.65$, $c_{pa} = 1.005$ kJ/kg K, $\gamma_{air} = 1.44$, $c_{pg} = 1.15$ kJ/kg K and $\gamma_{gases} = 1.33$.

Refer Fig. 21.36.

Now,
$$\frac{T_2}{T_1} = \left(\frac{p_2}{p_1}\right)^{\frac{\gamma-1}{\gamma}} = \left(\frac{3.3}{1.1}\right)^{\frac{1.4-1}{1.4}} = 1.369$$

$$\therefore \quad T_2 = 293 \times 1.369 = 401 \text{ K}$$

$$\eta_{compressor\ (\text{L.P.})} = 0.82 = \frac{T_2 - T_1}{T_2' - T_1} = \frac{401 - 293}{T_2' - 293}$$

$$\therefore \quad T_2' = \left(\frac{401 - 293}{0.82}\right) + 293 = 425 \text{ K}$$

Again,
$$\frac{T_4}{T_3} = \left(\frac{p_4}{p_3}\right)^{\frac{\gamma-1}{\gamma}} = \left(\frac{9.45}{3.15}\right)^{\frac{1.4-1}{1.4}} = 1.369$$

$$\therefore \quad T_4 = 300 \times 1.369 = 411 \text{ K}$$

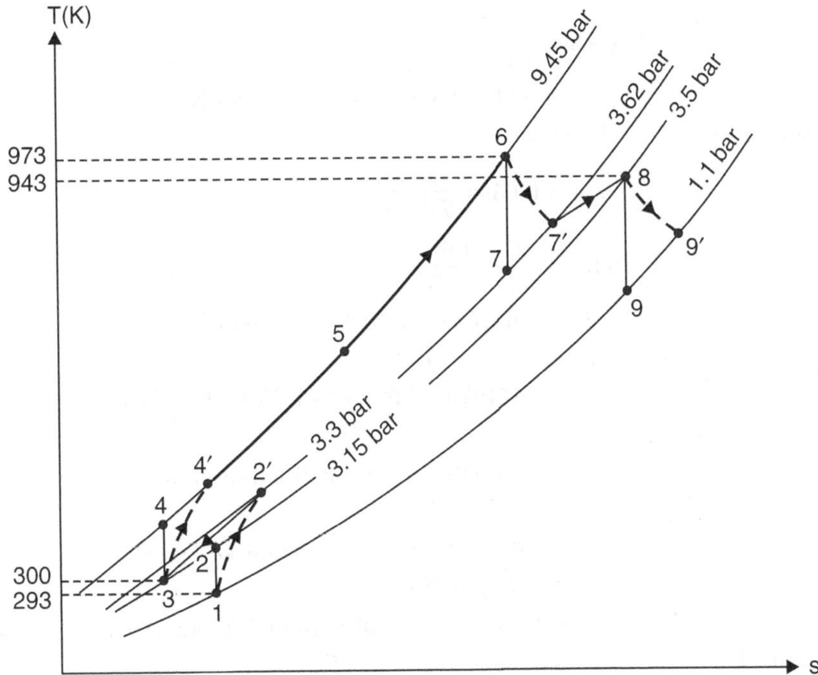

Fig. 21.36

Now, $\eta_{compressor\ (\text{H.P.})} = \dfrac{T_4 - T_3}{T_4' - T_3}$

$$0.82 = \frac{411 - 300}{T_4' - 300}$$

$\therefore$ $T_4' = \left(\dfrac{411 - 300}{0.82}\right) + 300 = 435 \text{ K}$

Similarly, $\dfrac{T_6}{T_7} = \left(\dfrac{p_6}{p_7}\right)^{\frac{\gamma-1}{\gamma}} = \left(\dfrac{9.45}{3.62}\right)^{\frac{1.33-1}{1.33}} = 1.268$

$\therefore$ $T_7 = \dfrac{T_6}{1.268} = \dfrac{973}{1.268} = 767 \text{ K}$

Also, $\eta_{turbine\ (\text{H.P.})} = \dfrac{T_6 - T_7'}{T_6 - T_7}$

$$0.85 = \frac{973 - T_7'}{973 - 767}$$

$\therefore$ $T_7' = 973 - 0.85(973 - 767) = 798 \text{ K}$

Again, $\dfrac{T_8}{T_9} = \left(\dfrac{p_8}{p_9}\right)^{\frac{\gamma-1}{\gamma}} = \left(\dfrac{3.5}{1.1}\right)^{\frac{1.33-1}{1.33}} = 1.332$

$\therefore$ $T_9 = \dfrac{T_8}{1.332} = \dfrac{943}{1.332} = 708 \text{ K}$

$$\eta_{turbine\ (\text{L.P.})} = \frac{T_8 - T_9'}{T_8 - T_9}$$

$$0.85 = \frac{943 - T_9'}{943 - 708}$$

$\therefore$ $T_9' = 943 - 0.85(943 - 708) = 743$ K.

Effectiveness of heat exchanger,

$$\varepsilon = 0.65 = \frac{T_5 - T_4'}{T_9' - T_4'}$$

i.e., $$0.65 = \frac{T_5 - 435}{743 - 435}$$

$\therefore$ $T_5 = 0.65(743 - 435) + 435 = 635$ K

$$W_{turbine\,(H.P.)} = c_{pg}\,(T_6 - T_7') = 1.15(973 - 798) = 201.25 \text{ kJ/kg of gas}$$

$$W_{turbine\,(L.P.)} = c_{pg}(T_8 - T_9') = 1.15(943 - 743) = 230 \text{ kJ/kg of gas}$$

$$W_{compressor\,(L.P.)} = c_{pa}(T_2' - T_1) = 1.005(425 - 293) = 132.66 \text{ kJ/kg of air}$$

$$W_{compressor\,(H.P.)} = c_{pa}(T_4' - T_3) = 1.005(435 - 300) = 135.67 \text{ kJ/kg of air}$$

Heat supplied $= c_{pg}(T_6 - T_5) + c_{pg}(T_8 - T_7')$
$= 1.15(973 - 635) + 1.15(943 - 798) = 555.45$ kJ/kg of gas

(*i*) **Overall efficiency $\eta_{overall}$:**

$$\eta_{overall} = \frac{\text{Net work done}}{\text{Heat supplied}}$$

$$= \frac{[W_{turbine\,(H.P.)} + W_{turbine\,(L.P.)}] - [W_{comp.\,(L.P.)} + W_{comp.\,(H.P.)}]}{\text{Heat supplied}}$$

$$= \frac{(201.25 + 230) - (132.66 + 135.67)}{555.45}$$

$$= \frac{162.92}{555.45} = 0.293 \text{ or } \mathbf{29.3\%.} \quad \textbf{(Ans.)}$$

(*ii*) **Work ratio :**

$$\textit{Work ratio} = \frac{\text{Net work done}}{\text{Turbine work}}$$

$$= \frac{[W_{turbine\,(H.P.)} + W_{turbine\,(L.P.)}] - [W_{comp.\,(L.P.)} + W_{comp.\,(H.P.)}]}{[W_{turbine\,(H.P.)} + W_{comp.\,(L.P.)}]}$$

$$= \frac{(201.25 + 230) - (132.66 + 135.67)}{(201.25 + 230)} = \frac{162.92}{431.25} = 0.377.$$

i.e., *Work ratio* = **0.377. (Ans.)**

(*iii*) **Mass flow rate, $\dot{m}$:**

Net work done = 162.92 kJ/kg.

Since mass of fuel is neglected, for 6000 kW, mass flow rate,

$$\dot{m} = \frac{6000}{162.92} = 36.83 \text{ kg/s}$$

i.e., *Mass flow rate* = **36.83 kg/s. (Ans.)**

21.8. JET PROPULSION

The *principle of jet propulsion* involves *imparting momentum to a mass of fluid in such a manner that the reaction of imparted momentum provides a propulsive force.* It may be achieved by expanding the gas, which is at high temperature and pressure, through a nozzle due to which a high velocity jet of hot gases is produced (in the atmosphere) that gives a propulsive force (in opposite direction due to its reaction). For jet propulsion the open cycle gas turbine is *most suitable.*

The propulsion system may be classified as follows :

1. **Air stream jet engines**, (Air-breathing engines)

(*a*) *Steady combustion systems* ; continuous air flow

(*i*) Turbo-jet (*ii*) Turbo-prop

(*iii*) Ram jet

(*b*) *Intermittent combustion system* ; intermittent flow

(*i*) Pulse jet or flying bomb.

2. **Self contained rocket engines** (Non-air breathing engines)

(*i*) Liquid propellent (*ii*) Solid propellent.

In **air stream jet engines** the *oxygen necessary for the combustion is taken from the surrounding atmosphere* whereas in a **rocket engine** *the fuel and the oxidiser are contained in the body of the unit which is to be propelled.*

Note. The turbo-jet and turbo-prop are modified forms of simple open cycle gas turbine. The ram jet and pulse jet are *athodyds* (aero-thermo-dynamic ducts) *i.e.,* straight duct type of jet engines having *no compressor and turbine wheels.*

In the past air propulsion was achieved by a "**Screw propeller**". In this system the total power developed by the turbine (full expansion) is used to drive the compressor and propeller. Fig. 21.37 shows the power plant for screw propeller. By controlling the supply of fuel in the combustion chamber the power supplied to the propeller can be controlled. *The rate of increase of efficiency of screw propeller is higher at lower speeds but its efficiency falls rapidly at higher speeds above the sonic velocity.*

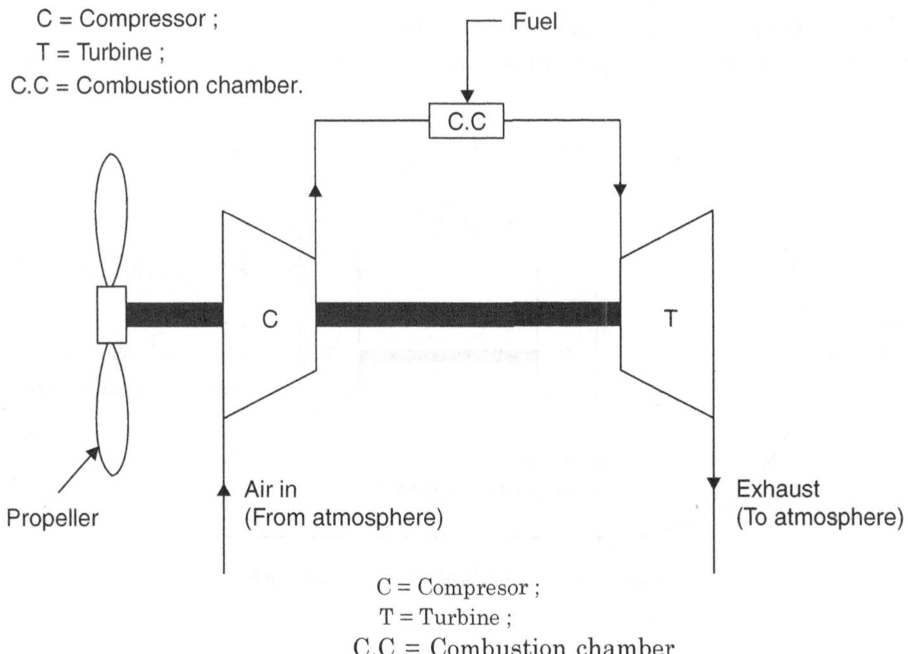

C = Compresor ;
T = Turbine ;
C.C = Combustion chamber

Fig. 21.37. Power plant for screw propeller.

21.8.1. Turbo-Jet

21.8.1.1. Description

Fig. 21.38 shows a turbo-jet unit.

- It consists of *diffuser* at entrance which slows down the air (entering at velocity equal to the plane speed) and part of the kinetic energy of the air stream is converted into pressure ; this type of compression is called as *ram compression.*
- The air is further compressed to a pressure of 3 to 4 bar in a rotary compressor (usually of axial flow type).
- The compressed air then enters the combustion chamber (C.C.) where fuel is added. The combustion of fuel takes place at sensibly constant pressure and subsequently temperature rises rapidly.
- The hot gases then enter the gas turbine where *partial expansion* takes place. The *power produced is just sufficient to drive the compressor, fuel pump and other auxiliaries.*
- The exhaust gases from the gas turbine which are at a higher pressure than atmosphere are expended in a nozzle and a very high velocity jet is produced which provides a forward motion to the air-craft by the jet reaction (Newton's third law of motion).

At higher speeds the turbo-jet gives higher propulsion efficiency. The turbo-jets are most suited to the air-crafts travelling *above 800 km/h.*

The overall efficiency of a turbo-jet is the product of the thermal efficiency of the gas turbine plant and the propulsive efficiency of the jet (nozzle).

Advantages of Turbo-jet engines

1. Construction much simpler (as compared to multi-cylinder piston engine of comparable power).
2. Engine vibrations absent.
3. Much higher speeds possible (more than 3000 km/h achieved).
4. Power supply is uninterrupted and smooth.
5. Weight to power ratios superior (as compared to that of reciprocating type of aero-engine).

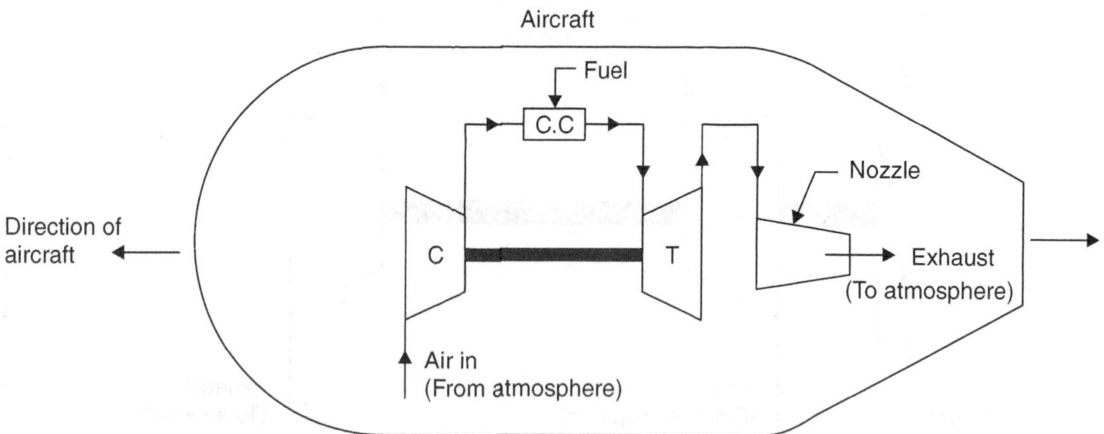

Fig. 21.38. Gas turbine plant for turbo-jet.

6. Rate of climb higher.
7. Requirement of major overhauls less frequent.
8. Radio interference much less.
9. Maximum altitude ceiling as compared to turbo-prop and conventional piston type engines.
10. Frontal area smaller.
11. Fuel can be burnt over a large range of mixture strength.

Disadvantages of turbo-jet engines

1. Less efficient.
2. Life of the unit comparatively shorter.
3. The turbo-jet becomes rapidly inefficient below 550 km/h.
4. More noisy (than a reciprocating engine).
5. Materials required are quite expensive.
6. Require longer strip since length of take-off is too much.
7. At take-off the thrust is low, this effect is overcome by boosting.

21.8.1.2. Basic Cycle for Turbo-jet Engine

The basic cycle for the turbo-jet engine is the *Joule or Brayton cycle* as shown is Fig. 21.39. The various processes are as follows :

Process 1-2 : The air entering from atmosphere is *diffused isentropically* from velocity C_1 down to zero (*i.e.,* C_2 = 0). This indicates that the diffuser has an efficiency of 100%, this is termed as *ram compression.*

Process 1-2′ is the *actual process.*

Process 2-3 : *Isentropic* compression of air.

Process 2′-3′ shows the *actual* compression of air.

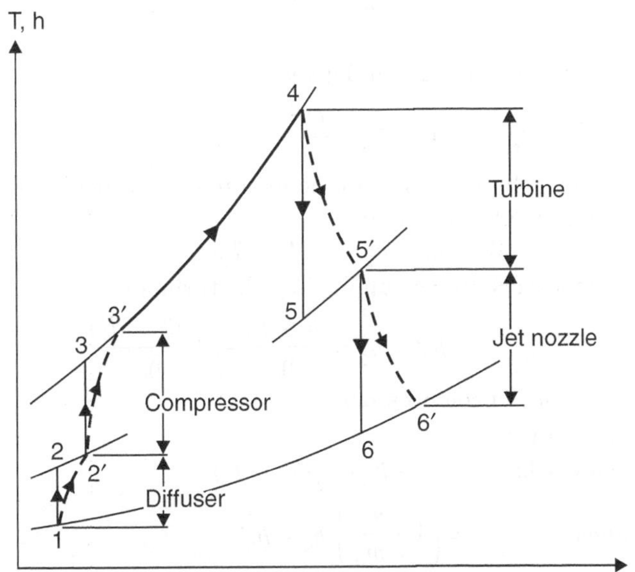

Fig. 21.39. *T-s* diagram of turbo-jet.

Process 3-4 : *Ideal* addition of heat at constant pressure $p_3 = p_4$

Process 3′-4 shows the *actual* addition of heat at constant process $p_3 = p_4$.

Process 4-5 : *Isentropic expansion* of gas in the *turbine.*

Process 4-5′ shows the *actual* expansion in the turbine.

Process 5-6 : *Isentropic expansion* of gas in the *nozzle.*

Process 5′-6′ shows the *actual expansion* of gas in the *nozzle.*

Consider *1 kg of working fluid flowing through the system.*

Diffuser :

Between states 1 and 2, the energy equation is given by :

$$\frac{C_a^2}{2} + h_1 + Q_{1-2} = \frac{C_2^2}{2} + h_2 + W_{1-2}$$

where C_a (= C_1) = Velocity of entering air from atmosphere.

In an ideal diffuser $C_2 = 0$, $Q_{1\text{-}2} = 0$ and $W_{1\text{-}2} = 0$.

$\therefore$ Enthalpy at state 2 is, $h_2 = h_1 + \frac{C_a^2}{2}$ kJ/kg

or
$$T_2 = T_1 + \frac{C_a^2}{2.c_p} \quad \text{...(21.7)} \quad (\because \; h = c_p.T)$$

Process 1-2′ shows actual process in diffuser.

Diffuser efficiency,
$$\eta_d = \frac{h_2 - h_1}{h_2' - h_1} = \frac{T_2 - T_1}{T_2' - T_1}$$

or
$$h_2' = h_1 + \frac{C_a^2}{2\,\eta_d}$$

or
$$T_2' = T_1 + \frac{C_a^2}{2\,.\,c_p\,.\,\eta_d} \quad \text{...(21.8)}$$

Compressor :

Energy equation between states 2 and 3 gives

$$h_2 + \frac{C_2^2}{2} + Q_{2-3} + W_c = h_3 + \frac{C_3^2}{2}$$

Assuming changes in potential and kinetic energies to be negligible, the ideal work expended in running the compressor is given as,

$$W_c = h_3 - h_2 = c_p(T_3 - T_2)$$

The actual compressor work (to be supplied by the turbine)

$$= h_3' - h_2 = \frac{h_3 - h_2}{\eta_c} = \frac{c_p\,(T_3 - T_2)}{\eta_c}$$

(where η_c = Isentropic efficiency of compressor)

Combustion chamber :

Ideal heat supplied per kg, $Q = h_4 - h_3 = c_p(T_4 - T_3)$

Actual heat supplied
$$= \left(1 + \frac{m_f}{m_a}\right) h_4 - h_3'$$

or
$$Q_a = c_{pg}\left(1 + \frac{m_f}{m_a}\right) T_4 - c_{pa}\,.\,T_3'$$

(where c_{pg} and c_{pa} are specific heats of gases and air at constant pressure respectively)

Turbine :

Between states 4 and 5, the energy equation is given by :

$$h_4 + \frac{C_4^2}{2} + Q_{4-5} = h_5 + \frac{C_5^2}{2} + W_t$$

If $Q_{4-5} = 0$, then turbine work,

$$W_t = (h_4 - h_5) + \frac{(C_4^2 - C_5^2)}{2}$$

If the change in kinetic energy is neglected, we have

$$W_t = (h_4 - h_5) = c_p(T_4 - T_5)$$

Actual turbine work $= h_4 - h_5' = c_p(T_4 - T_5') = c_p(T_4 - T_5) \times \eta_t$

(where η_t = Isentropic efficiency of turbine)

For the simplification, turbine work = compressor work

or $$c_p(T_4 - T_5') = c_p(T_4 - T_5)\,\eta_t = \frac{c_p\,(T_3 - T_2)}{\eta_c}$$

or $$T_5' = T_4 - (T_4 - T_5)\eta_t = T_4 - \frac{c_p\,(T_3 - T_2)}{\eta_c}$$

Jet nozzel :

Energy equation between states 5 and 6 gives

$$h_5 + \frac{C_5^2}{2} = h_6 + \frac{C_6^2}{2} \quad \text{...Ideal case}$$

$$h_5' + \frac{C_5'^2}{2} = h_6' + \frac{C_6'^2}{2} \quad \text{...Actual case}$$

If $C_5'^{\,2}$ is very less as compared to $C_6'^{\,2}$, we have

$$h_5' = h_6' + \frac{C_6'^2}{2}$$

or $$C_6' = \sqrt{2\,(h_5' - h_6')} = \sqrt{2\,\eta_n\,(h_5' - h_6)}$$

or $$C_6' = \sqrt{2\,\eta_n\,c_p\,(T_5' - T_6)} \quad \text{...(21.9)}$$

(where η_n = Nozzle efficiency)

Thermal efficiency (η_{th}) is given by :

$$\eta_{th} = \frac{(h_4 - h_6') - (h_3' - h_1)}{(h_4 - h_3')}$$

$$= \frac{(T_4 - T_6') - (T_3' - T_1)}{(T_4 - T_3')} \quad \text{...(21.10)}$$

21.8.1.3. Thrust, Thrust-power, Propulsive Efficiency and Thermal Efficiency.

Thrust (T)

Let C_a = Forward velocity of aircraft through air, m/s. Assuming the atmospheric air to be still the velocity of air, relative to the air craft, at entry to the aircraft will be C_a. It is called *velocity of approach of air.*

C_j = Velocity of jet (gases) relative to the exit nozzle/air craft ; m/s.

$\left[1 + \frac{\text{fuel}\ (m_f)}{\text{air}\ (m_a)}\right]$ = Mass of products leaving the nozzle for 1 kg of air.

Thrust is the force *produced due to change of momentum*

Now, absolute velocity of gases leaving aircraft = $(C_j - C_a)$

Absolute velocity of air entering the aircraft = **0**

$\therefore$ Change of momentum $= \left(1 + \frac{m_f}{m_a}\right)(C_j - C_a)$

Hence, thrust, $T = \left(1 + \frac{m_f}{m_a}\right)(C_j - C_a)$ N/kg of air/s ...(21.11)

$T = (C_j - C_a)$ N/kg of air/s, *neglecting mass of fuel* ...(21.12)

Thrust power (T.P.) :

It is defined as the rate at which work must be developed by the engine if the aircraft is to be kept moving at a constant velocity C_a *against friction force or drag.*

$\therefore$ Thrust power = Forward thrust × speed of aircraft

or $$\text{T.P.} = \left[\left(1 + \frac{m_f}{m_a}\right)(C_j - C_a)\right] C_a \text{ W/kg of air} \quad ...(21.13)$$

$= (C_j - C_a)\, C_a$ W/kg of air if *mass of fuel is neglected*

$$= \frac{(C_j - C_a)\, C_a}{1000} \text{ kW/kg of air} \quad ...(21.14)$$

Propulsive power (P.P.) :

The *energy required to change the momentum of the mass flow of gas represents the propulsive power.* It is expressed as the *difference between the rate of kinetic energies of the entering air and exit gases.*

Mathematically, $$\text{P.P.} = \Delta\text{K.E.} = \frac{\left(1 + \frac{m_f}{m_a}\right) C_j^2}{2} - \frac{C_a^2}{2} \text{ W/kg} \quad ...(21.15)$$

$= \frac{C_j^2 - C_a^2}{2}$ W/kg, *neglecting mass of fuel*

$$= \frac{C_j^2 - C_a^2}{2 \times 1000} \text{ kW/kg of air} \quad ...(21.16)$$

Propulsive efficiency ($\eta_{prop.}$) :

The *ratio of thrust power to propulsive power is called the propulsive efficiency of the propulsive unit.*

$$\eta_{prop.} = \frac{\text{Thrust power}}{\text{Propulsive power}} = \frac{\left[\left(1 + \frac{m_f}{m_a}\right)(C_j - C_a)\right] C_a}{\left[\frac{\left(1 + \frac{m_f}{m_a}\right) C_j^2}{2} - \frac{C_a^2}{2}\right]} = \frac{2\left[\left(1 + \frac{m_f}{m_a}\right)(C_j - C_a)\right] C_a}{\left[\left(1 + \frac{m_f}{m_a}\right) C_j^2 - C_a^2\right]} \quad ...(21.17)$$

Neglecting the mass of fuel,

$$\eta_{prop.} = \frac{2\,(C_j - C_a)\, C_a}{C_j^2 - C_a^2} = \frac{2\,(C_j - C_a)\, C_a}{(C_j + C_a)(C_j - C_a)}$$

or $$\eta_{prop.} = \frac{2\,C_a}{C_j + C_a} \quad \text{or} \quad \frac{2}{\left(\dfrac{C_j}{C_a} + 1\right)} \qquad ...(21.18)$$

From eqn. (21.18) it is evident that the *propulsive efficiency increases with an increase in aircraft velocity* C_a. $\eta_{prop.}$ becomes 100% when C_a approaches C_j ; thrust reduces to zero (Eqn. 21.12).

Thermal efficiency, (η_{th}) :

It is defined as the *ratio of propulsive work and the energy released by the combustion of fuel.*

or $$\eta_{th} = \frac{\text{Propulsive work}}{\text{Heat released by the combustion of fuel}} = \frac{\text{Increase in kinetic energy of the gases}}{\text{Heat released by the combustion of fuel}}$$

or $$\eta_{th} = \frac{\left(1 + \dfrac{m_f}{m_a}\right) C_j^2 - C_a^2}{2\left[\dfrac{m_f}{m_a} \times \text{calorific value}\right]} \qquad ...(21.19)$$

$$\simeq \frac{(C_j^2 - C_a^2)}{2 \times \left(\dfrac{m_f}{m_a}\right) \times \text{calorific value}} \qquad ...(21.20)$$

Overall efficiency (η_0) is given by :

$$\eta_0 = \eta_{th} \times \eta_{prop.} = \frac{(C_j^2 - C_a^2)}{2 \times \left(\dfrac{m_f}{m_a}\right) \times \text{calorific value}} \times \frac{2\,C_a}{C_j + C_a}$$

$$= \frac{(C_j - C_a)\,C_a}{\left(\dfrac{m_f}{m_a}\right) \times \text{calorific value}} \qquad ...(21.21)$$

For *maximum overall efficiency the aircraft velocity C_a is one half of the jet velocity C_j.*

The jet *efficiency* (η_{jet}) is defined as :

$$\eta_{jet} = \frac{\text{Final kinetic energy in the jet}}{\text{Isentropic heat drop in the jet pipe + Carry over from the turbine}}.$$

☞**Example 21.19.** *A turbo-jet engine consumes air at the rate of 60.2 kg/s when flying at a speed of 1000 km/h. Calculate :*

(i) Exit velocity of the jet when the enthalpy change for the nozzle is 230 kJ/kg and velocity co-efficient is 0.96.

(ii) Fuel flow rate in kg/s when air-fuel ratio is 70 : 1

(iii) Thrust specific fuel consumption

(iv) Thermal efficiency of the plant when the combustion efficiency is 92% and calorific value of the fuel used is 42000 kJ/kg.

(v) Propulsive power *(vi) Propulsive efficiency*

(vii) Overall efficiency.

Solution. Rate of air consumption, $\dot{m}_a = 60.2$ kg/s

Enthalpy change for nozzle, $\Delta h = 230$ kJ/kg

Velocity coefficient, $z = 0.96$

Air-fuel ratio $= 70 : 1$

Combustion efficiency, $\eta_{combustion} = 92\%$

Calorific value of fuel, C.V. $= 42000$ kJ/kg

Aircraft velocity, $C_a = \dfrac{1000 \times 1000}{60 \times 60} = 277.8$ m/s

(*i*) **Exit velocity of jet, C_j :**

$$C_j = z\sqrt{2\,\Delta h \times 1000}, \text{ where } \Delta h \text{ is in kJ}$$

$$= 0.96\sqrt{2 \times 230 \times 1000} = 651 \text{ m/s.}$$

i.e., **Exit velocity of jet = 651 m/s. (Ans.)**

(*ii*) **Fuel flow rate** :

Rate of fuel consumption, $\dot{m}_f = \dfrac{\text{Rate of air consumption}}{\text{Air-fuel ratio}}$

$= \dfrac{60.2}{70}$ **= 0.86 kg/s. (Ans.)**

(*iii*) **Thrust specific fuel consumption** :

Thrust is the force produced due to change of momentum.

Thrust produced $= \dot{m}_a (C_j - C_a)$, neglecting mass of fuel.

$= 60.2\,(651 - 277.8) = 22466.6$ N.

∴ Thrust specific fuel consumption

$$= \frac{\text{Fuel consumption}}{\text{Thrust}} = \frac{0.86}{22466.6}$$

= 3.828×10^{-5} kg/N of thrust/s. (Ans.)

(*iv*) **Thermal efficiency, $\eta_{thermal}$** :

$$\eta_{thermal} = \frac{\text{Work output}}{\text{Heat supplied}}$$

$$= \frac{\text{Gain in kinetic energy per kg of air}}{\text{Heat supplied by fuel per kg of air}}$$

$$= \frac{(C_j^2 - C_a^2)}{\left(\dfrac{m_f}{m_a}\right) \times \text{C.V.} \times \eta_{combustion} \times 1000}$$

$$= \frac{(651^2 - 277.8^2)}{2 \times \dfrac{1}{70} \times 42000 \times 0.92 \times 1000} = 0.3139 \text{ or } 31.39\%$$

i.e., Thermal efficiency **= 31.39%. (Ans.)**

(*v*) **Propulsive power :**

Propulsive power $= \dot{m}_a \times \left(\dfrac{C_j^2 - C_a^2}{2}\right) = \dfrac{60.2}{1000} \times \left(\dfrac{651^2 - 277.8^2}{2}\right)$ kW

= 10433.5 kW. (Ans.)

(*vi*) **Propulsive efficiency, $\eta_{prop.}$:**

$$\eta_{prop.} = \frac{\text{Thrust power}}{\text{Propulsive power}} = \frac{2\,C_a}{C_j + C_a} \quad ...(21.18)$$

$$= \frac{2 \times 277.8}{651 + 277.8} = 0.598 \text{ or } \mathbf{59.8\%}. \quad \textbf{(Ans.)}$$

(*vii*) **Overall efficiency, η_0 :**

$$\eta_0 = \frac{\text{Thrust work}}{\text{Heat supplied by fuel}} = \frac{(C_j - C_a)\,C_a}{\left(\dfrac{m_f}{m_a}\right) \times \text{C.V.} \times \eta_{combustion}} \quad ...(21.22)$$

$$= \frac{(651 - 277.8) \times 277.8}{\dfrac{1}{70} \times 42000 \times 0.92 \times 1000} = 0.1878 \text{ or } \mathbf{18.78\%}. \quad \textbf{(Ans.)}$$

Example 21.20. *The following data pertain to a turbo-jet flying at an altitude of 9500 m :*

Speed of the turbo-jet = *800 km/h*
Propulsive efficiency = *55%*
Overall efficiency of the turbine plant = *17%*
Density of air at 9500 m altitude = *0.17 kg/m³*
Drag on the plane = *6100 N*

Assuming calorific value of the fuels used as 46000 kJ/kg,
Calculate :

(*i*) *Absolute velocity of the jet.* (*ii*) *Volume of air compressed per min.*
(*iii*) *Diameter of the jet.* (*iv*) *Power output of the unit.*
(*v*) *Air-fuel ratio.*

Solution. *Given :* Altitude = 9500 m, $C_a = \dfrac{800 \times 1000}{60 \times 60} = 222.2$ m/s,

$\eta_{propulsive} = 55\%$, $\eta_{overall} = 17\%$; density of air at 9500 m altitude = 0.17 kg/m³ ; drag on the plane = 6100 N.

(*i*) **Absolute velocity of the jet, ($C_i - C_a$) :**

$$\eta_{propulsive} = 0.55 = \frac{2C_a}{C_j + C_a}$$

where, C_j = Velocity of gases at nozzle exit relative to the aircraft, and
C_a = Velocity of the turbo-jet/air-craft.

$$\therefore \quad 0.55 = \frac{2 \times 222.2}{C_j + 222.2}$$

i.e.,

$$C_j = \frac{2 \times 222.2}{0.55} - 222.2 = 585.8 \text{ m/s}$$

$\therefore$ Absolute velocity of jet = $C_j - C_a$ = 585.8 – 222.2 = 363.6 m/s.

(*ii*) **Volume of air compressed/min. :**

$$\text{Propulsive force} = \dot{m}_a\,(C_j - C_a)$$

$$6100 = \dot{m}_a\,(585.8 - 222.2)$$

$$\therefore \quad \dot{m}_a = 16.77 \text{ kg/s}$$

$\therefore$ Volume of air compressed/min. $= \dfrac{16.77}{0.17} \times 60 =$ **5918.8 kg/min. (Ans.)**

(*iii*) **Diameter of the jet, d :**

Now, $\frac{\pi}{4} d^2 \times C_j = 5918.8$

i.e., $\frac{\pi}{4} d^2 \times 585.8 = (5918.8/60)$

$\therefore$ $d = \left(\frac{5918.8 \times 4}{60 \times \pi \times 585.8}\right)^{1/2} = 0.463 \text{ m} = 463 \text{ mm}$

i.e., **Diameter of the jet = 463 mm. (Ans.)**

(*iv*) **Power output of the unit :**

Thrust power = Drag force × velocity of turbo-jet

$= 6100 \times 222.2$ N-m/s

$= \frac{6100 \times 222.2}{1000} = 1355.4$ kW

Turbine output $= \frac{\text{Thrust power}}{\text{Propulsive efficiency}} = \frac{1355.4}{0.55} =$ **2464.4 kW. (Ans.)**

(*v*) **Overall efficiency, η_0 :**

$$\eta_0 = \frac{\text{Heat equivalent of output}}{\dot{m}_f \times \text{C.V.}}$$

i.e., $$0.17 = \frac{2464.4}{\dot{m}_f \times 46000}$$

$\therefore$ $$\dot{m}_f = \frac{2464.4}{0.17 \times 46000} = 0.315 \text{ kg/s}$$

$\therefore$ Air-fuel ratio $= \frac{\text{Air used (in kg / s)}}{\text{Fuel used (in kg / s)}} = \frac{16.77}{0.315} = 53.24$

i.e., *Air-fuel ratio* = **53.24 : 1. (Ans.)**

Example 21.21. *In a jet propulsion unit air is drawn into the rotary compressor at 15°C and 1.01 bar and delivered at 4.04 bar. The isentropic efficiency of compression is 82% and the compression is uncooled. After delivery the air is heated at constant pressure until the temperature reaches 750°C. The air then passes through a turbine unit which drives the compressor only and has an isentropic efficiency of 78% before passing through the nozzle and expanding to atmospheric pressure of 1.01 bar with an efficiency of 88%. Neglecting any mass increase due to the weight of the fuel and assuming that R and γ are unchanged by combustion, determine :*

(*i*) *The power required to drive the compressor.*

(*ii*) *The air-fuel ratio if the fuel has a calorific value of 42000 kJ/kg.*

(*iii*) *The pressure of the gases leaving the turbine.*

(*iv*) *The thrust per kg of air per second.*

Neglect any effect of the velocity of approach.

Assume for air : R = 0.287 kJ/kg K, γ = 1.4.

Solution. *Given :* $T_1 = 15 + 273 = 288$ K,

$p_1 = 1.01$ bar, $p_2 = 4.04$ bar, $T_3 = 750 + 273 = 1023$ K,

$\eta_{compressor} = 82\%$, $\eta_{turbine} = 78\%$, $\eta_{nozzle} = 88\%$,

$R_{air} = 0.287$ kJ/kg K, $\gamma_{air} = 1.4$.

Refer Fig. 21.40.

$$\frac{T_2}{T_1} = \left(\frac{p_2}{p_1}\right)^{\frac{\gamma-1}{\gamma}} = \left(\frac{4.04}{1.01}\right)^{\frac{1.4-1}{1.4}} = 1.486$$

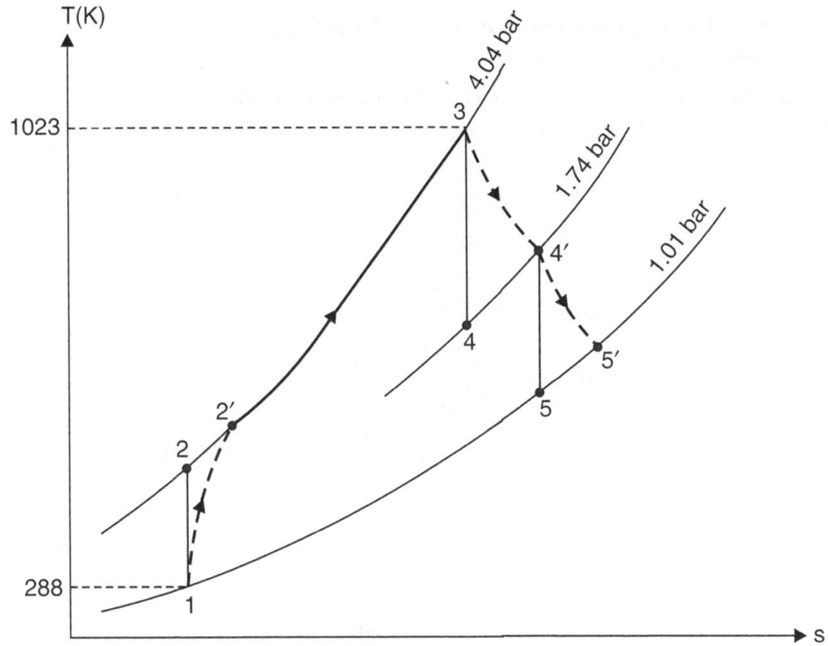

Fig. 21.40

$$\therefore \quad T_2 = 2.88 \times 1.486 = 428 \text{ K}$$

$$\eta_{compressor} = \frac{T_2 - T_1}{T_2' - T_1} \quad i.e., \quad 0.82 = \frac{428 - 288}{T_2' - 288}$$

$$\therefore \quad T_2' = \left(\frac{428 - 288}{0.82}\right) + 288 = 458.7 \text{ K}$$

$$c_p = R \times \left(\frac{\gamma}{\gamma - 1}\right) = 0.287 \times \frac{1.4}{(1.4 - 1)} = 1.004 \text{ kJ/kg K}$$

(*i*) **Power required to drive the compressor :**

Power required to drive the compressor (per kg of air/sec.)

$= c_p(T_2' - T_1) = 1.004(458.7 - 288) =$ **171.38 kW. (Ans.)**

(*ii*) **Air-fuel ratio :**

$$m_f \times C = (m_a + m_f) \times c_p \times (T_3 - T_2')$$

where, m_a = Mass of air per kg of fuel, and

= Air-fuel ratio.

$$\therefore \quad m_a = \frac{m_f \times C}{c_p\,(T_3 - T_2')} - m_f$$

$$\therefore \quad \frac{m_a}{m_f} = \frac{C}{c_p\,(T_3 - T_2')} - 1$$

$$= \frac{42000}{1.004(1023 - 458.7)} - 1 = 73.1$$

i.e., **Air-fuel ratio = 73.1 : 1. (Ans.)**

(*iii*) **Pressure of the gases leaving the turbine, p_4 :**

Neglecting effect of fuel on mass flow,

Actual turbine work = actual compressor work

i.e., $$c_p(T_2' - T_1) = c_p(T_3 - T_4')$$

or $$T_2' - T_1 = T_3 - T_4'$$

$\therefore$ $$458.7 - 288 = 1023 - T_4'$$

$\therefore$ $$T_4' = 852.3 \text{ K}$$

Also, $$\eta_{turbine} = \frac{T_3 - T_4'}{T_3 - T_4}$$

$$0.78 = \frac{1023 - 852.3}{1023 - T_4}$$

$\therefore$ $$T_4 = 1023 - \left(\frac{1023 - 852.3}{0.78}\right) = 804 \text{ K}$$

$\therefore$ $$\frac{T_4}{T_3} = \left(\frac{p_4}{p_3}\right)^{\frac{\gamma - 1}{\gamma}}$$

or $$\frac{p_4}{p_3} = \left(\frac{T_4}{T_3}\right)^{\frac{\gamma}{\gamma - 1}} = \left(\frac{804}{1023}\right)^{\frac{1.4}{1.4 - 1}} = 0.43$$

or $$p_4 = 4.04 \times 0.43 = \mathbf{1.74 \text{ bar.}} \quad \textbf{(Ans.)}$$

(*iv*) **Thrust per kg of air per second :**

$$\frac{T_4'}{T_5} = \left(\frac{p_4}{p_5}\right)^{\frac{\gamma - 1}{\gamma}} = \left(\frac{1.74}{1.01}\right)^{\frac{1.4 - 1}{1.4}} = 1.168$$

$\therefore$ $$T_5 = \frac{T_4'}{1.168} = \frac{852.3}{1.168} = 729.7 \text{ K}$$

$$\eta_{nozzle} = \frac{T_4' - T_5'}{T_4' - T_5}$$

$$0.88 = \frac{852 - T_5'}{852.3 - 729.7}$$

$\therefore$ $$T_5' = 852.3 - 0.88\ (852.3 - 729.7) = 744.4 \text{ K}$$

If C_j is the jet velocity, then

$$\frac{C_j^2}{2} = c_p(T_4' - T_5')$$

$\therefore$ $$C_j = \sqrt{2 \times c_p\ (T_4' - T_5')}.$$

$$= \sqrt{2 \times 1.004 (852.3 - 744.4) \times 1000} = 465.5 \text{ m/s}$$

$\therefore$ Thrust per kg per second = 1 × 465.5 = **465.5 N. (Ans.)**

Example 21.22. *A turbo-jet engine travels at 216 m/s in air at 0.78 bar and – 7.2°C. Air first enters diffuser in which it is brought to rest relative to the unit and it is then compressed in a compressor through a pressure ratio of 5.8 and fed to a turbine at 1110°C. The gases expand through the turbine and then through the nozzle to atmospheric pressure (i.e., 0.78 bar). The efficiencies of diffuser, nozzle and compressor are each 90%. The efficiency of turbine is 80%. Pressure drop in the combustion chamber is 0.168 bar. Determine :*

(*i*) *Air-fuel ratio ;*

(*ii*) *Specific thrust of the unit ;*

(*iii*) *Total thrust, if the inlet cross-section of diffuser is 0.12 m²*.

Assume calorific value of fuel as 44150 kJ/kg of fuel.

Solution. Refer Fig. 21.41.

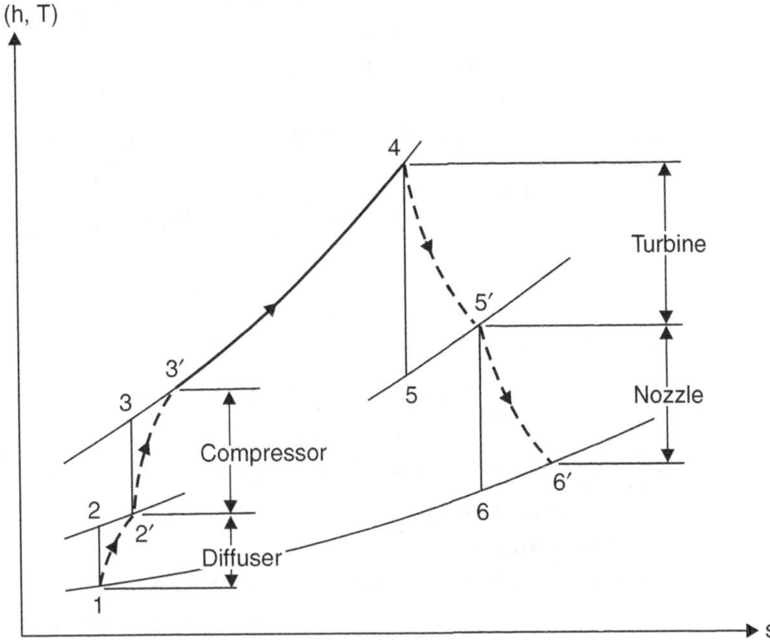

Fig. 21.41

Speed of the aircraft, $C_a = 216$ m/s

Intake air temperature, $T_1 = -7.2 + 273 = 265.8$ K

Intake air pressure, $p_1 = 0.78$ bar

Pressure ratio in the compressor, $r_p = 5.8$

Temperature of gases entering the gas turbine, $T_4 = 1110 + 273 = 1383$ K

Pressure drop in combustion chamber = 0.168 bar

$$\eta_d = \eta_n \; ; \; \eta_c = 90\% \; ; \; \eta_t = 80\%.$$

Calorific value of fuel, $C.V.$ = 44150 kJ/kg of coal

(*i*) **Air-fuel ratio :**

For ideal diffuser (*i.e.*, process 1-2) the energy equation is given by :

$$h_2 = h_1 + \frac{C_a^2}{2} \quad \text{or} \quad h_2 - h_1 = \frac{C_a^2}{2} \quad \text{or} \quad T_2 - T_1 = \frac{C_a^2}{2c_p}$$

or $$T_2 = T_1 + \frac{C_a^2}{2c_p} = 265.8 + \frac{216^2}{2 \times 1.005 \times 1000} = 289 \text{ K}$$

For actual diffuser (*i.e.*, process 1-2′),

$$\eta_d = \left(\frac{h_2 - h_1}{h_2' - h_1}\right) \quad \text{or} \quad h_2' - h_1 = \frac{h_2 - h_1}{\eta_d}$$

or $$h_2' = h_1 + \frac{h_2 - h_1}{\eta_d} = h_1 + \frac{C_a^2}{2\,\eta_d}$$

or $$T_2' = T_1 + \frac{C_a^2}{2\,c_p\,\eta_d} = 265.8 + \frac{216^2}{2 \times 1.005 \times 1000 \times 0.9} = 291.6 \text{ K}$$

Now $$\frac{T_2}{T_1} = \left(\frac{p_2}{p_1}\right)^{\frac{\gamma-1}{\gamma}} \text{ or } \frac{289}{265.8} = \left(\frac{p_2}{0.78}\right)^{\frac{1.4-1}{1.4}} \text{ or } (1.087)^{3.5} = \left(\frac{p_2}{0.78}\right)$$

or $$p_2 = 0.78 \times (1.087)^{3.5} = 1.044 \text{ bar}$$

Again, $$\frac{T_3}{T_2'} = (r_p)^{\frac{\gamma-1}{\gamma}} = (5.8)^{\frac{1.4-1}{1.4}} = 1.652 \text{ or } T_3 = 291.6 \times 1.652 = 481.7 \text{ K}$$

Also, $$\eta_c = \frac{T_3 - T_2'}{T_3' - T_2'} \text{ or } T_3' = T_2' + \frac{T_3 - T_2'}{\eta_c} = 291.6 + \frac{481.7 - 291.6}{0.9} = 502.8 \text{ K}$$

Assume $c_{pg} = c_{pa} = c_p$

Heat supplied $= (m_a + m_f)c_p\,T_4 - m_a c_p\,T_3' = m_f \times C$

or $$m_a\,c_p\,T_4 + m_f\,c_p\,T_4 - m_a\,c_p\,T_3' = m_f \times C$$

or $$m_a c_p (T_4 - T_3') = m_f (C - c_p\,T_4)$$

or $$\frac{m_a}{m_f} = \frac{C - c_p\,T_4}{c_p\,(T_4 - T_3')} = \frac{44150 - 1.005 \times 1383}{1.005(1383 - 502.8)} = 48.34$$

∴ Air-fuel ratio = **48.34. (Ans.)**

(*ii*) **Specific thrust of the unit :**

$$p_4 = p_3 - 0.168 = 5.8 \times 1.044 - 0.168 = 5.88 \text{ bar}$$

Assume that the turbine drives *compressor only* (and not accessories also as is the usual case)

∴ $$c_p(T_3' - T_2') = c_p(T_4 - T_5')$$

or $$T_3' - T_2' = T_4 - T_5' \text{ or } T_5' = T_4 - (T_3' - T_2')$$
$$= 1383 - (502.8 - 291.6) = 1171.8 \text{ K}$$

Also, $$\eta_t = \frac{T_4 - T_5'}{T_4 - T_5}$$

or $$T_5 = T_4 - \frac{T_4 - T_5'}{\eta_t} = 1383 - \frac{1383 - 1171.8}{0.8} = 1119 \text{ K}$$

Now, $$\frac{T_4}{T_5} = \left(\frac{p_4}{p_5}\right)^{\frac{\gamma-1}{\gamma}} = \left(\frac{5.88}{p_5}\right)^{\frac{1.4-1}{1.4}}$$

or $$\left(\frac{1383}{1119}\right)^{3.5} = \frac{5.88}{p_5} \text{ or } p_5 = 2.8 \text{ bar}$$

Again, $$\frac{T_5'}{T_6} = \left(\frac{p_5}{p_6}\right)^{\frac{\gamma-1}{\gamma}} = \left(\frac{2.8}{0.78}\right)^{\frac{1.4-1}{1.4}} = 1.44$$

or $$T_6 = \frac{T_5'}{1.44} = \frac{1171.8}{1.44} = 813.75 \text{ K}$$

and $$\eta_n = \frac{T_5' - T_6'}{T_5' - T_6} \quad \text{or} \quad T_6' = T_5' - \eta_n (T_5' - T_6)$$
$$= 1171.8 - 0.9(1171.8 - 813.75) = 849.5 \text{ K}$$

Velocity at the exit of the nozzle,

$$C_j = 44.72 \sqrt{h_5' - h_6'} = 44.72 \sqrt{c_p (T_5' - T_6')}$$
$$= 44.72 \sqrt{1.005(1171.8 - 849.5)} = 804.8 \text{ m/s}$$

Specific thrust $$= (1 + m_f) \times C_j = \left(1 + \frac{1}{48.34}\right) \times 804.8$$
$$= \textbf{821.45 N/kg of air/s.} \quad \textbf{(Ans.)}$$

(*iii*) **Total Thrust :**

Volume of flowing air, $V_1 = 0.12 \times 216 = 25.92 \text{ m}^3/\text{s}$

Mass flow, $$m_a = \frac{p_1 V_1}{RT_1} = \frac{0.78 \times 10^5 \times 25.92}{(0.287 \times 1000) \times 265.8} = 26.5 \text{ kg/s}$$

$\therefore$ Total thrust $= 26.5 \times 821.45 =$ **21768.4 N.** **(Ans.)**

☞**Example 21.23.** *The following data pertain to a jet engine flying at an altitude of 9000 metres with a speed of 215 m/s.*

Thrust power developed	*750 kW*
Inlet pressure and temperature	*0.32 bar, – 42°C*
Temperature of gases leaving the combustion chamber	*690°C*
Pressure ratio	*5.2*
Calorific value of fuel	*42500 kJ/kg*
Velocity in ducts (constant)	*195 m/s*
Internal efficiency of turbine	*86%*
Efficiency of compressor	*86%*
Efficiency of jet tube	*90%*

For air : $c_p = 1.005$, $\gamma = 1.4$, $R = 0.287$

For combustion gases, $c_p = 1.087$

For gases during expansion, $\gamma = 1.33$.

Calculate the following :

(*i*) *Overall thermal efficiency of the unit ;*

(*ii*) *Rate of air consumption ;*

(*iii*) *Power developed by the turbine ;*

(*iv*) *The outlet area of jet tube ;*

(*v*) *Specific fuel consumption is kg per kg of thrust.*

Solution. Refer Fig. 21.42.

Given : T.P. = 750 kW ; $p_1 = 0.32$ bar, $T_1 = -42 + 273 = 231$ K ; $T_3 = 690 + 273 = 963$ K ; $r_{pc} = 5.2$; $C = 42500$ kJ/kg ; $C_a = 215$ m/s, $C_4' = 19.5$ m/s, $\eta_c = 0.86$; $\eta_t = 0.86$; $\eta_{jt} = 0.9$.

Refer Fig. 21.42.

Let m_f = kg of fuel required per kg of air

Then, heat supplied per kg of air

$$= 42500\, m_f = (1 + m_f) \times 1.087(T_3 - T_2') \qquad \text{...}(i)$$

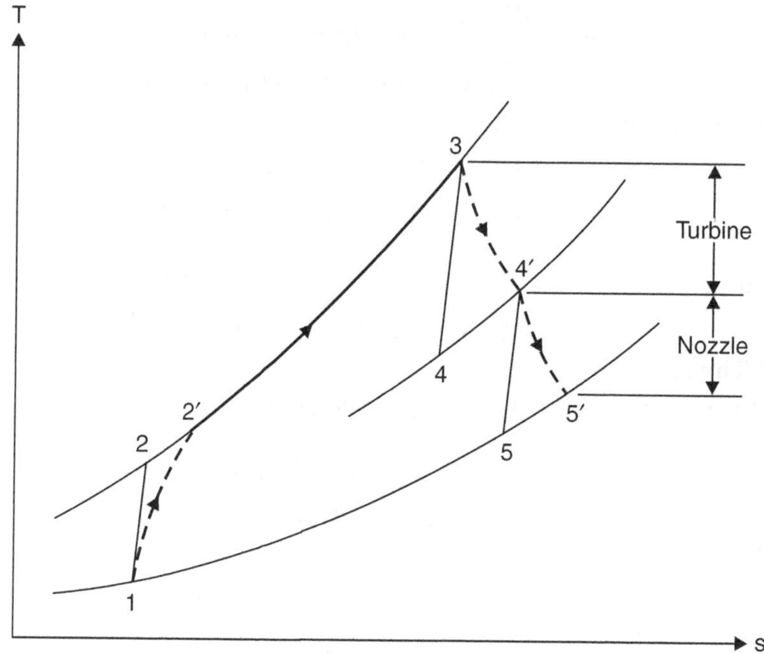

Fig. 21.42

Now, $$\frac{T_2}{T_1} = \left(\frac{p_2}{p_1}\right)^{\frac{\gamma-1}{\gamma}} = (5.2)^{\frac{1.4-1}{1.4}} = (5.2)^{0.2857} = 1.60$$

or $$T_2 = 231 \times 1.60 = 369.6 \text{ K}$$

Also, $$\eta_c = \frac{T_2 - T_1}{T_2' - T_1} \quad \text{or} \quad T_2' = T_1 + \frac{T_2 - T_1}{\eta_c} = 231 + \frac{369.6 - 231}{0.86} = 392.2 \text{ K}$$

Substituting the value of T_2' in eqn. (*i*), we get

$$42500\, m_f = (1 + m_f) \times 1.087(963 - 392.2) = 620.46(1 + m_f)$$

or $$42500\, m_f = 620.46 + 620.46\, m_f$$

or $$m_f = \frac{620.46}{(42500 - 620.46)} = 0.0148 = \text{fuel-air ratio}$$

∴ Air fuel ratio $$= \frac{1}{0.0148} = 67.56 : 1$$

The discharge velocity $C_j = C_5'$ cannot be determined from the thrust equation because the rate of air flow is not known. It may be determined from the expression of jet efficiency.

Jet efficiency, $$\eta_{jet} = \frac{\text{Final kinetic energy in the jet}}{\text{Isentropic heat drop in the jet pipe + Carry-over from the turbine}}$$

or $$\eta_{jet} = \frac{C_j^2/2}{c_{pg}(T_4' - T_5) + C_4'^2/2} \quad (\text{where } C_4' = 195 \text{ m/s}) \quad \ldots(i)$$

Since the turbine's work is to drive the compressor only, therefore,

$$c_{pa}(T_2' - T_1) = c_{pg}\left(1 + \frac{m_f}{m_a}\right)(T_3 - T_4')$$

or $$1.005(392.2 - 231) = 1.087(1 + 0.0148)(963 - T_4')$$

or $$T_4' = 963 - \frac{1.005(392.2 - 231)}{1.087(1 + 0.0148)} = 816.13 \text{ K}$$

Let r_{pt} = expansion pressure ratio in turbine *i.e.*, $r_{pt} = \dfrac{p_3}{p_4}$

r_{pj} = expansion pressure ratio in jet tube *i.e.*, $r_{pj} = \dfrac{p_4}{p_5}$

$$\therefore \quad r_{pt} \times r_{pj} = \frac{p_3}{p_4} \times \frac{p_4}{p_5} \simeq 5.2$$

Now, $$\eta_t = \frac{T_3 - T_4'}{T_3 - T_4} \quad \text{or} \quad T_4 = T_3 - \frac{T_3 - T_4'}{\eta_t}$$

$$= 963 - \frac{963 - 816.13}{0.86} = 792.2 \text{ K}$$

Also, $$\frac{T_3}{T_4} = \left(\frac{p_3}{p_4}\right)^{\frac{\gamma - 1}{\gamma}} = \left(\frac{p_3}{p_4}\right)^{\frac{1.33 - 1}{1.33}} \quad \text{or} \quad \frac{963}{792.2} = (r_{pt})^{0.248}$$

or $$r_{pt} = \left(\frac{963}{792.2}\right)^{\frac{1}{0.248}} = 2.197$$

$$\therefore \quad r_{pj} = \frac{p_4}{p_5} = \frac{5.2}{2.197} = 2.366$$

Thus, $$\frac{T_4'}{T_5} = (r_{pj})^{\frac{\gamma - 1}{\gamma}} = (2.366)^{\frac{1.33 - 1}{1.33}} = 1.238$$

or $$T_5 = \frac{T_4'}{1.238} = \frac{816.13}{1.238} = 659.23 \text{ K}$$

Substituting the values in eqn. (*i*), we get

$$0.9 = \frac{C_j^2 / 2}{1.087 \times 1000(816.13 - 659.23) + 195^2 / 2} = \frac{C_j^2 / 2}{189562.8}$$

$$\therefore \quad C_j = \sqrt{0.9 \times 189562.8 \times 2} = 584.13 \text{ m/s}$$

(*i*) **Overall efficiency, η_0 :**

$$\eta_0 = \frac{\left[\left(1 + \frac{m_f}{m_a}\right) C_j - C_a\right] C_a}{\left(\frac{m_f}{m_a}\right) \times C} = \frac{[(1 + 0.0148) \times 584.13 - 215] 215}{1000 \times 0.0148 \times 42500}$$

$$= 0.1291 \text{ or } \mathbf{12.91\%}. \quad \textbf{(Ans.)}$$

(*ii*) **Rate of air consumption, $\dot{m}_a$:**

Thrust power = Thrust × Velocity of the unit

$$750 = \left[\frac{\left\{\left(1 + \frac{m_f}{m_a}\right) C_j - C_a\right\} \dot{m}_a}{1000}\right] C_a$$

or $$750 = \frac{\{(1 + 0.0148) \times 584.13 - 215\} \times \dot{m}_a}{1000} \times 215 = 81.22\ \dot{m}_a$$

or $$\dot{m}_a = \frac{750}{81.22} = \mathbf{9.234\ kg/s.}\quad \textbf{(Ans.)}$$

(*iii*) **Power developed by the turbine, P_t :**

$$P_t = \dot{m}_a \left(1 + \frac{m_f}{m_a}\right) c_{pg}(T_3 - T_4')$$

$$= 9.234(1 + 0.0148) \times 1.087(963 - 816.13) = \mathbf{1496\ kW.}\quad \textbf{(Ans.)}$$

(*iv*) **The outlet area of jet tube, A_{jt} :**

Now, $$\frac{C_j^2 - C_4'^2}{2} = c_{pg}(T_4' - T_5')$$

or $$T_5' = T_4' - \frac{C_j^2 - C_4'^2}{2 \times c_{pg}}$$

$$= 816.13 - \frac{(584.13^2 - 195^2)}{2 \times 1.087 \times 1000} = 676.67\ \text{K}$$

Assume the exit pressure of the gases be equal to atmospheric pressure *i.e.*, 0.32 bar.

Density of exhaust gases, $$\rho = \frac{p_5'}{RT_5'} = \frac{0.32 \times 10^5}{0.29 \times 1000 \times 676.67} = 0.163\ \text{m}^3/\text{kg}$$

(Assuming $R = 0.29$ for the gases)

Also, discharge of jet area $$= A_{jt} \times C_j \times \rho = \dot{m}_a \left(1 + \frac{m_f}{m_a}\right)$$

or $$A_{jt} \times 584.13 \times 0.163 = 9.234(1 + 0.0148)$$

or $$A_{jt} = \mathbf{0.0984\ m^2.}\quad \textbf{(Ans.)}$$

(*v*) **Specific fuel consumption in kg per kg of thrust :**

Specific fuel consumption $$= \frac{0.0148 \times 9.234 \times 3600}{1000 \times (750 / 215)}$$

$$= \mathbf{0.141\ kg/thrust\text{-}hour.}\quad \textbf{(Ans.)}$$

21.8.2. Turbo-prop

Fig. 21.43 shows a turbo-prop system employed in aircrafts. Here the expansion of gases takes place *partly in turbine* (80%) and *partly* (20%) *in the nozzle.* The power developed by the turbine is consumed in running the compressor and the propeller. The propeller and jet produced by the nozzle give forward motion to the aircraft. The turbo-prob entails the advantages of turbo-jet (*i.e., low specific weight and simplicity in design*) and propeller (*i.e., high power for take-off and high propulsion efficiency at speeds below 600 km/h*). The overall efficiency of the turbo-prop is improved by providing the diffuser before the compressor as shown. The pressure rise takes place in the diffuser. This pressure rise takes place due to *conversion of kinetic energy of the incoming air* (equal to aircraft velocity) *into pressure energy by the diffuser*. This type of compression is known as "***ram effect***".

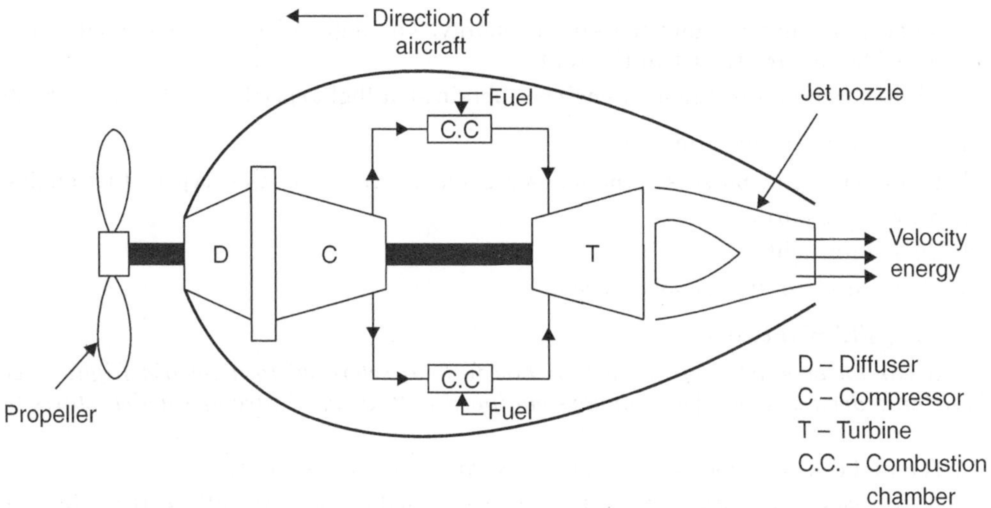

Fig. 21.43. Turbo-prop.

21.8.3. Ram-jet :

Ram jet is also called *athodyd, Lorin tube* or *flying stovepipe*. Ram jet engines have the capability to fly at *supersonic speeds*. Fig. 21.44 shows a schematic diagram of a ram jet engine (compressor and turbine are not necessary as the entire compression depends only on the ram compression).

- The ram jet engine consists of a *diffuser* (used for compression), *combustion chamber* and *nozzle*.
- The air enters the ram jet plant with *supersonic speed* and is slowed down to *sonic velocity* in the *supersonic diffuser*, consequently the pressure suddenly increases in the supersonic diffuser to the formation of shock wave. The pressure of air is further increased in the subsonic diffuser *increasing the temperature of the air above the ignition temperature*.
- In the combustion chamber, the fuel is injected through injection nozzles. The fuel air mixture is then ignited by means of a spark plug and combustion temperatures of the order of 2000 K are attained. The expansion of gases towards the diffuser entrance is restricted by pressure barrier at the after end of the diffuser and as a result the hot gases are constrained to move towards the nozzle and undergo expansion ; the pressure

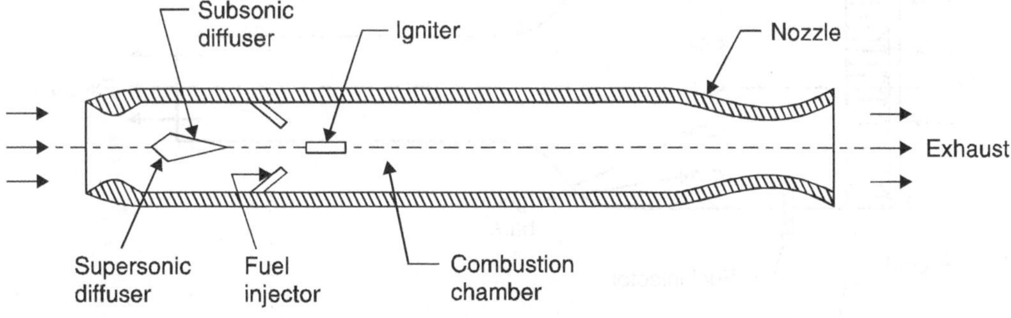

Fig. 21.44. Schematic diagram of a Ram-jet propulsion unit.

energy is converted into the kinetic energy. The high velocity gases leaving the nozzle provide forward thrust to the unit.

The best performance of ram jet engine is obtained at flight speed of 1700 km/h to 2000 km/h.

Advantages of ram-jet engine

The ram-jet engine possesses the following *advantages* over other types of jet engines :

1. No moving parts.
2. Light in weight.
3. Wide variety of fuels may be used.

Shortcomings/Limitations

1. It *cannot be started of its own.* It *has to be accelerated to a certain flight velocity by some launching device.* A ram-jet is *always equipped with a small turbo-jet which starts the ram jet.*

2. The fuel consumption is too large at low and moderate speeds.

3. For successful operation, the diffuser needs to be designed carefully so that kinetic energy associated with high entrance velocities is efficiently converted into pressure.

4. To obtain steady combustion, certain elaborate devices in form of flame holders or pilot flame are required.

21.8.4. Pulse-jet Engine

A *pulse-jet engine* is an intermittent combustion engine and it operates on a cycle similar to a reciprocating engine, whereas the turbo-jet and ram-jet engines are continuous in operation and are based on Brayton cycle. A pulse-jet engine like an athodyd, develops thrust by a high velocity of jet of exhaust gases without the aid of compressor or turbine. Its development is primarily due to the inability of the ram-jet to be self starting. Fig. 21.45 shows a schematic arrangement of a pulse jet propulsion unit.

- The incoming air is compressed by ram effect in the diffuser section and the grid passages which are opened and closed by V-shaped non-return valves.
- The fuel is then injected into the combustion chamber by fuel injectors (worked from the air pressure from the compressed air bottles). The combustion is then initiated by a spark plug (once the engine is operating normally, the spark is turned off and the residual flame in the combustion chamber is used for combustion).

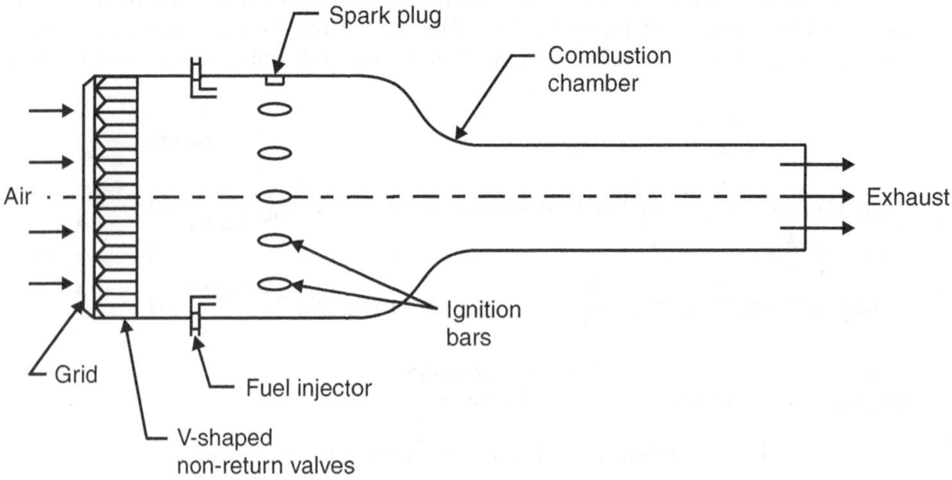

Fig. 21.45. Pulse-jet engine.

- As a result of combustion (of mixture of air and fuel) the temperature and pressure of combustion products increase. Because the combustion pressure is higher than the ram pressure, the non-return valves get closed and consequently the hot gases flow out of the tail pipe with a high velocity and in doing so give a forward thrust to the unit.
- With the escape of gases to the atmosphere, the static pressure in the chamber falls and the high pressure air in the diffuser forces the valves to open and fresh air is admitted for combustion during a new cycle.

Advantages :

1. Simple in construction and very inexpensive as compared to turbo-jet engine. Well adapted to pilotless aircraft.
2. Capable of producing static thrust and thrust in excess of drag at much low speeds.

Shortcomings :

1. High intensity of noise.
2. Severe vibrations.
3. High rate of fuel consumption and low thermodynamic efficiency.
4. Intermittent combustion as compared to continuous combustion in a turbo-jet engine.
5. The operating altitude is limited by air density consideration.
6. Serious limitation to mechanical valve arrangement.

21.8.5. Rocket Engines

Similar to jet propulsion, the thrust required for rocket propulsion is produced by the high velocity jet of gases passing through the nozzle. But the main difference is that in case of *jet propulsion the oxygen required for combustion is taken from the atmosphere and fuel is stored whereas for rocket engine, the fuel and oxidiser both are contained in a propelling body and as such it can function in vacuum.*

The rockets may be classified as follows :

1. **According to the type of propellents :**

(*i*) Solid propellent rocket

(*ii*) Liquid propellent rocket.

2. **According to the number of motors :**

(*i*) Single-stage rocket (consists of one rocket motor)

(*ii*) Multi-stage rocket (consists of more than one rocket motor).

Fig. 21.46 shows a simple type single stage liquid propellent (the fuel and the oxidiser are commonly known as propellents) rocket. It consists of a fuel tank *FT*, an oxidiser tank *O*, two pumps P_1, P_2, a steam turbine *ST* and a combustion chamber C.C. The fuel tank contains alcohol and oxidiser tank contains liquid oxygen. The fuel and the oxidiser are supplied by the pumps to the combustion chamber where the fuel is ignited by electrical means. The pumps are driven with the help of a steam turbine. Here the steam is produced by mixing a very concentrated hydrogen-peroxide with potassium permanganate. The products of combustion are discharged from the combustion chamber through the nozzle *N*. So the rocket moves in the opposite direction. In some modified form, this type of rocket may be used in missiles.

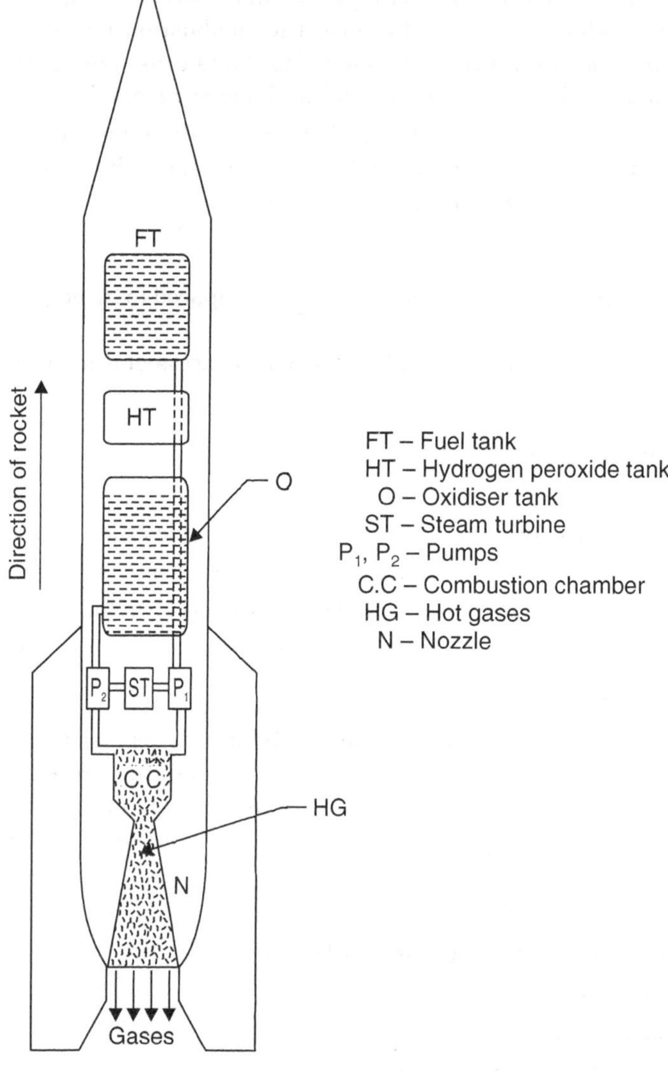

Fig. 21.46. Rocket.

21.8.5.1. Requirements of an ideal rocket propellent

An ideal rocket propellent should have the following *characteristics/properties* :

1. High heat value
2. Reliable smooth ignition
3. Stability and ease of handling and storing
4. Low toxicity and corrosiveness
5. Highest possible density so that it occupies less space.

21.8.5.2. Applications of rockets

The fields of application of rockets are as follows :

1. Long range artillary

2. Lethal weapons
3. Signalling and firework display
4. Jet assisted take-off
5. For satellites
6. For space ships
7. Research.

21.8.5.3. Thrust work, propulsive work and propulsive efficiency

In rocket propulsion, since air is self contained, the entry velocity relative to aircraft is zero. Neglecting the friction and other losses, we have the following formulae.

Thrust work $= C_j C_a$

Propulsive work $= C_j C_a + \dfrac{(C_j - C_a)^2}{2} = \dfrac{C_j^2 + C_a^2}{2}$

Rocket propulsive efficiency $= \dfrac{C_j C_a}{(C_j^2 + C_a^2)/2} = \dfrac{2\, C_j C_a}{C_j^2 + C_a^2} = \dfrac{2\left(\dfrac{C_a}{C_j}\right)}{1+\left(\dfrac{C_a}{C_j}\right)^2}$ ...(21.23)

HIGHLIGHTS

1. The gas turbines are mainly divided into two groups :
   (*i*) Constant pressure combustion gas turbine
      (*a*) Open cycle constant pressure gas turbine
      (*b*) Closed cycle constant pressure gas turbine.
   (*ii*) Constant volume combustion gas turbine.
2. Methods for improvement of thermal efficiency of open cycle gas turbine plant :
   (*i*) Intercooling (*ii*) Reheating (*iii*) Regeneration.
3. Types of jet propulsion systems :
   (*i*) Screw propeller (*ii*) Turbo-jet
   (*iii*) Turbo-prop (*iv*) Ram-jet.
4. Difference between jet propulsion and rocket propulsion :
   The main difference is that in case of jet propulsion the oxygen required for combustion is taken from the atmosphere and fuel is stored whereas for rocket engine the fuel and oxidiser both are contained in a propelling body and as such it can function in vacuum.
5. Classification of rockets :
   (*i*) According to the type of propellents :
      (*a*) Solid propellent rocket (*b*) Liquid propellent rocket.
   (*ii*) According to the number of motors :
      (*a*) Single-stage rocket (consists of one rocket motor)
      (*b*) Multi-stage rocket (consists of more than one rocket motor).

OBJECTIVE TYPE QUESTIONS

Choose the Correct Answer

1. Thermal efficiency of a gas turbine plant as compared to Diesel engine plant is
   (*a*) higher (*b*) lower
   (*c*) same (*d*) may be higher or lower.

2. Mechanical efficiency of a gas turbine as compared to internal combustion reciprocating engine is
(*a*) higher (*b*) lower
(*c*) same (*d*) unpredictable.

3. For a gas turbine the pressure ratio may be in the range
(*a*) 2 to 3 (*b*) 3 to 5
(*c*) 16 to 18 (*d*) 18 to 22.

4. The air standard efficiency of closed gas turbine cycle is given by (r_p = pressure ratio for the compressor and turbine)
(*a*) $\eta = 1 - \frac{1}{(r_p)^{\gamma - 1}}$ (*b*) $\eta = 1 - (r_p)^{\gamma - 1}$
(*c*) $\eta = 1 - \left(\frac{1}{r_p}\right)^{\frac{\gamma - 1}{\gamma}}$ (*d*) $\eta = (r_p)^{\frac{\gamma - 1}{\gamma}} - 1$.

5. The work ratio of closed cycle gas turbine plant depends upon
(*a*) pressure ratio of the cycle and specific heat ratio
(*b*) temperature ratio of the cycle and specific heat ratio
(*c*) pressure ratio, temperature ratio and specific heat ratio
(*d*) only on pressure ratio.

6. Thermal efficiency of closed cycle gas turbine plant increases by
(*a*) reheating (*b*) intercooling
(*c*) regenerator (*d*) all of the above.

7. With the increase in pressure ratio thermal efficiency of a simple gas turbine plant with fixed turbine inlet temperature
(*a*) decreases (*b*) increases
(*c*) first increases and then decreases (*d*) first decreases and then increases.

8. The thermal efficiency of a gas turbine cycle with ideal regenerative heat exchanger is
(*a*) equal to work ratio (*b*) is less than work ratio
(*c*) is more than work ratio (*d*) unpredictable.

9. In a two-stage gas turbine plant reheating after first stage
(*a*) decreases thermal efficiency (*b*) increases thermal efficiency
(*c*) does not effect thermal efficiency (*d*) none of the above.

10. In a two-stage gas turbine plant, reheating after first stage
(*a*) increases work ratio (*b*) decreases work ratio
(*c*) does not affect work ratio (*d*) none of the above.

11. In a two-stage gas turbine plant, with intercooling and reheating
(*a*) both work ratio and thermal efficiency improve
(*b*) work ratio improves but thermal efficiency decreases
(*c*) thermal efficiency improves but work ratio decreases
(*d*) both work ratio and thermal efficiency decrease.

12. For a jet-propulsion unit, ideally the compressor work and turbine work are
(*a*) equal (*b*) unequal
(*c*) not related to each other (*d*) unpredictable.

13. Greater the difference between jet velocity and aeroplane velocity
(*a*) greater the propulsive efficiency (*b*) less the propulsive efficiency
(*c*) unaffected is the propulsive efficiency (*d*) none of the above.

Answers

1. (*b*) 2. (*a*) 3. (*c*) 4. (*c*) 5. (*c*) 6. (*d*) 7. (*c*)
8. (*a*) 9. (*a*) 10. (*a*) 11. (*b*) 12. (*a*) 13. (*b*).

THEORETICAL QUESTIONS

1. What do you mean by the term 'gas turbine' ? How are gas turbines classified ?
2. State the merits of gas turbines over I.C. engines and steam turbines. Discuss also the demerits over gas turbines.
3. Describe with neat sketches the working of a simple constant pressure open cycle gas turbine.
4. Discuss briefly the methods employed for improvement of thermal efficiency of open cycle gas turbine plant.
5. Describe with neat diagram a closed cycle gas turbine. State also its merits and demerits.
6. Explain with a neat sketch the working of a constant volume combustion turbine.
7. Enumerate the various uses of gas turbines.
8. Write a short on fuels used for gas turbines.
9. Explain the working difference between propeller-jet, turbo-jet and turbo-prop.
10. State the fundamental differences between the jet propulsion and rocket propulsion.

UNSOLVED EXAMPLES

1. In an air standard gas turbine engine, air at a temperature of 15°C and a pressure of 1.01 bar enters the compressor, where it is compressed through a pressure ratio of 5. Air enters the turbine at a temperature of 815°C and expands to original pressure of 1.01 bar. Determine the ratio of turbine work to compressor work and the thermal efficiency when the engine operates on ideal Brayton cycle.

   Take : $\gamma = 1.4$, $c_p = 1.005$ kJ/kg K. **[Ans.** 2.393 ; 37.03%]

2. In an open cycle constant pressure gas turbine air enters the compressor at 1 bar and 300 K. The pressure of air after the compression is 4 bar. The isentropic efficiencies of compressor and turbine are 78% and 85% respectively. The air-fuel ratio is 80 : 1. Calculate the power developed and thermal efficiency of the cycle if the flow rate of air is 2.5 kg/s.

   Take $c_p = 1.005$ kJ/kg K and $\gamma = 1.4$ for air and $c_{pg} = 1.147$ kJ/kg K and $\gamma = 1.33$ for gases. $R = 0.287$ kJ/kg K. Calorific value of fuel = 42000 kJ/kg. **[Ans.** 204.03 kW/kg of air ; 15.54%]

3. A gas turbine has a pressure ratio of 6/1 and a maximum cycle temperature of 600°C. The isentropic efficiencies of the compressor and turbine are 0.82 and 0.85 respectively. Calculate the power output in kilowatts of an electric generator geared to the turbine when the air enters the compressor at 15°C at the rate of 15 kg/s.

   Take : $c_p = 1.005$ kJ/kg K and $\gamma = 1.4$ for the compression process, and take $c_p = 1.11$ kJ/kg K and $\gamma = 1.333$ for the expansion process. **[Ans.** 920 kW]

4. Calculate the thermal efficiency and the work ratio of the plant in example 3 (above), assuming that c_p for the combustion process is 1.11 kJ/kg K. **[Ans.** 15.8% ; 0.206]

5. The gas turbine has an overall pressure ratio of 5 : 1 and a maximum cycle temperature of 550°C. The turbine drives the compressor and an electric generator, the mechanical efficiency of the drive being 97%. The ambient temperature is 20°C and the isentropic efficiencies for the compressor and turbine are 0.8 and 0.83 respectively. Calculate the power output in kilowatts for an air flow of 15 kg/s. Calculate also the thermal efficiency and the work ratio.

   Neglect changes is kinetic energy, and the loss of pressure in combustion chamber.

   [Ans. 655 kW ; 12% ; 0.168]

6. Air is drawn in a gas turbine unit at 17°C and 1.01 bar and the pressure ratio is 8 : 1. The compressor is driven by the H.P. turbine and the L.P. turbine drives a separate power shaft. The isentropic efficiencies of the compressor, and the H.P. and L.P. turbines are 0.8, 0.85 and 0.83, respectively. Calculate the pressure and temperature of the gases entering the power turbine, the net power developed by the unit per kg/s of mass flow, the work ratio and the thermal efficiency of the unit. The maximum cycle temperature is 650°C.

   For the compression process take $c_p = 1.005$ kJ/kg K and $\gamma = 1.4$

   For the combustion process and expansion process, take

   $$c_p = 1.15 \text{ kJ/kg K and } \gamma = 1.333$$

   Neglect the mass of fuel. **[Ans.** 1.65 bar, 393°C ; 74.5 kW ; 0.201 ; 19.1%]

7. In a gas turbine plant, air is compressed through a pressure ratio of 6 : 1 from 15°C. It is then heated to the maximum permissible temperature of 750°C and expanded in two stages each of expansion ratio $\sqrt{6}$, the air being reheated between the stages to 750°C. A heat exchanger allows the heating of the compressed gases through 75 per cent of the maximum range possible. Calculate : (*i*) The cycle efficiency (*ii*) The work ratio (*iii*) The work per kg of air.

   The isentropic efficiencies of the compressor and turbine are 0.8 and 0.85 respectively.

   [Ans. (*i*) 32.75% (*ii*) 0.3852 (*iii*) 152 kJ/kg]

8. At the design speed the following data apply to a gas turbine set employing the heat exchanger : Isentropic efficiency of compressor = 75%, isentropic efficiency of the turbine = 85%, mechanical transmission efficiency = 99%, combustion efficiency = 98%, mass flow = 22.7 kg/s, pressure ratio = 6 : 1, heat exchanger effectiveness = 75%, maximum cycle temperature = 1000 K.

   The ambient air temperature and pressure are 15°C and 1.013 bar respectively. Calculate :

   (*i*) The net power output (*ii*) Specific fuel consumption

   (*iii*) Thermal efficiency of the cycle.

   Take the lower calorific value of fuel as 43125 kJ/kg and assume no pressure-loss in heat exchanger and combustion chamber. **[Ans.** (*i*) 2019 kW (*ii*) 0.4799 kg/kWh (*iii*) 16.7%]

9. In a gas turbine plant air at 10°C and 1.01 bar is compressed through a pressure ratio of 4 : 1. In a heat exchanger and combustion chamber the air is heated to 700°C while its pressure drops 0.14 bar. After expansion through the turbine the air passes through a heat exchanger which cools the air through 75% of maximum range possible, while the pressure drops 0.14 bar, and the air is finally exhausted to atmosphere. The isentropic efficiency of the compressor is 0.80 and that of turbine 0.85.

   Calculate the efficiency of the plant. **[Ans.** 22.76%]

10. In a marine gas turbine unit a high-pressure stage turbine drives the compressor, and a low-pressure stage turbine drives the propeller through suitable gearing. The overall pressure ratio is 4 : 1, and the maximum temperature is 650°C. The isentropic efficiencies of the compressor, H.P. turbine, and L.P. turbine are 0.8, 0.83, and 0.85 respectively, and the mechanical efficiency of both shafts is 98%. Calculate the pressure between turbine stages when the air intake conditions are 1.01 bar and 25°C. Calculate also the thermal efficiency and the shaft power when the mass flow is 60 kg/s. Neglect kinetic energy changes, and pressure loss in combustion. **[Ans.** 1.57 bar ; 14.9% ; 4560 kW]

11. In a gas turbine unit comprising L.P. and H.P. compressors, air is taken at 1.01 bar 27°C. Compression in L.P. stage is upto 3.03 bar followed by intercooling to 30°C. The pressure of air after H.P. compressor is 58.7 bar. Loss in pressure during intercooling is 0.13 bar. Air from H.P. compressor is transferred to heat exchanger of effectiveness 0.60 where it is heated by gases from L.P. turbine. The temperature of gases supplied to H.P. turbine is 750°C. The gases expand in H.P. turbine to 3.25 bar and are then reheated to 700°C before expanding in L.P. turbine. The loss of pressure in reheater is 0.1 bar. If isentropic efficiency of compression in both stages is 0.80 and isentropic efficiency of expansion in turbine is 0.85, calculate : (*i*) Overall efficiency (*ii*) Work ratio (*iii*) Mass flow rate when the gas power generated is 6500 kW. Neglect the mass of fuel.

    Take, for air : $c_p = 1.005$ kJ/kg K, $\gamma = 1.4$

    for gases : $c_{pg} = 1.15$ kJ/kg K, $\gamma = 1.3$. **[Ans.** (*i*) 16.17% (*ii*) 0.2215 (*iii*) 69.33 kg of air/sec.]

12. In a gas turbine installation, air is taken in L.P. compressor at 15°C 1.1 bar and after compression it is passed through intercooler where its temperature is reduced to 22°C. The cooled air is further compressed in H.P. unit and then passed in the combustion chamber where its temperature is increased to 677°C by

burning the fuel. The combustion products expand in H.P. turbine which runs the compressor and further expansion is continued in the L.P. turbine which runs the alternator. The gases coming out from L.P. turbine are used for heating the incoming air from H.P. compressor and then exhausted to atmosphere.

Taking the following data determine : (*i*) power output (*ii*) specific fuel consumption (*iii*) Thermal efficiency :

Pressure ratio of each compressor = 2, isentropic efficiency of each compressor stage = 85%, isentropic efficiency of each turbine stage = 85%, effectiveness of heat exchanger = 0.75, air flow = 15 kg/sec., calorific value of fuel = 45000 kJ/kg, c_p(for gas) = 1 kJ/kg K, c_p (for gas) = 1.15 kJ/kg K, γ (for air) = 1.4, γ (for gas) = 1.33.

Neglect the mechanical, pressure and heat losses of the system and fuel mass also.

[**Ans.** (*i*) 1849.2 kW (*ii*) 0.241 kg/kWh (*iii*) 33.17%]

13. A turbo-jet engine flying at a speed of 960 km/h consumes air at the rate of 54.5 kg/s. Calculate : (*i*) Exit velocity of jet when the enthalpy change for the nozzle is 200 kJ/kg and velocity co-efficient is 0.97, (*ii*) fuel flow rate in kg/s when air-fuel ratio is 75 : 1 (*iii*) Thrust specific fuel consumption (*iv*) Thermal efficiency of the plant when the combustion efficiency is 93% and calorific value of the fuel is 45000 kJ/kg, (*v*) Propulsive power (*vi*) Propulsive efficiency (*vii*) Overall efficiency.

[**Ans.** (*i*) 613.5 m/s (*ii*) 0.7267 kg/s (*iii*) 4.3×10^{-5} kg/N of thrust/s (*iv*) 25.44%.
(*v*) 8318 kW, (*vi*) 60.6%, (*vii*) 16.58%]

14. A turbo-jet has a speed of 750 km/h while flying at an altitude of 10000 m. The propulsive efficiency of the jet is 50% and overall efficiency of the turbine plant is 16%. The density of air at 10000 m altitude is 0.173 kg/m^3. The drag on the plant is 6250 N. The calorific value of the fuel is 48000 kJ/kg. Calculate : (*i*) Absolute velocity of the jet (*ii*) Volume of air compressed per minute (*iii*) Diameter of the jet (*iv*) Power output of the unit in kW (*v*) Air-fuel ratio. [**Ans.** (*i*) 417.3 m/s (*ii*) 5194 m^3/min (*iii*) 415 mm (*iv*) 2500 kW (*v*) 46.01]

22

Universities Examinations' Questions (Latest-selected) with Answers/Solutions

(A) Short Answer Questions.

(B) Conventional Questions with Answers/Solutions.

A. SHORT ANSWER QUESTIONS

Introduction to Internal Combustion Engines

Q. 22.1. What is a heat engine ?

Ans. Any type of engine or machine which derives heat energy from the combustion of fuel or any other source and converts this energy into mechanical energy is termed as a "*heat engine*".

Q. 22.2. What is an External Combustion (E.C.) engine ? Give examples.

Ans. An engine in which combustion of fuel takes place *outside* the cylinder is called an "*external combustion engine*".

Examples :

— Steam engine (In this case heat of combustion is employed to generate steam which is used to move a piston in cylinder).

— Steam turbine.

— Hot air engines.

— Closed cycle gas turbine.

Q. 22.3. What is an Internal Combustion (I.C.) engine ?

Ans. An engine in which combustion of fuel with oxygen of air occurs *within* the cylinder of the engine is called an "*internal combustion engine*".

Q. 22.4. Who invented the first fairly practical engine and when ?

Ans. The first fairly practical engine was invented by **J.J.E. Lenoir** which appeared on the scene *about* **1860.**

Q. 22.5. What are the advantages of a stirling engine ?

Ans. Low exhaust emission and multi-fuel capability.

Q. 22.6. Where are two-stroke petrol engines employed ?

Ans. Two-stroke engines are employed where *simplicity* and the *low cost* of the prime mover are primary considerations.

Q. 22.7. What are the fields of application of a two-stroke petrol engine ?

Ans. Mopeds, scooters, motor cycles, small electric generating sets, pumping sets, etc.

Q. 22.8. Where are two-stroke diesel engines used ?

Ans. These engines having very high power are usually employed for *ship propulsion* and generally have bores above 60 cm, uniflow with exhaust values or loop scavenged.

Q. 22.9. Give an example of a two-stroke diesel engine.

Ans. *Nordberg*-2 stroke, 12-cylinders, 80 cm bore and 155 cm stroke, develops 20000 kW at 120 r.p.m.

Q. 22.10. What is "mechanical efficiency" ?

Ans. The ratio of shaft energy to the energy available at the piston is called *mechanical efficiency.*

Q. 22.11. What is the function and material of an engine cylinder ?

Ans. The cylinder contains gas under pressure and guides the piston. It is made of hard grade cast iron and is usually cast in one piece.

Q. 22.12. What are inlet and exhaust manifolds ?

Ans. The system of pipes which connects the inlet ports of the various cylinders to a common intake pipe for the engine is called the *inlet manifold.* If the exhaust parts are similarly connected to a common exhaust system, this system of piping is called *exhaust manifold.*

Q. 22.13. What is the function of a piston ? Of which material(s) it is made of ?

Ans.

- A piston is fitted to each cylinder as a face to receive gas pressure and *transmit the thrust* to the connecting rod.
- Pistons are made of *cast iron* or *aluminium alloy* for lightness.

Q. 22.14. What is the function of a "spark plug" ?

Ans. The main function of a spark plug is to *conduct the high potential* from the ignition system into the combustion chamber. It provides the *proper gap* across which spark is produced by applying high voltage, to ignite the combustion chamber.

Q. 22.15. What is the function of a "simple carburettor" ?

Ans. The function of a carburettor is *to atomise and meter the liquid fuel* and mix it with the air as it enters the induction system of the engine, maintaining under all conditions of operation fuel-air proportions appropriate to these conditions.

Q. 22.16. What is the function of a "flywheel" ?

Ans. The function of a flywheel is to *store* the available mechanical energy when it is in excess of the load requirement and to *part with* the same when the available energy is less than that required by the load.

Q. 22.17. What is a "governor" ?

Ans.

- A governor may be defined as a *device for regulating automatically output* of a machine by regulating the supply of working fluid.
- Its function is to *control the fluctuations of engine speed due to changes of load.*

Q. 22.18. Differentiate among square, under square and over square engines.

Ans.

- An engine with $L = D$ is often called a *square* engine.
- If $L > D$ the engine is *under square.*
- If $L < D$ the engine is *over square.*

where L and D stand for stroke and bore of the engine.

Q. 22.19. What is "swept volume" ?

Ans. The volume swept through by the piston in moving between top dead centre and bottom dead centre is called *"swept volume* or *piston displacement"*.

Thus, when the piston is at B.D.C. (Bottom dead centre) :

Total volume = Swept volume + clearance volume.

Q. 22.20. What is "compression ratio" ? Explain briefly.

Ans. *Compression ratio* (r) is the ratio of cylinder volume to clearance volume.

Mathematically, $r = \dfrac{V_s + V_c}{V_c}$

where, V_s = swept volume, and V_c = clearance volume.

- The compression ratio varies from 5 : 1 to 11 : 1 (average value 7 : 1 to 9 : 1) in *S.I. engines* and from 12 : 1 to 24 : 1 (average value 15 : 1 to 18 : 1) in *C.I. engines.*
- Modern S.I. engines have compression ratios of 8 to 11, while C.I. engines have compression ratios in the range of 12 to 24. Engines with supercharges or turbocharges usually have lower compression ratios than naturally aspirated engines.

Q. 22.21. What is "piston speed" ?

Ans. The average speed of the piston is called *"piston speed"*. Piston speed = 2 LN, where L and N are the stroke length and engine speed (in r.p.m.) respectively.

Q. 22.22. What is a "smart engine" ?

Ans. A *"smart engine"* is an engine with *computer controls* that regulate operating characteristics such as air-fuel ratio, ignition timing, valve timing, exhaust control, intake timing, etc.

Q. 22.23. What do you mean by A/F and F/A ?

Ans.

- A/F (Air-fuel ratio) is the ratio of the air to mass of fuel input into engine.
- F/A (Fuel-air ratio) is the ratio of fuel to mass air input into the engine.

Q. 22.24. What is an 'Indicator diagram' ?

Ans. An *"indicator diagram"* is a graph between pressure and volume, the former being taken on vertical axis and the latter on the horizontal axis.

- This is obtained by an instrument known as *indicator.*

Q. 22.25. On which thermodynamic cycles do the S.I. engines and C.I. engines operate ?

Ans.

S.I. engines	*Otto cycle*
C.I. engines	*Diesel cycle*—For slow speed engines
	Dual cycle—For high speed engines

Q. 22.26. Give the applications of four-stroke and two-stroke cycle engines.

Ans. ***Engines***	***Applications***
Four-stroke cycle engines :	Cars, buses, trucks, tractors, industrial engines, aeroplanes, power generators, etc.
Two-stroke petrol engines :	Lawn movers, scooters, motor cycles (lubricating oil mixed with petrol)
Two-stroke diesel engine :	Used in very large sizes more than 60 cm bore, for *ship propulsion* because of low weight and compactness.

Air Standard cycles

Q. 22.27. What is a "cycle" ?

Ans. A *"cycle"* is defined as a repeated series of operations occurring in a certain order. In ideal cycle all accidental heat losses are prevented and the working substance is assumed to behave like a perfect working substance.

Q. 22.28. What is an "Air standard efficiency" ?

Ans. The efficiency of engine using air as the working medium is known as an *"Air standard efficiency"*. This efficiency is oftenly called *ideal efficiency*.

Q. 22.29. What is "Relative efficiency" ?

Ans. The ratio of actual thermal efficiency to air standard efficiency is termed as *"Relative efficiency"*.

Q. 22.30. Which cycle has the highest possible efficiency ?

Ans.

- *Carnot cycle* has the highest possible efficiency. It consists of four simple operations namely : Isothermal expansion, Adiabatic expansion, Isothermal compression and Adiabatic compression.
- Efficiency of cycle = $\dfrac{T_1 - T_2}{T_1}$, where T_1 and T_2 are temperatures (in **K**) of source and the sink respectively.

Q. 22.31. What is the expression for air standard efficiency of Otto cycle ?

Ans. $\eta_{otto} = 1 - \dfrac{1}{(r)^{\gamma-1}}$

where, r is the compression ratio and γ is the ratio of c_p and c_v $\left(i.e., \ \gamma = \dfrac{c_p}{c_v}\right)$.

Q. 22.32. What is the sequence of operations of an air standard Otto cycle ?

Ans. Adiabatic compression (of air) ; Heat addition at constant volume ;
Adiabatic expansion ; Heat rejection at constant volume.

Q. 22.33. The efficiency of an Otto cycle is 60% and γ = 1.5. What is the compression ratio ?

Ans. $\eta = 1 - \dfrac{1}{(r)^{\gamma-1}}$, or, $0.6 = 1 - \dfrac{1}{(r)^{1.5-1}}$ or $r = 6.25$.

Q. 22.34. What is the sequence of operations of a air standard Diesel cycle ?

Ans. Adiabatic compression (of air) ; Heat addition a constant pressure ; Adiabatic expansion; Heat rejection at constant volume.

Q. 22.35. What is the expression of air standard efficiency of a Diesel cycle ?

Ans.
$$\eta_{diesel} = 1 - \frac{1}{\gamma (r)^{\gamma-1}} \left[\frac{\rho^{\gamma} - 1}{\rho - 1} \right],$$

where, ρ is the *cut-off ratio* (It is the ratio of volume at cut-off and clearance volume).

Q. 22.36. What is the sequence of operations in an air standard dual combustion cycle ?

Ans. Adiabatic compression ; Addition of heat at constant volume ; Addition of heat at constant pressure ; Adiabatic expansion ; Rejection of heat at constant volume.

Q. 22.37. For a given compression ratio, give the comparison for air standard efficiencies of the Otto, Diesel and Dual combustion cycles.

Ans. For a given compression ratio,

$$\eta_{otto} > \eta_{dual} > \eta_{diesel}.$$

Q. 22.38. For the same compression ratio and the same heat input compare the air standard efficiencies of Otto, Diesel and Dual combustion cycles.

Ans. For the same compression ratio and the same heat input,

$$\eta_{otto} > \eta_{dual} > \eta_{diesel}.$$

Q. 22.39. Name the four operations used in an Atkinson cycle.

Ans. Two adiabatics, a constant volume and a constant pressure operations.

Q. 22.40. Name the four operations used in an Ericssion cycle.

Ans. Rejection of heat at constant pressure, isothermal compression, addition of heat at constant pressure and an isothermal expansion.

Q. 22.41. What is a Brayton cycle ?

Ans. *Brayton cycle* is a constant pressure cycle for a perfect gas. It is also called *Joule cycle.* An ideal gas turbine plant would perform the processes that make up a Brayton cycle.

Q. 22.42. What are the typical values of compression ratio and expansion ratio of a Miller cycle ?

Ans. Compression ratio is about 8 : 1, and expansion ratio is about 10 : 1

Fuel-air and Actual Cycles

Q. 22.43. What do you understand by 'fuel-air cycle approximation' ?

Ans. The theoretical cycle based on actual properties of the cylinder gases is called the 'Fuel-air cycle approximation'; it provides rough idea for comparison with the actual performance.

Q. 22.44. What factors are taken into considerations while making the fuel-air cycle calculations ?

Ans.

1. The actual composition of cylinder gases.
2. Increase of specific heats of gases with temperature increase.
3. Presence of gases such as CO_2, H_2 and O_2 at equilibrium condition.
4. The variation in the number of molecules present in the cylinder as the temperature and pressure change.

Q. 22.45. What is 'dissociation' ?

Ans.

- *Dissociation* (or chemical equilibrium loss) refers to disintegration of burnt gases at high temperature. It is a reversible process and increases with temperature.
- The dissociation, in general, lowers the temperature and consequently the pressures at the beginning of the stroke, this causes a loss of power and efficiency.

Q. 22.46. In which type of mixture dissociation is more severe ?

Ans. *Chemically correct mixture.* If the mixture is weaker, it gives temperatures lower than those required for dissociation to take place while if it is richer, during combustion it will give out CO and O_2 both of which suppress the dissociation of CO_2.

Q. 22.47. In which of the engines (S.I., C.I. engines) the dissociation has a more pronounced effect ?

Ans. The dissociation has a more pronounced effect in S.I. engines. In C.I. engines, the heterogeneous mixture of air and fuel tend to lower the temperature and hence the dissociation.

Q. 22.48. What is use of 'Chart for unburned mixture' ?

Ans. The chart for unburned mixture is employed for calculation of pressure, temperature, and energy of the cylinder contents before burning takes place.

Q. 22.49. Name the methods by which the problems involving variable specific heats can be solved.

Ans. The problems involving variable specific heats can be solved by the following methods:

(*i*) By integration of specific heat equation.

(*ii*) By enthalpy-entropy charts.

(*iii*) By gas tables.

Q. 22.50. How is the 'burning time loss' defined ?

Ans. The burning time loss or simply the time loss is defined as the loss of power due to time required for mixing the fuel with air and for complete combustion.

Q. 22.51. What are the factors on which the time required to complete the burning process in an actual cycle depend ?

Ans.

(*i*) Fuel-air ratio.

(*ii*) Fuel chemical structure and its ignition temperature.

(*iii*) The flame velocity and distance from the ignition point to the opposite side of the combustion chamber.

Q. 22.52. State whether flame speed in a rich mixture is low or high ; what is its effect on the efficiency ?

Ans. The flame speed in rich mixture is *low* and causes burning time loss leading to *lowering* of efficiency.

Q. 22.53. What is 'pumping loss' ?

Ans. The *pumping loss* is due to pumping gas from low pressure to higher exhaust pressure. It increases at part throttle because throttling causes reduction in suction pressure. It increases with increase in speed.

Q. 22.54. How is piston friction affected by changes in engine speed and mean effective pressure ?

Ans. The piston friction increases rapidly with engine speed and to small extent by increase in mean effective pressure.

Q. 22.55. How does engine efficiency vary with the load ?

Ans. The engine efficiency is maximum at full load and reduces with the decrease in load. It is due to the fact that direct heat loss, pumping loss and rubbing friction loss increase at lower loads.

Combustion in S.I. Engines

Q. 22.56. Define the term 'combustion'.

Ans. *Combustion* may be defined as a relatively rapid chemical combination of hydrogen and carbon in the fuel with the oxygen in the air, resulting in liberation of energy in the form of heat.

Q. 22.57. State the conditions which are necessary for combustion to take place.

Ans. The following conditions are necessary for combustion to take place :

(*i*) A combustible mixture.

(*ii*) Some means to initiate combustion.

(*iii*) Stabilization and propagation of flame in the combustion chamber.

Q. 22.58. List the factors which affect normal combustions in S.I. engines.

Ans. Induction pressure ; Engine speed ; Ignition timing ; Mixture strength ; Compression ratio ; Combustion chamber ; Fuel choice.

Q. 22.59. What are the two combustion abnormalities in S.I. engines ?

Ans.

(*i*) Pre or post ignition ;

(*ii*) Knock.

Q. 22.60. What are the factors on which duration of ignition lag depends ?

Ans. Fuel ; Mixture ratio ; Initial temperature and pressure ; Turbulence.

Q. 22.61. What are the factors which affect the ignition timings ?

Ans. Engine speed ; Mixture strength ; Part-load operation ; Type of fuel.

Q. 22.62. What is 'Pre-ignition' ?

Ans. *Pre-ignition* is the ignition of homogeneous mixture in the cylinder, before the timed ignition spark occurs, caused by the local overheating of the combustible mixture.

Q. 22.63. How is pre-ignition initiated ?

Ans. Pre-ignition is initiated by some *overheated projecting part* such as the sparking plug electrodes, exhaust valve head, metal corners in the combustion chamber, carbon deposits or protruding cylinder head gasket rim, etc.

However, pre-ignition is also caused by *persistent detonating pressure shock waves.*

Q. 22.64. What is the effect of pre-ignition on detonation in S.I. engines ?

Ans. Pre-ignition increases the tendency of detonation in the engines.

Q. 22.65. Define the term 'Detonation'.

Ans. A very sudden rise of pressure during combustion accompanied by metallic hammer like sound is called *detonation.*

Q. 22.66. What do you understand by 'Detonation zone' ?

Ans. The region in which detonation occurs is farthest removed from the sparking plug, and is named the 'detonation zone' and even with severe detonation this zone is *rarely more than one quarter the clearance volume.*

Q. 22.67. What minimum percentage of total mixture charge will be sufficient to produce a violent knock ?

Ans. As little as 5% of the total mixture charge when spontaneously ignited will be sufficient to produce a very violent knock.

Q. 22.68. Name the two general theories of knocking/detonation.

Ans.

(*i*) The auto-ignition theory ;

(*ii*) The detonation theory.

Q. 22.69. What is 'auto-ignition theory' ?

Ans. Auto-ignition refers to *initiation of combustion without the necessity of a flame.* The auto-ignition theory of knock assumes that the flame velocity is normal before the on-set of auto-ignition and that gas vibrations are created by a number of end-gas elements auto-igniting almost simultaneously.

Q. 22.70. What is 'detonation theory' ?

Ans. In this theory a true *detonating wave* formed by preflame reactions has been proposed as a machanism for explosive auto-ignition. Such a shock wave would travel through the chamber at about twice the sonic velocity and would compress the gases to pressures and temperatures where the reaction should be practically instantaneous.

Q. 22.71. What are the effects of detonation ?

Ans. Noise and roughness ; Mechanical damage ; Carbon deposits ; Increase in heat transfer ; Decrease in power output and efficiency ; Pre-ignition.

Q. 22.72. What are the methods by which detonation can be controlled ?

Ans. (*i*) Increasing engine r.p.m. ; (*ii*) Retarding spark ; (*iii*) Reducing pressure in the inlet manifold by throttling ; (*iv*) Making the ratio too lean or too rich, preferably latter ; (*v*) Water injection ; (*vi*) Use of high octane fuel.

Q. 22.73. What are the factors which affect detonation ?

Ans. (*i*) Fuel choice ; (*ii*) Induction pressure ; (*iii*) Engine speed ; (*iv*) Ignition timing ; (*v*) Mixture strength ; (*vi*) Compression ratio ; (*vii*) Combustible chamber design ; (*viii*) Cylinder cooling.

Q. 22.74. What is a 'Performance Number (PN)' ?

Ans. $\text{Performance Number (PN)} = \dfrac{\text{klimep of test fuel}}{\text{klimep of iso-octane}}$

where, *klimep* stands for knock limited indicated mean effective pressure.

Q. 22.75. What is 'Highest Useful Compression Ratio (HUCR)' ?

Ans. The *highest useful compression ratio* is the highest compression ratio employed at which a fuel can be used in a specified engine under specified set of operating conditions, at which detonation first becomes audible with both the ignition and mixture strength adjusted to give the highest efficiency.

Q. 22.76. What is 'swirl' ?

Ans. *Swirl* is the rotational flow of charge within the cylinder about its axis.

Q. 22.77. How is swirl generated ?

Ans. Swirl is generated by constructing the intake system to give a tangential component to the intake flow as it enters the cylinder. This is done by shaping and contouring the intake manifold, valve ports and even the piston face.

Q. 22.78. What is the function of 'swirl' ?

Ans. Swirl greatly *enhances the mixing of air and fuel* to give a homogeneous mixture in the very short time available for this in modern high speed engines. It is also a main mechanism for spreading of the flame front during the combustion process.

Q. 22.79. How are induction ports classified ?

Ans.

(*i*) Direct straight port ; (*ii*) Deflector wall port ;
(*iii*) Marked valve port ; (*iv*) Helical port.

Q. 22.80. What is 'swirl ratio' ?

Ans. *Swirl ratio* is the ratio of air rotational speed to crankshaft rotational speed.

Q. 22.81. What is the range of 'stroke-to-bore' ratio for various engines ?

Ans.

- For various engines the stroke-to-bore (L : D) ratio can range from 0.6 : 1 to 1.4 : 1.
- — When $L = D$, the $L : D$ ratio is said to be *square* ;
- — When $L < D$, the $L : D$ ratio is said to be *over square ;*
- — When $L > D$, the $L : D$ ratio is said to be *under square.*
- *"Over square"* engines are more suitable for saloon car petrol engines, whereas *"undersquare"* engines are better utilized for large diesel engines.

Combustion in C.I. Engines

Q. 22.82. For which reasons a C.I. engine is not much favoured in passenger cars ?

Ans. (*i*) Heavier weight ; (*ii*) Noise and vibration ; (*iii*) Smoke ; (*iv*) Odour.

Q. 22.83. What are the various phases of combustion in a C.I. engine ?

Ans. (*i*) Ignition delay period ; (*ii*) Period of rapid or uncontrolled combustion ; (*iii*) Period of controlled combustion ; (*iv*) After burning.

Q. 22.84. What do you understand by 'Period of physical delay' ?

Ans. The *period of physical delay* is the time between the beginning of injection and the attainment of chemical reaction condition. During this period, the fuel is atomised, vaporized, mixed with air, and raised in temperature.

Q. 22.85. List the factors which affect combustion in C.I. engines.

Ans. (*i*) Ignition quality of fuel ; (*ii*) Injection pressure of droplet size ; (*iii*) Injection advance angle ; (*iv*) Compression ratio ; (*v*) Intake temperature ; (*vi*) Jacket water temperature ; (*vii*) Intake pressure, supercharging ; (*viii*) Engine speed ; (*ix*) Engine size ; (*x*) Type of combustion chamber.

Q. 22.86. How much excess air (per cent) is used in most C.I. engines, and why ?

Ans. Most C.I. engines operate with at least *20% excess air* due to difficulty of introducing sufficient exposed oxygen to the fuel vapour in the given time available so that the combustion process can be completed before the exhaust valve opens. If the oxygen supply is *partially prevented from getting to the fuel vapour early enough during the power stroke then incomplete combustion, polluted exhaust gas and dark smoke will result.*

Q. 22.87. What should be the duration of the delay period in a C.I. engine ?

Ans. The delay period should be *as short as possible* since a long *delay period gives a more rapid rise in pressure and thus cause knocking.*

Q. 22.88. What do you mean by 'Diesel knock' ?

Ans. Diesel knock is the sound produced by the very rapid rate of pressure rise during the early part of the uncontrolled second phase of combustion. The primary cause of an excessively high pressure rise is the prolonged delay period.

Q. 22.89. How can the 'diesel knock' be controlled ?

Ans. The diesel knock can be controlled by *reducing the delay period.* The delay is reduced by the following :

(*i*) High charge temperature ; (*ii*) High fuel temperature ;
(*iii*) Good turbulence ; (*iv*) A fuel with a short induction period.

Q. 22.90. Name the various types of 'swirls' generated in the C.I. engines combustion chambers ?

Ans. (*i*) Induction swirl ; (*ii*) Compression swirl ; (*iii*) Combustion induced swirl.

Q. 22.91. List the four specific design of combustion chambers used in C.I. engines.

Ans. A. The *non-turbulent type* :

(*i*) Open or direct combustion chamber.

B. The *turbulent type* :

(*i*) Turbulent chamber ;
(*ii*) Pre-combustion chamber ;
(*iii*) Energy cell.

Q. 22.92. Name the modern starting aids of high speed engines.

Ans.

(*i*) Electric glow plugs (in the combustion chamber) ;
(*ii*) Manifold heaters (which ignite a small feed of fuel) ;
(*iii*) Injection of ether.

Air Capacity of Four-stroke Engines

Q. 22.93. What is air capacity (actual) ?

Ans. *Air capacity (actual)* is defined as the mass flow of fresh air through the engine per unit time.

Q. 22.94. What is an 'ideal air capacity' ?

Ans. The *ideal air capacity* corresponds to filling the displaced volume (*i.e.*, piston swept volume) with fresh mixture at inlet conditions.

Q. 22.95. Define the term 'volumetric efficiency' of an engine.

Ans. The *volumetric efficiency* of an engine is defined as the ratio of actual air capacity to the ideal air capacity. This is equal to the ratio of mass of air which enters or is forced into the cylinder in suction stroke to the mass of free air equivalent to the piston displacement at intake temperature and pressure conditions.

— Power output of an engine is proportional to volumetric efficiency provided the combustion is complete.

Q. 22.96. List the various factors which affect the volumetric efficiency.

Ans. (*i*) Fuel ; (*ii*) Heat transfer–high temperature ; (*iii*) Valve overlap ; (*iv*) Fluid friction losses ; (*v*) Choked flow ; (*vi*) Intake valve closure after B.D.C. ; (*vii*) Exhaust residual ; (*viii*) Exhaust gas recycle (EGR) ; (*ix*) Piston speed and engine size ; (*x*) Design of inlet and exhaust systems.

Two-stroke Engines

Q. 22.97. What is a common disadvantage for all the petrol as well as diesel two-stroke engines ?

Ans. The common *disadvantage* is greater cooling and lubrication requirements due to one power stroke in each revolution of crankshaft.

Q. 22.98. What are the two main disadvantages from which the two-stroke S.I. engines suffer ?

Ans.

(*i*) Loss of fuel ; (*ii*) Idling difficulty.

Q. 22.99. What are the reasons for use of two-stroke C.I. engines for marine propulsion ?

Ans.

(*i*) More uniform torque, the ideal requirement for the propeller.

(*ii*) More cooling is required in two-stroke engines, plenty of sea water is available for cooling.

(*iii*) C.I. engines have higher thermal efficiency.

(*iv*) Propeller imposes the condition that maximum power must be developed at about 100 r.p.m. Two-stroke engines may be made of slow speed, and with large displacement volume (over 60 cm bore) and of capacity 5000 kW and above. These slow speed engines can be *coupled directly to the propeller of the ship, without the necessity of gear reduction.*

Q. 22.100. Which type of two-stroke C.I. engine is mainly used for marine propulsion ?

Ans. Two-stroke C.I. opposed engine (cross-head type) is mainly used for marine propulsion.

Q. 22.101. State the reasons for use of two-stroke S.I. engines for low horse power two-wheelers.

Ans. In S.I. engines the charge consists of a mixture of air and fuel. During scavenging, both inlet and exhaust ports are open simultaneously for sometime. Some part of the fresh charge escapes with exhaust which results in higher fuel consumption and lower thermal efficiency.

- For *small two-wheeler engines the fuel economy is not a vital factor.* Here *light weight and initial cost* are the main considerations, which are the main characteristics of two-stroke S.I. engines.

Q. 22.102. What do you understand by "Scavenging process" ?

Ans. The process of clearing the cylinder after the expansion stroke is called *"scavenging process"*. The scavenging process is the replacement of combustion products in the cylinder from the previous stroke with fresh air charge to be burned in the next cycle.

Q. 22.103. What are the four distinct stages of scavenging process ?

Ans.

(*i*) Pre-blowdown ; (*ii*) Blowdown ;

(*iii*) Scavenging ; (*iv*) Additional charging.

Q. 22.104. Name the three theoretical scavenging processes.

Ans.

(*i*) Perfect scavenging ;

(*ii*) Perfect mixing ;

(*iii*) Short circuiting.

Q. 22.105. What is a 'short circuiting scavenging process' ?

Ans. In this process, the fresh charge coming from the scavenge manifold goes out of exhaust ports without removing combustion products/gases. It results in a dead loss and its occurrence must be checked/avoided.

Q. 22.106. Why more air input is required in a two-stroke cycle engine than in four stroke cycle engine, for the same power generation ?

Ans. Because some of the air is lost in the overlap period of the scavenging process.

Q. 22.107. What is "pressure loss coefficient" ?

Ans. *Pressure loss coefficient* is defined as the ratio between the main upstream and downstream pressures during the scavenging period and represents the pressure loss to which the scavenge air is subjected when it crosses the cylinder.

Q. 22.108. Name different scavenging systems/arrangements based on charge flow.

Ans.

(*i*) Uniflow scavenging ;
(*ii*) Loop or reverse scavenging ;
(*iii*) Cross scavenging.

Q. 22.109. What is crankcase scavenging ?

Ans. This type of scavenging is employed in the simplest type of two-stroke engine.

In this engine, the charge (fuel-air mixture in S.I. engine and air in C.I. engine) is compressed in the crankcase by the underside of the piston during the expansion stroke. There are three ports in this engine–intake port at the crankcase, transfer port and the exhaust port. The compressed charge passes through the transfer port into the engine cylinder flushing the products of combustion. This process is called scavenging, and this type of engine in called crankcase scavenged engine.

Q. 22.110. What are the demerits of crankcase scavenging ?

Ans.

(*i*) This system is *very uneconomical and inefficient in operation.*

(*ii*) Due to mixing of oil vapours from the crankcase with the scavenging air, *oil consumption is increased.*

Q. 22.111. Name the various types of scavenging pumps and blowers used in two-stroke engines.

Ans. (*i*) Piston type pump ; (*ii*) Roots blower ; (*iii*) Centrifugal blower.

Chemical Thermodynamics and Fuels

Q. 22.112. What is 'Stoichiometric mixture' ?

Ans. Stoichiometric (or chemically correct) mixture of air and fuel is one that *contains just sufficient oxygen for complete combustion of the fuel.*

— A *weak mixture* is one which has an *excess* of air.

— A rich mixture is one which has a *deficiency* of air.

Q. 22.113. What is 'Enthalpy of formation' ?

Ans. The *enthalpy of formation* is the increase in enthalpy when a compound is formed from its constituent elements in their material form and in a standard state.

Q. 22.114. Name the major fuel used for S.I. engines.

Ans. *Gasoline* (a mixture of various hydrocarbons such as paraffins, olefins, naphthenes, and aromatics).

Q. 22.115. What is 'Octane Number (ON) ?

Ans. The property of a fuel which describes how fuel will or will not self-ignite is called the *octane number* or *just octane.*

Q. 22.116. Name the two standard reference fuels used for determining octane number.

Ans. *Iso-octane* ($C_8 H_{18}$), and *Normal heptane* ($C_7 H_{16}$).

Q. 22.117. How is 'diesel knock' related to the cetane rating of the fuel ?

Ans. Higher the cetane rating of the fuel lesser is the propensity for diesel knock.

Q. 22.118. What is an ideal composition of CNG (Compressed Natural Gas) ?

Ans. Methane = 90% (minimum) ; Ethane content = 4% (max.) ; Propane content = 1.7% (max.) ; C_4 and higher = 0.7% (max.) ; C_6 and higher = 0.2% (max.) ; ($CO_2 + N_2$) = 0.2% (max.); Hydrogen = 0.1% (max.) ; CO = 0.1% (max.) ; O_2 = 0.5% (max.) ; Sulphur = 10% ppm (max.).

Fuel/Air Mixture Requirements

Q. 22.119. What do you mean by 'Steady running' ?

Ans. *Steady running* is defined as mean continuous operation at a required speed and power output with normal temperatures.

Q. 22.120. What does 'Transient operation' include ?

Ans. *'Transient operation'* includes *starting, warming up,* and *changing from one speed or load to another*, specially for automotive vehicle engines during acceleration and decelerations, and also *idling*.

Q. 22.121. Does best economy F/A ratio depend on speed ?

Ans. *No.* The best economy F/A ratio is *independent of speed.*

Q. 22.122. What is 'idling' ?

Ans. The no-load running mode of engine is called *idling.*

Q. 22.123. Does the richening of mixture improve combustion ?

Ans. *Yes.* The richening of mixture increases the probability of contact between fuel and air particles and thus improves combustion.

Q. 22.124. What is 'Maximum power range' ?

Ans. The maximum power range lies between 75% to 100% rated power.

Q. 22.125. What is the mixture requirement for maximum power ?

Ans. The mixture requirement for maximum power is a rich mixture of A/F about 14 : 1 or F/A $\simeq$ 0.7.

Q. 22.126. Why A/F ratios are lower in the multi-cylinder engines ?

Ans. To overcome maldistribution of air-fuel mixture in different cylinders.

Q. 22.127. What are 'Transient conditions' ?

Ans. *Transient conditions* are those conditions at which, speed, load, temperatures or pressures are abnormal or changing rapidly, like in starting of an engine, warming up of an engine, acceleration of vehicle and deceleration of vehicle.

Q. 22.128. What is the optimum amount of accelerating charge ?

Ans. The optimum amount of accelerating charge is that which gives *best power F/A ratios in the cylinder.*

Carburetion and Carburettors

Q. 22.129. What is 'Carburetion' ?

Ans. The process of preparing in the S.I. engine, a combustible fuel-air mixture outside the engine cylinder is *'carburetion'*. This complicated process is achieved in the induction system.

Q. 22.130. What is a 'Carburettor' ?

Ans. A *carburettor* is a device which atomises the fuel and mixes it with air. It is the most important part of the induction system.

Q. 22.131. What is the function of a throttle in a carburettor ?

Ans. It regulates the quantity of the mixture.

Q. 22.132. Name the factors which influence the process of carburetion.

Ans. (*i*) The engine speed ; (*ii*) The vaporisation characteristic of fuel ; (*iii*) The temperature of the incoming air ; (*iv*) The design of the carburettor.

Q. 22.133. How 'too high volatility' of petrol affect the working of S.I. engine ?

Ans. *Too high volatility* may form bubbles in the carburettor and fuel lines particularly when the engine temperatures are high, which interfere with the supply and metering of the fuel and may disturb the F/A ratio so seriously that engine may stop working.

Q. 22.134. What are the disadvantages of too low volatility of petrol ?

Ans. Too low volatility may cause petrol to condense on the cylinder walls, diluting and removing the lubricating oil film ; ultimately the petrol may reach the crankcase past the piston rings and dilute the engine oil. Condensation of petrol on cylinder wall also causes carbon deposits.

Q. 22.135. What is the purpose of a 'Float chamber' ?

Ans. It is meant for storage of fuel.

Q. 22.136. Name the devices/systems which are added to the simple carburettor for meeting the demand of the engine under all conditions of operation.

Ans. (*i*) Main metering system ; (*ii*) Idling system ; (*iii*) Power enrichment or economiser system ; (*iv*) Acceleration pump system ; (*v*) Choke.

Q. 22.137. Where is 'Auxiliary port carburettor' used ?

Ans. In aircraft—for altitude compensation.

Q. 22.138. What is a 'Choke' ?

Ans. A *Choke* is simply butterfly valve located between the entrance to the carburettor and the venturi throat.

Q. 22.139. What are the three basic forms of carburettors ?

Ans. (*i*) Updraught ; (*ii*) Downdraft ; (*iii*) Horizontal.

Q. 22.140. Give three examples of 'open choke type' carburettor.

Ans. (*i*) Zenith carburettor ; (*ii*) Solex carburettor ; (*iii*) Carter carburettor.

Q. 22.141. Give an example of 'Constant vacuum type' carburettor.

Ans. S.U. carburettor.

Q. 22.142. What are the leading characteristics of a 'Solex carburettor ?

Ans. (*i*) Easy starting ; (*ii*) Good performance ; (*iii*) Reliability.

Q. 22.143. What is the unique feature of 'Solex carburettor' ?

Ans. Bi-starter for cold starting.

Q. 22.144. What are the functions of a fuel injection system ?

Ans.

(*i*) To monitor the engine's operating variables ;

(*ii*) To transfer this information to a metering control ; then,

(*iii*) To discharge and atomise the fuel into the incoming air stream.

Fuel Injection Systems for C.I. Engines

Q. 22.145. What do you mean by the term 'Ignition delay' ?

Ans. The period between the start of injection and start of ignition is called the 'ignition delay' ; it is about 0.001 second for high speed engines and 0.002 second for low speed engines. *The lesser the relay period better is the engine performance.*

Q. 22.146. What is 'Solid or airless injection' ?

Ans. Injection of fuel directly into the combustion chamber without primary automatisation is termed as *solid injection.* It is also termed as *mechanical injection.*

Q. 22.147. Name the main types of modern fuel injection systems.

Ans. (*i*) Common-rail injection system ; (*ii*) Individual pump injection system ; (*iii*) Distributor system.

Q. 22.148. How are nozzle classified ?

Ans.

(*i*) Single hole nozzle ; (*ii*) Multi-hole nozzle ;

(*iii*) Circumferential nozzle ; (*iv*) Pintle nozzle ;

(*v*) Pintaux nozzle.

Q. 22.149. Name the three commonly used starting systems in large and medium size engines.

Ans. (*i*) Starting by an auxiliary engine ; (*ii*) Use of electric motors or self starters ; (*iii*) Compressed air system.

Ignition Systems (S.I. Engines)

Q. 22.150. What do you understand by the term 'Ignition' ?

Ans. '*Ignition*' is only a *pre-requisite of combustion.* It does not influence the gross combustion process. It is only a small scale phenomenon taking place within a specified small zone in the combustion chamber.

Q. 22.151. How much of spark energy is sufficient over entire range of operation for automotive engines ?

Ans. In normal practice, for automotive engines, the spark energy to the tune of 40 milli joules and duration of about 0.5 milli second is sufficient over entire range of operation.

Q. 22.152. Name the basic ignition systems which are in use.

Ans. (*i*) Battery ignition system (Conventional, transistor assisted) ; (*ii*) Magneto ignition system (Low tension, high tension) ; (*iii*) Electronic ignition system.

Q. 22.153. What is 'Magneto' ?

Ans. *'Magneto'* is a special type of ignition system with its own electric generator to provide the necessary energy for the system. It is an *efficient, reliable, self contained unit which is often preferred for aircraft engines because storage batteries are heavy and troublesome.*

Q. 22.154. What do you mean by the term 'Firing order' ?

Ans. *Firing order* is the order in which various cylinders of a multi-cylinder engine fire. The firing order is arranged to have power impulses equally spaced, and from the point of view of balancing.

Q. 22.155. What do you understand by the term 'Ignition timing' ?

Ans. *'Ignition timing'* is the correct instant for the introduction of spark near the end of compression stroke in the cycle. The ignition timing is fixed to obtain maximum power from the engine.

Q. 22.156. List the factors which affect ignition timings.

Ans. (*i*) Compression ratio ; (*ii*) Engine speed ; (*iii*) Mixture strength ; (*iv*) Combustion chamber design ; (*v*) Throttle opening ; (*vi*) Engine temperature ; (*vii*) Type of fuel.

Q. 22.157. What is the main function of a spark plug ?

Ans. The *main function of a spark plug* is to conduct the high potential from the ignition system into the combustion chamber. It provides the proper gap across which spark is produced by applying high voltage, to ignite the combustion chamber.

Q. 22.158. How does a cold plug differ from a hot plug ?

Ans. A *cold plug* has a short heat transfer path and a small area exposed to the combustion gases, as compared to a hot plug.

Engine Friction and Lubrication

Q. 22.159. What are the purposes of lubrication in I.C. engine ?

Ans.

(*i*) To reduce the rubbing action between different machine parts having relative motion with each other ;

(*ii*) To remove the heat generated inside the cylinder.

Q. 22.160. Define the term 'Engine friction'.

Ans. *'Engine friction'* is defined as the difference between the indicated power (I.P.) and brake power (B.P.)

Frictional power, F.P. = I.P. – B.P.

Q. 22.161. Name the losses which are included in the 'total engine friction loss'.

Ans. (*i*) Direct frictional losses ; (*ii*) Pumping loss ; (*iii*) Blowby losses ; (*iv*) Valve throttling losses ; (*v*) Combustion chamber pump loss ; (*vi*) Power loss to drive the auxiliaries.

Q. 22.162. How are 'Blowby losses' caused ?

Ans. *Blowby losses* are caused due to leakage of combustion products past the piston from the cylinder into the crankcase. These losses depend upon the inlet pressure and compression ratio. These losses increase directly with compression ratio but get reduced with an increase in engine speed.

Q. 22.163. Name the methods which are used to determine engine friction.

Ans. (*i*) From I.P. and B.P. measurements ; (*ii*) Morse test ; (*iii*) Willian's line method ; (*iv*) Motoring method ; (*v*) Deceleration method.

Q. 22.164. Define the term 'Lubrication' ?

Ans. *Lubrication* is the admittance of oil between two surfaces having relative motion.

Q. 22.165. What is 'Film lubrication' ?

Ans. It is that type of lubrication in which bearing surfaces are completely separated by a layer of film of lubricant and that the *frictional resistance arises only due to relative movements of the lubricant layers.*

Q. 22.166. What is 'Boundary lubrication' ?

Ans. Under the hydro-dynamic condition the oil film supports the load. If the oil film becomes thin enough so as not to support the load without occasional metal to metal contact the journal friction developed is called *boundary friction,* and the lubrication existing in this range is known as *boundary lubrication.*

Q. 22.167. What is 'Elasto-hydrodynamic lubrication' ?

Ans. When the load acting on the bearings is very high, the material itself deforms elastically against the pressure built up of the oil film. This type of lubrication is called *elasto-hydrodynamic lubrication* and it occurs between cams and followers, gear teeth, and rolling bearings when the contact pressures are extremely high.

Q. 22.168. What do you understand by the term 'Viscosity' of an oil ?

Ans. *Viscosity* is the ability of the oil to resist internal deformation due to mechanical stresses and hence it is a measure of the ability of the oil film to carry a load.

Q. 22.169. How is viscosity of an oil measured ?

Ans. Viscosity is measured by *viscosimeter.*

Q. 22.170. Name some important types of viscosimeters.

Ans. (*i*) Saybolt universal viscosimeter ; (*ii*) Red wood viscosimeter ; (*iii*) Engler viscosimeter ; (*iv*) Barbey viscosimeter.

Q. 22.171. What does high viscosity index indicate ?

Ans. A high viscosity index indicates relatively smaller changes in viscosity of the oil with the temperature.

Q. 22.172. Name the various lubrication systems used for I.C. engines.

Ans. (*i*) Wet sump lubrication system ; (*ii*) Dry sump lubrication system ; (*iii*) Mist lubrication system.

Q. 22.173. Which lubrication system is used in two stroke cycle engines ?

Ans. Mist lubrication system.

Engine Cooling

Q. 2.174. Why is the engine cooling required to keep the temperature of the engine low ?

Ans. To *avoid* loss of volumetric efficiency and hence power, engine seizure and danger of engine failure.

Q. 22.175. How much percent of total heat supplied in the fuel is removed by the cooling medium ?

Ans. Almost 25 to 35%.

Q. 22.176. How much percent of total heat supplied is carried away by lubricating oil and heat lost by radiation ?

Ans. 3 to 5%.

Q. 22.177. What is the mode of heat transfer from hot gases to the coolant ?

Ans. Heat transfer from hot gases to the coolant takes place by forced convection or by nucleate boiling when heat flux is high.

Q. 22.178. What are the two main methods of cooling of I.C. engines ?

Ans. (*i*) Air cooling ; (*ii*) Water/Liquid cooling.

Q. 22.179. Name the various methods used for circulating water around the engine cylinder.

Ans. (*i*) Thermo-system cooling ; (*ii*) Forced or pump cooling ; (*iii*) Cooling with thermostatic regulator ; (*iv*) Pressurised water cooling ; (*v*) Evaporative cooling.

Q. 22.180. What are the principal types of radiator cores ?

Ans. (*i*) Film type ; (*ii*) Fin and tube ; (*iii*) Pack type.

Supercharging of I.C. Engines

Q. 22.181. What is the purpose of supercharging ?

Ans. The purpose of supercharging is to *raise the volumetric efficiency* above that value which can be obtained by normal aspiration.

Q. 22.182. What is a 'Supercharger' ?

Ans. A supercharger is the apparatus used to *increase the air density*. It is merely a *compressor* which provides a denser charge to the engine, thereby enabling the consumption of a greater mass of charge with the same total piston displacement.

Q. 22.183. Which type of compressor is widely used as supercharger for reciprocating engines ?

Ans. Centrifugal type compressor.

Q. 22.184. What are the objects of supercharging ?

Ans. (*i*) To increase the power output for a given weight and bulk of the engine ; (*ii*) To compensate for loss of power due to altitude ; (*iii*) To obtain more power from an existing engine.

Q. 22.185. What is 'Boost pressure' ?

Ans. *Boost pressure* refers to the gauge pressure recorded when the air or mixture supply has passed through the supercharger.

Q. 22.186. What is 'pressure ratio' (in case of a supercharger) ?

Ans. *Pressure ratio* is the ratio of absolute pressure (boost pressure + atmospheric pressure) to that of the atmospheric pressure.

Q. 22.187. What are 'turbochargers' ?

Ans. *Turbochargers* are centrifugal compressors driven by the exhaust gas turbines.

Q. 22.188. What are the main types of turbocharging methods ?

Ans. (*i*) Constant pressure turbocharging ; (*ii*) Pulse turbocharging ; (*iii*) Pulse converter ; (*iv*) Two-stage turbocharger ; (*v*) Miller turbocharging ; (*vi*) Hyperbar turbocharging.

Testing and Performance of I.C. Engines

Q. 22.189. What does engine performance indicate ?

Ans. *'Engine performance'* is an indication of the *degree of success* with which it does its assigned job *i.e., convention of chemical energy contained in the fuel into the useful mechanical work.*

Q. 22.190. What are the basic parameters for the evaluation of engine performance ?

Ans. (*i*) Power and mechanical efficiency ; (*ii*) Mean effective pressure and torque ; (*iii*) Specific output ; (*iv*) Volumetric efficiency ; (*v*) Fuel-air ratio ; (*vi*) Specific fuel consumption ; (*vii*) Thermal efficiency and heat balance ; (*viii*) Exhaust smoke and other emissions ; (*ix*) Specific weight.

Q. 22.191. What is 'Indicated power (I.P)' ?

Ans. The total power developed by combustion of fuel in the combustion chamber is called *indicated power.*

Q. 22.192. What is 'Brake power' (B.P.) ?

Ans. The power developed by an engine at the output shaft is called the brake power.

Q. 22.193. What is 'Mechanical efficiency' ?

Ans. The ratio of B.P. to I.P. is called *mechanical efficiency* $\left(i.e., \eta_{mech.} = \frac{\text{B.P.}}{\text{I.P.}} \right)$.

Q. 22.194. What is 'Mean effective pressure' ?

Ans. *'Mean effective pressure (m.e.p.)'* is defined as hypothetical pressure which is thought to be acting on the piston throughout the power stroke.

Q. 22.195. What is 'Specific fuel consumption (s.f.c.)' ?

Ans. It is the *mass of fuel consumed per kW developed per hour,* and is a criterion of economical power production.

Q. 22.196. How can speed be measured ?

Ans. The speed may be measured by : (*i*) Revolution counters ; (*ii*) Mechanical tachometer ; (*iii*) Electrical tachometer.

Q. 22.197. Name the methods by which the fuel consumed by an engine can be measured.

Ans. (*i*) Fuel flow method ; (*ii*) Gravemetric method ; (*iii*) Continuous flow meters.

Q. 22.198. Name the methods by which air consumption can be measured.

Ans. (*i*) Air box method ; (*ii*) Viscous-flow air meter.

Q. 22.199. Name the smoke meters which are used for measurement of smoke.

Ans. (*i*) Bosch smoke meter ; (*ii*) Hatridge smoke meter ; (*iii*) PHS smoke meter.

Q. 22.200. Give the classification of dynamometers used for the measurement of B.P.

Ans. (*i*) Absorption dynamometer *e.g.,* Pony brake, Rope brake, Hydraulic brake, Fan brake, Electrical brake ; (*ii*) Transmission dynamometers.

Q. 22.201. What are the features of 'Rope brake dynamometer' ?

Ans. *"Rope brake dynamometer" is cheap and easily constructed but not very accurate because of changes in friction coefficient of rope with temperature.*

Q. 22.202. How is indicated power (I.P.) usually determined ?

Ans. *Indicated power* is usually determined with the help of a p-v diagram taken with the help of an *indicator.*

Q.22.203. What is the field of application of 'Morse test'?

Ans. Multi-cylinder engines' testing.

Q. 22.204. Name the methods which can be used to determine frictional power of an engine.

Ans. (*i*) Willian's line method (used for C.I. engines only) ; (*ii*) Morse test ; (*iii*) Motoring test ; (*iv*) Difference between I.P. and B.P.

Q. 22.205. Name the methods used for governing I.C. engines.

Ans. (*i*) Hit and miss method ; (*ii*) Quality governing ; (*iii*) Quantity governing.

Q. 22.206. In which type of engines 'Hit and miss method' of governing is commonly used ?

Ans. Gas engines.

Q. 22.207. For large engines which method of governing is preferred ?

Ans. Quantity governing.

Q. 22.208. What is the disadvantage of 'Quality governing method' ?

Ans. The ignition is not always satisfactory and thermal efficiency is reduced.

Q. 22.209. What are the three ways by which engine and exhaust noise can be reduced ?

Ans.

(*i*) *Passive.* Noise reduction is accomplished passively by correct design and the use of proper materials.

(*ii*) *Semi-active.* In semi-active noise abatement systems, 'hydraulics' are oftenly used.

(*iii*) *Active noise.* Active noise abatement is accomplished by generating 'antinoise' to cancel out engine exhaust noise.

Air Pollution from I.C. Engines and its Control

Q. 22.210. Define the term 'Air pollution'.

Ans. *Air pollution* can be defined as an addition to our atmosphere of any material which will have a deleterious effect on life upon our planet.

Q. 22.211. Name the major pollutants which are emitted from the exhaust due to incomplete combustion.

Ans. (*i*) Carbon monoxide (CO) ; (*ii*) Hydrocarbons (HC) ; (*iii*) Oxides of nitrogen (NO_x). Other products produced are acetylene, aldehydes, etc.

Q. 22.212. What are the products which are expelled from the exhaust when the combustion is complete ?

Ans. Water vapour and carbon dioxide (an inert gas-not directly harmful to humans).

Q. 22.213. What is 'Stoichiometric ratio' ?

Ans. The chemically correct air-fuel ratio by mass for complete combustion is known as *'stoichiometric ratio'*.

Q. 22.214. Name three main sources from which pollutants are emitted from S.I. engine.

Ans. (*i*) Crankcase ; (*ii*) The fuel system ; (*iii*) The exhaust system.

Q. 22.215. Name the main methods used for S.I. engine emission control.

Ans. (*i*) Modification in the engine design and operating parameters ; (*ii*) Treatment of exhaust products of combustion ; (*iii*) Modification of fuels.

Q. 22.216. What is a catalytic converter ?

Ans. A *catalytic converter* is a device which is placed in the vehicle exhaust system to reduce HC and CO by oxidising catalyst and NO by reducing catalyst.

Q. 22.217. Which is the best catalyst to control NO_x ?

Ans. Rhodium is the best catalyst to control NO_x but A/F ratio must be within a narrow range of 14.6 : 1 to 14.7 : 1.

Q. 22.218. What is the basic principle of crankcase blow-by control system ?

Ans. Recirculation of vapours back to the inlet manifold.

Q. 22.219. Name the methods by which NO_x emission can be reduced.

Ans. (*i*) Exhaust gas recirculation (EGR) ; (*ii*) Catalyst ; (*iii*) Water injection.

Q. 22.220. Name the factors on which the quantity of soot formed in a C.I. engine, depends.

Ans. (*i*) The local F/A ratios ; (*ii*) The type of fuel ; (*iii*) The pressure.

Q. 22.221. What is the cause of 'smoke' ?

Ans. Incomplete burning of fuel inside the combustion chamber.

Q. 22.222. What are the two major reasons for incomplete combustion ?

Ans. (*i*) Incorrect A/F ratio ; (*ii*) Improper mixing.

Q. 22.223. What are the two basic types of smoke meters for measuring smoke density ?

Ans. (*i*) Filter darkening type (*viz.* Bosch smoke meter) ; (*ii*) Light extinction type (*viz.* Hartridge smoke meter).

Q. 22.224. What are the methods which may be used for control of smoke ?

Ans.

- There is hardly any successful method to control the root/smoke except the engine has to run at lower load, *i.e., derating* and *maintain the engine at best possible condition.*
- The other methods which may be used for control of smoke are :

(*i*) Smoke suppress out additives ; (*ii*) Fumigation ; (*iii*) Catalytic mufflers.

Q. 22.225. Name the factors which affect odour production.

Ans. (*i*) F/A ratio ; (*ii*) Engine operation mode ; (*iii*) Engine type ; (*iv*) Fuel consumption ; (*v*) Odour suppressant additives.

Miscellaneous Engines

Q. 22.226. What is 'Dual-fuel operation' ?

Ans. *Dual-fuel operation* combines in a simple manner the possibility of operating a diesel engine on liquid fuels such as diesel oil or gas oil and on gaseous fuels such as natural gas, sewage gas and cook oven gas, etc.

— The engine can be switched from dual-fuel operation almost instantaneously in case of emergency.

Q. 22.227. What are the factors which affect combustion in a dual-fuel engine ?

Ans. (*i*) Pilot-fuel quantity ; (*ii*) Injection timing ; (*iii*) Cetane number of pilot-fuel ; (*iv*) Inlet temperature ; (*v*) Types of gaseous fuels ; (*vi*) Throttling ; (*vii*) Mixture strength.

Q. 22.228. What is a 'Multi-fuel engine' ?

Ans. A *'Multi-fuel engine'* is one which can operate satisfactorily (with substantially unchanged performance and efficiency) on a wide variety of fuels ranging from diesel oil, crude oil, IP-4 to lighter fuel like gasoline, and even normal lubricating oil.

Q. 22.229. What difficulties are associated with multi-fuel operation ?

Ans. (*i*) Tendency of vapour lock in the fuel pump while using lighter fuels ; (*ii*) Tendency of increased wear in the fuel pump due to lower lubricity of gasoline ; (*iii*) In view of differences in heating values and compressibility of fuels, different volumes of fuels need be injected.

Q. 22.230. What is a 'Stratified engine' ?

Ans. A '*Stratified engine*' is usually defined as a S.I. engine (stratified diesel engine has also been developed) in which the mixture in the zone of spark plug is very much richer than that in the rest of the combustion chamber *i.e.,* one which burns leaner overall fuel-air mixtures.

Q. 22.231. What do you understand by the term 'Charge stratification' ?

Ans. *Charge stratification* means providing different fuel-air mixture strengths at various places in the combustion chamber.

Q. 22.232. What are the main merits of stratified charge engine ?

Ans. The stratified charge engine combines the advantages of both *petrol engines* (very good full load power characteristics *e.g.,* high degree of air utilisation, high speed and flexibility) and *diesel engines* (good part-load characteristics) and at the same time avoids as far as possible their disadvantages.

Q. 22.233. What are the advantages of burning overall fuel-air mixtures ?

Ans. (*i*) Higher thermodynamic efficiency ; (*ii*) Reduced air pollution.

Q. 22.234. What is the basis for stirling engine ?

Ans.

— The alternate compression at low temperature and expansion at high temperature of a working fluid is the basis for the stirling engine.

— The working fluid is heated in a radically different manner. It burns fuel outside the engine itself, and continuously.

Q. 22.235. Which fuels can be used for a 'stirling engine' ?

Ans. Since the stirling engine is an external combustion engine, it possesses multi-fuel capacity. It can use any petroleum fraction such as *gasoline, diesel, methanol-gasoline blends,* etc., with no octane or cetane requirements. Thus, the stirling engine has the desirable characteristic of *adaptability to changing fuel availability.*

Q. 22.236. On which principle does Wankel engine work ?

Ans. Four-phase principle (The word *phase* corresponds to stroke of the reciprocating engine).

Q. 22.237. What are the characteric features of a variable compression Ratio (VCR) 'engine' ?

Ans. In the VCR engine a *'high compression ratio'* is employed for good stability and low load operation and a *'low compression ratio'* is used at full-load to allow the turbocharger to boost the inlet pressures without increasing the peak cycle pressure.

— The VCR concept is beneficial only at *part load* and the part-load efficiency of the diesel engine is higher than that of the gasoline engine.

Q. 22.238. How can variable compression ratio be obtained ?

Ans. (*i*) By changing the clearance volume ; (*ii*) By changing both the clearance and swept volume of the engine.

Q. 22.239. What are 'free-piston engine plants' ?

Ans. Free-piston engine plants are the conventional gas turbine plants with the difference that the *air compressor and combustion chamber are replaced by a free piston engine.*

B. CONVENTIONAL QUESTIONS WITH ANSWERS/SOLUTIONS

Theoretical Questions with Answers

Q. 22.1. *Explain various factors that influence the flame speed.*

Ans. Following are the factors that influence the flame speed :

1. **Fuel air ratio.** The composition of the working mixture influences the rate of combustion and the amount of heat evolved. *With hydrocarbon fuels the maximum flame velocities occur when mixture strength is 110% of stoichiometric (i.e., about 10% richer than stoichiometric). When the mixture is made leaner or is enriched still more, the velocity of flame diminishes.*

2. **Compression ratio.** A higher compression ratio increases the pressure and temperature of the working mixture and decreases the concentration of residual gases. These favourable conditions *reduce the ignition lag of combustion and hence less ignition advance is needed.* High pressures and temperatures of compressed mixtures also speed up the second phase of combustion. Total ignition angle is reduced.

3. **Intake temperature and pressure.** *Increase in intake temperature and pressure increases the flame speed.*

4. **Engine load.** *With increase in engine load the cycle pressure increases. Hence, the flame speed increases.*

5. **Turbulence.** Turbulence plays a very vital role in combustion phenomenon. The flame speed is very low in non-turbulent mixtures. *A turbulent motion of the mixture intensifies the processes of heat transfer and mixing of the burned and unburned portions in the flame front.* These two factors cause the velocity of turbulent flame to increase practically in proportion to the turbulence velocity.

6. **Engine speed.** *The higher the engine speed, the greater the turbulence inside the cylinder. For this reason the flame speed increases almost linearly with engine speed.*

Q. 22.2. ***What is a performance number ?***

Ans. The Octane scale, *i.e.* percentage of iso-octane in an iso-octane and *n*-heptane mixture fails to serve as soon as fuels more resistant than iso-octane are to be rated, such fuels are not uncommon. In order to extend the octane scale, the knock resistance of the fuel is measured in terms of Army and Navy *Performance Number (PN). It is the ratio of the knock limited indicated mean effective pressure (klimep) of the test fuel to the knock limited indicated mean effective pressure of iso-octane thus,*

$$\text{PN} = \frac{\text{klimep of test fuel}}{\text{klimep of iso-octane}} \times 100$$

and by definition the PN of iso-octane is 100.

Q. 22.3. ***What are the factors that limit the compression ratio that can be used in petrol engines ?***

Ans.

- For thermal efficiency to increase for any internal combustion engine, the compression ratio must increase. *In a practical engine the compression ratio is limited because of high material loading, high temperatures and fuel combustion problems.*
- In petrol engines the charge consists of petrol vapour and air. During compression the pressure and temperature both increase. The value of compression ratio is limited to the value that the temperature reached after compression or during the compression does not cause self ignition of fuel which causes knocking or pinking. On account of this fact the value of compression ratio is limited to 5 to 8 (in rare case 10) in petrol engines.
- *The addition of tetraethyl lead to the fuel helps in preventing knocking or pinking, thus enabling the higher compression ratios.* The use of tetraethyl lead is *now in disfavour because of atmospheric pollution and possible damage to health.* Research and development into fuel combustion techniques are *making the use of lead free petrol possible with increasing compression ratio.*

Q. 22.4. ***"Abnormal combustion knock produced by surface ignition in S.I. engines is more harmful than normal combustion knock". Justify this statement.***

Ans. In normal combustion knock, the unburnt charge auto-ignities before the flame started by spark travels across combustion chamber. Normal combustion knock with fuel of lower than the required octane value can be reduced and eliminated by retarding the ignition timing.

The *abnormal combustion knock is caused by surface ignition. Surface ignition is defined as the initiation of a flame front by a hot surface other than the spark.* Mostly surface ignition is due to carbon deposits. These deposits occupy space and so increase the compression ratio. An entirely distinct flame front is produced and the combustion becomes both erratict and uncontrollable. Surface ignition may occur before (pre-ignition) or after (post-ignition) normal ignition. Pre-ignition tends to raise temperature and pressure in chamber which cause temperature of hot spot to rise further and encourage still earlier pre-ignition. The cumulative effect tends to raise peak pressures and encourage the possibility of detonation. It also cause peak pressure to occur progressively earlier in cycle. Pre-ignition may advance these peak pressures to such a point that they occur before piston reaches T.D.C. on compression stroke. In such a case peak pressures will oppose piston movements during last part of compression stroke thus decreasing total output as well as rough engine operation.

Q. 22.5. ***Explain the following terms as applied to S.I. engines :***

(i) Pre-ignition; (ii) Auto-ignition ; (iii) Detonation.

Ans. (*i*) Pre-ignition.

- The increase in the rate of heat transfer to the walls may cause local overheating specially of the spark plug, which may reach a temperature high enough to *ignite the charge before the passage of spark.* This phenomenon is called **pre-ignition**.

- Pre-ignition may also be caused by *overheated exhaust valves or glowing carbon deposits in the combustion chamber.*

(ii) Auto-ignition.

- The analysis of knocking phenomenon (in S.I. engines) by high speed cinematography has led to two general theories—the *auto-ignition theory* and the *detonation theory.* Auto-ignition *refers to initiation of combustion without the necessity of a flame.* The auto-ignition theory of knock assumes that the flame velocity is *normal* before the onset of auto-ignition and that the gas vibrations are created by a number of end gas elements auto-igniting almost instantaneously.
- Auto-ignition does not occur immediately as the self-ignition temperature is reached. Some ignition delay period is required before the mixture becomes explosive. During the delay period some chemical reactions occur, which are called *preflame reactions,* because these reactions prepare the mixture for giving rise to a flame.

(iii) Detonation.

- In *detonation* a true pressure wave formed by preflame reactions is the mechanism for explosion. Such a shock wave would travel through the chamber at about twice the sonic velocity and would compress the gases to pressures and temperatures where the reactions should be practically instantaneous.

Q. 22.6. *How tetra-ethyl lead (T.E.L.) improves the quality of fuel for S.I. engine ?*

Ans. *Tetra-ethyl lead improves the quality of fuel by delaying auto-ignition and allowing it to occur only at a higher temperature.*

Q. 22.7. *What do you understand by octane number of 85 and cetane number of 75 ? What is H.U.C.R. ?*

Ans.

- A fuel of octane number of 85, gives the same knock intensity as 85% volume iso-octane plus 15% volume heptane, in a standard similar test.
- The cetane number of a fuel is the percentage by volume of cetane in a mixture of cetane and α-methyl naphthalene ($C_{10}H_7CH_3$) that has the same performance in the standard test engine as that of the fuel. Thus cetane number of 75 means the fuel has the same performance as of mixture of 75% cetane and 25% α-methyl naphthalene, both by volume, in the standard test engine.
- *Highest Useful Compression Ratio (H.U.C.R.).* The tendency of an engine to detonate increases as the compression ratio rises. By further increasing the compression ratio of the engine the detonation will, in time, becomes so severe that the power of the engine will commence to decrease due to overheating. *The compression ratio at which this occurs in a specified test engine, under specified operating conditions, is known as the Highest Useful Compression Ratio or H.U.C.R.*

  This method of classification is now little used as it compares fuels when producing a violent detonation which would not be tolerated in the normal running of any engine.

Q. 22.8. *Shape of the clearance volume controls the detonation in case of S.I. engine. Comment.*

Ans. Clearance volume has an effect on compression ratio,

Since compression ratio = $\dfrac{V_s + V_c}{V_c} = 1 + \dfrac{V_s}{V_c}$.

If clearance volume is reduced, compression ratio is increased which will increase chances of detonation in S.I. engines.

Q. 22.9. ***Discuss the effect of engine variables on ignition lag.***

Ans.

- *Ignition lag* is *not a period of inactivity but is a chemical process.*
- The ignition lag in terms of *crank angles is 10° to 20° and in terms of seconds 0.0015 seconds or so.*

Effects of engine variables on ignition lag :

(*i*) Fuel. Ignition lag depends on chemical nature of fuel. *The higher the self ignition temperature of fuel, longer the ignition lag.*

(*ii*) Mixture ratio. Ignition lag is *smallest for the mixture ratio which gives the maximum temperature.* This mixture ratio is somewhat *richer than the stoichiometric ratio.*

(*iii*) Initial temperature and pressure. *Ignition lag is reduced if the initial temperature and pressure are increased and the initial temperature and pressure can be increased by increasing the compression ratio.*

(*iv*) Turbulence. Ignition lag when expressed in degrees of crank rotation *increases linearly with engine speed.* Increasing the engine speed means increasing the turbulence.

Q. 22.10. ***Discuss the effects of the following variables on engine heat transfer : (i) Spark advance ; (ii) Engine output ; (iii) Pre-ignition and knocking.***

Ans. (*i*) Spark advance. *A spark advance more than the optimum as well as less than the optimum will result in increased heat rejection to the cooling system.* This is mainly due to the fact that the spark time other than MBT value (minimum spark advance for best torque) *will reduce the power output and thereby more heat is rejected.*

(ii) **Engine output.** Engines which are designed for high mean effective pressures or high piston speeds will reject less heat (on percentage basis). Less heat will be lost to coolant for the same indicated power in large engines.

(iii) **Pre-ignition and knocking.** Effect of pre-ignition is the same as advancing the ignition timing. *Large spark advance might lead to erratic running and knocking.* Though knocking causes large changes in local heat transfer conditions, the overall effect on heat transfer due to knocking appears to be negligible. However no authentic information is available regarding the effect of pre-ignition and knocking on engine heat transfer.

Q. 22.11. ***The higher compression ratio that can be used in an S.I. engine is limited by the detonation characteristics of the available fuel. Justify this statement.***

Ans. In normal combustion, the flame started by spark travels across the combustion chamber. As the flame front advances, it compresses the unburnt charge in last portion of combustion chamber. If this unburnt charge does not reach its critical temperature for auto-ignition, it will not auto-ignite and flame front will move across this unburnt charges in normal manner. In abnormal combustions called *detonation* the end charge auto-ignites before the flame front reaches it. In order to auto-ignite, the last unburnt portion of charge must reach above a certain critical temperature and remain at this temperature for a certain length of time. During this period certain chemical reactions take place which prepare the charge for auto-ignition. The time required in this preparation is called *ignition delay.*

In *order to limit detonation, we should not allow the unburnt charge to reach its critical temperature.* Increase in temperature of mixture reduces delay period of end charges and hence tendency of detonation increases. Increase in temperature is obtained by avoiding detonation, we should limit the compression ratio. Hence there is a critical compression ratio for a fuel for a given engine *setting above which knock occurs.* This *compression ratio is called highest useful compression ratio.*

Q. 22.12. *The retarding of spark timing in a S.I. engine will reduce detonation. Justify the statement.*

Ans. By retarding the spark from the optimized timing, *i.e.,* having the spark closer to T.D.C., the peak pressures are reached further down on the power stroke. This might reduce knocking, though this affect adversely the brake torque and power output of the engine.

For the same reason the spark should be retarded with low octane fuels.

Q. 22.13. *What action can be taken with regard to the following variables, in order to reduce the possibility of detonation in an S.I. engine ? Justify your answers by reasons :*

(i) Compression ratio ; (ii) Mass of charge induced ; (iii) Mixture inlet temperature ; (iv) Engine speed ; (v) Distance of flame travel.

Ans. The following actions are recommended for *reducing the possibility of detonation in S.I. engine.*

(i) **Compression ratio.** Increasing the compression ratio increases both the temperature and pressure. Increase in temperature reduces the delay period of end charge. Increase in temperature as well as increase in pressure both lead to greater collisions of molecular resulting in greater formation of chemical species responsible for knocking. Hence tendency to knock increases. Therefore in order to reduce possibility of detonation, it is essential to keep *compression ratio as low as possible* and *there is a critical compression ratio above which knock occurs.* This compression ratio is called *highest useful compression ratio (H.U.C.R.).*

(ii) **Mass of charge induced.** The volume of engine cylinder is same and if the *mass of charge is increased,* then density of mixture will be increased and increase in density mean better contact between fuel particles and ignition lag period will be reduced. When ignition lag is reduced, it means the chances of fuel in last portion of combustion chamber to auto-ignite are more, hence *knocking chances will be increased.* Hence *it is necessary to keep mass of charge induced as low as possible for the given volume of cylinder to reduce possibility of detonation.*

(iii) **Mixture inlet temperature.** Increasing the mixture inlet temperature reduces the delay period which in turn increases the chances of knocking. Hence the *mixture inlet temperature should be kept as low as possible.*

(iv) **Engine speed.** If the speed of engine is decreased, then turbulence of mixture will be reduced which will result in reduced flame speed. Also lower the engine speed, the longer is the absolute time for the flame to traverse the cylinder which increases the time available for pre-flame reactions. Hence tendency to knock is increased at lower speeds. Therefore, *engine speed should be increased to prevent detonation.*

(v) **Distance of flame travel.** If the distance of flame travel is increased, then there are chances that fuel in last portion of combustion chamber may auto-ignite before the flame may traverse the chamber. Hence *it is necessary that flame travel distance be reduced to prevent detonation.*

Q. 22.14. *"Auto-ignition is the cause of detonation." Justify the statement.*

Ans. Detonation is caused by the auto-ignition of the end gas. In S.I. engine, the ignition is initiated by a spark produced across the spark plug electrodes. The normal combustion takes place by flame propagation through the air-fuel mixture present in the combustion chamber. If conditions are such that the end charge reaches self-ignition temperature and completes the delay period before the arrival of flame front, it would auto-ignite and cause detonation.

Q. 22.15. *"Compressed natural gas (CNG) is preferable in S.I. engine than C.I. engine." Justify the statement.*

Ans. *CNG is highly knock-resistant, and does not pre-ignite easily.* It *mixes easily with air, and gives better manifold distribution.* These are desirable properties for its use in S.I. engine. On the other hand, its limitations for use in C.I. engine are that *it is not viscous enough to maintain lubrication of injector pump*, and *due to high self-ignition temperature of CNG it would need a very high compression ratio for self-ignition*. Hence *CNG is preferable in S.I. engine than in C.I. engine.*

Q. 22.16. *Explain the difference between :*

(i) Pre-ignition ; (ii) Auto-ignition; (iii) Detonation.

Ans. *(i)* Pre-ignition. In S.I. engines, the combustion during the normal working is initiated by an electric spark. The spark is timed to occur at a definite point just before the end of the compression stroke. The ignition of the charge should not occur before the spark is introduced in the cylinder. If the ignition starts due to any other source when the piston is still moving on the compression stroke, it is known as *pre-ignition. Pre-ignition will develop excessive pressure before the end of compression stroke tending to push the piston opposite to the direction in which it is moving, resulting in loss of power, violent thumping, stopping the engine or even mechanical damage.* Pre-ignition *may occur on account of persistent detonation, overheated spark plug points, overheated exhaust valve, incandescent carbon deposits on the surface of the cylinder or spark plug, or faulty timings of the spark plug.*

***(ii)* Auto-ignition.** It is one of the theories of knocking in S.I. engines. *Auto-ignition* refers to the *initiation of combustion without the necessity of a flame.* The auto-ignition theory of knocking assumes that the flame velocity is normal before the onset of auto-ignition, and that gas vibrations are initiated by a number of end gas elements autoigniting almost simultaneously. Auto-ignition does not occur immediately as the self-ignition temperature is reached. Some ignition delay period is required before the reaction becomes explosive. During the delay period some preflame reactions occur before giving rise to a flame The exact method of formation of the preflame reactions is not known.

***(iii)* Detonation.** The second theory of knocking in the S.I. engines is detonation. It *is the name given to the violent waves produced within the cylinder of an S.I. engine.* The noise produced is like that produced by a sharp ringing blow upon the metal of a cylinder.

The region in which the detonation occurs is far away from the spark plug, and is known as the *detonation zone*. After a spark is produced, there is a rise of pressure and temperature due to the combustion of the ignited fuel. This rise in temperature and pressure both combine to increase the velocity of the flame, compressing the unburnt portion of the charge of the detonation zone. Finally, the temperature in the detonation zone reaches such a high value that *chemical reaction occurs at a far greater rate than in the advancing flame. Before the flame completes its course* across the combustion chamber, the whole mass of remaining unburnt charge *ignites instantaneously.* This spontaneous ignition of a portion of the charge sets *rapidly moving high pressure waves* that hit cylinder walls with such violence that the cylinder walls gives out a loud pinking noise. *It is this noise that expresses or indicates detonation.*

Q. 22.17. *On what basis are S.I. engines fuels compared when they are better than iso-octane in anti-knock characteristics ?*

Ans. When S.I. engine fuels are better than iso-octane in anti-knock characteristics, the quality of the fuel is measured *in terms of Army and Navy performance number (PN).* It is the *ratio of the knock limited indicated mean effective pressure (klimep) of the test fuel to the knock limited imep of iso-octane, i.e.,*

$$PN = \frac{\text{klimep of test fuel}}{\text{klimep of iso-octone.}}$$

By above definition the PN of iso-octane is 100.

Another attempt to extend the octane scales is *Wiese method* :

$$ON = 100 + \frac{PN - 100}{3}$$

Some lead compounds, such as tetra-ethyl-lead (TEL), when added to iso-octane produce fuel of greater anti-knock quality than iso-octane. The anti-knock quality of fuel above 100 octane number *are also measured in terms of millilitres of TEL per US gallon of iso-octane.*

Q. 22.18. *Discuss the three basic requirements of a good S.I. engine combustion chamber.*

Ans. The three basic requirements of a good S.I. engine combustion chamber are

1. High power output with minimum octane requirements
2. High thermal efficiency
3. Smooth engine operation.

1. High power output, requires :

(*i*) *High compression ratio.* The compression ratio is limited by the phenomenon of detonation. It depends upon the design of combustion chamber and fuel quality. *Any change in design that improves the anti-knock characteristics of a combustion chamber permits the use of a higher compression ratio which should result in higher output and efficiency.*

(*ii*) *Small or no excess air.*

(*iii*) *A complete utilisation of the air – no dead pockets.*

(*iv*) *An optimum degree of turbulence.* Turbulence is induced by inlet flow configuration or squish. Turbulence induced by squish is preferable to inlet turbulence as the volumetric efficiency is not affected.

(*v*) *High volumetric efficiency.* This is achieved by having large diameter, valve timings and straight passage ways by streamlining the combustion chamber so that flow is with lesser pressure drop. This means more charge per stroke and proportionate increase in the power output.

2. High thermal efficiency requires :

(*i*) *High compression ratio.*

(*ii*) *Compact combustion chamber for small heat* loss during combustion, reduced flame travel and thus less combustion time loss.

3. Smooth engine operation requires :

(*i*) *Moderate rate of pressure rise during combustion.*

(*ii*) *Absence of detonation,* by proper location of spark plug and exhaust valve, by satisfactory cooling of spark plug points (to avoid pre-ignition) and of exhaust valve head, and by short distance of flame travel.

Q. 22.19. *Refer to Fig. 1 and answer the following :*

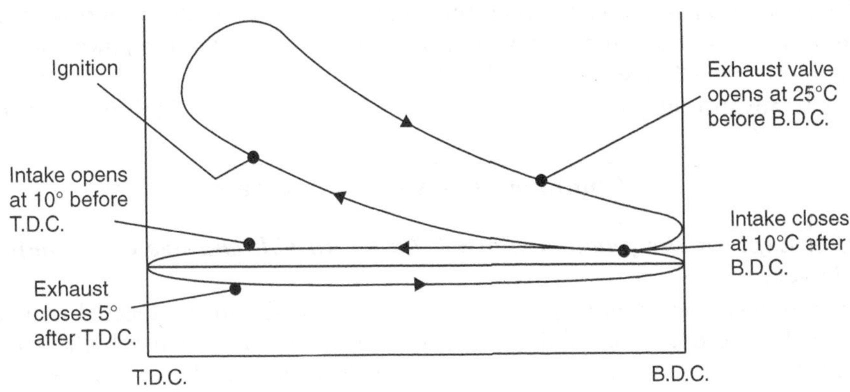

Fig. 1. Indicator diagram of S.I. engine.

(i) What is the value of the angle of overlap ? What effect on volumetric efficiency would you expect if the angle of valve overlap is increased leaving intake opening unaltered.

(ii) Re-sketch the same indicator diagram if the throttle is partly open.

(iii) If the throttle is partly opened, will you like to have the spark advanced or retarded ? Explain your answer.

Ans. (*i*) Angle of valve overlap

= 10° (inlet opening before T.D.C.) + 5° (exhaust closing after T.D.C.)

= 15°

The effect of valve overlap by closing the exhaust valve after T.D.C. reduces the back pressure, and *continues to drive out the combustion* products, leaving scope for admitting more charge. This effect *improves the volumetric efficiency*. However, *if valve overlap is increased beyond a certain limit, some fresh charge may be lost through the exhaust valve. This would decrease the charging efficiency*, and *also increase the fuel consumption (in S.I. engines).*

(*ii*) If throttle valve is partly open, the pumping loop will be larger and the power loop will be smaller (relative to that when the throttle is fully open). For the purpose of comparison, the indicator diagrams for full throttle open and part throttle are drawn below :

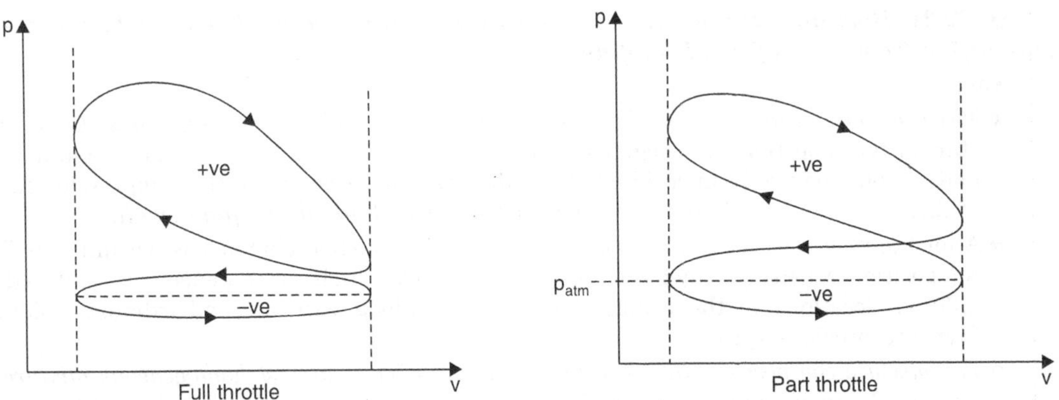

Fig. 2

(*iii*) The spark will have to be *advanced* on partly opened throttle. At part load the vacuum in the manifold is high, and as a result less air and fuel mixture enters the cylinder. Also the effect of exhaust dilution is more. Further the mixture is less highly compressed. Mixtures at lower compression ratios burn more slowly, and therefore to get maximum power (for that throttle opening), the ignition must take place earlier in the cycle.

The spark advance at part throttle is obtained by a "vacuum ignition advance" device.

Combustion on C.I. Engines

Q. 22.20. *Explain the phenomenon of knock in C.I. engines and compare it with S.I. engine knock.*

Ans. If the delay period is long, a large amount of fuel will be injected and accumulated in the chamber. The auto-ignition of this large amount of fuel may cause high rate of pressure rise and high maximum pressure which may cause *knocking* in diesel engines. A long delay period not only increases the amount of fuel injected by the moment of ignition, but also improves the homogenity of the fuel air mixture and its chemical preparedness for explosion-typeself ignition similar to detonation in S.I. engines.

Let us compare the phenomenon of detonation in S.I. engines with that of knocking in C.I. engines. Both are processes of auto-ignition subject to the ignition time lag characteristics of the fuel-air mixture. Differences in knocking phenomenon of the S.I. engine and the C.I. engine should also be carefully noted :

1. In the S.I. engine, the detonation occurs near the end of combustion whereas in the C.I. engine detonation occurs near the beginning of combustion.

2. The detonation in the S.I. engine is of a homogeneous charge causing very high rate of pressure rise and very high maximum pressure. In C.I. engine the fuel and air are imperfectly mixed and hence the rate of pressure rise is normally lower than that in the detonating park of the charge in the S.I. engine.

3. Since in the C.I. engine the fuel is injected into the cylinder only at the end of the compression stroke there is no question of pre-ignition or pre-mature ignition, as in the S.I. engine.

4. In the S.I. engine it is relatively easy to distinguish between knocking and non-knocking operations as the human ear easily finds the distinction.

5. Factors that tend to reduce detonation in the S.I. engine increase knocking in the C.I. engine.

Q. 22.21. *How does the mixture composition in the combustion chamber of a C.I. engine differ from that of a S.I. engine ?*

Ans.

- *Mixture composition in combustion chamber of C.I. engine is heterogeneous.* Before the start of fuel injection (few degrees before the T.D.C. position), the cylinder contains air, and combustion products residuals from the previous cycle. Since the compression ratios of diesel engines are high, the proportion of such *residuals is quite small.*
- After the injection starts, its distribution in the combustion chamber is not uniform. The composition consists of fuel droplets, and vaporized fuel, heterogeneously scattered at different locations in the combustion chamber, along with air and exhaust residuals from the previous cycle.
- *The mixture composition in S.I. combustion chamber is almost homogeneous mixture of air, petrol vapour and exhaust residuals from the previous cycle.* Since the compression ratios in petrol engines are quite low, and also since the mixture is throttled, the

proportion of *exhaust residuals in S.I. combustion chambers is quite high as compared to that in C.I. combustion chambers,* and *more so at lighter loads.*

Q. 22.22. *Explain briefly the phenomenon of "Diesel knock".*

Ans. Phenomenon of "Diesel knock." In C.I. engines the injection process takes place over a definite interval of time. Consequently, as the first few droplets to be injected are passing through the ignition lag period, additional droplets are being injected into the chamber. If *the ignition delay is longer*, the actual burning of the first few droplets is delayed and a greater quantity of fuel droplets gets accumulated in the chamber. When the actual burning commences, the additional fuel can cause too rapid a rate of pressure rise, as shown on pressure crank angle diagram above, resulting in a jamming of forces against the piston (as if struck by a hammer) and rough engine operation. If the ignition delay is quite long, so much fuel can accumulate that the rate of pressure rise is almost instantaneous. Such a situation produces the extreme pressure differentials and violent gas vibration known as knocking (diesel knock), and is evidenced by audible knock.

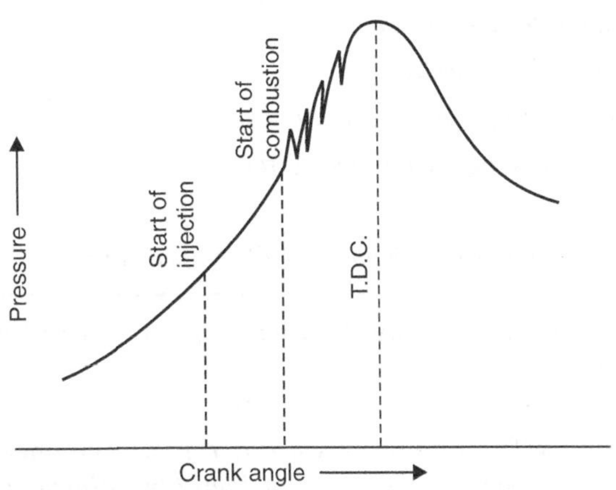

Fig. 3. Phenomenon of "Diesel knock."

Q. 22.23. *The factors that tend to increase detonation in S.I. engine tend to reduce knocking in C.I. engine.*

Discuss the above statement with reference to the following influencing factors :

(i) Compressions ratio ; (ii) Inlet temperature ; (iii) Inlet pressure ; (iv) Self-ignition temperature of fuel ; (v) Time lag of ignition of fuel ; (vi) rpm ; (vii) Combustion chamber wall temperature.

Ans. Detonation in S.I. engines and knocking in C.I. engines are fundamentally similar phenomenon. Both are processes of auto-ignition subject to the ignition time-lag characteristics of the fuel-air mixture. The *differences* in the two phenomena are as follows :

1. In the *S.I. engine,* the *detonation occurs near the end of the combustion process,* in the *C.I. engine* detonation *occurs near the beginning of combustion.*

2. In the *S.I. engine,* it is *relatively easy to distinguish between knocking and non-knocking operation as the human ear easily finds distinction.*

3. The *detonation in the S.I. engine is of homogeneous charge causing very high rate of pressure rise and very high maximum pressure. In the C.I. engine the fuel and air are imperfectly mixed and hence the role of pressure rise is normally lower than that in the detonating part of the charge in the S.I. engine.*

4. Since in C.I. engine the fuel is injected into the cylinder only at the end of compression stroke there is *no question of pre-ignition as in the S.I. engine.*

It is most important to care and note that *factors that tend to reduce detonation in S.I. engine increase knocking in the C.I. engine and vice-versa.* The detonation in the S.I. engine is *due to simultaneous auto-ignition of the last part of the charge.* To eliminate detonation in the S.I. engine we want to prevent altogether the auto-ignition of the last part of the average and therefore desire a *long delay period and high self ignition temperature of the fuel*. To eliminate knock in the C.I. engine, we want *auto-ignition as early as possible and therefore desire a short delay period and low self-ignition temperature of the fuel.*

Factors tending to reduce knocking in S.I. and C.I. engines

Factors	S.I. Engines	C.I. Engines
(*i*) *Compression ratio*	Low	High
(*ii*) *Inlet temperature*	Low	High
(*iii*) *Inlet pressure*	Low	High
(*iv*) *Self-ignition temperature of fuel*	High	Low
(*v*) *Time lag of ignition of fuel*	Long	Short
(*vi*) *r.p.m.*	High	Low
(*vii*) *Combustion chamber wall temperature*	Low	High

Q. 22.24. *Why does rate of pressure rise during combustion is limited to a certain value ?*

Ans.

- The rate of *pressure rise* is a very important aspect from engine design and operation point of view. It *considerably influences the maximum cycle pressure, the power output and the smooth running of the engine.*
- Higher rate of pressure rise during combustion *cause rough running of the engine because of vibration and jerks produced in the crankshaft rotation.* It also *tends to create a situation conducive to an undesirable occurrence known as knocking.* A higher rate of pressure rise near the end of compression stroke and beginning of power stroke would produce high peak pressure giving increased power output of the engine. But if the rate of pressure rise exceeds about 3 to 3.5 bar per degree of crank rotation, the running of engine becomes rough and noisy. Hence the rate of pressure rise is to be limited to a certain value.

Q. 22.25. *Discuss the influence of ignition delay on combustion processes in S.I. and C.I. engines. Explain how the presence of a knock inhibitor in fuel oil helps to change the ignition delay in C.I. engines.*

Ans. Effect of ignition delay on combustion processes in S.I. and C.I. engines :

S.I. engine. *The ignition delay affects the time required for formation of flame nucleus after passing the spark.* It is not much of a problem, as the spark timings can always be adjusted according to the ignition delay.

However, if the ignition delay is long, the end gases will have less tendency for auto-ignition, thereby inhibit knocking. On the other hand, if the ignition delay is short, a substantial part of end mixture will auto-ignite before the flame front reaches it. This will cause severe knocking, vibrations, and increased heat loss, and may also promote pre-ignition. Thus the *ignition delay in S.I. engine affects the last phase of combustion which may be abnormal combustion.*

C.I. Engines. The *ignition lag is the preparatory phase during which some fuel has already been admitted, but has not yet ignited.* It *is of extreme importance because of its effects on both the combustion rate and knocking.* It has *also influence on engine starting ability and the presence of smoke in the exhaust.*

— A *short ignition delay will give near constant pressure combustion,* and *low maximum temperatures and pressures. This will reduce the thermal efficiency though the engine will run smooth and knock free.*

— On the other end, a long *ignition delay will cause the rate of pressure rise per crank degree* $\left(\frac{dp}{d\theta}\right)$ *to be high. This will result in rough running of engine. If* $\frac{dp}{d\theta}$ *exceeds about 3.5 bar per crank degree, the engine will start knocking, which is not desirable.* Thus, the ignition delay in C.I. engines (which according to Ricardo is the *first stage of 'combustion'*) will affect the uncontrolled combustion, *i.e.*, the second stage of combustion.

Knock inhibitors :

The knock inhibitors like *amylnitrate, nitro-ether, boron and its compounds are added to diesel fuels to improve their cetane number.* These additives are highly explosive, and have high heating values. They combine with carbon and hydrogen to form a *BCH* type of compound.

Because of their high explosiveness, the addition of such compounds (in very small quantities) *reduces the ignition delay, and thus increase the cetane number of diesel fuels.*

Q. 22.26. *"The requirement of air motion and swirl in a C.I. engine combustion chamber is much more stringent than in an S.I. engine." Justify this statement ?*

Ans. Air motions are required in both S.I. and C.I. engines. *In S.I. engine, we call it turbulence* and in *C.I. engine, we call it swirl.* Turbulence which is required in S.I. engines implies *disordered air motion with no general direction of flow, to break up the surface of flame front and to distribute flame throughout an externally prepared combustible mixture.* Air swirl which is required in C.I. engines is *an orderly movement of whole body of air with a particular direction of flow to bring a continuous supply of fresh air to each burning droplet* and *sweep away the products of combustion which otherwise would suffocate it.*

If there is no turbulence in S.I. engines, the time occupied by each explosion would be so great as to make high speed internal combustion engines impracticable. Insufficient turbulence lowers the efficiency due to incomplete combustion of fuel. *In case of C.I. engines, it is impossible to inject fuel droplets so that they distribute uniformly throughout the combustion space, the fuel air mixture formed in combustion chamber is essentially heterogeneous.* Under these conditions, if the air within the cylinder were motionless, *only a small portion of fuel would find sufficient oxygen and even burning of this fuel would be slow or even choked.* So *it is essential to impart swirl to air so that a continuous supply of fresh air is brought to each burning droplet and the products of combustion are swept away.*

Q. 22.27. *The induction swirl in a C.I. engine helps in increasing indicated thermal efficiency. Justify this statement.*

Ans. Induction swirl is used in direct injection type engines, where the entire combustion space is directly above the piston, and hence the surface-to-volume ratio of the combustion chamber is low. Further, the compressed air and the combustion products do not have to pass through a neck narrow connecting passage. Also, the mean combustion temperatures are lower, and there is less turbulence. All these factors result in less heat losses, and thus the indicated thermal efficiency is increased.

Q. 22.28. *How C.I. engine combustion chambers are classified ? What type of swirl is used in these chambers ?*

Ans. *C.I. engine combustion chambers* are classified *on the basis of method of generating swirl*. There are four types of combustion chambers :

1. Open combustion chambers
2. Divided or turbulent swirl chambers
3. Pre-combustion chambers
4. Air cell combustion chambers.

1. **Open combustion chambers.** The method of swirl used in this type of combustion chamber is *induction swirl*. In induction swirl, flow of air is directed towards the cylinders during its entry. *An open combustion chamber is one in which combustion space is essentially a simple cavity with little restriction and there are no large differences in pressure between different parts of chamber during the combustions process.*

2. **Divided or Turbulent swirl chambers.** The method of generating swirl in the divided chamber is by compression. Compression swirl is one in which air is forced through a tangential passage into a separate swirl chamber during compression stroke. *A divided combustion chamber is defined as one in which the combustion space is divided into two or more distinct compartments between which there are restrictions or throats small enough so that considerable pressure differences occur between them during combustion process.*

3. **Pre-combustion chamber.** The type of swirl used is *combustion induced swirl.* A pre-combustion chamber consists of pre-combustion chamber or anti-chamber connected to the main chamber through a number of very small holes. Pre-combustion chamber contains 20 to 30% of clearance volume. Pre-combustion chamber has multi fuel capability without any modification in the injection system in the type of swirl used *i.e.* combustion induced swirl, swirl turbulence is created by use of initial pressure rise due to partial combustion.

4. **Air-cell combustion chamber.** The type of swirl used in this combustion chamber is combustion induced swirl. Here there is no organised air swirl. The advantages for this type of combustion chamber are ; the maximum pressure on the main chamber is fairly low and hence it gives smooth running and easy starting. This type of combustion chamber is most suitable for comparatively small engines of medium duty where a relatively high fuel consumption can be tolerated.

Q. 22.29. *In agriculture field, it is better to use C.I. engine than S.I. engine. Justify this statement.*

Ans.

- The diesel engine has high enthalpy efficiency and particularly at part loads.
- Due to high compression ratio the engine is robust in construction.
- There are less fire hazards with diesel oil, and the insurance premiums are considerably lower.
- Also the diesel fuel is much cheaper on the volume basis (per litre), and further cheaper on a mass basis (the density of diesel oil being about 12% more than that of petrol).
- The heating values of both the fuels on mass basis are comparable.
- All these features make a C.I. engine more suitable than S.I. engine in agricultural field. These are manufactured in varying power range from 15 to 50 kW for agricultural machineries.

Q. 22.30. *How can a diesel engine be converted to a CNG engine ?*

Ans. The CNG has a high self-ignition temperature (about 730°C), hence a very high compression ratio would be required if used in C.I. engine. To convert a diesel engine to a CNG engine the modifications necessary are :

1. *Arrangement to add the gaseous fuel to the air induced by the engine.* A supercharger at a pressure slightly above the atmospheric pressure may also be used.

2. *A pilot injector to be provided instead of the conventional injector. The pilot injector would inject a small quantity of liquid fuel* (about *5 per cent of that required at full load*) *to initiate the combustion.* This *quantity is to remain constant at all loads.*

Thus when CNG is used in a diesel engine, the engine would work like a diesel engine, to start the combustion, but it propagates by flame front in a manner similar to that in an S.I. engine.

Q. 22.31. ***"The maximum substitution of diesel by CNG in a C.I. engine is limited by the cetane characteristics of the available fuel." Justify this statement.***

Ans. The diesel engine is designed for fuel with a given ignition quality with regards to auto-ignition, rate of pressure rise, knocking, maximum pressure etc. (*i.e.* fuel of a specific cetane number). Also, it is known that a diesel engine can burn the gaseous fuel without much modification. Therefore, if the diesel (liquid) fuel is to be substituted by the gaseous fuel (CNG), for optimum performance the *maximum substitute of diesel by CNG is limited by the cetane characteristics of the available fuel.*

Q. 22.32. ***Write a short note on aids for starting C.I. engines under extreme cold climate.***

Ans. Cold starting aids for C.I. engines :

Many methods have been used in the past to achieve easy cold starting. Few of them are :

1. Injecting a small quantity of lubricating oil or fuel oil. This temporarily raises the compression ratio, and seals the piston rings and valves.

2. Provision of cartridges. These may be self-igniting or require lighting before insertion into the combustion chamber.

3. Pre-heating the engine cylinder by warm water.

4. Modifying valve timings for starting.

5. Starting as petrol engine by providing a carburettor and a spark plug. At starting compression ratio is reduced by providing an auxiliary chamber.

Modern starting aids :

1. Electric glow lamps in the combustion chambers.

2. Manifold heaters which ignite a small feed of fuel.

3. The injection, into the intake, of controlled amount of low ignition temperature liquids, usually ethyl-ether, with addition of other fuels.

Q. 22.33. ***How do you define volumetric efficiency of an I.C. engine ? How is it related to the power output of the engine ? How is the volumetric efficiency affected by speed and altitude ?***

Ans.

- The *volumetric efficiency* of an engine is defined *as the ratio of actual air capacity to the ideal air capacity.* This is equal to the *ratio of mass of air which enters or is forced into the cylinder in suction stroke to the mass of free air equivalent to the piston displacement at intake temperature and pressure conditions*

$$\eta_{vol} = \frac{\text{Mass of charge actually induced}}{\begin{array}{l}\text{Mass of charge represented by volume at}\\ \text{intake temperature and pressure conditions}\end{array}}$$

- **Power output** of an engine is *proportional to volumetric efficiency provided the combustion is complete.*
- *The volumetric efficiency of an engine is affected by many variables* such as *compression ratio, valve timing, induction and port design, mixture strength, latent heat of evaporation of the fuel, heating of the induced charge, cylinder temperature and the atmospheric conditions.*
  - At *higher speeds* on account of internal effects the charge inhaled decreases and consequently the *volumetric efficiency decreases.* At *lower speeds volumetric efficiency is nearly constant, at high speeds it falls rapidly.*

— At altitude the pressure and temperature both decrease. The effect of both is to reduce the density and consequently mass inhaled reduces. Thus *higher the altitude lower will be the volumetric efficiency.*

— With normal aspiration the volumetric efficiency is seldom above 80%, and to improve on this figure, *supercharging* is done. *Air is forced into the cylinder by a blower or fan which is driven by the engine.*

Q. 22.34. *Why the inlet valve be kept open for a few degree of crank angles even when the piston is on the compression stroke ? Assume that the engine under consideration is a high speed one.*

Ans. For high speed engines the cylinder pressure at the end of suction stroke may be below inlet pressure. Further, there may be inertia effect in the inlet pipe, due to high velocity of the charge. *In this case, delaying the inlet valve closing beyond the bottom dead centre* (*i.e.*, when the piston is on the compression stroke) *would allow more fresh charge to enter the cylinder.* This *would increase the volumetric efficiency. Higher the engine speed and longer the inlet pipe, longer the inlet valve be kept open when the piston is on the compression stroke* (to take advantage of inertia of incoming air in the inlet pipe).

Q. 22.35. *"4-stroke I.C. engine is always economical and less pollutant than 2-stroke I.C. engine." Justify this statement.*

Ans.

- In two-stroke engine the charge has to be compressed outside for scavenging and charging (this consumes some engine power). A part of this charge escapes directly through exhaust ports (short circuiting). Thus power spent in compressing this fraction of the charge is wasted. Particularly in S.I. engines the charge consists of air-fuel mixture. This loss of power and charge is absent in 4-stroke engine. Therefore 4-stroke engine is always economical than 2-stroke engine.
- Further the loss of charge increases HC in the exhaust in case of two-stroke engines, Hence 4-stroke engine is also less pollutant than 2-stroke engine.

Q. 22.36. *Define volumetric efficiency of an I.C. engine. What are the effects of resistance of the intake system, residual gas content and heat transfer to the incoming charge on volumetric efficiency ?*

Ans. *Volumetric efficiency* *of an I.C. engine is defined as the ratio of actual air capacity to ideal air capacity.*

It may also be defined as,

$$\text{Volumetric efficiency} = \frac{\text{Unit air charge}}{\begin{array}{c}\text{Mass of air that would fill the displacement of one}\\ \text{cylinder at inlet temperature and pressure}\end{array}}$$

Effects of various factors on volumetric efficiency :

1. Effect of resistance of intake system. The *mass of air inducted is relatively small because of fluid friction in the manifold part and because of throttling* (fluid friction) *across the slowly opening and slowly closing inlet valve.* However, these losses become smaller as the *piston speed increases* and unit air charge *could approach the limiting value.* The charge of air would correspond to filling the displacement volume with air at ambient temperature and pressure, that is, the volumetric efficiency would be 100 per cent. But with speed increase, the increasing velocities of air would lead to progressively greater fluid frictional losses, even though some ramming gain would arise from the momentum of incoming air column in inlet manifold. Thus the unit air charge would have constantly *declining characteristics, hence the volumetric efficiency.*

2. Residual gas content. At the end of exhaust stroke the exhaust gas, normally, has a higher pressure and temperature than the fresh charge (air for I.C. engine, air and fuel for S.I. engine) in the inlet manifold. When the inlet valve opens, a portion of residual expands into the intake manifold and is then drawn back into the cylinder as the piston descends on the intake stroke. It follows that *high exhaust pressures or early opening intake valves reduce the unit air charge because of this residual flowback with the effect of reducing volumetric efficiency. At high speeds, the volumetric efficiency is greatly reduced.*

3. Heat transfer to incoming charge. The cylinder and combustion chamber walls, in a real engine, are hot and confine the hot residual at the end of exhaust stroke. When the residual flows into the inlet manifold, the inlet valve and port are heated. The entering fresh charge is also heated, not only by walls, but also by the hot inlet valve and port. Thus the *volumetric efficiency is low since the inlet valve closed late, since residual is inducted with the fresh charge and since the fresh charge is heated.* The mixing process of cold charge with hot residual *does not reduce the volumetric efficiency since expansion of charge is almost compensated by construction of residual.*

— *With increase in load (at constant speed), unit air charge and volumetric efficiency decrease, since the temperatures of the walls are increased.*

Q. 22.37. ***What are the requirements of an ignition system for an I.C. engine ?***

Ans. The following are the *requirements of an ignition system* :

1. A source of electrical energy.
2. A device for transforming the low voltage from the source to a high voltage required to produce a spark across the spark plug electrodes.
3. A device for distributing the high voltage to the spark plug of each cylinder in every cycle for a multi-cylinder engine.
4. The timings should be accurately controlled depending upon load and speed of the engine.
5. Long duration of the spark, and sufficient energy to ensure ignition of the mixture.
6. Good performance at high speeds.
7. Be able to fire even when spark plugs are partially fouled.
8. Longer life of breaker points and spark plug.
9. Good starting when the breaker points open slowly at cranking speed.
10. Good reducibility of secondary voltage rise and maximum rise.

Q. 22.38. ***What is the difference between ignition timing and firing order ?***

Ans. Ignition timing is *the correct instant for the introduction of spark near the end of compression stroke in the cycle.* The ignition timing is fixed *to obtain maximum power from the engine.*

Firing order is *the order in which various cylinders of a multi-cylinder engine fire.* The firing order is arranged *to have power impulses equally spaced, and from the point of view of balancing.*

Q. 22.39. ***State the functions of an ignition coil and a condenser in the battery ignition system of a multi-cylinder S.I. engine.***

Ans. Function of ignition coil and condenser :

(*i*) **Ignition coil.** The *function of the ignition coil is to step up 6 to 12 volts of the battery to a high tension voltage (10000 to 20000 volts) sufficient to promote an electric spark across the electrodes of the spark plugs.* The ignition coil consists of two insulated conducting coils called the primary and secondary windings. The primary winding is connected to the battery, and the secondary winding is connected to spark plugs through the distributor. In order to boost the voltage, the primary winding has a few hundred turns of relatively thick wire, whereas the secondary winding consists of several thousand turns of very fine wire.

(*ii*) **Condenser.** The *function of condenser in the ignition system is to help the rapid collapse of the magnetic field and to store up the energy momentarily when the contact breaker points open, so that due to high voltage it may not jump between the breaker points.*

Without the condenser, the induced current would establish an arc across the contact points when they separate, and therefore the collapse of the field would be prolonged, and the voltage rise in the secondary coil would be slow. Meanwhile most of the energy stored in the magnetic field would be consumed in an arc across the contact breaker points (rather than arc across the spark-plug electrodes).

Q. 22.40. ***What is the main difference between the battery and electronic systems ?***

Ans. The main difference between the battery and electronic ignition systems is as follows :

- In **battery ignition system** *contact breaker is used for making and breaking the primary circuit of the ignition coil.* This making and breaking of the primary circuit is responsible for providing a high voltage across the spark plug electrodes. The contact breaker consists essentially of a fixed metal point against which another metal point bears. A cam driven by the engine shaft is arranged to open the breaker points whenever an electric discharge is required.
- In **electronic ignition systems** *electronic triggering is used to interrupt a circuit carrying a relatively high current.* It makes an *ideal replacement for the breaker points and the condenser.* Many variations of the electronic ignition system are available.
  — In one of the versions the *contact breaker and the cam assembly of the conventional battery ignition system are replaced by a magneto-pulse generating system which detects the distributor shaft position and sends electrical pulse to an electronic control module. The module switches off the flow of current to the primary coil, inducing a high voltage in the secondary winding, which is distributed to the spark plugs as in the conventional breaker system. The control module contains timing circuit which later closes the primary circuit so that the build up of the primary circuit current can occur for the next cycle.*

Thus the *main difference between the conventional (Battery) ignition system and electronic ignition system is* **that in the former the electrical circuit is made and break by mechanical devices, whereas in the later it is by electronic circuit.**

Q. 22.41. ***What are the differences between battery and magneto-ignition systems ?***

Ans.

- In **battery ignition system** *a 6-12 V battery* is used to provide primary voltage, and *a separate ignition coil is required to boost up this voltage* needed to operate the spark plug.
- **Magneto** *is a special type of ignition system with its own electric generator to provide the necessary energy for the system. It is mounted on the engine and replaces all the components of the battery ignition system except the spark plug.* The magneto when rotated by the engine is capable of producing a very high voltage and *does not need a battery as a source of external energy.*

The differences between the battery ignition system and the magneto ignition system are given in the table below :

Battery ignition system	Magneto ignition system
1. Current for primary circuit is obtained from the battery.	1. The required electric current is generated by the magneto.
2. Difficult to start the engine when the battery is discharged.	2. There is no problem of battery discharge.
3. Maintenance problems are more due to battery.	3. Maintenance problems are less, since there is no battery.
4. A good spark is available at the spark plug even at low speed.	4. During starting the quality of spark is poor due to low speed.
5. Efficiency of the system decreases with the reduction in spark intensity as engine speed rises.	5. Efficiency of the system improves as the engine speed rises due to high intensity spark.
6. Occupies more space	6. Occupies less space.
7. Commonly employed in cars and light commercial vehicles.	7. Mainly used in racing cars and two wheelers.

Q. 22.42. *"S.I. engines are generally not supercharged." Justify this statement.*

Ans. The factors which affect knocking in **S.I. engines** :

— Compression ratio ;

— Mixture strength ;

— Fuel characteristics (Octane number, ON) ;

— Initial pressure.

— In these engines the limit of supercharging is fixed mainly by *knocking,* because the knocking tendency of most fuels is increased by increasing the inlet pressure and temperature, or both. At the same ON requirement, if the charge density is increased the compression ratio has to be decreased considering the knock limits. Thus the *power by the supercharged engine is increased but at reduced thermal efficiency.*

— Further, supercharged S.I. engines are *usually to run on rich mixture, for maximum power.* This also results in a higher s.f.c.

Therefore, *S.I. engines are not generally supercharged, except to compensate for loss of power at high altitudes.*

Q. 22.43. *"Supercharging is preferred in diesel engine than petrol engine." Justify this statement.*

Ans.

- Supercharging increases the pressure and temperature of the charge at the end of compression. Thus *by supercharging the effective compression ratio is increased.* This reduces ignition delay in **C.I. engine**, thereby the combustion becomes smooth, the rate of pressure rise in second stage combustion is reduced, and the *tendency for knocking is avoided.*
- In **S.I. engine** the *short ignition delay promotes detonation.* To overcome this either the compression ratio is reduced (thereby thermal efficiency is also reduced), or a fuel of higher ON would have to be used. Both of these would *increase the operating cost.*

Hence, supercharging is preferred in diesel engine than in petrol engine.

Q. 22.44. *"Supercharging is essential for an aircraft engine." Justify this statement.*

Ans. Aircraft engines fly at high altitudes where the pressure and hence the density of atmospheric air falls down much below that at sea level. If the engine admits air by natural aspiration the power output of the engine will be very low. To maintain ground level power supercharging is essential for an aircraft engine.

Q. 22.45. *Explain the factors that limit the extent of supercharging of S.I. and C.I. engines.*

Ans.

- In *S.I. engine the main limiting factors are detonation and pre-ignition due to increase in temperature by excessive supercharging.*
- The *other limitations common to both types of engines which do not permit large increase in boost pressure are :*

(*i*) High boost pressures necessitate the cylinders to be designed to withstand the resulting high combustion pressures which will make the engine structure heavy.

(*ii*) High peak pressures and increase in bearing pressures will result in increase of friction losses.

(*iii*) Cooling system will have to handle extra heat, resulting in increase of temperatures of cylinder wall and piston top.

(*iv*) The power to drive the supercharger increase rapidly with the increase of pressure ratio. With excessive delivery pressures all the power gained by the engines may be absorbed by the supercharger. Delivery pressures upto about 1.5 bar are normally used in practice.

Q. 22.46. *What do you mean by supercharging of I.C. engines ? Explain, why supercharging is essential for the aircraft engines ?*

Ans. Meaning of Supercharging. Supercharging is the *process of supplying air/air-fuel mixture into an engine at a pressure higher than the pressure at which a naturally aspirated engine takes from atmosphere.* The increased pressure (and therefore increased density) results in increase in power developed by the engine of given displacement. The supercharging is done by a pressure boosting device, called *supercharger.* The supercharger may be driven directly by the engine, or by an exhaust driven turbine, called turbocharger.

The supercharging is done to

(*i*) Obtain more power from an engine of given displacement or

(*ii*) Compensate for loss of power at higher altitude, such as for mountains or for aircraft engines.

Supercharging is essential for aircraft engines because the aircraft engines loose power as the altitude increases due to reduction in atmospheric density.

Fuel-air Cycle

Q. 22.47. *Explain why a S.I. engine fails to operate if the air-fuel ratio is more than 20 : 1 while a C.I. engine can operate on an air-fuel ratio of even 50 : 1.*

Ans.

- In *S.I. engine* charge consists of air and fuel, and ignition takes place with the help of spark. *When the mixture is lean the fuel will not burn by spark.*
- However, in case of *C.I. engine* the *fuel is supplied in fine particles in the end of compression stroke, compressed air having high pressure and temperature and on coming in contact with this air it burns.* Hence *there is no problem of combustion of fuel even with the lean mixtures.*

Q. 22.48. ***'Air fuel ratio in a S.I. engine varies from 8 to 16 approximately while such variation in a C.I. engine is from 100 at no-load to 20 at full load.' Explain.***

Ans. In *S.I. engines, the combustion is homogeneous*. A flame nucleus is formed at the spark plug electrodes, and the flame propagates in a more or less homogeneous mixture of air and fuel. The ignition limits of air-fuel ratio are narrow, between about 8 : 1 to 18 : 1. The mixture proportion in S.I. engines *should be within this limit for the initiation and sustaining of the flame.* The carburettor supplies the mixture of air and fuel between this limit, depending upon the engine requirements of starting from cold, idling, normal running (maximum economy mixture), and maximum power (rich mixture).

On the other hand, in *C.I. engines the combustion is heterogeneous,* and the load control is by varying the quantity of fuel. There is no throttling of inlet air. A very small quantity of fuel is supplied by injector at starting and no load. As the load increases, the quantity of fuel is increased. At full load, for smoke free exhaust and best utilization of air, the air-fuel ratio reduces to about 20 : 1, depending upon the type of combustion chamber.

The ignition starts at several points in the combustion chamber at locations where the local mixture of air-fuel formed is between the ignition limits (irrespective of the overall air-fuel ratio in the cylinder being much higher than the limits of ignition).

Q. 22.49. ***What is the difference between air-standard cycles and fuel-air cycles ?***

Ans. In the air-standard cycle the working substance during the operation of the cycle is taken as air. It is considered as a perfect gas with constant composition and constant properties throughout the cycle. The fuel-air cycle, on the other hand, takes into account the following :

(*i*) *The actual composition of the cylinder gas.* The cylinder gases contain fuel, air, water vapour and the residual gases. The fuel-air ratio changes during the operation of the engine.

(*ii*) The variation in the specific heat and γ with temperature.

(*iii*) The effect of dissociation at high temperatures, and the presence of CO, H_2, H and O_2 at equilibrium conditions.

(*iv*) The variation in the number of molecules consequent on chemical reaction.

Causes of deviation of actual cycles from fuel-air cycles :

(*i*) The progressive combustion rather than the instantaneous combustion.

(*ii*) The heat transfer to and from the working medium during compression and expansion.

(*iii*) Loss of work on the expansion stroke due to early opening to the exhaust valve, and exhaust blowdown.

(*iv*) Gas-leakage, fluid friction etc.

Q. 22.50. ***(a) Make a comparative statement of operations and working media for air cycle, fuel-air cycle and actual cycle of SI engines.***

(b) Explain (i) Air-standard efficiency, (ii) Thermal efficiency, and (iii) Relative efficiency of an engine.

Ans. (*a*) **Comparison of operation and working media :**

(*i*) **Air cycle.** The working medium is air throughout the cycle. It is assumed to be an ideal gas with constant properties. The working medium does not leave the system, and performs cyclic processes. There are not inlet and exhaust strokes. The compression and expansion process are isentropic. The heat addition and rejection are instantaneous at T.D.C. and B.D.C. respectively, at constant volume.

(*ii*) **Fuel air cycle.** The cylinder gases contain fuel, air, water vapour and residual gases. The fuel-air ratio changes during the operation of the engine which changes the relative amounts of CO_2, water vapour etc. The variations in the values of specific heat and γ with temperature, the effects of dissociation, and the variations in the number of molecules before and after combustion are considered.

Besides taking the above factors into consideration, the following *assumptions* are commonly made for the operation :

1. No chemical change prior to combustion.
2. Charge always in equilibrium after combustion.
3. Compression and expansion processes-frictionless, adiabatic.
4. Fuel completely vaporized and mixed with air.
5. Burning taking place instantaneously, at constant volume, at T.D.C.

The fuel-air cycle gives a very good estimate of the actual engine with regards to efficiency, power output, peak pressure, exhaust temperature, etc.

(*iii*) **Actual cycle :**

1. The working substance is a mixture of air and fuel vapour, with the products of combustion left from the previous cycle.
2. The working substance undergoes change in the chemical composition.
3. Variation in specific heats with temperature takes place. Also the temperature and composition changes due to residual gases.
4. The combustion is progressive rather than instantaneous.
5. Heat transfer to and from the working medium to the cylinder walls takes place.
6. Exhaust blowdown losses take place, *i.e.* loss of work due to early opening of the exhaust valve.
7. Gas leakage and fluid friction are present.

The points (4) to (7) make the actual cycle differ from the fuel-air cycle.

(*b*) (*i*) **Air-standard Efficiency (η_{as}).** It is the efficiency of the idealised cycle, in which *air is assumed to be the working substance.* The compression and expansion processes are assumed frictionless and adiabatic. The addition of the heat is at constant volume (for Otto cycle), or at constant pressure (for Diesel cycle), and the heat rejection is at constant volume.

(*ii*) **Thermal Efficiency (η_{th}).** It is the efficiency of the actual engine, defined as the *actual work (Indicated or Brake), divided by the heat released by the combustion of fuel.*

Thus, $$\eta_{th} = \frac{\text{Power output, kW (Indicated or Brake)}}{\text{Fuel consumption (kg / s)} \times \text{C of fuel}}$$

Based on indicated power, it is indicated thermal efficiency, and based on brake power it is brake thermal efficiency.

(*iii*) **Relative Efficiency.** It is the *ratio of thermal efficiency and air-standard efficiency.* Generally relative efficiency is based on indicated thermal efficiency.

Two Stroke Engines

Q. 22.51. ***In a 2-stroke engine it is better to have deflector top type piston. Justify.***

Ans. The two stroke engines have a deflector top type piston to reflect the fresh charge towards the top of the cylinder, before flowing to the exhaust ports. This serves the double purpose of (*i*) *scavenging the upper part of the cylinder of combustion products,* and (*ii*) *preventing the fresh from flowing directly to the exhaust ports.*

Q. 22.52. ***Discuss briefly 'Mist lubrication system'.***

Ans.

- This system is used where crankcase lubrication is not suitable. In two stroke engine as the charge is compressed in the crankcase, it is not possible to have the lubricating oil in the sump. Hence mist lubrication is adopted in practice. In *such engines, the lubricating oil is mixed with the fuel, the usual ratio being 3% to 6%. The oil and the fuel mixture is inducted through the carburettor. The fuel is vaporized and oil in the form of mist goes via the crankcase into the cylinder. The oil which strikes the crankcase walls lubricates the main and connecting rod bearings, and the rest of the oil lubricates the piston, piston rings and the cylinder.*
- In some of the modern engines, the lubricating oil is directly injected into the carburettor and the quantity of oil is regulated. In this system the main bearings also receive oil from a separate pump.

Q. 22.53. ***Discuss the two main disadvantages of two-stroke cycle S.I. engine. How these disadvantages are avoided in the two-stroke cycle C.I. engines ?***

Ans.

- The *two-stroke S.I.* engine suffers from *two main disadvantages*—**fuel loss** and **idling difficulty.** The two-stroke C.I. engine does not suffer from these disadvantages, and hence C.I. engine is more suitable for two-stroke operation.
  - — If the fuel is supplied to the cylinders after the exhaust ports are closed, there will be no loss of fuel, and the indicated thermal efficiency of the two stroke engine will be as good as that of four-stroke engine. In an S.I. engine the scavenging is done with fuel-air mixture. In C.I. engine the fuel loss is avoided, as the fuel is injected near the end of compression stroke when the exhaust valve is closed.
  - — The two-stroke S.I. engine runs irregularly and may even stop at low speeds when the m.e.p is reduced to 2 bar. This is because large amount of residual gas is (more than in four-stroke engine) mixed with small amount of charge. At low speeds there may be backfiring due to slow burning rate. In C.I. engine there is no difficulty at idling because the fresh charge (air) is not reduced, and also backfiring is absent, as there is no fuel present in the inlet system.

Q. 22.54. ***Why do the two-stroke C.I. engines find wide use in marine propulsions ?***

Ans.

- Two-stroke C.I. engines find wide uses in marine propulsion for the following *reasons* :
  1. More uniform torque, the ideal requirement for the propeller.
  2. More cooling is required in two-stroke engines, plenty of sea water is available for cooling.
  3. C.I. engines have no loss of fuel in scavenging. Hence have higher thermal efficiency.
  4. *Propeller imposes the condition that maximum power must be developed at about 100 rpm. Two stroke engines may be made of slow speed, and with large displacement volume (over 60 cm bore), and of capacity 5000 kW and above. These slow speed engines can be coupled directly to the propeller of the ship, without the necessity of gear reduction.*
- *Two-stroke C.I. opposed piston engine (cross-head type) is mainly used for marine propulsion.*

Q. 22.55. *Why are two-stroke S.I. engines more commonly used in low horse power two wheelers ?*

Ans. When applied to S.I. engines, the two-stroke cycle engine has certain *disadvantages* which have restricted its use to *small low horse power engines.* In S.I. engines the charge consists of a mixture of air and fuel. During scavenging both, inlet and exhaust ports are open simultaneously for sometime. *Some part of the fresh charge escapes with exhaust which results in higher fuel consumption and lower thermal efficiency.* For small two-wheeler engines the **fuel economy is not a vital factor.** Here **light weight and low initial cost are the main consideration,** which are the *main characteristics of two-stroke S.I. engines.*

Q. 22.56. *Why is crankcase scavenging used only for low power engines ?*

Ans. Crankcase scavenging is *very uneconomical and inefficient in operation.*

— The amount of air which can be used for scavenging is less than the swept-volume of the cylinder. The amount of charge transferred through the port is only 40 to 50 per cent of cylinder volume, and hence it is not possible to scavenge the cylinder completely of the products of combustion. This results in low m.e.p., typical values being 3 to 4 bar, thus limiting the power output.

— A *further disadvantage* is that the oil vapour from the crankcase mixes with the scavenging air. This results in *high oil consumption.*

Because of these disadvantages, the crankcase scavenging is not preferred for high output two-stroke engine (where separate scavenging pump is used) and is used only for low power engines.

Q. 22.57. *Why are two-stroke diesel engines, for large power, more common than two-stroke S.I. engines ?*

Ans. In addition to *high fuel consumption,* the other drawback of two-stroke S.I. engine is the *lack of flexibility—the capacity to run with equal efficiency at any speed.* If the throttle is closed below the best-point, the amount of fresh mixture entering the cylinder is not enough to clear out all the exhaust, some of which remains in the cylinder *to dilute the charge.* This *results in irregular running of the engine.*

The two-stroke diesel engine does not suffer from these defects. There is *no loss of fuel with exhaust gases as the intake charge in diesel engines is air only.* The two-stroke diesel engine is therefore used quite widely for *large power output.* It has further *advantage* of a *higher output from the same size engine,* and *absence of complicated valve mechanism* (over the four-stroke engine).

Q. 22.58. *What is the reason that two-stroke engine is not used in car even though it develops theoretically twice power than that of four-stroke engine ?*

Ans. A majority of cars are fitted with S.I. engines *due to light weight and good pick up. The two-stroke S.I. engine is not used in cars as it suffers from two big disadvantages*—**fuel loss** *and* **idling difficulty**.

— In S.I. engines using carburettor, the scavenging is done with fuel air mixture, and only the fuel mixed with the retained air is used for combustion. Thus a part of fuel is lost with scavenging air, giving poor fuel economy.

— The two-stroke S.I. engine runs irregularly and even may stop at low speeds when m.e.p. is reduced to about 2 bar. This is due to large amount of residual gas (more than in four stroke engine) mixing with small amount of charge. At low speeds there may be *back firing* due to slow burning rate.

Both the above drawbacks may be avoided by using fuel injection. But this makes the system complicated, and the maintenance cost is also increased (fuel injection pump is the first to give trouble), and hence not suitable for car engine.

Fuels

Q. 22.59. *Briefly explain the chemical structure of petroleum.*

Ans.

- *Petroleum is basically a mixture of hydrocarbons compounds which are made up exclusively of carbon and hydrogen atoms.* In addition, it may contain small quantitatives of other compounds having sulphur, oxygen and hydrogen.
- In some petroleum, metallic compounds such as derivatives of vanadium, iron, nickel, arsenic etc. are also found.
- A variety of hydrocarbons differing widely in molecular structure ranging from simplest form such as methane or marsh gas to most intricate groups as in paraffin wax of bitumen, are present in petroleum.
- In a hydrocarbon, molecule, the carbon and hydrogen atoms may be linked in different ways and this *linking influences the chemical and physical properties of the different hydrocarbon groups.*

Q. 22.60. *Give the general chemical formulae of the following fuels :*

(i) Paraffin : (ii) Olefin ; (iii) Diolefin ; (iv) Naphthalene ; (v) Aromatic.

Also state their molecular arrangements and mention whether they are saturated or unsaturated.

Ans. Primary Hydrocarbon Families in Petroleum

Family	*General formula*	*Molecular arrangement*	*Saturated/ unsaturated*
(*i*) Paraffin	C_nH_{2n+2}	Chain	Saturated
(*ii*) Olefin	C_nH_{2n}	Chain	Unsaturated
(*iii*) Diolefin	C_nH_{2n-2}	Chain	Unsaturated
(*iv*) Naphthalene	C_nH_{2n}	Ring	Saturated
(*v*) Aromatic	C_nH_{2n-6}	Ring	Highly unsaturated

Q. 22.61. *What are five primary hydrocarbon families found in petroleum ? Which are chain types ? Which are ring types ? Which of primary families tends to be better S.I. engine fuel and C.I. engine fuel ?*

Ans. The five primary hydrocarbon families which are found in petroleum have been listed below :

(*i*) Paraffin	*Chain type*	C_nH_{2n+2}
(*ii*) Olefin	*Chain type*	C_nH_{2n}
(*iii*) Diolefin	*Chain type*	C_nH_{2n-2}
(*iv*) Naphthalene	*Ring type*	C_nH_{2n}
(*v*) Aromatic	*Ring type*	C_nH_{2n-6}

S.I. engine fuel. Paraffin series have the maximum tendency to knock whereas aromatic series have the minimum tendency to knock and Naphthene series comes in between the two.

Hence *for S.I. engine fuel, Aromatic family is the best fuel.*

C.I. engine fuel. Better fuel for C.I. engine will be one which will have higher value of cetane number and the more paraffins hydrocarbons are contained in fuel, higher will be its cetane number. Hence the *paraffinic hydrocarbons provide a better fuel for C.I. engine.*

Q. 22.62. *What are the different kinds of fuels used in an I.C. engine ?*

Ans. The different kinds of fuels used in I.C. engine are :

(*i*) Gasoline (*ii*) Diesel

(*iii*) LPG (*iv*) Benzol

(*v*) Alcohol

(*vi*) Gaseous fuels such as natural gas, produces gas, blast furnance gas and coke oven gas.

(*vii*) Ammonia.

Q. 22.63. *What are the important properties which S.I. engine fuel possess ?*

Ans. The following are the important properties which S.I. engine fuel should possess :

1. *Volatility. i.e.*, easily vapourise and mix with air, should not form vapour lock.
2. *Antiknock* quality, and should not preignite easily.
3. Absence of gum and varnish deposits.
4. Low sulphur contents, anti-corrosion and clean burning.
5. Ease of handling.
6. High calorific value.
7. Low cost and availability.

Q. 22.64. *Why volatility is an important quality of S.I. engine fuels ?*

Ans. The volatility of an S.I. engine fuel is an important quality because it *affects* the following in the engine :

(*i*) Starting (*ii*) Warm up

(*iii*) Normal operation (*iv*) Crankcase dilution

(*v*) Vapour lock (*vi*) Carburettor icing.

Q. 22.65. *Explain briefly the following in regard to a fuel :*

(i) Vapour lock characteristics; (ii) Crankcase dilution.

Ans. (*i*) **Vapour lock characteristics :**

- Vapour lock is a situation *where too lean a mixture is supplied to the engine*. The automotive fuel pump should handle both liquids and vapours. If the amount of fuel evaporated in the fuel system is very high the fuel pump is mainly pumping vapour and very little liquid will go to the engine. This results in very weak mixture which cannot maintain engine output.
- *Vapour lock causes uneven running of an engine, stalling while idling, when thoroughly heated irregular acceleration, difficult starting when hot, or momentary stalling when running.*
- *The vapour lock tendency of the gasoline is related to front end volatility.* The vapour liquid ratio (V/L) of a gasoline directly correlates with the degree of vapour lock likely to be experienced in the fuel system. At V/L ratio of 24 vapour lock may start, and at V/L ratio of 36 vapour lock may be very severe. Therefore, the volatility of the gasoline should be maintained as low as practicable to prevent this type of difficulty.

(*ii*) **Crankcase dilution :**

- If the tail-end evaporation has too high evaporation temperatures, this part of the fuel will not be completely vaporised and will be carried as fuel droplets into the combustion chamber.This liquid fuel gets past the piston rings into the crankcase where it dilutes the oil and decreases viscosity. It also washes away the lubricating oil film on the cylinder walls.
- *Crankcase dilution is more at low engine operating temperatures,* such as those encountered in cold weather, stop and go driving where the oil temperature never gets high enough to evaporate the diluent. The crankcase dilution will be significant in such cases.

- *The relative tendency of the fuels to cause the dilution of the lubricating oil lies in the order of their 90 per cent ASTM distillation temperature.* As long as the 90 per cent point, which indicates the boiling range of the 80–100 per cent fraction, lies near 180°C, dilution is not a danger.
- *Engines using heavy fuels, such as kerosene and distillate, may suffer from poor lubrication of pistons and rings because of excessive dilution.*

Q. 22.66. *While volatility of the fuel is a determining factor in the selection of fuels for S.I. engines, ignition quality of the fuel is the primary deciding factor for C.I. engines. Discuss briefly the statement.*

Ans.

- **Volatility** of *a liquid is its tendency to evaporate under a given set of conditions.* It is an extremely important characteristic of S.I. engine which affects engine performance and fuel economy characteristics.
  - — Cold starting of S.I. engine is improved if *front end volatility* is higher but it may lead to increased problems of hot starting and vapour lock.
  - — The *mid range* (20 to 80 per cent) *portion* should be volatile enough to give satisfactory air-fuel ratios under a variety of operating conditons.
  - — *Low tail end volatility* will help in good mixture distribution and hence good fuel economy.

In addition to above *volatility affects short and long trip economy, acceleration and power, warm up, smoothness, hot stalling, carburettor icing, dilution, deposits and spark plug fouling* etc.

Thus volatility of fuel is a determining factor in selection of fuel for S.I. engines.

- **Ignition delay**, the *time period between start of injection and start of combustion has a great influence on correct optimisation of diesel engine.*
  - — If it is *too long* the rate of pressure rise, once it starts, can become so rapid that *severe diesel knock and engine damage can occur.*
  - — If it is *too short* then there is not sufficient time for complete mixing and *smoking can result.*

Ignition quality of fuel is the primary deciding factor for C.I. engines because delay period affects rate of pressure rise and hence knocking (and this delay period depends on the ignition quality of fuel). If a fuel is of good ignition quality, then its delay period will be less and hence chances of knocking will be lessened. The diesel engineer aims at using a *fuel of good ignition quality which means a fuel of high cetane number.*

Q. 22.67. *Distinguish clearly between 'Octane Number' and 'Cetane Number'. What are their significances in rating of fuels for S.I. and C.I. engines ?*

Ans. Octane Number :

- The **octane number rating of the fuel** *is the percentage, by volume, of iso-octane in a mixture of iso-octane and normal heptane, which exactly matches the knocking intensity of a given fuel, in a standard engine under given standard operating conditions.*
- It is an expression which indicates the *ability of a fuel to resist knock in S.I. engine.* The *higher the octane number rating of a fuel, the greater will be its resistance to knock, and the higher will be the compression ratio which may be used without knocking.* Since power output and specific fuel consumptions are functions of compression ratio, it can be said that these are also functions of octane number rating.

Cetane Number :

- It is defined as the *percent by volume of cetane in a mixture of cetane and alphamethyl-naphthalene that produces the same ignition lag as the fuel being tested, in the same engine and under the same operating conditions.*

- *Knock* in a C.I. engine is *due to sudden ignition and abnormally rapid combustion of accumulated fuel in the combustion chamber.* Such a situation occurs because of an ignition lag in combustion of fuel between time of injection and actual burning. The *property of ignition lag is measured in terms of cetane number.*
  — *Higher cetane number means a lower delay period and smoother engine operation.*
  — Lower the cetane number, *lower are starting times at ambient temperatures. Lower the cetane number, higher are the hydrocarbon emissions and noise levels.*

Q. 22.68. *What are the reference fuels for 'Octane Number' ?*

Ans. Reference fuels for 'Octane Number' are iso-octane and normal heptane.

Q. 22.69. *What are the reference fuels for 'Cetane Number' ?*

Ans. Reference fuels for 'Cetane Number' are cetane and α-methyl naphthalene.

Q. 22.70. *What is performance number (PN) ?*

Ans. Performance number is *indicative of the maximum power which may be obtained with that fuel, without knock, relative to maximum power which may be obtained using iso-octane, also without knock.* Iso-octane is assigned a PN of 100.

Q. 22.71. *What is the significance of ASTM distillation curve ?*

Ans.

- The method of measuring volatility of gasoline is standardized by the American Society of Testing Materials (ASTM), and the *graphical representation of the results of the tests is referred to as the* **ASTM distillation curve.** Since gasoline is a mixture of different hydrocarbons, volatility depends on the *fractional composition of the fuel.*

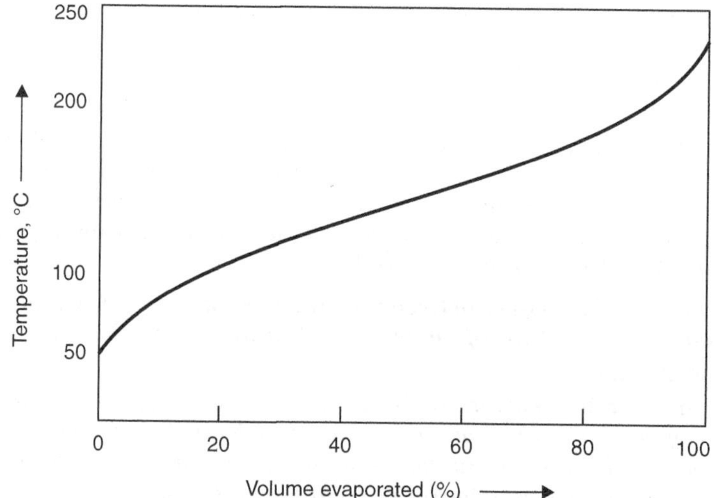

Fig. 4. Typical ASTM distillation curve for gasoline.

- The usual practice of measuring the fuel volatility is the distillation of fuel in a special device at atmospheric pressure and in the presence of its own vapour. The fraction that boils at a definite temperature is measured and plotted. The characteristic points are the temperature at which 10, 40, 50 and 90 per cent of the volume evaporates, as well as the temperature at which boiling of the fuel terminates. A typical ASTM distillation curve is plotted in Fig. 4.

Q. 22.72. ***Explain the effect of fuel viscosity on diesel engine performance.***

Ans. Fuel viscosity affects the atomisation and penetration of fuel into combustion chamber.

An increase in viscosity increases the mean droplet diameter which increases the delay period, and hence the knocking tendency. Higher fuel viscosity also reduces penetration. Hence the smoke free air-fuel ratio has to be kept high. This reduces the m.e.p. However, the lubrication of plunger-barrel assembly is dependent only on fuel viscosity, and on this account the diesel engine fuels are required to have some viscosity.

Q. 22.73. ***What qualities are desired in fuels to inhibit detonation ?***

Ans. S.I. engines :

The tendency of an engine to knock is very much affected by the properties of the fuel used. In general, *lower the self-ignition temperature of the fuel or greater its preflame reactivity, the greater the tendency to knock.*

- Octane number is the measure of the resistance to knock in S.I. engines. *Higher the octane number, lesser is the tendency to knock.*
- Fuels of paraffin series have the maximum, and of the aromatic series have the minimum tendency to knock. In aliphatic hydrocarbons, saturated compounds show, in general, lesser tendency to knock than the unsaturated hydrocarbons, with the exception of compounds like ethylene and acetylene.

  For *most hydrocarbons, a more compact molecular structure is associated with a lower tendency to detonate.*
- Thus the *best SI engine fuel is that having the highest octane number.*

C.I. engine :

- Knock in C.I. engine occurs because of a ignition lag in the combustion of fuel between the time of injection and the time of actual burning. As the ignition lag increases, the amount of fuel accumulated in the combustion chamber increases ; and when combustion actually takes place abnormal amount of energy is suddenly released causing an excessive rate of pressure rise which results in an audible knock. Hence *a good C.I. engine fuel should have a short ignition lag and will ignite more easily.*
- The present day rating of C.I. engine fuel is the cetane number, the *best fuel in general will have a cetane rating sufficiently high to avoid objectionable knock.*

NUMERICAL PROBLEMS WITH SOLUTIONS

Q. 22.1. *A diesel engine having compression ratio of 16 uses the fuel C_7H_{16}. The compression follows the law $pv^{1.4}$ = constant. If the engine uses 64% excess air and the temperature at the beginning of compression is 325 K, find the percentage of stroke at which combustion is completed.*

Assume $c_p = 1.0 + 21 \times 10^{-5}$ T kJ/kg K and T is in degrees' kelvin. C.V. = 44000 kJ/kg. Assume that air contains 23% by weight of O_2. **(MU)**

Solution. *Given :* $r = 16$; Excess air used = 64% ; T_1 = 325 K ; C.V. = 44000 kJ/kg.

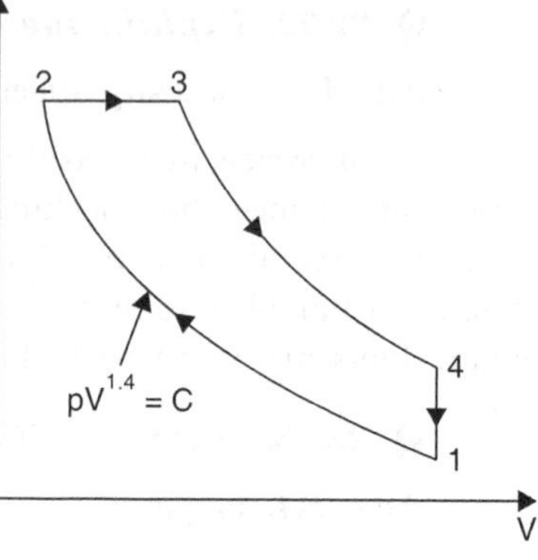

Fig. 5

Fig. 5 shows the cycle of operation. The minimum amount of air required to burn 1 kg of fuel can be calculated by using the following chemical equation :

$$C_7H_{16} + 11O_2 = 7CO_2 + 8H_2O$$
$$100 + 352 = 308 + 144$$

$\therefore$ 1 kg of fuel requires $\frac{352}{100}$ kg of O_2 and

$$\frac{352}{100} \times \frac{100}{23} \text{ kg of air} = 15.3 \text{ kg}$$

Actual air supplied = 15.3 × 1.64 = 25.092 kg/kg of fuel.

$$\% \text{ cut-off} = \frac{v_3 - v_2}{v_1 - v_2} = \frac{v_3 - v_2}{16v_2} = \frac{1}{16}\left(\frac{v_3}{v_2} - 1\right) \quad ...(i)$$

$$\left(\because \frac{v_1}{v_2} = r = 16\right)$$

From *compression process 1-2*, we have:

$$\frac{T_2}{T_1} = \left(\frac{v_1}{v_2}\right)^{\gamma - 1} = (16)^{1.4-1} \quad \text{or} \quad T_2 = 325 \times (16)^{0.4} = 985 \text{ K}$$

The heat supplied per kg of fuel = Heat given to the gases during the process 2-3.

$$\therefore \quad 44000 = \int_{T_2}^{T_3} \left(\frac{m_a}{m_f} + 1\right) c_p \, dT$$

$$44000 = \int_{T_2}^{T_3} (25.092 + 1)(1.0 + 21 \times 10^{-5} T) \, dT$$

or $$\frac{44000}{26.092} = \int_{985}^{T_3} (1.0 + 21 \times 10^{-5} \, T) \, dT$$

or $$1686.34 = (T_3 - 985) + \frac{21 \times 10^{-5}}{2} [T_3^2 - (985)^2]$$

or $$1686.34 = T_3 - 985 + 10.5 \times 10^{-5} \, T_3^2 - 101.87$$

or $$10.5 \times 10^{-5} \, T_3^2 + T_3 - 2762.7 = 0$$

or $$T_3 = \frac{-1 \pm \sqrt{1^2 + 4 \times 10.5 \times 10^{-5} \times 2762.7}}{2 \times 10.5 \times 10^{-5}} = \frac{-1 + 1.47}{2 \times 10.5 \times 10^{-5}} = 2238 \; K.$$

From *constant pressure process 2-3*, we have

$$\frac{v_2}{T_2} = \frac{v_3}{T_3} \quad \text{or} \quad \frac{v_3}{v_2} = \frac{T_3}{T_2} = \frac{2238}{985} = 2.272$$

Substituting this in eqn. (*i*), we have:

$$\therefore \quad \% \text{ cut-off} = \frac{1}{16}(2.272 - 1) = \mathbf{0.0795} \quad \textbf{or} \quad \mathbf{7.95\%.} \quad \textbf{(Ans.)}$$

Q. 22.2. *Calculate the diameter and length of the stroke of a diesel engine working on 4-stroke constant pressure cycle from the following data :*

Power developed (indicated)	*= 18.75 kW*
Speed of the engine	*= 200 r.p.m.*
Compression ratio	*= 14*
Fuel cut-off	$= \frac{1}{20}$ *th of the stroke*
Stroke-to-diameter ratio	*= 1.5*
Index of expansion	*= 1.3*
Index of compression	*= 1.35*

Assume the pressure and temperature of the air at the inlet as 1 bar and 40°C respectively.

(PTU)

Solution. *Given :* I.P. = 18.75 kW ; N = 200 r.p.m. ;

$$r = 14, \frac{v_3}{v_2} = \frac{v_s}{20}, \frac{L}{D} = 1.5 ;$$

$$p_1 = 1 \text{ bar} ; T_1 = 40 + 273 = 313 \text{ K}$$

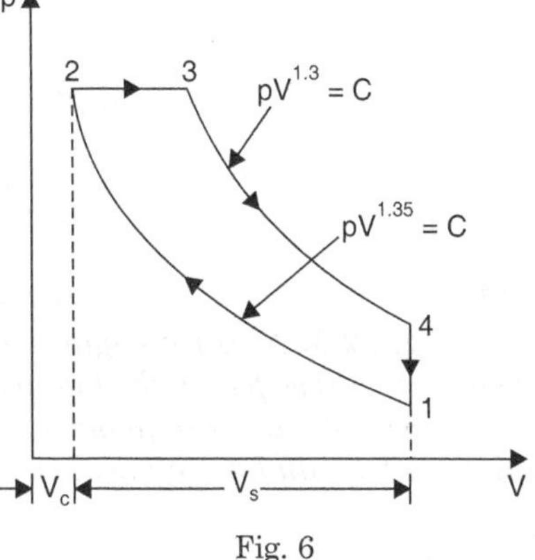

Fig. 6

Diameter and length of the stroke, D, L :

$$v_s = (r - 1)\, v_c$$

From the *compression process 1-2,* we have

$$p_1 v_1^{\,n} = p_2 v_2^{\,n}$$

or,
$$p_2 = p_1 \left(\frac{v_1}{v_2}\right)^n = 1 \times (r)^{1.35}$$

$$= 1 \times (14)^{1.35} = 35.26 \text{ bar}$$

$$p_2 = p_3 = 35.26 \text{ bar}$$

From *expansion process*, we have

$$p_3 v_3^{\,n} = p_4 v_4^{\,n}$$

$$p_4 = p_3 \left(\frac{v_3}{v_4}\right)^n = p_3 \left[\frac{v_c + \frac{v_s}{20}}{v_c + v_s}\right]^{1.3} = p_3 \left[\frac{v_c + \frac{1}{20}(r-1)v_c}{v_c + (r-1)v_c}\right]^{1.3}$$

$$= 35.26 \left[\frac{1 + \frac{1}{20}(14 - 1)}{1 + 13}\right]^{1.3} = 2.188 \text{ bar}$$

$$p_m = \frac{1}{v_s}\left[p_2(v_3 - v_2) + \frac{p_3 v_3 - p_4 v_4}{1.3 - 1} - \frac{p_2 v_2 - p_1 v_1}{1.35 - 1}\right]$$

$$v_1 = r v_c ; v_2 = v_c$$

$$v_3 = v_c + \frac{1}{20} v_s = v_c + \frac{1}{20}(r - 1)\, v_c = 1.65\, v_c ; v_4 = v_1 = r v_c = 14\, v_c$$

$$\therefore \quad p_m = \frac{1}{(r-1)\,v_c}\left[35.26\,(1.65\,v_c - v_c) + \frac{35.26 \times 1.65\,v_c - 2.188 \times 14 v_c}{1.3 - 1} - \frac{35.26\,v_c - 1 \times 14 v_c}{1.35 - 1}\right]$$

$$= \frac{1}{13}\left[35.26(1.65-1) + \frac{35.26 \times 1.65 - 2.188 \times 14}{0.3} - \frac{35.26 - 14}{0.35}\right]$$

$$= \frac{1}{13}(22.92 + 91.82 - 60.74) = 4.154 \text{ bar}$$

$$\text{I.P.} = \frac{p_m\,LANk \times 10}{6}, \text{ where } p_m \text{ is in bar}$$

$$= \frac{4.154 \times L \times \frac{\pi}{4} D^2 \times 200 \times \frac{1}{2} \times 10}{6}, \text{ where } L \text{ and } D \text{ are in metres}$$

$$18.75 = \frac{4.154 \times 1.5D \times \frac{\pi}{4} D^2 \times 200 \times \frac{1}{2} \times 10}{6}$$

$$D = \left(\frac{18.75 \times 6 \times 4 \times 2}{4.154 \times 1.5 \times \pi \times 200 \times 10}\right)^{1/3} = \mathbf{0.284\ m\ or\ 28.4\ cm.} \quad \textbf{(Ans.)}$$

and, $\quad L = 1.5D = 1.5 \times 0.284 = \mathbf{0.426\ m \quad or \quad 42.6\ cm.} \quad \textbf{(Ans.)}$

Q. 22.3. *A petrol engine uses a fuel of calorific value 42000 kJ/kg. The compression and expansion curves follow the law $pV^{1.3}$ = constant. At 25% and 75% of the stroke on the compression curve, the pressures are found to be 2 bar and 5.2 bar respectively. If the relative efficiency of the engine is 50% and mechanical efficiency is 75%, find the specific fuel consumption on B.P. basis.*

(PTU)

Solution. *Given :* C.V. = 42000 kJ/kg ; $\eta_{relative}$ = 50%, $\eta_{mech.}$ = 75%.

$pV^{1.3}$ = constant ; p_a = 2 bar ;

p_b = 5.2 bar

$$p_a V_a^{1.3} = p_b V_b^{1.}$$

$$V_a = V_c + 0.75\,V_s$$

$$V_b = V_c + 0.25\,V_s$$

$$V_s = (r-1)\,V_c$$

$$\therefore \quad p_a\,(V_c + 0.75\,v_s)^{1.3} = p_b\,(V_c + 0.25\,V_s)^{1.3}$$

or, $$p_a\,[V_c + 0.75\,(r-1)V_c]^{1.3} = p_b\,[V_c + 0.25\,(r-1)V_c]^{1.3}$$

$$\therefore \quad \left[\frac{1 + 0.75(r-1)}{1 + 0.25(r-1)}\right]^{1.3} = \frac{p_b}{p_a}$$

$$\frac{1 + 0.75(r-1)}{1 + 0.25(r-1)} = \left(\frac{p_b}{p_a}\right)^{1/1.3}$$

$$= \left(\frac{5.2}{2}\right)^{0.7692} = 2.0854$$

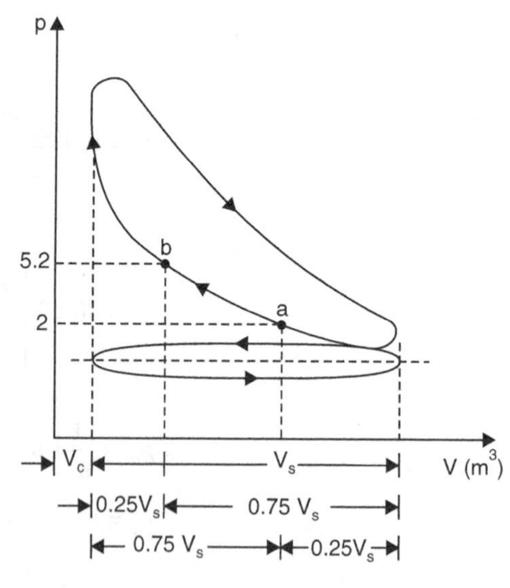

Fig. 7

$$\frac{1 + 0.75r - 0.75}{1 + 0.25r - 0.25} = 2.0854$$

$$\frac{0.25 + 0.75r}{0.75 + 0.25r} = 2.084$$

or, $$0.25 + 0.75r = 2.084(0.75 + 0.25\ r) = 1.563 + 0.521r$$

$\therefore$ $$r = \frac{(1.563 - 0.25)}{(0.75 - 0.521)} = 5.734$$

$$\eta_{air\text{-}standard} = 1 - \frac{1}{(r)^{\gamma-1}} = 1 - \frac{1}{(5.734)^{1.4-1}} = 0.5027 \text{ or } 50.27\%$$

$$\eta_{relative} = \frac{\eta_{th(I)}}{\eta_{air\text{-}standard}} \quad \text{or} \quad \eta_{th(I)} = 0.5 \times 0.5027 = 0.2513$$

$$\eta_{th(B)} = \eta_{th(I)} \times \eta_{mech.} = 0.2513 \times 0.75 = 0.1885 \text{ or } 18.85\%$$

Now, $$\eta_{th(B)} = \frac{\text{B.P.}}{\dot{m}_f \times \text{C.V}} = \frac{3600}{b.s.f.c. \times \text{C.V.}} \text{ kg/kWh}$$

$\therefore$ $$\textbf{b.s.f.c.} = \frac{3600}{\eta_{th(B)} \times \text{C.V}} = \frac{3600}{0.1885 \times 42000} = \mathbf{0.4547\ kg/kWh.} \quad \textbf{(Ans.)}$$

Q. 22.4. *A diesel engine contains 0.1 m³ of air at 0.98 bar and 30°C at the beginning of compression. The compression ratio is 15 and the volume at cut-off is 0.0125 m³. Determine for the corresponding air standard cycle :*

(i) The cut-off ratio ; *(ii) The per cent clearance ;*

(iii) The work done ; *(iv) The air standard efficiency.*

Take c_p = 1.005 kJ/kg K and γ = 1.4 **(DU)**

Solution. Fig. 8 shows the diesel cycle on *p-V* diagram.

Given : $V_1 = 0.1\ \text{m}^3$;

$T_1 = 30 + 273 = 303$ K;

$p_1 = 0.98$ bar.

Compression ratio, $$r = \frac{V_1}{V_2} = 15$$

$\therefore$ Clearance volume, $$V_2 = \frac{V_1}{15} = \frac{0.1}{15} = 0.006667\ \text{m}^3$$

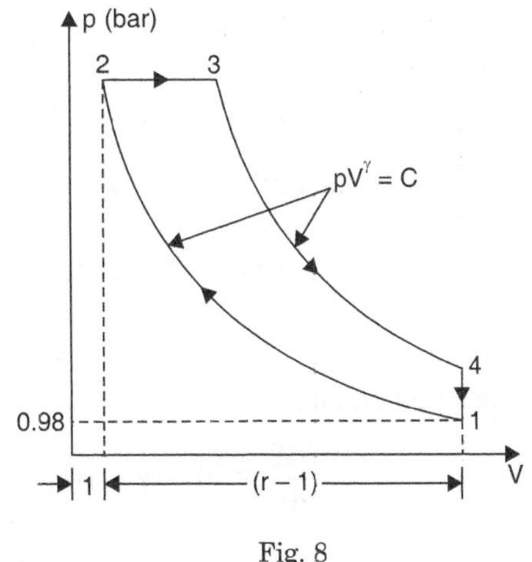

Fig. 8

(*i*) **Cut-off ratio,** $$\rho = \frac{V_3}{V_2} = \frac{0.0125}{0.006667} = \mathbf{1.875.} \quad \textbf{(Ans.)}$$

(*ii*) **The per cent clearance** $= \dfrac{1}{15} = 0.06667$ or **6.667%. (Ans.)**

(*iii*) **The work done, W :**

$$T_2 = T_1 \times (r)^{\gamma-1} = 303 \times (15)^{1.4-1} = 895.1 \text{ K}$$

$$T_3 = \rho \times T_2 = 1.875 \times 895.1 = 1678.3 \text{ K}$$

$$T_4 = T_3 \times \left(\frac{\rho}{r}\right)^{\gamma-1} = 1678.3 \times \left(\frac{1.875}{15}\right)^{1.4-1} = 730.5 \text{ K}$$

Heat supplied, $Q_{2-3} = c_p(T_3 - T_2) = 1.005(1678.3 - 895.1) = 787.1$ kJ/kg

Heat rejected, $Q_{4-1} = c_v\,(T_4 - T_1)$

$= 0.718\,(730.5 - 303) = 306.9$ kJ/kg

$$\left(\because \; c_v = \frac{c_p}{\gamma} = \frac{1.005}{1.4} = 0.718 \text{ kJ / kg K}\right)$$

$\therefore$ Work done, $W = Q_{2-3} - Q_{4-1} = 787.1 - 306.9 = 480.2$ kJ/kg

(*iv*) **The air standard efficiency,** $\eta_{\text{air-standard}}$ **:**

$$\therefore \quad \eta_{air\text{-}standard} = \frac{W}{Q_{2-3}} = \frac{480.2}{787.1} = \mathbf{0.61 \text{ or } 61\%.} \quad \textbf{(Ans.)}$$

Check :
$$\eta_{diesel} = 1 - \frac{\rho^\gamma - 1}{\gamma(r)^{\gamma-1}(\rho - 1)} = 1 - \frac{(1.875)^{1.4} - 1}{1.4 \times (15)^{1.4-1} \times (1.875 - 1)} = 0.61 \text{ or } 61\%$$

Q. 22.5. *In an engine working on the Otto cycle between given lower and upper limits of absolute temperature, T_1 and T_2 respectively, show that for maximum work to be done per kg, the ratio of compression is given by :*

$$r = \left(\frac{T_3}{T_1}\right)^{1.25}$$

where, γ *= Ratio of specific heats = 1.4.* **(NU)**

Solution. Fig. 9, shows the cycle for the Otto cycle.

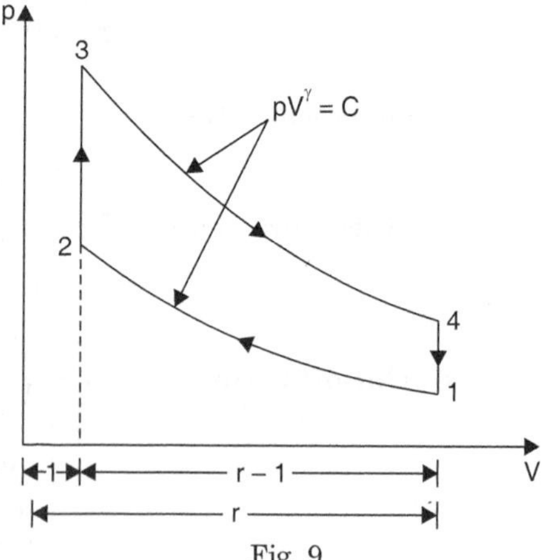

Fig. 9

$$T_2 = T_1 \times (r)^{\gamma-1} \; ; \; T_4 = T_3 \times \left(\frac{1}{r}\right)^{\gamma-1}$$

Work done, $W = c_v(T_3 - T_2) - c_v(T_4 - T_1)$

$$= c_v(T_3 - T_1 \times r^{\gamma-1}) - c_v \left\{T_3\left(\frac{1}{r}\right)^{\gamma-1} - T_1\right\}$$

For maximum work, *W* is differentiated with the variable *r* and equated to 0.

i.e.
$$\frac{dW}{dr} = [0 - (\gamma - 1)T_1(r)^{\gamma-2}] - [-T_3 \times (\gamma - 1)r^{-\gamma} - 0] = 0$$

or, $(\gamma - 1)\, T_1\,(r)^{\gamma-2} = (\gamma - 1)\, T_3\, r^{-\gamma}$

or $$\frac{T_3}{T_1} = \frac{(r)^{\gamma-2}}{(r)^{-\gamma}} = (r)^{2(\gamma-1)}$$

$\therefore$ $$r = \left(\frac{T_3}{T_1}\right)^{\frac{1}{2(\gamma-1)}} = \left(\frac{T_3}{T_1}\right)^{\frac{1}{2(1.4-1)}} = \left(\frac{T_3}{T_1}\right)^{1.25}$$ ...**Proved.**

Q. 22.6. *In a hypothetical air cycle, consisting of three processes, an adiabatic compression is followed by an isothermal expansion to the initial volume of compression. Finally a heat rejection process completes the cycle.*

(i) Draw the cycle on p-v and T-s diagrams and derive an expression for thermal efficiency of the cycle in terms of compression ratio r.

(ii) Also find the efficiency and m.e.p. of the cycle if the compression ratio is 14 and the induction conditions are 1 bar and 27°C. Take for air, c_p = 1.005 kJ/kg K and c_v = 0.718 kJ/kg K. **(PU)**

Solution. *Given :* $r = 14$; $p_1 = 1$ bar ; $T_1 = 27 + 273 = 300$ K ; $c_p = 1.005$ kJ/kg K ; $c_v = 0.718$ kJ/kg K.

(*i*) The *p-v* and *T-s* diagrams of the cycle are shown in Fig. 10

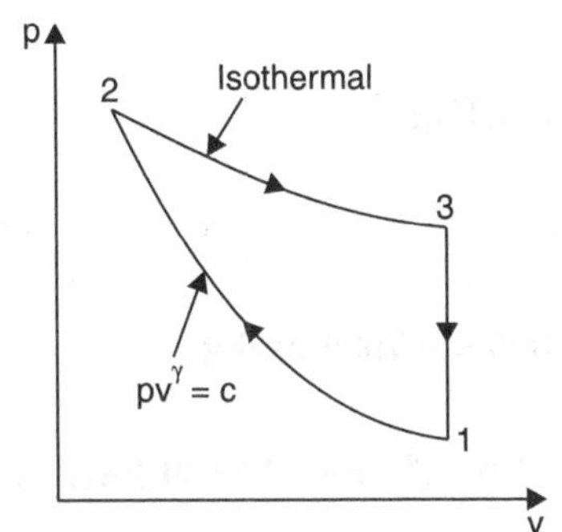

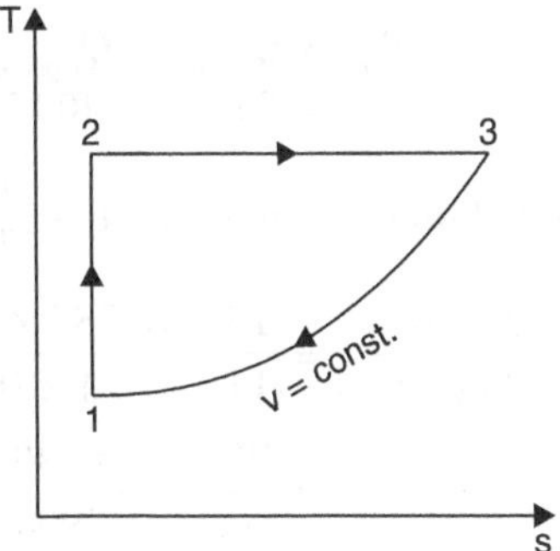

Fig. 10

Consider 1 kg of air.

Consider *adiabatic process 1-2 :*

$$\frac{T_2}{T_1} = \left(\frac{v_1}{v_2}\right)^{\gamma-1} \quad (r)^{\gamma-1} \quad \therefore \quad T_2 = T_1 (r)^{\gamma-1} \; ; \; Q_{1-2} = 0$$

Consider *isothermal process 2-3 :*

$$T_2 = T_3$$
$$Q_{2\text{-}3} = W = RT_2 \ln(r)$$

Consider *heat-rejection (constant volume) process 3-1 :*

$$Q_{3-1} = c_v (T_3 - T_1)$$

Efficiency, $$\eta = \frac{\text{Work done}}{\text{Heat added}} = \frac{Q_{2-3} - Q_{3-1}}{Q_{2-3}} = 1 - \frac{Q_{3-1}}{Q_{2-3}}$$

$$= 1 - \frac{c_v(T_3 - T_1)}{RT_2 \ln(r)} = 1 - \frac{c_v[T_1(r)^{\gamma-1} - T_1]}{RT_1(r)^{\gamma-1} \ln(r)}$$

$$= \mathbf{1 - \frac{c_v}{R\ln(r)}\left[1 - \frac{1}{(r)^{\gamma-1}}\right]} \quad \textbf{(Ans.)}$$

(ii) $Q_{added} = Q_{2-3} = RT_2 \ln(r)$

where, $R = c_p - c_v = 1.005 - 0.718 = 0.287$ kJ/kg K,

$$T_2 = T_1(r)^{\gamma-1} = 300 \times (14)^{1.4-1} = 300 \times 2.874 = 862 \text{ K} \quad \left(\because \gamma = \frac{c_1}{c_v} = \frac{1.005}{0.718} = 1.4 \right)$$

$\therefore$ $Q_{added} = 0.287 \times 862 \times \ln(14) = 652.88$ kJ/kg

$Q_{rejected} = c_v (T_3 - T_1) = 0.718(862 - 300) = 403.52$ kJ/kg

$$\eta = 1 - \frac{Q_{rejected}}{Q_{added}} = 1 - \frac{403.52}{652.88} = \mathbf{0.3819 \quad or \quad 38.19\% \quad (Ans.)}$$

$$\left[\text{Alternatively, } \eta = 1 - \frac{0.718}{0.287 \times \ln(14)} \left\{ 1 - \frac{1}{(14)^{1.4-1}} \right\} \right.$$

$$\left. = 1 - 0.9479 \times 0.652 = 0.3819 \text{ or } 38.19\% \right]$$

$W_{net} = Q_{added} - Q_{rejected}$

$= 652.88 - 403.52 = 249.36$ kJ/kg

Swept volume, $v_s = v_1 - v_2 = v_1 - \dfrac{v_1}{14} = \dfrac{13}{14} v_1$

or, $$\frac{13}{14}\left(\frac{RT_1}{p_1}\right) = \frac{13}{14} \times \frac{(0.287 \times 1000)}{1 \times 10^5} \times 300 = 0.7995 \text{ m}^3/\text{kg}$$

$\therefore$ $$\textbf{m.e.p.} = \frac{W_{net}}{v_s} = \frac{249.36}{0.7995} = \mathbf{311.89 \text{ kN/m}^2 \quad or \quad 3.1189 \text{ bar.} \quad (Ans.)}$$

Q. 22.7. *A single-cylinder, single-acting, 4-stroke gas engine develops 18.5 kW B.P. at 280 r.p.m. The atmospheric pressure and temperature are 1.036 bar and 20°C respectively. The diameter and stroke of the cylinder are 320 mm and 380 mm respectively. Volumetric efficiency = 71% at atmospheric condition. Calorific value of fuel = 18700 kJ/m³ at 15°C and 1.033 bar. A/F ratio by volume = 7.5 : 1. Broke thermal efficiency = 28.5%. Calculate :*

(i) The missing cycles per minute.

(ii) The air consumption in kg per minute assuming the volume of air during missed cycle is equal to the volume of mixture (gas and air) during firing cycle.

(iii) The gas consumption in kg/min.

Take R (for air and gas) = 287 Nm/kg K. **(MU)**

Solution. *Given :* $k = \dfrac{1}{2}$; B.P. = 18.5 kW ; N = 280 r.p.m. ; D = 320 mm = 0.32 m ; L = 380 mm = 0.38 m ; $\eta_{vol} = 0.71$; C.V. = 18700 kJ/m³ ; A/F ratio = 7.5 : 1 ; $\eta_{th(B)} = 28.5\%$; $R = 287$ J/kg K.

(i) **The missing cycles per minute :**

Stroke volume, $$V_s = \frac{\pi}{4} D^2 \times L = \frac{\pi}{4} \times 0.32^2 \times 0.38 = 0.03056 \text{ m}^3$$

Actual volume taken in $= V_s \times \eta_{vol} = 0.03056 \times 0.71$

$= 0.0217$ m³ at 1.036 bar and 27°C

The gas C.V. is given at 1.033 bar and 15°C, therefore, the actual volume of taken in at 1.033 bar and 15°C must be calculated.

∴ Actual volume inhaled at 1.033 bar and 15°C

$$= \frac{1.036}{1.033} \times \frac{288}{293} \times 0.0217 = 0.0214 \text{ m}^3$$

2.14 m^3 contains the air and gas in proportion of 7.5 : 1

∴ The gas taken in per working stroke $= \frac{0.0214}{8.5} = 0.002518 \text{ m}^3$

Heat supplied for each working stroke (firing stroke)

= Volume of gas × C.V. of gas

= 0.002518 × 18700 = 47.1 kJ

B.P. = Heat supplied per cycle × number of working cycles/sec. × $\eta_{th(B)}$

18.5 = 47.1 × number of working cycle/sec. × 0.285

∴ Number of working cycles/sec. $= \frac{18.5}{47.1 \times 0.285} = 1.38$/s or 83/min.

∴ **Missed cycles** $= \frac{N}{2} - 83 = \frac{280}{2} - 83 =$ **57. (Ans.)**

(*ii*) **Air consumption :**

Air consumption per minute

= Air during working (firing) cycles + air during non-firing cycles

$$= 0.0214 \times \frac{7.5}{8.5} \times 83 + 0.0214 \times 57 = 2.787 \text{ m}^3\text{/min}$$

Mass of this air is given by,

$$\mathbf{m_a} = \frac{pV}{RT} = \frac{1.033 \times 10^5 \times 2.787}{287 \times 288} = \mathbf{3.483 \text{ kg/min.}} \quad \textbf{(Ans.)}$$

(*iii*) **Gas consumption per minute :**

Gas consumption per minute = Gas used during firing cycles per minute

= 0.002518 × 83 = 0.209 m^3 at 1.033 bar and 15°C

Mass of this gas is also given by,

$$m_g = \frac{pV}{RT} = \frac{1.033 \times 10^5 \times 0.209}{287 \times 288} = \mathbf{0.2612 \text{ kg/min.}} \quad \textbf{(Ans.)}$$

Q. 22.8. *A 6-cylinder, 4-stroke direct injection oil engine has bore 140 mm and stroke 210 mm and it runs at 1600 r.p.m. It consumes 31 kg of fuel per hour. The calorific value of fuel is 42500 kJ/kg and its percentage composition by mass is carbon 86.2, hydrogen 13.5 and non-combustible 0.3. The percentage volumetric composition of dry exhaust gases is* CO_2 *= 7.0, CO = 0.1,* O_2 *= 11.2 and* N_2 *= 81.8. Barometric pressure is 755 mm Hg, room temperature 30°C and moisture in air is 0.02 kg/kg of air. If the indicated thermal efficiency and mechanical efficiency of the engine are 38% and 80% respectively, determine :*

(*i*) *The volumetric efficiency of the engine under rated condition ;*

(*ii*) *The rated output of the engine ;*

(*iii*) *The brake specific fuel consumption.* **(UPTU)**

Solution. *Given :* $n = 6$; D = 140 mm = 0.14 m ; L = 210 mm = 0.21 m ;

N = 1600 r.p.m ; fuel consumption = 31 kg/hour ; *C.V.* = 42500 kJ/kg ; C = 86.2% ;

H_2 = 13.5% ; non-combustible = 0.3% ; CO_2 = 7.0% ; CO = 0.1% ; O_2 = 11.2% ;

N_2 = 81.8% ; barometric pressure = 755 mm Hg, room temp., T = 30 + 273 = 303 K ;

moisture in air, m_{wv} = 0.02 kg/kg of air ; $\eta_{th(I)}$ = 38% ; $\eta_{mech.}$ = 80%

(*i*) **The volumetric efficiency, $\eta_{vol.}$:**

Fuel-air ratio, $$\frac{F}{A} = \frac{0.33}{C}\left[\frac{CO_2 + CO}{N_2}\right] = \frac{0.33}{0.862}\left[\frac{7 + 0.1}{81.8}\right] = 0.03323$$

or, $$\frac{A}{F} = 30 : 1$$

Now, $$\frac{F}{A} = \frac{31}{3600} \times \frac{1}{m_a} = 0.03323$$

or, $$m_a = \frac{31}{3600 \times 0.03323} = 0.259 \text{ kg/s}$$

Now, $$\frac{p_a}{p_i} = \frac{1}{1 + m_{wv} \times \frac{29}{18}} = \frac{1}{1 + 0.02 \times \frac{29}{18}} = 0.967$$

$$\rho_a = \frac{p_i}{RT} \times \frac{p_a}{p_i}$$

$$= \left(\frac{755}{760} \times 1.01325 \times 10^5\right) \times \frac{1}{287 \times 303} \times 0.967 = 1.119 \text{ kg/m}^3$$

$$\eta_{vol.} = \frac{\text{Mass of air taken in}}{\text{Mass equivalent to swept volume}}$$

$$= \frac{0.259}{6 \times \frac{\pi}{4} \times 0.14^2 \times 0.21 \times \frac{1600}{2 \times 60} \times 1.119} = \mathbf{0.895 \text{ or } 89.5\%} \quad \textbf{(Ans.)}$$

(*ii*) **The rated output of the engine, B.P. :**

$$\textbf{BP} = (m_f \times C.V.) \times (\eta_{th(I)} \times \eta_{mech.})$$
$$= (0.259 \times 0.03323) \times 42500 \times (0.38 \times 0.8) = \mathbf{111.2 \text{ kW.}} \quad \textbf{(Ans.)}$$

(*iii*) **The brake specific fuel consumption, b.s.f.c :**

$$\text{bsfc} = \frac{m_f / hour}{BP\,(\text{kg})} = \frac{31}{111.2} = \mathbf{0.279 \text{ kg/kWh.}} \quad \textbf{(Ans.)}$$

Q. 22.9. *A four-stroke, eight-cylinder engine of 9 cm bore and 8 cm stroke with a compression ratio 7 is tested at 4500 r.p.m. on a dynamometer which has 54 cm arm. During a 10 minutes test the dynamometer scale beam reading was 412.02 N and the engine consumed 4.4 kg of gasoline having a calorific value of 44000 kJ/kg. Air at 27°C and 1 bar was supplied to the carburettor at the rate of 6 kg/min. Calculate :*

(*i*) *The brake power developed ;* (*ii*) *bmep ;*
(*iii*) *bsfc ;* (*iv*) *Brake specific air consumption ;*
(*v*) *Brake thermal efficiency ;* (*vi*) *Volumetric efficiency ;*
(*vii*) *Air-fuel ratio.* **(Madras University)**

Solution. *Given :* $n = 8$; $D = 9$ cm $= 0.09$ m ; $L = 8$ cm $= 0.08$ m ; $r = 7$; $N = 4500$ r.p.m. ; dynamometer arm length = 54 cm = 0.54 m ; dynamometer scale beam reading = 412.02 N ; gasoline consumed in 10 minutes = 4.4 kg ; C.V. = 44000 kJ/kg ; temp. and pressure of air supplied = 27°C, 1 bar ; mass of air supplied to the carburettor = 6 kg/min.

(*i*) **The brake power developed BP :**

$$\text{BP} = \frac{2\pi NT}{60 \times 1000} = 2\pi \times \frac{4500}{60} \times \left(\frac{0.54 \times 412.02}{1000}\right) = \mathbf{104.85 \text{ kW.}} \quad \textbf{(Ans.)}$$

(*ii*) **b.m.e.p. :**

$$\text{B.P.} = \frac{np_{mb}LANk \times 10}{6}, \text{ where } p_{mb} \text{ is in bar}$$

or $$104.85 = \frac{8 \times p_{mb} \times 0.08 \times (\pi / 4) \times 0.09^2 \times 4500 \times \frac{1}{2} \times 10}{6} = 15.27\, p_{mb}$$

$\therefore$ $\mathbf{p_{mb} = 6.87}$ **bar. (Ans.)**

(*iii*) **b.s.f.c. :**

$$\text{b.s.f.c.} = \frac{(4.4/10) \times 60}{104.85} = \mathbf{0.2518 \ kg/kWh.} \quad \textbf{(Ans.)}$$

(*iv*) **Brake specific air consumption :**

Brake specific air consumption

$$= \frac{6 \times 60}{104.85} = \mathbf{3.4335 \ kg/kWh.} \quad \textbf{(Ans.)}$$

(*v*) **Brake thermal efficiency,** $\eta_{th(B)}$ **:**

$$\eta_{th(B)} = \frac{B.P.}{\dot{m}_f \times C.V} = \frac{104.85}{\frac{4.4}{(10 \times 60)} \times 44000} = \mathbf{0.3249} \quad \textbf{or} \quad \mathbf{32.49\%} \quad \textbf{(Ans.)}$$

(*vi*) **Volumetric efficiency,** $\eta_{vol.}$ **:**

Air consumption = 6 kg/min

Now, $$pv = mRT$$

$\therefore$ $$v = \frac{mRT}{p} = \frac{6 \times 287 \times (27 + 273)}{1 \times 10^5} = 5.166 \text{ m}^3/\text{min}$$

$$\text{Displacement volume} = n \times \frac{\pi}{4} \times D^2 \times L \times \frac{N}{2}$$

$$= 8 \times \frac{\pi}{4} \times (0.09)^2 \times 0.08 \times \frac{4500}{2} = 9.161 \text{ m}^3/\text{min}$$

$\therefore$ $$\eta_{vol} = \frac{\text{Air consumption}}{\text{Displacement volume}} = \frac{5.166}{9.161} = \mathbf{0.5639} \quad \textbf{or} \quad \mathbf{56.39\%} \quad \textbf{(Ans.)}$$

(*vii*) **Air-fuel ratio :**

$$\text{Air-fuel ratio} = \frac{\text{Brake specific air consumption}}{\text{b.s.f.c.}} = \frac{3.4335}{0.2518} = \mathbf{13.636.} \quad \textbf{(Ans.)}$$

Q. 22.10. *An automobile has three-litre S.I. V-6 engine which operates on a four-stroke cycle at 3600 r.p.m. The compression ratio is 9.5. The engine is square (i.e. bore = stroke). During a test, it is connected to a dynamometer which gives a brake output torque reading of 205 Nm at 3600 r.p.m. The air enters at 85 kPa and 60°C. The mechanical efficiency of engine is 85%. Calculate :*

(*i*) *Cylinder bore and stroke length ;*

(*ii*) *Clearance volume of one cylinder ;*

(*iii*) *BP and IP ;*

(*iv*) *Brake mean effective pressure.* **(DU)**

Solution. *Given :* $n = 6$; Displacement = 3 litres = 3000 cm³ ; $r = 9.5$; $L = D$;

$T = 205$ Nm ; $N = 3600$ r.p.m. ; $\eta_{mech.} = 85\%$.

(*i*) **Cylinder bore and length (D, L) :**

$$6 \times \frac{\pi}{4} D^2 \times L = 3000$$

or, $$6 \times \frac{\pi}{4} \times D^2 \times D = 3000 \quad \therefore \quad \mathbf{D = L = 8.6\ cm.} \quad \textbf{(Ans.)}$$

(*ii*) **Clearance volume of one cylinder, V_c :**

Swept volume per cylinder, $V_s = \frac{3000}{6} = 500\ cm^3$

$$r = \frac{V_s + V_c}{V_c} \quad \text{or} \quad 9.5 = \frac{500 + V_c}{V_c} = \frac{500}{V_c} + 1$$

$$\mathbf{V_c = 58.82\ cm^3.} \quad \textbf{(Ans.)}$$

(*iii*) **BP and IP :**

$$\mathbf{BP} = \frac{2\pi NT}{60 \times 1000} = \frac{2\pi \times 3600 \times 205}{60 \times 1000} = \mathbf{77.28\ kW.} \quad \textbf{(Ans.)}$$

$$\mathbf{IP} = \frac{BP}{\eta_{mech.}} = \frac{77.28}{0.85} = \mathbf{90.92\ kW.} \quad \textbf{(Ans.)}$$

(*iv*) **Brake mean effective pressure, p_{mb} :**

$$\text{BP} = \frac{np_{mb}\ LANk \times 10}{6}$$

or $$77.28 = \frac{6 \times p_{mb} \times (8.6/100) \times \frac{\pi}{4}\left(\frac{8.6}{100}\right)^2 \times 3600 \times \frac{1}{2} \times 10}{6} = 8.992\ p_{mb}$$

$\therefore$ $$\mathbf{p_{mb} = 8.59\ bar.} \quad \textbf{(Ans.)}$$

Q. 22.11. *A 9-cylinder, 4-stroke S.I. engine of bore 144 mm and stroke 187.5 mm has a compression ratio of 6.8 and develops brake power of 450 kW at 2000 r.p.m. The mixture supplied is 20% rich. The fuel lower C.V. is 44000 kJ/kg and it contains 86% carbon and 14% H_2. Assuming the volumetric efficiency of 75% at 15°C and 1.013 bar and mechanical efficiency of 90%, determine indicated thermal efficiency of the engine. With what standard performance would you compare this efficiency ?*

Air contains 23% O_2 by mass. **(Roorkee University)**

Solution. *Given :* $n = 9$; $k = \frac{1}{2}$; $D = 144$ mm $= 0.144$ m ; $L = 187.5$ mm $= 0.1875$ m ; $r = 6.8$; B.P. = 450 kW ; $N = 2000$ r.p.m. ; $C.V. = 44000$ kJ/kg ; $C = 86\%$; $H_2 = 14\%$; $\eta_{vol.} = 75\%$ at 15°C ; $\eta_{mech} = 90\%$

Indicated thermal efficiency $\eta_{th(I)}$:

Swept volume of the engine per minute,

$$V = n \times \frac{\pi}{4} D^2 \times L \times \eta_{vol.} \times \frac{N}{2}$$

$$= 9 \times \frac{\pi}{4} \times 0.144^2 \times 0.1875 \times 0.75 \times \frac{2000}{2} = 20.6\ m^3$$

Mass of this mixture at 15°C and 1.013 bar is given by

$$m = \frac{pV}{RT} = \frac{1.013 \times 10^5 \times 20.6}{287 \times (15 + 273)} = 25.25 \text{ kg (air + fuel) per min.}$$

Mass of air per kg of fuel for chemically correct mixture

$$= \frac{100}{23}\left[\frac{32}{12}C + 8H\right] = \frac{100}{23}\left[\frac{32}{12} \times 0.86 + 8 \times 0.14\right] = 14.84 \text{ kg}$$

Therefore, a mixture of air and fuel contains 1.2 kg fuel and 14.84 kg air [= 16.04 kg of mixture (since mixture is 20% rich)].

$\therefore$ Maximum mass of fuel supplied $= \frac{25.25}{16.04} = 1.574$ kg/min.

Now, $$\text{I.P.} = \frac{\text{B.P.}}{\eta_{mech}} = \frac{450}{0.90} = 500 \text{ kW}$$

$$\therefore \quad \eta_{th(I)} = \frac{\text{I.P.}}{\dot{m}_f \times \text{C.V.}} = \frac{500}{\frac{1.574}{60} \times 44000} = \mathbf{0.433 \quad or \quad 43.3\% \quad (Ans.)}$$

The performance of the engine can be compared with air standard efficiency of engine-cycle,

$$\eta_{air\text{-}standard} = 1 - \frac{1}{(r)^{\gamma-1}} = 1 - \frac{1}{(6.8)^{1.4-1}} = \mathbf{0.535 \quad or \quad 53.5\%.}$$

Q. 22.12. *A four-cylinder, four-stroke petrol engine of cylinder bore and stroke each equal to 77 mm has a compression ratio of 8.5 : 1. The relative efficiency is 50% when s.f.c. on I.P. is 0.28 kg/kWh. Determine :*

(i) The C.V. of the petrol in MJ/kg ;

(ii) The petrol consumption in kg/h.

Given that the i.m.e.p. is 950 kPa when the engine speed is 3000 r.p.m. Take γ for air = 1.4.

(PU)

Solution. *Given :* $N = 4$; $k = \frac{1}{2}$; $D = L = 77$ mm $= 0.077$ m ; $r = 8.5 : 1$

$\eta_{relative} = 50\%$; i.s.f.c. $= 0.28$ kg/kWh ; $p_{mi} = 950$ *kPa* $= 9.5$ bar ;

$N = 3000$ r.p.m. ; $\gamma = 1.4$

(*i*) **The C.V. of the petrol in MJ/kg :**

Air-standard efficiency of otto cycle,

$$\eta_{air\text{-}standard} = 1 - \frac{1}{(r)^{\gamma-1}} = 1 - \frac{1}{(8.5)^{1.4-1}} = 0.575$$

Also, $$\eta_{relative} = \frac{\eta_{th(I)}}{\eta_{air\text{-}standard}} \quad \text{or} \quad \eta_{th(I)} = 0.5 \times 0.575 = 0.2875$$

Now, $$\eta_{th(I)} = \frac{I.P.}{\dot{m}_f \times C.V.} = \frac{3600}{\frac{\dot{m}_f}{I.P.} \times C.V.} = \frac{3600}{0.28 \times C.V.}$$

$$\therefore \quad \textbf{C.V.} = \frac{3600}{0.28 \times 0.2875} = \mathbf{44720 \text{ kJ/kg} \quad or \quad 44.72 \text{ MJ/kg} \quad (Ans.)}$$

(*ii*) **The petrol consumption in kg/h :**

$$\text{IP} = \frac{n\, p_{mi}\, LANk \times 10}{6}$$

$$= \frac{4 \times 9.5 \times 0.077 \times \frac{\pi}{4}(0.077)^2 \times 3000 \times \frac{1}{2} \times 10}{6} = 34.06 \text{ kW}$$

$\therefore$ **Petrol consumption** = 0.28 × 34.06 = **9.537 kg/h. (Ans.)**

Q. 22.13. *In a trial of 4-stroke single-cylinder petrol engine of 150 mm diameter, 200 mm stroke, the net load of the dynamometer is 294.3 N at 450 mm radius at 2200 r.p.m. The fuel supply to the engine was at the rate of 0.16 kg/min with calorific value of 43.5 MJ/kg. The supply was stopped and the engine was rotated with motor which needed 4 kW to keep it running at the same speed. Calculate BP, IP, mechanical efficiency, brake thermal efficiency, bmep and imep.* **(PTU)**

Solution. *Given :* D = 150 mm = 0.15 m ; L = 200 mm = 0.2 m ; $(W - S)$ = 294.3 N ; R = 450 mm = 0.45 m ; N = 2200 r.p.m ; m_f = 0.16 kg/min ; C = 43.5 MJ/kg ; motoring power = 4 kW.

Brake power,
$$\text{B.P.} = \frac{(W - S)\pi DN}{60 \times 1000} = \frac{294.3 \times \pi \times (0.45 \times 2) \times 2200}{60 \times 1000}$$

= **30.51 kW. (Ans.)**

Indicated power, I.P. = B.P. + motoring power

= 30.51 + 4 = **34.51 kW. (Ans.)**

$$\eta_{\text{mech.}} = \frac{\text{B.P.}}{\text{I.P.}} = \frac{30.51}{34.5} = \mathbf{0.884} \quad \textbf{or} \quad \mathbf{88.4\%.} \quad \textbf{(Ans.)}$$

$$\eta_{th.(B)} = \frac{\text{B.P.}}{\dot{m}_f \times C} = \frac{30.51}{\left(\frac{0.16}{60}\right) \times (43.5 \times 10^3)} = \mathbf{0.263} \quad \textbf{or} \quad \mathbf{26.3\%} \quad \textbf{(Ans.)}$$

$$\text{BP} = p_{mb} \times L \times A \times \frac{N}{2 \times 60}$$

$$30.51 = p_{mb} \times 0.2 \times \frac{\pi}{4} \times 0.15^2 \times \frac{2200}{2 \times 60}$$

or,
$$\mathbf{p_{mb} = 470.87 \text{ kN/m}^2} \quad \textbf{or} \quad \mathbf{4.7087 \text{ bar.}} \quad \textbf{(Ans.)}$$

$$\mathbf{p_{mi}} = \frac{p_{mb}}{\eta_{mech}} = \frac{4.7087}{0.884} = \mathbf{5.326 \text{ bar.}} \quad \textbf{(Ans.)}$$

Q. 22.14. *A four-stroke, four-cylinder SI engine, 82 mm bore × 130 mm stroke, develops 28 kW of brake power at 1500 rpm with twenty per cent rich mixture. The volume of air sucked into the cylinder when measured at 20°C and 760 mm of Hg is 80 per cent of swept volume. The heating value of fuel is 42 MJ/kg and the theoretical air-fuel ratio is 14.8. If the mechanical efficiency of the engine is 85 per cent, find*

(*i*) *The indicated thermal efficiency, and*

(*ii*) *The brake mean effective pressure.* **(MDU)**

Solution. *Given :* D = 82 mm = 0.082 m; L = 132 mm = 0.132 m ; $B. P.$ = 28 kW ; N = 1500 r.p.m. ; T = 20 + 273 = 293 K ; p = 760 mm Hg, V = 0.8 V_s ; C = 42 MJ/kg, theoretical A/F ratio = 14.8, $\eta_{mech.}$ = 85 per cent.

(*i*) **The indicated thermal efficiency $\eta_{th\,(I)}$**

Indicated power, $IP = \frac{B.P.}{\eta_{mech.}} = \frac{28}{0.85} = 32.94$ kW

Swept volume, $V_s = \frac{\pi}{4} \times (0.082)^2 \times 0.132 = 6.971 \times 10^{-4}$ m³/stroke

$= 6.971 \times 10^{-4} \times \frac{1500}{2} \times 4$ m³/min = 2.0913 m³/min

Volume of air admitted at 293 K and 760 mm Hg (1.013 bar)

= 0.8 × 2.0913 = 1.673 m³/min

and its mass is $\frac{1.013 \times 10^5 \times 1.673}{287 \times 293} = 2.015$ kg/min $\left(\because \; m = \frac{pV}{RT}\right)$

Fuel consumption, $\dot{m}_f = \frac{2.015}{14.8 \times 60} = 2.269 \times 10^{-3}$ kg/s

$\therefore \quad \eta_{th.(I)} = \frac{I.P.}{\dot{m}_f \times C} = \frac{32.94}{2.269 \times 10^{-3} \times (42 \times 10^3)}$ = **0.3456 or 34.56% (Ans.)**

(*ii*) **The brake mean effective pressure, b.m.e.p. :**

Brake power (B.P.) $= (p_{mb} \times L \times A) \times \left[\frac{r.p.m.}{2 \times 60}\right] \times$ no. of cylinders

$$28 = p_{mb} \times 0.132 \times \frac{\pi}{4} \times (0.082)^2 \times \left(\frac{1500}{2 \times 60}\right) \times 4 = 0.03485\, p_{mb}$$

$\therefore \quad \mathbf{p_{mb}} = \frac{28}{0.03485}$ = **803.44 kN/m² or 8.0344 bar (Ans.)**

Q. 22.15. *At maximum power, a 6-cylinder engine of 112.5 mm bore and 125 mm stroke, when running at 2800 r.p.m. against a torque of 550 Nm consumed 658 kg of air per hour. Assuming air-fuel ratio by volume 40 : 1, calculate the volumetric efficiency of the engine on S.T.P. basis. Take the volume of fuel into account in calculations. Take heat value of air = 2990 kJ/kg.*

The best consumption on a weak mixture was 40.8 kg/h of a fuel whose heat value was 44400 kJ/kg when running at the same speed against a torque of 467.5 Nm. Also calculate the amount of fuel wasted in the case. **(GTU)**

Solution. *Given :* $n = 6$; D = 112.5 mm = 0.1125 m ; L = 125 mm = 0.125 m ; N = 2800 r.p.m. ; T = 550 Nm ; m_a = 6 kg/h ; A/F ratio = 40 : 1 ; C.V. of air = 2990 kJ/kg.

The volume efficiency of the engine at S.T.P. basis :

V_a (volume of air used at S.T.P./min.)

$$= \frac{mRT}{p} = \frac{(658/60) \times 287 \times 273}{1.013 \times 10^5} = 8.482 \text{ m}^3/\text{min}$$

V_{mix} (Volume of mixture) $= \frac{41}{40} \times 8.482 = 8.694$ m³/min, as A/F ratio = 40 : 1

V_s (swept volume) $= n \times \frac{\pi}{4} D^2 L \times \frac{N}{2}$

$$= 6 \times \frac{\pi}{4} \times (0.1125)^2 \times 0.125 \times \frac{2800}{2} = 10.44 \text{ m}^3/\text{min.}$$

$\eta_{\mathbf{vol.}} = \frac{V_{mix}}{V_s} = \frac{8.694}{10.44}$ = **0.8327 or 83.27% (Ans.)**

Loss of fuel :

$$\text{B.P. (output)} = \frac{2\pi NT}{60 \times 1000} = \frac{2\pi \times 2800 \times 550}{60 \times 1000} = 161.27 \text{ kW} = 9676 \text{ kJ/min}$$

$$\text{Input} = m_a \times \text{C.V. of air} = \frac{658}{60} \times 2990 = 32790 \text{ kJ/min}$$

$\therefore$ Combustion chamber efficiency, $\eta_{comb.} = \frac{\text{Output}}{\text{Input}} = \frac{9676}{32790} = 0.295$ or 29.5%

$$\text{Output of weak mixture} = \frac{467.5}{550} \times 9676 = 8224.6 \text{ kJ/min}$$

$\therefore$ The required input $= \frac{\text{Output}}{\eta_{comb}}$

$$= \frac{224.6}{0.295} = 27880 \text{ kJ/min} \quad (\text{as } \eta_{comb.} \text{ is constant})$$

$\therefore$ Amount of fuel required per minute

$$= \frac{27880}{44400} = 0.628 \text{ kg/min.}$$

Hence loss of fuel $= 40.8 - 0.628 \times 60 =$ **3.12 kg/h. (Ans.)**

Q. 22.16. *A 4-stroke, 6-cylinder diesel engine of bore 80 mm and stroke 100 mm has rated speed 2500 r.p.m. It is to be operated at an altitude of 4000 m where ambient pressure and temperature are expected to be 0.7 bar and – 5°C respectively. Estimate the probable loss of power as a percentage. If a supercharger of pressure ratio 2 is to be used as a corrective measure, determine the power required to operate the supercharger which has an adiabatic efficiency of 75%. Assume the following data :*

$\eta_{vol.}$ = 80%, c_p = 1 kJ/kg K ; ambient condition at sea level is 1 bar and 20°C. State any assumption made. **(MU)**

Solution. *Given :* $k = \frac{1}{2}$; $n = 6$; $D = 80$ mm $= 0.08$ m ; $L = 100$ mm $= 0.1$ m ; $N = 2500$ r.p.m. ; pressure ratio of the supercharger = 2 : $\eta_{adia.} = 75\%$; $\eta_{vol.} = 80\%$; $c_p = 1$ kJ/kg K.

Loss of power as a percentage :

Let the suffices 1 and 2 correspond to atmospheric condition and condition at an altitude of 4000 m respectively.

$$\rho_1 = \frac{p_1}{RT_1} = \frac{1 \times 10^5}{287 \times (20 + 273)} = 1.19 \text{ kg/m}^3$$

$$\rho_2 = \frac{p_2}{RT_2} = \frac{0.7 \times 10^5}{287 \times (-5 + 273)} = 0.91 \text{ kg/m}^3$$

The mass consumed by the engine is proportional to the density as the volume consumed per unit time is same assuming $\eta_{vol.}$, is also same under both atmospheric conditions.

$$\therefore \quad \frac{m_2}{m_1} = \frac{\rho_2}{\rho_1} = \frac{0.91}{1.19} = 0.765$$

The power developed by the engine is proportional to the air mass consumed by the engine

$$\therefore \quad \textbf{Loss of power} = \frac{m_1 - m_2}{m_1} = 1 - \frac{m_2}{m_1} = 1 - 0.765 = \mathbf{0.235} \quad \textbf{or} \quad \mathbf{23.5\%} \quad \textbf{(Ans.)}$$

The power required to run the supercharger :

The stroke volume of the 6-cylinder engine consumed per second,

$$V_s = n \times \frac{\pi}{4} \times D^2 L \times \frac{N}{2 \times 60}$$

$$= 6 \times \frac{\pi}{4} \times 0.08^2 \times 0.1 \times \frac{2500}{2 \times 60} = 0.06283 \text{ m}^3/\text{s}$$

V_{actual} (actual volume of air consumed) = $V_s \times \eta_{vol.}$ = 0.06283 × 0.8 = 0.0503 m³/s

When the supercharger is used, the pressure of air supplied to the engine is p_3.

$$\frac{p_3}{p_2} = 2 \quad \text{...(Given)}$$

$$p_3 = 2 \times 0.7 = 1.4 \text{ bar}$$

$$T_2 = -5 + 273 = 268 \text{ K} \quad \text{...(Given)}$$

$$\frac{T_3}{T_2} = \left(\frac{p_3}{p_2}\right)^{\frac{\gamma-1}{\gamma}} = (2)^{\frac{1.4-1}{1.4}} = 1.22$$

$$\therefore \quad T_3 = 268 \times 1.22 = 327 \text{ K}$$

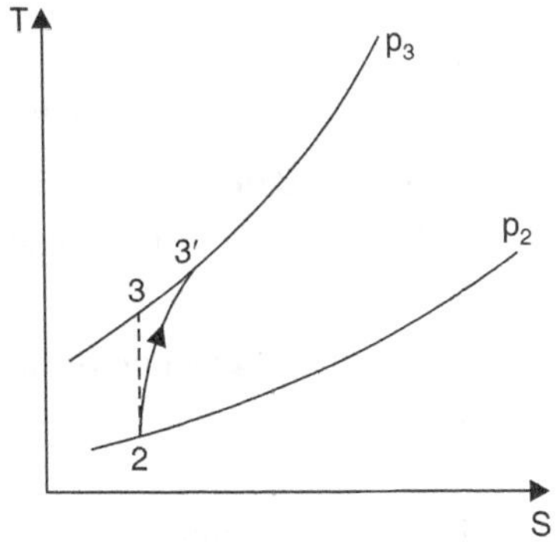

The adiabatic efficiency is given by,

$$\eta_{adia} = \frac{T_3 - T_2}{T_3' - T_2} = 0.75$$

Fig. 11

$$\therefore \quad T_3' = \frac{T_3 - T_2}{0.75} + T_2 = \frac{327 - 268}{0.75} + 268 = 346.7 \text{ K}$$

The mass of air consumed, when its temperature and pressure are 268 K and 0.7 bar respectively, is given by

$$m_{actual} = \frac{pV_{actual}}{RT} = \frac{0.7 \times 10^5 \times 0.0503}{287 \times 268} = 0.046 \text{ kg/s}$$

∴ Power required to run the supercharger

$$= m_{actual} \times c_p (T_3' - T_2)$$

$$= 0.046 \times 1 (346.7 - 268) = \mathbf{3.62 \ kW.} \quad \textbf{(Ans.)}$$

Q. 22.17. *A 4-stroke gas engine develops 3.5 kW B.P. at 160 r.p.m. and at full load. Assuming the following data, find the relative efficiency on I.P. basis and A : F ratio used :*

Volumetric efficiency *87%*
Mechanical efficiency *73.5%*
Clearance volume ...... *2100 cm³*
Swept volume ...... *9000 cm³*
Fuel consumption ...... *5 m³/h*
Calorific value of fuel ...... *18000 kJ/m³*
All working cycles are effective. **(PU)**

Solution. *Given :* $k = \frac{1}{2}$; B.P. = 3.5 kW ; N = 160 r.p.m. ; $\eta_{vol.}$ = 87% ; $\eta_{mech.}$ = 73.5% ; V_c = 2100 cm^3 ; V_s = 9000 cm^3 ; gas used = 5 m^3/h ; C = 18000 kJ/m^3.

Relative efficiency, η_{rel} :

Compression ratio, $$r = \frac{V_s + V_c}{V_c} = \frac{9000 + 2100}{2100} = 5.286$$

$$\eta_{air\text{-}standard} = 1 - \frac{1}{(r)^{\gamma-1}} = 1 - \frac{1}{(5.286)^{1.4-1}} = 0.4862 \quad \text{or} \quad 48.62\%$$

$$\text{I.P.} = \frac{\text{B.P.}}{\eta_{mech.}} = \frac{3.5}{0.735} = 4.762 \text{ kW}$$

$$\eta_{th(I)} = \frac{\text{I.P.}}{V_g \times C} = \frac{4.762}{\frac{5}{3600} \times 1800} = 0.19 \quad \text{or} \quad 19\%$$

$$\eta_{\textbf{relative}} = \frac{\eta_{th(I)}}{\eta_{air\text{-}standard}} = \frac{0.19}{0.4862} = \mathbf{0.39 \quad or \quad 39\% \quad (Ans.)}$$

A / F ratio :

Volume of mixture taken in per stroke

$$= \text{Swept volume} \times \eta_{vol.} \times = 9000 \times 0.87 = 7830 \text{ cm}^3$$

Volume of gas per working stroke

$$= \frac{5}{60} \times \frac{1}{(N/2)} = \frac{5}{60} \times \frac{1}{(160/2)} \times 10^6 \text{ cm}^3 = 1041.7 \text{ cm}^3$$

∴ Volume of air = 7830 – 1041.7 = 6788.3 cm^3

∴ A/F ratio by volume = $\frac{6788.3}{1041.7} \simeq$ **6.5 : 1 by volume. (Ans.)**

Q. 22.18. *A single-cylinder, four-stroke diesel engine running at 440 r.p.m. with cylinder displacement of 0.006 m^3 was arranged to draw air through a calibrated orifice in an air box, the pulsations being sufficiently damped by this procedure. The readings obtained were : barometer 736 mm Hg, air temperature 17°C ; depression in the air box 125 mm of water ; diameter of orifice 25 mm ; co-efficient of discharge 0.62.*

Calculate the volumetric efficiency of the engine referred to inlet conditions. **(PTU)**

Solution. *Given :* n = 1 ; N = 440 r.p.m. ; V_s = 0.006 m^3 ; barometer reading = 736 mm Hg ; air temperature, T = 17 + 273 = 290 K, diameter of orifice, d_0 = 25 mm = 0.025 m ; C_d = 0.62.

Volumetric efficiency, $\eta_{vol.}$:

The flow of air through orifice $= C_d \times A_0 \times \text{velocity}$

$$= C_d \times A_0 \times \sqrt{2g \times h_w \times \frac{\rho_w}{\rho_a}}$$

$$= 0.62 \times \frac{\pi}{4} \times (0.025)^2 \times \sqrt{2 \times 9.81 \times 0.125 \times \frac{1000}{1.181}}$$

$$= 0.01381 \text{ m}^3/\text{s}$$

$$\left[\text{where, } \rho_a = \frac{p_a}{RT_a} = \frac{(0.736 \times 1.336) \times 10^5}{287 \times (17 + 273)} = 1.181 \text{ kg/m}^3 \text{ (1 mHg = 1.336 bar)}\right]$$

Swept volume per sec. $= 0.006 \times \dfrac{440}{2 \times 60} = 0.022 \text{ m}^3/\text{s}$

$$\therefore \quad \eta_{vol.} = \frac{\text{Volume of air at inlet condition}}{\text{Swept volume}}$$

$$= \frac{0.01387}{0.022} = 0.6304 \quad \text{or} \quad \mathbf{63.04\%} \quad \textbf{(Ans.)}$$

Q. 22.19. *The air flow to a four-cylinder, four-stroke oil engine is measured by means of a 5 cm diameter orifice having coefficient of discharge of 0.6. During a test on the engine the following data were recorded : Bore = 10 cm ; stroke = 12 cm ; speed = 1200 rpm ; brake torque = 120 Nm ; fuel consumption = 5 kg/h ; calorific value of fuel = 42 MJ/kg ; pressure drop across orifice is 4.6 cm of water ; ambient temperature and pressure are 17°C and 1 bar respectively.*

Calculate : (i) The thermal efficiency on brake power basis ;

(ii) The brake mean effective pressure ;

(iii) The volumetric efficiency based on free-air condition. **(MDU)**

Solution. *Given :* $n = 4$; $d = 5$ cm $= 0.05$ m ; $C_d = 0.6$; $D = 10$ cm $= 0.1$ m ; $L = 12$ cm $= 0.12$ m ; $N = 1200$ r.p.m. ; $T = 120$ Nm ; $m_f = 5$ kg/h, $C = 42$ MJ/kg ; $\Delta p = 4.6$ cm of water ; $T = 17 + 273 = 290$ K, $p = 1$ bar.

(*i*) **Brake thermal efficiency, $\eta_{th(B)}$:**

$$\text{B.P.} = \frac{2\pi NT}{60 \times 1000} \text{ kW} = \frac{2\pi \times 1200 \times 120}{60 \times 1000} = 15.08 \text{ kW}$$

$$\eta_{th(B)} = \frac{\text{B.P.}}{\dot{m}_f \times C} = \frac{15.08}{(5/3600) \times 42 \times 10^3} = \mathbf{0.2585} \quad \textbf{or} \quad \mathbf{25.85\%} \quad \textbf{(Ans.)}$$

(*ii*) **The brake mean effective pressure, bmep :**

$$\text{BP} = n \times \left(\frac{p_{mb}\, LANk}{3}\right) \text{ kW, where } p_{mb} \text{ is in bar.}$$

$$\left(\because \quad k = \frac{1}{2} \text{ for four - stroke engine}\right)$$

$$15.08 = \frac{4 \times 10 \times p_{mb} \times 0.12 \times \frac{\pi}{4} \times 0.1^2 \times 1200 \times \left(\frac{1}{2}\right)}{6}$$

or, $\mathbf{p_{mb} = 4.0}$ **bar (Ans.)**

(*iii*) **The volumetric efficiency, η_{vol} :**

$$V_s = n \times \frac{\pi}{4} \times D^2 \times L \times \frac{1200}{60 \times 2}$$

$$= 4 \times \frac{\pi}{4} \times 0.1^2 \times 0.12 \times \frac{1200}{60 \times 2} = 0.0377 \text{ m}^3/\text{s}$$

Now, $\rho_a h_a = \rho_w h_w$

or, $$h_a = \frac{\rho_w h_w}{\rho_a} = \frac{1000 \times (4.6/100)}{\rho_a} = \frac{46}{\rho_a}$$

But $$\rho_a = \frac{p_a}{RT_a} = \frac{1 \times 10^5}{287 \times 290} = 1.2 \text{ kg/m}^3$$

$$\therefore \quad h_a = \frac{46}{1.2} = 38.333 \text{ m of air}$$

Volume rate of air at free-air conditions.

$$V_{actual} = C_d \times \frac{\pi}{4}(d)^2 \times \sqrt{2gh_a}$$

$$= 0.6 \times \frac{\pi}{4} \times (0.05)^2 \times \sqrt{2 \times 9.81 \times 38.333} = 0.0323 \text{ m}^3\text{/s}$$

$$\therefore \quad \eta_{vol.} = \frac{V_{actual}}{V_s} = \frac{0.0323}{0.0377} = \mathbf{0.8567 \quad or \quad 85.67\% \quad (Ans.)}$$

Q. 22.20. *Air consumption for four-stroke petrol engine is measured by means of a circular orifice of diameter 35 mm. The coefficient of discharge for the orifice is 0.6 and the differential pressure across the orifice is 140 mm of water. The ambient conditions are 21°C and 76 cm of Hg. The piston displacement volume is 1800 cm³. The engine develops 28 kW of brake power at 2500 r.p.m. and consumes 7.8 kg/h of fuel having C.V. of 44000 kJ/kg. Determine :*

(*i*) *The volumetric efficiency on the basis of air alone ;*

(*ii*) *The air-fuel ratio ;*

(*iii*) *Brake mean effective pressure ;*

(*iv*) *Brake thermal efficiency.*

Take for air R = 287 J/kg K. **(Madras University)**

Solution. *Given :* $k = \frac{1}{2}$; d_0 = 35 mm = 0.035 m ; C_d = 0.6 ; h_w = 140 mm of water ; T_a = 21 + 273 = 294 K ; p_a = 76 cm of Hg = $\frac{76}{100} \times 9.81 \times (13.6 \times 1000) \times 10^{-5}$ bar $\simeq$ 1.014 bar ; V_s = 1800 cm³ ; B.P. = 28 kW ; N = 2500 r.p.m., m_f = 7.8 kg/h ; C = 44000 kJ/kg ; R_a = 287 J/kg K

(*i*) **The volumetric efficiency, $\eta_{vol.}$:**

$$\rho_w h_w = \rho_a h_a$$

or
$$1000 \times \frac{140}{1000} = \frac{p_a}{R_a T_a} \times h_a$$

or
$$1000 \times \frac{140}{1000} = \frac{1.014 \times 10^5}{287 \times 294} \times h_a$$

$$\therefore \quad h_a = 116.5 \text{ m of air}$$

Air consumption = C_d × area of orifice × velocity

$$= 0.6 \times \frac{\pi}{4} \times (0.035)^2 \times \sqrt{2gh_a}$$

$$= 0.6 \times \frac{\pi}{4} \times (0.035)^2 \times \sqrt{2 \times 9.81 \times 116.5} = 0.0276 \text{ m}^3\text{/s}$$

Air

h_w = 140 mm

Fig. 12

Displacement volume per second = $\frac{1800}{10^6} \times \frac{2500}{2 \times 60}$ = 0.0375 m³/s

$$\therefore \quad \eta_{vol} = \frac{\text{Air consumption}}{\text{Displacement volume}} = \frac{0.0276}{0.0375} = \mathbf{0.736 \quad or \quad 73.6\% \quad (Ans.)}$$

(*ii*) **The air-fuel ratio :**

$$m_a = \rho_a \times (0.0276 \times 3600) = \frac{p_a}{R_a T_a} \times (0.0276 \times 3600)$$

$$= \frac{1.014 \times 10^5}{287 \times 294} \times (0.0276 \times 3600) = 119.4 \text{ kg/h}$$

$\therefore$ **Air-fuel ratio** $= \frac{m_a}{\dot{m}_f} = \frac{119.4}{7.8}$ = **15.3 (Ans.)**

(*iii*) **Brake mean effective pressure, p_{mb} :**

$$BP = \frac{p_{mb}\, LANk \times 10}{6}$$

or
$$28 = \frac{p_{mb} \times (1800 \times 10^{-6}) \times 2500 \times \frac{1}{2} \times 10}{6} = 3.75\, p_{mb}$$

($\because$ $LA = 1800 \times 10^{-6}$ m^3 ...Given)

$\therefore$ $\mathbf{p_{mb} = 7.47}$ **bar (Ans.)**

(*iv*) **Brake thermal efficiency, $\eta_{th(B)}$:**

$$\eta_{th(B)} = \frac{\text{B.P.}}{\dot{m}_f \times C} = \frac{28}{\left(\frac{7.8}{3600}\right) \times 44000} = \mathbf{0.2937} \text{ or } \mathbf{29.37\%} \text{ (Ans.)}$$

Q. 22.21. *A four-stroke petrol engine has six cylinders of 7.5 cm bore and 9 cm stroke. The engine is coupled to a brake which has a torque arm radius of 38 cm. At 3300 rev/min, with all cylinders operating the net brake load is 323 N. When each cylinder in turn rendered inoperative, the average net brake load produced at the same speed by the remaining five cylinders is 245 N. Estimate imep of the engine.* **(PTU)**

Solution. *Given :* $n = 6$, $D = 7.5$ cm $= 0.075$ m ; $L = 9$ cm $= 0.09$ m ; $R = 38$ cm $= 0.38$ m ; $N = 3300$ r.p.m. ; net brake load with all cylinders working = 323 N ; average net brake load produced when each cylinder in turn rendered inoperative = 245 N.

Indicated mean effective pressure, p_{mi} :

Indicated power,
$$\text{I.P.} = 6 \times \left(\frac{323 - 245}{1000}\right) \times 2\pi R \left(\frac{N}{60}\right) \text{ kJ/s(kW)}$$
$$= 6 \times \left(\frac{323 - 245}{1000}\right) \times 2\pi \times 0.38 \times \frac{3300}{60} = 61.46 \text{ kW}$$

Also,
$$\text{I.P.} = \frac{10n\, p_{mi}\, LANk}{6}, \text{ where } p_{mi} \text{ is in bar}$$

$\therefore$
$$61.46 = \frac{10 \times 6 \times p_{mi} \times 0.09 \times \frac{\pi}{4} \times 0.075^2 \times 3300 \times \frac{1}{2}}{6}$$

or $\mathbf{p_{mi} = 9.368}$ **bar. (Ans.)**

Q. 22.22. *The bore and stroke of a water-cooled, vertical, single-cylinder four-stroke diesel engine are 80 mm and 110 mm respectively. Find the mean effective pressure and torque developed by the engine if its rating is 4 kW at 1500 r.p.m.* **(MU)**

Solution. *Given :* $D = 80$ mm $= 0.08$ m ; $L = 110$ mm $= 0.1$ m ; I.P. = 4 kW ; $N = 1500$ r.p.m.

Mean effective pressure, p_m :

$$\text{I.P.} = p_m \times (L \times A) \times Nk$$

$$4 \times 1000 = p_m \times 0.11 \times \frac{\pi}{4} \times 0.08^2 \times \left(\frac{1500}{60} \times \frac{1}{2}\right)$$

$\therefore$ $p_m = \mathbf{578727\ N/m^2}$ **or 5.78 bar (Ans.)**

Torque developed, $T_{dev.}$

Also,
$$\text{I.P.} = \frac{2\pi N T_{dev.}}{60 \times 1000}$$

or,
$$4 \times 1000 = \frac{2\pi \times 1500 \times T_{dev.}}{60}$$

or, $T_{dev.} = \mathbf{25.46}$ **Nm (Ans.)**

Q. 22.23. *A 4-stroke petrol engine with a swept volume of 5 litres has a volumetric efficiency of 75 per cent when running at 3000 r.p.m. The engine is fitted with a carburettor which has a choke diameter of 35 mm. Considering single jet carburettor and neglecting the effects of compression, calculate the pressure and air velocity at the choke. The ambient conditions are 1 bar and 300 K. Table C_d (throat) = 0.85.* **(UPTU)**

Solution. *Given :* Swept volume = 5 litres = 5×10^{-3} m³ ; $\eta_{vol.}$ = 75% ; N = 3000 r.p.m. ; D_2 = 35 mm = 0.035 m ; p_1 = 1 bar, T_1 = 300 K ; C_d = 0.85

The velocity at the throat , $C_t (= C_2)$:

The actual volume taken in by the engine per second

$$= (5 \times 10^{-3} \times 0.75) \times \frac{3000}{2 \times 60}$$

$$= 0.09375 \text{ m}^3/\text{s}$$

Fig. 13

Also, $$0.09375 = \frac{\pi}{4} D_2^2 \times C_2$$

or $$C_2 = \frac{0.09375 \times 4}{\pi \times (0.035)^2} = \mathbf{97.44 \text{ m/s.}} \quad \textbf{(Ans.)}$$

The pressure at the throat, p_2 :

$$C_2 = C_d \sqrt{2 c_p T_1 [1 - \left(\frac{p_2}{p_1}\right)^{\frac{\gamma-1}{\gamma}}]}$$

$$97.44 = 0.85 \sqrt{2 \times (1.005 \times 1000) \times 300 \left[1 - \left(\frac{p_2}{1}\right)^{\frac{1.4-1}{1.4}}\right]}$$

$$\left(\frac{97.44}{0.85}\right)^2 = 2 \times (1.005 \times 1000) \times 300 \left[1 - \left(\frac{p_2}{1}\right)^{0.286}\right]$$

$$13141.25 = 603000 \left[1 - \left(\frac{p_2}{1}\right)^{0.286}\right]$$

$$0.9782 = \left(\frac{p_2}{1}\right)^{0.286}$$

$\therefore$ $$p_2 = (0.9782)^{\frac{1}{0.286}} = (0.9782)^{3.496} = \mathbf{0.926 \text{ bar.}} \quad \textbf{(Ans.)}$$

Q. 22.24. *A gas engine 250 mm bore and 500 mm stroke is designed to run at 250 r.p.m. and governed by but and miss governing. With fixed positions of gas cock and ignition ; the indicator diagrams gave the following values of meps :*

Firing cycle (positive loop) = 5 bar ; pumping loop = 0.2 bar ; non-firing cycle = 0.5 bar.

When the engine is developing 15 kW brake power, the explosions per minute were 110 and gas consumption was 0.1 m³/min. Determine the following :

(i) The pumping power of firing and non-firing cycles ;

(ii) True F.P. and mechanical efficiency of the engine ;

(iii) Explosions per minute at no load and gas consumption. **(NU)**

Solution. *Given :* D = 250 mm = 0.25 m ; L = 500 mm = 0.5 m ; N = 250 r.p.m. ; m.e.p.s. ; Firing cycle = 5 bar ; Pumping loop = 0.2 bar ; Non-firing cycle = 0.5 bar ; B.P. = 15 kW ; Explosions per min. = 110 ; Gas consumption = 0.1 m³/min.

(*i*) **The pumping power of firing and non-firing cycles :**

Firing cycle :

$$\text{IP (of positive loop)} = \frac{p_{mi} LANk \times 10}{6} = \frac{5 \times 0.5 \times \frac{\pi}{4} \times 0.25^2 \times 110 \times 10}{6}$$

[where p_{mi} is in bar and Nk = 110] = **22.5 kW. (Ans.)**

$$\text{IP (of pumping loop)} = \frac{0.2 \times 0.5 \times \frac{\pi}{4} \times 0.25^2 \times 110 \times 10}{6} = \mathbf{0.9\ kW.}\quad \textbf{(Ans.)}$$

Non-firing cycle :

$$\text{IP (of non-firing cycle)} = \frac{0.5 \times 0.5 \times \frac{\pi}{4} \times 0.25^2 \left(\frac{250}{2} - 110\right) \times 10}{6} = 0.306\ \text{kW}$$

$$\left(\text{Since non-firing cycle} = \frac{250}{2} - 110 = 15\right)$$

∴ Net IP developed by the engine

$$= 22.5 - 0.9 - 0.306 = 21.294\ \text{kW}$$

(*ii*) **True F.P. and $\eta_{mech.}$:**

$$\text{True FP} = \text{IP} - \text{BP} = 21.294 - 15 = \mathbf{6.294\ kW.}\quad \textbf{(Ans.)}$$

$$\eta_{mech.} = \frac{\text{BP}}{\text{IP}} = \frac{15}{21.294} = \mathbf{0.7044 \ \text{or}\ 70.44\%}\quad \textbf{(Ans.)}$$

(*iii*) **Explosion per minute at no load and gas consumption :**

At no load, the power developed by the engine must be equal to overcome the friction power. Assuming N_0 are the explosions per minute under no-load conditions, then we can write:

$$6.294 = \left[\frac{5 \times 0.5 \times \frac{\pi}{4} (0.25)^2\ N_0 \times 10}{6}\right] - \left[\frac{0.2 \times 0.5 \times \frac{\pi}{4} \times 0.25^2 \times N_0 \times 10}{6}\right] - \left[\frac{0.5 \times 0.5 \times \frac{\pi}{4}\ 0.25^2 \times \left(\frac{250}{2} - N_0\right) \times 10}{6}\right]$$

[It may be noted that engine always runs at 250 r.p.m. whether engine is fully loaded or at no load as the function of governor is to maintain constant speed of the engine irrespective of the load on engine.]

$$\therefore \quad 6.294 = \frac{\frac{\pi}{4} \times 0.25^2 \times 10}{6} [5 \times 0.5\ N_0 - 0.2 \times 0.5\ N_0 - 0.5 \times 0.5\ (125 - N_0)]$$

or, $$6.294 = 0.0818\ (2.5\ N_0 - 0.1\ N_0 - 31.25 + 0.25\ N_0)$$

or, $$76.944 = (2.65\ N_0 - 31.25)$$

$$\therefore \quad N_0 = \frac{76.944 + 31.25}{2.65} = 40.8 \text{ say } \mathbf{41\ explosions\ per\ min.}\quad \textbf{(Ans.)}$$

23

"GATE" and "UPSC" Examinations' Questions (Latest-selected) with Answers / Solutions

(A) Conventional Questions with Answers/Solutions
(B) Multiple-Choice Questions with Answers and "Explanations"

A. CONVENTIONAL QUESTIONS WITH ANSWERS/SOLUTIONS

Q. 23.1. A 16-cylinder diesel engine has a power output of 800 kW at 900 rpm. The engine works on 4-stroke cycle and has fuel consumption of 0.238 kg/kWh. The pressure in the cylinder at the beginning of injection is 32.4 bar and maximum cylinder pressure is 55 bar. The injector is set at 214 bar and maximum pressure at the injector is around 600 bar. The coefficient of discharge for the injector is 0.6. The specific gravity of the fuel is 0.86. Calculate the orifice area required per injector if the injection takes place over 10 degree crank rotation. ***(IES)***

Solution. *Given:* No. of cylinders, n = 16; Power output = 800 kW; N = 900 rpm; Specific fuel consumption = 0.238 kg/kWh; Maximum pressure in the injector = 600 bar; Initial pressure in the cylinder = 32.4 bar; Maximum pressure in the cylinder = 55 bar, injector set at 214 bar; Co-efficient of discharge, $C_d = 0.6\ \frac{\rho_t}{\rho_w}$; Injection takes place over 10° crank rotation.

Orifice area required per injector, A:

Initial pressure difference between injector and cylinder

$$= 214 - 32.4 = 181.6 \text{ bar}$$

Maximum pressure difference between injector and cylinder

$$= 600 - 55 = 545 \text{ bar}$$

$$\text{Mean pressure difference} = \frac{181.6 + 545}{2} = 363.3 \text{ bar}$$

For orifice,
$$m_f = AC_d\sqrt{2\Delta p \times \rho} \quad ...(i)$$

Now, mass flow per minute,

$$m_f = \frac{0.238 \times 800}{16 \times 60}$$

$= 0.1983$ kg/min

and, mass flow per cycle $= \dfrac{0.1983}{(900/2)}$

$= 4.407 \times 10^{-4}$ kg/cycle

Time for one revolution $= \dfrac{60}{900}$ s

Time for injection $= \dfrac{10°}{360°} \times \dfrac{60}{900}$

$= 1.852 \times 10^{-3}$ s

$\therefore$ Mass flow rate at injection $= \dfrac{4.407 \times 10^{-4}}{1.852 \times 10^{-3}} = 0.238$ kg/s

Substituting the values in equation (*i*), we get :

$$0.238 = A \times 0.6\sqrt{2 \times (363.3 \times 10^5) \times (0.86 \times 1000)}$$

$$= 1.5 \times 10^5$$

or,
$$A = \frac{0.238}{1.5 \times 10^5}$$

$= \mathbf{1.587 \times 10^{-6}\ m^2}$ or $\mathbf{1.587\ mm^2}$. **(Ans.)**

Q. 23.2. An automobile carburettor having its float chamber vented to the atmosphere is tested at sea level conditions in the factory without air cleaner. The main metering system of this carburetor is found to yield a fuel-air ratio of 0.065. The venturi throat pressure is 0.84 bar. The carburettor is now installed in an automobile and air cleaner is placed on the inlet to carburettor. The air flow rate with or without air filter is 230 kg/h. The pressure drop through the filter is found to be 0.035 bar at sea level conditions. Assuming Z = 0 and orifice coefficient to be constant, calculate:

(*i*) The venturi throat pressure with air cleaner.

(*ii*) Fuel-air ratio with air cleaner. Assuming air as incompressible. ***(IES)***

Solution. *Given*: Fuel-air ratio without air cleaner = 0.065; $p_{venturi} = 0.84$ bar; Air flow rate, $m_a = 230$ kg/h; Pressure drop through the filter, $\Delta p_{filter} = 0.035$ bar; $Z = 0$.

(*i*) Venturi throat pressure with air cleaner, $\mathbf{p_1}$:

Pressure difference between throat and inlet *without filter*,

$$\Delta p_{f_1} = 1.0135 - 0.84$$

$$= 0.1735 \text{ bar}$$

Let the pressure at throat be p_1 *with filter*, then pressure drop,

$$\Delta p_{f_2} = 1.0135 - 0.035 - p_1$$

$$= (0.9785 - p_1) \text{ bar}$$

We know that : $m = C_d A\sqrt{2\Delta p_f \rho_f}$

Here, $m_1 = m_2, A_1 = A_2$ and $\rho_1 = \rho_2$

$\therefore$
$$\Delta p_{f_1} = \Delta p_{f_2}$$

or
$$0.1735 \text{ bar} = (0.9785 - p_1) \text{ bar}$$

or,
$$p_1 = \mathbf{0.805\ bar.}\ \textbf{(Ans.)}$$

***(ii)* Fuel-air ratio with air cleaner,** $\left(\frac{\mathbf{F}}{\mathbf{A}}\right)_{\textbf{with filter}}$ **:**

$$\Delta p_{f_2} = 1.0135 - 0.805 = 0.2085 \text{ bar}$$

Now, $$\left(\frac{F}{A}\right)_{\text{with filter}} = \sqrt{\frac{\Delta p_{f_2}}{\Delta p_{f_1}}} \times \left(\frac{F}{A}\right)_{\text{without filter}}$$

$$= \sqrt{\frac{0.2085}{0.1735}} \times 0.065 = \mathbf{0.0713.} \quad \textbf{(Ans.)}$$

Q. 23.3. A six cylinder, four-stroke spark ignition engine of 10 cm × 12 cm (bore, stroke) with a compression ratio of 6 is tested at 4800 rpm on a dynamometer of arm 55 cm. During a 10 minute test, the dynamometer reads 45 kg and engine consumed 5 kg of petrol of C.V. 45 MJ/kg. The carburettor receives air at 29°C and 1 bar at the rate of 10 kg/min. Calculate:

(*i*) Brake power.

(*ii*) Brake mean effective pressure.

(*iii*) Brake specific fuel consumption.

(*iv*) Brake specific air consumption.

(*v*) Brake thermal efficiency.

(*vi*) Air fuel ratio. ***(GATE)***

Solution. *Given*: No. of cylinders, $n = 6$; Four-stroke engine, $k = \frac{1}{2}$; Bore = 10 cm = 0.1 m ; Stroke, $L = 12$ cm = 0.12 m ; Speed, $N = 4800$ rpm; Length of dynamometer arm, $l = 55$ cm = 0.55 m. Duration of the test = 10 min.; Petrol consumed, $m_f = 5$ kg, C.V. = 45 MJ/kg; Air temperature = 29°C, pressure = 1 bar.

(*i*) **Brake power, BP:**

$$\text{BP} = T \times \omega = \frac{T \times 2\pi N}{60}$$

Here, $$T = F \times l = 45\ (\times 9.81) \times 0.55$$
$$= 242.8 \text{ Nm}$$

$\therefore$ $$\text{BP} = \frac{242.8 \times 2\pi \times 4800}{60} (\times 10^{-3})$$
$$= \mathbf{122.04\ kW} \quad \textbf{(Ans.)}$$

(*ii*) **Brake mean effective pressure,** p_{mb}**:**

$$\text{BP} = \frac{n p_{mb} LANk}{60 \times 1000}$$

or $$122.04 = \frac{6 \times p_{mb} \times 0.12 \times \left(\frac{\pi}{4} \times 0.1^2\right) \times 4800 \times \frac{1}{2}}{60 \times 1000}$$

$$= 0.2262 \times 10^{-3} \times p_{mb}$$

or $$p_{mb} = \frac{122.04}{0.2262 \times 10^{-3}} = \mathbf{5.39\ bar.} \quad \textbf{(Ans.)}$$

(*iii*) **Brake specific fuel consumption, BSFC :**

$$\text{BSFC} = \frac{m_f}{\text{BP}} = \frac{\left(\frac{5}{10}\times 60\right)}{122.04}$$

$$= \mathbf{0.245\ kg/kWh.} \quad \textbf{(Ans.)}$$

(*iv*) **Brake specific air consumption, BSAC :**

$$\text{BSAC} = \frac{m_a}{\text{BP}} = \frac{10\times 60}{122.04}$$

$$= \mathbf{4.916\ kg/kWh.} \quad \textbf{(Ans.)}$$

(*v*) **Brake thermal efficiency, $(\eta_{th})_B$:**

$$(\eta_{th})_B = \frac{\text{BP}}{m_f \times \text{C.V.}} = \frac{122.04}{\left[\frac{5\times(45\times 10^3)}{10\times 60}\right]}$$

$$= \mathbf{0.3254} \text{ or } \mathbf{32.54\%} \quad \textbf{(Ans.)}$$

(*vi*) **Air-fuel ratio, A/F:**

$$\text{A/F} = \frac{10}{(5/10)} = \mathbf{20 : 1.} \quad \textbf{(Ans.)}$$

Q. 23.4. A sharp edged circular orifice of diameter 3.8 cm and coefficient of discharge as 0.6 is used to measure air consumption of a four-stroke petrol engine. Pressure drop through the orifice is 145 mm of water and barometer reads 75.5 cm of Hg. The compression ratio of the engine is 6 and the piston displacement volume is 2000 cm^3.

The temperature of air is taken to be 26°C. At 2600 rpm, the engine brake power recorded is 29.5 kW. The fuel consumption is 0.14 kg/min and the calorific value of fuel used is 43960 kJ/kg.

Calculate the following:

(*i*) Volumetric efficiency.

(*ii*) Air-fuel ratio.

(*iii*) Brake mean effective pressure.

(*iv*) Brake thermal efficiency.

(*v*) Relative efficiency. ***(IES)***

Solution. *Given*: Orifice diameter = 3.8 cm = 0.038 m; Coefficient of discharge, C_d = 0.6;

Four stroke cycle engine, *i.e.*, $k = \frac{1}{2}$; Pressure drop, Δp = 145 mm of H_2O = $\frac{145}{1000} \times 9810 = 1422$ N/m^2;

Barometer reading = 75.5 cm of Hg = $\frac{75.5}{100} \times 13.6 \times 9810 = 100729$ N/m^2 ; Compression ratio, r = 6; Stroke volume = 2000 cm^3 = 2000 × 10^{-6} m^3; Temperature of air = 26 + 273 = 299 K; Brake power, BP = 29.5 kW at 2600 rpm; Fuel consumption, m_f = 0.14 kg/min; C.V. = 43960 kJ/kg.

(*i*) **Volumetric efficiency, $\eta_{vol.}$:**

$$\eta_{vol.} = \frac{\text{Actual volume inhaled/cycle}}{\text{Swept volume}}$$

Now, $$\rho_{air} = \frac{100729}{287\times 299} = 1.174 \text{ kg/m}^3 \qquad \left[\because\ pv = m\text{RT or } p = \frac{m}{v}\text{RT} = \rho\text{RT or } \rho = \frac{p}{\text{RT}}\right]$$

and, Air inhaled in orifice metre per second

$$= C_d \times A \sqrt{2\Delta p \, \rho_{air}}$$

$$= 0.6 \times \frac{\pi}{4} \times (0.038)^2 \times \sqrt{2 \times 1422 \times 1.174}$$

$$= 0.0393 \text{ kg/s}$$

Swept volume $= 2000 \times 10^{-6} \times \dfrac{2600}{2 \times 60}$

$$= 0.0433 \text{ m}^3/\text{s}.$$

Actual mass of air inhaled/second, $\dot{m}_a = 0.0393$ kg/s

$\therefore$ Actual volume inhaled/second $= \dfrac{0.0393}{1.174} = 0.03348 \text{ m}^3/\text{s}$

Hence, $\eta_{vol.} = \dfrac{0.03348}{0.0433}$

$= \mathbf{0.7732}$ or **77.32%** **(Ans.)**

(*ii*) **Air-fuel ratio, A/F:**

$$\frac{A}{F} = \frac{\dot{m}_a}{\dot{m}_f} = \frac{0.0393}{(0.14/60)}$$

$= \mathbf{16.84.}$ **(Ans.)**

(*iii*) **Brake mean effective pressure, p_m:**

$$\text{BP} = \frac{p_m LANk}{60 \times 1000}$$

or, $29.5 = \dfrac{p_m \times (0.0433)}{1000}$

or, $p_m = \dfrac{29.5 \times 1000}{0.0433}$

$= 6.81 \times 10^5 =$ **6.81 bar.** **(Ans.)**

(*iv*) **Brake thermal efficiency, $(\eta_{th})_B$:**

$$(\eta_{th})_B = \frac{\text{BP}}{\dot{m}_f \times \text{C.V.}} = \frac{29.5 \times 10^3}{(0.14/60) \times 43960 \times 10^3}$$

$= \mathbf{0.2876}$ or **28.76%.** **(Ans.)**

(*v*) **Relative efficiency, $\eta_{relative}$:**

$$\eta_{relative} = \frac{\text{Actual thermal efficiency}}{\text{Air standard efficiency}}$$

Now, Air standard efficiency $= 1 - \dfrac{1}{(r)^{\gamma-1}}$

$$= 1 - \frac{1}{(6)^{1.4-1}}$$

$= 0.5116$ or 51.16%

$\therefore$ $\eta_{relative} = \dfrac{28.76}{51.16}$

$= \mathbf{0.5621}$ or **56.21%** **(Ans.)**

Q. 23.5. A turbo-jet engine consumes air at the rate of 60.2 kg/s when flying at a speed of 1000 km/h. Calculate :

(*i*) Exit velocity of the jet when the enthalpy change for the nozzle is 230 kJ/kg and velocity co-efficient is 0.96.

(*ii*) Fuel flow rate in kg/s when air-fuel ratio is 70 : 1.

(*iii*) Thrust specific fuel consumption.

(*iv*) Thermal efficiency of the plant when the combustion efficiency is 92% and calorific value of the fuel used is 42000 kJ/kg.

(*v*) Propulsive power.

(*vi*) Propulsive efficiency.

(*vii*) Overall efficiency. **(*IES*)**

Solution. *Given*: Rate of air consumption, $\dot{m}_a = 60.2$ kg/s

Enthalpy change for nozzle, $\Delta h = 230$ kJ/kg

Velocity coefficient, $z = 0.96$

Air-fuel ratio $= 70 : 1$

Combustion efficiency, $\eta_{combustion} = 92\%$

Calorific value of fuel, C.V. $= 42000$ kJ/kg

Aircraft velocity, $C_a = \dfrac{1000 \times 1000}{60 \times 60} = 277.8$ m/s

(*i*) **Exit velocity of jet, C_j:**

$$C_j = z\sqrt{2\Delta h \times 1000}\text{, where } \Delta h \text{ is in kJ}$$

$$= 0.96\sqrt{2 \times 230 \times 1000} = 651 \text{ m/s.}$$

i.e., **Exit velocity of jet** **= 651 m/s. (Ans.)**

(*ii*) **Fuel flow rate :**

Rate of fuel consumption,

$$\dot{m}_f = \frac{\text{Rate of air consumption}}{\text{Air-fuel ratio}}$$

$$= \frac{60.2}{70} = \mathbf{0.86\ kg/s.}\ \textbf{(Ans.)}$$

(*iii*) **Thrust specific fuel consumption :**

Thrust is the force produced due to change of momentum.

Thrust produced $= \dot{m}_a\,(C_j - C_a)$, neglecting mass of fuel.

$= 60.2\,(651 - 277.8) = 22466.6$ N.

∴ Thrust specific fuel consumption

$$= \frac{\text{Fuel consumption}}{\text{Thrust}} = \frac{0.86}{22466.6}$$

$$= \mathbf{3.828 \times 10^{-5}\ kg/N\ of\ thrust/s.}\ \textbf{(Ans.)}$$

(*iv*) **Thermal efficiency, $\eta_{thermal}$:**

$$\eta_{thermal} = \frac{\text{Work output}}{\text{Heat supplied}}$$

$$= \frac{\text{Gain in kinetic energy per kg of air}}{\text{Heat supplied by fuel per kg of air}}$$

$$= \frac{(C_j^2 - C_a^2)}{\left(\frac{m_f}{m_a}\right) \times \text{C.V.} \times \eta_{\text{combustion}} \times 1000}$$

$$= \frac{(651^2 - 277.8^2)}{2 \times \frac{1}{70} \times 42000. \times 0.92 \times 1000} = 0.3139 \text{ or } 31.39\%$$

i.e., **Thermal efficiency** = **31.39%** **(Ans.)**

(*v*) **Propulsive power:**

$$\text{Propulsive power} = \dot{m}_a \times \left(\frac{C_j^2 - C_a^2}{2}\right) = \frac{60.2}{1000} \times \left(\frac{651^2 - 277.8^2}{2}\right) \text{ kW}$$

= **10433.5 kW.** **(Ans.)**

(*vi*) **Propulsive efficiency, $\eta_{prop.}$:**

$$\eta_{\text{prop.}} = \frac{\text{Thrust power}}{\text{Propulsive power}} = \frac{2C_a}{C_j + C_a}$$

$$= \frac{2 \times 277.8}{651 + 277.8} = \mathbf{0.598} \text{ or } \mathbf{59.8\%.} \textbf{ (Ans.)}$$

(*vii*) **Overall efficiency, η_0:**

$$\eta_0 = \frac{\text{Thrust work}}{\text{Heat supplied by fuel}} = \frac{(C_j - C_a)C_a}{\left(\frac{m_f}{m_a}\right) \times \text{C.V.} \times \eta_{\text{combustion}}}$$

$$= \frac{(651 - 277.8) \times 277.8}{\frac{1}{70} \times 42000 \times 0.92 \times 1000} = \mathbf{0.1878 \text{ or } 18.78\%.} \textbf{ (Ans.)}$$

Q. 23.6. Calculate the diameter of the injector orifice of a six cylinder, 4-stroke C.I. engine using the following data:

Brake power = 250 kW
Engine speed = 1500 rpm
BSFC = 0.3 kg/kW h
Cylinder pressure = 35 bar
Injection pressure = 200 bar
Specific gravity of fuel = 0.88
Coefficient of discharge of the fuel orifice = 0.92
Duration of injection = 36° of crank angle. ***(IES)***

Solution. *Given*: Number of cylinders, n = 6; BP = 250 kW; N = 1500 rpm; BSFC = 0.3 kg/kWh; $p_{\text{cyl.}}$ = 35 bar; $p_{\text{inj.}}$ = 200 bar; Sp. gravity of fuel = 0.88; C_d (fuel orifice) = 0.92; Duration of injection = 36° of crank angle.

Diameter of the orifice, d :

Fuel consumption per hour = BSFC × power = 0.3 × 250 = 75 kg/h

Fuel consumption per cylinder $= \frac{75}{6} = 12.5$ kg/h = 0.208 kg/min

Fuel to be injected per cycle $= \frac{0.208}{(1500/2)} = 2.77 \times 10^{-4}$ kg

Velocity of injection, $V_{inj.} = C_d \sqrt{\frac{2(p_{inj.} - p_{cyl.})}{\rho_f}}$

$$= 0.92 \sqrt{\frac{2(200-35)\times 10^5}{(0.88 \times 1000)}} = 178.16 \text{ m/s}$$

Fuel injected per second in kg $= \frac{2.77 \times 10^{-4}}{\text{Time duration for injection}}$

$$= \frac{2.77 \times 10^{-4}}{\left(\frac{36°}{360°} \times \frac{60}{1500}\right)} = 0.06925 \text{ kg/s}$$

Volume of fuel injected per second,

$$Q = \frac{0.06925}{880} = 7.87 \times 10^{-5} \text{ m}^3\text{/s}$$

Now, $Q = \frac{\pi}{4} d^2 V_{inj.}$

or, $$d = \left(\frac{4Q}{\pi \times V_{inj}}\right)^{1/2} = \left(\frac{4 \times 7.87 \times 10^{-5}}{\pi \times 178.16}\right)^{1/2}$$

$= \mathbf{7.5 \times 10^{-4}}$ **m** or **0.75 mm. (Ans.)**

Q. 23.7. The following data are known for a four cylinder four-stroke petrol engine : Cylinder dimensions : 11 cm bore, 13 cm stroke, engine speed : 2250 rpm, brake power : 50 kW, friction power : 15 kW, fuel consumption rate : 10.5 kg/h, calorific value of fuel : 50,000 kJ/kg, air inhalation rate : 300 kg/h, ambient condition : 15°C, 1.03 bar.

Estimate :

(*i*) Brake mean effective pressure.

(*ii*) Volumetric efficiency.

(*iii*) Brake thermal efficiency.

(*iv*) Mechanical efficiency. ***(GATE)***

Solution. *Given*: $n = 4$; Four-stroke petrol engine, *i.e.*, $k = \frac{1}{2}$; $D = 11$ cm = 0.11 m; $L = 13$ cm = 0.13 m; $N = 2250$ rpm; BP = 50 kW; Friction power, FP = 15 kW; $m_f = 10.5$ kg/h; C.V. = 50000 kJ/kg; $m_a = 300$ kg/h; $T_{amb.} = 15 + 273 = 288$ kW, $p_{mb} = 1.03$ bar.

(*i*) **Brake mean effective pressure, p_{mb} :**

$$\text{BP} = \frac{p_{mb}\, LANk}{60 \times 1000}$$

or, $$50 = \frac{4 \times p_{mb} \times 0.13 \times \frac{\pi}{4} \times (0.11)^2 \times 2250 \times \frac{1}{2}}{60 \times 1000} = 9.266 \times 10^{-5}\, p_{mb}$$

or, $$p_{mb} = \frac{50}{9.266 \times 10^{-5}} = 5.326 \times 10^5 = \textbf{5.396 bar.} \quad \textbf{(Ans.)}$$

(*ii*) **Volumetric efficiency, $\eta_{vol.}$:**

Swept volume $$= \frac{\pi}{4} D^2 \times L \times \left(\frac{N}{2}\right) \times 4$$

$$= \frac{\pi}{4} \times 0.11^2 \times 0.13 \times \frac{2250}{2} \times 4 = 5.56 \text{ m}^3/\text{min}$$

Volume flow rate at ambient condition,

$$V_a = \frac{m_a R T_{amb}}{p_{amb.}} = \frac{(300/60) \times 287 \times 288}{1.03 \times 10^5} = 4.012 \text{ m}^3/\text{min.}$$

$$\therefore \quad \eta_{vol.} = \frac{V_a}{\text{Swept volume}} = \frac{4.012}{5.56} = \textbf{0.7216} \text{ or } \textbf{72.16\%} \quad \textbf{(Ans.)}$$

(*iii*) **Brake thermal efficiency, $(\eta_{th.})_B$:**

$$(\eta_{th})_B = \frac{\text{BP}}{m_f \times \text{C.V.}} = \frac{50}{\left(\frac{10.5}{3600}\right) \times 50000} = \textbf{0.3429} \text{ or } \textbf{34.29\%} \quad \textbf{(Ans.)}$$

(*iv*) **Mechanical efficiency, $\eta_{mech.}$:**

Indicated power, IP = BP + FP = 50 + 15 = 65 kW

$$\therefore \quad \eta_{mech.} = \frac{\text{BP}}{\text{IP}} = \frac{50}{65} = \textbf{0.7692} \text{ or } \textbf{76.92\%} \quad \textbf{(Ans.)}$$

Q. 23.8. A four-stroke diesel engine of 3000 CC capacity develops 14 kW per m^3 of free air induced per minute. When running at 3500 rev/min it has a volumetric efficiency of 85% referred to free air-conditions of 1.013 bar and 27°C. It is proposed to boost the power of the engine by supercharging by a blower (driven mechanically from the engine) of pressure ratio 1.7 and isentropic efficiency of 80%. Assuming that at the end of induction the cylinders contain a volume of charge equal to the swept volume, at the pressure and temperature of the delivery from the blower, estimate the increase in BP to be expected from the engine. Take overall mechanical efficiency as 80%. γ for air = 1.4, R = 0.287 kJ/kg K. ***(IES)***

Solution. *Given*: Four-stroke engine, $k = \frac{1}{2}$; Volume of cylinders = 3000 cm^3; Power developed = 14 kW per m^3 of free air induced per minute; Speed, N = 3500 rpm; $\eta_{vol.}$ = 85%; Pressure ratio of blower, (p_2/p_1) = 1.7; $\eta_{isen.}$ = 80%; $\eta_{mech.}$ = 80%; γ for air = 1.4, R = 0.287 kJ/kg K; $p_{amb.}$ = 1.013 bar; $T_{amb.}$ = 27 + 273 = 300 K.

Net increase in BP:

Swept volume per minute, $$V_s = (3000 \times 10^{-6}) \times \frac{3500}{2} = 5.25 \text{ m}^3/\text{min.}$$

Unsupercharged induced volume $= \eta_{vol.} \times V_s = 0.85 \times 5.25 = 4.46$ m^3/min.

Blower delivery pressure, $p_2 = 1.7 \times 1.013 = 1.722$ bar

Temperature after isentropic compression,

$$T_{2'} = T_1 \left(\frac{p_2}{p_1}\right)^{\frac{\gamma-1}{\gamma}} = 300 (1.7)^{\frac{1.4-1}{1.4}} = 349 \text{ K}$$

Also, $$\eta_{isen.} = 0.8 = \frac{T_{2'} - T_1}{T_2 - 300} = \frac{349 - 300}{T_2 - 300}$$

or, $$T_2 = 300 + \left(\frac{349 - 300}{0.8}\right) = 361 \text{ K}$$

The blower delivers 5.25 m³/min at 1.722 bar and 361 K, the equivalent volume at 1.013 bar and 300 K, V_1 is the calculate as follows:

$$\frac{p_1 V_1}{T_1} = \frac{p_2 V_2}{T_2} \quad \text{or} \quad V_1 = \frac{p_2 V_2 T_1}{p_1 T_2}$$

$$= \frac{1.722 \times 5.25 \times 300}{1.013 \times 3.61} = 7.42 \text{ m}^3/\text{min}$$

∴ Increase in the induced volume = $V_1 - V_2$

$$= 7.42 - 4.46 = 2.96 \text{ m}^3/\text{min}.$$

Increase in indicated power (IP) from air induced = 14 × 2.96 = 41.44 kW

Increase in indicated power (IP) due to increased induction pressure = $\Delta p \times V_s$

$$= [(1.722 - 1.013) \times 10^5] \times \frac{5.25}{60} \times 10^{-3} = 6.2 \text{ kW}$$

Total increase in IP = 41.44 + 6.2 = 47.64 kW

Increase in BP = $\eta_{mech.}$ × 47.64 = 0.8 × 47.64 = 38.11 kW

To find the net increase in BP, let us first calculate *power required to drive the blower.*

Mass of air delivered by the blower,

$$m_2 = \frac{p_2 V_2}{R T_2} = \frac{(1.722 \times 10^5) \times 5.25}{(0.287 \times 10^3) \times 361} = 8.726 \text{ m}^3/\text{min} \quad \text{or} \quad 0.145 \text{ kg/s}$$

∴ Power required by the blower

$$= \frac{m c_p \Delta T}{\eta} = \frac{0.145 \times 1.005 \times (361 - 300)}{0.8} = 11.11 \text{ kW}$$

∴ Net increase in BP = 38.11 – 11.11 = **27 kW. (Ans.)**

Q. 23.9. A liquid fuel C_7H_{16} is burned with 10% more air than the stoichiometric air. Assuming complete combustion, calculate :

(*i*) Mass of air supplied per kg of fuel, and

(*ii*) Volumetric analysis of the dry products of combustion. Assume air contains 21% O_2 by volume. *(IES)*

Solution. *Given*: Liquid fuel burned = C_7H_{16} (with 10% more air mass stoichiometric air); Air contains 21% O_2 by volume.

(*i*) **Mass of air supplied per kg of fuel burned, m_a:**

Equation for stoichiometric composition is given as:

$$C_7H_{16} + 11O_2 \longrightarrow 7\,CO_2 + 8H_2O \qquad ...(i)$$

When 10% more air is supplied, we have the following equation:

$$C_7H_{16} + 12.1\,O_2 + \left[12.1 \times \left(\frac{79}{21}\right)\right] N_2 = a\,CO_2 + b\,H_2O + c\,O_2 + d\,N_2 \qquad ...(ii)$$

Comparing eqns. (*i*) and (*ii*), we get:

C balance $\longrightarrow a = 7$;

H_2 balance $\longrightarrow b = 8$;

O_2 balance $\longrightarrow 2a + b + 2c$

$= 14 + 8 + 2c$

$= (12.1 \times 2) = [(2 \times 7) + 8 + 2c]$

or, $c = 1.1$,

N_2 balance $\longrightarrow d = 45.52$

∴ Mass of air for 1 mole of fuel

$= (12.1 \times 32) + (45.52 \times 28) = 1661.76$ kg of air

Mass of air from 100 kg of fuel = 1661.76 kg of air

∴ Mass of air for 1 kg of fuel $= \frac{1661.76}{100}$ = **16.62 kg of air. (Ans.)**

(*ii*) **Volumetric analysis of dry products of combustion:**

	Moles	***% Vol.***
$\mathbf{CO_2}$	**7**	**13.05**
$\mathbf{O_2}$	**1.1**	**2.05**
$\mathbf{N_2}$	**45.52**	**84.90**
	53.62	**100%** **(Ans.)**

Q. 23.10. Find the percentage increase in the efficiency of a diesel cycle having a compression ratio 'r' of 16 and cut-off ratio 'ρ' is 10% of the swept volume, if c_v decrease by 2%. Take c_v = 0.717 kJ/kg K and γ = 1.4. ***(GATE)***

Solution. *Given*: Compression ratio, $r = 16$, cut-off ratio, ρ = 10% of the swept volume; c_v = 0.717 kJ/kg K; $\gamma = 1.4$.

Percentage increase in efficiency if c_v decreases by 2%:

Refer to Fig. 1.

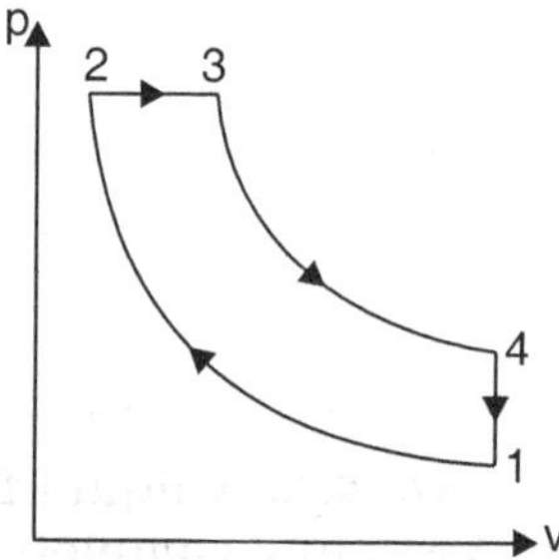

Fig. 1. Diesel cycle.

Air standard efficiency of diesel cycle is given by:

$$\eta_{diesel} = 1 - \frac{1}{\gamma (r)^{\gamma-1}} \left[\frac{\rho^r - 1}{\rho - 1} \right]$$

where, $r = \frac{V_1}{V_2}$; $\rho = \frac{V_3}{V_2}$, $\gamma = \frac{c_p}{c_v}$

Now, $c_p = \gamma \times c_v = 1.4 \times 0.717$
$= 1.0038$ kJ/kg K

$R = c_p - c_v = 1.0038 - 0.717$
$= 0.2868$ kJ/kg K

$$\therefore \quad \eta_{diesel} = 1 - \frac{1}{1.4(16)^{1.4-1}} \left[\frac{\rho^{1.4} - 1}{\rho - 1} \right]$$

where, $\rho = \frac{V_3}{V_2}$, $V_3 - V_2 = 0.1\,(V_1 - V_2)$

or, $$\left(\frac{V_3}{V_2} - 1 \right) = 0.1 \left(\frac{V_1}{V_2} - 1 \right) = 0.1\,(16 - 1) = 1.5$$

or, $\rho - 1 = 1.5$ or $\rho = 2.5$

Hence, $$\eta_{diesel} = 1 - \frac{1}{1.4(16)^{0.4}}\left[\frac{(2.5)^{1.4}-1}{2.5-1}\right] = 0.5905$$

When c_v decrease by 2%:

$$c_v' = (1 - 0.02) \times 0.717 = 0.70266 \text{ kJ/kg K}$$

$$R = 0.2868 \text{ kJ/kg K},$$

$$c_p' = 0.2868 + 0.70266 = 0.98946 \text{ kJ/kg K}$$

and, $$\gamma' = \frac{c_p'}{c_v'} = \frac{0.98946}{0.70266} = 1.4081$$

$\therefore$ $$\eta_{diesel} = 1 - \frac{1}{1.4081(16)^{0.4081}}\left[\frac{(2.5)^{1.4081}-1}{2.5-1}\right] = 0.5978$$

$\therefore$ Percentage increase in efficiency $= \frac{0.5978 - 0.5905}{0.5905} \times 100 =$ **1.236%** **(Ans.)**

Q. 23.11. A four-stroke single cylinder petrol engine mounted on a motor cycle was put to load test. The load measured on dynamometer was 30 kg with drum diameter and speed respectively at 900 mm and 2000 rpm. The engine consumed 0.15 kg of fuel in one minute, the calorific value of fuel being 43.5 MJ/kg. The fuel supply to the engine was stopped and was driven by a motor which needed 5 kW of power to keep it running at the same speed, the efficiency of the motor being 80%. The engine cylinder bore and stroke are respectively 150 mm and 200 mm. Calculate:

(*i*) brake power

(*ii*) indicated power

(*iii*) mechanical efficiency

(*iv*) brake thermal efficiency

(*v*) indicated thermal efficiency

(*vi*) brake mean effective pressure, and

(*vii*) indicate mean effective pressure. ***(IES)***

Solution. *Given*: Four-stroke cycle engine *i.e.*, $k = \frac{1}{2}$; $W = 30$ kg; Diameter of brake drum, $D_b = 900$ mm $= 0.9$ m; $N = 2000$ rpm; $m_f = 0.15$ kg/min.; C.V. = 43.5 MJ/kg; Cylinder bore, $D = 150$ mm $= 0.15$ m; Stroke length, $L = 200$ mm $= 0.2$ m; Power required to keep the engine running at same speed = frictional power, (FP) = 5 kW × η_{motor} = 5 × 0.8 = 4 kW

(*i*) **Brake power, BP:**

$$\text{BP} = \text{Brake torque} \times \text{angular speed} = T \times \omega = W \times r \times \frac{2\pi N}{60}$$

$$= \left[30 \times 9.81 \times \left(\frac{0.9}{2}\right)\right] \times \frac{2\pi \times 2000}{60} (\times 10^{-3}) = \mathbf{27.7 \text{ kW}} \quad \textbf{(Ans.)}$$

(*ii*) **Indicated power, IP:**

$$\text{IP} = \text{BP} + \text{FP} = 27.7 + 4 = \mathbf{31.7 \text{ kW}} \quad \textbf{(Ans.)}$$

(*iii*) **Mechanical efficiency, $\eta_{mech.}$:**

$$\eta_{mech.} = \frac{\text{BP}}{\text{IP}} = \frac{27.7}{31.7} = \mathbf{0.874} \text{ or } \mathbf{87.4\%} \quad \textbf{(Ans.)}$$

(*iv*) **Brake thermal efficiency, $(\eta_{th})_B$:**

$$(\eta_{th})_B = \frac{27.7}{\left[\frac{0.15}{60} \times (43.5 \times 10^3)\right]} = \mathbf{0.255} \text{ or } \mathbf{25.5\%} \quad \textbf{(Ans.)}$$

(*v*) **Indicated thermal efficiency, $(\eta_{th})_I$:**

$$(\eta_{th})_I = \frac{31.7}{\left[\frac{0.15}{60} \times (43.5 \times 10^3)\right]} = \mathbf{0.29} \text{ or } \mathbf{29\%} \textbf{ (Ans.)}$$

(*vi*) **Brake mean effective pressure, p_{mb}:**

$$\text{BP} = \frac{p_{mb}\, LANk}{60 \times 1000}$$

or,
$$27.7 = \frac{p_{mb} \times 0.2 \times \left(\frac{\pi}{4} \times 0.15^2\right) \times 2000 \times \frac{1}{2}}{60 \times 1000} = 5.89 \times 10^{-5}$$

or,
$$p_{mb} = \frac{27.7}{5.89 \times 10^{-5}} = 4.7 \times 10^5 = \mathbf{4.7\ bar.} \textbf{ (Ans.)}$$

(*vii*) **Indicated mean effective pressure, p_{mi}:**

$$\text{IP} = \frac{p_{mi}\, LANk}{60 \times 1000}$$

or,
$$31.7 = \frac{p_{mi} \times 0.2 \times \left(\frac{\pi}{4} \times 0.15^2\right) \times 2000 \times \frac{1}{2}}{60 \times 1000}$$
$$= 5.89 \times 10^{-5} \times p_{mi}$$

or,
$$p_{mi} = \frac{31.7}{5.89 \times 10^{-5}} = 5.38 \times 10^5 = \mathbf{5.38\ bar.} \textbf{ (Ans.)}$$

Q. 23.12. Explain clearly the terms preignition, detonation and diesel knock. ***(IES)***

Solution. (*i*) **Preignition:** Refer to Article **5.5.**

(*ii*) **Detonation:** Refer to Article **5.6.**

(*iii*) **Diesel knock:** Refer to Article **6.5.**

Q. 23.13. What is supercharging? Explain thermodynamic cycle and supercharging power. ***(IES)***

Solution. Refer to Article **16.2.**

Q. 23.14. Explain the principle of a Turbo-jet engine with a suitable figure.

Solution. Turbo-Jet engine :

Description : Fig. 2 shows a turbo-jet unit.

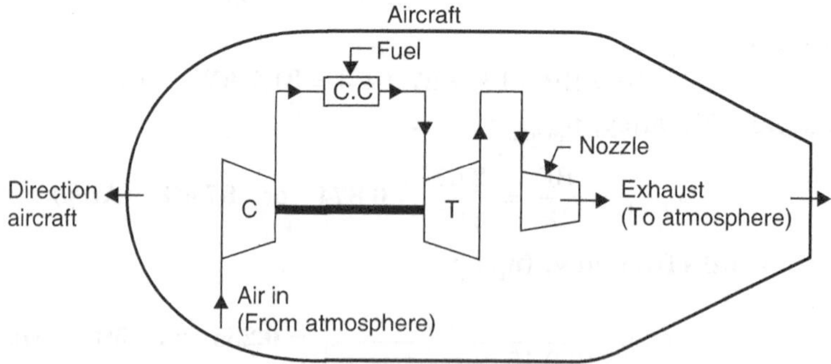

Fig. 2. Gas turbine plant for turbo-jet.

- It consists of *diffuser* at entrance which slows down the air (entering at velocity equal to the plane speed) and part of the kinetic energy of the air stream is converted into pressure; this type of compression is called as *ram compression.*
- The air is further compressed to a pressure of 3 to 4 bar in a rotary compressor (usually of axial flow type).
- The compressed air then enters the combustion chamber (C.C.) where fuel is added. The combustion of fuel takes place at sensibly constant pressure and subsequently temperature rises rapidly.
- The hot gases then enter the gas turbine where *partial expansion* takes place. The *power produced is just sufficient to drive the compressor, fuel pump and other auxiliaries.*
- The exhaust gases from the gas turbine which are at a higher pressure than atmosphere are expended in a nozzle and a very high velocity jet is produced which provides a forward motion to the aircraft by the jet reaction (Newton's third law of motion).

At higher speeds the turbo-jet gives higher propulsion efficiency. The turbo-jets are most suitable to the aircrafts travelling *above 800 km/h.*

The overall efficiency of a turbo-jet is the product of the thermal efficiency of the gas turbine plant and the propulsive efficiency of the jet (nozzle).

Advantages of Turbo-jet engines:

1. Construction much simpler (as compared to multi-cylinder piston engine of comparable power).
2. Engine vibrations absent.
3. Much higher speeds possible (more than 3000 km/h achieved).
4. Power supply is uninterrupted and smooth.
5. Weight to power ratios superior (as compared to that of reciprocating type of aero-engine).
6. Rate of climb higher.
7. Requirement of major overhauls less frequent.
8. Radio interference much less.
9. Maximum altitude ceiling as compared to turbo-prop and conventional piston type engines.
10. Frontal area smaller.
11. Fuel can be burnt over a large range of mixture strength.

Disadvantages of Turbo-jet engines

1. Less efficient.
2. Life of the unit comparatively shorter.
3. The turbo-jet becomes rapidly inefficient below 550 km/h.
4. More noisy (than a reciprocating engine).
5. Materials required are quite expensive.
6. Require longer strip since length of take-off is too much.
7. At take-off the thrust is low, this effect is overcome by boosting.

Q. 23.15. Make a detailed comparison of S.I. and C.I. engines. *(GATE)*

Solution. Comparison of S.I. and C.I. engines.

S.No.	Aspect	S.I. engines	C.I. engines
1.	*Thermodynamic cycle*	Otto cycle	Diesel cycle......For slow speed engines Dual cycle......For high speed engines
2.	*Fuel used*	Petrol	Diesel.
3.	*Air-fuel ratio*	10 : 1 to 20 : 1	18 : 1 to 100 : 1.
4.	*Compression ratio*	Upto 11 ; Average value 7 to 9 ; Upper limit of compression ratio fixed by *anti-knock quality of fuel.*	12 to 24 ; Average 15 to 18 ; Upper limit of compression ratio is limited by *thermal and mechanical stresses.*
5.	*Combustion*	Spark ignition	Compression ignition.
6.	*Fuel supply*	By carburettor......cheap method	By injection......explosive method.
7.	*Operating pressure* (*i*) *Compression Pressure* (*ii*) *Maximum pressure*	7 bar to 15 bar 45 bar to 60 bar	30 bar to 50 bar 60 bar to 120 bar
8.	*Operating speed*	High speed: 2000 to 6000 r.p.m.	Low speed: 400 r.p.m. Medium speed: 400 to 1200 r.p.m. High speed: 1200 to 3500 r.p.m.
9.	*Control of power*	Quantity governing......by throttle	Quality governing......by rack.
10.	*Calorific value*	44 MJ/kg	42 MJ/kg
11.	*Cost of running*	High	Low
12.	*Maintenance cost*	Minor maintenance required	Major overhaul required but less frequently.
13.	*Supercharging*	Limited by *denoted*. Used only in *aircraft engines.*	Limited by *blower power and mechanical and thermal stresses. Widely used.*
14.	*Two-stroke operation*	*Less suitable*, fuel loss in scavenging. But small two-stroke engines are used in mopeds, scooters and motorcycles due to their *simplicity and low cost.*	No fuel loss in scavenging. *More suitable.*
15.	*High powers*	No	Yes
16.	*Uses*	Mopeds, scooters, motorcycles, simple engine passenger cars, air-crafts etc.	Buses, trucks, locomotives, tractors, earth moving machinery and stationary generating plants.

Q. 23.16. Give the comparison of gasoline and diesel emission. *(IES)*

Solution. Refer to Article **18.7.**

Q. 23.17. Explain briefly various methods by which SI engine emission can be controlled. *(IES)*

Solution. Refer to Article **18.4.**

Q. 23.18. What is delay period in CI engines? What are the factors on which the delay period depends? Discuss any four of them.

Solution. Refer to Article **6.4.**

Q. 23.19. Discuss the advantages and disadvantages of using LPG in SI engine. *(IES)*

Solution. Advantages of using LPG in SI engine:

1. LPG is highly knock resistance and does not preignite easily.
2. Cheaper than gasoline.
3. Since it burns clearly, the residue and oil contamination is small.
4. Provider better manifold distribution and mixes easily with air.
5. Crankcase oil dilution is small.

Disadvantages:

1. LPG reduces volumetric efficiency due to its high heat to vaporisation.
2. Response to blending is poor.
3. More advantageous for engines of higher compression ratio.
4. Handling has to be done for pressure of about 18 bar.

Q. 23.20. Discuss the basic properties that a lubricant should process to meet the lubricant requirement of internal combustion engines. *(GATE)*

Solution. Refer to Article **14.5.3.**

B. MULTIPLE-CHOICE QUESTIONS WITH ANSWERS AND "*EXPLANATIONS*"

23.1. The order of values of thermal efficiency of Otto, Diesel and Dual cycle, when they have equal compression ratio and same heat input is given by

(*a*) $\eta_{otto} > \eta_{diesel} > \eta_{dual}$
(*b*) $\eta_{diesel} > \eta_{dual} > \eta_{otto}$
(*c*) $\eta_{dual} > \eta_{diesel} > \eta_{otto}$
(*d*) $\eta_{otto} > \eta_{dual} > \eta_{diesel}$.

Ans. (*d*)

Explanation: The order of values of thermal efficiency of otto, diesel and dual cycle, when they have equal compression ratio and same heat input is $\eta_{otto} > \eta_{dual} > \eta_{diesel}$.

23.2. Consider the following statements regarding knock rating of SI engine fuels:

1. Iso-octane is assigned a rating of zero octane number.
2. Normal heptane is assigned a rating of hundred octane number.
3. Iso-octane is assigned a rating of hundred octane number.
4. Normal heptane is assigned a rating of zero octane number.

Which of these statements are correct?

(*a*) **1 and 2**
(*b*) **2 and 3**
(*c*) **3 and 4**
(*d*) **4 and 1.**

Ans. (*c*)

Explanation. "*Iso-octane* (C_8H_{18})" has a *very high resistance to knock* and as such it is given an octane number of 100; and "*Normal heptane* (C_7H_{16})" which is *very prone to knock* is, therefore, given a *zero* value.

23.3. In spark ignition engines knocking can be reduced by:

(*a*) increasing the compression ratio

(*b*) increasing the cooling water temperature

(*c*) retarding the spark advance

(*d*) increasing the inlet air temperature.

Ans. (*c*)

Explanation. In S.I. engine knocking can be reduced by *retarding* the spark advance; the reason being that *burning gases are not compressed* and as such there is less rise in temperature of the end charge.

23.4. The tendency of knocking in CI engine is reduced by

(*a*) high self-ignition temperature of fuel

(*b*) decrease in jacket water temperature

(*c*) injection of fuel just before TDC

(*d*) decrease in injection pressure.

Ans. (*c*)

Explanation. If the fuel is injected just before TDC (top dead centre), surroundings will have high temperature, causing short delay period. All other factors (given above) increase the delay period of the fuel.

23.5. Consider the following statements relevant to the ignition system of SI engine:

1. **Too small a dwell angle will lead to the burning of condensers and contact points.**
2. **Too small a dwell angle will result in misfiring.**
3. **Too large a dwell angle will result in burning of condenser and contact points.**
4. **Too large a dwell will result in misfiring.**

Which of these statements are correct?

(*a*) 1 and 2 **(*b*) 2 and 3**

(*c*) 3 and 4 **(*d*) 4 and 1.**

Ans. (*b*)

Explanation. The period measured in degree of cam rotation during which the contact point remains closed is termed as '*dwell angle*'.

— Too small dwell angle will result in lower secondary voltage leading to poor spark or missing.

— Too large dwell angle leads to burning of condenser and contact points due to over saturation of winding.

23.6. If the performance of diesel engines of different sizes, cylinder dimensions and power rating are to be compared, which of the following parameters can be used for such comparison?

(*a*) Swept volume **(*b*) Air fuel ratio**

(*c*) Specific brake fuel consumption **(*d*) Volumetric efficiency.**

Ans. (*d*)

Explanation. The parameter which can be used for comparison of performance of the diesel engines is *'volumetric efficiency'* because it indicates the *inhaling capacity*. Higher the volumetric efficiency, the more air sucked in and consequently more fuel burnt.

23.7. Which one of the following action(s) increase(s) the knocking tendency in the SI engine?

(*a*) Increasing mixture strength beyond equivalence ratio (ϕ) = 1.4.

(*b*) Retarding the spark and increasing the compression ratio.

(*c*) Increasing the compression ratio and reducing engine speed.

(*d*) Increasing both mixture strength beyond equivalence (ϕ) = 1.4 and the compression ratio.

Ans. (*c*)

Explanation:

— The knocking tendency in SI engine is increased by the increase in *'compression ratio'*. Increase in compression ratio results in increase in temperature due to which knocking tendency increases.

— Reduced engine speed results in decrease in turbulence which increases the knocking tendency.

23.8. An engine using octane-air mixture has N_2, O_2, CO_2, CO and H_2O as constituents in the exhaust gas. Which one of the following can be conceded?

(*a*) Supply mixture is stoichiometric

(*b*) Supply mixture has incomplete combustion

(*c*) Supply mixture is rich

(*d*) Supply mixture is lean.

Ans. (*d*)

Explanation. When the supply mixture is *lean*, it means that an *excess* amount of air is used; subsequently there is presence of O_2 in the exhaust gases.

23.9. A four-stroke diesel engine when running at 2000 rpm has an injection duration of 1.5 ms. What is the corresponding duration of the crank angle in degrees?

(*a*) 18° **(*b*) 9°**

(*c*) 38° **(*d*) 15°.**

Ans. (*a*)

Explanation. N = 2000 rpm (same for two or four-stroke engine); Duration of injection = 1.5 ms; Crank angle, θ = ?

$$\text{Duration of injection} = \frac{\theta}{360} \times \frac{60}{N}$$

or, $$1.5 \times 10^{-3} = \frac{\theta}{360} \times \frac{60}{2000} \quad \text{or} \quad \theta = \frac{1.5 \times 10^{-3} \times 360 \times 2000}{60} = 18°.$$

23.10. Which of the following cannot be caused by a hot spark plug?

1. Pre-ignition **2. Post-ignition**

3. Detonation **4. Run-on-ignition**

Select the correct answer using the code given below:

(*a*) 1 and 4 **(*b*) 2 only**

(*c*) 2 and 3 **(*d*) 3 only.**

Ans. (*b*)

Explanation. A hot spark plug can cause *only pre-ignition* and *not post-ignition*.

23.11. Consider the following statements:

1. **Supercharging increases the power output and increases the volumetric efficiency.**
2. **Supercharging is more suitable for SI engines than CI engines.**
3. **The limits of supercharging for an SI engine is set by knock while that for a CI engine is set by thermal loading.**

Which of these statements are correct?

(*a*) 2 and 3 **(*b*) 1, 2 and 3**

(*c*) 1 and 3 **(*d*) 1 and 3.**

Ans. (*c*)

Explanation. Supercharging increases the power output and volumetric efficiency as well. Supercharging of "SI engines" is employed only for aircraft and racing car engines, due to the fact that *increase in supercharging pressure increases the tendency* to *detonate/knock and pre-ignite*. The supercharging limits for a "CI engine" (unlike SI engine where the limits of supercharging is due to combustion) is reached by *thermal loading*. The very high temperature of the piston and cylinder causes scuffing of piston rings and heavy liner wear.

23.12. Which one of the following fuels can be obtained by fermentation of vegetable matter?

(*a*) Benzene **(*b*) Diesel**

(*c*) Gasoline **(*d*) Alcohol.**

Ans. (*d*)

Explanation. Alcohol (fuel) can be obtained by fermentation of vegetable matter.

23.13. An engine produces 10 kW brake power while working with a brake thermal efficiency of 30%. If the calorific value of the fuel used is 40,000 kJ/kg, then what is the fuel consumption?

(*a*) 1.5 kg/hour **(*b*) 3.0 kg/hour**

(*c*) 0.3 kg/hour **(*d*) 1.0 kg/hour.**

Ans. (*b*)

Explanation. BP = 10 kW; $(\eta_{th})_B$ = 30%; C.V. = 40,000 kJ/kg; m_f = ?

Now, $(\eta_{th})_B = \dfrac{BP}{m_f \times C.V.}$ or $0.3 = \dfrac{10 \times 3600}{m_f \times 40000}$ or m_f = **3 kg/h**

23.14. A 40 kW engine has a mechanical efficiency of 80%. If the frictional power is assumed to be constant with load, what is the approximate value of the mechanical efficiency at 50% of the rated load?

(*a*) 45% **(*b*) 55%**

(*c*) 65% **(*d*) 75%.**

Ans. (*c*)

Explanation. BP = 40 kW; $\eta_{mech.}$ = 80%; $\eta_{mech.}$ at 50% of the rated load = ?

$$\eta_{mech.} = \frac{BP}{IP} \text{ or } 0.8 = \frac{40}{IP} \text{ or } IP = \frac{40}{0.8} = 50 \text{ kW}$$

Frictional power, FP = IP – BP = 50 – 40 = 10 kW

At **50% of the rated load:** $BP = \dfrac{40}{2} = 20$ kW

IP = BP + FP = 20 + 10 = 30 kW (since FP is assumed constant)

$$\therefore \qquad \eta_{mech.} = \frac{BP}{IP} = \frac{20}{30} = 0.666 \text{ or } 66.6\% \simeq \mathbf{65\%}$$

23.15. Consider the following statements:

In order to prevent detonation in a spark ignition engine, the charge away from the spark plug should have

1. Low temperature **2. Low density**

3. Long ignition delay

Which of these statements is/are correct?

(*a*) Only 1 **(*b*) Only 2**

(*c*) Only 3 **(*d*) 1, 2 and 3.**

Ans. (*d*)

Explanation. In a SI engine, in order to prevent detonation, the charge away from the spark plug should have *low temperature*, *low density* and *long ignition delay* (to avoid pre-ignition of end charge).

23.16. Consider the following statements:

1. For a diesel cycle, the thermal efficiency decreases as the cut-off ratio increases.

2. In a petrol engine the high voltage for spark is in the order of 1000 V.

3. The material for centre electrode in spark plug is carbon.

Which of these statements is/are correct?

(*a*) Only 1 **(*b*) Only 1 and 2**

(*c*) Only 2 and 3 **(*d*) 1, 2 and 3.**

Ans. (*a*)

Explanation. In case of a diesel cycle, the thermal efficiency decreases with the increase in cut-off ratio.

23.17. For the same indicated work per cycle, mean speed and permissible fluctuation of speed, what is the size of flywheel required for a multi-cylinder engine in comparison to a single-cylinder engine?

(*a*) Bigger

(*b*) Smaller

(*c*) Same

(*d*) Depends on thermal efficiency of the engine.

Ans. (*b*)

Explanation. The size of flywheel required for multi-cylinder engine in comparison to single-cylinder engine (for the same indicated power per cycle mean speed and permissible fluctuation of speed) will be *small* since the *turning moment will be more uniform.*

23.18. The tendency of petrol to detonate in terms of octane number is determined by comparison of fuel with which of the following?

(*a*) Iso-octane

(*b*) Mixture of normal heptane and iso-octane

(*c*) Alpha methyl naphthalene

(*d*) Mixture of methane and ethane.

Ans. (*b*)

Explanation. The tendency of petrol to detonate in terms of octane number is determined by the comparison of *mixture of normal heptane and iso-octane.* Iso-octane is given an octane member of *100* and normal heptane a *zero* value. Blends of these fuels define the knock resistance of *intermediate octane members*, and thus a blend of 10% *n*-heptane and 90% iso-octane by volume has an octane number of 90.

23.19. Where does mixing of fuel and air take place in case of diesel engine?

(*a*) Injection pump **(*b*) Injector**

(*c*) Engine cylinder **(*d*) Inlet manifold.**

Ans. (*c*)

Explanation. In case of a diesel engine *diesel is directly injected into the cylinder* where it gets mined with air.

23.20. Consider the following statements:

1. Supercharging increases the power output of an engine.

2. Supercharging increases the brake thermal efficiency considerably.

3. Supercharging helps scavenging of cylinders.

Which of these statements are correct?

(*a*) Only 1 and 2 **(*b*) Only 2 and 3**

(*c*) Only 1 and 3 **(*d*) 1, 2 and 3.**

Ans. (*d*)

Explanation. Supercharging *increases* the power out put of engine, *increases* the brake thermal efficiency and results in *better scavenging* of cylinders.

23.21. The purpose of employing supercharging for an engine is

(*a*) to provide forced cooling air

(*b*) to raise exhaust pressure

(*c*) to inject excess fuel for coping with higher load

(*d*) to supply an intake of air at a density greater than the density of the surrounding atmosphere.

And. (*d*)

Explanation. The purpose of supercharging is to *raise the volumetric efficiency* above that value which can be obtained by normal aspiration. One of the methods which might be utilized to increase the air consumption of an engine is to *increase the density of charge, such that a greater mass of charge is introduced into the same volume or same total piston displacement.*

23.22. Piston compression rings are made of which one of the following?

(*a*) Cast iron **(*b*) Bronze**

(*c*) Aluminium **(*d*) White metal.**

Ans. (*a*)

Explanation. The compression rings are usually made of '*cast iron*' of fine grain and high elasticity which is not affected by the working heat. Some rings are of alloy spring steel. They are split at one point so that they can be expanded and slipped over the end of the piston and into the ring grooves which have been cut in the piston.

23.23. The Cetane number of automotive diesel fuel used in India is in which one of the following ranges?

(*a*) 30–40 **(*b*) 41–50**

(*c*) 51–60 **(*d*) 61–70.**

Ans. (*b*)

Explanation. The cetane number (CN) of automotive diesel fuel (a measure of its ignition quality) used in India lies between 40–50.

- For higher engines the cetane number required is 50, for medium speed engines about 40 and for slow speed engines about 30.

23.24. The delay period in CI engine depends upon which of the following?

(*a*) Temperature and pressure in the cylinder at the time of injection

(*b*) Nature of the fuel mixture strength

(*c*) Relative velocity between the fuel injection and air turbulence pressure of residual gases

(*d*) All of the above.

Ans. (*d*)

Explanation. The delay period depends upon the following *factors*:

(*i*) Temperature and pressure in the cylinder at the time of injection; (*ii*) Nature of fuel mixture strength; (*iii*) Relative velocity between the fuel injection and air turbulence; (*iv*) Presence of residual gases; (*v*) Rate of fuel injection; (*vi*) To some extent the fineness of the fuel spray.

- The delay period should be *as short as possible* since a *long delay period* gives a more rapid rise in pressure and thus *causes knocking*.

23.25. Match List-I (SI engine Operational Mode with List-II (*A-F* Ratio Supplied by the Carburettor) and select the correct answer using the code given below the lists:

List I	List II
A. Idling	1. 3
B. Cruising	2. 10
C. Maximum Power	3. 13
D. Cold starting	4. 16
	5. 20

Codes:

	A	B	C	D
(*a*)	2	4	5	1
(*b*)	4	5	3	2
(*c*)	2	4	3	1
(*d*)	4	5	3	1

Ans. (*c*)

Explanation. Air-fuel ratio supplied by the carburettor during different operations are: Idling = 10; Cruising = 16; Maximum power = 13; Cold starting = 3.

23.26. What is the main objective of supercharging of the engine?

(*a*) To reduce the mass of the engine per brake power

(*b*) To reduce space occupied by engine

(*c*) To increase the power output of engine

(*d*) All of the above.

Ans. (*d*)

Explanation. The objectives of supercharging are:

(*i*) To increase the power output for a given weight and bulk of the engine ...... relates to aircraft, marine and automotive engines where weight and space are important; (*ii*) To compensate for loss of power due to altitude ...... relates to aircraft and other engines which are used at high altitudes; (*iii*) 20 obtain more power from an existing engine.

23.27. Which one of the following automobile exhaust gas pollutants is a major cause of photochemical smog?

(*a*) CO **(*b*) HC**

(*c*) NO_X **(*d*) SO_X.**

Ans. (*c*)

Explanation. NO_X pollutant in automobile exhaust gas is a major cause of photochemical smog.

23.28. Which one of the following is represented by the molecular structure of the paraffin family of hydrocarbon fuel given below, for an IC Engine?

```
   H  H  H  H
   |  |  |  |
H—C—C—C—C—H
   |  |  |  |
   H  H  H  H
```

(*a*) Ethane **(*b*) Propane**

(*c*) Butane **(*d*) Hexane.**

Ans. (*c*)

Explanation. C_4H_{10} is the formula for '*Butane gas*'. It is highly flamable, colourless and can be liquified easily.

23.29. In a SI engine, combustion stage-I takes 1 ms and combustion stage-II takes 1.5 ms when the engine runs at 1000 rpm. If stage-I time duration is independent of engine speed what will be the additional spark advance necessary when the engine speed is doubled?

(*a*) 0° **(*b*) 6°**

(*c*) 12° **(*d*) 24°.**

Ans. (*b*)

Explanation. The necessary advance time when the engine speed is 1000 rpm = 1.5 ms – 1 ms = 0.5 ms

When the engine speed is doubled (*i.e.*, 2 × 1000 = 2000 rpm), the necessary spark advance

$$\text{angle} = \frac{360° \times 2000}{60} \times (0.5 \times 10^{-3}) = 6°$$

23.30. Which one of the following set of materials is most commonly used in catalytic converters for CI engines?

(*a*) Platinum, palladium and rhodium **(*b*) Palladium, rhodium and ruthenium**

(*c*) Rhodium, ruthenium and platinum **(*d*) Ruthenium, platinum and palladium.**

Ans. (*a*)

Explanation. The following set of materials is most commonly used in catalytic converters for CI engines:

Platinum (for HC); Palladium (for CO); Rhodium (for NO_x).

23.31. The bore and stroke of the cylinder of a 6-cylinder engine working on an Otto-cycle are 17 cm and 30 cm respectively, total clearance volume is 9225 cm³, then what is the compression ratio?

(*a*) 7.8 **(*b*) 6.2**

(*c*) 15.8 **(*d*) 5.4.**

Ans. (*d*)

Explanation. D = 17 cm = 0.17 m; L = 30 cm = 0.3 m; V_c = 9225 cm³ = 9225 × 10^{-6} m³

$$V_c \text{ per cylinder} = \frac{9225 \times 10^{-6}}{6} \text{ m}^3 = 1537.5 \times 10^{-6} \text{ m}^3$$

$$\text{Swept volume,} \quad V_s = \frac{\pi}{4} D^2 L = \frac{\pi}{4} \times 0.17^2 \times 0.3 = 6.809 \times 10^{-3} \text{ m}$$

$$\text{Compression ratio,} \quad r = \frac{V_s + V_c}{V_c} = 1 + \frac{V_s}{V_c}$$

$$= 1 + \frac{6.809 \times 10^{-3}}{1537.5 \times 10^{-6}} = \mathbf{5.428}$$

23.32. Which of the following symptoms shows that the combustion is necessarily complete?

(*a*) Presence of free carbon in exhaust **(*b*) Presence of CO in exhaust**

(*c*) Presence of oxygen in exhaust **(*d*) Presence of nitrogen in exhaust.**

Ans. (*c*)

Explanation. Presence of oxygen in exhaust is the symptom which shows that the combustion is necessarily complete.

When C is present in exhaust, further reactions takes place as follows:

$2C + O_2 \rightarrow 2CO$ (Incomplete combustion): $2CO + O_2 \rightarrow 2CO_2$. Presence of nitrogen does not necessarily tell about the complete combustion because NO formation takes place at very high temperature.

23.33. Detonation in the SI engine becomes severe during

(*a*) Low load and high speed **(*b*) High load and high speed**

(*c*) Low load and low speed **(*d*) High load and low speed.**

Ans. (*d*)

Explanation. In the SI engine, detonation becomes *severe* during *high load and low speed* because at high load the temperature of cylinder and combustion chamber walls become high and hence temperature of end charge will also be more, leading to increase in the knocking tendency in the engine.

23.34. Consider the following:

1. EGR

2. Increased compression ratio.

3. Thermal reactor in the exhaust manifold.

4. Use of 5% lean mixture.

Which of these are used to control NO_X emission from engine exhaust?

(*a*) 1 only **(*b*) 1, 2 and 4 only**

(*c*) 2, 3 and 4 only **(*d*) 1, 2, 3 and 4.**

Ans. (*a*)

Explanation. The following are the three methods for reducing peak cycle temperature and thereby reducing NO_X emission:

1. EGR (Exhaust gas recirculation); 2. Catalyst; 3. Water injection.

23.35. The main object of Morse test is to find out

(*a*) Performance of a petrol engine

(*b*) Performance of a diesel engine

(*c*) Frictional power of an engine

(*d*) Indicated power of a multi-cylinder engine.

Ans. (*d*)

Explanation. Morse test is conducted to find out the indicated power of a multi-cylinder engine.

23.36. Consider the following statements regarding SI engine:

1. Specific output is higher than CI engine.

2. It can burn leaner mixtures than CI engine.

3. For a given volumetric capacity, its power output is lesser than CI engine.

4. Its exhaust pollution is higher relative to CI engine.

Which of these statements is/are correct?

(*a*) 1, 2, 3 and 4 **(*b*) 2 only**

(*c*) 2, 3 and 4 only **(*d*) 1 and 4 only.**

Ans. (*d*)

Explanation. Specific output and exhaust pollution of SI engine are higher than that of CI engine.

23.37. Pistons of diesel engines are usually cooled by

(*a*) Air **(*b*) Water**

(*c*) Lubricating oil **(*d*) Fuel oil.**

Ans. (*c*)

Explanation. Pistons of diesel engines are usually cooled by lubricating oil.

23.38. When a hydrocarbon fuel burns in excess air, Orsat analysis of products of combustion will show

(*a*) CO, CO_2 and N_2 **(*b*) CO_2, CO, N_2 and H_2O**

(*c*) CO_2, N_2, O_2 and H_2O **(*d*) CO_2, N_2 and O_2.**

Ans. (*d*)

Explanation. When excess air is used for burning hydrocarbon fuel, the products of combustion, as shown by Orsat apparatus, will be: CO_2, N_2 and O_2.

23.39. By supercharging the diesel engine, the possibility of knocking

(*a*) Decreases **(*b*) Increases**

(*c*) Remains constant **(*d*) None of the above.**

Ans. (*a*)

Explanation. Supercharging tends to decrease the possibility of knocking in a CI engine, whereas it tends to increase the possibility of detonation in SI engine.

23.40. In the SI engine, highest UBHC concentration is observed during

(*a*) Maximum load **(*b*) Acceleration**

(*c*) Deceleration **(*d*) Idling.**

Ans. (*c*)

Explanation. In the SI engine, highest unburnt hydrocarbon (UBHC) concentration is observed during *deceleration*.

23.41. An IC engine has a bore and stroke each equal to 2 units. The total area to calculate heat loss from the engine can be taken as

(*a*) 4π **(*b*) 5π**

(*c*) 6π **(*d*) 8π.**

Ans. (*a*)

Explanation. The total area to calculate heat loss from an IC engine = area of wall of engine = $\pi DL = \pi \times 2 \times 2 = 4\pi$.

23.42. In a petrol engine, the tendency for detonation increases with

(*a*) retarded spark timing **(*b*) running the engine at high speed**

(*c*) supercharging **(*d*) increasing the cooling rate.**

Ans. (*c*)

Explanation. In a petrol engine, the tendency for detonation increases with *supercharging,* since supercharging increases the temperature of the end charge which enhances the chances of detonation.

23.43. A two-stroke IC engine has a mean effective pressure of 6 bar. The speed of the engine is 1000 r.p.m. and the diameter of the piston and stroke are 110 mm and 140 mm respectively. What is the indicated power developed?

(*a*) 10.3 kW **(*b*) 13.3 kW**

(*c*) 15.3 kW **(*d*) 20.3 kW**

Ans. (*b*)

Explanation.

$$\text{IP} = \frac{np_{mi}LANk \times 10}{6} \text{ kW}$$

$$= \frac{1 \times 6 \times 0.14 \times \frac{\pi}{4} \times (0.11)^2 \times 1000 \times 1 \times 10}{6} = 13.3 \text{ kW}$$

23.44. A 4-cylinder four-stroke petrol engine develops 14.7 kW at 1000 r.p.m. The mean effective pressure is 5.5 bar. What are the bore and stroke of the engine, if the length of stroke is 1.5 times the bore?

(*a*) 60.6 mm, 70.6 mm **(*b*) 70.7 mm, 80.8 mm**

(*c*) 87.9 mm, 131.8 mm **(*d*) 90.6 mm, 111.7 mm.**

Ans. (*c*)

Explanation.

$$\text{IP} = \frac{mp_{mi}LANk \times 10}{6} \text{ kW} = \frac{4 \times 5.5 \times 1.5D \times \frac{\pi}{4} D^2 \times 1000 \times \frac{1}{2} \times 10}{6}$$

$$D = 0.0879 \text{ m} \quad \text{or} \quad 87.9 \text{ mm}, L = 131.8 \text{ mm}$$

23.45. A single-cylinder, four-stroke cycle oil engine is fitted with a rope brake. The diameter of the brake wheel is 600 mm and the rope diameter is 26 mm. The dead

load on the brake is 200 N and the spring balance reads 30 N. If the engine runs at 450 rpm what will be the brake power of the engine?

(*a*) 1.5 kW **(*b*) 2.5 kW**

(*c*) 3.5 kW **(*d*) 4.5 kW.**

Ans. (*c*)

Explanation.

$$BP = \frac{(W-S)\,\pi(D_b+d)\,N}{60\times1000}\ \text{kW}$$

$$= \frac{(200-30)\pi\,(0.6\times0.026)\times450}{60\times1000} = 2.5\ \text{kW}$$

23.46. A rope brake was used to measure the brake power of a single-cylinder, four-stroke cycle petrol engine. It was found that the torque due to brake load is 175 mN and the engine makes 500 r.p.m. What is the brake power developed by the engine?

(*a*) 6.2 kW **(*b*) 7.4 kW**

(*c*) 9.16 kW **(*d*) 12.2 kW.**

Ans. (*c*)

Explanation.

$$BP = \frac{2\pi NT}{60\times1000} = \frac{2\pi\times500\times175}{60\times1000} = 9.16\ \text{kW.}$$

23.47. Consider the following statements comparing IC engines and gas turbines:

1. Gas turbines are simple, compact and light in weight.

2. Complete expansion of working substance is possible in IC engines and not in gas turbines.

3. There is flexibility in the design of different components of gas turbines as different processes take place in different components.

4. Even low grade fuels can be burnt in gas turbines.

Of these statements

(*a*) 1, 2 and 3 are correct **(*b*) 1, 3 and 4 are correct**

(*c*) 2, 3 and 4 are correct **(*d*) 1, 2 and 4 are correct.**

Ans. (*a*)

Explanation. 1, 2 and 3 statements are correct. Statement 4 is incorrect because low grade fuel cannot be burnt in gas turbines. Thus choice (*a*) is correct.

23.48. Which of the following factors increase detonation in the SI engine?

1. Increased spark advance

2. Increased speed

3. Increased air-fuel ratio beyond stoichimetric strength

4. Increased compression ratio

Select the correct answer using the codes given below:

Codes:

(*a*) 1 and 3 **(*b*) 2 and 4**

(*c*) 1, 2 and 4 **(*d*) 1 and 4.**

Ans. (*d*)

Explanation. In SI Engines, detonation increases due to increase in spark advance and increase in compression ratio. Increase in speed and increase in air-fuel ratio beyond stoichiometric strength do not effect detonation much.

23.49. Consider the following statements:

I. The performance of an S.I. engine can be improved by increasing the compression ratio.

II. Fuels of higher octane number can be employed at higher compression ratio. Of these statements.

(*a*) both I and II are true **(*b*) both I and II are false**

(*c*) I is true but II is false **(*d*) I is false but II is true.**

Ans. (*d*)

Explanation. Statement I is false since the performance of an SI engine *cannot* be improved by increasing the compression. Statement II is true; since high octane number tends to suppress detonation, therefore, to some extent fuels of higher octane number will prove useful at higher compression ratio. Thus (*d*) is the correct choice.

23.50. Besides mean effective pressure, the data needed for determining the indicated power of engine would include

(*a*) Piston diameter, length of stroke and calorific value of fuel

(*b*) Piston diameter, specific fuel consumption and calorific value of fuel

(*c*) Piston diameter, length of stroke and speed of rotation

(*d*) Specific fuel consumption, speed of rotation and torque.

Ans. (*c*)

Explanation. The formula of indicated power (IP) involves $p_m\ LAN$; *i.e.*, IP depends upon mean effective pressure (p_m), length of stroke (L), piston diameter $\left(\text{Ares}, A = \frac{\pi}{4} D^2\right)$, and speed rotation ($N$). Thus (*c*) is the correct choice.

23.51. Which one of the following curves is a proper representation of pressure differential (*y*-axis) *vs* velocity of air (*x*-axis) at the throat of a carburettor?

(*a*)

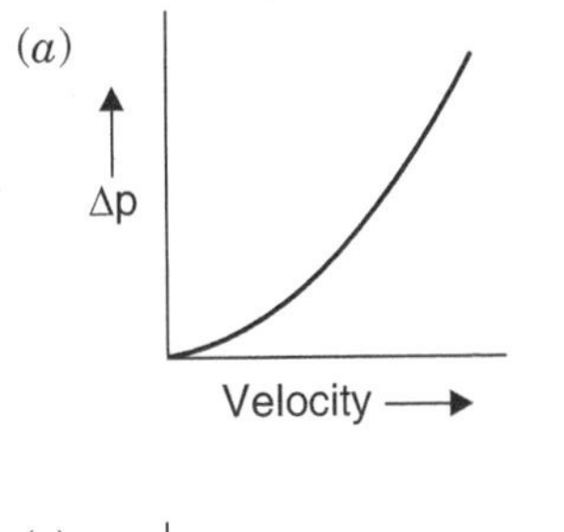

(*b*)

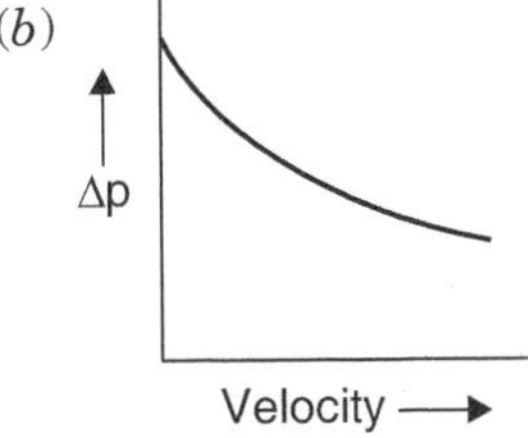

(*c*)

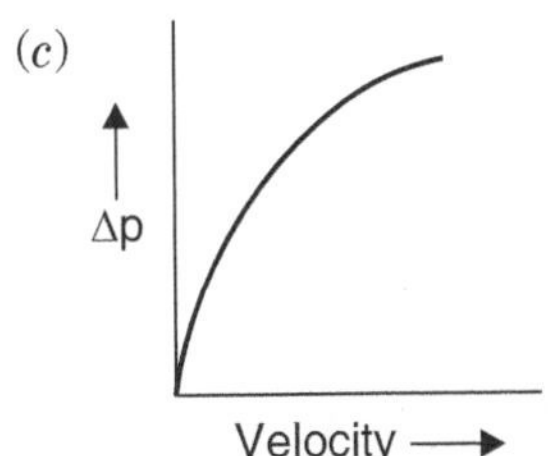

(*d*)

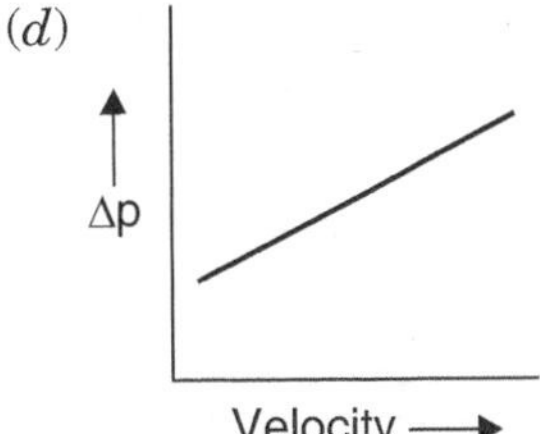

Ans. (*c*)

Explanation. $\Delta p \alpha\ v^2$, this relationship is shown in curve (*c*).

23.52. Match List I and List II and select the correct answer using the codes given below the lists:

List I *(Elements of a complete carburettor)*	**List II** *(Rich-mixture requirement)*
A. Idling system	**1. To compensate for dilution of charge**
B. Economiser	**2. For cold starting**
C. Acceleration pump	**3. For meeting maximum power range of operation**
D. Choke	**4. For meeting rapid opening of throttle**

Codes:

	A	B	C	D
(*a*)	1	2	3	4
(*b*)	1	3	4	2
(*c*)	2	3	4	1
(*d*)	4	1	2	3.

Ans. (*b*)

Explanation. Idling system compensates dilution of charge; **economiser** is used for meeting maximum power range of operation; **acceleration pump** for meeting rapid opening of throttle, and choke for cold starting thus (*b*) is the correct choice.

23.53. Which one of the following engines will have heavier flywheel than the remaining ones?

(*a*) 40 H.P. four-stroke petrol engine running at 1500 rpm

(*b*) 40 H.P. two-stroke petrol engine running at 1500 rpm

(*c*) 40 H.P. two-stroke diesel engine running at 750 rpm

(*d*) 40 H.P. four-stroke diesel engine running at 750 rpm.

Ans. (*a*)

Explanation. Because four-stroke engines require heavier flywheels as power stroke comes only once every four strokes and also petrol engine is running at the highest r.p.m.

Appendix

APPENDIX-I
GAS TABLES FOR AIR (SI UNITS)

$T(K)$	$h(kJ/kg)$	p_r	$u(kJ/kg)$	v_r	$\phi(kJ/kg\ K)$
200	199.97	0.3363	142.56	1707	1.29559
210	209.97	0.3987	149.69	1512	1.34444
220	219.97	0.4690	156.82	1346	1.39105
230	230.02	0.5477	164.00	1205	1.43557
240	240.02	0.6355	171.13	1084	1.47824
250	250.05	0.7329	178.28	979	1.51917
260	260.09	0.8405	185.45	887.8	1.55848
270	270.11	0.9590	192.60	808.0	1.59634
280	280.13	1.0889	199.75	738.0	1.63279
285	285.14	1.1584	203.33	706.1	1.65055
290	290.16	1.2311	206.91	676.1	1.66802
295	295.17	1.3068	210.49	647.9	1.68515
300	300.19	1.3860	214.07	621.2	1.70203
305	305.22	1.4686	217.67	596.0	1.71865
310	310.24	1.5546	221.25	572.3	1.73498
315	315.27	1.6442	224.85	549.8	1.75106
320	320.29	1.7375	228.42	528.5	1.76690
325	325.31	1.8345	232.02	508.4	1.78249
330	330.34	1.9352	235.61	489.4	1.79783
340	340.42	2.149	242.82	454.1	1.82790
350	350.49	2.379	250.02	422.2	1.85708
360	360.58	2.626	257.24	393.4	1.88543
370	370.67	2.892	264.46	367.2	1.91313
380	380.77	3.176	271.69	343.4	1.94001
390	290.88	3.481	278.93	321.5	1.96633
400	400.98	3.806	286.16	301.6	1.99194
410	411.12	4.153	293.43	283.3	2.01699
420	421.26	4.522	300.69	266.6	2.04142
430	431.43	4.915	307.99	251.1	2.06533
440	441.61	5.332	315.20	236.8	2.08870
450	451.80	5.775	322.62	223.6	2.11161
460	462.02	6.245	329.97	211.4	2.13407
470	472.24	6.742	337.32	200.1	2.15604
480	482.49	7.268	344.70	189.5	2.17760
490	492.74	7.824	352.08	179.7	2.19876
500	503.02	8.411	359.48	170.6	2.21952

T(K)	h(kJ/kg)	p_r	u(kJ/kg)	v_r	φ(kJ/kg K)
510	513.32	9.031	366.92	162.1	2.23993
520	523.63	9.684	374.36	154.1	2.25997
530	533.98	10.37	381.84	146.7	2.27967
540	544.35	11.10	389.34	139.7	2.29906
550	554.74	11.86	396.86	133.1	2.31809
560	565.17	12.66	404.42	127.0	2.33685
570	575.59	13.50	411.97	121.2	2.35531
580	586.04	14.38	419.55	115.7	2.37348
590	596.52	15.31	427.15	110.6	2.39140
600	607.02	16.28	434.78	105.8	2.40902
610	617.53	17.30	442.42	101.2	2.42644
620	628.07	18.36	450.09	96.92	2.44356
630	628.63	19.84	457.78	92.84	2.46048
640	649.22	20.64	465.50	88.99	2.47716
650	659.84	21.86	473.25	85.34	2.49364
660	670.47	23.13	481.01	81.89	2.50985
670	681.14	24.46	488.81	78.61	2.52589
680	691.82	25.85	496.62	75.50	2.54175
690	702.52	27.29	504.45	72.56	2.55731
700	713.27	28.80	512.33	69.76	2.57277
710	724.04	30.38	520.23	67.07	2.58810
720	734.82	32.02	528.14	64.53	2.60319
730	745.62	33.72	536.07	62.13	2.61803
740	756.44	35.50	544.02	59.82	2.63280
750	767.29	37.35	551.99	57.63	2.64737
760	778.18	39.27	560.01	55.54	2.66176
780	800.03	43.35	576.12	51.64	2.69013
800	821.95	47.75	572.30	48.08	2.71787
820	843.98	52.59	608.59	44.84	2.74504
840	866.08	57.60	624.95	41.85	2.77170
560	888.27	63.09	641.40	39.12	2.79783
880	910.58	68.98	657.95	36.61	2.82344
900	932.93	75.29	674.58	34.31	2.84856
920	955.38	82.05	691.28	32.18	2.87324
940	977.92	89.28	708.08	30.22	2.89748
960	1000.55	97.00	725.02	28.40	2.92128
980	1023.25	105.2	741.98	26.73	2.94468
1000	1046.04	114.0	758.94	25.17	2.96770

$T(K)$	$h(kJ/kg)$	p_r	$u(kJ/kg)$	v_r	$\phi(kJ/kg\ K)$
1020	1068.89	123.4	776.10	23.72	2.99034
1040	1091.85	133.3	793.36	22.39	3.01260
1060	1114.86	143.9	810.62	21.14	3.03449
1080	1137.89	155.2	827.88	19.98	3.05608
1100	1161.07	167.1	845.33	18.896	3.07732
1120	1184.28	179.7	862.79	17.886	3.09825
1140	1207.57	193.1	880.35	16.946	3.11883
1160	1230.92	207.2	897.91	16.064	3.13916
1180	1254.34	222.2	915.57	15.241	3.15916
1200	1277.79	238.0	933.33	14.470	3.17888
1220	1301.31	254.7	951.09	13.747	3.19334
1240	1324.93	272.3	968.95	13.069	3.21751
1260	1348.55	290.8	986.90	12.435	3.23638
1280	1372.24	310.4	1004.76	11.835	3.25510
1300	1395.97	330.9	1022.82	11.275	3.27345
1320	1419.76	352.5	1040.88	10.747	3.29160
1340	1443.60	375.3	1058.94	10.247	3.30959
1360	1467.49	399.1	1077.10	9.780	3.32724
1380	1491.44	424.2	1095.28	9.337	3.34474
1400	1515.42	450.5	1113.52	8.919	3.36200
1420	1539.44	478.0	1131.77	8.526	3.37901
1440	1563.51	506.9	1150.13	8.153	3.39586
1460	1587.63	537.1	1168.49	7.801	3.41247
1480	1611.79	568.8	1186.95	7.468	3.42892
1500	1635.97	601.9	1205.41	7.152	3.44516
1520	1660.23	636.5	1223.87	6.854	3.46120
1540	1684.51	672.8	1242.43	6.569	3.47712
1560	1708.83	710.5	1260.99	6.301	3.49276
1580	1733.17	750.0	1279.65	6.046	3.50829
1600	1757.57	791.2	1298.30	5.804	3.52364
1620	1782.00	834.1	1316.96	5.574	3.53879
1640	1806.46	878.9	1335.72	5.355	3.55381
1660	1830.96	925.6	1354.48	5.147	3.56867
1680	1855.50	974.2	1373.24	4.949	3.58335
1700	1880.1	1025	1392.7	4.761	3.5979
1750	1941.6	1161	1439.8	4.328	3.6336
1800	2003.3	1310	1487.2	3.944	3.6684
1850	2065.3	1475	1534.9	3.601	3.7023
1900	2127.4	1655	1682.6	3.295	3.7354

$T(K)$	$h(kJ/kg)$	p_r	$u(kJ/kg)$	v_r	$\phi(kJ/kg\ K)$
1950	2189.7	1852	1630.6	3.022	3.7677
2000	2252.1	2068	1678.7	2.776	3.7994
2050	2314.6	2303	1726.8	2.555	3.8303
2100	2377.4	2559	1775.3	2.356	3.8605
2150	2440.3	2837	1823.8	2.175	3.8901
2200	2503.2	3138	1872.4	2.012	3.9191
2250	2566.4	3464	1921.3	1.864	3.9474
2300	2629.68	3817	1994.25	1.729	4.78856
2350	2692.11	4197	2018.58	1.607	4.81586
2400	2756.54	4607	2088.71	1.495	4.84257
2450	2820.18	5048	2117.05	1.393	4.86883
2500	2886.61	5520	2166.42	1.300	4.89458
2550	2947.76	6030	2215.90	1.214	4.91986
2600	3011.69	6575	2265.51	1.135	4.94465
2650	3075.74	7159	2315.97	1.0623	4.96906
2700	3131.47	7782	2364.99	0.9957	4.99472
2750	3204.07	8448	2414.96	0.9342	5.01662
2800	3268.38	9159	2462.50	0.8104	5.03981
2850	3335.33	9916	2514.84	0.8248	5.06263
2900	3397.29	10722	2565.00	0.7762	5.08503
2950	3461.99	11579	2615.24	0.7312	5.10710
3000	3526.54	12490	2665.57	0.6893	5.12866
3050	3591.31	13458	2715.89	0.6404	5.15022
3100	3656.78	14483	2766.38	0.6143	5.17132
3150	3720.97	15569	2811.31	0.5806	5.92050
3200	3785.95	16720	2867.54	0.5493	5.21248
3250	3850.93	17939	2918.24	0.5120	5.232704
3300	3916.08	19224	2964.30	0.4926	5.22592
3350	3981.27	20582	3019.91	0.4671	5.27218
3400	4046.50	22016	3070.72	0.4432	5.29165
3450	4111.98	23295	3121.57	0.4250	5.31058
3500	4177.21	25123	3172.71	0.3998	5.32942

APPENDIX-II

GAS CONSTANTS AND SPECIFIC HEATS AT LOW PRESSURES AND 25°C

Gas	*M*	c_p *(kJ / kg-K)*	c_v *(kJ / kg-K)*	γ	*R* *(kJ / kg-K)*
Acetylene (C_2H_2)	26.036	1.6947	1.3753	1.232	0.3195
Air	28.97	1.0047	0.7176	1.4	0.287
Ammonia (NH_3)	17.032	2.089	1.5992	1.304	0.4882
Argon (A)	39.95	0.5208	0.3127	1.666	0.2081
Carbon dioxide (CO_2)	44.01	0.844	0.6552	1.288	0.1889
Corbon monoxide (CO)	28.01	1.0412	0.7444	1.399	0.2968
Chlorine (Cl_2)	70.914	0.4789	0.3617	1.324	0.1172
Ethane (C_2H_6)	30.068	1.7525	1.4761	1.187	0.2765
Ethylene (C_2H_4)	28.052	1.5297	1.2333	1.24	0.2764
Helium (He)	4.003	5.1954	3.1189	1.666	2.077
Hydrogen (H_2)	2.016	14.3136	10.190	1.4	4.125
Hydrazine (N_2H_4)	32.048	1.6453	1.3815	1.195	0.2594
Methane (CH_4)	16.043	2.1347	1.6164	1.321	0.5183
Neon (Ne)	20.183	1.0298	0.6179	1.666	0.4120
Nitrogen (N_2)	28.016	1.0399	0.7431	1.399	0.2968
Oxygen (O_2)	32	0.9185	0.6585	1.395	0.2598
Propane (C_3H_8)	44.094	1.6683	1.4799	1.127	0.1886
Sulphur dioxide (SO_2)	64.07	0.6225	0.4927	1.263	0.1298
Water vapour (H_2O)	18.016	1.8646	1.4033	1.329	0.4615
Xenon (Xe)	131.3	0.1582	0.0950	1.666	0.0633

APPENDIX-III

PHYSICAL PROPERTIES OF SELECTED FLUIDS

Temperature T (K)	*Density ρ (kg/m³)*	*Specific heat (c_p) (J/kg-K)*	*Thermal conductivity K (W/m-K)*	*Thermal diffusivity $\alpha \times 10^6$ (m^2/s)*	*Absolute viscosity $\mu \times 10^6$ ($N.s/m^2$)*	*Kinematic viscosity $\nu \times 10^6$ (m^2/s)*	*Prandtl number (Pr)*
Dry air at atmospheric pressure							
273	1.252	1011	0.0237	19.2	17.456	13.9	0.71
293	1.164	1012	0.0251	22.0	18.240	15.7	0.71
313	1.092	1014	0.0265	24.8	19.123	17.6	0.71
333	1.025	1017	0.0279	27.6	19.907	19.4	0.71
353	0.968	1019	0.0293	30.6	20.790	21.5	0.71
373	0.916	1022	0.0307	33.6	21.673	23.6	0.71
473	0.723	1035	0.0370	49.7	25.693	35.5	0.71
573	0.596	1047	0.0429	68.9	39.322	49.2	0.71
673	0.508	1059	0.0485	89.4	32.754	64.6	0.72
773	0.442	1076	0.0540	113.2	35.794	81.0	0.72
1273	0.268	1139	0.0762	240	48.445	181	0.74
Water at saturation pressure							
273	999.3	4226	0.558	0.131	1794	1.789	13.7
293	998.2	4182	0.597	0.143	993	1.006	7.0
313	992.2	4175	0.633	0.151	658	0.658	4.3
333	983.2	4181	0.658	0.159	472	0.478	3.00
353	971.8	4194	0.673	0.165	352	0.364	2.25
371	958.4	4211	0.682	0.169	278	0.294	1.75
393	963.4	4501	0.685	0.170	139	0.160	0.95
Refrigerant R12 (CCl_2F_2), saturated liquid							
223	1547	875.0	0.067	5.01	4.796	0.310	6.2
233	1519	884.7	0.069	5.14	4.238	0.279	5.4
243	1490	895.6	0.069	5.26	3.770	0.253	4.8
253	1461	907.3	0.071	5.39	3.433	0.235	4.4
263	1429	920.3	0.073	5.50	3.158	0.221	4.0
273	1397	934.5	0.073	5.57	2.990	0.214	3.8
283	1364	949.6	0.073	5.60	2.769	0.203	3.6
293	1330	965.9	0.073	5.60	2.633	0.198	3.5
303	1295	983.5	0.071	5.60	2.512	0.194	3.5
313	1257	1001.9	0.069	5.55	2.401	0.191	3.5
323	1216	1021.6	0.067	5.45	2.310	0.190	3.5
Unused engine oil, saturated liquid							
273	899.1	1796	0.147	911	3848	4280	471
293	888.2	1880	0.145	872	799	900	104
313	876.1	1964	0.144	834	210	240	28.7
333	864.1	2047	0.140	800	72.5	83.9	10.5
353	852.0	2131	0.138	769	32.0	37.5	4.90
373	840.0	2219	0.137	738	17.1	20.3	2.76
393	829.0	2307	0.135	710	10.3	12.4	1.75
413	816.9	2395	0.133	686	6.54	8.0	1.16
433	805.9	2483	0.132	663	4.51	5.6	0.84

APPENDIX-IV
HYDROCARBON FUELS

S. No.	Fuel group and general formula	Fuel	Molecular weight	Specific gravity	Boiling temp. at 1 atm, °C	Ignition temp. at 1 atm, °C	Constant pr. heating value kJ/kg		Critical compression ratio	A/F ratio	Octane rating research/Motor ml TEL/Sal			
							Higher	Lower			0	3	0	3
1	2	3	4	5	6	7	8	9	10	11	12	13	14	15
1.	Paraffins	Methane, CH_4	16	0.3	– 162	732	55010	49538	12.6	17.2	120		120	
	C_nH_{2n+2}	Ethane, C_2H_6	30	0.37	– 89	504	51565	47148	12.4	16.1	115		99	
		Propane, C_3H_8	44	0.5	– 42	—	49974	45979	12.2	15.7	112		97	
		Butane, C_4H_8	58	0.579	– 0.6	431	49128	45343	5.5	15.5	94	104	90	104
		Hexane, C_6H_{14}	86	0.659	69	261	48312	44736	3.3	15.2	25	65	26	65
		Heptane, C_7H_{16}	100	0.684	98	247	60633	44560	3.0	15.2	0	44	0	47
		Triptane, C_7H_{16}	100	0.690	81	454	47947	44435	14.4	15.2	112		101	116
		Iso-octane, C_8H_{18}	114	0.692	99.4	447	47813	44347	7.3	15.1	100	116	100	116
		Dodecane, $C_{12}H_{26}$	170	0.749	216		47474	44116						
		Hexadecane, (Cetane) $C_{16}H_{34}$	226	0.773	287		47269	43957						
2.	Napthnes	Cyclo-hexane, C_6H_{12}	84	0.779	81	270	46582	43442	4.9	14.8	84	97	78	87
	C_nH_{2n}	Cyclo-heptane, C_7H_{14}	98	0.816	119	—	46708	43534	3.4	14.8	39	60	41	65
		Cyclo-octane, C_8H_{16}	112	0.841	151	—	46725	43551		14.8	71		58	
3.	Aromatics	Benzene, C_6H_6	78	0.879	80	592	41835	40143		13.3	—	—	115	
	C_nH_{2n-6}	Toluene, C_7H_8	92	0.867	111	568	42437	40528	15	13.5	120	120	109	113
		Xylene, C_8H_{10}	106	0.864	139	563	42877	40805	15.5	13.7	118	120	115	120
4.	Alcohol	Methanol, CH_3OH	32	0.792	—	—	22726	20105		6.4	107			
		Ethanol, C_2H_5OH	46	0.785	—	—	29726	26992		9.0		102		

APPENDIX-V
DATA ON FUEL PROPERTIES

Fuel	*Formula (phase)*	*Molecular weight*	*Specific gravity*	*Heat of vaporization kJ/kg*	*Specific heat*		*Higher heating value kJ/kg*	*Lower heating value kJ/kg*	*LHV of stoich. mixture kJ/kg*	*(A/F)*	*(F/A)*	*Fuel octane rating*	+
					Liquid kJ/ kgK	*Vapour c_p kJ/kg K*						*RON*	*MON*
Commercial petroleum fuels													
Gasoline	$C_nH_{1.87n}$ (1)	110	0.72–0.78	305	2.4	1.7	47300	44000	2830	14.6	0.0685	92.98	80-90
Light diesel	$C_nH_{1.8n}$ (1)	170	0.84–0.88	270	2.2	1.7	44800	42500	2740	14.5	0.0690	—	—
Heavy diesel	$C_nH_{1.7n}$ (1)	200	0.82–0.95	230	1.9	1.7	43800	41400	2760	14.4	0.0697	—	—
Natural gas	$C_nH_{3.8n}$ N_{01n}(g)	18	0.79	—	—	2	50000	45000	2900	14.5	0.069	—	—
Pure hydrocarbons													
Methane	CH_4(g)	16.04	(0.72)	509	0.63	2.2	55500	50000	2720	17.23	0.0580	120	120
Propane	C_3H_8(g)	44.10	0.51 (20)	426	2.5	1.6	50400	46400	2750	15.67	0.0638	112	97
Isooctane	C_8H_{18}(I)	114.23	0.692	308	2.1	1.63	47800	44300	2750	15.13	0.0661	100	100
Cetane	$C_{16}H_{34}$(I)	226.44	0.773	358		1.6	47300	44000	2780	14.82	0.0675	—	—
Benzene	C_6H_6(I)	78.11	0.879	433	1.72	1.1	41900	40200	2820	13.27	0.0753	—	115
Toluene	C_7H_8(I)	92.14	0.867	412	1.68	1.1	42500	40600	2790	13.50	0.0741	120	109
Alcohols													
Methanol	CH_4O(I)	32.04	0.792	1103	2.6	1.72	22700	20000	2680	6.47	0.155	106	92
Ethanol	C_2H_6O(I)	46.07	0.785	840	2.5	1.93	29700	26900	2690	9.00	0.111	107	89
Other Fuels													
Carbon	C(s)	12.01	– 2	—	—	—	33800	33800	2700	11.51	0.0869		
Carbon monoxide	CO(g)	28.01	(1.25)	—	—	1.05	10100	10100	2910	2.467	0.405		
Hydrogen	H_2(g)	2.015	(0.0901)	—	—	1.44	142000	120000	3400	34.3	0.0292		

APPENDIX-VI
REQUIREMENTS FOR MOTOR GASOLINES (IS : 2796-1971)

S.No.	*Characteristics*	*Requirements*	
		83 Octane	*93 Octane*
(*i*)	Colour, visul	Orange	Red
(*ii*)	Copper-strip corrosion for 3 hours at 50°C	Not worse	No. 1
(*iii*)	Density at 150°C	Not limited but to be reported	
(*iv*)	Distillation :		
	(*a*) Initial boiling point	Not limited but to be reported	
	(*b*) Recovery up to 70°C, per cent by volume, *Min*	10	10
	(*c*) Recovery up to 125°C, per cent by volume, *Min*	50	50
	(*d*) Recovery up to 180°C, per cent by volume, *Min*	90	90
	(*e*) Final boiling point, *Max*	215°C	215°C
	(*f*) Residue, per cent by volume, *Max*	2	2
(*v*)	Octane number (Research method), *Min*	83	93
(*vi*)	Oxidation stability, in minutes, *Min*	360	360
(*vii*)	Residue on evaporation, mg/100 ml. *Max*	4.0	4.0
(*viii*)	Sulphur, total, per cent by weight, *Max*	0.25	0.20
(*ix*)	Lead content (as Pb), g/1, *Max*	0.56	0.80
(*x*)	Reid vapour pressure at 380°C, bar, *Max*	0.70	0.70

APPENDIX-VII
REQUIREMENT FOR DIESEL FUELS (IS : 1460-1974)

S.No.	*Characteristics*	*Requirements*	
		HSD	*LDO*
(*i*)	Acidity, inorganic	NIL	NIL
(*ii*)	Acidity, total mg of KOH/g (max)	0.50	—
(*iii*)	Ash, per cent by mass, max	0.01	0.02
(*iv*)	Carbon residue (Ramsbottom), per cent by mass (max)	0.20	1.50
(*v*)	Cetane number (min)	42	—
(*vi*)	Pour point (max)	6°C	12°C for winter 18°C for summer
(*vii*)	Copper strip corrosion for 3 hrs at 100°C	90	—
(*viii*)	Flash point :		
	(*a*) Abel, °C (min)	38	—
	(*b*) Pensky-Martens (closed), °C, (min)	—	66
(*ix*)	Kinematic viscosity, CS, at 38°C	20 to 7.5	2.5 to 15.7
(*x*)	Sediment, per cent by mass, (max)	0.05	0.10
(*xi*)	Total sulphur, per cent by mass, (max)	1.0	1.8
(*xii*)	Water content, per cent by volume, (max)	0.05	0.25
(*xiii*)	Total sediments, mg per 100 ml, (max)	1.0	—

APPENDIX-VIII
SPECIFICATIONS FOR DIESEL FUELS

Property		*BS 2869*	
		Class A1	*Class A2*
Viscosity, kinematic at 37.8°C (centistokes)	min.	1.5	1.5
Cetane number	max.	5.0	5.5
Carbon residue, Ramsbottom per cent by mass on 10 per cent residue	min.	50	45
Distillation, recovery at 350°C, per cent by volume	max.	0.2	0.2
Flashpoint, closed	min.	85	85.0
Pensky Martins °C	min.	56	56
Water content, per cent by volume	max.	0.05	0.05
Sediment, per cent	max.	0.01	0.01
Ash, per cent by mass	max.	0.01	0.01
Sulphur	max.	0.3	0.5
Copper corrosion test	max.	1	1
Cold filter plugging point (°C) max.	Summer (16 Mar./30 Sept.)	– 4	0
	Winter (1 Oct./15 March)	– 15	– 9

APPENDIX-IX
AVIATION TURBINE FUELS

Characteristics	*ASTM D1655*		*Mil-J-5624*		*Mil F-46005 Å*
Designation	Jet A	JP-1	JP-4 (up M = 2)	JP-5 M = 2 to 3	CITE-II
Flashpoint, °C (min-max)	43 – 46	43 (min)			
Freezing point, °C (max)	– 40	– 60	– 60	– 48	– 55
Gravity, API (min-max)	39 – 51	35 (max)	45 – 57	36 – 48	
Vapour pressure, Reid, icef/cm² cause (min-max)			0.14-0.21		0.21
Distillation, °C					
10% max	204	210		204	93
20% max			143		
50% max	232		188		163
90% max		254	243		
EP max	288	300		288	288
Heating value, lower, kcal/kg, min	10220	10167	10220	10167	
Sulphur, % by mass, max	0.3	0.2	0.4	0.4	0.4
Smoke point, mm (min)	25			20	
Aromatic, vol % max	20	20	25	25	25
Potential gum, mg/100 ml (max)	14	8	14	14	14

APPENDIX-X
SAE CLASSIFICATION OF LUBRICATING OILS

SAE viscosity number	*Viscosity units*	*Viscosity ranges*			
		18° C		*99°C*	
		Min.	*Max.*	*Min.*	*Max.*
5W	Centipoises	—	1200	—	—
	SUS	—	6000	—	—
10W	Centipoises	1200	2400	—	—
	SUS	6000*	12000	—	—
20W	Centipoises	2400	9600	—	—
	SUS	12000**	48000	—	—
20	Centistokes	—	—	5.7	9.6
	SUS	—	—	45	58
30	Centistokes	—	—	9.6	12.9
	SUS	—	—	58	70
40	Centistokes	—	—	12.9	16.8
	SUS	—	—	70	85
50	Centistokes	—	—	16.8	22.7
	SUS	—	—	85	110

* Minimum viscosity at – 18°C may be waived provided viscosity at 99° C is below 40 SUS.

** Minimum viscosity at – 18°C may be waived provided viscosity at 99°C is not below 45 SUS.

APPENDIX-XI
MAJOR CLASSES OF ENGINE OIL ADDITIVES AND THEIR PRIMARY FUNCTIONS

Additive class	*Function*
Detergent	Control of high-temperature deposits. If overbased, also acts as effective acid neutralizer.
Dispersent	Control of low-temperature sludge and varnish deposits.
Anti-wear	Reduce wear and prevent scoring, galling and seizure.
Anti-rust	Reduce rusting by acid neutralization of formation by protective film.
V.I. improver	Increase V.I. of oil, thereby reducing sensitivity of oil viscosity to temperature.
Pour point depresent	Reduce pour point of oil by interfering with wax crystallization.
Anti-foam	Reduce oil foaming by causing collapse of bubbles due to air entrainment.
Anti-oxidant	Reduce oil oxidation to protect alloy bearings against corrosive attack.

APPENDIX-XII
TYPICAL COMPOSITIONS OF LPG

Constituents	*% by weight*		
	I	*II*	*III*
Ethane	1.8	4.4	1.7
Propane	93.1	64.0	34.1
Propylene	2.8	28.7	62.1
Butane	2.3	2.9	2.1
Total	100.0	100.0	100.0

APPENDIX-XIII
TYPICAL COMPOSITION OF NATURAL GAS

	Constituents	*% by volume*
1.	Methane, CH_4	80
2.	Ethane	7
3.	Propane	6
4.	Isobutane	1.5
5.	Butane	2.5
6.	Pentane plus vapours	3.0
	Total	100.0

APPENDIX-XIV
GASEOUS FUELS

Type of gas	*Main Constituents*	*LHV = Kcal / m³ (Low heat value)*	*Theoretical air requirement (m³ air / m³ gas)*	*Ignition limits in % of air*	*Specific weights (kg / m³)*	*Methane number*
Hydrogen	H_2	2570	2.38	4 – 80	0.0889	0
Methane	CH_4	8550	9.50	5 – 15	0.7170	100.0
Ethane	C_3H_6	15370	16.90	3 – 14	1.3600	43.5
Propane	C_3H_8	22300	23.80	2.1 – 8.5	1.9600	35.0
Butane	C_4H_{10}	29500	32.00	1.5 – 8.5	2.6000	10.5
Blast-furnace gas	CO ; N_2 ; CO_2	700–950	0.74	12 – 70	1.2800	—
PRODUCER GAS						
Manufactured gas	CO ; H_2	100–1300	1.00	20 – 70	1.1000	33.0
Sewage gas (biogas)	CH_4 ; CO_2	5500	6.00	—	1.1000	130.0
Natural gas	CH_4 ; C_2H_6 ; C_3H_8 C_4H_{10} ; N_2	7000 – 9500	8 – 10.7	5 – 15	0.7600	90.0
Liquified gas (LPG)	C_3H_8 ; C_4H_{10}	23000 – 32000	—	1.5 – 20	—	10 – 35

1 Standard cubic metre (m³) equals 1 cubic metre of gas at 0°C and 760 mm of mercury.

APPENDIX-XV
GENERAL COMPOSITION OF BIOGAS PRODUCED FROM FARM WASTES

	Volume %
CH_4—Methane	54 to 70
CO_2—Carbon dioxide	27 to 45
H_2—Hydrogen	1 to 10
CO—Carbon monoxide	0 to 1
O_2—Oxygen	0 to 1
H_2S—Hydrogen sulphide	Traces

APPENDIX-XVI
PROPERTIES OF GOBAR GAS AS I.C. ENGINE FUEL

Calorific value	21500 kJ/m^3 19550 kJ/kg
Density	1.1 kg/m^3
Stoichiometric requirements for combustion (Air/Fuel Ratio)	6 m^3 Air/m^3 gas
Flammability limits	5 to 15% vol. in air-gas mixture
Knock resistance	130 methane number (Pure methane 100)
Ignition temperature	About 700°C

APPENDIX-XVII
TYPICAL VALVE TIMINGS FOR FOUR-STROKE SI ENGINES

Position	*Theoretical*	*Actual*	
		Low speed engine	*High speed engine*
Inlet valve opens (IVO) Inlet valve closes (IVC) Inlet valve is open for	TDC BDC 180°	10° b TDC 10° a BDC 200°	10° b TDC 60° a BDC 250°
Exhaust valve opens Exhaust valve closes Exhaust valve is open for	BDC TDC 180°	25° b BDC 5° a TDC 210°	55° b BDC 20° a TDC 255°
Valve overlap Spark.	Nil TDC	15° 15° b TDC	30° 30° b TDC

Note : Valve timing is different for different makes of engines.

b—before, a—after, TDC—Top dead centre, BDC—Bottom dead centre.

APPENDIX-XVIII
PORT TIMINGS FOR DIFFERENT TWO-STROKE ENGINES.

Engine type	*Exhaust*		*Lead angle deg.*	*Scavenge*		*Super-charge deg.*
	Opens	*Closes*		*Opens*	*Closes*	
1. Crankcase scavenging	60–76	60–67	10–20	45–60	45 – 60	0
2. Loop scavenging	60–76	60–76	10–20	45–60	45 – 60	0
3. Uniflow scavenging	80–90	38–59	20–40	38–55	38 – 55	0 – 70
4. Opposed piston	60–67	50–67	10–20	50–65	50 – 65	10 – 20

APPENDIX-XIX
APPLICATION OF I.C. ENGINES

Application	*Approx., engine power, kW*	*Predominant type SI or CI*	*Cycle*	*Cooling [A = Air W = water]*
1. Road vehicles				
(*i*) Mopeds, Scooters, Motor cycles	0.75 – 7	SI	2, 4	A
(*ii*) Small passenger car	15 – 75	SI	4	A, W
(*iii*) Large passenger car	75 – 200	SI	4	W
(*iv*) Light commercial	35 – 150	SI, CI	4	W
(*v*) Heavy commercial	120 – 300	CI	4	W
2. Railway locomotives	400 – 3000	CI	2, 4	W
3. Small aircraft				
(*i*) Helicopters	45 – 1500	SI	4	A
(*ii*) Airplanes	45 – 2700	SI	4	A
4. Marine				
(*i*) Motor boats	0.5 – 7.5	SI, CI	4	W
(*ii*) Ships	3500 – 22000	CI	2, 4	W
5. Off-road vehicles				
(*i*) Light vehicles (factory, airport etc.)	1.5 – 15	SI	2, 4	A, W
(*ii*) Agriculture	3 – 150	SI, CI	2, 4	A, W
(*iii*) Earth moving	40 – 750	CI	2, 4	W
(*iv*) Military	40 – 2000	CI	2, 4	A, W
6. Industrial, Stationary				
(*i*) Electric power	35 – 22000	CI	2, 4	W
(*ii*) Others	5 – 400	CI	2, 4	W
(*iii*) Gas Pipeline	750 – 5000	SI	2, 4	W
7. Home use-lawn movers	0.7 – 3	SI	2, 4	A

APPENDIX-XX
IGNITION SYSTEM CHARACTERISTICS

System (12 volt)	*Rise time* μ *sec*	*Arc duration* μ *sec*	*Energy mJ*	*Available voltage and drop off rpm*	
				kV	*rpm*
Conventional	80 – 200	1000 – 2000	20–60	20 – 25	2000
Transistor (TAC)	60 – 200	1000 – 3000	60–100	20 – 30	3000
Transistor (CD)*	1 – 100	5 – 300	5–100	15 – 30	8000
Magneto					
Low-tension	60^+	500	20	18 – 25	6000
High-tension	50^+	400	20	18 – 25	6000

* Eight cylinder engine, single coil.
+ At nominal speed (2500 rpm) and same capacitance load.

APPENDIX-XXI
COMPARATIVE IGNITION SYSTEM DATA OF SOME INDIAN AUTOMOBILES

		Hindustan Ambassador	*Fiat 1100*	*Jeep universal*	*Dodge / Fargo Model 89, M4 (Petrol type)*
1.	*Type of ignition*	Battery	Battery	Battery	Battery
2.	*Ignition coil*	Lucas Model LA 12	Prestolite	Prestolite 200713	Prestolite CAH 4001
3.	*Distributor*	Lucas DM 2	Lucas	Prestolite IAY-4401	IBR-4002
4.	*Control breaker gap*	0.36 – 0.40 mm	0.45 ± 0.03 mm	0.5 mm	0.457 – 0.559 mm
5.	*Type of automatic ignition advance*	Centrifugal and vacuum	Centrifugal	Centrifugal	Centrifugal and vacuum
6.	*Spark plug*	Mico Bosch HW 145 T 2 or KLG 50, 14mm	M14-1/225 14 mm 14 mm	Mico Bosch HW 145 T3 or KLG-TF-550,	Mico Bosch HW 145 T3 or KLG-TFS-50
7.	*Spark plug gap*	0.64 mm	0.5–0.6 mm	0.76 mm	0.7—10.81 mm

APPENDIX-XXII

RELATIVE HEAT LOSS DURING DIFFERENT PHASES OF THE CYCLE OF A SI ENGINE FOR A PARTICULAR DESIGN

Part of cylinder	*Fraction of total heat flow to coolant*	
	including piston friction	*not including piston friction*
Head and valve seat	0.5 to 0.55	0.57 to 0.63
Barrel	0.27 to 0.32	0.16 to 0.18
Exhaust port	0.17 to 0.22	0.20 to 0.25

APPENDIX-XXIII

RELATIVE HEAT LOSS TO COOLANT DURING DIFFERENT PHASES OF CYCLE

Phase	*Duration crankshaft degrees*	*Average relative area exposed*	*Estimated average temperature, K*	*Average temperature difference, °C*	*Heat loss*	
					Relative %	*% heat in fuel*
Inlet	180	0.7	320	– 110	– 2	– 0.3
Compression	150	0.7	470	35	1	0.2
Combustion	40	0.4	1600	1170	8	1.2
Expansion	125	0.7	2160	1720	40	6.0
Blow down	90	1.3	1820	1390	48	7.2
Exhaust	135	1.3	1150	720	5	0.7
				Total	100	15

APPENDIX-XXIV
EFFECT OF VARIABLES ON DELAY PERIOD

S.No.	*Increase in variable*	*Effect on delay period*	*Reason*
1.	Cetane number of fuel	Reduces	Reduces self-ignition temperature.
2.	Injection pressure	Reduces	Greater surface-volume ratio hence less physical delay.
3.	Injection advance angle	Increases	Pressures and temperatures lower when injection begins.
4.	Compression ratio	Reduces	Increases air temperature and pressure and reduces auto-ignition temperature.
5.	Intake temperature	Reduces	Increases air temperatures.
6.	Jacket water temperature	Reduces	Increases wall and hence air temperature.
7.	Fuel temperature	Reduces	Better vaporisation and increases chemical reaction.
8.	Intake pressure (supercharging)	Reduces	Increase in density reduces auto-ignition temperature.
9.	Speed	Reduces in milliseconds, increases in crank angle.	Less loss of heat more crank angle in a given time.
10.	Load (fuel-air-ratio)	Decreases	Operating temperature increases.
11.	Engine size	Little effect in milli seconds but crank angle decreases.	Low r.p.m.
12.	Type of combustion chamber	Lower for precombustion chamber	

APPENDIX-XXV
EFFECTS OF CHARACTERISTICS ON KNOCKING TENDENCY

	Characteristics	*Spark ignition engines*	*Compression ignition engines*
1.	Compression ratio	Low	High
2.	Inlet temperature	Low	High
3.	Inlet pressure	Low	High
4.	Ignition temp. of fuel	High	Low
5.	Time lag of fuel	Long	Short
6.	Speed	High	Low
7.	Cylinder size	Small	Large
8.	Combustion chamber wall temperature	Low	High

APPENDIX-XXVI
EXHAUST EMISSION STANDARDS FOR INDIAN VEHICLES

Refer table – 1 for exhaust emission standards for gasoline engines driven vehicles for two wheelers, three wheelers, passenger cars and Diesel engines driven vehicles of gross weight > 3.5 tonnes and also < 3.5 tonnes for Indian driving cycles.

Table-1

Category of vehicles and exhaust emissions	*Standards effective April 1991** **	*Standards effective April, 1996* ***	*Standards proposed April, 2000* ****
GASOLINE :			
Two Wheelers :			
(*a*) CO g/km	12 – 30	4.5	2.0
(*b*) HC g/km	8 – 12	—	—
(*c*) (HC + NO_x) g / km	—	3.6	1.5
Three Wheelers :			
(*a*) CO g / km	12 – 30	6.75	4.0
(*b*) HC g / km	8 – 12	—	—
(*c*) (HC + NO_x) g/km		5.40	1.5
Passenger Cars :			
(*a*) CO g/km	14.3 – 27.1	8.68 – 12.40	2.72
(*b*) HC g/km	2.0 – 2.9	—	—
(*c*) (HC + NO_x) g/km	—	3.00 – 4.36	0.97
DIESEL VEHICLES :			
A : Gross Vehicle Weight > 3.5 tonnes			
(*a*) CO g/kWh	14.0	11.2	4.5
(*b*) HC g/kWh	3.5	2.4	1.1
(*c*) NO_x g/kWh	18.0	14.4	8.0
(*d*) PM g/kWh	—	—	0.36
B : Gross Vehicle Weight < 3.5 tonnes			
(*a*) CO g/kWh	14.3 – 27.1	5.0 – 9.0	
(*b*) (HC + NO_x) g/kWh	2.7 – 6.9	2.0 – 4.0	

* For diesel vehicles, the standards were effective from April 1992.

** Based on reference mass of the vehicles and 'warm start' on Indian driving cycle.

*** Based on capacity of the engine and 'cold start' on Indian driving cycle.

**** Applicable to all categories of the engine and 'cold start' on Indian driving cycle.

Euro Norms at a Glance

The Euro norms for Petrol driven passenger cars, Diesel drawn passenger cars, Diesel light duty vehicles and Diesel heavy duty vehicles are given below :

- **Petrol Driven Passenger Cars :**

	1991/92		*1996*	*1998*	*1996*	*2000*		*2005*
	INDIA	EURO-I	INDIA	INDIA	EURO-II	INDIA	EURO-III	EURO-IV
CO g/km	14.3–27.1	2.72	8.68–12.4	4.34–6.20	2.2	2.72	2.3	1.0
HC g/km	2.0–2.9	—	—	—	—	—	0.20	0.1
NO_x g/km	—	—	—	—	—	—	0.15	0.08
HC + NO_x g/kg	—	0.97	3.4 – 4.36	1.5–2.18	0.57	0.97		

Note : 1. Norms for Passenger cars fitted with catalytic converter only.

2. In case of Euro III and Euro IV COP = type approved forms.

- **Diesel Driven Passenger Cars :**

	1991/92		*1996*		*2000*	
	INDIA	EURO-I	INDIA	EURO-II	INDIA	EURO-III
CO g/km	14.0 (g/kWh)	2.72	5.0–9.0	1.00	2.72–6.90	0.6
HC g/km	3.5 (g/kWh)	–	–	—	—	—
NO_x g/km	18.0 (g/kWh)	—	—	—	—	—
HC + NO_x g/km	—	0.97 (IDI) 1.36 (DI)	2.0–4.0	0.7	0.97–1.70	0.56
PM g/km	–	0.14 (IDI) 0.19 (DI)	— —	0.08 —	0.14 – 0.25 —	0.05 —

Note : 1. In case of Euro II and Euro III type approved forms = COP norms.

2. In India there is also option for testing with engine dynamometer (in g/kWh) as these come under diesel light duty vehicles GVW < 3.5 tonnes.

3. In European norms passenger cars refers to passenger cars with seating capacity less than 6 and GVW less than 2.5 tonnes.

- **Diesel Light Duty Vehicles < 3.5 tonnes :**

	1991/92		*1996*		*2000*
	INDIA	EURO-I	INDIA	EURO-II	INDIA
CO	14.0 (g/kWh)	2.72 – 6.90[2] (g/km)	11.2[1] g/km or 5.0–9.0[3]	1.0–1.5	4.5[1] (g/kWh) or 2.75 – 6.90[3] (g/km)
HC	3.5[1] (g/kWh)	—	2.4[1] g/kWh or HC + NO_x norms	—	1.1 g/kWh or HC + NO_x norms
NO_x	18 (g/kWh)	—	14.4[1] g/kWh)	—	8.0[1] g/kWh
HC + NO_x	—	0.97–1.7[3] (g/km)	2.0 – 4.0[3] (g/km)	0.7–1.3[2]	0.97 – 1.70[3] (g/km)
PM	—	0.14–0.7 (g/km)	—		0.61 kWh or 0.14 – 0.25 (g/km[3])

1 = 13 mode 2 = EDC + EUDC 3 = Indian driving cycle.

- **Diesel Heavy Duty Vehicles > 3.5 tonnes :**

	1991/92		*1996*		*2000*	
	INDIA	EURO-I	INDIA	EURO-II	INDIA	EURO-III
CO g/kWh	14.0	4.5	11.2	4.00	4.5	2.1
HC g/kWh	3.5	1.10	2.4	1.10	1.10	0.66
NO_x g/kWh	18.0	8.00	14.4	7.00	8.00	5.0
PM > 85 kWg/kWh	—	0.36	—	0.15	0.36	0.1
PM < 85 kWg/kWh	—	0.61	—	0.15	0.61	0.1

APPENDIX-XXVII
COMPARISON OF THERMAL AND CATALYTIC CONVERTER

Oxidising systems	*Catalytic reactors*	*Thermal reactors*
Possible Advantages		
Reduces HC and CO	80–90%	80–90%
Reduces NO	No	No
Reduces aldehydes	50% or more	50% or more
Use same design for all vehicles	Hopefully	Hopefully
Long life	Up to 50000 miles	Up to 100000 miles
Possible Disadvantages		
Cost	High	High
Volume	High	Higher
Engine mounting required	No	Yes
Weight added	Some	Significant
Container durability problem	Yes	Yes
Potential over temp. problem	Yes	Yes
Raises engine compartment temperature	Depends upon location	Yes
Requires non-leaded fuel	Probably	No
Requires air injection	Some do	Some do
Lowers fuel economy	No	Probably yes (depends Upon mixture Requirement)
Decreases power	Depends upon back Pressure	Depends upon back Pressure
Loss of catalytic material due to attrition	Yes	No
May emit other toxic material	Yes	No

APPENDIX-XXVIII
EMISSION CHARACTERISTICS OF VARIOUS ENGINES

Emission	*Four-stroke turbocharged*	*Two-stroke air scavenged*	*Four-stroke, normally aspirated*	
			Medium speed	*High speed*
1. HC	Low	High	Low	High
2. NO_x	High	Medium	Low	Low
3. RCHO	Medium	Low	Low	High
4. Smoke	Low	Low	High	High

APPENDIX-XXIX
INFLUENCE OF OPERATIONAL MODES ON EMISSION LEVELS

Engine exhaust constituent	*Concentration values as measured in exhaust gas*			
	Idle	*Acceleration*	*Partial load*	*Full load*
Two-cycle engine				
1. HC, ppm C	250	500	350	550
2. NO_x, ppm	180	1200	1100	1250
3. RCHO, ppm	17.0	9.3	1.0	5.5
4. Smoke, Hartridge unit	4	44	4	10
5. Odour, diesel intensity Turk	3.6	4.1	3.0	3.5
6. CO %	0.01	0.24	0.01	0.34
7. CO_2 %	0.83	5.42	3.79	5.29
Four cycle normally aspirated medium speed engine				
1. HC, ppm C	180	330	210	150
2. NO_x, ppm	330	920	590	780
3. RCHO, ppm	7.9	7.5	4.9	1.6
4. Smoke, Hartridge unit	4	44	4	10
5. Odour, diesel intensity (Turk)	3.6	4.1	3.4	3.5
6. CO %	0.02	0.08	0.04	0.26
7. CO_2%	2.56	3.40	5.33	6.68

APPENDIX-XXX
EXHAUST EMISSION FROM SELECTED ENGINES, mg/g FUEL

Engine	*CO*	*HC (as hexane)*	*NO_x*
1. SI engine, 50 kmph	241	6.1	16.3
2. SI engine, cold start	513	35	14.5
3. SI engine normal load	407	35	13
4. Regenerative gas turbine, 50 kmph	5.3	0.3	13.5
5. Gas turbine, cold start	5.0	1.1	—
6. Aircraft turbojet	3.3	0.4	5.5

Index

L

M

N

O

P

Q

S

T

V

W